Organosilicon Compounds

From Theory to Synthesis to Applications, Volume 1

Organosilicon Compounds

Theory and Experiment (Synthesis)

Edited by

Vladimir Ya. Lee

University of Tsukuba, Tsukuba, Japan

ELSEVIER

ACADEMIC PRESS

An imprint of Elsevier

Academic Press is an imprint of Elsevier
125 London Wall, London EC2Y 5AS, United Kingdom
525 B Street, Suite 1800, San Diego, CA 92101-4495, United States
50 Hampshire Street, 5th Floor, Cambridge, MA 02139, United States
The Boulevard, Langford Lane, Kidlington, Oxford OX5 1GB, United Kingdom

British Library Cataloguing-in-Publication Data
A catalogue record for this book is available from the British Library

Library of Congress Cataloging-in-Publication Data
A catalog record for this book is available from the Library of Congress

ISBN: 978-0-12-801981-8

For Information on all Academic Press publications
visit our website at https://www.elsevier.com/books-and-journals

Publisher: John Fedor
Acquisition Editor: Emily M. McCloskey
Editorial Project Manager: Susan E. Ikeda
Production Project Manager: Mohanapriyan Rajendran
Cover Designer: Matthew Limbert

Typeset by MPS Limited, Chennai, India

Contents

List of Contributors

Yitzhak Apeloig Technion-Israel Institute of Technology, Haifa, Israel
Antoine Baceiredo University of Toulouse, Toulouse, France
Olga A. Gapurenko Southern Federal University, Rostov-on-Don, Russia
Shintaro Ishida Tohoku University, Sendai, Japan
Takeaki Iwamoto Tohoku University, Sendai, Japan
Naokazu Kano University of Tokyo, Tokyo, Japan
Miriam Karni Technion-Israel Institute of Technology, Haifa, Israel
Tsuyoshi Kato University of Toulouse, Toulouse, France
Soichiro Kyushin Gunma University, Kiryu, Japan
Vladimir Ya. Lee University of Tsukuba, Tsukuba, Japan
Christoph Marschner Technische Universität Graz, Graz, Austria
Vladimir I. Minkin Southern Federal University, Rostov-on-Don, Russia
Ruslan M. Minyaev Southern Federal University, Rostov-on-Don, Russia
Yoshiyuki Mizuhata Kyoto University, Kyoto, Japan
Kohtaro Osakada Tokyo Institute of Technology, Yokohama, Japan
Akira Sekiguchi University of Tsukuba, Tsukuba, Japan
Makoto Tanabe Tokyo Institute of Technology, Yokohama, Japan
Norihiro Tokitoh Kyoto University, Kyoto, Japan
Boris Tumanskii Technion-Israel Institute of Technology, Haifa, Israel
Li-Wen Xu Hangzhou Normal University, Hangzhou, P. R. China; Lanzhou Institute of Chemical Physics, Chinese Academy of Sciences, Lanzhou, P. R. China

About the Author

Vladimir LEE earned his PhD from the N.D. Zelinsky Institute of Organic Chemistry Academy of Sciences in the USSR. Following two postdoctoral stints and a number of years working at universities and research institutes, he currently holds the position of Lecturer at the University of Tsukuba (Tsukuba, Japan). His career in the field of main group, mostly organosilicon, chemistry has spanned a period of over 30 years, with a predominant focus on the study (both experimental and theoretical) of the low-coordinate derivatives of the main group elements: cations, radicals, anions, carbenes, multiply-bonded species, aromatic compounds, clusters, and transition metal complexes. His work is documented in over 120 publications.

Preface

Organosilicon chemistry deals with organosilicon compounds, those compounds featuring carbon–silicon bonds that have been known since Friedel and Crafts landmark report of 1863 on the first organosilicon derivative, tetraethylsilane. By the present day, organosilicon reagents have become omnipresent, used in a wide variety of organic processes. Given the importance of these compounds, it comes as no surprise that organosilicon chemistry is repeatedly reviewed and discussed, and as the field continues to grow, there is a continual need for regular updates on its latest advances. That is the reason why we have attempted in this book to survey, analyze, and summarize the current state of affairs in the field of organosilicon chemistry, focusing on the most recent (published mostly after 2000) milestone advances and covering the literature up to the beginning of 2017.

As a reflection of the research and development process, and also in line with its title "*Organosilicon Compounds: From Theory to Synthesis to Applications*," the present book comprises three parts, *Theory*, *Experiment*, and *Applications*, made up of a total of 19 chapters, written by leading experts in their respective fields. Given the great number of chapters and a total of more than 1000 pages, the text has been broken up into 2 volumes: Volume 1 "*Theory and Experiment (Synthesis)*" covering computational and synthetic aspects of organosilicon chemistry, and Volume 2 "*Experiment (Physico-Chemical Studies) and Applications*" dealing with structural studies and the practical uses of organosilicon compounds.

1. The "THEORY" part of the book opens with "Nonclassical Organosilicon Compounds," a contribution from Olga Gapurenko, Ruslan Minyaev and Vladimir Minkin (Southern Federal University, Rostov-on-Don, Russian Federation). In this chapter, the authors discuss state-of-the-art computational approaches to the emerging field of nonstandard (so-called nonclassical) organosilicon compounds with unusual bonding situations (pyramidal structures featuring inverted tetrahedral silicon, sandwich complexes, hypercoordinate and high-spin systems) that classical bonding theory fails to adequately describe.

2. The "EXPERIMENT" part of the book deals with the most spectacular experimental advances in the field of synthetic and structural studies, and is divided into two large sections, *Synthesis* and *Physico-Chemical Studies*.
   a. The *Synthesis* section is further divided into three subsections, in accordance with the coordination number of the central silicon: normal tetracoordinate, low-coordinate, or hypercoordinate.

      In the first subsection "*Organosilicon Compounds of Tetracoordinate Silicon*," there are three contributions on the hot topics of transition metal complexes featuring silyl ligands, organosilicon clusters, and chiral organosilicon compounds. Makoto Tanabe and Kohtaro Osakada (Tokyo Institute of Technology, Yokohama, Japan), the authors of the chapter "Transition metal complexes of silicon," comprehensively cover those featuring single bonds between the silicon and transition metals (both late and early). The following chapter "Organosilicon Clusters" authored by Soichiro Kyushin (Gunma University, Kiryu, Japan) describes a variety of cage, polyhedral, and ring catenation compounds, as well as spirooligosilanes, and siliconoids: their synthesis, structural studies, and specific reactivity. In his chapter "Chiral Organosilicon Compounds," Li-Wen Xu (Hangzhou Normal University, Hangzhou, China) discusses advances in the synthesis of silicon-stereogenic compounds and their application in asymmetric synthesis.

      The focus of the second subsection, "*Organosilicon Compounds of Low-Coordinate Silicon*," is the flourishing field of studies on the silicon analogs of pivotal organic chemistry reactive intermediates, namely, silicon-centered cations, radicals, and anions, as well as silylenes, multiply-bonded silicon compounds, and silaaromatics. In the first contribution in this subsection, "Silicon-Centered Cations," Vladimir Lee and Akira Sekiguchi (University of Tsukuba, Tsukuba, Japan) provide a picture of the tricoordinate cations with the central six-valence-electron cationic silicon in the +IV oxidation state (so-called silylium ions): general synthetic approaches for their generation, structural studies (NMR spectroscopy, X-Ray crystallography), and application of silylium ions in organic synthesis. Discussing silyl radicals (in the chapter "Silicon-Centered Radicals"), Boris Tumanskii, Miriam Karni and Yitzhak Apeloig (Technion-Israel Institute of Technology, Haifa, Israel) focus on persistent and stable representatives with a particular emphasis on a group of polysilyl radicals and EPR spectroscopy as a major tool for identification and investigation of free radical species. Silyl anions is the subject of the next contribution "Silicon-Centered Anions" by Christoph Marschner (Graz University of Technology, Graz, Austria), in which he gives an overview of a variety of anionic silicon derivatives, including the rather promising functionalized silyl anions, silyl dianions, delocalized silyl anions, as well as compounds with a negative charge centered on the sp^2-Si atoms. Shintaro Ishida and Takeaki Iwamoto (Tohoku University, Sendai, Japan), in their contribution "Stable Silylenes and Their Transition Metal Complexes," further continue the story of the low-coordinate silicon derivatives, comprehensively covering recent progress in the emerging field of stable silylenes: their synthesis, structural studies, and use as ligands for transition metal complexes. One of the most challenging topics of contemporary organosilicon chemistry, namely, multiply-bonded derivatives, is overviewed by Antoine Baceiredo and Tsuyoshi Kato (Université de Toulouse, Toulouse, France) in their chapter "Multiple Bonds to Silicon," summarizing the latest developments in the field of doubly- and triply-bonded silicon compounds, both homonuclear and heteronuclear. The *Low-Coordinate Silicon* subsection is completed by a contribution from Yoshiyuki Mizuhata and Norihiro Tokitoh (Kyoto University, Kyoto, Japan) "Silaaromatics and Related Compounds," in which the authors discuss the peculiar properties of silaaromatic compounds, both neutral and charged, as well as those of their transition metal complexes.

The ***Synthesis*** section ends with the subsection "***Organosilicon Compounds of Hypercoordinate Silicon***." In his chapter "Penta- and Hexacoordinated Silicon (IV) Compounds," Naokazu Kano (University of Tokyo, Tokyo, Japan) classifies the title derivatives based on the type of element bound to the silicon, as well as the type of ligand at the silicon center.

b. Successful preparation of organosilicon compounds, as described in the above ***Synthesis*** section, requires subsequent structural studies, and overviews of such techniques are summarized in the ***Physico-Chemical Studies*** section of the experimental part of the book. It comprises four contributions dealing with the major state-of-the-art instrumental methods for assessing the structures of organosilicon derivatives. The first one, "X-Ray Crystallography of Organosilicon Compounds" written by Niepötter Benedikt and Dietmar Stalke (University of Göttingen, Göttingen, Germany), provides a fresh look into the nature of chemical bonding based on experimental and computational studies of electron density distribution. ^{29}Si NMR spectroscopy is a major tool for the study of organosilicon compounds in solutions, and Frank Uhlig (Graz University of Technology, Graz, Austria) in his contribution "^{29}Si NMR Spectroscopy" deals with the general features of this spectroscopic technique, emphasizing particular aspects of the range of silicon nuclei resonances, spin–spin coupling constants, and ^{29}Si NMR pulse techniques. In the next chapter "Thermochemistry of Organosilicon Compounds," Rosa Becerra and Robin Walsh (University of Reading, Reading, UK) review, update and evaluate enthalpies of formation along with bond dissociation enthalpies for a range of organosilicon compounds, with solid support from theoretical calculations. The chapter "UV-Photoelectron Spectroscopy of Organosilicon Compounds," by Anna Chrostowska and Clovis Darrigan (Université de Pau et des Pays de l'Adour, Pau, France), completes the physico-chemical section, describing the advantages of UV-photoelectron spectroscopy (UV-PES) as a powerful tool in the study of the electronic structure of organosilicon compounds (as a special bonus for nonspecialized readers, basic information on UV-PES fundamentals and theoretical methods for the evaluation of ionization energies are also provided).

3. As the culmination of all research efforts, the final part of the book "APPLICATIONS" reports on the practical uses of organosilicon compounds in synthetic chemistry directed towards the creation of new materials. In the first contribution "Hydrosilylation of the carbon—carbon multiple bonds," Bogdan Marciniec, Hieronim Maciejewski, and Piotr Pawluć (Adam Mickiewicz University, Poznań, Poland) give an account of recent developments in catalytic hydrosilylation of alkenes and alkynes as one of the most important methods for preparation of organosilicon compounds such as alkenylsilanes, particularly stressing the use of hydrosilylation for the production of silicones, ceramics, and nanocomposites. In the next chapter "Polysilanes" Julian Koe and Michiya Fujiki (International Christian University, Tokyo, Japan) comprehensively cover different aspects of the chemistry of polysilanes, discussing their structures and electronic properties, synthetic approaches, functionalization, and their high-end applications as chemical sensors and devices (photovoltaic devices, organic light-emitting diodes). Dimitris Katsoulis, Randall Schmidt, and Gregg Zank (Dow Corning Corporation, Midland, USA), in their chapter "Siloxanes and Silicones," discuss the production, specific properties, and ubiquitous application of silicones as one of the most popular silicon-based synthetic polymers, widely applied in various industrial fields (transportation, construction, fabrication of optoelectronic devices, healthcare applications). In the very final contribution of the ***Applications*** section, "Organosilicon Dendrimers and Irregular Hyperbranched Polymers," Aziz Muzafarov, Elena Tatarinova, Nataliya Vasilenko, and Galina Ignat'eva (A.N. Nesmeyanov Institute of Organoelement Compounds, Moscow, Russian Federation) present a complete picture of the silicon-based dendrimers and hyperbranched polymers, macromolecular nanoobjects that have found widespread application as functional materials for photonics, optoelectronics, ceramics, catalytic, and drug delivery systems.

As editor, I am very grateful to all the abovementioned authors whose excellent contributions will hopefully make this book a useful reference source and guide to contemporary organosilicon chemistry, further inspiring those working in the field. And though the book is first and foremost addressed to those working in the fascinating and challenging fields of organosilicon and organometallic chemistry, it could also be of interest to and useful for both main group and transition metal chemistry specialists, material science experts, and the rest of the chemical community.

Vladimir Ya. Lee
University of Tsukuba, Tsukuba, Japan
March 2017

PART I

Theory

CHAPTER 1

Nonclassical Organosilicon Compounds

Olga A. Gapurenko, Ruslan M. Minyaev and Vladimir I. Minkin
Southern Federal University, Rostov-on-Don, Russia

Chapter Outline

1.0 Abbreviations

AO	atomic orbital
Cp	cyclopentadienyl
Cp*	pentamethylcyclopentadienyl
DA	dihedral angle
DFT	density functional theory
MO	molecular orbital
NICS	nucleus independent chemical shift
PES	potential energy surface
TS	transition state

1.1 Pyramidal Structures

Molecular structures of main group elements with pyramidal shapes are among the most interesting chemical objects. Their thermodynamic stability depends on the orbital interaction between the frontier π-MOs of the basal ring and the valence AOs of the apical atom. Such type compounds may be described as *nido*-clusters, the structure of which complies with Wade's rule.[1,2] Analysis of the MO formation from the orbitals of molecular fragments (basal ring and the apical atom) predicts the stability of pyramidal species as the result of formation of four bonding MOs, which, regardless of the basal ring size, may be populated by no more than 8 electrons—the so-called, eight electron rule[3] (Fig. 1.1).

Organosilicon Compounds.
DOI: http://dx.doi.org/10.1016/B978-0-12-801981-8.00001-0

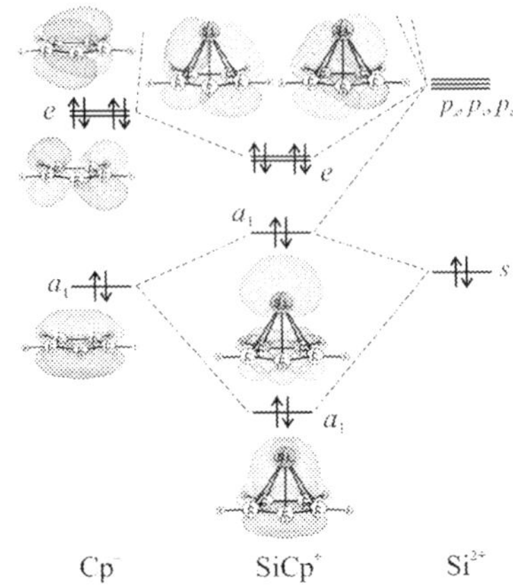

Figure 1.1
Schematic diagram of formation of MOs of pyramidal structures of group 14 elements on the example of the silicocenium ion $SiCp^+$: interaction of the Cp^- π-orbitals and valence AOs of an apical silicon atom.

Group 14 element pyramidal compounds based on the Cp and Cp* ligands have become known since the early 1970s.[4,5] They have η^5-coordination of the apical center and bear a positive charge as dictated by the 8-electron counting rule. A salt $[Cp^*Si]^+[B(C_6F_5)_4]^-$ of the pyramidal silicon cation was synthesized by Jutzi and coworkers in 2004.[6] The structure of the half-sandwich Cp^*Si^+ cation is characterized as a π-complex with η^5-coordinated Cp* ligand. The X-ray diffraction study showed the highly symmetrical regular pentagonal pyramidal structure of the Cp^*Si^+ cation with the Si–C distances about 2.15 Å. This geometry was well reproduced by the MP2/def2TZVPP calculations[7] of the cation, at which all C–Si bond lengths are equal to 2.141 Å.

Whereas Cp and Cp* are well-known ligands for the formation of stable half-sandwich compounds, much less is known on the ability of their full-silicon analogs, i.e., propensity of Si_5R_5 ligands to form stable η^5-coordinated species. Currently, there are only theoretical studies attempting to address this question. According to the early calculations,[8] full-silicon analog of ferrocene $(H_5Si_5)_2Fe$ has two planar η^5-coordinated Si_5-rings.[8] Nonplanarity of the bare $Si_5H_5^-$ anion is explained by the pseudo Jahn-Teller effect and is also eliminated by coordination with main group metal cations.[9] A bipyramidal complex $[Mg^{2+}\ (Si_5H_5)^-\ Mg^{2+}]$ contains planar Si_5H_5-rings with high degree of aromaticity: the $NICS(1)_{zz}$ value is -32.1.[9]

The calculations at the B3LYP/Def2TZVP level of theory predict a stable half-sandwich structure $Si[Si_5(SiMe_3)_5]^+$ **1** with a planar Si_5-ring and pyramidally-pentacoordinate apical silicon, corresponding to a true minimum on the PES.[10] According to the calculations, two complexes with $[AlCl_4]^-$ and $[B(C_6F_5)_4]^-$ counterions, $\mathbf{1[AlCl_4]^-}$ and $\mathbf{1[B(C_6F_5)_4]^-}$, preserve the pyramidal structures with pentacoordinated apical silicon atoms and almost planar Si_5-base. Tetrachloroaluminate complex **1[AlCl$_4$]$^-$** has a much more distorted basal ring [DAs = 2.5–9.1 degrees] than tetraarylborate complex **1[B(C$_6$F$_5$)$_4$]$^-$** (DAs = 0.1–2.4 degrees) that can be explained by the different nucleophilicity of the anions: less nucleophilic tetrakis(pentafluorophenyl)borate anion has less effect on the structure of the pyramidal cation **1**. This leads to the closely comparable bond lengths in **1[B(C$_6$F$_5$)$_4$]$^-$** and **1**: cyclic Si–Si bond lengths are 2.258 Å in **1** and 2.255–2.260 Å in **1[B(C$_6$F$_5$)$_4$]$^-$**, whereas in **1[AlCl$_4$]$^-$** these values are of larger range: 2.248–2.261 Å (Fig. 1.2). But in all cases cyclic Si–Si bond lengths are lying between the standard single (2.32 Å)[11] and double (2.14 Å)[11] bonds. Pyramidal Si–Si bonds in the bare cation **1** are 2.681 Å representing the average for the same bonds in **1[B(C$_6$F$_5$)$_4$]$^-$** and **1[AlCl$_4$]$^-$** (Table 1.1).

Natural charge (+0.15) on the apical silicon in the complex **1[B(C$_6$F$_5$)$_4$]$^-$** is very close to that in the cation **1** (+0.11), thus confirming little effect of the $[B(C_6F_5)_4]^-$ anion on the base properties of the silicon pyramidal system **1**. The same trend is observed for the Si–E bond indices of all representing structures: the WBI values of **1** and **1[B(C$_6$F$_5$)$_4$]$^-$** are very close (about 0.52) whereas the WBI values of **1[AlCl$_4$]$^-$** are remarkably smaller (0.44–0.50) (Table 1.1).

Another hypothetical pyramidal structure $Si[Ge_5(SiMe_3)_5]^+$ **2** with an apical silicon and planar Ge_5-basal ring also corresponds to an energy minimum on the PES.[10] Due to the comparable electronegativity of silicon and germanium atoms, the cation **2** has charge distribution and bond indices very close to those of **1** (Table 1.1) indicating the potential

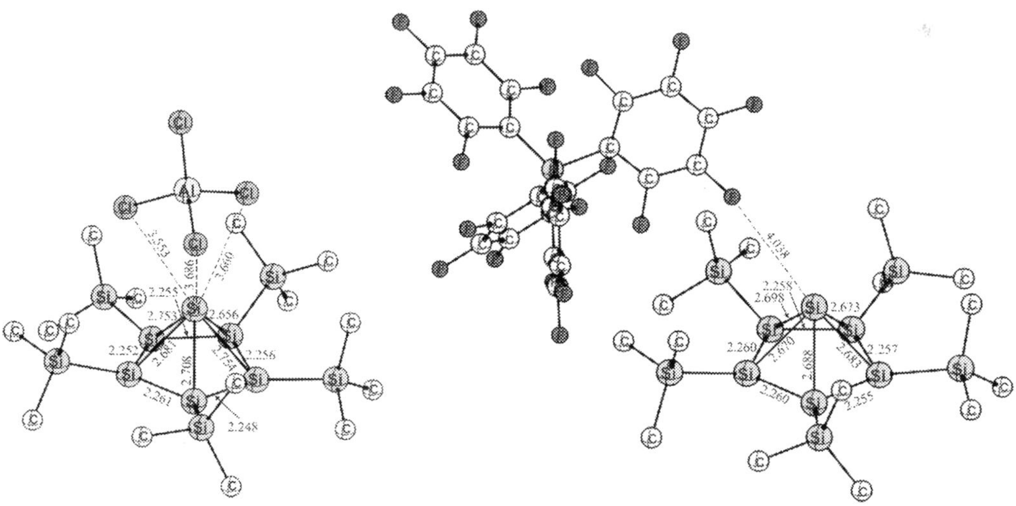

Figure 1.2
Optimized geometries of **1[AlCl$_4$]$^-$** and **1[B(C$_6$F$_5$)$_4$]$^-$** at the B3LYP/Def2TZVP level of theory.[10] Here and in all other figures: bond lengths are given in Å. Hydrogen atoms are omitted for clarity.

Table 1.1: Bond lengths (in Å), natural atomic charges (q), Wiberg bond indices (WBI), and NICS$(1)_{zz}$ for structures 1 and 2 (R = $SiMe_3$, E = Si or Ge) calculated at the B3LYP/Def2TZVP level of theory[10]

Structure	Si–E	E–E	q_{Si}	q_E	WBI (Si–E)	WBI (E–E)	NICS $(1)_{zz}$
1, $Si[Si_5R_5]^+$	2.681	2.258	+0.11	−0.07	0.52	1.14	−17.7
1$[AlCl_4]^-$	2.656–2.754	2.248–2.261	+0.26	−0.05÷ − 0.14	0.44–0.50	1.13–1.19	−18.0
1$[B(C_6F_5)_4]^-$	2.670–2.698	2.255–2.260	+0.15	−0.05÷ − 0.09	0.50–0.52	1.14–1.15	−16.4
2, $Si[Ge_5R_5]^+$	2.788	2.374	+0.14	−0.10	0.51	1.12	−16.1

possibility of stabilizing the electrically neutral complex **2[B(C_6F_5)$_4$]**$^-$, like in the case of structure **1**. Cations and complexes **1, 1[AlCl$_4$]**$^-$, **1[B(C_6F_5)$_4$]**$^-$, and **2** have negative NICS $(1)_{zz}$[12] for the basal Si_5 and Ge_5-rings with close values from −16.1 to −18.0 denoting the cyclic electron delocalization and aromatic character of the rings.

Another potential pyramidal configuration of silicon is feasible in the square-pyramidal structure, as is the case of its intensely studied carbon-analog, pyramidane C_5H_4 **3**, with a very unusual "umbrella" configuration of the apical atom.[13,14] Successful synthesis of "heavy" group 14 element pyramidanes **4** with apical Ge, Sn, and Pb and C_4, Si_4, and Ge_4-basal rings[15,16] raises hopes for the synthetic realization of pyramidanes with silicon at the apex. The DFT calculations (B3LYP/6-311 + G**, B3LYP/def2TZVP) of the realistic compounds **5–7** Si[E_4R_4] (E = C, Si, Ge; R = $SiMe_3$, $SiMe^tBu_2$) predict their stable square-pyramidal structures[15,16] (Scheme 1.1).

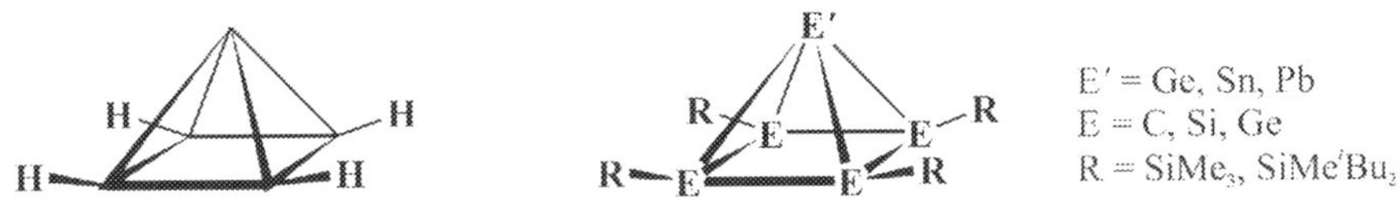

3 4

Scheme 1.1
Pyramidane **3** and its group 14 analogs **4**.

Symmetric square-pyramidal structure **5a** Si[$C_4(SiMe_3)_4$] corresponds to the TS on the PES, the structure of which is only 0.7 kcal mol^{-1} destabilized towards the nearest minimum **5**. Distortion of the basal Si_4-ring of **5** is negligible (DA = 0.4 degrees) as well as the differences in the geometry and other characteristics of **5** and **5a** (Table 1.2). Thus, compound **5** can be regarded as a synthetically quite realistic pyramidal structure (Fig. 1.3).

The previous calculations[17,18] of pentasilapyramidane Si_5H_4 and its related cation $Si_5H_5^+$ showed that these model compounds are not stable and correspond to third- and second-order saddle points on the PES, respectively. However, bulky substituents can stabilize the pyramidal structure of all-silicon pyramidane **6** Si[$Si_4(SiMe^tBu_2)_4$] to fit its structure to the energy minima.[16]

Table 1.2: Bond lengths (in Å), natural atomic charges (q), Wiberg bond indices (WBI), and NICS(1)$_{zz}$ for structures 5–7 (E = C, Si, or Ge, R^1 = SiMe$_3$, R^2 = SiMe^{t}Bu$_2$) calculated at the B3LYP/Def2TZVP level of theory[16]

Structure	Si–E	E–E	q_{Si}	q_E	WBI (Si–E)	WBI (E–E)	NICS (1)$_{zz}$
5, Si[C$_4$(R^1)$_4$] (C_2)	2.039, 2.043	1.484, 1.486	+0.71	−0.65, −0.64	0.49, 0.48	1.15	−13.0
5a, Si[C$_4$(R^1)$_4$] (TS, C_4)	2.042	1.485	+0.71	−0.64	0.48	1.15	−13.0
6, Si[Si$_4$(R^2)$_4$]	2.542	2.284	−0.12	−0.15	0.67	1.05	−10.7
7, Si[Ge$_4$(R^2)$_4$] (C_2)	2.456, 2.902	2.454	+0.02	−0.25, −0.12	0.81, 0.55	0.97	–
7a, Si[Ge$_4$(R^2)$_4$] (TS, C_4)	2.642	2.409	−0.09	−0.18	0.66	1.02	−11.0

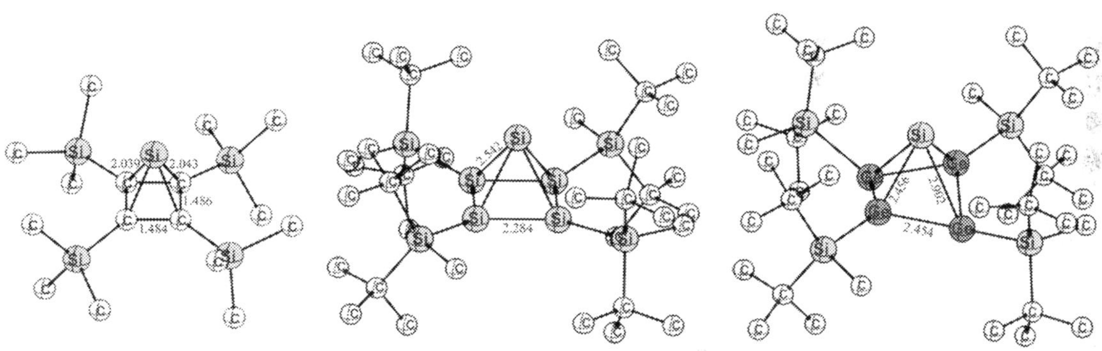

5, C_2 6, C_4 7, C_2

Figure 1.3
Optimized geometries of compounds **5–7** at the B3LYP/Def2TZVP level of theory.[16]

The calculations predicted stabilization of an effectively pyramidal structure **7** Si[Ge$_4$(SiMe^{t}Bu$_2$)$_4$] with Ge$_4$-basal ring. Highly symmetrical pyramidal system **7a** corresponds to the TS structure of the degenerate conformational rearrangement of two C_2 isomers **7** and **7′** (Scheme 1.2).[16] Although the basal DA in **7** is rather substantial (24.4 degrees), the energy barrier against the rearrangement is exceptionally low ($\Delta E = 2.1$ kcal mol^{-1}), accordingly the folding of the basal Ge$_4$-fragment is extremely fast being responsible for the observation of the averaged pyramidal geometry of **7** at ambient temperature. And indeed, it has been found that due to the very narrow energy gap between the pyramidal and folded structures the experimentally realized analog of **7**, pentagermapyramidane Ge[Ge$_4$(SiMe^{t}Bu$_2$)$_4$], crystallized in the two structurally distinct forms: square-pyramidal and folded.[19]

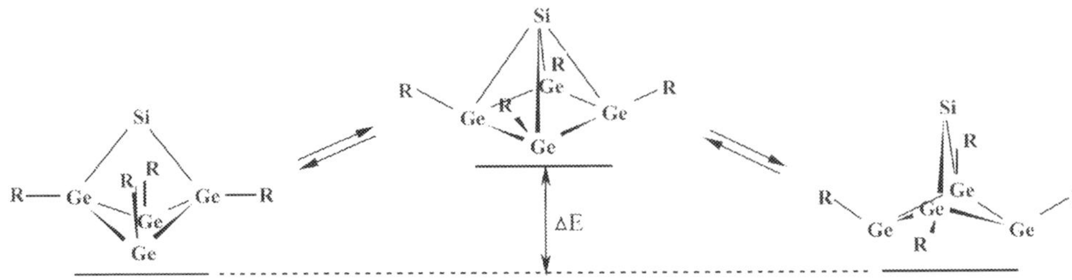

Scheme 1.2
Conformational rearrangement of two C_2 isomers **7**.

1.2 Sandwich Compounds

In 1951, two research groups independently reported on the synthesis of a $[H_5C_5]_2Fe$ compound, the structure of which caused intensive debates,[20,21] resulting in the establishment of the unusual sandwich structure of this compound,[22,23] named as ferrocene. A brief history of this discovery that led to the Nobel Prize being awarded to Wilkinson and Fischer is well described in a short essay by Laszlo and Hoffmann.[24] The multicenter bonding stabilizing the sandwich structure of ferrocene is primarily ensured by the orbital interactions between two vacant d_{xz} and d_{yz}-AO of Fe(II) ion with two highest occupied π-orbitals of cyclopentadienide ligands.[25] Overall, the valence bonding MOs of ferrocene and its structural analogs can be filled by 18-electrons which led to the formulation of the 18-electron rule governing stability of sandwich systems: the sum of the valence electrons of the metal atom and π-electrons of two cyclopolyenes must not exceed 18.[25,26]

Whereas the 18-electron rule regulates stability of transition metal sandwich complexes, in the case of similar main group compounds, in which the valence shells of the central atoms have no *d*-electrons, the 8-electron rule has been formulated stating that the sum of valence electrons of the central atom and π-electrons of two cyclopolyene rings must be equal or not exceed eight.[3,13,27] Thus, the sandwich structure of silicocene Cp_2Si **8** can be stabilized only for the rather unrealistic case of a hexacation. In the electrically neutral form with 14 π-electrons, an additional six electrons fill in the antibonding orbitals that leads to destabilization of the symmetrical sandwich structure and its distortion to a bent form.[28] The distortion can be eliminated by steric factors of Cp substituents. This electron-rich sandwich structure was successfully realized in the form of $Cp^*{}_2Si$[29,30] and $Cp^*[{}^iPr_5C_5]Si$[31]. Another possible way to stabilize the silicon sandwich center was demonstrated by the synthesis of bis(η^5-dicarbollide)silicon sandwich complex **9**[32,33] (Scheme 1.3).

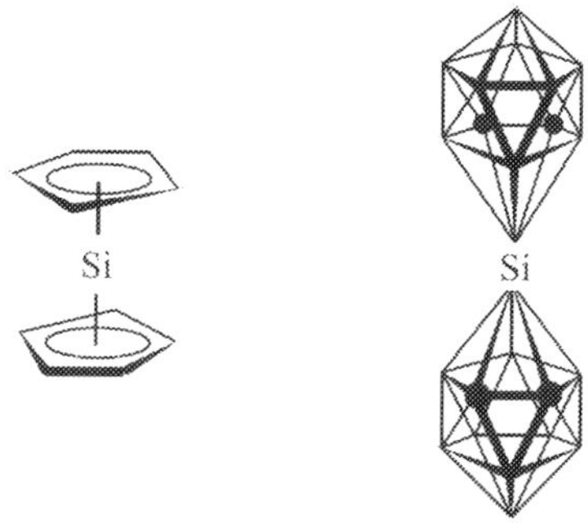

8 9

Scheme 1.3
Electron-rich silicon sandwich compounds. For **9**: black dots—CH groups; unmarked—BH groups.

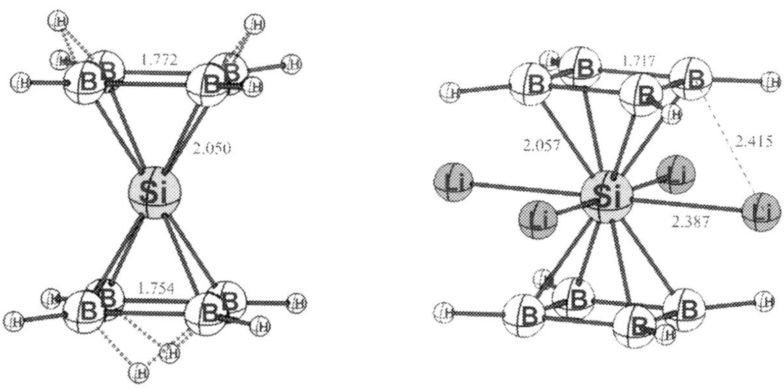

10, D_{2d} **11**, D_{4h}

Figure 1.4
The geometric characteristics of sandwich structures **10** and **11** calculated by the DFT/M062X/6-311++G(3df,3p) method.

The 8-electron rule allows to predict a series of stable sandwich structures with various atomic centers and different size cycles as the ligands. Of special interest are the compounds containing boron cycles, for which the required number of inner (skeletal) eight electrons may easily be adjusted with the help of bridging hydrogen atoms or counterions.[34]

Since the central silicon atom contains four valence electrons and π-orbitals of boron rings have no electrons, the stable silicon sandwiches must bear a positive 4+ charge that ought to be compensated by either four counterions (like sodium or lithium ions) or four bridging hydrogen atoms. Thus, to stabilize a silicon sandwich with $(BH)_n$ boron rings, the latter must be used as the $(HB)_nH_2$ species. And indeed, the DFT calculations have shown the compliance of structures **10** and **11** (Fig. 1.4) with local minima on the relevant PESs.

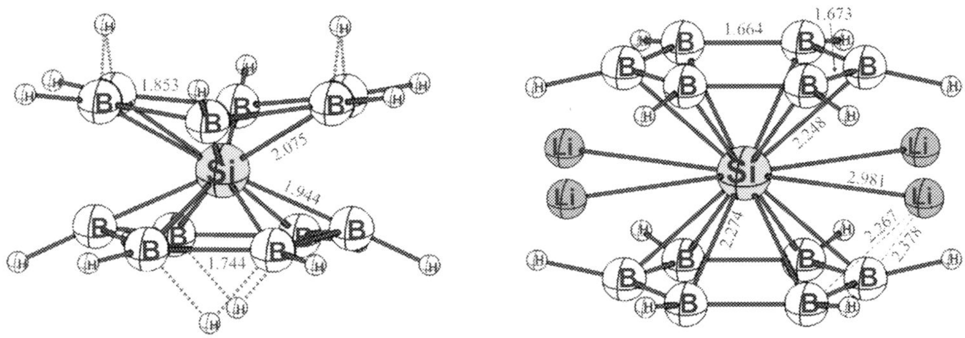

12, D_{2d} 13, D_{2h}

Figure 1.5
The geometric characteristics of sandwich structures **12** and **13** calculated by the DFT/B3LYP/6-311++G(3df,3p) method.

The silicon sandwiches **12** and **13** with six-membered boron cycles stabilized by bridging hydrogen atoms or counterions are depicted in Fig. 1.5.

Experimental detection and theoretical findings of a variety of sandwich-like alkali[26,35] and alkaline earth[4,36] metal compounds as well as some third and fourth row elements (Si, Ge, Sn, Pb, P, As, Bi)[37,38] confirm the validity of the 8-electron rule. Energetic and structural characteristics of new sandwich compounds of the second row elements **14** and **15** $X[Y_3]_2$ and $X[Y_4]_2$ ($X = B^{5-}$, C^{4-}, N^{3-}, O^{2-}; Y = C, Si) (Scheme 1.4) stabilized with lithium countercations were studied using the DFT [B3LYP/6-311 + G(d,p)] and *ab initio* [MP2(full)/6-311 + G(d,p)] methods.[38] Carbon sandwich systems were found to be thermodynamically more stable than their silicon analogs. The Bader topological analysis of electron density manifested the hypercoordinated nature of the central carbon atoms.

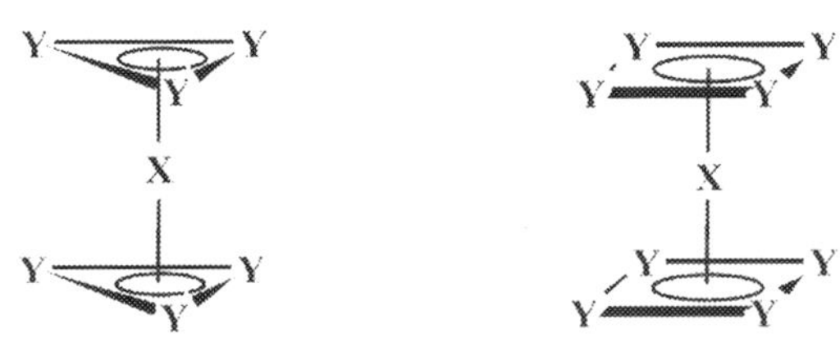

14 15

Scheme 1.4
Sandwich compounds $X[Y_3]_2$ and $X[Y_4]_2$ ($X = B^{5-}$, C^{4-}, N^{3-}, O^{2-}; Y = C, Si).[38]

Fig. 1.6 shows the correlation diagram for the formation of bonding MOs of the sandwich systems from the group orbitals of three-membered rings and the central atom. In Fig. 1.7, the geometric characteristics calculated for the sandwich structures obeying the 8-electron rule are shown.[38]

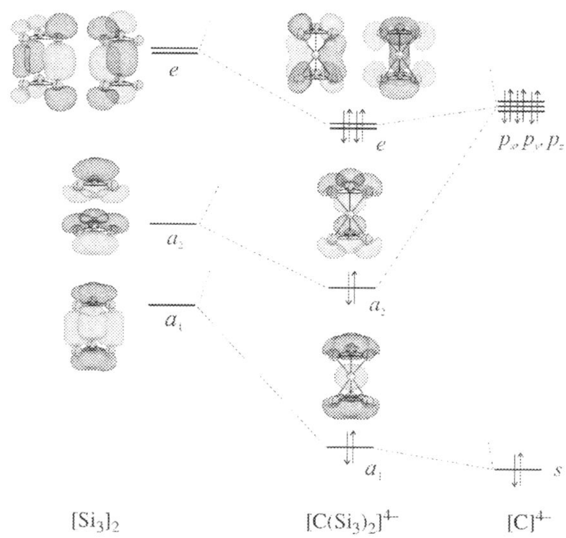

Figure 1.6
Schematic diagram for the formation of the principal bonding MOs of the sandwich systems **14** from the fragment orbitals of $[Si_3]_2$ and $[C]^{4-}$.

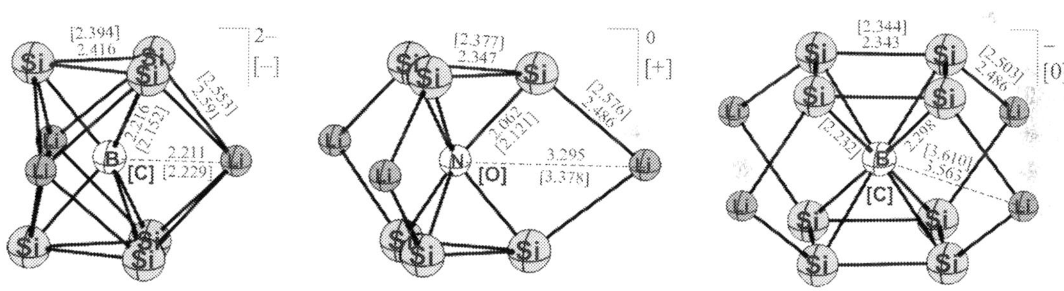

14, D_{3h} **14**, D_{3h} **15**, D_4

Figure 1.7
The geometric characteristics of systems **14** and **15** calculated at the B3LYP/6-311 + G(d,p) level of theory.[38]

Stable systems of this type obeying the octet rule[3,13,35] are exemplified[39] by molecules and ions $X[(SiH)_3]_2$ ($X = B^+$, C^{2+}).

Organosilicon cycles can be used for the formation of stable triple-decker sandwich systems of transition metals. DFT calculations (B3LYP/TZVP) by Koukaras and Zdetsis predicted a series of novel stable organometallic silicon-based multidecker sandwiches, including compounds **16** and **17** (Scheme 1.5), which exhibit similar, albeit not identical, properties to their thoroughly studied corresponding carborane-based structures.[40] The similarities between the two type of structures extend to their symmetry, structural, and electronic properties. For the silicon–carbon structures, depending on the transition metal used, the energy gaps can be adjusted within a wider range of values as a result of near-Fermi level states (frontier orbitals) with high contributions from the silicon atoms. The synthesis of linked silicon–carbon multidecker sandwiches could be further facilitated by the existence of additional linking regions present in these compounds compared to their corresponding carborane structures, and their spin dependence offers an added functionalization. In cases where the magnetic properties are of special interest, the silicon–carbon based structures exhibit higher spin polarization with less delocalization and are seemingly better suited as synthons for possible manufacturing of molecular magnetic materials.

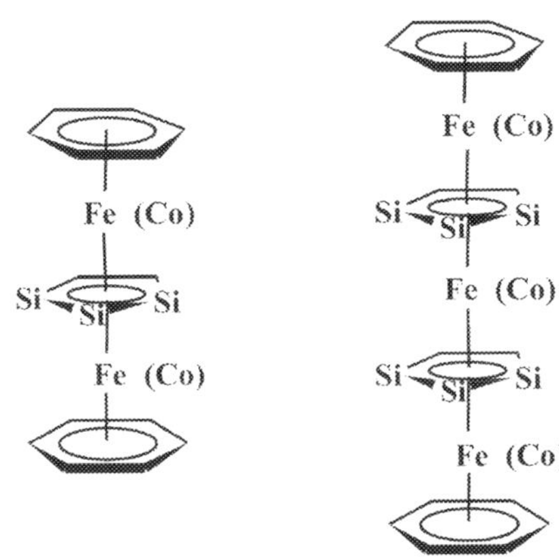

16 17

Scheme 1.5
Structures of $C_{14}Si_3Fe_2H_{14}$ **16** and $C_{16}Si_6Fe_3H_{16}$ **17** (Fe can be replaced by Co).

1.3 Hypercoordination and Hypervalency

Hypervalent interaction is a type of the donor–acceptor interaction occurring between the closed shell main group element centers X←Y, where Y stands for an electron-rich atom donating a lone electron pair.[41–45] The hypervalent bonding is often named as secondary or dative bond; and this interaction governs stereochemical courses of the nucleophilic S_N2 substitution reactions at silicon[46–50] centers, defines the molecular structures of compounds with spatially suitably fitted functionalities, secondary and tertiary structures, and biological activity, as well as modes of packing molecules in a crystal.

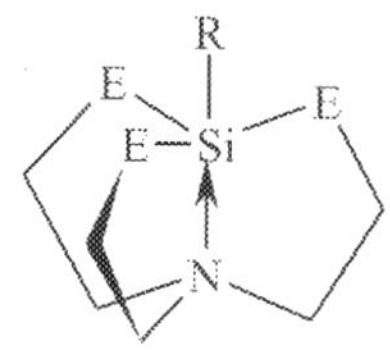

18, C_3

Scheme 1.6
Silatranes (E = CH_2, O; R = H, F).

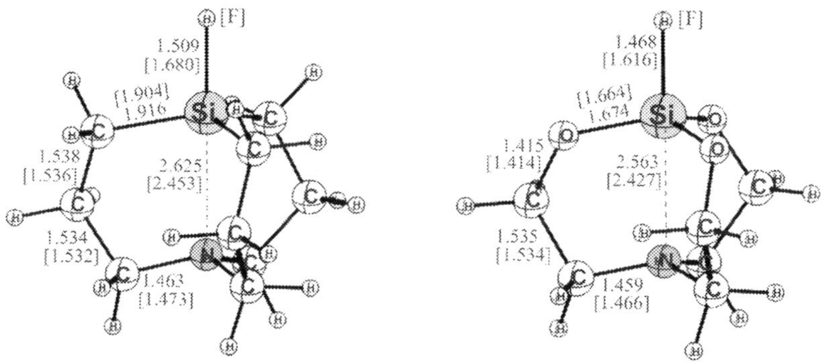

18a [18b] **18c [18d]**

Figure 1.8
Geometric parameters of silatranes **18**, calculated by the B3LYP/6-311 + G** method. Bond lengths for systems with R = F are given in square brackets.

The hypervalent attractive (O)N → Si interaction leads to decrease in the noncovalent contacts in trigonal-bipyramidal silicon complexes[43,44,50] including silatranes **18**,[50] so that these contacts become considerably shorter than the sum of van der Waals radii (Scheme 1.6). Examples of the structures of silatranes calculated by the quantum chemical methods[51] are given in Fig. 1.8.

The calculated Si···N distances are much shorter than the sum of van der Waals radii of silicon and nitrogen atoms (3.50 Å) and approach the length of an ordinary Si–N bond (1.8–1.9 Å). The calculated geometric parameters of structures **18a–d** are well consistent with the data of the gas-phase microwave measurements. For example, the experimental Si–N distance is 2.45 Å in methylsilatrane[52] and 2.324 Å in fluorosilatrane.[53] The calculated length of the Si–N contact is fairly similar to those found by the X-ray diffraction method: 2.146 Å in H-silatrane, 2.042 Å in fluorosilatrane, and 2.175 Å in methylsilatrane.[54,55] The data given in Fig. 1.8 indicate that the Si–N distance shortens on going from **18a** to **18b** and from **18c** to **18d**. Simultaneously, the angle of valence bond pyramidalization at the nitrogen atom (8–17 degrees) decreases, as compared to ammonia molecule (39 degrees).

The hypervalent Si···N distances strongly depend on the electronegativity of the axial substituent at the central Si atom.[50,51,56] The strength of the hypervalent N→Si bonds in silatranes **18** is very difficult to estimate because of the lack of a suitable open-chain structure without the N→Si contact. The strength of the hypervalent N→Si bonds can be estimated in systems **19a**,[56] for which the corresponding open-chain structures **19b** do exist, being estimated by the *ab initio* calculations[56] of bimolecular systems $H_3N \rightarrow SiY_3R$ as 8–10 kcal mol^{-1}. Comparable value of the hypervalent interaction has been revealed in the quasimonocyclic conformations **19a**. The calculations[48,51,56,57] showed that the formation of secondary N→X(R) (X = C, Si, Ge) bonds of the hypervalent type results in the considerable stabilization of the quasimonocyclic conformations **19a** as compared with the strain-free conformations **19b**. The energy of the X···N contact was found to be in the range ~1–8 kcal mol^{-1} (Scheme 1.7).

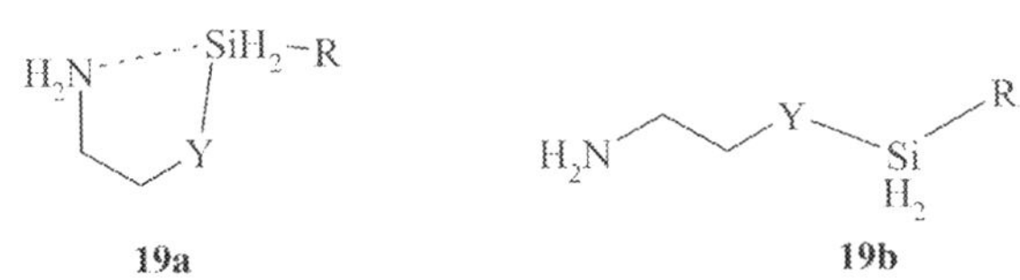

Scheme 1.7

Hypervalent systems **19a** and their open-chain isomers **19b** (Y = CH_2, O; R = H, F).

The hypervalent Y→X interaction is also a key factor determining the reaction mechanism of the bond-switching equilibration **20a**⇄**20b** (Scheme 1.8) occurring with extremely low activation energy barriers.[57,58]

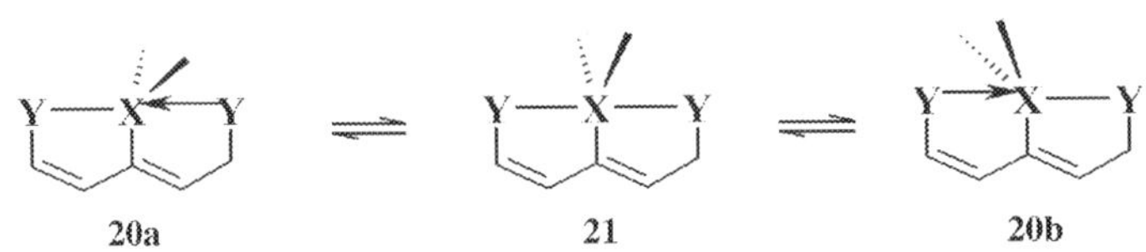

Scheme 1.8

Bond-switching equilibration between two degenerated hypervalent structures **20** (X = SiH_2, GeH_2, SnH_2; Y = NH, O, F^+).

Somewhat similar, although driven by different mechanism, rearrangement was recently reported for the case of the silylium ion **22** (Scheme 1.9). The NMR study detected fast rearrangement **22**→**23**→**24** occurring in solution, and DFT calculations showed that this process can proceed also in the gas phase.[59] The hypervalent F···Si···F interactions enable the low activation barrier (4.5 kcal mol^{-1}) of the reaction.

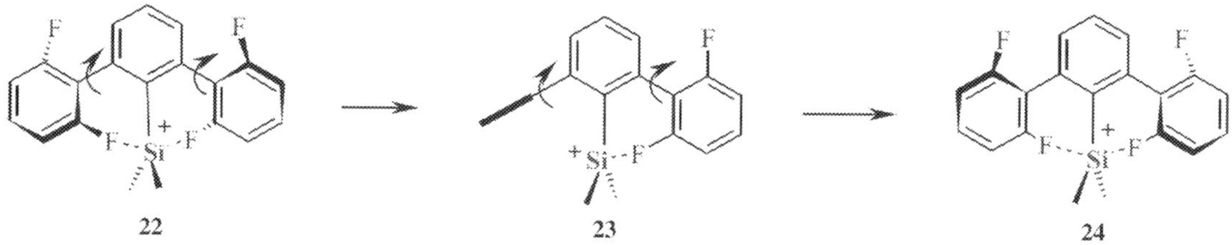

Scheme 1.9
Rotational rearrangement of silicon-containing hypercoordinated system.

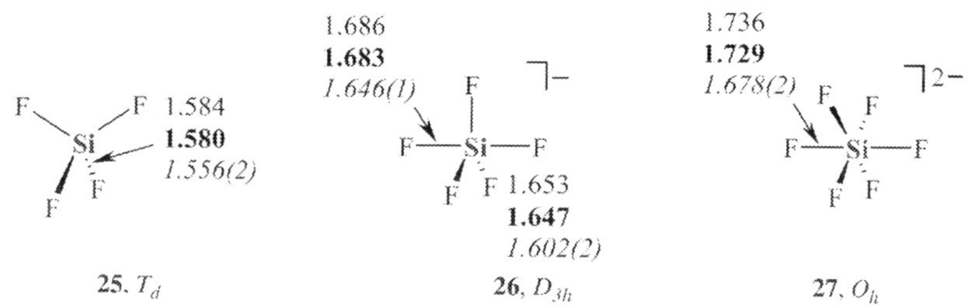

Figure 1.9
Geometric parameters of silicon fluorides SiF_n^{m-} ($n = 4,5,6$; $m = 0, -1, -2$), **25–27**, calculated by *ab initio* MP2(full)/6-311 + G** (boldface numbers) and the B3LYP/6-311 + G** methods. Experimental data are shown in italics (for SiF_4, electron diffraction data are given[60]; for SiF_5^-, X-ray data are given for $[PhCH_2NMe_3][SiF_5]$[61]; for SiF_6^{2-}, X-ray data are given for $K_2SiF_6 \cdot KNO_3$[62]).

In contrast to carbon, silicon is prone to the formation of sufficiently stable structures, such as SiF_4 **25**, and hypervalent SiF_5^- **26** and SiF_6^{2-} **27** (Fig. 1.9). With increase in the coordination number, the lengths of the Si–F bonds substantially increase from **25** to **26** to **27**. This trend is due to transition from classical two-center two electron bonds (2c–2e) in **25** to three-center two-electron bonds (3c–2e) in **26** and **27**. In the anionic **26**, three equatorial Si–F bonds are ordinary (2c–2e) bonds, whereas the axial Si–F bonds are (3c–2e) bonds. Accordingly, the axial bonds are weaker and longer than the equatorial ones. In the dianionic **27**, all bonds are (3c–2e) bonds, and therefore they are all longer than even the axial bonds in **26**.

Stabilization of bisphenoid structure SiF_5^- and octahedral structure SiF_6^{2-} with equal lengths of the apical bonds can be explained by the orbital diagram depicted in Fig. 1.10. The formation of the hypervalent bisphenoid structure of SiF_5^- is associated with two three-center orbitals $1a_1'$ and $2a_1''$, the former giving larger contribution to this effect. When fluoride anion approaches SiF_5^- **26**, analogous three-orbital interaction provides for the formation of a three-center four-electron bond to form the octahedral structure SiF_6^{2-} **27** with all bonds of the (3c–2e) type. Therefore, all Si–F bonds in the dianionic system **27** will be longer than the Si–F bonds in a neutral molecule **25** and anionic **26**. Thus, there

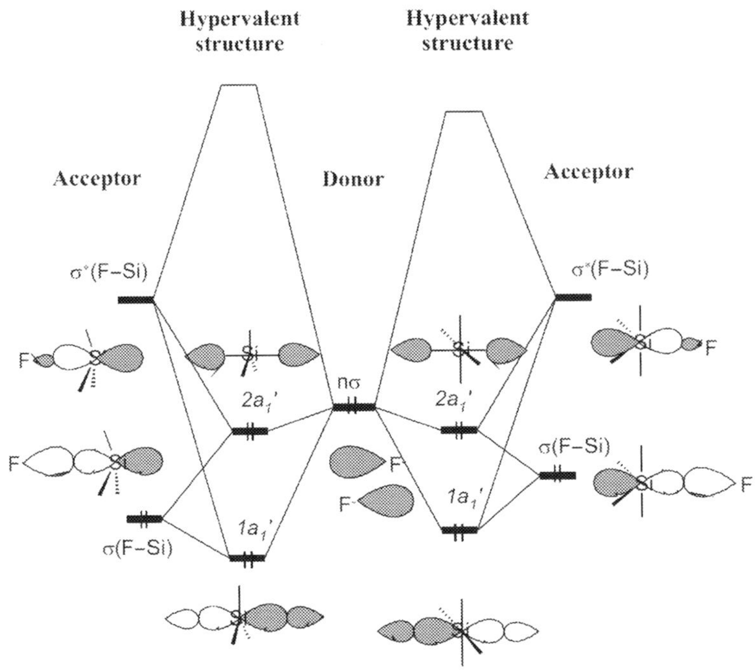

Figure 1.10
Three-orbital correlation diagram for the formation of three-center two-electron bonds in the anions SiF_5^- and SiF_6^{2-}.

is a resonance **27a** ↔ **27b** ↔ **27c** (Scheme 1.10), which averages all bonds and slightly strengthens these bonds (three-dimensional aromaticity). An evidence stems from the comparison of bond lengths of **25, 26**, and **27**. Whereas the difference between the bond lengths of **25** and **26** is about 0.1 Å, that between **26** and **27** is twice as short.

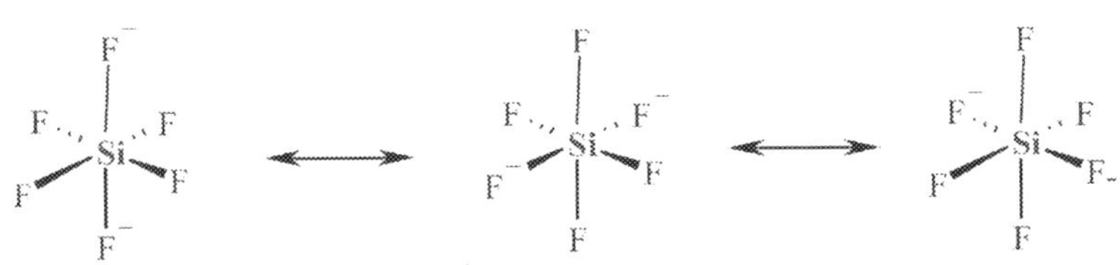

27a **27b** **27c**

Scheme 1.10
Resonance forms of SiF_6^{2-}.

It is interesting to trace the influence of a counterion on stabilization of the structures of hypervalent ionic complexes $Li^+SiF_5^-$ and $(Li_3F_2)^+SiF_5^-$ (Fig. 1.11).[63] Both of these complexes have bisphenoid structure SiF_5^-, wherein the first complex structure **28** is

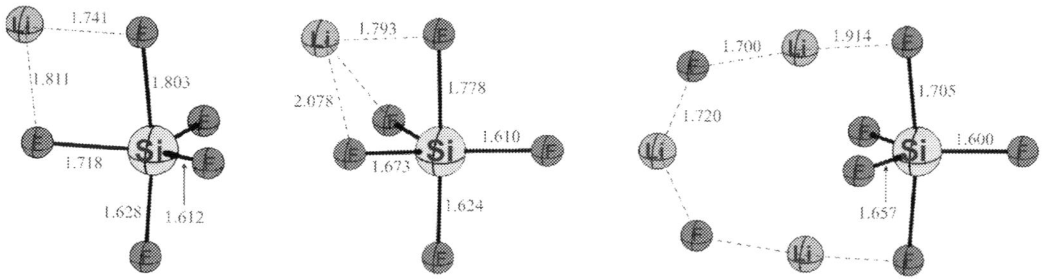

28 **28a** **29**

Figure 1.11
Geometric parameters of complexes **28** $Li^{+}SiF_5^{-}$ and **29** $(Li_3F_2)^{+}SiF_5^{-}$, calculated by the DFT (B3LYP/6-311 + G*) method.[63]

stabilized with unsymmetrical apical Si–F bonds and the second complex **29** retains the structure with symmetrical apical Si–F bonds. Thus, the structures of these complexes can be regarded as zwitterions $Li^{+}SiF_5^{-}$ and $(Li_3F_2)^{+}SiF_5^{-}$, respectively. The ionic structures of these complexes are evidenced by the distribution of electron density, showing that in these systems, all the negative charges are localized on SiF_5^{-} fragments.

The structure of the TS **28a** corresponds to the rotation of lithium atom around the apical Si–F bond with the barrier of 4.8 (DFT) and 4.3 (MP2) kcal mol^{-1}. Interestingly, the structure of the system **28** includes a fairly strong hypervalent interaction between the fluorine atom of the LiF fragment and the silicon atom. The structure of this complex can be represented by a valence scheme that implies donation of the lone electron pair of F^{-} to the σ*- orbital of the apical Si–F bond (Scheme 1.11).

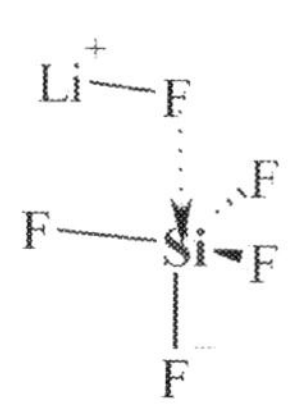

Scheme 1.11
Valence scheme of **28**.

The hypervalent (3c–2e) bonds with the central silicon in compounds **18–20, 28, 29** manifest ability of silicon to form the multicenter bonding networks that can be significantly extended in the compounds in which silicon appears in the nonstandard stereochemical coordinations. An interesting case is the planar tetracoordinate silicon molecules and ions MAl_4^{-} and MAl_4 (M = Si, Ge) (Scheme 1.12) which were theoretically designed[64] and experimentally detected[65] featuring tetracoordinate and hypercoordinate Si-centers.

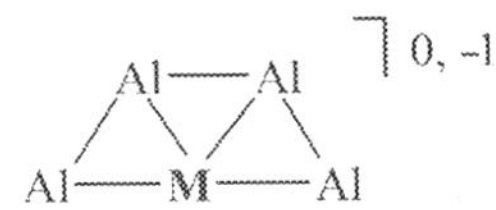

Scheme 1.12
Planar complexes (M = Si, Ge).

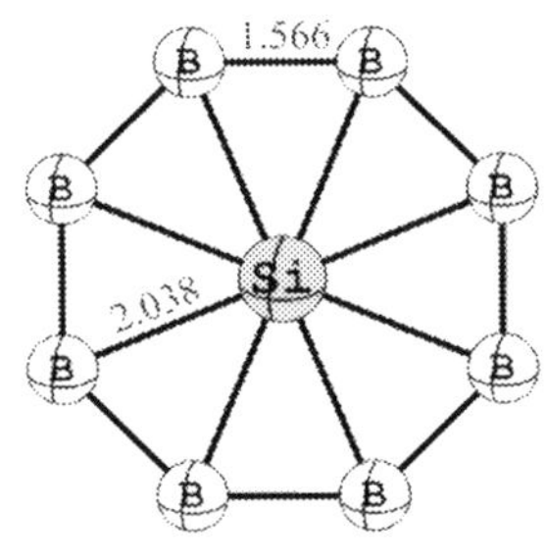

Figure 1.12
Geometrical parameters of **30** calculated by the DFT B3LYP/6-311G(2df) method.[67]

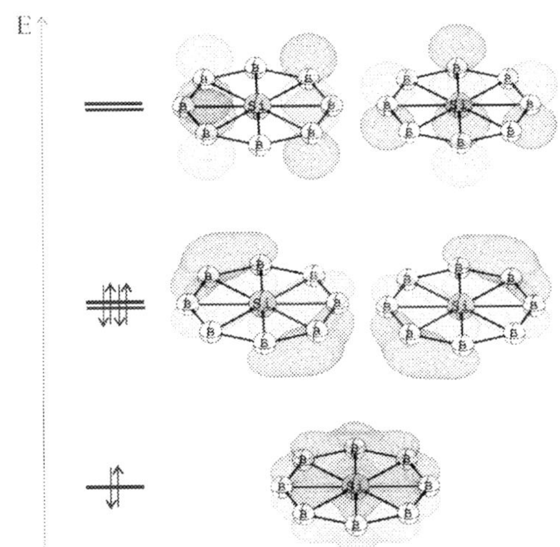

Figure 1.13
Valence π-MOs of B_8Si **30**.

A possibility of the occurrence of compounds with a planar octacoordinate silicon atom in a B_8Si molecular wheel had been simultaneous presented.[13,66,67] In a compound **30** with octacoordinated central silicon, the lengths of the BSi bonds were calculated to be 2.038 Å. These values slightly exceed the sum of the covalent radii of boron and silicon (1.98 Å), which may be explained as a consequence of the multicenter bonding of the central Si atom with the surrounding ligating boron atoms (Fig. 1.12). Compound **30** is an aromatic 6π-electron system with the three highest occupied π-orbitals (Fig. 1.13).

Li and coworkers performed DFT study of a series of S-shaped or cyclic $(B_nE_mSi)_2H_2$ molecules (E = B, C, Si; $n = 3-6$; $m = 1, 2$) containing double planar tetra-, penta-, and hexacoordinate silicon centers (Scheme 1.13).[68] The Si–B distances computed for **31** and **32** are close to those of SiB_8.[66,67]

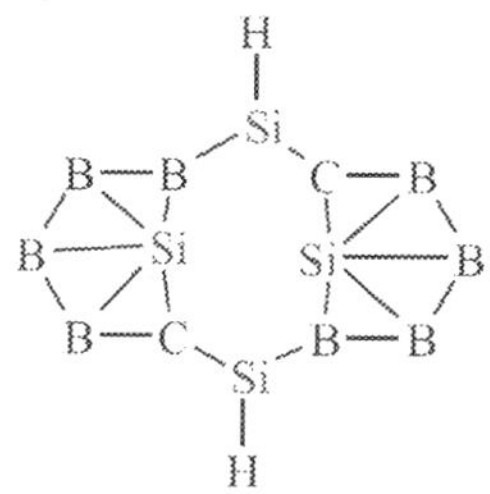

31

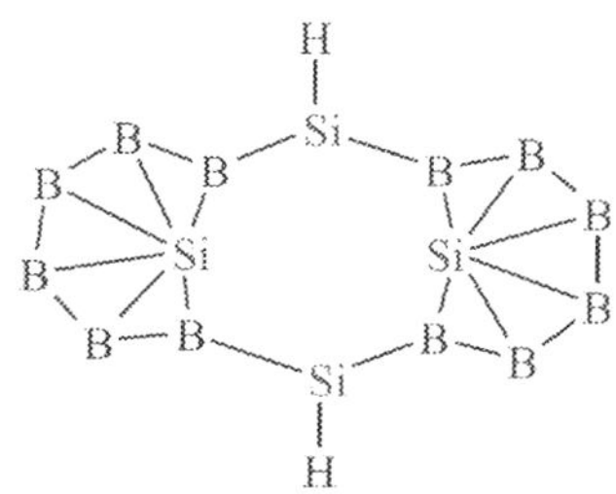

32

Scheme 1.13
Cyclic systems with hypercoordinated silicon.

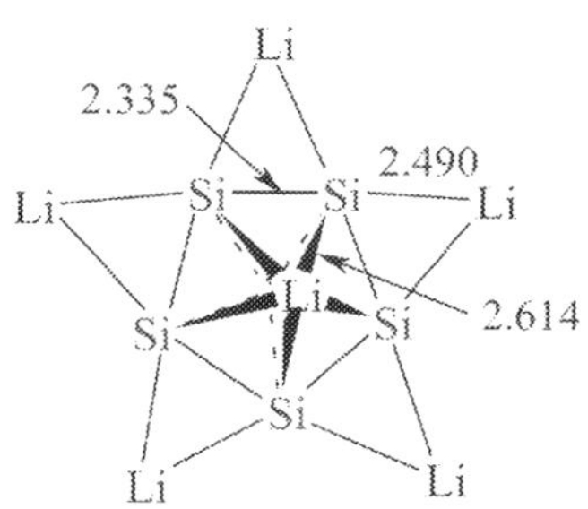

Figure 1.14
Geometry of perlithiated star-like system $Si_5Li_7^+$ calculated at the B3LYP/def2TZVPP level.[70]

A series of complexes possessing pentacoordinate and hexacoordinate planar silicon atoms in complexes $SiCu_5H_5$ and $SiCu_6H_6$ have been computationally studied and found to be the minima on the respective PESs.[69] Of special interest is the star-like structure shown in Fig. 1.14.

1.4 Silicon Analogs of Triangulenes

Rapid development of nanotechnology increases the interest in not only graphene,[71] but also in its analogs such as silicene, a prospective material for nanoelectronic devices,[72,73] and germanene. These two-dimensional structures arose considerable interest in studying other similar type structures, among them are the silicon-containing triangulenes representing an interesting example of zig-zag edged open-shell molecules with the expected unusual magnetic properties. Starting from phenalenyl **33**, the first member of the triangulene family, and gradually increasing molecular size, one can trace the changes in the structure and properties of these molecules depending on the composition. There are three variations of silicon containing triangulenes: full silicon analogs, hybrid half-silicon half-carbon species, and silicon–germanium systems (Scheme 1.14).

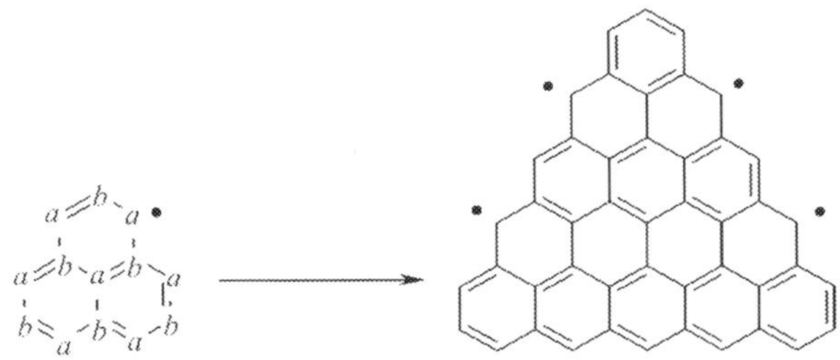

33, m = 2
S = ½

m = 5
S = 2

Scheme 1.14
Triangulene family (*m*, number of rings on one side of the triangle; S, spin state).

The full silicon analogs of triangulenes are nonplanar with chair-like conformation of the rings. Nevertheless silicon systems have the same trends in the structure and electronic properties as the planar carbon triangulenes. The most important feature is alternation of the spin density with its maximum absolute values located at the *a*-type atoms along the perimeter.[74]

A remarkable difference of atom positions (*a* and *b*) in the triangulene framework appears in carbon-silicon triangulenes where two possible versions are realized with alternating C and Si atoms. Both series [$a = \text{C}$, $b = \text{Si}$ (1); and $a = \text{Si}$, $b = \text{C}$ (2)] have planar structures when $m \leq 3$ but above this value the series (2) becomes nonplanar and looses the main triangulene properties. Thus, it is important that more electronegative atoms occupy *a*-position with maximum spin density to preserve the basic carbon triangulene properties.[75] Another computational study predicts planar structure for both series of carbon-silicon triangulenes.[76]

Structure of the nonplanar silicon-germanium hybrid systems does not depend so much on the arrangement of the atoms, probably due to comparable electronegativity of the Si and Ge atoms. All compounds [$a = \text{Si}$, $b = \text{Ge}$; and $a = \text{Ge}$, $b = \text{Si}$] when $m \leq 5$ have highly symmetrical structures and regular alternation of spin density[77] (Fig. 1.15).

Silicon triangulenes can be regarded as the analogs of the true carbon triangulenes and reproduce their peculiar magnetic properties. The calculated exchange spin coupling constants for all described systems have high positive values that mean ferromagnetic ordering of the unpaired electrons in these systems is much like that in the parent carbon triangulenes.[74,75,77] Therefore the silicon triangulenes may have the promising properties of nonmetallic magnetic materials.

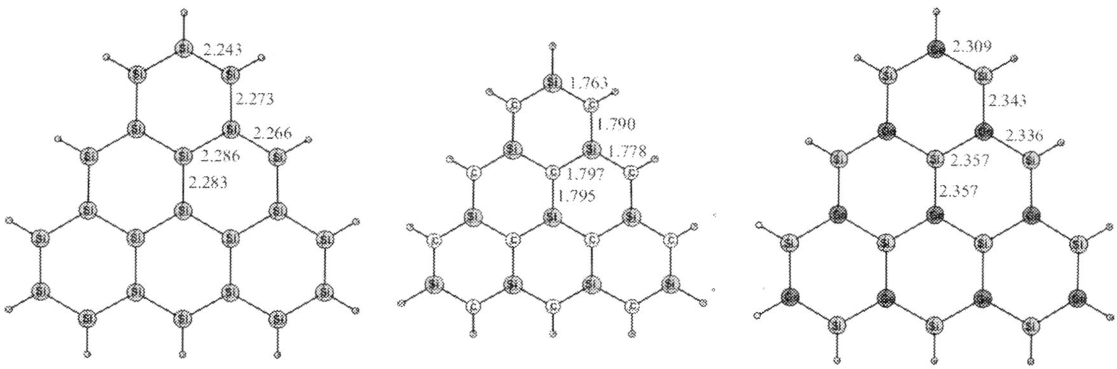

$Si_{22}H_{12}$ $C_{12}Si_{10}H_{12}$ $Si_{12}Ge_{10}H_{12}$

Figure 1.15
Optimized structures of silicon triangulenes ($m = 3$) at the UB3LYP/6-311 + G(d,p) level of theory.[74,75,77]

Acknowledgments

This work was supported by the basic part of internal SFedU grant (project No. 213.01–2014/005).

References

1. Wade, K. The Structural Significance of the Number of Skeletal Bonding Electron-Pairs in Carboranes, the Higher Boranes and Borane Anions, and Various Transition-Metal Carbonyl Cluster Compounds. *J. Chem. Soc. Chem. Commun.* **1971,** 792–793.
2. Welch, A. J. The Significance and Impact of Wade's Rules. *Chem. Commun.* **2013,** *49*, 3615–3616.
3. Minkin, V. I.; Minyaev, R. M.; Zhdanov, Yu. A. *Nonclassical Structures of Organic Compounds;* Mir: Moscow, 1987.
4. Jutzi, P.; Burford, N. Structurally Diverse π-Cyclopentadienyl Complexes of the Main Group Elements. *Chem. Rev.* **1999,** *99*, 969–990.
5. Jutzi, P.; Reumann, G. Cp* Chemistry of Main-Group Elements. *J. Chem. Soc. Dalton Trans.* **2000,** 2237–2244.
6. Jutzi, P.; Mix, A.; Rummel, B.; Schoeller, W. W.; Neumann, B.; Stammler, H.-G. The $(Me_5C_5)Si^+$ Cation: A Stable Derivative of HSi^+. *Science* **2004,** *305*, 849–851.
7. Berger, R. J. F.; Leszczyńska, K.; Jutzi, P. A Theoretical Study of the Half-Sandwich Silicon(II) Cation $[(Me_5C_5)Si]^+$ and of the Cationic Complex $[(Me_5C_5)Si\ (DME)]^+$. *Silicon* **2010,** *2*, 229–234.
8. Kudo, T.; Nagase, S. Theoretical Study of the Persilacyclopentadienyl Sandwich Compound, $(Si_5H_5)_2Fe$. *J. Mol. Struct. (THEOCHEM)* **1994,** *311*, 111–121.
9. Sergeeva, A. P.; Boldyrev, A. I. Flattening a Puckered Pentasilacyclopentadienide Ring by Suppression of the Pseudo Jahn-Teller Effect. *Organometallics* **2010,** *29*, 3951–3954.
10. Gapurenko, O. A.; Minyaev, R. M.; Minkin, V. I.; Lee, V. Ya.; Sekiguchi, A. Group 14 Element Cationic Pentagonal-Pyramidal Complexes $E^a[\eta^5\text{-}E^b{}_5(SiMe_3)_5]^+$ (E^a = Si – Pb, E^b = Si, Ge): A Quantum-Chemical Study. *Phosphorus, Sulfur, Silicon, Relat. Elem.* **2016,** *191*, 609–612.
11. Pyykkö, P.; Atsumi, M. Molecular Double-Bond Covalent Radii for Elements Li–E112. *Chem. Eur. J.* **2009,** *15*, 12770–12779.

12. Fallah-Bagher-Shaidaei, H.; Wannere, C. S.; Corminboeuf, C.; Puchta, R.; Schleyer, P. v. R. Which NICS Aromaticity Index for Planar π Rings is Best? *Org. Lett.* **2006,** *8*, 863–866.
13. Minkin, V. I.; Minyaev, R. M.; Hoffmann, R. Non-classical Structures of Organic Compounds: Unusual Stereochemistry and Hypercoordination. *Russ. Chem. Rev.* **2002,** *71*, 869–892.
14. Lewars, E. *Computational Chemistry (Introduction to the Theory and Applications of Molecular and Quantum Mechanics)*, 2nd ed.; Springer: Berlin, 2011.
15. Lee, V. Ya.; Ito, Y.; Sekiguchi, A.; Gornitzka, H.; Gapurenko, O. A.; Minkin, V. I.; Minyaev, R. M. Pyramidanes. *J. Am. Chem. Soc.* **2013,** *135*, 8794–8797.
16. Lee, V. Ya.; Gapurenko, O. A.; Ito, Y.; Meguro, T.; Sugasawa, H.; Sekiguchi, A.; Minyaev, R. M.; Minkin, V. I.; Herber, R. H.; Gornitzka, H. Pyramidanes: The Covalent form of the Ionic Compounds. *Organometallics* **2016,** *35*, 346–356.
17. Gapurenko, O. A.; Minyaev, R. M.; Minkin, V. I. Silicon Analogues of Pyramidane: A Quantum-Chemical Study. *Mendeleev Commun.* **2012,** *22*, 8–10.
18. Korkin, A. A.; Murashov, V. V.; Leszczynski, J.; Schleyer, P. v. R. Theoretical Study of Cyclic $Si_5H_5^+$ Structural Isomers. Are there Any Analogies with the Corresponding Carbon Species? *J. Phys. Chem.* **1995,** *99*, 17742–17747.
19. Lee, V. Ya.; Ito, Y.; Gapurenko, O. A.; Sekiguchi, A.; Minkin, V. I.; Minyaev, R. M.; Gornitzka, H. Pentagermapyramidane: Crystallizing the "Transition State" Structure. *Angew. Chem., Int. Ed.* **2015,** *54*, 5654–5657.
20. Kealy, T. J.; Pauson, P. L. A New Type of Organo-Iron Compound. *Nature* **1951,** *168*, 1039–1040.
21. Miller, S. A.; Tebboth, J. A.; Tremaine, J. F. Dicyclopentadienyliron. *J. Chem. Soc.* **1952,** 632–635.
22. Wilkinson, G.; Rosenblum, M.; Whiting, M. C.; Woodward, R. B. The Structure of Iron Bis-Cyclopentadienyl. *J. Am. Chem. Soc.* **1952,** *74*, 2125–2126.
23. Fischer, E. O.; Pfab, W. Z. Cyclopentadien-Metallkomplexe, ein neuer Typ Metallorganischer Verbindungen. *Naturforsch. B* **1952,** *7*, 377–379.
24. Laszlo, P.; Hoffmann, R. Ferrocene: Ironclad History or Rashomon Tale? *Angew. Chem., Int. Ed.* **2000,** *39*, 123–124.
25. Long, N. J. *Metallocenes: An Introduction to Sandwich Complexes;* Blackwell Science: London, 1998.
26. Togni, A.; Halterman, R. L., Eds. *Metallocenes: Synthesis, Reactivity, Applications;* Wiley-VCH Verlag GmbH: Weinheim, 1998.
27. Collins, J. B.; Schleyer, P. v. R. Sandwich-Type Molecules of First-Row Atoms. Instability of Bis-(η^3-cyclopropenyl)beryllium. *Inorg. Chem.* **1977,** *16*, 152–155.
28. Lee, T. J.; Rice, J. E. Silicocene, $(C_5H_5)_2Si$: A Highly Symmetric Sandwich Compound? *J. Am. Chem. Soc.* **1989,** *111*, 2011–2017.
29. Jutzi, P.; Kanne, D.; Krüger, C. Decamethylsilicocene – Synthesis and Structure. *Angew. Chem., Int. Ed.* **1986,** *25*, 164.
30. Kühler, T.; Jutzi, P. Decamethylsilicocene: Synthesis, Structure, Bonding and Chemistry. *Adv. Organomet. Chem.* **2003,** *49*, 1–34.
31. Jutzi, P.; Mix, A.; Neumann, B.; Rummel, B.; Stammler, H.-G. Novel *p*-Complexes of Divalent Silicon: Mixed Substituted Neutral Sandwich Compounds and the Half-Sandwich Cation $(^iPr_5C_5)Si^+$. *Chem. Commun.* **2006,** 3519–3521.
32. Schubert, D. M.; Rees, W. S., Jr.; Knobler, C. B. Studies of Metallacarborane Derivatives Containing Aluminium and Silicon. *Pure Appl. Chem.* **1987,** *59*, 869–878.
33. Schubert, D. M.; Rees, W. S., Jr.; Knobler, C. B.; Hawthorne, M. F. Synthesis, Structural Characterization, and Reactivity of the Bis(η^5-dicarbollide) Silicon Sandwich Compound *commo*-3,3'-Si(3,1,2-$SiC_2B_9H_{11})_2$. *Organometallics* **1990,** *9*, 2938–2944.
34. Gribanova, T. N.; Starikov, A. G.; Minyaev, R. M.; Minkin, V. I.; Siebert, M. R.; Tantillo, D. J. Sandwich Compounds of Transition Metals with Cyclopolyenes and Isolobal Boron Analogues. *Chem. Eur. J.* **2010,** *16*, 2272–2281.

35. Eiermann, M.; Halner, K. Spectroscopic Evidence for the Formation of a Bridged Lithium Dicyclopentadienyl Lithiate Complex. *J. Am. Chem. Soc.* **1992,** *114*, 135–140.
36. Williams, R. A.; Tesh, K. F.; Hanusa, T. P. Encapsulated Alkaline-Earth Metallocenes. Synthesis, Solution Behavior, and Solid-State Structures of Bis(tetraisopropylcyclopentadienyl)calcium and -barium, $[(C_3H_7)_4C_5H]_2Ca$ and $[(C_3H_7)_4C_5H]_2Ba$. *J. Am. Chem. Soc.* **1991,** *113*, 4843–4851.
37. Zaegel, F.; Galucci, J. C.; Meunier, P.; Gautheron, R.; Sivik, M. R.; Paquette, L. A. Preparation, X-Ray Crystallographic Analysis, and Stereoselective Electrophilic Capture of Coordinatively Isomeric exo-Lithium Complexes of Isodicyclopentadiene. *J. Am. Chem. Soc.* **1994,** *116*, 6466–6467.
38. Gapurenko, O. A; Minyaev, R. M.; Minkin, V. I. Theoretical Design of New Sandwich Compounds of Boron, Carbon, Nitrogen, and Oxygen. *Russ. J. Gen. Chem.* **2009,** *79*, 728–739.
39. Srinivas, G. N.; Hamilton, T. P.; Jemmis, E. D.; McKee, M. L.; Lammertsma, K. Will an η^3-Si_3H_3 Ligand form Sandwich Compounds with Main Group Elements? *J. Am. Chem. Soc.* **2000,** *122*, 1725–1728.
40. Koukaras, E. N.; Zdetsis, A. D. Multidecker Sandwiches of Silicon–Carbon Clusters. *Organometallics* **2009,** *28*, 4308–4315.
41. Minkin, V. I.; Minyaev, R. M. Cyclic Aromatic Systems with Hypervalent Centers. *Chem. Rev.* **2001,** *101*, 1247–1266.
42. Belogolova, E. F.; Sidorkin, V. F. Ab initio and DFT Study of the Structure of Pentacoordinated Silicon Compounds: Si-Substituted (O–Si)dimethyl(N-acetylacetamidomethyl)silanes. *J. Mol. Struct. (THEOCHEM)* **2004,** *668*, 139–145.
43. Voronkov, M. G.; Trofimova, O. M.; Bolgova, Y. I.; Chernov, N. F. Hypervalent Silicon-Containing Organosilicon Derivatives of Nitrogen Heterocycles. *Russ. Chem. Rev.* **2007,** *76*, 825–845.
44. Negrebetsky, V. V.; Tandura, S. N.; Baukov, Yu. I. Stereochemical Non-rigidity of Complexes of Hypercoordinate Group 14 Elements. *Russ. Chem. Rev.* **2009,** *78*, 21–51.
45. Minyaev, R. M.; Gribanova, T. N.; Minkin, V. I. Hyperbonding and Hypercoordination in Main-Group Chemistry. In *Comprehensive Inorganic Chemistry II;* Reedijk, J., Poeppelmeier, K., Eds.; Elsevier: Oxford, 2013; pp 109–132.
46. Pestunovich, V.; Kirpichenko, S.; Voronkov, M. Silatranes and their Tricyclic Analogs. In *The Chemistry of Organic Silicon Compounds;* Rappoport, Z., Apeloig, Y., Eds.; Vol. 2; John Wiley & Sons, Ltd: Chichester, UK, 1998; pp 1447–1537.
47. Chandrasekaran, A.; Day, R. O.; Holmes, R. R. A New Class of Silatranes: Structure and Dynamic NMR Behavior. *J. Am. Chem. Soc.* **2000,** *122*, 1066–1072.
48. Belogolova, E. F.; Sidorkin, V. F. Geometry of Intramolecular Silicon Complexes: An *Ab initio* Estimation of the Sensitivity to the Effect of Medium. (Aroyloxymethyl)trifluorosilanes. *Russ. Chem. Bull.* **2003,** *52*, 1472–1476.
49. Kost, D.; Kalikhman, I. Hypervalent Silicon Compounds. In *The Chemistry of Organic Silicon Compounds;* Rappoport, Z., Apeloig, Y., Eds.; Vol. 2; John Wiley & Sons, Ltd: Chichester, UK, 1998; pp. 1339–1445.
50. Milov, A. A.; Minyaev, R. M.; Minkin, V. I. Quantum-Chemical Study of X←N Intermolecular Coordination in $F_nH_{(4-n)}X\cdots NH_3$ (X = C, Si, Ge, Ti; n = 0–4) Bimolecular Complexes. *Russ. J. Inorg. Chem.* **2004,** *49*, 779–791.
51. Milov, A. A.; Minyaev, R. M.; Minkin, V. I. Hypervalent Intramolecular X←N (X = C, Si, Ge) Coordination in Atranes: Quantum-Chemical Study. *Russ. J. Org. Chem.* **2003,** *39*, 340–347.
52. Shen, Q.; Hilderbrandt, R. L. The Structure of Methyl Silatrane (1-Methyl-2,8,9-trioxa-5-aza-1-silabicyclo (3.3.3)undecane) as Determined by Gas Phase Electron Diffraction. *J. Mol. Struct.* **1980,** *64*, 257–262.
53. Forgács, G.; Kolonits, M.; Hargittai, I. The Gas-Phase Molecular Structure of 1-Fluorosilatrane from Electron Diffraction. *Struct. Chem.* **1990,** *1*, 245–250.
54. Greenberg, A.; Wu, G. Structural Relationships in Silatrane Molecules. *Struct. Chem.* **1990,** *1*, 79–85.
55. Hencsei, P. Evaluation of Silatrane Structures by Correlation Relationships. *Struct. Chem.* **1991,** *2*, 21–26.

56. Milov, A. A.; Minyaev, R. M.; Minkin, V. I. Quantum-Chemical Investigation of the Hyprevalent Intramolecular Coordination X←N (X = C, Si, Ge) in Quasimonocyclic Models of IVa Group Atranes. *Russ. J. Org. Chem.* **2004,** *40*, 261–268.
57. Minkin, V. I.; Minyaev, R. M.; Milov, A. A.; Gribanova, T. N. Intramolecular Y←O (Y = N, P, As, Sb, Bi) Coordination in Organopnictogen Compounds: An Ab initio and DFT Study. *Russ. Chem. Bull.* **2001,** *50*, 2028–2045.
58. Atsumi, T.; Abe, T.; Akiba, K.; Nakai, H. Theoretical Study of Hypervalent Bonds in 1,6-Diaza-1,6-dihydro- and 1,6-Dihydro-1,6-dioxapentalene Systems with a Heteroatom X at 6a Position (X = 14–16 Group Atoms). *Bull. Chem. Soc. Jpn.* **2010,** *83*, 892–899.
59. Romanato, P.; Duttwyler, S.; Linden, A.; Baldridge, K. K.; Siegel, J. S. Intramolecular Halogen Stabilization of Silylium Ions Directs Gearing Dynamics. *J. Am. Chem. Soc.* **2010,** *132*, 7828–7829.
60. Beagley, B.; Brown, D. P.; Freeman, J. M. The Si-F Length in SiF: A New Electron Diffraction Study. *J. Mol. Struct.* **1973,** *18*, 337–338.
61. Schrobilgen, D.; Krebs, R. Structural Chemistry of Pentacoordinated Silicon. Molecular Structures of the Pentafluorosilicate Anion and the Diphenyltrifluorosilicate Anion. *Inorg. Chem.* **1984,** *23*, 1378–1381.
62. Rissom, C.; Schmidt, H.; Voigt, W. Crystal Structure and Thermal Properties of a New Double Salt: $K_2SiF_6 \cdot KNO_3$. *Cryst. Res. Technol.* **2008,** *43*, 74–82.
63. Getmanskii, I. V.; Minyaev, R. M.; Minkin, V. I. Effect of Counterions on Hypervalent Interactions in Prereaction Complexes of S_N2 Reactions of Bisphenoid Compounds of the Second and Third Period Elements. *Russ. Chem. Bull.* **2012,** *61*, 2036–2048.
64. Ivanov, A. S.; Boldyrev, A. I. Reliable Predictions of Unusual Molecules. *Phys. Chem. Chem. Phys.* **2012,** *14*, 15943–15952.
65. Boldyrev, A. I.; Li, X.; Wang, L.-S. Experimental Observation of Pentaatomic Tetracoordinate Planar Si- and Ge-Containing Molecules: MAl_4^- and MAl_4. *Angew. Chem., Int. Ed.* **2000,** *39*, 3307–3310.
66. Wang, Z.-X.; Schleyer, P. v. R. Construction Principles of "Hyparenes": Families of Molecules with Planar Pentacoordinate Carbons. *Science* **2001,** *292*, 2465–2469.
67. Minyaev, R. M.; Gribanova, T. N.; Starikov, A. G.; Minkin, V. I. Octacoordinated Main-Group Element Centres in a Planar Cyclic B_8 Environment: An Ab initio Study. *Mendeleev Commun.* **2001,** *6*, 213–214.
68. Li, S.-D.; Guo, J.-C.; Miao, C.-Q.; Ren, G.-M. C_{2h} $(B_nE_mSi)_2H_2$ Molecules (E = B, C, Si; $n = 3 - 6$; $m = 1, 2$) Containing Double Planar Tetra-, Penta-, and Hexacoordinate Silicons. *J. Phys. Chem. A* **2005,** *109*, 4133–4136.
69. Li, S. D.; Ren, G. M.; Miao, C. Q. Hexacoordinate Planar Main Group Atoms Centered in Hexagonal Hydrocopper Complexes Cu_6H_6X (X = Si, P, As). *Inorg. Chem.* **2004,** *43*, 6331–6333.
70. Tiznado, W.; Perez-Peralta, N.; Islas, R.; Toro-Labbe, A.; Ugalde, J. M.; Merino, G. Designing 3-D Molecular Stars. *J. Am. Chem. Soc.* **2009,** *131*, 9426–9431.
71. Rao, C. N. R.; Sood, A. K., Eds. *Graphene: Synthesis, Properties, and Phenomena;* Wiley: Weinheim, 2012.
72. Kara, A.; Enriquez, H.; Seitsonen, A. P.; Lew Yan Voone, L. C.; Vizzini, S.; Aufray, B.; Oughaddou, H. A Review on Silicene: New Candidate for Electronics. *Surf. Sci. Rep.* **2012,** *67*, 1–18.
73. Tao, L.; Cinquanta, E.; Chiappe, D.; Grazianetti, C.; Fanciulli, M.; Dubey, M.; Molle, A.; Akinwande, D. Silicene Field-Effect Transistors Operating at Room Temperature. *Nat. Nanotechnol.* **2015,** *10*, 227–231.
74. Gapurenko, O. A.; Starikov, A. G.; Minyaev, R. M.; Minkin, V. I. Carbon and Silicon Triangulenes: Searching for Molecular Magnets. *Russ. Chem. Bull.* **2011,** *60*, 1517–1524.
75. Gapurenko, O. A.; Minyaev, R. M.; Starikov, A. G.; Minkin, V. I. Hybrid Carbon–Silicon Triangulenes. *Dokl. Chem.* **2013,** *448*, 23–28.
76. Zhang, S.; Zhou, J.; Li, X.; Wang, Q. Magnetism of Triangular Nanoflakes with Different Compositions and Edge Terminations. *J. Nanopart. Res.* **2012,** *14* (1171).
77. Gapurenko, O. A.; Starikov, A. G.; Minyaev, R. M.; Minkin, V. I. Germanium, Carbon-Germanium and Silicon-Germanium Triangulenes. *J. Comput. Chem.* **2015,** *36*, 2193–2199.

PART II

Experiment

SECTION 1

Synthesis

SECTION 1.1

Organosilicon Compounds of Tetracoordinate Silicon

CHAPTER 2

Transition Metal Complexes of Silicon (Excluding Silylene Complexes)

Makoto Tanabe and Kohtaro Osakada

Tokyo Institute of Technology, Yokohama, Japan

Chapter Outline

2.1 Introduction

Four-coordinating silicon is the most common state of the element in molecular compounds. Transition metal complex with a silyl (SiRR′R″) ligand is also included in this category because the bond between the metal and silicon has a covalent character. The coordination bond of the organosilyl group to the transition metals, however, shows a high reactivity, which is comparable with the metal alkyl bond. Reports of the transition metal complexes having silyl ligands started in the 1960s, and a number of the compounds were already reported. The chemistry of transition metal organosilyl complexes with metal–silicon bonds has close relevance to hydrosilylation of unsaturated organic compounds,[1] dehydroucoupling reactions,[2] and redistribution reactions[3] catalyzed by transition metal complexes. These reactions were applied to advanced materials, polymer synthesis, and biomedical chemistry.[4] The majority of transition metal silyl complexes were stabilized by phosphine, *N*-heterocyclic carbene, pyridine, cyclopentadienyl, and allyl ligands. The chemistry of the transition metal silyl complexes has been investigated to understand the above catalytic reactions, which involve Si–H bond-activating and Si–C bond-forming reactions.[5] Therefore, transition metal silyl complexes with monodentate organosilyl groups and their reactivity toward unsaturated organic compounds were well studied. The organosilyl ligands function as the electron-donating ligands and exhibit trans labile behavior.

Organosilicon Compounds.
DOI: http://dx.doi.org/10.1016/B978-0-12-801981-8.00002-2

This section mostly deals with recent reports on the complexes having transition metal–silicon single bonds. π-Coordination of an Si–H group to the metal involves bonding of the two atoms to the metal, although the transition metal provides a coordination site, and it may be classified as the complexes with the Si center having more than four coordinating atoms.

Some of these complexes are also included in this chapter because they are closely related to the simple silyl complexes and their reactions. Chemical properties of the silyl–transition metal complexes are also included in this section.

2.2 Late Transition Metal Complexes

Terminal Si-Ligands. In the 1960s, there were a number of examples of the late transition metal complexes with an $SiCl_3$ ligand because the strongly electron-withdrawing substituents were regarded as a π-acceptor ligand, which forms the stable M–Si bond. The treatment of late transition metal carbonyl complexes with $HSiCl_3$ afforded trichlorosilyl complexes $[M(SiCl_3)(CO)_5]$ (**1**: M = Mn, Re) or $[CpM(SiCl_3)(CO)_n]$ (**2**: M = Mo ($n = 3$), Fe ($n = 2$), Ni ($n = 1$); Cp = η^5-C_5H_5) (Scheme 2.1).[6] In contrast, the complexes having an SiH_3 group are very rare. The reaction of *trans*-$[PtH_2(PCy_3)_2]$ with SiH_4 was reported to form immediately *trans*-silyl(hydride)platinum complex **3** which was identified by X-ray crystallography.[7] Although the SiH_3 ligand appears to be sensitive to air and moisture, the platinum complex is significantly air stable due to the presence of the bulky PCy_3 ligand.

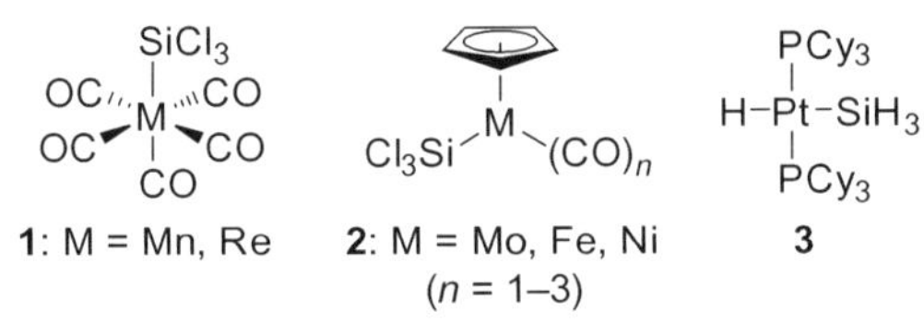

Scheme 2.1
Silyl complexes.

The silyl iron complex with $SiMe_3$ ligand, $[CpFe(SiMe_3)(CO)_2]$, was the first reported transition metal silyl complex.[8] The addition of primary silanes (H_3Si^tBu and $H_3SiCMe_2CMe_2H$) with the Fe complex on irradiation caused dimerization to give diiron complex **4** with the bridging silylene and carbonyl ligands (Scheme 2.2). A similar silyl exchange of $[Cp^*Fe(SiMe_3)(CO)_2]$ ($Cp^* = \eta^5$-C_5Me_5) with the bulky ligand yielded the mononuclear iron complex **5** through exchange of the silyl groups.[9]

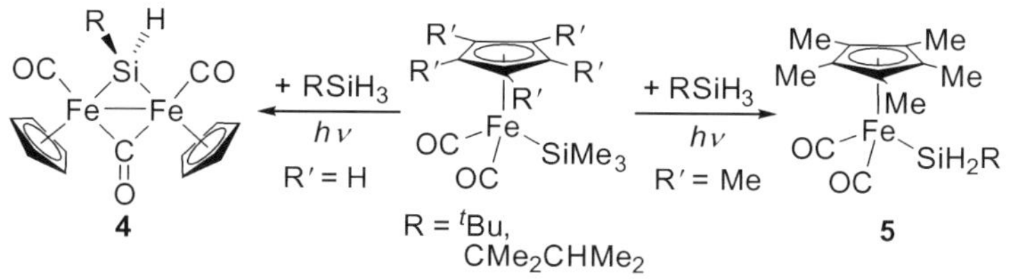

Scheme 2.2
Silyl ligand exchange.

Silyl polyhydride Re(VII) complexes **6**, obtained from $ReH_7(PPh_3)_2$ and $HSiR_3$, (R = Ph, Et etc.), have an unusual nine-coordinate tricapped trigonal-prismatic structure with two P and one Si atoms within the equatorial plane and the six hydride ligands at the positions out of the plane (Scheme 2.3).[10] The short Re–Si bond (2.474(4) Å) was attributed to the possible interaction of two hydride ligands between Re and Si atoms, which was described as η^3-H_2SiR_3 coordination. The organosilyl ligands tend to generate a coordinatively unsaturated site at the metal center due to large trans influence of the Si atom. The pentacoordianted silyl-rhodium(III) complexes **7** form a distorted square-pyramidal coordination around the Rh center with silyl ligand at the apical position, indicating the larger trans effect of the silyl atom than the hydride ligand.[11]

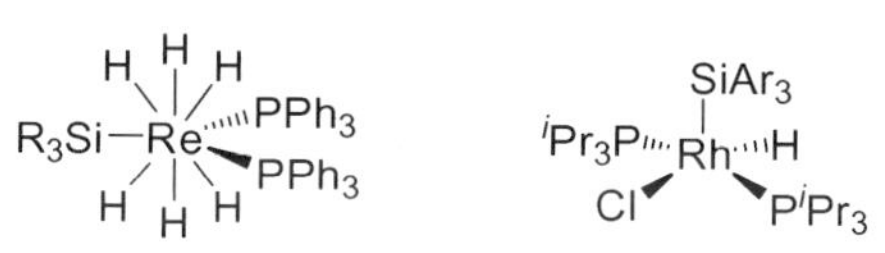

6
R = Ph, Et *etc.*

7
Ar = C_6H_5, 4-Me-C $_6H_4$ *etc.*

Scheme 2.3
Re and Rh silyl complexes.

Bis(silyl) and silyl(hydride) nickel complexes were known as unstable complexes. Lappert and Speier reported that the reactions of $[Ni(PPh_3)_4]$ with $Ph_{4-n}SiH_n$ ($n = 1-3$) produced pyrophoric and highly colored materials which were believed to contain paramagnetic tetrahedral Ni(II) complexes.[12] The nickel complexes were stabilized by the introduction of electron-withdrawing groups at the Si atom,[13] similar to complexes **1** and **2**. Radius reported synthesis of the silyl(hydride) and bis(silyl) nickel(II) complexes stabilized by strong electron-donating carbene ligands.[14] The dinickel(0) complex **8** with monodentate NHC ligands reacted with organosilanes to give stable silyl(hydride) complexes **9** and bis (silyl) complexes **10** (Scheme 2.4). The formation of the two different structures is dependent on the stoichiometry and the steric demand of the organosilanes used.

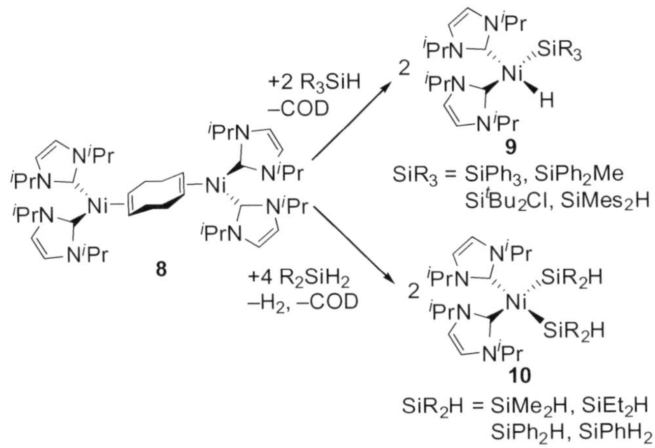

Scheme 2.4
Silyl Ni complexes with a NHC ligand.

Bidentate Si-Ligands. The bidentate silyl ligands with one or two M−Si bonds are expected to form thermodynamically stable complexes by the chelating effect. Disilametallacycles have been easily prepared from the double Si−H oxidative addition of the substrates to low-valent transition metal complexes or metathesis reactions of the corresponding dianion species with dihalide metal complexes (Scheme 2.5).[15]

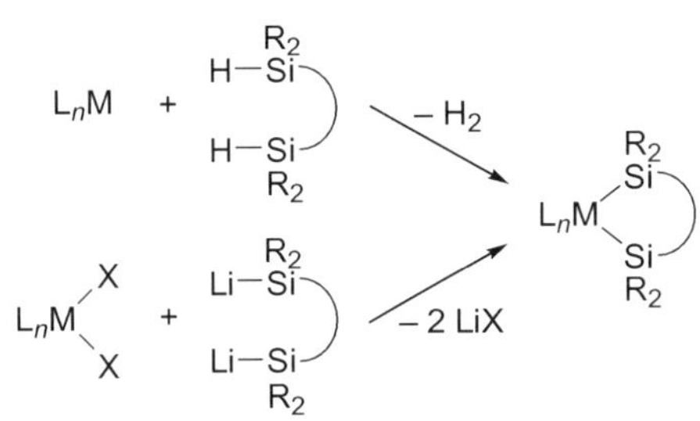

Scheme 2.5
Disilametallacycles.

Tanaka and coworkers developed dehydrogenative double silylations of 1,2-bis (dimethylsilyl)benzene with alkynes, yielding the six-membered ring compounds through two Si−C bond formation. They also successfully identified the corresponding disilaplatinacyclic intermediate {*cyclo*-[Pt($Me_2SiC_6H_4SiMe_2$)]($(PPh_3)_2$)} (**11**), followed by insertion of alkynes to give double-silylated products (Scheme 2.6).[16]

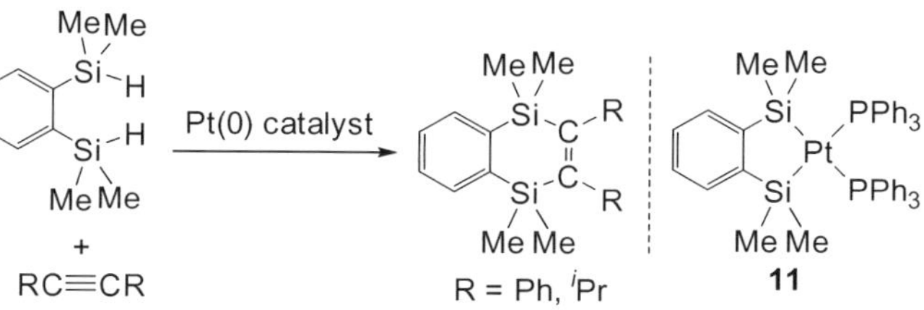

Scheme 2.6
Cyclodimerization of 1,2-disilylbenzene and alkynes.

The strongly electron-donating 1,2-disilylbenzene ligands provide stable transition metal complexes with high oxidation state.[17] The reactions of $H_3SiC_6H_4SiH_3$ with Ni(0), Pd(0), and Pt(0) complexes in 2:1 ratio yielded the tetrakis(silyl) metal(VI) complexes **12** with an octahedral coordination mode, coordinated by two chelating disilanyl ligands (Scheme 2.7).[18] Aromatic trisilane reacted with Ni(0) or Pt(0) complexes to form trisilyl(hydrido) nickel(IV) and -platinum(IV) complexes **13**.[19] The reaction with Pd(0) complex, however, did not produce the corresponding Pd(IV) complex, probably due to reductive elimination of the compound with an Si–Si bond.

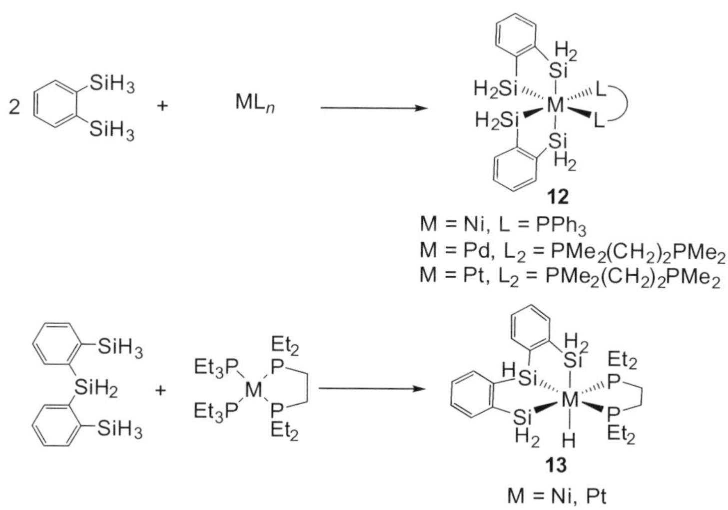

Scheme 2.7
Ni, Pd, and Pt complexes of di- and trisilylbenzene ligands.

Nagashima and Sunada reported iron disilyl complex **14** which was prepared from photoreaction of a 1:2 mixture of $[(\eta^4\text{-}C_6H_8)Fe(CO)_3]$ and 1,2-bis(dimethylsilyl)benzene (Scheme 2.8).[20] The complex has a distorted octahedral structure with two Fe–Si σ-bonds of a disilyl ligand and two η^2-(H–Si) coordination of the other ligand. In solution, the two disilyl ligands around the Fe center appear to be equivalent by NMR spectroscopy, indicating the hydride transfer is too rapid to distinguish the ligands even at −90 °C. The iron complex **14** catalyzed hydrosilylation of alkenes and C−H functionalization of arenes with high activity, which was attributed to the generation of unsaturated Fe species by dissociation of two weakly coordinated η^2-(H−Si) moieties. The ligand dissociation was caused by addition of CO or 2,6-dithiaheptane, yielding complex **15** in high yields.

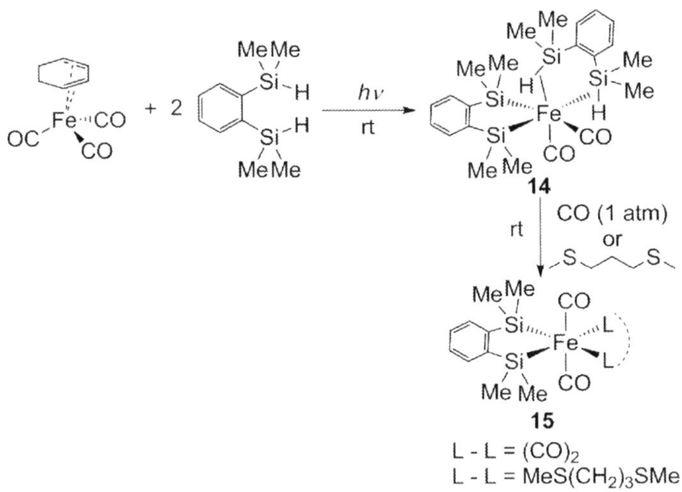

Scheme 2.8
Fe complexes with bis(silyl)benzene ligand.

Stobart developed systematic preparation of multidentate (phosphinoalkyl)silyl ligands, which bonded with late transition metals as the chelating ligand via coordination of the phosphorus atom and oxidative addition of the Si−H bond.[21] The double chelating Pt (II) complexes **16** were obtained from the reactions of (phosphinoethyl)silyl ligands with [Pt(cod)$_2$], [Pt(PPh$_3$)$_4$], or [PtCl$_2$(cod)] in the presence of NEt$_3$ (Scheme 2.9).[22] Ito and Murakami performed the mechanistic study on the Pd-catalyzed bissilylation and silylstannylation of unsaturated compounds and isolated the thermodynamically stable bis (silyl)- and silyl(stannyl)palladium intermediates.[23] Stoichiometric reaction of Pd$_2$(dba)$_3$ · CHCl$_3$ with ^{t}BuMe$_2$SiSnMe$_3$ did not form any products. The disilane and silylstannane with tethered PPh$_2$ groups reacted with the same Pd(0) complex, resulting in quantitative formation of stable Pd(II) complex **17** through the Si−Si and Si−Sn bond cleavage. The recent study of transition metal complexes with bidentate (phosphinoalkyl) silane ligands was reviewed.[24]

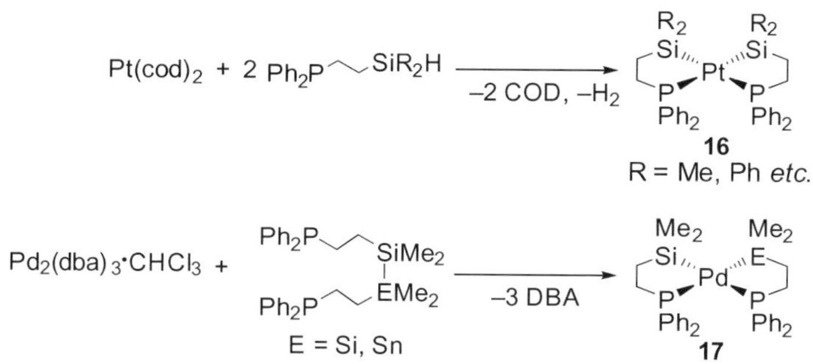

Scheme 2.9
Pd and Pt complexes of P, Si-chelating ligands.

Deng prepared Fe(II) and Co(II) complexes **18** with two strongly donating bidentate ligands, silyl anion and *N*-heterocyclic carbene, denoted as a CSi ligand,[25] which were formed via the formal insertion of silylene into a M–C bond of the ligands (Scheme 2.10). The latter cobalt complexes affect the hydrosilylation of alkenes with high activity and selectivity.

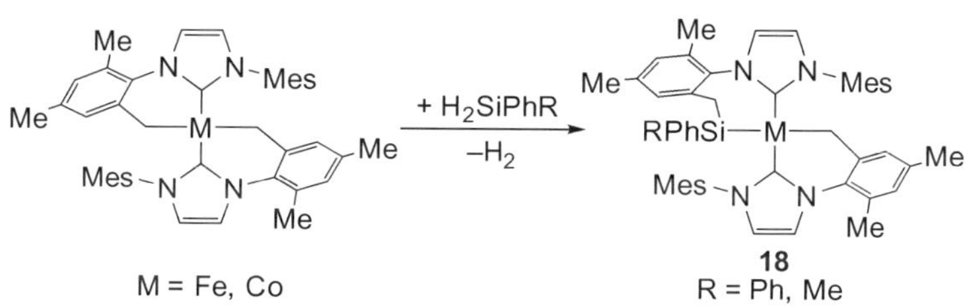

Scheme 2.10
Fe and Co complexes of silylated NHC ligands.

Tri- and Tetradentate Si-Ligands. The tridentate silyl pincer ligands have been actively studied by many groups. The complexes are coordinatively unsaturated metal species and effective for activation of small molecule and inert bond, as well as the catalytic reactions.

Incorporation of silyl groups into ancillary ligands is expected to enhance the catalytic performance of the metal center due to the electron-releasing properties and strong trans effect character. In contrast, the facile cleavage of the M–Si bonds has retarded such an application. In order to improve the properties, bi-, tri-, and tetradentate ligands including Si atom have been developed. There are many examples of the tridentate silyl pincer ligands with five electron donor, namely, PSiP,[26–28] NSiN,[29] and SSiS[30] (Scheme 2.11).

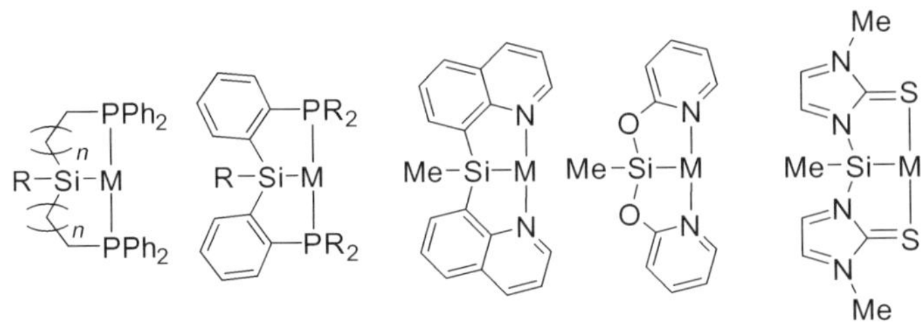

Scheme 2.11
Pincer ligands with coordinating Si atom.

The silyl pincer complexes exhibit remarkable reactivity for the inert bond activation, such as catalytic CO_2 fixation to allenes, boranes, and silanes,[31] and Si—C(sp^2) and Si—C(sp^3) bond-cleavage processes induced by chelating effects.[32] Complexes with the PSiP-pincer ligands show not only the interesting structure but also unique reactivity. Iwasawa reported the first preparation of η^2-(Si—H) Ni(0) and Pd(0) complexes **19** which are regarded as "frozen intermediates" of the oxidative addition of the Si—H bond in the catalytic reactions (Scheme 2.12).[33] A similar complex with more bulky PSiP ligands provided square-planar silyl(hydride) Ni(II) and Pd(II) complexes as the final product of oxidative addition of the Si—H bond. These complexes are believed to be in equilibrium between η^2-(Si—H)M(0) and silyl(hydride) Si—M(II)—H structures, formed via reversible Si—H oxidative addition and dissociation of the phosphine ligand.[31b,34] This structural change was proposed as the crucial step of the catalytic reactions.

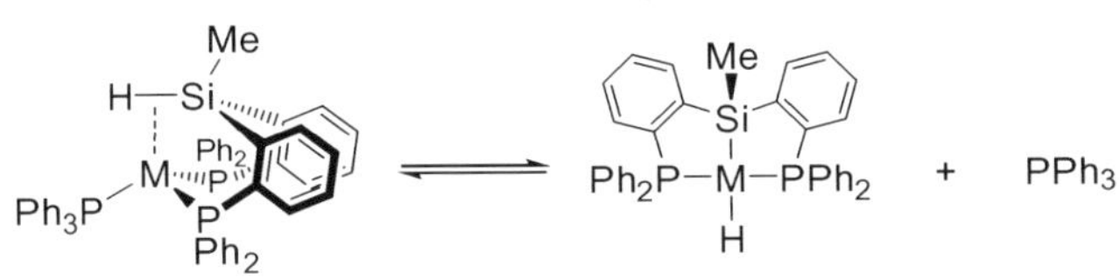

Scheme 2.12
Reaction of complex with a PSiP ligand.

A new PSiN-type mixed-donor pincer, composed of hard (N) and soft (P) donor atoms, was reported by Turculet's group.[35] They prepared a series of Ru, Rh, Ir, Pd, and Pt complexes with the PSiN pincer ligand. Studies involving the Pd and Pt complexes **20** disclosed hemilabile behavior of the amino group by ligand exchange of PMe_3. Treatment of **20′** with BPh_3 led to regeneration of **20**, together with formation of Me_3PBPh_3 (Scheme 2.13). Ozerov also developed a novel tridentate SiNN pincer ligand, composed of amido, quinoline, and silyl donors, and used it to synthesize the unique Ir(IV) complex **21** featuring also two boryl and one hydride ligands, whose structure was largely distorted from ideal octahedral coordination.[36] The Ir SiNN complex catalyzed dehydrogenative C–H(sp) borylation of terminal alkynes (RC≡C–H, R = Ph, Bu, etc.), yielding alkynylboronates (RC≡C–Bpin, pin = pinacolate) in high yield within minutes.

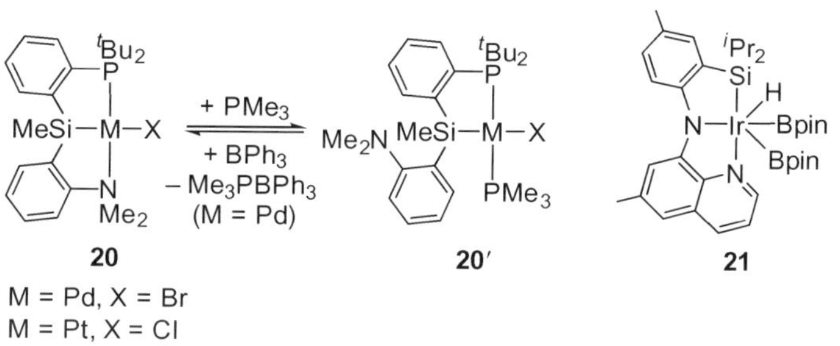

Scheme 2.13
Complexes of PSiN and SiNN ligands.

The SiP_3-type tetradentate scaffold, such as $(Ph_2PCH_2CH_2)_3Si$-, is suitable for formation of trigonal-bipyramidal (TBP) geometry around the metal center, having the three phosphine ligands at the equatorial positions. The five-coordinated Rh(I) complex **22** contained a SiP_3-type ligand with a strong trans-labilizing abilities which caused elongation of the Rh–CO bond (Scheme 2.14).[37,38] Nakazawa prepared similar Rh and Ir complexes **23** with tris(phosphino)silyl ligand ($[EP_3] = [(2\text{-}Ph_2PC_6H_4)_3E]^-$; E = Si, Ge, Sn), which has rigid *o*-phenylene group rather than the flexible ethylene tether.[39] Unno and Takeda reported octahedral Ir(III) complex **24** coordinated by a SiS_3-type ligand, leading to an octahedral configuration including chloride and hydride ligands.[40] The σ-electron donating properties of the Si atom resulted in elongation of the bond trans to the Si atom and provided coordinatively unsaturated complexes.

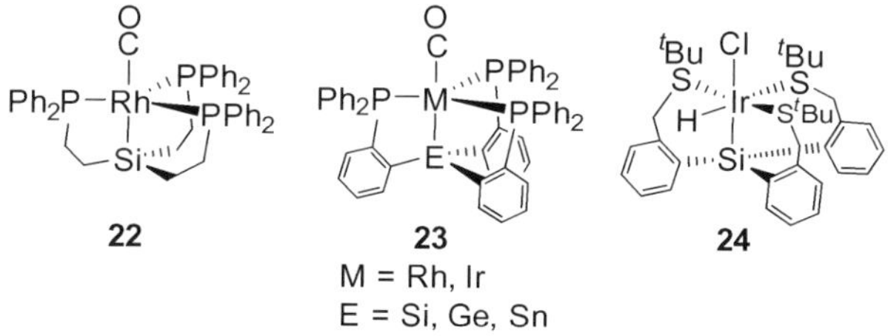

Scheme 2.14
Complexes of SiP_3 and SiS_3 ligands.

Peters and coworkers studied the chemical and magnetic properties of transition metal complexes with tetradentate tris(phosphino)silyl ligands ($[SiP_3] = [(2\text{-}R_2PC_6H_4)_3Si]^-$; R = Ph, ^{i}Pr). The ligands provided four-coordinate, trigonal-pyramidal complexes of a variety of late transition metals (Fe, Ru, Os, Co, Rh, Ir, Ni, Pd, Pt).[41–43] Low-valent iron species have been proposed to play a prominent role as the intermediate of enzymatic hydrogenase activity. The Fe[SiP_3] complexes bonded terminally with N_2 and CO ligands can be accessed in three formal iron oxidation states and spin states **25** (Fe(I), S = 1/2), **26** (Fe(0), S = 0), and **27** (Fe(II), S = 1) by electrochemical and chemical redox reactions (Scheme 2.15). Addition of electrophilic $ClSiMe_3$ and $Me_3Si(OTf)$ to the anionic Fe(0) complexes resulted in silylation of coordinated N_2 and CO ligands at the Fe[SiP_3] scaffold (**28** and **29**).[44]

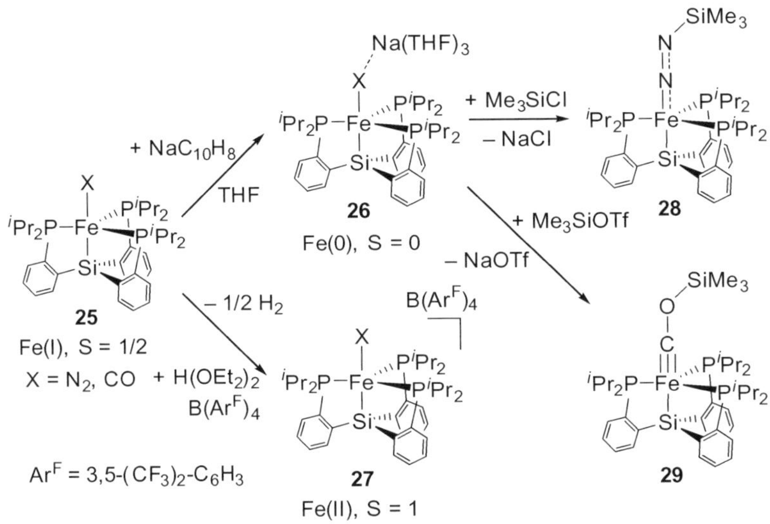

Scheme 2.15
Fe complexes with SiP_3 ligands.

Protonation of $[SiP_3]Pt{-}CH_3$ complex using $H(OEt_2)_2[B(Ar^F)_4]$ ($Ar^F = 3,5\text{-}(CF_3)_2\text{-}C_6H_3$) in toluene provided a toluene-coordinated cationic complex **30** through elimination of CH_4 and accompanying η^2-coordination of para C–H bond of the solvent (Scheme 2.16).[45] Similar protonation of $[SiP_3]Ni{-}CH_3$ complex under an N_2 atmosphere produced end-on coordinated $[SiP_3]Ni{-}N_2$ complex **31**. The strongly electrophilic metal centers induced C–H bond activations and N_2 fixation caused by the donation of these substrates to the empty d_z2 orbital.

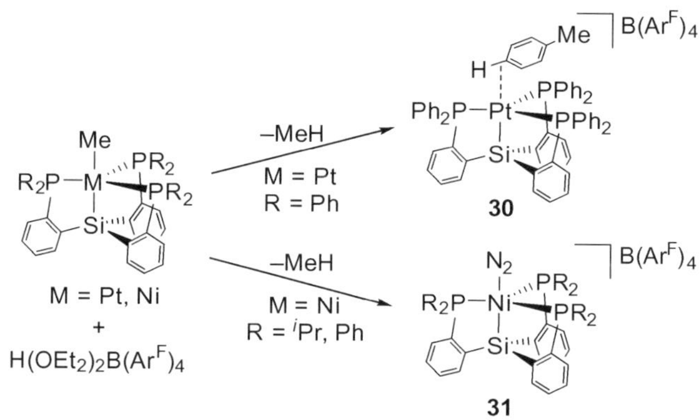

Scheme 2.16
Protonation of Ni and Pt complexes with SiP_3 ligands.

Agostic Interaction. Transition-metal silane σ-complexes,[46] which have the coordination of H–Si bond to a metal center through a three-center-two-electron bond, are of significant current interest due to their proposed intermediacy in catalytic transformation, such as hydrosilylation, dehydrogenative silylation, and dehydrocoupling polymerization. Since the report of the first silane σ-complex in 1969,[47] the activation of Si–H bonds has been the focus of intense research efforts. The majority of silane σ-complexes involves η^2-(H–Si) moieties. The bonding of the arene-coordinated Cr complex $[(\eta^6\text{-}C_6Me_6)Cr(SiHPh_2)(H)(CO)_2]$ (**32**)[48] and the isoelectronic Mn complex $[(\eta^5\text{-}C_5Me_5)Mn(SiHPh_2)(H)(CO)_2]$ (**33**)[49] was compared (Scheme 2.17). The structural and 29Si NMR spectroscopic data indicate stronger Si–H bond interaction of **32** than those of **33**, because of difference in the steric factor of the π-ligands. Therefore, reductive elimination of H_2SiPh_2 from **32**, induced by the addition of phosphine, occurs with a larger rate constant than **33** at the same temperature.

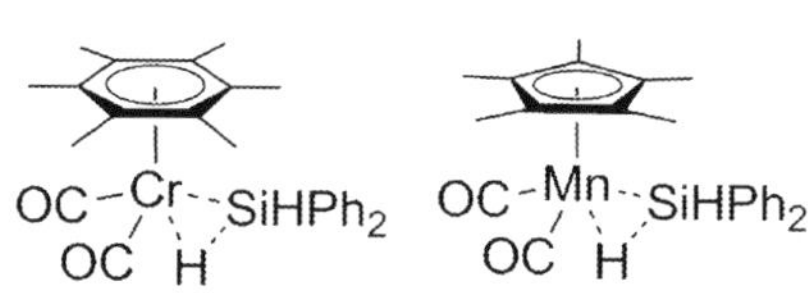

	32	33
M–Si	2.456(1) A	2.395(1) A
J(SiH)	70.8 Hz	65.4 Hz

Scheme 2.17
Cr and Mo complexes with Si–M–H bond.

Hillhouse found that one-electron oxidation of three-coordinate Ni(I) complex with a $SiHMes_2$ group by ferrocenium tetraarylborate $[Cp_2Fe][BAr^F{}_4]$ caused a partial 1,2-H migration from silicon to nickel giving a cationic Ni(II) complex **34** with a 3-center-2-electron (3c-2e) bond between Ni, Si, and bridging H atoms (Scheme 2.18).[50] Dinuclear Ni complexes with a Ni(I)–Ni(I) bond, which were stabilized by chelating naphthyridine-diimine ligand, reacted with H_2SiPh_2 to form an equilibrium mixture of the silane adduct **35** and free silane.[51] The Ni–H–Si bonding arrangement of **35**, formed by the double Si–H oxidative addition, is a similar 3c-2e bond noted in the Hillhouse's complex. These dinickel complexes demonstrated much higher catalytic activity for alkyne hydrosilylations with Ph_2SiH_2 than that of analogous mononuclear Ni catalysts with *N*-chelating ligands.

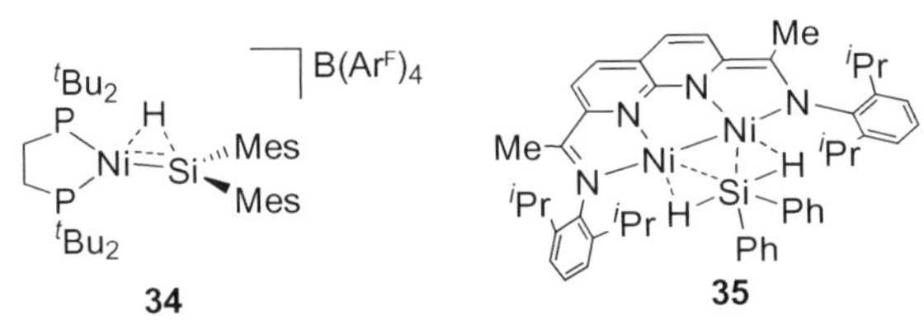

Scheme 2.18
Ni complexes with Si ligands.

The reactions of primary or secondary silanes with transition metals may cause two successive SiH activation reactions to form silylene complexes. The Si–H oxidative addition of an Si–H bond, followed by 1,2-hydrogen migration would form the $M = SiR_2$ species (Scheme 2.19, path A). The η^2-SiH coordination to transition metal is common but the coordination of two Si–H bonds results in two M–H–Si interactions, which is described as η^3-SiH_2 coordination mode (path B).

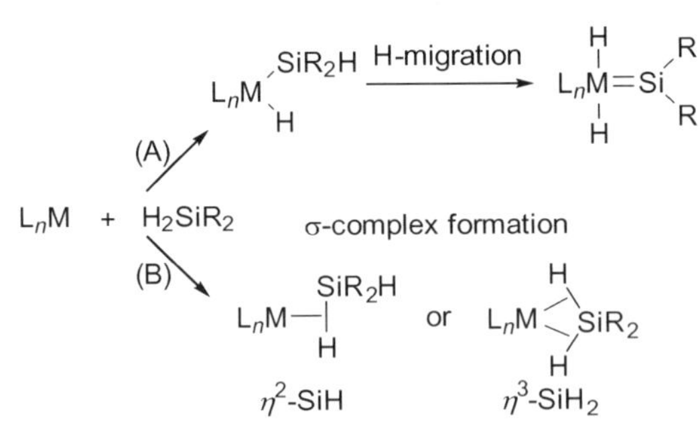

Scheme 2.19
Activation of R_2SiH_2 by metal complexes.

Peters reported the Fe complex with a tripodal tris(phosphino)borate ligand, MeFe[PhB($CH_2P^iPr_2)_3$], which reacted with primary aryl silanes yielding unusual Fe complex **36** with a η^3-H_2SiPhMe ligand via Si–H bond activation and methyl migration from Fe to Si (Scheme 2.20).[52] Very short Fe···Si distance (2.1280(7) Å) is suggested to reflect a potential contribution of the silylene trihydride structure **36′**. Tilley prepared several η^3-silane ruthenium σ-complexes **37**, obtained from the reaction of secondary silanes with [Ru{PhB($CH_2PPh_2)_3$}(μ-Cl)]$_2$. The highly electrophilic Si center interacted with Lewis base to form a hexacoodinate silicon center **38.**[53]

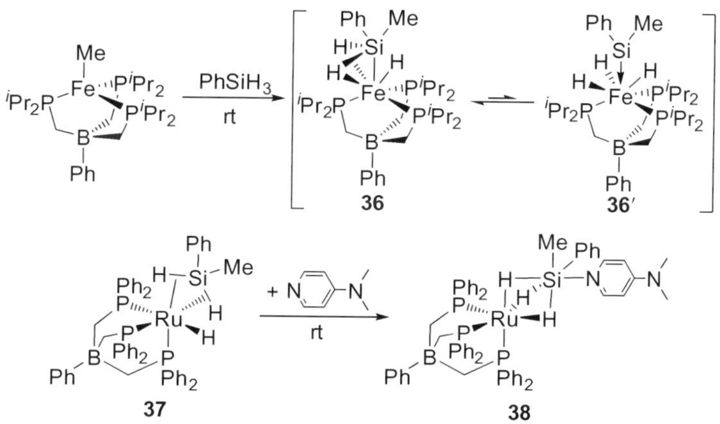

Scheme 2.20
Fe and Ru complexes with silane ligands.

Caulton found unique Si–C(sp^3) bond coordination of the PNP pincer ligands to Ni(II) center (PNP is the tridentate ligand [($^tBu_2PCH_2SiMe_2)_2N]^-$) (Scheme 2.21).[54] Abstraction of the Cl ligand of Ni complex by Na[B($Ar^F)_4$] provided coordinatively unsaturated and highly electrophilic Ni(II) center in **39**, featuring a transannular interaction of the Si–C σ-bond with the metal center. The DFT calculation supported that this Si–C bond donation is energetically preferred to any C–H agostic donation or coordination of any solvents.

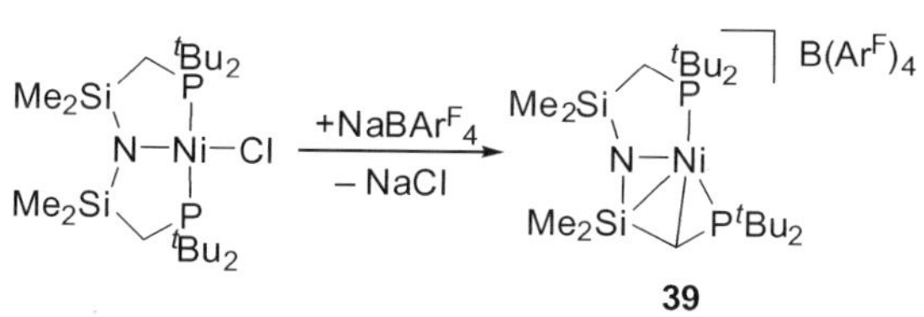

Scheme 2.21
Reaction of Ni–PNP complex.

Bridging Si-ligand. The bridging Si-ligands with metal–metal bonds coordinate to two or three metal centers with several bonding modes. The bridging silylene coordination (μ-SiR_2) is much more stable than that of the mononuclear silylene complexes (:SiR_2) due to effective π-back donation from the metal centers to electrophilic Si atom. The bridging silyl groups with Si–H bonds (SiR_2H or $SiRH_2$) also form a stable coordination via the M–Si single bonds and M–H–Si 3c-2e bonds. The bridging silyl ligands are thermally stable, but they exhibit fluxionality due to the facile cleavage and formation of the Si–H bond. There have been a number of reports of the dinuclear transition metal complexes with the bridging Si-ligands. The chemistry of these complexes has been reviewed.[55] The multinuclear complexes including more than three metals with bridging Si-ligands had been uncommon. The first preparation of a triangular triplatinum complex **40** with strong electron-releasing μ-$SiPh_2$ and PMe_3 ligands, involved the thermal reaction of $[Pt(PMe_3)_4]$ with H_2SiPh_2 in 1:1 ratio.[56] A similar Pt_3 complex bridged by silafluorenyl groups was reported by Braddock-Wilking.[57] Also, in the case of palladium and nickel, multinuclear complexes involving more than three metals were obtained. The Pd_4Si_3 complex **41** has a planar hexagonal core incorporating one Pd center, three Pd atoms, and three silylene ligands.[58] The short Pd_{cent}–Si bond distances of **41** (2.2521(8)–2.2674(8) Å) are comparable to the Pd = Si bond in a mononuclear Pd complex $[Pd\{SiC(SiMe_3)_2CH_2CH_2C(SiMe_3)_2\}_2]$ (2.263(1) Å).[59] Analogous Pt_3 and Pd_4 complexes with μ-germylene ligand were also produced.[60] The apical position of the central Pd atom in the planar tetrapalladium core **42** with disilyl ligands is available for potential coordination of the organic substrates.[61] Johnson reported systematic preparation of tetra- and pentanickel complexes **43** and **44** with a nonclassical 3c-2e bonding interaction of the primary Si–H bonds.[62] The Ni_4Si_3 core was composed of a Ni_3Si_3 ring with a chair conformation and the Ni atom bonded with the six atoms and with a P^iPr_3 ligand at the apical position. A part of the Ni_5Si_4 complex has a similar conformation to the Ni_4Si_3 complex. The central nickel was surrounded by four Ni atoms in a square-planar configuration and the four μ-Si ligands are included within the Ni_5 plane. The folding Pd_{11} nanosheet complex **45** is composed of two Pd_7 sheets with junction of the linear-shaped Pd_3 units which are bridged by two silylyne and four silylene ligands.[63] Although the bridging silylene ligand is five-coordinated similar to those of multinuclear Pd complexes **41** and **42**, the silylyne ligand features an unprecedented seven-coordinated structure having two apical and five equatorial Pd atoms (Scheme 2.22).

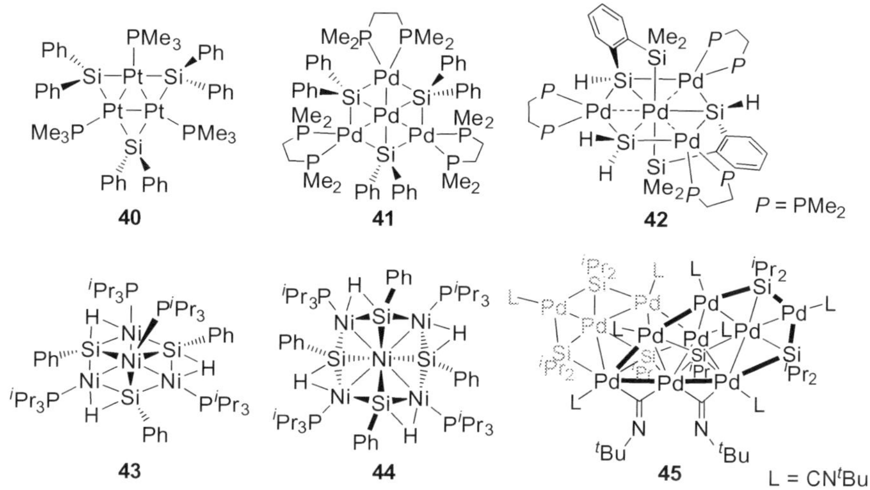

Scheme 2.22
Multinuclear Pt, Pd, and Ni complexes with μ-Si ligands.

2.3 Early Transition Metal Complexes

Metallocenes with Si-Ligands. Complexes of early transition metals with Si-ligands had been much less common until the mid-1980s, but their research took off at the period of discovery of efficient catalysis for dehydrogenative polycondensation of silanes catalyzed by Ti and Zr complexes having cyclopentadienyl (Cp and Cp* = C_5Me_5) ligands.[64] It prompted the study of chemical properties of group 4 transition metals having silyl ligands and auxiliary Cp (or Cp*) ligands.

The use of $Al(SiMe_3)_3(OEt_2)$ as the source of the silyl ligand led to isolation of the silyl complex of Ti(IV), $Cp_2TiCl(SiMe_3)$.[65] Zr and Hf analogs are obtained by a similar procedure, and the produced complexes, $Cp_2M(Cl)(SiMe_3)$ (M = Zr, Hf) **46**, are further transformed to the complexes having a chelating ligand **47** (Scheme 2.23).[65]

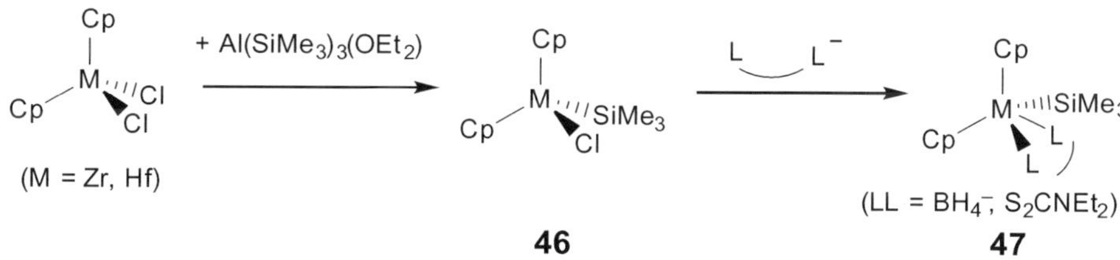

Scheme 2.23
Silyl complexes of Zr and Hf.

The use of potassium silyl anion also provides the group 4 metallocene derivatives having $Si(SiMe_3)_2(NEt_2)$ ligands, such as $Cp_2MCl\{Si(SiMe_3)_2(NEt_2)\}$ (M = Zr, Hf).[66]

Chloro(silyl)zirconium and hafnium complexes with Cp and Cp* ligands **48** undergo exchange of the secondary or tertiary silyl ligand on addition of $PhSiH_3$ (Scheme 2.24).[67]

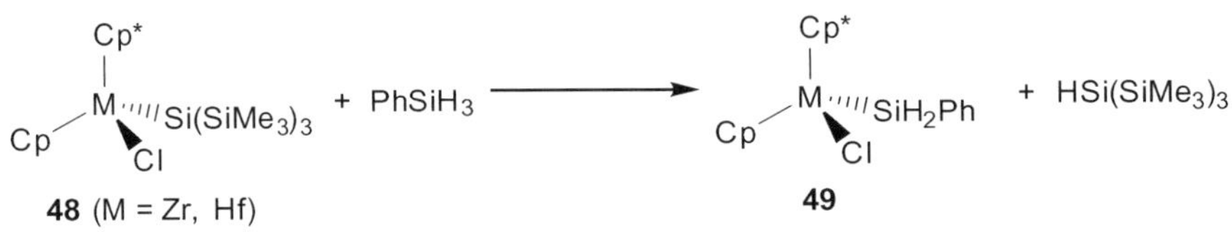

Scheme 2.24
Exchange and desilylation of silyl ligand of Zr and Hf complexes.

Hf complex Cp(Cp*)HfCl(SiH_2Ph) **49** reacts with H_3SiPh to yield both Cp(Cp*)HfCl(H) **50** and Cp(Cp*)HfCl(SiHPh-SiH_2Ph) **51** (Scheme 2.25).[68] Further addition of $PhSiH_3$ to the phenylsilyl complex results in the formation of the complex with higher oligosilyl ligand, Cp(Cp*)HfCl{$(SiHPh)_2SiH_2Ph$} **52** and a mixture of oligosilanes.

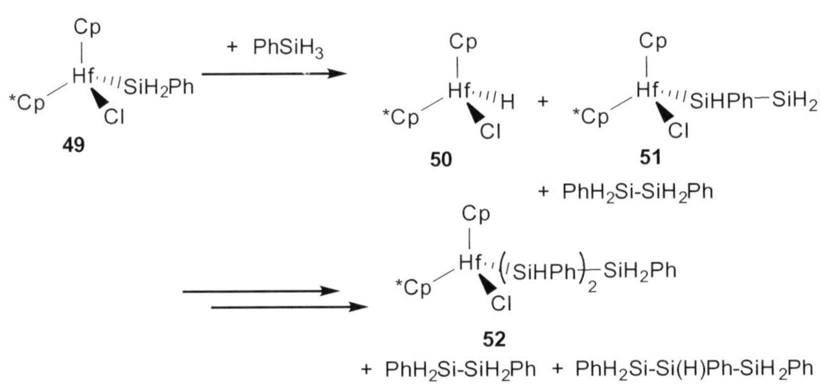

Scheme 2.25
Si–Si bond formation from the silyl ligands of Hf complex.

The mechanism based on the σ-bond metathesis was proposed to account for dehydrocoupling of phenylsilane to form the dimer to tetramer, as well as higher polymers, formulated as H-$(SiHPh)_n$-H, promoted by the Zr and Hf complexes. Details of the reactions as well as results of the kinetic study suggested that the σ-bond metathesis occurs in each Si–Si bond forming step of the reaction.

Silyl(hydride)hafnium complex Cp(Cp*)HfH{$Si(SiMe_3)_3$} was obtained from the reaction of $LiSi(SiMe_3)_3$ with Cp(Cp*)Hf(H)Cl.[69] The former complex undergoes exchange of the silyl ligand on addition of $SiH_2(SiMe_3)_2$ to form Cp(Cp*)HfH{SiH($SiMe_3)_2$}, and reaction of ethylene with the complex affords the five-membered ring metallacycle, Cp(Cp*)Hf{$(CH_2)_4$}.

Cationic silyl-hafnium complex, $[Cp_2Hf\{Si(tBu)_2Ph\}]^+[CH_3B(C_6F_5)_3]^-$ **54** was synthesized by addition of $B(C_6F_5)_3$ to $Cp_2Hf(CH_3)\{Si(tBu)_2Ph\}$ **53** (Scheme 2.26).[70] X-ray crystallography revealed the structure with the methyl group that bridges Hf and B atoms with a large Hf−C−B angle (162.0(9) degrees). The complex having a $SiH(Mes)_2$ ligand, $Cp_2Hf(CH_3)\{SiH(Mes)_2\}$ **55**, also reacts with $B(C_6F_5)_3$ to produce the cationic complex $[Cp_2Hf\{SiH(Mes)_2\}]^+[CH_3B(C_6F_5)_3]^-$ **56**, in which Si−H bond of the silyl ligand is π-bonded with the Hf center. Further addition of benzene or toluene causes σ-bond metathesis of the aromatic and aliphatic C−H bond to release $H_2Si(Mes)_2$ and forms the $[Cp_2Hf(CH_3)Ph][B(C_6F_5)_3]$ **57**.

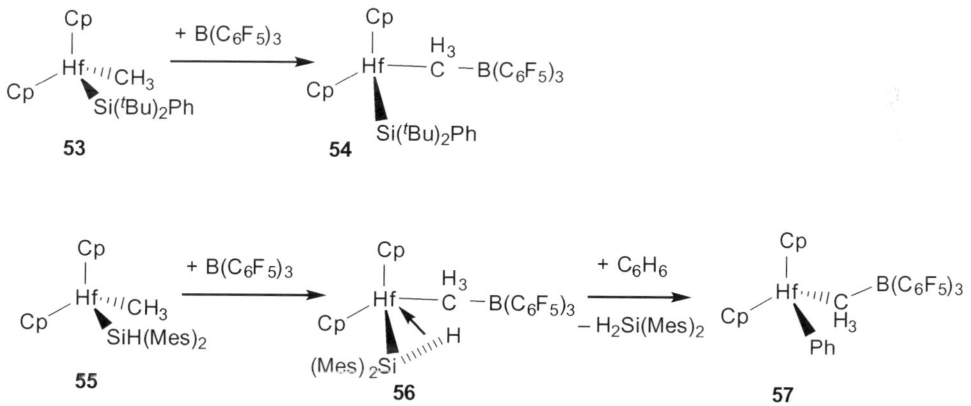

Scheme 2.26
Addition of Lewis acid to Hf complex with silyl ligand.

These results suggest that the cationic complexes undergo smooth σ-bond metathesis resulting in the functionalization of the aromatic and aliphatic hydrocarbons.

Reaction of H_3SiPh with Cp_2TiMe_2 affords dinuclear Ti(III) complex with bridging phenylsilyl ligands, $[Cp_2Ti(\mu\text{-}H{-}Si(H)Ph)]_2$ **58**, having a six-membered ring structure composed of two Ti, two Si, and two H atoms. The product is converted into a mixture of hydrido-bridged complex, $[Cp_2Ti(\mu\text{-}H)(\mu\text{-}H{-}Si(H)Ph)TiCp_2]$ **59** and poly(phenylsilane) (Scheme 2.27).[71] The complexes have two Ti(III) centers and show paramagnetic properties. Analogous reaction in the presence of PMe_3 and PEt_3 forms the mononuclear complex having a silyl ligand, $Cp_2Ti(SiH_2Ph)(PR_3)$ (R = Me, Et).[72] Use of SiH_4 as the source of the Si-ligand provides the complexes having Ti(III) centers which are bridged by SiH_3 ligands, **60** and **61**.[71,73]

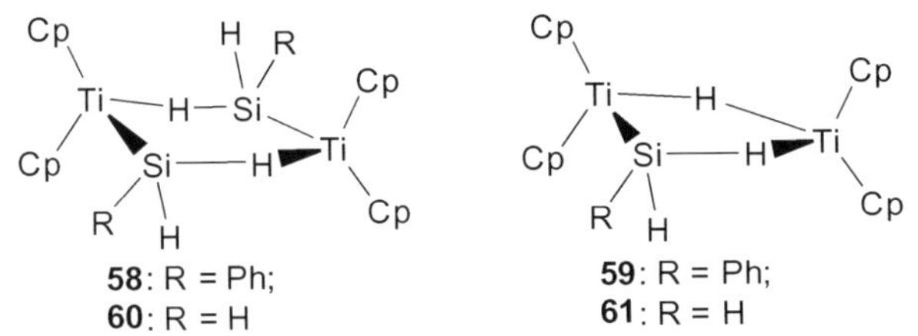

Scheme 2.27
Dinuclear Ti(III) complexes with bridging silyl ligands.

Both complexes **58** and **59** are regarded as the intermediates of the dehydrogenative coupling of primary organosilanes.

Cationic dinuclear Zr(IV) complex, $[Cp_2Zr(\mu\text{-}H{-}Si(H)Ph)]_2^{2+}$ **62**, was synthesized by using Cp_2ZrCl_2, BuLi, $B(C_6F_5)_3$, and H_3SiPh as the starting materials.[74] The reaction of BuLi with Cp_2ZrCl_2 at room temperature and further addition of $B(C_6F_5)_3$ yields Zr(III) complex $[Cp_2Zr]^+$ **63**. Harrod proposed the reactions summarized in Scheme 2.28 form the Si–Si bond under special conditions; dehydrogenative polycondensation of phenylsilane, catalyzed by $Cp_2ZrCl_2/B(C_6F_5)_3/BuLi$, would involve mononuclear Zr(IV) and Zr(III) intermediates.

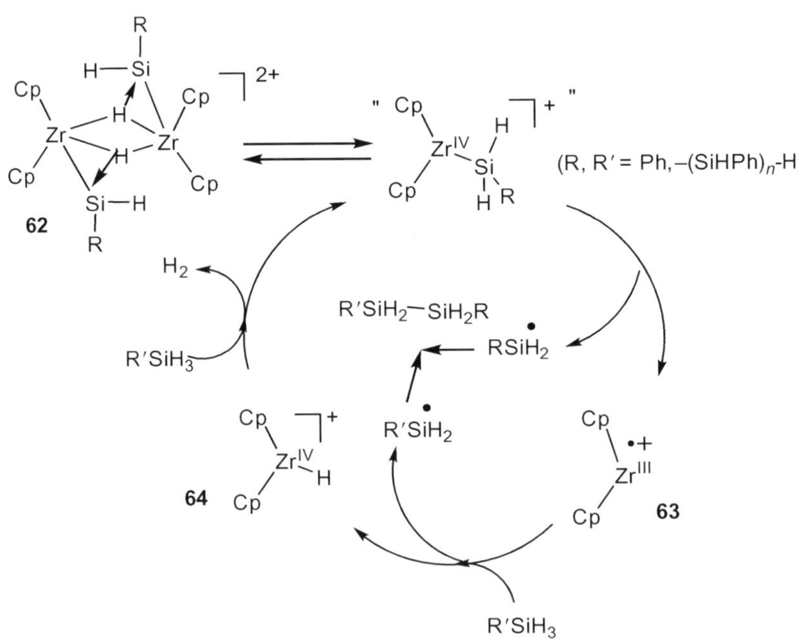

Scheme 2.28
A redox mechanism for dehydrocoupling of organosilanes.

Dinuclear Zr(IV) complex **62** is equilibrated with cationic mononuclear Zr(IV) complex which is split to a silyl radical and Zr(III) complex **63**. Further reaction of $PhSiH_3$ (or primary silane) with **63** affords another silyl radical and hydride zirconium(IV) complex **64**. The resulted complex easily regenerates the silylzirconium(IV) complex upon addition of primary silane. The Si–Si bond formation among $PhSiH_3$ and its oligomers PhH_2Si-$(SiHPh)_n$-SiH_2Ph occurs via coupling of the radical intermediates.

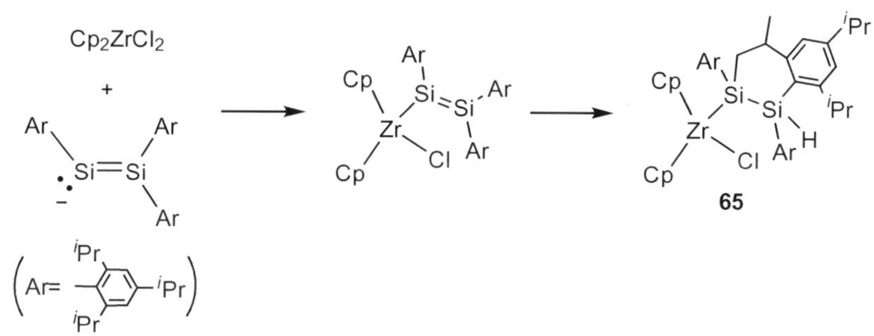

Scheme 2.29
Cyclization of disilenyl complex of Zr.

Addition of lithium salt of disilenide to Cp_2ZrCl_2 forms the zirconocene complex having a silylated vinyl ligand. Intramolecular C–H bond activation of the ligand forms the complex with a cyclic silyl ligand **65** (Scheme 2.29).[75]

Classification of simplified coordination mode of organosilanes is summarized in Scheme 2.30. A silyl ligand having Si–H bonds is able to form the σ-bonded complex as well as σ,π-bonded complex (Scheme 2.30A). The silane ligand also has two possible coordination modes, hydride(silyl) complexes with two σ-bonds, and a π-bonding between the Si–H group and the metal center (Scheme 2.30B).

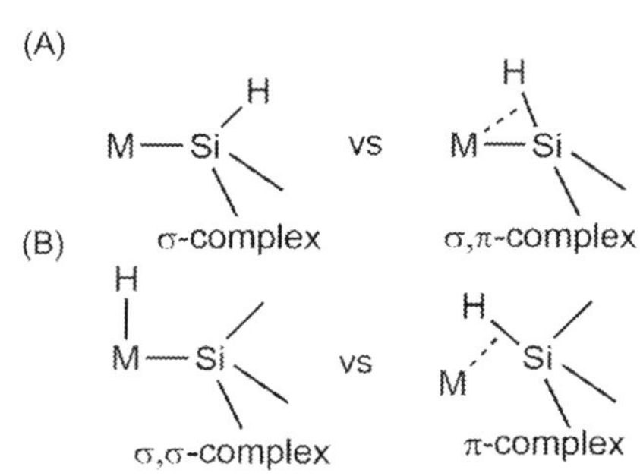

Scheme 2.30
Possible coordination modes of silyl and hydride ligands.

The preference of a coordination mode among those shown above highly depends on the nature of metal and auxiliary ligands. Such π-bonding of the Si–H group is observed in the Ti–silane complex, $Cp_2Ti(\eta^2\text{-}H{-}SiH_2Ph)(PMe_3)$, in which an Si–H group is coordinated with Ti center to form a Ti–Si–H triangle with Ti–H and Ti–Si distances of 1.81 and 2.60 Å, respectively.[76] Relative stability of the complex having Si–H–M bonding compared to that with a hydride and a silyl ligand vary depending on the substituents on the Si atom and auxiliary ligands. Zirconocene and hafnocene complexes with hydride and Ph_3Si ligands were characterized as the hydride complexes lacking any nonclassical coordination.[77] Titanocene complexes with PMe_3 coligand were prepared, and both the silyl-hydride complex **67** and Si–H–Ti complex **68** were obtained (Scheme 2.31).[78]

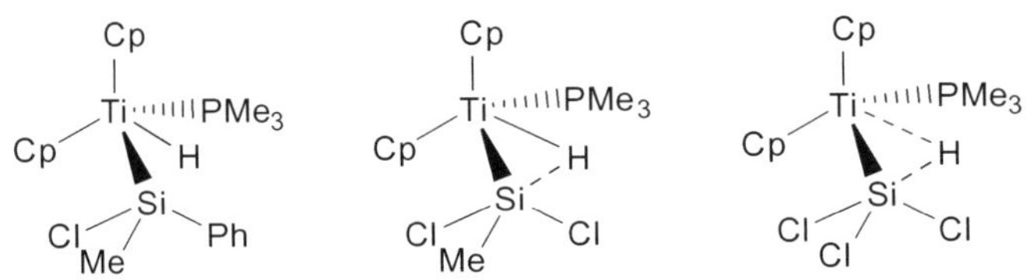

66 67 68

Scheme 2.31
Detailed coordination modes of silane and silyl ligands.

Use of chloro(methyl)phenylsilane affords complex **66** having both hydride and silyl ligands with no interaction between them. The complex with the dichloro(methyl)silane ligand involves interaction between the coordinated silicon atom and hydride ligand **67**. Complex **68** contains a trichlorosilane ligand, which is coordinated to the metal center via σ-bond of the silyl ligand and π-coordination of the Si–H group to the metal center. DFT calculations of the two latter complexes agree well with the proposed coordination of the silyl ligand. Detailed molecular orbital calculation study and the experimental results revealed that the back-bonding character of the coordination is enhanced by electron-withdrawing substituents at the Si atom.[79]

Titanocene complex of alkynyl silane **69** was also studied, both experimentally and computationally. The alkynyl silane has *C, C, Si, H*-η^4-coordination mode to the Ti center (Scheme 2.32). The bonding nature in the complex was monitored by spectroscopy, and ^{29}Si NMR and ^{1}H NMR (Si–H) signals were observed at −0.5 and −3.74 ppm at 303 K, while cooling the solution to 193 K caused a shift of these signals to 17.6 and −7.32 ppm, respectively.

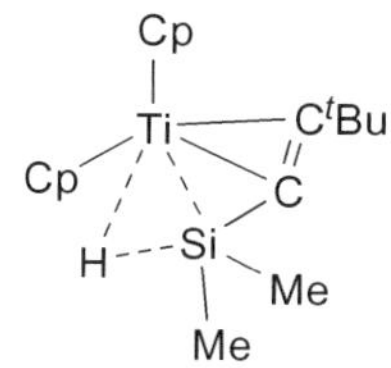

69

Scheme 2.32
Coordination of Ti to the vinylsilane ligand.

It was therefore suggested that the Ti–Si–H interaction mode varies depending on the temperature.[80]

Metallocenes of group 5 transition metals, Nb and Ta, also form silyl complexes. Trihydride complexes of Nb and Ta react with $PhMe_2SiH$ to yield the silyl complex via replacement of a hydride with the silyl group (Scheme 2.33).[81]

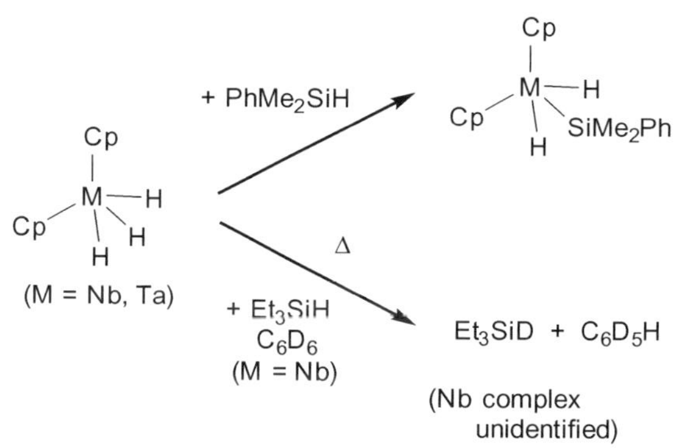

Scheme 2.33
Exchange of the hydride ligand with the silyl ligand.

Hydrogen–deuterium exchange was noted in the reaction using Et_3SiH at high temperature, suggesting that C–H (or C–D) and Si–H scrambling takes place at the Nb center. Similar bis(hydride)silyl complexes of niobocene were reported with the (trimethylsilyl) cyclopentadienyl group as the auxiliary ligands.[82]

Bis(silyl)hydride complexes of niobocene were prepared by the reaction of organosilanes with a Nb(III) complex, $Cp_2Nb(C_2H_4Ph)$.[83] Scheme 2.34 compares the coordination and structures of the complexes. The hydride ligand of complex **70** is located between the two Nb–Si bonds. The complex with $SiMe_2Cl$ ligands was characterized by X-ray crystallography and NMR spectroscopy, and also by neutron diffraction and detailed relaxation time measurement, which suggested contribution of interligand hypervalent interactions.[84] The complex having two different silyl ligands **71** and that with a chelating Si–N–Si ligand **72** were also synthesized and studied in detail, and the latter was shown to be free from the intramolecular hypervalent interactions.[85]

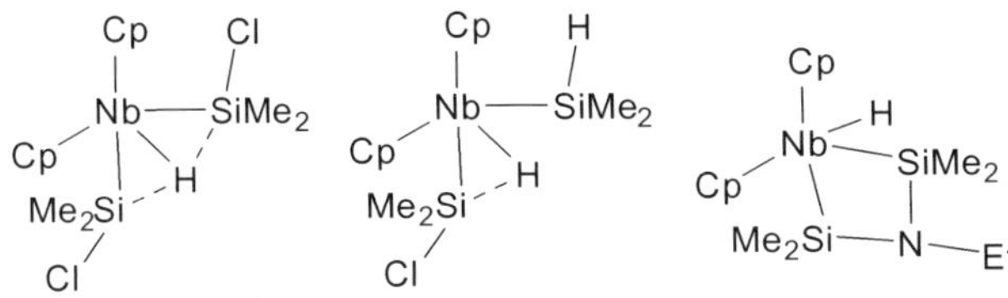

70 **71** **72**

Scheme 2.34
Coordination of Nb to organosilanes.

Tantalum complexes containing mono- and bis-silyl ligands as well as silyl complexes of Nb(IV) and Nb(III) were characterized.[86,87] Pentamethyldisilane and its methoxy derivative react with $Cp_2TaH(CH_2=CHCH_3)$ to yield the complex having a hydride and two silyl ligands **73, 74** and/or base-stabilized bis(silylene)tantalum complex **75**[88] (Scheme 2.35).

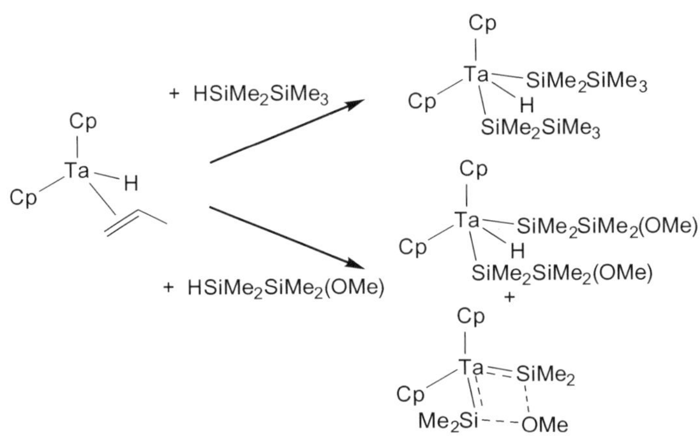

73
74
75

Scheme 2.35
Addition of disilanes to Ta complex.

Persilyl metallacycles of early transition metals have been reported.[89] Discovery of dipotassium derivatives of oligosilanes enabled their systematic synthesis. Oligosilanes capped with terminal tris(trimethylsilyl)silyl groups can be converted to the corresponding dianions via reductive desilylation (Scheme 2.36).[90] Reaction of the dianion, $[(Me_3Si)_2SiSiMe_2SiMe_2Si(SiMe_3)_2]^{2-}\cdot 2K^+$, with Cp_2MCl_2 (M = Zr, Hf) yields the corresponding five-membered ring metallacycle **76**.

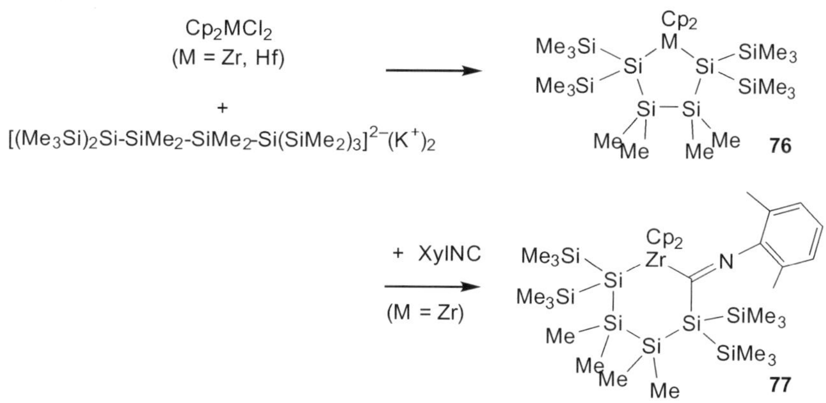

Scheme 2.36
Persilazirconacycle and its chemical properties.

The persilazirconacyclopentane undergoes insertion of nitrile into a Zr–Si bond to form the six-membered ring metallacycle containing Si and C as the coordinating atoms **77**. Reaction of silyl anions with potassium countercation with Cp_2ZrCl_2 and Cp_2HfCl_2 yielded the corresponding bis(silyl) complexes,[91] while bis(silyl)acetylene, having a linear structure, yielded dinuclear Zr complex rather than the metallacycle.[92] Dianion of 1,1′-bis(silyl) ferrocene also functions as the ligand precursor of Zr complex with the chelating bis(silyl) ligand.[93]

The analogous reaction of the dipotassium salts of the oligosilanes with Cp_2TiCl_2 does not yield the neutral titanacycle but the titanate **78**, which is isolated with potassium countercation complexed with the crown ether molecule (Scheme 2.37). Accompanied formation of a persilacyclobutane **79** was also noted.[94]

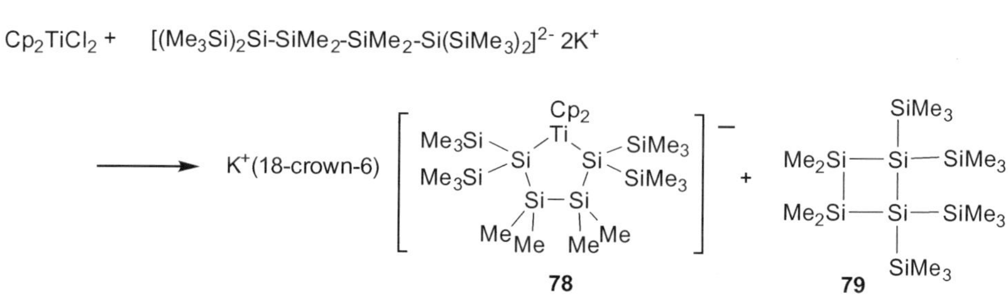

Scheme 2.37
Persilatitanacyclopentane.

1,2-Dianion of the disilane derivative reacts with Cp_2HfCl_2 to yield the isolable complex, formulated as $Cp_2Hf[(Me_3Si)_2Si\text{-}Si(SiMe_3)_2]$.[95] NMR and DFT studies revealed that the bond order between the Si atoms is less than unity, and that the Si–Si bonding has a character intermediate between the disilene ligand and bis(silylene) ligand, as shown in Scheme 2.38. It is contrasted with the bonding of the alkene complex of the same metal, whose bonding character is intermediate between the metalacycle and π-complex.

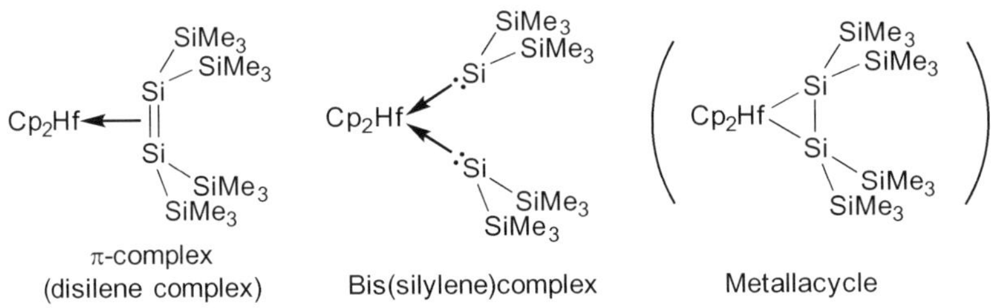

Scheme 2.38
Canonical structures of bis(silylene) complex of Hf.

Three-membered ring metallacyles, composed of Zr, Si, and N atoms **80**, were prepared and characterized (Scheme 2.39).[96]

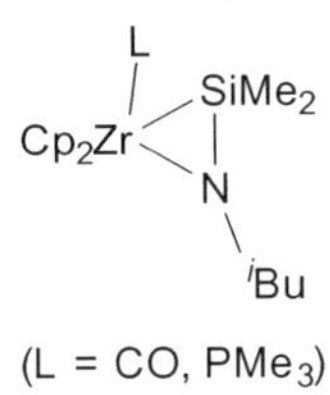

80

Scheme 2.39
Silazane complex of Zr.

Half-metallocenes and nonmetallocenes with Si-ligands. Half-zirconocenes and -hafnocenes having a piano stool structure were prepared using silyl anion, and were stabilized by coexistence of the imido ligand which functions as a three-electron donor.[97,98] Zirconium (IV) complex with permethylcyclopentadienyl and bulky silyl ligand, **81**, was reported (Scheme 2.40).

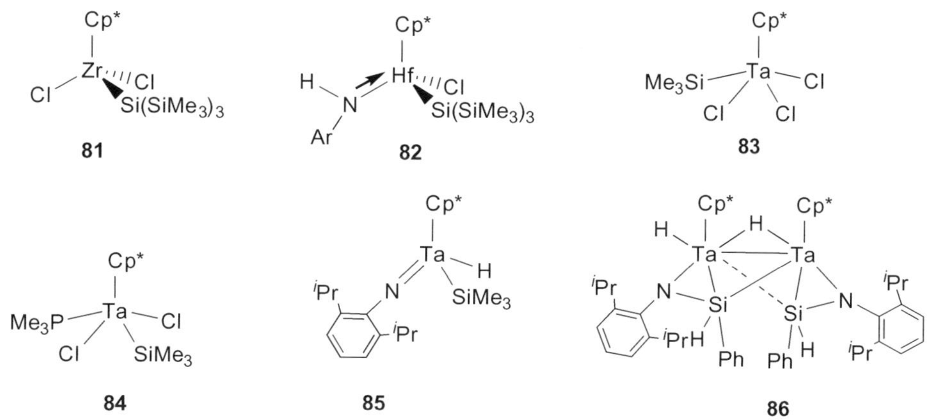

Scheme 2.40
Mono- and dinuclear complexes with silyl ligands.

Analogous mononuclear tantalum complexes with a pentamethylcyclopentadienyl ligand, $Cp^*Ta(SiMe_3)Cl_3$ **83** and $Cp^*Ta(SiMe_3)Cl_2(PMe_3)$ **84**, were isolated, and the complexes with auxiliary imido ligands were obtained as the mononuclear and dinuclear complexes **85** and **86**.[98]

Half-sandwich complexes of Nb and Ta with the hydride and silyl ligands have the issue of interligand interaction, similar to that of the niobocene derivatives. Nb and Ta complexes with imide and PMe_3 ligands react with $HSiClMe_2$ to form different products, complexes **87** and **88**, depending on the central metal (Scheme 2.41).[99] Detailed NMR studies revealed that the interaction does not directly correlate with the J(Si–H) values.[100]

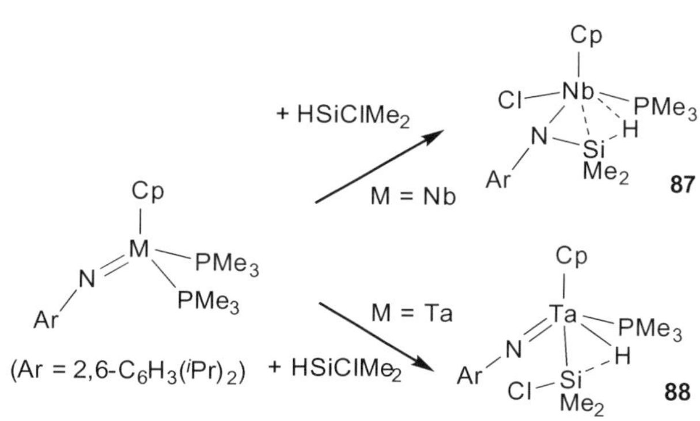

Scheme 2.41
Coordination of Ta and Nb to Si-compounds.

Vanadium is a coordinatively labile metal and the number of its isolated complexes is smaller compared to other transition metals, but half-sandwich complexes of V with silyl ligand **89** were reported (Scheme 2.42).[101] Methyl ligand attached to the cyclopentadienyl vanadium complex is replaced by silyl ligands on addition of the silanes.

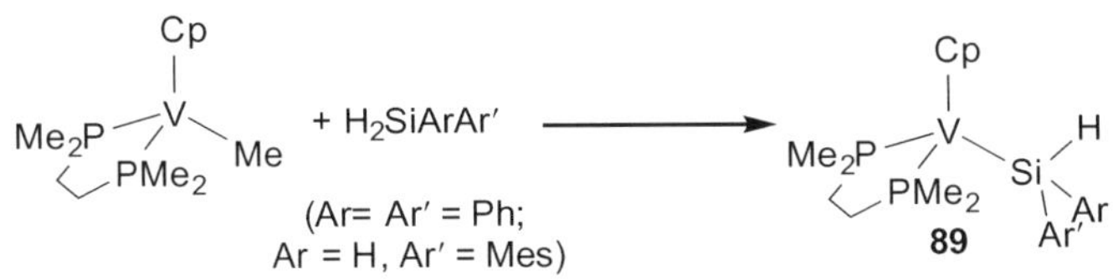

Scheme 2.42
Vanadium complex with a silyl ligand.

N-Heterocyclic carbene-stabilized silylene and amidinate-stabilized silylene are used as precursors of the silyl ligand to stabilize the V cyclopentadienyl complexes **90** and **91** (Scheme 2.43).[102]

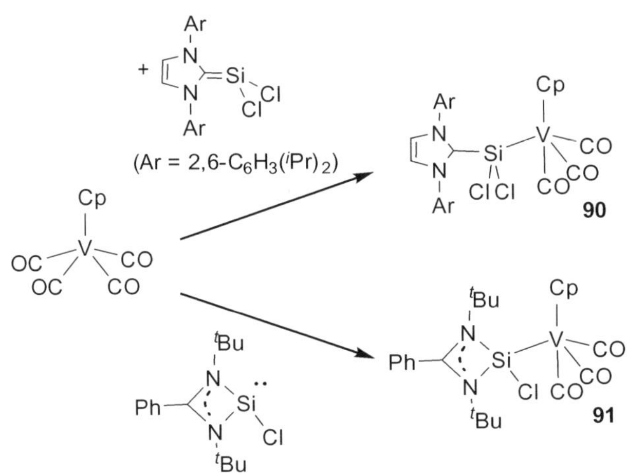

Scheme 2.43
Silylene coordination to the V complex.

The latter ligand was used to prepare titanocene derivatives with two silyl ligands.[103] Silyl complexes of group 4 transition metals without Cp-ligands were also prepared (Scheme 2.44). Reaction of silyl anions with halo complexes of these metals formed the silyl complexes with auxiliary alkyl ligands **92**, with alkoxy ligands **93**, and amide ligands **94**.[104]

Reaction of $K^+[Si(SiMe_3)_3]^-$ with $HfCl_4$(tmen) (tmen = *N,N,N′,N′*-tetramethylethylenediamine) forms the complex with a silyl ligand, $HfCl_3[Si(SiMe_3)_3]$

(tmen).[105] Bulky electron-donating *t*-butoxy ligands stabilize the Zr complex with Si $(SiMe_3)_3$ ligand.[106]

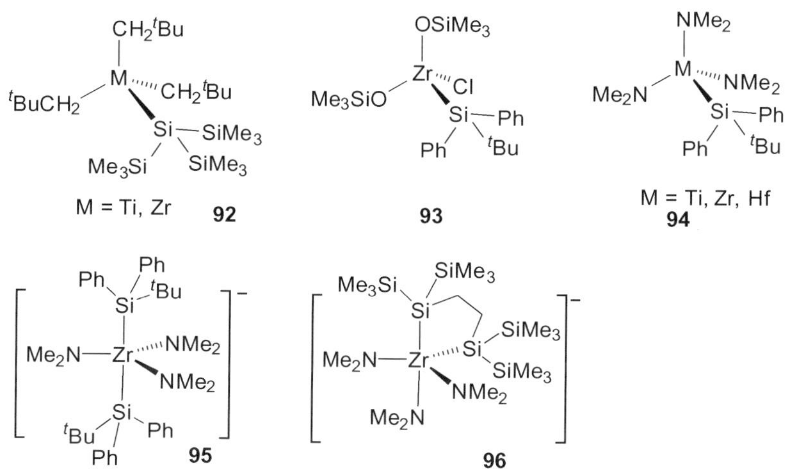

Scheme 2.44
Ti, Zr, Hf–Si complexes with verified structures.

The reaction of phenylsilane with $Me_2Ti(dmpe)_2$ (dmpe = 1,2-bis(dimethylphosphino) ethane) forms the Ti(0) complex with a trimer of phenylsilane as the ligand **97**. Si–H bonds of the terminal silyl groups are each coordinated to the titanium center via π-interaction.[107] The central Si–H bond reacts further with another equivalent of $PhSiH_3$ to form the branched tetrasilane as the ligand **98** (Scheme 2.45).

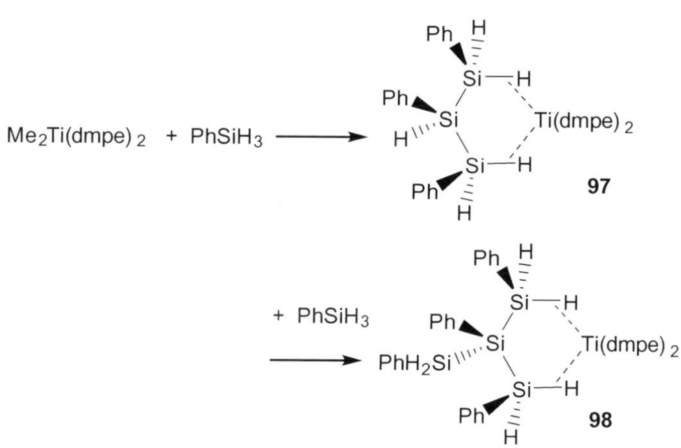

Scheme 2.45
Chelating coordination of trisilane in the Ti complex.

Other nonmetallocene complexes of Zr, Hf, and Ta were also reported.[108]

2.4 Conclusions

Transition metal silylene complexes as well as σ-bond metathesis of the complexes with organosilane ligands have been major research topics in organosilicon-transition metal chemistry. In the past few decades, nonclassical or bridging coordination of silyl ligands to the transition metals have been reported, as described in this chapter. Some of the new coordination bonds are unique to the silyl ligands, and that will motivate not only inorganic chemistry, but organic and catalytic chemistry as well.

References

1. (a) Speier, J. L. Homogenous Catalysis of Hydrosilation by Transition Metals. *Adv. Organomet. Chem.* **1979,** *17*, 407–447.
   (b) Troegel, D.; Stohrer, J. Recent Advances and Actual Challenges in Late Transition Metal Catalyzed Hydrosilylation of Olefins From an Industrial Point of View. *Coord. Chem. Rev.* **2011,** *255*, 1440–1459.
   (c) Roy, A. K. A Review of Recent Progress in Catalyzed Homogeneous Hydrosilation (Hydrosilylation). *Adv. Organomet. Chem.* **2007,** *55*, 1–59.
2. (a) Tilley, T. D. The Coordination Polymerization of Silanes to Polysilanes by a "σ-Bond Metathesis" Mechanism. Implications for Linear Chain Growth. *Acc. Chem. Res.* **1993,** *26*, 22–29.
   (b) Gauvin, F.; Harrod, J. F.; Woo, H.-G. Catalytic Dehydrocoupling: A General Strategy for the Formation of Element–Element Bonds. *Adv. Organomet. Chem.* **1998,** *42*, 363–405.
   (c) Corey, J. Y. Dehydrocoupling of Hydrosilanes to Polysilanes and Silicon Oligomers: A 30 Year Overview. *Adv. Organomet. Chem.* **2004,** *51*, 1–52.
3. Curtis, M. D.; Epstein, P. S. Redistribution Reactions on Silicon Catalyzed by Transition Metal Complexes. *Adv. Organomet. Chem.* **1981,** *19*, 213–255.
4. (a) Tilley, T. D. In *The Chemistry of Organic Silicon Compounds;* Patai, S., Rappoport, Z., Eds.; John Wiley & Sons: New York, 1989; p 1415. Chapter 24.
   (b) Tilley, T. D. In *The Silicon-Heteroatom Bond;* Patai, S., Rappoport, Z., Eds.; John Wiley & Sons: New York, 1991. Chapters 9 and 10, pp. 245 and 309.
5. (a) Corey, J. Y.; Braddock-Wilking, J. Reactions of Hydrosilanes With Transition-Metal Complexes: Formation of Stable Transition-Metal Silyl Compounds. *Chem. Rev.* **1999,** *99*, 175–292.
   (b) Corey, J. Y. Reactions of Hydrosilanes With Transition Metal Complexes and Characterization of the Products. *Chem. Rev.* **2011,** *111*, 863–1071.
6. (a) Jetz, W.; Graham, W. A. G. Chlorosilyl Derivatives of Transition Metals and Evidence for a Conformational Effect on Metal Carbonyl Stretching Bands. *J. Am. Chem. Soc.* **1967,** *89*, 2773–2775.
   (b) Jetz, W.; Graham, W. A. G. Trichlorosilyl Hydrides of Transition Metals. *J. Am. Chem. Soc.* **1969,** *91*, 3375–3376.
7. Ebsworth, E. A. V.; Marganian, V. M.; Reed, F. J. S.; Gould, R. O. Platinum Hydrides Containing Silyl or Germyl Ligands. Crystal Structure of *trans*-Hydridosilylbis(tricyclohexylphosphine)platinum(II). *J. Chem. Soc., Dalton Trans.* **1978,** 1167–1170.
8. Piper, T. S.; Lemal, D.; Wilkinson, G. A Silyliron Compound: An Fe–Si δ Bond. *Naturwissenschaften* **1956,** *43*, 129.
9. Kawano, Y.; Tobita, H.; Ogino, H. Photoreactions of Silyliron(II) Complexes $Cp^*Fe(CO)_2SiMe_3$ ($Cp^* = \eta^5$-C_5H_5, η^5-C_5Me_5) in the Presence of Trihydrosilanes. *J. Organomet. Chem.* **1992,** *428*, 125–143.

10. Luo, X.-L.; Baudry, D.; Boydell, P.; Charpin, P.; Nierlich, M.; Ephritikhine, M.; Crabtree, R. H. Reaction of Heptahydridobis(triphenylphosphine)rhenium With Silanes: Preparation and Characterization of the First Silyl Polyhydride Complexes, $ReH_6(SiR_3)(PPh_3)_2$ ($SiR_3 = SiPh_3$, $SiEt_3$, $SiHEt_2$). *Inorg. Chem.* **1990,** *29*, 1511–1517.
11. Osakada, K.; Koizumi, T.; Yamamoto, T. A New Series of Mono- and Dinuclear Hydridosilylrhodium(III) Complexes, $RhCl(H)(SiAr_3)L_2$ and $RhL(SiAr_3)H(\mu\text{-}Cl)(\mu\text{-}H)RhH(SiAr_3)L$ ($L = P(i\text{-}Pr_3)$). Preparation by Oxidative Addition of $HSiAr_3$ and Molecular Structures of the Complexes. *Organometallics* **1997,** *16*, 2063–2069.
12. Lappert, M. F.; Speier, G. Binuclear Organometallic Compounds. VIII. Oxidative Addition of Si–H or Si–Cl Compounds to Low Valent Fe, Co, or Ni Complexes Stabilised by Mono- or Bi-dentate Phosphines. *J. Organomet. Chem.* **1974,** *80*, 329–339.
13. (a) Kiso, Y.; Tamao, K.; Kumada, M. Silicon Hydrides and Nickel Complexes IV. Preparation of Silicon–Nickel Complexes by the Reaction of Silicon Hydrides With α-Alkyl–Nickel Complexes. *J. Organomet. Chem.* **1974,** *76*, 95–103.
    (b) Bierschenk, T. R.; Guerra, M. A.; Juhlke, T. J.; Larson, S. B.; Lagow, R. J. Synthesis of Trifluorosilyl Organometallic Complexes From Trifluorosilyl Radicals and Metal Atoms. *J. Am. Chem. Soc.* **1987,** *109*, 4855–4860.
    (c) Janikowski, S. K.; Radonovich, L. J.; Groshens, T. J.; Klabunde, K. J. Tricarbonylbis(trichlorosilyl) nickel. An Unusual Carbonyl Complex of Nickel(II). *Organometallics* **1985,** *4*, 396–398.
14. Zell, T.; Schaub, T.; Radacki, K.; Radius, U. Si–H Activation of Hydrosilanes Leading to Hydrido Silyl and Bis(silyl) Nickel Complexes. *Dalton Trans.* **2011,** *40*, 1852–1854.
15. Tanabe, M.; Osakada, K. Sila- and Germametallacycles of Late Transition Metals. *Organometallics* **2010,** *29*, 4702–4710.
16. Yamashita, H.; Tanaka, M. Transition Metal-Catalyzed Synthesis of Silicon Polymers. *Bull. Chem. Soc. Jpn.* **1995,** *68*, 403–419.
17. Shimada, S.; Tanaka, M. Group 10 Transition-Metal Complexes With Metal–Silicon Bonds Derived From 1,2-Disilylbenzenes and Bis(2-silylphenyl)silane. *Coord. Chem. Rev.* **2006,** *250*, 991–1011.
18. (a) Shimada, S.; Tanaka, M.; Shiro, M. The First Silylpalladium(IV) Complex. *Angew. Chem., Int. Ed. Engl.* **1996,** *35*, 1856–1858.
    (b) Shimada, S.; Rao, M. L. N.; Tanaka, M. Reaction of 1,2-Disilylbenzene With Bis[1,2-bis (dimethylphosphino)ethane]nickel(0). Isolation and Characterization of the First Silylnickel(IV) Complex. *Organometallics* **1999,** *18*, 291–293.
    (c) Shimada, S.; Tanaka, M.; Honda, K. Unusual Reactivity of 1,2-Disilylbenzene Toward Pt(0) Complexes. Isolation of the First $Pt^{IV}Si_4P_2$ and Dinuclear, Mixed-Valence $Pt^{II}Pt^{IV}Si_4P_4$ Complexes. *J. Am. Chem. Soc.* **1995,** *117*, 8289–8290.
19. Chen, W.; Shimada, S.; Tanaka, M.; Kobayashi, Y.; Saigo, K. Reaction of $[2\text{-}(SiH_3)C_6H_4]_2SiH_2$ With $Ni(Et_2PCH_2CH_2PEt_2)(PEt_3)_2$: Characterization of η^2-(Si – H)Ni and Ni^{IV} – H Complexes. *J. Am. Chem. Soc.* **2004,** *126*, 8072–8073.
20. (a) Sunada, Y.; Tsutsumi, H.; Shigeta, K.; Yoshida, R.; Hashimoto, T.; Nagashima, H. Catalyst Design for Iron-Promoted Reductions: An Iron Disilyl-Dicarbonyl Complex Bearing Weakly Coordinating η^2-(H–Si) Moieties. *Dalton Trans.* **2013,** *42*, 16687–16692.
    (b) Sunada, Y.; Soejima, H.; Nagashima, H. Disilaferracycle Dicarbonyl Complex Containing Weakly Coordinated η^2-(H–Si) Ligands: Application to C–H Functionalization of Indoles and Arenes. *Organometallics* **2014,** *33*, 5936–5939.
21. (a) Auburn, M. J.; Holmes-Smith, R. D.; Stobart, S. R.; Zaworotko, M. J.; Cameron, T. S.; Kumari, A. The Phosphinomethylsilyl Group as a Bifunctional Bridging Ligand. *X*-Ray Crystal Structure of Hexacarbonyl-bis(μ-diphenylphosphino-methyldimethylsilyl)diruthenium(II), and of its Reaction Product With Trifluoroacetic Acid, a Mononuclear Ruthenium(II) Complex Incorporating a Unique Co-orinated Silanol. *J. Chem. Soc., Chem. Commun.* **1983,** 1523–1525.

(b) Auburn, M. J.; Holmes-Smith, R. D.; Stobart, S. R. (Phosphinoalkyl)silyl Complexes. 3. "Chelate-Assisted" Hydrosilylation: Formation of Enantiomeric and Diastereoisomeric Iridium(III) Complexes With Chelating (Phosphinoethyl)silyl Ligands. *J. Am. Chem. Soc.* **1984,** *106*, 1314–1318.

(c) Auburn, M. J.; Grundy, S. L.; Stobart, S. R.; Zaworotko, M. J. Phosphinoalkylsilyl Complexes. 6. Isolation of a Silyl Complex of Iridium(I). Crystal and Molecular Structure of Dicarbonyl (triphenylphosphine) [[(Diphenylphosphino)ethyl]dimethylsilyl]iridium. *J. Am. Chem. Soc.* **1985,** *107*, 266–267.

22. Holmes-Smith, R. D.; Stobart, S. R.; Cameron, T. S.; Jochem, K. The Phosphinoethylsilyl Group as a Bifunctional Chelate. Asymmetric Induction at Platinum(II). *J. Chem. Soc., Chem. Commun.* **1981,** 937–939.

23. (a) Murakami, M.; Yoshida, T.; Ito, Y. Syntheses and Interconversion of [Bis(silyl)alkene]palladium(0) and Bis(silyl)palladium(II) Complexes: A Mechanistic Model for Palladium-Catalyzed Bis-Silylation. *Organometallics* **1994,** *13*, 2900–2902.

(b) Murakami, M.; Yoshida, T.; Kawanami, S.; Ito, Y. Synthesis, Structure, and Reaction of the First Thermally Stable *cis*-(Silyl)(stannyl)palladium(II) Complex. *J. Am. Chem. Soc.* **1995,** *117*, 6408–6409.

24. Okazaki, M.; Ohshitanai, S.; Iwata, M.; Tobita, H.; Ogino, H. Synthesis and Reactivity of (Phosphinoalkyl) silyl Complexes. *Corrd. Chem. Rev.* **2002,** *226*, 167–178.

25. (a) Mo, Z.; Liu, Y.; Deng, L. Anchoring of Silyl Donors on a *N*-Heterocyclic Carbene Through the Cobalt-Mediated Silylation of Benzylic C–H Bonds. *Angew. Chem., Int. Ed.* **2013,** *52*, 10845–10849.

(b) Ouyang, Z.; Deng, L. Iron(II) Complexes Featuring Bidentate N-Heterocyclic Carbene–Silyl Ligands: Synthesis and Characterization. *Organometallics* **2013,** *32*, 7268–7271.

26. (a) Brost, R. D.; Bruce, G. C.; Joslin, F. L.; Stobart, S. R. Phosphinoalkylsilyl Complexes. 12. Stereochemistry of the Tridentate Bis(diphenylphosphinopropyl)silyl (biPSi) Framework: Complexation that Introduces "Face Discrimination" at Coordinatively Unsaturated Metal Centers. X-Ray Crystal and Molecular Structures of $Pt[SiMe(CH_2CH_2CH_2PPh_2)_2]Cl$, $IrH[SiMe(CH_2CH_2CH_2PPh_2)_2]Cl$, and $RuH[SiMe(CH_2CH_2CH_2PPh_2)_2](CO)_2$. *Organometallics* **1997,** *16*, 5669–5680.

(b) Stobart, S. R.; Zhou, X.; Cea-Olivares, R.; Toscano, A. Activation of Water and of Dioxygen by a Bis (diphenylphosphinopropyl)silyl (biPSi) Complex of Ruthenium(II): Formation of Bis (diphenylphosphinopropyl)siloxo Cage Complexes. Concomitant Oxygen Atom Insertion into a Silicon–Carbon Bond. *Organometallics* **2001,** *20*, 4766–4768.

(c) Korshin, E. E.; Leitus, G.; Shimon, L. J. W.; Konstantinovski, L.; Milstein, D. Silanol-Based Pincer Pt (II) Complexes: Synthesis, Structure, and Unusual Reactivity. *Inorg. Chem.* **2008,** *47*, 7177–7189.

(d) Fang, H.; Choe, Y.-K.; Li, Y.; Shimada, S. Synthesis, Structure, and Reactivity of Hydridoiridium Complexes Bearing a Pincer-Type PSiP Ligand. *Chem. Asian J.* **2011,** *6*, 2512–2521.

(e) Wu, S.; Li, X.; Xiong, Z.; Xu, W.; Lu, Y.; Sun, H. Synthesis and Reactivity of Silyl Iron, Cobalt, and Nickel Complexes Bearing a [PSiP]-Pincer Ligand via Si–H Bond Activation. *Organometallics* **2013,** *32*, 3227–3237.

(f) Bernal, M. J.; Torres, O.; Martín, M.; Sola, E. Reversible Insertion of Carbenes into Ruthenium–Silicon Bonds. *J. Am. Chem. Soc.* **2013,** *135*, 19008–19015.

(g) Dixon, L. S. H.; Hill, A. F.; Sinha, A.; Ward, J. S. N-Heterocyclic Silyl Pincer Ligands. *Organometallics* **2014,** *33*, 653–658.

(h) Whited, M. T.; Deetz, A. M.; Boerma, J. W.; DeRosha, D. E.; Janzen, D. E. Formation of Chlorosilyl Pincer-Type Rhodium Complexes by Multiple Si–H Activations of Bis(phosphine)/Dihydrosilyl Ligands. *Organometallics* **2014,** *33*, 5070–5073.

(i) Nova, A.; Suh, H.-W.; Schmeier, T. J.; Guard, L. M.; Eisenstein, O.; Hazari, N.; Maseras, F. An Unusual Example of Hypervalent Silicon: A Five-Coordinate Silyl Group Bridging Two Palladium or Nickel Centers Through a Nonsymmetrical Four-Center Two-Electron Bond. *Angew. Chem., Int. Ed.* **2014,** *53*, 1103–1108.

27. (a) MacInnis, M. C.; MacLean, D. F.; Lundgren, R. J.; McDonald, R.; Turculet, L. Synthesis and Reactivity of Platinum Group Metal Complexes Featuring the New Pincer-Like Bis(phosphino)silyl

Ligand [κ^3-(2-$Ph_2PC_6H_4)_2SiMe]^-$ ([PSiP]): Application in the Ruthenium-Mediated Transfer Hydrogenation of Ketones. *Organometallics* **2007,** *26*, 6522–6525.
(b) MacLean, D. F.; McDonald, R.; Ferguson, M. J.; Caddell, A. J.; Turculet, L. Room Temperature Benzene C–H Activation by a New [PSiP]Ir Pincer Complex. *Chem. Commun.* **2008,** 5146–5148.
(c) Morgan, E.; MacLean, D. F.; McDonald, R.; Turculet, L. Rhodium and Iridium Amido Complexes Supported by Silyl Pincer Ligation: Ammonia N – H Bond Activation by a [PSiP]Ir Complex. *J. Am. Chem. Soc.* **2009,** *131*, 14234–14236.
(d) Mitton, S. J.; McDonald, R.; Turculet, L. Synthesis and Characterization of Neutral and Cationic Platinum(II) Complexes Featuring Pincer-Like Bis(phosphino)silyl Ligands: Si – H and Si – Cl Bond Activation Chemistry. *Organometallics* **2009,** *28*, 5122–5136.
(e) MacInnis, M. C.; McDonald, R.; Ferguson, M. J.; Tobisch, S.; Turculet, L. Four-Coordinate, 14-Electron Ru^{II} Complexes: Unusual Trigonal Pyramidal Geometry Enforced by Bis(phosphino)silyl Ligation. *J. Am. Chem. Soc.* **2011,** *133*, 13622–13633.

28. (a) Takaya, J.; Iwasawa, N. Hydrocarboxylation of Allenes With CO_2 Catalyzed by Silyl Pincer-Type Palladium Complex. *J. Am. Chem. Soc.* **2008,** *130*, 15254–15255.
(b) Takaya, J.; Iwasawa, N. Bis(o-phosphinophenyl)silane as a Scaffold for Dynamic Behavior of H – Si and C – Si Bonds With Palladium(0). *Organometallics* **2009,** *28*, 6636–6638.
(c) Takaya, J.; Kirai, N.; Iwasawa, N. Efficient Synthesis of Diborylalkenes From Alkenes and Diboron by a New PSiP-Pincer Palladium-Catalyzed Dehydrogenative Borylation. *J. Am. Chem. Soc.* **2011,** *133*, 12980–12983.
(d) Kirai, N.; Takaya, J.; Iwasawa, N. Two Reversible σ-Bond Metathesis Pathways for Boron–Palladium Bond Formation: Selective Synthesis of Isomeric Five-Coordinate Borylpalladium Complexes. *J. Am. Chem. Soc.* **2013,** *135*, 2493–2496.

29. (a) Kwok, W.-H.; Lu, G.-L.; Rickard, C. E. F.; Roper, W. R.; Wright, L. J. Nucleophilic Substitution Reactions at the Si–Cl Bonds of the Dichloro(methyl)silyl Ligand in Five- and Six-Coordinate Complexes of Ruthenium(II) and Osmium(II). *J. Organomet. Chem.* **2004,** *689*, 2511–2522.
(b) Sangtrirutnugul, P.; Tilley, T. D. Alkyl and Hydrido Complexes of Platinum(IV) Supported by the Bis (8-quinolyl)methylsilyl Ligand. *Organometallics* **2008,** *27*, 2223–2320.
(c) Yang, J.; Rosal, I. D.; Fasulo, M.; Sangtrirutnugul, P.; Maron, L.; Tilley, T. D. Nickel Complexes With Bis(8-quinolyl)silyl Ligands. An Unusual Ni_3Si_2 Cluster Containing Six-Coordinate Silicon. *Organometallics* **2010,** *29*, 5544–5550.
(d) Lalrempuia, R.; Iglesias, M.; Polo, V.; Miguel, P. J. S.; Fernández-Alvarez, F. J.; Pérez-Torrente, J. J.; Oro, L. A. Effective Fixation of CO_2 by Iridium-Catalyzed Hydrosilylation. *Angew. Chem., Int. Ed.* **2012,** *51*, 12824–12827.

30. (a) Hill, A. F.; Neumann, H.; Wagler, J. Bis(methimazolyl)silyl Complexes of Ruthenium. *Organometallics* **2010,** *29*, 1026–1031.
(b) Armbruster, F.; Meyer, J.; Baldes, A.; Burgos, P. O.; Fernández, I.; Breher, F. A Triangulo Palladium Cluster Consisting of μ^3-Capping Silyl Ligands. *Chem. Commun.* **2011,** *47*, 221–223.

31. (a) Mitton, S. J.; Turculet, L. Mild Reduction of Carbon Dioxide to Methane With Tertiary Silanes Catalyzed by Platinum and Palladium Silyl Pincer Complexes. *Chem. Eur. J.* **2012,** *18*, 15258–15262.
(b) Suh, H.-W.; Schmeier, T. J.; Hazari, N.; Kemp, R. A.; Takase, M. K. Experimental and Computational Studies of the Reaction of Carbon Dioxide With Pincer-Supported Nickel and Palladium Hydrides. *Organometallics* **2012,** *31*, 8225–8236.
(c) Suh, H.-W.; Guard, L. M.; Hazari, N. A Mechanistic Study of Allene Carboxylation With CO_2 Resulting in the Development of a Pd(II) Pincer Complex for the Catalytic Hydroboration of CO_2. *Chem. Sci.* **2014,** *5*, 3859–3872.

32. (a) Mitton, S. J.; McDonald, R.; Turculet, L. Nickel and Palladium Silyl Pincer Complexes: Unusual Structural Rearrangements That Involve Reversible Si–C(sp^3) and Si–C(sp^2) Bond Activation. *Angew. Chem., Int. Ed.* **2009,** *48*, 8568–8571.

(b) Kameo, H.; Ishii, S.; Nakazawa, H. Si–C Bond Cleavage by Hydride Complexes of Rhodium and Iridium: Comparison of Si–C(sp^2) and Si–C(sp^3) Activation. *Dalton Trans.* **2013,** *42*, 4663–4669.

33. Takaya, J.; Iwasawa, N. Reaction of Bis(o-phosphinophenyl)silane With $M(PPh_3)_4$ (M = Ni, Pd, Pt): Synthesis and Structural Analysis of η^2-(Si–H) Metal(0) and Pentacoordinate Silyl Metal(II) Hydride Complexes of the Ni Triad Bearing a PSiP-Pincer Ligand. *Dalton Trans.* **2011,** *40*, 8814–8821.

34. Suh, H.-W.; Balcells, D.; Edwards, A. J.; Guard, L. M.; Hazari, N.; Mader, E. A.; Mercado, A. B. Q.; Repisky, M. Understanding the Solution and Solid-State Structures of Pd and Pt PSiP Pincer-Supported Hydrides. *Inorg. Chem.* **2015,** *54*, 11411–11422.

35. Ruddy, A. J.; Mitton, S. J.; McDonald, R.; Turculet, L. "Hemilabile" Silyl Pincer Ligation: Platinum Group PSiN Complexes and Triple C–H Activation to Form a (PSiC)Ru Carbene Complex. *Chem. Commun.* **2012,** *48*, 1159–1161.

36. Lee, C.-I.; Zhou, J.; Ozerov, O. V. Catalytic Dehydrogenative Borylation of Terminal Alkynes by a SiNN Pincer Complex of Iridium. *J. Am. Chem. Soc.* **2013,** *135*, 3560–3566.

37. Joslin, F. L.; Stobart, S. R. Synthesis and Structure of a Silyl-Rhodium(I) Complex: Stabilization Attributable to Multisite "Anchoring" of a Silyl Fragment. *J. Chem. Soc., Chem. Commun.* **1989,** 504–505.

38. Hendriksen, D. E.; Oswald, A. A.; Ansell, G. B.; Leta, S.; Kastrup, R. V. Selective Rhodium-Catalyzed Hydroformylation With the Tri- and Tetraphosphine Ligands $(CH_3)_{1,0}Si(CH_2CH_2PPh_2)_{3,4}$. Formation of $Rh[Si(CH_2CH_2PPh_2)_3](CO)$ via CH_3-Si Bond Cleavage and Structure of this Rh(I)-Si Bonded Complex. *Organometallics* **1989,** *8*, 1153–1157.

39. (a) Kameo, H.; Ishii, S.; Nakazawa, H. Synthesis and Reactivity of Rhodium Complexes Bearing $[E(o\text{-}C_6H_4PPh_2)_3]$-Type Tetradentate Ligands (E = Si, Ge, and Sn). *Organometallics* **2012,** *31*, 2212–2218.
(b) Kameo, H.; Ishii, S.; Nakazawa, H. Synthesis of Iridium Complexes Bearing $\{o\text{-}(Ph_2P)C_6H_4\}_3E$ Type (E = Si, Ge, and Sn) Ligand and Evaluation of Electron Donating Ability of Group 14 Elements E. *Dalton Trans.* **2012,** *41*, 8290–8296.

40. Takeda, N.; Watanabe, D.; Nakamura, T.; Unno, M. Synthesis and Complexation of a New Tripodal Tetradentate Ligand, a Silyl Ligand Tethered With Three Thioether Moieties. *Organometallics* **2010,** *29*, 2839–2841.

41. (a) Mankad, N. P.; Whited, M. T.; Peters, J. C. Terminal $Fe^I–N_2$ and $Fe^{II}\cdots H–C$ Interactions Supported by Tris(phosphino)silyl Ligands. *Angew. Chem., Int. Ed.* **2007,** *46*, 5768–5771.
(b) Mankad, N. P.; Müller, P.; Peters, J. C. Catalytic N – N Coupling of Aryl Azides to Yield Azoarenes via Trigonal Bipyramid Iron–Nitrene Intermediates. *J. Am. Chem. Soc.* **2010,** *132*, 4083–4085.
(c) Lee, Y.; Kinney, R. A.; Hoffman, B. M.; Peters, J. C. A Nonclassical Dihydrogen Adduct of $S = {}^1/_2$ Fe(I). *J. Am. Chem. Soc.* **2011,** *133*, 16366–16369.
(d) Takaoka, A.; Mankad, N. P.; Peters, J. C. Dinitrogen Complexes of Sulfur-Ligated Iron. *J. Am. Chem. Soc.* **2011,** *133*, 8440–8443.

42. (a) Takaoka, A.; Mendiratta, A.; Peters, J. C. E – H Bond Activation Reactions (E = H, C, Si, Ge) at Ruthenium: Terminal Phosphides, Silylenes, and Germylenes. *Organometallics* **2009,** *28*, 3744–3753.
(b) Takaoka, A.; Gerber, L. C. H.; Peters, J. C. Access to Well-Defined Ruthenium(I) and Osmium(I) Metalloradicals. *Angew. Chem., Int. Ed.* **2010,** *49*, 4088–4091.

43. Takaoka, A.; Peters, J. C. A Homologous Series of Cobalt, Rhodium, and Iridium Metalloradicals. *Inorg. Chem.* **2012,** *51*, 16–18.

44. (a) Lee, Y.; Mankad, N. P.; Peters, J. C. Triggering N_2 Uptake via Redox-Induced Expulsion of Coordinated NH_3 and N_2 Silylation at Trigonal Bipyramidal Iron. *Nat. Chem.* **2010,** *2*, 558–565.
(b) Lee, Y.; Peters, J. C. Silylation of Iron-Bound Carbon Monoxide Affords a Terminal Fe Carbyne. *J. Am. Chem. Soc.* **2011,** *133*, 4438–4446.

45. (a) Tsay, C.; Mankad, N. P.; Peters, J. C. Four-Coordinate, Trigonal Pyramidal Pt(II) and Pd(II) Complexes. *J. Am. Chem. Soc.* **2010,** *132*, 13975–13977.

(b) Tsay, C.; Peters, J. C. Thermally Stable N_2 and H_2 Adducts of Cationic Nickel(II). *Chem. Sci.* **2012,** *3*, 1313–1318.

46. Crabtree, R. H. Transition Metal Complexation of σ Bonds. *Angew. Chem., Int. Ed. Engl.* **1993,** *32*, 789–805.
47. Hoyano, J. K.; Elder, M.; Graham, W. A. G. Hydrogen-Bridged Silicon-Rhenium Bonds. A Diphenylsilane Complex of Rhenium Carbonyl. *J. Am. Chem. Soc.* **1969,** *91*, 4568–4569.
48. Schubert, U.; Müller, J.; Alt, H. G. Hydrido Silyl Complexes. 9. Chromium- Hydrogen-Silicon Three-Center Bonding in $C_6Me_6(CO)_2Cr(H)SiHPh_2$. *Organometallics* **1987,** *6*, 469–472.
49. Schubert, U.; Scholz, G.; Müller, J.; Ackermann, K.; Wörle, B.; Stansfield, R. F. D. Hydrido-silyl-komplexe: VII. Strukturchemische und ^{29}Si-NMR- spektroskopische untersuchungen an $C_5R'_5(CO)(L)Mn(H)SiR_3$. Einfluss der substituenten R und R′ sowie des liganden L auf die Mn, H, Si- dreizentrenbindung. *J. Organomet. Chem.* **1986,** *306*, 303–326.
50. Iluc, V. M.; Hillhouse, G. L. Arrested 1,2-Hydrogen Migration From Silicon to Nickel Upon Oxidation of a Three-Coordinate Ni(I) Silyl Complex. *J. Am. Chem. Soc.* **2010,** *132*, 11890–11892.
51. Steiman, T. J.; Uyeda, C. Reversible Substrate Activation and Catalysis at an Intact Metal–Metal Bond Using a Redox-Active Supporting Ligand. *J. Am. Chem. Soc.* **2015,** *137*, 6104–6110.
52. Thomas, C. M.; Peters, J. C. An η^3-H_2SiR_2 Adduct of $[\{PhB(CH_2PiPr_2)_3\}Fe^{II}H]$. *Angew. Chem., Int. Ed.* **2006,** *45*, 776–780.
53. (a) Lipke, M. C.; Tilley, T. D. High Electrophilicity at Silicon in η^3-Silane σ-Complexes: Lewis Base Adducts of a Silane Ligand, Featuring Octahedral Silicon and Three Ru–H–Si Interactions. *J. Am. Chem. Soc.* **2011,** *133*, 16374–16377.
    (b) Lipke, M. C.; Tilley, T. D. Stabilization of $ArSiH_4^-$ and SiH_6^{2-} Anions in Diruthenium Si–H σ-Complexes. *Angew. Chem., Int. Ed.* **2012,** *51*, 11115–11121.
54. Fan, H.; Fullmer, B. C.; Pink, M.; Caulton, K. G. $[Ni\{N(SiMe_2CH_2PtBu_2)_2\}]^+$: Direct Observation of Transannular Si–C(sp^3) Bond Coordination. *Angew. Chem., Int. Ed.* **2008,** *47*, 9112–9114.
55. (a) Ogino, H.; Tobita, H. Bridged Silylene and Germylene Complexes. *Adv. Organomet. Chem.* **1998,** *42*, 223–290.
    (b) Osakada, K.; Tanabe, M. Platinum and Palladium Complexes With Metal–Silicon Bonds. New Bonding, Structures, and Chemical Properties. *Bull. Chem. Soc. Jpn.* **2005,** *78*, 1887–1898.
56. (a) Osakada, K.; Tanabe, M.; Tanase, T. A Triangular Triplatinum Complex With Electron-Releasing $SiPh_2$ and PMe_3 Ligands: $[\{Pt(\mu\text{-}SiPh_2)(PMe_3)\}_3]$. *Angew. Chem., Int. Ed.* **2000,** *39*, 4053–4055.
    (b) Tanaka, K.; Kamono, M.; Tanabe, M.; Osakada, K. Ring Expansion of Cyclic Triplatinum(0) Silylene Complexes Induced by Insertion of Alkyne Into a Si–Pt Bond. *Organometallics* **2015,** *34*, 2985–2990.
57. Braddock-Wilking, J.; Corey, J. Y.; Dill, K.; Rath, N. P. Formation and X-Ray Crystal Structure Determination of the Novel Triplatinum Cluster $[(Ph_3P)Pt(\mu\text{-}SiC_{12}H_8)]_3$ From Reaction of Silafluorene With $(Ph_3P)_2Pt(\eta^2\text{-}C_2H_4)$. *Organometallics* **2002,** *21*, 5467–5469.
58. Yamada, T.; Mawatari, A.; Tanabe, M.; Osakada, K.; Tanase, T. Planar Tetranuclear and Dumbbell-Shaped Octanuclear Palladium Complexes With Bridging Silylene Ligands. *Angew. Chem., Int. Ed.* **2009,** *48*, 568–571.
59. Watanabe, C.; Iwamoto, T.; Kabuto, C.; Kira, M. Fourteen-Electron Bis(dialkylsilylene)palladium and Twelve-Electron Bis(dialkylsilyl)palladium Complexes. *Angew. Chem., Int. Ed.* **2008,** *47*, 5386–5389.
60. (a) Tanabe, M.; Ishikawa, N.; Chiba, M.; Ide, T.; Osakada, K.; Tanase, T. Tetrapalladium Complex With Bridging Germylene Ligands. Structural Change of the Planar Pd_4Ge_3 Core. *J. Am. Chem. Soc.* **2011,** *133*, 18598–18601.
    (b) Tanabe, M.; Tanaka, K.; Omine, S.; Osakada, K. A Triangular Triplatinum(0) Complex With Bridging Germylene Ligands: Insertion of Alkynes Into the Pt–Ge Bond Rather Than the Pt–Pt Bond. *Chem. Commun.* **2014,** *50*, 6839–6842.
61. Shimada, S.; Li, Y.-H.; Choe, Y.-K.; Tanaka, M.; Bao, M.; Uchimaru, T. Multinuclear Palladium Compounds Containing Palladium Centers Ligated by Five Silicon Atoms. *Proc. Natl. Acad. Sci. USA* **2007,** *104*, 7758–7763.

62. Beck, R.; Johnson, S. A. Structural Similarities in Dinuclear, Tetranuclear, and Pentanuclear Nickel Silyl and Silylene Complexes Obtained via Si–H and Si–C Activation. *Organometallics* **2012,** *31*, 3599–3609.
63. Sunada, Y.; Haige, R.; Otsuka, K.; Kyushin, S.; Nagashima, H. A Ladder Polysilane as a Template for Folding Palladium Nanosheets. *Nat. Commun.* **2013,** *4*, 3014.
64. Aitken, C.; Harrod, J. F.; Samuel, E. Polymerization of Primary Silanes to Linear Polysilanes Catalyzed by Titanocene Derivatives. *J. Organomet. Chem.* **1985,** *279*, C11–C13.
65. (a) Rösch, L.; Altnau, G. Synthese und eigenschaften von koordiniertem und solvatfreiem tris (trimethylsilyl)aluminium. *J. Organomet. Chem.* **1980,** *195*, 47–53.
    (b) Tilley, T. D. Trimethylsilyl Derivatives of Bis(cyclopentadienyl)zirconium and -hafnium. Crystal Structure of $(\eta^5\text{-}C_5H_5)_2Zr(SiMe_3)(S_2CNEt_2)$. *Organometallics* **1985,** *4*, 1452–1457.
66. Zirngast, M.; Baumgartner, J.; Marschner, C. Preparation, Structure and Reactivity of $Et_2N(Me_3Si)_2SiK$. *Eur. J. Inorg. Chem.* **2008,** 1078–1087.
67. Woo, H.-G.; Heyn, R. H.; Tilley, T. D. σ-Bond Metathesis Reactions for d^0 Metal-Silicon Bonds That Produce Zirconocene and Hafnocene Hydrosilyl Complexes. *J. Am. Chem. Soc.* **1992,** *114*, 5698–5707.
68. Woo, H.-G.; Walzer, J. F.; Tilley, T. D. A σ-Bond Metathesis Mechanism for Dehydropolymerization of Silanes to Polysilanes by d^0 Metal Catalysts. *J. Am. Chem. Soc.* **1992,** *114*, 7047–7055.
69. Casty, G. L.; Lugmair, C. G.; Radu, N. S.; Tilley, T. D.; Walzer, J. F.; Zargarian, D. Synthesis and Reactivity of 16-Electron Hafnocene Silyl Hydride Complexes. *Organometallics* **1997,** *16*, 8–12.
70. Sadow, A. D.; Tilley, T. D. Cationic Hafnium Silyl Complexes and Their Enhanced Reactivity in σ-Bond Metathesis Processes With Si–H and C–H Bonds. *J. Am. Chem. Soc.* **2003,** *125*, 9462–9475.
71. Aitken, C. T.; Harrod, J. F.; Samuel, E. Identification of Some Intermediates in the Titanocene-Catalyzed Dehydrogenative Coupling of Primary Organosilanes. *J. Am. Chem. Soc.* **1986,** *108*, 4059–4066.
72. (a) Samuel, E.; Mu, Y.; Harrod, J. F.; Dromzee, Y.; Jeannin, Y. Synthesis and Properties of Some Novel Bis(cyclopentadienyl)silyltitanium(III) Complexes. *J. Am. Chem. Soc.* **1990,** *112*, 3435–3439.
    (b) Britten, J.; Mu, Y.; Harrod, J. F.; Polowin, J.; Baird, M. C.; Samuel, E. Crystal Structures and Conformational Analysis of Complexes of the Type $Cp_2Ti(SiHRR')PMe_3$ (R, R′ = H, Me, Ph): Relationships Between Calculated Molecular Structures and Observed Solid-State Structures. *Orgnometallics* **1993,** *12*, 2672–2676.
73. (a) Hao, L.; Lebuis, A.-M.; Harrod, J. F.; Samuel, E. Preparation and Characterization of Titanocene Silyl Hydrides $[Cp_2Ti(\mu\text{-}HSiH_2)]_2$ and $[Cp_2Ti(\mu\text{-}HSiH_2)(\mu\text{-}H)TiCp_2]$. *Chem. Commun.* **1997,** 2193–2194.
    (b) Hao, L.; Lebuis, A.-M.; Harrod, J. F. Synthesis and Structural Characterization of $Cp_2Ti(SiH_3)(PMe_3)$. *Chem. Commun.* **1998,** 1089–1090.
74. Dioumaev, V. K.; Harrod, J. F. Studies of the Formation and Decomposition Pathways for Cationic Zirconocene Hydrido Silyl Complexes. *Organometallics* **1997,** *16*, 2798–2807.
75. Nguyen, T.; Scheschkewitz, D. Activation of a Si = Si Bond by η^1-Coordination to a Transition Metal. *J. Am. Chem. Soc.* **2005,** *127*, 10174–10175.
76. Spaltenstein, E.; Palma, P.; Kreutzer, K. A.; Willoughby, C. A.; Davis, W. M.; Buchwald, S. L. Preparation and X-Ray Structure of $Cp_2Ti(Ph_2SiH_2)(PMe_3)$. *J. Am. Chem. Soc.* **1994,** *116*, 10308–10309.
77. Kreutzer, K. A.; Fisher, R. A.; Davis, W. M.; Spaltenstein, E.; Buchwald, S. L. Synthesis, Characterization and Reactivity of Zirconocene and Hafnocene Silyl Hydride Complexes. *Organometallics* **1991,** *10*, 4031–4035.
78. Ignatov, S. K.; Rees, N. H.; Tyrrell, B. R.; Dubberley, S. R.; Razuvaev, A. G.; Mountford, P.; Nikonov, G. I. Nonclassical Titanocene Silyl Hydrides. *Chem. Eur. J.* **2004,** *10*, 4991–4999.
79. Scherer, W.; Meixner, P.; Barquera-Lozada, J. E.; Hauf, C.; Obenhuber, A.; Brück, A.; Wolstenholme, D. J.; Ruhland, K.; Leusser, D.; Stalke, D. A Unifying Bonding Concept for Metal Hydrosilane Complexes. *Angew. Chem., Int. Ed.* **2013,** *52*, 6092–6096.
80. (a) Ohff, A.; Kosse, P.; Baumann, W.; Tillack, A.; Kempe, R.; Görls, H.; Burlakov, V. V.; Rosenthal, U. Novel *trans*-η^2-Alkyne Complexes of Titanocene With Strong Si-H-Ti Interactions. Syntheis, Spectral Characteristics, and X-Ray Crystal Structure. *J. Am. Chem. Soc.* **1995,** *117*, 10399–10400.

(b) Peulecke, N.; Ohff, A.; Kosse, P.; Tillack, A.; Spannenberg, A.; Kempe, R.; Baumann, W.; Burlakov, V. V.; Rosenthal, U. Si—H Activation in Titnocene and Zirconocene Complexes of Alkynylsilanes RC≡CSiMe_2H (R = ^{t}Bu, Ph, SiMe_3, SiMe_2H): A Model to Understand Catalytic Reactions of Hydrosilanes. *Chem. Eur. J.* **1998,** *4*, 1852–1861.

81. Curtis, M. D.; Bell, L. G.; Butler, W. M. C—H Activation. Synthesis of Silyl Derivatives of Niobocene and Tantalocene Hydrides, Their H/D Exchange Reactions With C_6D_6, and the Structure of Cp_2Ta $(H)_2$SiMe_2Ph. *Organometallics* **1985,** *4*, 701–707.

82. Antiñolo, A.; Carrillo, F.; Fajardo, M.; Otero, A.; Lanfranchi, M.; Pellinghelli, M. A. Synthesis and Characterization of New Silyl Niobocene Complexes. X-Ray Molecular Structure of d^0 Nb(η^5-C_5H_4Si$Me_3)_2(H)_2$(SiPh_2H). *Organometallics* **1995,** *14*, 1518–1521.

83. Nikonov, G. I.; Kuzmina, L. G.; Lemenovskii, D. A.; Kotov, V. V. Interligand Hypervalent Interaction in the Bis(silyl) Hydride Derivatives of Niobocene. *J. Am. Chem. Soc.* **1995,** *117*, 10133–10134.

84. (a) Bakhmutov, V. I.; Howard, J. A. K.; Keen, D. A.; Kuzmina, L. G.; Leech, M. A.; Nikonov, G. I.; Vorontsov, E. V.; Wilson, C. C. Combined Single Crystal Neutron Diffraction and Solution NMR Studies of Mono- and Bis(silyl) Substituted Niobocene Hydrides With Nonclassical Interligand Interactions. *J. Chem. Soc., Dalton Trans.* **2000,** 1631–1635.
(b) Nikonov, G. I.; Kuzmina, L. G.; Howard, J. A. K. Niobocene Hydrides With Functionalised Silyl Ligands. Tuning of Interligand Hypervalent Interactions (IHI) M-H → Si-X. *J. Chem. Soc., Dalton Trans.* **2002,** 3037–3046.

85. (a) Nikonov, G. I.; Vyboishchikov, S. F.; Kuzmina, L. G.; Howard, J. A. K. Serendipitous Syntheses and Structures of [Cp_2Nb(H){(Si$Me_2)_2(\mu$-NR)}]. *Chem. Commun.* **2002,** 568–569.
(b) Dorogov, K. Y.; Dumont, E.; Ho, N.-N.; Churakov, A. V.; Kuzmina, L. G.; Poblet, J.-M.; Schultz, A. J.; Howard, J. A. K.; Bau, R.; Lledos, A.; Nikonov, G. I. Neutron and X-Ray Diffraction Studies and DFT Calculations of Asymmetric Bis(silyl)niobocene Hydrides. *Organometallics* **2004,** *23*, 2845–2847.
(c) Dorogov, K. Y.; Yousufuddin, M.; Ho, N.-N.; Churakov, A. V.; Kuzmina, L. G.; Schultz, A. J.; Mason, S. A.; Howard, J. A. K.; Lemenovskii, D. A.; Bau, R.; Nikonov, G. I. Syntheses and Structures of Asymmetric Bis(silyl) Niobocene Hydrides. *Inorg. Chem.* **2007,** *46*, 147–160.

86. Jiang, Q.; Carroll, P. J.; Berry, D. H. Synthesis of Mono- and Bis(silyl) Complexes of Tantalum. *Organometallics* **1991,** *10*, 3648–3655.

87. Arnold, J.; Tilley, T. D.; Rheingold, A. L.; Gelb, S. J. Preparation and Reaction Chemistry of Trimethylsilyl Derivatives of Niobium. Redox Chemistry of (η^5-$C_5H_5)_2$Nb(SiMe_3)Cl and X-Ray Structures of (η^5-$C_5H_5)_2$Nb(SiMe_3)(η^2-C_2H_4) and [(η^5-$C_5H_5)_2$Nb(CH_2SiMe_3)Cl]PF_6. *Organometallics* **1987,** *6*, 473–479.

88. Koshikawa, H.; Okazaki, M.; Matsumoto, S.; Ueno, K.; Tobita, H.; Ogino, H. Synthesis and Structure of a Base-Stabilized Silyl(silylene)tantalum Complex. *Chem. Lett.* **2005,** *34*, 1412–1413.

89. Igonin, V. A.; Ovchinnikov, Y. E.; Dement'ev, V. V.; Shklover, V. E.; Timofeeva, T. V.; Frunze, T. M.; Struchkov, Y. T. Crystal Structures of Cycloheteropentasilanes (η^5-Cp$)_2$Ti(Si$Ph_2)_5$ and O(Si$Ph_2)_5$. *J. Organomet. Chem.* **1989,** *371*, 187–196.

90. Kayser, C.; Kickelbick, G.; Marschner, C. Simple Synthesis of Oligosilyl-α,ω-dipotassium Compounds. *Angew. Chem., Int. Ed.* **2002,** *41*, 989–992.

91. Kayser, C.; Frank, D.; Baumgartner, J.; Marschner, C. Reactions of Oligosilyl Potassium Compounds With Group 4 Metallocene Dichlorides. *J. Organomet. Chem.* **2003,** *667*, 149–153.

92. Fischer, R.; Frank, D.; Gaderbauer, W.; Kayser, C.; Mechtler, C.; Baumgartner, J.; Marschner, C. α,ω-Oligosilyl Dianions and Their Application in the Synthesis of Homo- and Heterocyclosilanes. *Organometallics* **2003,** *22*, 3723–3731.

93. Wagner, H.; Baumgartner, J.; Marschner, C. 1,1′-Oligosilylferrocene Compounds. *Organometallics* **2007,** *26*, 1762–1770.

94. (a) Zirngast, M.; Flörke, U.; Baumgartner, J.; Marschner, C. Oligosilylated Group 4 Titanocenes in the Oxidation State +3. *Chem. Commun.* **2009,** 5538–5540.

(b) Arp, H.; Zirngast, M.; Marschner, C.; Baumgartner, J.; Rasmussenn, K.; Zark, P.; Müller, T. Synthesis of Oligosilanyl Compounds of Group 4 Metallocenes With the Oxidation State +3. *Organometallics* **2012,** *31*, 4309–4319.

95. (a) Fischer, R.; Zimgast, M.; Flock, M.; Baumgartner, J.; Marschner, C. Synthesis of a Hafnocene Disilene Complex. *J. Am. Chem. Soc.* **2005,** *127*, 70–71.
(b) Zirngast, M.; Flock, M.; Baumgartner, J.; Marschner, C. Group 4 Metallocene Complexes of Disilenes, Digermenes, and a Silagermene. *J. Am. Chem. Soc.* **2009,** *131*, 15952–15962.

96. (a) Procopio, L. J.; Carroll, P. J.; Berry, D. H. η^2-Silanimine Complexes of Zirconocene: Synthesis, Structure, and Reactivity of $Cp_2Zr(\eta^2\text{-}SiMe_2=N^tBu)(PMe_3)$. *J. Am. Chem. Soc.* **1991,** *131*, 1870–1872.
(b) Procopio, L. J.; Carroll, P. J.; Berry, D. H. Structure and Reactivity of $Cp_2Zr(\eta^2\text{-}Me_2Si=N^tBu)(CO)$: An Usual Silanimine Carbonyl Complex With Extensive σ-π^* Back Bonding. *Polyhedron* **1995,** *14*, 45–55.

97. (a) Arnold, J.; Roddick, D. M.; Tilley, T. D.; Rheingold, A. L.; Geib, S. J. Preparation and Characterization of Tris(trimethylsilyl)silyl and Tris(trimethylsilyl)germyl Derivatives of Zirconium and Hafnium. X-Ray Crystal Structures of $(\eta^5\text{-}C_5Me_5)Cl_2HfSi(SiMe_3)_3$ and $(\eta^5\text{-}C_5Me_5)Cl_2HfGe(SiMe_3)_3$. *Inorg. Chem.* **1988,** *27*, 3510–3514.
(b) Burckhardt, U.; Casty, G. L.; Gavenonis, J.; Tilley, T. D. Neutral and Anionic Silyl Hydride Derivatives of the Tantalum Imido Fragment Cp*(DippN=)Ta ($Cp^*=\eta^5\text{-}C_5Me_5$; Dipp = $2,6\text{-}^iPr_2C_6H_3$). Reactivity of σ-Bonds and Intramolecular C–H Bond Activation Involving the Silyl Ligands. *Organometallics* **2002,** *21*, 3108–3122.

98. (a) Arnold, J.; Shina, D. N.; Tilley, T. D.; Arif, A. M. Preparation and Reaction Chemistry of Trimethylsilyl Derivatives of Tantalum. X-Ray Structures of d^0 $(\eta^5\text{-}C_5Me_5)Ta(SiMe_3)Cl_3$ and $d^1(\eta^5\text{-}C_5Me_5)Ta(SiMe_3)(PMe_3)Cl_2$. *Organometallics* **1986,** *5*, 2037–2044.
(b) Burckhardt, U.; Casty, G. L.; Tilley, T. D.; Woo, T. K.; Rothlisberger, U. Ditantalum Hydride Complexes With Bridging $(2,6\text{-}^iPr_2C_6H_3)NSiHPh$ Silanimine Ligands Resulting From $PhSiH_3$-Imido Ligand Coupling. A Combined Spectroscopic and Theoretical Investigation. *Organometallics* **2000,** *19*, 3830–3841.

99. (a) Nikonov, G. I.; Mountford, P.; Green, J. C.; Cooke, P. A.; Leech, M. A.; Blake, A. J.; Howard, J. A. K.; Lemenovskii, D. A. Contrasting Nonclassical Silicon-Hydrogen Interactions in Niobium and Tantalum Half-Sandwich Complexes: Si–H···M Agnostic Versus M–H···Si–Cl Interligand Hypervalent Interactions. *Eur. J. Inorg. Chem.* **2000,** 1917–1921.
(b) Nikonov, G. I.; Mountford, P.; Ignatov, S. K.; Green, J. C.; Leech, M. A.; Kuzmina, L. G.; Razuvaev, A. G.; Rees, N. H.; Blake, A. J.; Howard, J. A. K.; Lemenovskii, D. A. Surprising Diversity of Non-classical Silicon-Hydrogen Interactions in Half-Sandwich Complexes of Nb and Ta: M–H···Si–Cl Interligand Hypervalent Interaction (IHI) Versus Stretched and Unstretched β-Si–H···M Agostic Bonding. *Dalton Trans.* **2001,** 2903–2915.
(c) Nikonov, G. I.; Mountford, P.; Dubberley, S. R. Tantalizing Chemistry of the Half-Sandwich Silylhydride Complexes of Niobium: Identification of Likely Intermediates on the Way to Agostic Complexes. *Inorg. Chem.* **2003,** *42*, 258–260.
(d) Ignatov, S. K.; Rees, N. H.; Merkoulov, A. A.; Dubberley, S. R.; Razuvaev, A. G.; Mountford, P.; Nikonov, G. I. Silyl Hydrides of Tantalum Supported by Cyclopentadienyl-Imido Ligand Sets: Syntheses, X-Ray, NMR, and DFT Studies. *Organometallics* **2008,** *27*, 5968–5977.

100. Dubberley, S. R.; Ignatov, S. K.; Rees, N. H.; Razuvaev, A. G.; Mountford, P.; Nikonov, G. I. Are *J* (Si-H) NMR Coupling Constants Really a Probe for the Existence of Nonclassical H–Si Interactions? *J. Am. Chem. Soc.* **2003,** *125*, 642–643.

101. Shinohara, A.; McBee, J.; Waterman, R.; Tilley, T. D. Paramagnetic Vanadium Silyl Complexes: Synthesis, Structure, and Reactivity. *Organometallics* **2008,** *27*, 5717–5722.

102. (a) Ghadwal, R. S.; Azhakar, R.; Pröpper, K.; Holstein, J. J.; Dittrich, B.; Roesky, H. W. N-Heterocyclic Carbene Stabilized Dichlorosilylene Transition-Metal Complexes of V(I), Co(I), and Fe(0). *Inorg. Chem.* **2011,** *50*, 8502–8508.
(b) Azhakar, R.; Ghadwal, R. S.; Roesky, H. W.; Hey, J.; Stalke, D. Facile Access to Transition-Metal-Carbonyl Complexes With an Amidinate-Stabilized Chlorosilylene Ligand. *Chem. Asian J.* **2012,** *7*, 528–533.

103. Blom, B.; Driess, M.; Gallego, D.; Inoue, S. Facile Access to Silicon-Functionalized Bis-Silylene Titanium(II) Complexes. *Chem. Eur. J.* **2012,** *18*, 13355–13360.

104. (a) McAlexander, L. H.; Hung, M.; Li, L.; Diminnie, J. B.; Xue, Z.; Yap, G. P. A.; Rheingold, A. L. Alkyl-Silyl Complexes Free of Anionic π-Ligands. Synthesis and Characterization of $(Me_3ECH_2)_3MSi(SiMe_3)_3$. *Organometallics* **1996,** *15*, 5231–5235.
(b) Wu, Z.; Diminnie, J. B.; Xue, Z. Synthesis and X-Ray Crytstal Structures of a Chlorobis (trimethylsiloxy)zirconium Silyl Derivative, $(Me_3SiO)_2Zr(SiPh_2Bu^t)Cl \cdot 2THF$. *Organometallics* **1998,** *17*, 2917–2920.
(c) Wu, Z.; Diminnie, J. B.; Xue, Z. Synthesis and Characterization of Group 4 Amido Silyl Complexes Free of Anionic π-Ligands. *Inorg. Chem.* **1998,** *37*, 6366–6372.

105. Frank, D.; Baumgartner, J.; Marschner, C. First Successful Reaction of a Silyl Anion With Hafnium Tetrachloride. *Chem. Commun.* **2002,** 1190–1192.

106. Heyn, R. H.; Tilley, T. D. Tris(trimethyl)silyl Derivatives of Tri-*tert*-butoxyzirconium and Tri-*tert*-butoxyhafnium. X-Ray Crystal Structure of $(Me_3CO)_3ZrSi(SiMe_3)_3$. *Inorg. Chem.* **1989,** *28*, 1768–1769.

107. Spencer, M. D.; Shelby, Q. D.; Girolami, G. S. Titanium-Catalyzed Dehydrocoupling of Silanes: Direct Conversion of Primary Monosilanes to Titanium(0) Oligosilane Complexes With Agostic α-Si-H···Ti Interactions. *J. Am. Chem. Soc.* **2007,** *129*, 1860–1861.

108. (a) Wu, Z.; Xue, Z. Synthesis and Characterization of Tantalum Silyl and Disilyl Imido Complexes that do not Contain Anionic π-Ligands. *Organometallics* **2000,** *19*, 4191–4192.
(b) Wu, Z.; Cai, H.; Yu, X.; Blanton, J.; Diminnie, J. B.; Pan, H.-J.; Xue, Z.; Bryan, J. C. Synthesis of Tantalum(V) Amido Silyl Complexes and the Unexpected Formation of $(Me_2N)_3Ta(\eta^2\text{-}ONMe_2)[OSi(SiMe_3)_3]$ From the Reaction of $(Me_2N)_4Ta[Si(SiMe_3)_3]$ With O_2. *Organometallics* **2002,** *21*, 3973–3978.
(c) Yu, X.; Bi, S.; Guzei, I. A.; Lin, Z.; Xue, Z.-L. Zirconium, Hafnium, and Tantalum Amide Silyl Complexes: Their Preparation and Conversion to Metallaheterocyclic Complexes via γ-Hydrogen Abstraction by Silyl Ligands. *Inorg. Chem.* **2004,** *43*, 7111–7119.
(d) Qiu, H.; Cai, H.; Woods, J. B.; Wu, Z.; Chen, T.; Yu, X.; Xue, Z.-L. Disilyl Complexes of Zirconium, Hafnium, and Tantalum. Their Synthesis, Characterization, and Exchanges With Silyl Anions. *Organometallics* **2005,** *24*, 4190–4197.
(e) Yu, X.; Cai, H.; Guzei, I. A.; Xue, Z. Unusual Equilibria Involving Group 4 Amides, Silyl Complexes, and Silyl Anions via Ligand Exchange Reactions. *J. Am. Chem. Soc.* **2004,** *126*, 4472–4473.
(f) Cai, H.; Yu, X.; Chen, S.; Qiu, H.; Guzei, I. A.; Xue, Z. Amide-Silyl Ligand Exchanges and Equilibria Among Group 4 Amide and Silyl Complexes. *Inorg. Chem.* **2007,** *46*, 8071–8078.
(g) Chen, S. J.; Dougan, B. A.; Chen, X.-T.; Xue, Z. –L. Preparation of Zirconium Guanidinate Complexes From the Direct Insertion of a Carbodiimine and Aminolysis Using a Guanidine. Comparison of the Reactions. *Organometallics* **2012,** *31*, 3443–3446.

CHAPTER 3

Organosilicon Clusters

Soichiro Kyushin

Gunma University, Kiryu, Japan

Chapter Outline

Organosilicon Compounds.
DOI: http://dx.doi.org/10.1016/B978-0-12-801981-8.00003-4

3.1 Introduction

The chemistry of organosilicon clusters dates back to 1972, when West reported the formation of fused-ring oligosilanes with 8–13 silicon atoms.[1] After this pioneering work, organosilicon clusters began to be studied extensively in the 1980s. For example, Ishikawa and Kumada synthesized bicyclo[2.2.1]heptasilane and bicyclo[2.2.2]octasilane in 1982.[2] Boudjouk synthesized spiropentasilane in 1984.[3] Masamune reported bicyclo[1.1.0]-tetrasilane, bicyclo[1.1.1]pentasilane, tricyclo[2.2.0.0^{2,5}]hexasilane, and tetracyclo-[3.3.0.0^{2,7}.0^{3,6}]octasilane in 1985–90.[4–6] Matsumoto and Nagai reported a ladder oligosilane in 1987.[7] Persilapolyhedranes such as octasilacubane,[8] hexasilaprismane,[9] and tetrasilatetrahedrane[10] have been synthesized. Surprisingly, this important progress was made only within about 10 years. These studies established this area as one of the most rapidly developing research areas in organosilicon chemistry. Since then, the chemistry of organosilicon clusters has continued to be further developed.

This chapter reports the progress in the chemistry of organosilicon clusters. As previous studies have been summarized in many reviews,[11] this chapter deals with them briefly and focuses on the recent studies in detail. Organosilicon clusters described in this chapter include cage compounds, polyhedranes, ring catenation compounds, spirooligosilanes, and siliconoids. Although linear, branched, and monocyclic oligosilanes and polysilanes are not described here, some examples that are useful for understanding the properties of organosilicon clusters are included.

3.2 Synthesis

3.2.1 General Synthetic Methods: Si–Si Bond Formation

In contrast to a variety of methods of C–C bond formation, there have been used only a few methods of Si–Si bond formation. These methods are briefly discussed in this section.

3.2.1.1 Wurtz-type coupling of halosilanes with metals

Kipping used this method for the first time for the coupling of dichlorodiphenylsilane with sodium in 1921.[12] Since Gilman determined the structures of Kipping's phenyl-substituted cyclooligosilanes,[13] and Kumada synthesized many oligosilanes,[14] Wurtz-type coupling has been used as a major method of Si–Si bond formation. Monochlorosilanes and dichlorosilanes give symmetrically substituted linear and cyclic oligosilanes, respectively.[15,16] Wurtz-type coupling of two kinds of chlorosilanes has been used for the synthesis of linear, branched, and cyclic oligosilanes.[1,14,17]

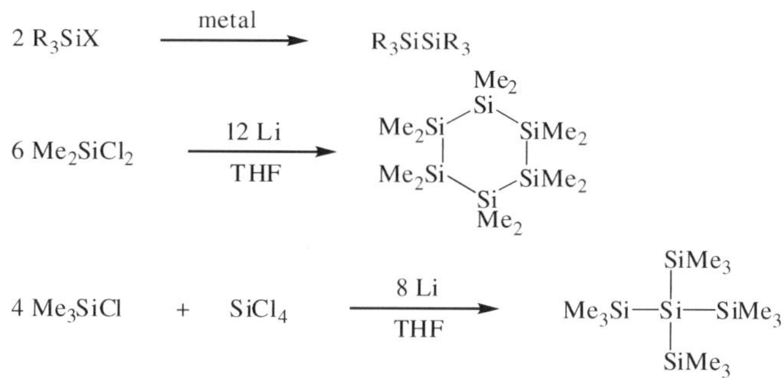

Wurtz-type coupling has been carried out by using various metals and reducing agents. Lithium, sodium, sodium/potassium, potassium, magnesium/magnesium bromide, lithium naphthalenide, and potassium graphite are usually used. Chlorosilanes are usually used as halosilanes. However, when the reaction time is very long or a reaction does not proceed, bromosilanes or iodosilanes are used instead.

3.2.1.2 Coupling of halosilanes with silyl anions

This method is an alternative way to oligosilanes. A Si–Si bond is formed by a reaction of silyl anion with halosilane. This method is useful for the synthesis of unsymmetrically substituted oligosilanes.

$$R_3SiM + R'_3SiX \rightarrow R_3SiSiR'_3$$

Silyllithium, silylsodium, and silylpotassium are used as silyl anions.[18] These silyl anions are prepared by the reactions of the corresponding chlorosilanes with lithium, sodium, and potassium. Silyllithium can also be prepared by the reactions of silylmercury compounds with lithium. Silyl-substituted silylpotassium can be prepared conveniently by the reactions of a silyl-substituted silanes with potassium *tert*-butoxide.[19] Recently, preparation of silylmagnesium reagents have been reported. Silylmagnesium reagents can be prepared by the corresponding silyllithiums with Grignard reagents.[20] Silylmagnesium reagents react with chlorooligosilanes without cleavage of Si–Si bonds. As silyllithium sometimes leads to formation of a complex mixture as a result of Si–Si bond cleavage, silylmagnesium is a useful reagent for synthesis of oligosilanes.

3.2.1.3 Rearrangement of oligosilanes with aluminum chloride

Silicon skeletons can be rearranged by aluminum chloride. This rearrangement was found by Ishikawa and Kumada[2,21,22] and studied by West[23] and Marschner.[24] A typical example is shown in the following equation, where dodecamethylcyclohexasilane is transformed into nonamethyltrimethylsilylcyclopentasilane in the presence of a catalytic amount of aluminum chloride almost quantitatively.[22a]

The reaction mechanism of this rearrangement was proposed by Ishikawa and Kumada.[22c] The mechanism is exemplified by decamethyltetrasilane **1**. Compound **1** is chlorinated by a trace amount of hydrogen chloride in the presence of aluminum chloride. The resulting 1-chlorotetrasilane reacts with **1** to give 2-chlorotetrasilane via intermolecular methyl–chlorine exchange. 2-Chlorotetrasilane is rearranged via the transition state **2** to more branched tetrasilane. The branched tetrasilane reacts with **1** to give the product **3** and 1-chlorotetrasilane.

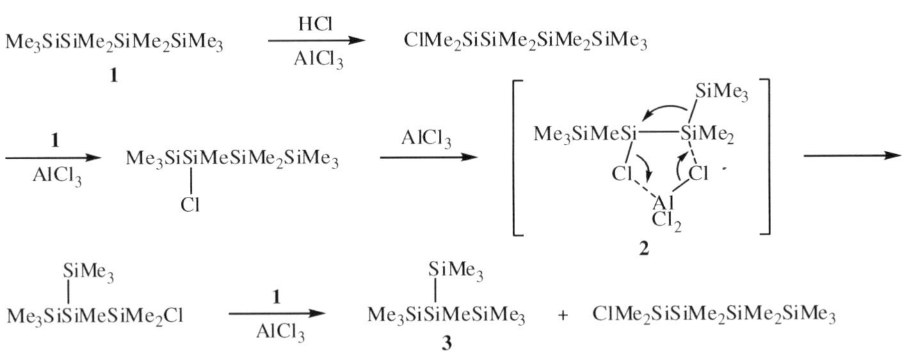

West studied rearrangement of **4** with Al(Fe)Cl_3 and found that rearrangement and redistribution of alkyl groups occur simultaneously.[23] He suggested that both reactions can be explained by the common intermediate **5**. This intermediate could exchange alkyl groups with uncomplexed silane or rearrange via intramolecular 1,2-Si–Si and Si–C bond shifts to give **6**.

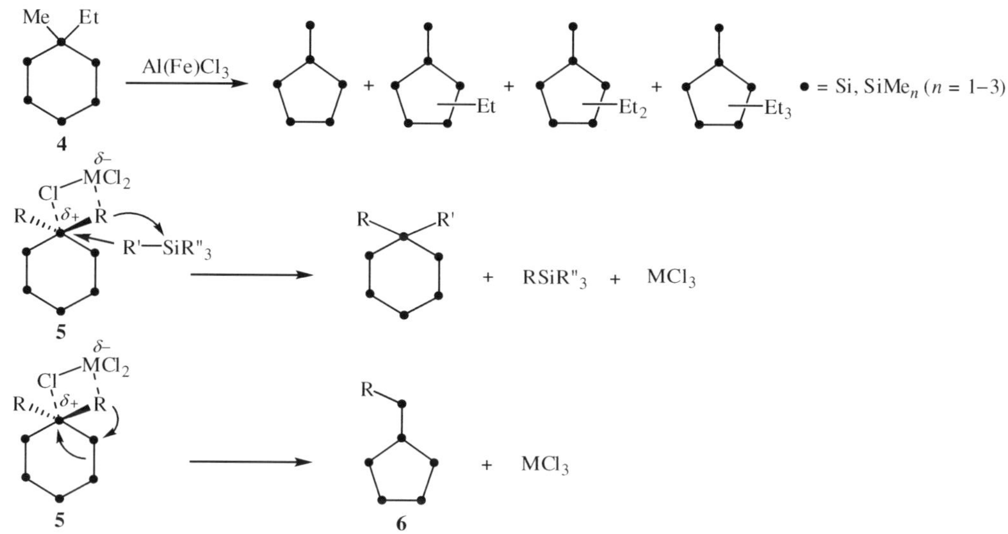

More recently, Müller and Marschner suggested a rearrangement mechanism via silyl and germyl cations to explain the rearrangement of **7** to **8** in the presence of Al(Fe)Cl_3. This mechanism is supported by theoretical calculations.[24b]

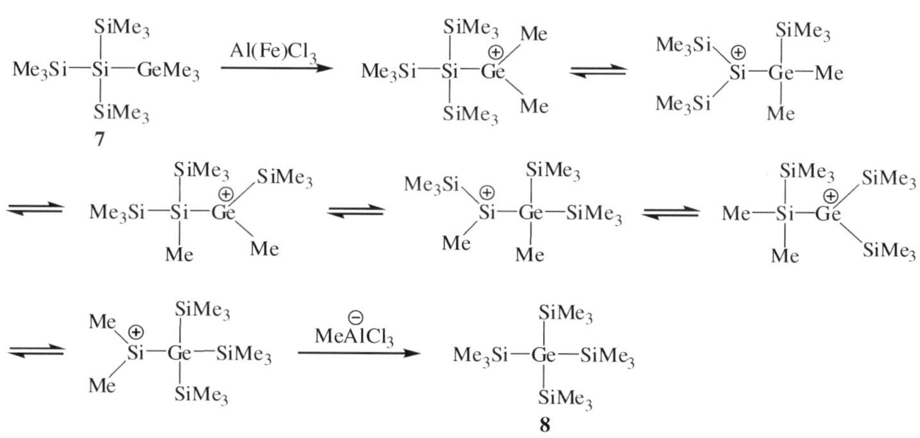

As a result of this rearrangement, more branched oligosilanes are formed. Although prediction of rearrangement products is not easy, novel oligosilanes which cannot be obtained by other methods can be synthesized by this method. For example, Marschner synthesized silyl-substituted decasilaadamantane by this rearrangement.[25] This result is explained in Section 3.2.3.

3.2.1.4 Oxidative coupling of silyl anion

Silyl anion has been known to donate an electron to electron acceptors. By recombination of the resulting silyl radical, a Si–Si bond is formed. This method is effective when silyl anion is readily available. 1,2-Dibromoethane and halosilanes are used as electron acceptors.

$$2\ R_3SiM \xrightarrow{\text{acceptor}} R_3SiSiR_3$$

3.2.1.5 Dehydrogenative coupling of hydrosilanes with transition metal catalysts

Formation of Si–Si bonds from hydrosilanes in the presence of transition metal catalysts is an attractive way for the synthesis of oligosilanes and polysilanes because many hydrosilanes are available as starting materials.

$$2\ R_3SiH \xrightarrow{\text{transition metal catalyst}} R_3SiSiR_3$$

Many kinds of transition metal catalysts have been employed.[26] This method is effective when methyl- and phenyl-substituted dihydro- and trihydrosilanes are used. By using this method, polysilane polymers have been synthesized.[27] However, this method has not yet been applied to synthesis of organosilicon clusters.

3.2.2 Early Work

West reported the coupling of dichlorodimethylsilane and trichloromethylsilane with sodium/potassium alloy in the presence of naphthalene.[1] The reaction mixture was separated to give fused-ring oligosilanes with 8–13 silicon atoms together with a large amount of polymers. Photolysis of **9** gave the silylene extrusion product **11**. Possible structures of these compounds were proposed. Later, the structure of **10** was determined by X-ray crystallography.[27] The structure of $Me_{18}Si_{10}$ is proposed to be octadecamethylbicyclo-[4.4.0]decasilane,[1,2] but Hengge pointed out that this structure cannot be distinguished from bi(nonamethylcyclopentasilanyl).[29]

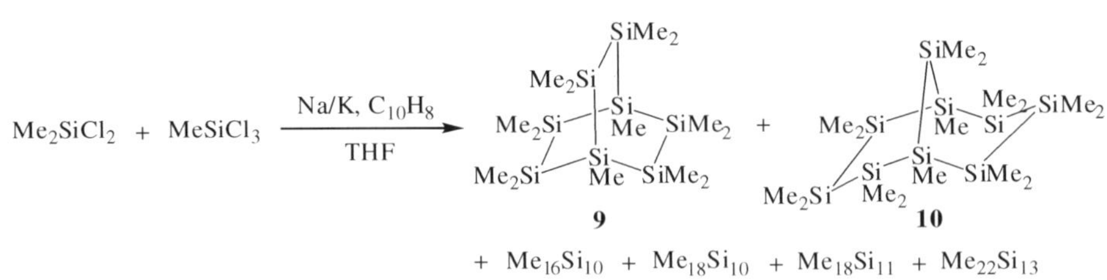

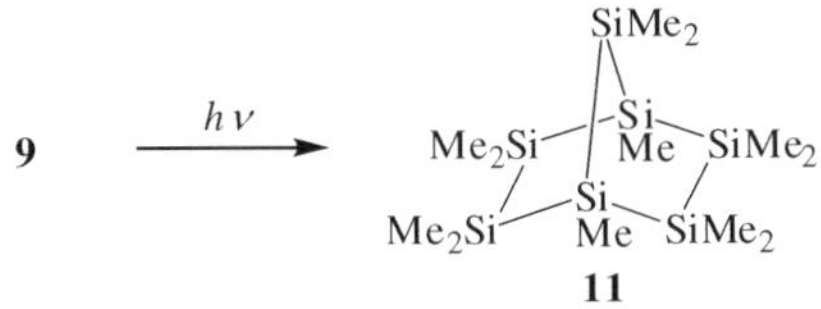

3.2.3 Cage Compounds

3.2.3.1 Bicyclo[1.1.1]pentasilanes and persilastaffanes

Masamune synthesized bicyclo[1.1.1]pentasilane **13a–c** by the reaction of **12** with lithium naphthalenide and subsequent reactions with dichlorosilanes.[5] The structure of **13a** was determined by X-ray crystallography.

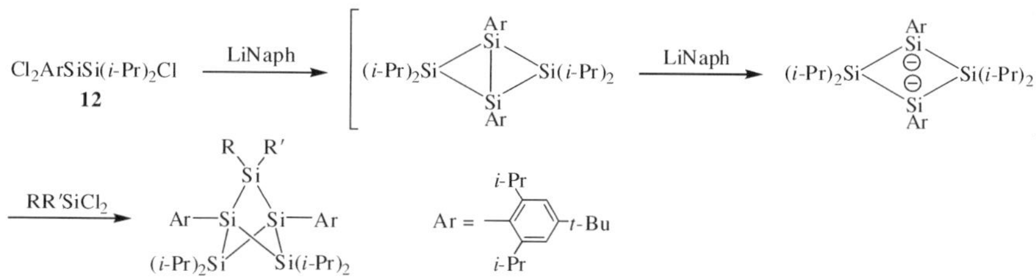

13a: R = R′= H
13b: R = Me, R′= H
13c: R = Ph, R′= H

Bicyclo[1.1.1]pentasilanes were also obtained by the bond cleavage of pentasila[1.1.1]-propellane (Section 3.6.4).

Iwamoto synthesized bicyclo[1.1.1]pentasilane **14** by the reaction of 1,1,3,3-tetrasilylcyclotetrasilane with potassium *tert*-butoxide and the subsequent reaction of 1,3-dipotassiocyclotetrasilane with dichlorodiisobutylsilane.[29] Compound **14** was converted into potassium bicyclo[1.1.1]pentasilanide **15** and 1-bromobicyclo[1.1.1]pentasilane **16**. The coupling of **15** with **16** gave decasila[2]staffane **17**. This compound was converted into silylpotassium **18**, and the reaction of **18** with **16** gave pentadecasila[3]staffane **19**.

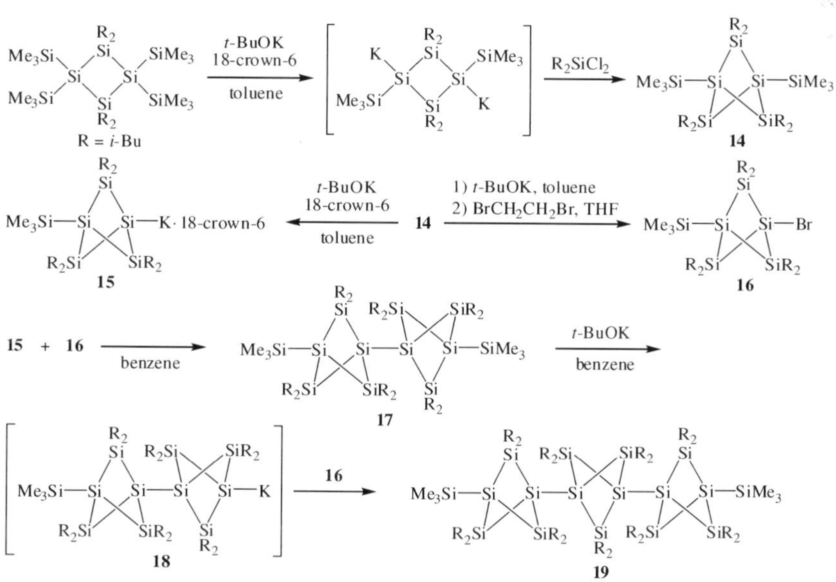

3.2.3.2 Bicyclo[2.2.1]heptasilanes and bicyclo[2.2.2]octasilanes

Bicyclo[2.2.1]heptasilane **11** and bicyclo[2.2.2]octasilane **9** were obtained by West as described above.[1] Ishikawa and Kumada have reported the formation of bicyclo[2.2.1]-heptasilanes **20** and **23** and bicyclo[2.2.2]octasilane **22** by the rearrangement of **9** and **21** with a catalytic amount of aluminum chloride and subsequent reactions with methylmagnesium bromide.[2]

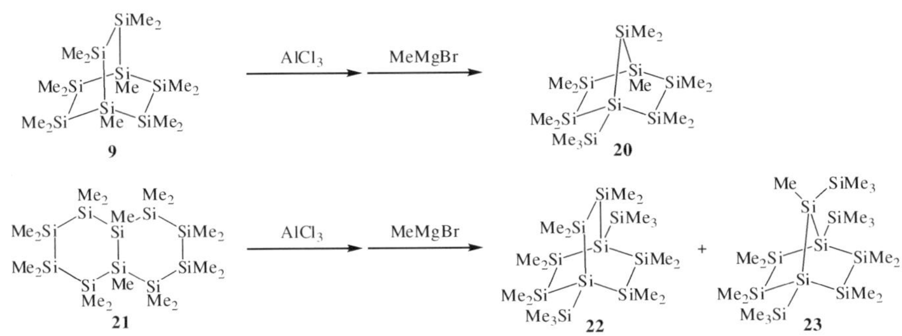

Compound 9 was also synthesized by the dimerization of tris(chlorodimethylsilyl)methylsilane with lithium.[30] The methyl groups at the bridgehead positions can be converted into chlorine and potassium atoms by using boron trichloride and potassium graphite, respectively.

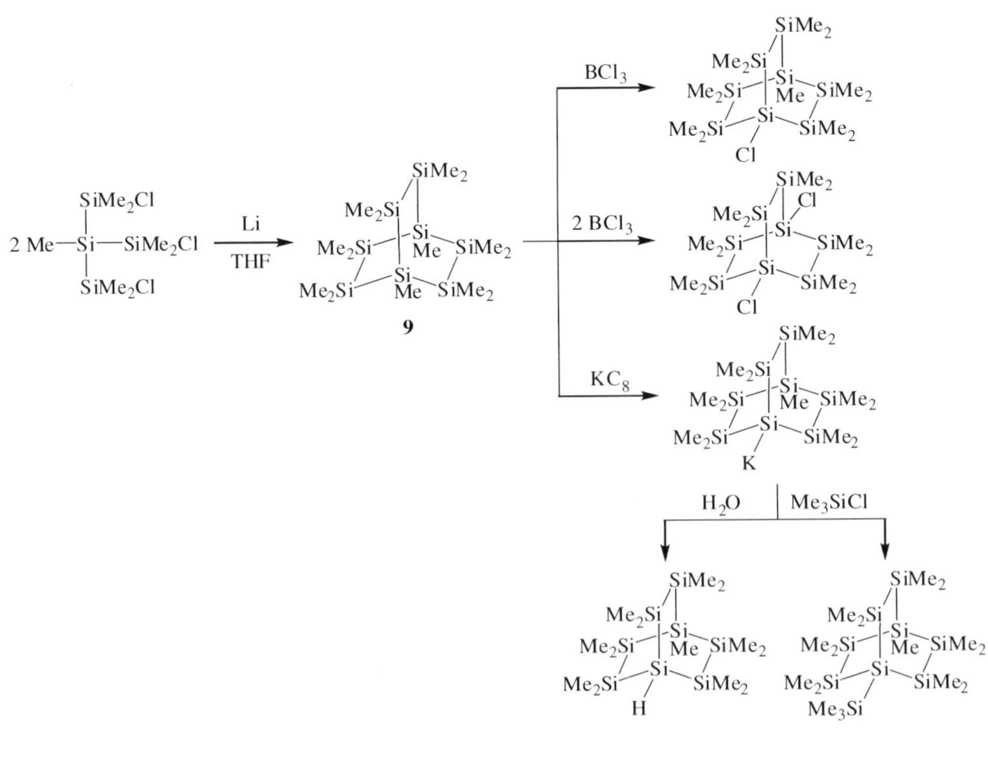

Marschner synthesized bicyclo[2.2.1]heptasilane **24** and bicyclo[2.2.2]octasilane **22** by the reaction of 1,4-dipotassiocyclohexasilane with dichlorodimethylsilane and 1,2-dihalodisilane, respectively.[31,32] The trimethylsilyl groups at the bridgehead positions of **22** were converted into hydrogen atoms, methyl groups, and other silyl groups via silylpotassiums **25a**,**b**.

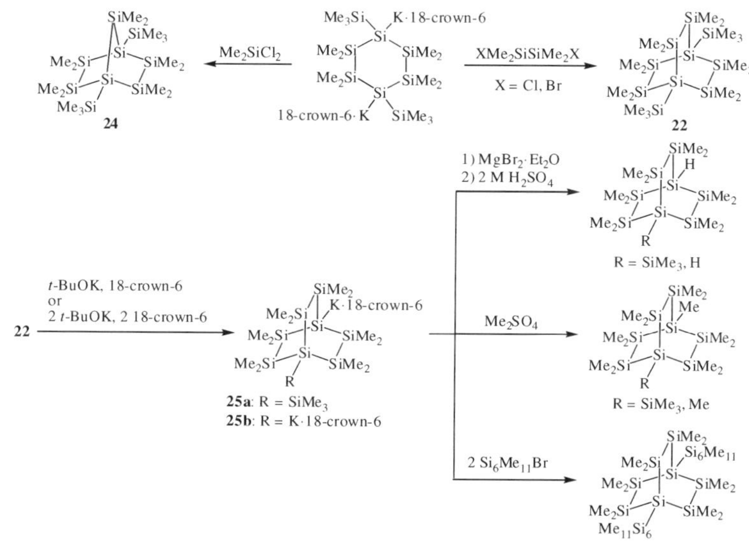

The reaction of **22** or **24** with potassium *tert*-butoxide followed by the coupling with α,ω-dichlorooligosilanes gave **26** or **27**, in which two bicyclo[2.2.1]heptasilane or bicyclo-[2.2.2]octasilane units are connected by an oligosilane chain.[33]

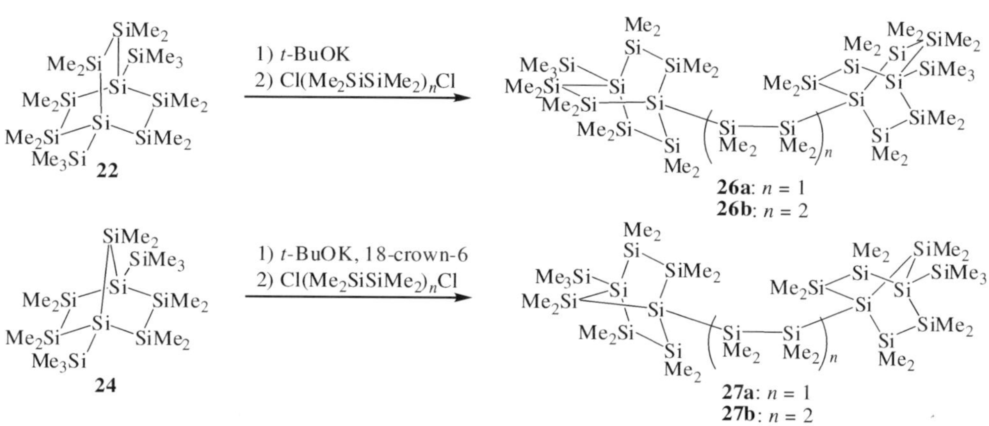

3.2.3.3 Bicyclo[3.3.1]nonasilanes and bicyclo[4.3.1]decasilane

Bicyclo[3.3.1]nonasilane **10** was obtained by West as described above.[1] Bicyclo[3.3.1]-nonasilane **28** and bicyclo[4.3.1]decasilane **29** were synthesized by the reactions of 1,3-dipotassiocyclohexasilane with 1,3-dichlorotrisilane and 1,4-dichlorotetrasilane, respectively.[34]

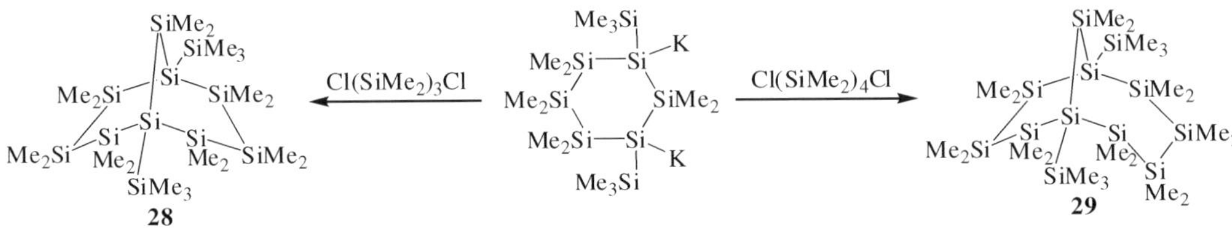

3.2.3.4 Tricyclo[2.1.0.0^{2,5}]pentasilanes

Compound **31** was synthesized by the reaction of $Ca^{2+} \cdot$ **30**$^{2-}$ with trichlorophenylsilane. This chlorosilane was transformed with potassium graphite into the rearranged silylpotassium **34**. This rearrangement may involve 1,3-silyl migration (**32** to **33**) and bridgehead-to-bridgehead 1,2-silyl migration (**33** to **34**). The reaction of **34** with chlorotrimethylsilane gave **35**.[35]

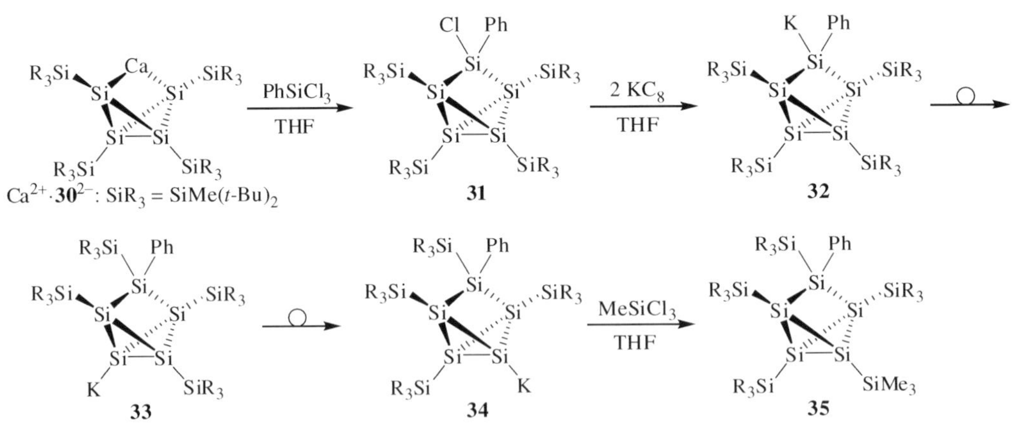

Other tricyclo[2.1.0.0^{2,5}]pentasilane derivatives **92**, **95**, and **165** are described in Sections 3.2.6, 3.2.7, and 3.6.4.

3.2.3.5 Tricyclo[2.2.0.0^{2,5}]hexasilanes

Tricyclo[2.2.0.0^{2,5}]hexasilanes **36a,b** and **37** were formed together with other products in the reduction of 1,1,2,2-tetrachlorodisilane with lithium naphthalenide. The X-ray structure of **37** has been reported.[6]

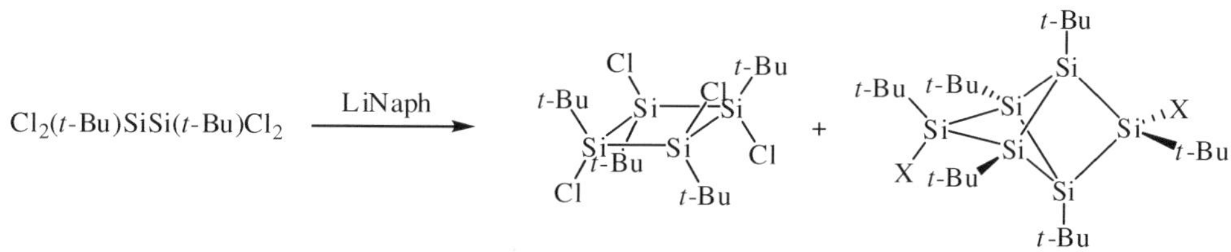

36a: X = Cl
36b: X = Me

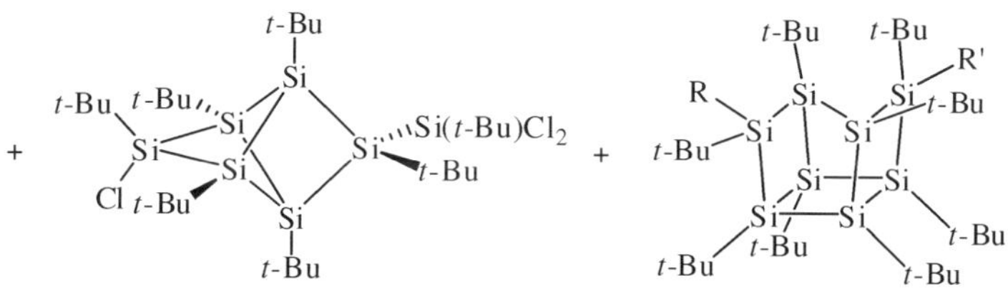

37

38a: R = R' = Cl
38b: R = Me, R' = Cl
38c: R = R' = Me

Other tricyclo[2.2.0.0^{2,5}]hexasilanes **162**, **164a–c**, and **184** are described in Sections 3.6.4 and 3.6.6.

3.2.3.6 Heptasilanortricyclene

Heptasilanortricyclene **40** was synthesized by the intramolecular coupling of the oligosilane dendrimer **39** with lithium naphthalenide.[36]

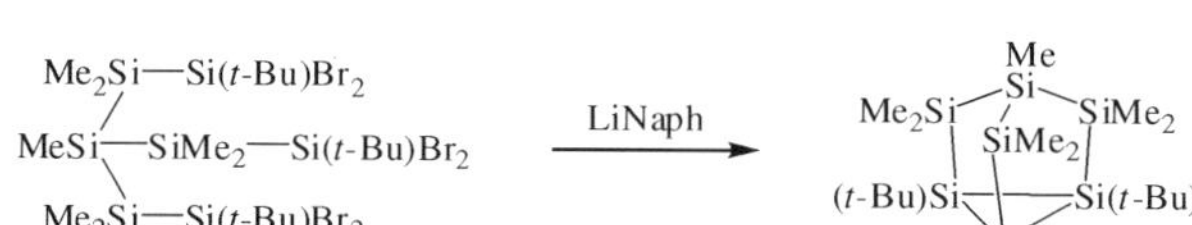

39 **40**

3.2.3.7 Decasilaadamantane

Rearrangement of tetradecasilane **41** with a catalytic amount of aluminum chloride gave tetrakis(trimethylsilyl)decasilaadamantane **42** in 78% yield.[25] Although this rearrangement pathway cannot be understood easily, Marschner explained that tricyclic precursor **41** is suitable for the construction of the tricyclic adamantane skeleton.

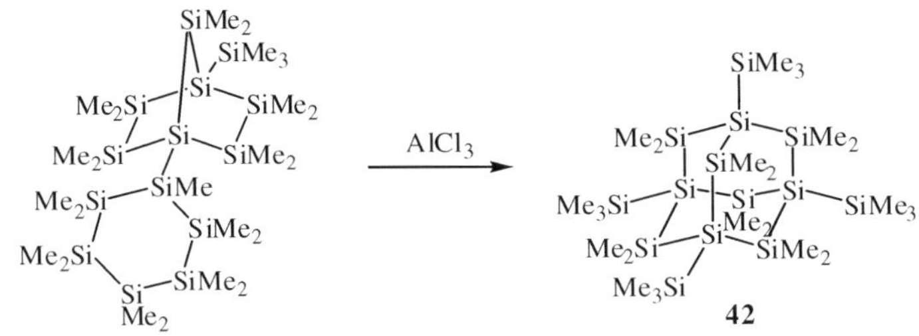

41 42

3.2.3.8 Tetracyclo[3.3.0.0^{2,7}.0^{3,6}]octasilanes

Tetracyclo[3.3.0.0^{2,7}.0^{3,6}]octasilanes **38a–c** were formed in the reduction of 1,1,2,2-tetrachlorodisilane with lithium naphthalenide as shown above.[6]

The reactions of octasilacubane **43** with phosphorus pentachloride, bromine, or iodine gave three stereoisomers of **44–46**.[37a,b] Compound **47** was formed in the reduction of **44–46** with sodium together with **43**.[37b]

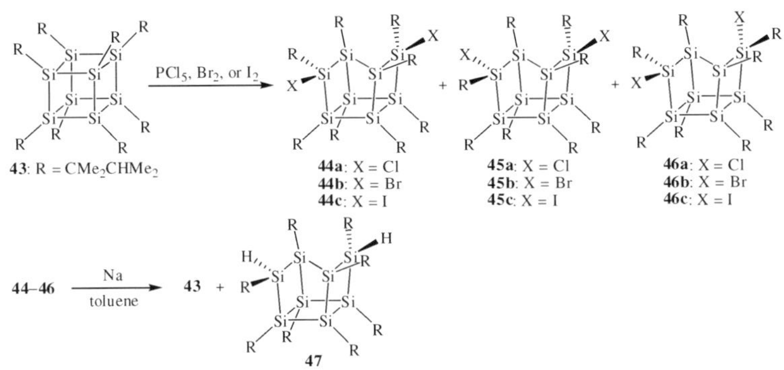

43: R = CMe$_2$CHMe$_2$

44a: X = Cl; **44b**: X = Br; **44c**: X = I

45a: X = Cl; **45b**: X = Br; **45c**: X = I

46a: X = Cl; **46b**: X = Br; **46c**: X = I

47

Compound **45b** was also obtained on irradiation of **43** with tetrabromomethane or 1,2-dibromoethane in the absence of oxygen.[37c] In the presence of oxygen, compound **43** reacts with tetrabromomethane without irradiation to give **45b** and monooxidation product **48**.

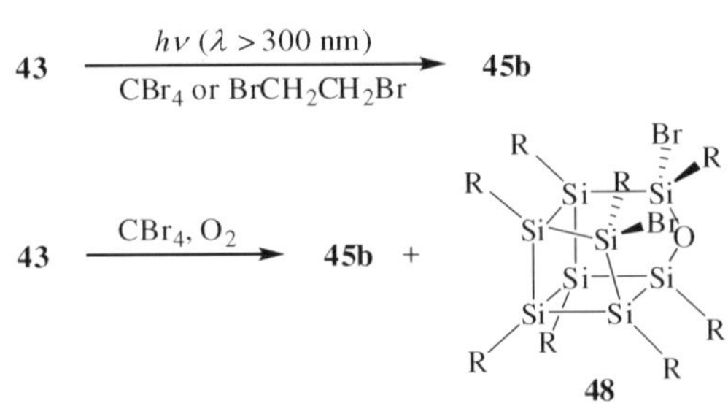

48

Another tetracyclo[3.3.0.0^{2,7}.0^{3,6}]octasilane **156** is described in Section 3.6.4.

3.2.4 Polyhedranes

3.2.4.1 Tetrasilatetrahedranes and bi(tetrasilatetrahedranyl)

Tetrasilatetrahedranes **49** and **50** were synthesized by the coupling of 2,2,3-3-tetrabromotetrasilane with tri-*tert*-butylsilylsodium in THF.[10,38]

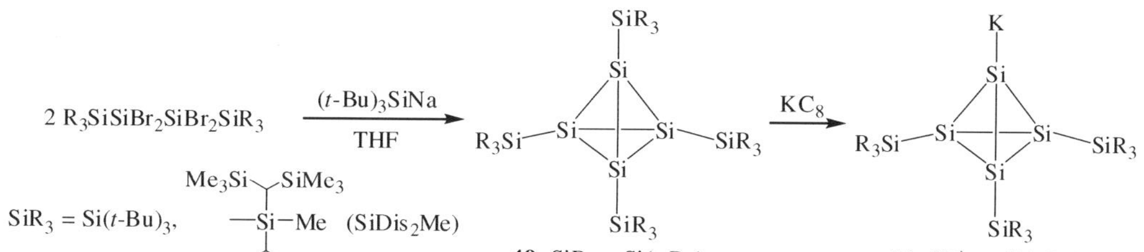

49: $SiR_3 = Si(t\text{-}Bu)_3$
50: $SiR_3 = SiDis_2Me$
51: $SiR_3 = Si(t\text{-}Bu)_3$
52: $SiR_3 = SiDis_2Me$

The reduction of **49** and **50** with potassium graphite gave tetrasilatetrahedranylpotassiums **51** and **52**, respectively.[38,39] The oxidative coupling of **51** with silicon tetrabromide gave bi-(tetrasilatetrahedranyl) **53**.[39]

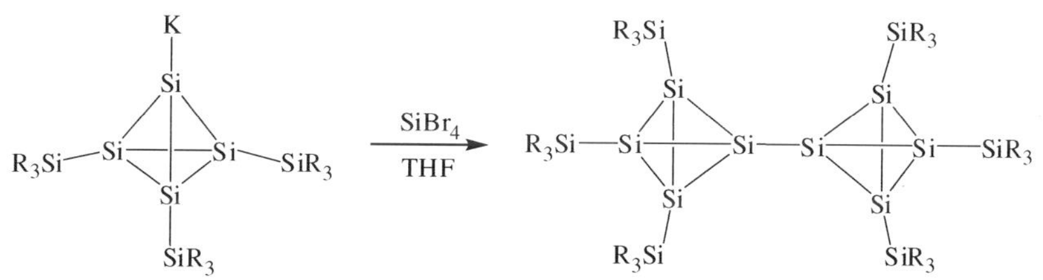

51: R = *t*-Bu
53: R = *t*-Bu

3.2.4.2 Hexasilaprismanes

Hexasilaprismane with 2,6-diisopropylphenyl (Dip) groups (**54**) was synthesized by the coupling of trichlorosilane or 1,1,2,2-tetrachlorodisilane with magnesium/magnesium bromide.[9] More recently, hexasilaprismane with 2,4,6-triisopropylphenyl (Tip) groups (**55**) was synthesized by the dimerization of 1,2,3-trichlorocyclotrisilane with magnesium.[40]

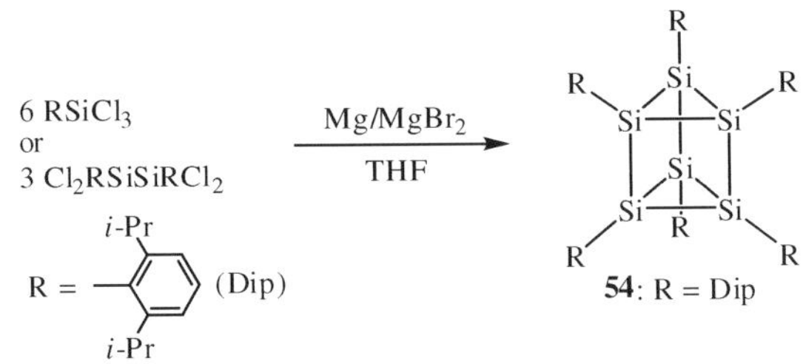

54: R = Dip

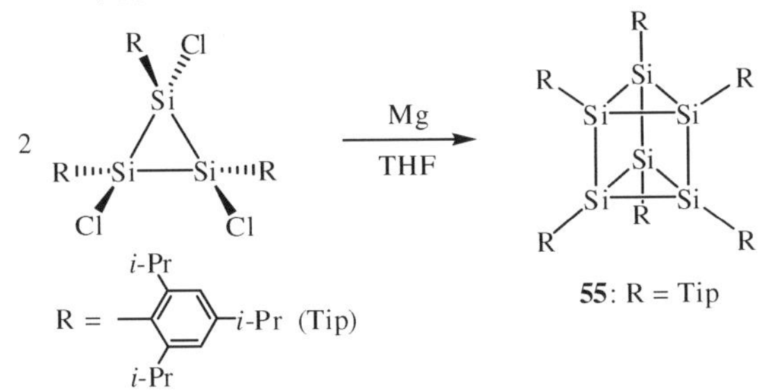

55: R = Tip

3.2.4.3 Octasilacubanes

Octasilacubanes with alkyl, aryl, and silyl substituents have been reported. Octasilacubane **56** was synthesized by the coupling of trichlorosilane or 1,2,3,4-tetrachlorocyclotetrasilane with sodium in the presence of 12-crown-4.[11d,41] Octasilacubane **43** was obtained by the coupling of trichlorosilane with sodium.[42] Octasilacubane **57** was synthesized by the coupling of trichlorosilane with magnesium/magnesium bromide.[43] Octasilacubane **58** was synthesized by the coupling of tribromosilane or 2,2,3,3-tetrabromotetrasilane with sodium.[8]

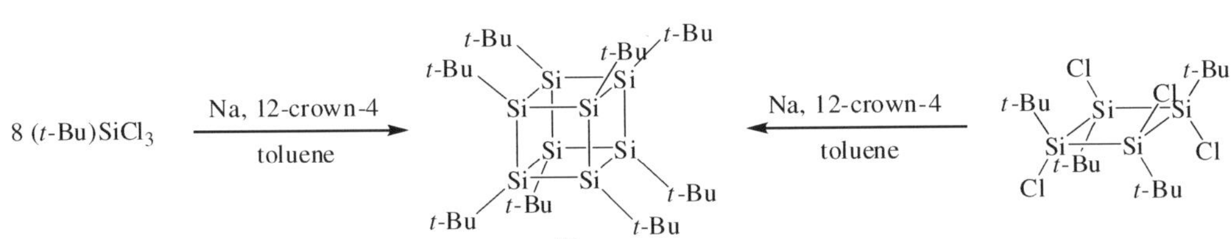

56

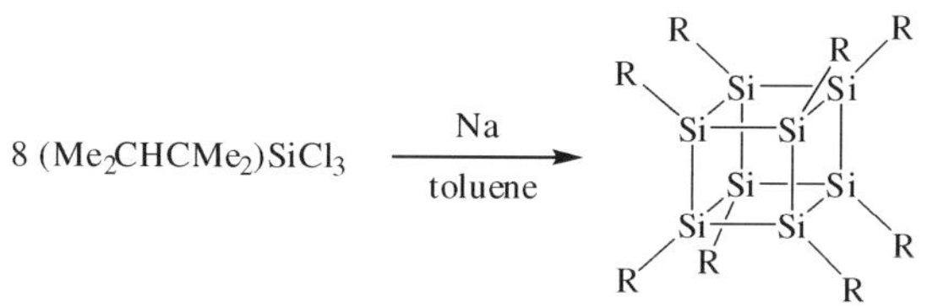

43: R = CMe_2CHMe_2

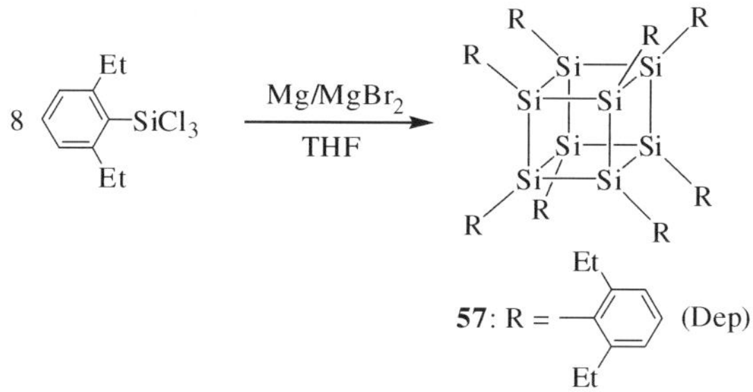

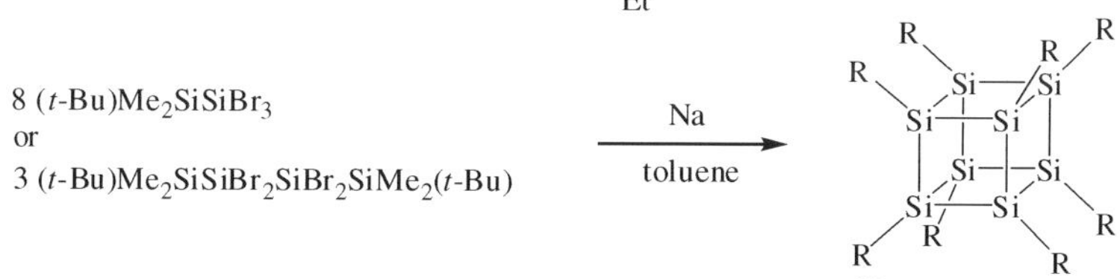

58: R = $SiMe_2(t\text{-}Bu)$

3.2.5 Ring Catenation Compounds

3.2.5.1 Bicyclo[1.1.0]tetrasilanes

Masamune synthesized bicyclo[1.1.0]tetrasilane **59** by intramolecular Si–Si bond formation of 1,3-dichlorocyclotetrasilane with lithium naphthalenide.[4a]

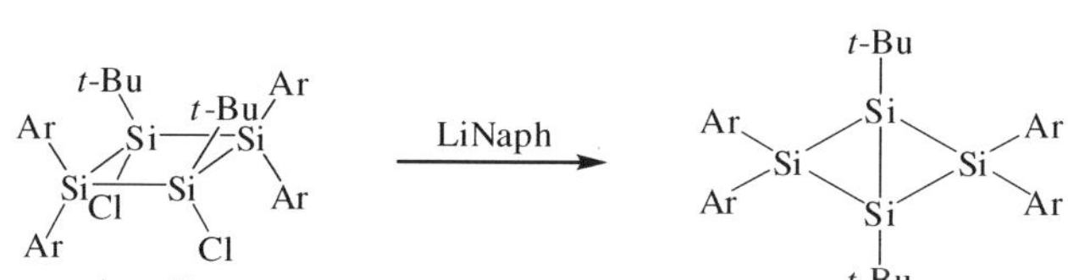

59

Bicyclo[1.1.0]tetrasilane **61** was obtained by the photolysis of **60** or **62** and the reaction of **63** with potassium graphite.[44]

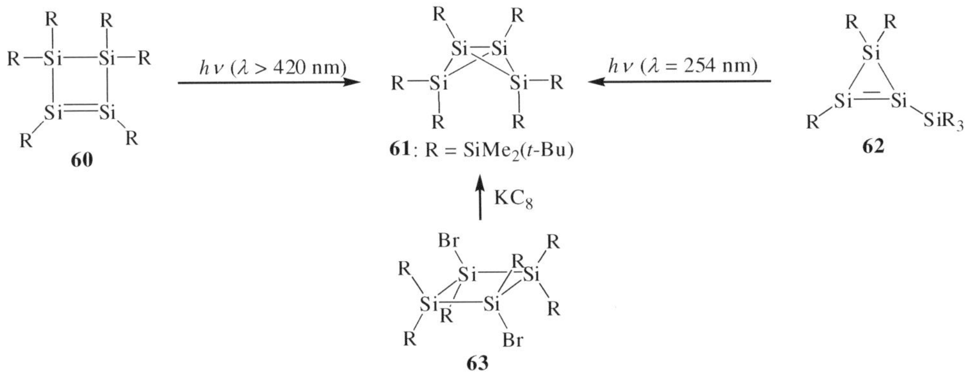

The reaction of 2,2,3-tribromotetrasilane with tri-*tert*-butylsilylsodium gave a mixture of **64** and **65** (10:1). These isomers were separated by preparative HPLC. Compound **65** was also obtained by the reaction of dianion of tetrasilatetrahedrane with methanol (Section 3.7.1).[45]

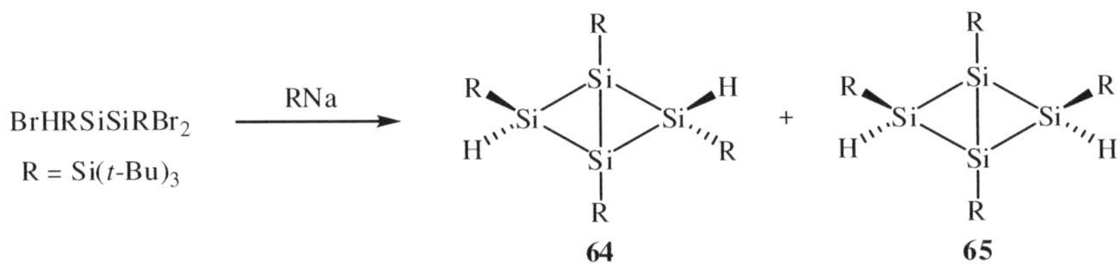

Several compounds containing the bicyclo[1.1.0]tetrasilane skeleton have been reported. 1,2,5,6-Tetrasilabenzobenzvalene **67** was synthesized by the reaction of the potassium salt of tetrasilacyclobutadiene dianion $2K^+ \cdot$ **66**.[2] with *o*-dibromobenzene[46]

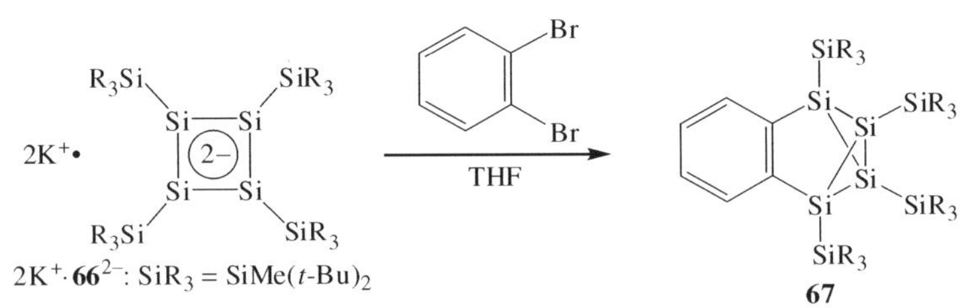

The reaction of **69** with boron triiodide gave iodonium-bridged bicyclo[1.1.0]tetrasilane **68**. The oxygen-bridged bicyclo[1.1.0]tetrasilane **70** was obtained on standing of a solution of **69** in C_6D_6 for one week.[47] Another method of obtaining **70** is described in Section 3.6.5.

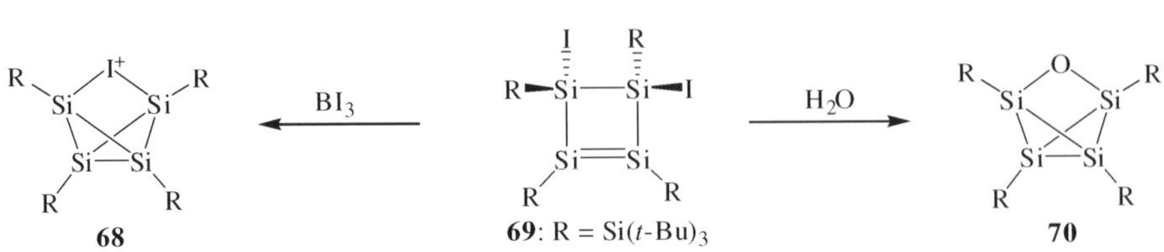

3.2.5.2 Tricyclo[3.1.0.0^{2,4}]hexasilane

Compound **71** was synthesized by the coupling of dibromosilane and 2,2,3,3-tetrabromotetrasilane with sodium.[48] The X-ray crystallographic analysis of **71** showed that this compound has an anti structure of the tricyclic silicon skeleton.

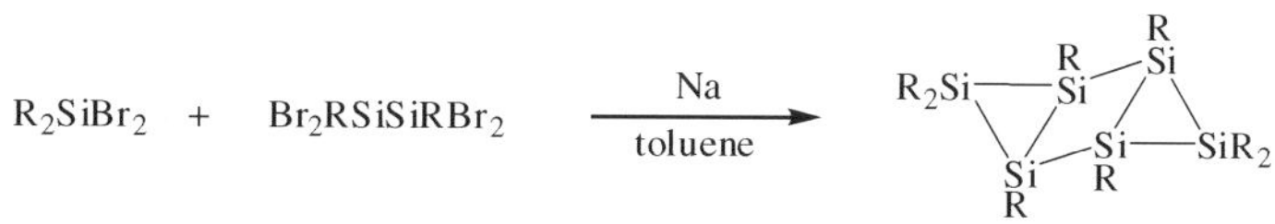

71: R = $SiMe_2(t\text{-}Bu)$

3.2.5.3 Tetracyclo[3.3.0.0^{1,3}.0^{5,7}]octasilane

Compound **72** was obtained by the dimerization of 1,1,3-trichlorocyclotetrasilane with potassium graphite.[49] The formation of this compound is explained by the dimerization of the cyclotetrasilane, intramolecular Si–Si bond formation, double 1,2-silyl migration, and intramolecular Si–Si bond formation.

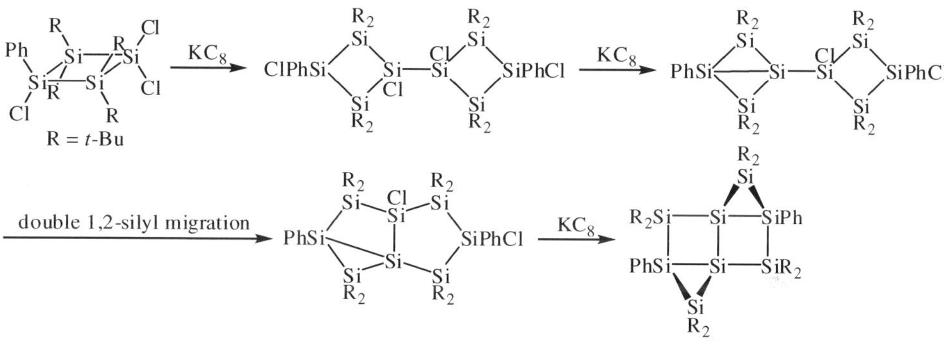

72

3.2.5.4 Pentacyclo[5.1.0.0^{1,6}.0^{2,5}.0^{3,5}]octasilane

Iwamoto has reported the reaction of disilenide with silicon tetrachloride. When the resulting product **73** was heated at 40°C, compound **74** was formed together with silene.[50]

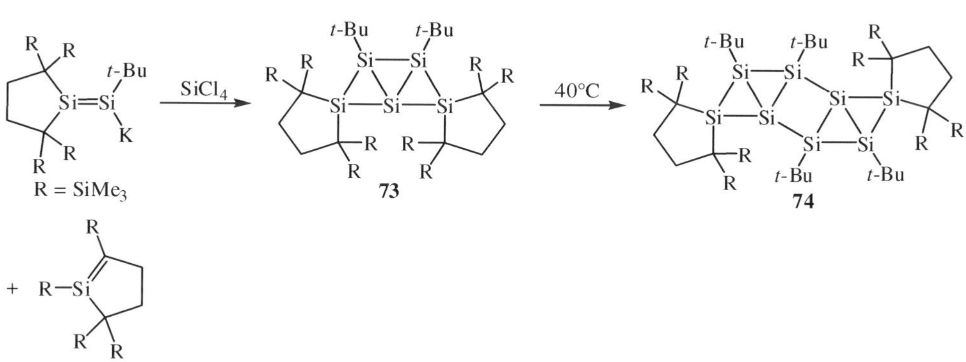

3.2.5.5 Ladder oligosilanes[51]

Matsumoto and Nagai synthesized ladder oligosilanes by the coupling of 1,2-dichlorodisilane and 1,1,2,2-tetrachlorodisilane with lithium in THF. Separation of the reaction mixture by recycling HPLC gave bicyclic to pentacyclic ladder oligosilanes.[7,51] Similar coupling of 1,1,4,4-tetrachlorotetrasilane and 1,1,2,2-tetrachlorodisilane gave tricyclic to octacyclic ladder oligosilanes.[52]

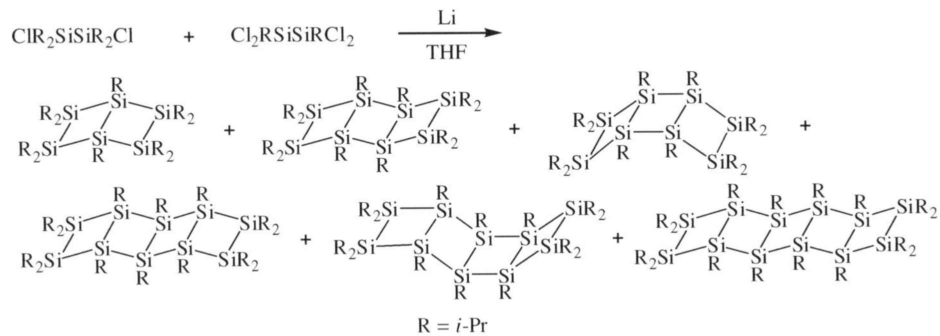

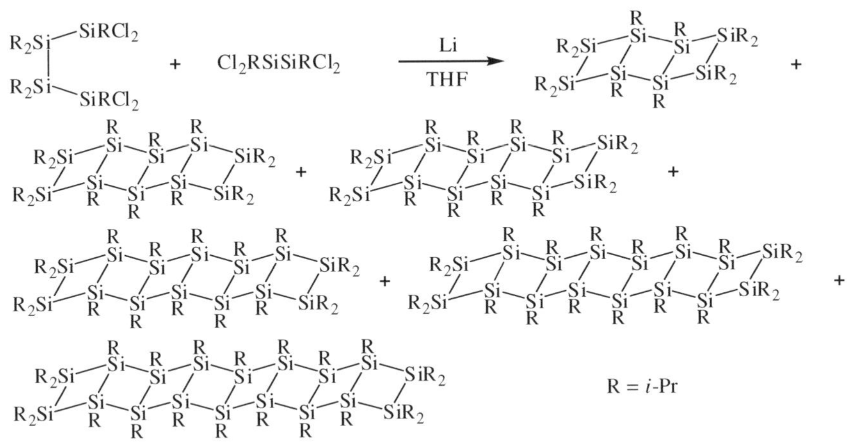

Bicyclic ladder oligosilane **75** was synthesized by intramolecular oxidative coupling of 1,4-dipotassiocyclohexasilane with 1,2-dibromoethane.[31,32]

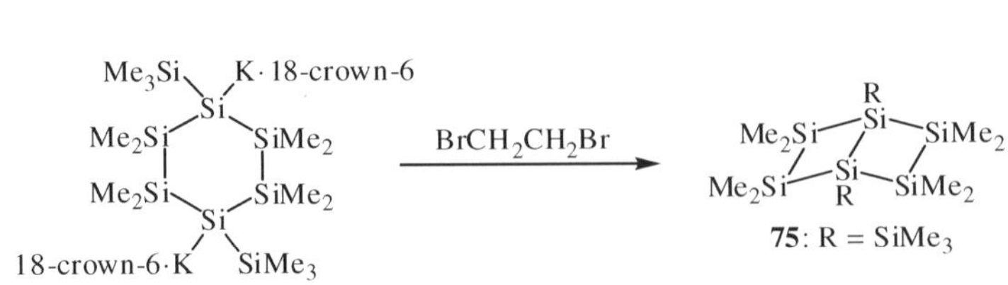

Another bicyclic ladder oligosilane **163** was obtained in the reaction of a siliconoid with bismuth trichloride.[40] This result is discussed in Section 3.6.4.

3.2.5.6 Tricyclo[5.3.0.0^{2,6}]decasilane

The reduction of 1,1-dibromocyclopentasilane with potassium graphite gave tricyclo-[5.3.0.0^{2,6}]decasilane **76**.[53] The initially generated silylene is rearranged to disilene, and the disilene dimerizes to **76**. The X-ray crystallographic analysis showed the anti arrangement of the two cyclopentasilane rings.

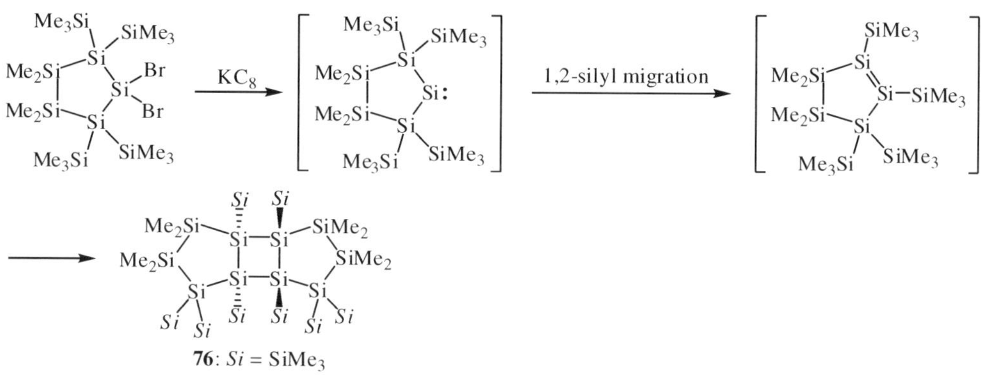

3.2.5.7 Bicyclo[3.3.0]octasilanes, bicyclo[4.2.0]octasilanes, and bicyclo[4.3.0]nonasilane

Bicyclo[3.3.0]octasilane **77** and bicyclo[4.3.0]nonasilane **79** were synthesized by the reactions of 1,2-dipotassiocyclopentasilane **78** with 1,3-dichlorotrisilane or 1,4-dichlorotetrasilane, respectively.[34]

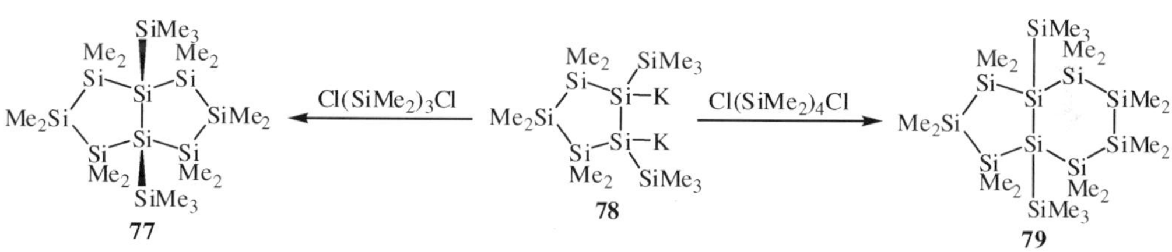

Other examples of bicyclo[3.3.0]octasilanes **142** and **145–147** and bicyclo[4.2.0]octasilanes **143** and **144** are discussed in Section 3.6.1.

3.2.5.8 Bicyclo[4.4.0]decasilane

As described above, West reported the formation of bicyclo[4.4.0]decasilane by the coupling of dichlorodimethylsilane and trichloromethylsilane with sodium/potassium.[1] Later, Hengge synthesized **80** by the reaction of 1,2-dichloro-1,1,2,2-tetramethyldisilane and trichloromethylsilane with lithium and determined the trans structure by X-ray crystallography.[29]

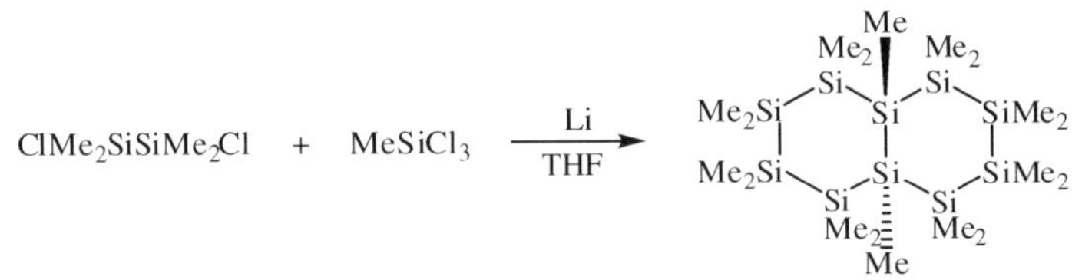

80

3.2.5.9 *Bicyclo[1.1.0]tetrasil-1(2)-ene and bicyclo[3.3.0]octasil-1(5)-ene: organosilicon clusters with a Si═Si double bond*

Bicyclo[1.1.0]tetrasil-1(2)-ene **81** · DMAP was obtained by heating **73** in the presence of 4-(*N*,*N*-dimethylamino)pyridine (DMAP).[50]

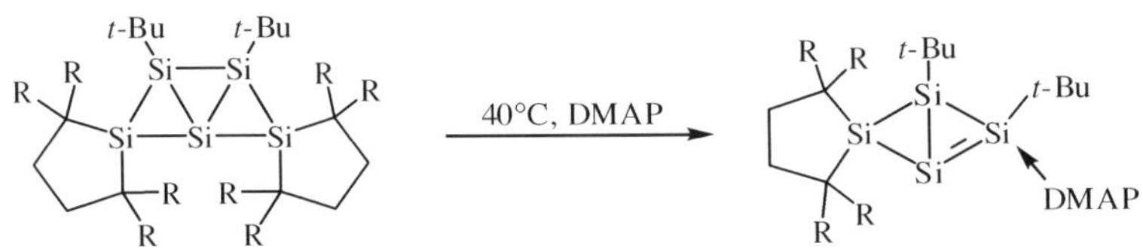

73: R = $SiMe_3$ **81**·DMAP

The coupling of 1,1-dichlorocyclotetrasilane **82** with lithium gave **84**.[54] The formation of **84** is explained by double 1,2-silyl migration of initially formed disilene **83**. Theoretical calculations of model compounds support this result: the energy of **83′** is 19.6 kcal mol^{-1} higher than that of **84′** at the B3LYP/6-31G(d) level.

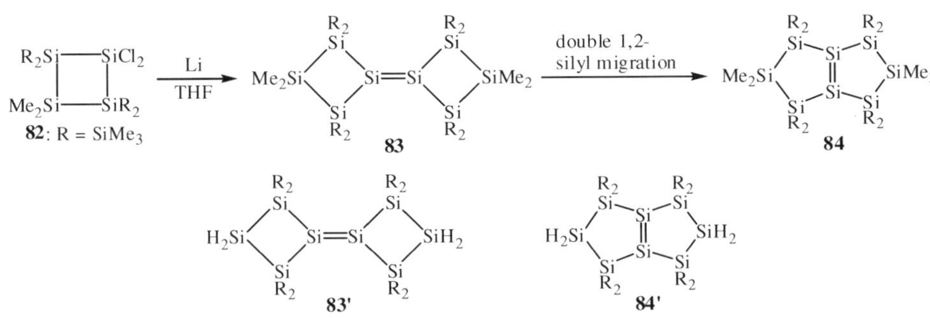

82: R = $SiMe_3$ **83** **84** **83'** **84'**

Other examples of organosilicon clusters with a Si═Si double bond (**101** and **102**) are described in Section 3.2.8.

3.2.6 *Spirooligosilanes*

The first spirooligosilane **85** was reported by Boudjouk in 1984.[3] This compound was synthesized by intramolecular coupling of tetrakis(bromodimethylsilyl)silane with lithium. Although **85** was not isolated, its structure was determined by the ^{1}H NMR spectrum and conversion into tetrasilylsilanes.

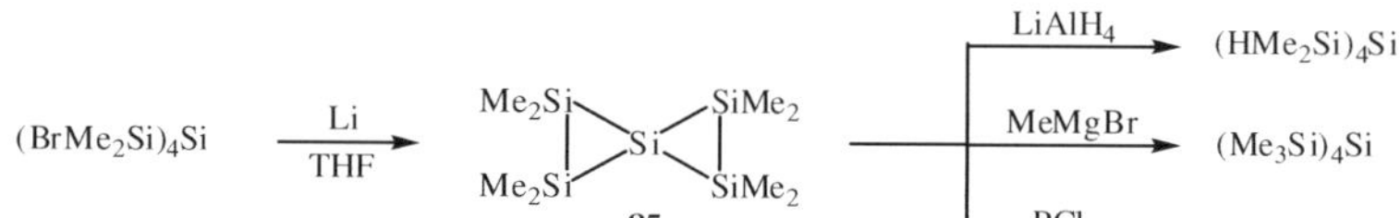

85

The coupling of 1,1-dibromocyclotetrasilane and 2,2-dibromotrisilane with potassium graphite gave spiro[3.2]hexasilane **86**.[55]

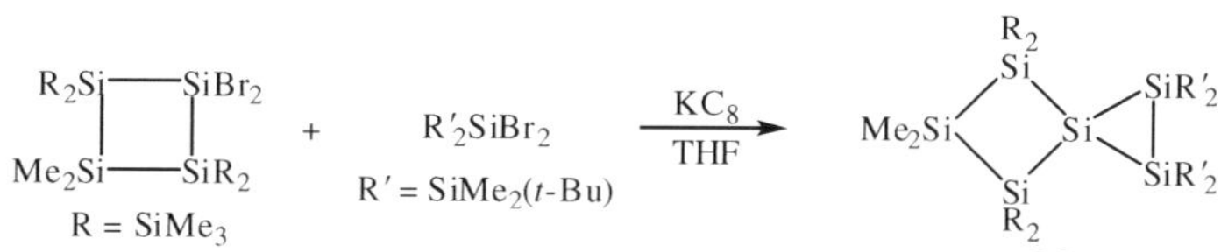

86

Spiro[4.4]nonasilane **88** and spiro[5.4]decasilane **90** were synthesized by the reactions of silylpotassium **87** or **89** with 1,4-difluorotetrasilane.[56] Nucleophilic attack of silylpotassium to the fluorosilane forms a Si–Si bond. The eliminated fluoride ion attacks a trimethylsilyl group, and the resulting silyl anion attacks the remaining fluorosilane moiety to give **88** and **90**. Spirosilane **88** was also formed by the reaction of tris(trimethylsilyl)-silylpotassium with 1,4-difluorotetrasilane, but this reaction was not so clean.

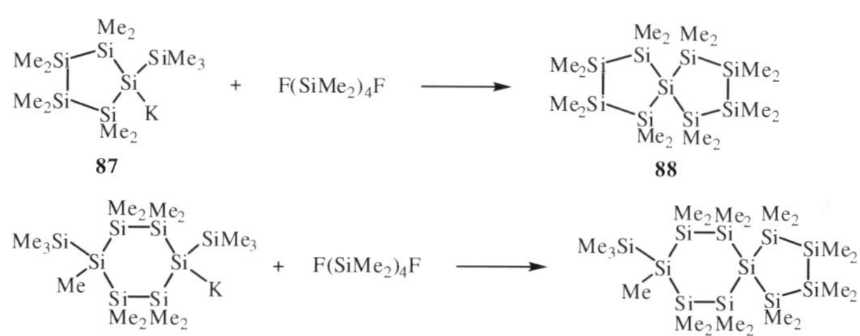

The reaction of tetrasilatetrahedranylpotassium **51** with silicon tetrachloride gave the spiro cluster **92** containing two bicyclo[1.1.0]tetrasilane units.[57] The formation of **92** is explained by the formation of dichlorobis(tetrasilatetrahedranyl)silane **91** and subsequent double 1,2-silyl migration.

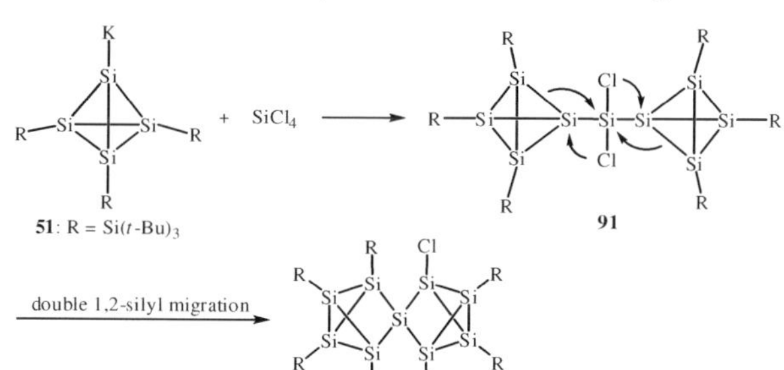

Spirooligosilane containing Si–Si double bonds has been reported by Kira.[58] Spiropentasiladiene **93** was isolated from the reaction mixture of the reduction of 1,1-dibromo-1-chlorodisilane with potassium graphite. In this reaction, 1-silylcyclotrisilene **62** was formed as a major product.[59] They found the formation of **93** after detailed analysis of the reaction mixture.

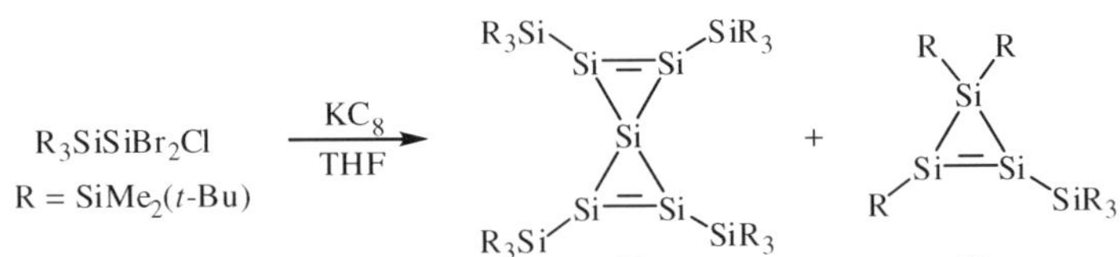

93 **62**

3.2.7 Siliconoids: Organosilicon Clusters Containing Unsubstituted Silicon Atoms

3.2.7.1 Si_8 cluster

Dimerization of 1,1,2,3,4-pentaiodocyclotetrasilane **94** with tri-*tert*-butylsilylsodium gave the organosilicon cluster **95**.[60] This cluster was also synthesized by the reaction of tetrasilatetrahedranylpotassium **51** with iodine monochloride.[39]

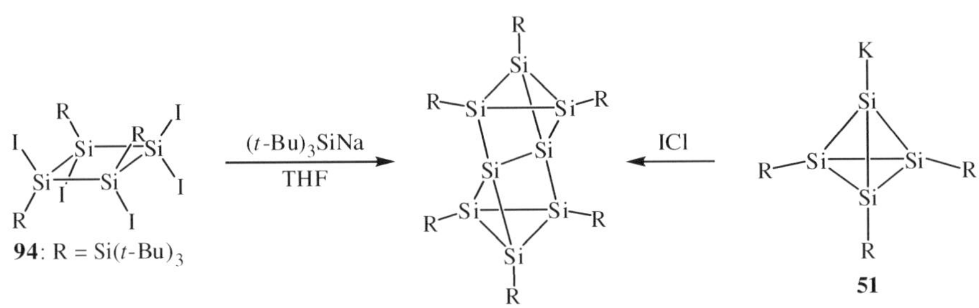

95

3.2.7.2 Pentasila[1.1.1]propellane

Breher has reported the synthesis of pentasila[1.1.1]propellane **96**.[61] This compound was synthesized by the coupling of hexachlorodisilane and 3 equiv of dichlorodimesitylsilane with lithium naphthalenide.

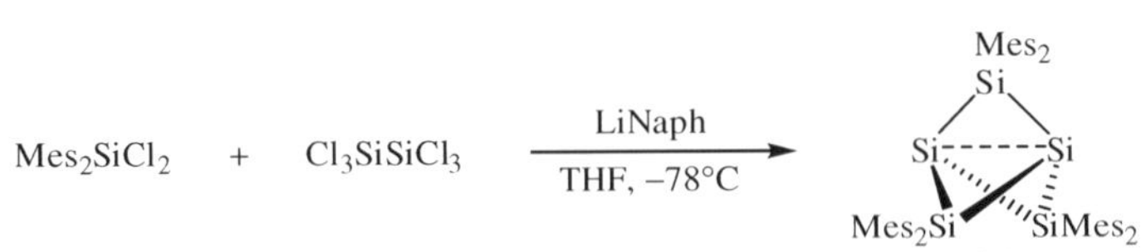

96

3.2.7.3 Tricyclo[2.1.0.0^{1,3}]pentasilane

Scheschkewitz has reported the highly strained compound **97** with fused three cyclotrisilane rings.[62] This compound was synthesized by the reaction of disilenide with silicon tetrachloride.

Another tricyclo[2.1.0.0^{1,3}]pentasilane **73** was synthesized by the similar way as described in Section 3.2.5.

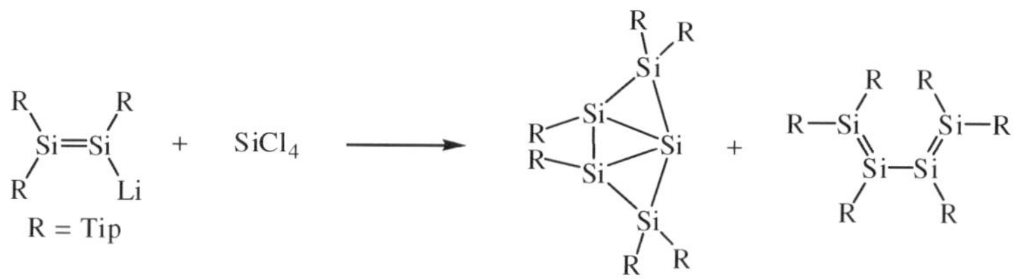

97

3.2.7.4 Isomers of hexasilabenzene

Scheschkewitz synthesized a novel isomer of hexasilabenzene (**98**) by dimerization of 1,1,2-trichlorocyclotrisilane with lithium naphthalenide.[63] This compound was also synthesized by the reaction of Cp*Si$^+$ with disilenide.[64] Thermolysis and photolysis of **98** lead to isomerization to **99**. Compound **99** has the silicon skeleton representing a global energy minimum among hexasilabenzene isomers (Si_6H_6).[65]

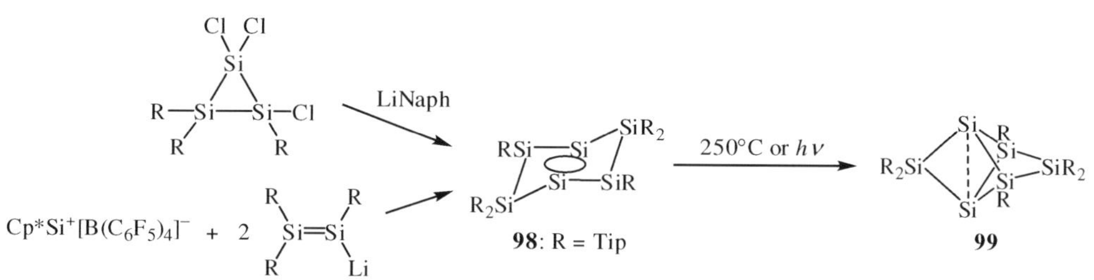

98: R = Tip **99**

3.2.8 Control of Oligomerization by the Ring Size of Cyclooligosilane Precursors[66]

Recently, oligomerization of 1,1,2,2-tetrachlorocyclooligosilanes with sodium was studied.[67–69] The oligomerization highly depends on the ring size of the starting compounds because four *tert*-butyl groups are pushed toward the tetrachlorodisilanylene moiety as the ring size increases. The Wurtz-type coupling of 1,1,2,2-tetrachlorocyclotetrasilane, -cyclopentasilane, and -cyclohexasilane gave the tetramer **100**, the trimer **101**, and the dimer **102**, respectively. These reactions provide novel silicon skeletons of cyclotetrasilane-fused octasilacuneane, cyclopentasilane-fused hexasilabenzvalene, and tetrasilane-bridged bicyclo[4.1.0]heptasil-1(6)-ene.

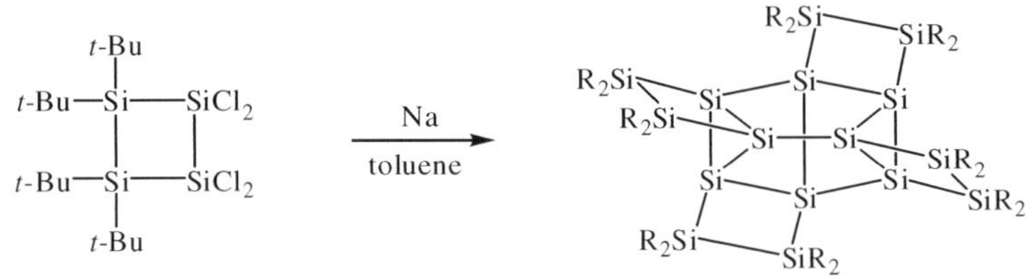

100: R = *t*-Bu

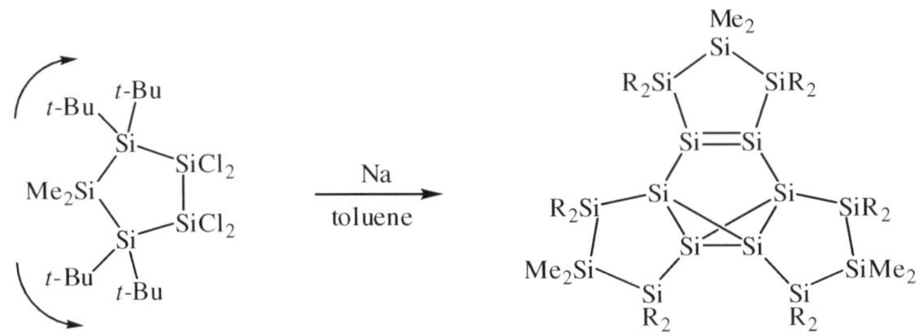

101: R = *t*-Bu

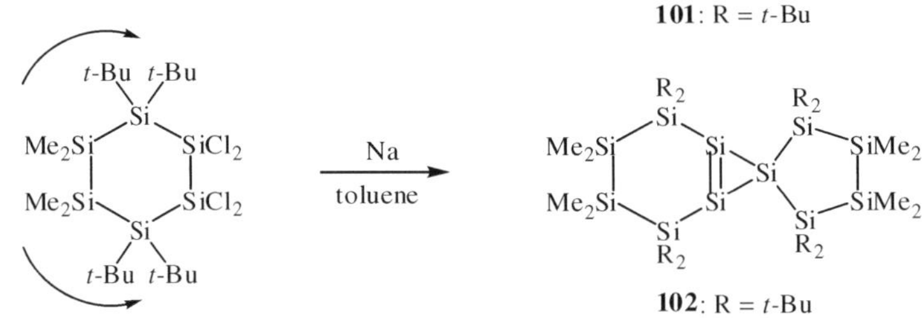

102: R = *t*-Bu

3.3 Structural Analysis by X-ray Crystallography and Temperature-Dependent ^1H NMR Spectroscopy

X-ray crystallography is the best way to determine and analyze structures of organosilicon clusters. Structures of many compounds have been reported so far. These studies revealed that silicon skeletons can adopt highly strained and unique structures. Temperature-dependent NMR measurement gives useful information on molecular dynamics. In this section, some of recent topics are described.

3.3.1 Bond-Stretch Isomers and Molecular Dynamics of Bicyclo[1.1.0]tetrasilanes[70]

Theoretical calculations predicted that bicyclo[1.1.0]tetrasilane has two bond-stretch isomers.[71] A short-bond isomer has a short bridgehead Si–Si bond (r), a small dihedral angle between two cyclotrisilane rings (ϕ), and a large Si–Si–R bond angle (θ). A long-bond isomer has large r, large ϕ, and small θ (Fig. 3.1). In the short-bond isomer, main lobes of the bridgehead silicon atoms interact with each other, and the Si–Si bond becomes short. In contrast, weak interaction between back lobes of the bridgehead silicon atoms results in the long-bond isomer. Recently, Kira demonstrated an important role of π–σ^* orbital mixing in these bond-stretch isomers.[72] The mixing between π-type orbitals at the

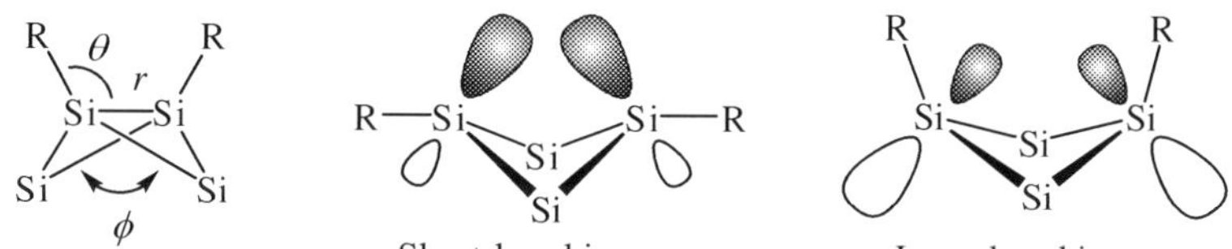

Figure 3.1
Schematic models of short-bond and long-bond isomers of bicyclo[1.1.0]tetrasilane.

bridgehead silicon atoms and σ* orbitals of the Si(bridgehead)−R(substituent) bonds stabilizes the structures of these isomers.

The structures of bicyclo[1.1.0]tetrasilanes **59** and **61** were analyzed by X-ray crystallography.[4b,44c] These compounds have small r, small ϕ, and large θ (Table 3.1). These values are in accord with those of the short-bond isomer of the model compound **103**, indicating that they are short-bond isomers.

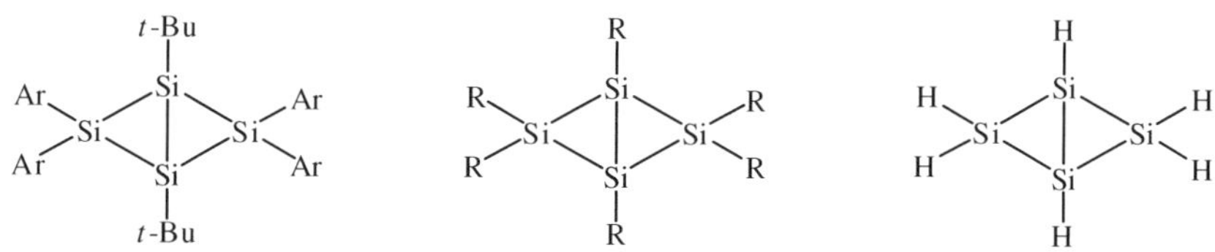

59: Ar = Dep **61**: R = $SiMe_2$(t-Bu) **103**

On the other hand, Kira has reported that 1,3-disilabicyclo[1.1.0]butane **104** is a long-bond isomer.[73] The X-ray crystallographic analysis of **104** showed large r, large ϕ, and small θ (Table 3.2). These values are similar to the calculated values of the long-bond isomer of the model compound **105**. These results clearly indicate that **104** is a long-bond isomer.

Table 3.1: Comparison of structural parameters of 59, 61, and 103

Compound	r/Å	ϕ/deg	θ/deg
59[4b]	2.373(3)	121.0	146.26, 147.48
61[44c]	2.367(1)	129.25(4)	149.14(4), 150.95(4)
103[a] (short-bond isomer)[71f]	2.380	120.0	146.1
103[b] (long-bond isomer)[44c]	2.859	142.2	92.20

[a]Calculated at the GVB/3-21G* level.
[b]Calculated at the B3LYP/6-31G(d) level.

Table 3.2: Comparison of structural parameters of 104 and 105[73]

Compound	r/Å	ϕ/deg	θ/deg
104	2.412(1)	141.1(1)	118.76(2)
105 (short-bond isomer)[a]	2.201	126.74	152.14
105 (long-bond isomer)[a]	2.448	141.48	106.93

[a]Calculated at the B3LYP/6-311 + G(d,p) level.

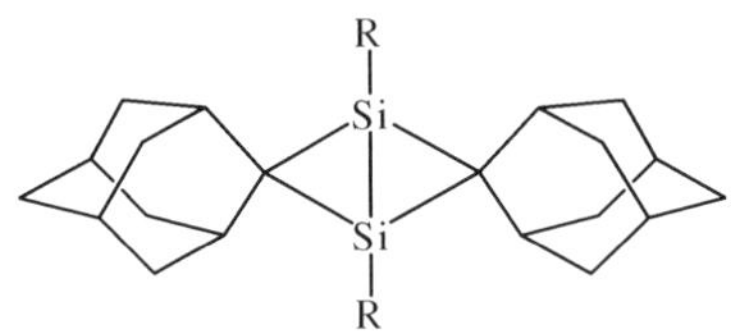

104: R = $SiMe_2(t\text{-}Bu)$

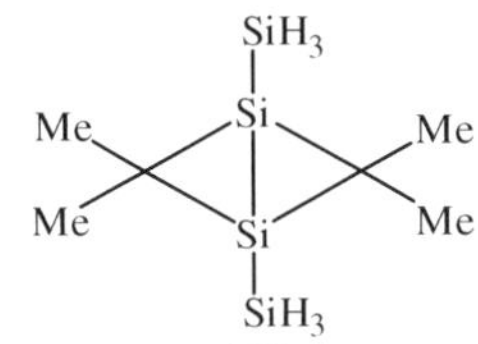

105

Inversion of the bicyclo[1.1.0]tetrasilane skeleton was observed by temperature-dependent ^{1}H NMR measurements.[4a] The ^{1}H NMR spectrum of **59** is temperature dependent at -50 to 120°C, showing flipping of the bicyclo[1.1.0]tetrasilane skeleton at the NMR timescale. The inversion mechanism via a planar intermediate or transition state in addition to the rapid rotation of the aryl rings was suggested.

59: Ar = Dep

The thermodynamic parameters of the inversion of bicyclo[1.1.0]tetrasilane **61** have been reported.[44c] Temperature-dependent ^{1}H NMR spectra of **61** were measured and analyzed by simulation of the observed spectra. The $\Delta H^{\ddagger}$ and $\Delta S^{\ddagger}$ values were estimated to be 6.67 ± 0.18 kcal mol^{-1} and -22.7 ± 0.8 cal mol^{-1} K^{-1}, respectively. The $\Delta H^{\ddagger}$ value is much smaller than the calculated value (18 kcal mol^{-1}) for the inversion of **103** through a planar transition state.[74] The small $\Delta H^{\ddagger}$ value and the remarkably large negative $\Delta S^{\ddagger}$ value may suggest a different flipping pathway of **61**.

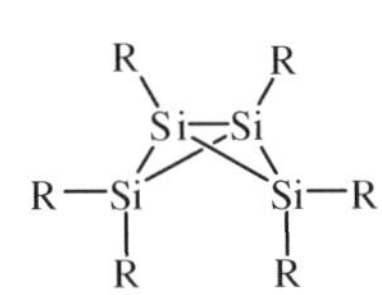

61: R = $SiMe_2(t\text{-}Bu)$
103: R = H

3.3.2 Bridgehead Si–Si Bonds of Pentasila[1.1.1]propellane and Related Compounds

The nature of the bridgehead Si–Si bond of pentasila[1.1.1]propellane **96** has been studied experimentally and theoretically by Breher.[61] The X-ray structure of **96** is shown in Fig. 3.2. The bond length between the two bridgehead silicon atoms is 2.636(1) Å. This value is close to the Si–Si bond length of hexa-*tert*-butyldisilane (2.697 Å),[75] which is the longest among values reported until now.

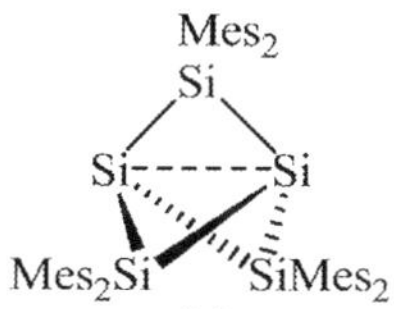

96

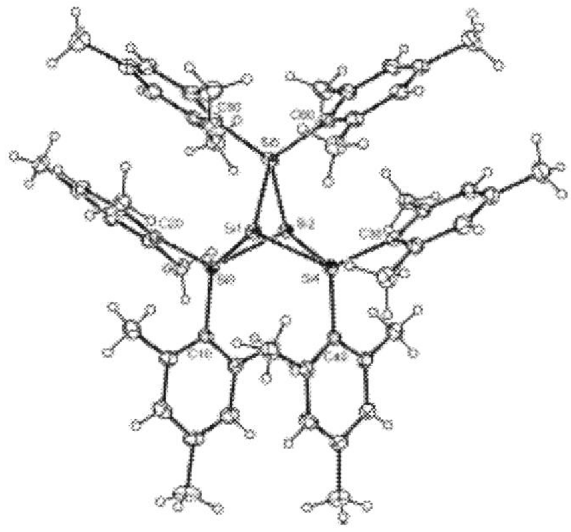

Figure 3.2
Molecular structure of **96**.

This unusually long Si–Si bond is expected to be considerably weak. The bond strength was estimated to be ~174 kJ mol^{-1} on the basis of thermodynamic considerations and theoretical calculations. This value is far smaller than those of normal disilanes (ca. 306–332 kJ mol^{-1}).[76]

A pentasila[1.1.1]propellane structure is included in compound **99**.[65] X-ray crystallographic analysis showed that the bond length between the two bridgehead silicon atoms is 2.7076(8) Å, which is significantly longer than that of **96** (Fig. 3.3). As two propellane blades are connected by the Si3 atom, the dihedral angle between these blades (96.7°) is smaller than other two values (131.64°).

The Si$_{11}$ cluster **106**, which can be regarded as a [3.2.2]propellane derivative, has a similar bridgehead Si–Si bond (Fig. 3.4).[40] The Si1–Si2 bond length (2.4976(9) Å) is considerably shorter than those of **96** and **99**, probably due to the staggered conformation around the Si1–Si2 bond in contrast to the eclipsed conformation of **96** and **99**.

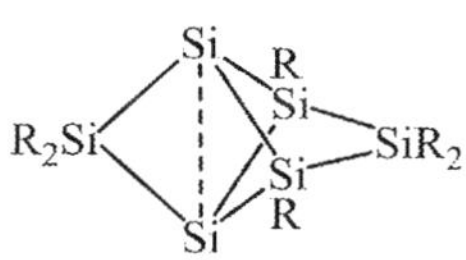

99: R = Tip

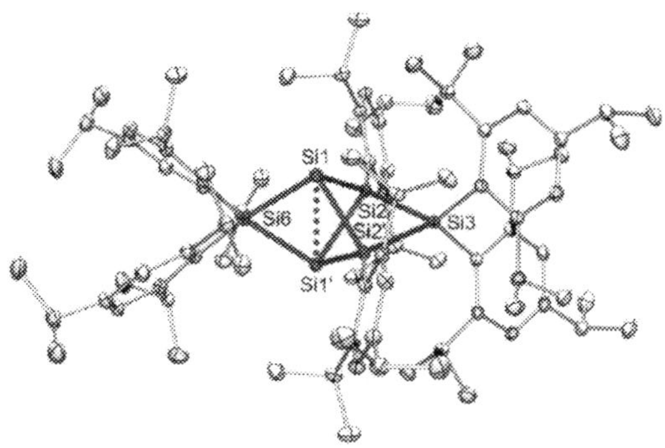

Figure 3.3
Molecular structure of **99**.

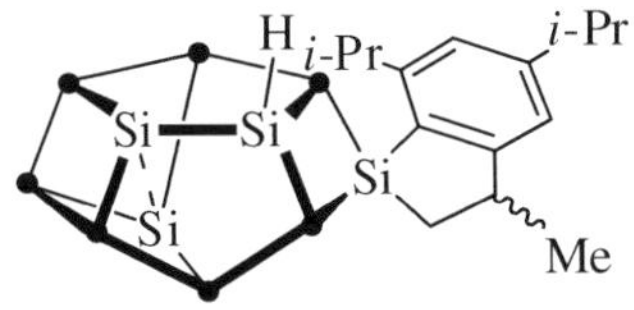

106: ● = SiTip

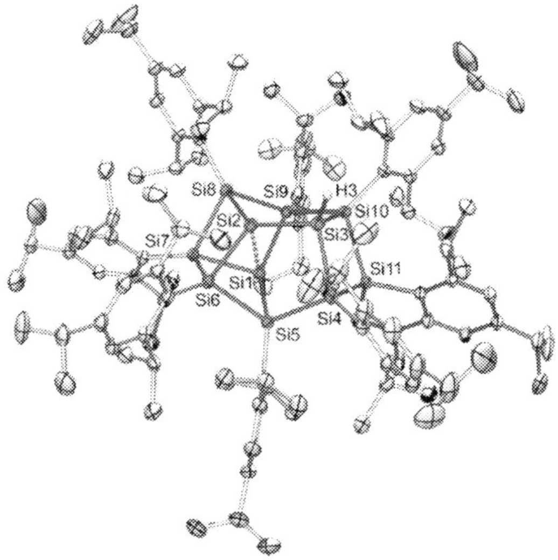

Figure 3.4
Molecular structure of **106**.

3.3.3 Trigonal Monopyramidal and Inverted Tetrahedral Structures of Silicon Atoms[11]

An ideal sp^3 silicon atom has a tetrahedral structure like an sp^3 carbon atom. Recent studies showed that silicon atoms can be deformed to highly strained structures, such as trigonal monopyramidal and inverted tetrahedral structures. These structures are easily distinguished by the sum of three bond angles around a silicon atom (Σ Si): this value of the trigonal monopyramidal structure is near 360°, and that of the inverted tetrahedral structure is less than 360°.

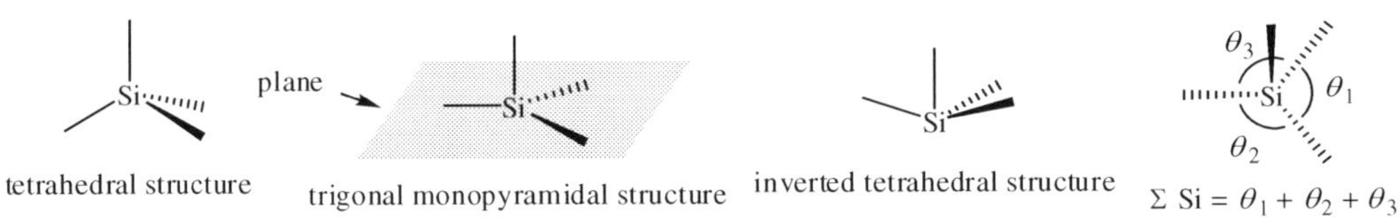

These structural deformations take place in some branched oligosilanes. For example, organosilanes with three bulky substituents and one small substituent take the trigonal monopyramidal structure to reduce steric hindrance among the bulky substituents.[77,78]

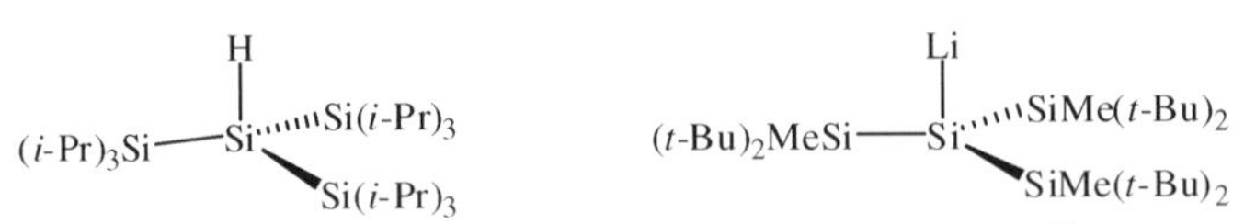

Σ Si = 354.30°,[77a]
353.7, 355.23°[77b]

Σ Si = 359.16°[78]

Some silicon atoms in organosilicon clusters suffer constriction by network structures. These silicon atoms are deformed and adopt trigonal monopyramidal and inverted tetrahedral structures. Examples reported until now are listed below.

3.3.3.1 Trigonal monopyramidal structures

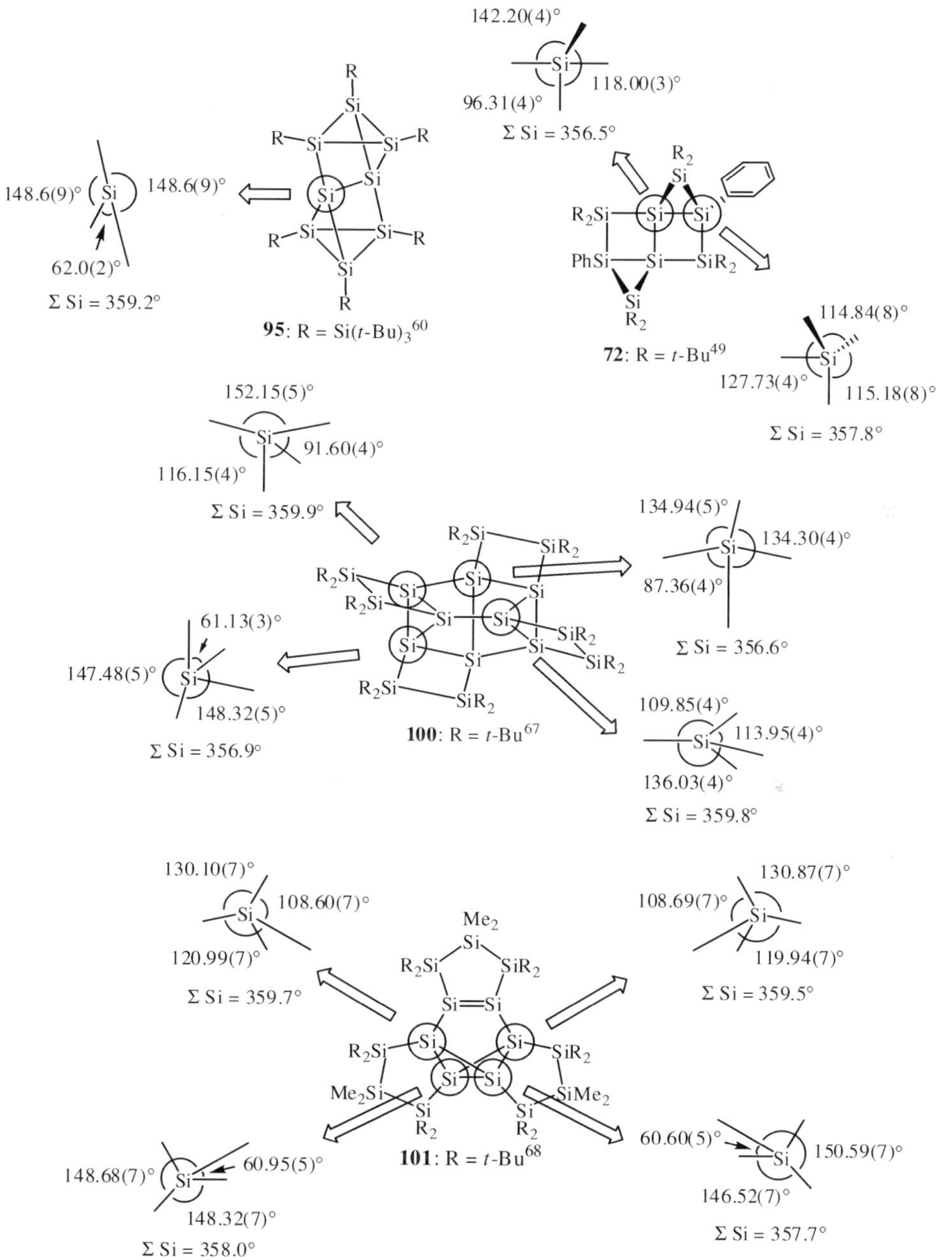

3.3.3.2 Inverted tetrahedral structures

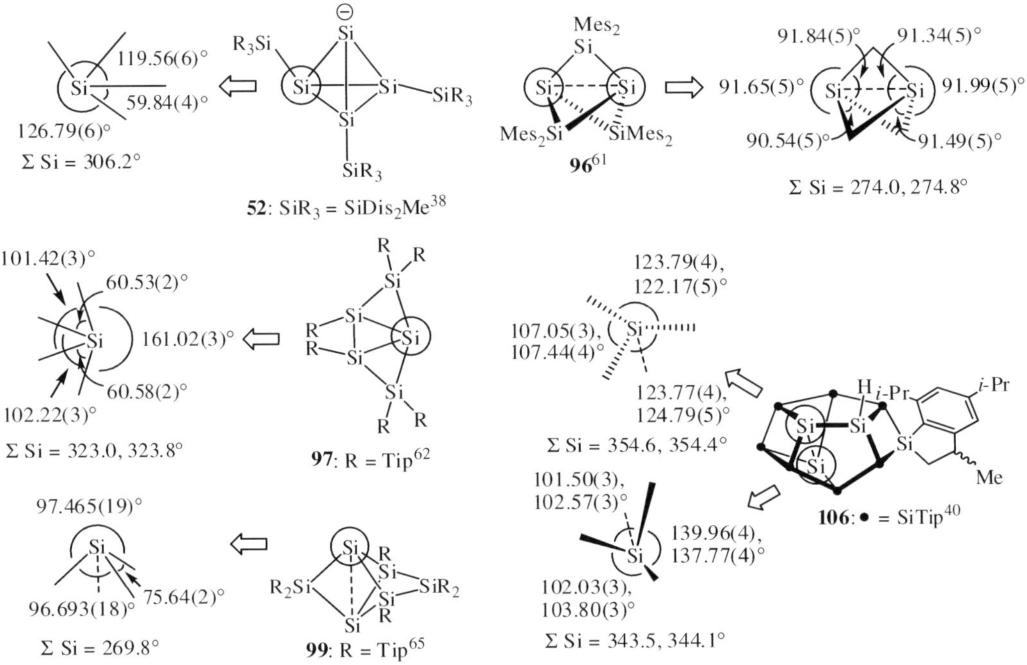

These compounds have highly rigid structures except for tetrasilatetrahedranylpotassium **52**. Compound **52** has unexpected structural features (Fig. 3.5).[38] The Si1–Si4 bond (2.7288 (15) Å) is much longer than other skeletal Si–Si bonds (2.2948(14)–2.3527(13) Å, av 2.322 Å). The Si4–Si15 bond is significantly tilted toward the Si1 atom, and the Si4 atom has an inverted tetrahedral structure with the Σ Si value of 306.2°. In addition, only one ^{29}Si signal of the tetrasilatetrahedrane skeleton was observed at −153.6 ppm, indicating that the four skeletal silicon atoms are equivalent on the NMR timescale. This result is explained by the migration of the silyl groups around the tetrasilatetrahedrane core.

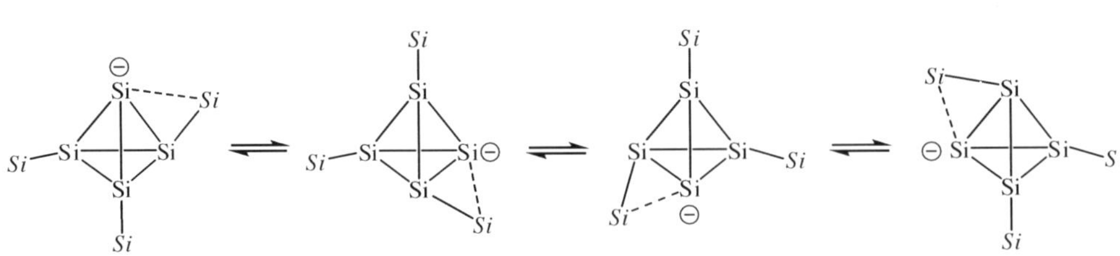

52: *Si* = $SiDis_2Me$

An organosilicon cluster containing an inverted tetrahedral tin atom has been reported quite recently.[79] Tetrasilastannapyramidane **107** has a nearly planar Si_4-base with the fold

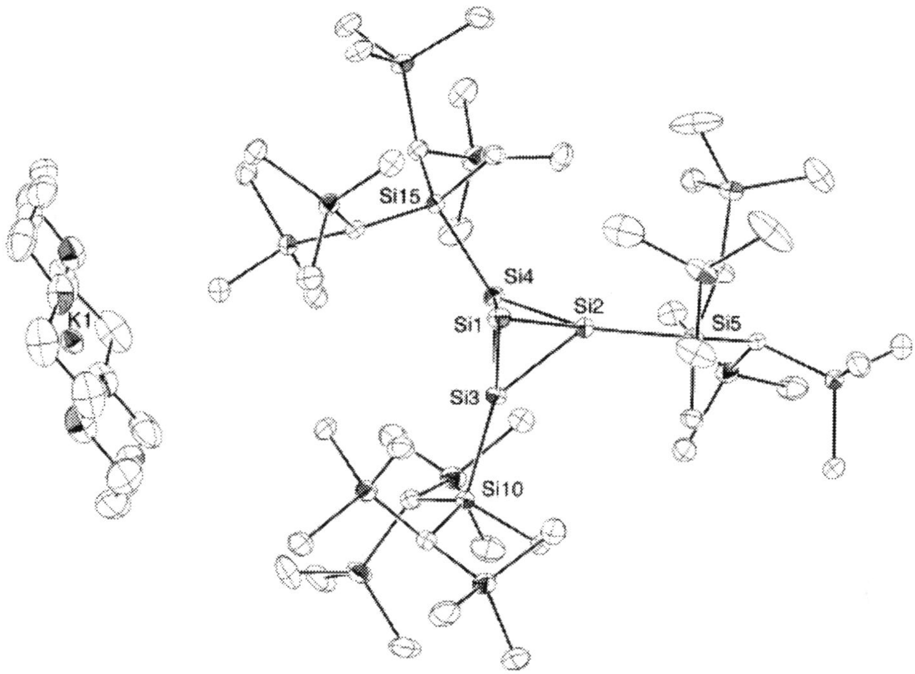

Figure 3.5
Molecular structure of **52**.

angle of 0.70°. The apical tin atom is connected with four silicon atoms with remarkably long Sn–Si bonds (2.763(4)–2.801(3) Å). The Si–Sn–Si bond angles are very small (48.2(1)–48.4(1)°).

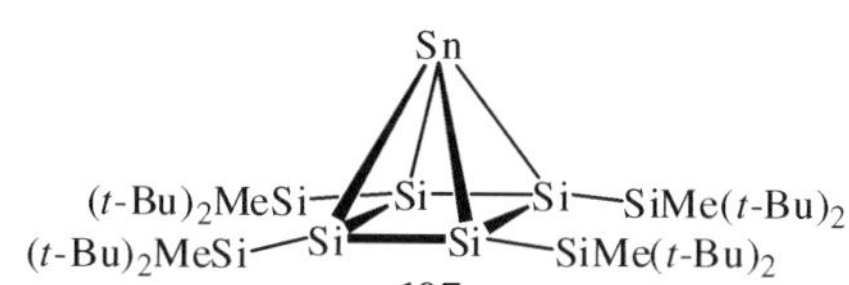

107

3.3.4 Tricyclic Isomer of Hexasilabenzene

The X-ray crystallographic analysis of **98** showed an unusual structure (Fig. 3.6).[63] This compound has a tricyclic chair-like structure, and the Si1 (and Si1′), Si2 (and Si2′) and Si3 (and Si3′) atoms bear no, two, and one Tip groups, respectively. The Si1–Si3–Si1′–Si3′ ring has a rhomboid structure. The Si1–Si3 (2.3275(5) Å) and Si1–Si3′ (2.3034(5) Å) bonds are remarkably short.

Theoretical calculations showed that its composition is best described by the resonance structures **108a** and **108b**.[63,80] Two π, two σ, and two nonbonding electrons are concerned with this resonance. The nucleus-independent chemical shift (NICS(0)) calculated at the

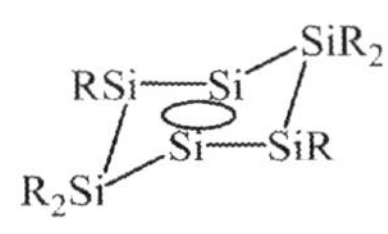

98: R = Tip

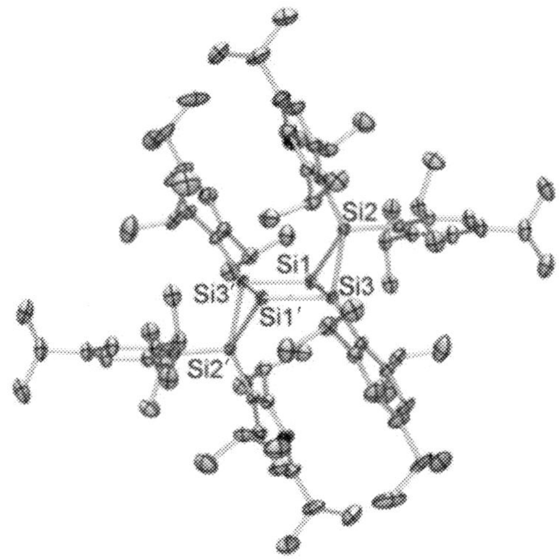

Figure 3.6
Molecular structure of **98**.

center of the central Si_4 ring is -23.8. Considering shielding effects by the σ framework, the NICS(0) value becomes -6.4, showing aromaticity by the six electrons. As this aromaticity is different from Hückel's aromaticity, Scheschkewitz proposed the term "dismutational aromaticity" to denote this effect.

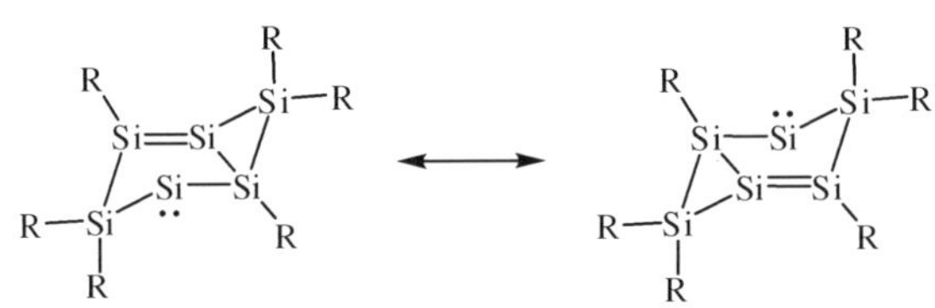

108a: R = Tip **108b**

3.4 Structural Analysis by ^{29}Si NMR Spectroscopy

Structures of organosilicon clusters cannot be determined when single crystals are not obtained or severe disorder of molecules inhibits structural analysis. In these cases, ^{29}Si NMR spectroscopy is an alternative way of structural determination of silicon skeletons.[81] ^{29}Si INEPT–INADEQUATE NMR spectroscopy is effective for this purpose. This method and other ^{29}Si NMR techniques which are helpful for structural analysis of organosilicon clusters are described in this section.

3.4.1 ^{29}Si INEPT–INADEQUATE NMR Spectroscopy

^{29}Si INEPT–INADEQUATE NMR spectroscopy was used for the first time by West to determine whether two silicon atoms are connected or separated.[82] When two silicon atoms are connected, a large $^{29}Si-^{29}Si$ coupling constant ($^1J_{Si-Si}$) is observed, but when they are separated, the $^{29}Si-^{29}Si$ coupling is not observed or observed with a relatively small $^nJ_{Si-Si}$ ($n \geq 2$) value. The ^{29}Si INEPT–INADEQUATE NMR spectra of **109–112** are shown in Fig. 3.7. The coupling constants of **109** and **111** are large, showing the presence of Si–Si bonds. On the other hand, the coupling constants of **110** and **112** are small, showing the that two silicon atoms are separated.

Later, 2D ^{29}Si INEPT–INADEQUATE NMR spectroscopy was developed to determine structures of organosilicon clusters by Kabe and Masamune.[83,84] They measured the 2D ^{29}Si INEPT–INADEQUATE NMR spectra of **37** and **38b**. In the ^{29}Si NMR spectrum of **37** in Fig. 3.8 (left), solid lines show correlations with large $^1J_{Si-Si}$ values, and dashed lines

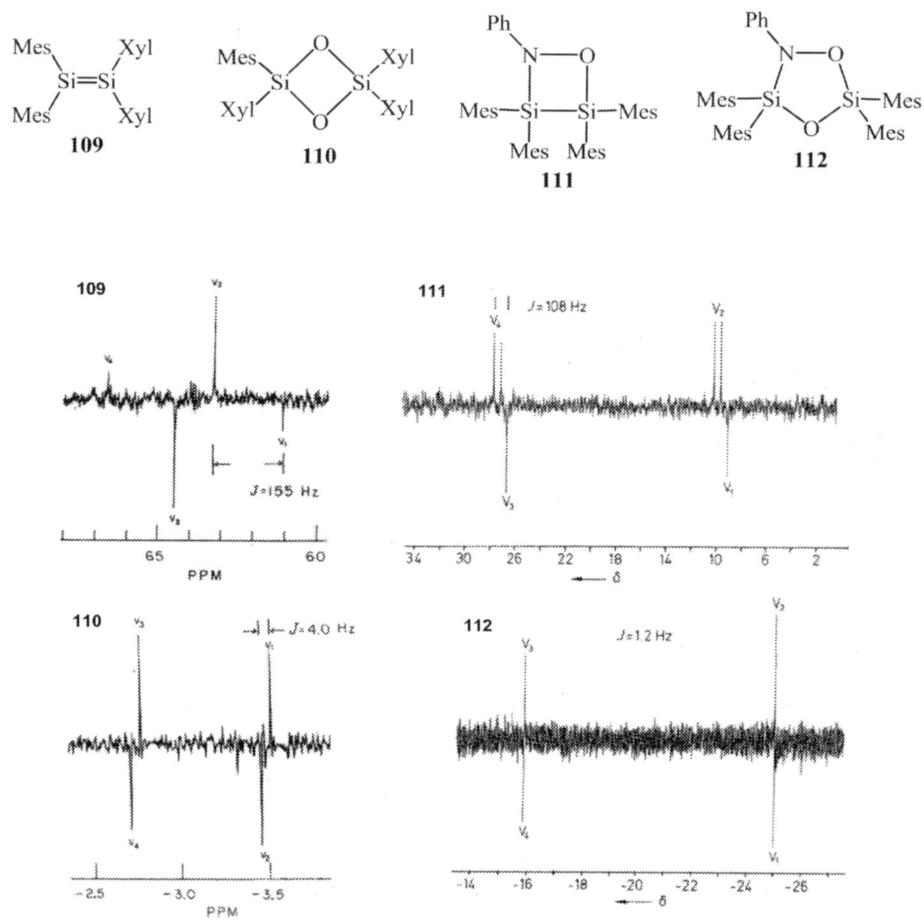

Figure 3.7
^{29}Si INEPT–INADEQUATE NMR spectra of **109–112**.

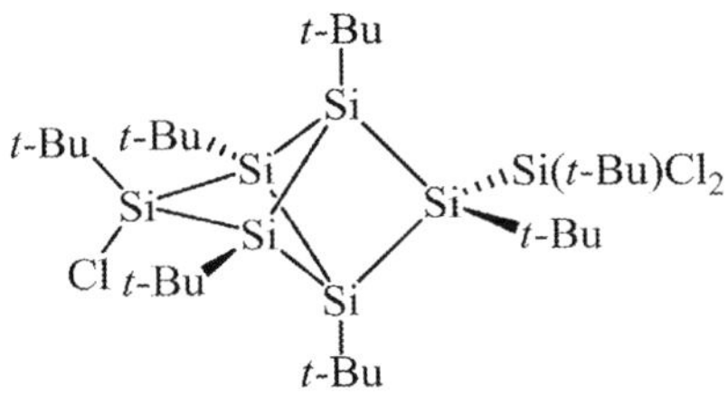

37

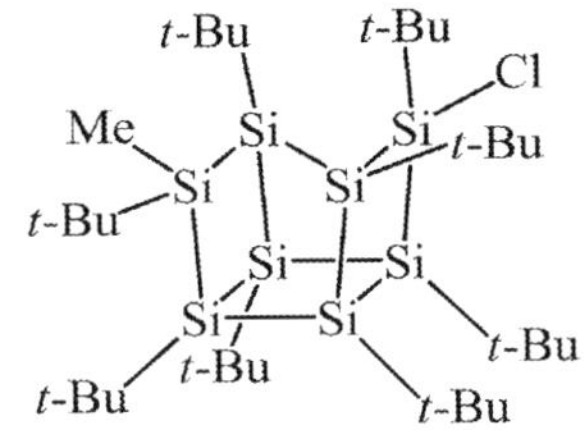

38b

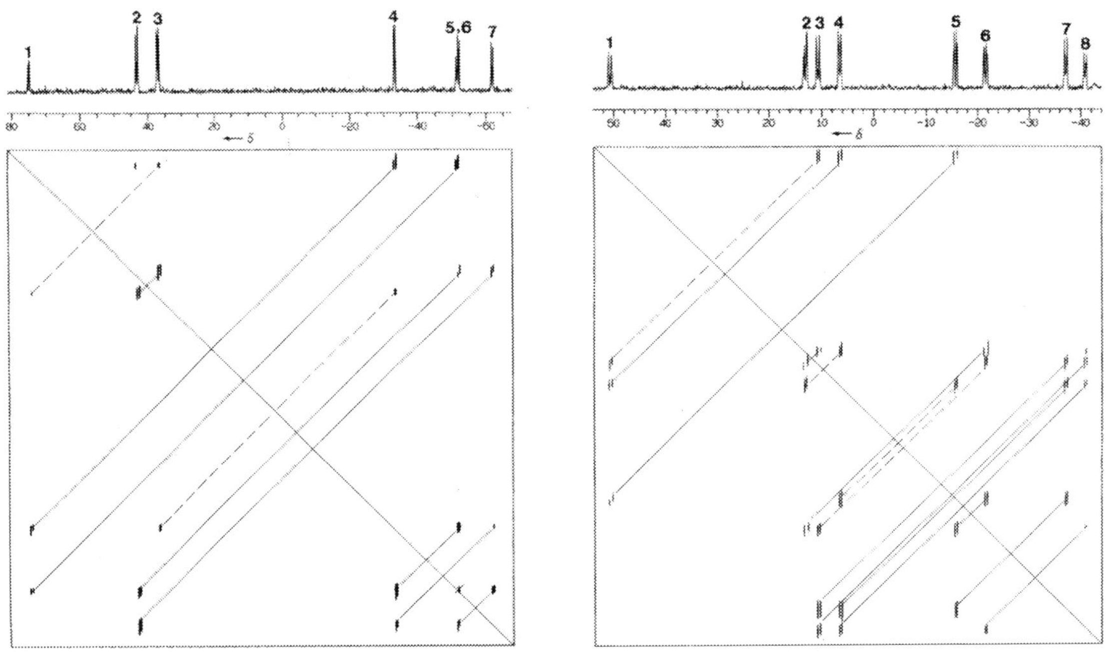

Figure 3.8
2D ^{29}Si INEPT–INADEQUATE NMR spectra of **37** in $CDCl_3$ (left) and **38b** in C_6D_6 (right).

show correlations with small $^{n}J_{Si-Si}$ ($n \geq 2$) values. The coupling constants are summarized in Table 3.3. From these data, the silicon skeleton **37** can be clearly determined.

Compound **38b** contains eleven Si–Si bonds. The 2D ^{29}Si INEPT–INADEQUATE NMR spectrum of **38b** shows eleven correlations ($^{1}J_{Si-Si}$) shown by solid lines and four correlations ($^{n}J_{Si-Si}$ ($n \geq 2$)) shown by dashed lines (Fig. 3.8 (right)).[83] In Table 3.3, the four $^{n}J_{Si-Si}$ ($n \geq 2$) values (24.1, 23.1, 22.5, and 19.9 Hz) are larger than two $^{1}J_{Si-Si}$ values (18.0 and 17.3 Hz). This problem makes the structural determination of **38b** partially obscure.

Table 3.3: $^{29}Si-^{29}Si$ coupling constants for 37 and 38b

Compound	^{1}J/Hz		^{n}J ($n \geq 2$)/Hz
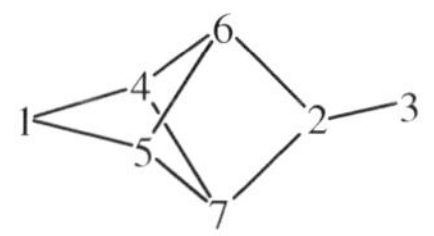 **37**	43.0 (2 − 3) 34.1 (2 − 6) 29.3 (1 − 5) 24.5 (4 − 6) −[a] (5 − 6)	36.1 (1 − 4) 32.2 (2 − 7) 27.4 (5 − 7) 23.4 (4 − 7)	13.7 (1 − 3) 16.6 (3 − 4)
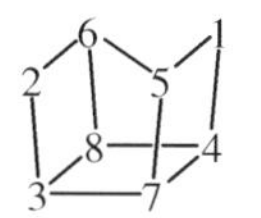 **38b**	42.2 (2 − 6) 41.0 (2 − 3) 30.8 (4 − 7) 28.3 (3 − 7) 25.7 (5 − 6) 17.3 (6 − 8)	41.8 (1 − 5) 33.4 (1 − 4) 28.9 (3 − 8) 25.7 (4 − 8) 18.0 (5 − 7)	24.1 (1 − 3) 23.1 (4 − 5) 22.5 (3 − 6) 19.9 (2 − 4)

[a]The chemical shifts of these ^{29}Si nuclei were too close to permit accurate measurement of the coupling constant between them.

These two examples show advantage and limitation of this method. The limitation can be overcome by theoretical calculations with the GIAO method. The GIAO calculations make assignment of ^{29}Si NMR signals possible. Combination of ^{29}Si INEPT–INADEQUATE NMR spectroscopy and GIAO calculations would give helpful methodology for structural determination of organosilicon clusters. Such an example has been reported by Hassler and Marschner.[36] They assigned the signals at −87.42 (Si1), −80.57 (Si3), and −14.59 (Si2)

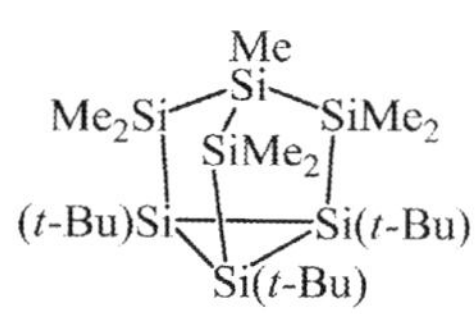

40

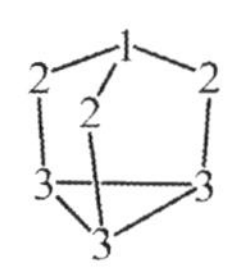

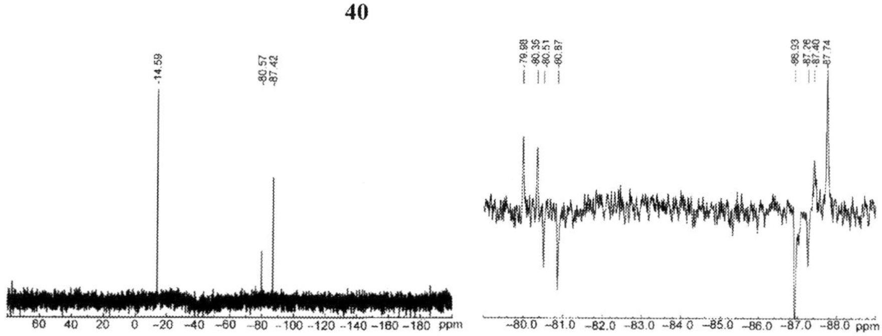

Figure 3.9

^{29}Si INEPT NMR spectrum (left) and a part of the ^{29}Si INEPT–INADEQUATE NMR spectrum (right) of **40**.

ppm in Fig. 3.9 (left) and ${}^{1}J_{\text{Si1-Si2}} = 52.7$, ${}^{1}J_{\text{Si2-Si3}} = 48.4$, ${}^{2}J_{\text{Si1-Si3}} = 9.6$, and ${}^{2}J_{\text{Si3-Si2'}} = 8.6$ Hz (Fig. 3.9 (right)). From these coupling constants, the structure of **40** was clearly confirmed.

3.4.2 2D ^{29}Si/^{1}H Correlation NMR Spectroscopy

Some ^{29}Si signals of organosilicon clusters, especially those of unsubstituted silicon atoms, cannot be observed easily by ^{29}Si NMR spectroscopy. To solve this problem, 2D ^{29}Si/^{1}H correlation NMR spectroscopy has been used by Scheschkewitz.[40,62,63,65] This technique is very effective to observe and assign such ^{29}Si NMR signals. Three examples are described in this section.

Compound **97** shows three signals at −124.8, −108.4, and 7.4 ppm in the ^{29}Si NMR spectrum. In the 2D ^{29}Si/^{1}H correlation NMR spectrum, only two signals at −108.4 and 7.4 ppm show cross-peaks with one and two Tip groups, respectively. These data support the structure of **97**.[62]

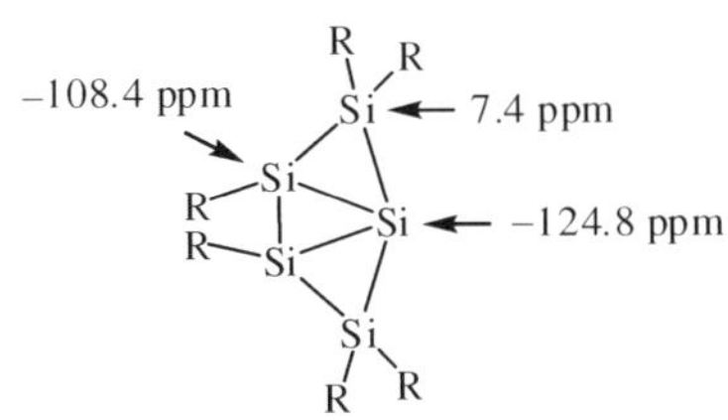

97: R = Tip

Similarly, ^{29}Si signals of **98** were assigned by the 2D ^{29}Si/^{1}H correlation NMR spectrum. Three signals at −89.3, −84.8, and 124.6 ppm showed correlation with one, two, and no Tip groups, respectively, supporting the structure of **98**.[63]

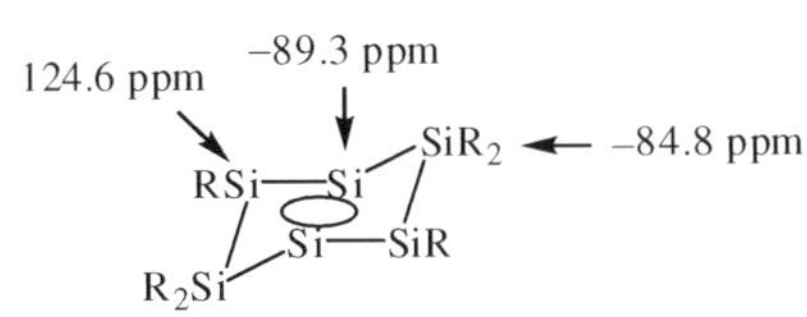

98: R = Tip

2D ^{29}Si/^{1}H correlation NMR measurement is useful to observe missing signals of **101** in the 1D ^{29}Si NMR spectrum.[68] This compound has the C_2 symmetry and is expected to show eight ^{29}Si signals in the ^{29}Si NMR spectrum. Unfortunately, only five signals assignable to the peripheral silicon atoms (Si4−Si8) were observed in the ^{29}Si NMR measurement using the DEPT technique. However, eight signals were clearly observed in the 2D ^{29}Si/^{1}H correlation NMR spectrum. These signals were assigned on the basis of the GIAO calculation as shown in Table 3.4.

Table 3.4: Calculated and observed ^{29}Si chemical shifts of 101

Nucleus	δ (calcd)/ppm[a]	δ (obs)/ppm[b]	
Si1	249.4	187.3	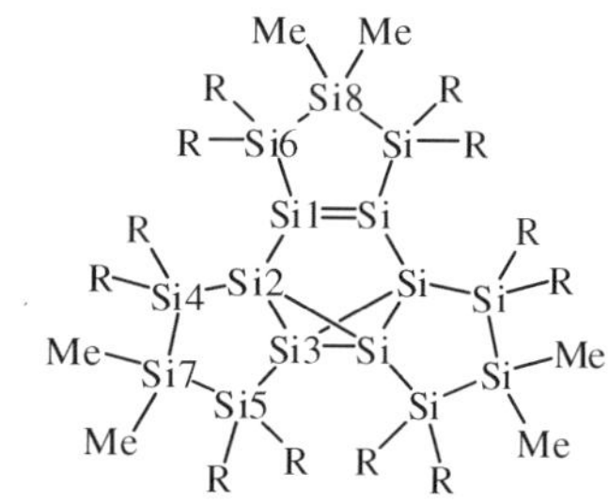 101: R = *t*-Bu
Si2	− 69.4	− 84.6	
Si3	− 20.9	− 54.2	
Si4	21.2	10.9	
Si5	27.9	17.8	
Si6	24.7	16.5	
Si7	− 8.8	− 21.4	
Si8	3.5	− 14.5	

[a]Calculated at the B3LYP/6-311 + G(2d,p) level.
[b]In C_6D_6 at 60°C.

3.4.3 Solid State ^{29}Si CP-MAS NMR Spectroscopy

As the number of silicon atoms in organosilicon clusters increases, their solubility tends to decrease. Large organosilicon clusters such as **100** are sparingly soluble in common organic solvents. This makes measurement of ^{29}Si NMR spectra in solution very problematic or impossible. Solid state ^{29}Si cross-polarization magic angle spinning (CP-MAS) NMR spectroscopy is an alternative method to solve this problem. This method gives satisfactory ^{29}Si NMR spectra with acceptable resolution. In Fig. 3.10, the ^{29}Si CP-MAS NMR spectrum of **100** is shown.[67]

As compound **100** has the C_2 symmetry, eight ^{29}Si signals were observed. These signals were assigned on the basis of a GIAO calculation of chemical shifts. Although there is a

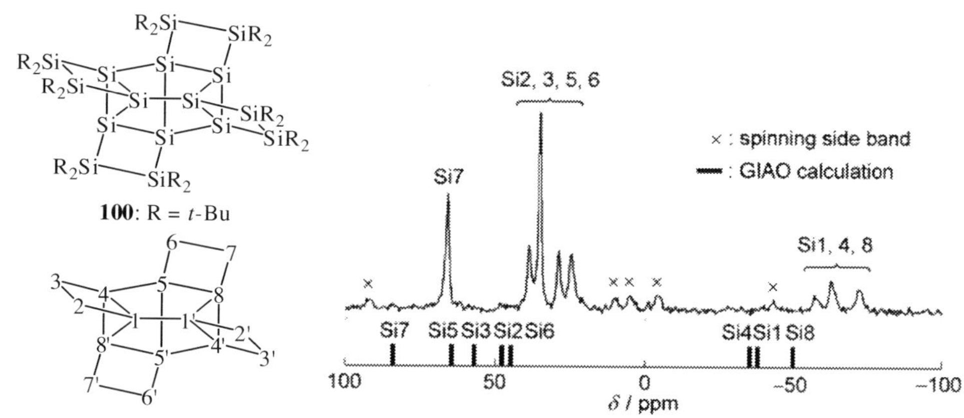

Figure 3.10
^{29}Si CP-MAS NMR spectrum of **100** at room temperature along with the chemical shifts calculated by the GIAO method at the B3LYP/6-311 + G(2d,p) level.

difference between the observed and calculated chemical shifts, the pattern of the calculated chemical shifts well reproduces the observed spectrum. An important point of this methodology is the rigid structure of **100** in the solid state. The rigid structure is suitable for assignment of signals by the GIAO calculation.

Combination of ^{29}Si CP-MAS NMR measurement and GIAO calculations is an effective way to analyze insoluble organosilicon compounds. For example, the detailed structure of poly(dimethylsilylene) has not been studied because of its well-known insolubility in almost all organic solvents. This problem has recently been overcome by using combination of ^{29}Si CP-MAS NMR measurement and GIAO calculations. Its end groups, degree of polymerization, and no-branched structure were clearly shown using this method.[85]

3.4.4 Unusual Downfield Shifts of ^{29}Si NMR Signals

In ^{29}Si NMR spectra of organosilicon clusters, unusual downfield shifts are sometimes observed. To explain the downfield shifts, the following effects have been discussed.

3.4.4.1 Ring current effect

Compound **99** shows four ^{29}Si signals at −274.2, 7.5, 14.8, and 174.6 ppm.[65] These signals were assigned by 2D $^{29}Si/^{1}H$ correlation NMR measurement and theoretical calculations of the model compound **113a** (−267.0, 20.4, 32.1, and 207.0 ppm). The signal at 174.6 ppm is the most deshielded among the ^{29}Si chemical shifts of tetracoordinate silicon atoms.

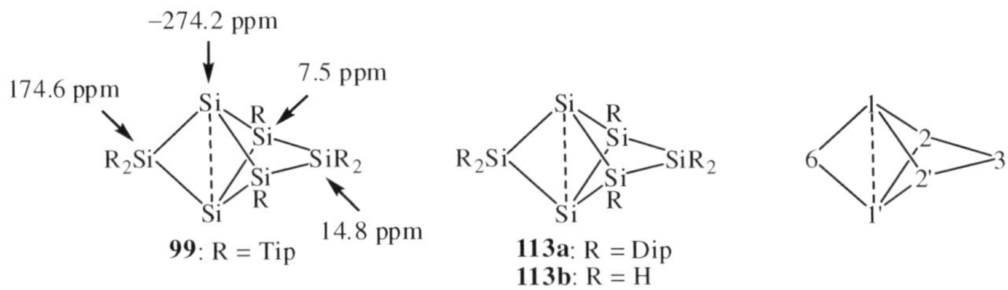

To understand this downfield shift, magnetically induced current density field vectors in **113b** were calculated with the gauge-including magnetically induced currents (GIMIC) method.[65] The results are shown in Fig. 3.11. Fig. 3.11(A) shows vectors in the Si6–Si1–Si1′–Si3 plane, and dominating current vortices are shown in Fig. 3.11(C). A diatropic current loop is present, including the Si1 and Si1′ atoms. The Si6 atom is surrounded by paratropic current vortex, which gives rise to the remarkable downfield shift at 174.6 ppm. The diatropic current branches around the Si2 and Si2′ atoms, as shown in Fig. 3.11(B). These two branch currents cancel the shielding effect at the Si2 and Si2′

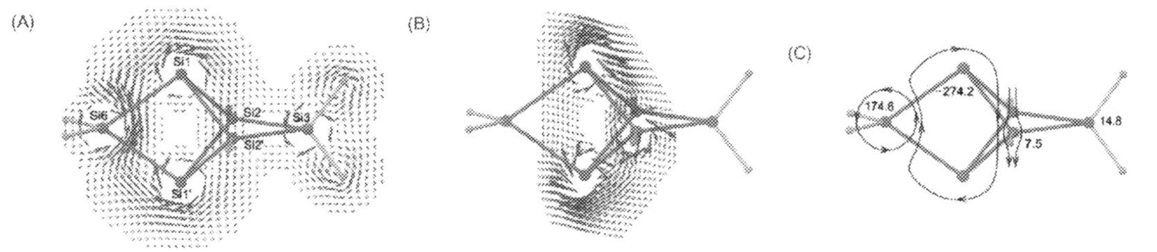

Figure 3.11
Magnetically induced current density field vectors of **113b** in the Si6–Si1–Si1′–Si3 plane (A) and in the Si1–Si1′–Si2′ plane (B). Dominating current vortices are shown in (C).

atoms. As a result, the signal of the Si2 and Si2′ atoms is observed at a normal position (7.5 ppm). The ring current on the silicon σ skeleton is a new concept.

3.4.4.2 Charge distribution effect

Lee, Frenking, and Sekiguchi reported silicon-containing organogermane clusters **114** and **115**.[86] Compounds **114** and **115** show the ^{29}Si NMR signals of the skeletal silicon atoms at 91.64 and 125.85 ppm, respectively. The authors explained that the reason for such atypical downfield shifted signals in both **114** and **115** is hard to rationalize considering only substituent effects; instead, they proposed that such extreme deshielding results from the unusual bonding situations in **114** and **115**, which yield a particular charge distribution of the electrons.

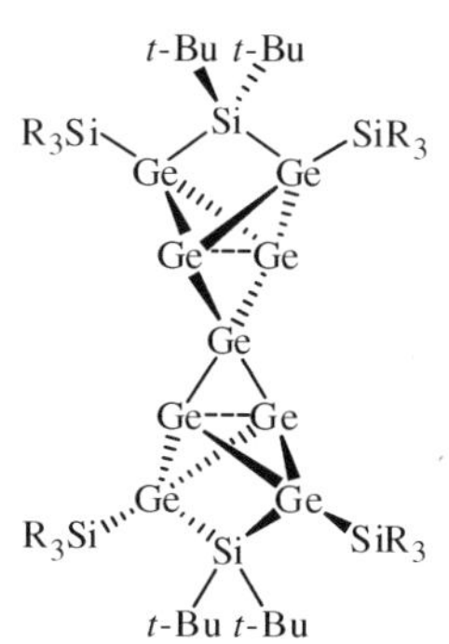

114: $SiR_3 = SiMe(t\text{-}Bu)_2$

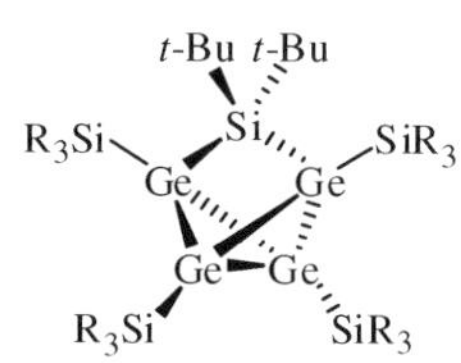

115: $SiR_3 = SiMe(t\text{-}Bu)_2$

3.4.4.3 Steric compression effect

Marschner compared the ^{29}Si chemical shifts of a series of trimethylsilyl-substituted cyclopentasilanes **116**–**119** (Table 3.5).[24a] The signals of quarternary silicon atoms (*Si*Si_4) were observed at similar positions of −132.0 (**116**), −130.8 (**117**), −129.6 (**118**), and

Table 3.5: ^{29}Si chemical shifts (δ/ppm) of cyclopentasilanes 116–119

Compound[a]	Me_3Si	a	b	c	d	e
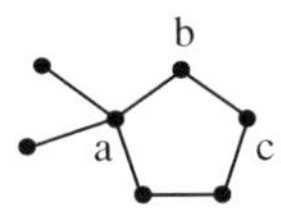 116	−7.6	−132.0	−31.6	−40.2		
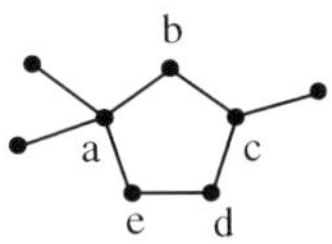 117	−7.2, −7.8, −9.6	−130.8	−24.7	−80.5	−31.1	−27.2
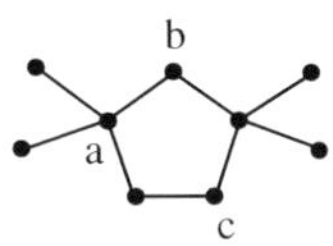 118	−7.6	−129.6	−19.9	−24.8		
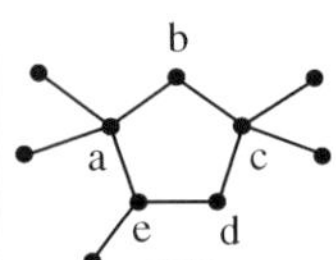 119	−6.8, −7.3, −8.3, −9.3, −10.4	−120.8	−14.8	−129.4	−16.7	−66.8

[a]The symbol • denotes $SiMe_n$ ($n = 0$–2).

−120.8 and −129.4 ppm (**119**), but tend to shift downfield as these molecules become more crowded. The signals of the tertiary silicon atoms ($MeSiSi_3$) shift downfield from −80.5 ppm (**117**) to −66.8 ppm (**119**). Similar tendency was observed for secondary silicon atoms (Me_2SiSi_2), where signals shift downfield in the order of **116** (−31.6 and −40.2 ppm), **117** (−24.7, −31.1, and −27.2 ppm), **118** (−19.9 and −24.8 ppm), and **119** (−14.8 and −16.7 ppm). The last data of **119** were observed remarkably downfield compared with that of decamethylcyclopentasilane (−42.1 ppm).[87] These downfield shifts are explained by the increased steric crowding by additional trimethylsilyl groups. Similar results were obtained in other examples of mono- and bicyclic oligosilanes.[34]

3.4.4.4 Consideration of the Ramsay's equation

As explained in Section 3.4.3, compound **100** shows eight ^{29}Si signals (Fig. 3.10). The signal at 66.4 ppm was assigned to the Si7 atom on the basis of a GIAO calculation.[67] This signal shifts to a significantly downfield region compared with the signals of other peripheral Si2, Si3, and Si6 atoms at around 30 ppm. The downfield shift was explained by Ramsey's equation,[88] where $\sigma_{N,uv}{}^{p}$ is the paramagnetic term of the nuclear magnetic shielding tensor of the nucleus N, c represents the speed of light, φ_k and φ_a denote wavefunctions of occupied and unoccupied molecular orbitals, ε_k and ε_a are their energies, $l_{O,u}$ is the angular momentum operator that represents the interaction of the external magnetic field with the electrons, and $l_{N,v} \cdot r_N^{-3}$ corresponds to the interaction of the electrons with the nuclear magnetic dipole field. This equation implies that when the energy difference between the occupied and unoccupied molecular orbitals is small, the signal shifts to a downfield region.

$$\sigma_{N,uv}{}^{p} = \frac{2}{c^2}\sum_{k}^{occ}\sum_{a}^{vac}\frac{\langle\varphi_k|l_{O,u}|\varphi_a\rangle\cdot\langle\varphi_a|l_{N,v}\cdot r_N^{-3}|\varphi_k\rangle}{e_k - e_a}$$

The HOMO and LUMO of **100** calculated at the B3LYP/6-31G(d) level are shown in Fig. 3.12. The σ orbitals of the Si1–Si4 (Si1′–Si4′), Si1–Si8′ (Si1′–Si8), and Si7–Si8 (Si7′–Si8′) bonds contribute mainly to the HOMO. The σ* orbitals of the silicon atom-to-substituent bonds of cyclotetrasilane rings contribute mainly to the LUMO. In both cases, the σ and σ* orbitals around the Si7 atom are largely concerned. Because

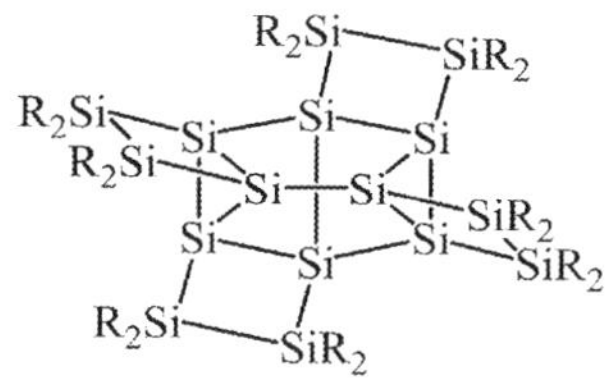

100: R = *t*-Bu

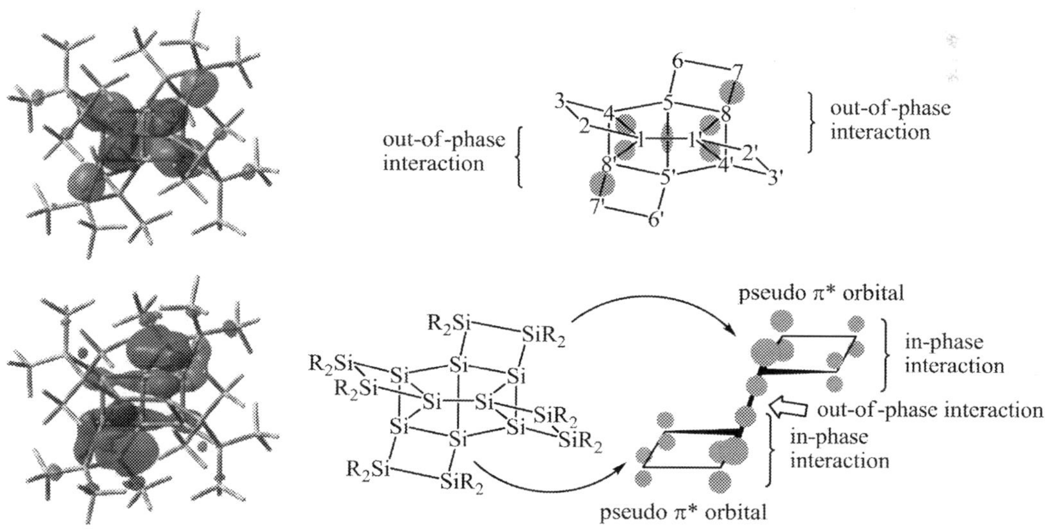

Figure 3.12
The HOMO (above) and LUMO (below) of **100** calculated at the B3LYP/6-31G(d) level.

of the smallest energy difference between the HOMO and the LUMO, the Si7 signal shows a remarkable downfield shift.

3.5 Electronic Properties

Electronic properties of organosilicon clusters have been studied by UV/Vis spectroscopy and theoretical calculations of molecular orbitals. The fundamentals of molecular orbitals of organosilicon clusters and recent topics of electronic properties of unique organosilicon clusters are described in this section.

3.5.1 Fundamentals of Molecular Orbitals of Organosilicon Clusters

In order to understand molecular orbitals of organosilicon clusters, the fundamentals of orbital interaction in silicon skeletons are considered. Fig. 3.13 shows σ conjugation between two Si–Si σ orbitals.[89] Interaction of two Si–Si σ orbitals gives two new orbitals. In-phase interaction gives a new orbital consisting of two lobes with the same phase sign. Out-of-phase interaction gives one with the different phase signs. These two orbitals are occupied by four electrons from two Si–Si σ bonds. As a result, the new orbital with the different phase signs becomes the HOMO. Similarly, the HOMOs of organosilicon clusters have lobes with alternate phase signs along Si–Si bonds due to σ conjugation. For example, the HOMO of **100** in Fig. 3.12 has lobes along the Si4–Si1–Si8′–Si7′ and Si4′–Si1′–Si8–Si7 bonds with alternate phase signs.

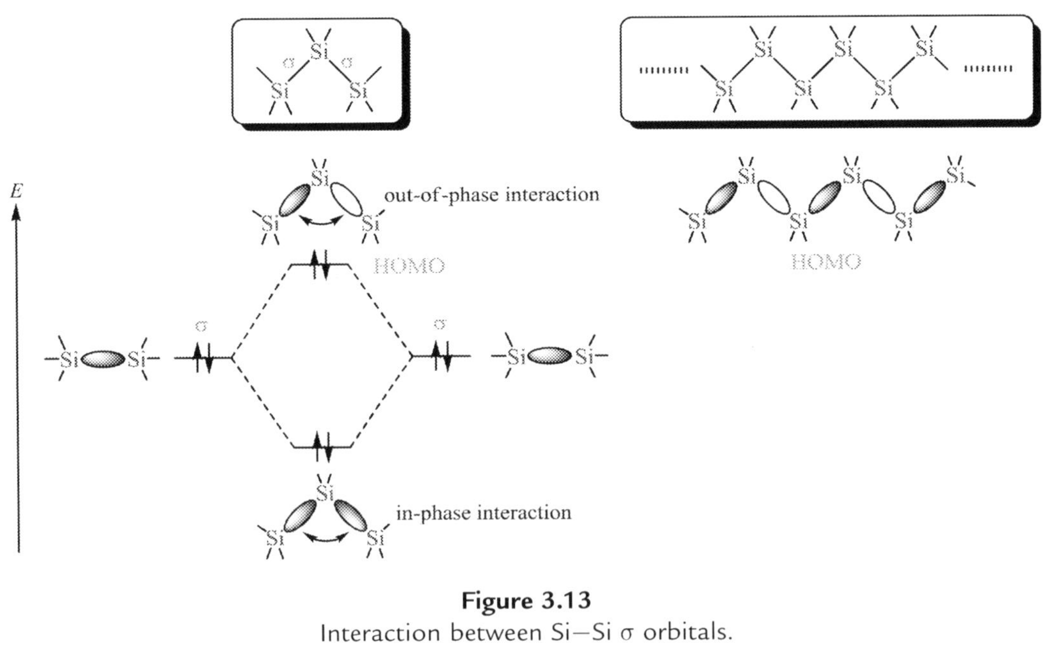

Figure 3.13
Interaction between Si–Si σ orbitals.

LUMOs of organosilicon clusters are formed by interaction of Si–Si and/or Si–C σ^* orbitals (Fig. 3.14). In-phase interaction of two Si–Si and/or Si–C σ^* orbitals gives a new orbital called a pseudo π^* orbital because two lobes interact with each other like a π orbital. The energy level of this pseudo π^* orbital is lower than those of Si–Si and/or Si–C σ^* orbitals because of the stabilization by in-phase interaction. LUMOs of organosilicon clusters are formed by this interaction among several Si–Si and/or Si–C σ^* orbitals. For example, the LUMO of **100** in Fig. 3.12 is formed by two sets of pseudo p* orbitals above and below the cyclotetrasilane rings.

In Fig. 3.15, the UV/Vis spectra of **100** and **120** are shown.[67] Although both compounds have 16 silicon atoms, their UV/Vis spectra are remarkably different. Ladder oligosilane **120** shows the lowest energy absorption band as distinctive band at 464 nm. This result indicates a small HOMO–LUMO energy gap compared with other electronic transitions. As this compound has a molecular long axis, σ conjugation among Si–Si bonds extends along this axis in the HOMO. Similarly, in-phase interaction among Si–C σ^* orbitals in the LUMO extends along the molecular long axis.[90] As a result, the energy levels of the HOMO and the LUMO are separated from those of other molecular orbitals (Fig. 3.16), and the lowest energy absorption band is observed as a distinctive band. On the other hand, compound **100** does not show a clear band above 250 nm, and absorption tails to ca. 570 nm. As this compound does not have a molecular long axis, the HOMO–LUMO

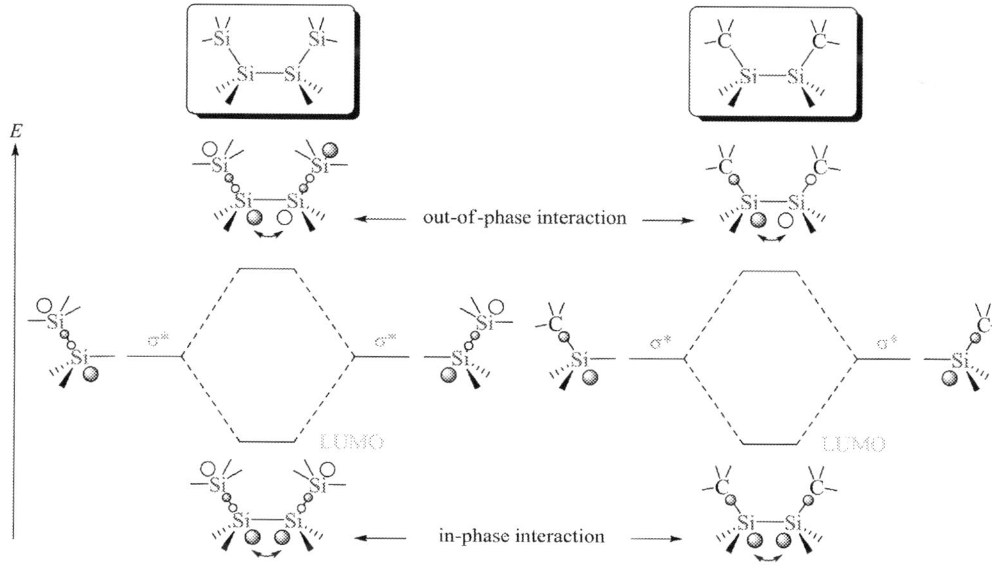

Figure 3.14
Interaction between Si–Si and Si–C σ^* orbitals.

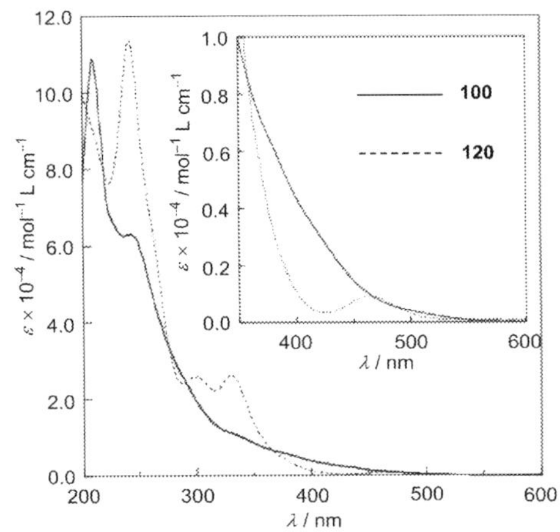

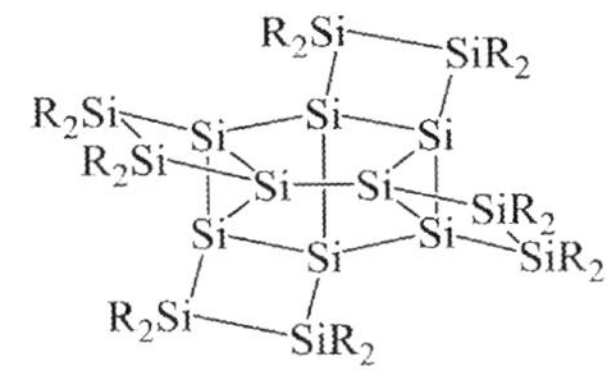

100: R = *t*-Bu

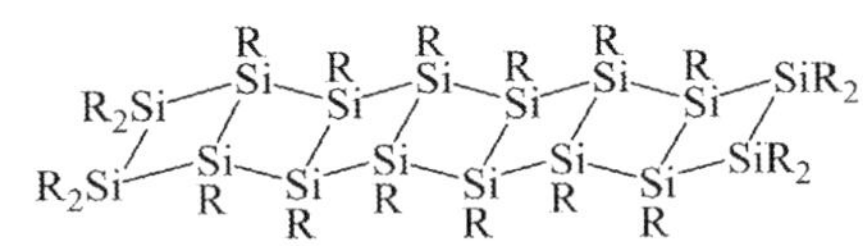

120: R = *i*-Pr

Figure 3.15
UV/Vis spectra of **100** in methylcyclohexane and **120** in hexane at room temperature.

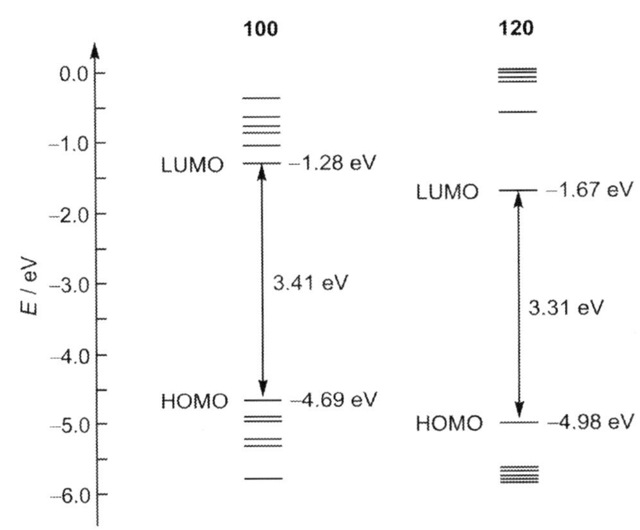

Figure 3.16
Comparison of the energy levels of the molecular orbitals of **100** and **120** calculated at the B3LYP/6-31G(d) level.

energy gap is close to those of other transitions (Fig. 3.16). As a result, absorption bands of many transitions overlap, and compound **100** does not show a clear band above 250 nm.

This situation is changed when an organosilicon cluster contains a Si=Si double bond. Compound **101** shows a distinctive lowest energy absorption band at 585 nm.[68] The $\pi-\pi^*$ transition of the Si=Si double bond has a small HOMO–LUMO energy gap compared to other transitions (Fig. 3.17).

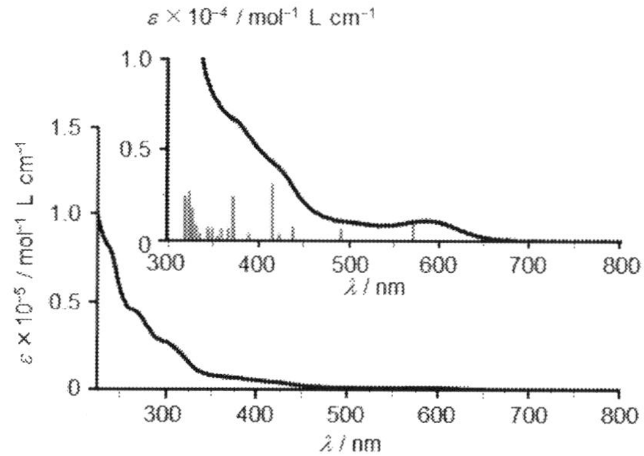

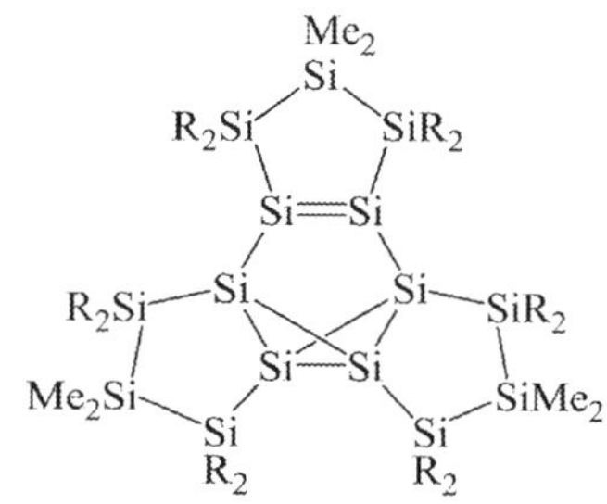

101: R = *t*-Bu

Figure 3.17

UV/Vis spectrum of **101** in hexane at room temperature with transitions calculated at the B3LYP/ 6-31 + G(2d,p) level (gray bars).

Marschner demonstrated a useful direction for understanding UV/Vis spectra of cyclic oligosilanes with silyl substituents.[34] They measured the UV spectra of **121**–**125** in Fig. 3.18 and assigned absorption bands on the basis of the data of linear oligosilanes Me-$(SiMe_2)_n$Me reported by Gilman.[91] The absorption bands of **121**–**125** at 242–250, 275–261, 273–278, 284, and 297 nm were assigned to pentasilane, hexasilane, heptasilane, octasilane, and decasilane chains. Compound **121** has hexasilane and heptasilane chains from one trimethylsilyl group along the ring to the other trimethylsilyl group. This compound shows absorption bands due to the hexasilane and heptasilane chains at 255 and 275 nm. Similarly, compounds **123**, **124**, and **125** show absorption bands due to the heptasilane, hexa- and octasilane, and penta- and decasilane chains, respectively. Compound **122** shows an absorption band due to the pentasilane chain at 242 nm but not the octasilane chains. This may be caused by the conformation of the cycloheptasilane ring affected by 1,3-interaction of the trimethylsilyl groups. In the case of **125**, the ring is large and flexible, and therefore, the effect of the 1,3-interaction is not large.

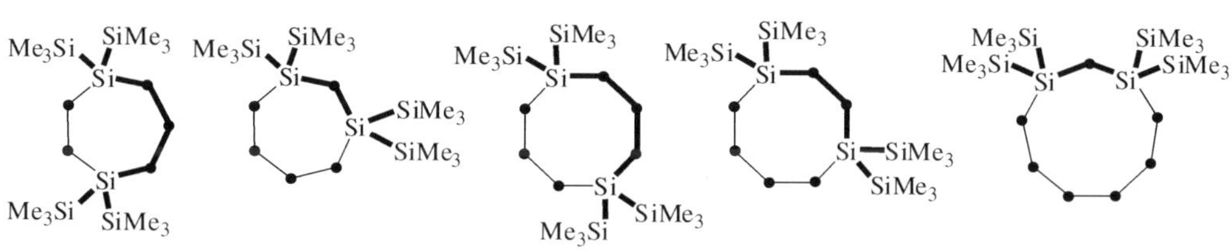

121 **122** **123** **124** **125**

• = $SiMe_2$

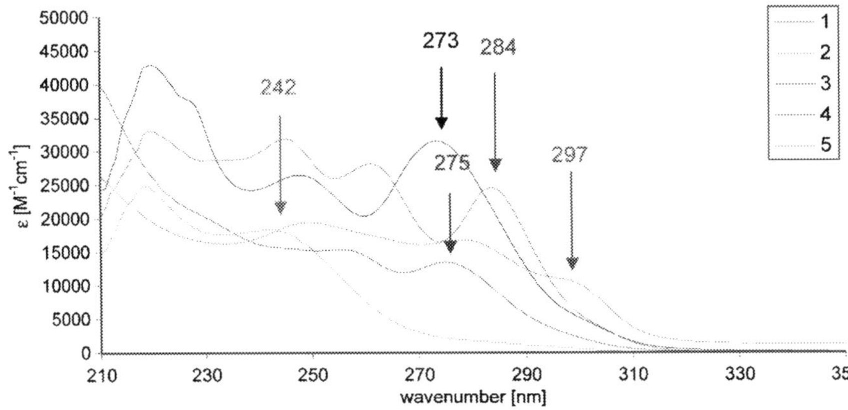

Figure 3.18
UV spectra of **121**–**125** with the lowest energy band for each compound marked.

3.5.2 Recent Topics

3.5.2.1 Spiropentasiladiene

Spiropentasiladiene **93** has been reported to have a twisted structure with the D_2 symmetry due to the steric hindrance among bulky silyl substituents.[58] The dihedral angle between the two cyclotrisilene rings is 78.26(0)°, and the torsional angle Si2–Si1–Si6–Si7 is 30.0(5)°. This compound shows four major absorption bands at 383 (ε 18,100 mol^{-1} L cm^{-1}), 428 (ε 11,700 mol^{-1} L cm^{-1}), 500 (ε 3640 mol^{-1} L cm^{-1}), and 560 nm (ε 2530 mol^{-1} L cm^{-1}) in the UV/Vis spectrum. The lowest energy absorption band is highly red-shifted compared to those of silyl-substituted cyclotrisilenes **62** (482 nm) and **126** (466 nm).[59,92]

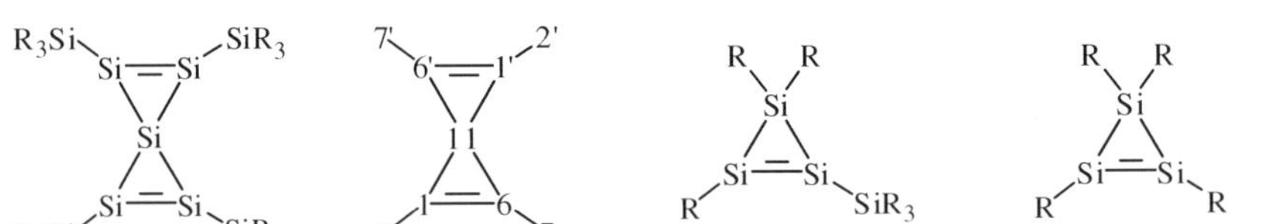

93: R = SiMe$_2$(*t*-Bu) **62**: R = SiMe$_2$(*t*-Bu) **126**: R = SiMe(*t*-Bu)$_2$

The large bathochromic shift is explained by the spiroconjugation. Molecular orbitals of model compounds are shown in Fig. 3.19. 3,3-Disilylcyclotrisilene **127** has π and π* orbitals. Two π orbitals of spiropentasiladiene **128** with the D_{2d} symmetry are degenerate, but two π* orbitals are highly splitted probably because of effective through-space interaction between the two π* orbitals. The energy levels of the two degenerated π orbitals are separated when the structure of **128** (D_{2d}) is deformed to **128** (D_2). As a result of the splitting of the π and π* orbitals, the energy gap between the π and π* orbitals becomes small, and **128** (D_2) shows a large bathochromic shift.

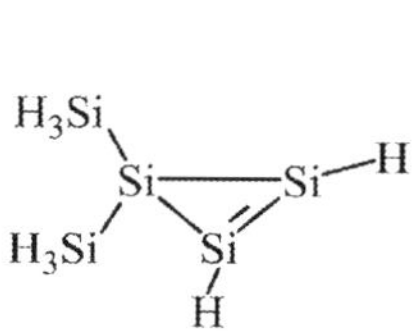

127 (C_{2v})

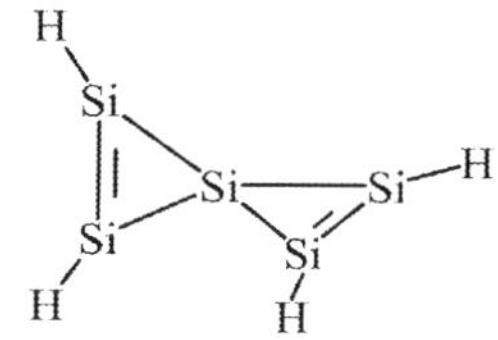

128 (D_{2d})

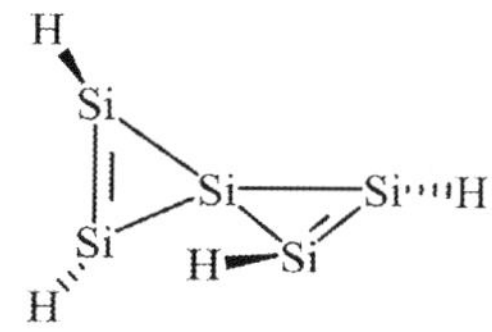

128 (D_2)

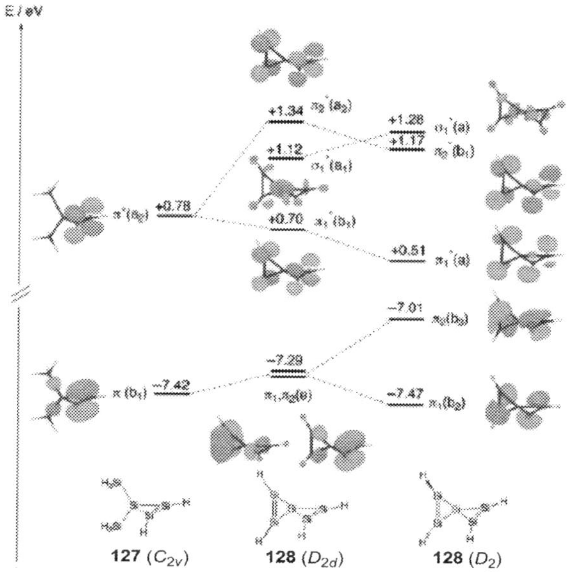

Figure 3.19
Schematic molecular orbitals and energy levels of **127** (C_{2v}), **128** (D_{2d}), and **128** (D_2) calculated at the HF/6-311++G(3df, 2p) level.

3.5.2.2 Pentasila[1.1.1]propellane and a hexasilabenzene isomer with a global minimum energy

In Fig. 3.20, the UV/Vis spectra of pentasila[1.1.1]propellane **96** is shown.[61] This compound shows absorption bands at 325 and 396 nm. These bands fit the transitions of the model compound **129** calculated by a TD-DFT method.[93] The band at 325 nm corresponds to the HOMO−1 to LUMO transition, and the band at 396 nm is due to the HOMO to LUMO transition. The major contribution to the HOMO−1 and LUMO is the bridgehead Si−Si σ and σ* orbitals, respectively. The HOMO is formed by σ conjugation among the six bridge Si−Si σ orbitals.

Compound **96** shows a very weak absorption band at 546 nm. The TD-DFT calculation showed that this band is assigned to the HOMO−1 to LUMO transition from the ground state to the triplet excited state. As this type of absorption is not usually observed, the bridgehead Si−Si bond of **96** shows a very unique electronic property.

A related compound **99** shows the lowest energy absorption band at 473 nm (ε 700 mol^{-1} L cm^{-1}).[65] This absorption cannot be ascribed to the forbidden transition from the ground state

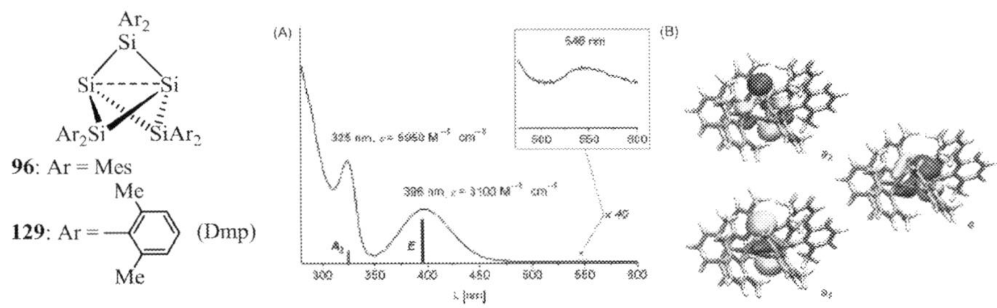

Figure 3.20
(A) UV/Vis spectra of **96** in THF with transitions of **129** calculated at the B3LYP/def2-TZVP level. (B) The HOMO−1 (a_1), the HOMO (e), and the LUMO (a_2) of **129**.

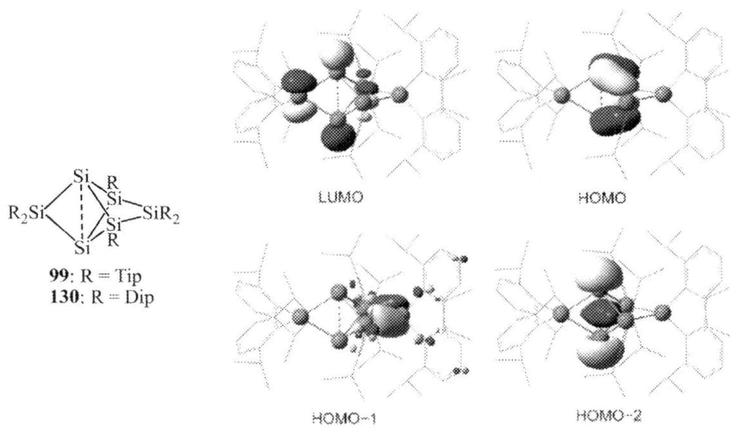

Figure 3.21
Frontier orbitals of **130** at isovalue 0.05 calculated at the B3LYP/6-31G(d) level.

to the triplet excited state because of the relatively large molecular extinction coefficient. Instead, a TD-DFT calculation of the model compound **130** showed that the band is due to the HOMO to LUMO transition (Fig. 3.21). The HOMO is located at the Si−Si bonds at two bridges. The HOMO−2 and LUMO are formed mainly by the bridgehead Si−Si σ and σ* orbitals, respectively. The disappearance of the absorption due to the transition from the ground state to the triplet excited state may be due to the rigid structure caused by an additional bridge and thus less vibronic coupling or possibly the tailing of the intense band at 473 nm.

3.5.2.3 Persilastaffanes

Persilastaffanes **14**, **17**, and **19** show intense absorption band I and weak absorption band II in the UV spectra (Fig. 3.22).[30] The bands I and II shift bathochromically as the number of bicyclo[1.1.1]pentasilane units increases.

The frontier orbitals of the model compounds **131**−**133** are shown in Fig. 3.23. The HOMOs are two degenerate Si−Si σ orbitals at the bridges (σ_{cage}). The σ_{axis} orbital is formed mainly

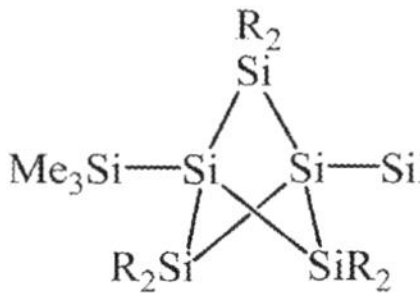

14: R = *i*-Bu

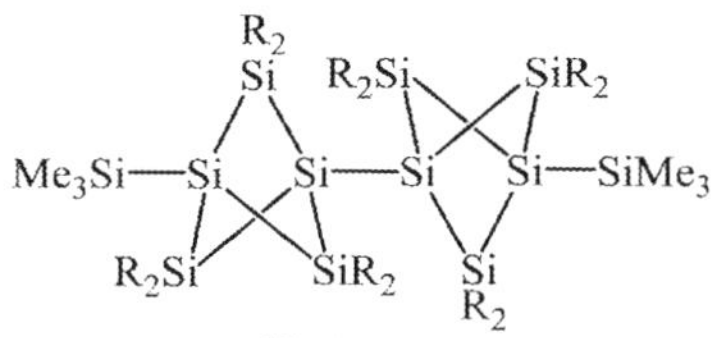

17: R = *i*-Bu

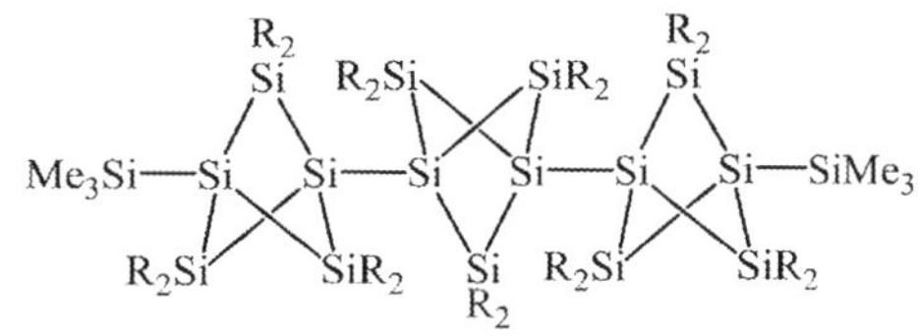

19: R = *i*-Bu

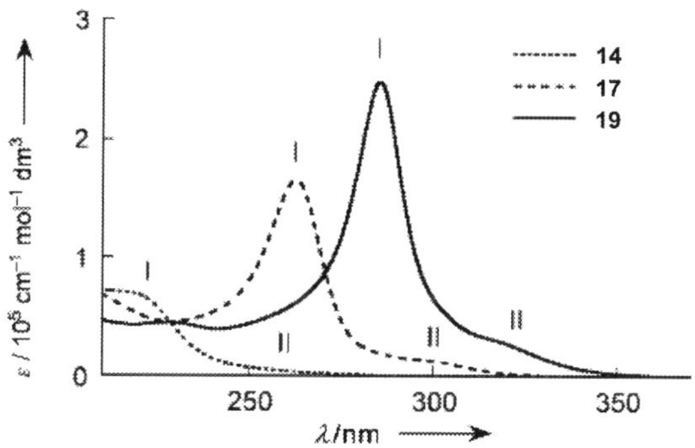

Figure 3.22
UV spectra of **14**, **17**, and **19** in hexane at room temperature.

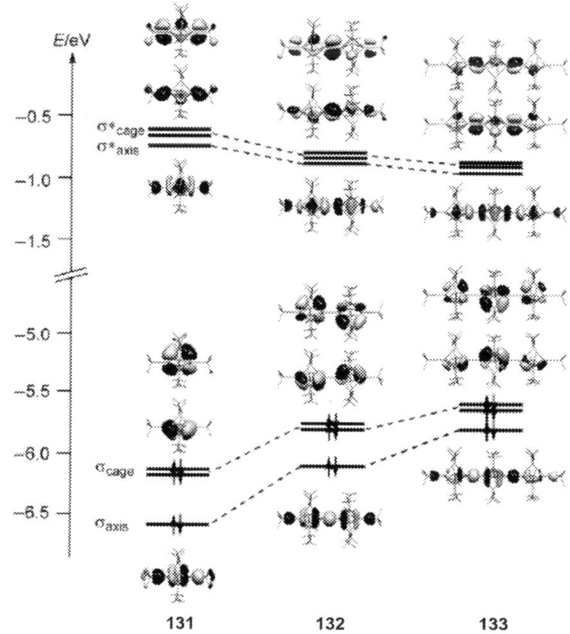

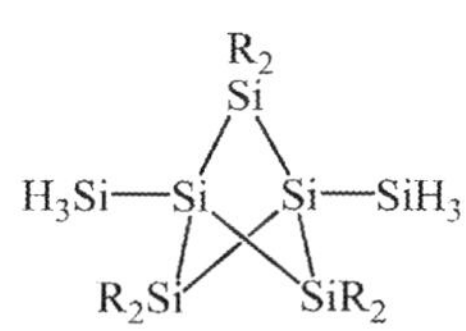

131: R = Me

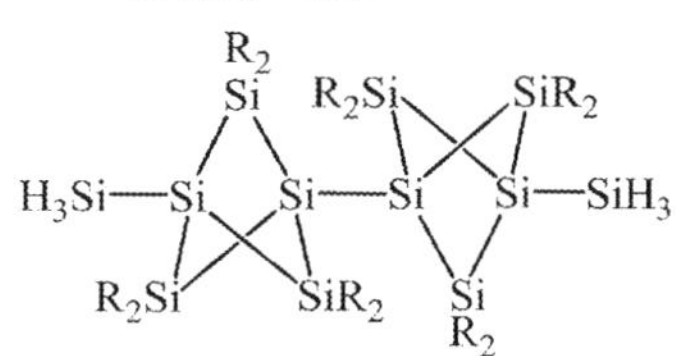

132: R = Me

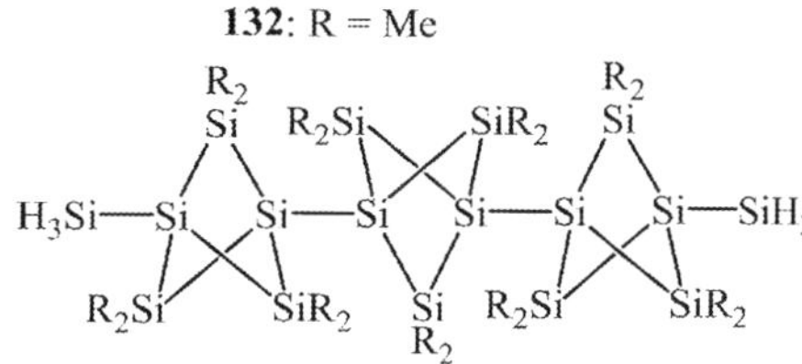

133: R = Me

Figure 3.23
Frontier Kohn–Sham orbitals of **131**–**133** calculated at the B3LYP/def2-TZVP//B3LYP/6-31G(d) level.

by the bridgehead Si–Si σ orbitals. Similarly, the σ^*_{axis} (LUMO) and σ^*_{cage} orbitals are formed by the bridgehead and bridge Si–Si σ* orbitals, respectively. According to theoretical calculations, the bands I and II were assigned to the σ_{axis} to σ^*_{axis} transition and the σ_{cage} to σ^*_{cage} transition with a significant contribution of the σ_{cage} to σ^*_{axis} transition, respectively. The remarkable bathochromic shifts of the bands I and II are explained by the delocalization of these orbitals over the persilastaffanes and the decrease of transition energies.

3.6 Reactions

Various types of reactions of organosilicon clusters have been reported, including Si–Si bond cleavage, rearrangement of silicon skeletons, isomerization, oxidation, photochemical reactions, etc. These reactions are described in this section.

3.6.1 Si–Si Bond Cleavage

The bridgehead Si–Si bonds of bicyclo[1.1.0]tetrasilanes **59** and **61** have been reported to be cleaved by water and chlorine.[4a,94] The reaction of a mixture of bicyclo[1.1.0]tetrasilanes **64** and **65** with iodine gave two kinds of bond cleavage products. Cyclotrisilanes **134** and **135** and cyclotetrasilanes **137–139** were formed by cleavage of the Si–Si bonds at the 1,2-positions and the bridgehead positions, respectively.[45]

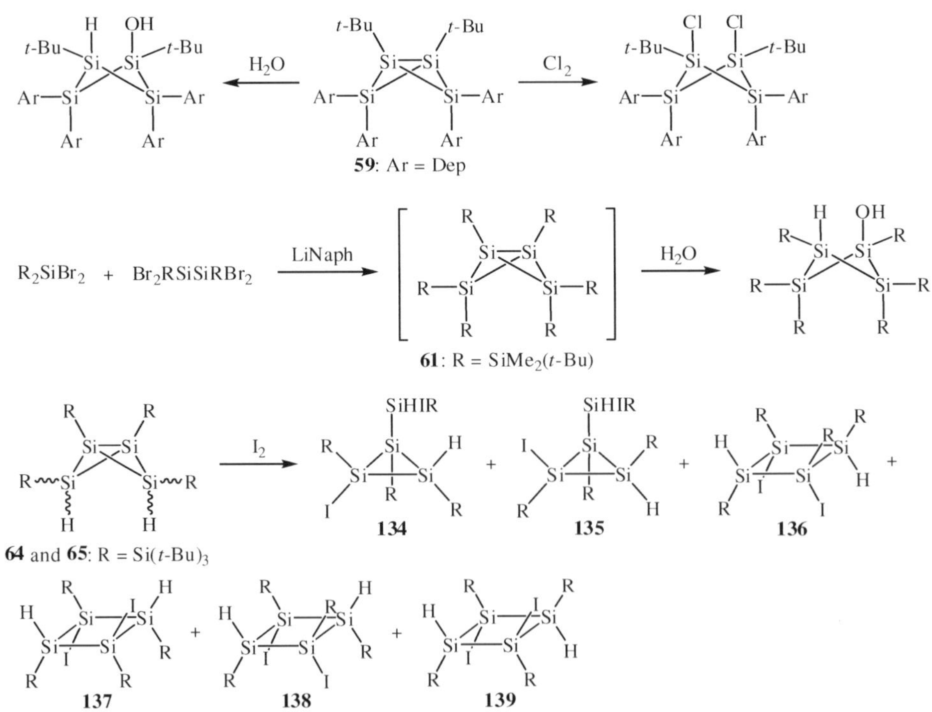

Similarly, the bridgehead Si–Si bond of the bicyclic ladder oligosilane **140** is cleaved by a palladium complex, hydrochloric acid, and hydrobromic acid.[95]

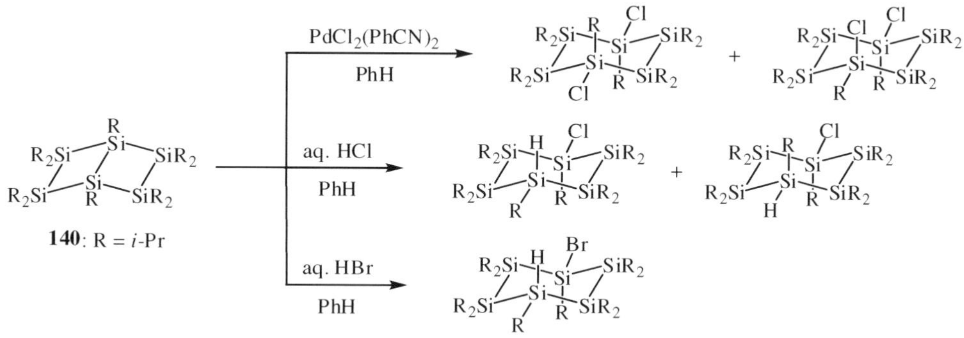

The Si–Si bond cleavage of the tricyclic ladder oligosilane **141** is cleaved by a palladium complex. The rearranged product **142** and the simple bond cleavage product **143** were formed. When phosphorus pentachloride was used, **144** and **143** were obtained.[96]

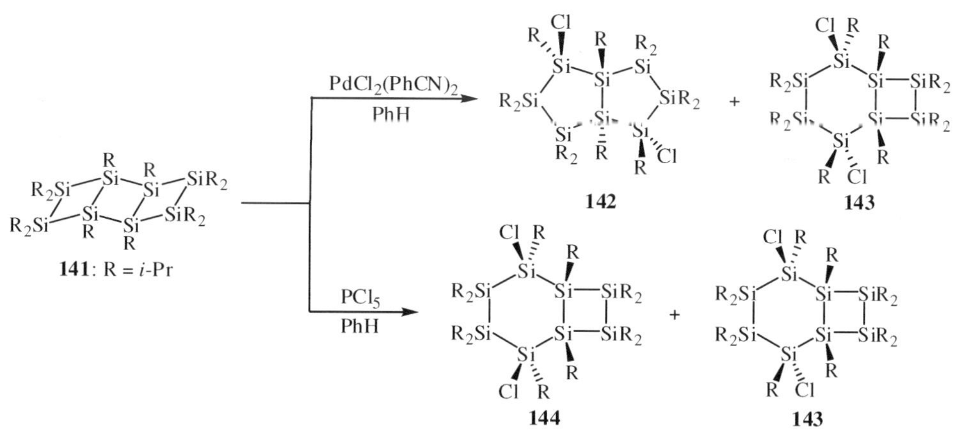

The bridgehead Si═Si double bond of **84** is cleaved by water, carbon tetrachloride, and 9,10-phenanthrenequinone to give **145**, **146**, and **147**, respectively.[54]

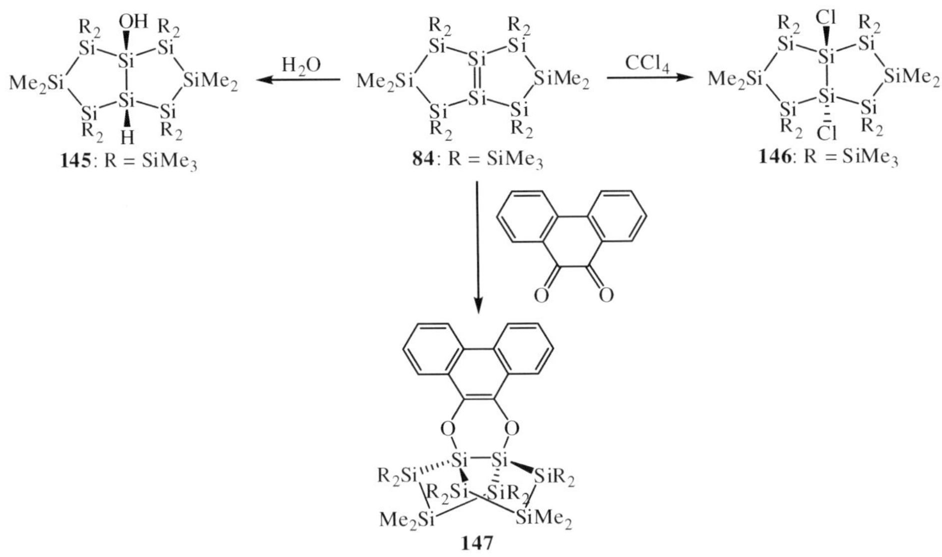

The reaction of tetrasilatetrahedrane **49** with iodine gave 3,4-diiodocyclotetrasilene **69**.[47,60] This compound was further iodinated to 1,1,2,3,4-pentaiodocyclotetrasilane **94** in benzene at room temperature to and to *trans*-$[(t\text{-Bu})_3Si]_2Si_4I_6$ in boiling benzene.[60]

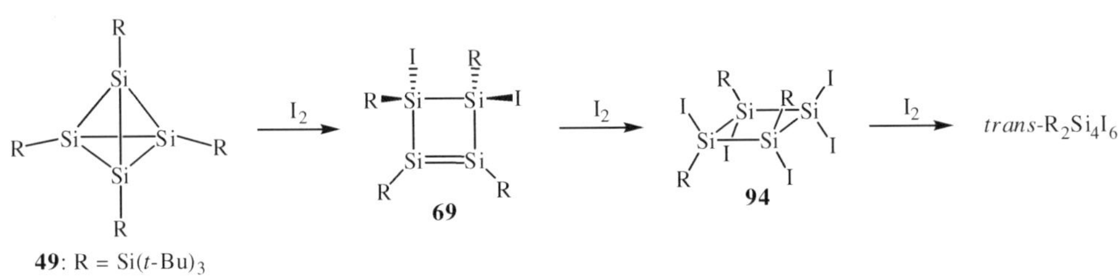

The Si–Si bond cleavage of octasilacubane **43** has already been described in Section 3.2.3.

1,2-Dipotassiocyclopentasilane **78** was formed by the reaction of bicyclo[3.3.0]-pentasilaneoctasilane **77** with potassium *tert*-butoxide. The initially formed cyclopentasilanylpotassium with a trisilanyl chain is attacked by potassium *tert*-butoxide to give **78** with elimination of 1,3-di-*tert*-butoxytrisilane.[34]

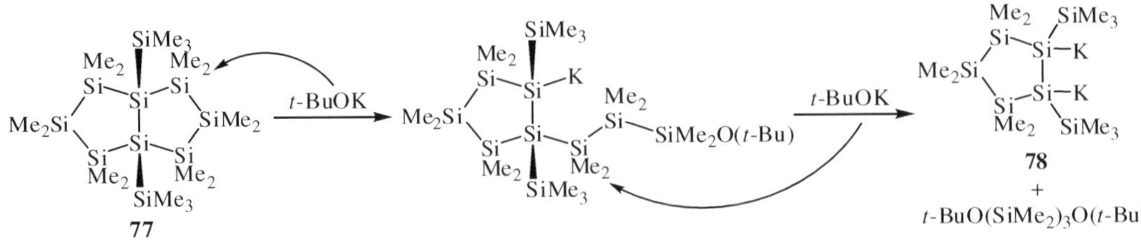

3.6.2 Rearrangement of Silicon Skeletons

Rearrangement of silicon skeletons with aluminum chloride is an important reaction of organosilicon clusters. Details of this rearrangement were described in Sections 3.2.1 and 3.2.3.

3.6.3 Isomerization

Kira has reported interconversion of bicyclo[1.1.0]tetrasilane **61** and cyclotetrasilene **60**.[44a] Compound **61** isomerizes to **60** thermally, and photolysis of **60** gives **61**.

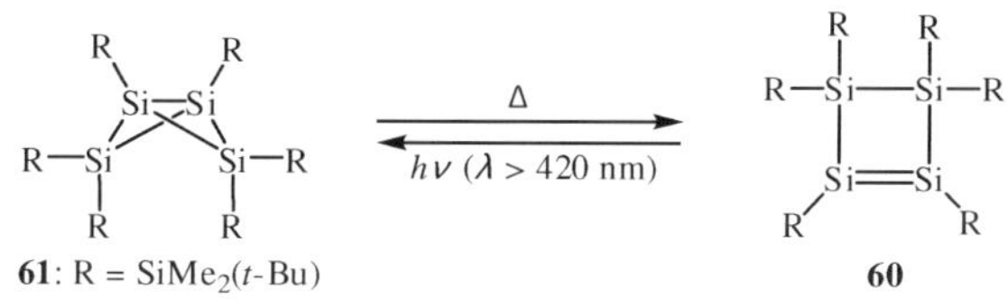

Thermodynamic parameters of the thermal isomerization from **61** to **60** were determined: $\Delta H^{\ddagger} = 16.5$ kcal mol^{-1}, $\Delta S^{\ddagger} = -20.8$ cal mol^{-1} K^{-1}.[44a] The large negative $\Delta S^{\ddagger}$ value shows that the transition state is significantly restricted. Two plausible reaction mechanisms were postulated.[94] In the path A, the isomerization of **61***, in which the bridgehead substituents are labeled as R*, gives **60*** via 1,2-silyl migration. The distribution of two R* groups on the double-bond and saturated silicon atoms (%2R*) should be 50 and 50%, respectively. On the other hand, skeletal isomerization gives **60***, in which %2R* is 100% on the double-bond silicon atoms (path B). The isomerization was carried out by using **61**-d_{12}. The %2R* values in **60**-d_{12} were 47 and 53% on the double-bond and saturated silicon atoms, respectively. This result supports the 1,2-silyl migration mechanism.

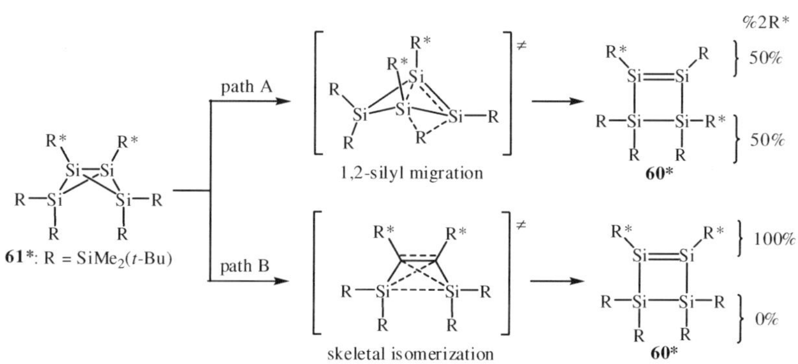

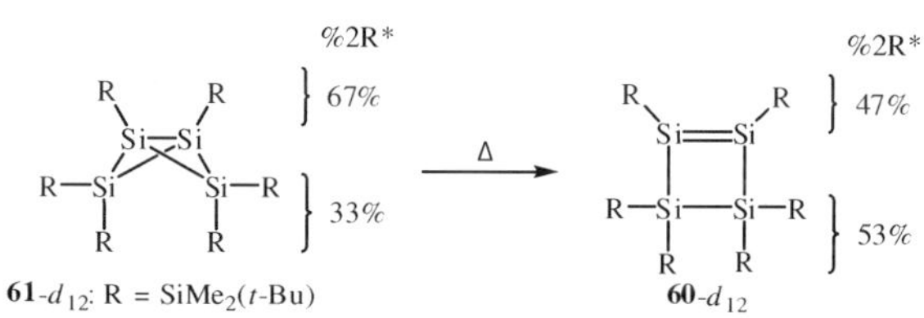

Wiberg observed isomerization of **65** to **64**.[45] Pure sample of compound **65** gave a 10:1 mixture of **64** and **65** on standing at room temperature for 2 weeks. The biradical **148** was suggested as a possible intermediate.

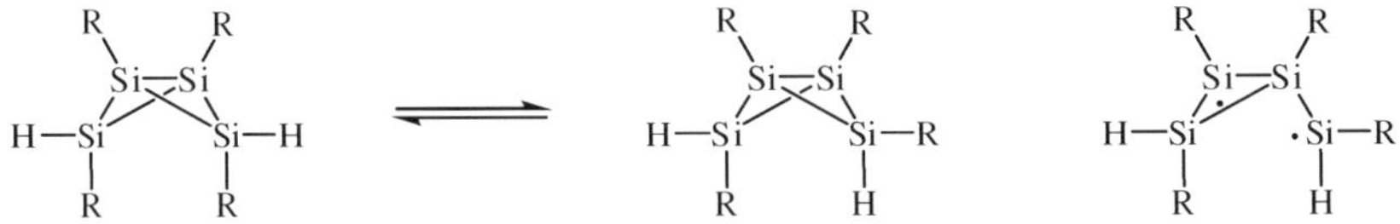

65: R = Si(*t*-Bu)$_3$ **64** **148**

The tricyclic ladder oligosilane **149** has been reported to isomerize to the more stable anti isomer **141** at 220°C.[97] This isomerization can be regarded as inversion of the bicyclo[2.2.0]hexasilane moiety. The thermodynamic parameters of this isomerization were measured: $E_a = 42.3 \pm 0.4$ kcal mol^{-1}, $\log A = 15.1 \pm 0.2$ s^{-1}, $\Delta H^{\ddagger} = 41.4 \pm 0.4$ kcal mol^{-1}, $\Delta S^{\ddagger} = 7.6 \pm 0.8$ cal mol^{-1} K^{-1}. Although a reaction mechanism is not clear, the large activation energy implies that the Si–Si bond at the bridgehead positions may be cleaved to give a biradical intermediate as reported in the isomerization of *syn*-tricyclo[4.2.0.0^{2,5}]octane to its anti isomer.[98]

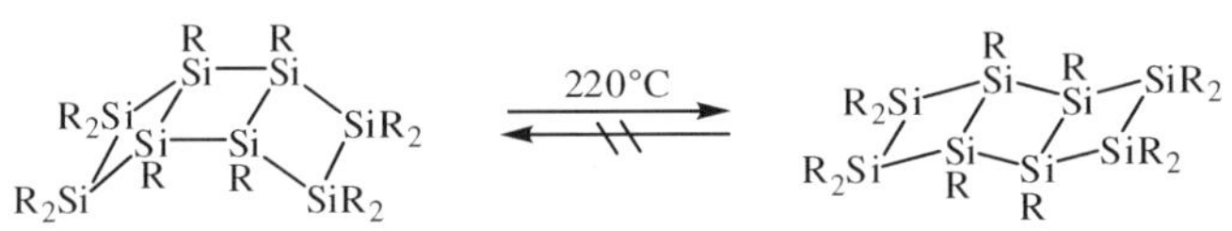

149: R = *i*-Pr **141**

Elimination of DMAP from the **81** · DMAP complex with triphenylborane gave **74** and **150** in 72 and 8% yields, respectively. On standing the mixture of **74** and **150**, **150** isomerized to **74**. These results are explained by isomerization of silylene **151** to disilene **81**, and dimerization of **81** to **74**. The isomerization between **81** and **151** was also supported by the reactions of **150** with DMAP and silicon tetrachloride to give **81** · DMAP and **152**, respectively.[50]

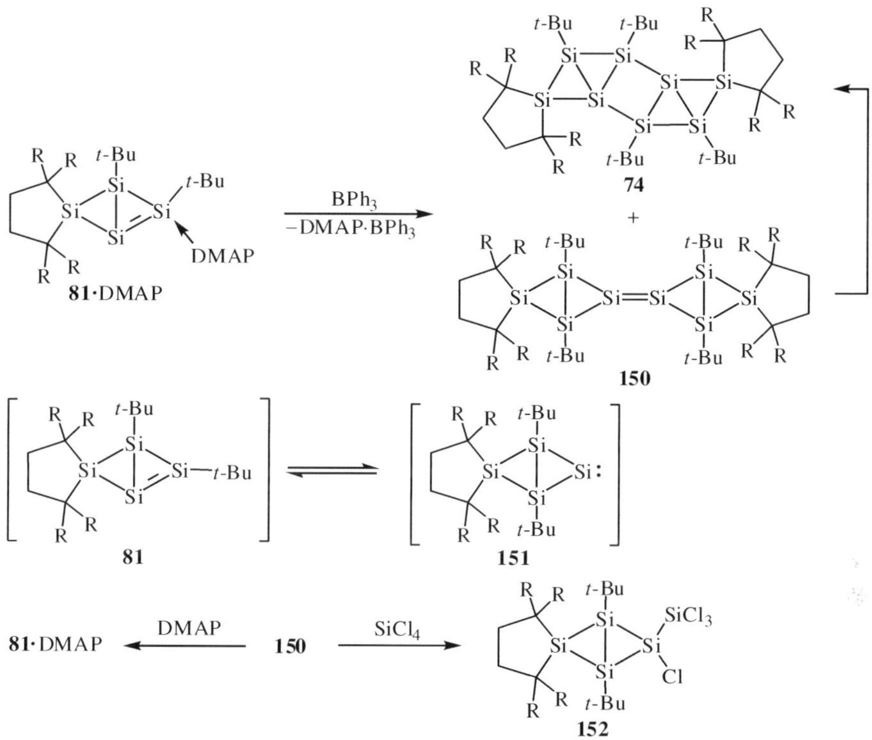

3.6.4 Unique Reactivity of Siliconoids

Halogenation of organosilicon cluster **95** proceeds via several types of Si−Si bond cleavage.[99] Chlorination of **95** with carbon tetrachloride gave **153** and **154**, while bromination with carbon tetrabromide gave two molecules of **155**. Compounds **156** and **157** were formed by chlorination with tin(IV) chloride and iodination with iodine, respectively.

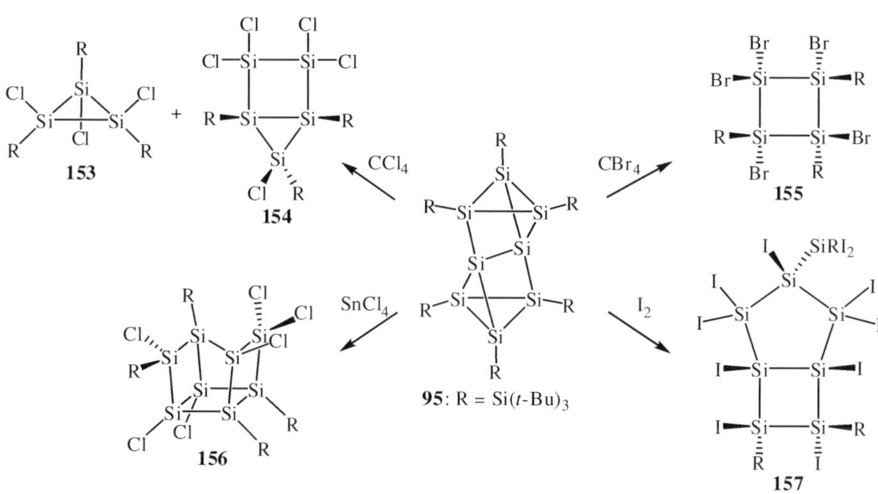

Although detailed reaction mechanisms are obscure, the formation of **153** and **154** is explained by chlorinative cleavage of four Si–Si bonds. In the bromination of **95**, bromotetrasilatetrahedrane **158** is formed via isomerization of **95** to **53** and cleavage of the central Si–Si bond with bromine. Further bromination of **158** gives **155** through **159**.

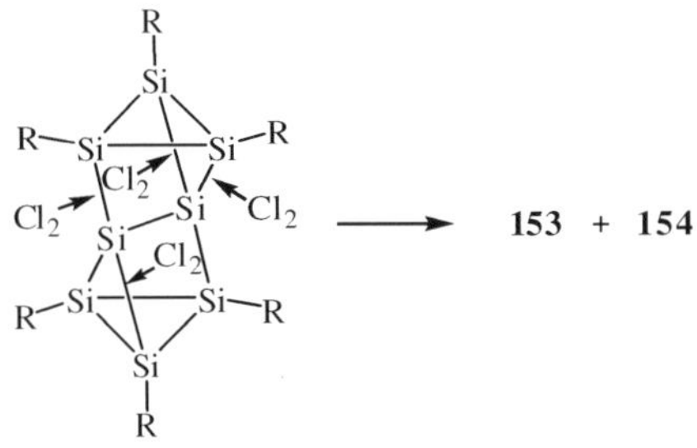

95: R = Si(*t*-Bu)$_3$

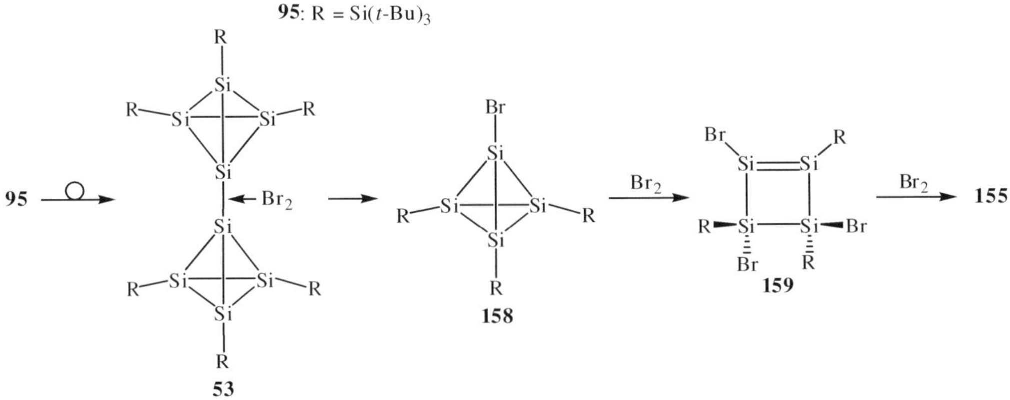

Chlorinative cleavage of the central Si–Si bond of **95** and skeletal rearrangement give octasilacubane **160**. Chlorination of two silicon atoms and skeletal rearrangement give **161**. Compound **156** is formed by replacement of two R groups by chlorine atoms. The mechanism of the formation of **157** is obscure, but it may be formed directly from **95**.[99]

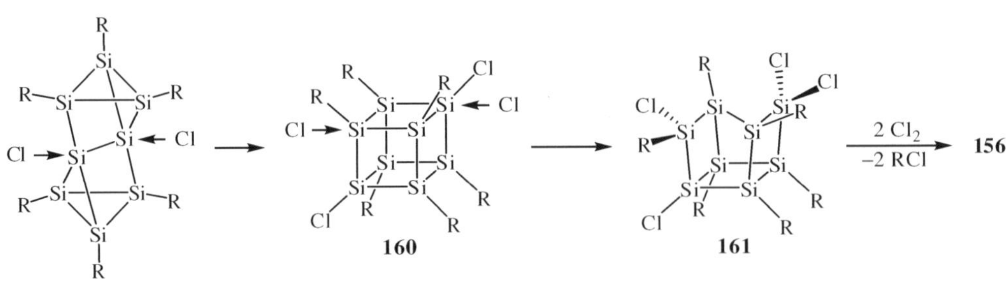

95: R = Si(*t*-Bu)$_3$

The long and weak bridgehead Si–Si bond of **96** shows both closed-shell and radical-type reactivity. The reactions of **96** with H_2O, PhSH, PhOH, Me_3SnH, and 9,10-dihydroanthracene gave bicyclo[1.1.1]pentasilane derivatives.[61]

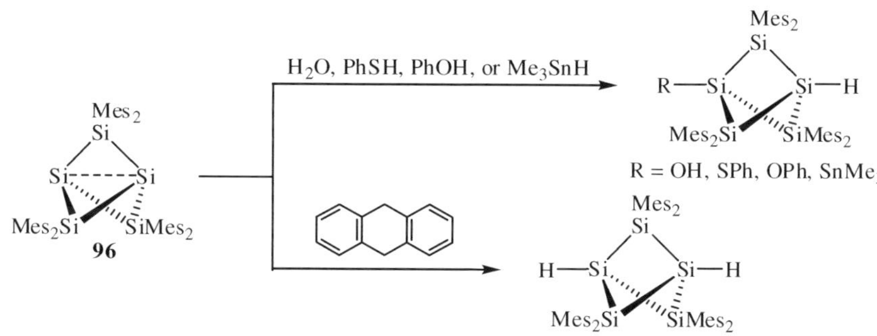

As described in Section 3.2.7, isomerization of the tricyclic isomer of hexasilabenzene **98** to **99** occurs thermally and photochemically.[65] By detailed study of the thermal isomerization, compound **106** was isolated as a by-product.[40] This compound is a dimerization product of **98**, but an $SiTip_4$ unit was removed from the exact dimerization product $Si_{12}Tip_{12}$ during the dimerization. As a result, three unsubstituted silicon vortices are formed. Two vortices form a Si–Si bond (Section 3.3.2), and the remaining one is inserted into a C–H bond of a Tip group.

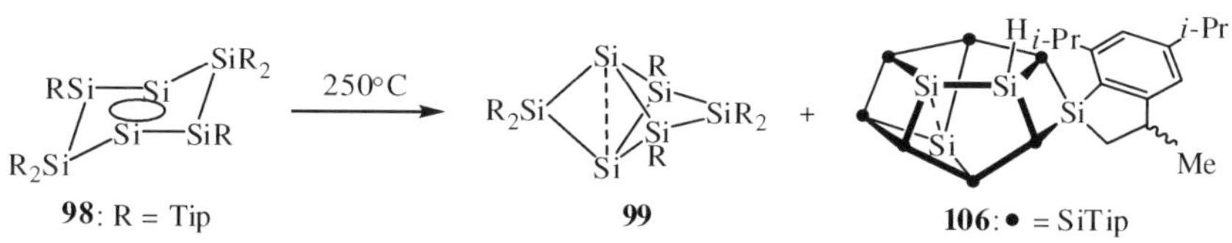

Chlorination of **98** with bismuth trichloride gave the dichlorinated compound **162** as a major product together with **163**, **164a**, and **99**.[40] Although the reaction mechanism is obscure, the presence of subvalent bismuth compounds might play an important role. The silicon skeleton of the dichlorinated compound **162** was rearranged to **165** thermally at 230°C.

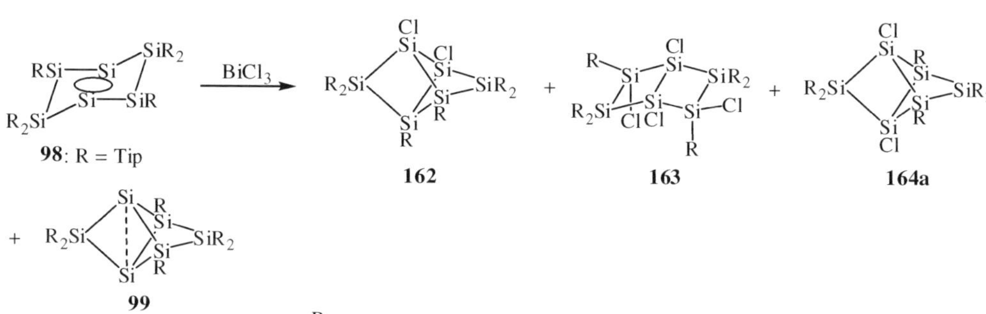

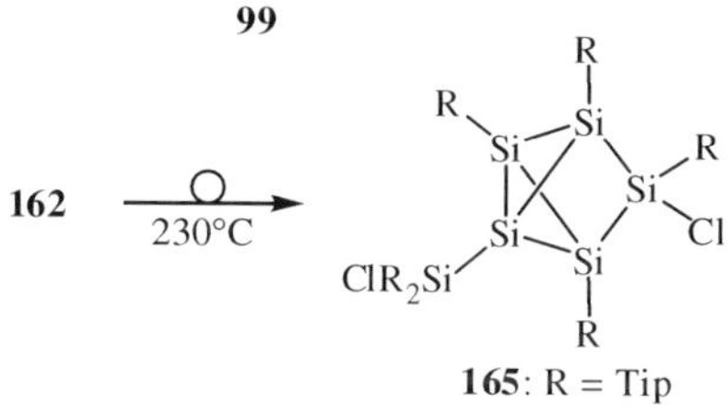

Halogenation of **99** with bromine or iodine at room temperature leads to cleavage of the bridgehead Si–Si bond to give the dihalogenated products **164b,c**.[65] When **99** was treated with an excess amount of iodine at 110°C, hexaiodocyclopentasilane **166** was obtained by contraction.[40]

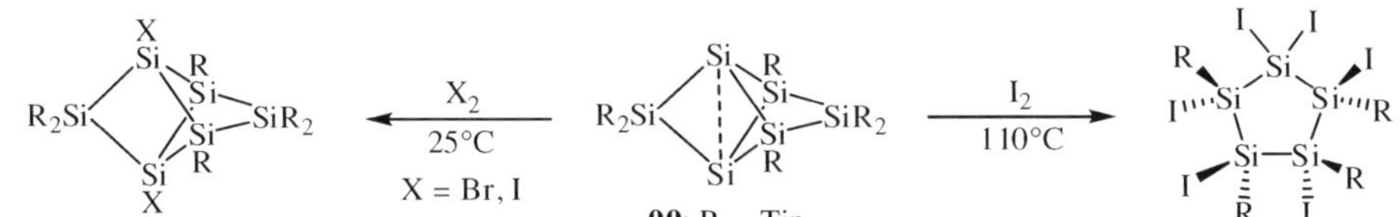

164b: X = Br
164c: X = I
99: R = Tip
166

Although the mechanisms of most reactions described here are not easy to understand, it is clear that siliconoids have the possibility of showing unprecedented reactivity.

3.6.5 Oxidation

Ladder oligosilanes have been reported to be oxidized with *m*-chloroperbenzoic acid (MCPBA) to give domino oxidation products.[100b] Oxidation of the tricyclic ladder oligosilane **141** with three equivalents of MCPBA gave the trioxidized compound **167**. Three oxygen atoms are inserted into one of two oligosilane main chains selectively. During the oxidation, mono- and dioxidation products were not detected. This result is explained by that once a Si–Si bond is oxidized, rapid oxidation of adjacent Si–Si bonds takes place successively. Similarly, **168** and **170** gave domino oxidation products **169** and **171** by four and five equivalents of MCPBA, respectively. However, in the oxidation of the bicyclic ladder oligosilane **140**, the monooxidized products **172** and **173** were formed along with the domino oxidation product **174**.[100a]

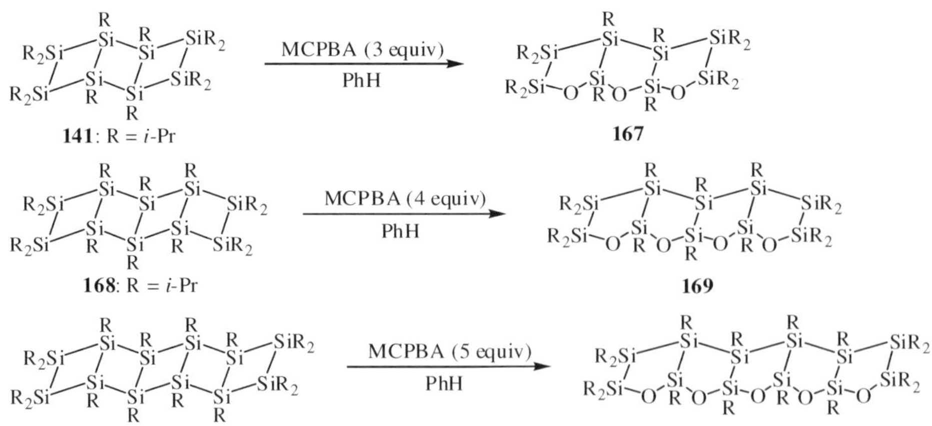

141: R = *i*-Pr
167
168: R = *i*-Pr
169
170: R = *i*-Pr
171

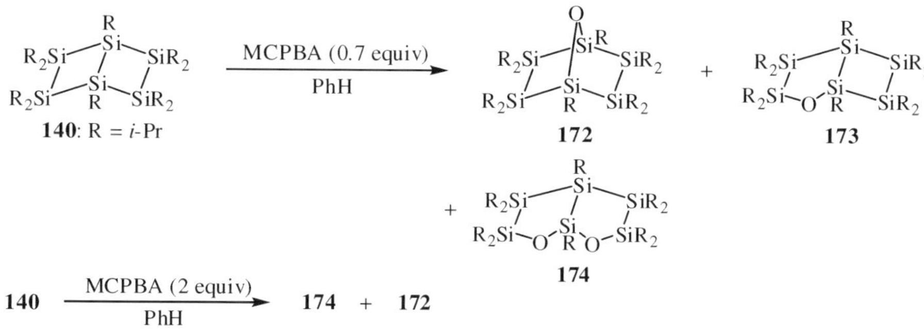

When an excess amount of MCPBA was used, all Si—Si bonds of ladder oligosilanes were oxidized, and ladder oligosiloxanes were obtained.[101]

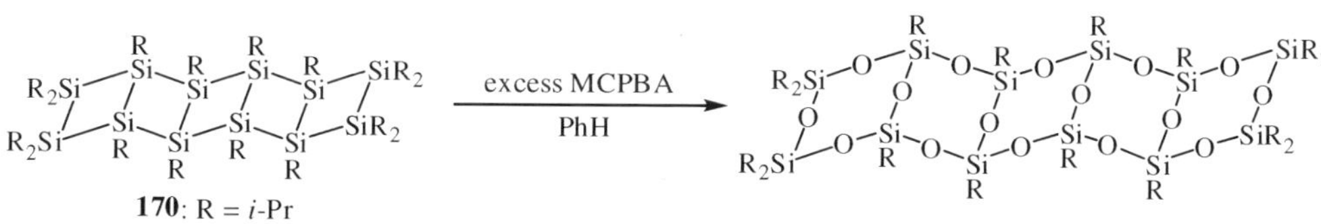

Oxidation of octasilacubanes has been studied.[102] Mono- and dioxidized compounds **175** and **176** were obtained by irradiation of **43** and DMSO in benzene, while oxidation of **58** with 14 equivalents of MCPBA gave silsesquioxane **177**.

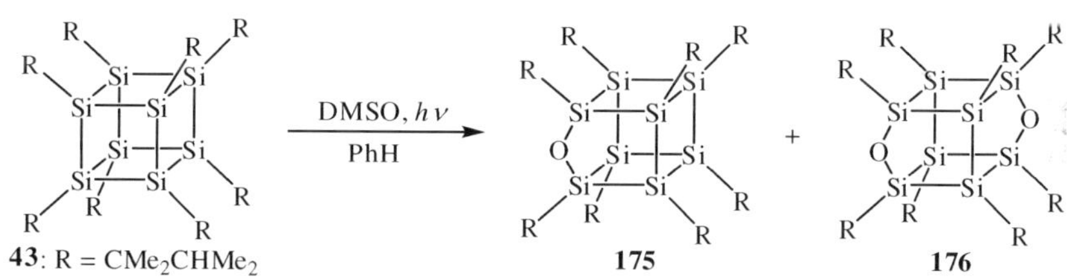

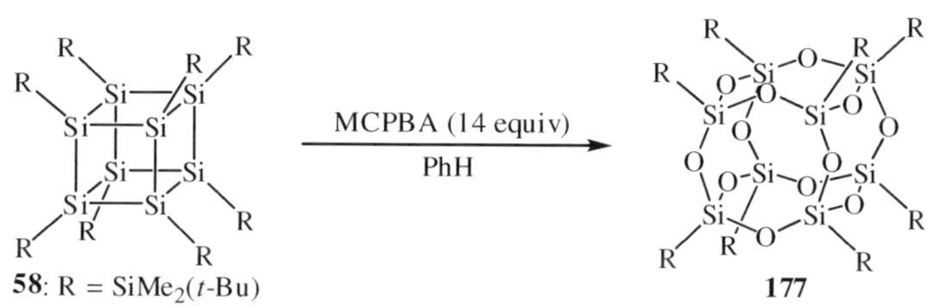

Monooxidation of tetrasilatetrahedrane **49** has been reported.[103] The reaction of **49** with two equivalents of $Ph_3C^+ \cdot TFPB$ (TFPB = tetrakis[3,5-bis(trifluoromethyl)phenyl]borate) gave

178^{+} · TFPB. This compound was treated with triethylamine to give **70**. A possible reaction mechanism may involve two-electron oxidation of **49** with Ph_3C^{+} · TFPB and cleavage of a Si–Si bond to give dication, which reacts with residual water in the reaction mixture to give **178**$^{+}$.

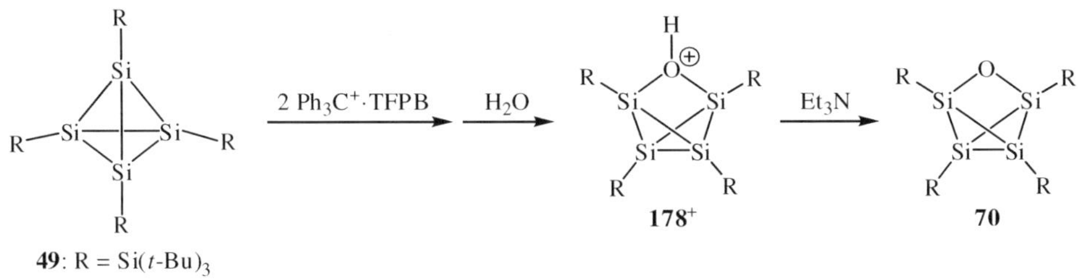

Autooxidation of **49** and **59** gave dioxidation products. On standing **49** in air at room temperature for months, the dioxide **179** was obtained mainly.[45] Bicyclo[1.1.0]tetrasilane **59** reacts with oxygen to give the dioxide **180**.[4a] It is noted that adjacent Si–Si bonds are oxidized selectively in these reactions.

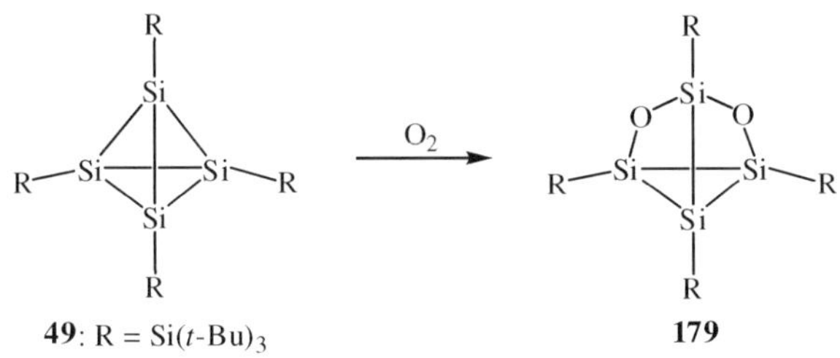

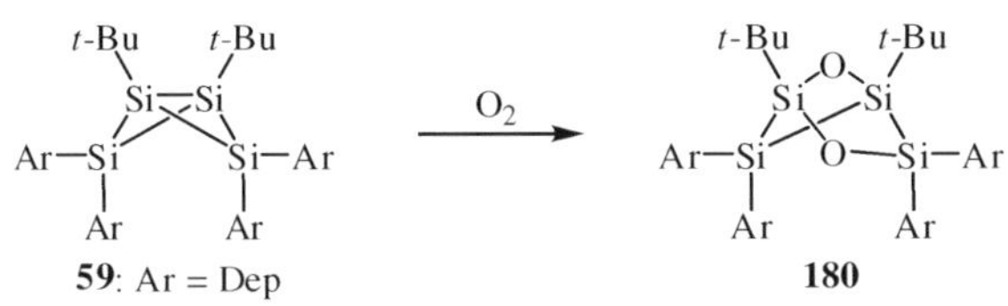

3.6.6 Photochemical Reactions

Irradiation of hexasilaprismane **54** with light ($\lambda = 340-380$ nm) has been reported to give hexasila-Dewar benzene **181**.[11d,104,105] Compound **54** was regenerated on irradiation of **181** with light ($\lambda > 460$ nm) or thermally. The thermodynamic parameters of this thermal reversion was estimated as $E_a = 13.7$ kcal mol^{-1}, $\Delta H^{\ddagger} = 13.2$ kcal mol^{-1}, and $\Delta S^{\ddagger} = -17.8$ cal mol^{-1} K^{-1}.

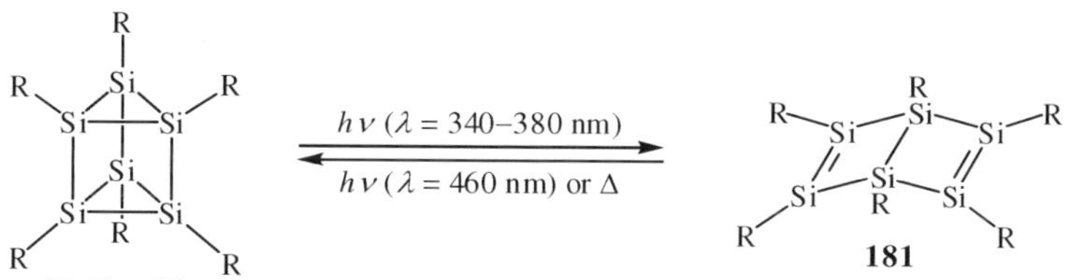

54: R = Dip

181

Photolysis of octasilacubane **43** in 3-methylpentane at 77K gave three new absorption bands at 340, 470, and 700 nm.[106] Similar absorption bands were observed in the transient absorption spectra in cyclohexane at room temperature. Theoretical studies on possible intermediates showed that these absorption bands can be ascribed to **182**.

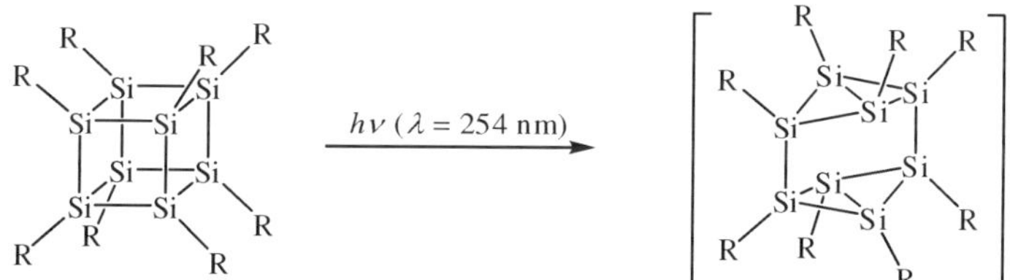

43: R = CMe_2CHMe_2

182

Photolysis of **71** gave its isomer **184**.[48] This isomerization is considered to proceed via the cyclohexasiladiene intermediate **183**. Formal symmetry-allowed photochemical $[_{\sigma}2_s + _{\sigma}2_s]$ ring opening and symmetry-allowed thermal $[_{\pi}2_s + _{\pi}2_a]$ ring closure are compatible with the Woodward–Hoffmann rule.

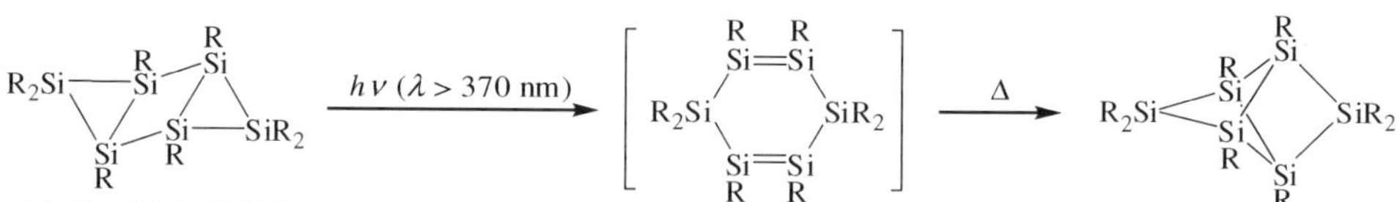

71: R = $SiMe_2(t\text{-Bu})$

183

184

Photolysis of the tricyclic ladder oligosilanes **141** and **185** in the presence of methanol and 2,3-dimethyl-1,3-butadiene gave **187a,b** and **188a,b**, respectively. Photolysis of **141** and **185** in the presence of anthracene gave **189** and **190**, respectively.[107] Two Si–Si bonds of **141** and **185** are cleaved photochemically, and the resulting cyclotetrasilenes were trapped by methanol, 2,3-dimethyl-1,3-butadiene, and anthracene. When photolysis of **141** was carried out in the presence of C_{60} and carbon disulfide, the cyclotetrasilene intermediate **186a** was trapped by carbon disulfide to give carbene **191**, and addition of **191** to C_{60} gave **192**.[108]

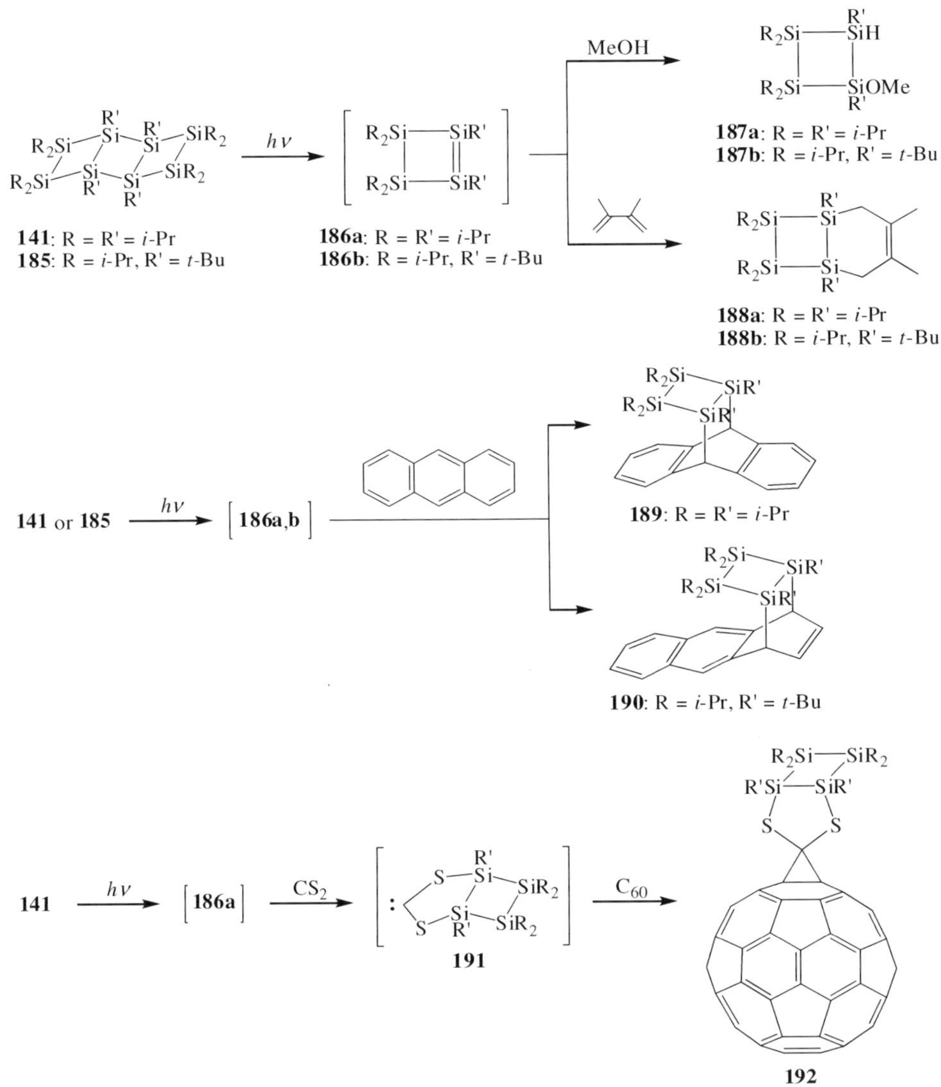

3.7 Ionic Organosilicon Clusters

3.7.1 Anions

The following anionic organosilicon clusters have already been discussed in this chapter. These compounds are useful building blocks for constructing organosilicon clusters.[29,31,38,39] The dynamic behavior of **52** was discussed in Section 3.3.3.

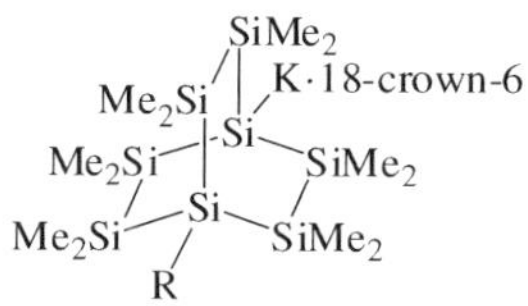

25a: R = $SiMe_3$
25b: R = K · 18-crown-6

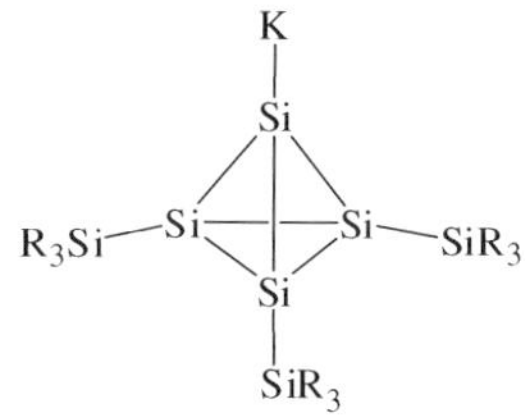

51: SiR_3 = $Si(t\text{-}Bu)_3$
52: SiR_3 = $SiDis_2Me$

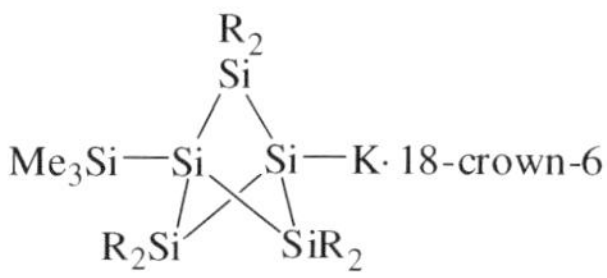

15: R = *i*-Bu

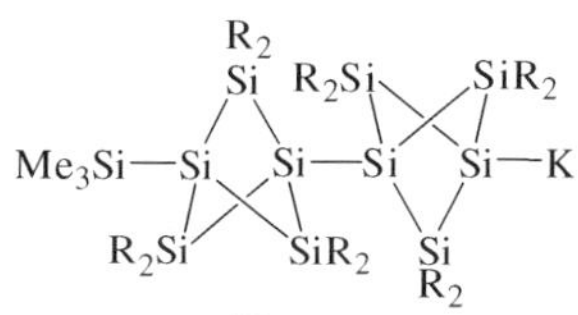

18: R = *i*-Bu

The calcium salt of bicyclo[1.1.0]tetrasilanide (Ca^{2+} · $\mathbf{30}^{2-}$) was obtained by the transmetallation of 2 K^+ · $\mathbf{66}^{2-}$ with calcium iodide.[109] This anionic organosilicon cluster was used for the synthesis of tricyclo[2.1.0.0^{2,5}]pentasilane **35** (Section 3.2.3)[35] and silylene complexes.[109,110]

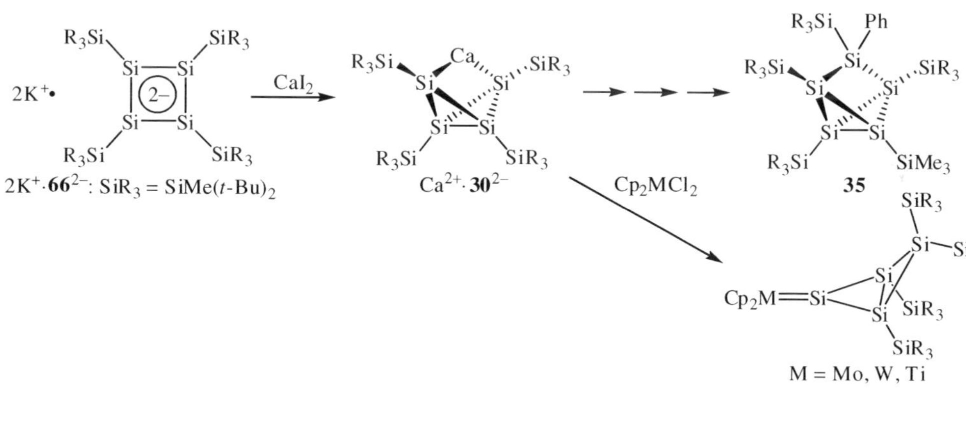

The reduction of tetrasilatetrahedrane **49** with sodium naphthalenide gave dianion, the structure of which was supposed to be **193** or 2Na^+ · $\mathbf{194}^{2-}$.[11i,45] Bicyclo[1.1.0]tetrasilane **65** was obtained by methanolysis of this dianion. On the other hand, the reaction of this dianion with dimethyl sulfate gave cyclotetrasilene **195**.

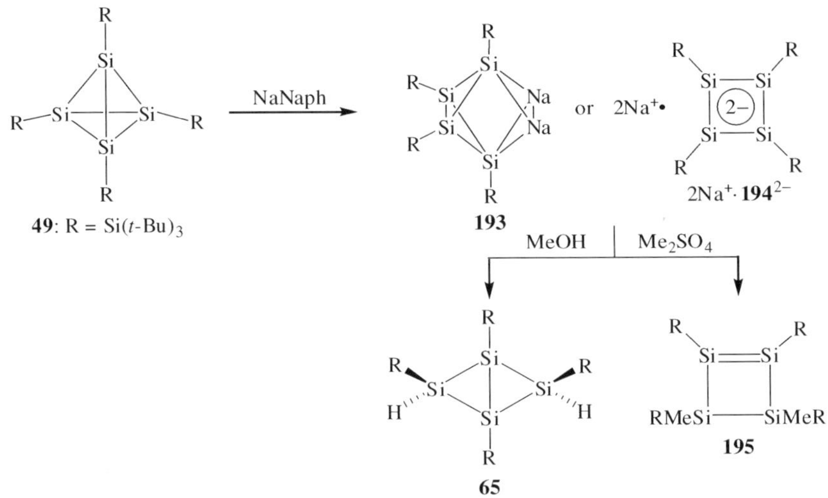

3.7.2 Radical Anions

In 1965, West reported the generation of a radical anion by the reduction of dodecamethylcyclohexasilane.[111] As dodecamethylcyclohexasilane contains only σ bonds, this radical anion is a σ-type species in contrast to radical anions of aromatic compounds. This radical anion was observed only at low temperatures, but later, he showed that the radical anion **196**$^{\bullet-}$ persists for several days at room temperature and for several weeks at 0°C.[112] Radical anions of oligosilanes were not isolated for a long time. Recently, the radical **197**$^{\bullet-}$ was isolated, and its structure was determined by X-ray crystallography.[113] The cyclotetrasilane ring of **197**$^{\bullet-}$ has a planar structure, and the Si–Si bond lengths (2.346(2)–2.347(2) Å) are slightly shorter compared to Si–Si bond lengths which are found in cyclopolysilanes (av. 2.372 Å) or for cyclotetrasilanes (2.363–2.445 Å).

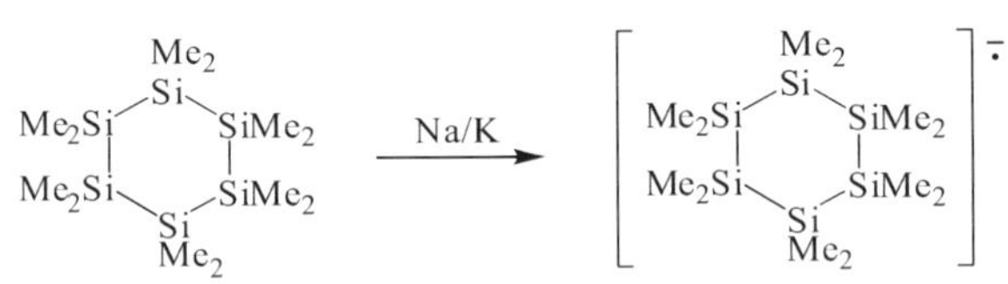

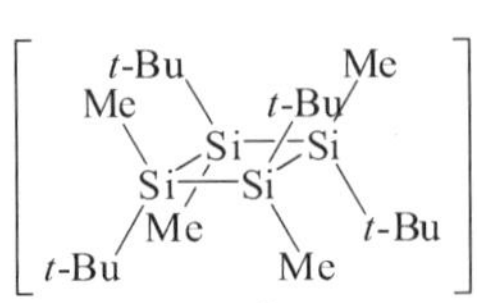

196$^{\bullet-}$

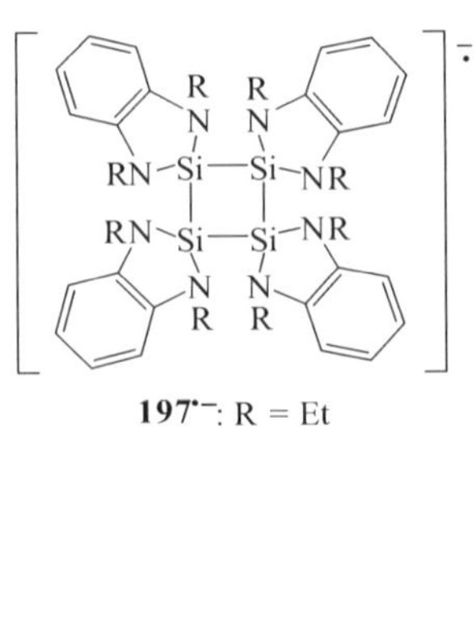

197$^{\bullet-}$: R = Et

Radical anions of organosilicon clusters have recently been reported. The radical anions of ladder oligosilanes were generated by the reduction of the corresponding ladder oligosilanes with potassium.[114] These radical anions were found to become more stable as the number of cyclotetrasilane rings increases. For example, the EPR signals of $\mathbf{140}^{\bullet-}$ observed at −70°C gradually decreased as the temperature was raised above −10°C. However, those of $\mathbf{170}^{\bullet-}$ persisted for several months at room temperature. The increased stability is explained by effective delocalization of an unpaired electron over the ladder oligosilane skeleton.

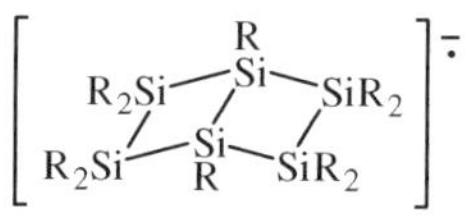

$\mathbf{140}^{\bullet-}$: R = *i*-Pr

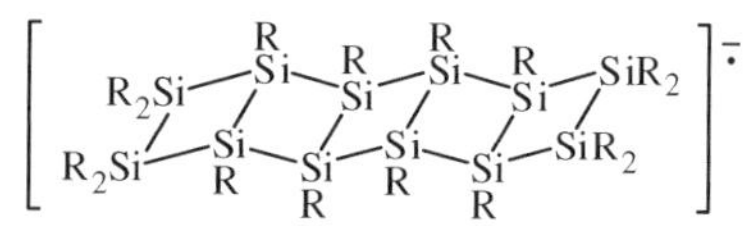

$\mathbf{170}^{\bullet-}$

Recently, the isolable radical anion of octasilacubane **56** was studied in detail.[115] The EPR spectrum of $\mathbf{56}^{\bullet-}$ showed satellites due to the tertiary ^{13}C nuclei of the eight *tert*-butyl groups. The SOMO of $\mathbf{56}^{\bullet-}$, which is formed by in-phase interaction among the eight σ* orbitals of the Si–C bonds, explains the satellites (Fig. 3.24).

The X-ray crystallography of $\mathbf{56}^{\bullet-}$ showed several structural features (Fig. 3.25). The Si–Si bond lengths of $\mathbf{56}^{\bullet-}$ (2.367(3)–2.380(3) Å, av. 2.373 Å) are shorter than those of **56**.[116] On the other hand, the Si–C bond lengths of $\mathbf{56}^{\bullet-}$ (1.938(5)–1.964(8) Å, av. 1.946 Å) are longer than those of **56**. The long Si–C bonds of $\mathbf{56}^{\bullet-}$ are explained by the presence of an unpaired electron in the Si–C σ* orbitals (Fig. 3.24). In the inner space of the octasilacubane skeleton, in-phase interaction among the eight Si–C σ* orbitals works effectively. The resulting lobe can be regarded as a cluster of pseudo π* orbitals between neighboring two silicon atoms (Section 3.5.1). The presence of an unpaired electron in this bonding orbital makes the Si–Si bonds short.

As mentioned in Section 3.5.1, π and π* orbitals play a predominant role in the organosilicon cluster containing a Si=Si double bond (**101**). As shown in Fig. 3.26, the

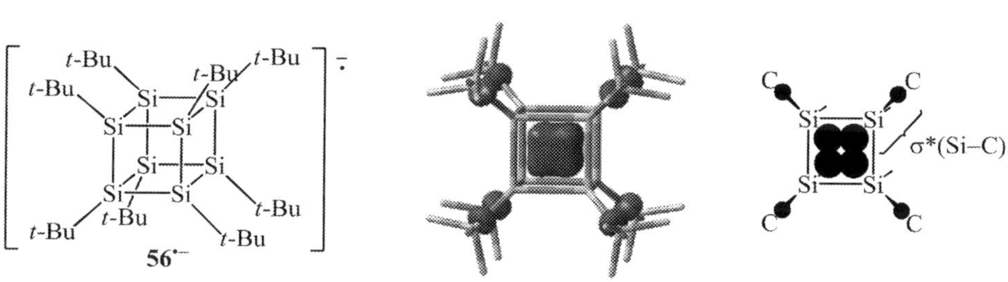

Figure 3.24
Schematic representation of the SOMO of $\mathbf{56}^{\bullet-}$ calculated at the UB3LYP/6-31G(d) level.

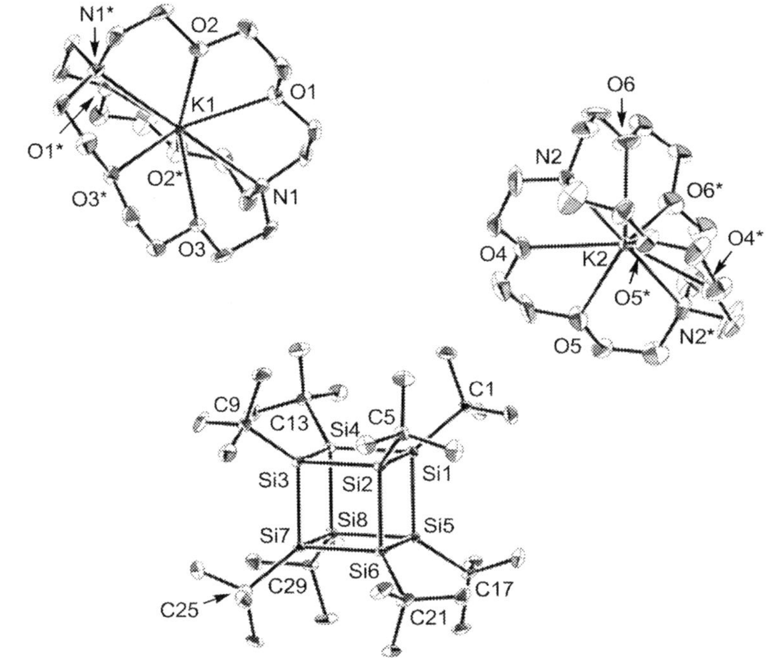

Figure 3.25
Structure of [K([2.2.2]cryptand)]$^{+}$**56**$^{\cdot-}$. The closest ion pair is shown.

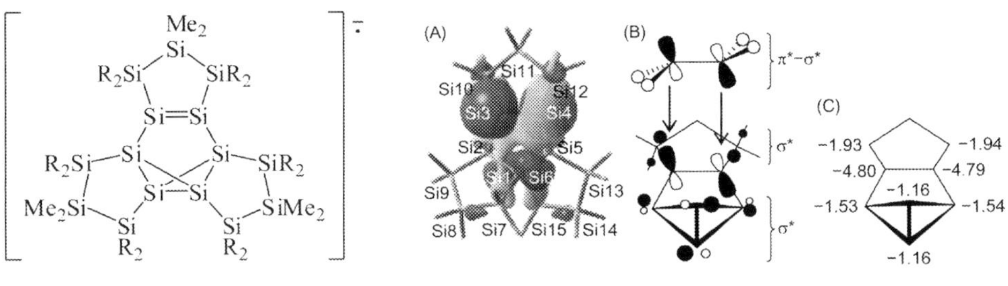

101$^{\bullet-}$: R = *t*-Bu
198$^{\bullet-}$: R = Me

Figure 3.26
(A) The SOMO, (B) schematic representation of the SOMO, and (C) selected isotropic splitting constants (mT) of **198**$^{\cdot-}$ calculated at the UB3LYP/6-311 + G(2df,p)//UB3LYP/6-31 + G(d) level.

SOMO of the model radical anion **198**$^{\bullet-}$ is formed mainly by the interaction among the $\pi^* - \sigma^*$ orbital of the Si=Si double bond (main), the four σ^* orbitals of the Si–Si bonds of the bicyclo-[1.1.0]tetrasilane moiety, and the σ^* orbitals of the Si10–C(*t*-Bu) and Si12–C (*t*-Bu) bonds. This theoretical result is in accord with the EPR spectrum of **101**$^{\bullet-}$, in which the satellites due to the ^{29}Si nuclei of the Si=Si double bond were observed with a much larger splitting constant than the others.[117]

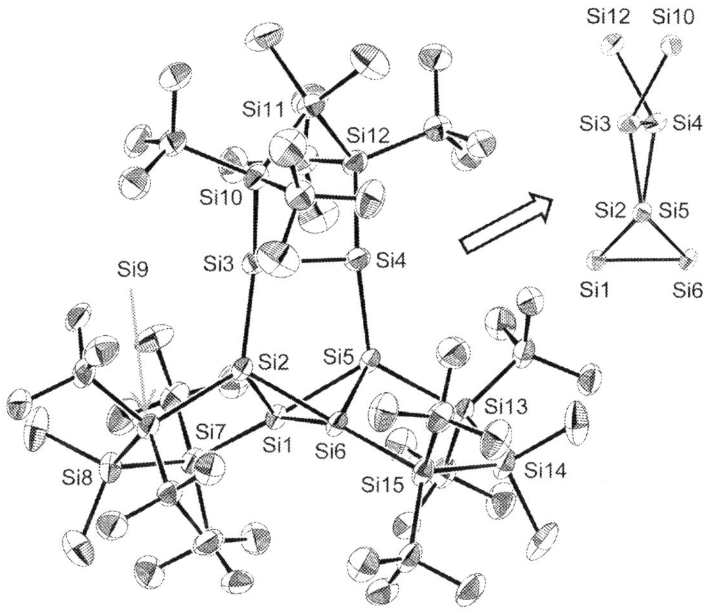

Figure 3.27
Molecular structure of $\mathbf{101}^{\bullet -}$.

Structural deformation of the Si=Si double bond of $\mathbf{101}^{\bullet -}$ was shown by X-ray crystallography (Fig. 3.27).[117] The Si=Si double bond of $\mathbf{101}^{\bullet -}$ has a highly *trans*-bent structure with the bend angles of 65.1 and 65.5°. These values are much larger than those of neutral **101** (42.5, 36.5°).[68] Theoretical calculations of various related compounds showed that the highly *trans*-bent structure of $\mathbf{101}^{\bullet -}$ is due to the hexasilabenzvalene structure. This result is in accord with a theoretical study,[118] but remarkably different from the structural deformation of the radical anion of $[(t\text{-Bu})_2\text{MeSi}]_2\text{Si}{=}\text{Si}[\text{SiMe}(t\text{-Bu}_2)]_2^{\bullet -}$.[119] This radical anion has a highly twisted structure with the twist angle of 88°.

3.7.3 Cations

Cationic organosilicon clusters have not yet been studied well. As explained in Section 3.6.5, Sekiguchi has reported that a dication was generated by two-electron oxidation of tetrasilatetrahedrane **49** with $\text{Ph}_3\text{C}^+ \cdot$ TFPB. The resulting dication may be the 2,4-dication of bicyclo[1.1.0]tetrasilane $\mathbf{199}^{2+}$ or the dication of tetrasilacyclobutadiene $\mathbf{194}^{42+}$.[103]

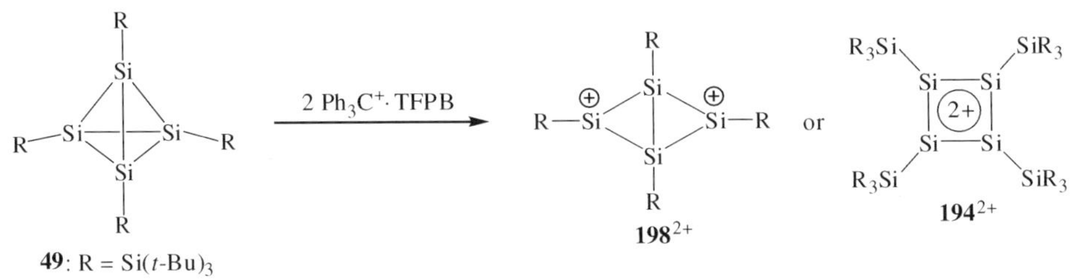

3.8 Conclusions

As described in this chapter, the chemistry of organosilicon clusters has been rapidly developing. It is surprising that a variety of organosilicon clusters have been synthesized from monosilanes by using only a few methods of Si–Si bond formation. The chemistry of organosilicon clusters is expected to develop from studies on small and simple clusters to those on large and complicated ones. This may be accomplished more easily with more advanced synthetic methods and strategies. Another future prospect is application of organosilicon clusters to materials science. Although studies on this research area have not yet been reported, they are expected to be a topical subject in the near future.

References

1. West, R.; Indriksons, A. Cyclic Polysilanes. VI. Bicyclic and Cage Permethylcyclopolysilanes. *J. Am. Chem. Soc.* **1972**, *94*, 6110–6115.
2. Ishikawa, M.; Watanabe, M.; Iyoda, J.; Ikeda, H.; Kumada, M. Aluminum Chloride Catalyzed Skeletal Rearrangement of Permethylated Cyclic Polysilanes. *Organometallics* **1982**, *1*, 317–322.
3. Boudjouk, P.; Sooriyakumaran, R. The Synthesis of the First Spiropentasilane, Octamethylspiropentasilane. *J. Chem. Soc. Chem. Commun.* **1984**, 777–778.
4. (a) Masamune, S.; Kabe, Y.; Collins, S.; Williams, D. J.; Jones, R. Tetrasilabicyclo[1.1.0]butane System: Preparation and Characterization of 1,3-Di-*tert*-butyl-2,2,4,4-tetrakis(2,6-diethylphenyl)tetrasilabicyclo-[1.1.0]butane. *J. Am. Chem. Soc.* **1985**, *107*, 5552–5553.
   (b) Jones, R.; Williams, D. J.; Kabe, Y.; Masamune, S. The Tetrasilabicyclo[1.1.0]butane System: Structure of 1,3-Di-*tert*-butyl-2,2,4,4-tetrakis(2,6-diethylphenyl)tetrasilabicyclo[1.1.0]butane. *Angew. Chem. Int. Ed. Engl.* **1986**, *25*, 173–174.
5. Kabe, Y.; Kawase, T.; Okada, J.; Yamashita, O.; Goto, M.; Masamune, S. A Bicyclo[1.1.1]pentasilane Derivative. Synthesis, Molecular Structure, and Comments on Structural Homology. *Angew. Chem. Int. Ed. Engl.* **1990**, *29*, 794–796.
6. Kabe, Y.; Kuroda, M.; Honda, Y.; Yamashita, O.; Kawase, T.; Masamune, S. Reductive Oligomerization of 1,2-Di-*tert*-butyl-1,1,2,2-tetrachlorodisilane: The Tricyclo[$2.2.0.0^{2,5}$]hexasilane and Tetracyclo-[$3.3.0.0^{2,7}.0^{3,6}$]octasilane Systems. *Angew. Chem. Int. Ed. Engl.* **1988**, *27*, 1725–1727.
7. Matsumoto, H.; Miyamoto, H.; Kojima, N.; Nagai, Y. The First Bicyclo[2.2.0]hexasilane System: Synthesis of Decaisopropylhexasilabicyclo[2.2.0]hexane. *J. Chem. Soc. Chem. Commun.* **1987**, 1316–1317.
8. Matsumoto, H.; Higuchi, K.; Hoshino, Y.; Koike, H.; Naoi, Y.; Nagai, Y. The First Octasilacubane System: Synthesis of Octakis(t-butyldimethylsilyl)pentacyclo[$4.2.0.0^{2,5}.0^{3,8}.0^{4,7}$]octasilane. *J. Chem. Soc. Chem. Commun.* **1988**, 1083–1084.

9. Sekiguchi, A.; Yatabe, T.; Kabuto, C.; Sakurai, H. The "Missing" Hexasilaprismane: Synthesis, X-ray Analysis, and Photochemical Reactions. *J. Am. Chem. Soc.* **1993**, *115*, 5853–5854.
10. Wiberg, N.; Finger, C. M. M.; Polborn, K. Tetrakis(tri-*tert*-butylsilyl)-*tetrahedro*-tetrasilane $(tBu_3Si)_4Si_4$: The First Molecular Silicon Compound with a Si_4 Tetrahedron. *Angew. Chem. Int. Ed. Engl.* **1993**, *32*, 1054–1056.
11. (a) West, R.; Carberry, E. Permethylpolysilanes: Silicon Analogs of Hydrocarbons. *Science* **1975**, *189*, 179–186.
 (b) West, R. Polysilanes. In *The Chemistry of Organic Silicon Compounds;* Patai, S., Rappoport, Z., Eds.; Wiley: Chichester, 1989; pp 1207–1240.
 (c) Tsumuraya, T.; Batcheller, S. A.; Masamune, S. Strained-Ring and Double Bond Systems Consisting of the Group 14 Elements Si, Ge, and Sn. *Angew. Chem. Int. Ed. Engl.* **1991**, *30*, 902–930.
 (d) Sekiguchi, A.; Sakurai, H. Cage and Cluster Compounds of Silicon, Germanium, and Tin. *Adv. Organomet. Chem.* **1995**, *37*, 1–38.
 (e) Hengge, E.; Janoschek, R. Homocyclic Silanes. *Chem. Rev.* **1995**, *95*, 1495–1526.
 (f) Sekiguchi, A.; Nagase, S. Polyhedral Silicon Compounds. In *The Chemistry of Organic Silicon Compounds;* Rappoport, Z., Apeloig, Y., Eds.; Vol. 2; Wiley: Chichester, 1998; pp 119–152.
 (g) Hengge, E.; Stüger, H. Recent Advances in the Chemistry of Cyclopolysilanes. In *The Chemistry of Organic Silicon Compounds;* Rappoport, Z., Apeloig, Y., Eds.; Vol. 2; Wiley: Chichester, 1998; pp 2177–2216.
 (h) Sekiguchi, A.; Lee, V. Ya. Cage Compounds of Heavier Group 14 Elements. In *The Chemistry of Organic Germanium, Tin and Lead Compounds;* Rappoport, Z., Ed.; Vol. 2; Wiley: Chichester, 2002; pp 935–962.
 (i) Wiberg, N.; Power, P. P. Molecular Cages and Clusters of the Heavier Group 14 Elements (E = Si, Ge, Sn or Pb) of Formula E_nR_m ($n \geq$ m). In *Molecular Clusters of the Main Group Elements;* Driess, M., Nöth, H., Eds.; Wiley-VCH: Weinheim, 2004; pp 188–208.
 (j) Schnepf, A. Metalloid Group 14 Cluster Compounds: An Introduction and Perspectives to This Novel Group of Cluster Compounds. *Chem. Soc. Rev.* **2007**, *36*, 745–758.
 (k) Iwamoto, T.; Ishida, S. Silicon Compounds with Inverted Geometry around Silicon Atoms. *Chem. Lett.* **2014**, *43*, 164–170.
 (l) Kyushin, S. Organosilicon Synthesis for Construction of Organosilicon Clusters.. In *In Efficient Methods for Preparing Silicon Compounds;;* Roesky, H. W., Ed.; Academic Press: London2016; pp 455–498.
12. Kipping, F. S.; Sands, J. E. Organic Derivatives of Silicon. Part XXV. Saturated and Unsaturated Silicohydrocarbons, Si_4Ph_8. *J. Chem. Soc.* **1921**, *119*, 830–847.
13. Gilman, H.; Schwebke, G. L. Organic Substituted Cyclosilanes. *Adv. Organomet. Chem.* **1964**, *1*, 89–141.
14. For a review, see Kumada, M.; Tamao, K. Aliphatic Organopolysilanes. *Adv. Organomet. Chem.* **1968**, *6*, 19–117.
15. Gilman, H.; Shiina, K.; Aoki, D.; Gaj, B. J.; Wittenberg, D.; Brennan, T. Hexasubstituted Disilanes from Chlorosilanes and Lithium in Tetrahydrofuran. *J. Organomet. Chem.* **1968**, *13*, 323–328.
16. Laguerre, M.; Dunogues, J.; Calas, R. One-Step Synthesis of Dodecamethylcyclohexasilane. *J. Chem. Soc. Chem. Commun.* **1978**, 272.
17. (a) Gilman, H.; Smith, C. L. Tetrakis(trimethylsilyl)silane. *J. Am. Chem. Soc.* **1964**, *86*, 1454.
 (b) Gilman, H.; Smith, C. L. Tetrakis(trimethylsilyl)silane. *J. Organomet. Chem.* **1967**, *8*, 245–253.
18. (a) Tamao, K.; Kawachi, A. Silyl Anions. *Adv. Organomet. Chem.* **1995**, *38*, 1–58.
 (b) Lickiss, P. D.; Smith, C. M. Silicon Derivatives of the Metals of Groups 1 and 2. *Coord. Chem. Rev.* **1995**, *145*, 75–124.
 (c) Belzner, J.; Dehnert, U. Alkaline and Alkaline Earth Silyl Compounds – Preparation and Structure. In *The Chemistry of Organic Silicon Compounds;* Rappoport, Z., Apeloig, Y., Eds.; Vol. 2; Wiley: Chichester, 1998; pp 779–825.

(d) Sekiguchi, A.; Lee, V. Ya; Nanjo, M. Lithiosilanes and Their Application to the Synthesis of Polysilane Dendrimers. *Coord. Chem. Rev.* **2000**, *210*, 11–45.
(e) Lee, V. Ya; Sekiguchi, A. Silicon-, Germanium-, and Tin-Centered Cations, Radicals, and Anions. In *Reviews of Reactive Intermediate Chemistry;* Platz, M. S., Moss, R. A., Jones, M., Jr., Eds.; Wiley: Hoboken, 2007; pp 47–120.
(f) Lee, V. Ya.; Sekiguchi, A. *Heavy Analogs of Carbanions: Si-, Ge-, Sn- and Pb-Centered Anions.* In *Organometallic Compounds of Low-Coordinate Si, Ge, Sn and Pb. From Phantom Species to Stable Compounds;* , Wiley: Chichester, 2010; pp 89–138.

19. Marschner, C. A New and Easy Route to Polysilanylpotassium Compounds. *Eur. J. Inorg. Chem.* **1998**, 221–226.
20. (a) Kawachi, A.; Tamao, K. Mixed Reagent (Aminosilyl)lithium/*i*-PrMgBr for the Synthesis of Functionalized Oligosilanes. *J. Organomet. Chem.* **2000**, *601*, 259–266.
    (b) Shibano, Y.; Sasaki, M.; Tsuji, H.; Araki, Y.; Ito, O.; Tamao, K. Conformation Effect of Oligosilane Linker on Photoinduced Electron Transfer of Tetrasilane-Linked Zinc Porphyrin–[60]Fullerene Dyads. *J. Organomet. Chem.* **2007**, *692*, 356–367.
21. For a review, see Kumada, M. Some Recent Studies of the Skeletal Transformations of Organopolysilanes. *J. Organomet. Chem.* **1975**, *100*, 127–137.
22. (a) Ishikawa, M.; Kumada, M. Ring Contraction of Cyclohexasilanes to Silylcyclopentasilanes and the Preparation of Monofunctional Nonamethylcyclopentasilanes. *J. Chem. Soc. Chem. Commun.* **1969**, 567–568.
    (b) Ishikawa, M.; Kumada, M. Aluminum Chloride-Catalysed Skeletal Rearrangement of Permethylated Linear Polysilanes. *J. Chem. Soc. Chem. Commun.* **1970**, 157.
    (c) Ishikawa, M.; Iyoda, J.; Ikeda, H.; Kotake, K.; Hashimoto, T.; Kumada, M. Aluminum Chloride Catalyzed Skeletal Rearrangement of Permethylated Acyclic Polysilanes. *J. Am. Chem. Soc.* **1981**, *103*, 4845–4850.
23. Blinka, T. A.; West, R. Metal Halide Catalyzed Rearrangements of Alkylcyclosilanes. *Organometallics* **1986**, *5*, 128–133.
24. (a) Wagner, H.; Wallner, A.; Fischer, J.; Flock, M.; Baumgartner, J.; Marschner, C. Rearrangement of Cyclic Silanes with Aluminum Trichloride. *Organometallics* **2007**, *26*, 6704–6717.
    (b) Wagner, H.; Baumgartner, J.; Müller, T.; Marschner, C. Shuttling Germanium Atoms into Branched Polysilanes. *J. Am. Chem. Soc.* **2009**, *131*, 5022–5023.
    (c) Wagner, H.; Baumgartner, J.; Marschner, C. Rearrangement/Fragmentation Reactions of Oligosilanes with Aluminum Chloride. *Organometallics* **2011**, *30*, 3939–3954.
25. Fischer, J.; Baumgartner, J.; Marschner, C. Synthesis and Structure of Sila-Adamantane. *Science* **2005**, *310*, 825.
26. (a) Harrod, J. F. Catalyzed Dehydrogenative Coupling as a Route to Novel Organometallic Polymers. . NATO ASI Series In *Transformation of Organometallics into Common and Exotic Materials: Design and Activation;* Laine, R. M., Ed.; , Vol. 141; Martinus Nijhoff Publishers: Dordrecht, 1988; pp 103–115.
    (b) Corey, J. Y. Dehydrogenative Coupling Reactions of Hydrosilanes. In *Advances in Silicon Chemistry;* Larson, G. L., Ed.; , Vol. 1; JAI: Greenwich, 1991; pp 327–387.
    (c) Gauvin, F.; Harrod, J. F.; Woo, H. G. Catalytic Dehydrocoupling: A General Strategy for the Formation of Element–Element Bonds. *Adv. Organomet. Chem* **1998**, *42*, 363–405.
    (d) Gray, G. M.; Corey, J. Y. Synthesis of Polysilanes by New Procedures: Part 2. Catalytic Dehydropolymerisation of Hydrosilanes. In *Silicon-Containing Polymers;* Jones, R. G., Ando, W., Chojnowski, J., Eds.; Kluwer: Dordrecht, 2000; pp 401–418.
    (e) Corey, J. Y. Dehydrocoupling of Hydrosilanes to Polysilanes and Silicon Oligomers: A 30 Year Overview. *Adv. Organomet. Chem* **2004**, *51*, 1–52.
27. Stallings, W.; Donohue, J. Crystal and Molecular Structure of Hexadecamethylbicyclo[3.3.1]nonasilane. *Inorg. Chem.* **1976**, *15*, 524–529.

28. Jenkner, P. K.; Hengge, E.; Czaputa, R.; Kratky, C. Synthesis, Properties and Characterization of Octadecamethylbicyclo[4.4.0]decasilane. *J. Organomet. Chem.* **1993**, *446*, 83–90.
29. Iwamoto, T.; Tsushima, D.; Kwon, E.; Ishida, S.; Isobe, H. Persilastaffanes: Design, Synthesis, Structure, and Conjugation between Silicon Cages. *Angew. Chem. Int. Ed.* **2012**, *51*, 2340–2344.
30. Setaka, W.; Hamada, N.; Kira, M. Novel Synthesis and Bridgehead Functionalization of Permethylbicyclo [2.2.2]octasilane. *Chem. Lett* **2004**, *33*, 626–627.
31. Fischer, R.; Konopa, T.; Ully, S.; Baumgartner, J.; Marschner, C. Route Si_6 Revisited. *J. Organomet. Chem.* **2003**, *685*, 79–92.
32. Wallner, A.; Hölbling, M.; Baumgartner, J.; Marschner, C. Structural and Spectroscopic Studies of Silylated Cyclo- and Bicyclosilanes. *Silicon Chem.* **2005**, *3*, 175–185.
33. Wallner, A.; Hlina, J.; Konopa, T.; Wagner, H.; Baumgartner, J.; Marschner, C.; Flörke, C. Cyclic and Bicyclic Methylpolysilanes and Some Oligosilanylene-Bridged Derivatives. *Organometallics* **2010**, *29*, 2660–2675.
34. Zirngast, M.; Baumgartner, J.; Marschner, C. Synthesis of Cyclic and Bicyclic Polysilanes of Variable Ring Sizes. *Organometallics* **2008**, *27*, 6472–6478.
35. Lee, V. Ya.; Yokoyama, T.; Takanashi, K.; Sekiguchi, A. Pentasilatricyclo[2.1.0.$0^{2,5}$]pentane and Its Anion. *Chem. Eur. J.* **2009**, *15*, 8401–8404.
36. Fischer, R.; Baumgartner, J.; Kickelbick, G.; Hassler, K.; Marschner, C. Adamantanes, Nortricyclenes, and Dendrimers with Extended Silicon Backbones. *Chem. Eur. J.* **2004**, *10*, 1021–1030.
37. (a) Unno, M.; Higuchi, K.; Ida, M.; Shioyama, H.; Kyushin, S.; Matsumoto, H.; Goto, M. Ring-Opening Reaction of Octakis(1,1,2-trimethylpropyl)octasilacubane. Chlorination with PCl_5 Leading to Stereoisomeric 4,8-Dichlorooctakis(1,1,2-trimethylpropyl)tetracyclo[3.3.0.$0^{2,7}$.$0^{3,6}$]octasilanes. *Organometallics* **1994**, *13*, 4633–4640.
    (b) Unno, M.; Shioyama, H.; Ida, M.; Matsumoto, H. Reductive Dehalogenation of 4,8-Dihalooctakis-(1,1,2-trimethylpropyl)tetracyclo[3.3.0.$0^{2,7}$.$0^{3,6}$]octasilanes with Sodium: Formation of Octasilacubane. *Organometallics* **1995**, *14*, 4004–4009.
    (c) Unno, M.; Masuda, H.; Matsumoto, H. Photo-Initiated Bromination of Octakis(1,1,2-trimethylpropyl) octasilacubane With Tetrabromomethane. *Silicon Chem.* **2002**, *1*, 377–381.
38. Ichinohe, M.; Toyoshima, M.; Kinjo, R.; Sekiguchi, A. Tetrasilatetrahedranide: A Silicon Cage Anion. *J. Am. Chem. Soc.* **2003**, *125*, 13328–13329.
39. Klapötke, T. M.; Vasisht, S. K.; Fischer, G.; Mayer, P. A Reactive Si_4 Cage: $K(SitBu_3)_3Si_4$. *J. Organomet. Chem.* **2010**, *695*, 667–672.
40. Abersfelder, K.; Russell, A.; Rzepa, H. S.; White, A. J. P.; Haycock, P. R.; Scheschkewitz, D. Contraction and Expansion of the Silicon Scaffold of Stable Si_6R_6 Isomers. *J. Am. Chem. Soc.* **2012**, *134*, 16008–16016.
41. Furukawa, K.; Fujino, M.; Matsumoto, N. Cubic Silicon Cluster. *Appl. Phys. Lett.* **1992**, *60*, 2744–2745.
42. Matsumoto, H.; Higuchi, K.; Kyushin, S.; Goto, M. Octakis(1,1,2-trimethylpropyl)octasilacubane: Synthesis, Molecular Structure, and Unusual Properties. *Angew. Chem. Int. Ed. Engl.* **1992**, *31*, 1354–1356.
43. Sekiguchi, A.; Yatabe, T.; Kamatani, H.; Kabuto, C.; Sakurai, H. Preparation, Characterization, and Crystal Structures of Octasilacubanes and Octagermacubanes. *J. Am. Chem. Soc.* **1992**, *114*, 6260–6262.
44. (a) Kira, M.; Iwamoto, T.; Kabuto, C. The First Stable Cyclic Disilene: Hexakis(trialkylsilyl)-tetrasilacyclobutene. *J. Am. Chem. Soc.* **1996**, *118*, 10303–10304.
    (b) Iwamoto, T.; Tamura, M.; Kabuto, C.; Kira, M. Photochemical Isomerization and X-ray Structure of Stable [Tris(trialkylsilyl)silyl]cyclotrisilene. *Organometallics* **2003**, *22*, 2342–2344.
    (c) Ueba-Ohshima, K.; Iwamoto, T.; Kira, M. Synthesis, Structure, and Facile Ring Flipping of a Bicyclo-[1.1.0]tetrasilane. *Organometallics* **2008**, *27*, 320–323.
45. Wiberg, N.; Auer, H.; Wagner, S.; Polborn, K.; Kramer, G. Disilene $R^*XSi{=}SiXR^*$ ($R^*{=}SitBu_3$) mit Siliciumgebundenen H-und Hal-Atomen X: Bildung, Isomerisierung, Reaktionen. *J. Organomet. Chem.* **2001**, *619*, 110–131.

46. Takanashi, K.; Lee, V. Ya.; Ichinohe, M.; Sekiguchi, A. 1,2,5,6-Tetrasilabenzobenzvalene: A Valence Isomer of 1,2,3,4-Tetrasilanaphthalene. *Chem. Lett.* **2007**, *36*, 1158−1159.
47. Wiberg, N.; Auer, H.; Nöth, H.; Knizek, J.; Polborn, K. Diiodotetrasupersilylcyclotetrasilene $(t\text{Bu}_3\text{Si})_4\text{Si}_4\text{I}_2$ − A Molecule Containing an Unsaturated Si_4 Ring. *Angew. Chem. Int. Ed.* **1998**, *37*, 2869−2872.
48. Iwamoto, T.; Uchiyama, K.; Kabuto, C.; Kira, M. Synthesis of Tricyclo[3.1.0.0^{2,4}]hexasilane and Its Photochemical Isomerization to Tricyclo[2.2.0.0^{2,5}]hexasilane. *Chem. Lett.* **2007**, *36*, 368−369.
49. Tsurusaki, A.; Koganezono, M.; Otsuka, K.; Ishida, S.; Kyushin, S. Two Pentasilahousanes Fused Together. *Chem. Eur. J.* **2014**, *20*, 9263−9266.
50. Iwamoto, T.; Akasaka, N.; Ishida, S. A Heavy Analogue of the Smallest Bridgehead Alkene Stabilized by A Base. *Nat. Commun.* **2014**, *5*, 6353/1−7.
51. (a) Matsumoto, H.; Kyushin, S.; Unno, M.; Tanaka, R. Syntheses, Structures, and Properties of Ladder Oligosilanes and Ladder Oligogermanes. *J. Organomet. Chem.* **2000**, *611*, 52−63.
    (b) Kyushin, S.; Matsumoto, H. Ladder Polysilanes. *Adv. Organomet. Chem.* **2003**, *49*, 133−166.
52. Kyushin, S.; Ueta, Y.; Tanaka, R.; Matsumoto, H. Hexa-, Hepta-, and Octacyclic Ladder Polysilanes. *Chem. Lett.* **2006**, *35*, 182−183.
53. Xiao, X.-Q.; Zhao, H.; Xu, Z.; Lai, G.; He, X.-L.; Li, Z. Reduction of the Dihalocyclopentasilanes. A Quest for a Homocyclic Silylene. *Chem. Commun.* **2013**, *49*, 2706−2708.
54. Kobayashi, H.; Iwamoto, T.; Kira, M. A Stable Fused Bicyclic Disilene as a Model for Silicon Surface. *J. Am. Chem. Soc.* **2005**, *127*, 15376−15377.
55. Iwamoto, T.; Furiya, Y.; Kobayashi, H.; Isobe, H.; Kira, M. Synthesis and Facile Ring Expansion of Silylenecyclotetrasilane. *Organometallics* **2010**, *29*, 1869−1872.
56. Hlina, J.; Mechtler, C.; Wagner, H.; Baumgartner, J.; Marschner, C. Multiple Silyl Exchange Reactions: A Way to Spirooligosilanes. *Organometallics* **2009**, *28*, 4065−4071.
57. Klapötke, T. M.; Vasisht, S. K.; Mayer, P. Spirocycle $(\text{Si}t\text{Bu}_3)_6\text{Si}_9\text{Cl}_2$: The First of Its Kind among Group 14 Elements. *Eur. J. Inorg. Chem.* **2010**, 3256−3260.
58. Iwamoto, T.; Tamura, M.; Kabuto, C.; Kira, M. A Stable Bicyclic Compound with Two Si=Si Double Bonds. *Science* **2000**, *290*, 504−506.
59. Iwamoto, T.; Kabuto, C.; Kira, M. The First Stable Cyclotrisilene. *J. Am. Chem. Soc.* **1999**, *121*, 886−887.
60. Fischer, G.; Huch, V.; Mayer, P.; Vasisht, S. K.; Veith, M.; Wiberg, N. $\text{Si}_8(\text{Si}t\text{Bu}_3)_6$: A Hitherto Unknown Cluster Structure in Silicon Chemistry. *Angew. Chem. Int. Ed.* **2005**, *44*, 7884−7887.
61. Nied, D.; Köppe, R.; Klopper, W.; Schnöckel, H.; Breher, F. Synthesis of a Pentasilapropellane. Exploring the Nature of a Stretched Silicon−Silicon Bond in a Nonclassical Molecule. *J. Am. Chem. Soc.* **2010**, *132*, 10264−10265.
62. Scheschkewitz, D. A Molecular Silicon Cluster with a "Naked" Vertex Atom. *Angew. Chem. Int. Ed.* **2005**, *44*, 2954−2956.
63. Abersfelder, K.; White, A. J. P.; Rzepa, H. S.; Scheschkewitz, D. A Tricyclic Aromatic Isomer of Hexasilabenzene. *Science* **2010**, *327*, 564−566.
64. Leszczyńska, K.; Abersfelder, K.; Majumdar, M.; Neumann, B.; Stammler, H.-G.; Rzepa, H. S.; Jutzi, P.; Scheschkewitz, D. The Cp^*Si^+ Cation as a Stoichiometric Source of Silicon. *Chem. Commun.* **2012**, *48*, 7820−7822.
65. Abersfelder, K.; White, A. J. P.; Berger, R. J. F.; Rzepa, H. S.; Scheschkewitz, D. A Stable Derivative of the Global Minimum on the Si_6H_6 Potential Energy Surface. *Angew. Chem. Int. Ed.* **2011**, *50*, 7936−7939.
66. For a review, see Kyushin, S.; Ishida, S.; Tsurusaki, A.; Otsuka, K. Effect of Ring Sizes of Cyclooligosilanes on Construction of Organosilicon Clusters. *J. Synth. Org. Chem. Jpn.* **2014**, *72*, 1290−1297.
67. Ishida, S.; Otsuka, K.; Toma, Y.; Kyushin, S. An Organosilicon Cluster with an Octasilacuneane Core: A Missing Silicon Cage Motif. *Angew. Chem. Int. Ed.* **2013**, *52*, 2507−2510.
68. Tsurusaki, A.; Iizuka, C.; Otsuka, K.; Kyushin, S. Cyclopentasilane-Fused Hexasilabenzvalene. *J. Am. Chem. Soc.* **2013**, *135*, 16340−16343.

69. Tsurusaki, A.; Kamiyama, J.; Kyushin, S. Tetrasilane-Bridged Bicyclo[4.1.0]heptasil-1(6)-ene. *J. Am. Chem. Soc.* **2014**, *136*, 12896–12898.

70. (a) Rohmer, M.-M.; Bénard, M. Bond-Stretch Isomerism in Strained Inorganic Molecules and in Transition Metal Complexes: A Revival? *Chem. Soc. Rev.* **2001**, *30*, 340–354.
    (b) Grützmacher, H.; Breher, F. Odd-Electron Bonds and Biradicals in Main Group Element Chemistry. *Angew. Chem. Int. Ed.* **2002**, *41*, 4006–4011.
    (c) Breher, F. Stretching Bonds in Main Group Element Compound – Borderlines between Biradicals and Closed-Shell Species. *Coord. Chem. Rev.* **2007**, *251*, 1007–1043.

71. (a) Schleyer, P. v. R.; Sax, A. F.; Kalcher, J.; Janoschek, R. Bond Stretch Isomerism in Bicyclo[1.1.0]-tetrasilanes. Contrasts between Strained Silicon and Carbon Ring Systems. *Angew. Chem. Int. Ed. Engl.* **1987**, *26*, 364–366.
    (b) Schoeller, W. W.; Dabisch, T.; Busch, T. Bond Stretch Isomerism in the Silicon Analogues of Bicyclo-[1.1.0]butane and of [1.1.1]propellane. Consequence of Orbital Nonhybridization. *Inorg. Chem.* **1987**, *26*, 4383–4389.
    (c) Nagase, S.; Kudo, T. An *Ab Initio* Comparative Study of Bicyclo[1.1.0]tetrasilane and Bicyclo[2.2.0]-hexasilane. *J. Chem. Soc., Chem. Commun.* **1988**, 54–56.
    (d) Boatz, J. A.; Gordon, M. S. Theoretical Studies of Bond Stretch Isomerism in Silabicyclobutanes. *J. Phys. Chem.* **1989**, *93*, 2888–2891.
    (e) Kudo, T.; Nagase, S. The Heavier Group 14 Analogues of the Bicyclo[1.1.0]butane Radical Cation. A Theoretical Study. *J. Phys. Chem.* **1992**, *96*, 9189–9194.
    (f) Boatz, J. A.; Gordon, M. S. Bond-Stretch Isomerism in Tetrasilabicyclo[1.1.0]butane Derivatives. *Organometallics* **1996**, *15*, 2118–2124.

72. Kira, M. Importance of $\pi-\sigma^*$ Orbital Mixing in Bonding and Structure of Bicyclo[1.1.0]tetrasilane and Related Compounds. *Organometallics* **2014**, *33*, 644–652.

73. Iwamoto, T.; Yin, D.; Kabuto, C.; Kira, M. The First 1,3-Disilabicyclo[1.1.0]butane with Long-Bridge Silicon–Silicon Bond. *J. Am. Chem. Soc.* **2001**, *123*, 12730–12731.

74. Collins, S.; Dutler, R.; Rauk, A. The Structure and Barrier to Inversion of Tetrasilabicyclo[1.1.0]butane. Comparison to Bicyclo[1.1.0]butane. *J. Am. Chem. Soc.* **1987**, *109*, 2564–2569; **1987**, *109*, 6217 (correction).

75. Wiberg, N.; Schuster, H.; Simon, A.; Peters, K. Hexa-*tert*-butyldisilane – The Molecule with the Longest Si–Si Bond. *Angew. Chem. Int. Ed. Engl.* **1986**, *25*, 79–80.

76. Becerra, R.; Walsh, R. Thermochemistry. In *The Chemistry of Organic Silicon Compounds;* Rappoport, Z., Apeloig, Y., Eds.; Vol. 2; Wiley: Chichester, 1998; pp 153–180.

77. (a) Kyushin, S.; Sakurai, H.; Matsumoto, H. Highly Planar Silane $[(i\text{-Pr})_3Si]_3SiH$ and Silyl Radical $[(i\text{-Pr})_3Si]_3Si\cdot$. *Chem. Lett.* **1998**, 107–108.
    (b) Gaspar, P. P.; Beatty, A. M.; Chen, T.; Haile, T.; Lei, D.; Winchester, W. R., et al. Tris (triisopropylsilyl)silane and the Generation of Bis(triisopropylsilyl)-silylene. *Organometallics* **1999**, *18*, 3921–3932.

78. Nakamoto, M.; Fukawa, T.; Lee, V. Ya.; Sekiguchi, A. Nearly Planar Nonsolvated Monomeric Silyl- and Germyllithiums as a Result of an Intramolecular CH–Li Agostic Interaction. *J. Am. Chem. Soc.* **2002**, *124*, 15160–15161.

79. Lee, V. Ya.; Gapurenko, O. A.; Ito, Y.; Meguro, T.; Sugasawa, H.; Sekiguchi, A., et al. Pyramidanes: The Covalent Form of the Ionic Compounds. *Organometallics* **2016**, *35*, 346–356.

80. Kratzert, D.; Leusser, D.; Holstein, J. J.; Dittrich, B.; Abersfelder, K.; Scheschkewitz, D.; Stalke, D. An Experimental Charge Density Study of Two Isomers of Hexasilabenzene. *Angew. Chem. Int. Ed.* **2013**, *52*, 4478–4482.

81. (a) Blinka, T. A.; Helmer, B. J.; West, R. Polarization Transfer NMR Spectroscopy for Silicon-29: The INEPT and DEPT Techniques. *Adv. Organomet. Chem* **1984**, *23*, 193–218.
    (b) Williams, E. A. NMR Spectroscopy of Organosilicon Compounds. In *The Chemistry of Organic Silicon Compounds;* Patai, S., Rappoport, Z., Eds.; Wiley: Chichester, 1989; pp 511–554.

(c) Takeuchi, Y.; Takayama, T. 29Si NMR Spectroscopy of Organosilicon Compounds. In *The Chemistry of Organic Silicon Compounds;* Rappoport, Z., Apeloig, Y., Eds.; , Vol. 2; Wiley: Chichester, 1998; pp 267–354.

(d) Brook, M. A. *Organosilanes: Where to Find Them, What to Call Them, How to Detect Them. Silicon in Organic, Organometallic, and Polymer Chemistry;* Wiley: New York, 20003–26.

82. (a) Yokelson, H. B.; Millevolte, A. J.; Adams, B. R.; West, R. Bonding in 1,3-Cyclodisiloxanes: ^{29}Si NMR Coupling Constants in Disilenes and 1,3-Cyclodisiloxanes. *J. Am. Chem. Soc.* **1987**, *109*, 4116–4118.

(b) Gillette, G. R.; Maxka, J.; West, R. Synthesis of the Novel Ring Systems 1,2,3,4-Oxazadisiletidine and 1,3,4,2,5-Dioxazadisilolidine. *Angew. Chem. Int. Ed. Engl.* **1989**, *28*, 54–55.

83. Kuroda, M.; Kabe, Y.; Hashimoto, M.; Masamune, S. ^{29}Si–^{29}Si Long Range Couplings in Strained Cyclic Polysilanes. *Angew. Chem. Int. Ed. Engl.* **1988**, *27*, 1727–1730.

84. 1D and 2D ^{29}Si INADEQUATE NMR spectroscopy was used to determine the structures of cyclic oligosilanes. See Hengge, E.; Schrank, F. ^{29}Si-Double-Quantum Coherence Spectroscopy (INADEQUATE). An Efficient Method for the Structure Elucidation of Silicon Frameworks. *J. Organomet. Chem.* **1989**, *362*, 11–16.

85. Kyushin, S.; Ichikawa, K.; Koyama, Y.; Shiraiwa, H.; Ichikawa, H.; Okamura, K.; Suzuki, K. Studies on the Detailed Structure of Poly(dimethylsilylene). *Organometallics* **2014**, *33*, 6298–6304.

86. Ito, Y.; Lee, V. Ya.; Gornitzka, H.; Goedecke, C.; Frenking, G.; Sekiguchi, A. Spirobis(pentagerma[1.1.1]-propellane): A Stable Tetraradicaloid. *J. Am. Chem. Soc.* **2013**, *135*, 6770–6773.

87. Stanislawski, D. A.; West, R. ^{29}Si and ^{13}C NMR Spectra of Permethylpolysilanes. *J. Organomet. Chem.* **1981**, *204*, 295–305.

88. (a) Ramsey, N. F. The Internal Diamagnetic Field Correction in Measurements of the Proton Magnetic Moment. *Phys. Rev.* **1950**, *77*, 567.

(b) Ramsey, N. F. Magnetic Shielding of Nuclei in Molecules. *Phys. Rev.* **1950**, *78*, 699–703.

(c) Ramsey, N. F. Dependence of Magnetic Shielding of Nuclei upon Molecular Orientation. *Phys. Rev.* **1951**, *83*, 540–541.

(d) Ramsey, N. F. Chemical Effects in Nuclear Magnetic Resonance and in Diamagnetic Susceptibility. *Phys. Rev.* **1952**, *86*, 243–246.

(e) Kaupp, M. Interpretation of NMR Chemical Shifts. In *Calculation of NMR and ESR Parameters. Theory and Applications;* Kaupp, M., Bühl, M., Malkin, V. G., Eds.; Wiley-VCH: Weinheim, 2004; pp 293–306.

(f) Karni, M.; Apeloig, Y.; Takagi, N.; Nagase, H. S. Ab Initio and DFT Study of the ^{29}Si NMR Chemical Shifts in RSi≡SiR. *Organometallics* **2005**, *24*, 6319–6330.

89. (a) Bock, H. Fundamentals of Silicon Chemistry: Molecular States of Silicon-Containing Compounds. *Angew. Chem. Int. Ed. Engl.* **1989**, *28*, 1627–1650.

(b) Bock, H.; Solouki, B. Photoelectron Spectra of Silicon Compounds. In *The Chemistry of Organic Silicon Compounds;* Patai, S., Rappoport, Z., Eds.; Wiley: Chichester, 1989; pp 555–653.

(c) Miller, R. D.; Michl, J. Polysilane High Polymers. *Chem. Rev.* **1989**, *89*, 1359–1410.

(d) Kira, M.; Miyazawa, T. Mechanistic Aspects of the Photochemistry of Organosilicon Compounds. In *The Chemistry of Organic Silicon Compounds;* Rappoport, Z., Apeloig, Y., Eds.; Vol. 2; Wiley: Chichester, 1998; pp 1311–1337.

(e) Hill, M. S. Homocatenation of Metal and Metalloid Main Group Elements. *Struct. Bond.* **2010**, *136*, 189–216.

90. Kosa, M.; Karni, M.; Apeloig, Y. Theoretical Study of Ladder Polysilanes. *Organometallics* **2007**, *26*, 2806–2814.

91. Gilman, H.; Atwell, W. H.; Schwebke, G. L. Ultraviolet Properties of Compounds Containing the Silicon–Silicon Bond. *J. Organomet. Chem.* **1964**, *2*, 369–371.

92. Ichinohe, M.; Matsuno, T.; Sekiguchi, A. Synthesis, Characterization, and Crystal Structure of Cyclotrisilene: A Three-Membered Ring Compound with a Si–Si Double Bond. *Angew. Chem. Int. Ed.* **1999**, *38*, 2194–2196.

93. Nied, D.; Klopper, W.; Breher, F. Pentagerma[1.1.1]propellane: A Combined Experimental and Quantum Chemical Study on the Nature of the Interactions between the Bridgehead Atoms. *Angew. Chem. Int. Ed.* **2009**, *48*, 1411–1416.

94. Iwamoto, T.; Kira, M. A Mechanistic Study of Thermal and Photochemical Isomerization between Hexasilyltetrasilabicyclo[1.1.0]butane and Hexasilyltetrasilacyclobutene. *Chem. Lett.* **1998**, 277–278.

95. (a) Kyushin, S.; Yamaguchi, H.; Okayasu, T.; Yagihashi, Y.; Matsumoto, H.; Goto, M. Selective Si–Si Bond Cleavage in Decaisopropylbicyclo[2.2.0]hexasilane. A Route to Sterically Hindered 1,4-Dichlorocyclohexasilanes. *Chem. Lett.* **1994**, 221–224.
(b) Meguro, A.; Sakurai, H.; Kato, K.; Kyushin, S.; Matsumoto, H. Selective Si–Si Bond Cleavage of Decaisopropylbicyclo[2.2.0]hexasilane with Hydrobromic Acid and Hydrochloric Acid. *Chem. Lett.* **2001**, 1212–1213.

96. Kyushin, S.; Sakurai, H.; Yamaguchi, H.; Matsumoto, H. Ring-Opening Reactions of *anti*-Dodecaisopropyltricyclo[4.2.0.0^{2,5}]octasilane. Formation of Novel Bicyclo[3.3.0]octasilane and Bicyclo-[4.2.0]octasilane Systems. *Chem. Lett.* **1996**, 331–332.

97. Kyushin, S.; Yagihashi, Y.; Matsumoto, H. Kinetic Study on Thermal syn–anti Isomerization of Dodecaisopropyltricyclo[4.2.0.0^{2,5}]octasilane. *J. Organomet. Chem.* **1996**, *521*, 413–415.

98. (a) Martin, H.-D.; Eisenmann, E.; Kunze, M.; Bonačic-Koutecký, V. Die C_8H_{12}-Energiehyperfläche Thermolyse von *syn*- und *anti*-Tricyclo[4.2.0.0^{2,5}]octan. Experimentelle und Theoretische Untersuchungen. *Chem. Ber.* **1980**, *113*, 1153–1179.
(b) Walsh, R.; Martin, H.-D.; Kunze, M.; Oftring, A.; Beckhaus, H.-D. Small Rings. Part 32. The Gas Phase Kinetics, Mechanism, and Energy Hypersurface for the Thermolyses of *syn*- and *anti*-Tricyclo-[4.2.0.0^{2,5}]octane. *J. Chem. Soc. Perkin Trans.* **1981**, 1076–1083.

99. Wiberg, N.; Vasisht, S.-K.; Fischer, G.; Mayer, P.; Huch, V.; Veith, M. Reactivity of the Unusually Structured Silicon Cluster Compound $Si_8(Si\textit{t}Bu_3)_6$. *Z. Anorg. Allg. Chem.* **2007**, *633*, 2425–2430.

100. (a) Kyushin, S.; Sakurai, H.; Yamaguchi, H.; Goto, M.; Matsumoto, H. Structure and Unusual Electronic Spectra of Decaisopropyl-7-oxabicyclo[2.2.1]heptasilane. *Chem. Lett.* **1995**, 815–816.
(b) Kyushin, S.; Tanaka, R.; Arai, K.; Sakamoto, A.; Matsumoto, H. Domino Oxidation of Ladder Oligosilanes: Formation of Novel Ladder Frameworks Containing Oligosiloxane and Oligosilane Chains. *Chem. Lett.* **1999**, 1297–1298.

101. Unno, M.; Tanaka, R.; Tanaka, S.; Takeuchi, T.; Kyushin, S.; Matsumoto, H. Oligocyclic Ladder Polysiloxanes: Alternative Synthesis by Oxidation. *Organometallics* **2005**, *24*, 765–768.

102. (a) Unno, M.; Yokota, T.; Matsumoto, H. Oxaoctasilahomocubane and Dioxaoctasilabishomocubane: Novel Silicon Ring Systems. *J. Organomet. Chem.* **1996**, *521*, 409–411.
(b) Unno, M.; Matsumoto, T.; Mochizuki, K.; Higuchi, K.; Goto, M.; Matsumoto, H. Structure and Oxidation of Octakis(*tert*-butyldimethylsilyl)octasilacubane. *J. Organomet. Chem.* **2003**, *685*, 156–161.

103. Ichinohe, M.; Takahashi, N.; Sekiguchi, A. Formation and Structure of Protonated Tetrasilatetrahedrane-Monooxide, $(\textit{tert}\text{-}Bu_3Si)_4Si_4OH^+$. *Chem. Lett.* **1999**, 553–554.

104. Lee, V. Ya.; Sekiguchi, A.; Ichinohe, M.; Fukaya, N. Stable Aromatic Compounds Containing Heavier Group 14 Elements. *J. Organomet. Chem.* **2000**, *611*, 228–235.

105. Sekiguchi, A.; Yatabe, T.; Doi, S.; Sakurai, H. Cage Compounds of Si and Ge: Synthesis and Structures. *Phosphorus Sulfur Silicon Relat. Elem.* **1994**, *93–94*, 193–196.

106. Horiuchi, H.; Nakano, Y.; Matsumoto, T.; Unno, M.; Matsumoto, H.; Hiratsuka, H. Electronic Structure and Photochemical Reaction Intermediates of Octakis(1,1,2-trimethylpropyl)octasilacubane. *Chem. Phys. Lett.* **2000**, *322*, 33–40.

107. Kyushin, S.; Meguro, A.; Unno, M.; Matsumoto, H. Photolysis of *anti*-Dodecaalkyltricyclo[4.2.0.0^{2,5}]-octasilane: Generation and Reactions of Cyclotetrasilene. *Chem. Lett.* **2000**, 494–495.

108. Nikawa, H.; Nakahodo, T.; Tsuchiya, T.; Wakahara, T.; Rahman, G. M. A.; Akasaka, T., et al. *S*-Heterocyclic Carbene with a Disilane Backbone. *Angew. Chem. Int. Ed.* **2005**, *44*, 7567–7570.

109. Takanashi, K.; Lee, V. Ya.; Yokoyama, T.; Sekiguchi, A. Base-Free Molybdenum and Tungsten Bicyclic Silylene Complexes Stabilized by a Homoaromatic Contribution. *J. Am. Chem. Soc.* **2009**, *131*, 916–917.
110. Lee, V. Ya.; Aoki, S.; Yokoyama, T.; Horiguchi, S.; Sekiguchi, A.; Gornitzka, H.; Guo, J. D.; Nagase, S. Toward a Silicon Version of Metathesis: From Schrock-Type Titanium Silylidenes to Silatitanacyclobutenes. *J. Am. Chem. Soc.* **2013**, *135*, 2987–2990.
111. Husk, G. R.; West, R. The Radical Anion of Dodecamethylcyclohexasilane. *J. Am. Chem. Soc.* **1965**, *87*, 3993–3994.
112. Helmer, B. J.; West, R. Synthesis and Properties of the Geometrical Isomers of *tert*-Butylmethylcyclosilanes, $(t\text{-BuMeSi})_4$ and $(t\text{-BuMeSi})_5$. *Organometallics* **1982**, *1*, 1458–1463.
113. Gehrhus, B.; Hitchcock, P. B.; Zhang, L. An Isolable Radical Anion and Dianion of a Cyclotetrasilane: Synthesis and Structure of $[\mathrm{Si}\{1,2\text{-}(\mathrm{NEt})_2\mathrm{C_6H_4}\}]_4^{-\bullet}$ and $[\mathrm{Si}\{1,2\text{-}(\mathrm{NEt})_2\mathrm{C_6H_4}\}]_4^{2-}$. *Angew. Chem. Int. Ed.* **2004**, *43*, 1124–1126.
114. Kyushin, S.; Miyajima, Y.; Matsumoto, H. Observation of Highly Stable Radical Anions of Ladder Oligosilanes. *Chem. Lett.* **2000**, 1420–1421.
115. Otsuka, K.; Matsumoto, N.; Ishida, S.; Kyushin, S. An Isolable Radical Anion of an Organosilicon Cluster Containing Only σ Bonds. *Angew. Chem. Int. Ed.* **2015**, *54*, 7833–7836.
116. (a) Tachibana, H.; Goto, M.; Matsumoto, M.; Kishida, H.; Tokura, Y. Crystal Structure and Optical Properties of Polymorphic Octasilacubane. *Appl. Phys. Lett.* **1994**, *64*, 2509–2510.
    (b) Furukawa, K.; Fujino, M.; Matsumoto, N. Superlattice Structure of Octa-*tert*-butylpentacyclo-$[4.2.0.0^{2,5}.0^{3,8}.0^{4,7}]$octasilane Found by Reinvestigation of X-ray Structure Analysis. *J. Organomet. Chem.* **1996**, *515*, 37–41.
117. Tsurusaki, A.; Kyushin, S. The Radical Anion of Cyclopentasilane-Fused Hexasilabenzvalene. *Chem. Eur. J.* **2016**, *22*, 134–137.
118. (a) Schoeller, W. W.; Busch, T. Electronic Structure of the Disilenyl Radical Anion. *J. Phys. Org. Chem.* **1994**, *7*, 251–255.
    (b) Kira, M. Distortion Modes of Heavy Ethylenes and Their Anions: $\pi-\sigma^*$ Orbital Mixing Model. *Organometallics* **2011**, *30*, 4459–4465.
119. Sekiguchi, A.; Inoue, S.; Ichinohe, M.; Arai, Y. Isolable Anion Radical of Blue Disilene $(^t\mathrm{Bu_2MeSi})_2\mathrm{Si{=}Si(SiMe_2{}^t\mathrm{Bu_2})_2}$ Formed upon One-Electron Reduction: Synthesis and Characterization. *J. Am. Chem. Soc.* **2004**, *126*, 9626–9629.

CHAPTER 4

Chiral Organosilicon Compounds

Li-Wen Xu[1,2]

[1]Hangzhou Normal University, Hangzhou, P. R. China [2]Lanzhou Institute of Chemical Physics, Chinese Academy of Sciences, Lanzhou, P. R. China

Chapter Outline

4.1 Introduction

Chirality is widely represented in a diverse array of life forms, and the synthesis of chiral organic compounds has triggered a heightened interest in the search for functional molecules. However, in contrast to the carbon-based enantioenriched organic compounds, chiral organosilicon compounds are not available in nature.[1] Therefore the synthesis of silicon-stereogenic organosilicon compounds is expected to be more difficult than that for the carbon-stereogenic organic compounds because of the lack of the silicon-stereogenic sources in nature. However, it will offer new opportunities and research topics in the development of such exciting construction of Si-centered stereochemistry.

Since the first example of silicon-stereogenic silanes reported by Kipping in 1907,[2] a quest for the synthesis of chiral tetrasubstituted silanes had remained pending over the next half century.[3] Although several optically active silicon-stereogenic silanes had been obtained through resolution procedures from the 1960s to the 1980s, the synthesis of chiral organosilicon compounds had been recognized as a highly challenging project at that time. Fortunately, the great contribution of Sommer's and Corriu's groups in the field of the silicon-centered stereochemistry laid the foundation of chiral organosilicon chemistry.[4] To our

Organosilicon Compounds.
DOI: http://dx.doi.org/10.1016/B978-0-12-801981-8.00004-6

delight, much attention has been paid by organic chemists very recently to the development of novel silicon-stereogenic silanes and the application of corresponding organosilicon compounds in organic synthesis. In this chapter, the author would like to provide a comprehensive summary of the synthesis of silicon-stereogenic compounds and outline briefly the application of such chiral organosilicon compounds in asymmetric synthesis.

4.2 Synthesis of Chiral Organosilicon Compounds

4.2.1 Optical Resolution or Kinetic Resolution

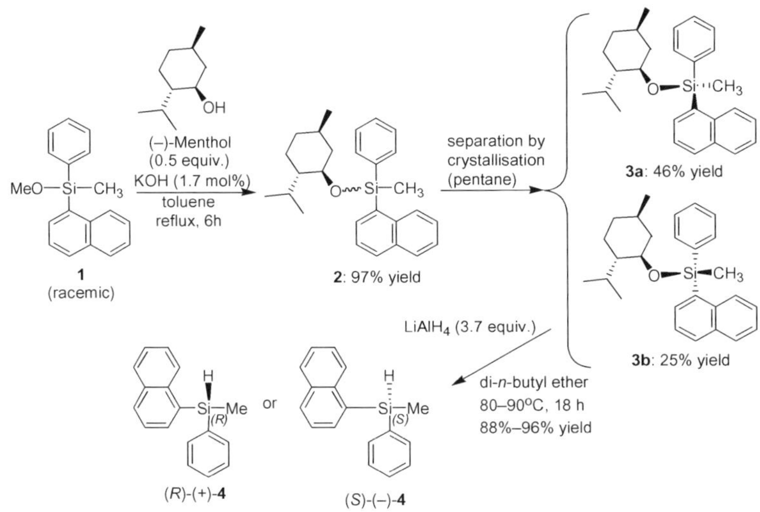

Scheme 4.1
The synthesis of silicon-stereogenic silane **4**.

Early in the 1960s, the synthesis of silicon-stereogenic silanes through the optical resolution of diastereomers proved to be possible and practical in terms of optical purity. A representative example of the preparation of functional chiral silane by Si–O coupling reaction was reported by Sommer and coworkers, in which they applied the (–)-menthol as a chiral reagent in the formation and separation of a diastereomer mixture (Scheme 4.1).[5] Subsequently, such a resolution method has been proved to be useful in the preparation of optically active arylsilane (Scheme 4.2, **5**–**7**).[6]

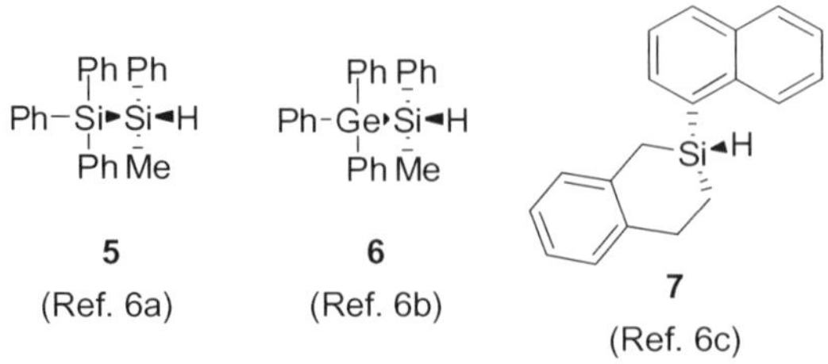

Scheme 4.2
Preparation of optically pure aryl hydrosilanes by kinetic resolution.

It has been reported that the introduction of (−)-menthoxy group to a silane molecule generally required high temperature (e.g., reflux in toluene) in the presence of strong base (KOH). Thus many unavoidable by-products usually formed because of unexpected side reactions. Therefore the development of mild reaction conditions for the silylation of chiral alcohol was valuable. Early in 1970, Holt et al. developed a mild Pd/C-catalyzed silylation at room temperature.[7] Subsequently in the 1970s, Corriu and coworkers also developed various silicon-stereogenic silanes (Scheme 4.3, chiral silanes **8–10**), which provided the structural diversity of optically active silanes for the further studies of silicon-centered stereochemistry.[8]

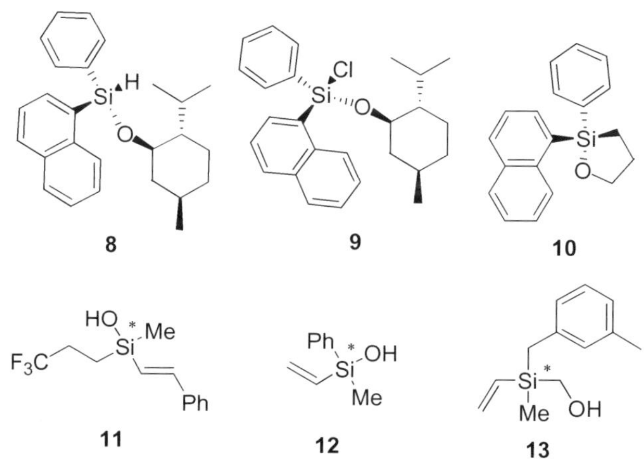

Scheme 4.3

Representative silicon-stereogenic silanes prepared before 1999.

Since the 1990s, direct resolution with the aid of chiral auxiliary or direct separation of diastereoisomer mixtures by modern chiral chromatographic technology have also been reported as practical methods. For example, in 1999, Mori and coworkers[9] reported the synthesis and resolution of silicon-stereogenic silanols with alkene group via chiral HPLC with Chiralpak AD column, in which the chiral separation is successful because the enantiomeric excess of the desired product could be achieved to be >99%*ee* (Scheme 4.3, compounds **11** and **12**). Similarly, the chiral (hydroxymethyl)methyl(3-methylbenzyl) vinylsilane **13** could be also obtained successfully by chromatographic resolution (HPLC) using a chiral tool (Scheme 4.3, compound **13**).[10]

The classical resolution of chlorosilane based on silylation reaction of chiral alcohol, such as (−)-menthol, was proved to be effective in the synthesis of silicon-stereogenic silanes. As shown in Scheme 4.4, Oestreich et al. reported a practical approach to the synthesis of chiral organosilicon compounds, in which the direct silylation reaction of racemic cyclic chlorosilane with the potassium alkoxides of (*S*)-1-phenylethanol or (−)-menthol led to the formation of separable diastereoisomers, and subsequent separation by flash chromatography resulted in the isolation of silicon-stereogenic silanes with good diastereoselectivity. For example, the desired organosilicon compound **20a** or **20b** could be obtained in excellent enantiomeric excess and good yield (Scheme 4.4).[11]

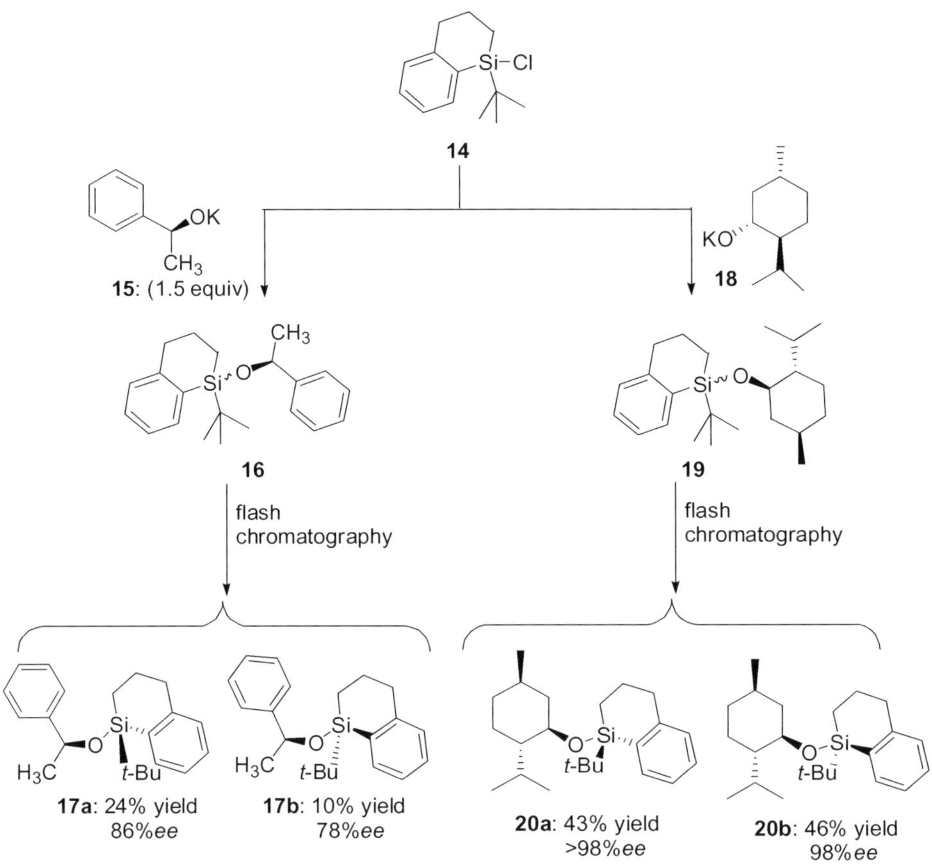

Scheme 4.4
Preparation of silicon-stereogenic silanes by chiral alcohols.

Except for the (−)-menthol and similar compounds as an auxiliary, a number of commercially available compounds with chiral functional groups, such as (+)-tartaric acid, chiral amino acid, have also been successfully employed as a chiral auxiliary. For example, in 1980, Terunuma et al. reported an example of optical resolution of the α-amino organosilanes with (+)-tartaric acid (Scheme 4.5),[12] and it was found that the chiral α-amino organosilanes **23** could be converted into various derivatives via hydrolysis or nucleophilic addition reaction.[13] Similarly to Terunuma's work, Tacke and coworkers have also described the application of *O,O′*-di-*p*-toluoyltartaric acid as resolution agents in the preparation of optically active (2-aminoethyl) cyclohexyl(hydroxymethyl)phenylsilanes (≥97%*ee*).[14]

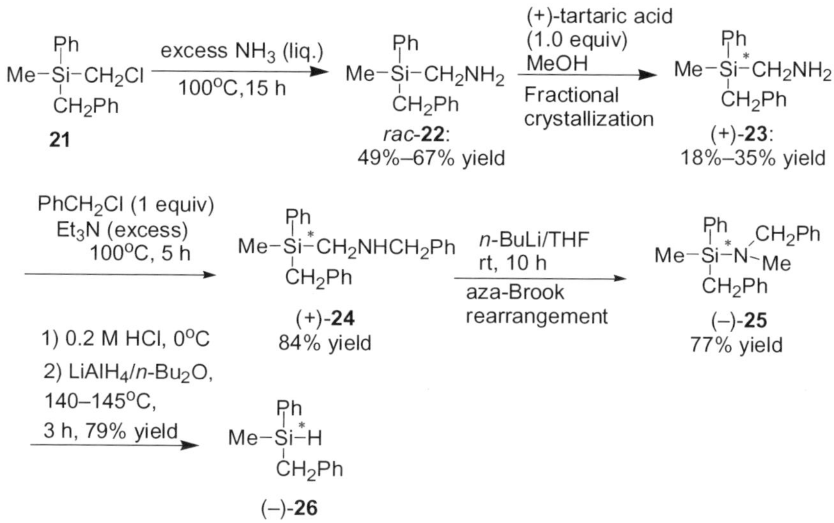

Scheme 4.5
Preparation of silicon-stereogenic silanes by chiral tartaric acid.

In 1999, Schaumann and coworkers[15] investigated the chiral amino alcohols as a chiral auxiliary to synthesize chiral silanes (Scheme 4.6). In fact, the direct resolution of simple *tert*-butyl(methyl)phenylmenthoxysilanes by crystallization or chiral column chromatography was proved to be quite difficult. In this context, Schaumann and coworkers reported a good example for the synthesis of simple silicon-stereogenic silanes. The authors succeed in the preparation (*R*)- and (*S*)-*tert*-butyl(methyl)phenylsilanes **32** (98%*ee*) starting from racemic *tert*-butyl(methyl)phenylsilyl chloride, in which the key step was the introduction of readily available chiral 2-amino-1-butanol to the silane and subsequent crystallization (Scheme 4.6).

Schaumann et al.[16] also reported an interesting example for the synthesis of silicon-stereogenic silanes in 2000, in which they found that the enantiomerically pure 1,1′-binaphthalenyl-2,2′-diol (BINOL, **33**) was a good auxiliary for the resolution of functional silanes (Scheme 4.7). In this approach, the BINOL-derived organosilicon compound **34** could be prepared easily by the simple silylation of racemic *tert*-butyl (methyl)phenylsilyl chloride and lithium derivative of (*R*)-BINOL, and the resulting diastereoisomeric mixture **34** was separable by preparative HPLC with Nucleosil 50−5 column. Then the subsequent reduction of chiral siloxane **34** led to the formation of silicon-stereogenic silane **35** in good yield and excellent enantiomeric excess (>98%*ee*).

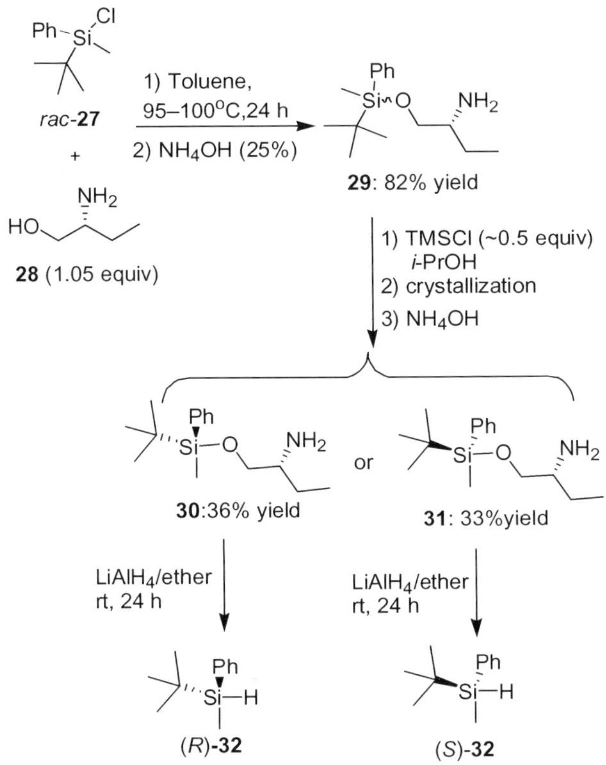

Scheme 4.6

Preparation of silicon-stereogenic silanes by chiral amino alcohols.

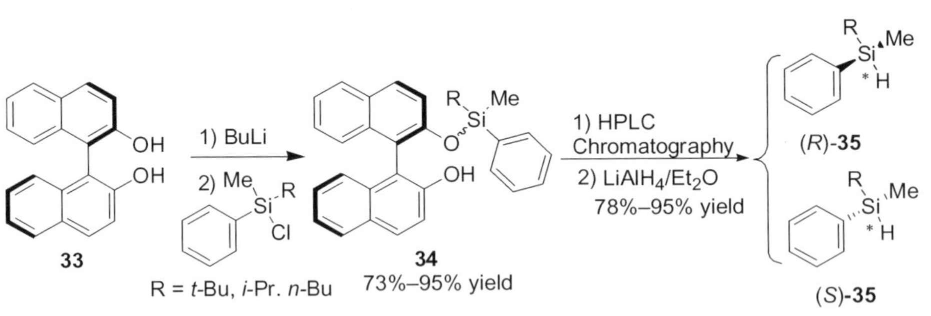

Scheme 4.7

BINOL-involved synthesis of silicon-stereogenic silanes **35**.

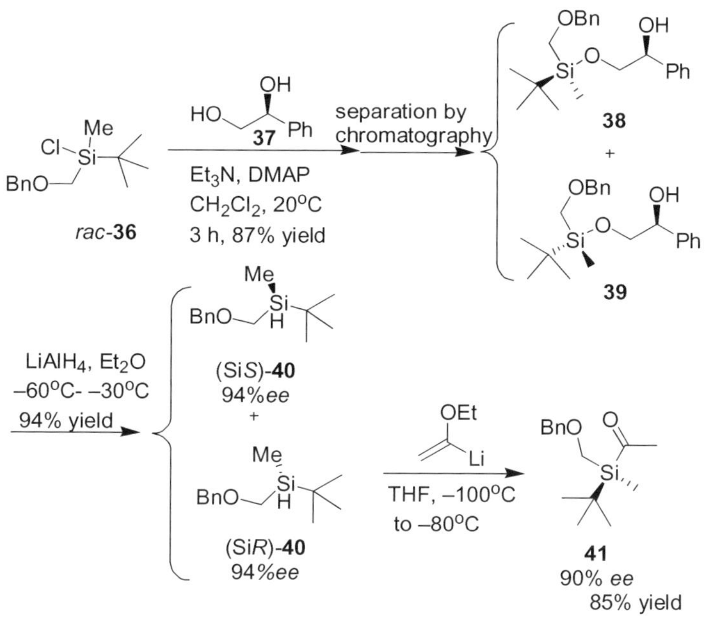

Scheme 4.8

The synthesis of silicon-stereogenic hydrosilane and silicon-stereogenic acylsilane.

In 2004, Bienz and coworkers developed a useful route to the preparation of optically active hydrosilanes with enantiomerically pure diol ((*S*)-1-phenylethane-1,2-diol, **37**). As shown in Scheme 4.8, the diastereomeric isomers **38** or **39** could be separated easily by flash column chromatography, and then the subsequent reduction resulted in the formation of silicon-stereogenic hydrosilane with excellent enantiomeric excess (94%*ee*).[17] In this work, the same authors also reported the further synthesis of silicon-stereogenic acylsilane from chiral hydrosilane **38** or **39**, in which the desired product (**41**) could be obtained with retention of configuration (90%*ee*) through the substitution reaction with 1-ethoxyvinyllithium and subsequent hydrolysis.[17]

Although kinetic resolution is not a new concept in organic chemistry, its powerful potential has still been in prospect in the synthesis of chiral organosilicon compounds. However, the synthetic limit of the method is the right choice of chiral reagent and necessity of stopping the reaction at the right time.[18] For example, Corriu et al.[19] have reported the kinetic resolution of bifunctional organosilanes by selective methanolysis of chloro-α-naphthylphenylsilane. However, the enantioselectivity is quite low. In this context, Yamamoto et al.[20] reported in 1993 that the cyclohex-1-enylcyclohexylmethylsilanol could be prepared by Sharpless epoxidation in the presence of a titanium-based catalyst system

(Scheme 4.9). As shown in Scheme 4.9, when cyclohex-1-enyl-cyclohexylmethylsilanol **45** was used in this reaction, a desired product, $(R)_{Si}$-**46**, was obtained with >99%*ee*, and $K_{rel} > 11$ at 71% conversion (K_{rel} values refer to "relative rates of kinetic resolution"). The diastereoisomer ratio of the epoxide product could be obtained with good diastereoselectivity (95:5).

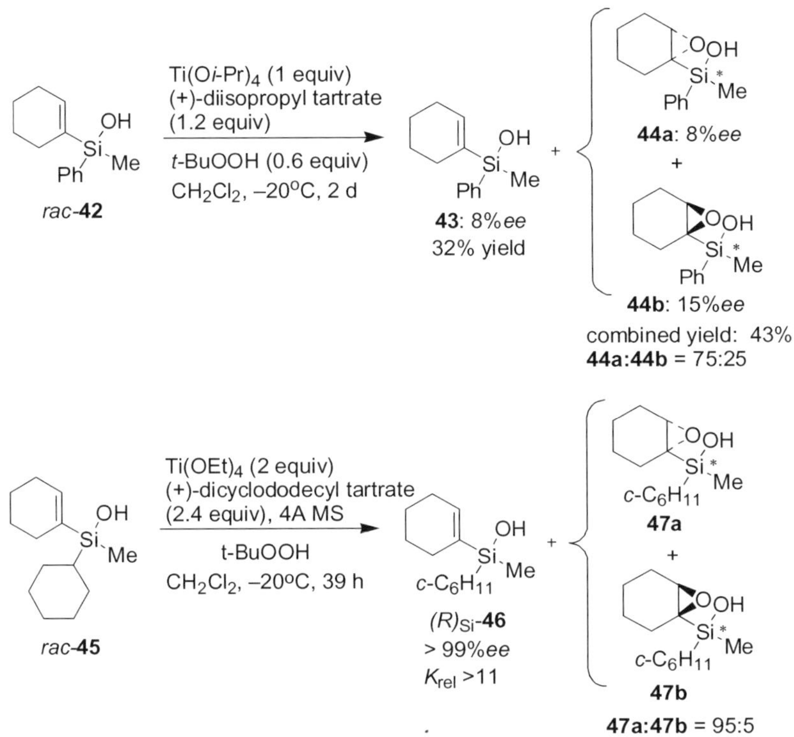

Scheme 4.9

Kinetic resolution of cyclohex-1-enylcyclohexylmethylsilanol by Sharpless epoxidation (4A MS: molecular sives 4 Å).

In 2006, Oestreich et al.[21] reported a useful kinetic resolution for the preparation of chiral silane via copper-catalyzed dehydrogenative silylation (Scheme 4.10). Although this process proceeded with only moderate diastereoselectivity (76:24 *dr*), the corresponding diastereoisomer mixture **51** containing chiral auxiliary could be separated by flash column chromatography, enabling further isolation of the optically pure silicon-stereogenic compounds in promising isolated yields and diastereoselectivity (>98:2 *dr*). And then the reduction of latter compounds with DIBAL-H gave silicon-stereogenic hydrosilane with excellent enantiomeric excess (up to 96%*ee*).

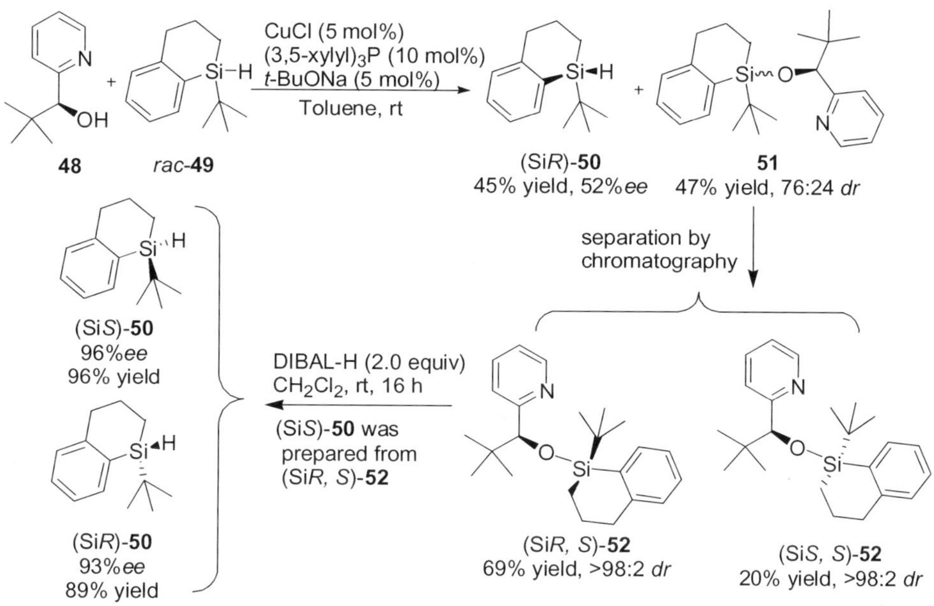

Scheme 4.10
Preparation of optically pure silanes with the aid of pyridine-derived alcohol.

4.2.2 Desymmetrization of Functional Organosilicon Compounds With Chiral Reagents

Although considerable efforts have been devoted to the preparation of silicon-stereogenic silanes through optical resolution and kinetic resolution, the resolution method basically requires suitable crystallization behavior, and suffers from rather low yields and a limited substrate scope. More efficient and atom-economic approaches promoted by readily available catalysts or chiral sources would be a highly appealing alternative and is arguably valuable in practical organic synthesis. Whereas the preparation of silicon-stereogenic silanes by the optical resolution or kinetic resolution has been widely studied in the past 50 years, the synthesis of optically pure silanes through desymmetrization o functional organosilicon compounds is a potentially useful synthetic approach.

In this field, an important early example was revealed in 1966 by Klebe et al. with chiral auxiliaries (Scheme 4.11).[22] In this pioneering work, the authors developed a silylation reaction of *N*-phenyl amino acid (**54**) with bis(*N*-methyl-acetamido)methylphenylsilane (**53**), which led to the facile formation of 2-silaoxazolidones **55** in the form of two diastereoisomers (about 2:1 *dr*). The subsequent exchange reaction with different alcohols, including phenol, accompanied with isomerization and crystallization, gave the optically active silanes **56**.[23]

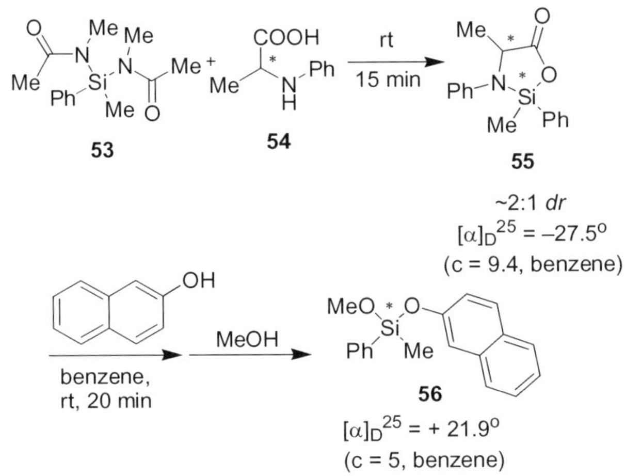

Scheme 4.11

The synthesis of silicon-stereogenic siloxane by desymmetrization of functional organosilicon compounds [c: concentration (g/100 mL)].

Early in 1987, it was reported that the silyl ether of symmetric chiral diol could be applied as a C_2 chiral auxiliary for the synthesis of chiral silane.[24] For example, Masuda and coworkers reported that the (*R*)-methylphenylpropylsilane **60** was obtained with promising enantioselectivity (up to 98%*ee*) by the desymmetric nucleophilic substitution of Grignard reagents with chiral 5,6-dimethoxy-1,3,2-dioxasilacycloheptane derivatives, followed by reduction (Scheme 4.12). However, the reaction of chiral siloxane **57** with most Grignard reagents or lithium reagents was not good in terms of diastereoselectivity/enantioselectivity (Scheme 4.12). For example, the reaction of **57** with 1-NpLi only gave a nearly racemic product (−)-**60** in poor yield (10%) and low enantioselectivity (2.8%*ee*).[24b]

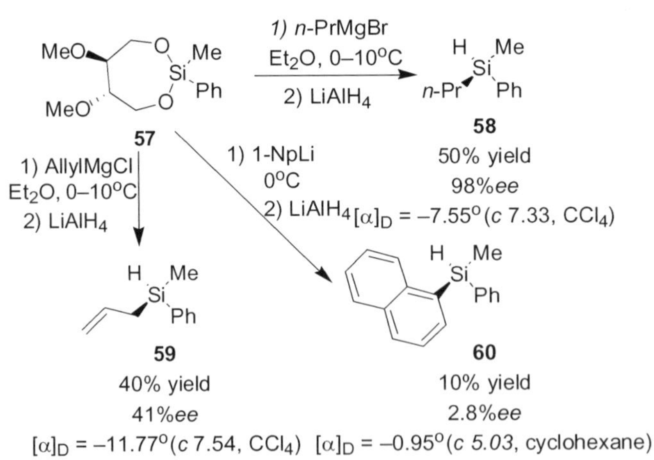

Scheme 4.12

The synthesis of silicon-stereogenic silanes with desymmetrization of symmetric siloxane with Grignard reagents [c: concentration (g/100 mL)].

Similarly to Masuda's strategy, in 2008, Tomooka and coworkers[25] developed an improved strategy with desymmetric nucleophilic substitution of lithium reagents with siloxane **61** for the stereoselective construction of silicon-stereogenic organosilicon compound. In this stereoselective base-mediated substitution reaction, bisoxazolin was found to be an ideal chiral base/auxiliary to induce good enantioselectivity (Scheme 4.13, up to 84%*ee*). And various substituted silanols **64** were prepared in excellent yields (up to 99%), albeit the enantiopurity is moderate.

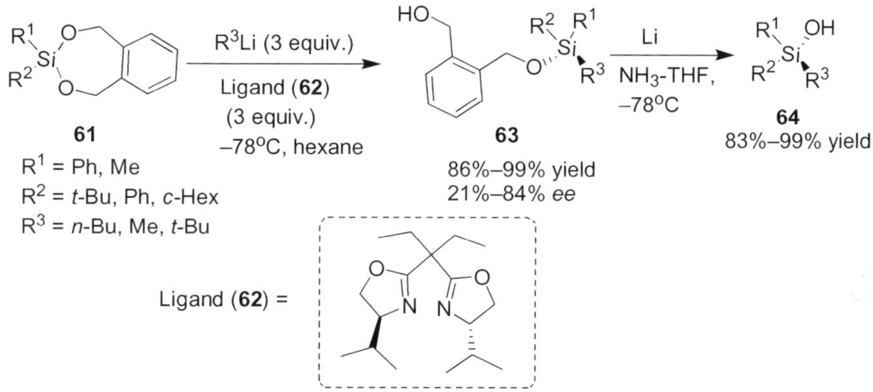

Scheme 4.13
Synthesis of silicon-stereogenic silanes by enantioselective substitution of dialkoxysilane.

In 1999, Tamao and coworkers[26] reported a new approach to optically active (1-naphthyl) phenylmethylfluorosilane (**70**) with the aid of a chiral secondary amine and its lithium salt (**66**). The synthetic method was initiated by desymmetrization of difluorosilane by amination (Scheme 4.14) and then successfully completed on the basis of epimerization of diastereomeric mixtures of fluorosilane (Scheme 4.15). The subsequent methylation and fluorination led to the stereospecific formation of the enantiomerically pure (1-naphthyl) phenylmethylfluorosilane (**70**).

Important progress in the enantioselective synthesis of silicon-stereogenic silanes by desymmetrization with the aid of chiral auxiliary was achieved by Tomooka and coworkers[27] in 2000, when they developed a novel procedure for the synthesis of chiral silanol via montmorillonite K10 clay-promoted intramolecular 1,4-aryl migration from silicon to carbon in the molecule of *tert*-butyldiarylsiloxy cyclic hemiaminal compound in the presence of an alcohol (Scheme 4.16). The mechanism most likely involved the intramolecular Friedel–Crafts reaction through a possible *N*-acyliminium ion intermediate to form the *β*-silyl cation intermediate. The key intermediate was then consumed by nucleophilic attack of alcohol to the silicon, which led to the diastereoselective formation of the desired 1,4-phenyl migration product **73**. This novel approach featured high diastereoselectivity (95%*de*) and good yields.

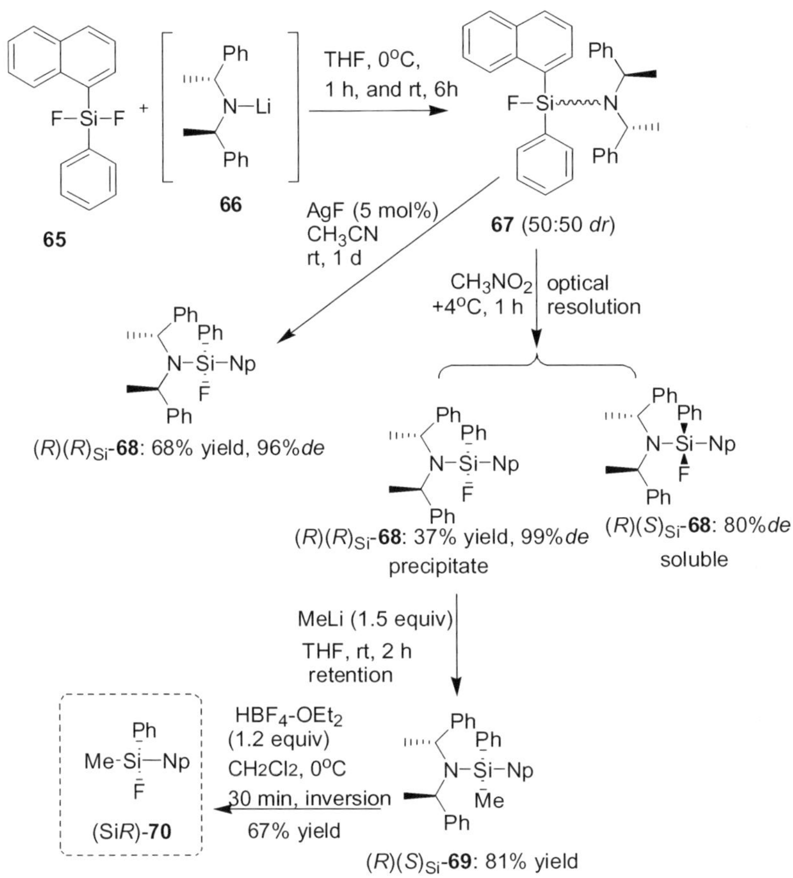

Scheme 4.14
The synthesis of optically active (1-naphthyl)phenylmethylfluorosilane.

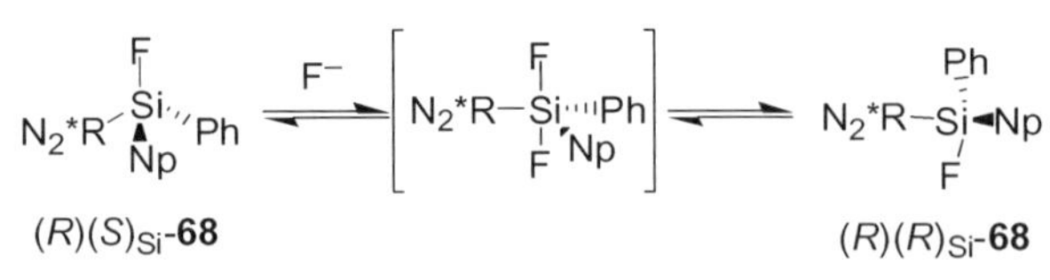

Scheme 4.15
Epimerization of diastereomeric mixtures of fluorosilane.

Similarly to Corriu's strategy of Rh-catalyzed dehydrogenative silylation with chiral alcohol, Leighton and coworkers[28] also developed a highly diastereoselective copper-catalyzed silylation of chiral alcohol **75** with dihydrosilane **76** in 2003. The reported method featured the high diastereoselectivity (up to 97/3 *dr*) by use of chiral alcohol and in

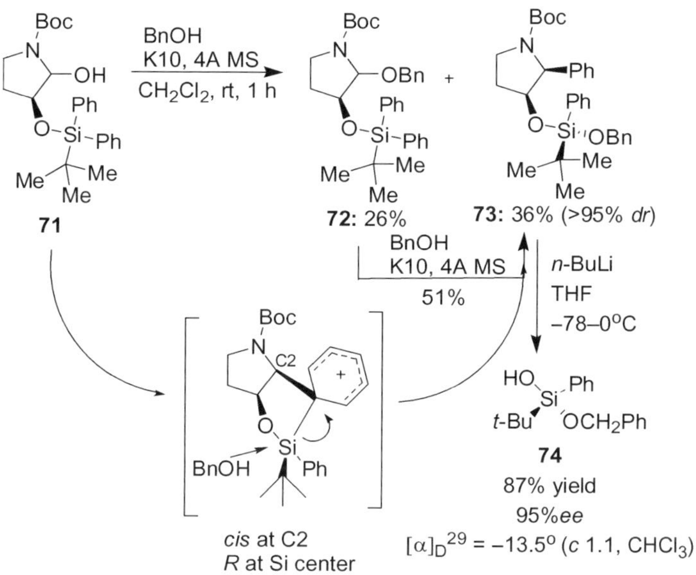

Scheme 4.16
The synthesis of silicon-stereogenic organosilicon compound **74** by intramolecular 1,4-aryl migration [Boc: *tert*-butyloxycarbonyl; Bn: benzyl; 4A MS: molecular sieves 4 Å].

the presence of chiral phosphine ligand **77** (Scheme 4.17). Notably, the phosphine ligand was proved to be an important factor in this reaction and it was found that the (*R*,*R*)-2,4-bis (diphenylphosphino)pentane (BDPP) and its analogs were effective phosphine ligands with varied diastereoselectivities (90:10–97:3 *dr*).

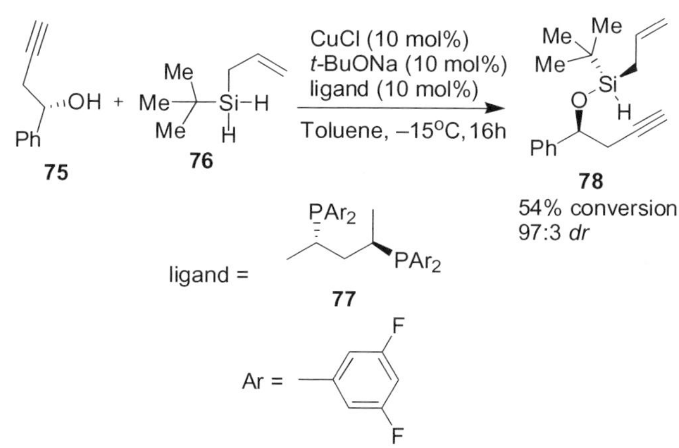

Scheme 4.17
The desymmetrization of dihydrosilane by Cu-catalyzed dehydrogenative silylation with chiral alcohol.

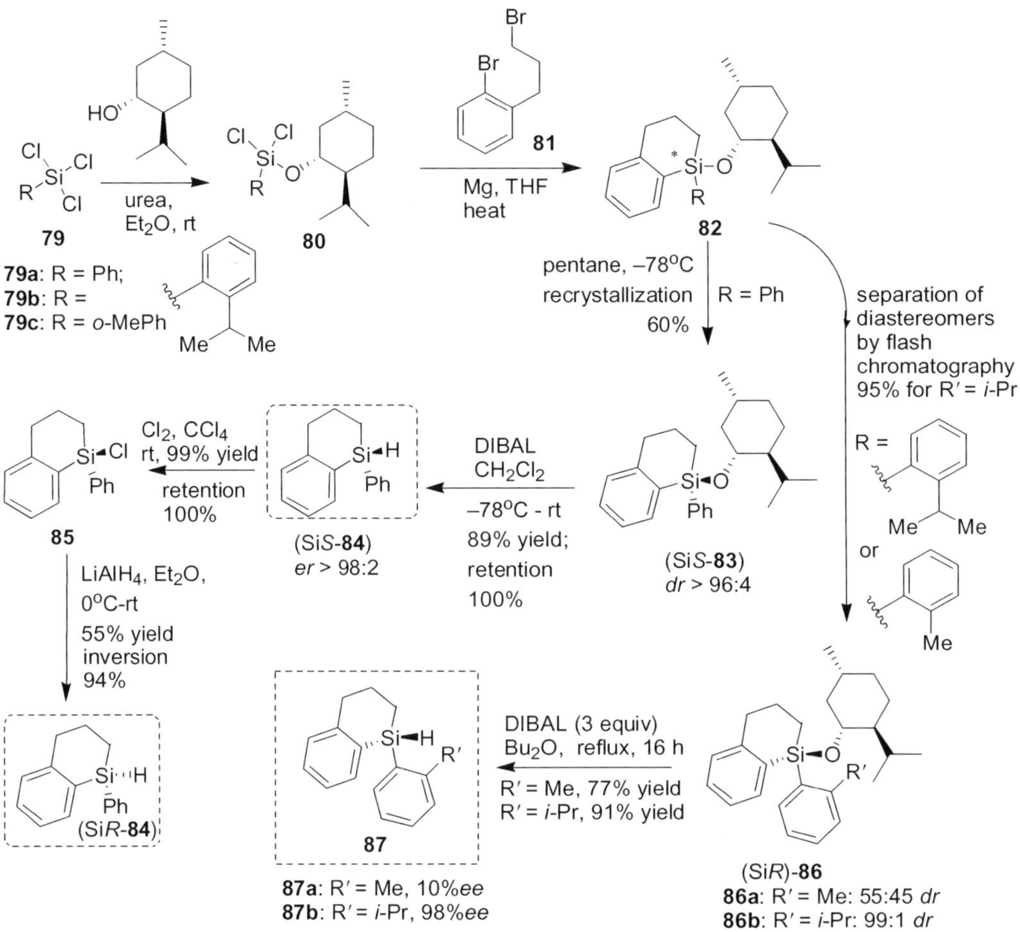

Scheme 4.18
The synthesis of silicon-stereogenic hydrosilanes by (−)-menthol.

In 2003, Oestreich and coworkers[29] also successfully developed a novel method by which the construction of the chiral cyclic silanes could be achieved through the Grignard addition reaction of dibromide **81** and a chiral dichlorosilane **80** under mild reaction conditions. Notably, this methodology relies on the enantiopure dichlorosilanes, such as (−)-menthyloxy-substituted dichlorosilanes **80** (Scheme 4.18).

Similar to Klebe's approach,[22,23] an improved study with chiral amino alcohols instead of amino acids was completed by Wada et al. in 2009.[30] As shown in Scheme 4.19, several optically active silanes (compounds **90** and **91**) were obtained by the desymmetrization of bifunctional organosilicon compound **89** with chiral amino alcohol **88**. And then the nucleophilic addition reaction of Grignard reagents, and subsequent reduction with diisobutylaluminum hydride (DIBAL), led to the desired silicon-stereogenic hydrosilane with high levels of diastereoselectivity and enantioselectivity.

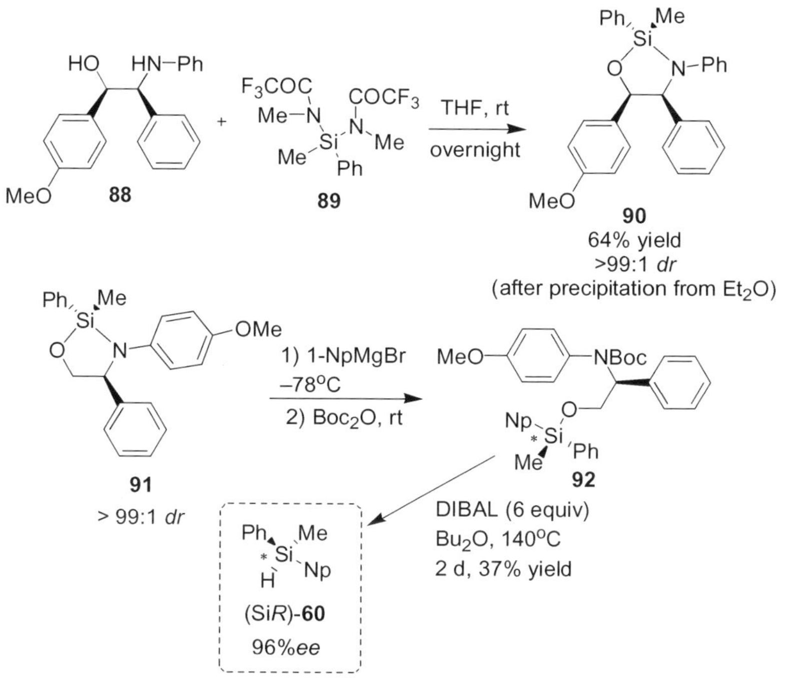

Scheme 4.19
The synthesis of silicon-stereogenic silane initiated by desymmetrization of bifunctional organosilicon compound with chiral amino alcohol [Boc: *tert*-butyloxycarbonyl].

In 2014, on the basis of the transformation of Si–OMe groups in methoxysilanes to Si–NR_2 functions,[31] Strohmann and Bauer developed a novel access to N,O-functionalized silicon-stereogenic organosilicon compounds by selective Si–N/Si–O exchange, providing chiral organosilicon compounds **96** or **98** with good to excellent optical purities (Schemes 4.20 and 4.21).[32] The selective exchange of the Si–N bond for a second Si–O bond is highly stereoselective. Notably, on the basis of determination of the absolute configuration by X-ray crystallography, the authors confirmed that the Si–N/Si–O exchange reaction proceeds with inversion of configuration at the stereogenic silicon center.

Very recently, Strohmann and Bauer also reported a convenient stereocontrolled route to N,O-functionalized silicon-stereogenic organosilicon compounds **101/102** with an additional aminomethyl function (Scheme 4.22).[33] Similarly to their previous strategy,[32] owing to the presence of three donor functions, the asymmetric substitution by *t*BuLi occurred with high stereochemical control of the nucleophilic attack. Later, the same authors presented a conclusive mechanism for the stereocontrolled nucleophilic substitution reactions on such chiral aminodimethoxysilanes suggesting that stereomutations are of fundamental importance for the design of the silicon-center chirality.[34]

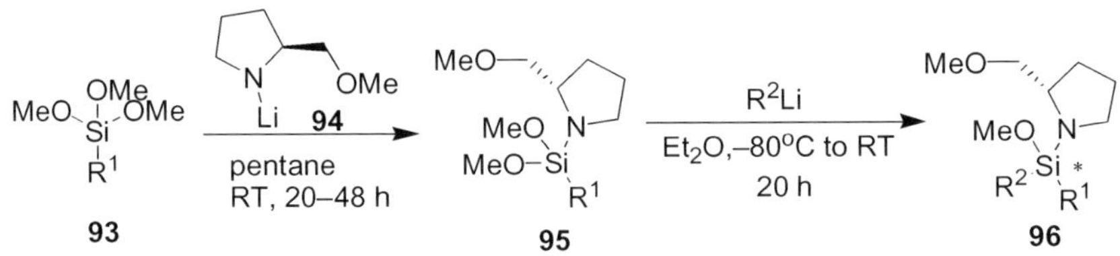

R^1	R^2	yield (%)	d.r.
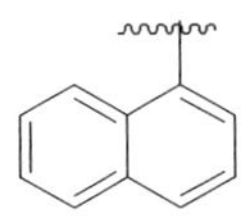	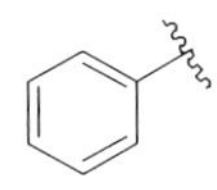	93	88:12
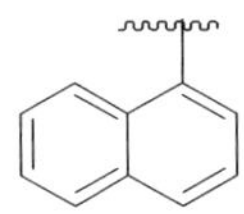	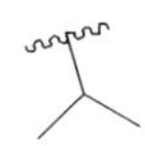	91	95:5
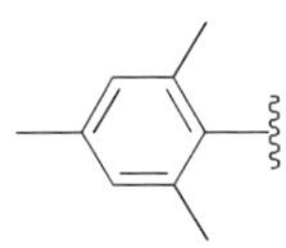	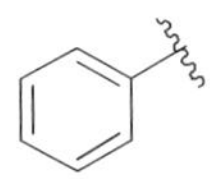	81	>99:1
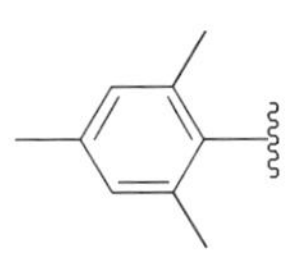	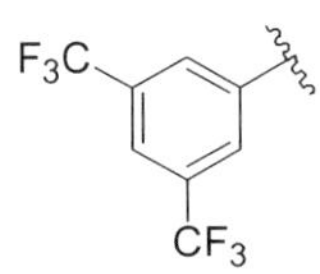	84	99:1

Scheme 4.20
Diastereoselective synthesis of silicon-stereogenic N,O-functionalized organosilanes.

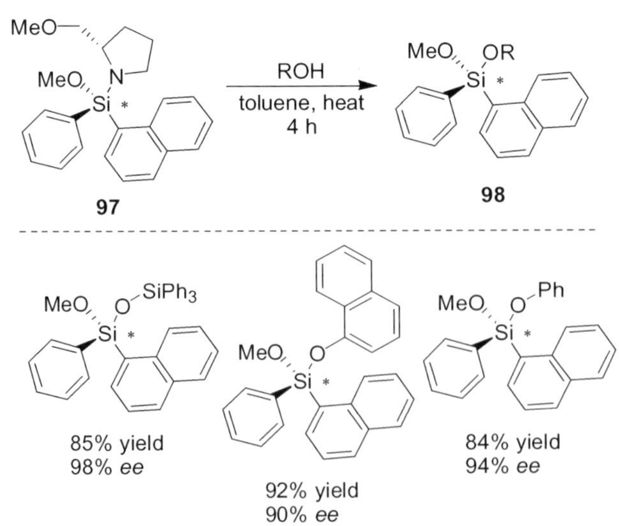

Scheme 4.21
Stereoselective synthesis of highly enantiomerically enriched alkoxy-functionalized siloxanes.

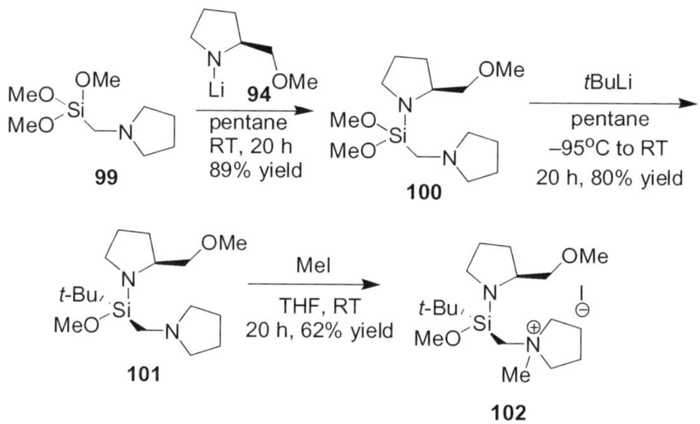

Scheme 4.22
Synthesis of stereochemically pure functionalized organosilicon compounds by substrate-controlled desymmetrization.

4.2.3 Asymmetric Catalysis: Enzymatic Method

Over the past decades, tremendous research efforts have also been made to establish catalytic asymmetric synthesis of enantiomerically pure compounds by enzymatic methods.[35] Since the first example of the enzymatic synthesis of silicon-stereogenic silane in 1985,[36] various enantiomeric excess organosilicon compounds have been synthesized by esterases and lipases.[37] However, the achievement of high level enantioselectivity was not an easy task 30 years ago. Tacke and coworkers greatly contributed to the stereoselective synthesis of chiral silanes by biocatalytic reduction in the 1980s.[38] For example, in 1988, Tacke et al.[38c] found that the achiral acetyldimethylphenylsilane could be reduced enantioselectively by resting free cells of the yeast *Trigonopsis variabilis* (DSM 70714) and bacterium *Corynebacterium dioxydans* (ATC 21766), giving optically active (*R*)-(1-hydroxyethyl)dimethylphenylsilane. Subsequently, Tacke et al. continued to investigate the enzymatic transformation with a new type of silicon-containing organosilicon compound (**103**). This bioconversion approach resulted in the formation of α-silyl substituted secondary alcohols with high enantioselectivity, which represented an efficient method for the preparation of such functional chiral silanes containing alcohol groups. As shown in Scheme 4.23, for the biotransformation with *Trigonopsis variabilis* (DSM 70714), the chiral silane **104** or **105** could be obtained in quite high yields and excellent enantiomeric excess after crystallization.[39]

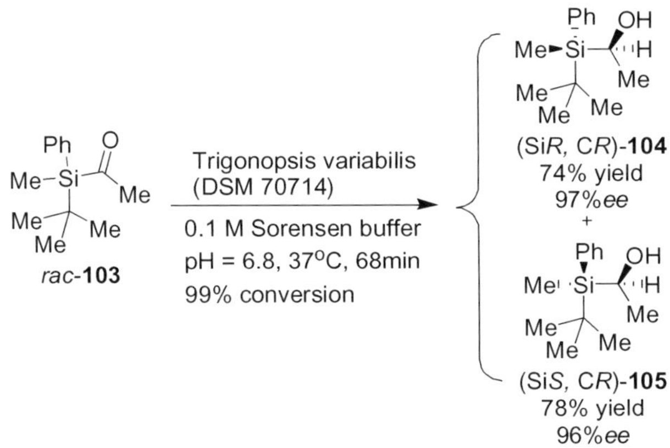

Scheme 4.23
Catalytic synthesis of silicon-stereogenic silanes by enzymatic method.

Among reported methods, the use of lipases constitutes a particularly useful enzymatic approach for the construction of silicon-stereogenic silanes. In this regard, Blanco and Djerourou[40] reported in 1991 that the enzymatic transesterifications could be carried out successfully for the lipases from *Candida cyclidracea (LCC)* and *Chromobacterium viscosum (LCV)*-catalyzed desymmetrization of symmetric 2-sila-1,3-propanediol **106** (Scheme 4.24). The lipase-catalyzed desymmetrization led to the stereoselective formation of monoester **107** with promising enantiomeric excess (up to 75%*ee*).

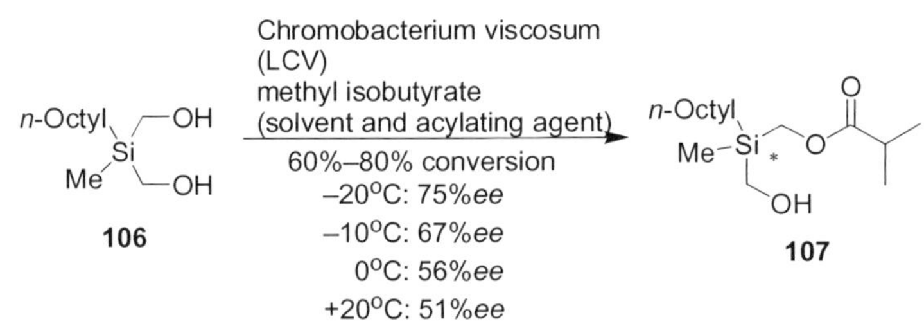

Scheme 4.24
Lipase-promoted desymmetrization of 2-sila-1,3-propanediol.

In 1994, a similar strategy of esterification catalyzed by hydrolase was reported by Tanaka and coworkers.[41] In this work, a commercially available crude *papain* from *papaya latex* was found to be effective and exhibited a good level of enantioselectivity (up to 92%*ee*) with 60% conversion for the esterification of silicon-containing alcohol with 5-phenylpentanoic acid in organic media.

In 2013, Xu and coworkers reported the catalytic synthesis of functional silicon-stereogenic silanes through *Candida antarctica* Lipase B (CAL-B)-catalyzed remote desymmetrization of silicon-centered diols, in which a new family of silicon-containing diols was used for the stereoselective synthesis of chiral functional silanes through enzymatic desymmetrization (Scheme 4.25).[42] The lipase-catalyzed remote desymmetrization was proved to be a conceptually new synthetic methodology in the construction of optically active silicon-stereogenic organosilicon compounds **109** (up to 90:10 *er*).

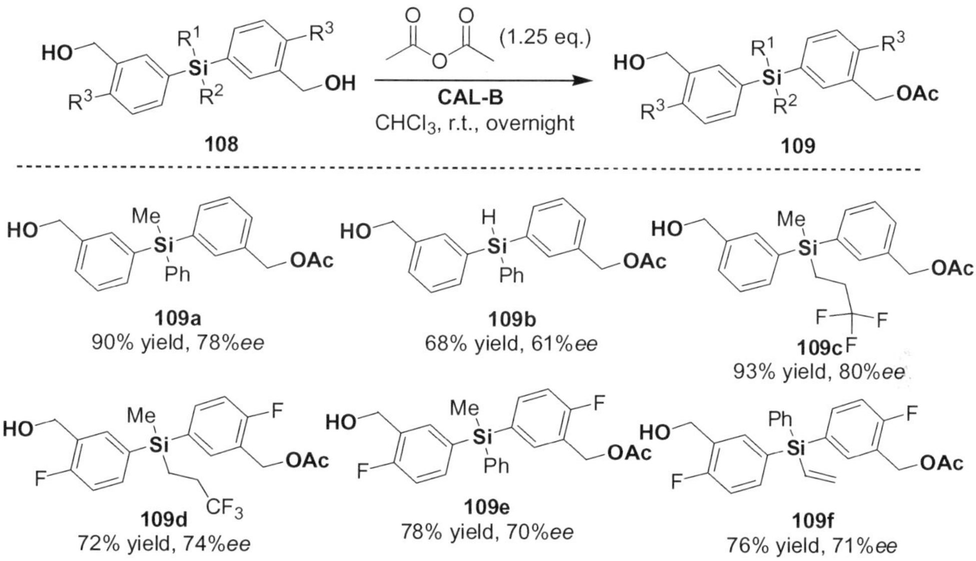

Scheme 4.25
Synthesis of silicon-stereogenic silanes by remote desymmetrization of prochiral silicon-containing diols.

4.2.4 Transition Metal-Catalyzed Synthesis of Silicon-Stereogenic Silanes

Despite the significant efforts that have been put into finding new synthetic methodologies in the past century, the catalytic synthesis of silicon-stereogenic silanes is still in its infant stage. In particular, whereas the preparation of chiral silanes has received much attention in the past 50 years, the synthesis of optically active silicon-stereogenic silanes and their functional derivatives was not an easy task and remained restricted to the optical resolution and kinetic resolution.[43] Fortunately, significant success has been achieved in recent years. Nevertheless, the exploration of novel and efficient catalytic methods with high level of enantioselectivity remains one of the principal challenges of asymmetric catalysis and organosilicon chemistry. In addition, although considerable research efforts have been dedicated to the evaluation of chiral metal catalysts, only a limited number of reports have described catalytic asymmetric synthesis of chiral silanes.

A solution to this synthetic problem has been shown to be based on the selective desymmetrization reaction,[44] especially with dihydrosilanes, and dependence on the catalytic efficiency of transition metal catalysts with chiral ligand. Probably the earliest examples of catalytic asymmetric synthesis of silicon-stereogenic silanes was reported by Corriu and coworkers in 1970s.[45] In this work, the authors used Rh-DIOP (**111**) complex as a catalyst to develop a promising method for enantioselective dehydrogenative hydrosilylation and desymmetrization of dihydrosilanes with alcohols (Scheme 4.26). Although the optical yields of the corresponding product was only up to 19%*ee* by using achiral alcohols and up to 56%*ee* with chiral (−)-menthol as a reagent, it was found that the hydrosilylation could be carried out smoothly with various dihydrosilanes in the presence of Rh/DIOP (**111**) complex. Thus, Corriu et al. provided the first example to demonstrate the catalytic ability of chiral systems for the catalytic synthesis of chiral silanes.

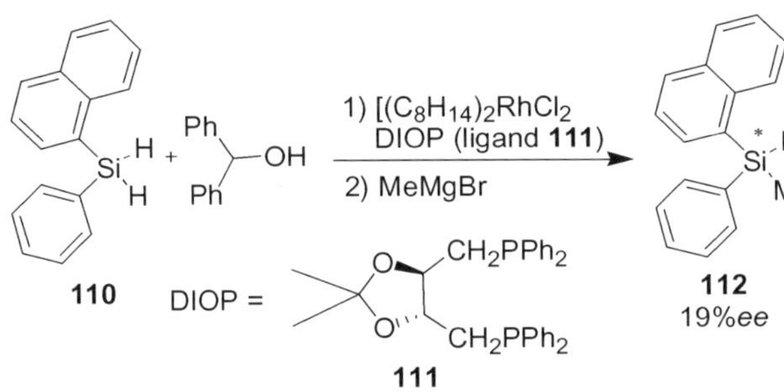

Scheme 4.26
Catalytic desymmetrization of dihydrosilanes with alcohols.

Corriu et al.[46] and other groups made important contributions in the 1960s to 1970s in the field of the asymmetric synthesis of silicon-stereogenic silanes using chiral reagents. For example, Corriu and Moreau[47] reported that the selective alcoholysis of dihydrosilanes led to the facile synthesis of silicon-stereogenic hydrosilanes **112** or **114** in moderate enantioselectivity (48%*ee* or 40%*ee*, respectively) by chiral alcohols or amino alcohols, (−)-menthol or (−)-ephedrine (Scheme 4.27). Although Corriu's method was not practical in the preparation of optically active silicon-stereogenic silanes, the Rh-catalyzed dehydrogenative silylation opened a novel pathway to the construction of silicon-centered stereochemistry by desymmetrization.

Almost at the same time, Corriu et al.[48] and Kumada et al.[49] independently reported the catalytic asymmetric hydrosilylation of ketones with dihydrosilanes for the stereoselective construction of silicon-stereogenic silanes. Unfortunately, the enantiomeric excesses of products were not good in the presence of phosphine ligands, including P-stereogenic phosphine ligand. For example, the rhodium-DIOP complex could promote the desymmetric hydrosilylation in moderate enantioselectivity (32%−40%*ee*).[48] However, the chiral monophosphine ligand **117** (but with only 79%*ee* for the (*R*)-($PhCH_2$)MePhP ligand) resulted in low enantioselectivity (Scheme 4.28, 7%−28%*ee*).[49]

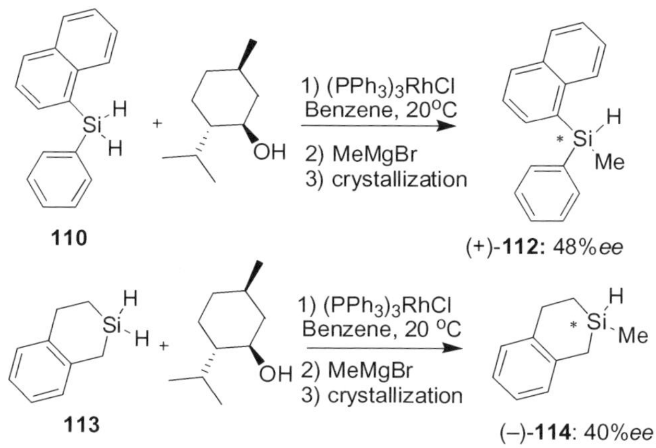

Scheme 4.27
The desymmetrization of dihydrosilanes by Rh-catalyzed dehydrogenative silylation with chiral alcohol.

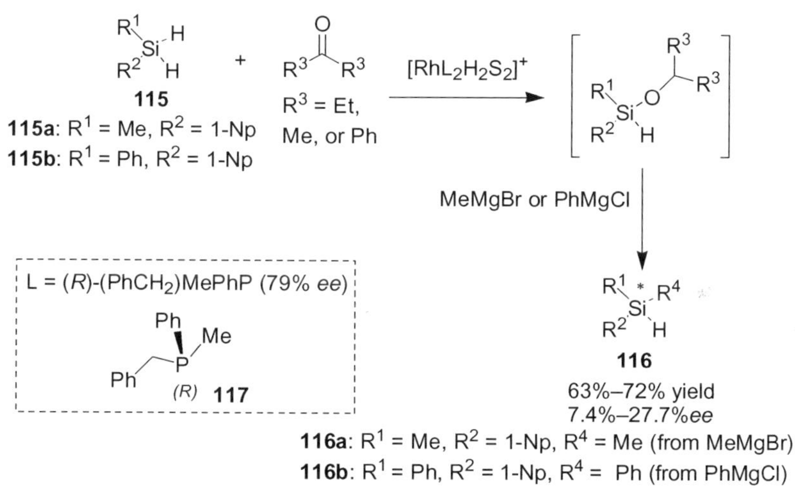

Scheme 4.28
Enantioselective desymmetrization of dihydrosilanes with ketones.

Despite Corriu-Kumada's catalytic desymmetrization of dihydrosilanes that were reported in 1970s, a limited progress was achieved in the synthesis of silicon-stereogenic silanes with high level of enantioselectivity. The main obstacle is probably the high activity of transition metals in the decomposition and nonselective hydrosilylation when using dihydrosilanes as a substrate. In 1994, Takaya et al.[50] found that rhodium(I)-BINAP or rhodium(I)-CyBINAP complexes were excellent catalysts for the hydrosilylation of symmetric ketones, which led to successful stereoselective formation of silicon-stereogenic silanes with excellent enantioselectivities (up to > 99%*ee* in the presence of CyBINAP ligand). As shown in Scheme 4.29, when CyBINAP was used as a chiral ligand, enantiomeric excesses of the corresponding products **118** in the hydrosilylation reaction were higher than those obtained with BINAP. In addition, it is worthy of note that the absolute configuration of the products obtained with (*R*)-CyBINAP are opposite to those with (*R*)-BINAP. It is also especially noteworthy that usually very low *ee*'s were obtained after prolonged reaction times because probably racemization of the products was induced in the presence of the catalyst system.

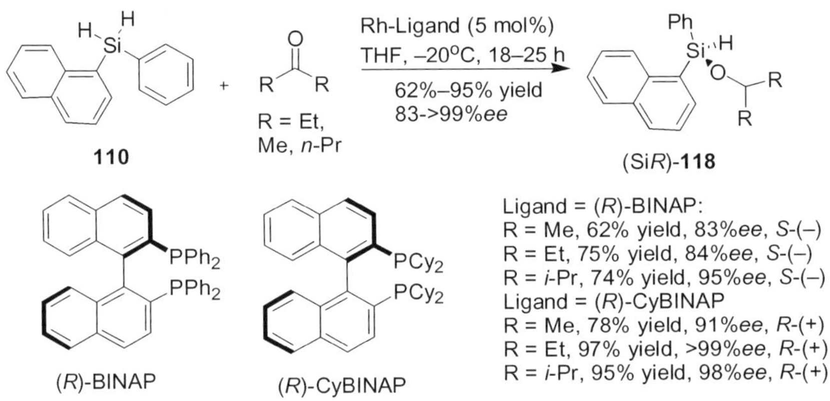

Scheme 4.29
Enantioselective desymmetrization of dihydrosilanes with ketones.

In 1996, a striking methodology was reported by Tamao and coworkers,[51] in which the authors developed a highly efficient enantioselective intramolecular hydrosilylation of bis (alkenyl)dihydrosilane (**119**) by Rh catalyst in the presence of chiral phosphine ligands. The asymmetric intramolecular hydrosilylation of **119** proceeded sequentially to form spirosilane in two steps. As shown in Scheme 4.30, very high enantioselectivities and chemoselectivities were achieved in this reaction by use of the silicon-containing diphosphines **120a–d**, which resulted in 98%–99%*ee* and good yields for the corresponding spirosilane products **121** (Scheme 4.30).

In 2010, inspired by enantioselective Si–H insertion reaction with metal carbenoid,[52] Katsuki and coworkers[53] described a novel strategy to construct functional silicon-stereogenic organosilicon compounds via catalytic asymmetric Si–H insertion of dihydrosilanes with α-diazocarbonyl compounds by chiral iridium(III)-salen complex (Scheme 4.31). In this work, Katsuki and coworkers emphasized the desymmetrization of prochiral dihydrosilanes **121** when Ir(salen) complex (**124**) was used as a chiral catalyst. Fortunately, only one of the four possible diastereomers **123** was produced exclusively, thus the excellent diastereo- and stereoselectivity was achieved in this reaction.

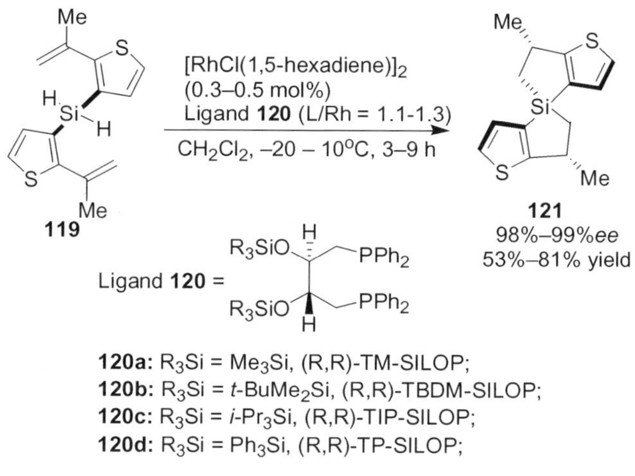

Scheme 4.30

Synthesis of chiral silane via asymmetric intramolecular hydrosilylation of bis(alkenyl)silane.

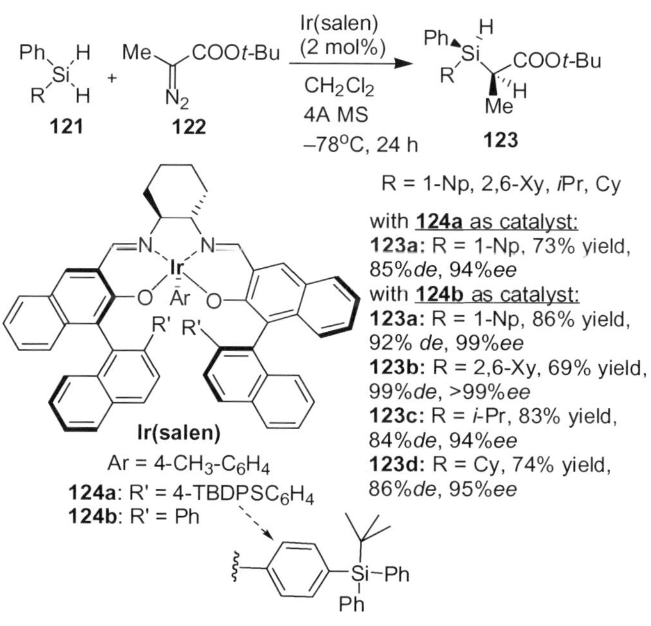

Scheme 4.31

Desymmetrization of prochiral dihydrosilane with α-diazocarbonyl compound.

Perhaps one of the most exciting recent studies on enantioselective synthesis of silicon-stereogenic silanes was published by the group of Shintani and Hayashi.[54] As an important example in this field, Hayashi and coworkers reported in 2011 that the use of a chiral palladium as catalyst for the enantioselective desymmetrization of silacyclobutanes led to a highly efficient desymmetrization reaction for the construction of silicon-stereogenic

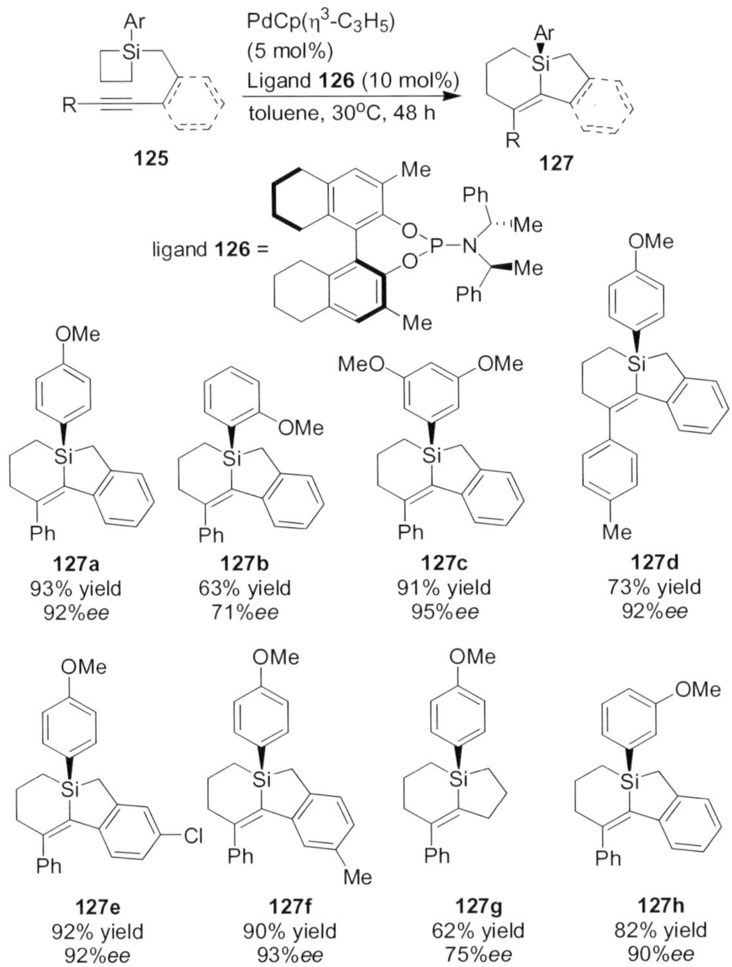

Scheme 4.32
Palladium-catalyzed desymmetrization of silacyclobutanes.

silacycles.[55] This intramolecular desymmetrization reaction was found to proceed via oxidative addition of a silicon–carbon bond of silacyclobutane to palladium(0), following intramolecular insertion of the alkyne substrate, and subsequent reductive elimination along with regeneration of palladium(0) to give chiral silacycles possessing a tetraorganosilicon stereocenter. However, the detailed mechanism is unclear at present. Among the tested phosphines serving as chiral ligands in the palladium complex in this work, BINOL-derived ligand **126** was found to be highly effective on the enantioselective induction in the desymmerization reaction of silacyclobutanes. As shown in Scheme 4.32, most silacycle products **127** with a silicon stereogenic center were obtained in excellent enantioselectivities (up to 95%*ee*) and good yields (62%–93% yields).

On the basis of their previous striking work of intramolecular desymmetrization of silacyclobutanes **128**,[55] Shintani and Hayashi further demonstrated in 2012 the potential of desymmetrization with intermolecular palladium-catalyzed desymmetrization of silacyclobutanes with alkynes.[56] With the chiral palladium catalyst system ($PdCp(\eta^3\text{-}C_3H_5)$ and ligand **126**), they developed an intermolecular cycloaddition reaction starting from 1-aryl-1-alkylsilacyclobutane **128** and dimethyl acetylenedicarbonylate **129**. The desymmetrization reaction occured with high level of enantioselectivity (Scheme 4.33). For example, the corresponding product **130a** could be obtained in 92%*ee* and 95% yield when the reaction temperature was suitable (10°C). Then the experimental results of substrate scope showed that various 1-alkyl-1-arylsilacyclobutanes **128** were tolerated for the ring-opening desymmetrization (Scheme 4.33), in which various silicon-stereogenic 1-sila-2-cyclohexenes **130** were formed with uniformly excellent enantioselectivities (90%–94%*ee*) and high yields (90%–95% yields). Thus, this synthetic approach featured readily available starting materials and high enantioselectivities.

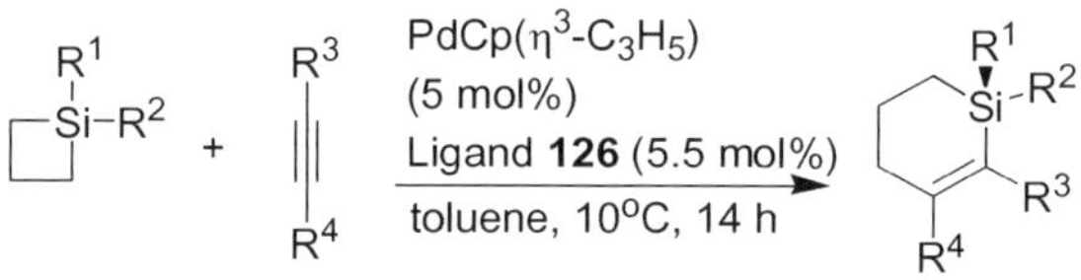

128 **129** **130**

130a: R^1 = 4-$MeOC_6H_4$, R^2 = Me, R^3 = R^4 = CO_2Me
95% yield, 92%*ee*
130b: R^1 = 4-$MeOC_6H_4$, R^2 = Me, R^3 = R^4 = CO_2Et
90% yield, 90%*ee*
130c: R^1 = 4-$CF_3C_6H_4$, R^2 = Me, R^3 = R^4 = CO_2Me
95% yield, 92%*ee*
130d: R^1 = 4-$MeOC_6H_4$, R^2 = $CH_2CH=CH_2$, R^3 = R^4 = CO_2Me
91% yield, 94%*ee*

Scheme 4.33
Palladium-catalyzed asymmetric synthesis of Si-stereogenic silacyclohexenes.

In 2012, Hayashi and Shintani[57] made a significant breakthrough in the synthesis of silicon-stereogenic silanes through catalytic desymmetrization of functional organosilicon compounds, in which the enantioselective C–H bond functionalization of 2-(arylsilyl)aryl triflates was successfully achieved. As shown in Scheme 4.34, such C–H activation-based cross-coupling transformation proceeded smoothly with catalytic $Pd(OAc)_2$ in the presence of 2 equivalents of Et_2NH in toluene. Similarly to previous reports on the C–H bond activation and intramolecular arylation reported by Shimizu et al.,[58] Hayashi and coworkers tested the chiral phosphine ligands in this work, and found that the use of Josiphos-type (R,S_p)-**132** resulted in good yields and excellent enantioselectivities for the

Pd-catalyzed intramolecular C−H bond activation and intramolecular arylation reaction of prochiral aryl triflates containing a (*tert*-butyl)diphenylsilyl group.[59] Under the optimized reaction conditions, various substituted benzosiloles with a broad range of groups on the aromatic ring were obtained with up to 98%*ee* by using Josiphos-type ligands (Scheme 4.34).

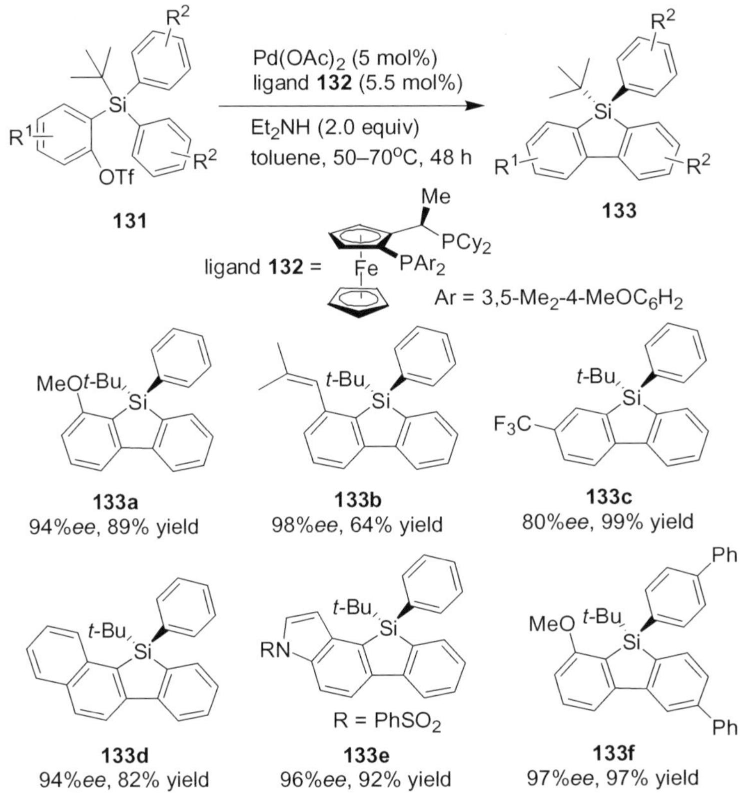

Scheme 4.34
Palladium-catalyzed asymmetric synthesis of Si-stereogenic dibenzosiloles.

Hayashi and Shintani also developed a rhodium-catalyzed asymmetric synthesis of silicon-stereogenic silanes in 2012.[60] The authors performed an interesting enantioselective transmetalation, in which the hydroxyl-tethered tetraorganosilanes bearing two identical aryl groups worked as the transmetalating reagents in rhodium-catalyzed transmetalation-based silylation (Scheme 4.35). With $[Rh(OH)(coe)_2]_2$ (coe: cyclooctene) and ligand **135** ((*S,S*)-Me-Duphos) in hand, they observed that good enantioselectivity for various substituted dibenzooxasilines **136** containing Si-stereogenic moiety can be achieved within the substrate scope investigation (71%−92%*ee*).

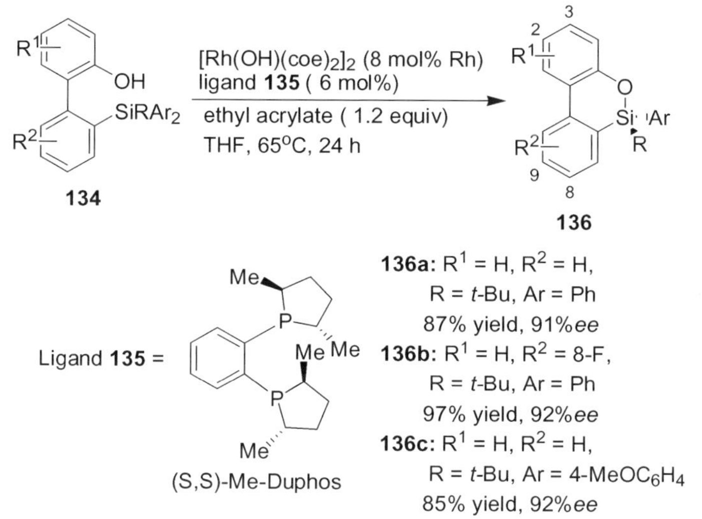

Scheme 4.35
Rhodium-catalyzed asymmetric synthesis of Si-stereogenic dibenzooxasilines.

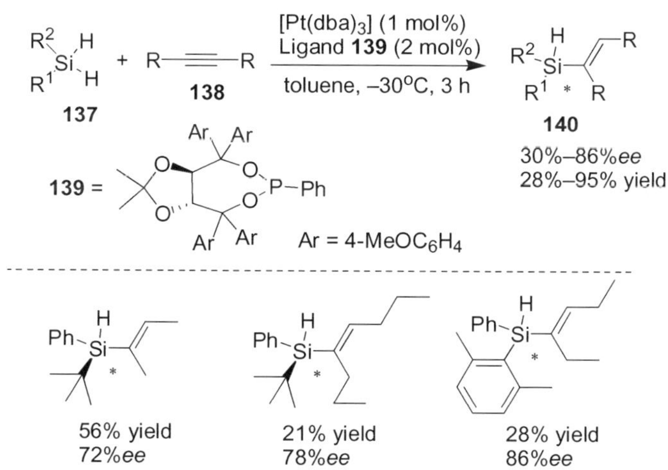

Scheme 4.36
Catalytic desymmetrization of dihydrosilanes by Pt-catalyzed hydrosilylation [dba: dibenzylideneacetone].

Recently, Tomooka and coworkers have developed a catalytic desymmetrization of dihydrosilanes based on Pt-catalyzed hydrosilylation of alkynes with TADDOL-based (TADDOL: $\alpha,\alpha,\alpha',\alpha'$-tetraaryl-1,3-dioxolane-4,5-dimethanol) phosphonite ligands (Scheme 4.36).[61] This new approach allows for the catalytic synthesis of chiral alkenylhydrosilane

140 in an enantioselective fashion with high atom efficiency. Despite this method affording a multifunctionalized nonracemic alkenylhydrosilane in only moderate to good enantioselectivities (30%–86%*ee*), it is really a synthetically valuable procedure.

Although palladium-catalyzed silicon-carbon bond-forming reaction through Si–H/C–I exchange demonstrated the potential use of hydrosilanes as silylating reagents 20 years ago,[62] limited progress has been made in catalytic desymmetrization of dihydrosilanes. In this regard, Yamanoi and Nishihara successfully developed a special desymmetrziation approach by silicon-carbon bond-forming cross-coupling reaction, in which an enantioselective silylation of arenes with dihydrosilanes was achieved in the presence of a $Pd_2(dba)_3$-phosphoramidite ligand to afford the optically active tertiary silanes with promising *ee* values (Scheme 4.37, 11%–77%*ee* for 13 examples).[63]

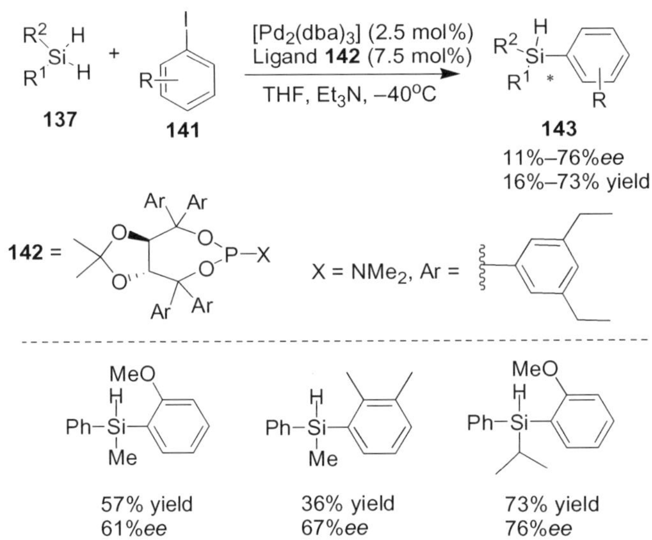

Scheme 4.37
Desymmetrization of dihydrosilanes by catalytic asymmetric arylation of secondary silanes.

In the same year, Tobisu and Chatani reported a rhodium-catalyzed coupling reaction of 2-(dimethylisopropylsilyl)phenylboronic acid **144** with internal alkynes, which provides a facile approach for the synthesis of 2,3-disubstituted benzosilole derivatives. On the basis of Rh-catalyzed cross-coupling reaction, the authors found that various functional groups are compatible in this reaction. Sequential twofold coupling enables modular synthesis of asymmetrically substituted 1,5-dihydro-1,5-disila-s-indacene, a π-extended molecule of interest in organic electronics. As shown in Scheme 4.38, the potential use of the desymmetrization reaction in catalytic asymmetric synthesis of silicon-stereogenic benzosiloles is demonstrated by Tobisu and Chatani.[64]

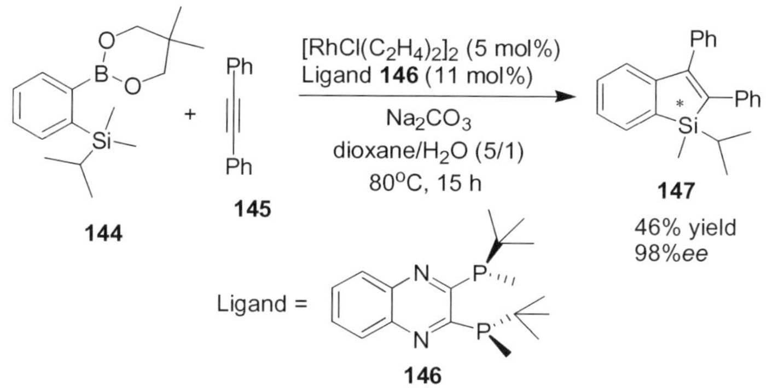

Scheme 4.38
Rhodium-catalyzed carbon-silicon bond activation for asymmetric synthesis of silicon-stereogenic benzosilole derivative.

A striking example was reported by Kuninobu and coworkers in 2013.[65] The authors demonstrated a facile and atom-economical synthesis of a quaternary silicon-containing spirosilabifluorene derivative **149** from bis(biphenyl)silane **148** by catalytic desymmetrization, double dehydrogenative silylation using rhodium catalyst. More importantly, the authors succeeded in the synthesis of silicon-stereogenic spirosilabifluorene derivatives **149a**–**e** with good enantioselectivity (Scheme 4.39, 70%–81%*ee*).

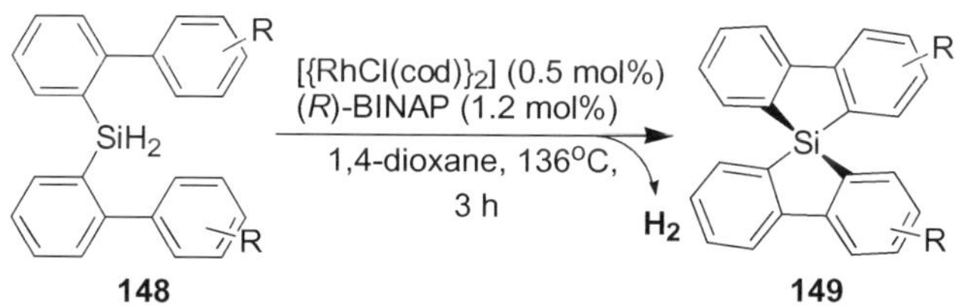

149a, R = 4-OMe, 95% yield, 81%*ee*;
149b, R = 4-*t*-Bu, 94% yield, 78%*ee*;
149c, R = 4-CF_3, 90% yield, 75%*ee*;
149d, R = 4-Ph, 90% yield, 70%*ee*;
149e, R = 2-OMe, 73% yield, 77%*ee*

Scheme 4.39
Synthesis of silicon-stereogenic spirosilabifluorene derivatives by catalytic silylation.

In 2015, Shintani and Nozaki reported a rhodium-catalyzed enantioselective synthesis of silicon-stereogenic dibenzosiloles through a [2 + 2 + 2] cycloaddition of silicon-containing prochiral triynes with internal alkynes. As shown in Scheme 4.40, high yields and enantioselectivities could be achieved by employing an axially chiral monophosphine ligand (up to 96%*ee*).[66] In 2015, similarly to the rhodium-catalyzed [2 + 2 + 2] cycloaddition, the same authors have also developed rhodium-catalyzed intramolecular alkynylsilylation of

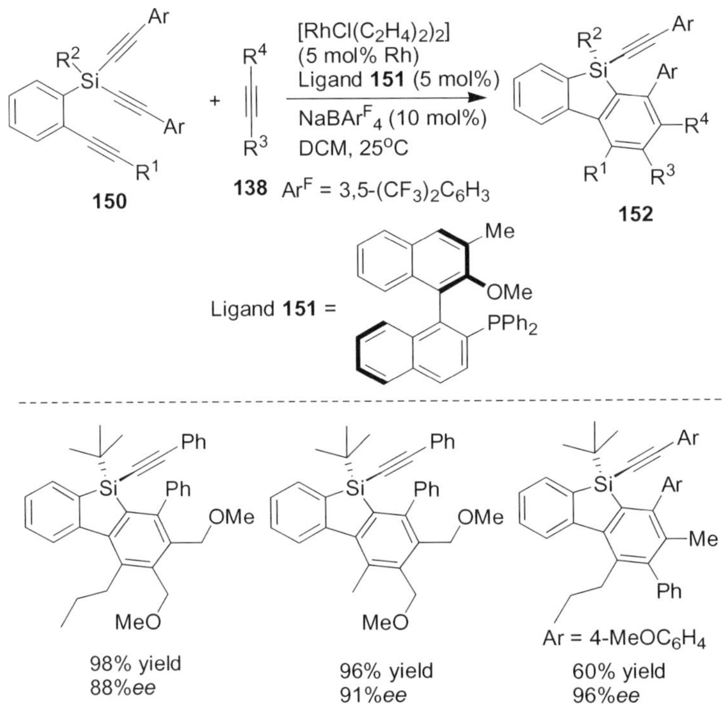

Scheme 4.40
Rhodium-catalyzed asymmetric synthesis of silicon-stereogenic dibenzosiloles by enantioselective [2 + 2 + 2] cycloaddition (DCM: dichloromethane).

alkynes under mild conditions.[67] It is recognized as the first alkynylsilylation of alkynes via the cleavage of a C(sp)−Si bond by transition-metal catalysts. Although the substrate scope is quite limited, this work still represents an important progress in this field.

4.2.5 Other Synthetic Methods by Transformation of Chiral Organosilicon Compounds

In the 1960s, Sommer and coworkers greatly contributed to the stereochemistry of chiral silanes. They found that many functional chiral organosilicon compounds with various substituents could be obtained by successive stereoselective transformation of functional groups on chiral organosilicon molecules, including chlorination, substitution reactions, or coupling with alkyllithium reagents and chiral hydrosilanes.[68] These findings laid the groundwork for chiral organosilicon chemistry. Subsequently, on the basis of Sommer's findings, other groups have successfully prepared functional silicon-stereogenic silanes containing vinyl, dienyl, and other reactive groups by the nucleophilic substitution reaction with organolithium reagents.[69]

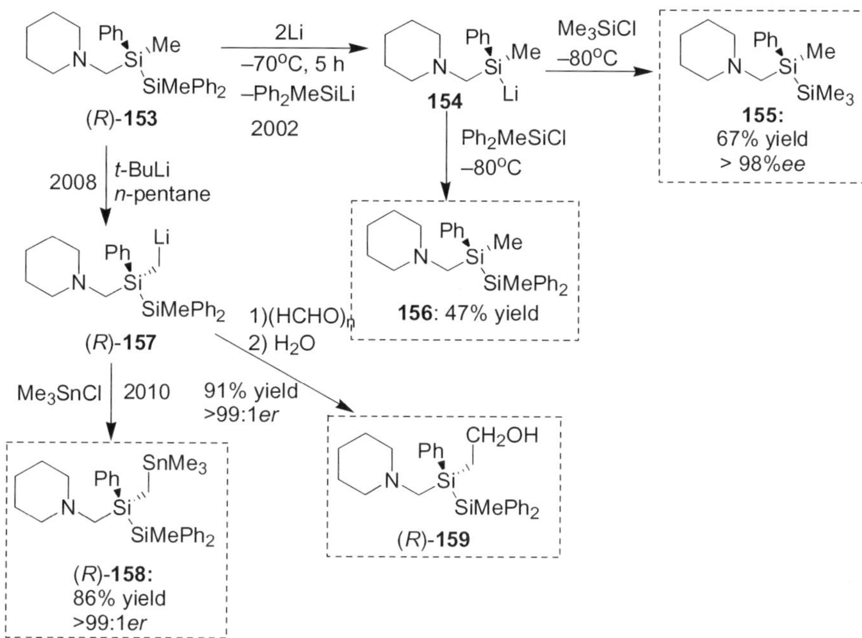

Scheme 4.41

The synthesis of silicon-stereogenic organosilicon compounds by metalation.

On the basis of the early studies of silicon-based stereochemistry and synthetic methods, in 2002 Strohmann et al.[70] expanded Sommer's method[71] to prepare the enantiomerically enriched silyllithium **154** or **157**, useful chiral reagents for the introduction of silyl groups to the synthesis of functional organosilicon compounds (Scheme 4.41). In this work, the retention of configuration at the silicon center was observed if a transmetalation step occurred with $MgBr_2(thf)_4$ in the reaction. It was found that the chiral silyllithium reagent **154** bearing silicon-stereogenic center could be transformed into various organosilicon compounds (Scheme 4.41).[72] Similarly to conversion of chiral organosilicon compound **154**, Strohmann et al.[73] also reported the direct α-lithiation of tertiary amine-substituted silane **157** as an efficient method for the preparation of silicon-stereogenic silane derivatives with functional groups, such as $-CH_2OH$ (**159**), $-SnMe_3$ (**158**).

In 2005, Oestreich and coworkers[74] found that the structural modification of chiral organosilicon compounds could prevent racemization by lithium chloride (t $< -40°C$) as well as dimerization ($t < -100°C$), which allows the authors to prepare an asymmmetrically substituted silyl anion with silicon-centered chirality by reductive metalation of a silyl chloride. In fact, the transformation of optically pure silyllithiums bearing stereogenic silicon center is a useful approach to introduce a chiral silicon center to the novel organosilicon compounds containing silicon-silicon bonds. In 2000–04, Kawakami and coworkers[75] reported several important examples of the silicon-based transformation of configurationally stable silyllithiums derived from chiral hydrosilanes. Especially in 2004, as shown in Scheme 4.42, the authors found that the stereochemistry of the transformation of chiral chlorosilane (*S*)-**161** (>99%*ee*) upon addition of achiral methyldiphenylsilyllithium **160** is largely different from that of the reaction of achiral methyldiphenylchlorosilane **163** with optically active silyllithium **162** (90%*ee*). In this work, Kawakami and coworkers[75e] also reported the selective functionalization of silicon-stereogenic disilanes by the cleavage of silicon-naphthyl bonds with bromine (Scheme 4.43). The regioselective cleavage reaction of the silicon-naphthyl bond of (*R*)-**165** (95%*ee*) upon its bromination afforded the functional organosilicon compound (*R*)-**166** in 68% yield and 84%*ee*.

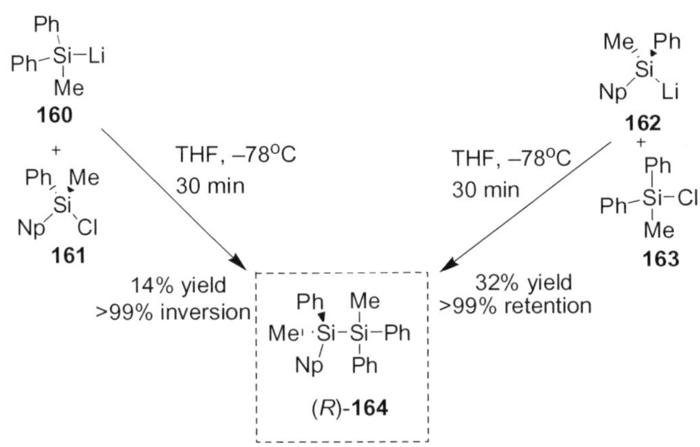

Scheme 4.42
Synthesis of silicon-stereogenic organodisilane by metalation.

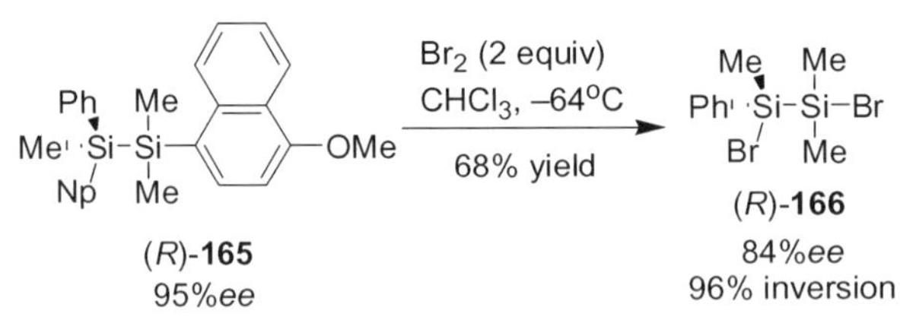

Scheme 4.43
Bromination of silicon-stereogenic organodisilane to form novel organosilicon compounds.

In 2010, Tomooka and coworkers[76] developed a new transformation based on reductive metalation of chiral chlorosilane, which provided a facile procedure to the synthesis of chiral silacarboxylic acids and their ester derivatives (Scheme 4.44). For example, during the synthesis of silicon-stereogenic silacarboxylic acid (**171**) and its ester derivative (**172**), five different functional chiral silanes have been prepared from the chiral silanol through five-step transformation, such as methylation with MeI, reduction with $LiAlH_4$, oxidation and chlorination with CCl_4, reductive metalation and then carbonylation with CO_2, and esterification by Mitsunobu reaction.[76] These results opened up the possibility for the preparation and synthetic use of complex and biologically active organosilicon compounds in enantiomerically enriched forms.

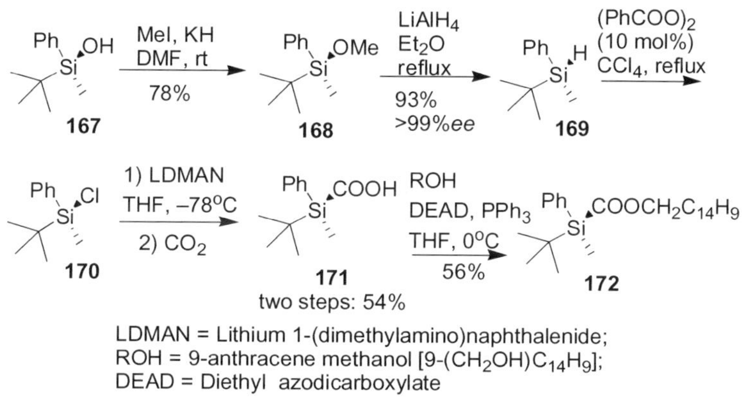

Scheme 4.44
The synthesis and application of silicon-stereogenic silanes.

The selective oxidation of chiral hydrosilanes is useful but a challenging reaction in organosilicon chemistry. There are few reports on the investigation of oxidation of chiral organosilicon compounds. As an important transformation of chiral hydrosilane, the oxidation of silicon-stereogenic hydrosilanes can potentially provide the optically active silanols, key intermediates in the preparation of useful building blocks of silicon-containing drug candidates in organosilicon chemistry and pharmacology.[77] Although the direct stereospecific oxidation of enantioenriched hydrosilanes is in prospect, the direct synthesis of silicon-stereogenic silanols was quite difficult.[78] For example, Adam and Curci have tried to develop a synthetic method to obtain the target through the oxidation of optically active silane with dioxirane under mild condition (Scheme 4.45).[79] In this work, it was found that the chiral hydrosilane **173** could be quantitatively and stereospecifically converted into the corresponding silanol **174**. In analogy to previous report, the same authors also found that the methyltrioxorhenium (MTO)-catalyzed oxidation of chiral silanes to silanols proceeded smoothly with urea/hydrogen peroxide adduct (UHP) as oxygen source, and high conversion and excellent selectivities were achieved to form the chiral silanol **174** (94%*ee*).[80]

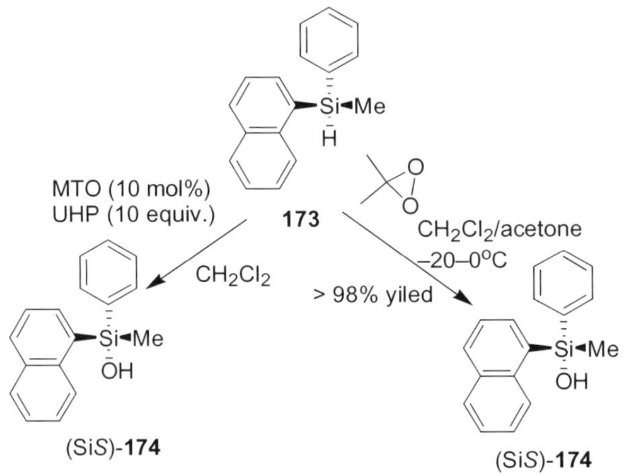

Scheme 4.45
The synthesis of silicon-stereogenic silanol by oxidation.

In the field of organosilicon chemistry, Brook and retro-Brook rearrangements are the important reactions. Especially, the retro-Brook reaction, in which the trialkylsilyl group shifts from the oxygen atom of a silyl ether to a carbanion, has been recognized as useful transformation for the preparation of functional organosilanes.[81] For example, Tomooka and coworkers[82] utilized the retro-Brook rearrangement of a simple allyloxysilane **175** as facile approach to prepare a new silicon-stereogenic organosilicon compound **176** (Scheme 4.46). Notably, the retro-Brook rearrangement proceeded with retention of configuration at the silicon center.

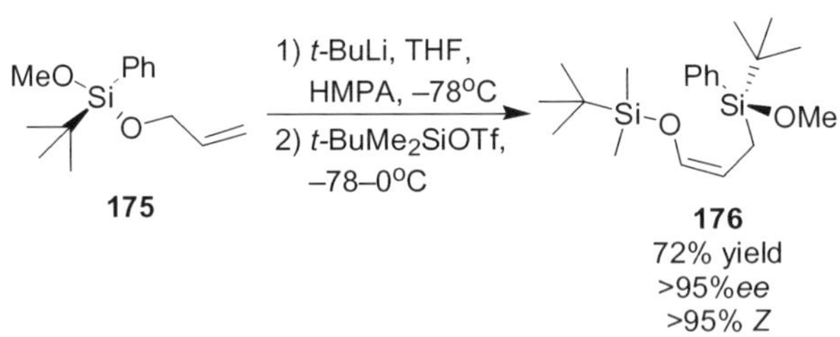

Scheme 4.46
The synthesis of novel silicon-stereogenic compound by retro-Brook rearrangement.

4.3 The Application of Chiral Organosilicon Compounds

Functional silicon compounds are generally known to be highly useful molecules in material chemistry and organic synthesis. Modern organic synthesis has witnessed a wide application of organosilicon compounds in organic synthesis.[83] In the past decades, the application of chiral silicon systems, especially silicon-stereogenic silanes, has also attracted much attention and special interest due to their wide application as chiral auxiliaries in stereoselective synthesis and asymmetric catalysis. For example, the application of silicon-stereogenic silanes as a chiral reagent or auxiliary in organometalic synthesis,[84] and polymer chemistry[85] has offered extremely useful methods in the field of organic synthesis and advanced materials. In addition, the concept of silicon-to-carbon chirality transfer of silicon-stereogenic silanes has been proved to be an extraordinarily useful approach for the synthesis of exciting compounds owing to the high level of stereoselectivity in silicon-involving organic transformations. Early in 1992, Chan and Wang[86] have reviewed the utilization of chiral organosilicon compounds as chiral auxiliaries or reagents in asymmetric synthesis. The silicon-stereogenic organosilicon moiety was proved to be very interesting because of selective controlling of both the course of the reaction and the stereoselectvity. Thus, silicon-stereogenic silanes could be used in the asymmetric reduction of carbonyl compounds as chiral reducing agent,[87] asymmetric nucleophilic addition to chiral acylsilanes or α-silyl thione,[88] selective epoxidation of chiral alkenylsilane,[89] catalytic transformation of silyl enol ethers bearing silicon-stereogenic center,[90] carbonylation of chiral α-silylcarbanions,[91] and intermolecular Si–H insertion reaction.[92] These reactions were intensively studied for more than 20 years. Notably, the selective transformation of silicon-stereogenic silanes, such as metalation, oxidation, and rearrangement, etc., facilitated the synthesis of novel and downstream functional chiral silanes and materials bearing stereogenic silicon atom.

Herein, we chose to provide representative examples of the application of silicon-stereogenic silanes in asymmetric synthesis, reported within the recent two decades. At the beginning, e.g., in the 1990s, Bienz and coworkers reported the synthesis of a series of silicon-stereogenic acylsilanes and related stereoselective preparation of α,δ-silylated γ,δ-unsaturated carboxylic acids,[93] and (−)-(*R*)-phenyl 2-phenylpropyl ketone.[94] As shown in Scheme 4.47, the [(benzyloxy)methyl](*tert*-butyl)methylsilyl group could be used as an excellent chiral auxiliary to affect the diastereoselective conjugate additions of organocuprates to enones, and the corresponding (−)-(*R*)-phenyl 2-phenylpropyl ketone **179** was formed with almost no loss of optical purity in the multistep reaction sequence involving the conjugate addition of cuprate, hydrolysis, and removal of the silicon group (Scheme 4.47). In 2002, the same group anchored the silicon-stereogenic auxiliary to the α- or β-hydroxyketones as a chiral silyl ethers **180** bearing a ketone functionality. The subsequent 1,2-addition of Grignard reagents to the chiral ketone **180** afforded the corresponding products **181a** and **181b** with promising diastereoselectivities (up to 92:2 *dr*) (Scheme 4.48).[95]

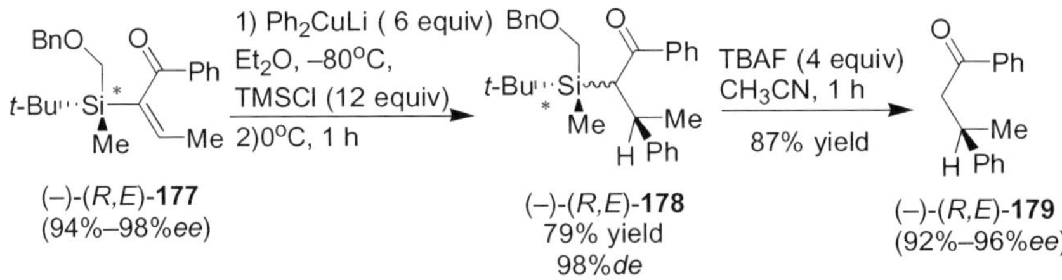

Scheme 4.47

The synthesis of chiral ketone from silicon-stereogenic enone **177**.

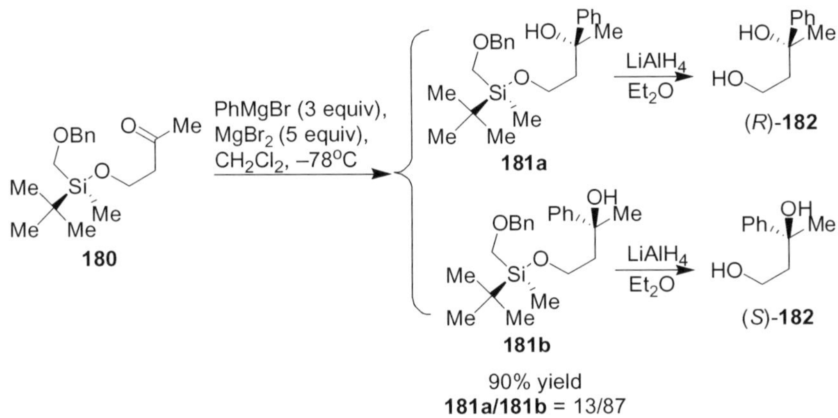

Scheme 4.48

The synthesis of chiral tertiary alcohols.

In the 21st century, a growing interest in the application of chiral organosilicon compounds in chiral synthesis has emerged. In 2002, Mori et al.[96] reported an example on the chirality transfer from silicon to carbon via stereoselective Simmons–Smith cyclopropanation of chiral alkenylsilanols (Scheme 4.49). The effect of substituent in silicon-stereogenic silanol was notable, resulting in the moderate (84:16 *dr*) selectivity for the Simmons–Smith cyclopropanation when a 3,3,3-trifluoropropyl group was employed on the chiral silanol. Fortunately, higher selectivity was achieved in the reaction of **183b**, bearing a cyclohexyl substitutent, and a single diastereomer was formed (Scheme 4.49). In the last step, the oxidative transformation of **184b** gave cyclopropanol **185** with good enantioselectivity (97%*ee*) under Tamao's oxidation conditions[97].

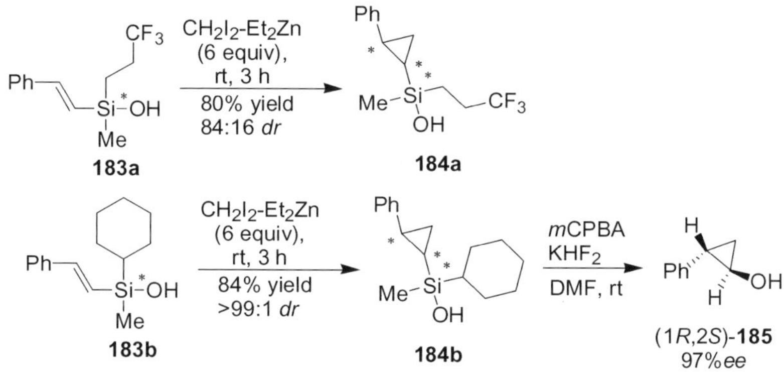

Scheme 4.49
The synthesis of chiral cyclopropanol derivative.

In 2003, Leighton et al.[98] studied the diastereomerical transformation of the silicon-stereogenic silane **186** to alcohols via the silylformylation-allylsilylation reaction under the optimized conditions ($[(PhO)_3P]_2Rh(CH_3COCH_3)_2]BF_4$, 900 psi, Co, benzene, 60°C) (psi: pounds per square inch), and the authors found that the resulting 1,5-*syn*-diol **187** could be obtained in promising yields without the loss of enantiomeric excess in all cases (Scheme 4.50).

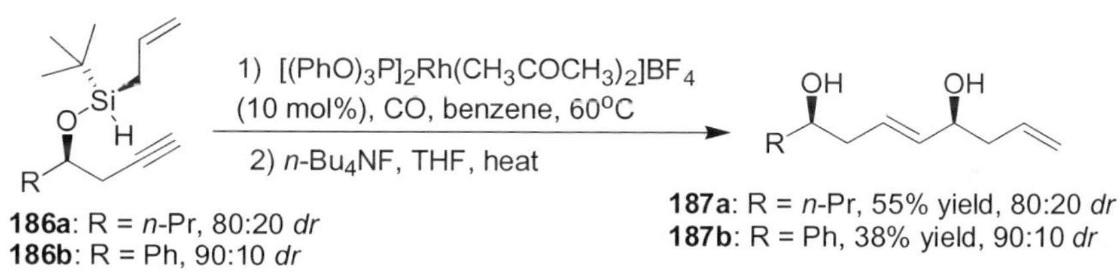

Scheme 4.50
Silylformylation-allylsilylation reactions of diastereomerically enriched silanes.

Since 2003, the concept, called as the chirality transfer from silicon to carbon of silicon-stereogenic moiety in organic synthesis, has been systematically forwarded by Oestreich and coworkers. With their novel cyclic chiral hydrosilanes in hand, they investigated asymmetric silylation of alcohols and hydrosilylation of olefins for the synthesis of corresponding chiral alcohols and silanes.[99] As shown in Scheme 4.51, the palladium-catalyzed hydrosilylation of norbornene derivative with chiral cyclic silanes was proved to be quite good in the transferring of chirality from silicon to carbon.[100] In recent years, the chirality transfer with silicon-stereogenic silane **193** or **194** was found to be exceptionally high (Scheme 4.52, 98%–99%*ct, ct = chirality transfer*).[101] The structural features of the

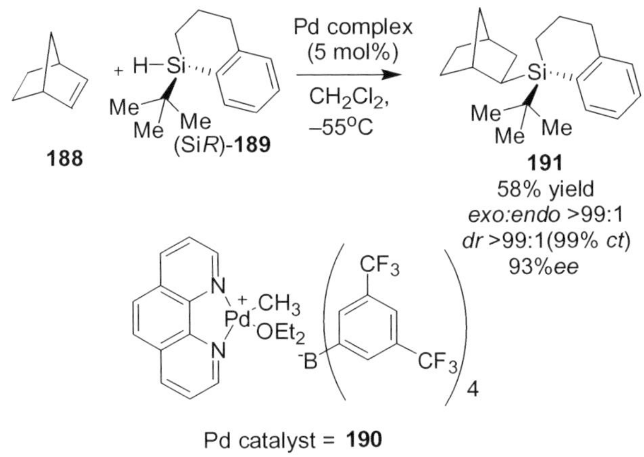

Scheme 4.51
The chirality transfer from silicon to carbon via catalytic asymmetric hydrosilylation.

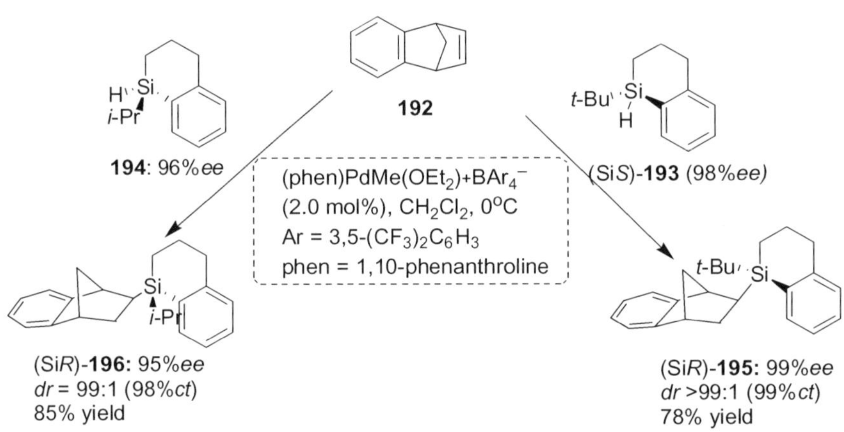

Scheme 4.52
The chirality transfer from silicon to carbon via catalytic asymmetric hydrosilylation.

silicon-stereogenic silane unit is essential for the achieving high diastereoselectivities in catalytic asymmetric hydrosilylation reaction. Also reported by Oestreich and coworkers, remarkably high stereocontrol was found in the double hydrosilylation of norbornadiene with silicon-stereogenic silanes.[102]

The application of the silicon-stereogenic silanes in the kinetic resolution of secondary alcohols and the asymmetric dehydrogenative silylation of alcohols were also attractive, and a great deal of interest in developing stereoselective processes in these reactions has emerged.[103] For example, the rhodium complex catalyzed dehydrogenative silylation of alcohols manifested remarkably high levels of diastereoselectivity, and its wide application to the kinetic resolution of functionalized alcohols containing pyridine moiety was proved to be practical because of extremely high stereoslectivity (up to 99%*ee*) with good selectivity factors (Scheme 4.53, s > 50, up to 900) ($s = \boldsymbol{K}_{rel}$, that refers to "relative rate of kinetic resolution"). As shown in Scheme 4.53, on the basis of copper-catalyzed silylation, kinetic resolution of similar nitrogen-donor-functionalized secondary and tertiary alcohols proceeded with the good enantioselectivity (70%–94%*ee*).[104]

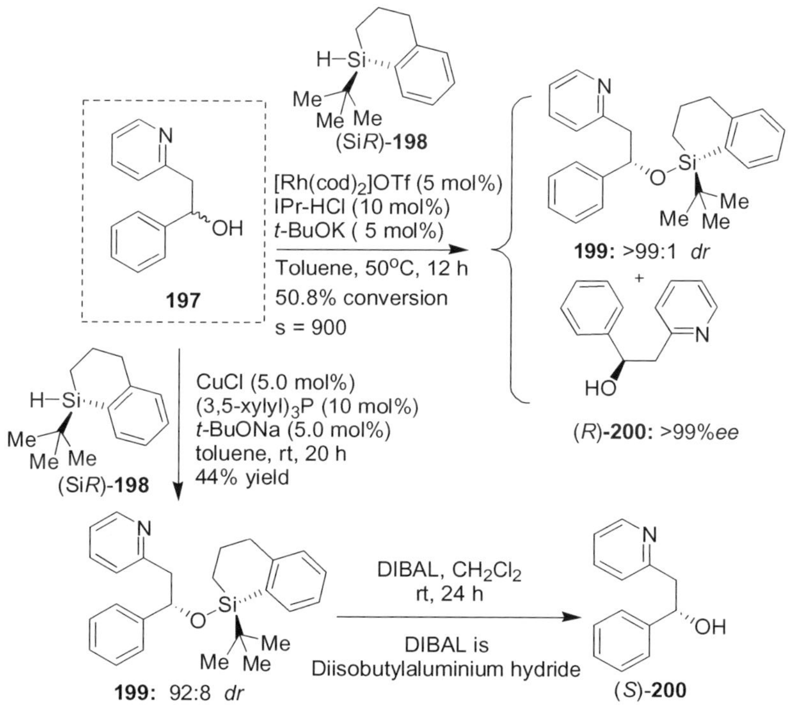

Scheme 4.53
Dehydrogenative silylation of alcohols with stereogenic silanes.

In 2008, Oestreich et al. reported that the chiral silane could be used as a reducing agent for the synthesis of chiral alcohol.[105] However, the enantioselectivity was not good for the $B(C_6F_5)_3$-catalyzed hydrosilylation of prochiral acetophenone using a silicon-stereogenic cyclic silane (Scheme 4.54).

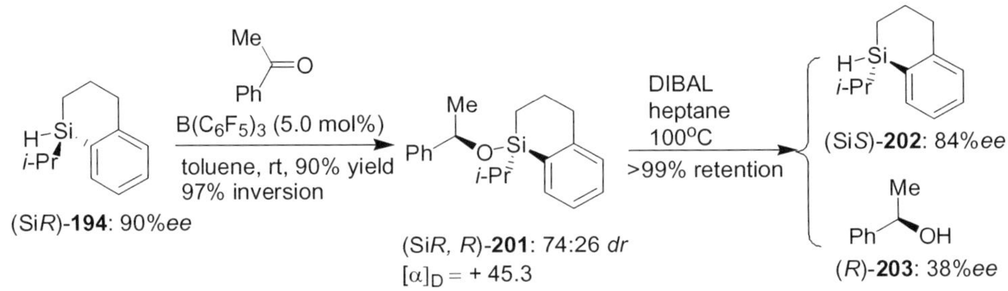

Scheme 4.54

Catalytic asymmetric hydrosilylation of ketone with silicon-stereogenic silane.

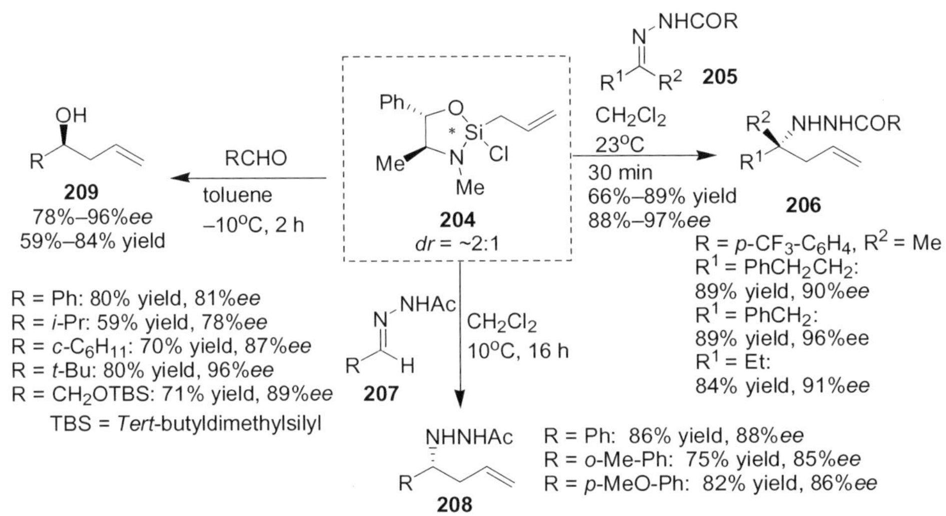

Scheme 4.55

Allylation of aldehydes and imines with silicon-stereogenic reagents.

Notably, the chiral organosilicon compounds could also be utilized as chiral reagents or chiral silicon-based Lewis acids. In this respect, Leighton and coworkers greatly contributed to the development of silicon-stereogenic reagents and chiral silicon-based Lewis acids for the catalytic allylation reaction. Since 2002, Leighton et al.[106] discovered a series of allylation of aldehydes, ketones, and imines with the chiral organosilicon compound **204** synthesized from pseudoephedrine and allytrichlorosilane as chiral reagent (Scheme 4.55). Although the chiral allylsilane was not optically pure (2:1 mixture of diastereomers) and the role of silicon-stereogenic moiety is unclear, the reactivity of the chiral silane reagent in these reactions is quite interesting. As proposed by the authors, hypercoordinate silicon species were believed to be formed as a reactive intermediate thus enabling the excellent diastereoselectivity in these allylation reactions. Meanwhile, Leighton and coworkers have expanded the use of these pseudoephedrine-derived chiral silanes to enantioselective Lewis acid-promoted transformations, such as Friedel–Crafts alkylation of benzoylhydrazones,[107] [3 + 2]acylhydrazone-enol cycloadditions,[108] Diels-Alder cycloadditions,[109] and Mannich reactions of aliphatic ketone-derived hydrazones.[110]

4.4 Summary

Over the past decades, a remarkable number of chiral organosilicon compounds, silicon-stereogenic silanes, have been synthesized with good stereoselectivity, which provide new approaches to chiral organosilicon compounds. However, despite significant effort having been put into finding new synthetic methodologies to access silicon-stereogenic silanes, the problem of their synthesis is as difficult as it is important. The story of synthesis and application of chiral organosilicon compounds with unprecedented reactivity will continue to expand with the discovery of new reactions in organic synthesis and asymmetric catalysis.

References

1. (a) Tacke, R.; Merget, M.; Bertermann, R.; Bernd, M.; Beckers, T.; Reissmann, T. Syntheses and Properties of Silicon- and Germanium-Containing α-Amino Acids and Peptides: A Study on C/Si/Ge Bioisosterism. *Organometallics* **2000,** *19*, 3486–3497.
   (b) Tacke, R.; Handmann, V. I.; Bertermann, R.; Burschka, C.; Penka, M.; Seyfried, C. Sila-Analogues of High-Affinity, Selective σ Ligands of the Spiro[indane-1,4′-piperidine] Type: Syntheses, Structures, and Pharmacological Properties. *Organometallics* **2003,** *22*, 916–924.
   (c) Corriu, R. J. P. Synthesis and Characterization of Alkylene-bridged Silsesquicarbodiimide Hybrid Xerogels. *J. Organomet. Chem.* **2003,** *686*, 127–133.
2. (a) Kipping, F. S., XXII Organic Derivatives of Silicon. Part II. The Synthesis of Benzylethylpropylsilicol, Its Sulphonation, and the Resolution of the *DL*-sulphonic Derivative Into its Optically Active Components. *J. Chem. Soc., Trans.* **1907,** 209–240.
   (b) Kipping, F. S., XLIV Organic Derivatives of Silicon. Part VI. The Optically Active Sulphobenzylethylpropylsilicyl Oxides. *J. Chem. Soc.* **1908,** 457–473.
3. Sommer, L. H.; Frye, C. L. Introduction to Enzymology. *J. Am. Chem. Soc.* **1959,** *81*, 1013–1014.
4. (a) Sommer, L. H. *Stereochemsitry, Mechanism and Silicon;* McGraw-Hill: New York, 1965.
   (b) Corriu, R. J. P.; Guérin, C. Nucleophilic Displacement at Silicon: Recent Developments and Mechanistic Implications. *Adv. Organomet. Chem.* **1982,** *20*, 265–312.
   (c) Corriu, R. J. P.; Guérin, C.; Moreau, J. J. E. Stereochemistry at silicon. *Top. Stereochem.* **1984,** *15*, 43–62.
   (d) Corriu, R. J. P.; Guerin, C.; Moreau, J. J. E. In The *Chemistry of* Organic Silicon Compounds*;* Patai, S., Rappoport, Z., Eds.; John Wiley & Sons: New York, 1989; pp 305–370.
5. Sommer, L. H.; Frye, C. L.; Parker, G. A.; Michael, K. W. Stereochemistry of Asymmetric Silicon. I. Relative and Absolute Configurations of Optically Active α-Naphthylphenylmethylsilanes. *J. Am. Chem. Soc.* **1964,** *86*, 3271–3276.
6. (a) Sommer, L. H.; Rosborough, K. T. Stereochemistry of Asymmetric Silicon. XVII. Synthesis, Resolution, and Stereochemistry of the 1,2,2,2-Tetraphenyl-1-Methyldisilane System. *J. Am. Chem. Soc* **1969,** *91*, 7067–7076.
   (b) Corriu, R. J. P.; Lanneau, G. F.; Ould-Kada, S. Methylphenyltriphenylgermylsilanes Fonctionnels Optiquement Actifs: I. Synthese Et Stereochimie. *J. Organomet. Chem.* **1983,** *248*, 23–37.

(c) Corriu, R. J. P.; Masse, J. *Bull. Soc. Chim. Fr.* **1969,** 3491.

(d) Citron, J. D. Preparation, Resolution and Dynamic Stereochemistry of the 7H-7-Methyl-7-Silabenz[d,e] anthracene System. *J. Organomet. Chem.* **1975,** *86*, 359–367.

(e) Roark, D. N.; Sommer, L. H. Dramatic Stereochemistry Crossover to Retention of Configuration With Angle-Strained Asymmetric Silicon. *J. Am. Chem. Soc.* **1973,** *95*, 969–971.

(f) Sommer, L. H.; Michael, K. W.; Korte, K. W. Stereochemistry of Asymmetric Silicon. VI. Synthesis, Stereochemistry, and Absolute Configurations of New Optically Active Monofunctional Organosilicon Systems. *J. Am. Chem. Soc.* **1967,** *89*, 868–875.

(g) Sommer, L. H.; Silverman, M. A. Stereochemistry of Asymmetric Silicon. XXII. Preparation and Properties of Optically Active Perfluorophenyl Compounds. *J. Org. Chem.* **1973,** *38*, 636–639.

7. Holt, A.; Jarvie, A. W. P.; Jervis, G. J. The Preparation and Determination of Optical Purity of Methylphenyl-1-naphthyl-(−)-menthoxysilanes. *J. Organomet. Chem.* **1970,** *21*, 75–77.

8. (a) Corriu, R. J. P.; Lanneau, G. F.; Lead, M. Action des organométalliques saturés ou aryliques sur les organosilanes asymétriques bifonctionnels. Synthèse de nouveaux silanes chiraux monofonctionnels. *J. Organomet. Chem.* **1974,** *64*, 79–91.

(b) Breliere, C.; Corriu, R. J. P.; Royo, G. Chemical and Stereochemical Behaviour of Bifunctional Enantiomeric Silicon Compounds: I. Study of α-Naphthylferrocenyl-fluorosilane and -Chlorofluorosilane. *J. Organomet. Chem.* **1978,** *148*, 107–118 and references cited therein.

9. Mori, A.; Toriyama, F.; Kajiro, H.; Hirabayashi, K.; Nishihara, Y.; Hiyama, T. Synthesis and Optical Resolution of Novel Chiral Silanols. *Chem. Lett.* **1999,** 549–550.

10. (a) Tacke, R.; Kosub, U.; Wagner, S. A.; Bertermann, R.; Schwarz, S.; Merget, S., et al. Enzymatic Transformation and Liquid-Chromatographic Separation as Methods for the Preparation of the (*R*)- and (*S*)-Enantiomers of the Centrochiral Hydridogermanes *p*-$XC_6H_4(H)Ge(CH_2OAc)CH_2OH$ (X = H, F). *Organometallics* **1998,** *17*, 1687–1699.

(b) Tacke, R.; Schmid, T.; Hofmann, M.; Tolasch, T.; Francke, W. Sila-linalool as a Pheromone Analogue: A Study on C/Si Bioisosterism. *Organometallics* **2003,** *22*, 370–372.

(c) Schmid, T.; Daiss, J. O.; Ilg, R.; Surburg, H.; Tacke, R. Enantiopure Chiral Derivatives of the Fragrance Materials Majantol and Sila-majantol: A Bioisosteric Carbon/Silicon Switch With Drastic Effects on the Sensory. *Organometallics* **2003,** *22*, 4343–4346.

11. Rendler, S.; Auer, G.; Keller, M.; Oestreich, M. Preparation of a Privileged Silicon-Stereogenic Silane: Classical Versus Kinetic Resolution. *Adv. Synth. Catal.* **2006,** *348*, 1171–1182.

12. Terunuma, D.; Murakami, K.; Kokubo, M.; Senda, K.; Nohira, H. Optical Resolutions and Configurations of Benzylmethylphenylsilylmethylamine and Its Derivatives. *Bull. Chem. Soc. Jpn.* **1980,** *53*, 789–794.

13. Terunuma, D.; Senda, K.; Sanazawa, M.; Nohira, H. The Configurations of Benzylmethyl(o- or p-tolyl) silylmethylamine and the Stereochemistry of the Reactions of Benzylmethylmethoxyphenylsilane With Some Organometallic Agents. *Bull. Chem. Soc. Jpn.* **1982,** *55*, 924–927.

14. (a) Tacke, R.; Reichel, D.; Kropfgans, M.; Jones, P. G.; Mutschler, E.; Gross, J., et al. Biological Recognition of Enantiomeric Silanes: Syntheses and Antimuscarinic Properties of Optically Active (2-Aminoethyl)cyclohexyl(hydroxymethy l)phenylsilanes and Related Quaternary Ammonium Derivatives. *Organometallics* **1995,** *14*, 251–262.

(b) Tacke, R.; Kornek, T.; Heinrich, T.; Burschka, C.; Penka, M.; Pülm, M. Syntheses and Pharmacological Characterization of Achiral and Chiral Enantiopure C/Si/Ge-Analogous Derivatives of the Muscarinic Antagonist Cycrimine: A Study on C/Si/Ge Bioisosterism. *J. Organomet. Chem.* **2001,** *640*, 140–165.

15. Jankowski, P.; Schaumann, E.; Wicha, J.; Zarecki, A.; Adiwidjaja, G. Facile Synthesis of Enantiomerically Pure *tert*-Butyl(methyl)phenylsilanes. *Tetrahedron: Asymmetry* **1999,** *10*, 519–526.

16. Jankowski, P.; Schaumann, E.; Wicha, J.; Zarecki, A.; Adiwidjaja, G.; Asztemborska, M. Preparation of Optically Active *n*-Butyl(methyl)phenyl-, *tert*-Butyl(methyl)phenyl- and Isopropyl(methyl)phenylsilanes From the Corresponding Silyl Chlorides Using 2,2′-Dihydroxy-1,1′-Binaphthyls as Resolving Agents. *Chem. Commun.* **2000,** 1029–1030.

17. Trzoss, M.; Shao, J.; Bienz, S. Preparation of a 'Si-centered' Chiral Auxiliary by Resolution. *Tetrahedron: Asymmetry* **2004,** *15*, 1501–1505.
18. (a) Holt, A.; Jarvie, A. W. P.; Jervis, G. J. Kinetic Resolution of an Organosilane. *Tetrahedron Lett.* **1968,** 4087–4088.
    (b) Holt, A.; Jarvie, A. W. P.; Jervis, G. J. Preparation and Spectroscopic Properties of Some New Asymmetric Organosilanes. *J. Chem. Soc. Perkin Trans.* **1973,** *2*, 114–122.
19. Corriu, R. J. P.; Lanneau, G. F. Preparation de deux nouveaux organosilanes asymetriques bifonctionnels induction asymetrique dans la menthanolyse d'un chlorosilane. *Tetrahedron Lett.* **1971,** 2771–2774.
20. Yamamoto, K.; Kawanami, Y.; Miyazawa, M. Kinetic Resolution of (±)-Cyclohex-1-enylsilanols by the Sharpless Asymmetric Epoxidation. *J. Chem. Soc. Chem. Commun.* **1993,** 436–437.
21. Rendler, S.; Auer, G.; Keller, M.; Oestreich, M. Preparation of a Privileged Silicon-Stereogenic Silane: Classical Versus Kinetic Resolution. *Adv. Synth. Catal.* **2006,** *348*, 1171–1182.
22. Klebe, J. F.; Finkbeiner, H. The Asymmetric Synthesis of an Optically Active Difunctional Silane. *J. Am. Chem. Soc.* **1966,** *88*, 4740–4741.
23. Klebe, J. F.; Finkbeiner, H. Optically Active Silicon in 2-Siloxazolidones-5. An Asymmetric Synthesis. *J. Am. Chem. Soc.* **1968,** *90*, 7255–7261.
24. (a) Kobayashi, K.; Kato, T.; Masuda, S. Asymmetric Synthesis of Silicon Compounds Using Chiral 5,6-Dimethoxy-1,3,2-dioxasilacycloheptane Derivatives. *Chem. Lett.* **1987,** 101–104.
    (b) Kobayashi, K.; Kato, T.; Unno, M.; Masuda, S. Asymmetric Synthesis of Organosilicon Compounds Using a C2 Chiral Auxiliary. *Bull. Chem. Soc. Jpn.* **1997,** *70*, 1393–1401.
25. Igawa, K.; Takada, J.; Shimono, T.; Tomooka, K. Enantioselective Synthesis of Silanol. *J. Am. Chem. Soc.* **2008,** *130*, 16132–16133.
26. Kawachi, A.; Maeda, H.; Mitsudo, K.; Tamao, K. Optical Resolution and Epimerization of Fluorosilane Having an Optically Active Amino Group: A New, Convenient Access to Optically Active Silicon Compounds. *Organometallics* **1999,** *18*, 4530–4533.
27. (a) Tomooka, K.; Nakazaki, A.; Nakai, T. A Novel Aryl Migration From Silicon to Carbon: An Efficient Approach to Asymmetric Synthesis of α-Aryl β-Hydroxy Cyclic Amines and Silanols. *J. Am. Chem. Soc* **2000,** *122*, 408–409.
    (b) Nakazaki, A.; Usuki, J.; Tomooka, K. Stereoselective 1,4-Phenyl Migration From Silicon to Carbon in α-Siloxy Cyclic Acetal Systems: A Concise Synthesis of 1,2-cis-Phenyl C-Glycoside and Enantioenriched Silanol. *Synlett* **2008,** 2064–2068.
28. Schmidt, D. R.; O'Malley, S. J.; Leighton, J. L. Catalytic Asymmetric Silane Alcoholysis: Practical Access to Chiral Silanes. *J. Am. Chem. Soc.* **2003,** *125*, 1190–1191.
29. Oestreich, M.; Schmid, U. K.; Auer, G.; Keller, M. A Convergent Method for the Synthesis of Highly Enantiomerically Enriched Cyclic Silanes With Silicon-Centered Chirality. *Synthesis (Mass)* **2003,** 2725–2739.
30. Oka, N.; Nakamura, M.; Soeda, N.; Wada, T. Stereocontrolled Synthesis of Tertiary Silanes Via Optically Pure 1,3,2-Oxazasilolidine Derivatives. *J. Organomet. Chem.* **2009,** *694*, 2171–2178.
31. Bauer, J. O.; Strohmann, C. One-Step Conversion of Methoxysilanes to Aminosilanes: A Convenient Synthetic Strategy to N,O-Functionalised Organosilanes. *Chem. Commun.* **2012,** *48*, 7212–7214.
32. Bauer, J. O.; Strohmann, C. Stereoselective Synthesis of Silicon-Stereogenic Aminomethoxysilanes: Easy Access to Highly Enantiomerically Enriched Siloxanes. *Angew. Chem. Int. Ed.* **2014,** *53*, 720–724.
33. Bauer, J. O.; Strohmann, C. From An α-Functionalized Silicon-Stereogenic N,O-Silane to a Monomeric and Tetracoordinate *t*BuLi Adduct With Lithium-Centered Chirality. *Angew. Chem. Int. Ed.* **2014,** *53*, 8167–8171.
34. Bauer, J. O.; Strohmann, C. Stereocontrol in Nucleophilic Substitution Reactions at Silicon: The Role of Permutation in Generating Silicon-Centered Chirality. *J. Am. Chem. Soc.* **2015,** *137*, 4304–4307.
35. (a) Ghanem, A.; Aboul-Enein, H. Y. Lipase-Mediated Chiral Resolution of Racemates in Organic Solvents. *Tetrahedron: Asymmetry* **2004,** *15*, 3331–3351.
    (b) Garcia-Urdiales, E.; Alfonso, I.; Gotor, V. Enantioselective Enzymatic Desymmetrizations in Organic Synthesis. *Chem. Rev.* **2005,** *105*, 313–354.

36. Sheldrick, W. S.; Linoh, H.; Tacke, R.; Lambrecht, G.; Moser, U.; Mutschler, E. Sila–Pharmaca. Part 32. Crystal and Molecular Structures of the (*R*)-Enantiomer and the Racemate of the Antimuscarinic Agent (Cyclohexyl)phenyl[2-(pyrrolidin-1-yl)ethyl]silanol (Sila–Procyclidine). *J. Chem. Soc., Dalton Trans.* **1985,** 1743–1746.
37. Smith, R. J.; Pietzsch, M.; Waniek, T.; Syldatk, C.; Bienz, S. Enzymatic Synthesis of Enantiomerically Enriched *D*- and *L*-3-Silylated Alanines by Deracemization of DL-5-Silylmethylated Hydantoins. *Tetrahedron: Asymmetry* **2001,** *12*, 157–165.
38. (a) Tacke, R.; Becker, B. Sila-Substitution of Drugs and Biotransformation of Organosilicon Compounds. *Main Group Met. Chem.* **1987,** *10*, 169–197.
    (b) Syldatk, C.; Andree, H.; Stoffregen, A.; Wagner, F.; Stumpf, B.; Ernst, L., et al. Enantioselective Reduction of Acetyldimethylphenylsilane by *Trigonopsis variabilis* (DSM 70714). *Appl. Microbiol. Biotechnol.* **1987,** *27*, 152–158.
    (c) Syldalk, C.; Stoffregen, A.; Wuttke, F.; Tacke, R. Enantioselective Reduction of Acetyldimethylphenylsilane: A screening With Thirty Strains of Microorganisms. *Biotechnol. Lett.* **1988,** *10*, 731–734.
    (d) Fritsche, K.; Syldatk, C.; Wagner, F.; Hengelsberg, H.; Tacke, R. Enzymatic Resolution of Rac-1,1-Dimethyl-1-Sila-Cyclohexan-2-OL by Ester Hydrolysis or Transesterification Using a Crude Lipase Preparation of Candida Cylindracea. *Appl. Microbiol. Biotechnol.* **1989,** *31*, 107–111.
    (e) Tacke, R.; Hengelsberg, H.; Zilch, H.; Stumpf, B. Enantioselective Microbial Reduction of 1,1-Dimethyl-1-Sila-Cyclohexan-2-One With Growing Cells of the Yeast Kloeckera Corticis (ATCC 20109). *J. Organomet. Chem.* **1989,** *379*, 211–216.
39. (a) Tacke, R.; Brakmann, S.; Wuttke, F.; Fooladi, J.; Syldatk, C.; Schomburg, D. Stereoselective Microbial Reduction of Racemic Acetyl(*t*-Butyl)methylphenylsilane by Trigonopsis Variabilis (DSM 70714) and Cornebacterium Dioxydans (ATCC 2176). *J. Organomet. Chem.* **1991,** *403*, 29–41.
    (b) Tacke, R.; Wuttke, F.; Henke, H. Zur Stereochemie Der Mikrobiellen Reduktion Von *Rac*-acetyl(*t*-butyl)methylphenylsilan Mit *Trigonopsis variabilis* (DSM 70714) und *Corynebacterium dioxydans* (ATCC 21766): Aufklärung der absoluten konfiguration der biotransformationsprodukte (Si*R*,C*R*)- und (Si*S*,C*R*)-*t*-butyl(l-hydroxyethyl)methylphenylsilan. *J. Organomet. Chem.* **1992,** *424*, 273–280.
    (c) Huber, P.; Bratovanov, S.; Bienz, S.; Syldatk, C.; Pietzsch, M. Chiral Silicon Groups as Auxiliaries for Enantioselective Synthesis: Access to Optically Active Silanes by Biotransformation and the Enantiospecific Preparation of (*R*)-(+)-1-Phenylethanol. *Tetrahedron: Asymmetry* **1996,** *7*, 69–78.
    (d) Tacke, R.; Heinrich, T. Syntheses of Enantiopure Si-Centrochiral Silicon-Based Muscarinic Antagonists Using an Enantioselective Enzymatic Esterification as the Key Step. *Silicon Chem.* **2002,** *1*, 35–39.
40. Djerourou, A. H.; Blanco, L. Synthesis of Optically Active 2-Sila-1,3-Propanediols Derivatives by Enzymatic Transesterification. *Tetrahedron Lett.* **1991,** *32*, 6325–6326.
41. Fukui, T.; Kawamoto, T.; Tanaka, A. Enzymatic Preparation of Optically Active Silylmethanol Derivatives Having a Stereogenic Silicon Atom by Hydrolase-Catalyzed Enantioselective Esterification. *Tetrahedron: Asymmetry* **1994,** *5*, 73–82.
42. Lu, X.; Li, L.; Yang, W.; Jiang, K.; Yang, K. F.; Zheng, Z. J., et al. Catalytic Synthesis of Functional Silicon-Stereogenic Silanes Through *Candida antarctica* Lipase B Catalyzed Remote Desymmetrization of Silicon-Centered Diols. *Eur. J. Org. Chem.* **2013,** 5814–5819.
43. Xu, L. W.; Li, L.; Lai, G. Q.; Jiang, J. X. The Recent Synthesis and Application of Silicon-Stereogenic Silanes: A Renewed and Significant Challenge in Asymmetric Synthesis. *Chem. Soc. Rev.* **2011,** *40*, 1777–1790.
44. (a) García-Urdiales, E.; Alfonso, I.; Gotor, V. Enantioselective Enzymatic Desymmetrizations in Organic Synthesis. *Chem. Rev.* **2005,** *105*, 313–354.
    (b) De Villegas, M. D. D.; Gálvez, J. A.; Etayo, P.; Badorrey, R.; López-Ram-de-Víu, P. Recent Advances in Enantioselective Organocatalyzed Anhydride Desymmetrization and Its Application to the Synthesis of Valuable Enantiopure Compounds. *Chem. Soc. Rev.* **2011,** *40*, 5564–5587.
    (c) Xu, L. W. Desymmetrization Catalyzed by Transition-Metal Complexes: Enantioselective Formation of Silicon-Stereogenic Silanes. *Angew. Chem. Int. Ed.* **2012,** *51*, 12932–12934.

45. (a) Corriu, R. J. P.; Moreau, J. J. E. Synthese asymetrique d'alcoxysilanes catalysee par des complexes du rhodium. *Tetrahedron Lett.* **1973,** *14*, 4469–4472.
    (b) Corriu, R. J. P.; Moreau, J. J. E. Asymmetric Synthesis at Silicon: II. Alcoholysis of Prochiral Organosilicon Compounds Catalysed by Rhodium Complexes. *J. Organomet. Chem.* **1976,** *120*, 337–346.
    (c) Corriu, R. J. P.; Moreau, J. J. E. Alcoolyse selective d'organosilanes catalysee par un complexe du rhodium. *J. Organomet. Chem.* **1976,** *114*, 135–144.
46. (a) Corriu, R. J. P.; Masse, J. P. Reactions stereospecifiques de lithiens sur un bicyclosilane asymetrique. *Tetrahedron Lett.* **1968,** *9*, 5197–5200.
    (b) Richter, W. J. Eine neue asymmetrische synthese von optisch aktiven silanen. *J. Organomet. Chem.* **1979,** *169*, 9–17.
    (c) Corriu, R. J. P.; Guérin, C. Nucleophilic Displacement at Silicon Sterochemistry and Mechanistic Implications. *J. Organomet. Chem.* **1980,** *198*, 231–340.
47. (a) Corriu, R. J. P.; Moreau, J. J. E. Synthses asymmetric d'alcoxysilanes catalyses par des complexes du rhodium. *Tetrahedron Lett.* **1973,** 4469–4472.
    (b) Corriu, R. J. P.; Moreau, J. J. E. Asymmetric Synthesis at Silicon: II. Alcoholysis of Prochiral Organosilicon Compounds Catalysed by Rhodium Complexes. *J. Organomet. Chem.* **1976,** *120*, 337–346.
48. (a) Corriu, R. J. P.; Moreau, J. J. E. Asymmetric Hydrosilylation of Ketones Catalysed by a Chiral Rhodium Complex. *J. Organomet. Chem.* **1974,** *64*, C51–C54.
    (b) Corriu, R. J. P.; Moreau, J. J. E. Asymmetric Synthesis at Silicon: I. Hydrosilylation of Carbonyl Compounds Catalysed by a Chiral Phosphine—Rhodium Complex: Stereochemistry and Mechanism. *J. Organomet. Chem.* **1975,** *85*, 19–33.
49. Hayashi, T.; Yamamoto, K.; Kumada, M. Asymmetric Synthesis of Bifunctional Organosilicon Compouns via Hydrosilylation. *Tetrahedron Lett.* **1974,** 331–334.
50. Ohta, T.; Ito, M.; Tsuneto, A.; Takaya, H. Asymmetric Synthesis of Silanes With a Stereogenic Centre at Silicon via Hydrosilylation of Symmetric Ketones With Prochiral Diaryl Silanes Catalysed by binap-Rh^{I} Complexes. *J. Chem. Soc. Chem. Commun.* **1994,** 2525–2526.
51. Tamao, K.; Nakamura, K.; Ishii, H.; Yamaguchi, S.; Shiro, M. Axially Chiral Spirosilanes Via Catalytic Asymmetric Intramolecular Hydrosilation. *J. Am. Chem. Soc.* **1996,** *118*, 12469–12470.
52. Zhang, Y. Z.; Zhu, S. F.; Wang, L. X.; Zhou, Q. L. Copper-Catalyzed Highly Enantioselective Carbenoid Insertion Into Si–H Bond. *Angew. Chem. Int. Ed.* **2008,** *47*, 8496–8498 and references cited therein.
53. Yasutomi, Y.; Suematsu, H.; Katsuki, T. Iridium(III)-Catalyzed Enantioselective Si – H Bond Insertion and Formation of an Enantioenriched Silicon Center. *J. Am. Chem. Soc.* **2010,** *132*, 4510–4511.
54. Shintani, R. Recent Advances in the Transition-Metal-Catalyzed Enantioselective Synthesis of Silicon-Stereogenic Organosilane. *Asian J. Org. Chem.* **2015,** *4*, 510–514.
55. Shintani, R.; Moriya, K.; Hayashi, T. Palladium-Catalyzed Enantioselective Desymmetrization of Silacyclobutanes: Construction of Silacycles Possessing a Tetraorganosilicon Stereocenter. *J. Am. Chem. Soc.* **2011,** *133*, 16440–16443.
56. Shintani, R.; Moriya, K.; Hayashi, T. Palladium-Catalyzed Desymmetrization of Silacyclobutanes With Alkynes: Enantioselective Synthesis of Silicon-Stereogenic 1-Sila-2-Cyclohexenes and Mechanistic Considerations. *Org. Lett.* **2012,** *14*, 2902–2905.
57. Shintani, R.; Otomo, H.; Ota, K.; Hayashi, T. Palladium-Catalyzed Asymmetric Synthesis of Silicon-Stereogenic Dibenzosiloles Via Enantioselective C–H Bond Functionalization. *J. Am. Chem. Soc.* **2012,** *134*, 7305–7308.
58. (a) Shimizu, M.; Mochida, K.; Hiyama, T. Modular Approach to Silicon-Bridged Biaryls: Palladium-Catalyzed Intramolecular Coupling of 2-(Arylsilyl)aryl Triflates. *Angew. Chem., Int. Ed.* **2008,** *47*, 9760–9764.
    (b) Shimizu, M.; Mochida, K.; Katoh, M.; Hiyama, T. Synthesis and Photophysical Properties of 2-Donor-7-Acceptor-9-Silafluorenes: Remarkable Fluorescence Solvatochromism and Highly Efficient Fluorescence in Doped Polymer Films. *J. Phys. Chem. C* **2010,** *114*, 10004–10014.

59. Shintani, R.; Otomo, H.; Ota, K.; Hayashi, T. Palladium-Catalyzed Asymmetric Synthesis of Silicon-Stereogenic Dibenzosiloles Via Enantioselective C–H Bond Functionalization. *J. Am. Chem. Soc.* **2012,** *134*, 7305–7308.
60. Shintani, R.; Maciver, E. E.; Tamakuni, F.; Hayashi, T. Rhodium-Catalyzed Asymmetric Synthesis of Silicon-Stereogenic Dibenzooxasilines Via Enantioselective Transmetalation. *J. Am. Chem. Soc.* **2012,** *134*, 16955–16958.
61. Igawa, K.; Yoshihiro, D.; Ichikawa, N.; Kokan, N.; Tomooka, K. Catalytic Enantioselective Synthesis of Alkenylhydrosilanes. *Angew. Chem. Int. Ed.* **2012,** *51*, 12745–12748.
62. Kunai, A.; Sakurai, T.; Toyoda, E.; Ishikawa, M.; Yamamoto, Y. Versatile Method for the Synthesis of Iodosilanes. *Organometallics* **1994,** *13*, 3233–3236.
63. Kurihara, Y.; Nishikawa, M.; Yamanoi, Y.; Nishihara, H. Synthesis of Optically Active Tertiary Silanes via Pd-Catalyzed Enantioselective Arylation of Secondary Silanes. *Chem. Commun.* **2012,** *48*, 11564–11566.
64. Onoe, M.; Baba, K.; Kim, Y.; Kita, Y.; Tobisu, M.; Chatani, N. Rhodium-Catalyzed Carbon–Silicon Bond Activation for Synthesis of Benzosilole Derivatives. *J. Am. Chem. Soc.* **2012,** *134*, 19477–19488.
65. Kuninobu, Y.; Yamauchi, K.; Tamura, N.; Seiki, T.; Takai, K. Rhodium-Catalyzed Asymmetric Synthesis of Spirosilabifluorene Derivatives. *Angew. Chem. Int. Ed.* **2013,** *52*, 1520–1522.
66. Shintani, R.; Takagi, C.; Ito, T.; Naito, M.; Nozaki, K. Rhodium-Catalyzed Asymmetric Synthesis of Silicon-Stereogenic Dibenzosiloles by Enantioselective [2 + 2 + 2] Cycloaddition. *Angew. Chem. Int. Ed.* **2015,** *54*, 1616–1620.
67. Shintani, R.; Kurata, H.; Nozaki, K. Rhodium-Catalyzed Intramolecular Alkynylsilylation of Alkynes. *Chem. Commun.* **2015,** *51*, 11378–11381.
68. (a) Sommer, L. H.; Citron, J. D. Stereochemistry of Asymmetric Silicon. VII. Silicon–Nitrogen Bond. *J. Am. Chem. Soc.* **1967,** *89*, 5797–5801.
    (b) Sommer, L. H.; Lyons, J. E. Stereospecific Substitution Reactions of Optically Active R_3Si^*H Catalyzed by Palladium and Nickel. *J. Am. Chem. Soc.* **1967,** *89*, 1521–1522.
    (c) Sommer, L. H.; Korte, W. D. Stereochemistry of Asymmetric Silicon. VIII. Stereochemistry Crossover and Leaving Group Basicity in Organometallic Coupling Reactions. *J. Am. Chem. Soc.* **1967,** *89*, 5802–5806.
    (d) Sommer, L. H.; McLick, J. Stereochemistry of Asymmetric Silicon. IX. Silicon–Sulfur Bond. *J. Am. Chem. Soc.* **1967,** *89*, 5806–5812.
    (e) Sommer, L. H.; Bauman, D. L. Stereochemistry of Asymmetric Silicon. XVIII. Hydrogen-Halogen Exchange of R_3SiH With Trityl Halides. *J. Am. Chem. Soc.* **1969,** *91*, 7076–7078.
69. (a) Daniels, R. G.; Paquette, L. A. Silanes in Organic Synthesis. XVIII. Preparation and Reactivity of Optically Active Vinyl- and Dienylsilanes. *Organometallic* **1982,** *1*, 1449–1453.
    (b) Larson, G. L.; Prieto, J. A.; Oritiz, E. The Preparation and Chemistry of (r)-(1-Naphthyl) phenylmethylsilylmethyllithium: Stereochemistry at Silicon in the Elimination of β-Hydroxysilanes. *Tetrahedron* **1988,** *44*, 3781–3790.
    (c) Omote, M., et al. Stereospecific Formation of Optically Active Trialkylsilyllithiums and Their Configurational Stability. *J. Organomet. Chem.* **2000,** *611*, 20–25.
    (d) Nanjo, M., et al. Convenient Access to Optically Active Silyl- and Germyllithiums: Synthesis, Absolute Structure, and Reactivity. *Tetrahedron Lett.* **2005,** *46*, 8945–8947.
70. Strohmann, C.; Hörnig, J.; Auer, D. Synthesis of a Highly Enantiomerically Enriched Silyllithium Compound. *Chem. Commun.* **2002,** 766–767.
71. (a) Sommer, L. H.; Lyons, J. E.; Fujimoto, H. A Stereochemistry of Asymmetric Silicon. XV. Stereospecific Hydrosilation and Exchange Reactions of $R_3SiH(D)$ Catalyzed by Group VIII Metal Centers. *J. Am. Chem. Soc.* **1969,** *91*, 7051–7061.
    (b) Sommer, L. H.; Mason, R. Optically Active R_3Si^*Li From Lithium Metal Cleavage of an Optically Active Disilane. *J. Am. Chem. Soc.* **1965,** *87*, 1619–1620.

72. (a) Strohmann, C.; Däschlein, C.; Kellert, M.; Auer, D. A Highly Enantiomerically Enriched Lithiosilane by Selective Cleavage of a Silicon–Phenyl Bond With Lithium. *Angew. Chem. Int. Ed.* **2007,** *46*, 4780–4782.

(b) Strohmann, C. C.; Däschlein, C. Synthesis of a Highly Enantiomerically Enriched Silagermane and Selective Cleavage of the Si – Ge Bond With Lithium. *Organometallics* **2008,** *27*, 2499–2504.

73. (a) Ott, H.; Däschlein, C.; Leusser, D.; Seibel, T.; Stalke, D.; Strohmann, C. Structure/Reactivity Studies on an α-Lithiated Benzylsilane: Chemical Interpretation of Experimental Charge Density. *J. Am. Chem. Soc.* **2008,** *130*, 11901–11911.

(b) Däschlein, C.; Gessner, V. H.; Strohmann, C. Preparation of "Si-Centered" Chiral Silanes by Direct α-Lithiation of Methylsilanes. *Chem. Eur. J* **2010,** *16*, 4048–4062.

74. Oestreich, M.; Auer, G.; Keller, M. On the Mechanism of the Reductive Metallation of Asymmetrically Substituted Silyl Chlorides. *Eur. J. Org. Chem.* **2005,** 184–195.

75. (a) Omote, M.; Tokita, T.; Shimizu, Y.; Imae, I.; Shirakawa, E.; Kawakami, Y. Stereospecific Formation of Optically Active Trialkylsilyllithiums and Their Configurational Stability. *J. Organomet. Chem.* **2000,** *611*, 20–25.

(b) Oh, H. S.; Imae, I.; Kawakami, Y. Evaluation of Absolute Configuration of Naphthylphenyl-Substituted Oligosilanes by CD Exciton Chirality Method. *Chirality* **2003,** *15*, 231–237.

(c) Oh, H. S.; Imae, I.; Kawakami, Y.; Raj, S. S.; Tamane, T. Synthesis, Stereochemistry and Chiroptical Properties of Naphthylphenyl-Substituted Optically Active Oligosilanes With α,ω-Chiral Silicon Centers. *J. Organomet. Chem.* **2003,** *685*, 35–43.

(d) Suzuki, K.; Kawakami, Y. Selective Formation of Both Enantiomerically Pure Disilanes From One Fluorosilane Antipode With Silyllithium. *Organometallics* **2003,** *22*, 2367–2369.

(e) Suzuki, K.; Kawakami, Y.; Velmurugan, D.; Yamane, T. Stereoselective Synthesis of Optically Active Disilanes and Selective Functionalization by the Cleavage of Silicon-Naphthyl Bonds With Bromine. *J. Org. Chem.* **2004,** *69*, 5383–5389.

76. Igawa, K.; Kokan, N.; Tomooka, K. Asymmetric Synthesis of Chiral Silacarboxylic Acids and Their Ester Derivatives. *Angew. Chem. Int. Ed.* **2010,** *49*, 740–743.

77. (a) Tacker, R.; Linoh, H., Ernst, L.; Moser, U.; Mutschler, E.; Sarger, S., et al. Sila-Pharmaka, 371) Darstellung und Eigenschaften der Enantiomere der Antimuskarinika Sila-Procyclidin und Sila-Tricyclamol-iodid: Optisch aktive Silanole mit Silicium als Chiralitätszentrum. *Chem. Ber* **1987,** *120*, 1229–1237.

(b) Yamamoto, K.; Kawanami, Y.; Miyazawa, M. Kinetic Resolution of (±)-Cyclohex-1-enylsilanols by the Sharpless Asymmetric Epoxidation. *J. Chem. Soc. Chem. Commun.* **1993,** 436–437.

78. Cavicchioli, M.; Montanari, V.; Resnati, G. Oxyfunctionalization Reactions by Perfluoro cis-2,3-Dialkyloxaziridines. Enantioselective Conversion of Silanes Into Silanols. *Tetrahedron Lett.* **1994,** *35*, 6329–6330.

79. Adam, W.; Mello, R.; Curci, R. *O*-Atom Insertion into Si–H Bonds by Dioxiranes: A Stereospecific and Direct Conversion of Silanes Into Silanols. *Angew. Chem. Int. Ed.* **1990,** *29*, 890–891.

80. Adam, W.; Mitchell, C.; Saha-Möller, C. R.; Weichold, O. Host – Guest Chemistry in a Urea Matrix: Catalytic and Selective Oxidation of Triorganosilanes to the Corresponding Silanols by Methyltrioxorhenium and the Urea/Hydrogen Peroxide Adduct. *J. Am. Chem. Soc.* **1999,** *121*, 2097–2103.

81. Tomooka, K. In *The Chemistry of Organolithium Compounds;* Rappoport, Z., Marek, I., Eds.; , Vol. 2; Wiley: Chichester, 2004; pp 749–828.

82. Nakazaki, A.; Nakai, T.; Tomooka, K. Asymmetric Retro-[1,4] Brook Rearrangement and Its Stereochemical Course at Silicon. *Angew. Chem. Int. Ed.* **2006,** *45*, 2235–2238.

83. (a) Paquette, L. A. Silicon-Mediated Organic Synthesis. *Science* **1982,** *217*, 793–800.

(b) Studer, A. S_Hi Reactions at Silicon as Unimolecular Chain Transfer Steps in the Formation of Cyclic Alkoxysilanes. *Angew. Chem. Int. Ed.* **1998,** *37*, 462–465.

(c) Brook, M. A. *Silicon in Organic, Organometallic, and Polymer Chemistry;* John Wiley & Sons: New York, 2000.

(d) Coelhol, P. J.; Blanco, L. Highly Diastereoselective Intramolecular Diels–Alder Reaction of Chiral Silatrienes. *Tetrahedron* **2003,** *59*, 2451–2456.
(e) Godineau, E.; Schenk, K.; Landais, Y. Synthesis of Fused Piperidinones Through a Radical-Ionic Cascade. *J. Org. Chem.* **2008,** *73*, 6983–6993.
(f) Recent Advances in Organosilicon Chemistry Directed Towards Organic Synthesis. In *Tetrahedron, Special Issue;* Langer, P. *Tetrahedron, Special Issue*, 65; ; 2009. ; pp 5445–5614.
(g) Xu, L. W.; Li, L.; Shi, Z. H. Asymmetric Synthesis With Silicon-Based Bulky Amino Organocatalysts. *Adv. Synth. Catal.* **2010,** *352*, 243–279.

84. Sakaba, H.; Hirata, T.; Kabuto, C.; Kabuto, K. Synthesis, Structure, and Dynamic Behavior of Tungsten Dihydride Silyl Complexes $Cp^*(CO)_2W(H)_2(SiHPhR)$ (R = Ph, H, Cl). *Organometallics* **2006,** *25*, 5145–5150.
85. (a) Oishi, M.; Kawakami, Y. A Stereoselective Approach to Optically Active Bifunctional 1,3-Dimethyl-1,3-Diphenyldisiloxanes. *Org. Lett.* **1999,** *1*, 549–552.
(b) Oishi, M.; Kawakami, Y. Synthesis of Stereoregular and Optically Active Polysiloxanes Containing 1,3-Dimethyl-1,3-Diphenyldisiloxane as a Constitutional Unit. *Macromolecules* **2000,** *33*, 1960–1963.
(c) Oishi, M.; Moon, J. Y.; Janvikul, W.; Kawakami, Y. Synthesis of Poly(methylphenylsiloxane) Rich in Syndiotacticity by Rh-Catalysed Stereoselective Cross-Dehydrocoupling Polymerization of Optically Active 1,3-Dimethyl-1,3-Diphenyldisiloxane Derivatives. *Polym. Int.* **2001,** *50*, 135–143.
(d) Park, S. Y.; Uenishi, K.; Kawakami, Y. Induced Circular Dichroism to Optically Active Poly{[methyl (1-naphthyl)silylene](*o*-phenylene)methylene}s Assisted by Supramolecular Complexation With β-Cyclodextrin. *Macromol. Chem. Phys.* **2005,** *206*, 1199–1205.
86. Chan, T. H.; Wang, D. Chiral Organosilicon Compounds in Asymmetric Synthesis. *Chem. Rev.* **1992,** *92*, 995–1006.
87. (a) Fry, J. L.; McAdam, M. A. Stereochemical Studies of Fluoride Ion-Assisted Reductions of Ketones by Organosilicon Hydrides. *Tetrahedron Lett.* **1984,** *25*, 5859–5862.
(b) Corriu, R. J. P. In *Frontiers of Organosilicon Chemistry;* Bassindale, A. R., Gaspar, P. P., Eds.; Royal Society of Chemistry: Cambridge, 1991; p 185.
88. (a) Brook, A. G.; Limburg, W. W. Stereochemistry of α-Silylcarbinol Rearrangements. II. The Absolute Configuration of Asymmetric Silanes. *J. Am. Chem. Soc.* **1963,** *85*, 832–833.
(b) Brook, A. G.; Pascoe, J. D. Stereochemistry at Asymmetric Carbon in the Silylcarbinol to Alkoxysilane Rearrangement. *J. Am. Chem. Soc.* **1971,** *93*, 6224–6227.
(c) Bienz, S.; Chapeaurouge, A. Chiral Acylsilanes in Organic Synthesis. Diastereoselective 1,2-Additions to Racemic Alkoxymethyl-Substituted Acylsilanes. *Helv. Chim. Acta* **1991,** *74*, 1477–1488.
(d) Bonnini, B. F.; Mazzanti, G.; Zani, P.; Maccagnani, G. Asymmetric Induction by Chiral Silicon. Synthesis and Electrophilic Reactivity Towards Dienes and Carbanions of *R*-(−)-Methyl-α-Naphthyl-Phenylsilyl Phenyl Thioketone. *J. Chem. Soc. Chem. Commun.* **1988,** 365–367.
(e) Bonnini, B. F.; Maccagnani, G.; Masiero, S.; Mazzanti, G.; Zani, P. Chiral Silicon in Asymmetric Synthesis. Addition of Organometallic Reagents to a Silyl Thione Chiral at Silicon. *Tetrahedron Lett* **1989,** *30*, 2677–2680.
(f) Bonini, B. F.; Masiero, S.; Mazzanti, G.; Zani, P. 1,2-Diastereoselection Induced by Chiral Silicon in the Addition of Grignard Reagents to Acylsilanes. *Tetrahedron Lett.* **1991,** *32*, 6801–6804.
89. Walkup, R. D.; Obeyeskere, N. U. Silicon-Functionalized Silyl Enol Ethers. Part 2. Silyl Enol Ethers Bearing Stereogenic Silicon Atoms and Chiral Alkoxy Groups: The Effect of These Groups Upon the Facial Selectivity of the Epoxidation of an Enol Double Bond. *J. Org. Chem.* **1988,** *53*, 920–923.
90. (a) Brook, A. G.; Duff, J. M.; Anderson, D. G. Optically Active Silyl- and Germylmethyllithium Reagents and the Stereochemistry of Carbene Insertions Into the Silicon-Hydrogen and Germanium-Hydrogen Bonds. *J. Am. Chem. Soc.* **1970,** *92*, 7567–7572.
(b) Larson, G. L.; Sandoval, S.; Cartledge, F.; Froonczek, F. R. (-Haloethyl)silanes Optically Active at Carbon and at Silicon. Synthesis and Absolute Configuration. *Organometallics* **1983,** *2*, 810–815.

91. Larson, G. L.; Torres, E. Asymmetric Induction by Chiral Silicon Groups. *J. Organomet. Chem.* **1985,** *293*, 19–27.
92. (a) Sommer, L. H.; Michael, K. W.; Fujimoto, H. Stereochemistry of Asymmetric Silicon. Stereospecific Platinum-Catalyzed Hydrosilation of 1-Octene With Optically Active R_3Si^*H. *J. Am. Chem. Soc.* **1967,** *89*, 1519–1521.
    (b) Stang, P. J.; Learned, A. E. Stereochemistry and Mode of Intermolecular Silicon-Hydrogen Unsaturated Carbene Insertions. *J. Am. Chem. Soc.* **1987,** *109*, 5019–5020.
    (c) Stang, P. J.; Learned, A. E. Axial Chirality by Asymmetric Induction. Diastereomeric Allene Formation Via Silicon as a Chiral Auxiliary. *J. Org. Chem.* **1989,** *54*, 1779–1781.
93. (a) Chapeaurouge, A.; Bienz, S. Chiral Acylsilanes in Organic Synthesis. Part 2. The Role of the Solvent, the Organometallic Reagent, and the Nature of the Substrate for the Diastereoselectivity of 1,2-Additions to Racemic Alkoxymethyl-Substituted Acylsilanes. *Helv. Chim. Acta* **1993,** *76*, 1876–1889.
    (b) Enev, V.; Stojanova, D.; Bienz, S. Silicon-Directed Stereoselective Preparation and Ireland Ester-Enolate Rearrangement of o-Acyl-Substituted α-Silylated Ally1 Alcohols. *Helv. Chim. Acta* **1996,** *79*, 391–404.
94. Bratovanov, S.; Bienz, S. A Chiral (Alkoxy)methyl-Substituted Silicon Group as an Auxiliary for the Stereoselective Cuprate Addition to α,β-Unsaturated Ketones: Synthesis of (−)-(R)-Phenyl 2-Phenylpropyl Ketone. *Tetrahedron: Asymmetry* **1997,** *8*, 1587–1603.
95. Trzoss, M.; Shao, J.; Bienz, S. A Chiral Silyl Ether as Auxiliary for the Asymmetric Nucleophilic Addition to α- and β-Silyloxy Carbonyl Compounds. *Tetrahedron* **2002,** *58*, 5885–5894.
96. Yamamura, Y.; Toriyama, F.; Kondo, T.; Mori, A. Chirality Transfer From Silicon to Carbon Via Diastereoselective Simmons–Smith Cyclopropanation of Chiral Alkenylsilanols. *Tetrahedron: Asymmetry* **2002,** *13*, 13–15.
97. Tamao, K.; Ishida, N.; Tanaka, T.; Kumada, M. Silafunctional Compounds in Organic Synthesis. Part 20. Hydrogen Peroxide Oxidation of the Silicon–Carbon Bond in Organoalkoxysilanes. *Organometallics* **1983,** *2*, 1694–1696.
98. Schmidt, D. R.; O'Malley, S. J.; Leighton, J. L. Catalytic Asymmetric Silane Alcoholysis: Practical Access to Chiral Silanes. *J. Am. Chem. Soc.* **2003,** *125*, 1190–1191.
99. Oestreich, M. Chirality Transfer From Silicon to Carbon. *Chem. Eur. J.* **2006,** *12*, 30–37.
100. Oestreich, M.; Rendler, S. "True"Chirality Transfer From Silicon to Carbon: Asymmetric Amplification in a Reagent-Controlled Palladium-Catalyzed Hydrosilylation. *Angew. Chem. Int. Ed.* **2005,** *44*, 1661–1664.
101. Rendler, S.; Oestreich, M.; Butts, C.; Lioyd-Jones, G. C. Intermolecular Chirality Transfer From Silicon to Carbon: Interrogation of the Two-Silicon Cycle for Pd-Catalyzed Hydrosilylation by Stereoisotopochemical Crossover. *J. Am. Chem. Soc.* **2007,** *129*, 502–503.
102. (a) Rendler, S.; Oestreich, M. Conformational Rigidity of Silicon-Stereogenic Silanes in Asymmetric Catalysis: A Comparative Study. *Beilstein J. Org. Chem.* **2007,** *3*, 1–5.
    (b) Rendler, S.; Fröhlich, R.; Keller, M.; Oestreich, M. Enantio- and Diastereotopos Differentiation in the Palladium(II)-Catalyzed Hydrosilylation of Bicyclo[2.2.1]alkene Scaffolds With Silicon-Stereogenic Silanes. *Eur. J. Org. Chem.* **2008,** 2582–2591.
103. (a) Oestreich, M. Silicon-Stereogenic Silanes in Asymmetric Catalysis. *Synlett* **2007,** 1629–1643.
    (b) Rendler, S.; Oestreich, M. Kinetic Resolution and Desymmetrization by Stereoselective Silylation of Alcohols. *Angew. Chem. Int. Ed.* **2007,** *47*, 248–250.
    (c) Weickgenannt, A.; Oestreich, M. Potassium *tert*-Butoxide-Catalyzed Dehydrogenative Si-O Coupling: Reactivity Pattern and Mechanism of an Underappreciated Alcohol Protection. *Chem. Asian J.* **2009,** *4*, 406–410.
    (d) Weickgenannt, A.; Mewald, M.; Oestreich, M. Asymmetric Si–O Coupling of Alcohols. *Org. Biomol. Chem.* **2010,** *8*, 1497–1504.

(e) Klare, H. F. T.; Oestreich, M. Chiral Recognition With Silicon-Stereogenic Silanes: Remarkable Selectivity Factors in the Kinetic Resolution of Donor-Functionalized Alcohols. *Angew. Chem. Int. Ed.* **2007,** *46*, 9335–9338.

104. (a) Rendler, S.; Auer, G.; Oestreich, M. Kinetic Resolution of Chiral Secondary Alcohols by Dehydrogenative Coupling With Recyclable Silicon-Stereogenic Silanes. *Angew. Chem. Int. Ed.* **2005,** *44*, 7620–7624.

(b) Karatas, B.; Rendler, S.; Fröhlich, R.; Oestreich, M. Kinetic Resolution of Donor-Functionalised Tertiary Alcohols by Cu–H-Catalysed Stereoselective Silylation Using a Strained Silicon-Stereogenic Silane. *Org. Biomol. Chem.* **2008,** *6*, 1435–1440.

(c) Rendler, S.; Plefka, O.; Karatas, B.; Auer, G.; Fröhlich, R.; Mück-Lichtenfeld, C., et al. Stereoselective Alcohol Silylation by Dehydrogenative Si–O Coupling: Scope, Limitations, and Mechanism of the Cu–H-Catalyzed Non-Enzymatic Kinetic Resolution With Silicon-Stereogenic Silanes. *Chem. Eur. J* **2008,** *14*, 11512–11528.

(d) Weickgenannt, A.; Mewald, M.; Muesmann, T. W. T.; Oestreich, M. Catalytic Asymmetric Si–O Coupling of Simple Achiral Silanes and Chiral Donor-Functionalized Alcohols. *Angew. Chem. Int. Ed.* **2010,** *49*, 2223–2226.

105. (a) Rendler, S.; Oestreich, M. Conclusive Evidence for an S_N2-Si Mechanism in the $B(C_6F_5)_3$-Catalyzed Hydrosilylation of Carbonyl Compounds: Implications for the Related Hydrogenation. *Angew. Chem. Int. Ed.* **2008,** *47*, 5997–6000.

(b) Hog, D. T.; Oestreich, M. $B(C_6F_5)_3$-Catalyzed Reduction of Ketones and Imines Using Silicon-Stereogenic Silanes: Stereoinduction by Single-Point Binding. *Eur. J. Org. Chem.* **2009,** 5047–5056.

106. (a) Kinnaird, J. W. A.; Ng, P. Y.; Kubota, K.; Wang, X.; Leighton, J. L. Strained Silacycles in Organic Synthesis: A New Reagent for the Enantioselective Allylation of Aldehydes. *J. Am. Chem. Soc.* **2002,** *124*, 7920–7921.

(b) Berger, R.; Rabbat, P. M. A.; Leighton, J. L. Toward a Versatile Allylation Reagent: Practical, Enantioselective Allylation of Acylhydrazones Using Strained Silacycles. *J. Am. Chem. Soc.* **2003,** *125*, 9596–9597.

(c) Kubota, K.; Leighton, J. L. A Highly Practical and Enantioselective Reagent for the Allylation of Aldehydes. *Angew. Chem. Int. Ed.* **2003,** *42*, 946–947.

(d) Berger, R.; Duff, K.; Leighton, J. L. Enantioselective Allylation of Ketone-Derived Benzoylhydrazones: Practical Synthesis of Tertiary Carbinamines. *J. Am. Chem. Soc.* **2004,** *126*, 5686–5687.

(e) Burns, N. Z.; Hackman, B. M.; Ng, P. Y.; Powelson, I. A.; Leighton, J. L. The Enantioselective Allylation and Crotylation of Sterically Hindered and Functionalized Aryl Ketones: Convenient Access to Unusual Tertiary Carbinol Structures. *Angew. Chem. Int. Ed.* **2006,** *45*, 3811–3813.

(f) Feske, M.; Santanilla, A. E.; Leighton, J. L. Asymmetric Allylation, Crotylation, and Cinnamylation of N-Heteroaryl Hydrazones. *Org. Lett.* **2010,** *12*, 688–691.

107. Shirakawa, S.; Berger, R.; Leighton, J. L. Enantioselective Friedel – Crafts Alkylations With Benzoylhydrazones Promoted by a Simple Strained Silacycle Reagent. *J. Am. Chem. Soc.* **2005,** *127*, 2858–2859.

108. Shirakawa, S.; Lombardi, P. J.; Leighton, J. L. A Simple and General Chiral Silicon Lewis Acid for Asymmetric Synthesis: Highly Enantioselective [3 + 2] Acylhydrazone – Enol Ether Cycloadditions. *J. Am. Chem. Soc.* **2005,** *127*, 9974–9975.

109. Kubota, K.; Hamblett, C. L.; Wang, X.; Leighton, J. L. Strained Silacycle-Catalyzed Asymmetric Diels–Alder Cycloadditions: the First Highly Enantioselective Silicon Lewis Acid Catalyst. *Tetrahedron* **2006,** *62*, 11397–11401.

110. Notte, G. T.; Leighton, J. L. A New Silicon Lewis Acid for Highly Enantioselective Mannich Reactions of Aliphatic Ketone-Derived Hydrazones. *J. Am. Chem. Soc.* **2008,** *130*, 6676–6677.

SECTION 1.2

Organosilicon Compounds of Low-Coordinate Silicon

CHAPTER 5

Silicon-Centered Cations

Vladimir Ya. Lee and Akira Sekiguchi
University of Tsukuba, Tsukuba, Japan

Chapter Outline

5.0 List of Abbreviations

BBN Borabicyclo[3.3.1]nonane
CP-MAS Cross polarization magic angle spinning
Fc Ferrocenyl
FLP Frustrated Lewis pair
Mes 2,4,6-Trimethylphenyl
MMA Methyl methacrylate

Organosilicon Compounds.
DOI: http://dx.doi.org/10.1016/B978-0-12-801981-8.00005-8

NHC	*N*-heterocyclic carbene
NHI	*N*-heterocyclic imine
NMR	Nuclear magnetic resonance
OTf	Triflate ion, $CF_3SO_3^-$
Ter	2,6-bis(2,4,6-trimethylphenyl)phenyl
THF	Tetrahydrofuran
WBI	Wiberg bond index

5.1 Introduction

Silylium ions R_3Si^+, the silicon analogs of carbenium ions R_3C^+ that are also isoelectronic and isolobal to boranes R_3B, are tricoordinate (in ideal case, trigonal-planar), electron-deficient species and extraordinarily powerful Lewis acids due to the electronic sextet and vacant *p*-orbital of their silicon centers. However, the properties of silylium ions are sharply different from those of carbenium ions, why so? Because silicon is not carbon. Silicon, which shares the same group 14 of the Periodic Table with its lighter homolog, carbon, nevertheless has properties distinctly different from those of carbon. These major distinctions between the carbon and silicon are electronegativity (2.55 vs 1.90: Pauling electronegativity scale), polarizability, and size: silicon is remarkably more electropositive, more polarizable, and larger. Intuitively, such intrinsic properties of silicon should favor easier generation and stabilization of silylium ions R_3Si^+, as compared to analogous carbenium ions R_3C^+. And this comes true in the gas phase where a vast number of silylium ion derivatives were successfully generated. However, in the condensed phase (that is in the solution) the situation is totally reversed and carbenium ions appear by far easier to be stabilized than the corresponding silylium ions. The inherent instability of the latter species is undoubtedly kinetic in its origin, resulting from the enormous electrophilicity of silylium ions, which in solution (or in the solid state) may interact with almost every available nucleophile, σ-, π-, or *n*-donor. Because of such extreme electrophilicity of silylium ions, greatly exceeding that of carbenium ions, even weak Lewis bases (e.g., typically inert solvents or counteranions) can readily coordinate to them, forming corresponding silylium ion complexes in which the cationic silicon center is not trigonal-planar anymore but tetrahedral instead. This problem was finally solved upon the introduction of the counteranions of exceptionally low nucleophilicity (such as borates and carboranes) and the use of nonpolar aromatic hydrocarbons (such as benzene or toluene) as the reaction solvents. The critical choice of the substituents at the cationic silicon is of paramount importance and is governed by two major demands: (1) steric bulkiness required for the effective protection of the cationic center by keeping away from it both counterions and solvent molecules (kinetic stabilization); (2) electron donating abilities essential for the stabilization of the positive charge (thermodynamic stabilization). When all these requirements, i.e., the right choice of substituents, counteranions, and reaction solvents, were satisfied, then the truly cationic derivatives were successfully synthesized.

In this chapter, we will briefly describe the whole story of silylium ions, including their synthetic approaches, structural studies (NMR [nuclear magnetic resonance], X-ray), their stable representatives (both stabilized and "free"), and finally discuss the most recent topic of silylium ions, namely, their application in organic synthesis. Since the chemistry of silylium ions has been repeatedly and comprehensively reviewed over the past two decades,[1] in this overview we will specifically focus on the latest advances in the field, mostly published after 2010. Moreover, in this chapter we will specifically deal with the chemistry of the low-coordinate six-valence-electrons silylium ion species with the cationic silicon in the + IV oxidation state, whereas four-valence-electrons silyliumylidene derivatives $RSi:^{+}$,[2] featuring + II oxidation state of the silicon center, will not be covered in this overview. Likewise, hypercoordinate (penta- and hexacoordinate) silicon cations,[3] as well as silicon-centered radical-cations,[4] will also be beyond the scope of this contribution. Discussing silylium ions in the condensed phase, we will leave numerous examples of their gas-phase generation and reactions outside the framework of this overview and refer our readers to the original contributions.[5]

5.2 Synthesis of Silylium Ions R_3Si^+

To succeed in the generation and subsequent isolation of silylium ions, one should overcome the major problem of the intrinsic kinetic instability of the latter species. This problem is associated with enormous electrophilicity of silylium ions and thus their exceptionally high reactivity (as compared to carbenium ions) toward practically any available nucleophiles, including solvents and counteranions that are otherwise considered as quite innocent. Now, for the generation of silylium ion derivatives it is commonly accepted that the best solvents are arenes (i.e., benzene, toluene) or halogenated arenes. For the least coordinating counteranions, there are two most prominent choices: fluorinated perarylborates (such as $[B(C_6F_5)_4]^-$), halogenated *closo*-carboranes (such as $[HCB_{11}H_5Br_6]^-$), and perhalogenated *closo*-boranes (such as $[B_{12}Cl_{12}]^{2-}$).[1a,d,6]

There are several major general synthetic routes for the synthesis of stable (and in most cases isolable) silylium ion derivatives R_3Si^+, distinguished from each other by the starting material used. Below the advantages and disadvantages of each approach will be briefly discussed.

5.2.1 From Silyl Hydrides R_3Si-H

This method, taking advantage of greater electronegativity of hydrogen and thus favorable polarization of the $Si^{\delta+}-H^{\delta-}$ bond, is by far the most popular and now the most frequently used for the selective and fast generation of silylium ions, often referred to as Bartlett–Condon–Schneider "***hydride-transfer reaction***"[7] (Scheme 5.1).

Scheme 5.1
Generation of silylium ions from silyl hydrides.

The whole process, involving oxidation of the silyl hydride R_3SiH with a powerful electrophile (trityl cation Ph_3C^+ is the most frequently used Lewis acid), is driven by the relative strength of the breaking and forming bonds: stronger E–H versus weaker Si–H. The abovementioned fluorinated perarylborates, halogenated *closo*-carborates, and perhalogenated *closo*-borates are the counteranions of choice for such hydride-transfer reactions.

The evident advantage of this method is its simplicity and ready availability of the Lewis acid reagents with which the reaction proceeds very fast (e.g., with $[Ph_3C]^+[B(C_6F_5)_4]^-$ as a Lewis acid reagent and aromatic hydrocarbon as a solvent, the reaction typically proceeds in a matter of minutes). On the other hand, the important steric bulkiness of the trityl cation precludes its direct approach toward the silyl hydrides with the sterically demanding substituents at the Si atom, which makes $[Ph_3C]^+$ ineffective for generation of such silylium ions (e.g., Mes_3Si^+ which was prepared by the alternative, so-called "allyl leaving group approach," procedure, see Section 5.2.2).

A variety of silylium ion derivatives, intra- or intermolecularly stabilized by coordination to n/π-donors, counteranions, or nucleophilic solvents, is readily available by this method (Scheme 5.2), the most prominent examples are triethylsilylium ion (**A**)[1a,c,d,8] and triisopropylsilylium ion (**B**).[1a,c,d,9]

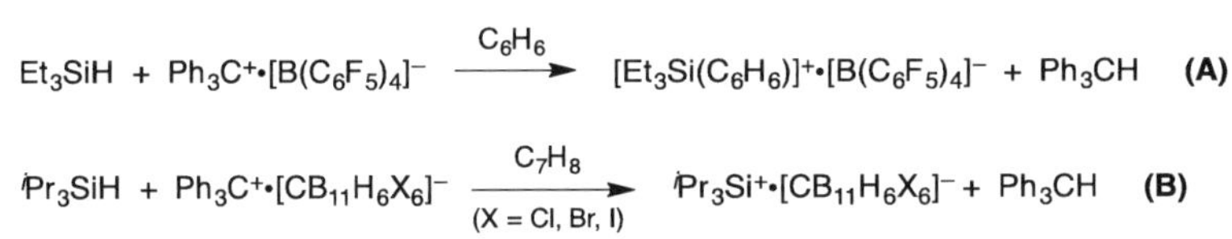

Scheme 5.2
Generation of silylium ions by the "hydride-transfer reaction".

The recent most spectacular advances in the generation of stabilized silylium ions were all achieved by application of this method (see Section 5.4.1).

5.2.2 From Silanes $R_3Si–CR_3$

This method is not as common as the above-described hydride-transfer reaction (Scheme 5.1), and found only a limited application in the generation of silylium ions requiring the use of very powerful electrophiles (such as carbenium or silylarenium ions) to break rather strong Si–C bond (Scheme 5.3).

Scheme 5.3
Generation of silylium ions from silanes.

However, this approach was the only one available for the synthesis and isolation of the very first stable silylium ion, namely trimesitylsilylium Mes_3Si^+, free from any intra- or intermolecular interactions with donors and therefore least electronically perturbed. As mentioned above, the common hydride-transfer reaction failed in this case because of the very high steric demand of the three mesityl substituents at the silicon center. By the so-called "***allyl leaving group approach,***" Lambert et al. generated Mes_3Si^+ by the treatment of allyltrimesitylsilane $H_2C=CH-CH_2-SiMes_3$ with $[Et_3Si(C_6H_6)]^+{\bullet}[B(C_6F_5)_4]^-$ to form the intermediate β-silyl substituted carbenium ion $Et_3Si-H_2C-CH^+-CH_2-SiMes_3$.[10] The latter species then underwent selective breaking of its $CH_2-SiMes_3$ bond driven by the formation of more favorable silylium ion Mes_3Si^+ and accompanied by liberation of allyltriethylsilane $H_2C=CH-CH_2-SiEt_3$ as a side product (Scheme 5.4).

$$H_2C{=}CH{-}CH_2{-}SiMes_3 + [Et_3Si(C_6H_6)]^+{\bullet}[B(C_6F_5)_4]^- \xrightarrow{C_6H_6}$$

$$[Et_3Si{-}CH_2{-}CH^+{-}CH_2{-}SiMes_3]{\bullet}[B(C_6F_5)_4]^- \longrightarrow$$

$$Mes_3Si^+{\bullet}[B(C_6F_5)_4]^- + H_2C{=}CH{-}CH_2{-}SiEt_3$$

Scheme 5.4
Generation of Mes_3Si^+ ion by the "allyl leaving group approach".

5.2.3 From Disilanes $R_3Si-SiR_3$

This method is also not very common (Scheme 5.5).

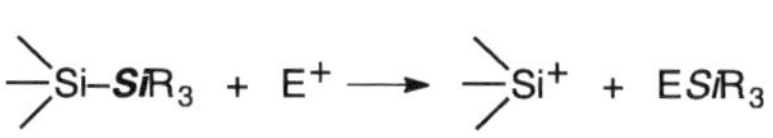

Scheme 5.5
Generation of silylium ions from disilanes.

For example, oxidative cleavage of the central Si–Si bond with strong Lewis acid was observed only for sterically overcrowded disilanes, such as hexa-*tert*-butyldisilane $^tBu_3Si-Si^tBu_3$. Thus, oxidation of the latter with $Ph_3C^+{\bullet}[B(3,5\text{-}(F_3C)_2C_6H_3)_4]^-$ in the presence of pivalonitrile $^tBu-C{\equiv}N$ resulted in the formation of the corresponding tri-*tert*-butylsilylium nitrile complex (Scheme 5.6).[11]

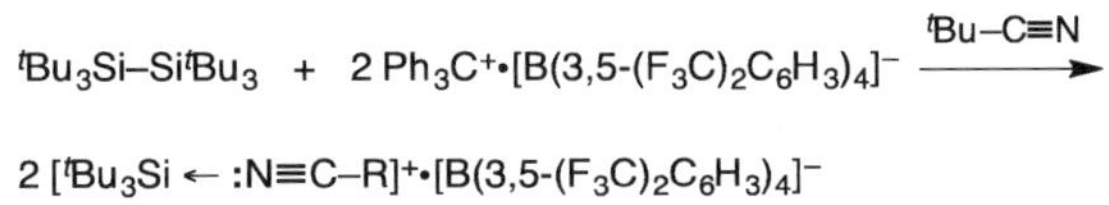

Scheme 5.6
Generation of the Tri-*tert*-butylsilylium nitrile complex.

5.2.4 From Silyl Halides $R_3Si{-}X$

Ionization of the carbon–halogen bond C–X is one of the most general methods for the generation of stable carbenium ions R_3C^+ in organic chemistry. In sharp contrast, this synthetically appealing approach does not work equally well for generation of silylium ions R_3Si^+ because of two major obstacles: (1) rather strong Si–X bonds of the silyl halide precursor R_3SiX; (2) great affinity of the developing cation R_3Si^+ toward the halide leaving group X^- due to the much higher halophilicity of silicon compared to that of carbon (Scheme 5.7).

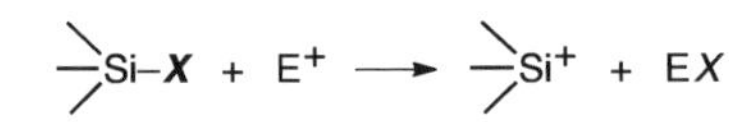

Scheme 5.7
Generation of silylium ions from silyl halides.

Accordingly, silyl cations generated in this way feature only a *partial* positive charge on their silicon centers, being better classified as the strongly polarized donor–acceptor complexes rather than the free silylium ions (Scheme 5.8).[12]

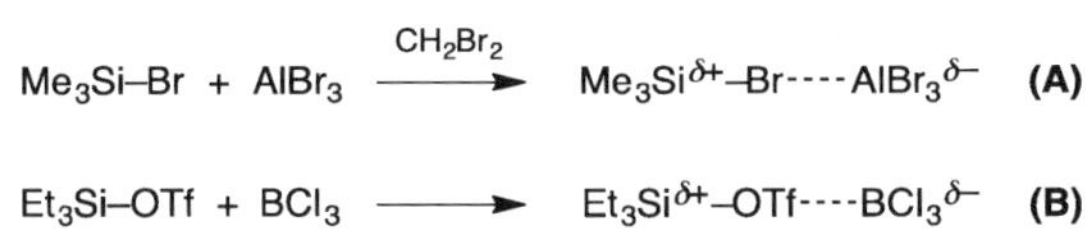

Scheme 5.8
Donor-acceptor complexes from silyl halides and silyl triflates.

5.2.5 From Low-Coordinate Organosilicon Derivatives

Although the low-coordinate organosilicon species, such as silylenes R_2Si: and silyl radicals $R_3Si•$, might be prospective as the selective red-ox precursors for silylium ions by oxidative addition of cationic species to silylenes or one-electron oxidation of silyl radicals, the scope of this seemingly attractive approach is notably restricted by the very limited availability of stable silylenes and silyl radicals. Accordingly, there are only very few examples of the synthesis of silylium ions (or related silicon-centered cations) based on this synthetic methodology, with some of them being discussed below.

5.2.5.1 *From silylenes R_2Si:*

Upon oxidative addition of the strong Lewis acid E^+ to silylene >Si:, the coordination number of the central silicon increases from 2 (in the starting silylene) to 3 (in the resulting product) (Scheme 5.9). The product of this reaction is silylium ion $>Si^+{-}E$ strongly stabilized by intramolecular electron donation.

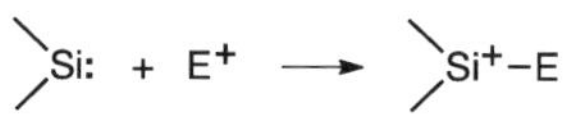

Scheme 5.9
Generation of silylium ions from silylenes.

A representative example of such synthetic approach is the reaction of nucleophilic *N*-heterocyclic silylene with strongly electrophilic silylated arenium ion (as synthetic equivalent of silylium ion) forming silaimidazolium ion $\mathbf{1^+ \bullet [B(C_6F_5)_4]^-}$ (Scheme 5.10).[13]

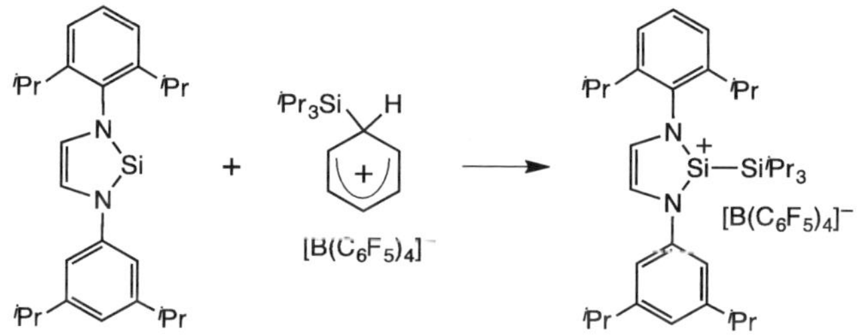

$1^+ \bullet [B(C_6F_5)_4]^-$

Scheme 5.10
Generation of silaimidazolium ion $\mathbf{1^+ \cdot [B(C_6F_5)_4]^-}$

Reaction of decamethylsilicocene $(\eta^5\text{-}Me_5C_5)_2Si$: with catechol gives rise to a silyl cation in the form of protonated decamethylsilicocene (Scheme 5.11).[14]

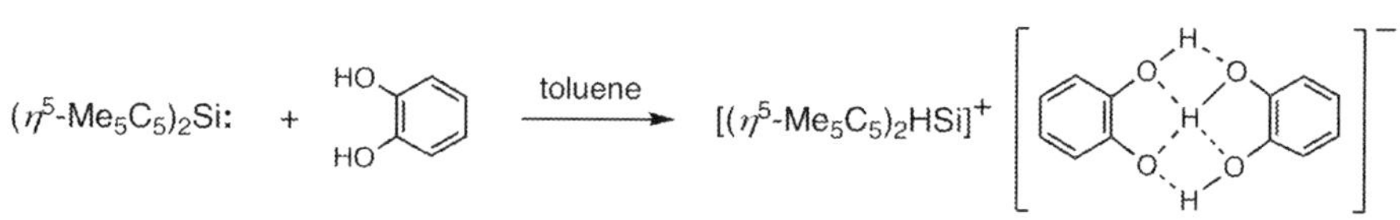

Scheme 5.11
Formation of protonated decamethylsilicocene.

5.2.5.2 From silyl radicals $R_3Si\cdot$

This synthetic approach involving oxidation of the isolable free silyl radicals $R_3Si\bullet$ with powerful Lewis acids (typically, Ph_3C^+ as a selective "one-electron transfer" reagent) is a latest development in the generation of stable silylium ions (Scheme 5.12). The advantages of this method are the remarkable ease of the experimental procedure and lack of the side products (apart from the readily separable Ph_3CH), whereas its evident drawback is restricted availability of the stable silyl radicals whose number is limited to just a few reported examples.[1b,15] Nevertheless, with the progress that can be anticipated in the field of stable silyl radicals, such a straightforward method could become a very attractive alternative for the traditional methods of generation of silylium ions.

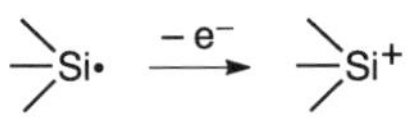

Scheme 5.12
Generation of silylium ions from stable silyl radicals.

A representative example of such selective transformation of free silyl radical to "free" silyl cation is the oxidation of the stable cyclotetrasilenyl radical **2•**[16] to the homoaromatic cyclotetrasilenylium ion $\mathbf{3^+\bullet[B(C_6F_5)_4]^-}$ [17] (Scheme 5.13).

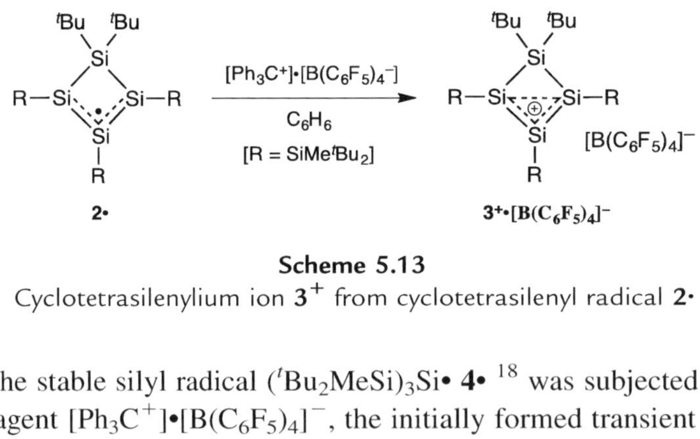

2• $\mathbf{3^+\bullet[B(C_6F_5)_4]^-}$

Scheme 5.13
Cyclotetrasilenylium ion $\mathbf{3^+}$ from cyclotetrasilenyl radical **2•**

However, when the stable silyl radical $(^tBu_2MeSi)_3Si\bullet$ **4•** [18] was subjected to its oxidation with the same reagent $[Ph_3C^+]\bullet[B(C_6F_5)_4]^-$, the initially formed transient silylium ion derivative $[(^tBu_2MeSi)_3Si^+]\bullet[B(C_6F_5)_4]^-$ $\mathbf{5^+\bullet[B(C_6F_5)_4]^-}$ underwent unavoidable isomerization via the fast 1,2-Me migration (as the silicon version of the Wagner–Meerwein rearrangement) from the substituent Si to the cationic Si center forming a novel silylium ion salt $[(^tBu_2MeSi)_2MeSi{-}^tBu_2Si^+]\bullet[B(C_6F_5)_4]^-$ $\mathbf{6^+\bullet[B(C_6F_5)_4]^-}$ (Scheme 5.14).[19] The original silylium ion $(^tBu_2MeSi)_3Si^+$ $\mathbf{5^+}$ can be stabilized and isolated as its nitrile complex $[(^tBu_2MeSi)_3Si \leftarrow : N{\equiv}C{-}CH_3]^+$ $\mathbf{7^+}$ by the reaction with acetonitrile (Scheme 5.14).[19]

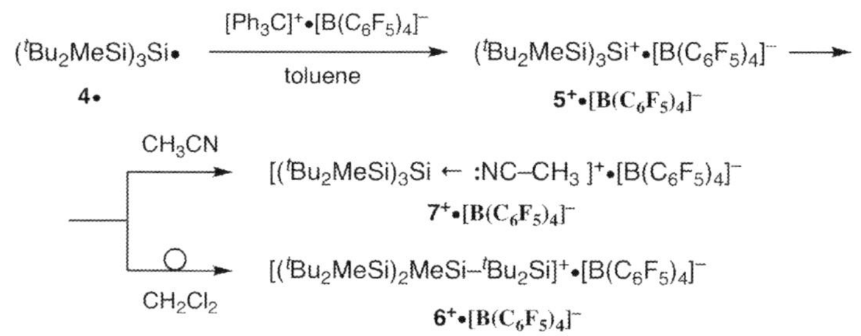

Scheme 5.14

Generation and subsequent isomerization of the transient silylium ion $({}^tBu_2MeSi)_3Si^+$ $\mathbf{5}^+$.

5.3 Structural Assessment of Silylium Ions

^{29}Si NMR spectroscopy and X-ray crystallography are the major instrumental techniques in the toolbox of the modern organosilicon chemist to disclose the decisive structural information on silylium ions. Of these two methods, ^{29}Si NMR spectroscopy is particularly diagnostic for the identification and characterization of silylium ions in solution (although the ^{29}Si NMR Cross polarization magic angle spinning (CP-MAS) technique also provides very important information about the anisotropy of the ^{29}Si NMR chemical shift tensors), whereas X-ray diffraction analysis is of indispensible importance for the structural assessment of silylium ion derivatives in their crystalline form. Overall, the ^{29}Si NMR chemical shift of the silicon center as well as its deviation from planarity are commonly applied as qualitative measures of the degree of cationic character of R_3Si^+ species.

5.3.1 ^{29}Si NMR Spectroscopy

The cationic silicon center of tricoordinate silylium ion derivatives R_3Si^+ is diagnostically strongly deshielded (as compared to their neutral tetracoordinate counterparts R_4Si), resonating at significantly lower field. Such enormous deshielding is caused by the remarkably small energy gap between the interacting occupied σ(Si–C) and low-energy lying vacant $3p$(Si) orbitals that are mixing in the applied magnetic field.[1a,c,d] This results in the increase in paramagnetic contribution, and finally, in the overall deshielding of the cationic ^{29}Si nucleus.[20] Thus, in triarylsilylium ions Ar_3Si^+ the resonances of the cationic silicon centers (independently on the NMR solvent used) were observed at +216 to +230 ppm in the aromatic hydrocarbons.[1d] For example, in the first genuine representative of the truly "free" tricoordinate silylium ion, namely Mes_3Si^+, the cationic silicon resonated at +225.5 ppm.[10]

The above-described tendency is even more pronounced in the tris(silyl)silylium ions $(R_3Si)_3Si^+$, in which the substituents are electropositive silyl groups. In the latter compounds, the σ(Si–Si) orbitals energy levels are notable elevated, resulting in the further decrease in the energy separation and consequently in the extreme deshielding of the cationic center. Although stable "free" tricoordinate tris(silyl)silylium ion derivatives are yet to be synthesized, theoretical calculations predicted for them extraordinary deshielding amounting to unprecedented +920 ppm for $(Me_3Si)_3Si^+$.[21] Such a low-field extreme for the tris(silyl)silylium ions has never been experimentally realized, and one of the closest approaches is the homoaromatic cyclotetrasilenylium ion **3^+•[B(C_6F_5)$_4$]$^-$** featuring an extraordinarily deshielded central silicon atom resonating at +315.7 ppm (for comparison, the terminal silicon atoms of the homoaromatic fragment were found at significantly higher fields at +77.3 ppm).[17] To a great surprise, structurally similar cycloterasilenylium ion derivative **8^+•[B(C_6F_5)$_4$]$^-$**, which differs from the above-described homoaromatic congener **3^+•[B(C_6F_5)$_4$]$^-$** only by substituents at the bridgehead silicon atoms, ^{t}Bu versus ^{t}Bu$_2$MeSi, manifested the sharply distinct properties of the allylic group lacking 1,3-orbital interaction diagnostic of the homoaromatic system.[22] Thus, as is expected for the classical allylic system, the most deshielded in **8^+** were the terminal silicon atoms (+286.8 ppm) while the central silicon atom of the allylic group resonated at higher field at +183.8 ppm.

"Free" trialkylsilylium ions R_3Si^+ were calculated to resonate at higher fields, e.g., Me_3Si^+ at +404 ppm, whereas experimental values for their coordinated analogs were found at remarkably lower frequencies (+138 ppm for $[Me_3Si^+][C_2H_5CB_{11}F_{11}]^-$, in which cationic Si center is coordinated to one of the F atoms of the *closo*-carborane anion).[23]

Coordination of any electron donors (*n*-, σ-, or π-), both intramolecular (by substituents) and intermolecular (by solvents and counteranions), results in the immediate decrease in the amount of positive charge at the silicon center and its notable shielding. Thus, the calculated range of the resonances for the tricoordinate silylium ions spans over a very broad area depending on the nature of substituents: from high-field (+40 ppm for $(Me_2N)_3Si^+$)[24] to exceptionally low-field (+572 ppm for $(Me_2B)_3Si^+$[25] and +920 ppm for $(Me_3Si)_3Si^+$.[21]

5.3.2 X-ray Crystallography

Crystallographic data are now available for a number of silylium ion structures with the vast majority of them representing species in which the cationic silicon center is stabilized (intra- or intermolecularly) by coordination of nucleophiles. In the field of tricoordinate silylium ions, the major breakthroughs in their X-ray structural assessment were achieved in 1993 and 2002. Thus, the first milestone discoveries were accomplished in 1993, when the groups of Lambert and Reed reported the crystal structures of tri(alkyl)silylium ion derivatives, namely, $[Et_3Si(toluene)]^+$•$[B(C_6F_5)_4]^-$ [26] and $[i\text{-}Pr_3Si]^+$•$[CB_{11}H_6Br_6]^-$ [27] derivatives. Both compounds were readily accessible by the hydride transfer reaction of hydrosilanes (Et_3SiH or $^iPr_3SiH$) using Ph_3C^+ as a powerful Lewis acid reagent and counteranions of particular low nucleophilicity (borate $[B(C_6F_5)_4]^-$ or carborane $[CB_{11}H_6Br_6]^-$) (Scheme 5.2). Although being claimed as closely approaching the "free" silylium ions, both $[Et_3Si(toluene)]^+$•$[B(C_6F_5)_4]^-$ and $[i\text{-}Pr_3Si]^+$•$[CB_{11}H_6Br_6]^-$ were not that "free," featuring notable coordination of the cationic center to either solvent (toluene) or counteranion ($[CB_{11}H_6Br_6]^-$), respectively, resulting in an appreciable charge transfer from the Si to the C (or Br) atom. Therefore, these original claims were later disproved based on the structural and spectroscopic evidences. Indeed, for the truly "free" silylium ions with their diagnostic trigonal-planar geometry and sp^2-hybridized Si center, one can expect 360 degrees (for the sum of the bond angles around the Si atom) and several hundreds ppm in the low-field region (calculated for the ^{29}Si NMR chemical shift: *vide supra*). In contrast, the cationic Si centers in both Et_3Si^+ and $^iPr_3Si^+$ were notably pyramidalized (342 degrees and 351 degrees) and their resonances were observed at remarkably high-fields (+92.3 ppm and +109.8 ppm). Accordingly it was finally agreed that the Et_3Si^+ is best described as arenium (Wheland σ-complex) rather than silylium ion, whereas $^iPr_3Si^+$ can be viewed as an intermediate between the bromonium and silylium ions.[1a,c,28]

The very first "free" silylium ion derivative, trimesitylsilylium tetrakis(pentafluorophenyl) borate $[Mes_3Si]^+[B(C_6F_5)_4]^-$, was prepared by Lambert et al. by the "*allyl leaving group approach*"[10b] in 1997 (Scheme 5.4).[10a] The resonance of the cationic silicon in $[Mes_3Si]^+[B(C_6F_5)_4]^-$, was observed at +225.5 ppm, which value is very close to those calculated for the "free" $[Mes_3Si]^+$ ion of +230.1 (GIAO/HF) and +243.9 ppm (GIAO/DFT).[29] Structural determination of the $[Mes_3Si]^+$ ion was enabled by the exchange of the borate anion $[B(C_6F_5)_4]^-$ with the *closo*-carborane anion $[CB_{11}HMe_5Br_6]^-$ and isolation of $[Mes_3Si]^+$•$[CB_{11}HMe_5Br_6]^-$ salt (Fig. 5.1).[30] The latter derivative revealed a truly "free" silylium ion structure, lacking any observable interactions with either counteranion or

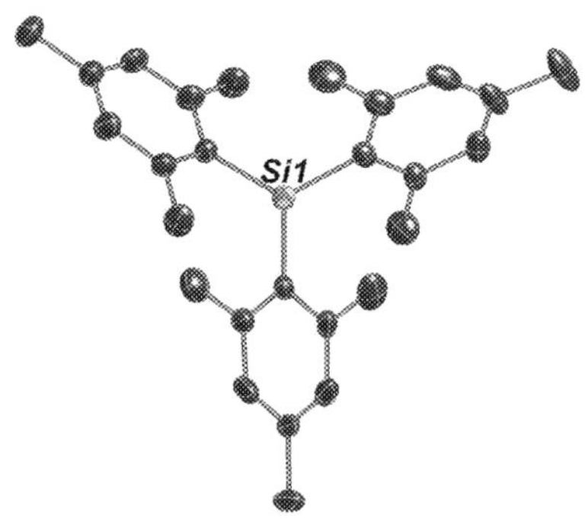

Figure 5.1
Crystal structure of the trimesitylsilylium ion derivative $[Mes_3Si]^+\cdot[CB_{11}HMe_5Br_6]^-$ (hydrogen atoms and $[CB_{11}HMe_5Br_6]^-$ counteranion are not shown).

solvent molecules, and featuring trigonal-planar geometry and low-field resonance for the cationic center: 359.9 degrees (for the sum of the bond angles) and +225.5 ppm in solution/ + 226.7 ppm in the solid state (for the ^{29}Si NMR resonance).

In the follow-up report, Müller et al. reported the crystal structure of the second "free" tri (aryl)silylium ion derivative, tris(pentamethylphenyl)silylium dodecachloro-*closo*-dodecaborane $[Ar_3Si^+]_2[B_{12}Cl_{12}]^{2-}$ (Ar = 2,3,4,5,6-Me_5C_6) $\mathbf{[9^+]_2{\bullet}[B_{12}Cl_{12}]^{2-}}$, prepared by the hydride transfer reaction involving also substituents exchange[20]:

$$3\ Ar_2(R)Si-H\ +\ Ph_3C^+ \rightarrow 2\ Ar_3Si^+\ +\ R_3Si-H$$

Like in $[Mes_3Si]^+$, the cationic Si center in $\mathbf{9^+}$ manifested ideal trigonal-planar geometry (360.0 degrees for the sum of the bond angles) and characteristic low-field ^{29}Si NMR resonance (+216.2 ppm in solution/ + 223.5 ppm in the solid state).

5.4 Stable Silylium Ions

5.4.1 Inter- and Intramolecularly Stabilized Silylium Ions

A number of inter- or intramolecularly stabilized silylium ions, in which the coordination number of the central silicon is expanded from 3 ("free" cations) to 4, or even 5 and 6 (stabilized cations) due to the coordination of external or internal nucleophile, have been reported since 2000. Because most of them were comprehensively covered in the recent reviews,[1a,c,d] in this chapter we will deal only with the latest advances in the field, the very spectacular examples of the donor-stabilized silylium ion derivatives, published mostly after 2010.

5.4.1.1 Oxidation of the Si–H bond

The overwhelming majority of the recently reported stabilized silylium ions were generated by the classical hydride transfer reaction (see Section 5.2.1 and Scheme 5.1), which is by far the most popular and most frequently used generation method, taking advantage of its straightforward synthetic procedure and the ready availability of the starting silyl hydride reagents. Several of the most remarkable examples of these silylium ion derivatives prepared by this method will be briefly described below.

Müller et al. reported a series of intramolecularly stabilized silylium ions as well as silyl-substituted vinyl cations and arenium ions, all being prepared by the hydride transfer reaction using $Ph_3C^+{\bullet}[B(C_6F_5)_4]^-$ (Scheme 5.15).[31] Among them are: (1) nitrilium complex of the transient 7-silanorbornadien-7-ylium ion $\mathbf{[10 \leftarrow : N{\equiv}C{-}CD_3]^+{\bullet}[B(C_6F_5)_4]^-}$ [31a,h]; (2) 2-silanorbornyl cation $\mathbf{11^+{\bullet}[B(C_6F_5)_4]^-}$ [31b,g,h]; silyl cation $\mathbf{12^+{\bullet}[B(C_6F_5)_4]^-}$ featuring a three-center two-electron Si–H–Si bond[31d,e]; β-disilacyclohexenylidenesubstituted vinyl cations **13^{+}•[B(C_6F_5)$_4$]$^{-}$** [31c,e,k]; bissilylated arenium ions **14^{+}•[B(C_6F_5)$_4$]$^{-}$** [31e,f]; hydrogen-bridged bis(silyl)cations **15^{+}•[B(C_6F_5)$_4$]$^{-}$** [31i]; aryl-bridged bis(silyl)cations **16^{+}•[B(C_6F_5)$_4$]$^{-}$**.[31j]

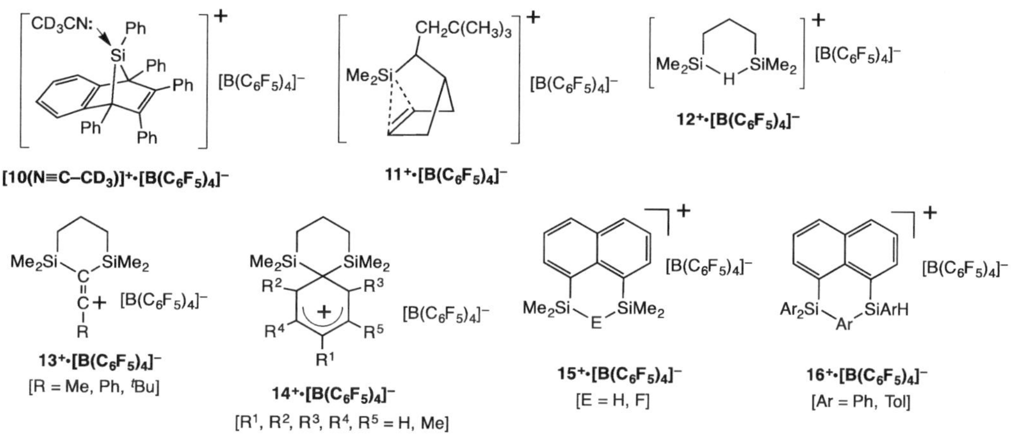

Scheme 5.15

Stabilized silylium ions prepared by the hydride transfer reaction.

Further increasing the steric bulk of substituent at the Si atom by using very demanding 2,6-bis(2,4,6-trimethylphenyl)phenyl (Ter) group, Müller et al. were able to stabilize the cationic dibenzosilanorbornadienylium isolated in the form of its [B(C_6F_5)$_4$]$^{-}$ salt **17^{+}•[B(C_6F_5)$_4$]$^{-}$** (Scheme 5.16).[2e] The ^{29}Si NMR resonance of the cationic silicon center was found at surprisingly high-field (compared to other triarylsilylium ions: +216 to +230 ppm, *vide supra*), practically independent of the NMR solvent used: +1.3 ppm (in benzene-d_6), +1.2 ppm (in toluene-d_8), and +2.2 ppm (in chlorobenzene-d_5). Such substantial shielding of the cationic silicon is caused by the intramolecular π-donation from the flanking aryl groups, further reinforced by the homoconjugation of the π-system of the dibenzonorbornadiene part and the Si cationic center. As is typical for silylium ions, upon addition of acetonitrile **17^{+}•[B(C_6F_5)$_4$]$^{-}$** readily formed the corresponding nitrilium complex [**17**←: N≡C–CH_3]$^{+}$•[B(C_6F_5)$_4$]$^{-}$ whose structure was confirmed by X-ray crystallography.

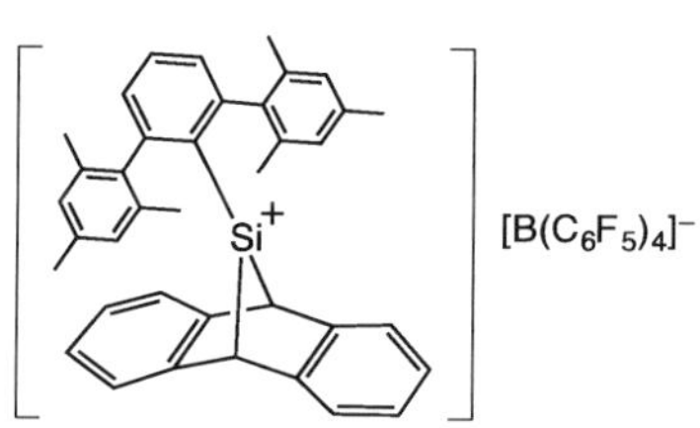

[17]$^{+}$•[B(C_6F_5)$_4$]$^{-}$

Scheme 5.16

Ter-substituted dibenzosilanorbornadienylium ion 17^{+}

Similar stabilization through the intramolecular π-donation from the flanking aryl groups experienced cationic silylium center in $\mathbf{18^+\bullet[B(C_6F_5)_4]^-}$ (Scheme 5.17).[32] Both steric and electronic [3p(Si)$-\pi$(aryl) interaction] stabilization provided by the extraordinarily bulky aryl substituent enable isolation of the $\mathbf{18^+\bullet[B(C_6F_5)_4]^-}$ as a room temperature stable derivative. Such stabilization occurred at the expense of the degree of the silylium ion character in $\mathbf{18^+}$ that was remarkably diminished. This was clearly manifested in the relatively high-field ^{29}Si NMR chemical shifts of $\mathbf{18^+}$ (+58.6 to +80.1 ppm), and in a significant departure of the cationic silicon center from planarity and accordingly, its notable pyramidalization (sum of the bond angles around Si = 346.1 degrees). Nevertheless, being sterically highly protected, cation **18** does not show observable interactions with either solvent molecules or counteranions.

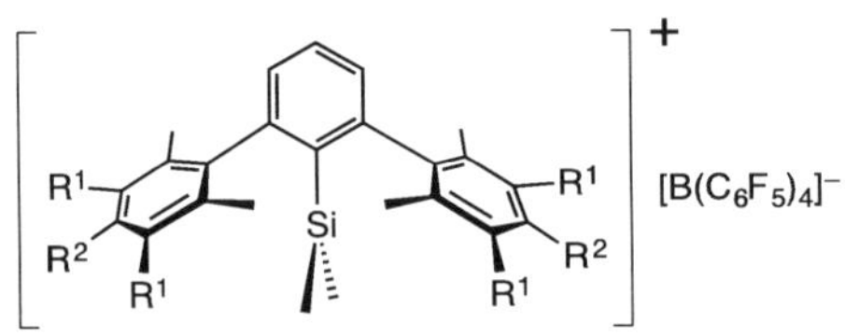

$\mathbf{18^+\bullet[B(C_6F_5)_4]^-}$ [R^1 = H, Me; R^2 = H, Me]

Scheme 5.17
Intramolecularly π-Stabilized Silylium Ion $\mathbf{18^+}$.

Interestingly, in a modification of $\mathbf{18^+}$, featuring halogen substituents in the *ortho*-positions of the flanking aryl groups, namely silylium ion $\mathbf{19^+\bullet[B(C_6F_5)_4]^-}$, the silicon centers were slightly more deshielded and their resonances were observed at +89 and +91 ppm, which values were still far from those expected for the free silylium ions (Scheme 5.18).[33] X-ray analysis of $\mathbf{19^+}$ (X = F) clearly showed that the cationic Si is effectively stabilized by the electron donation from the two F substituents occupying the apical positions of the silicon trigonal-bipyramidal geometry.

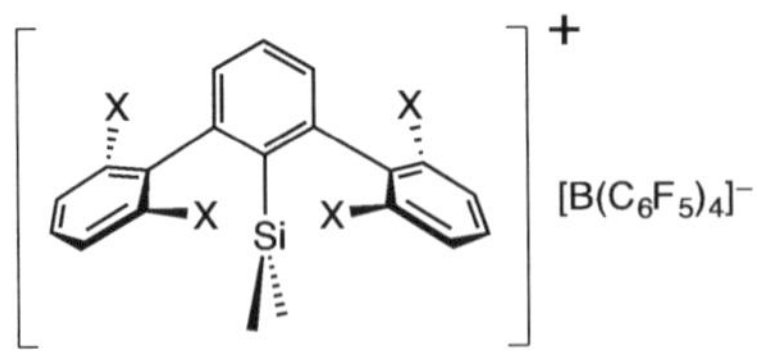

$\mathbf{19^+\bullet[B(C_6F_5)_4]^-}$ [X = F, Cl]

Scheme 5.18
Intramolecularly π Stabilized Silylium Ion $\mathbf{19^+}$ with the *ortho*-Halogen Substituents of the Flanking Aryl Groups.

The same authors also prepared a series of structurally similar silylium ion derivatives $\mathbf{20}^+$ featuring flanking aryl groups with different *ortho*-substituents: *ortho*-methyl (like in $\mathbf{18}^+$) and *ortho*-F (like in $\mathbf{19}^+$).[34] Both stabilization modes of the silylium ion through the coordination of either *ortho*-F or π-electron-rich aromatic ring were observed in the crystal structure of $\mathbf{20}^+$, and the delicate balance between these two coordination modes was discussed.

A novel synthetically rather attractive route to trarylsilylium ions was recently proposed by Müller et al.[20,35] As a feasible alternative to a classical *allyl leaving group* approach, this new method based on the standard *hydride transfer reaction* synthetic protocol involves oxidation of diaryl(methyl)silanes $Ar_2(Me)SiH$ with $Ph_3C^+{\bullet}[B(C_6F_5)_4]^-$, surprisingly forming triarylsilylium ions Ar_3Si^+ (along with Me_3SiH and Ph_3CH as side products) as a result of the substituents scrambling reaction. Like in the genuine Mes_3Si^+ ion,[10] Ar_3Si^+ exhibited diagnostic low-field resonances of their cationic silicon atoms observed at +216–230 ppm, essentially independent of the NMR arene solvent used thus indicating the lack of significant interactions between the silylium center and the solvent molecules.

In the several examples of Wagner–Meerwein-type rearrangements of germapolysilanes, the isomerization cascades of intermediate germylium ions to intermediate silylium ions were proposed.[36]

The ferrocene-stabilized silylium ion $[Fc(^tBu)MeSi]^+{\bullet}[B(C_6F_5)_4]^-$ (Fc = ferrocenyl), $\mathbf{21^+{\bullet}[B(C_6F_5)_4]^-}$ was readily prepared, and its crystal structure $2[Fc(^tBu)MeSi]^+{\bullet}[B_{12}Cl_{12}]^{2-}$ was later revealed after the exchange of counteranion from $[B(C_6F_5)_4]^-$ to $[B_{12}Cl_{12}]^{2-}$ (Scheme 5.19).[37] The positive charge in $\mathbf{21}^+$ is still largely maintained on the Si atom, which can be seen in its almost planar geometry and its remarkable deshielding (^{29}Si NMR: +114.5 ppm). Neither the solvent nor the counteranion notably coordinate to the cationic silicon in the solid state.

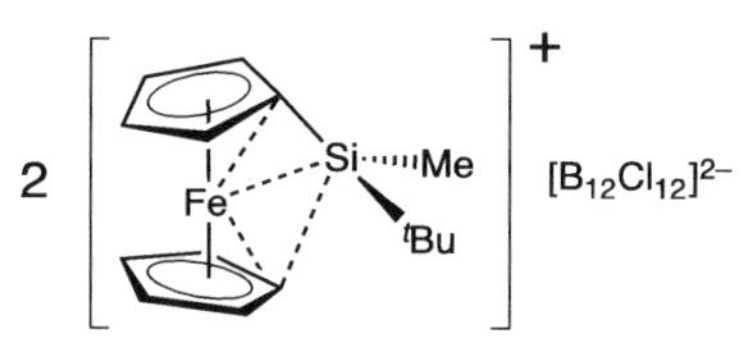

$\mathbf{[21]_2^+{\bullet}[B_{12}Cl_{12}]^{2-}}$

Scheme 5.19
Ferrocene-stabilized silylium ion $\mathbf{21}^+$

Further extending the range of intramolecularly stabilized silylium ion derivatives, the same authors synthesized and spectroscopically characterized several 1,3-dithiolane-/1,3-dithiane-stabilized silylium ions, in which the cationic silicon centers were reversibly coordinated by a sulfur *n*-donor.[38] Sulfur-stabilized silylium ions with chiral binaphthyl backbones were also smoothly generated by conventional hydride transfer reaction of the corresponding silyl hydrides and Ph_3C^+•$[B(C_6F_5)_4]^-$.[39]

A remarkable trimethylsilylium ion, as a simplest trialkylsilylium ion derivative, was stabilized and isolated as its *closo*-carborane salt Me_3Si^+•$[RCB_{11}F_{11}]^-$ (R = H, Et) by the hydride transfer reaction between Ph_3C^+•$[RCB_{11}F_{11}]^-$ and Me_3SiH.[23] Interestingly, although Me_3Si^+•$[RCB_{11}F_{11}]^-$ displayed a conductivity characteristic for ionic liquids, in contrast to most of them, Me_3Si^+•$[RCB_{11}F_{11}]^-$ is extremely reactive. In its solid state, Me_3Si^+•$[RCB_{11}F_{11}]^-$ manifested coordination of the cationic Me_3Si^+ unit to one of the fluorine atoms of the carborane counteranion, resulting in an insignificant pyramidalization (354.4 degrees) at the Si center and stretching of the coordinating B–F bonds of the carborane cage (compared with the noncoordinating B–F bonds). Likewise, in the liquid phase, Me_3Si^+•$[RCB_{11}F_{11}]^-$ also featured some degree of the anion coordination that can be seen in its ^{29}Si NMR resonance observed at +138 ppm. This value, even being low-field shifted compared with that of Me_3Si^+•$[B(C_6F_5)_4]^-$ (+84.8 ppm),[8b] was still by far smaller than the value of 400 ppm calculated for the "free" Me_3Si^+ cation.

In several years, the crystal structures of the related $[Me_3Si(arene)]^+$•$[B(C_6F_5)]^-$ salts, which revealed significant cation–anion interactions, were also reported.[40] Thus, in $[Me_3Si(benzene)]^+$•$[B(C_6F_5)]^-$ derivative, the silicon atom was remarkably pyramidalized with the sum of the bond angles around it of 341.7–343.1 degrees (for different sets of crystals). Reacting $[Me_3Si]^+$•$[B(C_6F_5)]^-$ with Me_3Si–X (X = F, Cl, Br, I), the same authors synthesized a whole series of bissilylated halonium salts $[Me_3Si–X–SiMe_3]^+$•$[B(C_6F_5)]^-$ that can be regarded as (Me_3Si–X–$[Me_3Si]^+$) solvent complexes.[41] In these complexes, the $[Me_3Si]^+$ unit has almost completely lost its silylium ion character (strong deviation from planarity) thus forming a covalently bound tetracoordinate Si center (like in Me_3Si–O–$SiMe_3$).

Homologous trialkylsilylium ions, namely $[R_3Si^+]_2$•$[B_{12}Cl_{12}]^{2-}$ (R = Me, Et, ^{i}Pr), were also readily available by the standard hydride transfer reaction procedure from $[Ph_3C^+]_2$•$[B_{12}Cl_{12}]^{2-}$ and 2 equivalents R_3SiH in 1,2-difluorobenzene.[42] Both CP-MAS ^{29}Si NMR spectroscopy (+117–126 ppm for the cationic Si atom) and X-ray crystallography (348–350 degrees around the silicon center) revealed notable coordination of the silylium ions to $[B_{12}Cl_{12}]^{2-}$ anion (via silicon–chlorine contacts).

An interesting silylium zwitterion [$\mathbf{Me_2Si}^+$–CH_2–$(CB_{11}Cl_{11})^-$] was synthesized by the treatment of the precursor $[^nBu_4N]^+$•[$\mathbf{HMe_2Si}$–CH_2–$CB_{11}Cl_{11}]^-$ (in which the silicon moiety is connected to the anionic carborane core by a covalent –CH_2– linker) with Ph_3C^+•$[B(C_6F_5)_4]^-$.[43] The silicon center in [$\mathbf{Me_2Si}^+$–CH_2–$(CB_{11}Cl_{11})^-$], stabilized by the weak coordination of a proximal chlorine atom (Si••••Cl = 2.304(8) Å), maintained significant silylium character as supported by its only modest pyramidalization (351 degrees) and relatively low-field resonance (+ 137.4 ppm, in the solid-state ^{29}Si NMR spectrum).

In the silylium ion $\mathbf{22^{+}{\bullet}[B(C_6F_5)_4]^{-}}$ stabilized by polyagostic Si−H•••Si interactions, there is an interesting hydride transfer between the two Si^{α}-centers assisted by the agostic bonding with the Si^{α}−H bonds, which can be seen in its ^{1}H- and ^{29}Si NMR spectra (Scheme 5.20).[44a] Such H-bridged silylium ion structure, featuring two $Si^{\alpha} \leftarrow H{-}Si^{\beta}$ agostic interactions, was further proved by the subsequent DFT computations.[44b] The cationic Si^{α}-centers were calculated to be essentially planar (358.9 degrees), and the three-center two-electron $Si^{\alpha}{-}H^{\alpha}{-}Si^{\alpha}$ bonding was characterized by the substantially larger Wiberg bond order (WBI) (0.426) than that for the $Si^{\alpha} \leftarrow H{-}Si^{\beta}$ agostic interaction (0.182). Accordingly, the electron occupancy of the Si^{β}−H bond was reduced to 1.806 caused by the electron density transfer to the adjacent cationic Si^{α}-centers.

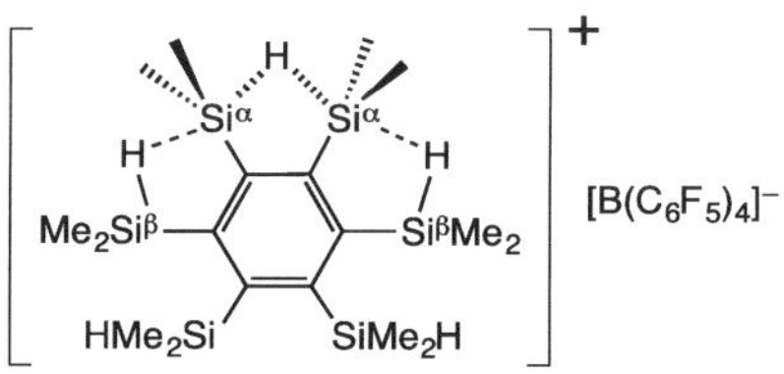

$\mathbf{22^{+}{\bullet}[B(C_6F_5)_4]^{-}}$

Scheme 5.20
Silylium ion $\mathbf{22^{+}}$ stabilized by polyagostic Si−H•••Si interactions.

Even highly constrained bridgehead silylium ion salt $\mathbf{23^{+}{\bullet}[CHB_{11}Cl_{11}]^{-}}$ was generated under the standard hydride transfer reaction conditions (Scheme 5.21).[45] The crystal structure analysis showed coordination of the silylium ion to the aromatic solvent (benzene, toluene, bromobenzene) with no direct contacts between the cationic and anionic portions of the molecule, which formulation was further supported by the observation of the ^{29}Si NMR chemical shift of the silylium ion at +105 ppm (in C_6D_5Br).

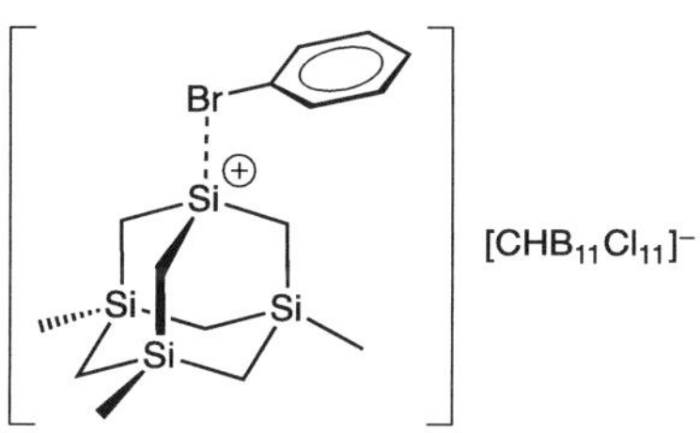

$\mathbf{23^{+}{\bullet}[CHB_{11}Cl_{11}]^{-}}$

Scheme 5.21
Bridgehead silylium ion $\mathbf{23^{+}}$.

5.4.1.2 Oxidation of the Si—C bond

Manners et al. developed a novel approach for the generation of silylium ions taking advantage of the inherently high strain of their sila[1]ferrocenophanes. They cleaved a strained Si—C bond in the latter compounds with $[H(OEt_2)]^+\bullet\{B[3,5\text{-}(CF_3)_2\text{-}C_6H_3]_4\}^-$, thus smoothly generating transient Tetrahydrofuran (THF)-solvated ferrocenyl-substituted silylium ion species $\mathbf{24^+\bullet\{B[3,5\text{-}(CF_3)_2\text{-}C_6H_3]_4\}^-}$ that can be stabilized upon coordination of a stronger Lewis base (such as pyridine) forming isolable pyridinium ion derivatives $\mathbf{25^+\bullet\{B[3,5\text{-}(CF_3)_2\text{-}C_6H_3]_4\}^-}$ (Scheme 5.22).[46] Reflecting coordination of an extra ligand (pyridine), the silicon center in $\mathbf{25^+\bullet\{B[3,5\text{-}(CF_3)_2\text{-}C_6H_3]_4\}^-}$ is markedly pyramidalized with the sum of the bond angles around Si of 337.7 degrees.

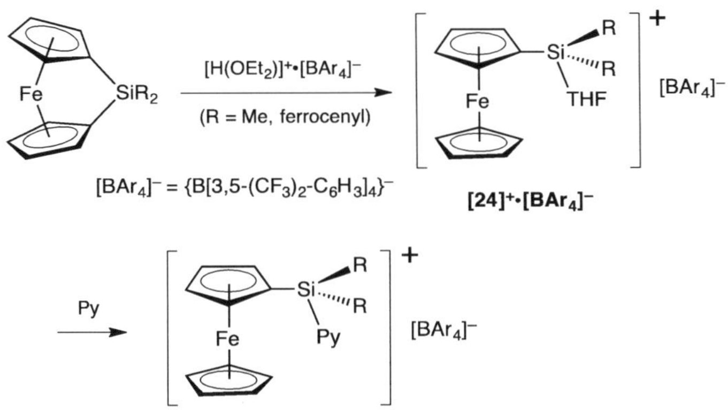

[24]+•[BAr4]−

[25]+•[BAr4]−

Scheme 5.22
Ferrocenyl-substituted silylium ion derivatives $\mathbf{24^+}$ and $\mathbf{25^+}$.

In an alternative approach, Oestreich et al. prepared already reported (*vide supra*), intramolecularly stabilized ferrocenyl-substituted silylium ion $[Fc(^tBu)MeSi]^+\bullet[B(C_6F_5)_4]^-$ (Fc = ferrocenyl), $\mathbf{21^+\bullet[B(C_6F_5)_4]^-}$, employing ready elimination of the cyclohexa-2,5-dien-1-yl group from the corresponding precursor upon its treatment with $Ph_3C^+\bullet[B(C_6F_5)_4]^-$ (the method coined as the "*cyclohexadienyl leaving group approach*," in analogy with the original Lambert's "*allyl leaving group approach*")[10] (Scheme 5.23).[47] The thus generated silylium ion $\mathbf{21^+}$ can be alternatively stabilized intermolecularly by coordination of the appropriate Lewis base, e.g., Ph_2S.

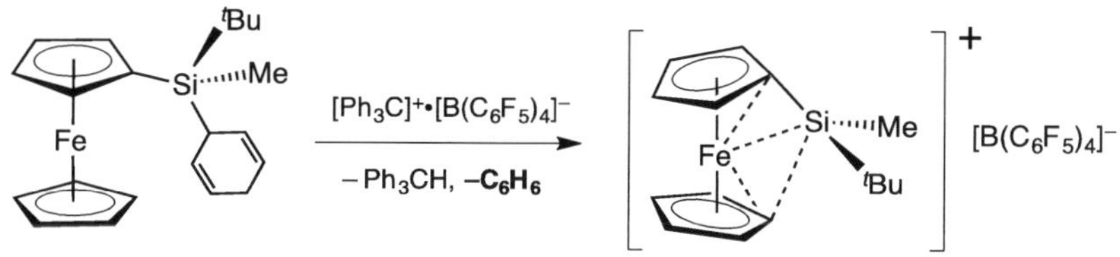

[21]+•[B(C6F5)4]−

Scheme 5.23
Cyclohexadienyl leaving group approach for the generation of ferrocenyl-substituted silylium ion **21**$^+$.

5.4.1.3 Other methods

Protonating *N*-heterocyclic silacyclopropene **26** with $[H(OEt_2)]^+{\bullet}[B(C_6F_5)_4]^-$, Driess et al. generated *n*-donor supported silacyclopropenylium ion $\mathbf{27^+{\bullet}[B(C_6F_5)_4]^-}$ featuring the cationic spiro-Si center shared by the six-membered ring N-heterocyclic unit and three-membered ring cyclopropene moiety (Scheme 5.24).[48] The highly electrophilic silylium center in **27**$^+$ is effectively stabilized by the intramolecular electron donation from the adjacent *N* atom.

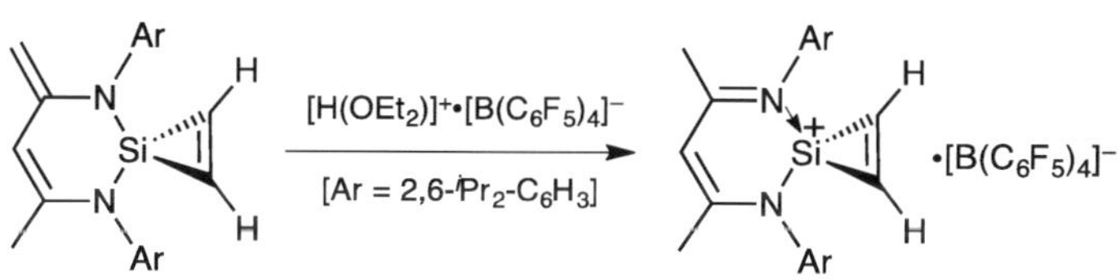

26 [27]+•[B(C6F5)4]−

Scheme 5.24
n-Donor-supported silacyclopropenylium ion derivative **27**$^+$.

N-heterocyclic carbenes (NHC) can also effectively stabilize the otherwise transient silylium ions, as was demonstrated by Driess et al. who reported the crystal structures of $[NHC \rightarrow SiH_3]^+$**•OTf**$^-$ and $[NHC \rightarrow SiH_3 \leftarrow NHC]^+$**•OTf**$^-$ complexes representing the first NHC adducts of the parent silylium ion H_3Si^+ (NHC = 1,3-bis(2,6-diisopropylphenyl) imidazole-2-ylidene).[49] The geometry of the cationic silicon center in the former and in the latter complexes was distorted tetrahedral and distorted trigonal-bipyramidal, respectively.

Likewise, *N*-heterocyclic imines (NHI) are also capable of stabilizing silylium ions. Thus, $[Me_2Si^+{-}N = C(N^tBu{-}CH = CH{-}N^tBu)]$ can be stabilized by the intermolecular coordination of Me_3SiNI^tBu (NI^tBu = bis(*tert*-butyl)imidazolin-2-imino) and isolated in the form of its $[MeB(C_6F_5)_3]^-$ salt.[50]

Other recently reported remarkable examples of the donor-stabilized silylium ion derivatives include: (1) tetrasilacyclobutadiene dication $[\mathbf{28}]^{2+}\bullet 2[Cp^*{}_2Zr_2Cl_7]^-$ [51]; (2) disilenyl cation $[\mathbf{29}]^+[OTf]^-$.[52]

In the former compound $[\mathbf{28}]^{2+}\bullet 2[Cp^*{}_2Zr_2Cl_7]^-$, unexpectedly formed upon the reaction of *N*-heterocyclic chloro silylene with Cp^*ZrCl_3 in the molar ratio 3:2, the planar rhombic Si_4-core consists of two *N*-donor stabilized silylium units (Si1) and two silylene-like moieties (Si2) (Scheme 5.25).[51] The four-coordinate cationic silicon centers Si1 in $[\mathbf{28}]^{2+}$ resonated at relatively high-field at +53.4 ppm because of the electron donation from the amidinato ligand. Based on the experimental and computational studies, the major contributor to the overall structure of $[\mathbf{28}]^{2+}$ is the charge localized resonance form with both positive charges sitting at the Si1 atoms. The following theoretical treatment of the above tetrasilacyclobutadiene dication classified it as a *nonaromatic* compound because it lacks a conjugated π-ring system required by the Hückel's rule.[51b]

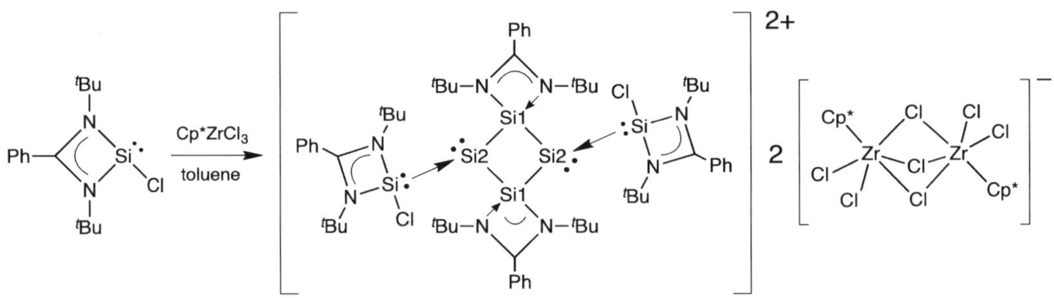

$[\mathbf{28}]^{2+}\bullet 2[Cp^*{}_2Zr_2Cl_7]^-$

Scheme 5.25
Tetrasilacyclobutadiene dication derivative $[\mathbf{28}]^{2+}$.

In the disilenyl cation $[\mathbf{29}]^+[OTf]^-$, formed upon the reaction of the NHC-complex of a stable disilyne[53] with MeOTf, the cationic NHC-substituted Si1 atom resonated at +54.0 ppm, the Si1–Si2 bond length of 2.192(2) Å was typical for the Si = Si double bonds, and the geometry around both doubly-bonded silicon atoms was nearly planar (Scheme 5.26).[52] From the experimental and computational results, the overall structure of $[\mathbf{29}]^+$ is best described as a hybrid of the two resonance extremes: (1) NHC-stabilized disilenyl cation with a positive charge at the Si1 atom; and (2) imidazolium-substituted disilene with a positive charge delocalized over the NHC-moiety.

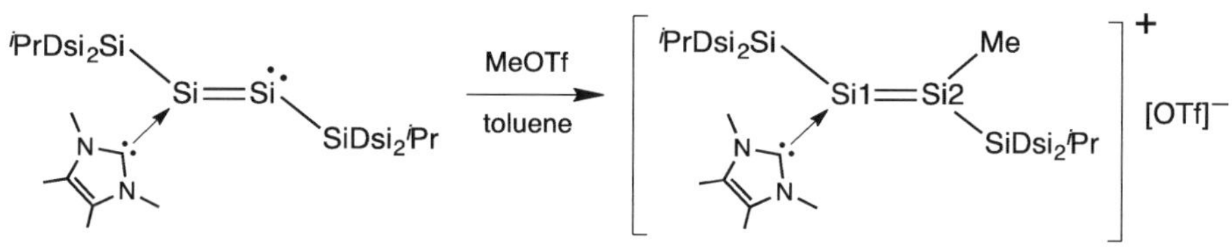

[29]+•[OTf]−

Scheme 5.26
Disilenyl cation derivative **29**$^{+}$.

5.4.2 "Free" Silylium Ions

The search for the "free" tricoordinate, truly trigonal-planar, silylium ions was a long-sought goal that was pursued by experimental silicon chemists for several decades. And to achieve the goal, it was first of all mandatory to solve the problem of extreme electrophilicity of silylium ions that can coordinate practically every available n-, π-, or σ-donor, even such typically innocent solvents as benzene or toluene or otherwise inert counteranions as perchlorate ion. Finally, the silylium ion problem[28a] was solved by a combination of several factors critical for stabilization and isolation of "free" silylium ions: (1) use of counteranions of particularly low nucleophilicity (such as $[B(C_6F_5)_4]^-$, $[CB_{11}H_6Br_6]^-$, or $[B_{12}Cl_{12}]^{2-}$); (2) application of nonpolar aromatic hydrocarbons (benzene, toluene) as solvents; and (3) smart choice of substituents at the cationic silicon center that must be bulky enough (for its kinetic stabilization) and electron donating (for its thermodynamic stabilization).

5.4.2.1 Trigonal-planar silylium ions

The tricoordinate silylium ions of the type R_3Si^+ (R = alkyl, aryl, silyl) with their trigonal-planar geometry, the least electronically perturbed by coordination of either counteranions or solvents, are the most challenging synthetic targets. However, no examples of the "free" tri (alkyl)- or tri(silyl)silylium ions were reported to date, and for the tri(aryl)silylium ion salts only two representatives that are unequivocally structurally characterized are currently known: these are $[Mes_3Si]^+\bullet[CB_{11}HMe_5Br_6]^-$ and $[Ar_3Si^+]_2[B_{12}Cl_{12}]^{2-}$ (Ar = 2,3,4,5,6-Me_5C_6).

The former, $[Mes_3Si]^+$, was actually the very first and most remarkable "free" silylium ion ever reported, prepared by the *allyl leaving group* approach (see Section 5.2.2; and for its spectroscopic and structural characteristics, see Sections 5.3.1 and 5.3.2).[30]

The other example of the "free" tri(aryl)silylium, namely tris(pentamethylphenyl)silylium ion $[Ar_3Si^+]_2[B_{12}Cl_{12}]^{2-}$ (Ar = 2,3,4,5,6-Me_5C_6) $\mathbf{[9^+]_2\bullet[B_{12}Cl_{12}]^{2-}}$, was reported a decade later by Müller et al. (see Section 5.3.2).[20] Although prepared by the standard hydride transfer reaction procedure, this method involved unusual substituents scrambling that enabled formation of the tri(aryl)silylium ion derivatives (for its spectroscopic and structural characteristics, see Section 5.3.2).

5.4.2.2 Delocalized silylium ions

A family of the cyclic organosilicon compounds, in which the positive charge is delocalized over the entire ring (or part of it) resulting in the overall aromatic stabilization of the cationic system, was prepared during the past two decades (Scheme 5.27). Although in these compounds the tricoordinate cationic silicon centers are not truly trigonal-planar (as in the silylium ions described above in the Section 5.4.2.1), they still enjoy the "freedom" from coordination of any sort of nucleophiles, neither solvents nor counteranions, as was evidenced by their X-ray crystallographic analysis.

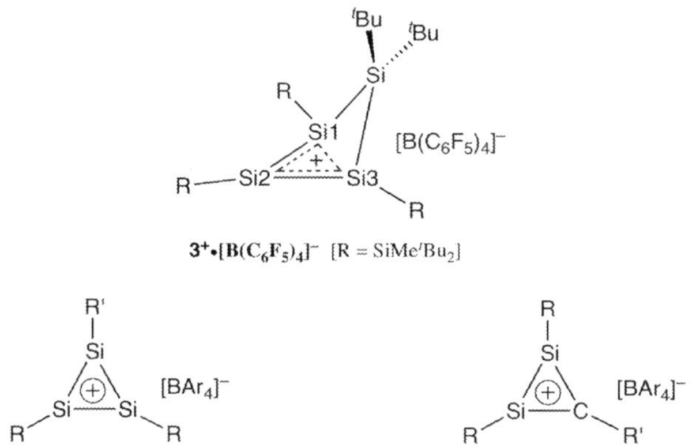

3^+•[B(C_6F_5)$_4$]$^-$ [R = SiMe^{t}Bu$_2$]

30^+•[BAr$_4$]$^-$ [R = Si^{t}Bu$_3$; R' = SiMe^{t}Bu$_2$; Ar = C_6F_5, 2,3,5,6-F_4-C_6H, 4-(^{t}BuMe$_2$Si)-C_6F_4]

31^+•[BAr$_4$]$^-$ [R = Si^{t}Bu$_3$; R' = 1-adamantyl; Ar = C_6F_5, 4-(^{t}BuMe$_2$Si)-C_6F_4]

Scheme 5.27
Delocalized silylium ions.

The first member of this family, homoaromatic cyclotetrasilenylium ion **3^+•[B(C_6F_5)$_4$]$^-$**, was prepared by the Me group abstraction from the corresponding neutral precursor cyclotrisilene by [Et_3Si(benzene)]$^+$•[B(C_6F_5)$_4$]$^-$ (see also Section 5.2.5.2) (Scheme 5.27).[17]

As a classical manifestation of homoaromatic system, the central silicon Si2 in **3^+** was extraordinarily deshielded (+315.7 ppm) whereas the terminal Si1 and Si3 were found at a much higher field (+77.3 ppm). Again, in accord with its homoaromatic composition, the geometry around the cationic silicon atoms (Si1, Si2, and Si3) in **3^+** was perfectly planar with the Si1–Si2 and Si2–Si3 bond lengths of 2.240(2) Å and 2.244(2) Å, respectively, being intermediate between those of the typical single and double bond lengths. The interatomic Si1•••Si3 distance of 2.692(2) Å was strikingly short, being only 15% longer than a standard Si–Si single bond and reasonably pointing to a homoaromatic nature of the Si1–Si3 interaction.

As mentioned above in the Section 5.3.1, structurally similar "free" cyclotetrasilenylium ion derivative $\mathbf{8^{+}\bullet[B(C_6F_5)_4]^{-}}$, differing from the above-described homoaromatic ion $\mathbf{3^{+}\bullet[B(C_6F_5)_4]^{-}}$ by only bridgehead substituents (^{t}Bu vs $^{t}Bu_2$MeSi), is not homoaromatic anymore representing instead a typical allylic system showing no signs of 1,3-orbital interaction.[22]

The 2π-electron aromatic cyclotrisilenylium $\mathbf{30^{+}\bullet[BAr_4]^{-}}$ [54a] and disilacyclopropenylium $\mathbf{31^{+}\bullet[BAr_4]^{-}}$ [54b] were both prepared by oxidative elimination of one of the silyl substituents from the corresponding neutral precursor by $[Ph_3C]^{+}\bullet[BAr_4]^{-}$ (see Section 5.2.3) (Scheme 5.27). Both cations were entirely "free," lacking any observable interactions with either solvents or counteranions both in the solid state and in solution, and their cyclopropenylium-type 2π-electron aromatic delocalization was supported by the X-ray and NMR data. Thus, in the solid state $\mathbf{30^{+}}$ forms nearly equilateral triangle with internal Si–Si–Si bond angles of nearly 60 degrees (av.) and endocyclic Si–Si bond lengths of 2.217(3) Å (av.), the latter value being intermediate between the typical Si–Si single and Si = Si double bond lengths. In solution, both $\mathbf{30^{+}\bullet[BAr_4]^{-}}$ and $\mathbf{31^{+}\bullet[BAr_4]^{-}}$ diagnostically manifested exceptional low-field resonances for their cationic atoms: +284.6 and 288.1 ppm (^{29}Si NMR for $\mathbf{30^{+}\bullet[BAr_4]^{-}}$); +208.2 ppm (^{29}Si NMR for $\mathbf{31^{+}\bullet[BAr_4]^{-}}$); and +253.7 ppm (^{13}C NMR for $\mathbf{31^{+}\bullet[BAr_4]^{-}}$).

Some closely related "free" cationic silaaromatic compounds were also reported albeit their crystal structures were not revealed. They include:

1. 6π-electron silatropylium ion (stable below −50°C) prepared by the hydride transfer reaction and featuring low-field resonance of its cationic Si atom (+142.9 ppm);[55]
2. 2-silaimidazolium ions prepared by either Cl abstraction by $[Et_3Si(benzene)]^{+}\bullet[B(C_6F_5)_4]^{-}$ (stable below −10°C)[56] (see Section 5.2.4) or by the reaction of *N*-heterocyclic silylenes with $[Et_3Si(benzene)]^{+}\bullet[B(C_6F_5)_4]^{-}/[^{i}Pr_3Si(benzene)]^{+}\bullet[B(C_6F_5)_4]^{-}$ [13] (see Section 5.2.5.1).

5.5 Application of Silylium Ions in Organic Synthesis

The reactivity of silylium ions is greatly dominated by their extraordinarily high Lewis acidity,[57] which makes them very useful reagents in organic synthesis and particularly in catalytic processes.[58] Although stoichiometric reactions involving participation of silylium ions are well known, their use as the advanced catalysts for a broad range of organic chemistry transformations is by far the most important practical application of silylium ion derivatives.

5.5.1 Diels–Alder Reactions

One of the most challenging catalytic applications of silylium ions in the C–C bond formation is Diels–Alder reaction. The very first development in this area was published in 2005 by Sawamura et al.,[59] who reported an excellent catalytic performance of the $[Et_3Si(toluene)]^+\bullet[B(C_6F_5)_4]^-$ (in comparison to the commonly used Me_3SiOTf and Et_3SiNTf_2) in Diels–Alder reactions, e.g., in the reaction of 1,3-cyclohexadiene and methyl acrylate giving practically quantitative yield of the cycloadduct (97%).

Further advance in the use of silylium ions as catalysts of Diels–Alder reactions was achieved by Oeistreich et al., who successfully employed their ferrocene-stabilized silylium ion $[Fc(^tBu)MeSi]^+\bullet[B(C_6F_5)_4]^-$ (Fc = ferrocenyl), $\mathbf{21^+\bullet[B(C_6F_5)_4]^-}$ [37a,60] (Section 5.4.1) as a superb catalyst of the [4 + 2] cycloadditions, involving a variety of deactivated 1,3-dienes (cyclohexadiene, butadiene) and dienophiles (α,β-unsaturated ketones) and proceeding with excellent *endo*-selectivity at low temperatures (−30 to −78°C).

Likewise, sulfur-stabilized silylium ions[38] can also serve as potent catalysts in Diels–Alder reactions at the temperatures as low as 0°C. Sulfur-stabilized silylium ions with chiral binaphthyl backbones[39] were also capable of catalyzing Diels–Alder reactions, involving both cyclohexadiene and pentadiene as a diene.

An in situ generated silylium ion–carbanion pair was proved to catalyze the enantioselective Diels–Alder reaction of cinnamates and cyclopentadiene with excellent enantiomeric (up to 97:3) and diastereomeric ($>$ 20:1) ratios.[61]

5.5.2 Friedel–Crafts Reactions

Recently, Oestreich et al. reported a general Friedel–Crafts-type procedure for the catalytic preparation of mono- and difunctionalized dibenzosiloles from readily available *ortho*-silylated biphenyls. This intramolecular electrophilic aromatic substitution, involving catalytically generated silicon electrophiles, results in the C–H bond silylation and formation of the desired products of up to 99% yields within minutes.[62] Likewise, catalytic S_EAr C–H silylation of 2- and 3-substituted pyridines with hydrosilanes in the presence of a transition metal catalyst, splitting the Si–H bond into a hydride and a silylium electrophile, forms the corresponding 5-silylated pyridines.[63] Further developing their findings, the same authors reported also S_EAr C–H silylation of electron-rich heteroarenes (such as indoles), promoted by the stabilized silylium ions formed upon the reaction of hydrosilanes with strong Brønsted acid ($[H(OEt_2)_2]^+[BAr^F{}_4]^-$).[64]

Friedel–Crafts variation of the intramolecular aryl coupling is initiated by the catalytic amount of silylium ion and enabled by the formation of stronger Si–F bonds (compared to the C–F bonds), resulting in the formation of a range of polycyclic arenes and graphene fragments.[65] As a representative example of such silylium-promoted Friedel–Crafts arylation, one can mention the addition of $[^{i}Pr_3Si]^+[CB_{11}H_6Cl_6]^-$ to 1-(2-fluorophenyl) naphthalene affecting the activation of the C–F bond and formation of a new arene-arene bond, thus leading to the fluoranthene product. In a similar vein, the same authors reacted 2-fluoro-2′-methylbihenyl with a catalytic amount (5 mol. %) of $[^{i}Pr_3Si]^+[CB_{11}H_6Cl_6]^-$ and stoichiometric amount of Me_2SiMes_2 thus making a C–F/C–H activation couple, in which silylium ion serves as a catalytic initiator and silane serves as Brønsted base and pro-Lewis acid catalyst.[66] In this case, silylium ion-promoted C–F activation of the substrate leads to formal C–H activation and the formation of new C_{aryl}–C_{alkyl} bond in the final product, fluorene.

Schulz et al. found that their solvent-coordinated silylium ion salt $[Me_3Si(arene)]^+[B(C_6F_5)_4]^-$ can effectively catalyze Friedel–Crafts disproportionation of *tert*-butylbenzene forming a mixture of 1,3-di-*tert*-butylbenzene (8.4%), 1,4-di-*tert*-butylbenzene (64.4%), and 1,3,5-tri-*tert*-butylbenzene (27.2%).[40]

5.5.3 Hydrodefluorination Reactions

The catalytic exchange of C–F bonds with C–H bonds in fluoroalkanes (hydrodefluorination) is a highly challenging process that recently attracted a great deal of attention. The stimulating report on the first room temperature catalytic hydrodefluorination of aliphatic C–F bonds was published by Ozerov et al. in 2005.[67a] They demonstrated that $[Et_3Si]^+[B(C_6F_5)_4]^-$ interact with trifluorotoluene derivatives $ArCF_3$ (Ar = C_6H_5, *p*-C_6H_4F, *p*-C_6H_4Cl, *p*-C_6H_4Br, *m*-C_6H_4F, *o,p*-$C_6H_3Cl_2$) and 1-fluoropentane at room temperature forming Et_3SiF and carbenium ion R^+ in the initial step, followed by the reaction of the latter with the hydride source (Et_3SiH) regenerating silylium ion catalyst Et_3Si^+ and completing the catalytic cycle (Scheme 5.28).

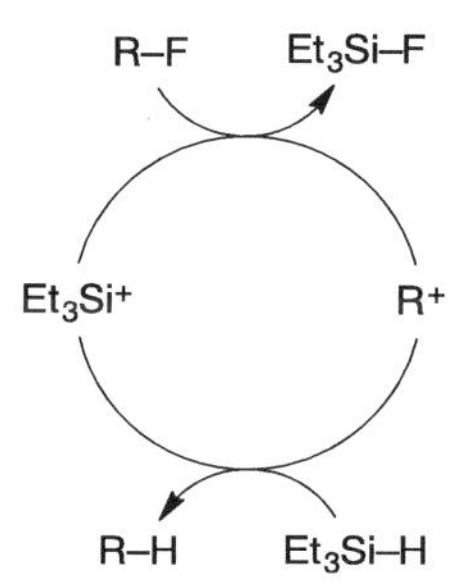

Scheme 5.28
Hydrodefluorination catalytic cycle.

Using carborane instead of borate anion, Ozerov et al. developed a novel catalytic system Et_3SiH and $[Ph_3C]^+[CHB_{11}H_5Cl_6]^-$ (as precursors for $[Et_3Si]^+[CHB_{11}H_5Cl_6]^-$) for hydrodefluorination of fluoroalkyl groups in $C_6F_5CF_3$, $C_6H_5CH_2CH_2CF_3$, and $CF_3CF_2CF_2CF_2CH_2CH_3$.[67b] In the presence of excess Et_3SiH, $C(sp^3)-F$ bonds catalytically activated and transformed into C–H bonds at room temperature with very high turnover numbers (up to 2700).

In the subsequent study, the same authors extended the hydrodefluorination process to more general hydrodehalogenation of C–F, C–Cl, and C–Br bonds with trialkylsilanes as stoichiometric reagents: $X_3C-Hal + R_3Si-H \rightarrow X_3C-H + R_3Si-Hal$.[67c]

In a similar approach, Müller et al. reacted fluorodecane or benzyl trifluoride (in the presence of equimolar amount of Et_3SiH) with a catalytic amount of the hydrogen (or fluorine)-bridged silylium ion derivative $\mathbf{15^+\bullet[B(C_6F_5)_4]^-}$ (Scheme 5.15) to give the corresponding alkanes as the hydrodefluorination products.[31i,68]

5.5.4 Activation of Small Molecules (CO_2, H_2)

Reduction of CO_2 by silanes R_3SiH ($R = Et$, ^{i}Pr) with stoichiometric amounts of $[Ph_3C]^+[B(C_6F_5)_4]^-$ in chlorobenzene forms (after hydrolysis) formic acid HCOOH and methanol CH_3OH. In benzene, the reaction of CO_2 with the preformed $[Et_3Si(C_6H_6)]^+[B(C_6F_5)_4]^-$ results in the formation (after hydrolysis) of benzoic acid PhCOOH.[69]

Interestingly, triarylsilylium borates $[Ar_3Si]^+[B(C_6F_5)_4]^-$ form frustrated Lewis pairs (FLPs) with sterically hindered phosphanes, which are capable of irreversible cleavage of dihydrogen at ambient conditions forming Ar_3SiH.[35,70] Moreover, silylium ion/phosphane FLPs, $(Me_5C_6)_3Si^+/PR_3$ (R = *tert*-butyl or cyclohexyl), are able to bind CO_2 forming silylacylphosphonium salt $(Me_5C_6)_3Si-O-C(O)-PR_3{}^+$.

Silylium ions can also pair with NHC forming silylium ion/NHC FLPs that are capable of fixation of CO_2, ^{t}BuNCO, and Ph–C≡C–C≡C–Ph.[71]

Nitrogen stabilized silylium ions, as novel intramolecular N/Si^+ FLP, showed excellent catalytic activity in the high-yield hydroboration of CO_2 to the methoxide level with 9-BBN (9-Borabicyclo[3.3.1]nonane), catecholborane, and pinacolborane.[72]

5.5.5 Hydrosilylation Reactions

Several examples of silylium ion catalyzing hydrosilylation reactions were recently published. Oestreich et al. reported the reduction of carbonyl compounds $R^1-C(O)-R^2$ with Fc(^{t}Bu)MeSiH catalyzed by in situ generated silylium ion $[Fc(^tBu)MeSi]^+[B(C_6F_5)_4]^-$ $\mathbf{21^+\bullet[B(C_6F_5)_4]^-}$ (from Fc(^{t}Bu)MeSiH and $[Ph_3C]^+[B(C_6F_5)_4]^-$) and resulted in the formation of hydrosilylation product $Fc(^tBu)MeSi-O-CHR^1R^2$.[73]

Silylium ions can also catalyze both 1,4-hydrosilylation of methyl methacrylate (MMA) with R_3SiH to generate the silyl ketene acetal initiator in situ and subsequent living polymerization of MMA.[74]

Intramolecular chain hydrosilylation of diorganyl[2-(trimethylsilylethynyl)phenyl]silanes with catalytic amount of $[Ph_3C]^+[B(C_6F_5)_4]^-$ as an initiator afforded the corresponding benzosiloles.[75] The reaction involves in situ generated silylium ions as chain carriers.

5.5.6 Miscellaneous Reactions of Silylium Ions

Apart from their above-described catalytic performance, silylium ions can also:

1. react with various alkenes forming 2-silanaphthalenes as the dehydrogenative annulation products[77];
2. initiate ring-opening polymerization of cyclic pentacoordinate sila[1]ferrocenophanes to give poly(ferrocenylsilanes) with pentacoordinate silicon moiety in the polymer backbone[78];
3. promote room temperature ring-opening polymerization of cyclic chlorophosphazene trimer[79];
4. promote reduction of imines to the corresponding amines with hydrosilanes.[80]

Catalytic amount of $[Me_3Si]^+$ ions causes isomerization of bis(trimethylsilyl) diazomethane $(Me_3Si)_2CNN$ to give bis(trimethylsilyl)aminoisonitrile $(Me_3Si)_2NNC$, which then undergoes unusual trimerization forming exclusively 4-diazenyl-3-hydrazinylpyrazol.[81]

5.6 Summary

Although the chemistry of silylium ion derivatives has notably progressed since the year 2000, it is still not as mature as the field of carbenium ions. If initially the first experimental efforts were dealing with the stabilization of otherwise transient species (by the smart choice of substituents, reaction solvents, and counteranions), then the attention of silicon chemists was focused on the isolation of "free" silylium ions lacking any observable coordination of their cationic centers with either internal or external nucleophiles. This culminated in 2002 when the Lambert and Reed groups jointly published the crystal structure of the first ever reported "free" trimesitylsilylium ion Mes_3Si^+, isolated as $[B(C_6F_5)_4]^-$ salt. Since then, after the silylium ion problem was finally resolved, the interest of the organosilicon community moved again, this time to the side of the practical use of silylium ion derivatives. Given exceptionally high electrophilicity of silylium ions, one can expect their most prospective utilization as very strong Lewis acidic catalysts in organic synthesis. This was proved successful, and as was discussed in the Section 5.5, silyium ions (either preformed or in situ generated) very successfully catalyze a number of organic transformations, including classical Diels–Alder, Friedel–Crafts reactions, hydrosilylation, and small molecules activation.

Looking to the future, we can expect further developments in the synthesis of stable silylium ions, e.g., the isolation of the still elusive trigonal-planar tris(silyl)silylium ion derivatives $(R_3Si)_3Si^+$. The application of silylium ions as powerful Lewis acidic catalysts also requires further developments, including the expansion of the scope of its synthetic applicability. Given the very high current research activity in the field, one can anticipate major breakthroughs to be achieved in the near future.

References

1. (a) Müller, T. Cations of Group 14 Organometallics. *Adv. Organomet. Chem.* **2005,** *53*, 155–215.
   (b) Lee, V. Ya; Sekiguchi, A. Stable Silyl, Germyl and Stannyl Cations, Radicals, and Anions: Heavy Versions of Carbocations, Carbon Radicals, and Carbanions. *Acc. Chem. Res.* **2007,** *40*, 410–419.
   (c) Lee, V. Ya; Sekiguchi, A. *Organometallic Compounds of Low-Coordinate Si, Ge, Sn and Pb;* Wiley: Chichester, 2010. Chapter 1, pp. 1–43.
   (d) Müller, T. Silylium Ions. *In Functional Molecular Silicon Compounds I (Series "Structure and Bonding")* **2014,** *155*, 107–162.
2. (a) Jutzi, P.; Mix, A.; Rummel, B.; Schoeller, W. W.; Neumann, B.; Stammler, H.-G. The $(Me_5C_5)Si^+$ Cation: A Stable Derivative of HSi^+. *Science* **2004**, *305*, 849–851. See also: Jutzi, P. The Pentamethylcyclopentadienylsilicon(II) Cation: Synthesis, Characterization, and Reactivity. *Chem. Eur. J.* **2014,** *20*, 9192–9207.
   (b) Driess, M.; Yao, S.; Brym, C.; van Wüllen, C. Low-Valent Silicon Cations With Two-Coordinate Silicon and Aromatic Character. *Angew. Chem. Int. Ed.* **2006,** *45*, 6730–6733.
   (c) Müller, T. Stability, Reactivity, and Strategies for the Synthesis of Silyliumylidenes, RSi: $^+$. A Computational Study. *Organometallics* **2010,** *29*, 1277–1283.
   (d) Xiong, Y.; Yao, S.; Inoue, S.; Irran, E.; Driess, M. The Elusive Silyliumylidene $[ClSi:]^+$ and Silathionium $[ClSi=S]^+$ Cations Stabilized by Bis(Iminophosphorane) Chelate Ligand. *Angew. Chem. Int. Ed.* **2012,** *51*, 10074–10077.
   (e) Gerdes, C.; Saak, W.; Haase, D.; Müller, T. Dibenzosilanorbornadienyl Cations and Their Fragmentation Into Silyliumylidenes. *J. Am. Chem. Soc.* **2013,** *135*, 10353–10361.
   (f) Yeong, H.-X.; Xi, H.-W.; Li, Y.; Lim, K. H.; So, C.-W. A Silyliumylidene Cation Stabilized by an Amidinate Ligand and 4-Dimethylaminopyridine. *Chem. Eur. J.* **2013,** *19*, 11786–11790.
   (g) Breit, N. C.; Szilvási, T.; Suzuki, T.; Gallego, D.; Inoue, S. From a Zwitterionic Phosphasilene to Base Stabilized Silyliumylidene-Phosphide and Bis(silylene) Complexes. *J. Am. Chem. Soc.* **2013,** *135*, 17958–17968.
   (h) Ahmad, S. U.; Szilvási, T.; Inoue, S. A Facile Access to a Novel NHC-Stabilized Silyliumylidene Ion and C–H Activation of Phenylacetylene. *Chem. Commun.* **2014,** *50*, 12619–12622.
   (i) Ahmad, S. U.; Szilvási, T.; Irran, E.; Inoue, S. An NHC-Stabilized Silicon Analogue of Acylium Ion: Synthesis, Structure, Reactivity, and Theoretical Studies. *J. Am. Chem. Soc.* **2015,** *137*, 5828–5836.
   (j) Li, Y.; Chan, Y.-C.; Li, Y.; Purushothaman, I.; De, S.; Parameswaran, P.; So, C.-W. Synthesis of a Bent 2-Sialallene With a Perturbed Electronic Structure From a Cyclic Alkyl(amino) Carbene-Diiodosilylene. *Inorg. Chem.* **2016,** *55*, 9091–9098.
3. Chult, C.; Corriu, R. J. P.; Reye, C.; Young, J. C. Reactivity of Penta- and Hexacoordinate Silicon Compounds and Their Role as Reaction Intermediates. *Chem. Rev.* **1993,** *93*, 1371–1448.

4. (a) Bock, H.; Solouki, B. Organosilicon Radical Cations. *Chem. Rev.* **1995,** *95*, 1161–1190.
   (b) Bock, H.; Kaim, W. Organosilicon Radical Cations. *Acc. Chem. Res.* **1982,** *15*, 9–17.
5. (a) Schwarz, H. Positive and Negative Ion Chemistry of Silicon-Containing Molecules in the Gas Phase. In *The Chemistry of Organic Silicon Compounds;* Patai, S., Rappoport, Z., Eds.; , Chapter 7; Wiley: Chichester, 1989; pp 445–510.
   (b) Goldberg, N.; Schwarz, H. Gas-Phase Ion Chemistry of Silicon-Containing Molecules. In *The Chemistry of Organic Silicon Compounds;* Rappoport, Z., Apeloig, Y., Eds.; , Vol. 2; Wiley: Chichester, 1998; Part 2, Chapter 18, pp 1105–1142.
   (c) Fornarini, S. Ion-Molecule Reactions of Silicon Cations. In *The Chemistry of Organic Silicon Compounds;* Rappoport, Z., Apeloig, Y., Eds.; , Vol. 3; Wiley: Chichester, 2001; Chapter 18, pp 1027–1057.
   (d) Kochina, T. A.; Vrazhnov, D. V.; Sinotova, E. N.; Voronkov, M. G. Silylium Ions. *Russ. Chem. Rev.* **2006,** *75*, 95–110.
   (e) Kochina, T. A.; Vrazhnov, D. V. Silylium Cations and Their Analogs: Radiochemical Approach (A Review). *Glass Phys. Chem.* **2009,** *35*, 443–454.
   (f) Kochina, T. A., et al. Nuclear-Chemical Generation of Si and Ge-Centered Cations. In *Physical Organic Chemistry;* Burley, K. T., Ed.; , Chapter 4; Nova Science Publishers: Hauppauge, NY, 2010; pp 145–169.
   (g) Kochina, T. A.; Vrazhnov, D. V.; Ignatyev, I. S.; Sinotova, E. N.; Voronkov, M. G. Nucleogenic Silylium Cations and Their Analogs. *J. Organomet. Chem.* **2011,** *696*, 1331–1340.
6. Reed, C. A. Carboranes: A New Class of Weakly Coordinating Anions for Strong Electrophiles, Oxidants, and Superacids. *Acc. Chem. Res.* **1998,** *31*, 133–139.
7. (a) Bartlett, P. D.; Condon, F. E.; Schneider, A. Exchange of Halogen and Hydrogen Between Organic Halides and Isoparaffins in the Presence of Aluminum Halides. *J. Am. Chem. Soc.* **1944,** *66*, 1531–1539.
   (b) Corey, J. Y. Generation of a Silicenium Ion in Solution. *J. Am. Chem. Soc.* **1975,** *97*, 3237–3238.
8. (a) Lambert, J. B.; Zhang, S. Tetrakis(pentafluorophenyl)borate: A New Anion for Silylium Cations in the Condensed Phase. *J. Chem. Soc., Chem. Commun.* **1993,** 383–384.
   (b) Lambert, J. B.; Zhang, S.; Ciro, S. M. Silyl Cations in the Solid and in Solution. *Organometallics* **1994,** *13*, 2430–2443.
9. Xie, Z.; Manning, J.; Reed, R. W.; Mathur, R.; Boyd, P. D. W.; Benesi, A.; Reed, C. Approaching the Silylium (R_3Si^+) Ion: Trends With Hexahalo (Cl, Br, I) Carboranes as Counterions. *J. Am. Chem. Soc.* **1996,** *118*, 2922–2928.
10. (a) Lambert, J. B.; Zhao, Y. The Trimesitylsilylium Cation. *Angew. Chem. Int. Ed. Engl.* **1997,** *36*, 400–401.
    (b) Lambert, J. B.; Zhao, Y.; Wu, H.; Tse, W. C.; Kuhlmann, B. The Allyl Leaving Group Approach to Tricoordinate Silyl, Germyl, and Stannyl Cations. *J. Am. Chem. Soc.* **1999,** *121*, 5001–5008.
11. Ichinohe, M.; Fukui, H.; Sekiguchi, A. Synthesis of ${}^tBu_3E^+$ Nitrile Complexes by Oxidative Cleavage of ${}^tBu_3E{-}E{}^tBu_3$ (E = Si, Ge, Sn). *Chem. Lett.* **2000,** *29*, 600–601.
12. (a) Olah, G. A.; Field, L. D. Organometallic Chemistry. 19. Correlation of Silicon-29 and Carbon-13 NMR Chemical Shifts in Analogous Compounds. Application in Attempted Preparation of Silicenium Ions. *Organometallics* **1982,** *1*, 1485–1487.
    (b) Olah, G. A.; Laali, K.; Farooq, O. Organometallic Chemistry. 21. Silyl Trifluoromethanesulfonate (Triflate)-Boron Trichloride (Tribromide) Complexes. *Organometallics* **1984,** *3*, 1337–1340.
13. Schäfer, A.; Schäfer, A.; Müller, T. Silaimidazolium and Silaimidazolidinium Ions. *Dalton Trans.* **2010,** *39*, 9296–9303.
14. (a) Jutzi, P.; Bunte, E.-A. $[(\text{-}Me_5C_5)_2SiH]^+$ – A Novel Silyl Cation. *Angew. Chem. Int. Ed. Engl.* **1992,** *31*, 1605–1607.
    (b) Müller, T.; Jutzi, P.; Kühler, T. The Nature of Protonated Decamethylsilicocene, $(Me_5C_5)_2Si^+H$. *Organometallics* **2001,** *20*, 5619–5628.

15. (a) Power, P. P. Persistent and Stable Radicals of the Heavier Main Group Elements and Related Species. *Chem. Rev.* **2003,** *103*, 789–810.
   (b) Lee, V. Ya; Sekiguchi, A. Si-, Ge-, and Sn-Centered Free Radicals: From Phantom Species to Grams-Order-Scale Materials. *Eur. J. Inorg. Chem.* **2005,** 1209–1222.
   (c) Lee, V. Ya; Nakamoto, M.; Sekiguchi, A. Making Stable Radicals of Heavy Elements of Groups 14 and 13: The Might of Silyl Substitution. *Chem. Lett.* **2008,** *37*, 128–133.
   (d) Lee, V. Ya; Sekiguchi, A. *Organometallic Compounds of Low-Coordinate Si, Ge, Sn and Pb*, Wiley: Chichester, 2010. Chapter 2, pp. 45–88.
16. Sekiguchi, A.; Matsuno, T.; Ichinohe, M. Cyclotetrasilenyl: The First Isolable Silyl Radical. *J. Am. Chem. Soc.* **2001,** *123*, 12436–12437.
17. Sekiguchi, A.; Matsuno, T.; Ichinohe, M. The Homocyclotrisilenylium Ion: A Free Silyl Cation in the Condensed Phase. *J. Am. Chem. Soc.* **2000,** *122*, 11250–11251.
18. Sekiguchi, A.; Fukawa, T.; Nakamoto, M.; Lee, V. Ya; Ichinohe, M. Isolable Silyl and Germyl Radicals Lacking Conjugation With π-Bonds: Synthesis, Characterization, and Reactivity. *J. Am. Chem. Soc.* **2002,** *124*, 9865–9869.
19. Nakamoto, M.; Fukawa, T.; Sekiguchi, A. Silylium Ion to Silylium Ion Rearrangement Caused by 1,3-Methyl Migration. *Chem. Lett.* **2004,** *33*, 38–39.
20. Schäfer, A.; Reissmann, M.; Jung, S.; Schäfer, A.; Saak, W.; Brendler, E.; Müller, T. Synthesis of Silylium and Germylium Ions by a Substituent Exchange Reaction. *Organometallics* **2013,** *32*, 4713–4722.
21. Ottosson, C.-H.; Cremer, D. Nature of the $Si(SiMe_3)_3^+$ Cation in Aromatic Solvents. *Organometallics* **1996,** *15*, 5495–5501.
22. Inoue, S.; Ichinohe, M.; Yamaguchi, T.; Sekiguchi, A. A Free Silylium Ion: A Cyclotetrasilenylium Ion With Allylic Character. *Organometallics* **2008,** *27*, 6056–6058.
23. Küppers, T.; Bernhard, E.; Eujen, R.; Willner, H.; Lehmann, C. W. $[Me_3Si][R\text{-}CB_{11}F_{11}]$ – Synthesis and Properties. *Angew. Chem. Int. Ed.* **2007,** *46*, 6346–6349.
24. Pidun, U.; Stahl, M.; Frenking, G. The Silaguanidinium Cation and the Search for a Stable Silylium Cation in Condensed Phases. *Chem. Eur. J.* **1996,** *2*, 869–876.
25. Ottosson, C.-H.; Szabó, K. J.; Cremer, D. The Tris(9-borabicyclo[3.3.1]nonyl)silylium Cation: A Suggestion for a Weakly Coordinated Silylium Cation in Solution. *Organometallics* **1997,** *16*, 2377–2385.
26. Lambert, J. B.; Zhang, S.; Stern, C. L.; Huffman, J. C. Crystal Structure of a Silyl Cation With No Coordination to Anion and Distant Coordination to Solvent. *Science* **1993,** *260*, 1917–1918.
27. Reed, C. A.; Xie, Z.; Bau, R.; Benesi, A. Closely Approaching the Silylium Ion (R_3Si^+). *Science* **1993,** *262*, 402–404.
28. (a) Reed, C. A. The Silylium Ion Problem, R_3Si^+. Bridging Organic and Inorganic Chemistry. *Acc. Chem. Res.* **1998,** *31*, 325–332.
   (b) Lee, V. Ya; Sekiguchi, A. Silicon-, Germanium-, and Tin-Centered Cations, Radicals, and Anions. In *Reviews of Reactive Intermediate Chemistry;* Moss, R. A., Platz, M. S., Jones, M., Jr., Eds.; , Chapter 2; Wiley: Hoboken, 2007; pp 47–120.
29. Müller, T.; Zhao, Y.; Lambert, J. B. Computational Evidence for a Free Silylium Ion. *Organometallics* **1998,** *17*, 278–280.
30. Kim, K.-C.; Reed, C. A.; Elliott, D. W.; Mueller, L. J.; Tham, F.; Lin, L.; Lambert, J. B. Crystallographic Evidence for a Free Silylium Ion. *Science* **2002,** *297*, 825–827.
31. (a) Schuppan, J.; Herrschaft, B.; Müller, T. The Elusive 7-Silanorbornadien-7-ylium: Synthesis and Characterization of Nitrilium and Oxonium Ions Deriving From 2,3-Benzo-7-silanorbornadien-7-ylium. *Organometallics* **2001,** *20*, 4584–4592.
   (b) Steinberger, H.-U.; Müller, T.; Auner, N.; Maerker, C.; Schleyer, P. V. R. The 2-Silanorbornyl Cation: An Internally Stabilized Silyl Cation. *Angew. Chem. Int. Ed. Engl.* **1997,** *36*, 626–628.
   (c) Müller, T., et al. Unusually Stable Vinyl Cations. *Angew. Chem. Int. Ed.* **2000,** *39*, 3074–3077.
   (d) Müller, T. A Silyl Cation With a Three-Center Si–H–Si Bond. *Angew. Chem. Int. Ed.* **2001,** *40*, 3033–3036.

(e) Müller, T. A Model System for the Generation of Silyl Cationic Species of Different Reactivity and Stability. In Organosilicon Chemistry *V;* Auner, N., Weis, J., Eds.; Wiley–VCH: Weinheim, 2003; pp 34–44.

(f) Meyer, R.; Werner, K.; Müller, T. Persistent Bissilylated Arenium Ions. *Chem. Eur. J.* **2002,** *8*, 1163–1172.

(g) Auner, N.; Fearon, G.; Weis, J. Fascinating Organosilicon Compounds. Introduction. In Organosilicon Chemistry *III;* Auner, N., Weis, J., Eds.; Wiley–VCH: Weinheim, 1998; pp 1–16.

(h) Auner, N.; Müller, T.; Ostermeier, M.; Schuppan, J.; Steinberger, H.-U. Intramolecularly π-Stabilized Silyl Cations. In Organosilicon Chemistry *IV;* Auner, N., Weis, J., Eds.; Wiley–VCH: Weinheim, 2000; pp 127–139.

(i) Panisch, R.; Bolte, M.; Müller, T. Hydrogen- and Fluorine-Bridged Disilyl Cations and Their Use in Catalytic C – F Activation. *J. Am. Chem. Soc.* **2006,** *128*, 9676–9682.

(j) Panisch, R.; Bolte, M.; Müller, T. Structure and Bonding in Bissilylated Arenium Ions. *Organometallics* **2007,** *26*, 3524–3529.

(k) Müller, T.; Juhasz, M.; Reed, C. A. The X-ray Structure of a Vinyl Cation. *Angew. Chem. Int. Ed.* **2004,** *43*, 1543–1546.

32. Duttwyler, S.; Do, Q.-Q.; Linden, A.; Baldridge, K. K.; Siegel, J. S. Synthesis of 2,6-Diarylphenyldimethylsilyl Cations: Polar-π-Distribution of Cation Character. *Angew. Chem. Int. Ed.* **2008,** *47*, 1719–1722.

33. Romanato, P.; Duttwyler, S.; Linden, A.; Baldridge, K. K.; Siegel, J. S. Intramolecular Halogen Stabilization of Silylium Ions Directs Gearing Dynamics. *J. Am. Chem. Soc.* **2010,** *132*, 7828–7829.

34. (a) Romanato, P.; Duttwyler, S.; Linden, A.; Baldridge, K. K.; Siegel, J. S. Competition Between π-Arene and Lone-Pair Halogen Coordination of Silylium Ions? *J. Am. Chem. Soc.* **2011,** *133*, 11844–11846.

(b) Ducos, P.; Liautard, V., et al. Chiral Memory in Silylium Ions. *Chem. Eur. J.* **2015,** *21*, 11573–11578.

35. (a) Schäfer, A.; Reissmann, M.; Schäfer, A.; Saak, W.; Haase, D.; Müller, T. A New Synthesis of Triarylsilylium Ions and Their Application in Dihydrogen Activation. *Angew. Chem. Int. Ed.* **2011,** *50*, 12636–12638.

(b) Schäfer, A., et al. Synthesis of Silylium and Germylium Ions by a Substituent Exchange Reaction. *Organometallics* **2013,** *32*, 4713–4722.

36. (a) Wagner, H.; Baumgartner, J.; Müller, T.; Marschner, C. Shuttling Germanium Atoms Into Branched Polysilanes. *J. Am. Chem. Soc.* **2009,** *131*, 5022–5023.

(b) Albers, L.; Meshgi, M. A.; Baumgartner, J.; Marschner, C.; Müller, T. Wagner–Meerwein-Type Rearrangements of Germapolysilanes – A Stable Ion Study. *Organometallics* **2015,** *34*, 3756–3763.

(c) Albers, L.; Baumgartner, J.; Marschner, C.; Müller, T. Cationic Si–H–Si Bridges in Polysilanes: Their Detection and Targeted Formation in Stable Ion Studies. *Chem. Eur. J.* **2016,** *22*, 7970–7977.

37. (a) Klare, H. F. T.; Bergander, K.; Oestreich, M. Taming the Silylium Ion for Low-Temperature Diels–Alder Reactions. *Angew. Chem. Int. Ed.* **2009,** *48*, 9077–9079.

(b) Müther, K.; Fröhlich, R.; Mück-Lichtenfeld, C.; Grimme, S.; Oestreich, M. A Unique Transition Metal-Stabilized Silicon Cation. *J. Am. Chem. Soc.* **2011,** *133*, 12442–12444.

(c) Müther, K.; Hrobárik, P.; Hrobárikova, V.; Kaupp, M.; Oestreich, M. The Family of Ferrocene-Stabilized Silylium Ions: Synthesis, ^{29}Si NMR Characterization, Lewis Acidity, Substituents Scrambling, and Quantum-Chemical Analyses. *Chem. Eur. J.* **2013,** *19*, 16579–16594.

(d) Schmidt, R. K.; Klare, H. F. T.; Fröhlich, R.; Oestreich, M. Planar Chiral, Ferrocene-Stabilized Silicon Cations. *Chem. Eur. J.* **2016,** *22*, 5376–5383.

38. Rohde, V. H. G.; Pommerening, P.; Klare, H. F. T.; Oestreich, M. Intramolecularly Sulfur-Stabilized Silicon Cations as Lewis Acid Catalysts. *Organometallics* **2014,** *33*, 3618–3628.

39. Rohde, V. H. G.; Müller, M. F.; Oestreich, M. Intramolecularly Sulfur-Stabilized Silicon Cations With Chiral Binaphthyl Backbones: Synthesis of Three Different Motifs and Their Application in Enantioselective Diels–Alder Reactions. *Organometallics* **2015,** *34*, 3358–3373.

40. Ibad, M. F.; Langer, P.; Schulz, A.; Villinger, A. Silylium–Arene Adducts: An Experimental and Theoretical Study. *J. Am. Chem. Soc.* **2011,** *133*, 21016–21027.
41. Lehmann, M.; Schulz, A.; Villinger, A. Bissilylated Halonium Ions: $[Me_3Si-X-SiMe_3][B(C_6F_5)_4]$ (X = F, Cl, Br, I). *Angew. Chem. Int. Ed.* **2009,** *48*, 7444–7447.
42. Kessler, M.; Knapp, C.; Sagawe, V.; Scherer, H.; Uzun, R. Synthesis, Characterization, and Crystal Structures of Silylium Compounds of the Weakly Coordinating Dianion $[B_{12}Cl_{12}]^{2-}$. *Inorg. Chem.* **2010,** *49*, 5223–5230.
43. Ramírez-Contreras, R.; Bhuvanesh, N.; Zhou, J.; Ozerov, O. V. Synthesis of a Silylium Zwitterion. *Angew. Chem. Int. Ed.* **2013,** *52*, 10313–10315.
44. (a) Khalimon, A. Y.; Lin, Z. H.; Simionescu, R.; Vyboishchikov, S. F.; Nikonov, G. I. Persistent Silylium Ions Stabilized by Polyagostic Si–H•••Si Interactions. *Angew. Chem. Int. Ed.* **2007,** *46*, 4530–4533.
    (b) Tussupbayev, S.; Nikonov, G. I., et al. Dynamics of Si–H–Si Bridges in Agostically Stabilized Silylium Ions. *J. Phys. Chem. A* **2009,** *113*, 1199–1209.
45. Wehmschulte, R. J.; Laali, K. K.; Borosky, G. L.; Powell, D. R. Synthesis and Structure of the First Bridgehead Silylium Ion. *Organometallics* **2014,** *33*, 2146–2149.
46. (a) MacLachlan, M. J.; Bourke, S. C.; Lough, A. J.; Manners, I. Ring-Opening Protonolysis of Strained Silicon-Containing Rings: A New Approach to Ions With Silylium Character. *J. Am. Chem. Soc.* **2000,** *122*, 2126–2127.
    (b) Bourke, S. C.; MacLachlan, M. J.; Lough, A. J.; Manners, I. Ring-Opening Protonolysis of Sila[1] ferrocenophanes as a Route to Stabilized Silylium Ions. *Chem. Eur. J.* **2005,** *11*, 1989–2000.
47. Simonneau, A.; Biberger, T.; Oestreich, M. The Cyclohexadienyl-Leaving-Group Approach toward Donor-Stabilized Silylium Ions. *Organometallics* **2015,** *34*, 3927–3929.
48. Yao, S.; Xiong, Y.; van Wüllen, C.; Driess, M. From an N-Heterocyclic Silacyclopropene to Donor-Supported Silacyclopropenylium Cations. *Organometallics* **2009,** *28*, 1610–1612.
49. Xiong, Y.; Yao, S.; Driess, M. Coordination of *N*-Heterocyclic Carbene to H_2SiX_2 (X = Cl, OTf) and H_3SiOTf (OTf = OSO_2CF_3): Synthesis of Donor-stabilized Parent Silylium Salts With Four- and Five-Coordinate Silicon Atoms. *Z. Naturforsch.* **2013,** *68b*, 445–452.
50. Ochiai, T.; Szilvási, T.; Inoue, S. Facile Access to Stable Silylium Ions Stabilized by *N*-Heterocyclic Imines. *Molecules* **2016,** *21*, 1155 Available from: http://dx.doi.org/10.3390/molecules21091155.
51. (a) Inoue, S.; Epping, J. D.; Irran, E.; Driess, M. Formation of a Donor-Stabilized Tetrasilacyclobutadiene Dication by a Lewis Acid Assisted Reaction of an *N*-Heterocyclic Chloro Silylene. *J. Am. Chem. Soc.* **2011,** *133*, 8514–8517.
    (b) Yang, Y.-F.; Cheng, G.-J.; Zhu, J., et al. Silicon-Containing Formal 4–Electron Four-Membered Ring Systems: Antiaromatic, Aromatic, or Nonaromatic? *Chem. Eur. J.* **2012,** *18*, 7516–7524.
52. Yamaguchi, T.; Asay, M.; Sekiguchi, A. $[[(Me_3Si)_2CH]_2{}^iPrSi(NHC)Si = Si(Me)Si^iPr[CH(SiMe_3)_2]_2]^+$: A Molecule With Disilenyl Cation Character. *J. Am. Chem. Soc.* **2012,** *134*, 886–889.
53. Yamaguchi, T.; Sekiguchi, A.; Driess, M. An *N*-Heterocyclic Carbene – Disilyne Complex and Its Reactivity Toward $ZnCl_2$. *J. Am. Chem. Soc.* **2010,** *132*, 14061–14063.
54. (a) Ichinohe, M.; Igarashi, M.; Sanuki, K.; Sekiguchi, A. Cyclotrisilenylium Ion: The Persilaaromatic Compound. *J. Am. Chem. Soc.* **2005,** *127*, 9978–9979.
    (b) Igarashi, M.; Ichinohe, M.; Sekiguchi, A. Air-Stable Disilacyclopropene With a Si = C Bond and Its Conversion to Disilacyclopropenylium Ion: Silicon–Carbon Hybrid 2–π-Electron Systems. *J. Am. Chem. Soc.* **2007,** *129*, 12660–12661.
55. (a) Nishinaga, T.; Izukawa, Y.; Komatsu, K. The First Cyclic–π-Conjugated Silylium Ion: The Silatropylium Ion Annelated With Rigid σ-Frameworks. *J. Am. Chem. Soc.* **2000,** *122*, 9312–9313.
    (b) Nishinaga, T.; Izukawa, Y.; Komatsu, K. The First Silatropylium Ion Stabilized by Rigid σ-Frameworks: Preparation, Properties, and Some Reactions. *Tetrahedron* **2001,** *57*, 3645–3656.
56. Ishida, S.; Nishinaga, T.; West, R.; Komatsu, K. Generation and Aromaticity of 2-Silaimidazolium Ion, a New–π-Conjugated Silylium Ion. *Chem. Commun.* **2005,** 778–780.

57. For the quantitative estimation of the Lewis acidity of silylium ions, see: Grossekappenburg, H.; Reissmann, M.; Schmidtmann, M.; Müller, T. Quantitative Assessment of the Lewis Acidity of Silylium Ions. *Organometallics* **2015,** *34*, 4952–4958.
58. (a) Schulz, A.; Villinger, A. "Tamed" Silylium Ions: Versatile in Catalysis. *Angew. Chem. Int. Ed.* **2012,** *51*, 4526–4528.
    (b) Stahl, T.; Klare, H. F. T.; Oestreich, M. Main-Group Lewis Acids for C–F Bond Activation. *ACS Catal.* **2013,** *3*, 1578–1587.
59. Hara, K.; Akiyama, R.; Sawamura, M. Strong Counteranion Effects on the Catalytic Activity of Cationic Silicon Lewis Acids in Mukaiyama Aldol and Diels–Alder Reactions. *Org. Lett.* **2005,** *7*, 5621–5623.
60. (a) Schmidt, R. K.; Müther, K.; Mück-Lichtenfeld, C.; Grimme, S.; Oestreich, M. Silylium Ion-Catalyzed Challenging Diels–Alder Reactions: The Danger of Hidden Proton Catalysis With Strong Lewis Acids. *J. Am. Chem. Soc.* **2012,** *134*, 4421–4428.
    (b) Nödling, A. R.; Müther, K., et al. Ferrocene-Stabilized Silicon Cations as Catalysts for Diels–Alder Reactions: Attempted Experimental Quantification of Lewis Acidity and ReactIR Kinetic Analysis. *Organometallics* **2014,** *33*, 302–308.
    (c) Klare, H. F. T.; Oestreich, M. Silylium Ions in Catalysis. *Dalton Trans.* **2010,** *39*, 9176–9184.
61. Gatzenmeier, T.; van Gemmeren, M.; Xie, Y.; Höfler, D.; Leutzsch, M.; List, B. Asymmetric Lewis Acid Organocatalysis of the Diels–Alder Reaction by a Silylated C–H Acid. *Science* **2016,** *351*, 949–952.
62. Omann, L.; Oestreich, M. A Catalytic S_EAr Approach to Dibenzosiloles Functionalized at Both Benzene Cores. *Angew. Chem. Int. Ed.* **2015,** *54*, 10276–10279.
63. Wübbolt, S.; Oestreich, M. Catalytic Electrophilic C–H Silylation of Pyridines Enabled by Temporary Dearomatization. *Angew. Chem. Int. Ed.* **2015,** *54*, 15876–15879.
64. (a) Chen, Q.-A.; Klare, H. F. T.; Oestreich, M. Brønsted Acid-Promoted Formation of Stabilized Silylium Ions for Catalytic Friedel–Crafts C–H Silylation. *J. Am. Chem. Soc.* **2016,** *138*, 7868–7871.
    (b) Bähr, S.; Oestreich, M. Electrophilic Aromatic Substitution With Silicon Electrophiles: Catalytic Friedel–Crafts C–H Silylation. *Angew. Chem. Int. Ed.* **2017,** *56*, 52–59.
65. Allemann, O.; Duttwyler, S.; Romanato, P.; Baldridge, K. K.; Siegel, J. S. Proton-Catalyzed, Silane-Fueled Friedel–Crafts Coupling of Fluoroarenes. *Science* **2011,** *332*, 574–577.
66. Allemann, O.; Baldridge, K. K.; Siegel, J. S. Intramolecular C–H Insertion *vs.* Friedel–Crafts Coupling Induced by Silyl Cation-Promoted C–F Activation. *Org. Chem. Front.* **2015,** *2*, 1018–1021.
67. (a) Scott, V. J.; Çelenligil-Çetin, R.; Ozerov, O. V. Room-Temperature Catalytic Hydrodefluorination of C (sp^3)–F Bonds. *J. Am. Chem. Soc.* **2005,** *127*, 2852–2853.
    (b) Douvris, C.; Ozerov, O. V. Hydrodefluorination of Perfluoroalkyl Groups Using Silylium-Carborane Catalysts. *Science* **2008,** *321*, 1188–1190.
    (c) Douvris, C.; Nagaraja, C. M.; Chen, C.-H., et al. Hydrodefluorination and Other Hydrodehalogenation of Aliphatic Carbon–Halogen Bonds Using Silylium Catalysis. *J. Am. Chem. Soc.* **2010,** *132*, 4946–4953.
68. (a) Lühmann, N.; Panisch, R.; Müller, T. A Catalytic C–C Bond-Forming Reaction Between Aliphatic Fluorohydrocarbons and Arylsilanes. *Appl. Organomet. Chem.* **2010,** *24*, 533–537.
    (b) Kordts, N.; Borner, C.; Panisch, R.; Saak, W.; Müller, T. Hydrogen-Bridged Digermyl and Germylsilyl Cations. *Organometallics* **2014,** *33*, 1492–1498.
    (c) Meier, G.; Braun, T. Catalytic C–F Activation and Hydrodefluorination of Fluoroalkyl Groups. *Angew. Chem. Int. Ed.* **2009,** *48*, 1546–1548.
69. Schäfer, A.; Saak, W.; Haase, D.; Müller, T. Silyl Cation Mediated Conversion of CO_2 Into Benzoic Acid, Formic Acid, and Methanol. *Angew. Chem. Int. Ed.* **2012,** *51*, 2981–2984.
70. (a) Reissmann, M.; Schäfer, A.; Jung, S.; Müller, T. Silylium Ion/Phosphane Lewis Pairs. *Organometallics* **2013,** *32*, 6736–6744.
    (b) Schäfer, A.; Reissmann, M.; Schäfer, A., et al. Dihydrogen Activation by a Silylium Silylene Frustrated Lewis Pair and the Unexpected Reaction of a Protonated Silylene. *Chem. Eur. J.* **2014,** *20*, 9381–9386.

71. Silva Valverde, M. F.; Theuergarten, E.; Bannenberg, T.; Freytag, M.; Jones, P. G.; Tamm, M. Frustrated N-Heterocyclic Carbene–Silylium Ion Lewis Pair. *Dalton Trans.* **2015,** *44*, 9400–9408.
72. von Wolff, N.; Lefevre, G.; Berthet, J.-C.; Thuéry, P.; Cantat, T. Implications of CO_2 Activation by Frustrated Lewis Pairs in the Catalytic Hydroboration of CO_2: A View Using N/Si^+ Frustrated Lewis Pairs. *ACS Catal.* **2016,** *6*, 4526–4535.
73. Müther, K.; Oestreich, M. Self-Regeneration of a Silylium Ion Catalyst in Carbonyl Reduction. *Chem. Commun.* **2011,** *47*, 334–336.
74. Xu, T.; Chen, E. Y.-X. Silylium Dual Catalysis in Living Polymerization of Methacrylates via *In Situ* Hydrosilylation of Monomer. *J. Polym. Sci., Part A: Polym. Chem.* **2015,** *53*, 1895–1903.
75. Arii, H.; Nakabayashi, K.; Mochida, K.; Kawashima, T. Intramolecular Chain Hydrosilylation of Alkynylphenylsilanes Using a Silyl Cation as a Chain Carrier. *Molecules* **2016,** *21*, 999 Available from: http://dx.doi.org/10.3390/molecules21080999.
76. For the reactivity of trialkylsilyl carborane reagents, see: Reed, C. A. H^+, $CH_3{}^+$, and R_3Si^+ Carborane Reagents: When Triflates Fail. *Acc. Chem. Res.* **2010,** *43*, 121–128.
77. (a) Arii, H.; Yano, Y.; Nakabayashi, K.; Yamaguchi, S.; Yamamura, M.; Mochida, K., et al. Regioselective and Stereospecific Dehydrogenative Annulation Utilizing Silylium Ion-Activated Alkenes. *J. Org. Chem.* **2016,** *81*, 6314–6319.
    (b) Arii, H.; Kurihara, T.; Mochida, K.; Kawashima, T. Silylium Ion-Promoted Dehydrogenative Cyclization: Synthesis of Silicon-Containing Compounds Derived From Alkynes. *Chem. Commun.* **2014,** *50*, 6649–6652.
78. Hatanaka, Y.; Okada, S.; Minami, T.; Goto, M.; Shimada, K. Synthesis, X-ray Structure, and Ring-Opening Polymerization of Pentacoordinate Silicon-Bridged [1]Ferrocenophane. *Organometallics* **2005,** *24*, 1053–1055.
79. Zhang, Y.; Huynh, K.; Manners, I.; Reed, C. A. Ambient Temperature Ring-Opening Polymerization (ROP) of Cyclic Chlorophosphazene Trimer $[N_3P_3Cl_6]$ Catalyzed by Silylium Ions. *Chem. Commun.* **2008,** 494–496.
80. Müther, K.; Mohr, J.; Oestreich, M. Silylium Ion Promoted Reduction of Imines With Hydrosilanes. *Organometallics* **2013,** *32*, 6643–6646.
81. Ibad, M. F.; Langer, P.; Reiss, F.; Schulz, A.; Villinger, A. Catalytic Trimerization of Bis-silylated Diazomethane. *J. Am. Chem. Soc.* **2012,** *134*, 17757–17768.

CHAPTER 6

Silicon-Centered Radicals

Boris Tumanskii, Miriam Karni and Yitzhak Apeloig

Technion-Israel Institute of Technology, Haifa, Israel

Chapter Outline

Organosilicon Compounds.
DOI: http://dx.doi.org/10.1016/B978-0-12-801981-8.00006-X

6.1 Introduction

Silyl radicals are of significant importance in diverse areas, such as organic synthesis, silicon chemistry, and material sciences.[1] In organic synthesis, silyl radicals found use in a variety of applications. For example, they abstract efficiently halogen atoms from organic halides (Scheme 6.1A), and are therefore attractive alternatives to toxic reducing reagents such as organotin compounds.[1d,f,i] Silyl radicals are also used as catalysts; for example, they catalyze the cyclotrimerization of acetylenes[2] (Scheme 6.1B). Silyl radicals are also efficient photoinitiators for polymerization[3] and they are intermediates in the synthesis and photodegradation of polysilanes, an important class of polymers (Scheme 6.1C).[4]

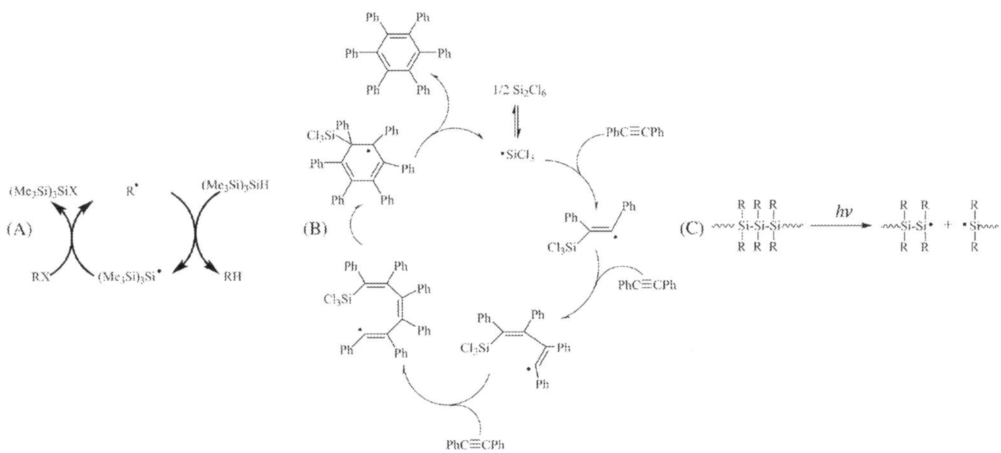

Scheme 6.1
Silyl radicals as (A) Reducing agents in organic synthesis, (B) As catalyst for alkyne cyclotrimerization, and (C) As intermediates in photodegradation of polysilanes.

Hydrosilation of unsaturated systems is a highly important industrial process[1f,5] which is usually conducted using a transition metal as a catalyst.[6] However, from economic and ecological considerations, a metal-free radical hydrosilation is highly desirable. The carbon-centered radical formed after initial silyl radical addition to an alkene is not effective in abstracting hydrogen from a trialkylsilane. This is due to the relatively strong Si–H bond in the trialkylsilanes. To overcome this problem the following three methods are mainly used in metal-free radical hydrosilylations: (1) application of tris(trimethylsilyl) silane which has a relatively weak Si–H bond[1i] resulting from stabilization of the silyl radical by the silyl substituents; (2) using thiyl radicals in conjunction with silicon hydrides[7]; and (3) application of silylated cyclohexadiene as a radical transfer hydrosilylating reagent (Scheme 6.2).[8]

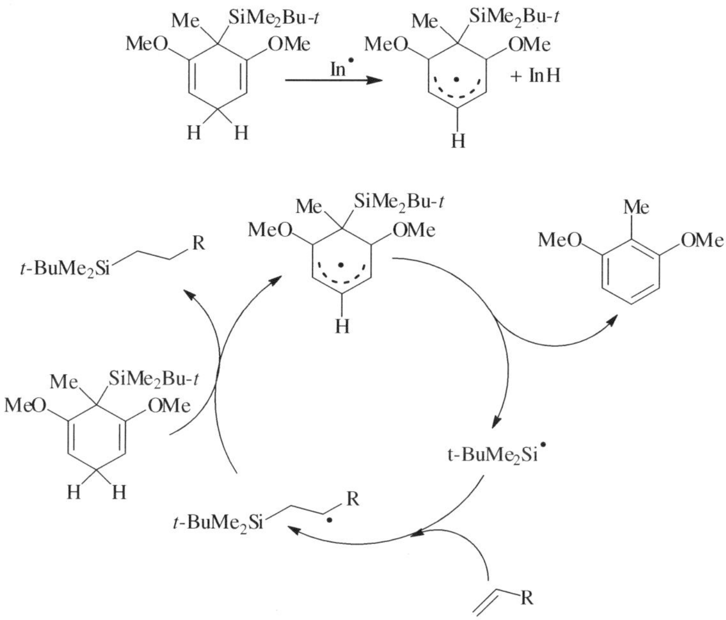

Scheme 6.2
Silylated cyclohexadiene as a radical transfer hydrosilylating reagent.[8]

Silyl radicals are often generated on silicon surfaces[9]—one of the technologically most important materials in the electronic industry. For example, significant recent attention is directed towards the synthesis of organic monolayers on silicon surfaces which can be modified for specific technological requirements,[10] as demonstrated in Scheme 6.3 for H–Si(111) surfaces.[11] Such surfaces undergo radical-activated reactions involving silyl radicals with a variety of terminal olefins to yield densely packed monolayers.[10]

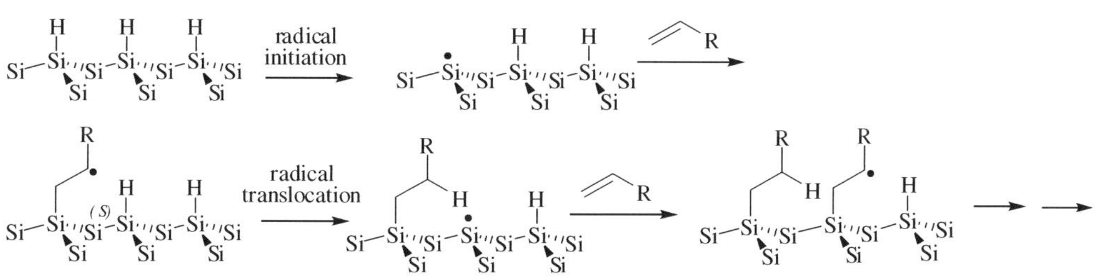

Scheme 6.3
Silyl radical activated olefination of a H–Si(111) surface.

The most exciting development in this field is the recent synthesis, isolation, and characterization of stable silyl radicals and their heavy congeners. This development is fundamentally interesting and opens many new technological opportunities.[1g,12]

Known stable organic radicals serve as "building blocks" for new materials which possess a unique combination of magnetic, electrochemical, and photochemical properties.[13] Stable aromatic organic radicals, such as dendritic and polymeric radicals of the triphenylmethyl type, have already been applied as organic magnets,[14] and as new paramagnetic materials.[15] Polynitroxyl-type radicals are applied in organic radical polymer batteries.[16] Fullerene-based stable radicals can potentially be used in electron-spin quantum computing.[17] Stable silyl radicals can in principle play a similar role. Indeed, recently high-power electrochemical energy storage systems based on the stable persilyl-substituted silyl radicals and other group 14 element radicals as anode materials were reported[12d] and they are discussed below.

The major focus of this review is on persistent and stable silyl radicals and in particular on stable polysilyl radicals which are the majority of this group. We define as a stable radical a radical that can be isolated, and handled as a pure compound. Radicals that are relatively long-lived and can be observed using conventional spectroscopic methods, but can not be isolated as pure compounds, are classified as "persistent."[18] The term "persistent" is subjective and we use it for radicals which have a lifetime longer than several hours. Spectral characteristics and reactivity data of short-lived alkyl-, alkoxy-, phenyl-, and halogen-substituted silyl radicals were previously reviewed comprehensively and the reader is referred to previous reviews for more details.[1c,f,9]

The most important spectroscopic tool for identifying and studying fundamental properties of free radicals is Electron Paramagnetic Resonance (EPR).[19] For the benefit of the nonexpert reader we therefore include a short introduction describing the basics of EPR spectroscopy.

6.2 Fundamentals of EPR Spectroscopy

6.2.1 Principles and Spectroscopy Techniques

EPR spectroscopy, also referred to as Electron Spin Resonance was discovered by the Russian physicist Yevgeny Zavoisky at Kazan State University in 1945.[20] EPR monitors directly unpaired electrons, that is the paramagnetism of free radicals, ions of variable valence, and of triplet state biradicals. In addition, EPR enables to study in detail the lifetime, structures, and other physical and chemical properties of these species.

EPR spectroscopy is based on the spin states of the unpaired electron and its associated magnetic moment. When an electron is placed within an applied magnetic field, H, the degeneracy of the two possible spin states of the electron is elevated resulting in the Zeeman effect. According to Eq. 6.1, where M_S is the magnetic quantum number equal to either $+1/2$

or $-1/2$, β is Bohr magneton and g is a proportionality coefficient usually termed the g-factor, the energy difference between the spin states in the presence of an external field is $g\beta$H. In the lower energy state the electron's magnetic moment, μ, is aligned with the magnetic field and a higher energy state occurs where μ is aligned opposite to the magnetic field. The two states are labeled by the projection of the electron spin, M_S, on the direction of the magnetic field, where $M_S = -1/2$ is the parallel state, and $M_S = +1/2$ is the antiparallel state (Fig. 6.1).

$$E = g\beta M_S H; \quad \text{i.e.,} \quad E = +1/2\, g\beta H \quad \text{or} \quad E = -1/2 g\beta H \tag{6.1}$$

The electron distribution between the two possible states obeys the Boltzmann law, that is $n+/n- = \exp(-\Delta E/kT)$ where $n+$ and $n-$ are the numbers of electrons at the upper and lower levels respectively, ΔE is the difference between the energies of the levels, k is the Boltzmann constant and T is the absolute temperature. According to Plank's law, electromagnetic radiation will be absorbed if $\Delta E = h\nu = g\beta H$, where ΔE is the difference in energy of the two states, h is Plank's constant, and ν is the radiation frequency.

In the absence of a magnetic field ($H = 0$) $n+ = n-$. When a magnetic field is applied, some of the electrons will absorb the alternating field energy and are elevated to the upper level. Simultaneously, an induced transition will occur from the upper to the lower level, leading to emission of energy. When equilibrium is achieved, the electron population of the lower level is slightly higher than that of the upper one, enabling a net absorption of high-frequency energy and giving rise to an EPR absorption. The radiation used in EPR spectroscopy is in the gigahertz range (X-diapason). In practice the frequency of the radiation is held constant while the magnetic field is varied in order to obtain an absorption spectrum. Absorption occurs when the magnetic field "tunes" to the two spin states so that their energy difference ΔE is equal to $h\nu$ (Fig. 6.1).

The EPR spectrum is characterized by the following parameters: g-factor, signal width, shape, and its hyperfine coupling constants ($hfccs$). As mentioned above, an EPR

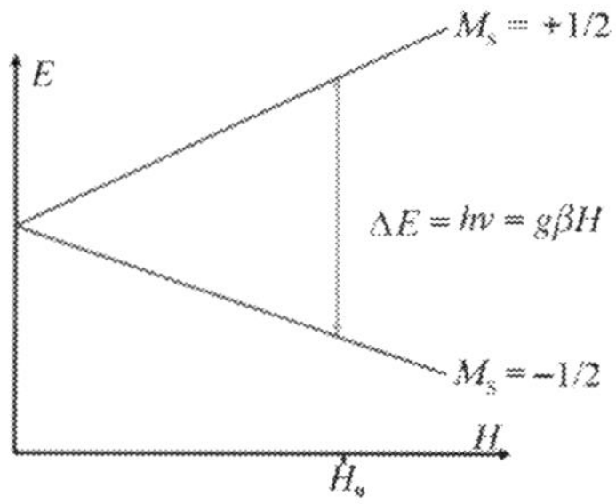

Figure 6.1
Energy levels in a magnetic field H of a system with one unpaired electron; β is Bohr magneton, g is a proportionality coefficient usually termed the g-factor.

spectrum is obtained by holding the frequency of radiation constant and varying the magnetic field.

As spectra can be obtained at a variety of frequencies, the field for resonance does not provide unique identification of the radicals. The proportionality *g*-factor, however, yields more useful information, because it is independent of the instrument used.

$$g = h\nu/\beta H_0 \tag{6.2}$$

For a free electron, the proportionality *g*-factor is 2.00232. For organic radicals, the *g* value is typically quite close to that of a free electron with values ranging from 1.999 to 2.014 (see Table 6.1). For transition metal compounds, large variations occur due to spin-orbit coupling which results in g-values ranging from 1.4 to 4.0. The *g*-factors of several types of organic and organoelement radicals are presented in Table 6.1. Note that alkyl-substituted silyl radicals, $R_3Si^•$ (R = alkyl), have *g*-values around 2.003, and silyl-substituted silyl radicals $(R_3Si)_3Si^•$ around 2.005.

6.2.1.1 The EPR spectrometer

The majority of EPR spectrometers operate at approximately 9.2–9.4 GHz, which corresponds to about 32 mm (microwave range). The sample may be irradiated continuously (i.e., continuous wave, abbreviated cw) or by pulses. The sample is placed in a resonant cavity which is located in the middle of an electromagnet. The absorption lines are detected when the separation of the energy levels is equal (or very close) to the radiation frequency, such that the source of microwave radiation and detector are contained within a microwave

Table 6.1: *g*-Factors of several types of organic and organoelement radicals

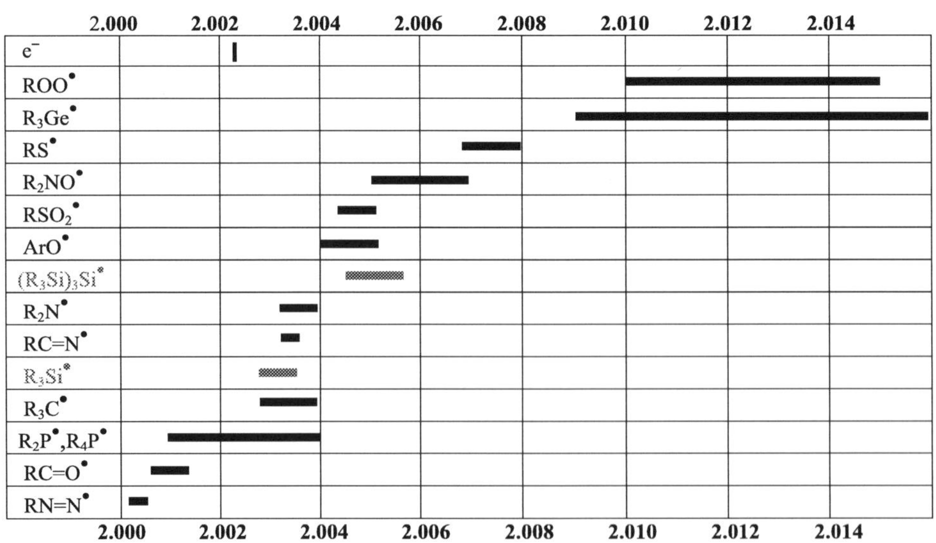

bridge control. Additionally, other components, such as an attenuator, field modulator, and amplifier are also included to enhance the performance of the instrument. A schematic diagram of a typical EPR spectrometer is shown in Fig. 6.2.

To minimize the noise from the microwave diode in steady state measurements, a magnetic field modulation scheme with phase sensitive detection is usually employed. As a result, the detected signal appears as a first derivative (Fig. 6.3). For example, in the Bruker spectrometer there is choice between 10.0 and 100 kHz modulation frequency.

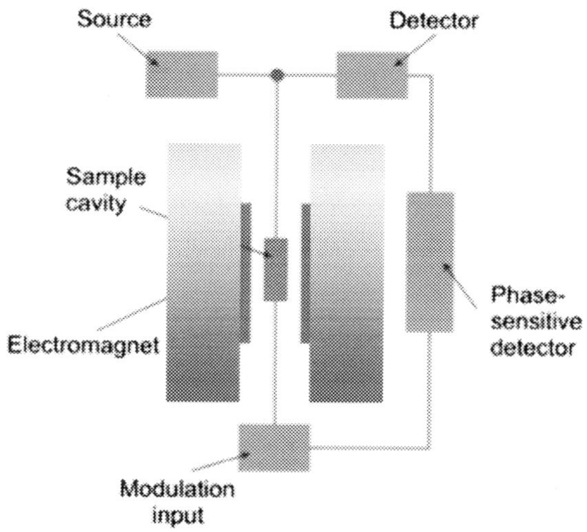

Figure 6.2
A schematic diagram of a typical EPR spectrometer.

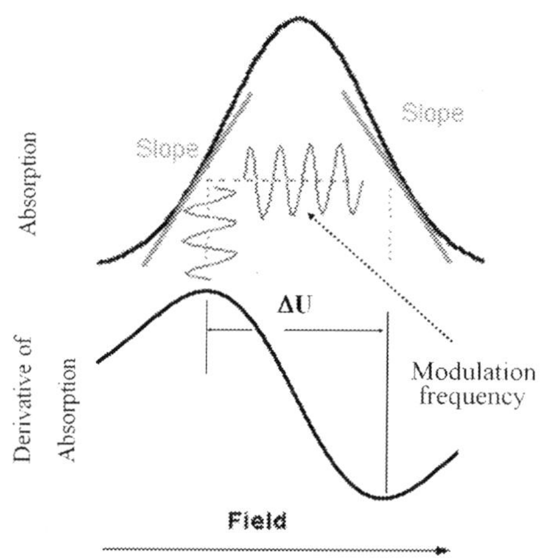

Figure 6.3
A magnetic field modulation scheme.

6.2.2 Line Width

The line width (ΔU) is defined as the distance between the points at which the absorption curve slope is maximal. Since the first derivative of the EPR spectra is recorded, the line width is measured as the distance between extremes of the experimental curve (Fig. 6.3). The line width gives the uncertainty in the excited state energy. The uncertainty is related by the Heisenberg relation ($\Delta E\tau \approx h$) to the mean lifetime of the spin state considered. In frequency units, the line width is $\Delta U = \Delta E/h = 1/\tau$ (τ is called the relaxation time). If τ is very small, then the line is strongly broadened and will not be observed experimentally. On the other hand if τ is too large, then the system that has absorbed the energy will not have time to reach its Boltzman equilibrium. In this case the microwave irradiation will equalize the energy level populations and decrease absorption intensity, this effect is called saturation. At a very high saturation the resonance may be unobservable. Paramagnetic complexes containing heavy atoms have short relaxation times (10^{-8} to 10^{-9} s) and short spin state lifetimes leading to relatively broad spectrum lines. Organic radicals including silyl radicals usually have long relaxation times (10^{-5} to 10^{-7} s) and thus narrow resonance lines (Fig. 6.4).

6.2.3 Line Shape

The line shape reflects the absorption intensity as a function of the magnetic field. In the solid state where usually there is no spin exchange interaction with the lattice, each of the

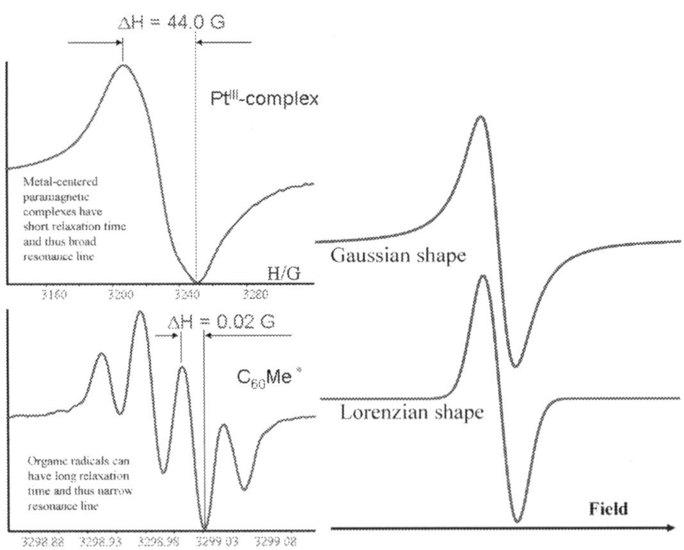

Figure 6.4
Examples of EPR spectra of paramagnetic species with different line width (left) and shape (right).

spins is in the local magnetic field built up by the neighboring spins. When this local magnetic field obeys Gauss distribution the EPR line will have a Gaussian shape (Fig. 6.4). In liquids, the local fields are averaged to zero, so a Lorentzian signal shape appears in cases where there is a high concentration of paramagnetic species. Also in the solid phase, the exchange interaction causes the central region of the line to have a Lorentzian shape (Fig. 6.4). Whether the Gaussian shape transforms to the Lorentzian shape depends on the exchange frequency. It is important to consider the line shape when simulating the EPR spectra.

6.2.4 Hyperfine Coupling

In addition to the applied magnetic field, unpaired electrons are also sensitive to their local environments. Frequently the nuclei of the atoms in a molecule or complex have a magnetic moment, which produces a local magnetic field at the measured unpaired electron. The resulting interaction between the unpaired electron and the nuclei is called the hyperfine interaction. Hyperfine coupling interactions provide important information about the studied paramagnetic compound including information about the number and identity of interacting nuclei in the compound as well as their distance from the unpaired electron helping in the assignment of the measured radical. As a result of the hyperfine coupling interaction, Eq. (6.1) is expanded to:

$$E = g\beta H_o M_S + aM_S m_I \tag{6.3}$$

where a is the *hfcc*, m_I is the nuclear spin quantum number of the neighboring nucleus, and M_S is the electron-spin quantum number. The coupling of the unpaired electron with its neighboring nuclei causes a splitting of the EPR signal. The splitting patterns that are observed in EPR spectra are determined by the same rules that apply to NMR spectroscopy. However, in EPR spectra it is more common to see coupling to nuclei with spins greater than ½. The number of lines which result from the coupling is given by Eq. (6.4), where N is the number of interacting equivalent nuclei and I is the spin ($n/2$).

$$\text{Number of lines} = 2NI + 1 \tag{6.4}$$

It is important to note that this formula only determines the number of lines in the spectrum, not their relative intensities. Coupling to a single nucleus with spin $n/2$ produces $(n + 1)$ lines of equal intensity. For example, coupling to a single vanadium nucleus ($I = 7/2$) will result in a spectrum of eight lines all of equal intensity. Coupling to N equivalent nuclei, each with spin ½ generates $(N + 1)$ lines, (i.e., $2NI + 1 = 2(N)(1/2) + 1 = N + 1$). However, since there are multiple nuclei interacting, the relative intensities of the lines follow a binomial distribution. For example, when an unpaired electron is coupled with two equivalent protons (spin = 1/2), the possible nuclear spin orientations will correspond to the following four projections: $-1(\uparrow\uparrow)$; $0\ (\uparrow\downarrow)$ and $(\downarrow\uparrow)$; and $+1\ (\downarrow\downarrow)$. As projection 0 is encountered twice as frequently as $+1$ and -1, the intensity of the central line will be twice more intense than the terminal ones (Fig. 6.5).

The value of the isotropic *hfcc* with nucleus *N*, *a*(*N*), is expressed by Eq. (6.5) where N = nucleus type, μ_0 is the vacuum permeability, μ_N is the nuclear magneton, g_N is the nucleus *g*-factor, and $\rho_N(0)$ is the net spin density (i.e., the difference between the α-spin and β-spin electron densities) at nucleus *N*.

$$a(N) = 2/3\mu_0 g_N \mu_N \rho_N(0) \tag{6.5}$$

Since the *s* orbital is the only orbital which does not have a node at the nucleus, the unpaired electron must reside in an orbital having an *s*-orbital contribution in order to contribute to *a*(*N*). The $2/3\mu_0\ g_N\ \mu_N\ \rho_N\ (0)$ term is constant for a specific nucleus and can be positive or negative. For example the magnetogyric ratio of ^{29}Si is negative, the signs of $a_{Si}(^{29}Si)$ are also negative. However, as the spin density at the nucleus, $\rho_N(0)$ can be positive (spin $\alpha > \beta$) or negative (spin $\alpha < \beta$), the sign of the *hfcc* depends on both the sign of the magnetic parameters of the specific nucleus and on the sign of the spin density at that nucleus.

It is important to note that ordinary EPR experiments yield only the absolute values of *hfcc*'s; therefore, to obtain the sign of *hfcc*'s and to obtain information on the mechanism which contributes to the *hfcc*, quantum mechanical calculations have to be performed. For main-group element radicals, two main mechanisms can contribute to $\rho_N(0)$ and therefore to the isotropic value of the *hfcc*[19a]: (1) The direct contribution (delocalization or

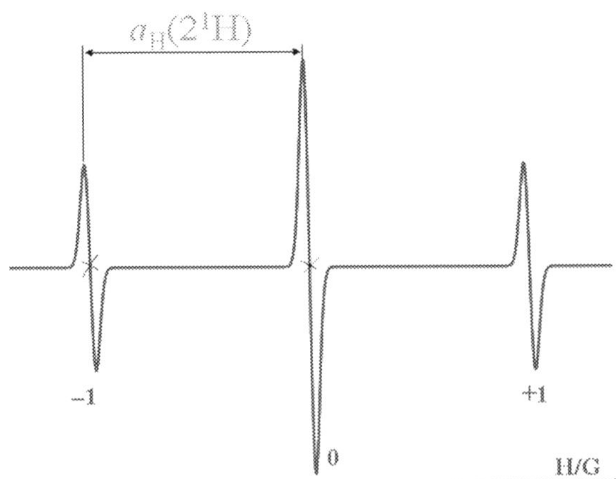

Figure 6.5
Theoretical EPR spectrum of an hypothetical radical containing two equivalent protons.

conjugation), which is due to spin density that arises from the spin occupation in the singly occupied molecular orbital (SOMO) having an s orbital contribution. This contribution is always positive. (2) The spin-polarization contribution (indirect), which results from a slightly different interaction of the unpaired α electron with the paired α and β electrons of inner shell orbitals or with those of the neighboring bonds. A positive spin density is induced at the nucleus where the direct spin density resides, and a negative spin density is induced at the adjacent atoms (spin polarization) (Fig. 6.6). The spin-polarization contribution is usually smaller than the direct contribution. The interplay of these two contributing mechanisms strongly depends on the radical geometry (e.g., degree of pyramidality of the radical center). In planar radicals, such as $H_3C^•$, so-called π radicals, the electron spin is located in a p orbital and therefore the main contribution to $^1H_\alpha$ *hfcc* is by spin polarization and the sign of $a(^1H_\alpha)$ is negative (Fig. 6.6A). In pyramidal radicals, for example $F_3C^•$, the SOMO has a significant s character at the C nucleus, and therefore the main contribution to $^{19}F_\alpha$ *hfcc* is by the direct mechanism and the sign of $a(^{19}F_\alpha)$is positive (Fig. 6.6B).

6.2.5 EPR of Triplet Biradicals

Biradicals are species with two unpaired electrons which occupy different orbitals. The ground state of a biradical can be either singlet (S = 0) or triplet (S = 1). Since Hund's rule applies to molecules with two electrons in singly occupied orbitals, the triplet state is usually more stable. The exchange interaction between the two paramagnetic fragments is an important characteristic of biradicals. When the interaction between the two radical centers is weak, the EPR spectrum of the biradical is simply the combination of the spectra of the independent radical centers. However, when the interaction is strong, the ground state multiplicity plays an important role. Fig. 6.7 shows schematically the energy levels for a triplet biradical with different symmetries.

(A) (B)

Figure 6.6

Examples of (A) A negative spin-density on the ^1H nuclei in planar $H_3C^•$ radical and (B) A positive spin-density on the ^{19}F nuclei in pyramidal $F_3C^•$ radical.

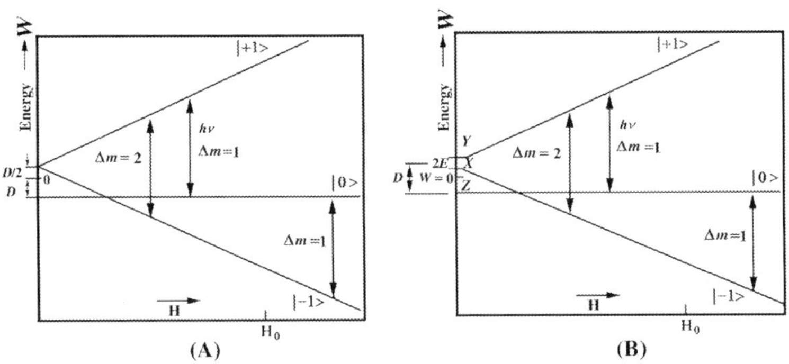

Figure 6.7

Energy levels as a function of the external magnetic field for a triplet biradical ($S = 1$, and $H \|Z$), (H is the external magnetic field and Z is the internal axis of the biradical): (A) With axial symmetry, where the X and Y axes are equivalent and the zero-field parameter $E = 0$. (B) Without symmetry, where X and Y axes are not equivalent and $E \neq 0$. Δm is the energy required for the transition between spin states; W is the energy of the spin levels; $H_0 = h\nu/g\beta$, D, and E are zero-field splitting parameters.

Fig. 6.7 demonstrates, that for a system with two unpaired electrons, the degeneracy of the energy of the spin state can be removed, even in the absence of an external magnetic field. This phenomenon is called zero-field splitting and parameters D and E measure the zero-field interaction, or the dipole–dipole coupling interaction. D and E can be extracted from the experimental EPR spectrum.

Experimental observation of a triplet state is possible only in the solid state or in a frozen glass. In liquid solution rotation of the triplet species leads to averaging of the dipole–dipole interaction to zero. The D (cm^{-1}) values are usually given as absolute magnitude, assuming that the spin–spin dipole interaction $|D'| = D/g\beta$(Gauss) is approximately given by Eq. (6.6), where g is the g-factor, β is Bohr magneton, and r is the average distance between the two unpaired electrons. Fig. 6.8 shows the calculated EPR spectrum for a biradical with axial symmetry (the X and Y axes are equivalent). If $E = 0$, the measured $|D'|$ value between line z and the spectrum center must be equal to the measured $|D'|$ value between the x,y-lines, otherwise $E \neq 0$. For example, a spectrum similar to that in Fig. 6.8, was observed for triplet planar hexasilylbenzene dianion[21] or for triplet coronene dication.[22]

$$|D'| = D/g\beta = (3/2)g^2\beta^2 r^3 \tag{6.6}$$

Another characteristic feature of triplet biradicals is the $\Delta M_S = 2$ transition also called the "half-field" transition. $\Delta M_S = 2$ refers to the transition between the $M_S = -1$ and $M_S = +1$ spin-quantum numbers. This "forbidden" transition is observed because the magnetic dipolar interaction admixes (at most orientations of the biradical relative to the external

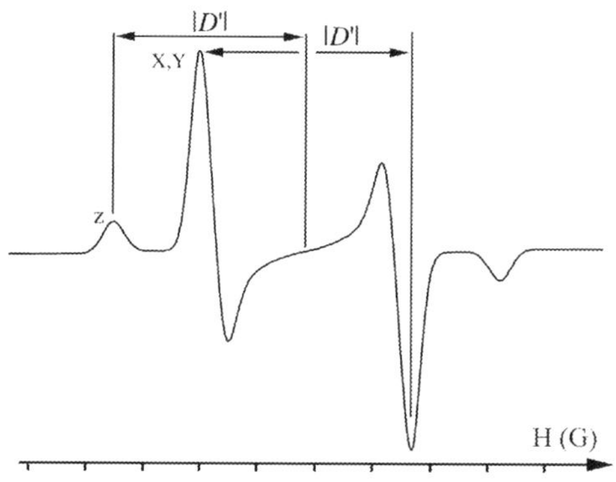

Figure 6.8
Theoretical EPR spectrum for a randomly oriented triplet system with axial symmetry, where the X and Y axes are equivalent in the region of $\Delta M_S = 1$ transition.

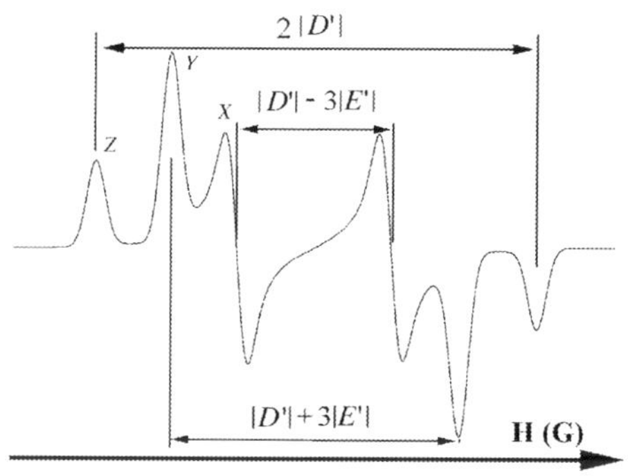

Figure 6.9
Theoretical EPR spectrum for a randomly oriented triplet biradical system with nonequivalent X and Y symmetry axes in the region of the $\Delta M_S = 1$ transition.

field) components with $M_S = \pm 1/2$ into each of the $M_S = -1$ and $M_S = 1$ levels. The amplitudes are of the order D/H, D being the dipole–dipole coupling parameter and H the external field. The amplitude of $\Delta M_S = 2$ transitions do not depend on the orientations of the biradical in the external field and the "half-field" region signal usually is a single absorption peak. The position of the "half-field" signal (H_{min}) is related to the electronic quadrupole splitting parameters D and E, according to Eq. (6.7),[19a] where h is Plank constant, v is the frequency of the instrument, β_o is Bohr's magneton, and g is the g-factor.

$$H_{min} = (2g_o\beta_o)^{-1}\{(hv)^2 - 4[(D^2/3)+E^2]\}^{1/2} \tag{6.7}$$

When the biradical has nonequivalent X and Y symmetry axes and $E \neq 0$, the $\Delta M_S = 1$ transition is characterized by three pairs of peaks with separations of $2|D'|$, $|D'| + 3|E'|$, and $|D'| - 3|E'|$, as shown in Fig. 6.9. For example, a spectrum similar to that in Fig. 6.9 was observed for a triplet aminyl diradical.[23]

In summary, the observation by EPR spectroscopy of the $\Delta M_S = 1$ and $\Delta M_S = 2$ transitions allows for unambiguous identification of the triplet states of biradicals, and to provide important information about their electronic properties by determining the parameters of zero-field splitting (D, E), their symmetry, and by obtaining a good estimate of the distance between the paramagnetic centers (Eq. 6.6).

6.2.6 Simulation of EPR Spectra

An indispensable step in the process of data analysis is obtaining a reliable and accurate simulation of the expected EPR spectrum for a particular paramagnetic system. In general, if the simulated spectrum does not fit the observed EPR spectrum this indicates that the suggested *hfcc* constants are wrong. There are several popular programs for simulating an EPR spectrum; for example, "EasySpin" toolbox for simulating and fitting a wide range of EPR spectra (included in the MATLAB (matrix laboratory) software system). The simpler "SimFonia" program of Bruker is a fast and friendly gateway to CW-EPR simulations on a Windows platform. These are fast and efficient algorithms based on perturbation theory that allow prompt feedback for simulations of both solution and powder EPR spectra.

6.3 Silyl Radicals—General Introduction

Silyl radicals are highly reactive species which were observed mainly as intermediates in radical propagating reactions. Similarly to carbon-centered radicals, great efforts were made to isolate and identify persistent and stable silyl radicals in order to gain more information on their properties. The results of these efforts are reviewed below.

We discuss first the factors which affect the stability, electronic properties, and structures of silyl radicals.

6.3.1 Controlling the Stability of Silyl Radicals

6.3.1.1 Thermodynamic stability—electronic effects

Understanding the effect of substituents on the thermodynamic and kinetic stability of silyl radicals is crucial for predicting their kinetic stability and their lifetime. However, the available data, either experimental or theoretical, is quite limited, in contrast to the extensively studied alkyl radicals (see e.g., refs. [18,24]). In the following section we discuss the electronic effects of α-substituents on silyl radicals and compare them with the corresponding substituted methyl radicals. The available experimental and theoretical data is presented in Tables 6.2 and 6.3. The experimental data was mainly derived from two sources: (1) the comprehensive NIST database;[24j] and (2) Handbook of Bond Dissociation

Table 6.2: Experimental[a] and calculated[b] radical stabilization energies (RSE) (ΔE, kcal mol^{-1}) derived from isodesmic Eqs. (6.8a) and (6.8b)

	ΔE of Eqs. (6.8a) and (6.8b) [c]			
	E = Si		E = C	
Radical	**Experiment**	**Theory**	**Experiment**	**Theory**
$(H_3C)H_2E^{\bullet}$	−1.0[d]	−1.0[e]	4.0[f]; 4.5[d,e]	3.8[f]; 3.9[e]
$(H_3C)_3E^{\bullet}$	−2.9[g]	−2.0 (−1.4)[h]	9.5[i]	6.8[j]
$PhH_2E^{\bullet}$	−0.4[d,e]	0.4[d,e]	14.6[f]; 16.5[d,e]	14.1[f]; 13.7[e]; 14.6[j]
$Ph_3E^{\bullet}$	3.2[g]	3.1[k]	24.1[l]	24.7[m]
$LiH_2E^{\bullet}$	—	12.0[n]	—	9.4[n]
$(H_3Si)H_2E^{\bullet}$	2.9[o]	2.8[k,o]	—	2.8[p]
$(H_3Si)_3E^{\bullet}$	—	6.3[h]	—	—
$(Me_3Si)H_2E^{\bullet}$	—	2.6[e]	5.0[e]	2.9[q]; 3.0[e]
$(Me_3Si)_3E^{\bullet}$	8.1[g]; 8.0[k]	7.2[h]; 8.8[k]	—	7.4[q]
$FH_2E^{\bullet}$	—	−2.6[e];	3.7[e]; 3.5[f]	3.8[e]; 3.5[f]; 3.1[j]
$F_3E^{\bullet}$	−11.5[g,d]	−7.7[h]; − 7.6[k]; − 9.7[r]	−2.4[d]	−1.0[m]
$ClH_2E^{\bullet}$	−1.3[i]	−0.2[e]; − 0.2[r]	5.5[f]; 5.7[e]	5.0[f]; 4.9[e]
$Cl_3E^{\bullet}$	−2.6[g]; 0.4[d]	0.8[k]; − 1.3[r]	11.2[d]	10.1[m]
$BrH_2E^{\bullet}$	—	1.9[h]; 0.6[r]	3.4[d]	3.4[f]
$Br_3E^{\bullet}$	1.3[s]; −12.0[d,s]	6.3[h],2.3[r]	9.6[d]	11.7[t]
$(H_2N)H_2E^{\bullet}$	—	1.0[e] (2.7[u])[e]	11.1[e] (14.0[u])[e]	12.0[e] (12.3[t])[e]; 10.7[j]
$(H_2N)_3E^{\bullet}$	—	−2.9[k]	—	11.2[q]
$(H_2P)H_2E^{\bullet}$	—	−2.9[v]	—	5.6[j]
$(CH_3O)H_2E^{\bullet}$	—	−2.8[e]	8.9[e]; 8.4[f]	8.7[e,f]; 7.6[j]
$(CH_3O)_3\ E^{\bullet}$	—	−3.9[k]	—	—
$(CH_3S)H_2E^{\bullet}$	—	4.5[h]	—	9.9[j,m]
$(CH_3S)_3\ E^{\bullet}$	4.7[d]	6.1[k]	—	—

[a]The experimental energies were derived from E—H bond dissociation energies or from the relevant standard heats of formation.[25a]
[b]For the level of calculations see the cited references.
[c]Positive values indicate stabilization by R.
[d]Ref. [25].
[e]Ref. [28].
[f]Ref. [24i].
[g]Ref. [29].
[h]Ref. [27a].
[i]Ref. [24j].
[j]Ref. [24g].
[k]Ref. [30].
[l]Ref. [31].
[m]Ref. [26g].
[n]Ref. [27b].
[o]ΔE derived from data in ref. [29] (exp.) and ref. [27a] (calc.).
[p]Ref. [24g], see also ref. [27b].
[q]From calculated bond dissociation free energies, ref. [32].
[r]Ref. [33].
[s]An uncertainly of up to ca. 10 kcal mol^{-1} in the experimental evaluation of the heat of formation of $Br_3Si^{\bullet}$ [34] leads to the large uncertainty in the estimated ΔE(8b).
[t]Ref. [35].
[u]For $((H_3C)_2N)H_2Si^{\bullet}$.
[v]Ref. [36].

Table 6.3: Experimental and calculated relative thermodynamic stabilities of substituted silyl versus methyl radicals calculated by Eq. (6.9)

	ΔE (kcal mol^{-1}) of Eq. (6.9) [a]	
Radical	**Experiment**[b]	**Theory**
$H_3E^{\bullet}$	−13.2[c]	−12.6[c]
$(H_3C)_3E^{\bullet}$	−0.8	0.5[d]
$Ph_3E^{\bullet}$	7.7	9.0[b]
$(H_3Si)_3E^{\bullet}$	—	−15.3[e]
$(Me_3Si)_3E^{\bullet}$	—	−16.1[e]
$F_3E^{\bullet}$	−4.1	−3.9 to −5.9[b]
$Cl_3E^{\bullet}$	0.6 to −2.4	−1.2
$(H_2N)_3E^{\bullet}$	—	1.5[b]

[a]Negative values imply that $R_3Si^{\bullet}$ is more stable than $R_3C^{\bullet}$.
[b]Derived from the experimental or calculated values given in Table 6.2 for Eq. (6.8b), E = C, Si and BDEs of CH_4 and SiH_4: ΔE(Eq. 6.9) = ΔE(Eq. 6.8b (E = C)) − ΔE(Eq. 6.8b(E = Si)) + ΔE(Eq. 6.9(R = H)).
[c]Ref. [27c].
[d]Apeloig et al., unpublished results at B3LYP/6−31 G(d).
[e]Ref. [38].

Energies in Organic Compounds.[25] We have also used data reported in the references that are cited in Table 6.2. The reported theoretical values are at different levels of calculations. For details the reader is referred to the original papers cited in Table 6.2. The accuracy of the variety of computational methods used to estimate the thermodynamic stability of organic radicals has been thoroughly discussed in references.[24c−e,26]

The thermodynamic stability of a substituted radical $R_nH_{3-n}E^{\bullet}$ is determined from the homolytic E−H bond dissociation energy (BDE) of the corresponding $R_nH_{4-n}E$. Substituents that reduce the BDE stabilize the radical, and vice versa. The thermodynamic stability of $RH_2E^{\bullet}$ relative to $H_3E^{\bullet}$, reflecting the effect of substituent R on the radical stability, is determined by the isodesmic hydrogen transfer Eq. (6.8a) (Eq. 6.8b for $R_3E^{\bullet}$).[26g] These energies are referred to as the *radical stabilization energy* (RSE) and provide the difference between the E−H BDE in EH_4 and H_3ER. The RSE energies are based on the relative BDEs of $R_nH_{4-n}E$ and the reference compound H_4E and thus provide *relative radical stabilities.* However, it should be noted that the trends given by these reactions depend not only on the effect of the substituents on the radical but also on their effect on the energy of the closed shell precursor $R_nH_{4-n}E$, due to interactions that are present in the closed shell precursor but not in the radical.[24e] The energies of isodesmic Eqs. (6.8a) and (6.8b) for E = Si and C and R = Me, Ph, Li, SiH_3, $SiMe_3$, F, Cl, Br, OMe, SMe, NH_2, and PH_2 are collected in the Table 6.2. Positive values of Eqs. (6.8a and 6.8b) indicate that R is stabilizing the radical.

$$RH_2E^{\bullet} + EH_4 \rightarrow REH_3 + H_3E^{\bullet} (E = C,\ Si) \tag{6.8a}$$

$$R_3E^{\bullet} + EH_4 \rightarrow R_3EH + H_3E^{\bullet}(E = C, Si) \quad (6.8b)$$

$$R_3C^{\bullet} + R_3SiH + \rightarrow R_3CH + R_3Si^{\bullet} \quad (6.9)$$

The thermodynamic stabilizing/destabilizing effects of the substituents are mainly attributed to their ability to interact with and delocalize the unpaired electron.[24h,i,26g,27a,b] For carbon-centered radicals, π-electron containing substituents, (e.g., allyl, phenyl) are highly stabilizing. Alkyl or silyl substituents having "pseudo π-type orbitals" are also stabilizing but to a smaller extent. For example, the calculated RSE of $PhH_2C^{\bullet}$, $MeH_2C^{\bullet}$, and $(Me_3Si)H_2C^{\bullet}$ are: 13.7−14.6, 3.9, and 2.8 kcal mol^{-1}, respectively (Table 6.2). Adjacent lone-pair orbitals have also a significant stabilizing effect; e.g., RSE of $(H_2N)H_2C^{\bullet}$ is 12.0 kcal mol^{-1}, (Table 6.2). The stabilizing effect of heteroatoms bearing lone-pairs decreases across the first row of the periodic table as a result of a decrease in their lone-pair donation ability and an increase in their σ-withdrawing ability;[24h,i] e.g., the RSE (calculated values, kcal mol^{-1}) follow the order: $(H_2N)H_2C^{\bullet}$ (10.7−12.3) > $(CH_3O)H_2C^{\bullet}$ (7.6−8.7) > $FH_2C^{\bullet}$ (3.1−3.8). Heteroatoms of the second row of the Periodic Table are in general more stabilizing than first row substituents, due to their higher donor ability (higher lying lone-pair orbitals) and lower electronegativity[24h] e.g., RSE (calculated, kcal mol^{-1}) of $ClH_2C^{\bullet}$ (5.0) > RSE of $FH_2C^{\bullet}$ (3.1−3.8) and RSE of $(CH_3S)H_2C^{\bullet}$ (9.9) > RSE of $(CH_3O)H_2C^{\bullet}$(7.6−8.7) (Table 6.2).

Examination of the data in Table 6.2 shows that silyl radicals are significantly less stabilized or more destabilized than the analogous carbon-centered radicals (Table 6.2). Substitution of $H_3Si^{\bullet}$ with one methyl group has a small destabilizing effect of 1 kcal mol^{-1}. Three methyl groups destabilize the silyl radical by 2.9 kcal mol^{-1}.[29] A single phenyl substituent has a minor effect (~ 0.5 kcal mol^{-1}) on the stability of $H_3Si^{\bullet}$, and $Ph_3Si^{\bullet}$ is stabilized by only ca. 3 kcal mol^{-1} relative to $H_3Si^{\bullet}$, in sharp contrast to $Ph_3C^{\bullet}$ which is stabilized by ca. 24 kcal mol^{-1} more stable than $CH_3^{\bullet}$. This is a clear demonstration of the fact that π-conjugation which is very effective for alkyl radicals is not effective for stabilizing silyl radicals.

The electropositive silyl substituent has a significant stabilizing effect on silyl radicals, and $(R_3Si)_3Si^{\bullet}$ (R = H, Me) is stabilized by ~7 − 9 kcal mol^{-1} relative to $H_3Si^{\bullet}$, similar to the effect on $H_3C^{\bullet}$. A β-silyl substitution is calculated to have no effect on the thermodynamic stability of silyl radicals.[37] The electropositive Li substituent has a large calculated stabilizing effect of 12 kcal mol^{-1}.[27b] On the other hand, the electronegative fluorine destabilizes the silyl radical and $F_3Si^{\bullet}$ is by 7.6−9.7 kcal mol^{-1} (11 kcal mol^{-1} experimental) less stable than $H_3Si^{\bullet}$ (Table 6.2). Chlorine substitution has a small effect on the stability of $H_3Si^{\bullet}$, while three bromine substituents stabilize the silyl radical by 2−6 kcal mol^{-1} (calculated), (1−12 kcal mol^{-1} exp., see comments in Table 6.2). The lone-pair donors, methoxy, amino, and phosphine substituents slightly destabilize silyl radicals in contrast to their stabilizing effect on the methyl radical. In contrast, a MeS substituent stabilizes the silyl radical (Table 6.2). In summary, for silyl radicals the substituents' stabilizing/destabilizing effect follows the order (Table 6.2): Li (strongly stabilizing) $>>$ SiR_3 $>$ MeS $>$ phenyl $>$ Br $>$ Cl $>$ NH_2 (slightly destabilizing) $\approx$ $PH_2 \approx Me \approx F$ (moderately destabilizing). Thus, in contrast to methyl radicals which can be stabilized significantly electronically by π-withdrawing or π-donor substituents or by σ-withdrawing electronegative substituents, silyl radicals are not significantly stabilized by electronic effects. An exception are electropositive substituents, e.g., Li, but such substituted silyl radicals are less accessible experimentally.

A comparison of the thermodynamic stability of trisubstituted silyl radicals and the corresponding trisubstituted alkyl radicals is of interest. Isodesmic Eq. (6.9) and Table 6.3 compares the RSEs of $R_3Si^{\bullet}$ to those of $R_3C^{\bullet}$; a positive value indicates that $R_3C^{\bullet}$ is more stable than $R_3Si^{\bullet}$. The discussion below is based on the theoretical calculated values, as more values are available. When both experimental and calculated valued are known, they are generally in good agreement.

The parent $H_3Si^{\bullet}$ radical is thermodynamically more stable than $H_3C^{\bullet}$ by $\sim$13 kcal mol^{-1}.[27c] However, as the effect of substituents on $H_3C^{\bullet}$ and $H_3Si^{\bullet}$ is very different (compare the energies of Eqs. 6.8a and 6.8b for E = C and E = Si, Table 6.2), the relative stabilities of substituted silyl and methyl radicals vary significantly depending on R. For example, three methyl groups stabilize the methyl radical by as much as 6.8 kcal mol^{-1} (calc., 9.5 kcal mol^{-1} [24j] (exp.),[24g] Table 6.2), in contrast to a destabilization of ca. 2 kcal mol^{-1} (calc.) of the silyl radical. These opposing substituent effects place $Me_3C^{\bullet}$ and $Me_3Si^{\bullet}$ at nearly the same thermodynamic stability (Eq. 6.9 for R = Me is nearly thermoneutral, Table 6.3). Similarly, the π-conjugating phenyl substituents in $Ph_3C^{\bullet}$ have a large stabilizing effect of 24 kcal mol^{-1} on the methyl radical while their effect on H_3Si is small, and thus $Ph_3C^{\bullet}$ is by 9.0 kcal mol^{-1} (calc.) more stable than $Ph_3Si^{\bullet}$ (Table 6.3). The electropositive silyl substituent has a similar stabilizing effect on both silyl and methyl radicals and therefore $(R_3Si)_3Si^{\bullet}$ is more stable than $(R_3Si)_3C^{\bullet}$ (R = H, Me) by $\sim$15–16 kcal mol^{-1} (Table 6.3). An amino group has a significant stabilizing effect on a methyl radical but a small destabilizing effect on the analogous silyl radical, and thus $(H_2N)_3Si^{\bullet}$ and $(H_2N)_3C^{\bullet}$ have a similar thermodynamic stability. In conclusion, depending on the substituents, a silyl radical can be either more stable or less stable thermodynamically than the corresponding methyl radical.

6.3.1.2 *Kinetic stabilization—steric effects*

Simple aliphatic carbon-[39] and silicon-centered radicals[40] are in general highly reactive intermediates and have very short lifetimes. For example, $MeH_2Si^{\bullet}$, $Me_2HSi^{\bullet}$, and $Me_3Si^{\bullet}$ were observed by EPR spectroscopy in solution only at low temperature (150–200K) and their half lifetimes ($\tau_{1/2}$) are at an estimated range of 0.005–0.01 s.[40] Bulky alkyl, aryl, and silyl substituents reduce the reactivity of simple C-centered radicals prolonging their lifetimes making them relatively long-lived.[18,41] Several stable carbon-centered radicals, stabilized by conjugation, e.g., triarylmethyl, were isolated.[42] One example of a stable carbon-centered silyl-substituted radical lacking resonance stabilization was also reported.[43] In contrast, triarylsilyl radicals are not stable enough to be isolated and their half-lifetimes are estimated to be 0.1–1 s.[44] However, a stable triarylgermyl radical was recently isolated.[45]

Kinetic stabilization of silyl radicals is effectively achieved by the introduction of bulky substituents. For example, $(Me_3Si)_3Si^•$ has a half-life of 0.1–1 s,[1c] $[(Me_3Si)_2CH]_3Si^•$ has a half-life of 10 min[46] and $(i\text{-}Pr_3Si)_3Si^•$ has a half-life of 3 days.[47] The use of even larger substituents, such as $t\text{-}Bu_2MeSi$, led recently to the isolation and structural characterization of stable silyl radicals and other group-14-centered radicals.[1e,g,12a,c,48] For example, $(t\text{-}Bu_2MeSi)_3Si^•$ is stable at room temperature for several years in an inert medium.[12a,c] These large silyl substituents sterically protect the radical center from undergoing typical radical reactions, such as dimerization, hydrogen abstraction, and disproportionation, and prolong their lifetime.

Another important factor controlling the lifetime of $R_3E^•$ is the identity of central atom E.[1c,18,39–41,44,49] The heavier the E atom, the longer is the E–E bond in the dimer formed by recombination of two $R_3E^•$ radicals and consequently the larger the substituents required to create sufficient steric repulsion which would inhibit dimerization and thus stabilize kinetically the radical. In $R_3C{-}CR_3$, the C–C distance is 1.5–1.6 Å[50] and six $R = Me_3Si$ substituents cause steric repulsion which is large enough to hinder dimerization, making $(Me_3Si)_3C^•$ persistent with a lifetime of several days at 298K.[18,41] In sharp contrast, the analogous Si-centered radical $(Me_3Si)_3Si^•$, which is thermodynamically more stable than $(Me_3Si)_3C^•$ (Table 6.3), is highly reactive with $\tau_{1/2}$ of only 0.1–1 s at 298K.[1c,40b] This large difference in the lifetimes between $(Me_3Si)_3C^•$ and $(Me_3Si)_3Si^•$ is due primarily to the fact that the Si–Si bond distance in the corresponding dimer (2.4 Å),[51] is much longer than that of a C–C bond and therefore the Me_3Si substituents are not large enough to suppress the dimerization of $(Me_3Si)_3Si^•$. Thus, the high kinetic stability of $(Me_3Si)_3C^•$ imposed by the substituents' steric effects overrides its lower inherent thermodynamic stability compared with that of $(Me_3Si)_3Si^•$, and higher inherent exothermicity of dimerization for $R_3C^•$ compared to $R_3Si^•$ (the C–C bond is much stronger than a Si–Si bond (90.2 kcal mol^{-1} for $H_3C{-}CH_3$ vs 76.7 kcal mol^{-1} for $H_3Si{-}SiH_3$[50])). The relationship between the dissociation energy of the central Si–Si bond in $R_3Si{-}SiR_3$ and the kinetic stability of the corresponding $R_3Si^•$ radicals is discussed in more detail in Section 6.6.3.

6.3.2 Structure

The methyl radical, $H_3C^•$, is planar.[52] In contrast, $H_3Si^•$ has a pyramidal C_{3v} structure (where the sum of the bond angles around the radical center ($\sum\theta$) is 332.6 degrees) and the inversion barrier is 5.3 kcal mol^{-1}.[53] $Me_3Si^•$ is slightly more pyramidal than $H_3Si^•$

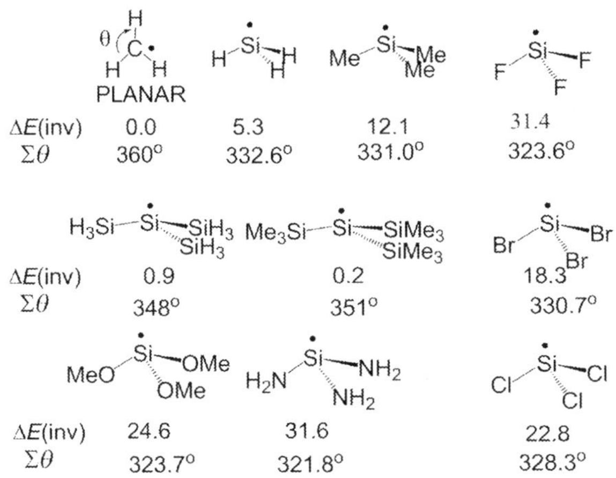

Figure 6.10

Calculated (UB3LYP/6−31 + G(d)) degree of pyramidality ($\sum\theta$) and inversion barriers ΔE (kcal mol^{-1}) of several silyl radicals (Y. Apeloig, et al., unpublished results).

($\sum\theta = 331.1$ degrees), but its inversion barrier is significantly higher, 12.1 kcal mol^{-1} (Apeloig et al., unpublished results). Calculations predict that α-trisubstituted silyl radicals $R_3Si^•$ (R = H_3C, H_2N, OH, F, H_3Si, H_2P, HS, Cl) are all pyramidal at the silicon radical center (Fig. 6.10).[54] The degree of pyramidality is larger for substituents that possess lone-pair electrons ($\sum\theta$ of 321−327 degrees). Similar observations were found for carbon-centered radicals.[24i] On the other hand, silyl substituents (and other electropositive substituents) decrease the degree of pyramidality and the inversion barrier; e.g., in $(H_3Si)_3Si^•$ $\sum\theta = 348$ degrees and the inversion barrier is only 0.9 kcal mol^{-1} and in $(Me_3Si)_3Si^•$ $\sum\theta = 351$ degrees and the inversion barrier is 0.2 kcal mol^{-1} (Fig. 6.10).

Larger silyl substituents decrease the degree of pyramidality and the inversion barrier even further, and X-ray crystallography analysis shows that $(t\text{-}Bu_2Me)_3Si^•$ is planar.[12c] More calculations on the structures and magnetic properties of various short-lived silyl radicals are available, and the interested reader is referred to references[49b,54,55] for more information. Accurate quantum mechanical calculations on the electronic and structural and environmental effects on the EPR parameters of a large variety of organic radicals can be found in ref. [56].

6.3.3 EPR Spectra of Silyl Radicals

The EPR spectrum of silyl radicals consists of a central singlet signal characterized by its *g*-factor which results from the unpaired electron residing on the nonmagnetic ^{28}Si nuclei (95.33% natural abundance). This central signal is split by the central ^{29}Si nuclei ($I = ½$, 4.67% abundance) and by other magnetic elements present in the substituents. Substituents, bonded directly to the silicon carrying the unpaired electron (α-position) affect significantly the EPR parameters of silyl radicals; the $a_{Si}(^{29}Si_\alpha)$ *hfccs* vary from 55.5 G for

$(i\text{-}Pr_3Si)_3Si^{\bullet}$[47] to 416 G for $Cl_3Si^{\bullet}$[57] and to 498 G for $F_3Si^{\bullet}$.[58] These large variations in a (^{29}Si) *hfccs* are attributed to changes in the geometry of the radical center, affecting its hybridization, and to the electronic effects of the α-substituents. Based on theoretical calculations it was suggested that an increase in the degree of pyramidality at the radical center and a higher electronegativity of the α-substituents lead to a larger contribution of the 3s-orbital to the SOMO and therefore to a larger $a_{Si}(^{29}Si)$.[49b,54] Electronegative substituents (e.g., F) increase the degree of pyramidalization at the radical center and thus increase $a_{Si}(^{29}Si)$[58] while electropositive substituents (e.g., silyl groups, metals) lower the degree of pyramidality decreasing $a_{Si}(^{29}Si)$ (Fig. 6.10).[49b,54] The latter effects are clearly evident in the spectra of the silyl radicals shown in Fig. 6.11, where silyl groups replace

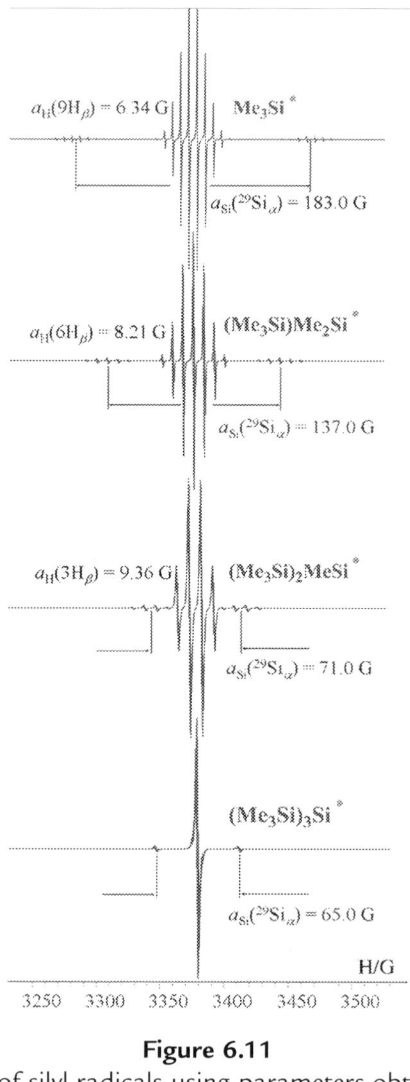

Figure 6.11
Simulation of EPR spectra of silyl radicals using parameters obtained experimentally.[1c,40,60]

consecutively the Me groups in $Me_3Si^{\bullet}$. In addition, bulky substituents reduce the degree of pyramidality at the radical center as a result of reduced steric repulsion between the substituents in more planar structures. Thus, the larger are the substituents the smaller is $a_{Si}(^{29}Si_{\alpha})$. This is clearly exemplified in the EPR data presented in Tables 6.4 and 6.5.

Theoretical quantum-mechanical calculations are very helpful in characterizing and studying silyl radicals. Thus, the calculated geometry and EPR spectrum are crucial for the correct assignment of a measured EPR spectrum to a specific radical and thus for its identification. In addition, the calculations allow to evaluate the role of the different effects that determine the magnetic properties of a given radical. The importance of theory to radical research led in recent years to numerous computational studies using a variety of theoretical approaches and levels of computational sophistication.[12a,49b,54–56,59]

Many of the recent studies of silyl radicals concentrated on silyl-substituted silyl radicals, some of which are stable, and these are discussed below in more detail.

Table 6.4: EPR parameters, degree of pyramidality and half-life time (at 298K) of bis(silyl) substituted silyl radicals

Radical	$a_{Si}(^{29}Si_{\alpha})$ G	$a_X(X)$,[a] G	g	$\Sigma\theta$ (Si)[b]	Half-life Time	Refs.
$H(tBu_2MeSi)_2Si^{\bullet}$ (**1**)	78.0[c]	11.7 ($^1H_{\alpha}$)	2.005	354.1	~ 1 s	[49b]
$H(iPr_3Si)_2Si^{\bullet}$ (**2**)	72.7	12.4 ($^1H_{\alpha}$)	2.005	352.1	~ 1 s	[49b]
$H(tBuMe_2Si)_2Si^{\bullet}$ (**3**)	79.8[c]	11.0 ($^1H_{\alpha}$)	2.005	350.4	~ 1 s	[49b]
$H[(Me_3Si)Me_2Si]_2Si^{\bullet}$ (**4**)	81.5	9.6 ($^1H_{\alpha}$)	2.005	348.9	~ 1 s	[49b]
$H[(Me_3Si)_3Si]_2Si^{\bullet}$ (**5**)	83.7	10.1 ($^1H_{\alpha}$), 4.5 ($^{29}Si_{\beta}$), 7.4 ($^{29}Si_{\gamma}$)	2.005	350.2	~ 1 s	[49b]
$Me[(Me_3Si)_3Si]_2Si^{\bullet}$ (**6**)	90.0	9.28 ($^1H_{\beta}$), 0.44 ($^1H_{\gamma}$)	2.0045	–	~ 1 s	[1f]
$F(tBu_2MeSi)_2Si^{\bullet}$ (**7**)	110.0	31.9 ($^{19}F_{\alpha}$)	2.0032	344.3	~ 3 s	[49b]
$Ph(tBu_2MeSi)_2Si^{\bullet}$ (**8**)	72.0	6.5 ($^{29}Si_{\beta}$)	2.0044	357.1	Persistent	[59]
$Mes(tBu_2MeSi)_2Si^{\bullet}$ (**9**)	69.2	7.4 ($^{29}Si_{\beta}$)	2.0045	359.9	Stable	[59]
3,5-$tBu_2C_6H_3(tBu_2MeSi)_2Si^{\bullet}$ (**10**)	70.2	7.0($^{29}Si_{\beta}$)	2.0053	360.0[d]	Stable	[12b]
4-$PhC_6H_4(tBu_2MeSi)_2Si^{\bullet}$ (**11**)	69.7	6.5 ($^{29}Si_{\beta}$)	2.0044	–	Stable	[12b]
4-$tBuC_6H_4(tBu_2MeSi)_2Si^{\bullet}$ (**12**)	69.9	7.0 ($^{29}Si_{\beta}$)	2.0045	–	Stable	[12b]
$(tBuHN)(tBuMe_2Si)_2Si^{\bullet}$ (**13**)	91.7	2.2 ($^{14}N_{\alpha}$)	2.0036	340.4	~ 3 min	[59]
$Me(iPr_3Si)_2Si^{\bullet}$ (**14**)	73.5	6.0 ($^{29}Si_{\beta}$), 9.5 ($^1H_{\beta}$)	2.0047	354.8	~ 5 min	[59]
1-Ad$(tBuMe_2Si)_2Si^{\bullet}$ (**15**)	79.1	14.5 ($^1H_{\beta}$), 4.0 ($^{29}Si_{\beta}$)	2.0045	346.6	Persistent	[59]

[a]Atom X responsible for the quoted *hfcc* is given in parenthesis.
[b]Calculated sum of bond angles at the radical center (in deg.)[49b,59]
[c]Values are extrapolated from a lower temperature experiment using a correction of 0.1 G/10K.
[d]Determined by X-ray crystallography.[12b]
Source: Adapted with permission from Tumanskii, B.; Karni, M.; Apeloig, Y. Persistent and Stable Silyl Radicals. In Encyclopedia of Radicals in Chemistry, Biology and Materials; *John Wiley & Sons, Ltd., 2012, Copyright 2012 Wiley-VCH Verlag GmbH & Co KGaA.*

Table 6.5: EPR parameters, half-life times (at 298K) and calculated degree of pyramidality of tris(silyl)silyl radicals

Radical	$a_{Si}(^{29}Si_{\alpha})$, G	$a_X(X)$,[a] G	g	$\Sigma\theta$ (Si)[b]	Half-life Times ($\tau_{1/2}$)	Refs.
$(Me_3Si)_3Si^{\bullet}$ (**18**)	63.8	7.1($^{29}Si_{\alpha}$), 0.43($^{1}H_{\gamma}$)	2.0053	351	~ 1 s	[1c,40b]
$(EtMe_2Si)_3Si^{\bullet}$ (**19**)	62.8	7.1 ($^{29}Si_{\beta}$), 0.37 (Me $^{1}H_{\gamma}$), 0.14 (Et $^{1}H_{\gamma}$)	2.0060	—	~ 1 s	[64]
$(Et_2MeSi)_3Si^{\bullet}$ (**20**)	60.3	7.3 ($^{29}Si_{\beta}$), 3.2 ($^{13}C_{\gamma}$), 0.27 (Et $^{1}H_{\gamma}$), 0.15 (Me $^{1}H_{\gamma}$)	2.0060	—	~ 1 s	[64]
$(Et_3Si)_3Si^{\bullet}$ (**21**)	57.2	7.9 ($^{29}Si_{\beta}$), 3.0 ($^{13}C_{\gamma}$), 0.12 ($^{1}H_{\gamma}$)	2.0063	—	~ 1 s	[64]
$(Me_3SiMe_2Si)_3Si^{\bullet}$ (**22**)	57.2	7.4 ($^{29}Si_{\beta}$), 4.0 ($^{13}C_{\gamma}$), 0.1 ($^{1}H_{\gamma}$)	2.0060	—	~ 1 s	[37,65]
$(t\text{-}Bu_2HSi)(Me_3Si)_2Si^{\bullet}$ (**23**)	61.6	7.9 ($^{29}Si_{\beta}$)	2.0056	—	~ 1 s	[66]
$(i\text{-}Pr_3Si)_3Si^{\bullet}$ (**24**)	55.5	8.1 ($^{29}Si_{\beta}$), 2.2 ($^{13}C_{\gamma}$)	2.0061	360	Persistent	[47]
$(t\text{-}BuMe_2Si)_3Si^{\bullet}$ (**25**)	56.8	8.0 ($^{29}Si_{\beta}$), 0.1 (tBu ^{1}H), 0.3 (Me ^{1}H)	2.0060	358.3	~ 1 s	[66,67]
$(t\text{-}Bu_2MeSi)_3Si^{\bullet}$ (**26**)	58.0	8.0 ($^{29}Si_{\beta}$), 0.14 (tBu ^{1}H), 0.21 (Me ^{1}H), 2.8 ($^{13}C_{\gamma}$)	2.0056	360	Stable	[12a,12c]
$[(Me_3Si)_3Si]_2(t\text{-}Bu_2MeSi)Si^{\bullet}$ (**27**)	63.9	8.0 ($^{29}Si_{\beta}$),	2.0053	360	Persistent	[63]
$(t\text{-}Bu_3Si)(t\text{-}Bu_2MeSi)_2Si^{\bullet}$ (**28**)	60.5	8.0 ($^{29}Si_{\beta}$)	2.0053	360	Stable	[59]
$(t\text{-}Bu_2MeSi)_2HSi(t\text{-}Bu_2MeSi)_2Si^{\bullet}$ (**29**)	59.3	7.3 ($^{29}Si_{\beta}$), 10.4 ($^{29}Si_{\gamma}$)	2.0051	359.8	Stable	[12a]
$(i\text{-}Pr_3Si)_2HSi(i\text{-}Pr_3Si)_2Si^{\bullet}$ (**30**)	57.1	7.5 ($^{29}Si_{\beta}$), 10.0 ($^{29}Si_{\gamma}$), 1.1 ($^{1}H_{\beta}$)	2.0054	360	Persistent	[12a]
$(i\text{-}Pr_3Si)_2HSi(t\text{-}Bu_2MeSi)_2Si^{\bullet}$ (**31**)	58.3	7.3 ($^{29}Si_{\beta}$), 10.5 ($^{29}Si_{\gamma}$), 0.65 ($^{1}H_{\beta}$)	2.0051	—	Persistent	[59]
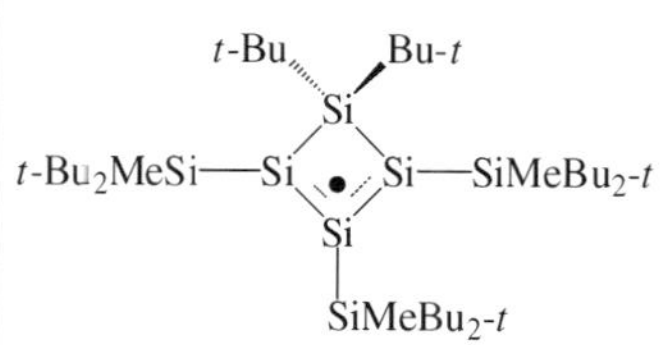 (**32**)	39.0 (2^{29}Si)	15.5 ($^{29}Si_{\beta}$)	2.0058	—	Stable	[68]

[a]Atom X responsible for the quoted *hfcc* is given in parenthesis.
[b]Calculated sum of the bond angles (deg.) around the radical center.[59]
Source: Adapted with permission from Tumanskii, B.; Karni, M.; Apeloig, Y. Persistent and Stable Silyl Radicals. In Encyclopedia of Radicals in Chemistry, Biology and Materials; *John Wiley & Sons, Ltd., 2012, Copyright 2012 Wiley-VCH Verlag GmbH & Co KGaA.*

6.4 Silyl Substituted Silyl Radicals

6.4.1 Mono-Silyl Substituted Silyl Radical

Only one example of a mono-silyl substituted silyl radical $(Me_3Si)Me_2Si^{\bullet}$, was reported.[40] $a(^{29}Si_{\alpha}) = 137.0$ G (Fig. 6.11) which is significantly smaller than the corresponding constant for $Me_3Si^{\bullet}$ ($a(^{29}Si_{\alpha}) = 183.0$ G; Fig. 6.11). This significant decrease in $a(^{29}Si_{\alpha})$ results from the neighboring electropositive silyl-substituent. Introduction of one silyl substituent does not provide enough kinetic stabilization and $(Me_3Si)Me_2Si^{\bullet}$ is short-lived ($\tau_{1/2} \sim 0.1$ s).

6.4.2 Bis(Silyl)-Substituted Silyl Radicals

A much wider class of bis(silyl) silicon-centered radicals is known. The EPR parameters, degree of pyramidality, and the kinetic stability of both short-lived and stable bis(silyl)silyl radicals $X(R_3Si)_2Si^{\bullet}$, **1**–**15**, where X = H, Me, 1-Ad, aryl, F, and amino are collected in Table 6.4.

6.4.2.1 $H(R_3Si)_2Si^{\bullet}$

Hydrogen-substituted bis(silyl)-substituted silyl radicals $H(R_3Si)_2Si^{\bullet}$ (**1**–**5**) were generated by hydrogen abstraction (Eq. 6.10) and were comprehensively studied by EPR spectroscopy and DFT (density functional theory) calculations (Table 6.4).[49b]

$$t\text{-BuOOBu-}t \xrightarrow{h\nu} t\text{-BuO}^{\bullet} \xrightarrow{R_2SiH_2} t\text{BuOH} + \underset{\mathbf{(1-5)}}{R_2HSi^{\bullet}} \qquad (6.10)$$

Two bulky silyl substituents at the Si-radical center, even as large as t-Bu_2MeSi, are not sufficient to prevent their fast dimerization, and the half-life of radicals **1**–**5** is around $\tau_{1/2} \sim 1$ s (Table 6.4). This is true also for the methyl-substituted radical **6**. The most reactive radical in the series, $H(Me_3Si)_2Si^{\bullet}$, was characterized only by its *hfcc* with $^1H_{\alpha}$ (11.0 G).[61] The calculated geometries of $H(R_3Si)_2Si^{\bullet}$ and their experimental *hfccs* show that they are slightly pyramidal; e.g., in **1** $\sum\theta(Si) = 354.1$ degrees (Table 6.4).

In contrast to silyl radicals **1**–**5** which have relatively short lifetimes, the analogous C-centered radicals with two bulky t-Bu_2MeSi substituents, $H(t\text{-}Bu_2MeSi)_2C^{\bullet}$ (**16**), or with one t-Bu_2MeSi group and one bulky alkyl substituent, i.e., $(t\text{-}Bu_2MeSi)(1\text{-Ad})HC^{\bullet}$ (**17**), are persistent at 240K.[49b] Evidently, these substituents are sufficiently large to prevent dimerization of C-centered radicals, but not of the corresponding silyl radicals. In agreement, radical **17** decays via a pseudo-first-order kinetics, which is consistent with an

H-abstraction decay mechanism.[49b] Thus, the decay mechanism is different for Si-centered radicals and C-centered radicals carrying the same silyl substituents; i.e., H-abstraction for C-centered radicals versus dimerization for Si-centered radicals. This difference in behavior is due to two main factors[49b]: (1) C–C bonds are much shorter than Si–Si bonds and thus the steric size of the substituents is much more effective in determining the dimerization rate of C-centered radicals; (2) The C–H bonds produced by H-abstraction are significantly stronger than Si–H bonds (e.g., 104.8 kcal mol^{-1} for H_3C–H vs 90.3 kcal mol^{-1} for H_3Si–H),[50] thus favoring H-abstraction for C-centered radicals.

6.4.2.2 $R'(R_3Si)_2Si^{\bullet}$ (R' = aryl)

The important properties of aryl-substituted radicals **8**–**12** are summarized in Table 6.4.[12a,b,59] $R'(R_3Si)_2Si^{\bullet}$, where R_3Si = tBu$_2$MeSi and R' = phenyl (**8**) or 2,4,6-trimethylphenyl (mesityl) (**9**) were synthesized according to Eq. (6.11). Radical **8** is persistent at room temperature but could not be isolated, while radical **9**, carrying the larger mesityl group, was successfully isolated.[59]

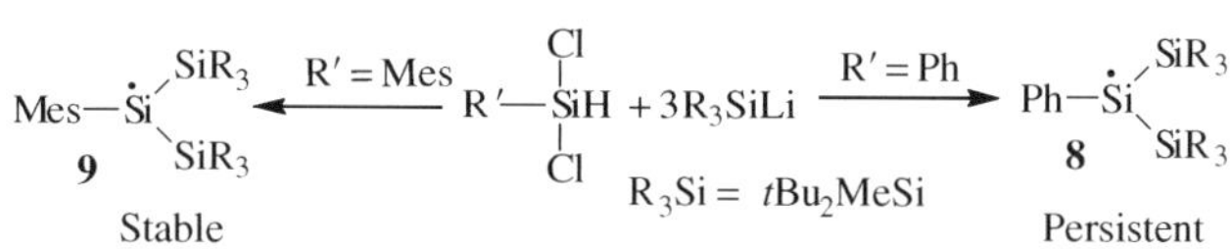

(6.11)

Calculations (UB3LYP/6–31 + G(d)) for **8** and **9** predict a planar Si-radical center (Table 6.4) and show that steric interactions force the aryl ring to be nearly perpendicular to the radical's molecular plane thus preventing conjugation between the radical's SOMO, and the aromatic π orbital of the aryl substituent.[59]

Stable radicals $R'(R_3Si)_2Si^{\bullet}$, where R_3Si = tBu$_2$MeSi and R' = 3,5-tBu$_2$C$_6$H$_3$ (**10**), 4-PhC$_6$H$_4$ (**11**) and 4-tBuC$_6$H$_4$ (**12**) were recently synthesized according to Eq. (6.12) and were isolated.[12b]

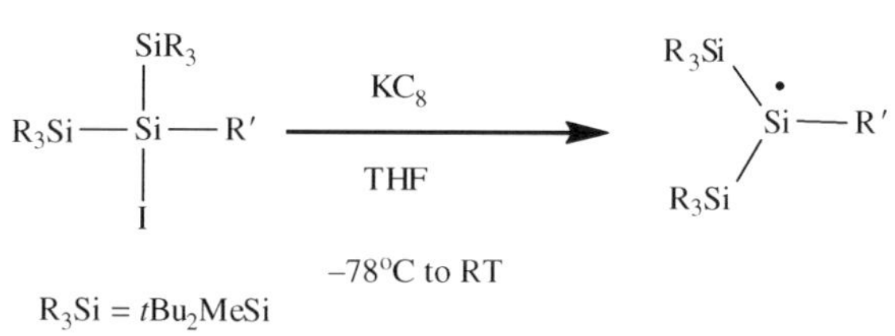

(6.12)

The molecular structure of silyl radical **10** was determined by X-ray crystallography (Fig. 6.12). The geometry around the Si radical center is planar ($\sum\theta$ = 360 degrees) and the bond angles are: $\angle$C1-Si1-Si2 = 117.0 degrees, $\angle$C1-Si1-Si2' = 117.0 degrees, and

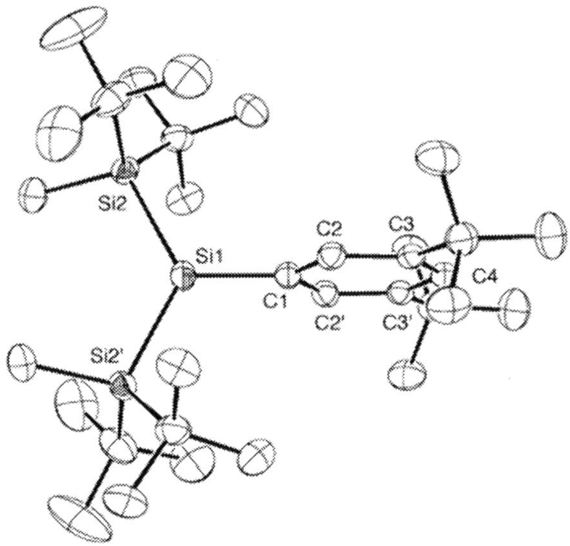

Figure 6.12
ORTEP view of the X-ray structure of 3,5-$tBu_2C_6H_3(tBu_2MeSi)_2Si^{\bullet}$ (**10**). *Adapted with permission from Taira, K.; et al., Isolable Aryl-Substituted Silyl Radicals: Synthesis, Characterization, and Reactivity. Chem. Eur. J.* ***2014****, 20 (30), 9342–9348, Copyright 2014 Wiley-VCH Verlag GmbH & Co KGaA.*

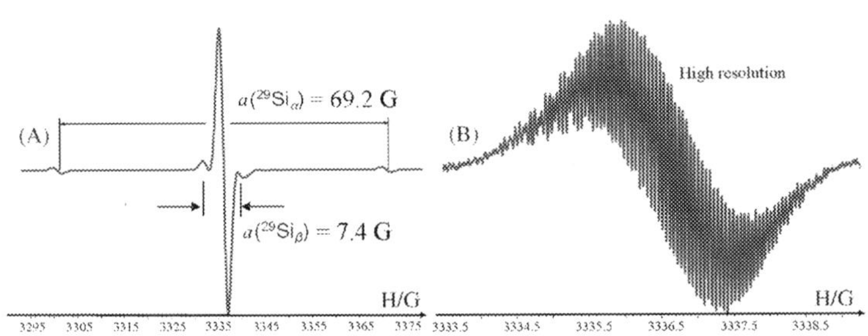

Figure 6.13
(A) EPR spectrum of radical **8** at 298K; (B) Expanded central line at high resolution at 350K.[59]

∠Si2-Si1-Si2' 126.0 degrees. The Si2-Si1-C1-C2 torsional angle is 88 degrees, indicating that the SOMO orbital is perpendicular to the π-system of the aryl substituent, preventing conjugation,[12b] in agreement with the calculated geometries of **8** and **9**.[59]

The EPR spectra of the aryl substituted radicals **8**–**12** show a *hfcc* of ~70 G with $^{29}Si_{\alpha}$ and 6.5–7.4 G with $^{29}Si_{\beta}$[12b,59] (Table 6.4). At high resolution, interactions with the *tert*-butyl hydrogens and methyl groups are observed for **8** and **10** respectively, see for example the EPR spectrum of **8** (Fig. 6.13).[59]

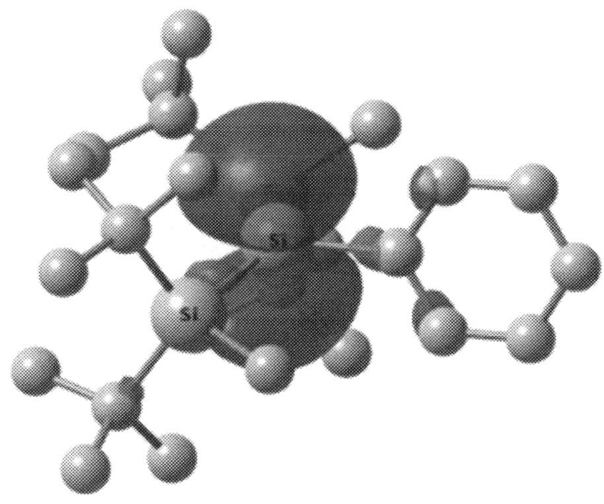

Figure 6.14
DFT calculated spin density of **8** (at 0.002 au contour level).[59]

An interesting feature of the spectra of aryl-substituted silyl radicals (**8**–**12**) is the absence of splitting from the aryl hydrogens which results from the absence of spin density delocalization into the perpendicular aromatic ring in agreement with the theoretical prediction that the aryl ring is perpendicular to the SOMO (Fig. 6.14). In contrast, such splitting were observed in the EPR spectra of the short-lived ($\tau_{1/2} \sim 1$ s) trimesitylsilyl[62] and tris(3,5-di-tert-butylphenyl)silyl[44] radicals. For $Ph_3Si^{\bullet}$, calculations predict that the phenyl rings are rotated by ~60 degrees relative to the radical's molecular plane, allowing some conjugation between the aryl's π-orbital and the SOMO and resulting in small spin density delocalization into the rings.[59]

An interesting peculiarity of radical **8**, is that it exists in equilibrium with its silene-type dimer (Eq. 6.13), which is formed similarly to the dimer of Gomberg's radicals ($Ph_3C^{\bullet}$).[12b]

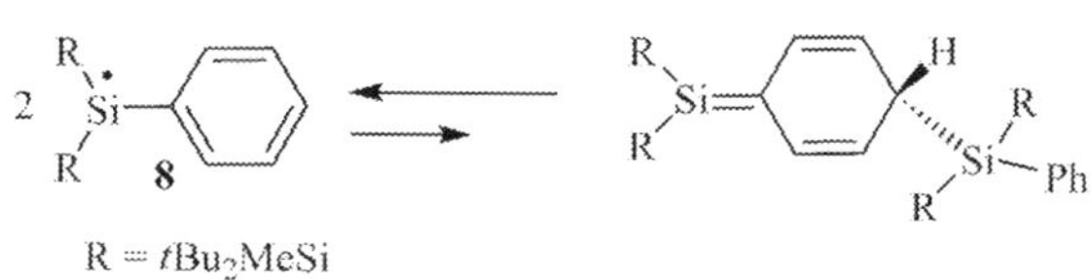

(6.13)

6.4.2.3 Amino-substituted silyl radical

(t-BuHN)(t-BuMe$_2$Si)$_2$Si$^{\bullet}$(**13**) was generated by photolysis of [(tBuMe$_2$Si)$_2$(tBuHN) Si]$_2$Hg.[59,63] Radical **13** is significantly less stable than the analogous aryl (**8**–**12**) and 1-adamantyl (**15**) substituted silyl radicals and its half-life time is only 2–3 min.[59] The EPR spectrum of **13** is characterized by $a_{Si}(^{29}Si_{\alpha}) = 91.7$ G and $a_N(^{14}N) = 2.2$ G. The relatively high $a_{Si}(^{29}Si_{\alpha})$ is consistent with the highly pyramidal ($\sum\theta$(Si) = 340.4 degrees) calculated structure.[59,63]

6.5 Tris(Silyl)-Substituted Silyl Radicals

This group comprises most of the known persistent and stable silyl radicals (Table 6.5). Their stabilization stems both from the highly thermodynamic stabilizing electronic effect of three silyl substituents (Table 6.2) and from the kinetic steric stabilizing effect of the bulky substituents that are used.

6.5.1 EPR Parameters

The EPR parameters, the degree of pyramidality, and the half-life time of several persistent and stable tris(silyl) substituted silyl radicals (**18**–**32**) are presented in Table 6.5. The EPR spectra of tris(silyl) substituted silyl radicals are characterized by a central signal at *g*-factor in the range of 2.0053–2.0065 which is split by hyperfine interactions with neighboring atoms. The *hfcc* of the central silicon α-^{29}Si of 55–64 G is much larger than the *hfcc* with the β-^{29}Si (7–9 G) or the γ-^{29}Si nuclei of the substituents (6–10 G). Additional splitting from the substituents' protons is generally very small (0.1–0.4 G) but observable in some cases (Table 6.5). The *g*-factor is not sensitive to the substitution pattern. The kinetic stability of the silyl radicals depends on the size of substituents at the radical center. Stable radicals are obtained when tBu$_2$MeSi-, or more bulky substituents are used (Table 6.5).

Sekiguchi and coworkers prepared in 2002 the first stable tris(silyl)silyl radical [(t-Bu$_2$MeSi)$_3$Si$^\bullet$] (**26**) by oxidation of (tBu$_2$MeSi)$_3$SiNa (Eq. 6.14) and isolated **26** as air-sensitive yellow needles.[12c] Radical **26** was later synthesized in higher yield by Apeloig et al. using reaction (Eq. 6.15).[12a]

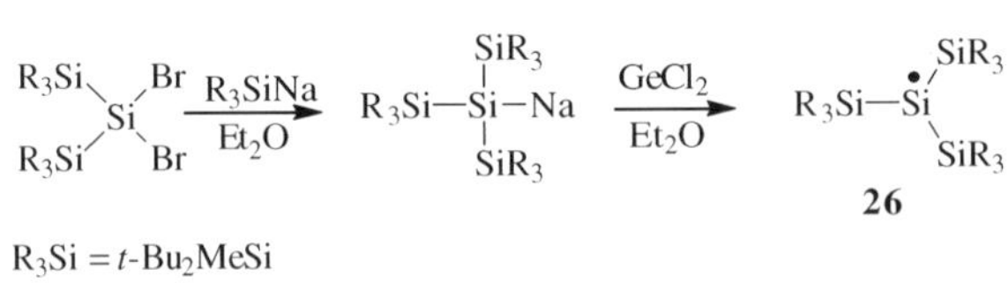

26 (6.14)

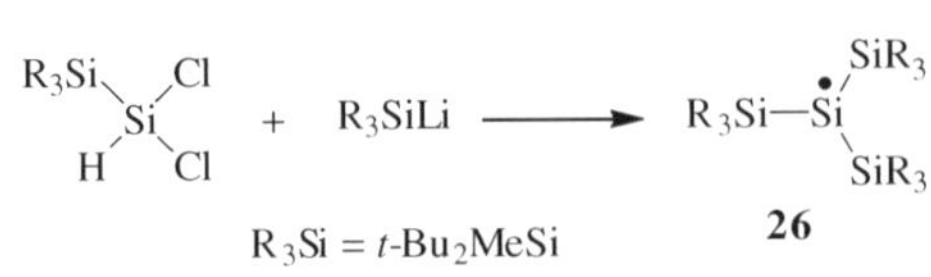

26 (6.15)

The EPR spectrum of radical **26** in pentane is shown in Fig. 6.15 and it is characteristic also to other tris(silyl)silyl radicals.[12a] It consists of a central singlet signal at $g = 2.0056$

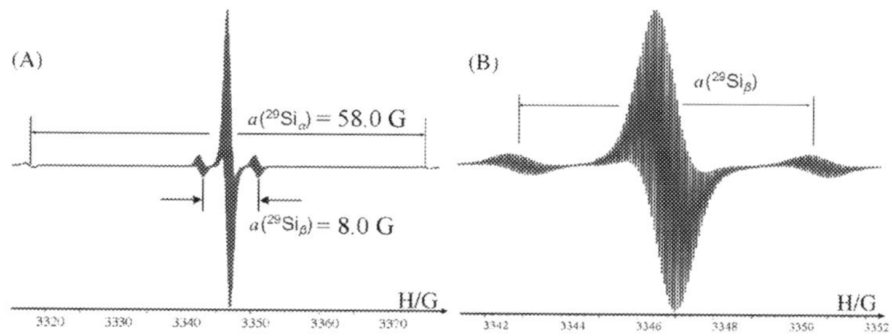

Figure 6.15
(A) High-resolution EPR spectrum of (t-$Bu_2MeSi)_3Si^•$ in pentane; (B) Expanded central multiplet. *Adapted by permission from Molev, G. et al., Isolable Photoreactive Polysilyl Radicals. Journal of the American Chemical Society* ***2009****, 131 (33), 11698–11700, Copyright 2009 American Chemical Society.*

that results from the unpaired electron residing on the nonmagnetic ^{28}Si nuclei, a doublet splitting with *hfcc* of 58.0 G assigned to the coupling of the unpaired electron with the $^{29}Si_\alpha$ nucleus, and a doublet splitting having a higher intensity and a smaller *hfcc* value (8.0 G) assigned to the *hfcc* with three $^{29}Si_\beta$ nuclei (Fig. 6.15A). Coupling with the methyl and t-butyl protons is also observed (the number of interacting nuclei is given in parenthesis): $a_H(54\,^1H) = 0.14$ G; $a_H(9\,^1H) = 0.21$ G (Fig. 6.15B).[12a] An important observation is that $a_{Si}(^{29}Si_\alpha)$ is linearly dependent on the temperature in the range of 180–340K, so that raising the temperature increases a_{Si} ($^{29}Si_\alpha$) by 1.73 G/100K.[59] This temperature dependence is due to the out-of-plane vibrations of the substituents and is similar to that observed in planar alkyl radicals.[69] This temperature dependence should be taken into consideration when assigning EPR signals to other silyl radicals.

6.5.2 X-ray Crystallography

The X-ray structure of radical **26** is shown in Fig. 6.16.[12c] **26** has a trigonal-planar geometry around the Si radical center ($\sum\theta = 360$ degrees). Steric effects dictate or contribute to many of the structural details, such as the planarity of the radical center and the clockwise arrangement in the Si1–Si2–Si3–Si4 plane of all methyl substituents at the peripheral Si atoms, thus minimizing the steric repulsion of the bulky tBu groups. The Si–Si bond lengths (average 2.42 Å) are slightly longer than the normal values of 2.30–2.40 Å.[70] The X-ray structures of radicals **28**[63] and **29**,[12a] which also have a planar radical center, are similar to those of radical **26**.

The allyl-type cyclotetrasilenyl radical (**32**) (Fig. 6.17) has an almost planar array of four silicon atoms, of which three (Si1, Si2, Si3) are trigonally coordinated.[68a] The sum of bond angles around each of these silicon atoms (Si1–Si3) is 360.0, 359.1, and 356.2 degrees, respectively. Si1 and Si2 have a planar geometry, but Si3 is slightly pyramidalized. The

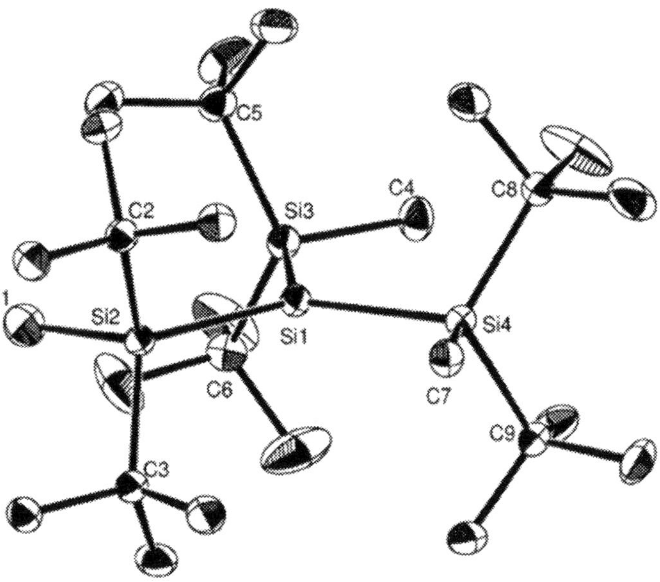

Figure 6.16

ORTEP view of the X-ray structure of $(tBu_2Me)_3Si^{\bullet}$ (**26**). *Adapted with permission from Sekiguchi, A.; Fukawa, T.; Nakamoto, M.; Lee, V.Ya.; Ichinohe, M. Isolable Silyl and Germyl Radicals Lacking Conjugation With π-Bonds: Synthesis, Characterization, and Reactivity. J. Am. Chem. Soc.* ***2002****, 124 (33), 9865–9869, Copyright 2002 American Chemical Society.*

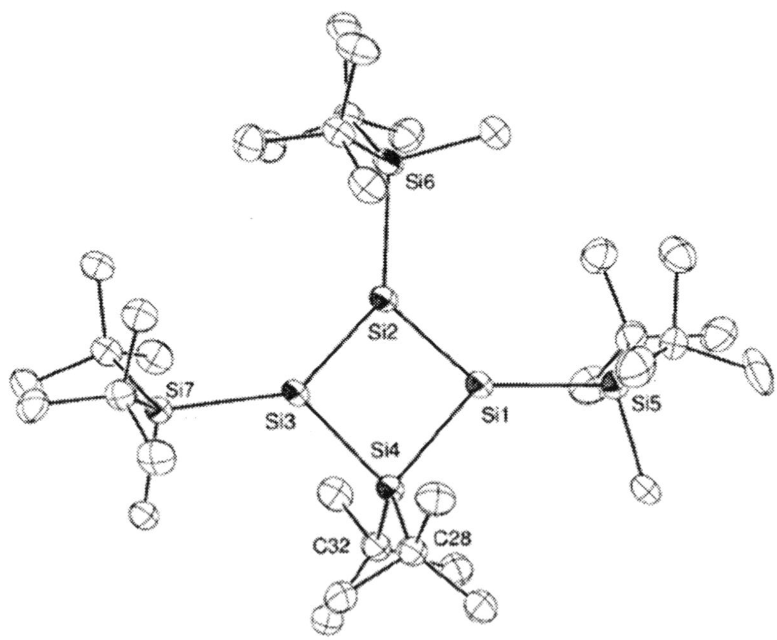

Figure 6.17

ORTEP view of the X-ray structure of cyclotetrasilenyl radical **32**. *Adapted with permission from Sekiguchi, A.; Matsuno, T.; Ichinohe, M. Cyclotetrasilenyl: The First Isolable Silyl Radical. J. Am. Chem. Soc.* ***2001****, 123 (49), 12436–12437, Copyright 2001 American Chemical Society.*

Si–Si bond lengths of 2.226 Å and 2.263 Å are intermediate between those of a Si = Si double bond (2.174 Å) and a Si–Si single bond (2.349–2.450 Å) in the four-membered ring of hexakis(tert-butyldimethylsilyl)cyclotetrasilene,[71] supporting the occurrence of allylic delocalization of the unpaired electron.[68a] As expected for an allyl-type radical,[19a] the largest spin density (positive) in **32** is located on the terminal Si1 and Si3 atoms (40.7 and 37.4 G), and the spin density (negative) on the Si2 atom is only 15.5 G.

6.5.3 Reactions

The two major decay routes of silyl radicals are (1) dimerization to the corresponding disilane and (2) hydrogen abstraction from the solvent or from the substituents of another silyl radical molecule. In the absence of scavengers, trapping reagents, or efficient H-donors, dimerization is the major decay route of silyl radicals and thus determines their lifetime.

6.5.3.1 Dimerization

The rate of dimerization depends on the dissociation energy of the central Si–Si bond (a thermodynamic factor) and on the size of the substituents and the steric effects which they exert (a kinetic factor). To demonstrate these effects, we discuss below the kinetic stability of several very reactive ($\tau_{1/2} \sim 0.1-1$ s) $R_3Si^{\bullet}$ radicals (**18, 22, 23, 25**, Table 6.5) toward dimerization to the corresponding oligosilanes $R_3Si-SiR_3$ (**18a, 22a, 23a**, and **25a**, respectively, Scheme 6.4).[66]

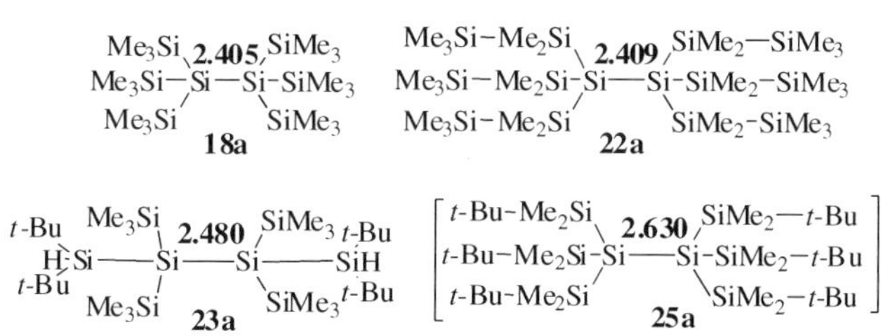

Scheme 6.4
Central Si–Si bond lengths (Å) in polysilanes **18a, 22a, 23a**, and **25a**.

The rate constants for the dimerization of radicals **22, 23**, and **25** to produce the corresponding disilanes, shown as the backward reaction of Eq. (6.16), were measured using EPR spectroscopy.[66] These radicals were generated photolytically ($\lambda \geq 300$ nm), **22** from disilane **22a** (Eq. 6.16) and radicals **23** and **25** from the corresponding disilylmercury compounds (Eq. 6.17).

$$R_3Si - SiR_3 \underset{k_{dim}}{\overset{h\nu \text{ or } \Delta}{\rightleftarrows}} 2\ R_3Si^{\bullet} \tag{6.16}$$

$$R_3Si - Hg - SiR_3 \xrightarrow{h\nu} 2\ R_3Si^{\bullet} + Hg \tag{6.17}$$

In the kinetic measurements, the light was turned off after a few seconds and the decrease in radical concentration [C] as a function of time was measured by following the decrease in the intensity of the radical's EPR signal, as shown in Fig. 6.18A for radical **23**. Good linear curves of 1/[C] versus time were obtained in all cases (e.g., Fig. 6.18B). The resulting second-order reaction rate constants (k_{dim}) at 290K are given in Table 6.6.[66] The dimerization rates in Table 6.6 cover a range of 500,000 and they are strongly dependent on the radical's substituents. The most reactive radical is **18** because the Me_3Si substituents are too small to effectively slow down its dimerization. The small steric repulsions in dimer **18a** are reflected in its relatively short Si–Si bond of 2.405 Å.[51] Substitution of the three Me_3Si substituents in **18** by three Me_3SiMe_2Si groups in **22** has a significant effect on k_{dim} which for radical **22** is reduced by a factor of 20,000 relative to radical **18** (Table 6.6). However, substitution of three Me_3Si substituents by Me_3SiMe_2Si substituents has only a minor effect on the central Si–Si bond distance which is 2.409 Å in **22a** (Scheme 6.4).[66] When one of the Me_3Si groups in **18** is substituted by the bulkier t-Bu_2HSi substituent (i.e., radical **23**), the rate of dimerization decreases further and is slower by a factor of 5×10^4 relative to radical **18**. In this case, the central r(Si–Si) distance in its dimer **23a** is

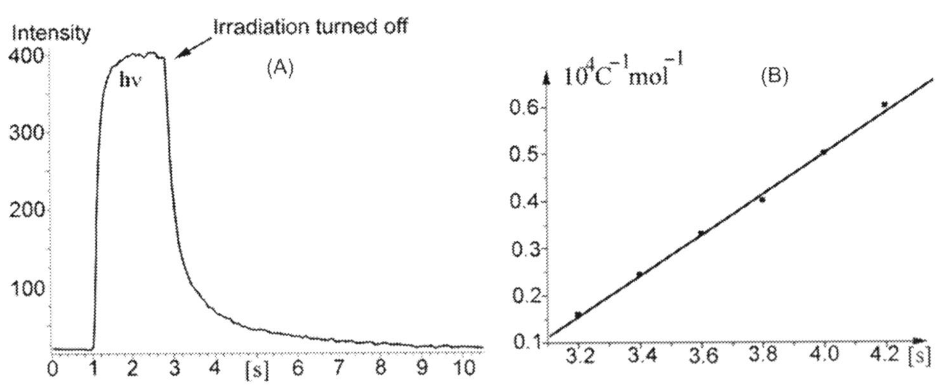

Figure 6.18

EPR measurements of k_{dim} of radical **23**: (A) Decay curve of the EPR signal at 290K; (B) Plot of 1/[C] versus time. *Reproduced by permission from Kravchenko, V.; Bravo-Zhivotovskii, D.; Tumanskii, B.; Botoshansky, M.; Segal, N.; Molev, G.; et al., Kinetic Stabilization of Polysilyl Radicals. In* Organosilicon Chemistry IV: From Molecules to Materials; *Auner, N., Weis, J., Eds. Wiley-VCH Verlag GmbH: Weinheim, 2005; pp 48–57, Copyright 2005 Wiley-VCH Verlag GmbH & Co. KGaA.*

Table 6.6: Kinetic data for dimerization of silyl radicals 18, 22, 23, and 25

Radical	k_{dim}[a]	Relative k_{dim}	r(Si–Si) in dimer (Å)
18	$\sim 1 \cdot 10^8$	$5 \cdot 10^5$	2.405
22	$5 \cdot 10^3$	25	2.409
23	$2 \cdot 10^3$	10	2.480
25	$2 \cdot 10^2$	1	2.630[b]

[a]Second order rate constant in $M^{-1} \cdot sec^{-1}$.
[b]Calculated.
Source: Adapted with permission from Tumanskii, B.; Karni, M.; Apeloig, Y. Persistent and Stable Silyl Radicals. In Encyclopedia of Radicals in Chemistry, Biology and Materials; *John Wiley & Sons, Ltd., 2012, Copyright 2012 Wiley-VCH Verlag GmbH & Co KGaA.*

elongated considerably to 2.480 Å.[66] Radical **25** with three *t*-$BuMe_2Si$ substituents is the least reactive and its dimerization rate is slower by a factor of 25 than that of radical **22** and 5×10^5 times slower than that of **18**. In **25a** (which has not yet been isolated) the calculated (ONIOM, MP2/6–31G(d):B3LYP/6–31G(d):HF/3–21G) central Si–Si bond distance is 2.630 Å,[66] significantly longer than that of **23a**.

DFT calculations show that β-alkyl or β-silyl substitution have a negligible thermodynamic effect on the stability of silyl radicals[37,53a] and therefore the thermodynamic stabilities of radicals **25** and **18** are expected to be nearly the same. The reason that **25** reacts 5×10^5 times more slowly than $(Me_3Si)_3Si^{\bullet}$ is therefore kinetic in origin and is due to the effective steric protection of the radical center of **25** by the three *t*-$BuMe_2Si$ substituents.[66] Similarly, radical **22** reacts 2×10^4 more slowly than **18**.[66] The effective protection of the radical center of **22** toward dimerization was explained by a protective substituents' "umbrella" around the silyl radical center.[66]

To obtain silyl radicals which are stable at room temperature, the size of the substituents has to be increased beyond that in **25**, as is the case for radical **26** which is substituted with three large *t*-Bu_2MeSi substituents.

The enthalpies (ΔH) of the homolytic cleavage of the central Si–Si bond in disilanes can be obtained from temperature-dependent EPR measurements of the intensity of the radical's signal using Eq. (6.18), in which C is the concentration of the radical, T is the absolute temperature, and A is a constant.

$$\mathrm{Ln}C^2 = \frac{\Delta H}{RT} - A \tag{6.18}$$

An example is shown in Fig. 6.19 for radical **23**.[66]

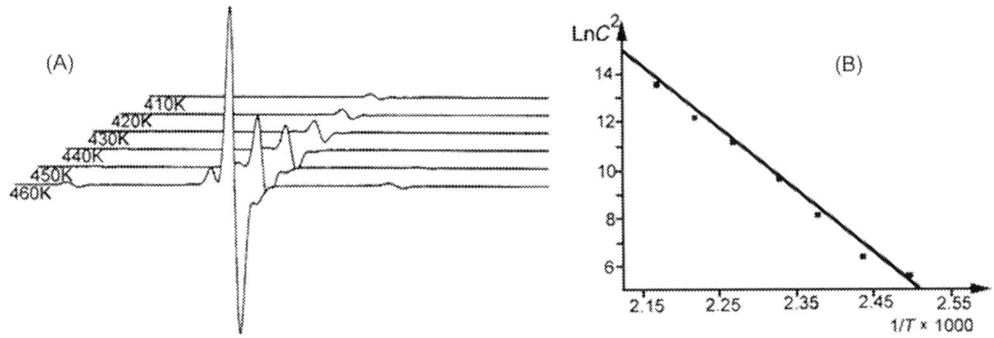

Figure 6.19

(A) Temperature dependence of the intensity of the EPR signal of radical **23**; (B) Plot of LnC^2 versus $1/T$ for radical **23**. *Reproduced by permission from Kravchenko, V.; Bravo-Zhivotovskii, D.; Tumanskii, B.; Botoshansky, M.; Segal, N.; Molev, G.; et al., Kinetic Stabilization of Polysilyl Radicals. In* From Organosilicon Chemistry VI: From Molecules to Materials*; Auner, N., Weis, J., Eds. Wiley-VCH Verlag GmbH: Weinheim, 2005; pp 48–57, Copyright 2005 Wiley-VCH Verlag GmbH & Co. KGaA.*

Temperature-dependent EPR measurements (Fig. 6.19) yield ΔH values of 56 and 58 kcal mol^{-1} for the cleavage of the central Si–Si bond in **22a** and **23a**, respectively. These values are smaller by 9–11 kcal mol^{-1} than for $Me_3Si–SiMe_3$ (67 kcal mol^{-1}).[72] The weaker central Si–Si bond in **22a** and **23a** reflects the thermodynamic stabilization of the corresponding silyl radicals (**22** and **23**, respectively) by the α-silyl substituents relative to methyl groups (see also Table 6.2). The similar Si–Si bond cleavage energy in **22a** and **23a** reflects the fact that β-silyl substitution has practically no effect on the thermodynamic stability of the silyl radicals, in agreement with theoretical calculations,[37,66] and the fact that the β-silyl groups do not raise the ground state energy of **22a** due to steric congestion as also reflected in its short central Si–Si bond (Scheme 6.4).[37,66] It was similarly estimated that ΔH for the cleavage of the long central Si–Si bond of congested **25a** is only about 8 kcal mol^{-1}.[66] In agreement, at relatively high concentrations of **25a** ($\sim 10^{-2}$ to 10^{-3} M, hexane), obtained by oxidation of the corresponding anion, radical **25** was observed at 290K due to the **25a**$\rightleftharpoons$**25** equilibrium.[66] The half-life of **25** at 298K is ~1 day and it decays by hydrogen abstraction, not via dimerization.[66]

6.5.3.2 Photochemistry

$(t\text{-}Bu_2MeSi)_3Si^{\bullet}$ (**26**) is stable upon UV–Visible irradiation.[12a] In contrast, the stable radical **29** (yellow solution in hexane) decays rapidly ($\tau_{1/2} \sim 100$ s) upon irradiation at $\lambda > 400$ nm, yielding a 1:1 ratio of the *corresponding* disilene (blue colored) and the disilane (Eq. 6.19).[12a] Formally this is a disproportionation reaction.

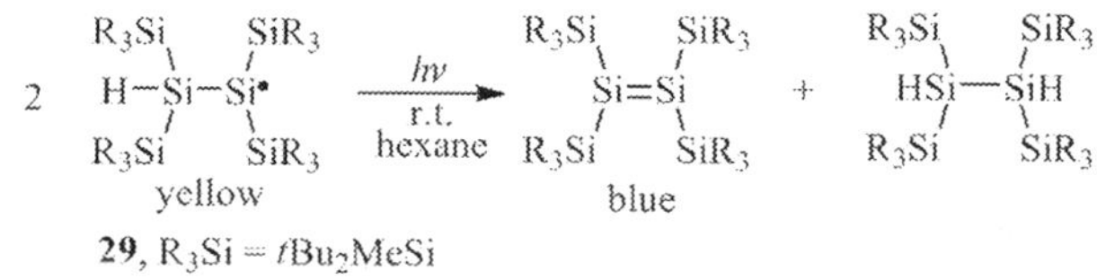

(6.19)

The photoreactivity of **29** is due to its β-hydrogen. Upon irradiation, **29** is photoexcitated (Scheme 6.5, step A), and then the photoexcited radical accepts an electron from another radical molecule forming an ion pair within a solvent cage (Scheme 6.5, step B) followed by proton transfer from the cation to the anion to yield the observed products (Scheme 6.5, step C).[12a] Strong support for this mechanism is the fact that $(t\text{-}Bu_2MeSi)_3Si^{\bullet}$ (**26**), which lacks β-hydrogens and is photostable, becomes photoreactive ($\tau_{1/2} \sim 100$ s), yielding $(t\text{-}Bu_2MeSi)_3SiH$ in the presence of hydrogen donors such as 2-propanol or triethylsilane.[12a]

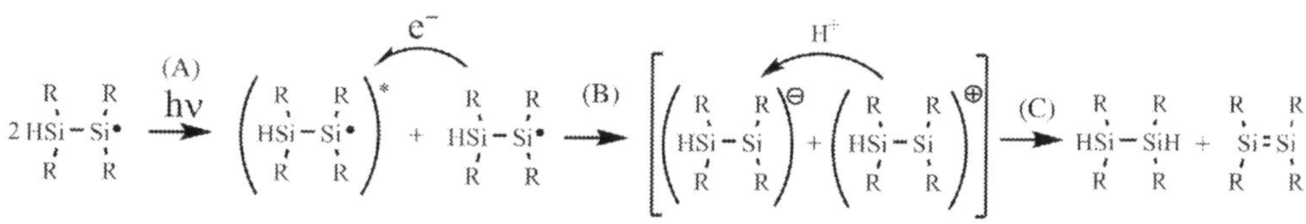

Scheme 6.5

Proposed mechanism for the photolysis of radical **29**.

6.5.3.3 Oxidation, reduction, and ionization

$(t\text{-}Bu_2MeSi)_3E^{\bullet}$ (E = Si, Ge, Sn) radicals undergo both oxidation and reduction processes, yielding the corresponding cationic and anionic derivatives.[1h] Their first oxidation and reduction potentials in tetrahydrofuran (THF), as measured by cyclic voltammetry, are the following: for the Si-centered radical $E_P(ox.) = 0.8$ V, $E_{1/2}(red.) = -1.43$ V; for the Ge-centered radical $E_P(ox.) = 0.6$ V, $E_{1/2}(red.) = -1.46$ V; and for the stable Sn-centered radical $E_P(ox.) = 0.2$ V, $E_{1/2}(red.) = -1.41$ V.[73]

UV photoelectron spectroscopy gives the following first ionization potentials for $(t\text{-}Bu_2MeSi)_3E^{\bullet}$ (E = Si, Ge, Sn): 6.15 eV for E = Si; 6.0 eV for E = Ge; and 5.8 eV for E = Sn. These ionization energies correspond to the removal of an electron from the SOMO of the radicals and they decrease in the order Si > Ge > Sn, in agreement with the trend in their oxidation potentials and the ionization energies of the isolated atoms (Si (8.15 eV), Ge (7.90 eV), Sn (7.34 eV)).[74]

Oxidation of radical **29** with HgF_2 yields the corresponding flourosilane (**29a**) (Scheme 6.6, path A), and its reduction with lithium metal in hexane or with $t\text{-}Bu_2MeSiLi$ in THF quantitatively yields the corresponding lithiosilane (**29b**) (Scheme 6.6, path B).[12a] Oxidation of anion **29b** by $GeCl_2$ yields radical **29** (Scheme 6.6, path C).[12a]

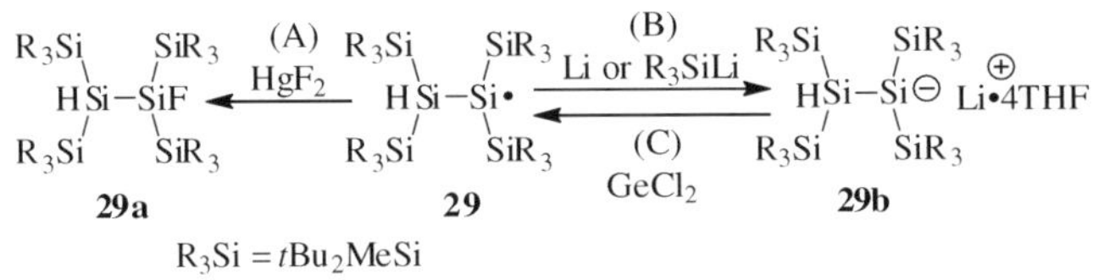

Scheme 6.6
Oxidation and reduction reactions of radical **29.**

6.5.4 Silyl Radicals in Batteries

Recently, in an exciting development, high-power electrochemical energy storage devices based on the stable persilyl-substituted free radicals of group 14 $(t\text{-}Bu_2MeSi)_3E^{\bullet}$ [E = Si, Ge, and Sn], as anode materials have been developed.[12c,12d,48d] The choice of these radicals is based on their low reduction potentials (below −1.6 V vs Ag/Ag^+)[73] and the reversibility of the redox process (Eq. 6.20). These properties make these radicals suitable candidates for electrochemical capacitor (EC) anode materials.

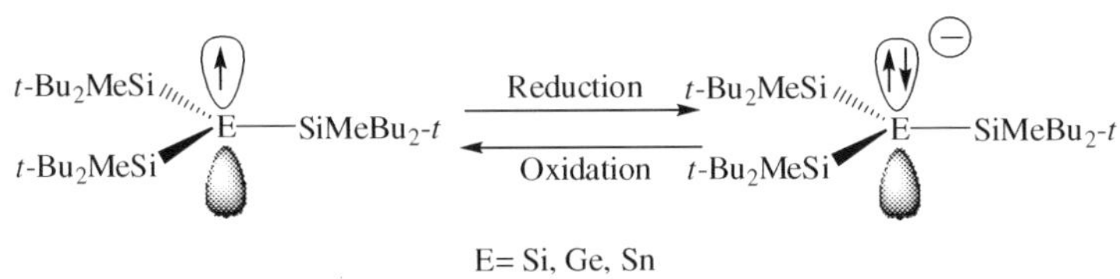

(6.20)

Furthermore, given the long distance of 9.08 Å between the Si radical centers in the crystal, one can expect that these radicals will readily, and with only slight structural changes in the electrode, interact with other ionic species that can be introduced into the available space by electrochemical reduction.

The battery prototype shown schematically in Fig. 6.20, showed a remarkable cycle stability without significant loss of power density, in comparison with similar characteristics of known organic radical batteries, the dual carbon cell, and the EC.[12d] Particularly important is the fact that these novel electrochemical energy storage systems employing stable heavy group 14 element radicals are lithium-free. The development of electrochemical energy storage devices based on stable $R_3E^{\bullet}$ is yet another example of how basic research can develop into practical applications.

6.5.5 Conformational Analysis of Stable Silyl Radicals in Solution

The preferred conformations of silyl radicals in solution can be determined by a combination of EPR spectroscopy and molecular orbital calculations. For demonstration, we

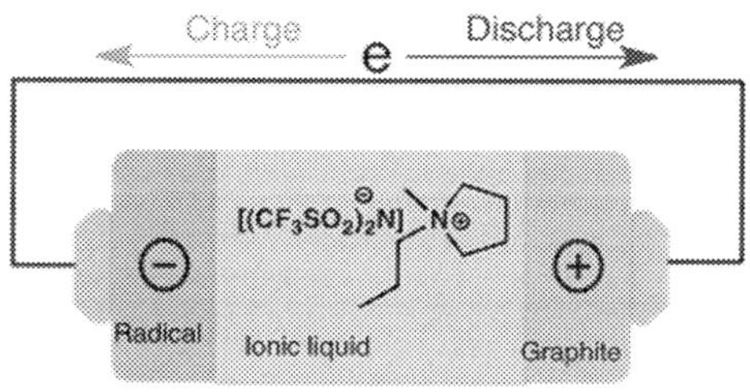

Figure 6.20
Schematic representation of a battery based on stable persilyl-substituted radicals of group 14 elements. *Adopted by permission from Maruyama, H.; Nakano, H.; Nakamoto, M.; Sekiguchi, A. High-Power Electrochemical Energy Storage System Employing Stable Radical Pseudocapacitors. Angew. Chem. Int. Ed.* ***2014****, 53 (5), 1324–1328, Copyright 2005 Wiley-VCH Verlag GmbH & Co. KGaA.*

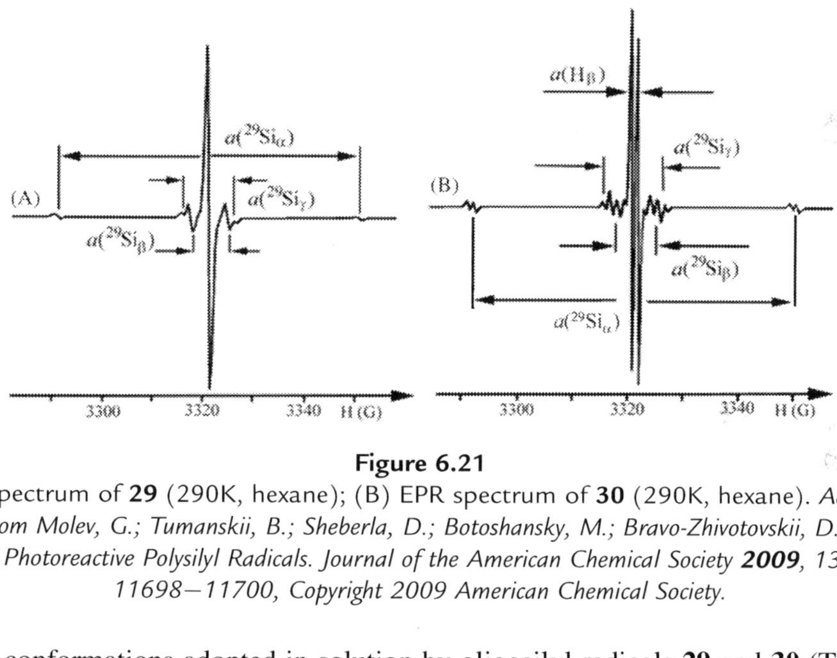

Figure 6.21
(A) EPR spectrum of **29** (290K, hexane); (B) EPR spectrum of **30** (290K, hexane). *Adapted with permission from Molev, G.; Tumanskii, B.; Sheberla, D.; Botoshansky, M.; Bravo-Zhivotovskii, D.; Apeloig, Y., Isolable Photoreactive Polysilyl Radicals. Journal of the American Chemical Society* ***2009****, 131 (33), 11698–11700, Copyright 2009 American Chemical Society.*

discuss the conformations adopted in solution by oligosilyl radicals **29** and **30** (Table 6.5). Radical **29** was isolated and its structure was determined by X-ray crystallography. In the solid state, the radical center of **29** is planar ($\sum\theta = 359.8$ degrees),[12a] similar to the stable radicals **26**[12c] and **28**.[63] Calculations predict a planar Si center also for radical **30**.[12a] The EPR spectrum in hexane solution of **29** and of **30**, both having a β-hydrogen, is shown in Fig. 6.21. Both radicals exhibit the expected splitting with the α, β, and γ ^{29}Si nuclei of 59.3, 7.3, and 10.4 G, respectively, for **29** (Fig. 6.21A, Table 6.5), and 57.1, 7.5, and 10.0 G, respectively, for **30** (Fig. 6.21B). The g-factor and the $a_{Si}(^{29}Si_\alpha)$ and $a_{Si}(^{29}Si_\beta)$ values of **29** and of **30** are typical for planar silyl radicals (Table 6.5). There is one important difference in the EPR spectra of the two radicals. The EPR spectrum of **30**

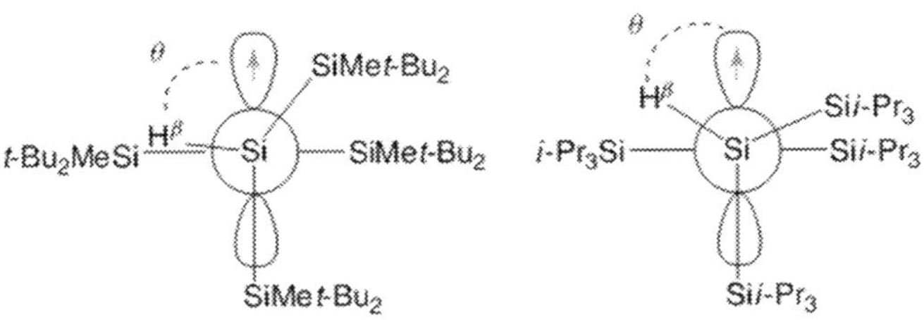

29 **30**

Figure 6.22
Newman projections of the preferred conformations of radicals **29** and **30.**

exhibits a doublet arising from coupling with the β-hydrogen with $a_H(^1H_\beta) = 1.1$ G (Fig. 6.21B),[12a] which is missing in the spectrum of **29**. This difference allows the determination of the conformation adopted by these radicals in hexane solution. Theoretically, the *hfcc* splitting constant by H_β depends on the angle θ between the $H_\beta{-}Si_\beta$ bond and the principal axis of the Si(3p) orbital occupied by the unpaired electron, according to Eq. (6.21), where B_0 and B are constants.[19a] a_H (H_β) can therefore serve as a sensitive conformational probe.[19a,61,75]

$$a(H_\beta) = B_0 + B\cos^2\theta \tag{6.21}$$

In **29** (Fig. 6.21A), where a doublet splitting is not observed (a_H (H_β) ≤ 0.4 G), θ is close to 90 degrees. In contrast, in **30**, a_H (H_β) $= 1.1$ G (Fig. 6.21B) and therefore $\theta \neq 90$ degrees. DFT calculations (UB3LYP/TZVP//UB3LYP/6−31G(d)) support the above conclusion, yielding the conformations shown in Fig. 6.22 where in **29** $\theta = 82$ degrees and no spin density is found on H_β, while in **30**, $\theta = 63$ degrees and some spin density resides on H_β. Solving Eq. (6.21), using the measured a_H (H_β) for **30**, the absence of a_H (H_β) for **29**, and the calculated θ angles, gives $B = 5.4$ G and $B_0 \sim 0$ G. Thus, in branched polysilyl radicals the maximum value of a_H (H_β), when θ = 0 degrees, is 5.4 G. Based on this value the conformation of other silyl radicals can be evaluated. For example in radical **31** where a_H (H_β) is 0.65 G, θ = 70 degrees.[63]

6.6 Silicon-Centered Bi- and Triradicals

6.6.1 Triplet Silyl Biradical

The first stable triplet silyl biradical (**33**) was synthesized recently by Sekiguchi and coworkers by the reductive dehalogenation of a *m*-bis(halosilyl)benzene (Eq. 6.22) and was isolated and structurally characterized by X-ray crystallography.[76]

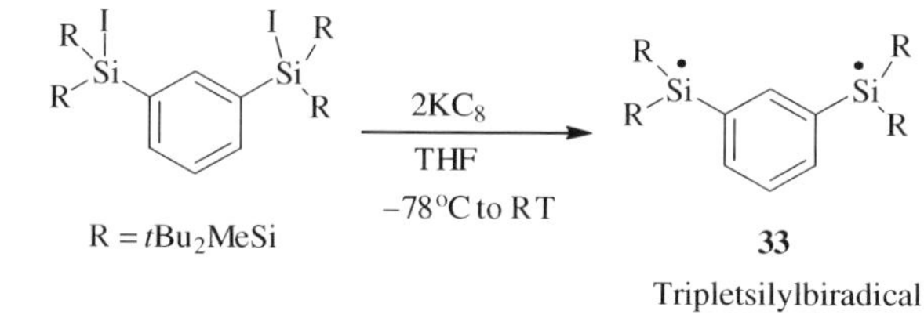

(6.22)

The X-ray structure (Fig. 6.23) of **33** shows that the two half-filled 3p(Si) orbitals and the aromatic ring π-orbitals are orthogonal to each other, similarly to **8**–**12**. This conformation does not allow formation of a C = Si π bond. Both Si-radical centers have a nearly trigonal-planar structure ($\sum\theta(Si) = 358.6$ degrees). The r(Si1-C2) and r(Si1#-C2#) distances of 1.911 Å are slightly longer than typical Si–C(aryl) single bonds (1.879 Å).[76]

The EPR spectrum of **33** in the $\Delta M_S = 1$ transition region consists of six lines and is characteristic for triplet species with $|D'| = 138.0$ G and $|E'| = 17.2$ G (Fig. 6.24). The characteristic signal in the $\Delta M_S = 2$ transition region was also observed.[76]

In contrast to *m*-bis(halosilyl)benzene which yields the triplet biradical **33**, the isomeric *p*-bis(halosilyl)benzene reacts with KC_8 to yield a diamagnetic quinodimethane species featuring two exocyclic Si=C double bonds (Eq. 6.23).[76]

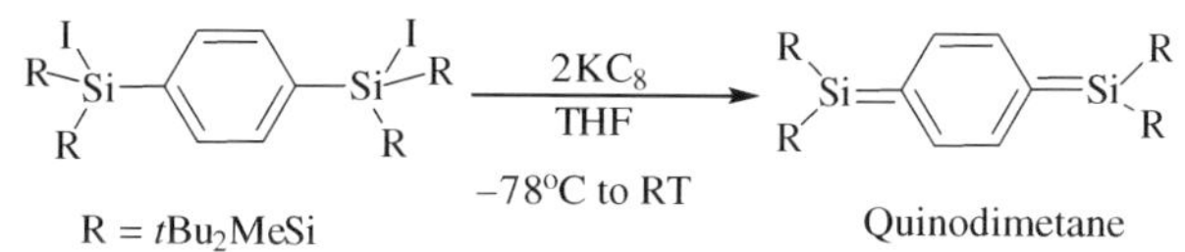

(6.23)

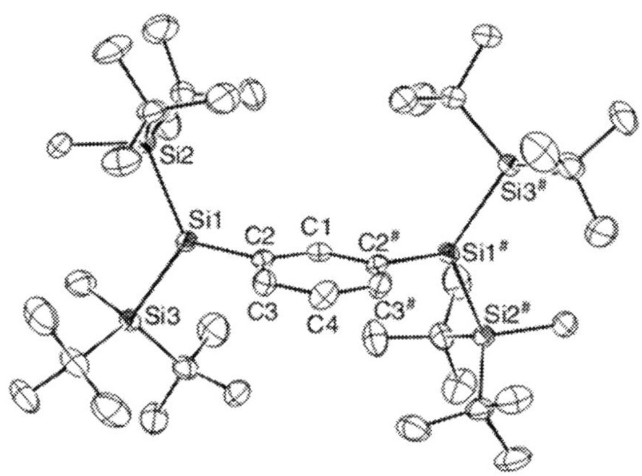

Figure 6.23

ORTEP view of the X-ray structure of biradical **33**. *Adapted with permission from Nozawa, T.; Nagata, M.; Ichinohe, M.; Sekiguchi, A., Isolable p- and m-[(tBu$_2$MeSi)$_2$Si]$_2$C$_6$H$_4$: Disilaquinodimethane vs Triplet Bis (silyl radical). J. Am. Chem. Soc.* ***2011****, 133 (15), 5773–5775, Copyright 2011 American Chemical Society.*

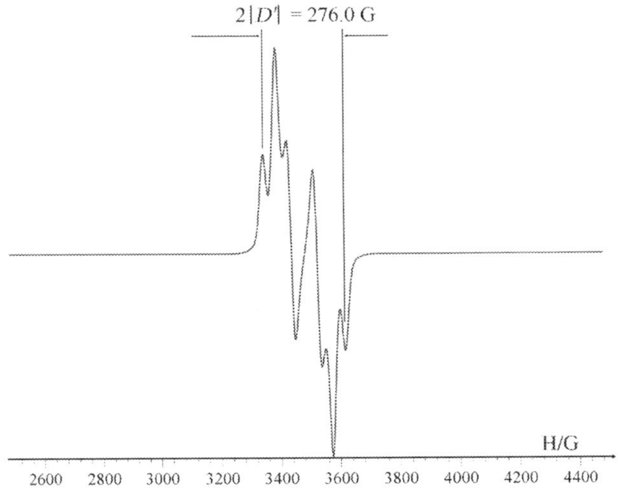

Figure 6.24
Simulation of the EPR spectrum of triplet biradical **33** using experimental parameters obtained at 80K .[76]

Biradicals $[Li(tBu_2MeSi)_2Si^{\bullet}]_2$ which result from aggregation of silyllitium radicals are discussed in Section 6.7.3.

6.6.2 Si-Centered Triradical

Reductive deiodination of 1,3,5-tris[2,2-di-*tert*-butyl-1-(di-*tert*-butylmethylsilyl)-1-iodo-2-methyldisilanyl]benzene (**34**) with KC_8 (Eq. 6.24) results in the formation of isolable silyl triradical 1,3,5-$[(tBu_2MeSi)_2Si]_3C_6H_3$ (**35**).[77] EPR spectroscopy of triradical **35** shows that it has a quartet ground state, with characteristic $\Delta m = 1$, $\Delta m = 2$, and $\Delta m = 3$ transitions.[19a,77]

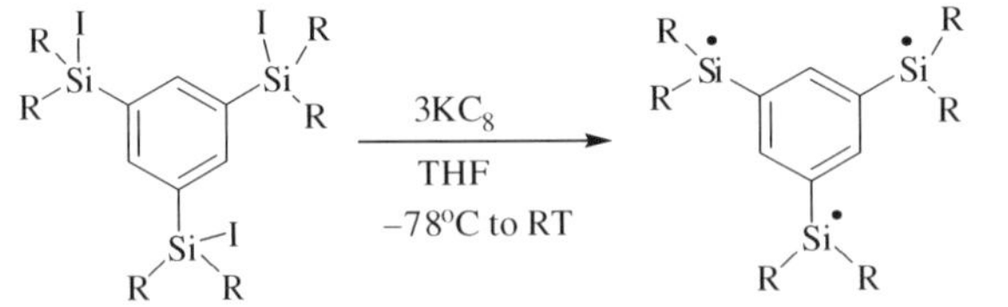

34

35

R = tBu$_2$MeSi

Quartet silyl triradical

(6.24)

The molecular structure of **35**, determined by X-ray crystallography, is shown in Fig. 6.25. The geometry around the three Si radical centers is nearly trigonal-planar, ($\sum\theta$ = (Si1) = 359.9, Si4 = 357.9, and Si7 = 357.8 degrees). A difference between the structures of

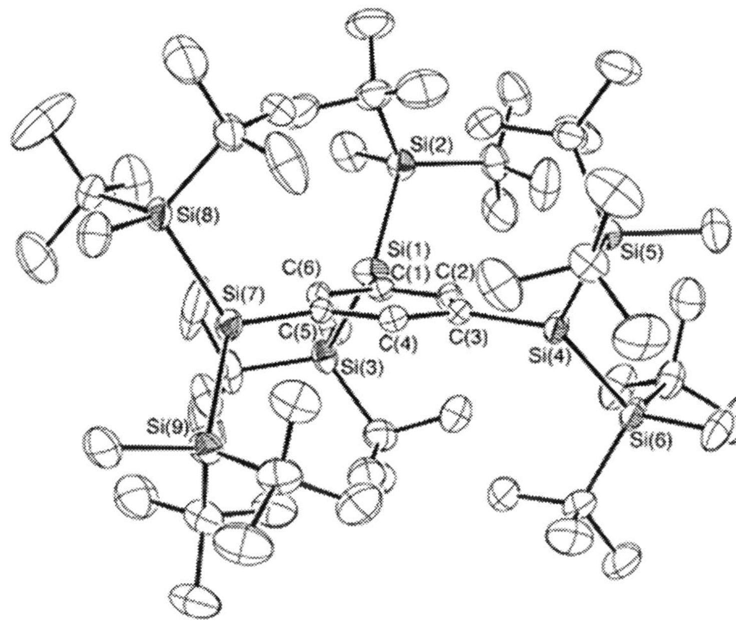

Figure 6.25
ORTEP view of the X-ray structure of triradical **35**. *Adapted with permission from Nozawa, T.; Ichinohe, M.; Sekiguchi, A., 1,3,5-[(tBu$_2$MeSi)$_2$Si]$_3$C$_6$H$_3$: Isolable Si-centered Triradical with a High-spin Quartet Ground State. Chem. Lett.* ***2015****, 44 (1), 56–57, Copyright 2015 Chemical Society of Japan.*

triradical **35** and biradical **33** was observed in the conformation of the aryl group and (tBu$_2$MeSi)$_2$Si units relative to the radical center. In biradical **33**, the 3p(Si) orbital is nearly perpendicular (87 degrees) to the π-system of the aromatic ring; in contrast, in **35**, these dihedral angles are 62–70 degrees to minimize the steric repulsion between the bulky substituents, allowing some 3p(Si)-π conjugation.[77]

6.6.3 Thermally Accessible Triplet State of the Highly Twisted Tetrakis(di-tert-butylmethylsilyl)disilene

Rotation around the π-C=C double bond, leaving the σ-(C=C) bond intact, is one of the most fundamental processes in chemistry. Highly twisted tetrasubstituted olefins, R$_2$C=CR$_2$ exhibit chemical behavior which suggests that they possess some diradical character; however, generally diradicals were not observed spectroscopically. Disilenes, R$_2$Si=SiR$_2$, have a significantly lower barrier to rotation around the Si=Si π-double bond (15–25 kcal mol^{-1}), than around C=C π-bonds in olefins (ca. 65 kcal mol^{-1}). Consequently, when carrying bulky substituents, disilenes can have highly twisted geometries. For example, tetrakis(di-*tert*-butylmethylsilyl)disilene (**36**)[78] has a highly twisted Si=Si double bond, with an average

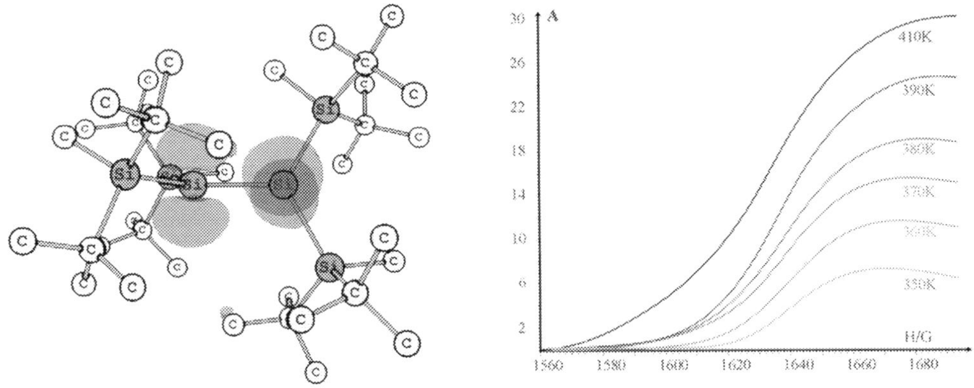

Figure 6.26
(A) A schematic description of the optimized structure of triplet biradical **37** and the Mulliken spin density calculated at the UBP86-D3/TZVP level of theory. Hydrogens are omitted for clarity. (B) Integration curves of the half-field absorption signal (A) of a powder sample of **36** at 350–410K. *Reproduced by permission from Kostenko, A.; Tumanskii, B.; Karni, M.; Inoue, S.; Ichinohe, M.; Sekiguchi, A.; et al. Observation of a Thermally Accessible Triplet State Resulting from Rotation around a Main-Group Bond. Angew. Chem. Int. Ed.* ***2015****, 54 (41), 12144–12148, Copyright 2015 Wiley-VCH Verlag GmbH & Co. KGaA.*

Si–Si=Si–Si torsional angle (Ω) of 56 degrees. This raises the possibility to access thermally a 1,2-disilyl biradical having a triplet state in which the Si=Si π-orbital is rotated by ca. 90 degrees—a species unprecedented in organic or organosilicon chemistry (Eq. 6.25).

An EPR study of a powder sample of **36** at 350–410K shows a half-field transition ($\Delta M_S = \pm 2$) at $H_{min} = 1634.0$ G. The calculated values (at the ROB3LYP-TZVP//UPB86-D3/TZVP level of theory) of the zero-field splitting parameters D and E for the optimized geometry of the nearly perpendicular rotated triplet **37**, correspond to a half-field transition at $H_{min} = 1632.1$ G, in excellent agreement with the experimental value, supporting the thermal formation of the triplet diradical **37**. The calculated spin density in **37** resides almost entirely in the two nearly orthogonal 3p orbitals of the central Si atoms (Fig. 6.26A).[79]

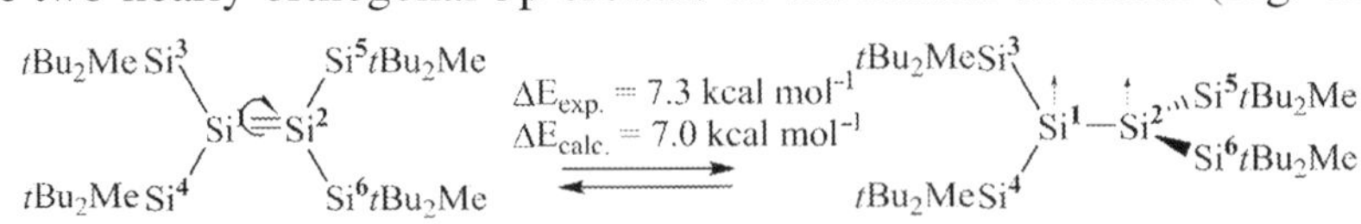

(6.25)

36, singlet average Ω = 55.7°

37, triplet average Ω = 75.2°

The energy gap separating singlet **36** and triplet **37** of only 7.4 kcal mol^{-1} determined experimentally (Fig. 6.26B), suggests that at temperatures above 350K (Fig. 6.26B) **36** should have a significant triplet diradical contribution.

6.7 Silicon-Centered Anion-Radicals

One-electron reduction of low-coordinated organosilicon compounds (e.g., disilenes, disilynes, silylenes) by alkali metals yields the corresponding anion-radical salts. These species can exist as contact ion pairs, solvent-separated ion pairs, or free ions with weak coulombic interaction with the cation, depending on the solvent. For simplicity, we use the term anion-radicals for all these charged species in strongly solvating media (e.g., THF, dimethoxymethane (DME), crown ethers). In contrast, in poor solvating media (e.g., hexane, toluene), a tight contact ion pair with significant Si−M bonding is formed and we classify the species as a metal-substituted silyl radical.

6.7.1 Reduction of Multiply-Bonded Silicon-Compounds

6.7.1.1 Disilenes

Anion-radicals of stable disilenes have been studied extensively and the ones that are persistent or stable are collected in Table 6.7. Weidenbruch et al. were the first to report the EPR spectra of disilene anion-radicals, those of tetramesityldisilene (**38**) and of tetra-*tert*-butyldisilene (**39**) (Table 6.7), which were generated by the multiple reduction of the

Table 6.7: EPR parameters and stability of persistent silyl anion-radicals

Radical	$a_{Si}(^{29}Si_\alpha)$ G	$a_X(X)$,[a] G	*g*	Stability	Refs.
$[Mes_2Si = SiMes_2]^{\bullet-}$ (**38**)	49.6	–	2.0031	Persistent	[80]
$[tBu_2Si - SiBut_2]^{\bullet-}$ (**39**)	33.6	–	2.0035	Persistent	[80]
$[(iPr_3Si)_2Si = Si(SiPri_3)_2]^{\bullet-}$ (**40**)	24.5	–	2.0062	Persistent	[81]
$[(tBuMe_2Si)_2Si = Si(SiMe_2But)_2]^{\bullet-}$ (**41**)	29.2	–	2.0058	Persistent	[81]
$[(iPr_2MeSi)_2Si = Si(SiMePri_2)_2]^{\bullet-}$ (**42**)	31.8	2.5 ($^{29}Si_\beta$)	2.0063	Persistent	[81]
$[(tBu_2MeSi)_2Si = Si(SiMeBut_2)_2]^{\bullet-}$ (**43**)	24.5	–	2.0061	Stable	[78]
[b]$[(Dsi)_2iPrSi{\equiv}SiPri(Dsi)_2]^{\bullet-}$ (**44**)	39.2	22.4 ($^{29}Si_\beta$), 2.3 ($^1H_\lambda$)	1.9996	Stable	[82]
[c]$[(tBu_2MeSi)_2Si = PMes^*]^{\bullet-}$)THF) (**45**)	50.0	54.0 ($^{31}P_\alpha$)	2.0083	Persistent	[83]
[c]$[(tBu_2MeSi)_2Si = PMes^*]^{\bullet-}$ (benzene) (**46**)	65.0	15.0 ($^{31}P_\alpha$)	2.0076	Persistent	[83]
$(tBu_2MeSi)_2Si^{\bullet-}$ (DME) (**47**)	29.1	10.2 ($^{29}Si_\beta$)	2.0074	Stable	[84]
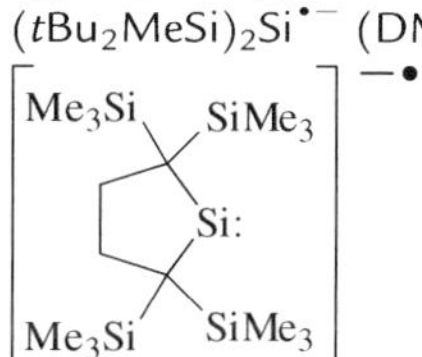 (**48**)	29.9	13.0 ($^{29}Si_\lambda$), 16.6 ($^{29}Si_\lambda$)	2.0077	Persistent	[85]

[a]Atom X responsible for the quoted *hfcc* is given in parenthesis.
[b]Dsi = $CH(SiMe_3)_2$.
[c]Mes* = 2,4,6-*t*-Bu_3-C_6H_2.
Source: Adapted with permission from Tumanskii, B.; Karni, M.; Apeloig, Y. Persistent and Stable Silyl Radicals. In Encyclopedia of Radicals in Chemistry, Biology and Materials; *John Wiley & Sons, Ltd., 2012, Copyright 2012 Wiley-VCH Verlag GmbH & Co KGaA.*

corresponding 1,2-dichlorodisilanes as the source of the disilenes and which were further reduced to the anion-radicals.[80] Reduction of tetrakis(trialkylsilyl)disilenes $R_2Si{=}SiR_2$ ($R{=}SiPri_3$, $SiMe_2Bu$-*t*, $SiMePri_2$) by potassium in DME at room temperature produces the corresponding persistent disilene anion radicals **40**–**42**, respectively (Table 6.7).[81] Based on their *hfccs*, which are in the range 24–32 G (Table 6.6) about one-half of the *hfccs* of the corresponding tris(trialkylsilyl)silyl radicals which exhibit $a(^{29}Si){=}55.5–63.8$ G (Table 6.5), the authors concluded that the silicon atoms in **40**–**42** are nearly planar.[81]

Reduction of $(t\text{-}Bu_2MeSi)_2Si{=}Si(SiMeBu\text{-}t_2)_2$ with *t*-BuLi in THF (Eq. 6.26) yields the isolable anion radical **43** which crystallizes from THF as the Li·4THF salt.[78] The molecular structure of **43** was determined by X-ray crystallography, and it is shown in Fig. 6.27.[78] The central Si1–Si2 bond length in **43** of 2.341 Å is 3.6% longer than that of precursor disilene (2.260 Å) and it is in the typical range of Si–Si single bonds. The twist angle around the central Si–Si bond (i.e., the angles between the Si2–Si1–Si4 and Si5–Si2–Si6 planes) is 88 degrees, indicating that on reduction the π (Si=Si) bond is broken, consistent with the extra electron entering the π^* (Si=Si) orbital. The Si2 atom is planar ($\sum\theta(Si) = 359.9$ degrees), whereas Si1 is slightly pyramidal ($\sum\theta(Si) = 352.7$ degrees), indicating that Si2 has a radical character while Si1 has a silyl anion character.[78] The EPR spectrum of **43** (in 2-methyltetrahydrofuran at r.t.) shows a strong signal with $g = 2.0061$, accompanied by a pair of satellite lines with a splitting at 24.5 G due to

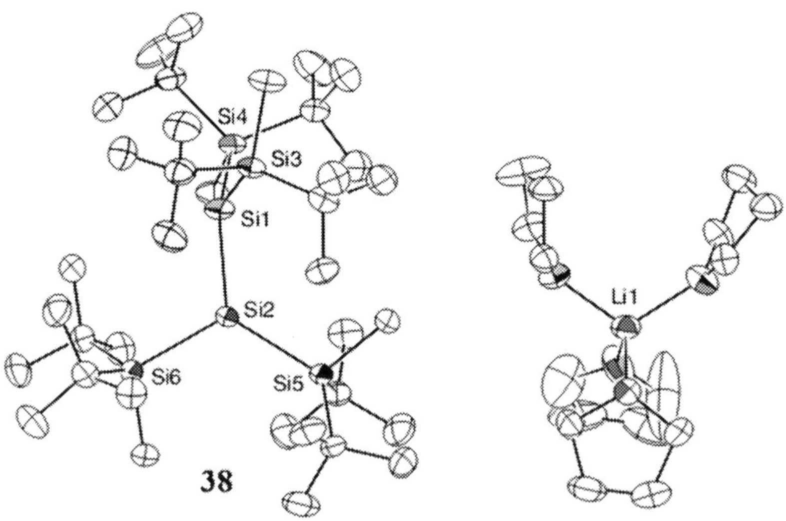

Figure 6.27
ORTEP view of the X-ray structure of **43**. *Adapted with permission from Sekiguchi, A.; Inoue, S.; Ichinohe, M.; Arai, Y., Isolable Anion Radical of Blue Disilene* ($tBu_2MeSi)_2SiSi(SiMetBu_2)_2$ *Formed upon One-Electron Reduction: Synthesis and Characterization. J. Am. Chem. Soc.* ***2004****, 126 (31), 9626–9629,. Copyright 2004 American Chemical Society.*

coupling of the unpaired electron with two magnetically equivalent silicon atoms, indicating that the spin is delocalized over both the Si1 and Si2 nuclei.[78]

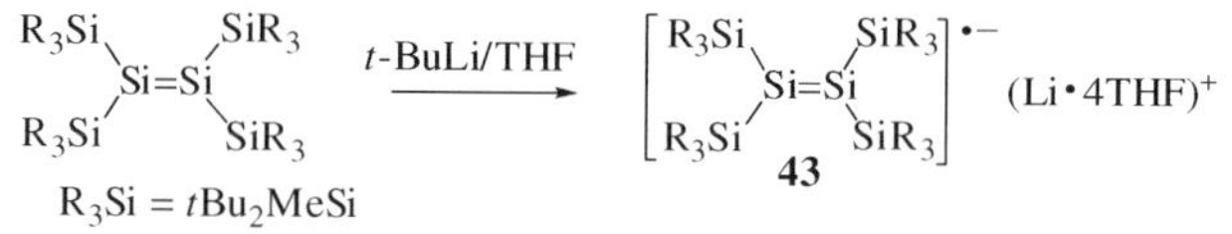

(6.26)

6.7.1.2 Disilynes

Reaction of the stable triply bonded disilyne **44a** with an equivalent amount of KC_8 in dimethyl ether produces the disilyne anion-radical **44**, which was isolated as dark brown crystals in 63% yield (Eq. 6.27).[82]

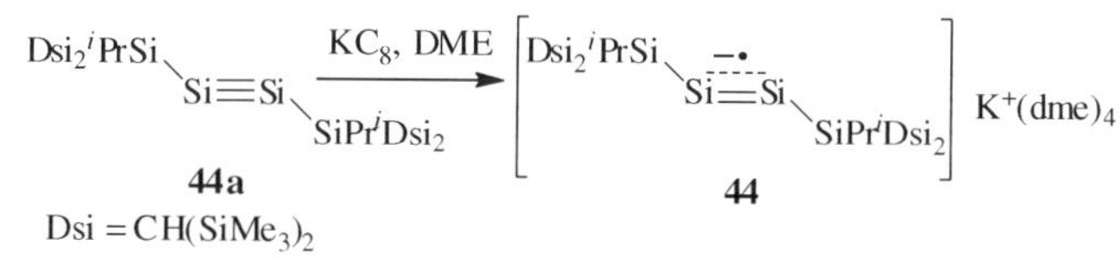

(6.27)

X-ray crystallography (Fig. 6.28) reveals that **44** is a free anion radical with a Si1–K · 4DME distance of 11 Å. The central Si–Si bond length of 2.1728(14) Å is 5% longer than in disilyne **44a** (2.0622(9) Å).[86] This bond elongation is consistent with the notion that the additional electron enters the in-plane antibonding π^* orbital of **44a**. The Si4–Si2–Si1 and Si3–Si1–Si2 bond angles of 112.8 and 113.9 degrees respectively, are significantly smaller (i.e., the degree of bending of the substituents is larger) than the corresponding angles of 137.4 degrees in disilyne **44a**.[86] The fact that the two Si–Si–Si

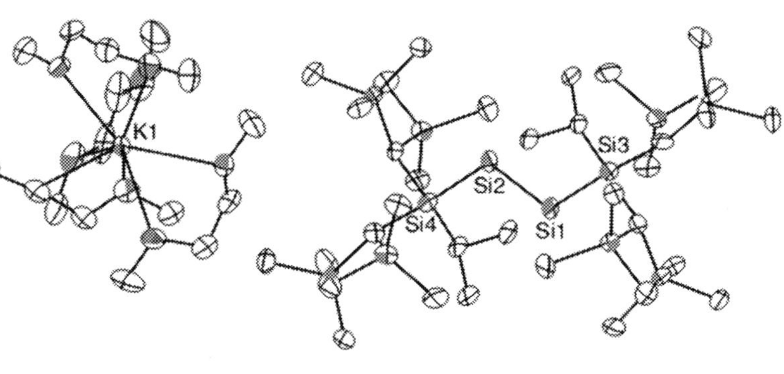

Figure 6.28
ORTEP view of the X-ray structure of the disilyne anion-radical **44**. *Adapted with permission from Kinjo, R.; Ichinohe, M.; Sekiguchi, A. An Isolable Disilyne Anion Radical and a New Route to the Disilenide Ion upon Reduction of a Disilyne. J. Am. Chem. Soc.* ***2007****, 129 (1), 26–27, Copyright 2007 American Chemical Society.*

bending angles in anion-radical **44** are essentially equal indicates that the unpaired electron is equally delocalized between the two central silicon atoms (Si1 and Si2).[82]

The EPR spectrum of **44** shows a triplet signal with a g-value of 1.9996. The triplet splitting (2.3 G) arises from the interaction of the unpaired electron with the two protons of the isopropyl groups. The central signal is accompanied by two pairs of satellite signals, due to coupling with the two magnetically equivalent Si1 and Si2 nuclei ($a_{Si}(^{29}Si1) = a_{Si}(^{29}Si2) = 39.2$ G) and with two Si_β nuclei, ($a_{Si}(2^{29}Si_\beta) = 22.4$ G) (Table 6.6).[82]

6.7.1.3 Phosphasilene

Reduction of phosphasilene $R_2Si = PR$ (**45a**) with KC_8 yields the anion-radical **45** (Eq. 6.28), which is persistent at room temperature and was characterized by EPR spectroscopy.[83] In THF, the EPR spectrum of **45** shows a doublet resulting from the coupling of the unpaired electron to the phosphorus nucleus ($a_P(^{31}P) = 54.0$ G), accompanied by a pair of silicon satellites ($a_{Si}(^{29}Si) = 50.0$ G). The EPR spectrum changes dramatically when the solvent is changed to the nonpolar benzene. $a_P(^{31}P)$ decreases significantly to 15.0 G, whereas $a_{Si}(^{29}Si)$ increases to 65.0 G (Table 6.6). This change can be explained by a localized electronic structure of the anion-radical in benzene (**46**), where the unpaired electron is localized on Si_α and the negative charge is mainly localized on P, and a more delocalized structure in the polar THF.[83]

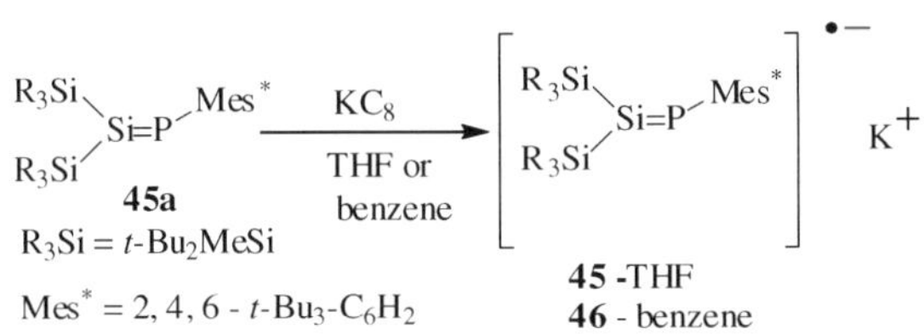

(6.28)

6.7.1.4 Radical-anions of silanone

As stable silanones are not known, the EPR spectra of their anion-radicals $[(Me_3Si)_3CSi(R) = O]^{\bullet -}$ (**49–54**) (Table 6.8) were observed indirectly when photolytically generated t-butoxyl radicals abstract an hydrogen atom from potassium silaenolate **49a–54a** in t-butyl alcohol/di-t-butyl peroxide as solvent (Eq. 6.29).[87] The large *hfccs* to ^{29}Si in anion-radicals **49–54** of 186–165 G indicate that the unpaired electron occupies the $Si(3sp^3)$-$O(2p)$ π^* orbital,[88] centered principally on the silicon atom which is strongly pyramidal (resonance structure **A**, below). The magnitude of coupling to the α-hydrogen (H_α) in **49** (37.8 G) is larger than that in $Me_2HSi^\bullet$ [$a_H(^1H_\alpha) = 17.0$ G][40b] and similar to that observed for silyl radicals carrying electronegative groups (e.g., $H_2FSi^\bullet$, $a_H(^1H_\alpha) = 34.6$ G).[58] Hyperfine coupling to the β-protons (**50–53**) has the following features: in anion-radical **50** three methyl groups give rise to a binominal quartet as expected, with $a_H(3H_\beta) = 4.25$ G.

Table 6.8: EPR parameters of silanone radical-anions[87]

Radical	$a_{Si}(^{29}Si_\alpha)$ G	$a_X(X)$, [a] G	*g*
$[(Me_3Si)_3CSi(H)=O]^{\bullet-}$ (**49**)	186.0	37.8 ($^1H_\alpha$); 0.13(27^{1}H)	2.0028
$[(Me_3Si)_3CSi(Me)=O]^{\bullet-}$ (**50**)	172.1	4.25 (3$^1H_\beta$)	2.0027
$[(Me_3Si)_3CSi(Et)=O]^{\bullet-}$ (**51**)	164.7	8.7 (1$^1H_\beta$)[b]	2.0026
$[(Me_3Si)_3CSi(Bu)=O]^{\bullet-}$ (**52**)	164.8	8.4 (1$^1H_\beta$)	2.0026
$[(Me_3Si)_3CSi(Ph)=O]^{\bullet-}$ (**53**)	—	0.11 (27^{1}H)	2.0026
$[(Me_3Si)_3CSi(F)=O]^{\bullet-}$ (**54**)	—	95.8 (1$^{19}F_\alpha$)	2.0019

[a]Atom X responsible for the quoted *hfcc* is given in parenthesis.
[b]Coupling is observed to only one of the two methylene protons.

The methylene hydrogens in the ethyl or butyl side chains of **51** and **52**, respectively, give a doublet splitting rather than the anticipated triplet, with $a_H(1\ H_\beta) = 8.70$ and 8.44 G. The nonequivalence of the methylene protons could be the result of a restricted rotation about the $Si-CH_2R$ bond.[87]

In conclusion, the EPR spectra indicate that the main contributing resonance structure to these anion-radicals is **A** in which the spin is localized mainly on the silicon atom and the negative charge on oxygen atom.

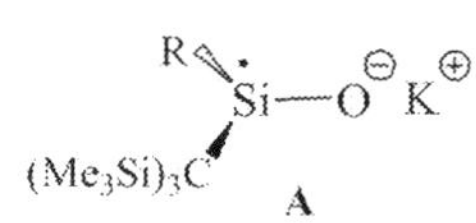

A

The radical-anions **49**–**54** are relatively short-lived under the experimental conditions, but the decay products and their decay mechanisms are not known.

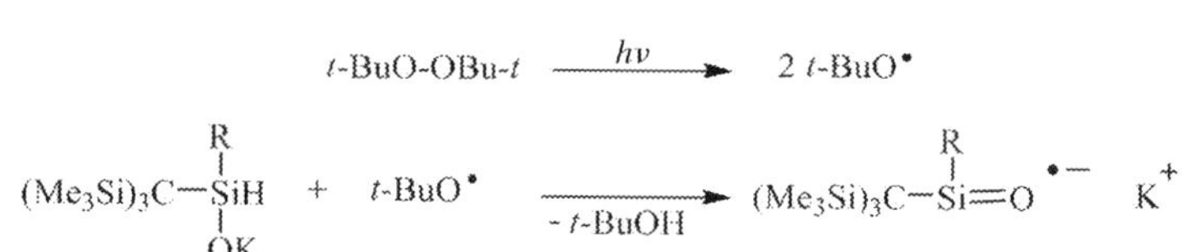

(6.29)

R = H (**49a**), Me (**50a**), Et (**51a**), Bu (**52a**), Ph (**53a**), F (**54a**)

6.7.2 Reduction of Silylenes

Several stable and persistent anion-radicals of silylenes have been reported.[84] Addition of crown-ethers in THF or DME to 1,2-dimetallodisilane **47a** (M = Li, Na) leads to cleavage of the central Si–Si bond, yielding the corresponding stable silylene anion-radicals (Eq. 6.30).[84] The EPR parameters of **47** (Table 6.7) are independent of the counter alkali cations (Li, Na), indicating that they are free ion pairs.

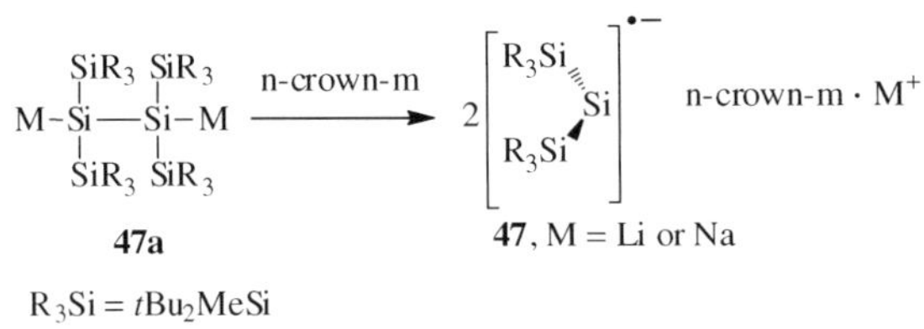

(6.30)

One-electron reduction of stable silylene (**48a**) by alkali metals yields the corresponding anion-radical **48** (Eq. 6.31).[85] The EPR parameters of **48** in DME (Table 6.6) are $a_{Si}(^{29}Si_{\alpha}) = 29.9$ G, $a_{Si}(^{29}Si_{\gamma}) = 13.0$ G and 16.6 G, $g = 2.0077$, and they are independent of the counter alkali cations (Li, Na, K, Cs, Rb), indicating that in DME **48** is free anion-radical.[85]

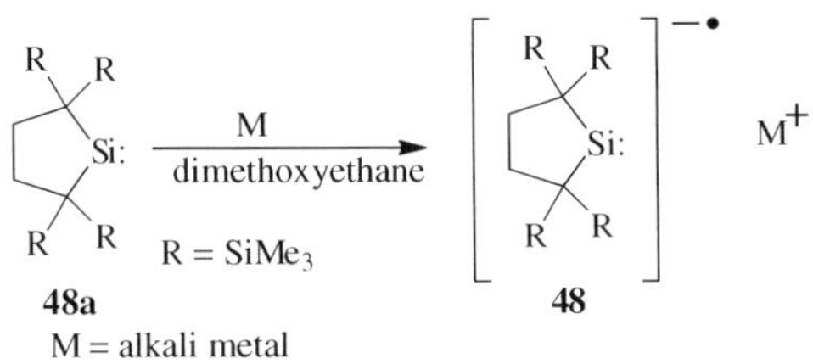

(6.31)

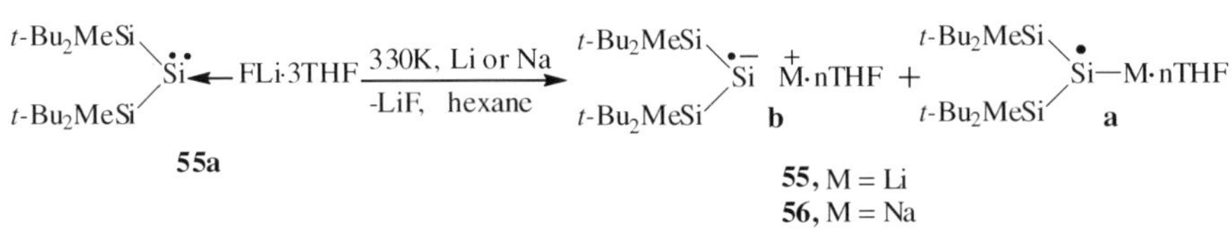

(6.32)

Li- or Na-substituted silyl anion radicals **55** and **56** were observed when silylenoid **55a** was stirred at 330K in hexane with lithium or sodium powder, respectively (Eq. 6.32).[89] The EPR spectrum of **55** shows the expected quartet splitting of all signals by ^{7}Li ($I = 3/2$) and **56** shows quartet splitting of the signals by ^{23}Na ($I = 3/2$) (Table 6.7).[89] The $a_{Si}(^{29}Si_{\alpha})$ values of anion radicals **47, 48, 55**, and **56** (Tables 6.7 and 6.9) are the smallest and their g-values are the largest among related polysilyl radicals, indicating that the SOMO of these anion-radicals has a nearly pure 3p character.[84,85,89] Comparison of the EPR spectra of **47, 55**, and **56** in hexane and DME provides interesting information about the degree of their association. In DME, $[(t\text{-}Bu_2MeSi)_2Si]^{\bullet-}\ Na^{+}$ (DME) **47** has $a_{Si}(^{29}Si_{\alpha}) = 29.1$ G, which is smaller than that of $(t\text{-}Bu_2MeSi)_2LiSi^{\bullet}$ (**55**) or $(t\text{-}Bu_2MeSi)_2NaSi^{\bullet}$ (**56**) in hexane, where $a_{Si}(^{29}Si_{\alpha}) = 33.0\text{–}34.0$ G (Table 6.9).[89,90] In **56**, $a_{Na}(^{23}Na) = 2.8$ G, while in **47** coupling with ^{23}Na is not observed (Table 6.7).[84] This data is consistent with the description of these species as being covalently bound (structure **a** in Eq. 6.32) (i.e., metallosilyl radicals) in nonpolar solvents and separated anion-radical ion-pairs (structure **b** in Eq. 6.32) in better solvating media such as ethers.

West et al. reported that the reduction of stable saturated *N*-heterocyclic silylene with Na/K in THF (Eq. 6.33) gives the corresponding 1,2-dihydrodisilane, which they suggested is formed via fast dimerization of the corresponding silylene anion-radical.[91]

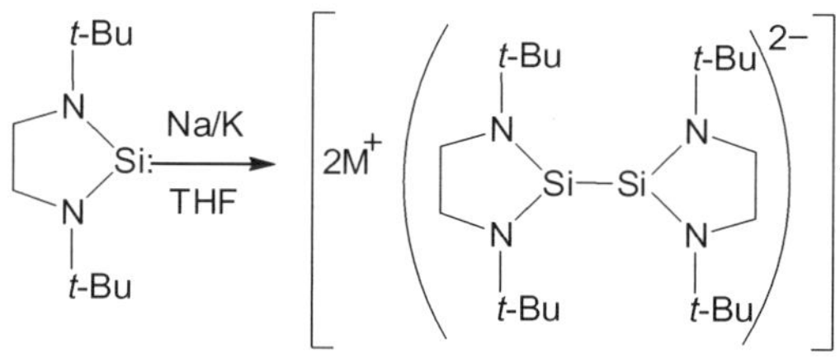

(6.33)

6.7.3 Alkali Metal- and Mercury-Substituted Silyl Radicals

Silyl radicals monosubstituted with Li (**55**) or Na (**56**) were discussed in Section 6.7.2.

6.7.3.1 Aggregated silyllithium radicals

Aggregated lithiosilanes **57a** and **58a** are photolytic precursors to the interesting silyllithium aggregated silyl radicals **57** and biradical **58** respectively (Scheme 6.7).[90]

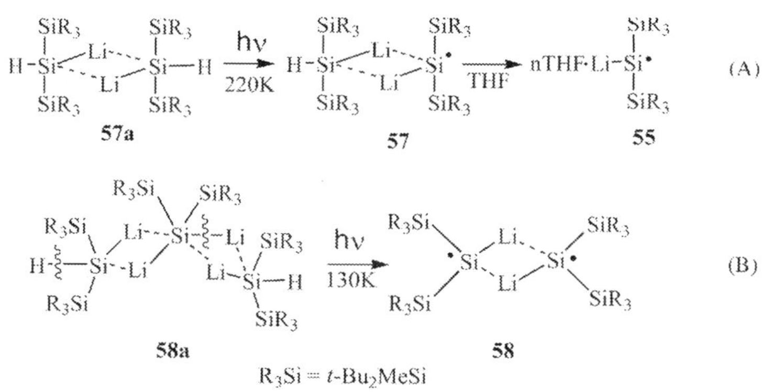

Scheme 6.7
Synthesis of silyllithium aggregated silyl radical **57** and biradical **58**.

UV irradiation of an hexane solution of silyllithium dimer **57a** at 220K yields radical **57** exhibiting a septet EPR signal (Fig. 6.29A), which results from the splitting of the main line ($g = 2.007$) by two equivalent ^{7}Li nuclei ($I = 3/2$) with $a_{\mathrm{Li}}(2^{7}\mathrm{Li})$ of 4.17 G. The hyperfine structure of the EPR signal and the cyclic structure of precursor **57a** allows the confident assignment of the observed signal to cyclic **57** (Scheme 6.7A), **57** is the first known silyl radical that is part of an aggregated silyllithium skeleton.[90] Addition of THF to a hexane solution of **57** cleaves its cyclic Si_2Li_2 skeleton (Scheme 6.7A) yielding the stable lithium-substituted silicon-centered radical **55** exhibiting a quartet EPR signal ($a_{\mathrm{Si}}(^{29}\mathrm{Si}_{\alpha}) = 34.0$ G, $a_{\mathrm{Si}}(2^{29}\mathrm{Si}_{\beta}) = 9.7$ G, $a_{\mathrm{Li}}(^{7}\mathrm{Li}) = 1.6$ G, $g = 2.007$) (Fig. 6.29B, Table 6.9).[90]

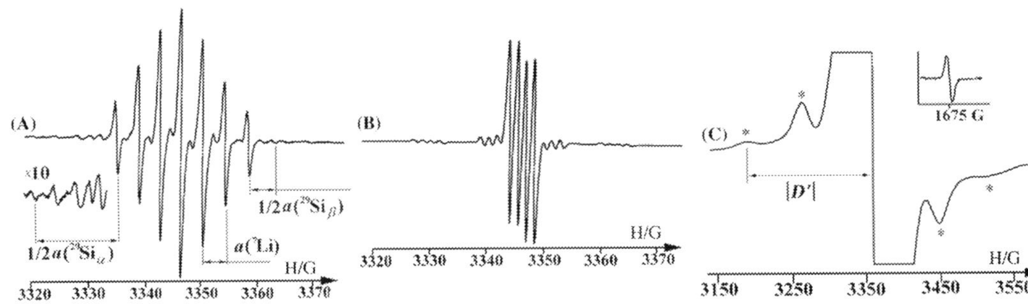

Figure 6.29

(A) EPR spectrum of **57** observed during UV irradiation of **57a** in hexane at 220K. (B) EPR spectrum of silyl radical **55** observed after addition of THF to **57** in hexane at 300K. (C) EPR spectrum of triplet bis(silyl) radical **58** (marked by "*") during UV irradiation of **58a** at 130K. The $\Delta M_S = 2$ absorption is shown in the insert. *Reproduced by permission from Bravo-Zhivotovskii, D.; Ruderfer, I.; Melamed, S.; Botoshansky, M.; Tumanskii, B.; Apeloig, Y., Nonsolvated, Aggregated 1,1-Dilithiosilane and the Derived Silyl Radicals. Angew. Chem. Int. Ed.* ***2005****, 44 (5), 739–743, Copyright 2005 Wiley-VCH Verlag GmbH & Co. KGaA.*

Table 6.9: EPR parameters of persistent alkali metal and mercury substituted silyl radicals

Radical	$a_{Si}(^{29}Si_\alpha)$ G	$a_{Si}(X)$,[a] G	g	Stability	Refs.
$(t\text{-}Bu_2MeSi)_2Si^\bullet - Li{\bullet}nTHF$ (hexane) (**55**)	34.0	9.7 ($^{29}Si_\beta$), 1.6 (^7Li)	2.0075	Persistent	[90]
$(t\text{-}Bu_2MeSi)_2Si^\bullet - Na{\bullet}nTHF$ (hexane) (**56**)	33.0	9.0 ($^{29}Si_\beta$), 2.8 (^{23}Na)	2.0074	Persistent	[89]
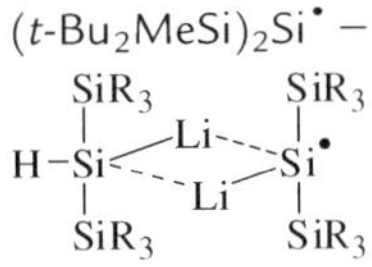 (**57**)	30.0	9.0 ($^{29}Si_\beta$), 4.17 (^7Li)	2.0073	Persistent	[90]
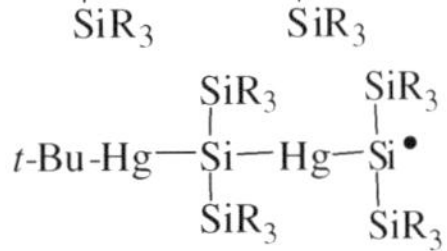 (**59**)	56.0	105.3 (^{199}Hg)	1.983	Persistent	[92]
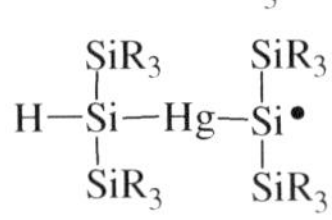 (**60**)	56.0	104.9 (^{199}Hg)	1.984	Persistent	[92]
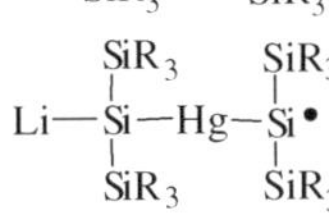 (**61**)	56.0	63.4 (^{199}Hg)	1.984	Persistent	[92]

[a]Atom X responsible for the quoted *hfccs* is given in parenthesis.

Source: Adapted with permission from Tumanskii, B.; Karni, M.; Apeloig, Y. Persistent and Stable Silyl Radicals. In Encyclopedia of Radicals in Chemistry, Biology and Materials; *John Wiley & Sons, Ltd., 2012, Copyright 2012 Wiley-VCH Verlag GmbH & Co KGaA.*

UV irradiation of a frozen hexane solution (130K) of silyl lithium aggregate **58a** (Scheme 6.7B) produces a characteristic EPR signal of a triplet biradical[19a] with four side-bands from $\Delta M_S = 1$ transitions and a signal at half-field at 1675 G resulting from the $\Delta M_S = 2$ transition (Fig. 6.29C).[90] The zero-field splitting parameter $|D'|$ of 172 G indicates a substantial interaction between the two radical centers with a calculated distance (based on $|D'|$[19a]) of 4.35 Å, which is similar to that between the skeletal Si atoms in **58a** (4.57 Å).[90] The suggestion that the two unpaired electrons are centered on the silicon atoms of the cyclic $\{[(R_3Si)_2Si]_2Li_2\}$ fragment with a geometry similar to that in **58a** leads to the assignment of the structure of the observed triplet bis(silyl)radical to **58**.[90]

6.7.3.2 Hg-substituted silyl radicals

Photolysis ($\lambda > 300$ nm) for 10 s of silyl mercurials **59a** (Eq. 6.34) and **60a** (Eq. 6.35) or oxidation of **61a** (Eq. 6.36) yields EPR spectra with g-factors of about 1.983 (Table 6.9).[92] These g-factors are significantly smaller (upfield-shifted) relative to those of all other silyl radicals, which are in the range of $g = 2.004 - 2.007$. These EPR spectra (Fig. 6.30), exhibiting the small *g*-factor (Table 6.9), are assigned to the mercury-substituted silyl radicals **59**–**61**. This assignment is supported by the observation of isotropic hyperfine interactions of the unpaired electron with the ^{199}Hg and ^{29}Si nuclei with satellite lines' intensities corresponding to their natural abundances, and by the significantly upfield-shifted *g*-factor of 1.990 of the carbon analog $ClHgH_2C^•$.[93]

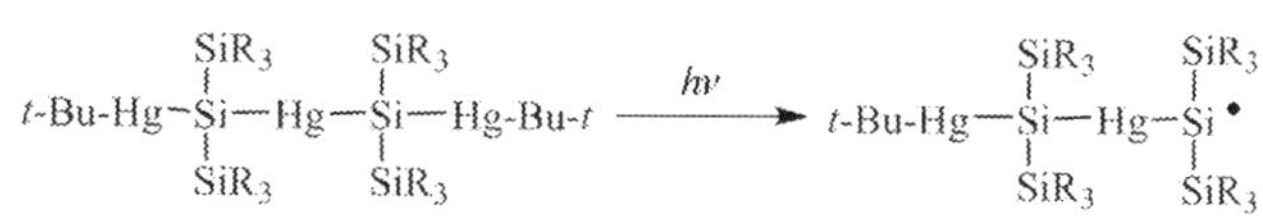

(6.34)

59a $R_3Si = t\text{-}Bu_2MeSi$ **59**

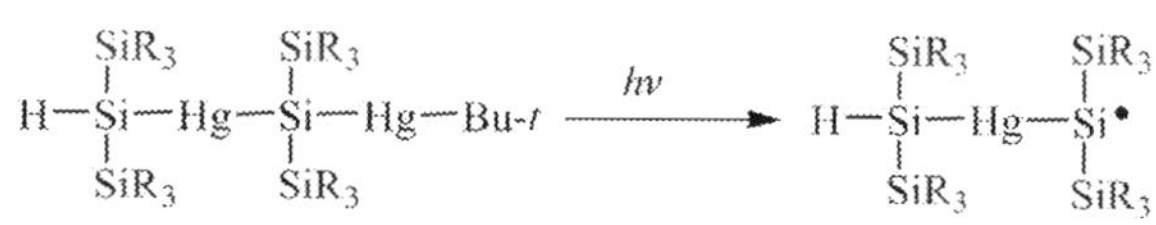

(6.35)

60a **60**

$R_3Si = t\text{-}Bu_2MeSi$

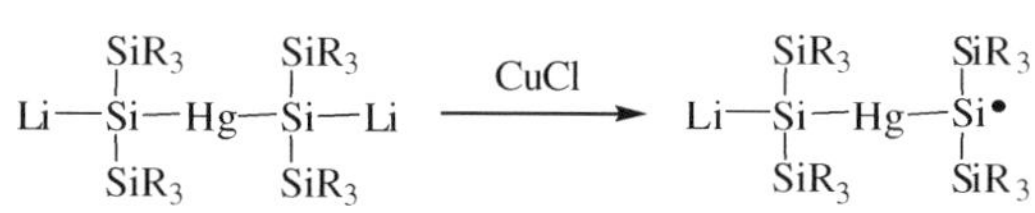

(6.36)

61a **61**

$R_3Si = t\text{-}Bu_2MeSi$

The magnitude of $a_{Hg}(^{199}Hg)$ of 63–105 G in radicals **59**–**61** (Table 6.9) indicates that only 0.5%–0.8% of the spin density resides in a 6s-type orbital of the mercury atom and that the spin of these radicals is localized on the central silicon atom. This estimate is based on the fact that a full spin residing in the 6 s orbital results in $a_{Hg}(^{199}Hg)$ of ~ 12000 G.[19a] This huge *hfcc* value implies that even small changes in the distribution of the spin density indicate meaningful changes in $a_{Hg}(^{199}Hg)$.

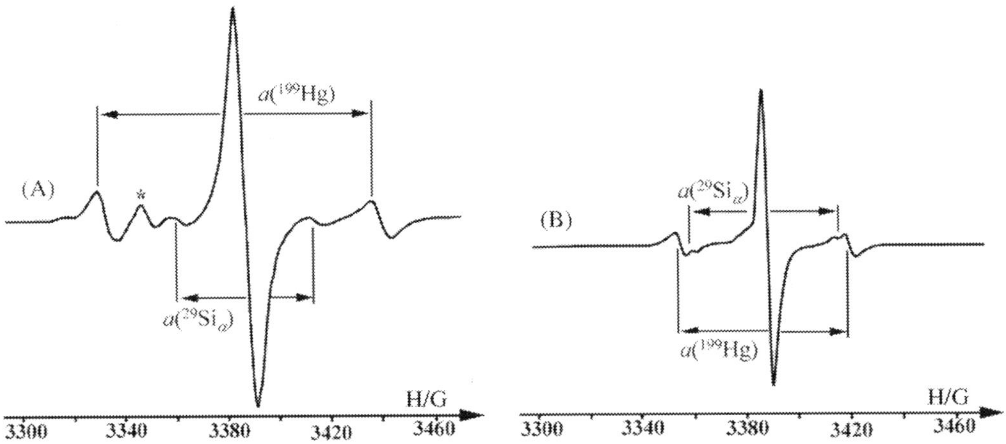

Figure 6.30

EPR spectra of (A) Radical **59** obtained after UV irradiation ($\lambda = 300$ nm) of a hexane solution of **59a** at 200K; line of a nonassigned silyl radical is marked by an asterisk. (B) Radical **61** in hexane at 290K. *Adapted with permission from Bravo-Zhivotovskii, D.; Ruderfer, I.; Yuzefovich, M.; Kosa, M.; Botoshansky, M.; Tumanskii, B.; et al. Mercury-Substituted Silyl Radical Intermediates in Formation and Fragmentation of Geminal Dimercury Silyl Compounds. Organometallics.* ***2005****, 24 (11), 2698–2704, Copyright 2005 American Chemical Society.*

6.8 Transition Metal Substututed Silyl Radicals

The recent synthesis and isolation of stable silylenes[94] (and other group-14 metallylenes[95]) has opened new opportunities for the synthesis of novel transition metal substituted silyl radicals. Such radicals, most of them persistent, were recently produced by the addition of transition metal-centered radicals to stable *N*-heterocyclic silylenes,[96] germylene,[97] and carbene (Scheme 6.8).[98] These novel paramagnetic species **62–68** were studied by EPR spectroscopy (Table 6.10) and DFT calculations, revealing strong dependence of the spin density distribution on the nature of the main group E atom and of the transition metal M. Two types of bonds form on addition of transition metal fragments to divalent E atoms of N-heterocyclic metallylenes: coordination (A) or covalent (B) (Scheme 6.8).

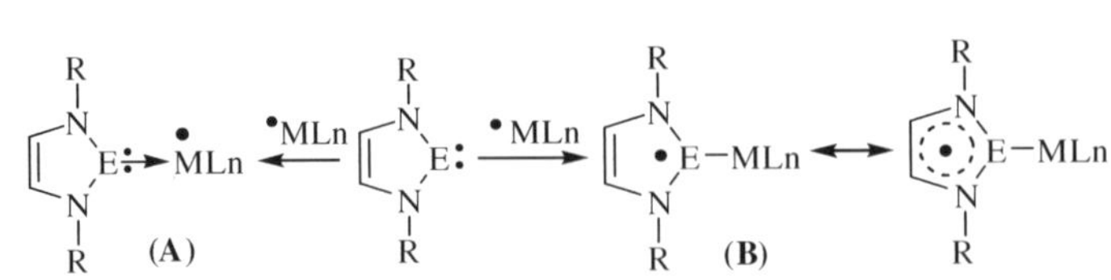

Scheme 6.8

Formation of two bond types on the addition of transition metal-centered radicals to the divalent center E of *N*-heterocyclic metallylenes.

Table 6.10: EPR parameters of persistent transition metal substituted silyl radicals obtained by addition to silylenes

Radical	$a_{Si}(^{29}Si_{\alpha})$ G	$a_X(X)$,[a] G	g[b]	Stability	Refs.
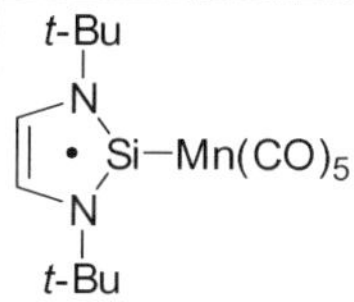 (**62**)	b	5.7 (^{14}N), 5.7 (^{1}H), 8.9 (^{55}Mn)	2.0027	Persistent	[96]
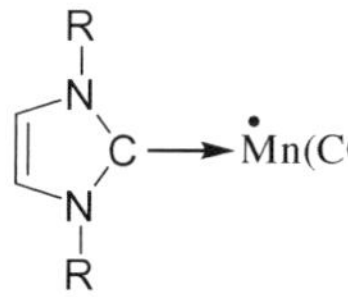 (**63**)	b	24.0 (^{55}Mn)	2.023	Persistent	[98]
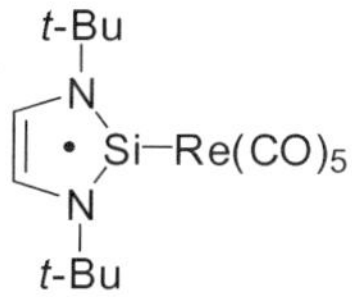 (**64**)	b	7.1 (^{14}N), 4.7 (^{1}H), 36.5 ($^{185,187}Re$)	2.0039	Persistent	[96]
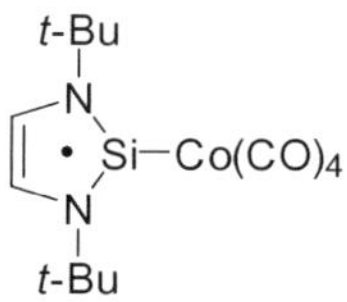 (**65**)	b	5.3 (^{14}N), 5.3 (^{1}H), 6.4 (^{59}Co)	2.003	Persistent	[97]
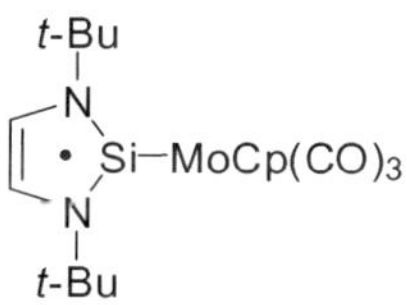 (**66**)	b	5.7 (^{14}N), 5.7 (^{1}H)	2.0027	Persistent	[96]
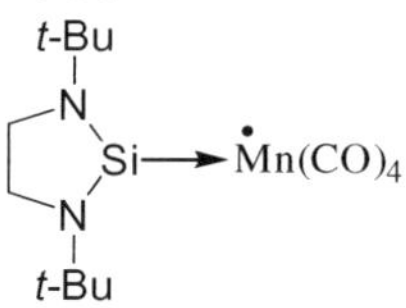 (**67**)	b	13.6 (^{55}Mn)	2.028	Persistent	[59]
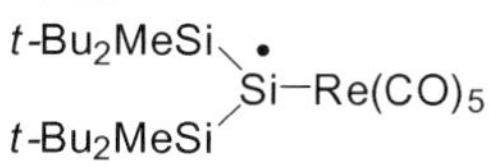 (**68**)	59.6 59.0	40.0 ($^{185,187}Re$) 38.8 ($^{185,187}Re$)	2.013 2.0127	Persistent	[49b]
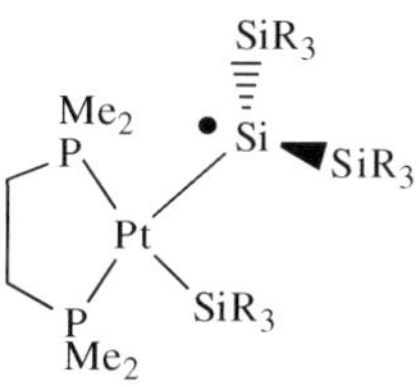 (**69**) $R_3Si = t BuMe_2Si$	b	54.6(^{195}Pt), 60.3 (^{31}P), 7.1(^{31}P)	2.023	Persistent	[99]

[a]Atom X responsible for the quoted *hfccs* is given in parenthesis.
[b]Satellite lines of $^{29}Si_{\alpha}$ are not observed due to low signal/noise ratio.
Source: Adapted with permission from Tumanskii, B.; Karni, M.; Apeloig, Y. Persistent and Stable Silyl Radicals. In Encyclopedia of Radicals in Chemistry, Biology and Materials; *John Wiley & Sons, Ltd., 2012, Copyright 2012 Wiley-VCH Verlag GmbH & Co KGaA].*

EPR spectroscopy combined with DFT calculations of the produced metallo-silyl radical adducts allow the evaluation of the nature of the E–M bond in these paramagnetic species. For example, the reaction of $(CO)_5Mn^{\bullet}$ (produced by photolysis of $Mn_2(CO)_{10}$) with unsaturated silylene **62a** yields radical **62** (Eq. 6.37) which has a half-life of several weeks at room temperature. Radical **62** is characterized by a relatively small splitting of the main line by ^{55}Mn, ($a_{Mn}(^{55}Mn) = 8.9$ G) and by $a_N(2^{14}N) = a_H(2\ H) = 5.7$ G. The observed splitting by two magnetically equivalent ^{14}N nuclei and by two proton nuclei (Table 6.10) clearly indicates that the spin is delocalized over the entire heterocyclic moiety (type B in Scheme 6.8).[59] Similarly, delocalized silyl radicals **64–66** of type B in Scheme 6.8 were observed in reactions of Re-, Co-, and Mo-centered radicals with silylene **62a**.[96,97] In contrast, the reaction of the analogous carbene with $(CO)_5Mn^{\bullet}$ results in the substitution of one of the CO ligands, yielding a persistent Mn-centered radical (**63**) ($a_{Mn}(^{55}Mn) =$ 24.0 G).[98] The large a_{Mn} (^{55}Mn) value and the lack of splitting by the nitrogen and proton nuclei indicate that this radical is of type A in Scheme 6.8, with the spin localized mainly on the manganese.[98] Similar reactions of a variety of transition metal-centered radicals with the saturated silylene **67a** (Eq. 6.38) do not lead to radical adducts, except with $(CO)_5Mn^{\bullet}$ yielding a persistent Mn-centered radical (**67**) ($a_{Mn}(^{55}Mn) = 13.6$ G, $g = 2.028$).[59]

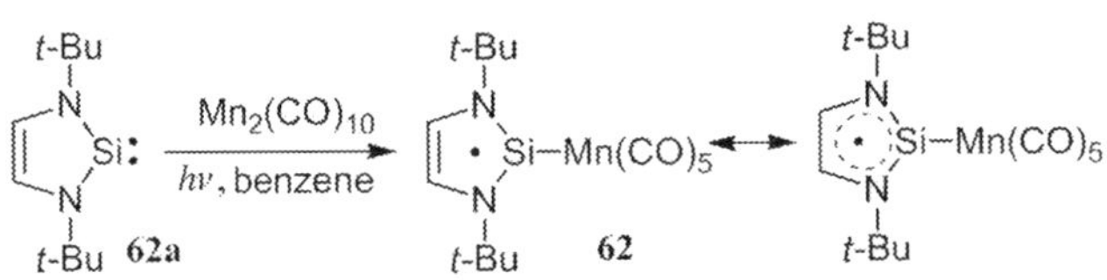

(6.37)

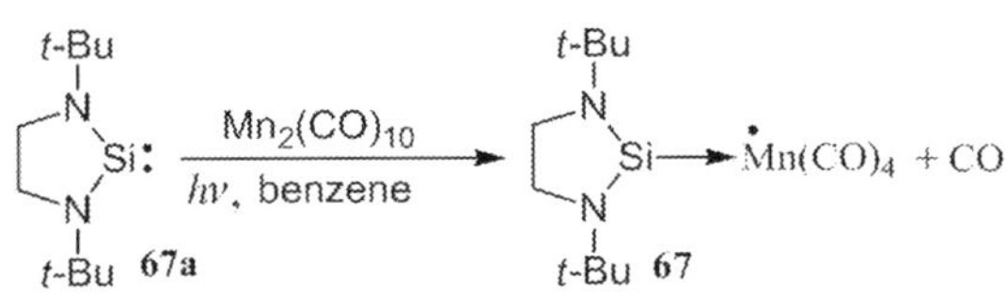

(6.38)

In contrast to transition metal-substituted silyl radicals derived from heterocyclic silylene **62a,** which are usually persistent, radical adducts of silylenes **62a** and **67a** with phosphorus-centered radicals, such as $(i\text{-PrO})_2(O)P^{\bullet}$, are highly reactive and they are not observed by EPR.[100]

A persistent Re-substituted acyclic Si-centered radical **68** is produced by UV irradiation of a 1:1 mixture of silylenoid (**55a**)[89] and $Re_2(CO)_{10}$ in toluene (39).[49b]

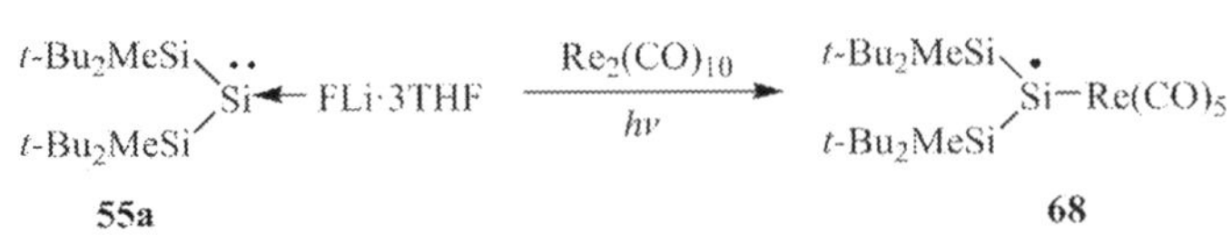

(6.39)

The EPR spectrum of the irradiated mixture of reaction 6.39 reveals the presence of two rotamers of radical **68** (**68a** and **68b**) (Fig. 6.31) in the ratio of about 30:70. The intensity

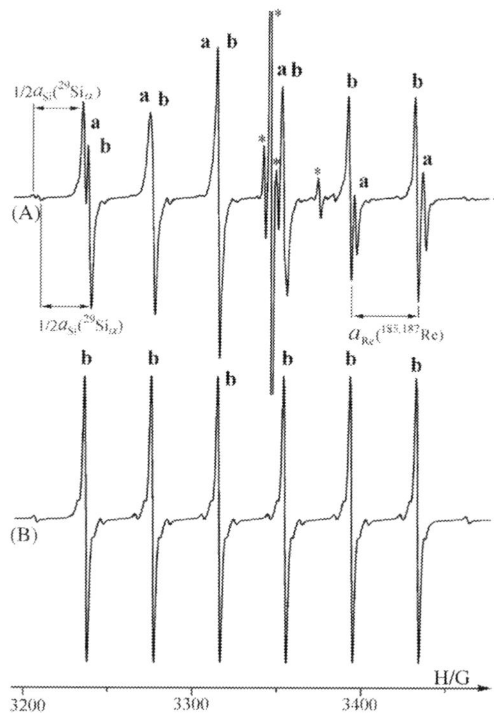

Figure 6.31

EPR spectra of silyl radical **68:** (A) Experimental spectrum showing a superposition of the two rotamers of **68**. Signals of $(t\text{-}Bu_2MeSi)_3Si^{\bullet}$ (a byproduct) are marked by an asterisk. (B) Simulated spectrum of rotamer **68b**. *Adapted with permission from Sheberla, D.; et al., Electronic Structure of Bis(silyl)carbon-, Bis(silyl)silicon-, and Bis(silyl)germanium-Centered Radicals $(R_3Si)_2XE^{\bullet}$ (E = C, Si, Ge; X = H, Re(CO)$_5$, F): EPR and DFT Studies. Organometallics **2010**, 29 (21), 5596–5606, Copyright 2010 American Chemical Society.*

ratio of the signals of the two rotamers does not depend on the irradiation time or the temperature (in the range of 230–320K). According to DFT calculations, in the more stable rotamer, the *t*-butyl groups of the silyl substituents are in *anti*-arrangement and in the less stable rotamer the *t*-butyl groups are in a *syn* arrangement, the *anti*-rotamer being more stable by 0.6 kcal mol^{-1}.[49b]

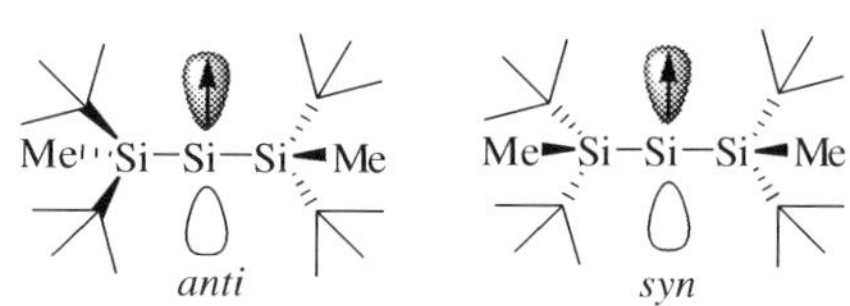

68a **68b**

A persistent Pt-substituted acyclic Si-centered radical **69** was produced by UV irradiation of a cyclic silylmercury complex of platinum (**69a**) in toluene solution (Eq. 6.40).[99]

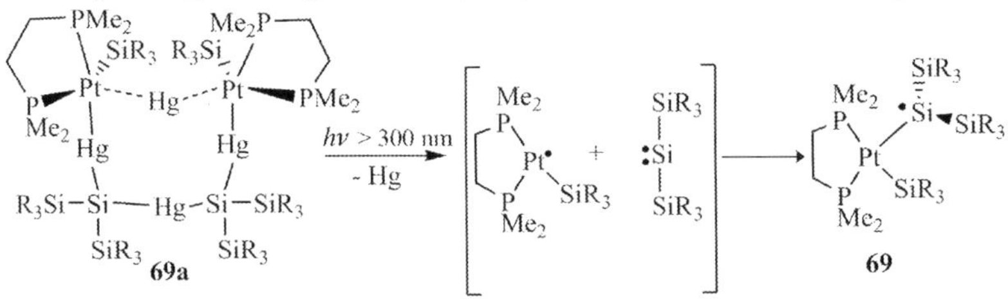

(6.40)

$R_3Si = tBuMe_2Si$

The EPR spectrum **69** is shown in Fig. 6.32. It is characterized by three major parameters: (1) a significantly downfield-shifted *g*-factor (2.023) relative to other silyl radicals, which usually have *g*-factors in the range of 2.003–2.007; (2) isotropic hyperfine interaction of the unpaired electron with two magnetically nonequivalent ^{31}P nuclei $a(^{31}P_1) = 60.3$ G and $a(^{31}P_2) =$ 7.1 G, each appearing as a doublet of doublets; and (3) isotropic hyperfine interaction of the unpaired electron with the magnetic isotope ^{195}Pt [$a(^{195}Pt) = 54.6$ G]. The two very different

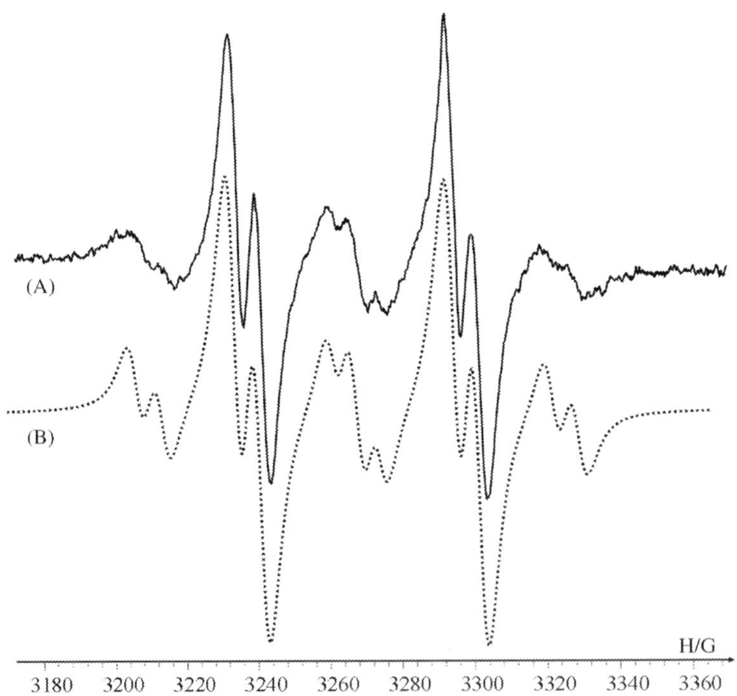

Figure 6.32
(A) EPR spectra of radical **69** in toluene at 300K ($g_{iso} = 2.023$, $a_1(^{31}P) = 60.3$ G, $a_2(^{31}P) = 7.1$ G, $a(^{195}Pt) = 54.6$ G); (B) Simulated spectrum, using the *hfc* constants observed at 300K.

$a(^{31}P)$ indicate their very different degree of interaction with the unpaired electron. The larger $a(^{31}P)$ of 60.3 G is assigned to the phosphorus atom located *cis* to the trivalent silicon atom, and the smaller $a(^{31}P)$ of 7.4 G to the *trans* phosphorus atom.[99]

6.9 Conclusions

The recent developments in the study of silyl radicals have been spectacular. Two decades ago, only short-lived silyl radicals were known and only a few were spectroscopically characterized. Since then, many new types of persistent and stable silyl radicals have been synthesized, isolated, and characterized, including by EPR spectroscopy and X-ray crystallography. These exciting developments were reviewed in this chapter. Silyl radicals are of increasing interest not only to chemists but also to physicists and material scientists. We therefore believe that the chemistry of silyl radicals in general and of persistent and stable silyl radicals in particular will evolve in many new directions in the future.

Acknowledgments

Our research on silyl radicals was supported by the Israel Science Foundation (ISF), by the USA–Israel Binational Science Foundation (BSF), by the Deutsche-Israel Project (DIP), by the German-Israel Foundation for Scientific Research (GIF) and by the Minerva Foundation in Munich. BT is grateful to the Center for Absorption in Science, Ministry of Immigrant Absorption, State of Israel, for financial support. The authors are grateful to Dr. Dmitry Bravo-Zhivotovskii and Arseni Kostenko for helpful discussions.

References

1. (a) Tumanskii, B.; Karni, M.; Apeloig, Y. *Persistent and Stable Silyl Radicals. Encyclopedia of Radicals in Chemistry, Biology and Materials;* John Wiley & Sons, Ltd., 2012.
   (b) Chivers, T.; Konu, J. Stable and Persistent Radicals of Group 13-17 Elements. In *Comprehensive Inorganic Chemistry II;* Poeppelmeier, J. R., Ed.; . *2nd ed.* Elsevier: Amsterdam, 2013; pp 349–373.
   (c) Chatgilialoglu, C. Structural and Chemical Properties of Silyl Radicals. *Chem. Rev.* **1995,** *95* (5), 1229–1251.
   (d) Chatgilialoglu, C.; Schiesser, C. H. In *The Chemistry of Organic Silicon Compounds;* Rappoport, Z., Apeloig, Y., Eds.; , Vol. 3; John Wiley & Sons, Ltd.: Chichester, 2001; pp 341–390.
   (e) Power, P. P. Persistent and Stable Radicals of the Heavier Main Group Elements and Related Species. *Chem. Rev.* **2003,** *103* (3), 789–810.
   (f) Chatgilialoglu, C. *Organosilanes in Radical Chemistry;* John Wiley & Sons, Ltd., 2004.
   (g) Lee, V. Ya; Sekiguchi, A. Si-, Ge-, and Sn-Centered Free Radicals: From Phantom Species to Grams-Order-Scale Materials. *Eur. J. Inorg. Chem.* **2005,** *2005* (7), 1209–1222.
   (h) Lee, V. Ya, et al. In *Reviews of Reactive Intermediates Chemistry;* Moss, R. A., Platz, M. S., Jr, M. J., Eds.; Wiley-VCH Verlag GmbH: Hoboken, NJ, 2007.
   (i) Chatgilialoglu, C. $(Me_3Si)_3SiH$: Twenty Years After Its Discovery as a Radical-Based Reducing Agent. *Chem. Eur. J.* **2008,** *14* (8), 2310–2320.
   (j) Lee, V. Ya; Nakamoto, M.; Sekiguchi, A. Making Stable Radicals of Heavy Elements of Groups 13 and 14: The Might of Silyl Substitution. *Chem. Lett.* **2008,** *37*, 128–133.

2. (a) Zhu, Z.; Wang, C.; Xiang, X.; Pi, C.; Zhou, X. DyI_2 Initiated Mild and Highly Selective Silyl Radical-Catalyzed Cyclotrimerization of Terminal Alkynes and Polymerization of MMA. *Chem. Commun.* **2006,** *19*, 2066–2068.
(b) Yang, J.; Verkade, J. G. Disilane-Catalyzed Cyclotrimerization of Acetylenes. *J. Am. Chem. Soc.* **1998,** *120* (27), 6834–6835.
3. (a) Lalevée, J.; Blanchard, N.; El-Roz, M.; Graff, B.; Allonas, X.; Fouassier, J. P. New Photoinitiators Based on the Silyl Radical Chemistry: Polymerization Ability, ESR Spin Trapping, and Laser Flash Photolysis Investigation. *Macromolecules* **2008,** *41* (12), 4180–4186.
(b) Lalevée, J., et al. Overview of Radical Initiation. *Encyclopedia of Radicals in Chemistry, Biology and Materials;* John Wiley & Sons, Ltd., 2012.
4. (a) Kira, M.; Badugu, R. *Handbook of Photochemistry and Photobiology* **2003,** *Vol. 1*, 83–124.
(b) Birot, M.; Pillot, J.-P.; Dunogues, J. Comprehensive Chemistry of Polycarbosilanes, Polysilazanes, and Polycarbosilazanes as Precursors of Ceramics. *Chem. Rev.* **1995,** *95* (5), 1443–1477.
5. Chatgilialoglu, C. Organosilanes as Radical-Based Reducing Agents in Synthesis. *Acc. Chem. Res.* **1992,** *25* (4), 188–194.
6. (a) Widenhoefer, R. A.; DeCarli, M. A. Tandem Cyclization/Hydrosilylation of Functionalized 1,6-Dienes Catalyzed by a Cationic Palladium Complex. *J. Am. Chem. Soc.* **1998,** *120* (15), 3805–3806.
(b) Hydrosilylation. *A Comprehensive Review on Recent Advances;* Springer, 2009.(c)Marciniec, B. Ed. Catalytic Aspects of Hydrosilylation. In Comprehensive Handbook on Hydrosilylation; Pergamon: Amsterdam, **1992**; Vol. Chapter 2, 8–98.
7. Roberts, B. P. Polarity-Reversal Catalysis of Hydrogen-Atom Abstraction Reactions: Concepts and Applications in Organic Chemistry. *Chem. Soc. Rev.* **1999,** *28* (1), 25–35.
8. Amrein, S.; Timmermann, A.; Studer, A. Radical Transfer Hydrosilylation/Cyclization Using Silylated Cyclohexadienes. *Org. Lett.* **2001,** *3* (15), 2357–2360.
9. Chatgilialoglu, C.; Timokhin, V. I. Chapter 2 – Silyl Radicals in Chemical Synthesis. In *Advances in Organometallic Chemistry;* Anthony, F. H., Mark, J. F., Eds.; , Vol. 57; Academic Press, 2008; pp 117–181.
10. Cicero, R. L.; Linford, M. R.; Chidsey, C. E. D. Photoreactivity of Unsaturated Compounds With Hydrogen-Terminated Silicon(111). *Langmuir* **2000,** *16* (13), 5688–5695.
11. Rijksen, B.; Caipa Campos, M. A.; Paulusse, J. M. J.; Zuilhof, H. *Silicon Radical Surface Chemistry. Encyclopedia of Radicals in Chemistry, Biology and Materials;* John Wiley & Sons, Ltd., 2012.
12. (a) Molev, G.; Tumanskii, B.; Sheberla, D.; Botoshansky, M.; Bravo-Zhivotovskii, D.; Apeloig, Y. Isolable Photoreactive Polysilyl Radicals. *J. Am. Chem. Soc.* **2009,** *131* (33), 11698–11700.
(b) Taira, K., et al. Isolable Aryl-Substituted Silyl Radicals: Synthesis, Characterization, and Reactivity. *Chem. Eur. J.* **2014,** *20* (30), 9342–9348.
(c) Sekiguchi, A.; Fukawa, T.; Nakamoto, M.; Lee, V. Ya; Ichinohe, M. Isolable Silyl and Germyl Radicals Lacking Conjugation With π-Bonds: Synthesis, Characterization, and Reactivity. *J. Am. Chem. Soc.* **2002,** *124* (33), 9865–9869.
(d) Maruyama, H.; Nakano, H.; Nakamoto, M.; Sekiguchi, A. High-Power Electrochemical Energy Storage System Employing Stable Radical Pseudocapacitors. *Angew. Chem. Int. Ed.* **2014,** *53* (5), 1324–1328.
13. (a) Ratera, I.; Veciana, J. Playing With Organic Radicals as Building Blocks for Functional Molecular Materials. *Chem. Soc. Rev.* **2012,** *41* (1), 303–349.(b)Stable Free Radicals: Fundamentals and Applied Aspects of Odd-Electron Compounds. John Wiley & Sons Ltd.: West Sussex, 2010.
14. Rajca, A. From High-Spin Organic Molecules to Organic Polymers With Magnetic Ordering. *Chem. Eur. J.* **2002,** *8* (21), 4834–4841.
15. Castellanos, S.; Velasco, D.; López-Calahorra, F.; Brillas, E.; Julia, L. Taking Advantage of the Radical Character of Tris(2,4,6-trichlorophenyl)methyl to Synthesize New Paramagnetic Glassy Molecular Materials. *J. Org. Chem.* **2008,** *73* (10), 3759–3767.
16. Nakahara, K.; Iriyama, J.; Iwasa, S.; Suguro, M.; Satoh, M.; Cairns, E. J. Al-laminated Film Packaged Organic Radical Battery for High-Power Applications. *J. Power Sources* **2007,** *163* (2), 1110–1113.

17. Wesenberg, J. H.; Ardavan, A.; Briggs, G. A. D.; Morton, J. J. L.; Schoelkopf, R. J.; Schuster, D. I., et al. Quantum Computing With an Electron Spin Ensemble. *Phys. Rev. Lett.* **2009,** *103* (7), 070502.
18. Griller, D.; Ingold, K. U. Persistent carbon-centered radicals. *Acc. Chem. Res.* **1976,** *9* (1), 13–19.
19. (a) Weil, J. A.; Bolton, J. *Electron Paramagnetic Resonance*, 2nd ed.;; John Wiley & Sons, Inc: Hoboken, NJ, 2007.
 (b) Walton, J. C. *Analysis of Radicals by EPR. Encyclopedia of Radicals in Chemistry, Biology and Materials;* John Wiley & Sons, Ltd., 2012.
 (c) Gerson, F.; Huber, W. *Electron Spin Resonance Spectroscopy of Organic Radicals;* Wiley-VCH: Weinheim, 2003.
20. Zavoisky, E. *J. Phys. U.S.S.R.* **1945,** *9*, 245.
21. Ebata, K.; Setaka, W.; Inoue, T.; Kabuto, C.; Kira, M.; Sakurai, H. Planar Hexasilylbenzene Dianion With Thermally Accessible Triplet State. *J. Am. Chem. Soc.* **1998,** *120* (6), 1335–1336.
22. Krusic, P. J.; Wasserman, E. Coronene Dication: A Thermally Accessible Triplet. *J. Am. Chem. Soc.* **1991,** *113* (6), 2322–2323.
23. Rajca, A.; Shiraishi, K.; Pink, M.; Rajca, S. Triplet (S = 1) Ground State Aminyl Diradical. *J. Am. Chem. Soc.* **2007,** *129* (23), 7232–7233.
24. (a) Poutsma, M. L. The Radical Stabilization Energy of a Substituted Carbon-Centered Free Radical Depends on Both the Functionality of the Substituent and the ordinarily of the Radical. *J. Org. Chem.* **2011,** *76* (1), 270–276.
 (b) Wiberg, K. B.; Petersson, G. A. A Computational Study of RXHn X-H Bond Dissociation Enthalpies. *J. Phys. Chem. A* **2014,** *118* (12), 2353–2359.
 (c) Wodrich, M. D.; Corminboeuf, C.; Wheeler, S. E. Accurate Thermochemistry of Hydrocarbon Radicals via an Extended Generalized Bond Separation Reaction Scheme. *J. Phys. Chem. A* **2012,** *116* (13), 3436–3447.
 (d) Chan, B.; Radom, L. BDE261: A Comprehensive Set of High-Level Theoretical Bond Dissociation Enthalpies. *J. Phys. Chem. A* **2012,** *116* (20), 4975–4986.
 (e) Wodrich, M. D.; McKee, W. C., et al. On the Advantages of Hydrocarbon Radical Stabilization Energies Based on R – H Bond Dissociation Energies. *J. Org. Chem.* **2011,** *76* (8), 2439–2447.
 (f) Menon, A. S.; Henry, D. J.; Bally, T.; Radom, L. Effect of Substituents on the Stabilities of Multiply-Substituted Carbon-Centered Radicals. *Org. Biomol. Chem.* **2011,** *9* (10), 3636–3657.
 (g) Coote, M. L.; Lin, C. Y.; Beckwith, A. L. J.; Zavitsas, A. A. A Comparison of Methods for Measuring Relative Radical Stabilities of Carbon-Centered Radicals. *Phys. Chem. Chem. Phys.* **2010,** *12* (33), 9597–9610.
 (h) Bernardi, F.; Epiotis, N. D.; Cherry, W.; Schlegel, H. B.; Whangbo, M.-H.; Wolfe, S. A Molecular Orbital Interpretation of the Static, Dynamic, and Chemical Properties of CH_2X Radicals. *J. Am. Chem. Soc.* **1976,** *98* (2), 469–578.
 (i) Henry, D. J.; Parkinson, C. J.; Mayer, P. M.; Radom, L. Bond Dissociation Energies and Radical Stabilization Energies Associated With Substituted Methyl Radicals. *J. Phys. Chem. A* **2001,** *105* (27), 6750–6756.
 (j) NIST standard Reference Data Base Number 69, http:webbook.nist.gov and references therein. 2011.
25. (a) Luo, Y. R. *Handbook of Bond Dissociation Energies in Organic Compounds;* CRC Press: Florida, 2003.
 (b) Luo, Y.-R., Comprehensive Handbook of Chemical Bond Energies CRC Press: 2007.
26. (a) Sengupta, A.; Ramabhadran, R. O.; Raghavachari, K. Breaking a Bottleneck: Accurate Extrapolation to "Gold Standard" CCSD(T) Energies for Large Open Shell Organic Radicals at Reduced Computational Cost. *J. Comput. Chem.* **2016,** *37* (2), 286–295.
 (b) Sengupta, A.; Raghavachari, K. Prediction of Accurate Thermochemistry of Medium and Large Sized Radicals Using Connectivity-Based Hierarchy (CBH). *J. Chem. Theory Comput.* **2014,** *10* (10), 4342–4350.
 (c) McKee, W. C.; Schleyer, P. v R. Correlation Effects on the Relative Stabilities of Alkanes. *J. Am. Chem. Soc.* **2013,** *135* (35), 13008–13014.

(d) Chan, B.; Radom, L. Hierarchy of Relative Bond Dissociation Enthalpies and Their Use to Efficiently Compute Accurate Absolute Bond Dissociation Enthalpies for C-H, C-C, and C-F Bonds. *J. Phys. Chem. A* **2013,** *117* (17), 3666–3675.

(e) Gonthier, J. F., et al. Quantification of "Fuzzy" Chemical Concepts: A Computational Perspective. *Chem. Soc. Rev.* **2012,** *41* (13), 4671–4687.

(f) Zavitsas, A. A.; Rogers, D. W.; Matsunaga, N. Shortcomings of Basing Radical Stabilization Energies on Bond Dissociation Energies of Alkyl Groups to Hydrogen. *J. Org. Chem.* **2010,** *75* (16), 5697–5700.

(g) Zipse, H. Radical Stability—A Theoretical Perspective. In *Radicals in Synthesis I;* Gansäuer, A., Ed.; , Vol. 263; Springer Berlin Heidelberg, 2006; pp 163–189.

27. (a) Wu, Y.-D.; Wong, C.-L. Substituent Effect on the Dissociation Energy of the Si-H Bond: A Density Functional Study. *J. Org. Chem.* **1995,** *60* (4), 821–828.

(b) Coolidge, M. B.; Borden, W. T. Ab Initio Calculations of the Effects of Substituents on the Stabilization of Silyl Radicals Versus Methyl Radicals. *J. Am. Chem. Soc.* **1988,** *110* (7), 2298–2299.

(c) Parthiban, S.; Martin, J. M. L. Assessment of W1 and W2 Theories for the Computation of Electron Affinities, Ionization Potentials, Heats of Formation, and Proton Affinities. *J. Chem. Phys.* **2001,** *114* (14), 6014–6029.

28. Fu, Y.; Yu, T.-Q.; Wang, Y.-M.; Liu, L.; Guo, Q.-X. α-Substituent Effects on Si–H, P–H and S–H Bond Dissociation Energies. *Chin. J. Chem.* **2006,** *24* (3), 299–306.

29. Beccerra, R.; Walsh, A. R. In *The Chemistry of Organic Silicon Compounds;* Rappoport, Z., Apeloig, A. Y., Eds.; , Vol. 2; John Wiley & Sons: New York, 1998; pp 153–180.

30. Wang, H.-J.; Fu, Y. Designing New Free-Radical Reducing Reagents: Theoretical Study on Si–H Bond Dissociation Energies of Organic Silanes. *J. Mol. Struct.: THEOCHEM* **2009,** *893* (1–3), 67–72.

31. Bordwell, F. G.; Cheng, J.; Ji, G. Z.; Satish, A. V.; Zhang, X. Bond Dissociation Energies in DMSO Related to the Gas Phase Values. *J. Am. Chem. Soc.* **1991,** *113* (26), 9790–9795.

32. Lin, C. Y.; Peh, J.; Coote, M. L. Effects of Chemical Structure on the Thermodynamic Efficiency of Radical Chain Carriers for Organic Synthesis. *J. Org. Chem.* **2011,** *76* (6), 1715–1726.

33. Wang, L.; He, Y.-L. Halogenated Silanes, Radicals, and Cations: Theoretical Predictions on Ionization Energies, Structures and Potential Energy Surfaces of Cations, Proton Affinities, and Enthalpies of Formation. *Int. J. Mass. Spectrom.* **2008,** *276* (1), 56–76.

34. (a) Hildenbrand, D. L.; Lau, K. H.; Sanjurjo, A. Experimental Thermochemistry of the SiCl and SiBr Radicals; Enthalpies of Formation of Species in the Si-Cl and Si-Br Systems. *J. Phys. Chem. A* **2003,** *107* (28), 5448–5451.

(b) Ponomarev, D.; Takhistov, V.; Slayden, S.; Liebman, J. Thermochemistry of Organic, Elementorganic and Inorganic Species. Part XX. Enthalpies of Formation for Free Radicals of Main Group Elements' Halogenides. *J. Mol. Struct.* **2008,** *876* (1–3), 15–33.

35. Marshall, P.; Srinivas, G. N.; Schwartz, M. A Computational Study of the Thermochemistry of Bromine- and Iodine-Containing Methanes and Methyl Radicals. *J. Phys. Chem. A.* **2005,** *109* (28), 6371–6379.

36. Nyulaszi, L.; Belghazi, A.; Szetsi, S. K.; Veszpremi, T.; Heinicke, J. Substituent Effect of Second Row Elements on Silyl Centers. *J. Mol. Struct.: THEOCHEM* **1994,** *119* (1), 73–81.

37. Apeloig, Y.; Bravo-Zhivotovskii, D.; Yuzefovich, M.; Bendikov, M.; Shames, A. I. Polysilyl Radicals: EPR Study of the Formation and Decomposition of Star Polysilanes. *Appl. Magn. Reson.* **2000,** *18* (3), 425–434.

38. Longshaw, A. I.; Carland, M. W.; Krenske, E. H.; Coote, M. L.; Sherburn, M. S. Tris(trimethylsilyl) methane Is Not an Effective Mediator of Radical Reactions. *Tetrahedron. Lett.* **2007,** *48* (32), 5585–5588.

39. Fessenden, R. W.; Schuler, R. H. Electron Spin Resonance Studies of Transient Alkyl Radicals. *J. Chem. Phys.* **1963,** *39* (9), 2147–2195.

40. (a) Krusic, P. J.; Kochi, J. K. Electron Spin Resonance of Organosilyl Radicals in Solution. *J. Am. Chem. Soc.* **1969,** *91* (14), 3938–3940.

(b) Bennett, S. W.; Eaborn, C.; Hudson, A.; Jackson, R. A.; Root, K. D. J. An Electron Spin Resonance Study of Some Silyl Radicals. *J. Chem. Soc. A* **1970,** 348–351.

41. Mendenhall, G. D.; Griller, D.; Lindsay, D.; Tidwell, T. T.; Ingold, K. U. Kinetic Applications of Electron Paramagnetic Resonance Spectroscopy. XIII. Di-*tert*-butylmethyl and Related Radicals. *J. Am. Chem. Soc.* **1974,** *96* (8), 2441–2447.
42. (a) Hicks, R. G. What's New in Stable Radical Chemistry? *Org. Biomol. Chem.* **2007,** *5* (9), 1321–1338.
 (b) Armet, O.; Veciana, J.; Rovira, C.; Riera, J.; Castaner, J.; Molins, E., et al. Inert Carbon Free Radicals. 8. Polychlorotriphenylmethyl Radicals: Synthesis, Structure, and Spin-Density Distribution. *J. Phys. Chem.* **1987,** *91* (22), 5608–5616.
43. Apeloig, Y.; Bravo-Zhivotovskii, D.; Bendikov, M.; Danovich, D.; Botoshansky, M.; Voronkov, M., et al. Synthesis and X-ray Molecular Structure of the First Stable Organic Radical Lacking Resonance Stabilization. *J. Am. Chem. Soc.* **1999,** *121* (35), 8118–8119.
44. Sakurai, H.; Umino, H.; Sugiyama, H. Electron Spin Resonance Studies on Tris(3,5-di-tert-butylphenyl) silyl and Germyl Radicals. *J. Am. Chem. Soc.* **1980,** *102* (22), 6837–6840.
45. Drost, C.; Griebel, J.; Kirmse, R.; Lönnecke, P.; Reinhold, J. A Stable and Crystalline Triarylgermyl Radical: Structure and EPR Spectra. *Angew. Chem. Int. Ed.* **2009,** *48* (11), 1962–1965.
46. Hudson, A.; Lappert, M. F.; Lednor, P. W. Subvalent Group 4B Metal Alkyls and Amides. Part 4. An Electron Spin Resonance Study of Some Long-Lived Photochemically Synthesised Trisubstituted Silyl, Germyl, and Stannyl Radicals. *J. Chem. Soc., Dalton Trans.* **1976,** *22*, 2369–2375.
47. Kyushin, S.; Sakurai, H.; Matsumoto, H. Highly Planar Silane $[(i\text{-}Pr)_3Si]_3SiH$ and Silyl Radical $[(i\text{-}Pr)_3Si]_3Si$. *Chem. Lett.* **1998,** *27* (2), 107–108.
48. (a) Lee, V. Ya; Sekiguchi, A. *Organometallic Compounds of Low-Coordinate Si, Ge, Sn and Pb;* John Wiley & Sons, Ltd., 2010415–431.
 (b) Becker, M.; Förster, C.; Franzen, C.; Hartrath, J., et al. Persistent Radicals of Trivalent Tin and Lead. *Inorg. Chem.* **2008,** *47* (21), 9965–9978.
 (c) Förster, C.; Klinkhammer, K. W.; Tumanskii, B.; Krüger, H.-J.; Kelm, H. Stable Mononuclear Lead (III) Compound: A Lead-Centered Radical. *Angew. Chem. Int. Ed.* **2007,** *46* (7), 1156–1159.
 (d) Sekiguchi, A.; Fukawa, T.; Lee, V. Ya; Nakamoto, M. Tin-Centered Radical and Cation: Stable and Free. *J. Am. Chem. Soc.* **2003,** *125* (31), 9250–9251.
49. (a) Cotton, J. D.; Cundy, C. S.; Harris, D. H.; Hudson, A.; Lappert, M. F.; Lednor, P. W. Photochemical Synthesis and Electron Spin Resonance Characterisation of Stable Trivalent Metal Alkyls (Si, Ge, Sn) and Amides (Ge and Sn) of Group IV Elements. *J. Chem. Soc., Chem. Commun.* **1974,** *16*, 651–652.
 (b) Sheberla, D., et al. Electronic Structure of Bis(silyl)carbon-, Bis(silyl)silicon-, and Bis(silyl) germanium-Centered Radicals $(R_3Si)_2XE^{\bullet}$ (E = C, Si, Ge; X = H, $Re(CO)_5$, F): EPR and DFT Studies. *Organometallics* **2010,** *29* (21), 5596–5606.
50. Lide, D. R. *CRC Handbook of Chemistry and Physics;* CRC Press/Taylor and Francis: Boca Raton, FL, 2010.
51. Fronczek, F. R.; Lickiss, P. D. Structure of the Highly Crowded Alkyne Di[tris(trimethylsilyl)methyl] acetylene and the Octasilane Hexakis(trimethylsilyl)disilane. *Acta Crystallogr., Sect. C* **1993,** *49* (2), 331–333.
52. Bickelhaupt, F. M.; Ziegler, T.; Schleyer, P. V. R. $CH_3^{\bullet}$ Is Planar Due to H – H Steric Repulsion. Theoretical Study of $MH_3^{\bullet}$ and MH_3Cl (M = C, Si, Ge, Sn). *Organometallics* **1996,** *15* (5), 1477–1487.
53. (a) Apeloig, Y. In *The Chemistry of Organic Silicon Compounds;* Patai, S., Rappoport, Z., Eds.; John Wiley & Sons, Ltd.: Chichester, 1989; pp 57–226.
 (b) Karni, M.; Apeloig, Y.; Kapp, J.; Schleyer, P. R. In *The Chemistry of Organic Silicon Compounds;* Rappoport, Z., Apeloig, Y., Eds.; , Vol. 3; John Wiley & Sons, Ltd.: Chichester, 2001; pp 1–163.
 (c) Apeloig et al. unpublished results.
54. Guerra, M. Theoretical Study of the Effect of α-Substituents on the Silicon-29 Hyperfine Splitting Constant in Silyl Radicals: An Electronic Rather Than Structural Effect. *J. Am. Chem. Soc.* **1993,** *115* (25), 11926–11929.
55. (a) Hermosilla, L.; Calle, P.; García de la Vega, J. M.; Sieiro, C. Density Functional Theory Predictions of Isotropic Hyperfine Coupling Constants. *J. Phys. Chem. A* **2005,** *109* (6), 1114–1124.

(b) Hermosilla, L.; Calle, P.; García de la Vega, J. M., et al. Theoretical Isotropic Hyperfine Coupling Constants of Third-Row Nuclei (^{29}Si, ^{31}P, and ^{33}S). *J. Phys. Chem. A* **2005,** *109* (33), 7626–7635.

56. (a) Improta, R.; Barone, V. Interplay of Electronic, Environmental, and Vibrational Effects in Determining the Hyperfine Coupling Constants of Organic Free Radicals. *Chem. Rev.* **2004,** *104* (3), 1231–1254.
(b) Barone, V.; Cimino, P.; Stendardo, E. Development and Validation of the B3LYP/N07D Computational Model for Structural Parameter and Magnetic Tensors of Large Free Radicals. *J. Chem. Theory Comput.* **2008,** *4* (5), 751–764.
(c) Barone, V.; Cimino, P. Accurate and Feasible Computations of Structural and Magnetic Properties of Large Free Radicals: The PBE0/N07D Model. *Chem. Phys. Lett.* **2008,** *454* (1–3), 139–143.

57. Roncin, J. Etude par Résonance Paramagnétique Electronique des Radicaux Libres Produits par Irradiation γ à 77°K de Dérivés Globulaires du Méthane et du Silane. *Mol. Cryst.* **1967,** *3* (1), 117–144.

58. Merritt, M. V.; Fessenden, R. W. ESR Spectra of the Fluorinated Silyl Radicals. *J. Chem. Phys.* **1972,** *56* (5), 2353–2357.

59. Sheberla, D., [Ph.D. Thesis]. Technion-Israel Institute of Technology: Haifa, Israel, 2011.

60. Chatgilialoglu, C.; Guerrini, A.; Lucarini, M. The Trimethylsilyl Substituent Effect on the Reactivity of Silanes. Structural Correlations Between Silyl Radicals and Their Parent Silanes. *J. Org. Chem.* **1992,** *57* (12), 3405–3409.

61. Jackson, R. A.; Rhodes, C. J. Configurations of Silicon-Centred Radicals: The Bis(trimethylsilyl)silyl radical. *J. Organomet. Chem.* **1987,** *336* (1–2), 45–48.

62. Gynane, M. J. S.; Lappert, M. F.; Riley, P. I.; Rivière, P.; Rivière-Baudet, M. Triaryl-Silyl, -Germyl, and -Stannyl Radicals $^{\cdot}MAr_3$ (M = Si, Ge, or Sn and Ar = 2,4,6-$Me_3C_6H_2$) and $^{\cdot}Ge(2,6\text{-}Me_2C_6H_3)_3$: Synthesis and ESR studies. *J. Organomet. Chem.* **1980,** *202* (1), 5–12.

63. Kaushansky, A., [M.S. Thesis]. Technion-Israel Institute of Technology: Haifa, Israel, 2013.

64. Kyushin, S.; Sakurai, H.; Betsuyaku, T.; Matsumoto, H. Highly Stable Silyl Radicals $(Et_nMe_{3-n}Si)_3Si\bullet$ ($n = 1 - 3$). *Organometallics* **1997,** *16* (25), 5386–5388.

65. Azinović, D.; Bravo-Zhivotovskii, D.; Bendikov, M.; Apeloig, Y.; Tumanskii, B.; Vepřek, S. Spectroscopic Studies of the Role of Silyl Radicals in Photolysis of Polysilanes. *Chem. Phys. Lett.* **2003,** *374* (3–4), 257–263.

66. Kravchenko, V.; Bravo-Zhivotovskii, D.; Tumanskii, B.; Botoshansky, M.; Segal, N.; Molev, G., et al. Kinetic Stabilization of Polysilyl Radicals. In *From Organosilicon Chemistry VI: From Molecules to Materials;* Auner, N., Weis, J., Eds.; Wiley-VCH Verlag GmbH: Weinheim, 2005; pp 48–57.

67. Kira, M.; Obata, T.; Kon, I.; Hashimoto, H.; Ichinohe, M.; Sakurai, H., et al. Persistent Tris(*t*-butyldimethylsilyl)silyl Radical and Its New Generation Methods. *Chem. Lett.* **1998,** *27* (11), 1097–1098.

68. (a) Sekiguchi, A.; Matsuno, T.; Ichinohe, M. Cyclotetrasilenyl: The First Isolable Silyl Radical. *J. Am. Chem. Soc.* **2001,** *123* (49), 12436–12437.
(b) Matsuno, T.; Ichinohe, M.; Sekiguchi, A. Cyclotetrasilenide Ion: A Reversible Redox System of Cyclotetrasilenyl Cation, Radical, and Anion. *Angew. Chem. Int. Ed.* **2002,** *41* (9), 1575–1577.

69. Griller, D.; Preston, K. F. Vibrational Analysis by Electron Paramagnetic Resonance Spectroscopy. The Isopropyl Radical. *J. Am. Chem. Soc.* **1979,** *101* (8), 1975–1979.

70. Kaftory, M.; Kapon, M.; Botoshansky, M. In *The Chemistry of Organic Silicon Compounds;* Rappoport, Z., Apeloig, Y., Eds.; John Wiley & Sons, Inc: New York, 1998; pp 181–265.

71. (a) Wiberg, N.; Auer, H.; Nöth, H.; Knizek, J.; Polborn, K. Diiodotetrasupersilylcyclotetrasilene $(^tBu_3Si)_4Si_4I_2$-A Molecule Containing an Unsaturated Si_4 Ring. *Angew. Chem. Int. Ed.* **1998,** *37* (20), 2869–2872.
(b) Kira, M.; Iwamoto, T., et al. The First Stable Cyclic Disilene: Hexakis(trialkylsilyl) tetrasilacyclobutene. *J. Am. Chem. Soc.* **1996,** *118*, 10303.

72. Davidson, I. M. T.; Stephenson, I. L. The Silicon-Silicon Bond Dissociation Energy in Hexamethyldisilane. *J. Chem. Soc. A* **1968,** *0*, 282–287.

73. Becker, J. Y.; Lee, V. Ya; Nakamoto, M.; Sekiguchi, A.; Chrostowska, A., et al. Electrochemical Properties and Computations of Stable Radicals of the Heavy Group 14 Elements (Si, Ge, and Sn). *Chem. Eur. J.* **2009,** *15* (34), 8480–8484.
74. Chrostowska, A.; Dargelos, A.; Graciaa, A.; Baylère, P.; Lee, V. Ya; Nakamoto, M., et al. Electronic Structure of Stable Radicals of the Heavy Group 14 Elements: UV-Photoelectron Spectroscopy Characterization. *Organometallics* **2008,** *27* (13), 2915–2917.
75. (a) Stanislawski, D. A.; Buchanan, A. C.; West, R. Nuclear Magnetic Resonance and Electron Spin Resonance Observation of Hindered Rotation in Sym-tetra-*tert*-butyldisilane and the Corresponding Disilanyl Radical. *J. Am. Chem. Soc.* **1978,** *100* (25), 7791–7794.
    (b) Sakurai, H.; Kira, M.; Sato, M. A Study of Electron Spin Resonance Spectra of the 1,1,2,2-Tetramethyldisilanyl Radical. *Chem. Lett.* **1974,** *3* (11), 1323–1326.
76. Nozawa, T.; Nagata, M.; Ichinohe, M.; Sekiguchi, A. Isolable p- and m-$[(t\mathrm{Bu_2MeSi})_2\mathrm{Si}]_2\mathrm{C_6H_4}$: Disilaquinodimethane vs Triplet Bis(silyl radical). *J. Am. Chem. Soc.* **2011,** *133* (15), 5773–5775.
77. Nozawa, T.; Ichinohe, M.; Sekiguchi, A. 1,3,5-$[({}^t\mathrm{Bu_2MeSi})_2\mathrm{Si}]_3\mathrm{C_6H_3}$: Isolable Si-centered Triradical With a High-Spin Quartet Ground State. *Chem. Lett.* **2015,** *44* (1), 56–57.
78. Sekiguchi, A.; Inoue, S.; Ichinohe, M.; Arai, Y. Isolable Anion Radical of Blue Disilene $(\mathrm{tBu_2MeSi})_2\mathrm{SiSi}(\mathrm{SiMetBu_2})_2$ Formed Upon One-Electron Reduction: Synthesis and Characterization. *J. Am. Chem. Soc.* **2004,** *126* (31), 9626–9629.
79. Kostenko, A.; Tumanskii, B.; Karni, M.; Inoue, S.; Ichinohe, M.; Sekiguchi, A., et al. Observation of a Thermally Accessible Triplet State Resulting From Rotation Around a Main-Group Bond. *Angew. Chem. Int. Ed.* **2015,** *54* (41), 12144–12148.
80. Weidenbruch, M.; Kramer, K.; Schäfer, A.; Konrad Blum, J. Siliciumverbindungen mit starken intramolekularen sterischen Wechselwirkungen, 17. Disilenyl-Anionen $[\mathrm{R_2Si=SiR_2}]^{-\cdot}$ und andere langlebige Silicium-zentrierte Radikale. *Chem. Ber.* **1985,** *118* (1), 107–115.
81. Kira, M.; Iwamoto, T. Stable Cyclic and Acyclic Persilyldisilenes. *J. Organomet. Chem.* **2000,** *611* (1–2), 236–247.
82. Kinjo, R.; Ichinohe, M.; Sekiguchi, A. An Isolable Disilyne Anion Radical and a New Route to the Disilenide Ion Upon Reduction of a Disilyne. *J. Am. Chem. Soc.* **2007,** *129* (1), 26–27.
83. Lee, V. Ya; Kawai, M.; Sekiguchi, A.; Ranaivonjatovo, H.; Escudié, J., et al. Phosphasilene and Phosphagermene and Their Anion-Radicals. *Organometallics* **2009,** *28* (15), 4262–4265.
84. (a) Inoue, S.; Ichinohe, M.; Sekiguchi, A. Isolable Silylene Anion Radical: Structural Characteristics in the Solid State and in Solution. *J. Am. Chem. Soc.* **2007,** *129* (19), 6096–6097.
    (b) Inoue, S.; Ichinohe, M.; Sekiguchi, A. Isolable Alkali-Metal-Substituted Silyl Radicals $({}^t\mathrm{Bu_2MeSi})_2\mathrm{SiM}$ (M = Li, Na, K): Electronically and Sterically Accessible Planar Silyl Radicals. *Organometallics* **2008,** *27* (7), 1358–1360.
85. Ishida, S.; Iwamoto, T.; Kira, M. Radical Anion of Isolable Dialkylsilylene. *J. Am. Chem. Soc.* **2003,** *125* (11), 3212–3213.
86. Sekiguchi, A.; Kinjo, R.; Ichinohe, M. A Stable Compound Containing a Silicon-Silicon Triple Bond. *Science* **2004,** *305* (5691), 1755–1757.
87. Davies, A. G.; Eaborn, C.; Lickiss, P. D.; Neville, A. G. EPR Spectra of Tris(trimethylsilyl)methyl (hydroxy)silyl Radicals, $(\mathrm{Me_3Si})_3\mathrm{CSi(R)OH}^{\cdot}$ and of Tris(trimethylsilyl)methylsilanone Radical Anions, $(\mathrm{Me_3Si})_3\mathrm{CSi(R)=O}^{-\cdot}$ (R = H, Me, Et, Bu, Ph, F). *J. Chem. Soc., Perkin Trans. 2* **1996,** (2), 163–169.
88. Davies, A. G.; Neville, A. G. A Study by Electron Spin Resonance Spectroscopy of the Di-*t*-butylsilanone Radical Anion. *J. Organomet. Chem.* **1992,** *436* (3), 255–263.
89. Molev, G.; Bravo-Zhivotovskii, D.; Karni, M.; Tumanskii, B.; Botoshansky, M.; Apeloig, Y. Synthesis, Molecular Structure, and Reactivity of the Isolable Silylenoid With a Tricoordinate Silicon. *J. Am. Chem. Soc.* **2006,** *128* (9), 2784–2785.
90. Bravo-Zhivotovskii, D.; Ruderfer, I.; Melamed, S.; Botoshansky, M.; Tumanskii, B.; Apeloig, Y. Nonsolvated, Aggregated 1,1-Dilithiosilane and the Derived Silyl Radicals. *Angew. Chem. Int. Ed.* **2005,** *44* (5), 739–743.

91. Haaf, M.; Schmedake, T. A.; Paradise, B. J.; West, R. Synthesis and Reactivity of the Stable Silylene *N*, *N′*-di-tert-butyl-1,3-diaza-2-sila-2-ylidene. *Can. J. Chem.* **2000,** *78* (11), 1526−1533.
92. Bravo-Zhivotovskii, D.; Ruderfer, I.; Yuzefovich, M.; Kosa, M.; Botoshansky, M.; Tumanskii, B., et al. Mercury-Substituted Silyl Radical Intermediates in Formation and Fragmentation of Geminal Dimercury Silyl Compounds. *Organometallics* **2005,** *24* (11), 2698−2704.
93. Kerr, C. M. L.; Wargon, J. A.; Williams, F. ^{201}Hg Quadrupole Interaction in the Electron Spin Resonance of the CH_2HgCl Radical. *J. Chem. Soc., Faraday Trans. 2* **1976,** *72* (0), 552−556.
94. Haaf, M.; Schmedake, T. A.; West, R. Stable Silylenes. *Acc. Chem. Res.* **2000,** *33* (10), 704−714.
95. (a) Arduengo, A. J. Looking for Stable Carbenes: The Difficulty in Starting Anew. *Acc. Chem. Res.* **1999,** *32* (11), 913−921.
    (b) Kühl, O. *N*-Heterocyclic Germylenes and Related Compounds. *Coord. Chem. Rev.* **2004,** *248* (5−6), 411−427.
96. Tumanskii, B.; Pine, P.; Apeloig, Y.; Hill, N. J.; West, R. Radical Reactions of a Stable *N*-Heterocyclic Silylene: EPR Study and DFT Calculation. *J. Am. Chem. Soc.* **2004,** *126* (25), 7786−7787.
97. Tumanskii, B.; Pine, P.; Apeloig, Y.; Hill, N. J.; West, R. Radical Reactions of a Stable N-Heterocyclic Germylene: EPR Study and DFT Calculation. *J. Am. Chem. Soc.* **2005,** *127* (23), 8248−8249.
98. Tumanskii, B.; Sheberla, D.; Molev, G.; Apeloig, Y. Dual Character of Arduengo Carbene−Radical Adducts: Addition Versus Coordination Product. *Angew. Chem. Int. Ed.* **2007,** *46* (39), 7408−7411.
99. Molev, G., [Ph.D. Thesis]. Haifa, Israel, 2009.
100. Sheberla, D.; Tumanskii, B.; Tomasik, A. C.; Mitra, A.; Hill, N. J.; West, R., et al. Different Electronic Structure of Phosphonyl Radical Adducts of *N*-Heterocyclic Carbenes, Silylenes and Germylenes: EPR Spectroscopic Study and DFT Calculations. *Chem. Sci.* **2010,** *1* (2), 234−241.

CHAPTER 7

Silicon-Centered Anions

Christoph Marschner

Technische Universität Graz, Graz, Austria

Chapter Outline

7.0 List of Abbreviations

Ad	1-Adamantyl
AIBN	Azoisobutyronitrile
Ar	Aryl
BTMSA	Bis(trimethylsilyl)acetylene
^{n}Bu	*n*-Butyl
^{t}Bu	*tert*-Butyl
DME	1,2-Dimethoxyethane
Dsi	Bis(trimethylsilyl)methyl
ee	Enantiomeric excess
er	Enantiomeric ratio
Et	Ethyl
HMPA	Hexamethylphosphoric triamide
LiDBB	Lithium 4,4-di-*tert*-butylbiphenylide
Me	Methyl

Organosilicon Compounds.
DOI: http://dx.doi.org/10.1016/B978-0-12-801981-8.00007-1

Mes	Mesityl
Naph	Naphtyl
Ph	Phenyl
^{i}Pr	Isopropyl
RT	Room temperature
THF	Tetrahydrofuran
thex	1,1,2-Trimethylpropyl
Tip	2,4,6-Triisopropylphenyl
Tol	Tolyl
To^M	Tris(4,4-dimethyl-2-oxazolinyl)phenylborate
Tsi	Tris(trimethylsilyl)methyl
XRD	X-ray diffraction
xs	Excess

7.1 Introduction

The expression "silicon-centered anions" usually refers to so called silyl anions or silanides, which are negatively charged three-coordinate species, isoelectronic to carbanions. To retain electroneutrality they require a counterion, typically a positively charged metal atom. Irrespective of nature and strength of interaction between these two ions, the ion pair will be referred to as silyl anion or silanide throughout the following article. These compounds comprise the main part of the review. Similar to carbanions which are important building blocks for organic synthesis, silyl anions are preeminent reagents for organosilicon chemistry, and therefore their chemistry has been the subject of a number of excellent reviews.[1–7] Although silyl anions are well-established synthons, the current article is mainly concerned with their formation and preparation, particularly focusing on synthetic methods. Some examples of silyl anion reactivity will be given (in particular when reactions lead to the formation of other silyl anions) but an exhaustive description of the diverse chemistry done with silyl anions would be clearly beyond the scope of this article.

Certain silicon-centered anions do not meet the crude definition given above for silyl anions in such a way that they are either lower (1,1-dianions, silenides, or disilenides) or higher (organosilicates) coordinate species. While 1,1-dianions, silenides, and disilenides exhibit typical silanide reactivity and are thus discussed together with other silyl anions, the higher coordinate silicon-centered anions (silicates) will be treated at the very end of the review.

7.2 General Synthetic Methods for the Preparation of Silyl Anions

For carbanions arguably the most versatile method of preparation is the use of strong bases to deprotonate hydrocarbons thus forming the desired compounds. This is of course only possible if the precursor is sufficiently acidic. Otherwise synthetic efforts to generate carbanions typically start from alkyl halides, employing Grignard, Barbier, or Ziegler reaction type conditions, which yield organo magnesium, zinc, or lithium compounds.[8]

For the generation of silyl anions both of the methods mentioned for carbanion synthesis are associated with some more or less severe drawbacks. Deprotonation of silyl hydrides with strong bases is known, but its use is limited to cases where the formed silyl anion is strongly stabilized by phenyl or silyl substituents. The major problem of this approach is that due to the lower electronegativity of silicon the polarization of the Si–H bond is reversed to that of the C–H bond. Hydrogen atoms attached to silicon therefore have partial hydridic character, which impedes facile deprotonation. Also metalation of silyl halides with magnesium or lithium using the Grignard or Ziegler protocols cannot be considered as generally applicable. Since silyl halides are excellent electrophiles, the generation of highly nucleophilic silyl anions in their presence usually leads to effective Wurtz type coupling reactions.

Still the most regularly employed method for the synthesis of silyl anions is the disilane cleavage reaction, which often is carried out as a one-pot reaction starting from silyl halides, which react with elemental alkali metal (usually lithium) in a Wurtz type reaction to a disilane which in a second step is cleaved by an excess of alkali metal to generate a metalated silyl compound (silyl anion). This method was pioneered by Gilman and coworkers.[9,10] It is usually carried out in ethereal solvents and typically requires at least one phenyl group attached to silicon (Scheme 7.1). Although facile access to $PhMe_2SiLi$,[11] $Ph_2MeSiLi$,[12] and Ph_3SiLi[13] is possible by this method, it needs to be pointed out that even minor variations such as the exchange of the phenyl group of $PhMe_2SiLi$ against *m*- or *p*-tolyl cause the reaction to fail.[14] Synthesis of *o*-$TolMe_2SiLi$ was, however, reported recently by reaction of *o*-$TolMe_2SiCl$ with lithium.[15]

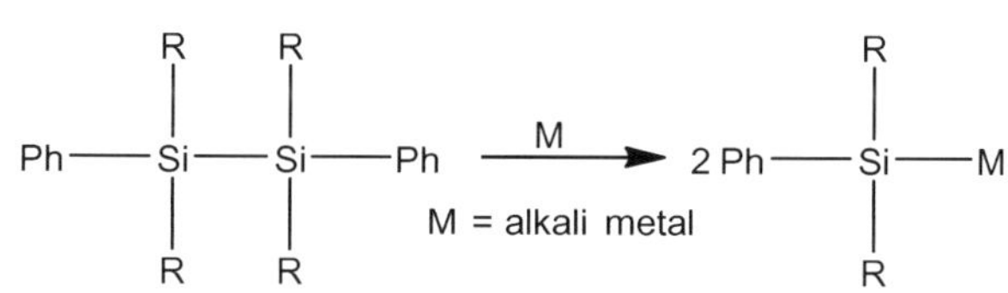

Scheme 7.1
Disilane cleavage with alkali metals.

Another preparative method that also makes use of Si–Si bond cleavage is the treatment of disilanes or oligosilanes with strong nucleophiles such as methyllithium or potassium alkoxides. Of course this method is not as atom-economic as the alkali metal disilane cleavage as it delivers only one equivalent of silyl anion per disilane compared to two equivalents from the alkali metal disilane cleavage. However, the silyl abstraction method with strong nucleophiles offers access to a wider range of nonstabilized and less symmetric silyl anions (Scheme 7.2). In the case of hexaalkyldisilanes, such as $Me_3SiSiMe_3$, reaction with methyllithium only proceeds in the presence of a strong donor solvent (usually mixtures of THF [tetrahydrofuran] and HMPA [hexamethylphosphoric triamide]).[16]

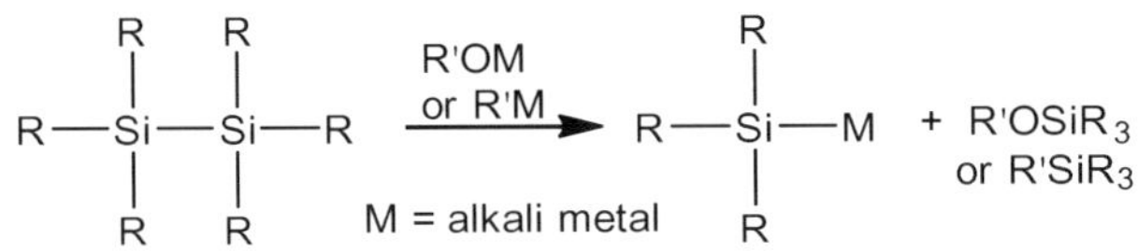

Scheme 7.2
Disilane cleavage with strong nucleophiles.

Another widespread method for the synthesis of silyl anions is the reaction of alkali metals with silyl-mercury compounds. The latter are obtained by the Vyazankin reaction,[17] that is the conversion of silyl hydrides with dialkyl mercury compounds to disilylmercury compounds (Scheme 7.3). Synthesis of silyl anions via the silyl mercury route provides easy access to base-free silanides.[18]

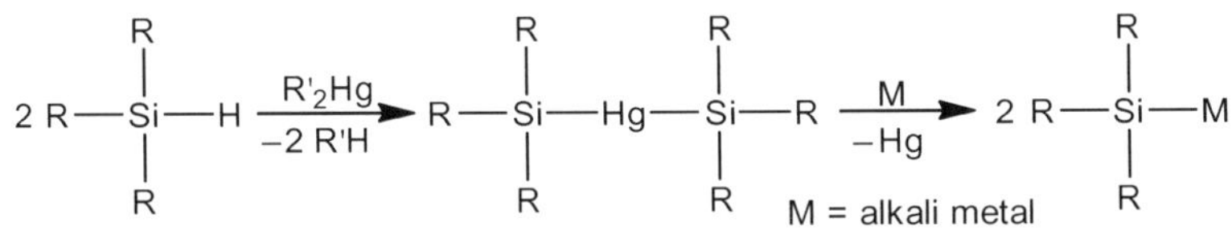

Scheme 7.3
Transmetalation of silyl mercury compounds with alkali metals.

The disilane cleavage reaction with alkali metals is facilitated by the fact that two stable anionic compounds are formed. This concept can be extended to Si–C bond cleavage reactions with alkali metals for cases where in addition to a silyl anion also a stable carbanion forms. Typical examples are the cleavage of $R_3Si{-}CH_2Ph$[19] or $R_3Si{-}Ph$[20] bonds with lithium leading to R_3SiLi and $PhCH_2Li$ or PhLi, respectively.

While most of the silanides covered in this review have alkali metal counterions, also compounds with alkaline earth or even group 12 counterions can be considered as silanides. Quite similar to what is known for carbanions, the presence of less electropositive counterions has a moderating effect on silanide reactivity. Nucleophilicity and in particular reduction potential of magnesium or zinc silanides is less pronounced, which often goes along with enhanced selectivity and cleaner reactions. Since Grignard or Barbier type reactions of silyl halides with magnesium or zinc are usually not working well, metathesis of alkali metal silanides with salts of a less electropositive metal is usually the most convenient and reliable synthetic access to group 2 and 12 silanides (Scheme 7.4).

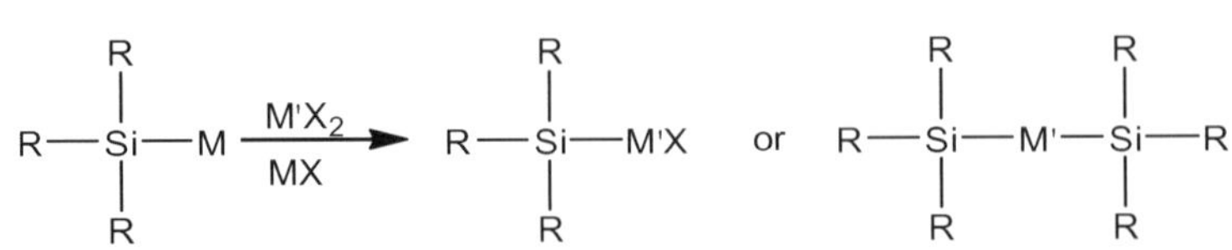

Scheme 7.4
Metathesis of alkali metal silanides with salts of less electropositive metals.

7.3 Synthesis of Different Silyl Anions

The following main section covers reported preparations of various silyl anions since the year 2000. In several cases also references are made to earlier seminal contributions. These earlier references are by no means complete and their choice should not be taken as a scientific assessment but arguably reflects the author's own research interests. A classification of silyl anions was done according to their substituent pattern. Treatment of alkylated and arylated compounds is followed by a short excursion to chiral silyl anions. The next section deals with the large class of oligosilanyl anions, followed by the description of functionalized silyl anions, which is subdivided into hydrogen-, halogen-, amino-, and alkoxy-substituted compounds. The substituents referred to are usually attached to the negatively charged silicon atom, leading to ambiphilic (silylenoid) reactivity. The section about functionalized compounds concludes with organic functional groups (such as vinyl and alkynyl) attached to silicon and is followed by a larger section on silyl dianions. Short sections on sila-enolates, silole anions, and silenyl and disilenyl anions follow before the final section on hypercoordinate anions (silicates).

Silyl anions typically have metal counterions. Compounds treated in this chapter contain counterions of the relatively electropositive metals of groups 1, 2, and 12. Silyl anions with tetraalkylammonium counterions are not stable (as the R_4N^+ unit serves as an alkylating agent) but such counterions are frequently found for silicates.

7.3.1 Alkylated Silyl Anions

Associated with the difficulties of generating peralkylated silyl anions from respective disilanes, their synthesis has not been very well investigated. The prototypical Me_3SiLi is available from the reaction of $Me_3SiSiMe_3$ with MeLi in THF/HMPA[16] or from the reaction of $(Me_3Si)_2Hg$ with lithium in THF.[21] The very bulky $^tBu_3Si$ group (supersilyl) is sufficiently large that no Wurtz type coupling occurs upon reaction of the respective halides with alkali metals. Reaction of $^tBu_3SiBr$ with alkali metals thus provides convenient access to $^tBu_3SiM$ (M = alkali metal).[22] Conversion of $^tBu_3SiNa$ to alkaline-earth metal supersilanides $(^tBu_3Si)_2Be$, $(^tBu_3Si)_2Mg$, and $^tBu_3SiMgBr(THF)$ was achieved by reaction with $BeCl_2$ and $MgBr_2$, respectively.[23]

7.3.2 Arylated Silyl Anions

As mentioned, for the cleavage of disilanes it is important to obtain reasonably stabilized anions, which is most easily achieved by attachment of a phenyl group to the silicon atom. For this reason $PhMe_2SiLi$ is arguably the most popular silyllithium compound, while the structurally similar Me_3SiLi, which requires working either with HMPA or $^tBu_2Hg$, is encountered less frequently.

Synthesis of $PhMe_2SiLi$ usually is carried out in situ by reaction of $PhMe_2SiCl$ with lithium in THF as described meticulously by Fleming and coworkers.[11] The compound has enjoyed tremendous popularity in organic synthesis, especially when converted to a cuprate,[24] as a very soft silyl nucleophile. Its reaction with $ZnCl_2$ was shown to give $(PhMe_2Si)_2Zn$, while reactions with MeZnCl and MeMgI gave $Me(PhMe_2Si)Zn \cdot LiCl$ and $Me(PhMe_2Si)Mg \cdot LiI$, all of which were tested as copper-mediated soft silyl transfer reagents.[25] Related zincates obtained from the reaction of $PhMe_2SiLi$ with R_2Zn (R = Me, Et)[26] or by reaction of 3 equivalents $PhMe_2SiLi$ with $ZnCl_2$ were also studied.[27] Formation of $PhEt_2SiLi$ is also straightforward from $PhEt_2SiCl$[28] or $(PhEt_2Si)_2$.[29] Ph_3SiK prepared by reaction of $Ph_3SiSiPh_3$ with potassium in THF was used to determine the pK_a of Ph_3SiH by reaction with various arylmethanes to be 35.1.[30]

The synthesis of (2-methoxyphenyl)dimethylsilyllithium which is considered as a superior alternative to $PhMe_2SiLi$, with regard to both the electron-donating properties of the silyl group and the Tamao–Fleming oxidation of sensitive substrates, was achieved by reaction of the respective disilane with methyllithium in THF/HMPA in a ratio of 4/1.[14]

The formation of $^tBu_2PhSiNa$[31] was reported from $^tBu_2PhSiBr$ by reaction with sodium in heptane at 80°C and also the respective lithium and potassium compounds were obtained this way.[32]

Starting from a diphenylalkylsilylfluoride (**1**) Sieburth and coworkers prepared a silyllithium compound (**2**) by treatment with lithium in THF. Although no spectroscopic characterization of the anion was provided, subsequent derivatization proved its formation (Scheme 7.5).[33]

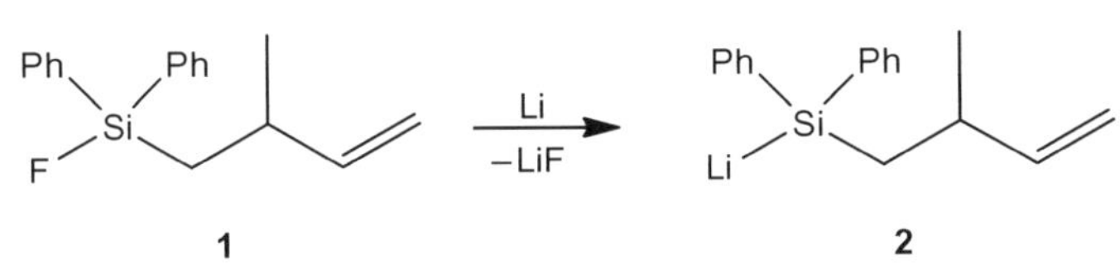

Scheme 7.5
Formation of a lithium silanide via Si–F bond metalation.

While Si–Si bond cleavage seems the most straightforward way to obtain Ph_3SiM (M = Li, Na, K), an alternative access to crown ether adducts of these anions was reported recently by reaction of $Ph_3SiSiMe_3$ with either $Me_3SiCH_2Li \cdot$ 12-crown-4, $NaO^tBu \cdot$ 15-crown-5, or $KO^tBu \cdot$ 18-crown-6.[34] $Ph_3SiK \cdot (THF)$ was further reacted with CaI_2 to give $(Ph_3Si)_2Ca \cdot (THF)_4$.[35]

In the course of studies concerning the preparation of enantiomerically enriched or pure silyl anions Strohmann and coworkers have investigated also cleavage of Si–C bonds.[19] 9-Methylfluorenyl (**3**) and diphenylmethyl (**4**) substituted organosilanes were found to be cleaved with lithium to give one equivalent of a silyllithium compound and the respective organolithium compound each (Scheme 7.6).[19] Subsequently, it was shown that also Si–Ph bonds (**5**) can be cleaved selectively with lithium (Schemes 7.6 and 7.10).[20]

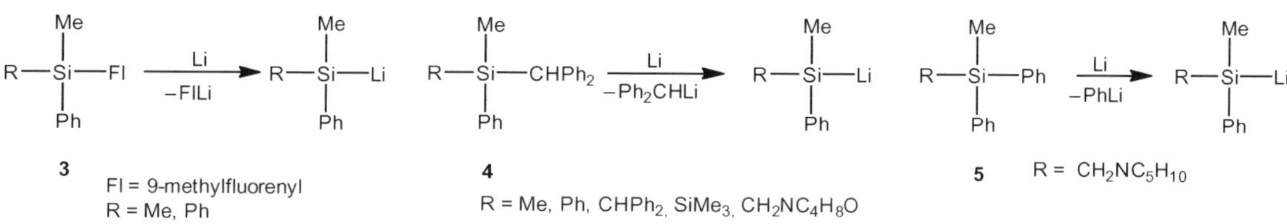

Scheme 7.6
Silyl anion formation involving Si–C cleavage with alkali metals.

In a concluding study the competition between Si–Si and Si–C cleavage with lithium in functionalized oligosilanes was investigated.[36] Aromatic substituents at the silicon atoms proved to be necessary for the viability of any cleavage reaction, as these were found to stabilize the radical anion after electron transfer to the corresponding oligosilane. Yet, selective cleavage reactions do not depend on the number of attached aryl groups.[36]

7.3.3 Chiral Silyl Anions

An interesting aspect of silyl anions is that, given three different substituents, they are chiral. However, in a similar sense as for carbanions the configurational stability of chiral silyl anions is low. There are only a few experimental studies concerning inversion barriers of silyl anions,[37,38] but together with theoretical considerations[39] a range of values between 17 kcal mol^{-1} $[(H_3Si)_3Si^-]$ and 48 kcal mol^{-1} $[(H_3C)_3Si^-]$ seems realistic for alkali metal or free silanides. There have been a few attempts to prepare enantiomerically enriched or enantiomerically pure lithiosilanes. Besides early reports by Sommer[40] and Corriu,[41] more recent work was carried out by Kawakami,[42] Oestreich,[43] and especially Strohmann and coworkers.[44–46]

The reactions of chiral disilanes (**6**) and stannylsilanes (**8**) with lithium and methyllithium, respectively, at low temperature showed that the chirality at the silicon atom is maintained in the formed silyl anions (**7**, **9**) (Scheme 7.7).[42]

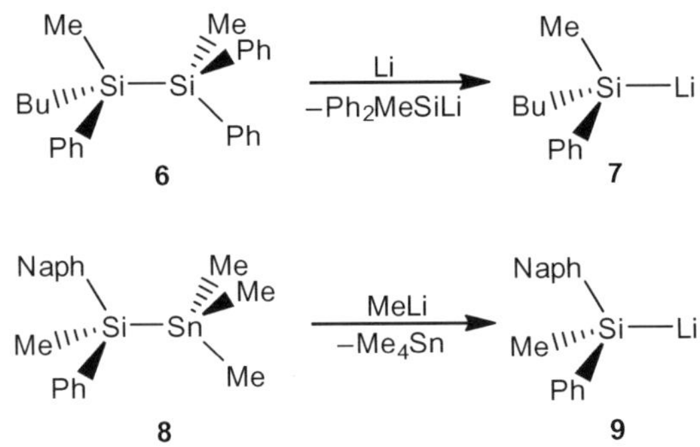

Scheme 7.7
Cleavage of chiral disilanes and stannylsilanes to chiral lithium silanides.

Oestreich and coworkers studied the stereochemical course of the reductive metalation of silyl chlorides with silicon-centered chirality, with LiDBB (lithium 4,4-di-*tert*-butylbiphenylide). They identified two major detrimental factors for stereoselection during the silyl anion formation. The inevitable formation of lithium chloride causes racemization of silyl chlorides and in addition the possible disilane formation was also found to be not stereoselective. However, these issues can be resolved by design of a sterically encumbered silyl chloride (**10**), which is not racemized by lithium chloride and does not engage in disilane formation at temperatures below $-100°C$. Thus the first generation of an asymmetrically substituted silyl anion (**11**) by reductive metalation of silyl chloride **10** with silicon-centered chirality was achieved (Scheme 7.8).[43]

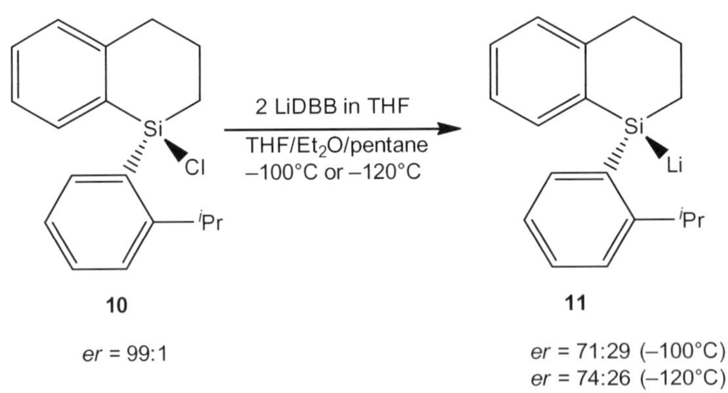

Scheme 7.8
Formation of a chiral lithium silanide via Si–Cl metalation.

Strohmann and coworkers utilized disilane cleavage of an enantiomerically pure asymmetric disilane (**12**) with lithium at $-70°C$. Trapping of the obtained lithiosilane (**13**) with $PhMe_2SiCl$ gave the starting material with enantiomeric excess (*ee*) >98% indicating stereoselective silanide formation. Isolated solutions of the lithiomethylphenyl(1-piperidinylmethyl)silane **13** racemized within a few hours at temperatures between 0 and 20°C. However, after transmetalation with $MgBr_2 \cdot (THF)_4$ at −70°C no significant racemization of the resulting magnesium silanide **14** was observed at room temperature (Scheme 7.9).

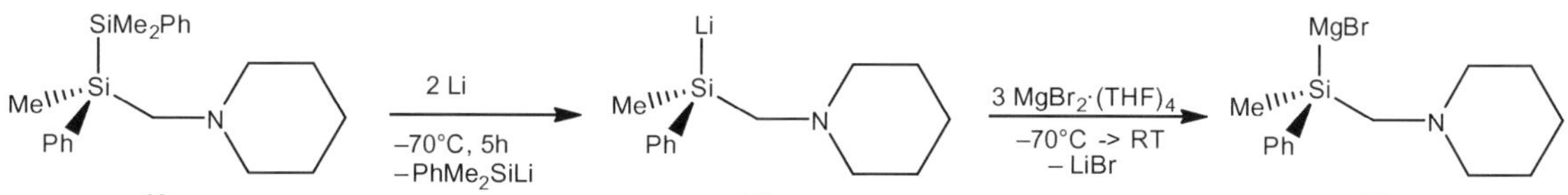

12 13 14

Scheme 7.9
Formation of a chiral lithium silanide via Si–Si bond cleavage.

In a subsequent study the synthetic method was refined in a way that not a Si–Si but the Si–Ph bond of disilane **15** was cleaved. The reaction proceeded at −50°C in a stereoselective manner to give chiral silanide **16** (Scheme 7.10).[45]

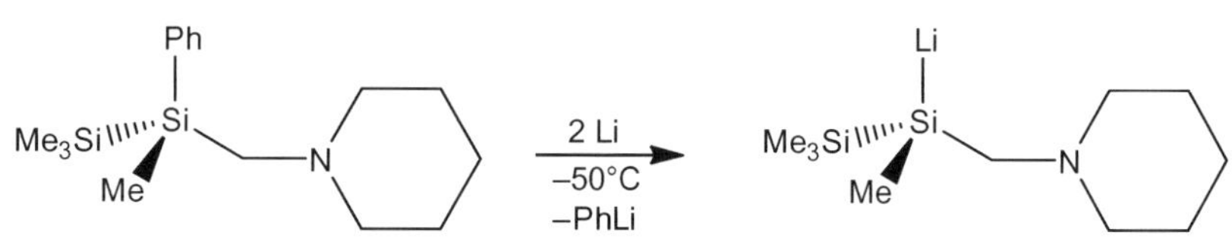

15 16

Scheme 7.10
Formation of a chiral lithium silanide via Si–Ph bond cleavage.

Also the stereospecifity of the reaction of enantiomerically pure lithiosilanes with aliphatic and aromatic halides was investigated. It was found that both the nature of the organic group as well as of the corresponding halide of the used electrophile have an influence on product composition and enantiomeric ratios. For cases of alkyl chlorides, experimental and quantum chemical studies support the preference of an S_N2 mechanism favoring retention of configuration at silicon. Alkyl bromides and aryl halides tend to react under inversion via a halide–lithium ate complex.[46]

Chiral lithium silanides are not necessarily required to have three different substituents at silicon. This was shown by the use of (−)-sparteine which serves as a donor for $PhMe_2SiLi$,[47] $Ph_2MeSiLi$,[48] Et_2NPh_2SiLi,[47] and $(Et_2N)_2PhSiLi$.[48] Nuclear magnetic resonance (NMR)-spectroscopic and structural studies clearly point out the inequivalence of the two identical groups at the silicon atom. No examples of transfer of the (−)-sparteine induced stereogenic information at the silicon center to electrophiles were reported.

Employing a similar approach Nanjo, Mochida, and coworkers succeeded in the optical resolution of ^{t}BuMePhSiLi, prepared by cleavage of (^{t}BuMePhSi)$_2$ with lithium in THF, using again (−)-sparteine as a chiral ligand. The absolute structure of the sparteine−silanide complex was confirmed by X-ray diffraction (XRD).[49] Hydrolysis reactions with H_2O or EtOH gave the corresponding hydrosilane ^{t}BuMePhSiH in quantitative yields with high enantiomeric excess (95%*ee*). In the same study also the germyllithium compound, ^{t}BuMePhGeLi was optically resolved using (−)-sparteine. Reactions of this compound with benzyl halides gave the expected coupling product ^{t}BuMePhGeCH$_2$Ph in low to moderate yields with *ee* of 96%, 68%, and 23% for benzyl fluoride, chloride, and bromide, respectively.[49]

7.3.4 Oligosilanyl Anions

The chemistry of oligosilanyl anions dates back to Gilman's seminal work[50,51] which described the reaction of oligosilanes such as $(Me_3Si)_4Si$ (**17**) with methyllithium (Scheme 7.11). The easily accessible $(Me_3Si)_3SiLi$[52] (**18**) has become quite popular in main group and transition metal chemistry.[53] However, the methyllithium reaction has a tendency to cleave internal Si−Si bonds before peripheral ones (Scheme 7.11).[51,54] For instance hexakis(trimethylsilyl)disilane (**20**) is cleaved with methyllithium to afford $(Me_3Si)_3SiLi$ (**18**) and $(Me_3Si)_3SiMe$ (Scheme 7.11).[51,54] Conversely, use of KO^{t}Bu as metalating reagent yields selective trimethylsilyl group cleavage to give pentakis(trimethylsilyl)disilanylpotassium (**21**) (Scheme 7.11).[55] This way not only $(Me_3Si)_3SiK$ (**19**)[56] and $(Me_3Si)_3SiSi(SiMe_3)_2K$ (**21**) can be obtained but also numerous other complex oligosilanyl anions are accessible (see below).[5]

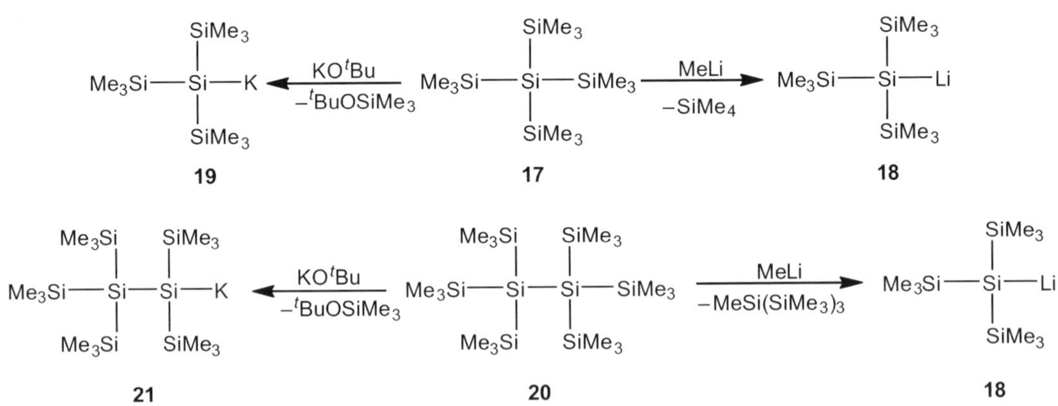

Scheme 7.11
Comparison of MeLi and KO^{t}Bu in reactions with $(Me_3Si)_4Si$ and $(Me_3Si)_3SiSi(SiMe_3)_3$.

Exchange of one or more of the Me_3Si groups of $(Me_3Si)_4Si$ (**17**) against other silyl groups can easily be accomplished for ^{t}BuMe$_2$Si,[57] thexMe$_2$Si,[57] ^{i}Pr$_3$Si,[57] Ph$_3$Si,[58] Me$_3$SiSiMe$_2$,[59] (Me$_3$Si)$_3$Si,[56] (Me$_3$Si)$_2$PhSi,[57] (Me$_3$Si)$_2$PhCCSi,[60] (Me$_3$Si)$_3$SiSiMe$_2$,[56] (Me$_3$Si)$_3$Si(SiMe$_2$)$_2$,[61] and cyclo-Si_6Me_{11}.[62] Reactions of these derivatives with KO^{t}Bu give the respective anions in the same selective manner as was found for **17** (Scheme 7.12).[57–62]

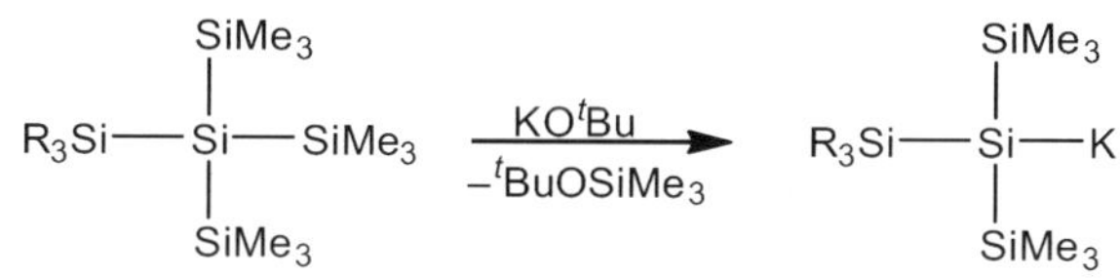

Scheme 7.12
Formation of various isotetrasilanylpotassium compounds.

Replacement of a Me_3Si group of $(Me_3Si)_4Si$ (**17**) against alkyl groups (Me,[57] Et,[56] ^{i}Pr,[57]) or Ph[57] is also possible and reactions of these oligosilanes with KO^{t}Bu lead to the expected silyl anions in clean reactions. Reaction of octamethyltrisilane, where formally two Me_3Si groups of $(Me_3Si)_4Si$ (**17**) are exchanged for methyl groups, with KO^{t}Bu is not clean[56] but when instead of methyl groups trimethylsilylmethyl substituents were used the clean formation of the pentaalkyldisilanylpotassium compound **23** was observed via the reaction of **22** with KO^{t}Bu (Scheme 7.13).[63]

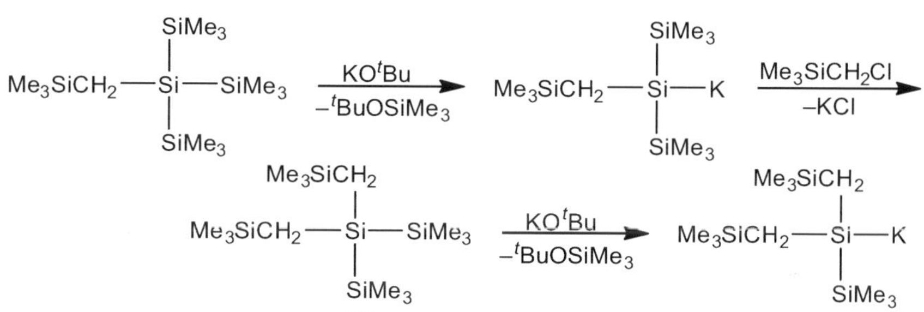

Scheme 7.13
Formation of successively mono- and dialkylated oligosilanylpotassium compounds.

The trimethylsilyl cleavage method with KO^{t}Bu yields even complex silanides such as $(Me_3Si)_2MeSiSiMe_2SiMe(SiMe_3)K$[64] cleanly. It was shown that the reaction of $(Me_3Si)_4Si$ (**17**) is not restricted to KO^{t}Bu but works also with RbO^{t}Bu and CsO^{t}Bu.[65] When carried out in DME (1,2-dimethoxyethane) at elevated temperature the reaction works even with NaO^{t}Bu.[66] Although no facile conversion of $(Me_3Si)_4Si$ (**17**) with LiO^{t}Bu was observed, the various accessible potassium silanides can be reacted with LiBr to obtain lithium silanides, and in very much the same way magnesium[67,68] and zinc silanides[69,70] were obtained by reaction of potassium silanides with $MgBr_2 \cdot Et_2O$ and $ZnBr_2$ or $ZnCl_2$, respectively.

The dimethylsilyl substituted potassium silanide $(Me_2HSi)_3SiK$ (**24**) was obtained by reaction of $(Me_2HSi)_4Si$ with KO^{t}Bu in benzene.[70,71] Reactions of **24** and **19** with To^{M}ZnCl or To^{M}MgBr (To^M = tris(4,4-dimethyl-2-oxazolinyl)phenylborate) gave monomeric magnesium or zinc silyl compounds (**25**) (Scheme 7.14).[70]

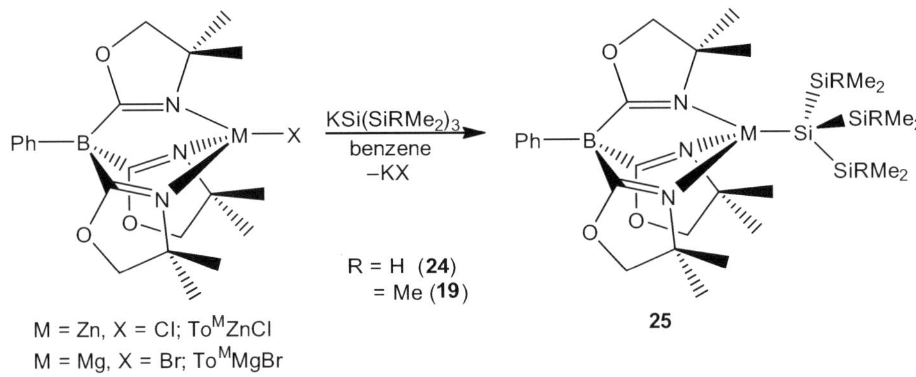

Scheme 7.14
Formation of oligosilanyl zinc and magnesium compounds.

The two-coordinate compound $[(Me_2HSi)_3Si]_2Zn$, which was generated by reaction of $ZnCl_2$ with two equivalents of $(Me_2HSi)_3SiK$ (**24**), underwent facile reductive elimination to $(Me_2HSi)_3Si{-}Si(SiHMe_2)_3$ and elemental zinc.[71]

7.3.4.1 Cyclic oligosilanyl anions

The reaction of KO^{t}Bu with dodecamethylcyclohexasilane (**26**) in diglyme is interesting as it leads to the formation of mainly undecamethylcyclohexasilanylpotassium (**27**). What seems to be a simple methyl abstraction is actually a more complex redistribution reaction with varying amounts of nonamethylcyclopentasilanylpotassium (**28**) and ^{t}BuOSiMe$_3$ being the side products (Scheme 7.15).[72]

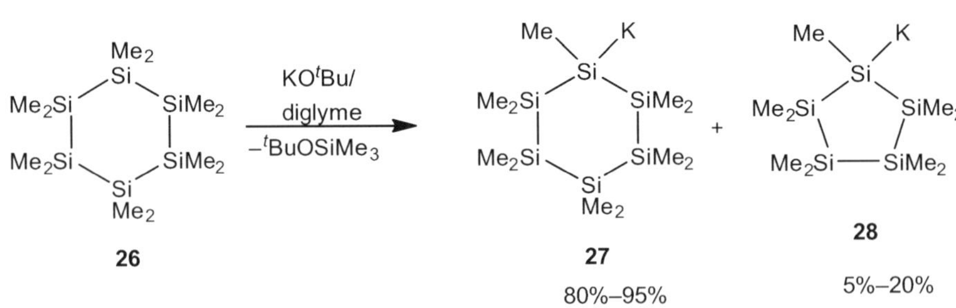

Scheme 7.15
Formation of undecamethylcyclohexasilanylpotassium.

Reactions of KO^{t}Bu with trimethylsilyl substituted cyclosilanes, however, are very clean. Several examples of silylated cyclotetrasilanyl (**29**),[73–75] cyclopentasilanyl (**30**),[59] and cyclohexasilanyl (**31**)[76] potassium compounds obtained this way were reported (Fig. 7.1).

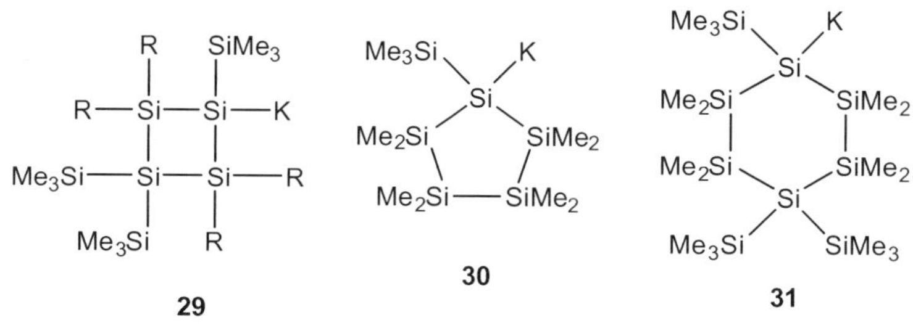

29 R = Me, $SiMe_3$

30

31

Figure 7.1
Other examples of cyclosilanylpotassium compounds.

Another cyclopentasilanylpotassium compound was obtained in an attempt to generate a five-membered homocyclic disilylated silylene in the reaction of a geminal dichlorosilylene unit (**32**) with potassium graphite (KC_8) (Scheme 7.16).[77] The mechanism for the formation of compound **33** and in particular the origin of the additional Me_3Si group is still unclear.[77]

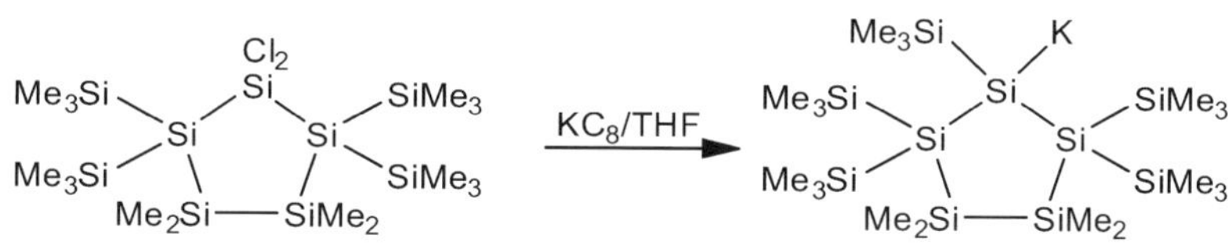

32

33

Scheme 7.16
Attempted synthesis of cyclic silylene giving a potassium cyclopentasilanyl compound.

An interesting example of cyclotrisilanyllithium compounds (**35**) was reported to be formed during the reduction of chlorosilyl disilenes **34** with lithium (Scheme 7.17).[78]

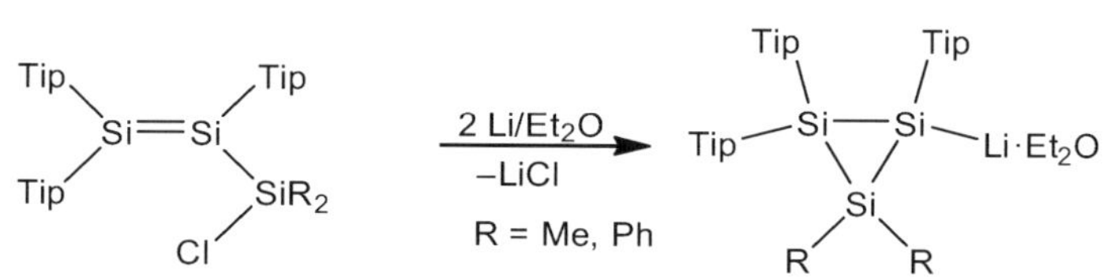

34

35

Scheme 7.17
Lithium-induced transformation of disilenylchlorosilanes to lithium cyclotrisilanides.

In an approach to zwitterionic silyl anions the reaction of $(MeOMe_2Si)_4Si$ (**36**) with MO^tBu (M = Li, Na, K) yielded silyl anions (**37**) where the countercation is effectively coordinated by the methoxy groups (Scheme 7.18).[79]

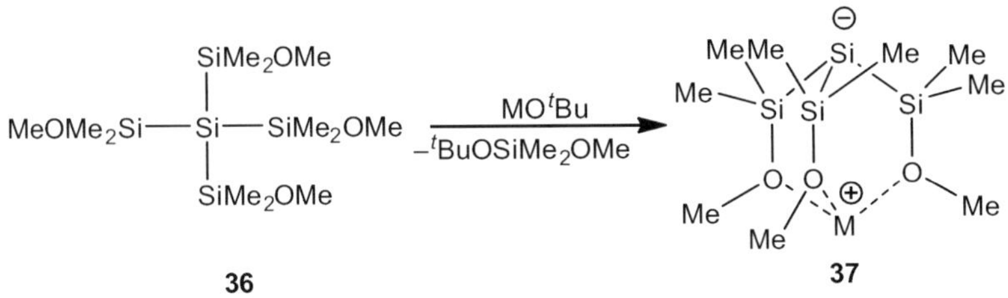

Scheme 7.18
Formation of zwitterionic silanides from tetrakis(methoxydimethylsilyl)silane.

An even more effective charge separation was achieved when the reaction was carried out with $(MeOCH_2CH_2OMe_2Si)_4Si$ (**38**).[79] The coordination of the counterion is so strong that the respective potassium silanide (**39**) reacts with $ZnCl_2$ only as a Lewis base to give a base adduct (**40**) without potassium chloride formation (Scheme 7.19).[79]

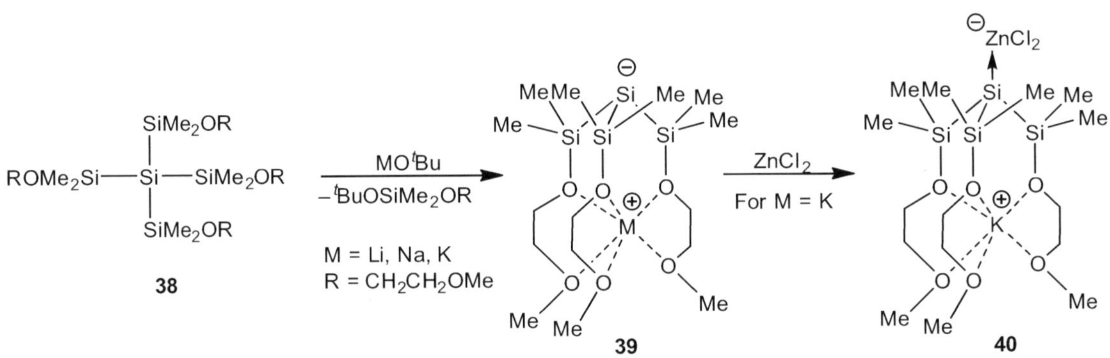

Scheme 7.19
Zwitterionic silanides as Lewis bases.

The formation of an oligosilanyl anion (**42**) was also observed in the reaction of diaminosilylene **41** with $(Me_3Si)_3SiLi$ (**18**), by insertion of the silylene into the Si–Li bond of **18** (Scheme 7.20).[80,81]

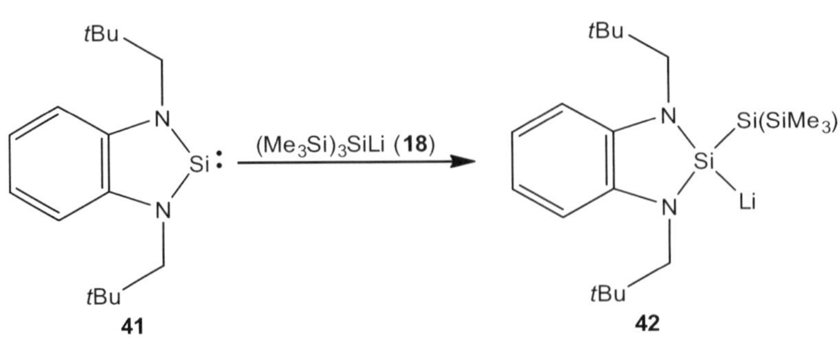

Scheme 7.20
Insertion of an N-heterocyclic silylene into a Si–Li bond.

Formation of $(Me_3SiMe_2Si)_3SiLi$ was reported by Apeloig and coworkers by reaction of $(Me_3SiMe_2Si)_3SiH$ (obtained from the reaction of 3 Me_3SiMe_2SiLi with $HSiCl_3$) with $^tBu_2Hg$ and subsequent treatment with lithium either in hexane or THF.[82] The analogous $(Me_3SiMe_2Si)_3SiK$ was prepared by reaction of $(Me_3SiMe_2Si)_4Si$ (formed by reaction of $(Me_3Si)_3SiK$ (**19**) with excess Me_3SiMe_2SiF) with KO^tBu.[59]

Realizing that the steric bulk of the $^tBu_3Si$ group is not sufficient to prohibit dimerization of disilyne, Wiberg and coworkers set out to study the class of disupersilylsilyl groups. A key reaction in these studies was the formation disupersilylsilanides: $(^tBu_3Si)_2RSiM$ (**44**).[83] These were prepared: (1) by metalation of $(^tBu_3Si)_2RSiBr$ (R = H, Me) (**43**) with Li, Na, or K in heptane, benzene, or THF (Scheme 7.21); (2) by activated metalation of $(^tBu_3Si)_2SiR_2$ with lithium naphthalenide in THF at room temperature (R = F), at −100°C (R = Cl), and at −130°C (R = Br); (3) by transmetalation between $(^tBu_3Si)_2RSiBr$ (R = H, Me) (**43**) and $^tBu_3SiNa$ in THF; (4) by insertion of the silylene $(^tBu_3Si)PhSi$: (from $(^tBu_3Si)PhSiHCl$ and $^tBu_3SiNa$) into the Si–Na bond of $^tBu_3SiNa$ in THF.[83] Group 12 disupersilylsilanides (**45**) were obtained by reaction of the alkali metal silanides with 0.5 equivalents of $ZnCl_2$, CdI_2, or $HgCl_2$ (Scheme 7.21).[84]

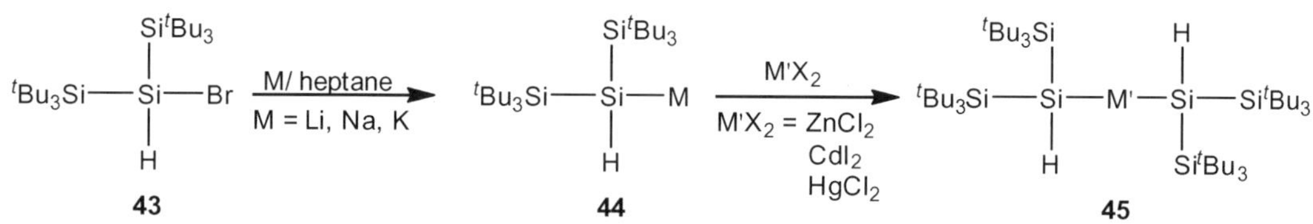

Scheme 7.21
Formation of hydrosilanides by metal halogen exchange followed by transmetalation.

The formation of a disupersilyldisilanide was accomplished by reaction of $(^tBu_3Si)H_2SiSiBrH(Si^tBu_3)$ with $^tBu_3SiNa$ yielding $(^tBu_3Si)H_2SiSiHNa(Si^tBu_3)\cdot(THF)_3$ and $^tBu_3SiBr$.[85]

With the intention to generate a radical anion, Apeloig and coworkers irradiated $[Li(^iPr_3Si)_2Si]_2Hg$ (**46**). However, instead of the expected $[(^iPr_3Si)_2SiLi]^\bullet$ the hydrosilanide $H(^iPr_3Si)_2SiLi$ (**47**) was obtained, which was also formed upon treatment of **46** with lithium (Scheme 7.22).[86]

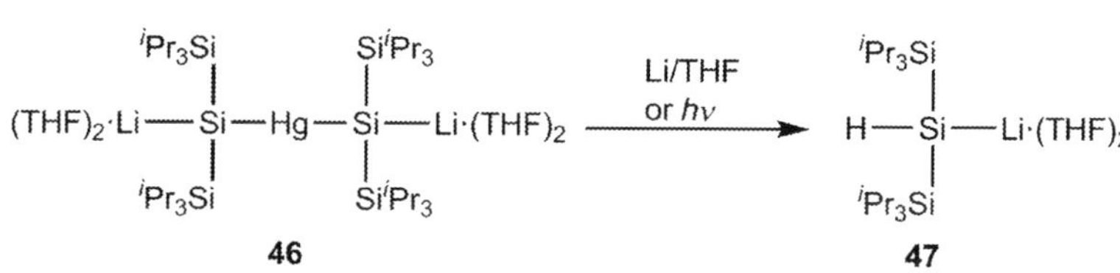

Scheme 7.22
Hydrosilanide formation from a lithiated silylmercury compound.

In cases, where the formed silyl anion is well stabilized, deprotonation of silyl hydrides can be feasible. Kira and coworkers have studied this for a number of compounds of the type: R_2SiH_2 (**48**).[87] Although high yielding in some cases, this method is extremely dependent on base and substitution pattern of the hydrosilane. Reaction of $(^tBuMe_2Si)_2SiH_2$ (**48a**) with ^tBuLi gave the expected $(^tBuMe_2Si)_2Si(H)Li$ (**49a**) (Scheme 7.23) while the conversion with ^nBuLi led to quantitative $(^tBuMe_2Si)_2Si(H)^nBu$ formation. $(Me_3Si)_2SiH_2$ (**48b**) and $(^iPr_2MeSi)_2SiH_2$ (**48c**) were also found to give the hydrosilyllithium species (**49b,c**), whereas $(^iPr_3Si)_2SiH_2$ did not react at all. $(^tBuMe_2Si)TolSiH_2$ (**48d**) reacts with ^tBuLi to give a mix of $(^tBuMe_2Si)TolSi(H)Li$ (**49d**) (21%) and $(^tBuMe_2Si)TolSi(H)^tBu$ (67%), though **49d** can be obtained cleanly when $^iPr_2NLi$ is used as base (Scheme 7.23). All attempts to achieve deprotonation of dialkylsilanes or diarylsilanes gave exclusively hydride displacement.[87]

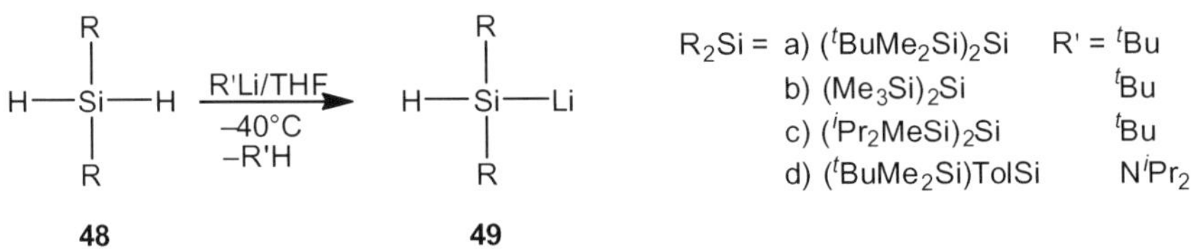

Scheme 7.23
Hydrooligosilanide formation by deprotonation with ^tBuLi or $^iPr_2NLi$.

The geometry of base-free oligosilanyl lithium compounds such as $(^tBu_2MeSi)_3SiLi$ (**51**) can be dominated by intramolecular CH–Li agostic interactions.[88] The compound can be prepared by lithiation of the corresponding radical (**50**) (Scheme 7.24). Addition of THF destroys the agostic interaction and causes formation of a separated ion pair (**52**) with loss of the planar geometry (Scheme 7.24).

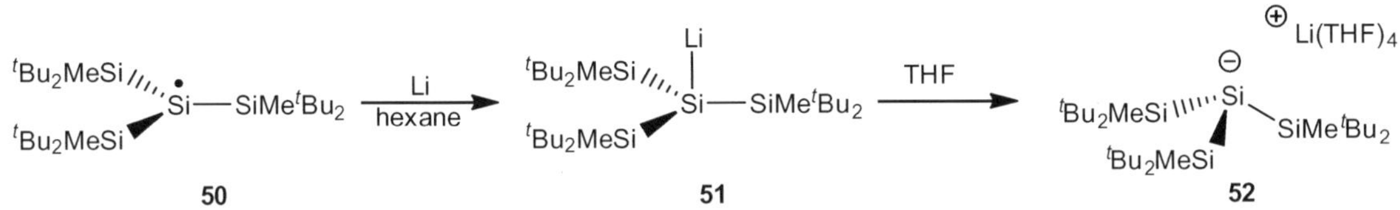

Scheme 7.24
Lithiation of an oligosilanyl radical to a planar oligosilanide.

Conceptually similar is the formation of the cyclotetrasilenide **54** which is formed by reaction of a cyclotetrasilenylium ion (**53**) with lithium, sodium, or potassium graphite. The structure of the obtained cyclotetrasilenide **54** shows interaction of the alkali metal with all three low valent silicon atoms (Scheme 7.25).[89]

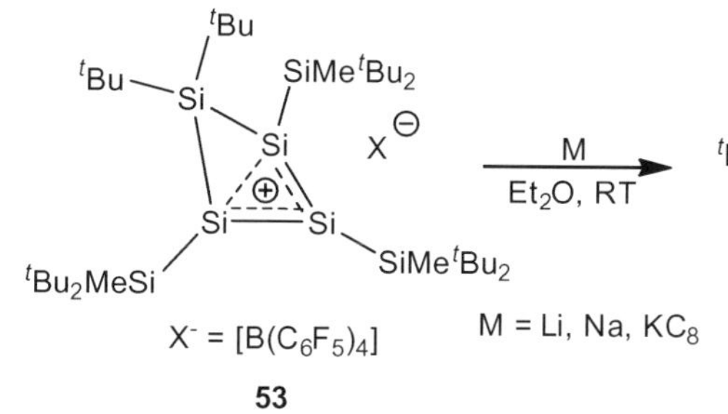

Scheme 7.25
Metalation of an oligosilenyl cation to a cyclotetrasilenide.

Marschner and coworkers found an interesting reaction sequence that starts out with $(Me_3Si)_3SiK$ (**19**) which undergoes reaction with excess of silyl fluorides. In the course of the reaction the anionic character of the compound is maintained and trimethylsilyl groups are exchanged for other groups such as $PhMe_2Si$ or Me_3SiMe_2Si (Scheme 7.26 top). In a final step salt elimination leads to a neutral tetrasilylated silane **55**.[59] The reaction can also be used to obtain spirocyclic oligosilanes (**56**) (Scheme 7.26 bottom).

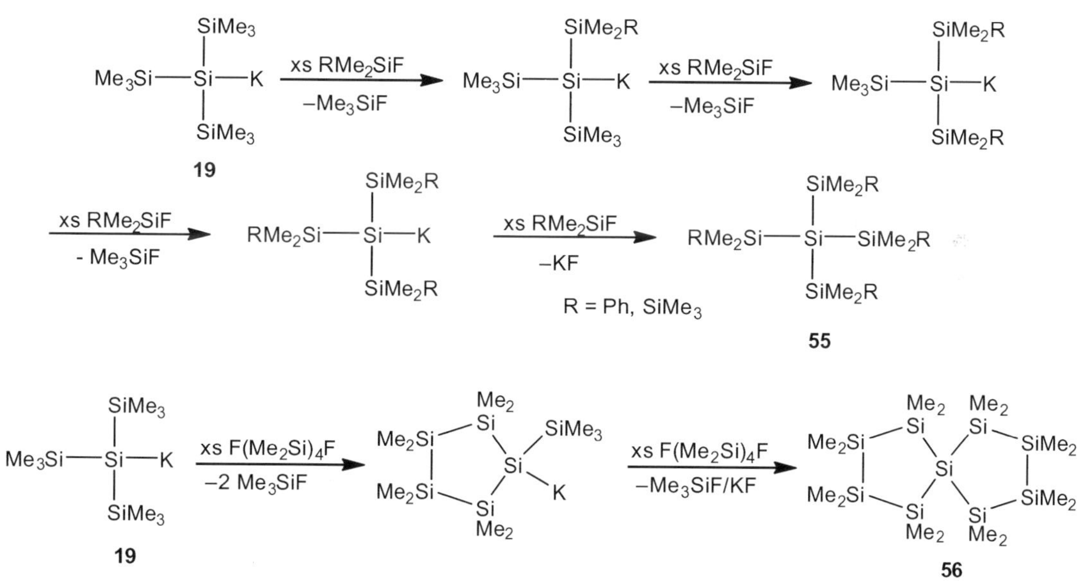

Scheme 7.26
Subsequent silyl exchange of a potassium oligosilanide with silyl fluorides leading to branched (top) or spirocyclic (bottom) oligosilanes.

It is known that electropositive substituents lower the singlet–triplet gap of silylenes. Replacing one of two silyl groups in a disilylated silylene by an alkali metal thus generates a triplet ground state. Starting out from a bis(tri-*tert*-butylsilyl)silacyclopropene (**57**) Sekiguchi and coworkers cleaved one of the exocyclic Si–Si bonds by treatment with alkali metals. The formed silacyclopropenyl anion **58** was then photolytically converted into a metallasilylsilylene (**59**), which was shown EPR spectroscopically to possess a triplet ground state (Scheme 7.27).[90]

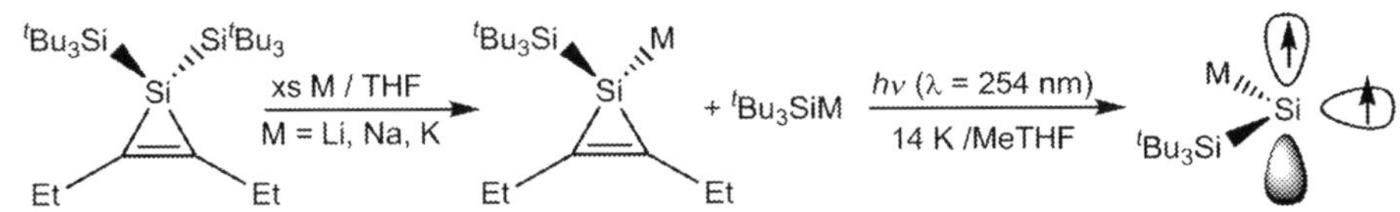

57 58 59

Scheme 7.27
Metalation of a silacyclopropene and photolytic conversion to a metallasilylsilylene.

Extending their silyl mercury chemistry Apeloig and coworkers have shown that oligosilyl hydrides can be converted to the respective oligosilyl zinc compounds by reaction with Et_2Zn or $^tBu_2Zn$ (Scheme 7.28).[91] The reaction is effectively catalyzed either by a small amount of $^tBu_2Hg$ or by AIBN (azoisobutyronitrile). This way $(Me_3SiMe_2Si)_2Zn$ and $[(Me_3Si)_3Si]_2Zn$ were obtained in excellent yields from the respective hydrosilanes Me_3SiMe_2SiH and $(Me_3Si)_3SiH$. Reaction of $(^tBuMe_2Si)_2Si(H)Cl$ (**60**) with $^tBu_2Zn$ (5 mol% $^tBu_2Hg$) gave $[(^tBuMe_2Si)_2(Cl)Si]_2Zn$ (**61**), which could be converted to the geminal dimetalated species **62** by reaction with lithium (Scheme 7.28).[91]

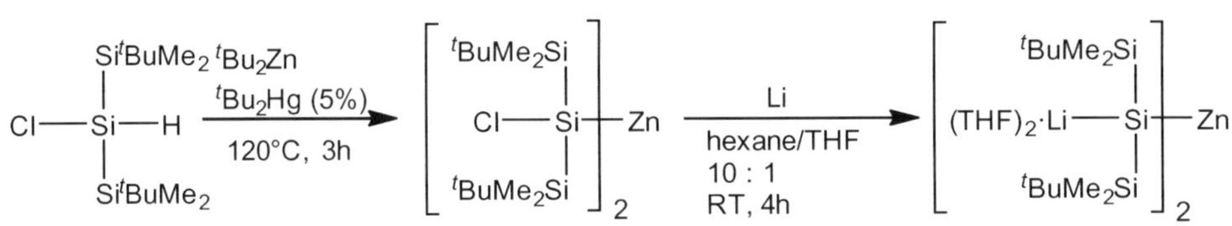

60 61 62

Scheme 7.28
Zincation of a hydrochlorosilane and subsequent lithiation to a geminal dimetalated silane.

Sekiguchi and coworkers observed the rearrangement of a lithium disilenide (**64**), obtained by the addition of methyllithium to disilyne **63**, to a lithium disilacyclopropanylsilylhydrosilanide (**65**). The reaction seems to involve intramolecular deprotonation of a $CH(SiMe_3)_2$ group. The thus formed carbanion adds to the disilene unit generating the final product **65** (Scheme 7.29).[92]

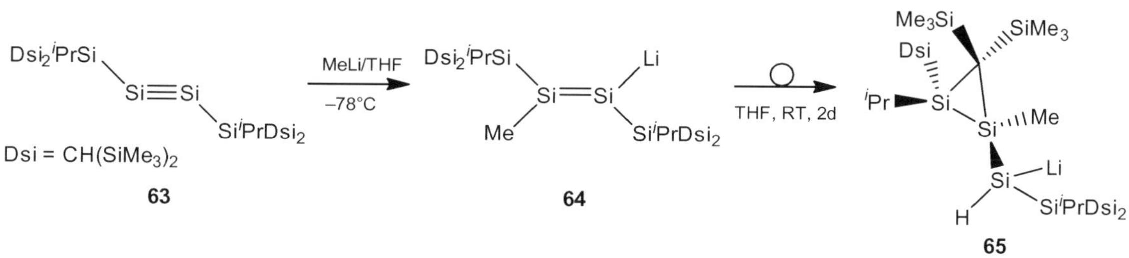

Scheme 7.29
Rearrangement of a disilenide to a disilacyclopropanyl substituted hydrosilanide.

The (η^3-disilaallyl)lithium compound **67** was obtained in the reaction of lithium disilenide **66** with tris(3,5-di-*tert*-butylphenyl)acetaldehyde (Scheme 7.30). Single crystal XRD analysis and NMR spectra show that the compound has a π-allyl-type structure, with an interaction between the Li^+ ion and the π-electrons of the silene moiety in the solid state and in toluene solution (**67a**), whereas no such interaction is present in THF (**67b**) (Scheme 7.30).[93]

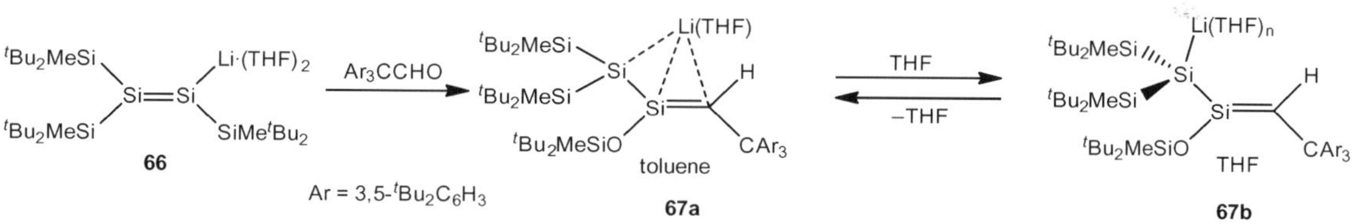

Scheme 7.30
Reaction of a disilenide with a triarylacetaldehyde forming a (η^3-disilaallyl)lithium compound.

7.3.4.2 Polycyclic and cage type oligosilanyl anions

Besides the numerous examples of silicon Zintl anions[6,94–96] a few instances of polycyclic polysilane anions have been reported. In the reaction of the tetrasilatetrahedrane $(Dsi_2MeSiSi)_4$ **68a** with 2 equivalents KC_8 selectively an exocyclic Si–Si bond is cleaved and the tetrasilatetrahedranide **69** is formed in addition to Dsi_2MeSiK (Scheme 7.31).[97]

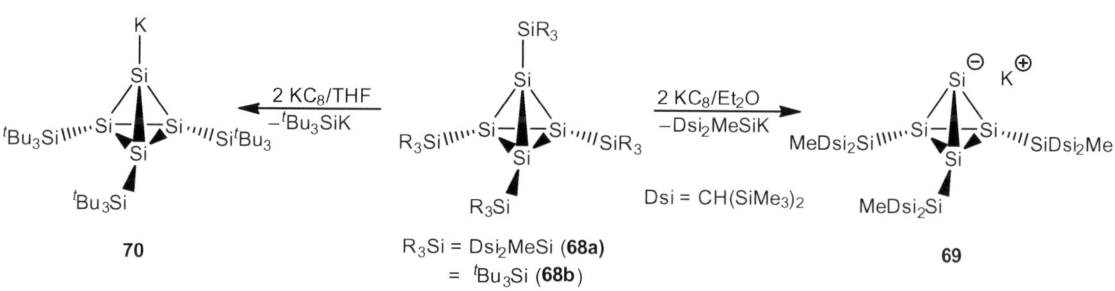

Scheme 7.31
Silyl group abstraction from tetrasilylated tetrasilatetrahedranes with KC_8 yielding respective silanides.

A very similar reaction occurs when Wiberg's tetrasilatetrahedrane ($(^tBu_3SiSi)_4$ (**68b**) is subjected to KC_8 using slightly different reaction conditions. Again the respective cage anion **70** is formed (Scheme 7.31).[98]

Utilizing silylpotassium chemistry Iwamoto et al. were able to prepare persilastaffanes.[99] Starting from a 1,1,3,3-tetrakis(trimethylsilyl)cyclotetrasilane (**71**) a 1,3-dipotassium compound (**72**) is obtained by treatment with two equivalents of KO^{t}Bu (Scheme 7.32). By reaction with $^tBu_2SiCl_2$ the persilastaffane **73** with trimethylsilyl groups attached to the bridgehead positions was generated. Further addition of KO^{t}Bu to **73** caused formation of silyl anion **74** which was converted to the respective silyl bromide **75** with $Br(CH_2)_2Br$ (Scheme 7.32). A persilastaffane dimer **76** was obtained from the coupling of **74** with **75**. The dimer **76** could again be converted to a silyl anion (**77**) with KO^{t}Bu, allowing even the synthesis of a persilastaffane trimer (**78**) (Scheme 7.32).[99]

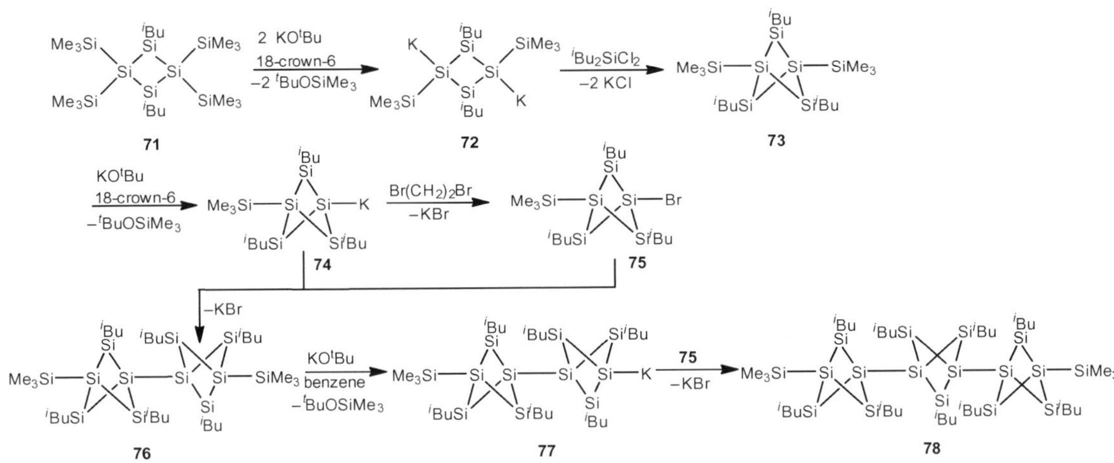

Scheme 7.32
Synthesis of persilastaffanes involving several oligosilanylpotassium intermediates.

The reaction of the chlorinated pentasilatricyclo[2.1.0.0^{2,5}]pentane **79** with potassium graphite was reported to lead to the formation of silyl anion **80**. Since the negative charge in **80** is not located at the same silicon atom bearing the chloride in **79**, a rearrangement of another initially formed silyl anion to **80** can be assumed (Scheme 7.33).[100]

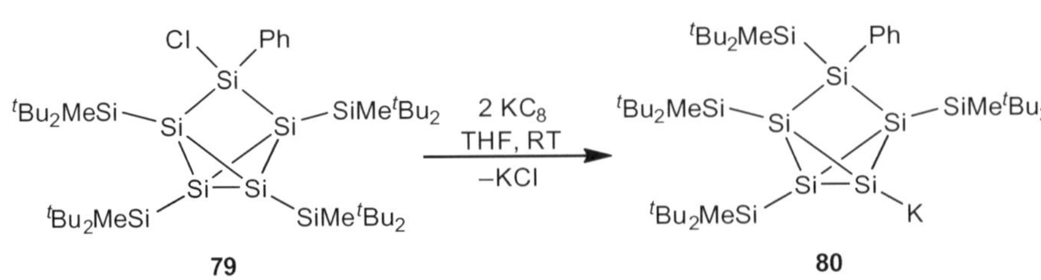

Scheme 7.33
Synthesis of a metalated cage-oligosilane.

7.3.5 Functionalized Anions

While much of the chemistry concerning alkylated, arylated, and silylated silyl anions was studied early on, the recent two decades have seen the advent of functionalized silyl anions. Arguably the most interesting ones of those are α-functionalized anions. By attaching electronegative substituents with leaving group character, silylenoid compounds with ambiphilic properties can be obtained. But also the introduction of other heteroatoms and even organic functional groups has been achieved and led to compounds with interesting structural and chemical properties.

7.3.5.1 Hydrogen substituted anions

As outlined above (Scheme 7.21) the reaction of $(^tBu_3Si)_2Si(H)Br$ with alkali metals was found to proceed to hydrosilanides[83,101] which were also converted to the respective group 12 silanides (Scheme 7.21).[84] Also Kira's study concerning hydrosilyllithium formation by deprotonation (Scheme 7.23) was already discussed above.[87]

Also mentioned above (Scheme 7.22) Bravo-Zhivotovskii et al. found that the reaction of $(THF)_2 \cdot Li(^iPr_3Si)_2Si\text{-}Hg\text{-}Si(Si^iPr_3)_2Li \cdot (THF)_2$ (**46**) with lithium gave instead of the expected geminal dilithiosilane the hydrosilanide $H(^iPr_3Si)_2SiLi \cdot (THF)_2$ (**47**).[86] Related, although less clean, chemistry was observed in the reaction of $(^tBu_2MeSi)_2SiH_2$ (**81**) with $^tBu_2Hg$ and further with elemental lithium.[102] A mixture of dimeric hydrosilanide **82** (**82**$_2$) in addition to **82** cocrystallized with the geminal silyl dianion **83b** (**82**$_2$ · **83b**) was obtained (Scheme 7.34).

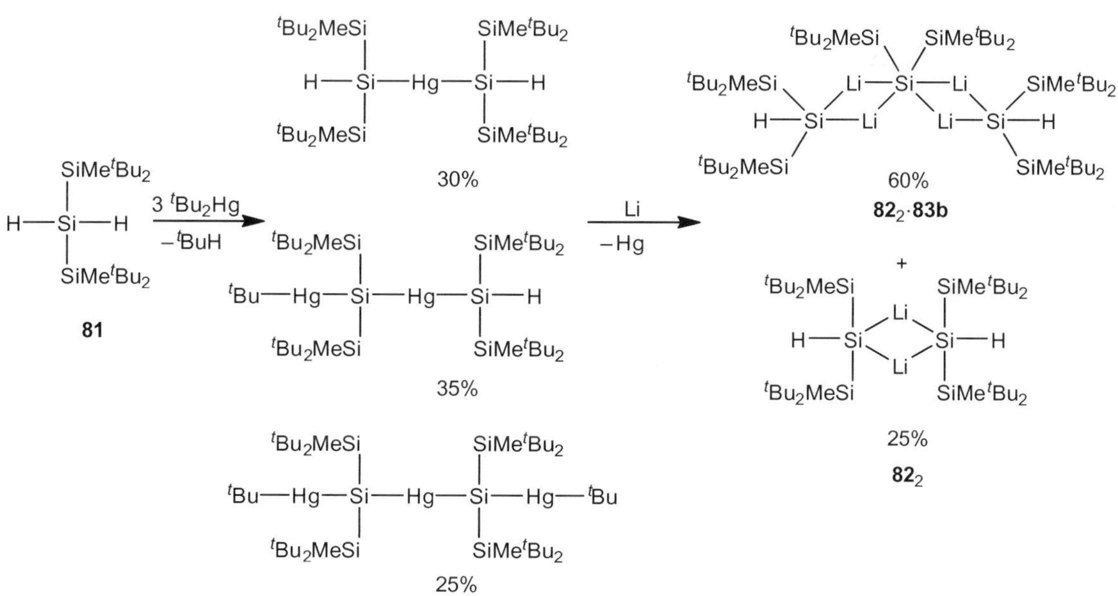

Scheme 7.34
Mercuration of $(^tBu_2MeSi)_2SiH_2$ and subsequent lithiation.

A synthesis of the hydrosilylzinc compound **84** was recently reported by reaction of a carbene-stabilized dialkylzinc complex with phenylsilane (Scheme 7.35).[103]

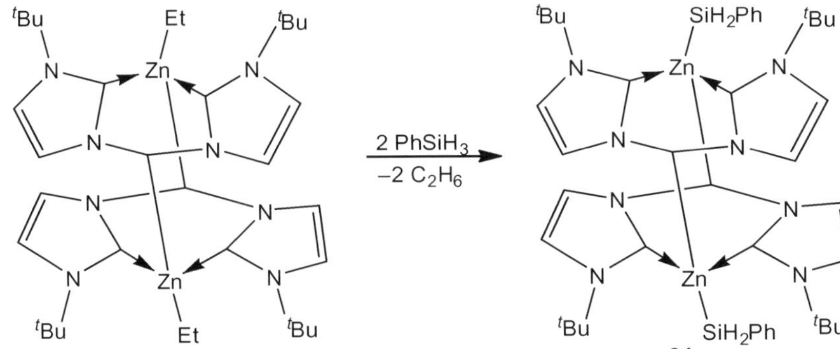

84

Scheme 7.35
Formation of a zinc phenylsilanide in the reaction of a dialkylzinc complex with $PhSiH_3$.

The formation of H_3SiK by reaction of Na/K alloy with SiH_4 is a long known reaction.[104,105] H_3SiK was found to be a promising hydrogen storage material and therefore alternative synthetic access is of interest.[106,107] Given the hazards associated with the use of SiH_4 the recent report of the formation of H_3SiK by hydrogenolysis of Ph_3SiK can be considered a substantial improvement.[108] Alternatively, also the hydrogenolysis of SiK was used to obtain H_3SiK.[106] The reaction of SiH_4 with Na/K alloy does not necessarily stop with the formation of H_3SiK but can be directed further to $(H_3Si)_3SiK$.[109] An alternative synthetic access to $(H_3Si)_3SiK$ was reported recently via the reaction of $(H_3Si)_4Si$ with KO^tBu.[110]

7.3.5.2 Halogen substituted anions

In the course of Wiberg's exhaustive studies on silanes containing the supersilyl group ($^tBu_3Si$), reactions of $(^tBu_3Si)_2SiX_2$ (X = F, Cl, Br) with lithium naphthalenide (LiNp) were found to give $(^tBu_3Si)_2Si(X)Li$, which due to its steric bulk is not undergoing a condensation reaction.[83]

The tris(trimethylsilyl)methyl group (Tsi) is also very bulky and was used by Lee and coworkers to obtain the halide substituted silylene Tsi(Br)Si: by reaction of $TsiSiBr_3$ with 2 equivalents of lithium naphthalenide.[111] However, spectroscopic and reactivity studies indicated that the formed compound might be more appropriately described as $TsiSiBr_2Li$.[112] Depending on reaction conditions the compound is able to react either as nucleophile, electrophile, or silylene. Treatment with MesLi gives silylenoid TsiMesSiBrLi and further reaction with $MgBr_2$ gives the magnesium silylenoid TsiMesSiBrMgBr.[113] Reaction of $TsiSiBr_3$ with 4 equivalents LiNp led to the formation of the stable halodilithiosilane $TsiSiBrLi_2$.[114] The compound might be considered as the lithium bromide adduct of a lithiosilylenoid. This assignment is supported by the fact that the compound does not react as an electrophile for instance with MeLi (MeLi/TMEDA) or MesLi but acts as a dinucleophile with Me_3SiBr to $(Me_3Si)_2TsiSiBr$.[114] Reaction of $(Me_3Si)_3SiF$ (**85a**) with KO^tBu was found to give in a clean reaction $K(Me_3Si)_2SiSi(SiMe_3)_2F$ (**86a**) (Scheme 7.36).[115] It was assumed that the reaction occurs through the transient formation and subsequent dimerization of $(Me_3Si)_2(F)SiK$ (**87a**). Replacement of one Me_3Si group of **85a** by $^iPr_3Si$ allowed the NMR spectroscopic detection of $(Me_3Si)(^iPr_3Si)(F)SiK$ (**87b**) as an intermediate (Scheme 7.36).[116]

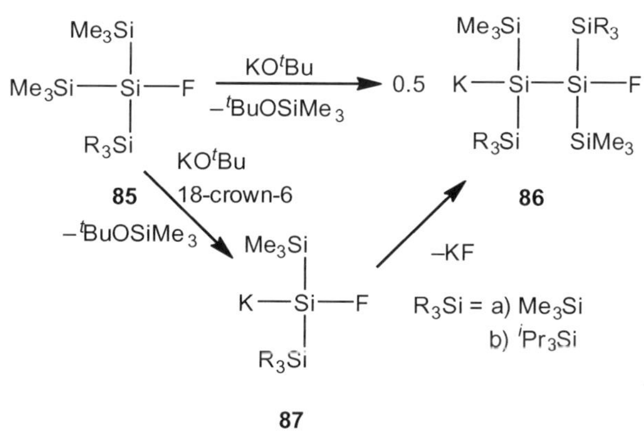

Scheme 7.36
Formation of a β-fluorosilanide via condensation of fluorosilylenoid.

Increasing the steric bulk even further Apeloig and coworkers obtained the stable silylenoid $(^tBu_2MeSi)_2SiFLi \cdot (THF)_3$ which features a tricoordinate silicon atom with the lithium coordinating to fluoride.[117] Compound $[(^tBuMe_2Si)_2(Cl)Si]_2Zn$ (**61**) described above (Scheme 7.28) can of course also be considered as silylenoid.[91]

7.3.5.3 N-substituted anions

N-Substituted silyl anions were reported by Tamao and Kawachi as first examples of silylenoid compounds.[118–120] These compounds were found to display distinct NMR spectroscopic signatures[120,121] but with respect to reactivity behaved very much like other silyl anions without much electrophilic properties. No tendency for self-condensation was thus observed.

The picture of aminosilyl anions behaving very much like ordinary silyl anions holds also for amino substituted oligosilanides. These can be obtained using the trimethylsilyl cleavage protocol with KO^{t}Bu.[122] Mono- and bis(diethylamino) substituted silylpotassium compounds (**88**) were reported (Scheme 7.37).

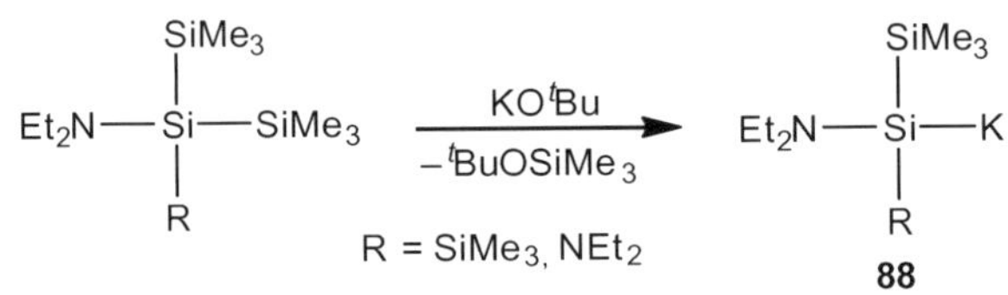

Scheme 7.37
Synthesis of amino substituted potassium oligosilanides by silyl abstraction with KO^{t}Bu.

Reaction of the silylaminochlorosilanes [($Me_3Si)_2$N]RPhSiCl (R = Me, Ph, N($SiMe_3)_2$) (**89**) with lithium in THF at low temperature afforded the silanides [($Me_3Si)_2$N]RPhSiLi (**90**), which rearranged to lithium amides [Li(Me_3Si)N]RPhSiSiMe_3 (**91**) when the temperature was raised (Scheme 7.38). The reaction of the hydrochlorosilane [($Me_3Si)_2$N]HPhSiCl with lithium in THF at −78°C yielded a mixture of [($Me_3Si)_2$N]HPhSiLi and [($Me_3Si)_2$N] HPhSiSiHPhLi.[123]

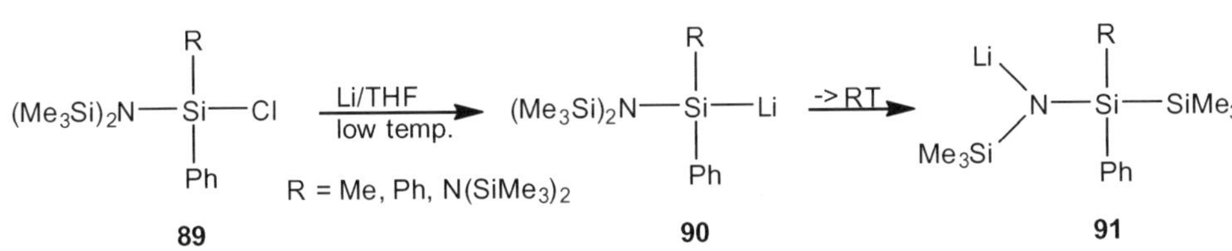

Scheme 7.38
Synthesis of silylamino substituted silanides at low temperature followed by silyl group migration.

As outlined above in the oligosilanide section, Gehrhus, Lappert, and coworkers showed that silylene insertion into the Si–Li bond of a silyllithium compound is a facile process (Scheme 7.20). The used N-heterocyclic silylene **41** causes silylenoid character of the formed anion **42** with two Si–N bonds.[80,81] In a similar way Driess and coworkers demonstrated insertion of their N-heterocyclic silylene **92** into C–Li, C–Zn, C–Al, and H–Al bonds of MeLi, Me_2Zn, Me_3Al, and $H_3Al \cdot NMe_3$, respectively. The thus formed compounds **93–96** can also be considered as silylenoids (Scheme 7.39).[124]

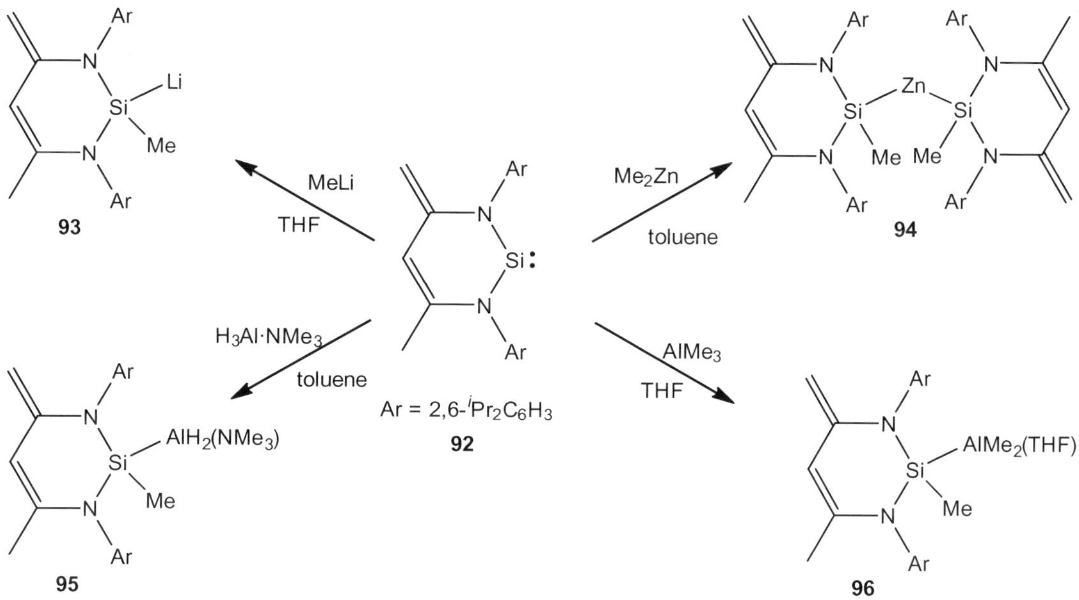

Scheme 7.39
Metalation reactions of the Driess silylene.

Zinc silylenoids were also obtained by Oestreich and coworkers by reacting Et_2NPh_2SiLi with zinc chloride.[125] In the course of studies on the synthesis of polypyrazolylsilanes and disilanes, Breher and coworkers converted the tetrapyrazolylsilane **97** to the zwitterionic tris (3,5-dimethylpyrazolyl)silanide **98**, which either exists as a base-free polymeric chain or as THF adduct (Scheme 7.40).[126]

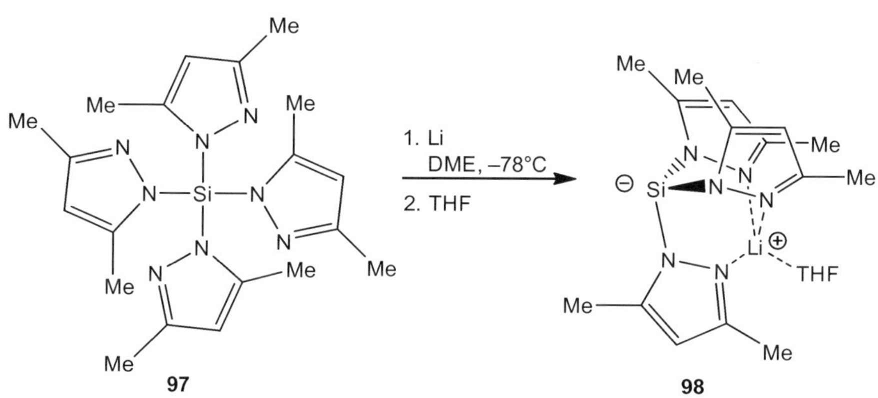

Scheme 7.40
Si–N Cleavage reaction of a tetrapyrazolylsilane with lithium giving a zwitterionic silanide.

Related chemistry was also studied by Krempner and coworkers, who attached four pyrazolyl units to a neopentasilane (**99**) which upon reaction with KO^{t}Bu in THF in the presence of 18-crown-6 gave the respective silanide **100** (Scheme 7.41).[127]

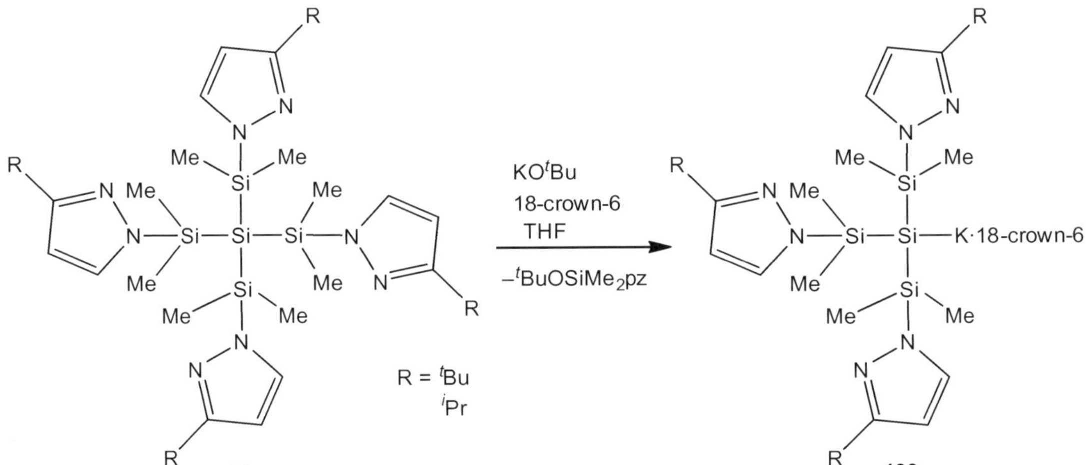

99 **100**

Scheme 7.41
KO^{t}Bu induced silyl abstraction yielding a zwitterionic isotetrasilanide.

7.3.5.4 O- and S-substituted anions

The first example of an alkoxysilyl anion (^{t}BuO)Ph$_2$SiLi (**103a**) was reported by Tamao and Kawachi to be formed either by transmetalation of a stannylalkoxysilane (**101**) with *n*-butyl lithium or by lithiation of a chloroalkoxysilane (**102a**) (Scheme 7.42).[128,129] It can be kept monomeric by complexation with 12-crown-4 or otherwise exhibits pronounced ambiphilic

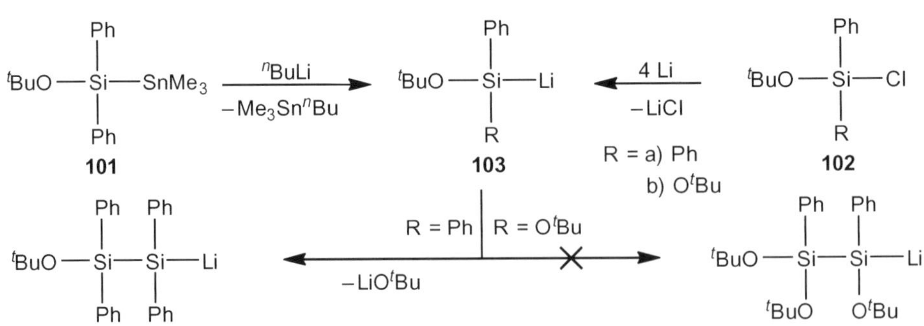

101 **103** **102** **104**

Scheme 7.42
The surprisingly different reactivity of lithium mono- and dialkoxysilanides.

behavior and undergoes dimerization to the respective 2-alkoxydisilanyl lithium compound **104** (Scheme 7.42).[128] The analogous dialkoxysilanide $(^{t}BuO)_2PhSiLi$ (**103b**) was prepared by reaction of $(^{t}BuO)_2PhSiCl$ (**102b**) with lithium, however, it shows no self-condensation tendency (Scheme 7.42).[129]

The alkoxysilanide $(MeO)(Me_3Si)_2SiK$ (**106**) was prepared by reaction of $MeOSi(SiMe_3)_3$ (**105**) with $KO^{t}Bu$.[130] Depending on whether the reaction was carried out with or without 18-crown-6 either the silylenoid **106** or the condensation product **107** was obtained (Scheme 7.43).[130]

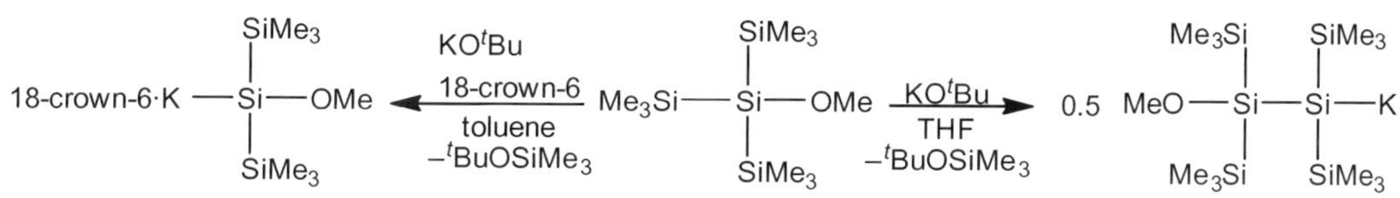

106 **105** **107**

Scheme 7.43
Different reactivity of methoxybis(trimethylsilyl)silylpotassium with and without crown-ether.

Synthesis, stability, and reactivity of substituted lithium trimethylsiloxysilanides was studied in some detail by Popowski and coworkers.[131,132] Reaction of chlorosilanes $(Me_3SiO)RR'SiCl$ with lithium in THF at low temperature gave either the silyllithium derivatives $(Me_3SiO)RR'SiLi$, the condensation products $(Me_3SiO)(RR'Si)_2Li$, or trimethylsilylated silanides $Me_3SiRR'SiLi$.[131,132] The formation of $(Me_3SiO)RR'SiLi$ occured not only by direct metalation of the chlorosilanes but, depending on the substituent pattern, also by cleavage of previously formed disilanes $(Me_3SiO)(RR'Si)_2(OSiMe_3)$.[131,132]

Reactions of the N-heterocyclic silylene **41** with either NaOMe or with $NaCH(SiMe_3)(SiMe_2OMe)$, which is known to extrude sodium methoxide, proceeded to give dimeric sodium silylenoids **108** (Scheme 7.44).[133]

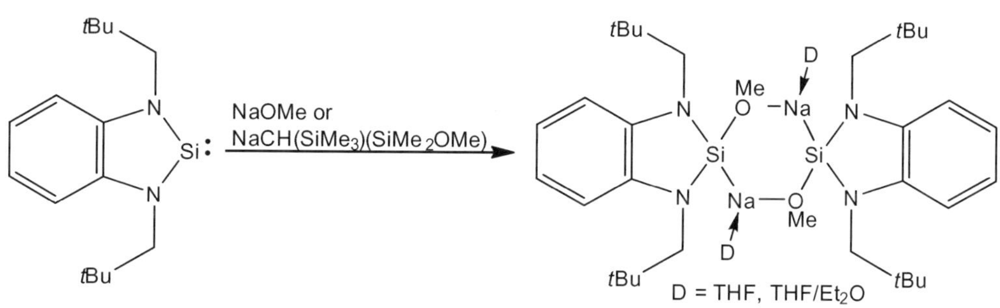

41 **108**

Scheme 7.44
Silylenoid formation by addition of NaOMe to an N-heterocyclic silylene.

Besides aminosilanides, which behave like ordinary silyl anions, and alkoxysilanides, which display ambiphilic behavior and can undergo self-condensation, Tamao and Kawachi also introduced sulfur-substituted silyl anions (**110**) obtained from the lithiation of a stannylsilane (**109**).[134] These compounds exhibit yet another reactivity pattern and show α-elimination of lithium thiolate to give a free silylene (**111**), which can be trapped with various reagents (Scheme 7.45).

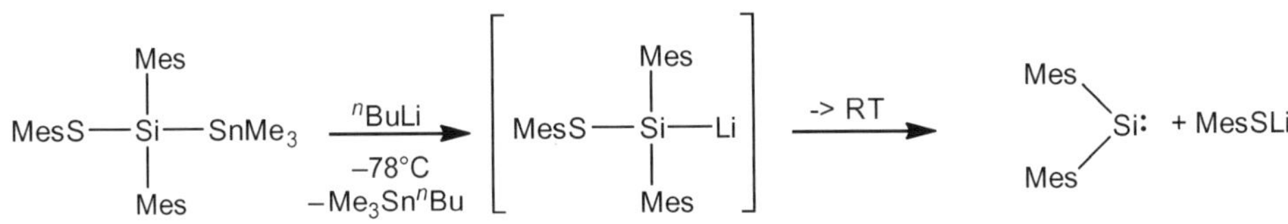

Scheme 7.45
Formation of a sulfur-substituted lithium silanide which decomposes to a silylene by α-elimination of lithium thiolate.

7.3.5.5 *Other functionalized anions*

Alkynyl-substituted oligosilanides **113** were reported to be formed from bis[tris(trimethylsilyl)silyl]acetylene (**112**),[135] or other tris(trimethylsilyl)silylated alkynes in the reaction with KO^{t}Bu (Scheme 7.46).[60]

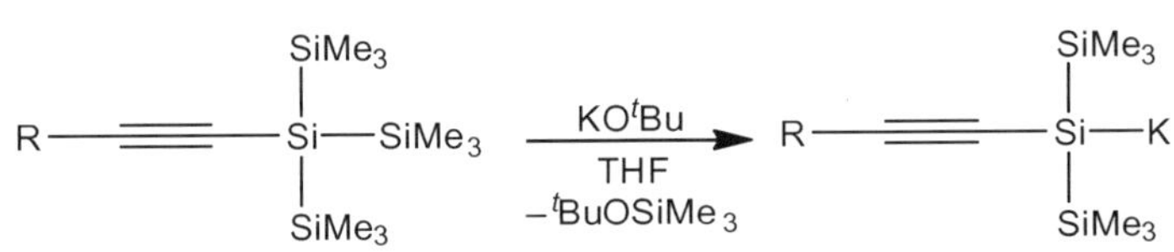

Scheme 7.46
Facile formation of alkynylated oligosilanides using trimethylsilyl cleavage with KO^{t}Bu.

Reactions of KO^{t}Bu with substrates with two different types of trimethylsilylated silyl groups provide some insight into the ease of silanide formation. Reaction of the 1,3-propynylidene bridged compound **114** proceeded selectively to the alkynyl substituted silanide **115** (Scheme 7.47). Conversely, the reaction of phenylethynylpentakis(trimethylsilyl)disilane (**116**) with KO^{t}Bu led to a mixture of the two possible silylpotassium compounds **117** and **118** indicating that the stabilization derived from alkynyl substituents is similar to that of silyl groups (Scheme 7.47).[60]

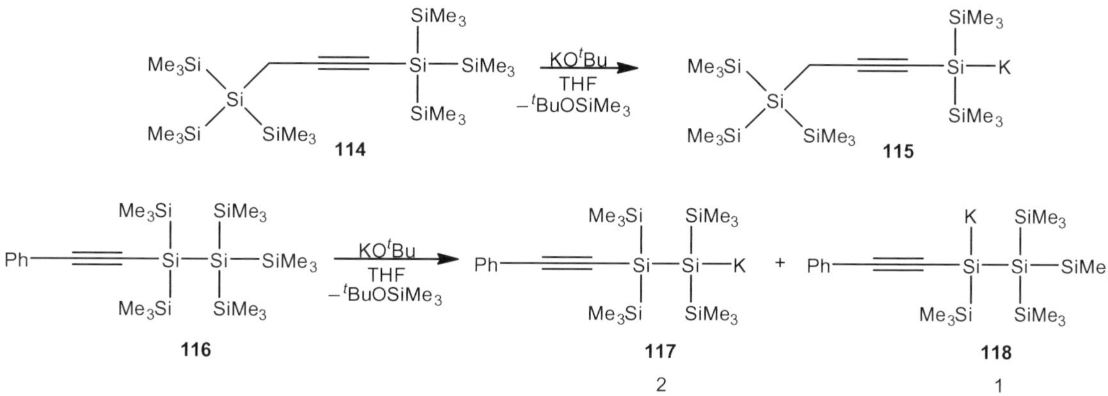

Scheme 7.47
Reactions of alkynylated oligosilanes with KO^{t}Bu providing insight into the different stabilization capacity of silyl, alkynyl, and alkyl substituents.

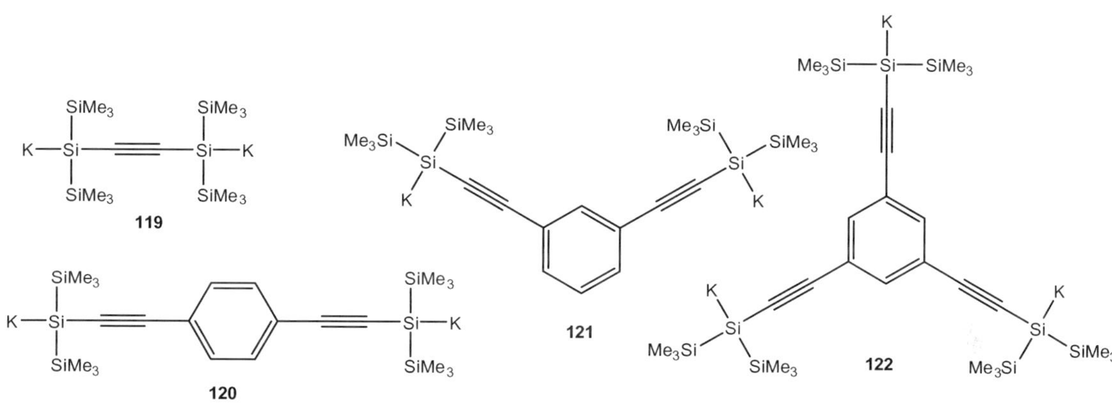

Figure 7.2
Examples di- and trimetalated alkynylsilanes.

Compounds where tris(trimethylsilyl)silyl groups are separated either by a single alkynylene unit or by bis- or tris(alkynylated) phenyl rings can be treated with KO^{t}Bu to give di- (**119**, **120**, **121**) or even trimetalated (**122**) silanes (Fig. 7.2).[136]

Kira and coworkers prepared a number of alkynylsilyllithium compounds (**124**) by lithiation reactions of the corresponding hydrosilanes (**123**) with ^{t}BuLi in THF. The crystal structure of (^{t}BuMe$_2$Si)$_2$(Me$_3$SiCC)SiLi revealed that it exists as a syn dimer with a planar four-membered ring composed of two lithium and two anionic silicon atoms with one lithium atom being coordinated by the two alkynyl groups and the other one by a THF molecule (Scheme 7.48).[137]

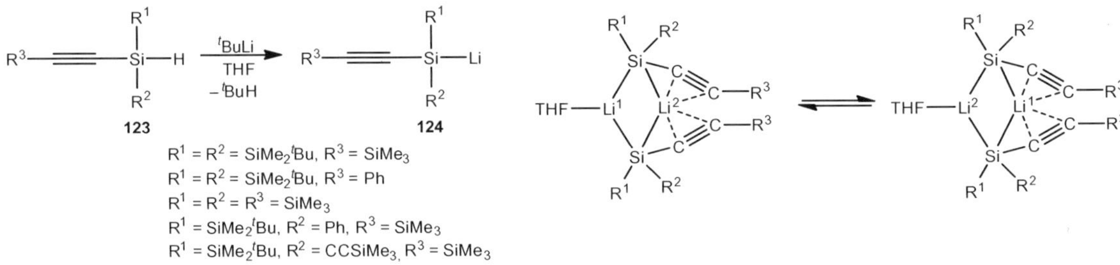

Scheme 7.48
Deprotonation of alkynylated hydrosilanes with ^{t}BuLi (left side). Coordination motif of the lithium alkynylsilanide (right side).

The number of reported vinylated silyl anions is very small. Marschner and coworkers demonstrated that vinylated tris(trimethylsilyl)silanes **125** undergo facile reaction with KO^{t}Bu to form the respective vinylated silylpotassium compounds **126**.[138] The few examples include also the formation of the vinylene bridged 1,4-silyl dianion **127** (Scheme 7.49).[138]

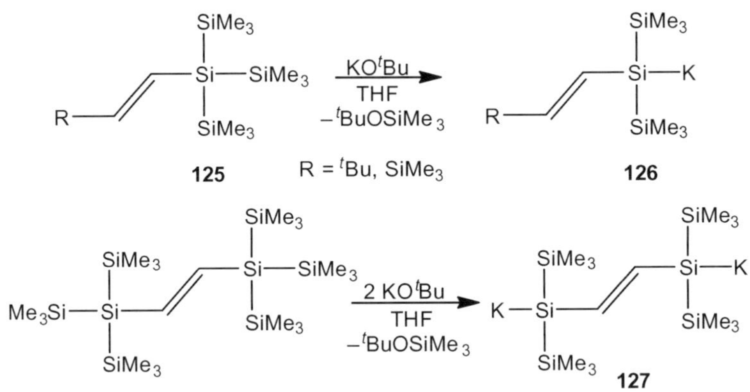

Scheme 7.49
Formation of vinylated oligosilanides using trimethylsilyl cleavage with KO^{t}Bu.

Another interesting case of a divinylene bridged 1,4-dianion (**129**) was reported by Sekiguchi and coworkers, who reacted a 1,4-disilyl-1,4-disilabicyclo[2.2.0]-hexa-2,5-diene (**128**) with excess lithium in THF.[139] In the solid state structure the six-membered ring of **129** features a boat conformation with the two lithium atoms bridging the anionic silicon atoms (Scheme 7.50).[139]

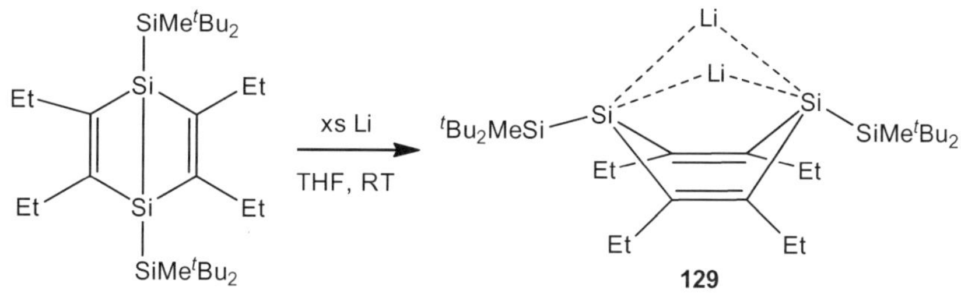

Scheme 7.50
Si–Si–Cleavage with lithium to a 1,4-divinylene bridged disilanide.

Lithium 3-sila-β-diketiminates **130** were obtained by reaction between $(Me_3Si)_3SiLi$ (**18**) and appropriate nitriles PhCN/ArCN (Ar = 2,6-$Me_2C_6H_3$).[140] Due to the reluctance of silicon to form double bonds, the 3-sila-β-diketiminates (**130**) display a higher degree of charge localization at the 3-position than previously noted for the respective carbon analogs (Scheme 7.51).[140]

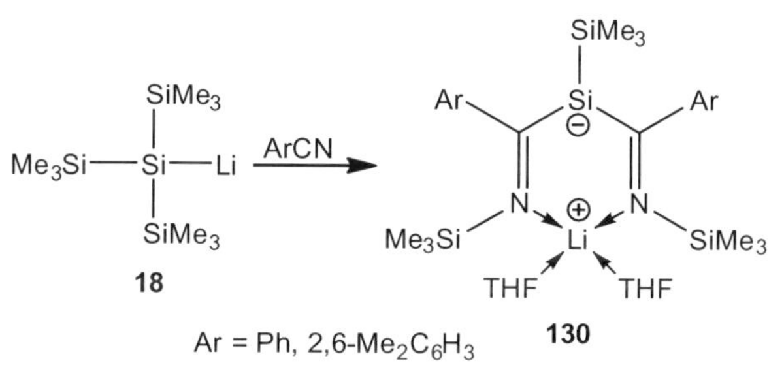

Scheme 7.51
Addition of $(Me_3Si)_3SiLi$ to aromatic nitriles yields 3-sila-β-diketiminates.

7.3.6 Silyl Dianions

As silyl anions are versatile building blocks for organic and organometallic synthesis, double charged compounds are even more interesting for the synthesis of either unsaturated or cyclic compounds. The arguably most interesting of these silyl dianions are geminal (1,1) and vicinal (1,2) dianions.

7.3.6.1 Geminal silyl dianions

A marvelously simple access to geminal 1,1-dianionic silanes was developed by Sekiguchi and coworkers reacting silacyclopropenes **131** with lithium to form the respective dilithiosilane and bis(trimethylsilyl)acetylene (Scheme 7.52). The thus obtained

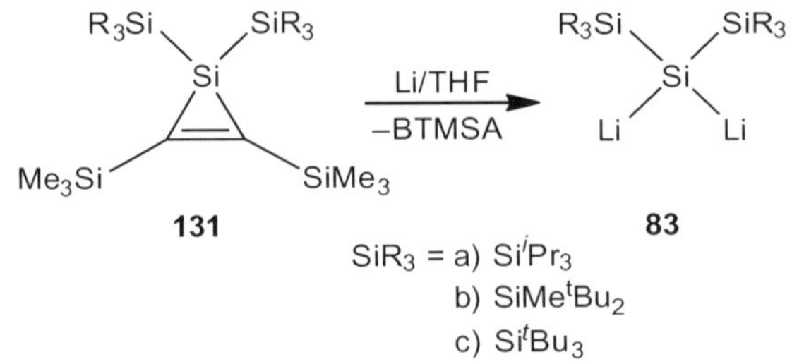

Scheme 7.52
Metalation of a silacyclopropene to a geminal dilithiosilane.

$(^i\mathrm{Pr_3Si})_2\mathrm{SiLi_2}$ (**83a**),[141] $(^t\mathrm{Bu_2MeSi})_2\mathrm{SiLi_2}$ (**83b**),[142] and $(^t\mathrm{Bu_3Si})_2\mathrm{SiLi_2}$ (**83c**)[143] were used as starting materials for a number of compounds containing double bonds to silicon.[143–149]

Kira and coworkers found that $(^i\mathrm{Pr_3Si})_2\mathrm{SiLi_2}$ (**83a**) is also formed in the reaction of tetrakis (triisopropylsilyl)disilene (**132**) with lithium (Scheme 7.53).[150]

Scheme 7.53
Metalation of a tetrasilylated disilene yielding a geminal dilithiosilane.

As outlined above Apeloig and coworkers obtained the geminal disilanide $(^t\mathrm{Bu_2MeSi})_2\mathrm{SiLi_2}$ (**83b**) as a mixture with the respective hydrosilanide **82** (Scheme 7.34).[102] The coordination motive of this mixture can be changed by addition of THF to give an unusual trimeric dianionic structure **83b**$_3$ (Scheme 7.54).[151] Changing the silyl substituent from $^t\mathrm{Bu_2MeSi}$ to

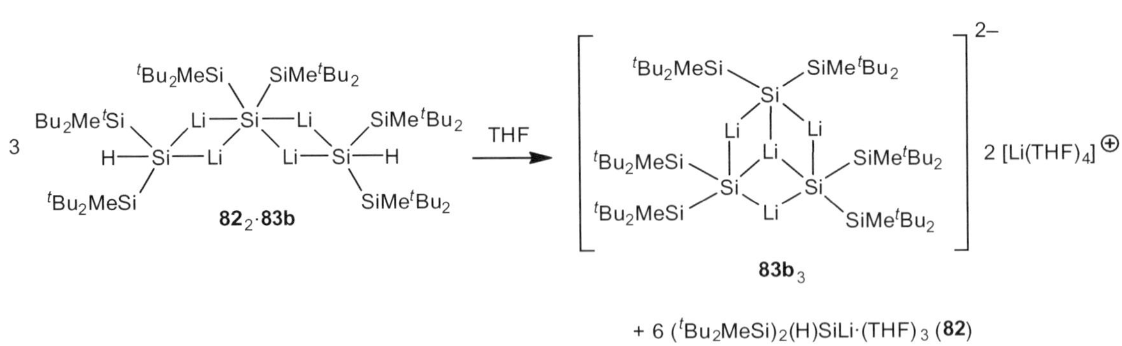

Scheme 7.54
Solvent-induced change of coordination motif from a mixed aggregate of geminal dianion and hydrosilanide to a dianionic trimeric structure.

^{i}Pr$_3$Si gives a cleaner reaction and conversion of [^{t}BuHg(^{i}Pr$_3$Si)$_2$Si]$_2$Hg with lithium in hexane gave the mixed aggregate [(^{i}Pr$_3$Si)$_2$SiLi$_2$]$_2$ · (^{t}BuLi)$_2$.[152] Treatment of the latter in hexane at 60°C results in decomposition to ^{t}BuLi and an insoluble powder of [(^{i}Pr$_3$Si)$_2$SiLi$_2$]$_n$, which upon addition of THF gives the known (^{i}Pr$_3$Si)$_2$SiLi$_2$ (**83a**).[152]

Reacting (^{t}BuMe$_2$Si)$_2$SiLi$_2$ (**133**) (obtained from the reaction of (^{t}BuMe$_2$Si)$_2$Si(HgSiMe$_2{}^t$Bu)$_2$ with lithium in hexane)[153] with ^{t}BuMgCl · MgCl$_2$ the dimeric 1,1-dimagnesiosilane **134** was obtained (Scheme 7.55).[153] Reaction with excess Grignard reagent gave the di-Grignard compound **135** and reaction of **134** with ZnCl$_2$ gave the trimeric 1,1-dizinciosilane **136** (the lithium chloride in **136** originates from incomplete separation of LiCl during the synthesis of **134**) (Scheme 7.55).[153]

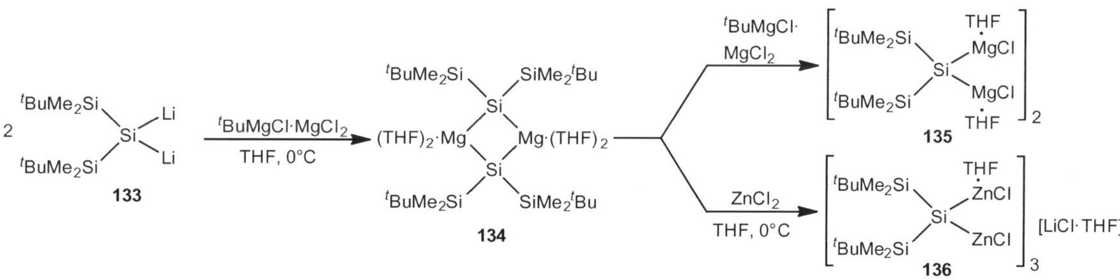

Scheme 7.55
Conversion of a geminal disilanide to 1,1-dimagnesiosilanes or a 1,1-dizinciosilane.

As outlined already above [(^{t}BuMe$_2$Si)$_2$(Cl)Si]$_2$Zn (**61**) was treated with lithium to give the geminal dimetalated [(^{t}BuMe$_2$Si)$_2$(Li)Si]$_2$Zn (**62**) (Scheme 7.28).[91] Formation of the stable halodilithiosilane TsiSiBrLi$_2$ by reaction of TsiSiBr$_3$ with 4 equivalents LiNp was mentioned above in the section concerned with halosilyl anions.[114]

7.3.6.2 Vicinal dianions

The first 1,2-dilithiodisilane was reported by Belzner et al.[154] Starting out from a cyclotrisilane bearing 2-(dimethylaminomethyl)phenyl substituents (**137**) reaction with excess lithium in THF yields the respective 1,2-dilithiodisilane **138** (Scheme 7.56).

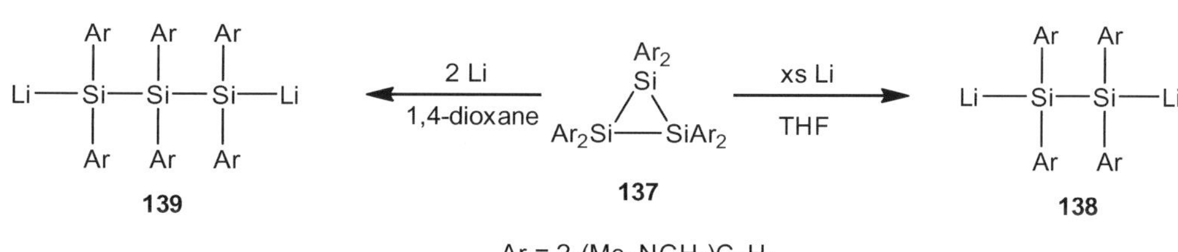

Scheme 7.56
Trisilirane ring-opening with lithium either to a 1,2-dilithiodisilane or a 1,3-dilithiotrisilane.

Interestingly reaction of the same starting material with two equivalents of lithium in 1,4-dioxane gave the 1,3-dilithiotrisilane **139**, which can be regarded as a plain ring-opening product (Scheme 7.56).[154]

Reaction of hexakis(trimethylsilyl)disilane (**20**) with two equivalents KO^{t}Bu and 18-crown-6 in benzene gives the 1,2-dipotassium compound **140** (Scheme 7.57).[73] If one of the central silicon atoms of **20** is replaced with germanium (**141**), the reaction still works but selectively the germanide **142** is formed in the first step before the dianion **143** is formed (Scheme 7.57).[155]

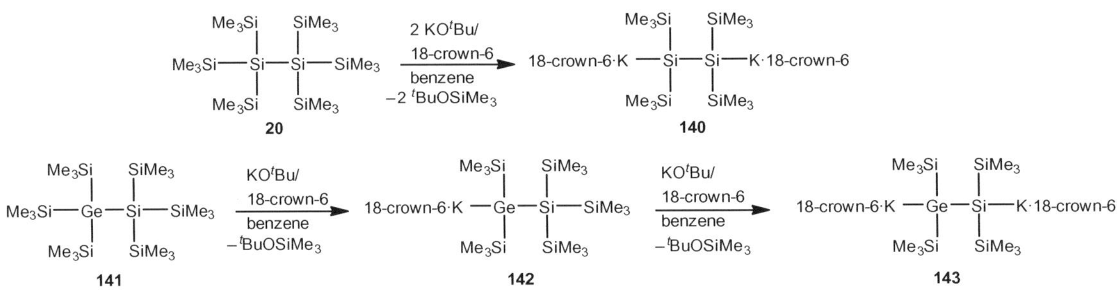

Scheme 7.57
Vicinal dianion formation by double silyl group abstraction with KO^{t}Bu.

The same type of chemistry can also be used with cyclic starting materials such as 1,1,2,2-tetrakis(trimethylsilyl)hexamethylcyclopentasilane, which upon reaction with two equivalents KO^{t}Bu/18-crown-6 in benzene gives the respective cyclic 1,2-dianion.[156]

As outlined above Kira and coworkers found that reaction of tetrakis(triisopropylsilyl)disilene **132** with lithium gave geminal dilithiosilane **83a** (Scheme 7.53).[150]
However, changing the silyl substituent on the disilene to either ^{t}BuMe$_2$Si or ^{i}Pr$_2$MeSi (**144**) causes the formation of the respective 1,2-dilithiodisilanes **145** instead (Scheme 7.58).[150]

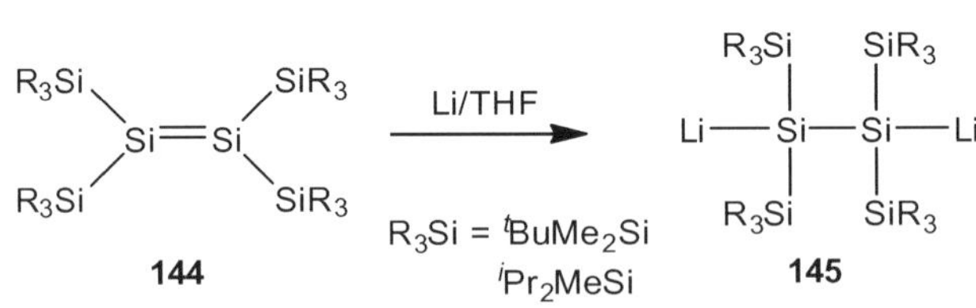

Scheme 7.58
Lithiation of tetrasilyldisilenes to 1,2-dilithiotetrasilyldisilanes.

Sekiguchi and coworkers found that reaction of another tetrasilylated disilene, namely, tetrakis(di-*tert*-butylmethylsilyl)disilene (**146**), with alkali metal naphthalenides again gives the respective 1,2-dianion (**147**) (Scheme 7.59).[157,158] However, depending on reaction conditions the 1,2-dianion can react further either to disilenides **66**[157] or to silylene radical anions **148** (Scheme 7.59).[158]

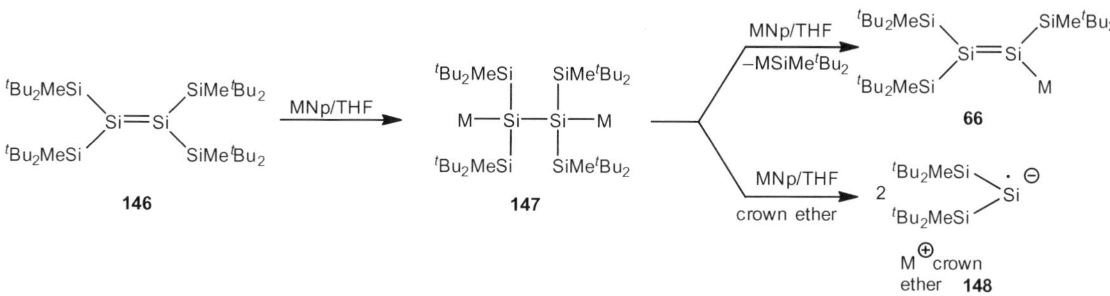

Scheme 7.59
Metalation of a tetrasilyldisilene to a 1,2-metallatetrasilyldisilane and further to either a disilenide or a disilylated silylene radical anion.

7.3.6.3 1,3-Dianions

As outlined above Belzner and coworkers obtained 1,3-dilithiotrisilane **139** by Si–Si cleavage of a trisilirane (Scheme 7.56).[154] The acyclic 1,3-dipotassium dianionic **150a** compound was obtained by double trimethylsilyl abstraction with KO^{t}Bu from the respective bridged α,ω-bis[tris(trimethylsilyl)silyl]dimethylsilane **149a** (Scheme 7.60).[61,135]

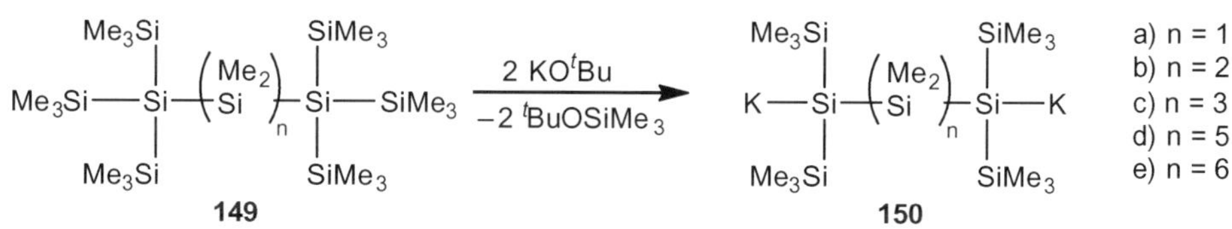

Scheme 7.60
α,ω-Oligosilandiide formation by double trimethylsilyl abstraction with KO^{t}Bu.

Transmetalation of 1,3-dianion **150a** with $MgBr_2 \cdot Et_2O$ gave a cyclic magnesacyclotetrasilane **151**,[68] whereas the reaction with $ZnCl_2$ gave the dimeric bicyclic chlorodizincate **152** (Scheme 7.61).[159]

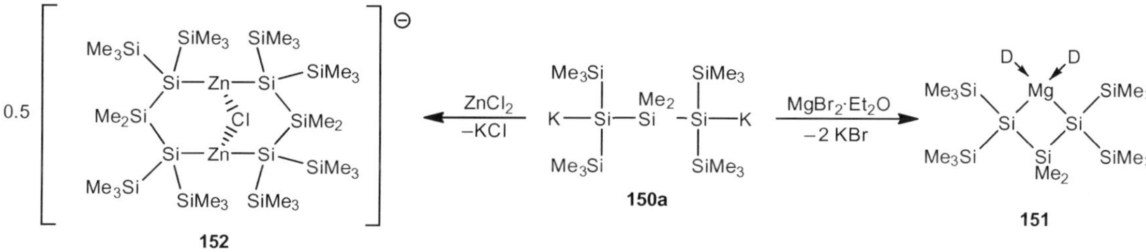

Scheme 7.61
Transmetalation of a 1,3-dipotassiosilane with magnesium bromide and zinc chloride.

Dianion formation using silyl group abstraction with KO^{t}Bu works also for cyclic oligosilanes. From octakis(trimethylsilyl)cyclotetrasilane **153** the dipotassium

1,3-*trans*-cyclotetrasilandiide **154** was obtained (Scheme 7.62).[73,74] Similarly, Iwamoto et al. used a 1,3-dianionic cyclotetrasilane **72** as a central intermediate for the synthesis of persilastaffanes (Scheme 7.32).[99]

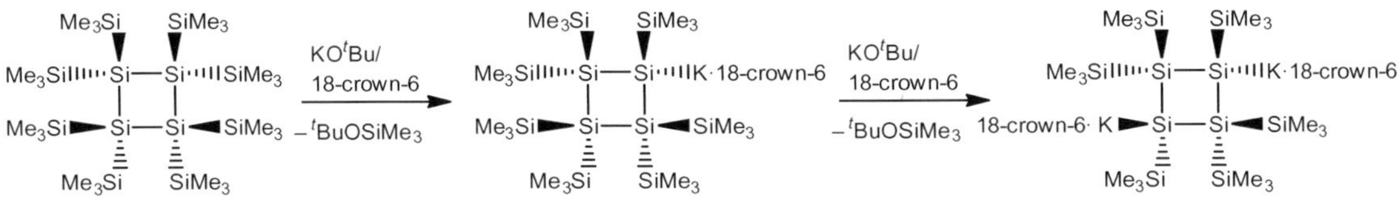

153 154

Scheme 7.62

Mono and dimetalation of octakis(trimethylsilyl)cyclotetrasilane with KO^{t}Bu.

A related compound with *tert*-butyl groups was reported by Kyushin and coworkers, who metalated hexa-*tert*-butyl-1,3-dibromocyclotetrasilane **155** with potassium in benzene to obtain the respective 1,3-dipotassiumcyclotetrasilane **156** (Scheme 7.63).[160]

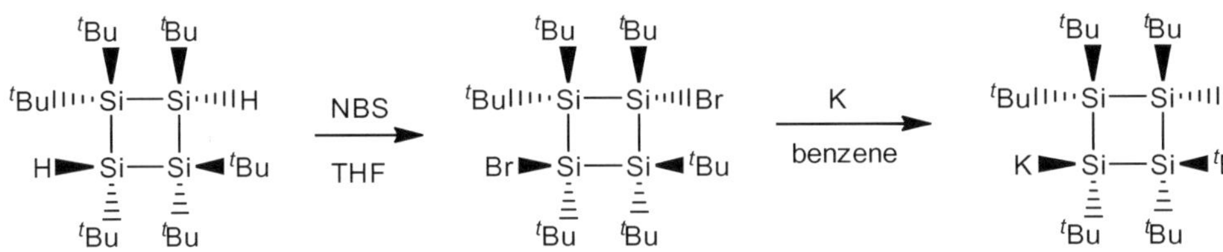

155 156

Scheme 7.63

Direct metalation of a 1,3-dibromocyclotetrasilane with potassium.

Employing the silyl abstraction reaction with KO^{t}Bu 1,1,3,3-tetrakis(trimethylsilyl) hexamethylcyclopentasilane **157** gave the 1,3-*trans*-cyclopentasilandiide **158** (Scheme 7.64).[68,159,161] In a similar way as shown for the acyclic **150a** (Scheme 7.61)

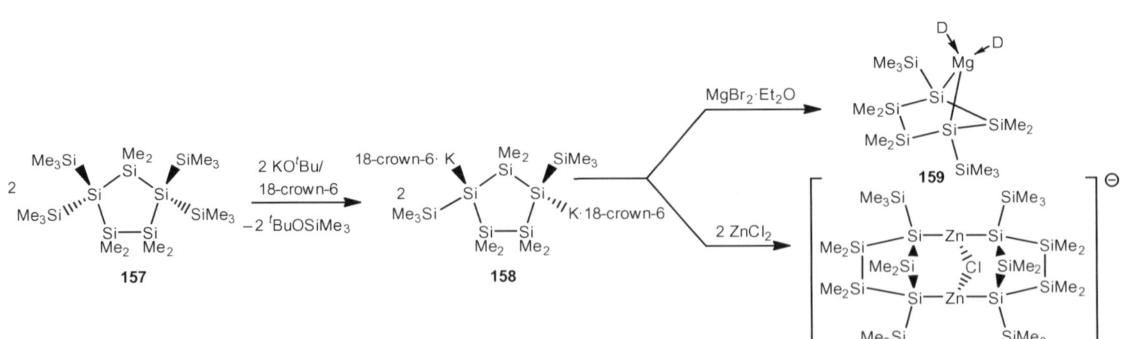

157 158 159 160

Scheme 7.64

1,3-*trans*-Cyclopentasilandiide formation and subsequent transmetalation with $MgBr_2$ and $ZnCl_2$.

reactions with $MgBr_2 \cdot Et_2O$ and $ZnCl_2$ gave the bicyclic disilylmagnesium compound **159**[68] or the dimeric tetracyclic chlorodizincate **160** (Scheme 7.64).[159]

Extension of the silyl abstraction reaction with KO^{t}Bu to a cyclohexasilane gave the 1,3-*trans*-cyclohexasilandiide **162** from the cyclohexasilane **161** (Scheme 7.65).[156]

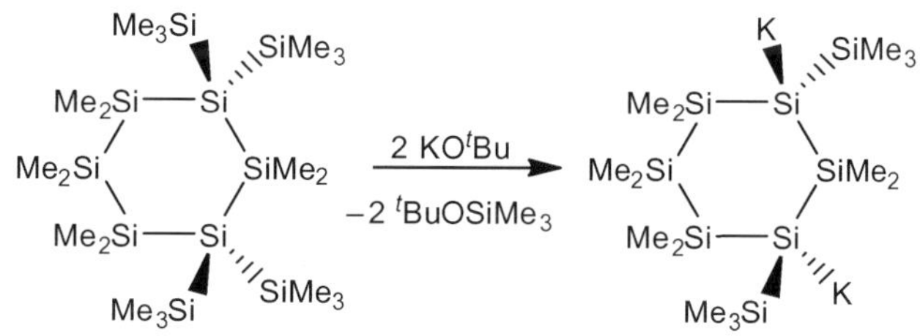

161 **162**

Scheme 7.65
1,3-*trans*-Cyclohexasilandiide formation.

Similar to a previously reported procedure which described a tetrasilyl-1,3-disila-2,4-digermabicyclo[1.1.0]butane-2,4-diide,[162] Lee, Sekiguchi, and coworkers prepared the related calcium tetrasilabicyclo[1.1.0]butane-2,4-diide **164** by reaction of the dipotassium tetrasilacyclobutadiene dianion $[(^tBu_2MeSi)_4Si_4]^{2-} \cdot 2K^+$ (**163**) with calcium iodide (Scheme 7.66).[163]

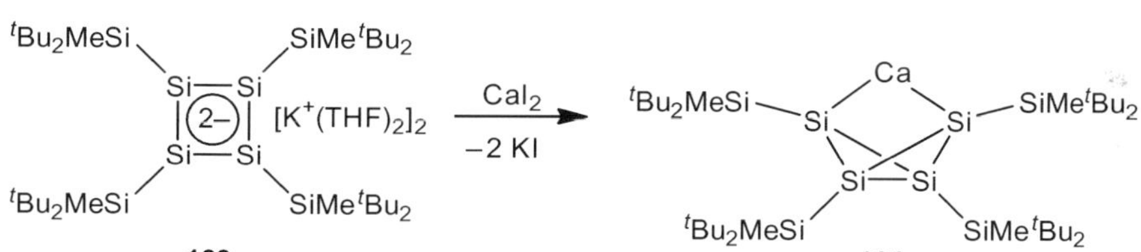

163 **164**

Scheme 7.66
Isomerization of the tetrasilacyclobutadiene dianion upon transmetalation.

7.3.6.4 1,4-Dianions

First examples of 1,4-dilithiotetrasilanes were obtained via cyclotetrasilane cleavage with lithium.[164] As outlined above 1,4-dipotassiumtetrasilane **150b** can obtained by double trimethylsilyl abstraction with KO^{t}Bu from the respective 1,2-bis[tris(trimethylsilyl)silyl] tetramethyldisilane **149b** (Scheme 7.60).[61,135] The analogous 1,4-dilithium compound **165** was reported by Apeloig and coworkers, who could show that for **149b** the reaction with methyllithium does not lead to the cleavage of internal Si–Si bonds (Scheme 7.67).[54]

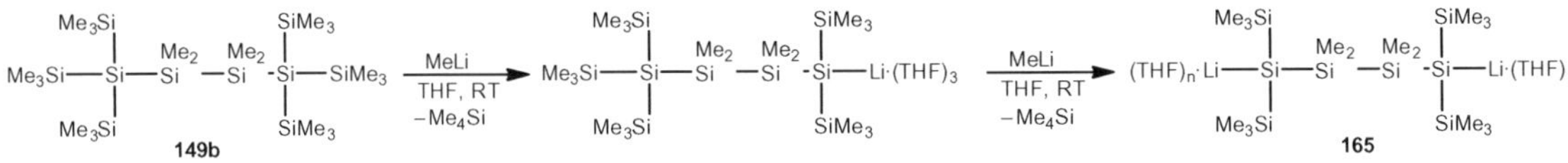

Scheme 7.67
1,4-Dilithiotetrasilane formation by reaction of 1,2-bis[tris(trimethylsilyl)silyl]tetramethyldisilane with methyllithium.

Reaction of 1,1,4,4-tetrakis(trimethylsilyl)octamethylcyclohexasilane **166** with 2 equivalents of KO^{t}Bu led to the clean formation of dipotassium 1,4-*trans*-cyclohexasilandiide (Scheme 7.68) **167**.[76] Transmetalation with $MgBr_2$ gave the 1,4-*cis*-cyclohexasilandiide **168** (Scheme 7.68). Reaction of **167** with 1,2-dichlorotetramethyldisilane gave bicyclo[2.2.2] octasilane **169**, which again can be converted to the 1,4-dianion **170** by reaction with 2 equivalents of KO^{t}Bu (Scheme 7.68).

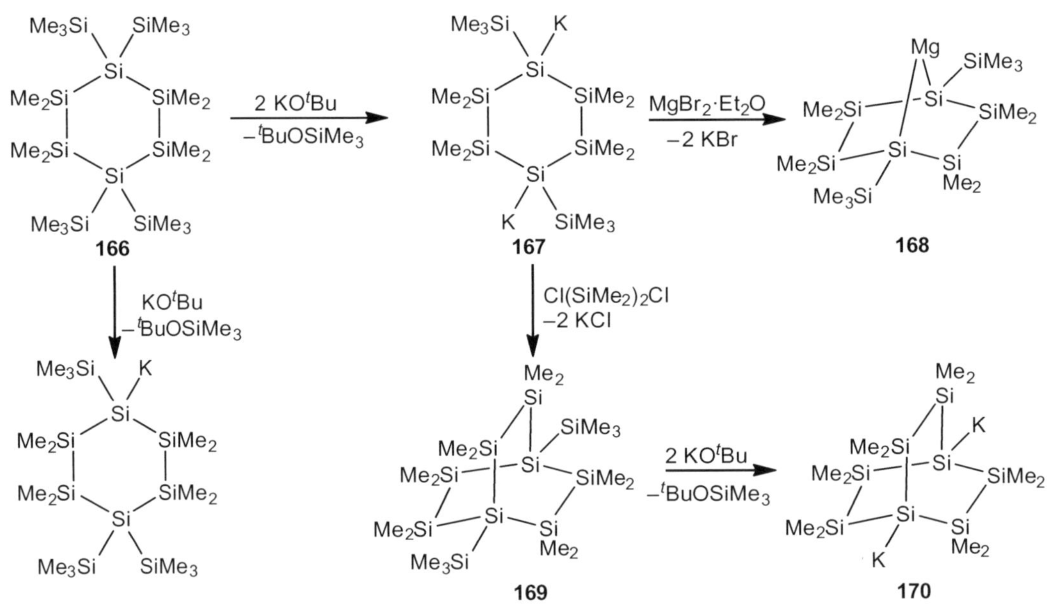

Scheme 7.68
1,4-*trans*-Cyclohexasilandiide formation, subsequent conversion to a bicyclo[2.2.2]octasilane which is again easily metalated at the bridgehead positions.

Using the ferrocene oligosilanyl dianion **171** (formation described below in Scheme 7.72 in more detail) the tetrasilaferrocenophane **172** was prepared.[165] This compound was converted into a 1,4-*trans*-tetrasilanyldiide **173** (Scheme 7.69) by action of two equivalents KO^{t}Bu.[165]

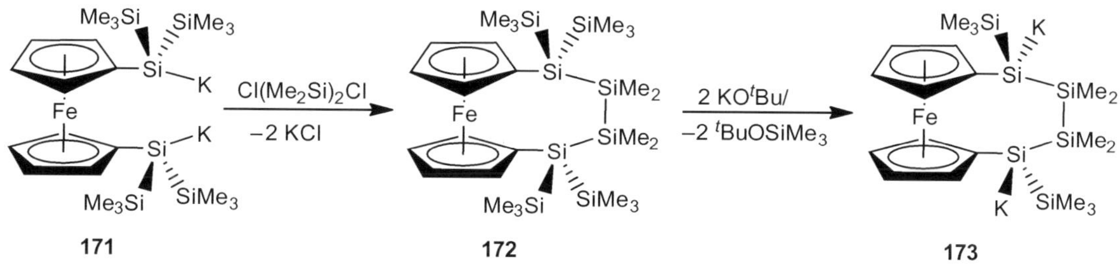

Scheme 7.69
Synthesis of a tetrasilaferrocenophane and subsequent conversion to a 1,4- tetrasilanyldiide.

7.3.6.5 Dianions with longer spacer units

The first synthetic attempts to obtain oligosilanyldianions with longer spacer units employed alkali metal-mediated ring-opening of arylated cyclosilanes such as perphenylated cyclopenta-, or -hexasilanes leading to α,ω-dimetalated perphenylpenta-, and hexasilanes.[166,167] More recent attempts include α,ω-oligosilandiide formation by double trimethylsilyl abstraction with KO^{t}Bu as outlined above (Scheme 7.60).[61,135]

7.3.6.6 Bridged dianions

The dilithium salt of two bridged silaindenes **176** was prepared from the reaction of the dilithium silaindene salt **174** (Scheme 7.73) with 1,2-bis(3-butyl-1-chloro-2-phenyl-1-silaindenyl)ethane (**175**) (Scheme 7.70). It was demonstrated that silole dianions can perform stepwise reactions and that when a quenching reagent has remote electrophilic centers, these can be attacked separately maintaining single charges on the silicon centers.[168]

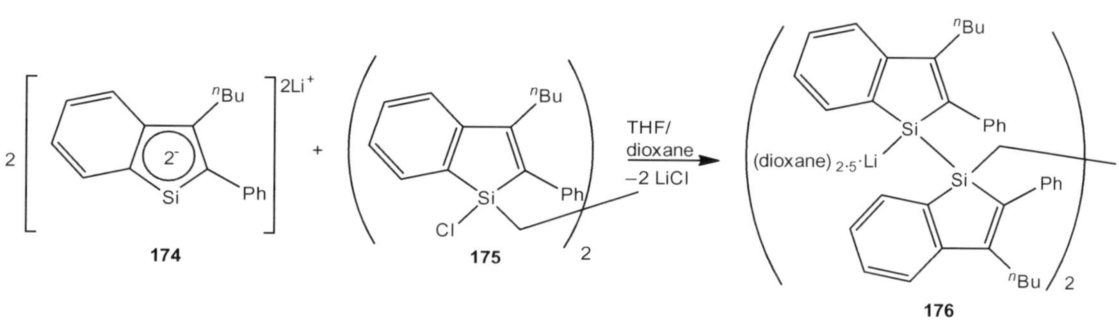

Scheme 7.70
Formation of a bridged disilaindenyl dianion via reaction of a dilithium silaindene with a silaindene containing spacer unit.

Reactions of alkylene bridged tris(trimethylsilyl)silyl units **177** with KO^{t}Bu were studied independently by Marschner's group[169,170] and Xue and coworkers for alkylene spacers ranging from methylene to 1,4-butylene (Scheme 7.71).[171] Partial hydrolysis of the obtained dianions **178** was found to lead to cyclization reactions where trimethylsilane was released and cyclic silyl anions **179** were formed (Scheme 7.71).[169,170]

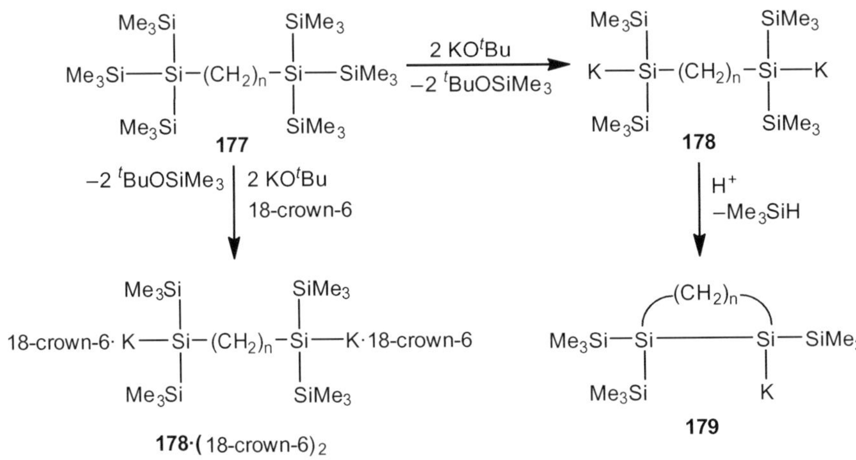

Scheme 7.71
Formation of alkylene bridged oligosilanyl dianions and partial hydrolysis leading to ring closure.

Simple access to the ferrocene bridged potassium oligosilanyldiide **171** is provided by the reaction of 1,1′-bis[tris(trimethylsilyl)silyl]ferrocene **180** with two equivalents of KO^{t}Bu (Scheme 7.72).[165]

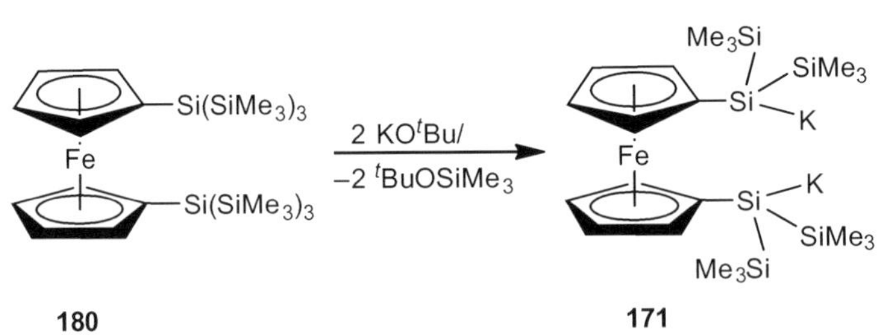

Scheme 7.72
Preparation of a ferrocene bridged oligosilanyl dianion.

7.3.7 Delocalized Silyl Anions

Synthesis and properties of silole anions and dianions have been studied by a number of groups. Boudjouk and coworkers have extended previous work[172] to silaindenyl systems generating dilithium and disodium salts[173] of 3-butyl-2-phenyl-1-silaindene (**174**) from the respective dichlorosilaindene **181** (Scheme 7.73).

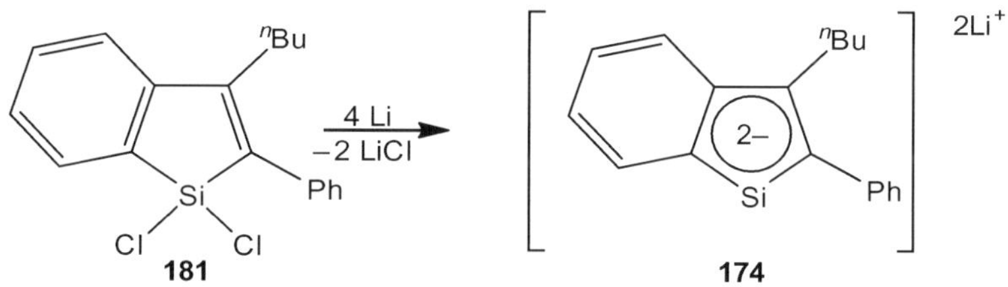

Scheme 7.73
Silaindenyl dianion formation by reduction of the dichlorosilaindene with lithium.

Extending earlier work on silole dianions,[174] Dysard and Tilley demonstrated selective formation of the trimethylsilylated silole anion **183** by reaction of 1,1-bis(trimethylsilyl) tetramethyl-1-silacyclopentadiene (**182**) with benzyl potassium (Scheme 7.74).[175] Transmetalation of **183** with magnesium bromide provided the magnesium silole salt **184** (Scheme 7.74).[175]

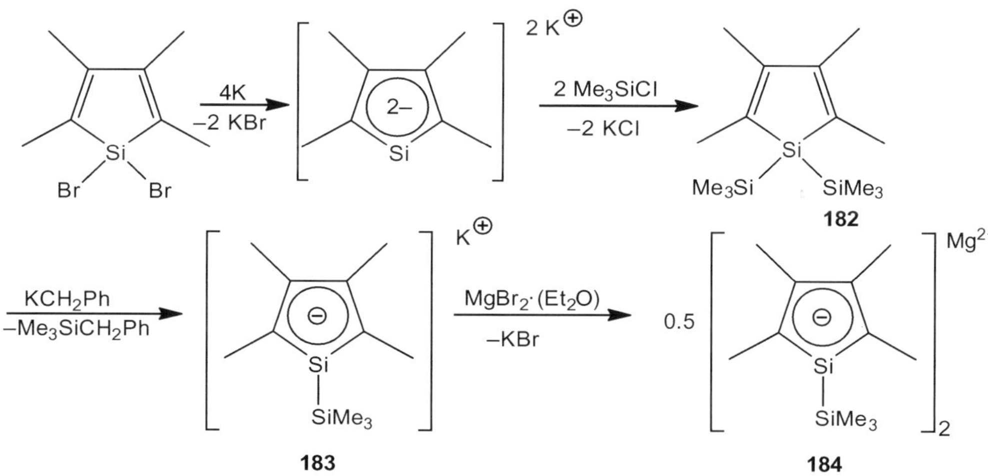

Scheme 7.74
Selective anion formation from bis(trimethylsilyl)tetramethylsilole and its subsequent transmetalation.

Pietschnig et al. obtained silafluorenyl anion **187** by reacting terphenyl trifluorosilane $Tip_2C_6H_3SiF_3$ (**185**) with excess sodium (Scheme 7.75). The formation of **187** was explained by initial C–C bond insertion of a transiently formed fluorosilylene and subsequent reduction of the thus formed fluorinated silole **186**.[176]

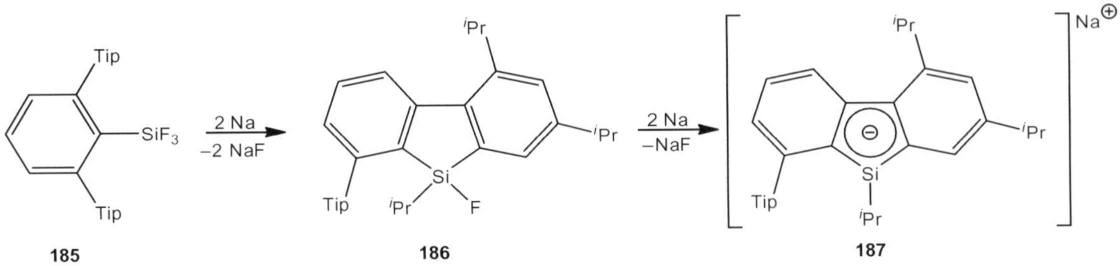

Scheme 7.75
Silafluorenyl anion formation via silylene C–C bond insertion and successive reduction.

1-Silafluorene dianion **189** was synthesized by West and coworkers using potassium reduction of 1,1-dichloro-1-silafluorene **188** in THF (Scheme 7.76). Its crystal structure showed equal C–C bond lengths in the silole ring and C–C bond length alternation in the six-membered benzene rings, suggesting aromatic delocalization of electrons in the silole ring.[177]

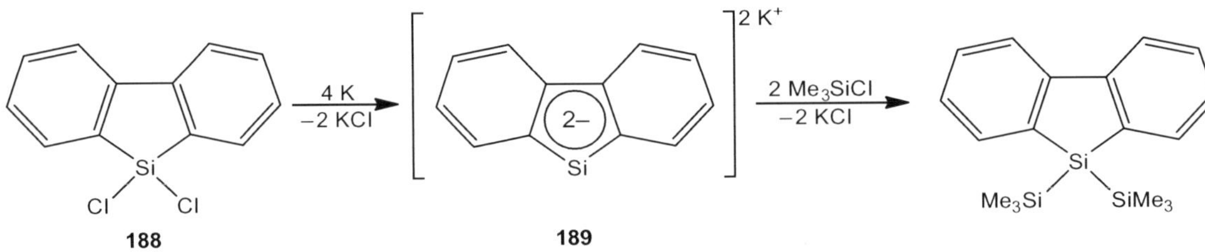

Scheme 7.76
Silafluorenyl dianion formation via reduction of a dichlorosilafluorene with potassium.

The trimethylsilyl substituted tetraphenylsilole anion **191** was obtained from the reaction of the silole dianion $[Ph_4C_4Si]^{2-} \cdot 2Li^+$ (**190**) with trimethylchlorosilane (Scheme 7.77). It can be alkylated to give **192** using methyl iodide, ethyl iodide, and isobutyl bromide (Scheme 7.77).[178]

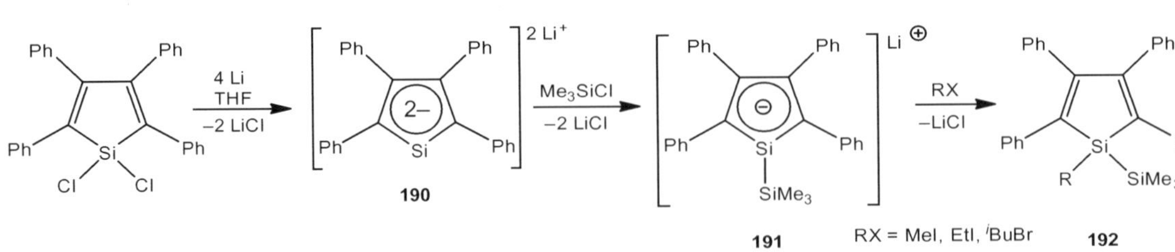

Scheme 7.77
Trimethylsilyltetraphenylsilole anion formation by selective silylation of the dianion.

Lee, Sekiguchi, and coworkers prepared the first examples of heavier group 14 element cyclobutadiene dianions consisting of silicon and germanium atoms. Reduction of dichlorodisiladigermetene **193** with KC_8 in THF yielded tetrakis(di-*tert*-butylmethylsilyl)-1,2-disila-3,4-digermacyclobutadiene dianion **194** (Scheme 7.78).[179] The analogous all silicon species **163**, was obtained similarly by the reduction of the tetrasilylated tetrabromocyclotetrasilane **195** with KC_8 in THF (Scheme 7.78).[179] Both compounds exhibit folded four-membered rings with two η^2-1,3-coordinated potassium cations above and below the ring. Although **163** and **194** have very similar geometries in the solid state, the ^{29}Si NMR spectra suggest that for **194** in solution the negative charges are located at the more electronegative Ge atoms, making the two endocyclic Si atoms doubly bonded. In contrast to this, compound **163** is not displaying typical disilene resonances.[179]

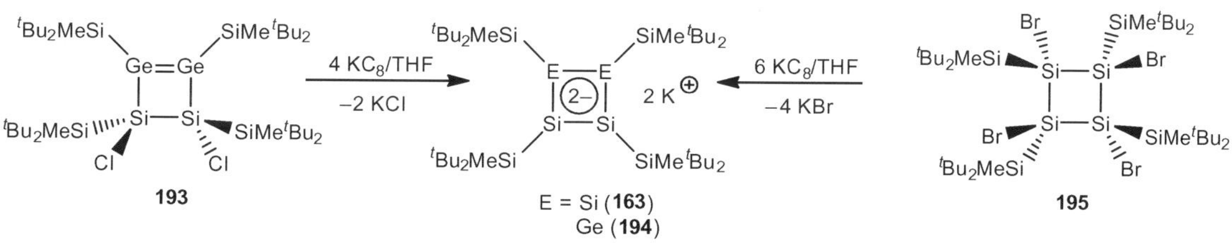

Scheme 7.78

Preparation of heavier group 14 element cyclobutadiene dianions with either four silicon atoms or two silicon and two germanium atoms in the ring.

Also the formation of cyclopentadienyl anions with heavy group 14 elements was addressed by Lee, Sekiguchi, and coworkers. Reduction of 1,2-disila-3-germacyclopenta-2,4-diene **196** with KC_8 in THF occurred with the formation of ^{t}Bu$_2$MeSiK and the potassium salt of the 1,2-disila-3-germacyclopentadienide, which upon treatment with LiBr gave corresponding Li derivative **197** (Scheme 7.79). Single crystal XRD analysis and ^{29}Si NMR spectroscopy suggested aromatic character of **197** which was also confirmed by a nucleus-independent chemical shift (NICS) value of −12.[180,181]

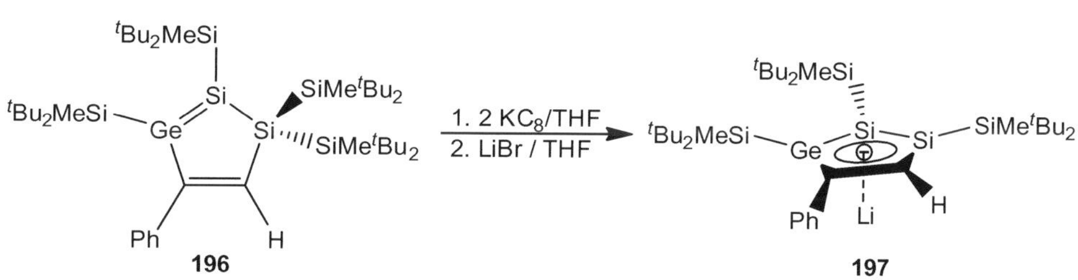

Scheme 7.79

Formation of a lithium 1,2-disila-3-germacyclopentadienide.

In a related study the reduction of the nonconjugated 1,2,3-trisilacyclopenta-1,4-diene **198** with KC_8 proceeded to give a potassium 1,2,3-trisilacyclopentadienide (Scheme 7.80). Again transmetalation with LiBr gave the corresponding aromatic lithium salt **199** (Scheme 7.80).[182]

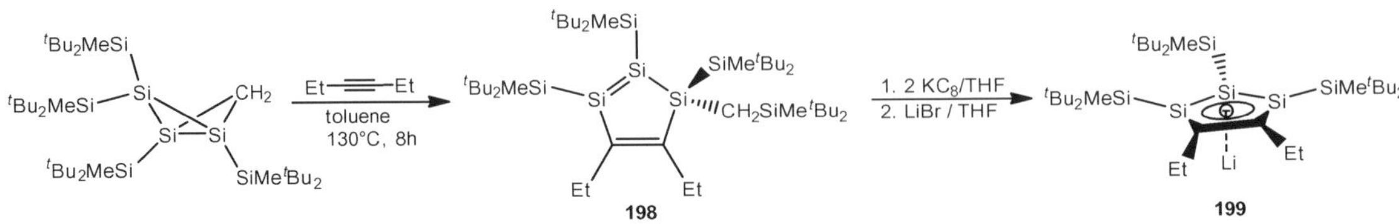

Scheme 7.80
Formation of a lithium 1,2,3-trisilacyclopentadienide.

7.3.8 Sila-Enolates

Early work on the chemistry of sila-enolates by Apeloig and coworkers[183] and in particular by Ishikawa, Ohshita, and coworkers[184–187] has shown that reaction of organolithium compounds with acyl tris(trimethylsilyl)silanes leads to the facile formation of lithium sila-enolates. However, these compounds proved to be very reactive and thus characterization was mainly carried out NMR spectroscopically. A stable potassium silenolate was reported to form by reaction of pivaloyltris(trimethylsilyl)silane **200** with KO^{t}Bu (Scheme 7.81).[188] Crystal structure analysis and theoretical calculations[189] revealed that the silenolate **201** represents the keto form rather than the silenol tautomer with a Si–C bond distance of 1.926(3) Å.[188]

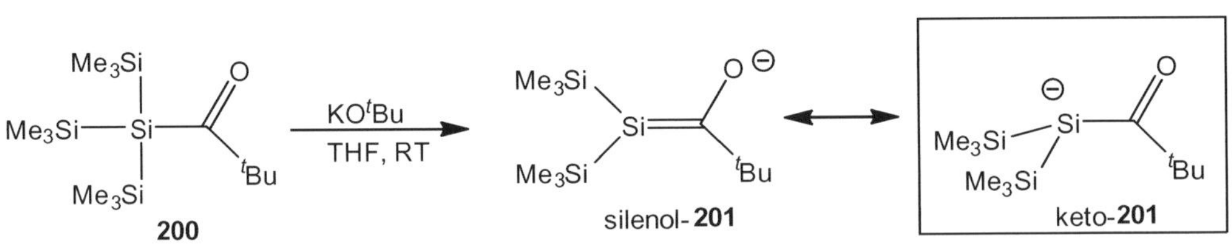

Scheme 7.81
Keto tautomer of a silenolate obtained by reacting pivaloyltris(trimethylsilyl)silane with KO^{t}Bu.

Apeloig and coworkers were able to prepare and characterize the silenol tautomer of silenolate **203** with a Si–C bond distance of 1.822(7) Å. The metalation of the compound was achieved by metal halogen exchange of bromoacylsilane **202** with ^{t}BuMe$_2$SiLi

(Scheme 7.82). As a consequence of this strategy, substitution is the major reaction pathway leading to trisilylacylsilane **204** as main product (Scheme 7.82).[190]

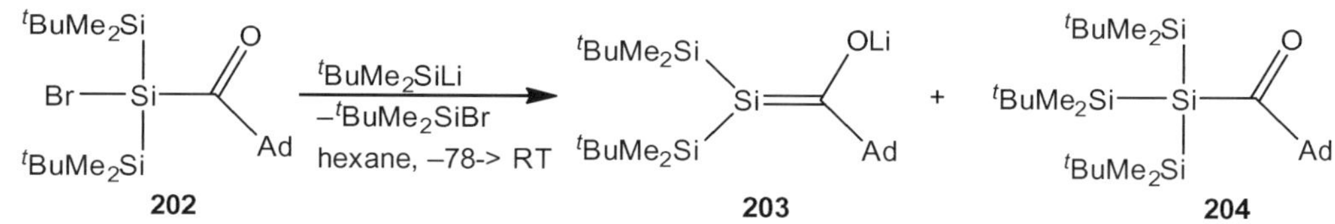

Scheme 7.82
Silenol tautomer of a silenolate obtained by metal halogen exchange of a bromoacylsilane.

In a recent study of exocyclic silenolates Stueger and coworkers showed that the question of whether the keto or silenol form of the structure is preferred strongly depends on solvation effects.[191] Starting from cyclic acylsilanes **205** it was found that the coordination possibilities of the potassium counterion determine the preference for either the keto- (**206**) or silenol (**207**) form (Scheme 7.83). This has not only structural and spectroscopic consequences but leads also to different reactivity. Reaction of the respective silenolates **206** or **207** with ^{i}Pr$_3$SiCl led either to acylsilane **208** or the silenol silyl ether **209** (Scheme 7.83).

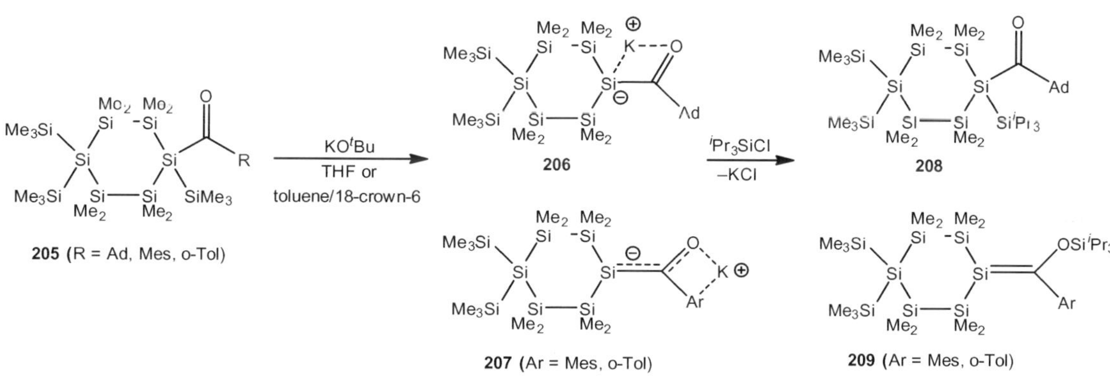

Scheme 7.83
Formation of either the keto or enol form of a silenolate depending on counterion coordination preferences.

Ohshita et al. reported that reactions of bis(acyl)trisilanes **211** (formed by acylation of silenolate **210**) with $(Me_3Si)_3SiLi$ (**18**) in THF gave acylsilenolates **212**, which upon treatment with alkyl halides, were converted to the Si-alkylated bis(acyl)silanes **213** (Scheme 7.84).[192]

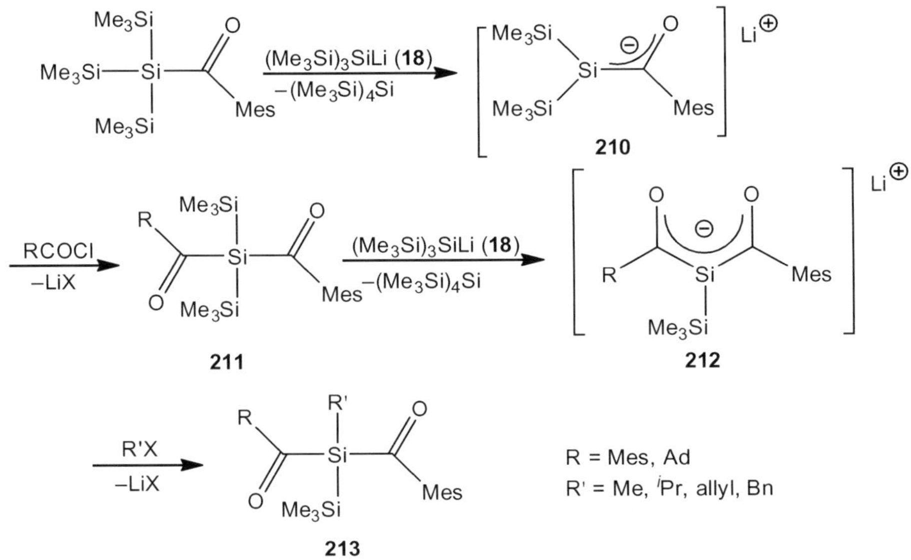

Scheme 7.84
Acylsilenolates by metalation of bis(acyl)trisilanes with $(Me_3Si)_3SiLi$.

7.3.9 Silenyl and Disilenyl Anions

While there now have been a substantial number of disilenides reported (see below) there are only two examples of silenides, both reported by Apeloig and coworkers. These compounds (**215**) were obtained by addition of THF to the aggregates of lithium silenolate with silanides $[(^tBu_2MeSi)_2Si=C(OLi)(1\text{-}Ad)]\cdot(R_3SiLi)$ ($R_3Si = {}^tBu_2MeSi$, $^tBuMe_2Si$) (**214**) (Scheme 7.85). The authors suggest that the silenides **215** were formed by elimination of $^tBu_2MeSiOLi$ to form transient silyne – silylidene intermediates which were trapped by the aggregated R_3SiLi species.[193]

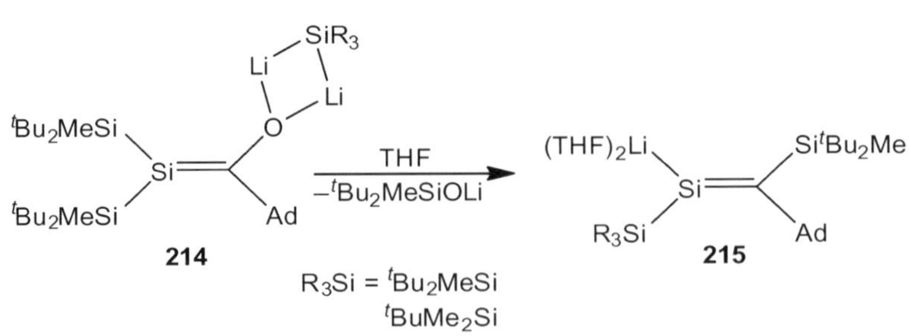

Scheme 7.85
Rearrangement of a lithium silenolate/silanide mixed aggregate to a lithium silenide.

First examples of isolated disilenyl anions,[194] were reported in 2004 independently by Scheschkewitz[195] and Sekiguchi's group.[196] Previous indication of the selective formation of the disilenyl lithium $Tip_2Si = SiTip(Li)$ (**217**) (Tip = 2,4,6-$^iPr_3C_6H_2$) was given by Weidenbruch et al.[197] in their spectacular synthesis of a tetrasilabutadiene. Scheschkewitz developed an improved protocol reducing Tip_2SiCl_2 (**216**) with an excess of lithium powder in DME at elevated temperature to obtain the disilenide **217** in 51% yield (Scheme 7.86).[195]

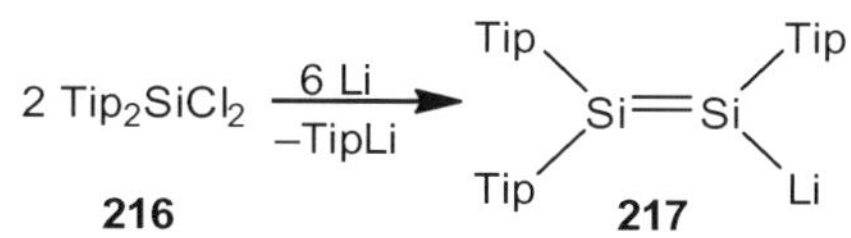

Scheme 7.86
Selective generation of a lithium disilenide from a diaryldichlorosilane.

In the course of these studies the unusual α,ω-dianionic unsaturated oligosilane **219** was prepared. Reaction of $Tip_2Si = SiTip(Li)$ (**217**) with $TipSiCl_3$ yielded a dichlorosilyldisilene **218**, which upon reaction with activated magnesium Mg* in THF gave the magnesium salt of a trisilene-1,3-diide **219** (Scheme 7.87).[198]

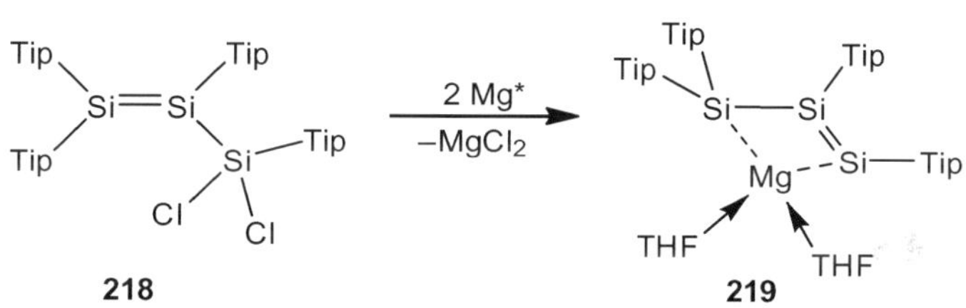

Scheme 7.87
Synthesis of a magnesium salt of a trisilene-1,3-diide.

Sekiguchi's first approach to the disilenyl lithium compound **66** was less straightforward as it involved the initial synthesis of the tetrasilabutadiene **220** which was converted to the disilenide species **66** by reaction with either ^tBuLi[196] or KC_8[199] (Scheme 7.88).

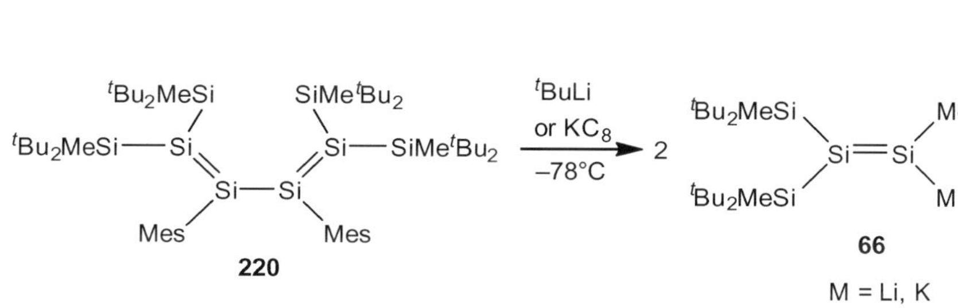

Scheme 7.88
Disilenide formation by Si–Si cleavage of a tetrasilabutadiene.

The metalating step in this reaction is actually the cleavage of a Si–Si single bond and accordingly a similar result could be achieved utilizing a simpler system, namely the reaction of tetrakis(di-*tert*-butylmethylsilyl)disilene (**146**) with alkali metal naphthalenides (Scheme 7.89).[157]

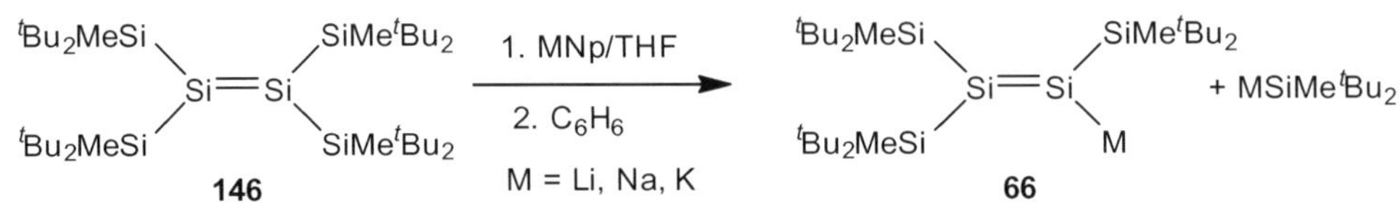

Scheme 7.89
Disilenide formation by Si–Si cleavage of a tetrasilyldisilene.

Following Sekiguchi's seminal synthesis of disilyne **63**,[200] reactivity studies have shown that reaction with methyllithium led to the methyl-substituted disilenide **64** (Scheme 7.29).[92,201] Change of solvent to DME caused the formation of the separate ion pair disilenide **221** (Scheme 7.90).[201]

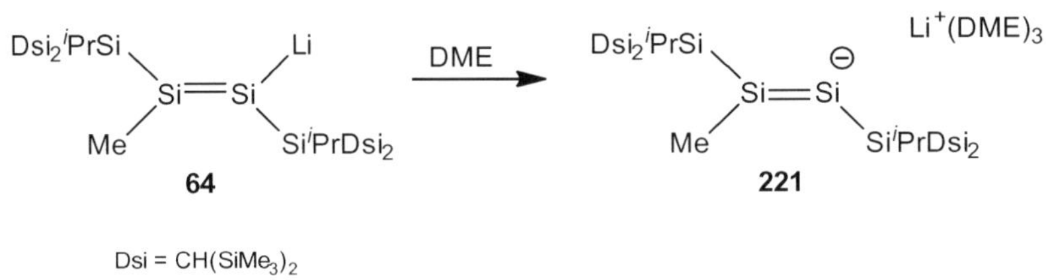

Scheme 7.90
Ion pair separation of a lithium disilenide.

Reaction of disilyne **63** with potassium graphite led to the formation of the isolable disilyne anion radical **222** (Scheme 7.91).[201]

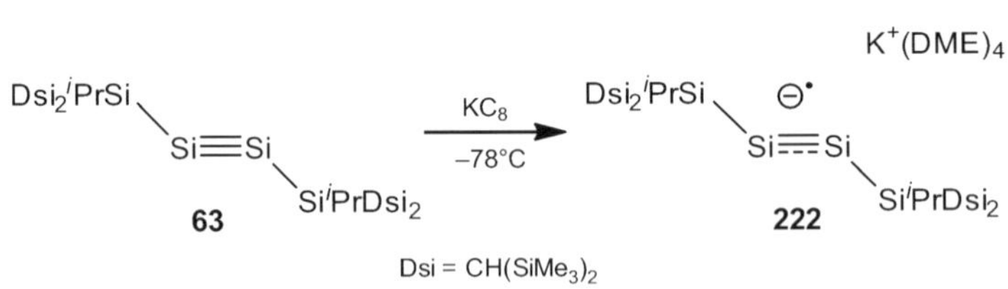

Scheme 7.91
Potassium graphite reduction of a disilyne to an anion radical.

Starting out from Kira's first isolable dialkylated silylene **223**,[202] reaction with ^{t}BuSiCl$_3$ proceeded with insertion in one of the Si–Cl bonds. Reaction of the thus obtained trichlorodisilane **224** with potassium graphite gave the trialkyl-substituted disilenide **225** (Scheme 7.92).[203]

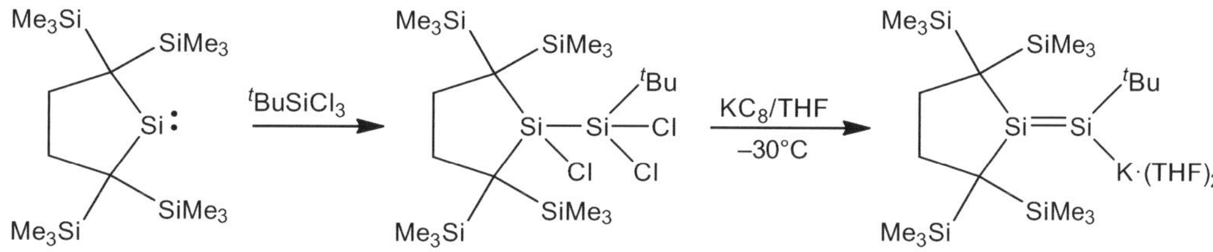

223 224 225

Scheme 7.92
Formation of a potassium disilenide by reductive dechlorination of a trichlorodisilane.

An interesting cyclic disilenide was reported by Lee, Sekiguchi, and coworkers. The addition of 12-crown-4 to the above described lithium 1,2,3-trisilacyclopentadienide **199** (Scheme 7.80) clearly shows the decisive role of the Li^+ counterion in the stabilization of the aromatic derivative. Upon addition of the crown ether the cyclic disilenide **226** was isolated which formed as the result of a 1,2-migration of the silyl substituent and exists as a separated ion pair (Scheme 7.93). Although isomeric to the aromatic 1,2,3-trisilacyclopentadienide **199** the formed disilenide **226** features localized Si = Si and C = C bonds with the anionic charge residing on one of the sp^2-Si atoms.[182]

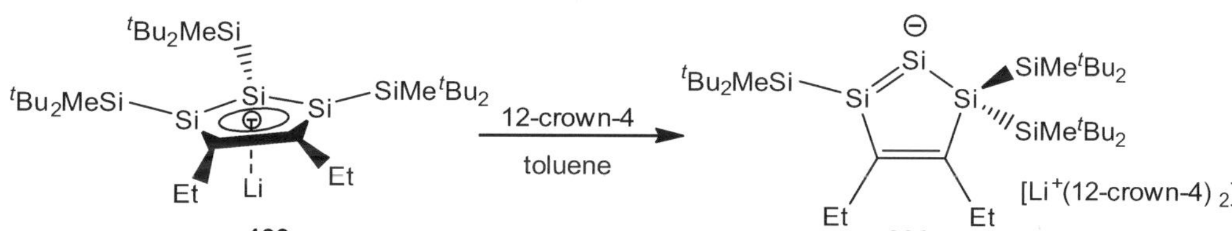

199 226

Scheme 7.93
Cyclic disilenide formation upon charge separation of a lithium 1,2,3-trisilacyclopentadienide.

Sekiguchi and coworkers also reported the unusual reaction of disilenide **66** with adamantanone to give silanide-substituted silene **227** (Scheme 7.94). The choice of the carbonyl compound proved to be crucial as the reaction did not proceed in cases of less hindered ketones. It seems likely that **227** formed by addition of the resulting ^{t}Bu$_2$MeSiOLi to a 1,2-disilaallene intermediate produced in a sila-Peterson-type reaction.[204]

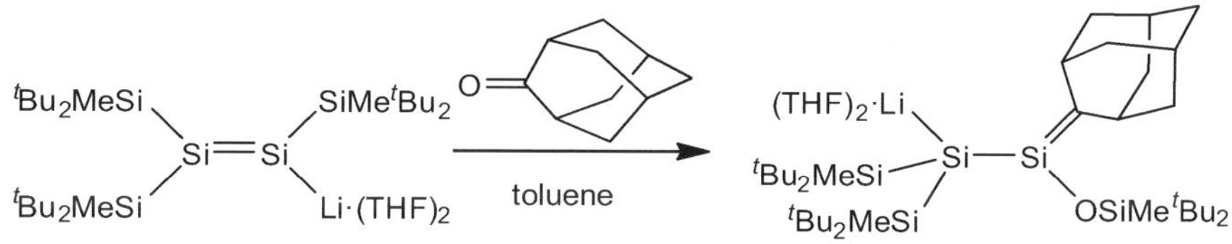

66 227

Scheme 7.94
Formation of a silanide-substituted silene via a sila-Peterson reaction.

7.3.10 Hypercoordinate Anions (Silicates)

Besides the large number of low-coordinate silyl anions there is also the class of hypercoordinated silicon compounds bearing a negative charge. These compounds, usually referred to as silicates, can be envisioned to be adducts of anions such as lithium alkyls or alkoxides with a neutral tetravalent silane. Such species are also known for carbon but in general their nature is transient as transition states in nucleophilic substitution reactions. For silicon the situation can be different as it is more electropositive and in addition has a much larger coordination sphere. The fields of zwitterionic pentacoordinate silicates[205] and pentaorganosilicates[206] have been subjects to recent reviews.

7.3.10.1 Zwitterionic silicates

Recent research on zwitterionic pentasilicates was largely conducted by Tacke and coworkers.[207–211] The general synthetic strategy to obtain these compounds usually employs either a neutral silane with an attached amino group or already a silicate with a covalently connected ammonium group (**228, 230**). The formation of the zwitterionic pentasilicates is then usually achieved by addition of bidentate ligands which engage in covalent bonds with the silicon atom (Scheme 7.95). The negative charge of **229** and **231**, located at the silicon atom, is counter-balanced by the attached ammonium group.

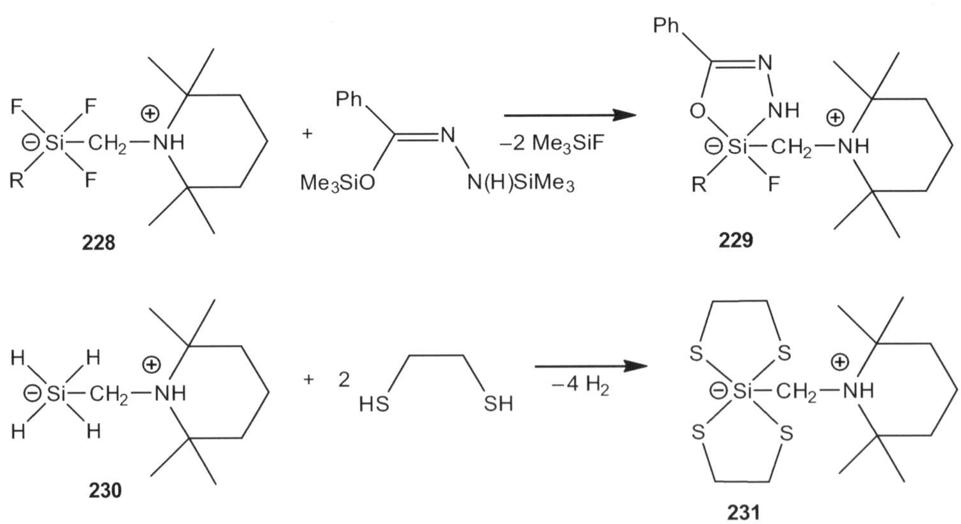

Scheme 7.95
Typical examples for the formation of zwitterionic pentasilicates.

The choice of recently used bidentate dianionic ligands includes hydrazonates,[207] 1,2-ethanedithiolates and 1,2-benzodithiolates,[208,210] α-amino acids and α-hydroxycarboxylic acids,[209] 2-mercaptoacetic acid,[210] and 1,2-diolates.[211] This chemistry is, however, not restricted to zwitterionic examples. Reactions of α-hydroxycarboxylic acids with $Si(OMe)_4$ in the presence of morpholine were found to give octahedrally coordinated silicates **232** (Scheme 7.96).[212]

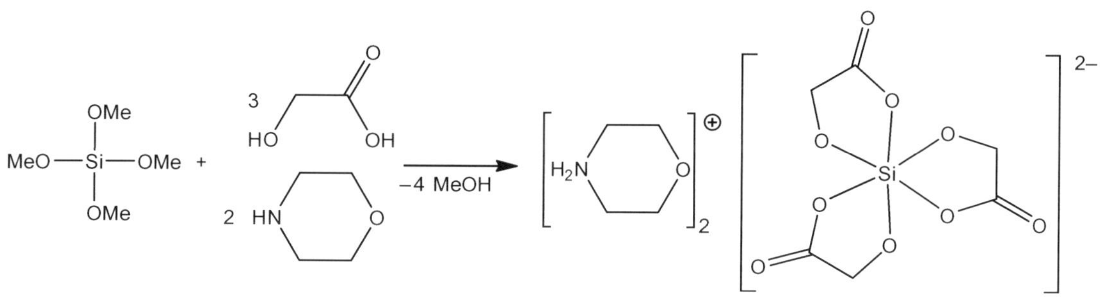

232

Scheme 7.96
Formation of a hexacoordinated silicate by reaction with a bidentate dianionic ligand.

Attempts to prepare a tris[citrato(2−)]silicate dianion by reaction of $Si(OMe)_4$ with citric acid led to the formation of the hexacoordinate silicate dianion **233** with two tridentate citrato(3−) ligands instead (Scheme 7.97).[213]

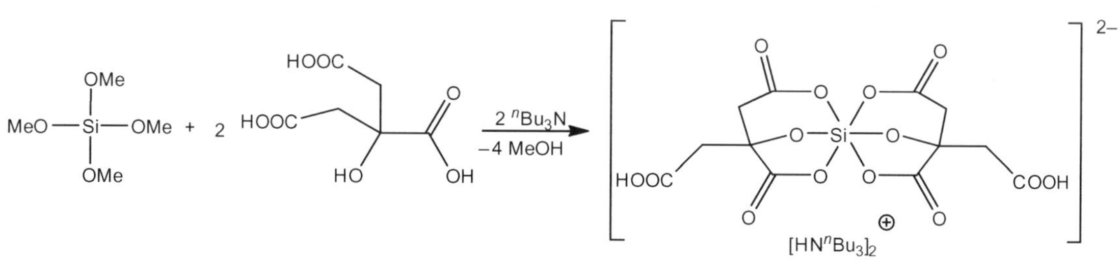

233

Scheme 7.97
Formation of a hexacoordinate silicate dianion with tridentate ligands.

Reaction of $(MeO)_3SiCH_2NH_2$ (**234**) and related compounds with (*R,R*)-tartaric acid gave the optically active zwitterionic λ^5Si,λ^5Si'-disilicate **235** containing two pentacoordinate (formally negatively charged) silicon atoms and two ammonium units (Scheme 7.98).[214]

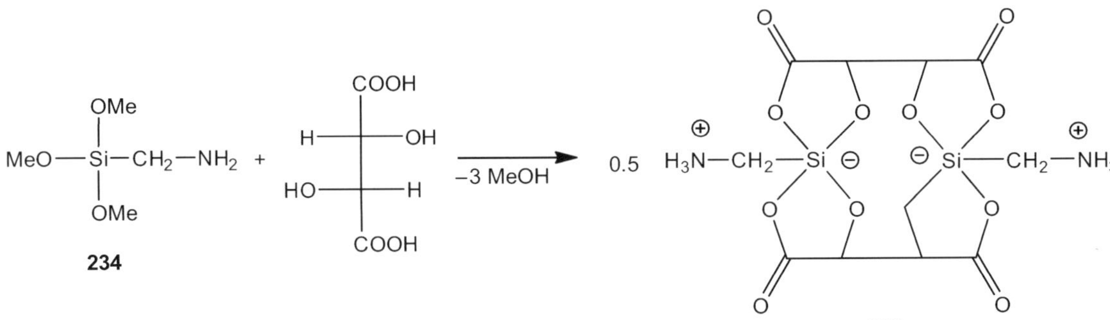

234

235

Scheme 7.98

Formation of an optically active zwitterionic disilicate.

Related work was done with carbohydrates. Five-coordinate phenylsilicates were formed in reactions of $PhSi(OMe)_3$ with some monosaccharides in methanol in the presence of a stoichiometric amount of [K(18-crown-6)]OR (R = Me, ^{t}Bu).[215] For the cases of D-fructose and D-ribose only a single five-coordinate species in methanol was detected by ^{29}Si NMR spectroscopy. For the fructose example the silicon central atom in the solid state structure of the formed silicate is part of two chelate rings, with the ligands being β-D-fructofuranose-O^2,O^3 dianions.[215]

Recently silicate formation of some spirosilanes was reported to serve as fluoride sensors. The reactions between compound **236** and different fluoride sources result in a silicate (**237**) with structural alterations which can be monitored and quantified by UV and fluorescence spectroscopy (Scheme 7.99). Maximum fluorescence emission is shifted from 290 nm for the neutral silane **236** to 311 nm for the fluorosilicate **237**.[216]

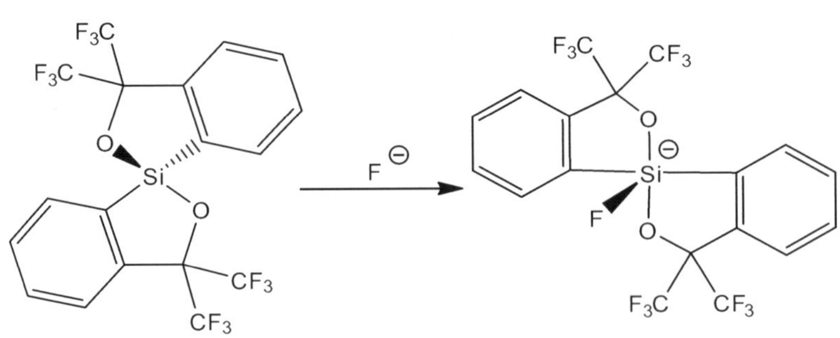

236

237

Scheme 7.99

Use of a spirocyclic silane acting as fluoride sensor via formation of a fluorosilicate.

7.3.10.2 Pentaorganosilicates

Stable silicates with five carbon substituents are extremely rare and for a long time were thought to be not stable at ambient temperature. However, starting from spirocylic silanes **238**, addition of methyllithium was found to give stable pentacoordinate anions **239** (Scheme 7.100). These possess trigonal bipyramidal geometry and undergo configurational isomerization in solution, which was studied in some detail.[217–219]

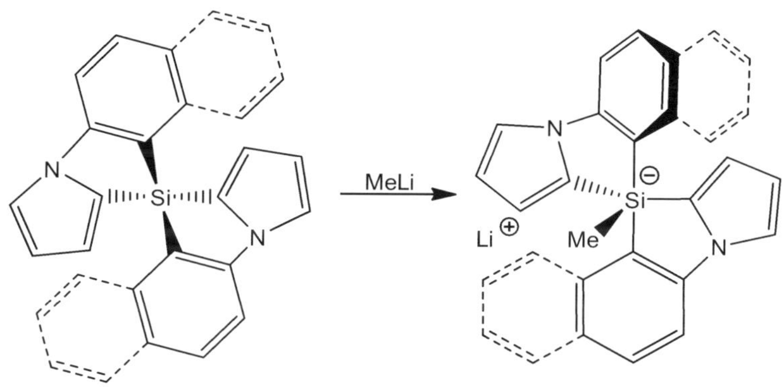

238 **239**

Scheme 7.100
Synthesis of stable pentaorganosilicates.

7.3.10.3 Hydridosilicates

First examples of hydridosilicates were reported by Corriu and coworkers, who studied the reactions of trialkoxysilanes $(RO)_3SiH$ with various nucleophiles.[220] In the course of the reaction of 10-phenyl-10-germa-9-silatriptycene with phenyllithium in THF/HMPA Bickelhaupt's group showed that the formation of 9,10-diphenyl-10-germa-9-silatriptycene proceeded via an intermediate hydridosilicate, which was identified by ^{29}Si NMR spectroscopy. Modeling this reaction, the attack of phenyllithium on diphenylsilane revealed the formation of the stable hydridosilicate **240Li** (Scheme 7.101).[221]

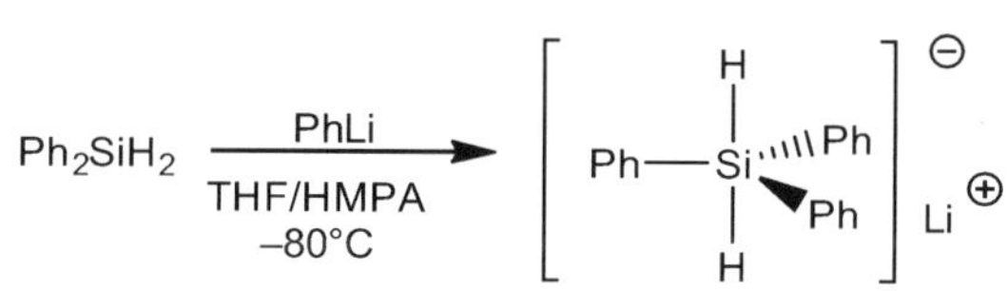

240Li

Scheme 7.101
Formation of a lithium dihydridosilicate by the reaction of Ph_2SiH_2 and PhLi in THF/HMPA.

Later Steed, McGrady and coworkers showed that reaction of Ph_2SiH_2 with potassium metal in DME in the presence of 18-crown-6 leads to the formation the same silicate **240K** with a different counterion $[K(18\text{-crown-6})]^+[Ph_3SiH_2]^-$ (Scheme 7.102),[222] which can also be prepared by reaction of triphenylsilane with KH (Scheme 7.102).[223] The crown ether complexing the alkali metal countercation, provides greater solubility of the complex and stabilization. The structure of **240K** in the solid state was studied by single crystal XRD analysis. The silicate acts as a good hydride transfer reagent to a variety of substrates. In its reaction with $(^i PrO)_3SiH$ the formation of $[(^i PrO)_3SiH_2]^-$ was detected.[223] An example of a cationic calcium hydride cluster showed that the nature of the counterion of $[Ph_3SiH_2]^-$ can also be substantially more complex.[224]

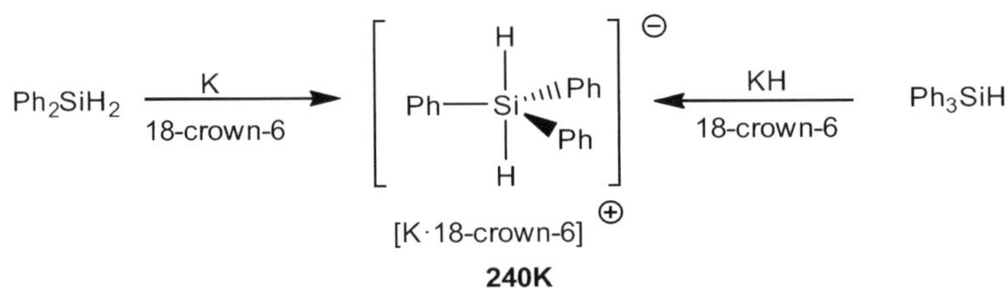

Scheme 7.102
Formation of a potassium dihydridosilicate using either $PhSiH_3$/KH or Ph_2SiH_2/K and 18-crown-6.

A transient involvement of hydridosilicates **242** and **243** was postulated for the reactions of 1,4-dilithio-1,3-butadienes **241** with $PhSiH_3$. Depending on the solvent (either Et_2O or THF) either siloles **244** or 3-silacyclopentenes **245** were obtained (Scheme 7.103).[225]

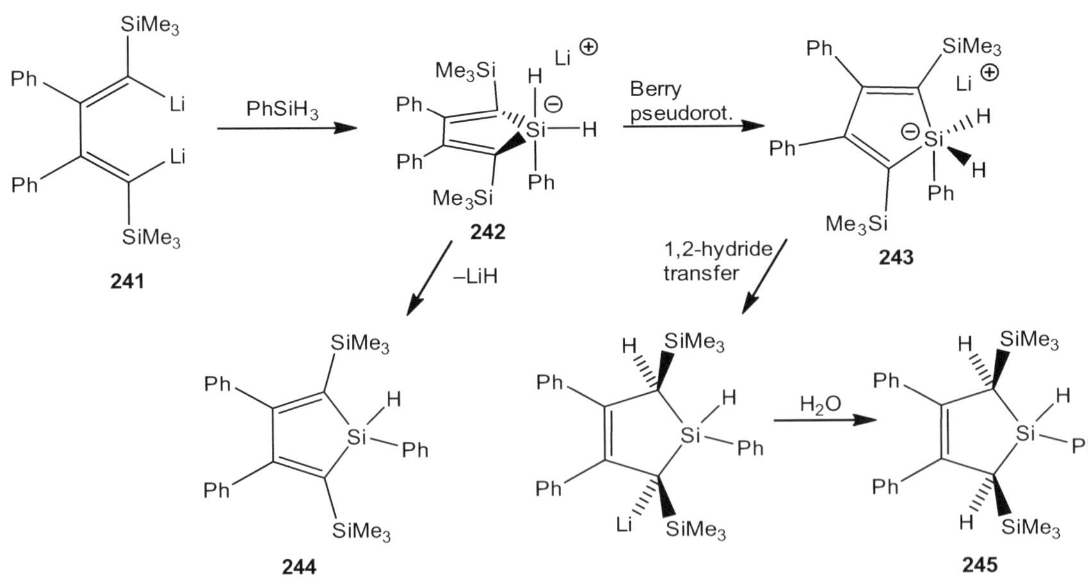

Scheme 7.103
Formation of either siloles or 3-silacyclopentenes involving hydridosilicate intermediates.

7.4 Conclusion and Outlook

The chemistry of simple silyl anions has become a mature field. Reliable synthetic methods for their preparation are available and have been adopted by a wide audience of chemists. Since the year 2000, therefore, many instances of more sophisticated silanides have been reported. In particular functionalized and structurally unusual compounds were studied by a number of research groups. Stable examples of silylenoids, compounds which display ambiphilic properties and are likely intermediates in the Wurtz type synthesis of polysilanes, could be prepared and studied. Also, several silanides containing one or more Si–Si double bonds are known now. Despite the typical problems associated with the characterization of extremely sensitive compounds, structural data about silanides is now very abundant and also their spectroscopic properties are well understood.

As silyl anions are typically very reactive, they are attractive synthetic building blocks to create bonds of silicon to electrophilic centers. This makes it very likely that their synthesis and study will remain a central part of organosilicon chemistry. Silanides with novel and unusual properties and features will very likely emerge both from research targeted at this field but also as a by-product from studies concerned with the chemistry of structurally challenging and unusual organosilicon chemistry in general.

References

1. Tamao, K.; Kawachi, A. Silyl Anions. *Adv. Organomet. Chem.* **1995,** *38*, 1–58.
2. Lickiss, P. D.; Smith, C. M. Silicon Derivatives of the Metals of Groups 1 and 2. *Coord. Chem. Rev.* **1995,** *145*, 75–124.
3. Lerner, H.-W. Silicon Derivatives of Group 1, 2, 11 and 12 Elements. *Coord. Chem. Rev.* **2005,** *249*, 781–798.
4. Sekiguchi, A.; Lee, V. Ya; Nanjo, M. Lithiosilanes and Their Application to the Synthesis of Polysilane Dendrimers. *Coord. Chem. Rev.* **2000,** *210*, 11–45.
5. Marschner, C. Preparation and Reactions of Polysilanyl Anions and Dianions. *Organometallics* **2006,** *25*, 2110–2125.
6. Präsang, C.; Scheschkewitz, D. Silyl Anions. In *Functional Molecular Silicon Compounds II;* Scheschkewitz, D., Ed.; Springer: Heidelberg, 2013; pp 1–47.
7. Lee, V. Ya; Sekiguchi, A. *Heavy Analogs of Carbanions: Si-, Ge-, Sn- and Pb-Centered Anions.* In *Organometallic Compounds of Low-Coordinate Si, Ge, Sn and Pb;* John Wiley & Sons, Ltd., 201089–138.
8. Carey, F. A.; Sundberg, R. J. *Organometallic Compounds of Group I and II Metals. Advanced Organic Chemistry Part B: Reactions and Synthesis;* Springer: Boston, MA, 1990.
9. Gilman, H.; Lichtenwalter, G. D. Cleavage of Symmetrically Substituted Disilanes by Lithium in Tetrahydrofuran. *J. Am. Chem. Soc.* **1958,** *80*, 608–611.
10. George, M. V.; Peterson, D. J.; Gilman, H. Preparation of Silyl- and Germylmetallic Compounds. *J. Am. Chem. Soc.* **1960,** *82*, 403–406.
11. Fleming, I.; Roberts, R. S.; Smith, S. C. The Preparation and Analysis of the Phenyldimethylsilyllithium Reagent and Its Reaction With Silyl Enol Ethers. *J. Chem. Soc., Perkin Trans.* **1998,** *1*, 1209–1214.
12. Chen, H.-M.; Oliver, J. P. Silylcupration of Acetylenes. *J. Organomet. Chem.* **1986,** *316*, 255–260.

13. Woo, H. G.; Freeman, W. P.; Tilley, T. D. Synthesis, Characterization, and Reactivity of Triphenylsilyl, Triphenylgermyl, and Triphenylstannyl Derivatives of Zirconium and Hafnium. *Organometallics* **1992,** *11*, 2198–2205.
14. Lee, T. W.; Corey, E. J. (2-Methoxyphenyl)dimethylsilyl Lithium and Cuprate Reagents Offer Unique Advantages in Multistep Synthesis. *Org. Lett.* **2001,** *3*, 3337–3339.
15. Yamamoto, Y.; Matsubara, H.; Murakami, K.; Yorimitsu, H.; Osuka, A. Activator-Free Palladium-Catalyzed Silylation of Aryl Chlorides With Silylsilatranes. *Chem. Asian. J.* **2015,** *10*, 219–224.
16. Still, W. C. Conjugate Addition of Trimethysilyllithium. A Preparation of 3-Silyl Ketones. *J. Org. Chem.* **1976,** *41*, 3063–3064.
17. Razuvaev, G. A.; Vyazankin, N. S. Organosilicon and Organogermanium Derivatives With Silicon-Metal and Germanium-Metal Bonds. *Pure. Appl. Chem.* **1969,** *19*, 353–374.
18. Nanjo, M.; Sekiguchi, A.; Sakurai, H. Lithiosilanes With Si-Si Bonds: Synthesis and ^{6}Li, ^{7}Li, and ^{29}Si NMR Studies. *Bull. Chem. Soc. Jpn.* **1998,** *71*, 741–747.
19. Hörnig, J.; Auer, D.; Strohmann, C. *Syntheses of Silyllithium Reagents Starting From Tetraorganosilanes.* In *Organosilicon Chemistry V;* Wiley-VCH, 2003150–154.
20. Strohmann, C.; Schildbach, D.; Auer, D. Selective Si-C Bond Cleavage as Synthetic Entry to a Functionalized Lithiosilane. *J. Am. Chem. Soc.* **2005,** *127*, 7968–7969.
21. Hengge, E.; Holtschmidt, N. Darstellung Unsymmetrischer Disilane Mittels Trimethylsilyllithium. *J. Organomet. Chem.* **1968,** *12*, P5–P7.
22. Wiberg, N.; Amelunxen, K.; Lerner, H.-W.; Schuster, H.; Nöth, H.; Krossing, I., et al. Donor-Free and Donor-Containing Supersilyl Alkali Metal $^{t}Bu_3SiM$: Synthesis, Characterization, Structure. *J. Organomet. Chem.* **1997,** *542*, 1–18.
23. Lerner, H.-W.; Scholz, S.; Bolte, M.; Wiberg, N.; Nöth, H.; Krossing, I. Synthesis and Structures of Alkaline-Earth Metal Supersilanides: tBu$_3$SiMX and tBu$_3$Si-M-SitBu$_3$ (M = Be, Mg; X = Cl, Br). *Eur. J. Inorg. Chem.* **2003,** 666–670.
24. Crump, R. A. N. C.; Fleming, I.; Hill, J. H. M.; Parker, D.; Reddy, N. L.; Waterson, D. The Diastereoselectivity of Electrophilic Attack on Trigonal Carbon Adjacent to A Stereogenic Centre: Diastereoselective Alkylation and Protonation of Open-Chain Enolates Having a Stereogenic Centre Carrying a Silyl Group at the β Position. *J. Chem. Soc., Perkin Trans.* **1992,** *1*, 3277–3294.
25. Oestreich, M.; Weiner, B. Copper-Catalyzed Conjugate Addition of a Bis(triorganosilyl) Zinc and a Methyl (triorganosilyl) Magnesium. *Synlett* **2004,** 2139–2142.
26. Uchiyama, M.; Kobayashi, Y.; Furuyama, T.; Nakamura, S.; Kajihara, Y.; Miyoshi, T., et al. Generation and Suppression of 3-/4-Functionalized Benzynes Using Zinc Ate Base (TMP-Zn-ate): New Approaches to Multisubstituted Benzenes. *J. Am. Chem. Soc.* **2008,** *130*, 472–480.
27. Yonehara, M.; Nakamura, S.; Muranaka, A.; Uchiyama, M. Regioselective Silylzincation of Phenylallene Derivatives. *Chem. Asian. J.* **2010,** *5*, 452–455.
28. Tang, Z.; Mathieu, B.; Tinant, B.; Dive, G.; Ghosez, L. The Search for Tolerant Lewis Acid Catalysts. Part 2: Enantiopure Cycloalkyldialkylsilyl Triflimide Catalysts. *Tetrahedron* **2007,** *63*, 8449–8462.
29. Ishikawa, M.; Hatano, T.; Hasegawa, Y.; Horio, T.; Kunai, A.; Miyai, A., et al. Polymeric Organosilicon Systems. 12. Synthesis and Anionic Ring-Opening Polymerization of 1,2,5,6-Tetrasilacycloocta-3,7-diynes. *Organometallics* **1992,** *11*, 1604–1618.
30. Buncel, E.; Venkatachalam, T. K. Carbanion Mechanisms. XXI. Solution Acidity of Triphenylsilane. *J. Organomet. Chem.* **2000,** *604*, 208–210.
31. Lerner, H.-W.; Scholz, S.; Bolte, M. The Sodium Silanide tBu$_2$PhSiNa: Synthesis, Properties, Structure Analysis—A Synthetic Pathway to Introduce the tBu$_2$PhSi Ligand. *Z. Anorg. Allg. Chem.* **2001,** *627*, 1638–1642.
32. Lerner, H.-W.; Scholz, S.; Bolte, M.; Wagner, M. Alkali metal di-*tert*-butylphenylsilanides—tBu$_2$PhSiM (M = Li, Na, K)—Syntheses, Structures, and Reactivity. *Z. Anorg. Allg. Chem.* **2004,** *630*, 443–451.
33. Sen, S.; Purushotham, M.; Qi, Y.; Sieburth, S. M. Efficient Asymmetric Synthesis of Silanediol Precursors From 1,5-Dihydrosiloles. *Org. Lett.* **2007,** *9*, 4963–4965.

34. Leich, V.; Lamberts, K.; Spaniol, T. P.; Okuda, J. Crown Ether Adducts of Light Alkali Metal Triphenylsilyls: Synthesis, Structure and Hydrosilylation Catalysis. *Dalton. Trans.* **2014,** *43*, 14315–14321.
35. Okuda, J.; Spaniol, T. P.; Leich, V.; Maron, L. Hydrosilylation Catalysis by Earth Alkaline Metal Silyl: Synthesis, Characterization, and Reactivity of Bis(triphenylsilyl)calcium. *Chem. Commun.* **2014,** *50*, 2311–2314.
36. Däschlein, C.; Strohmann, C. The Competition Between Si–Si and Si–C Cleavage in Functionalised Oligosilanes: Their Reactivity With Elemental Lithium. *Dalton. Trans.* **2010,** *39*, 2062–2069.
37. Lambert, J. B.; Urdaneta-Perez, M. Diastereotopic Groups in Silyllithium and Germyllithium Compounds. Slow Inversion about Silicon and Germanium. *J. Am. Chem. Soc.* **1978,** *100*, 157–162.
38. Fischer, R.; Marschner, C. *Experimental Determination of the Inversion Barriers of Oligosilyl Anions.* In *Organosilicon Chemistry V;* Wiley-VCH, 2003190–194.
39. Flock, M.; Marschner, C. Silylanions: Inversion Barriers and NMR Chemical Shifts. *Chem. Eur. J.* **2002,** *8*, 1024–1030.
40. Sommer, L. H.; Mason, R. Optically Active R_3Si^*Li From Lithium Metal Cleavage of an Optically Active Disilane. *J. Am. Chem. Soc.* **1965,** *87*, 1619–1620.
41. Colomer, E.; Corriu, R. On the Behaviour of an Optically Active Cobalt—Silicon Bond. Evidence for the Formation of Analogues of Silyl-Grignard Reagents. *J. Organomet. Chem.* **1977,** *133*, 159–168.
42. Omote, M.; Tokita, T.; Shimizu, Y.; Imae, I.; Shirakawa, E.; Kawakami, Y. Stereospecific Formation of Optically Active Trialkylsilyllithiums and Their Configurational Stability. *J. Organomet. Chem.* **2000,** *611*, 20–25.
43. Oestreich, M.; Auer, G.; Keller, M. On the Mechanism of the Reductive Metallation of Asymmetrically Substituted Silyl Chlorides. *Eur. J. Org. Chem.* **2005,** 184–195.
44. Strohmann, C.; Hörnig, J.; Auer, D. Synthesis of a Highly Enantiomerically Enriched Silyllithium Compound. *Chem. Commun.* **2002,** 766–767.
45. Strohmann, C.; Däschlein, C.; Kellert, M.; Auer, D. A Highly Enantiomerically Enriched Lithiosilane by Selective Cleavage of a Silicon–Phenyl Bond With Lithium. *Angew. Chem. Int. Ed.* **2007,** *46*, 4780–4782.
46. Däschlein, C.; Bauer, S. O.; Strohmann, C. Mechanistic Insights into the Reaction of Enantiomerically Pure Lithiosilanes and Electrophiles: Understanding the Differences between Aryl and Alkyl Halides. *Eur. J. Inorg. Chem.* **2011,** 1454–1465.
47. Strohmann, C.; Däschlein, C.; Auer, D. Crystal Structures of the Chiral Lithiosilanes [(Li*s*)-$PhMe_2SiLi \cdot THF \cdot (-)$-Sparteine] and [$Ph_2(NEt_2)SiLi \cdot (-)$-Sparteine]. *J. Am. Chem. Soc.* **2006,** *128*, 704–705.
48. Däschlein, C.; Strohmann, C. Structural Studies on (–)-Sparteine-Coordinated Lithiosilanes. *Eur. J. Inorg. Chem.* **2009,** 43–52.
49. Nanjo, M.; Maehara, M.; Ushida, Y.; Awamura, Y.; Mochida, K. Convenient Access to Optically Active Silyl- and Germyllithiums: Synthesis, Absolute Structure, and Reactivity. *Tetrahedron Lett.* **2005,** *46*, 8945–8947.
50. Gilman, H.; Holmes, J. M.; Smith, C. L. Branched-Chain Methylated Polysilanes Containing a Silyl-lithium Group. *Chem. Ind.* **1965,** 848–849.
51. Gilman, H.; Harrell, R. L., Jr. Hexakis(trimethylsilyl)disilane: A Highly Branched and Symmetrical Organopolysilane. *J. Organomet. Chem.* **1967,** *9*, 67–76.
52. Gutekunst, G.; Brook, A. Tris(trimethylsilyl)silyllithium · 3 THF: A Stable Crystalline Silyllithium Reagent. *J. Organomet. Chem.* **1982,** *225*, 1–3.
53. Kornev, A. N. The Tris(trimethylsilyl)silyl Group in Organic, Coordination and Organometallic Chemistry. *Russ. Chem. Rev.* **2004,** *73*, 1065–1089.
54. Apeloig, Y.; Korogodsky, G.; Bravo-Zhivotovskii, D.; Blaser, D.; Boese, R. The Syntheses and Molecular Structure of a Branched Oligosilyl Anion With a Record of Nine Silicon Atoms and of the First Branched Oligosilyl Dianion. *Eur. J. Inorg. Chem.* **2000,** 1091–1095.

55. Marschner, C. A New and Easy Route to Polysilanylpotassium Compounds. *Eur. J. Inorg. Chem.* **1998,** 221–226.
56. Klinkhammer, K. W.; Schwarz, W. Synthesis of Tris(trimethylsilyl)silyl Potassium, Rubidium and Cesium and the Molecular Structures of Two Toluene Solvates. *Z. Anorg. Allg. Chem.* **1993,** *619*, 1777–1789.
57. Kayser, C.; Fischer, R.; Baumgartner, J.; Marschner, C. Tailor-Made Oligosilyl Potassium Compounds. *Organometallics* **2002,** *21*, 1023–1030.
58. Klink, R.; Schrenk, C.; Schnepf, A. $Si(SiMe_3)_2SiPh_3$ – A Ligand for Novel Sub-Valent Tin Cluster Compounds. *Dalton. Trans.* **2014,** *43*, 16097–16104.
59. Hlina, J.; Mechtler, C.; Wagner, H.; Baumgartner, J.; Marschner, C. Multiple Silyl Exchange Reactions: A Way to Spirooligosilanes. *Organometallics* **2009,** *28*, 4065–4071.
60. Mechtler, C.; Zirngast, M.; Baumgartner, J.; Marschner, C. Synthesis and Reactions of Alkynyl Oligosilanes. *Eur. J. Inorg. Chem.* **2004,** 3254–3261.
61. Kayser, C.; Kickelbick, G.; Marschner, C. Simple Synthesis of Oligosilyl-α,ω-dipotassium Compounds. *Angew. Chem. Int. Ed.* **2002,** *41*, 989–992.
62. Englich, U.; Graschy, S.; Hengge, E.; Herrmann, U.; Marschner, C.; Mechtler, C., et al. On Some Reactions of Undecamethylcyclohexasilanyl-Substituted Silanes. *J. Organomet. Chem.* **2000,** *598*, 202–210.
63. Wallner, A.; Hlina, J.; Wagner, H.; Baumgartner, J.; Marschner, C. Conformational Control of Polysilanes: the Use of CH_2-Spacers in the Silicon Backbone. *Organometallics* **2011,** *30*, 3930–3938.
64. Dzambaski, A.; Baumgartner, J.; Flock, M.; Hassler, K. Conformation Control of Oligosilanes by Trimethylsilyl Groups: Dodecamethyl-, Undecamethyl-2-trimethylsilyl- and Decamethyl-2, 4-bis (trimethylsilyl)-*n*-Pentasilane Studied by Raman Spectroscopy and Quantum Chemical Calculations. *Silicon* **2010,** *1*, 225–237.
65. Jenkins, D. M.; Teng, W.; Englich, U.; Stone, D.; Ruhlandt-Senge, K. Heavy Alkali Metal Tris (trimethylsilyl)silanides: A Synthetic and Structural Study. *Organometallics* **2001,** *20*, 4600–4606.
66. Marschner, C. to be published.
67. Farwell, J. D.; Lappert, M. F.; Marschner, C.; Strissel, C.; Tilley, T. D. The First Structurally Characterised Oligosilylmagnesium Compound. *J. Organomet. Chem.* **2000,** *603*, 185–188.
68. Gaderbauer, W.; Zirngast, M.; Baumgartner, J.; Marschner, C.; Tilley, T. D. Synthesis of Polysilanylmagnesium Compounds. *Organometallics* **2006,** *25*, 2599–2606.
69. Yan, K.; Upton, B. M.; Zhu, J.; Ellern, A.; Sadow, A. D. Nucleophilicity of Neutral versus Cationic Magnesium Silyl Compounds. *Organometallics* **2013,** *32*, 6834–6843.
70. Mukherjee, D.; Lampland, N. L.; Yan, K.; Dunne, J. F.; Ellern, A.; Sadow, A. D. Divergent Reaction Pathways of Tris(oxazolinyl)borato Zinc and Magnesium Silyl Compounds. *Chem. Commun.* **2013,** *49*, 4334–4336.
71. Lampland, N. L.; Ellern, A.; Sadow, A. D. β-SiH Rich Zinc Silyl Compounds: Reductive Elimination and β-Hydrogen Abstraction. *Inorg. Chim. Acta* **2014,** *422*, 134–140.
72. Uhlig, F.; Gspaltl, P.; Trabi, M.; Hengge, E. Zur Darstellung Von Permethylierten Cyclosilanylalkalimetall-Derivaten. *J. Organomet. Chem.* **1995,** *493*, 33–40.
73. Fischer, R.; Konopa, T.; Baumgartner, J.; Marschner, C. Small Cyclosilanes: Syntheses and Reactions toward Mono- and Dianions. *Organometallics* **2004,** *23*, 1899–1907.
74. Reinke, H.; Krempner, C. *Synthesis, Structure and Reactivity of Novel Oligosilyl Anions*. In *Organosilicon Chemistry V;* Wiley-VCH, 2003217–221.
75. Wallner, A.; Hlina, J.; Konopa, T.; Wagner, H.; Baumgartner, J.; Marschner, C., et al. Cyclic and Bicyclic Methylpolysilanes and Some Oligosilanylene-Bridged Derivatives. *Organometallics* **2010,** *29*, 2660–2675.
76. Fischer, R.; Konopa, T.; Ully, S.; Baumgartner, J.; Marschner, C. Route Si_6 Revisited. *J. Organomet. Chem.* **2003,** *685*, 79–92.
77. Xiao, X.-Q.; Zhao, H.; Xu, Z.; Lai, G.; He, X.-L.; Li, Z. Reduction of the Dihalocyclopentasilanes. A Quest for a Homocyclic Silylene. *Chem. Commun.* **2013,** *49*, 2706–2708.

78. Abersfelder, K.; Scheschkewitz, D. Syntheses of Trisila Analogues of Allyl Chlorides and Their Transformations to Chlorocyclotrisilanes, Cyclotrisilanides, and a Trisilaindane. *J. Am. Chem. Soc.* **2008,** *130*, 4114–4121.

79. Li, H.; Hung-Low, F.; Krempner, C. Synthesis and Structure of Zwitterionic Silylborates and Silylzincates With Pendant Polydonor Arms. *Organometallics* **2012,** *31*, 7117–7124.

80. Gehrhus, B.; Hitchcock, P. B.; Lappert, M. F.; Slootweg, J. C. The Diverse Reactions of the Silylene Si $[(NCH_2{}^tBu)_2C_6H_4\text{-}1,2]$ With $Li[Si(SiMe_3)_3](thf)_3$ and $K[N(SiMe_3)_2]$. *Chem. Commun.* **2000,** 1427–1428.

81. Cai, X.; Gehrhus, B.; Hitchcock, P. B.; Lappert, M. F.; Slootweg, J. C. The Stable Silylene Si $[(NCH_2{}^tBu)_2C_6H_4\text{-}1,2]$: Insertion Into Li-C or Li-Si Bonds of Lithium Alkyls LiR or $[LiSi(SiMe_3)_3(THF)_3]$ [R = Me, ^{t}Bu or $CH(SiMe_3)_2$]. *J. Organomet. Chem.* **2002,** *643–644*, 272–277.

82. Apeloig, Y.; Yuzefovich, M.; Bendikov, M.; Bravo-Zhivotovskii, D.; Blaser, D.; Boese, R. Branched Star-Type Polysilyllithium Compounds: The Effects of β-Silyl Substitution and of Complexation on Their Molecular Structure. *Angew. Chem., Int. Ed.* **2001,** *40*, 3016–3020.

83. Wiberg, N.; Niedermayer, W.; Nöth, H.; Warchhold, M. Alkali Metal Disupersilylsilanides $MSiXR^*_2$ ($R^* = SitBu_3$; X = H, Me, Ph, Hal): Synthesis, Characterization, Structures, Reactions. *J. Organomet. Chem.* **2001,** *628*, 46–56.

84. Wiberg, N.; Niedermayer, W.; Lerner, H.-W.; Bolte, M. Disupersilylsilanide $M(SiHR^*_2)_2$ der Zinkgruppenmetalle (M = Zn, Cd, Hg; $R^* = SitBu_3$): Synthesen, Charakterisierung und Strukturen. *Z. Anorg. Allg. Chem.* **2001,** *627*, 1043–1047.

85. Wiberg, N.; Auer, H.; Wagner, S.; Polborn, K.; Kramer, G.; Disilenes, R. *XSi:SiXR* ($R^* = SitBu_3$) With Silicon-Bound H- and Hal Atoms as X: Formation, Isomerization, and Reaction. *J. Organomet. Chem.* **2001,** *619*, 110–131.

86. Bravo-Zhivotovskii, D.; Yuzefovich, M.; Sigal, N.; Korogodsky, G.; Klinkhammer, K.; Tumanskii, B., et al. The Synthesis of the First Compound With Li-Si-Hg Bonding: $[\{Li(iPr_3Si)_2Si\}_2Hg]$—A Source for the $[Li(iPr_3Si)_2Si]^{\bullet}$ Radical. *Angew. Chem., Int. Ed.* **2002,** *41*, 649–651.

87. Iwamoto, T.; Okita, J.; Kabuto, C.; Kira, M. Sila-Metalation Route to Hydrido(trialkylsilyl)silyllithiums. *J. Am. Chem. Soc.* **2002,** *124*, 11604–11605.

88. Nakamoto, M.; Fukawa, T.; Lee, V. Ya; Sekiguchi, A. Nearly Planar Nonsolvated Monomeric Silyl- and Germyllithiums as a Result of an Intramolecular CH − Li Agostic Interaction. *J. Am. Chem. Soc.* **2002,** *124*, 15160–15161.

89. Matsuno, T.; Ichinohe, M.; Sekiguchi, A. Cyclotetrasilenide Ion: A Reversible Redox System of Cyclotetrasilenyl Cation, Radical, and Anion. *Angew. Chem., Int. Ed.* **2002,** *41*, 1575–1577.

90. Sekiguchi, A.; Tanaka, T.; Ichinohe, M.; Akiyama, K.; Gaspar, P. P. Tri-*tert*-butylsilylsilylenes With Alkali Metal Substituents (^{t}Bu$_3$Si)SiM (M = Li, K): Electronically and Sterically Accessible Triplet Ground States. *J. Am. Chem. Soc.* **2008,** *130*, 426–427.

91. Dobrovetsky, R.; Kratish, Y.; Tumanskii, B.; Botoshansky, M.; Bravo-Zhivotovskii, D.; Apeloig, Y. Radical Activation of Si-H Bonds by Organozinc and Silylzinc Reagents: Synthesis of Geminal Dizinciosilanes and Zinciolithiosilanes. *Angew. Chem., Int. Ed.* **2012,** *51*, 4671–4675.

92. Yamaguchi, T.; Ichinohe, M.; Sekiguchi, A. Addition of Methyllithium to Disilyne RSi≡SiR ($R = SiiPr[CH(SiMe_3)_2]$), Giving a Disilenyllithium, and Its Unexpected Isomerization to a Disilacyclopropylsilyllithium. *New J. Chem.* **2010,** *34*, 1544–1546.

93. Tanaka, H.; Inoue, S.; Ichinohe, M.; Sekiguchi, A. An (η^3-disilaallyl)lithium Derivative: Interconversion of π-Allyl-Type Bonding and η^1 Coordination to a Silyllithium Fragment. *Organometallics* **2009,** *28*, 6625–6628.

94. Ugrinov, A.; Sevov, S. C. Derivatization of Deltahedral Zintl Ions by Nucleophilic Addition: $[Ph - Ge_9 - SbPh_2]^{2-}$ and $[Ph_2Sb - Ge_9 - Ge_9 - SbPh_2]^{4-}$. *J. Am. Chem. Soc.* **2003,** *125*, 14059–14064.

95. Sevov, S. C.; Goicoechea, J. M. Chemistry of Deltahedral Zintl Ions. *Organometallics* **2006,** *25*, 5678–5692.

96. Scharfe, S.; Kraus, F.; Stegmaier, S.; Schier, A.; Fässler, T. F. Zintl Ions, Cage Compounds, and Intermetalloid Clusters of Group 14 and Group 15 Elements. *Angew. Chem. Int. Ed.* **2011,** *50*, 3630–3670.

97. Ichinohe, M.; Toyoshima, M.; Kinjo, R.; Sekiguchi, A. Tetrasilatetrahedranide: A Silicon Cage Anion. *J. Am. Chem. Soc.* **2003,** *125*, 13328–13329.
98. Klapoetke, T. M.; Vasisht, S. K.; Fischer, G.; Mayer, P. A Reactive Si_4 Cage: $K(SitBu_3)_3Si_4$. *J. Organomet. Chem.* **2010,** *695*, 667–672.
99. Iwamoto, T.; Tsushima, D.; Kwon, E.; Ishida, S.; Isobe, H. Persilastaffanes: Design, Synthesis, Structure, and Conjugation Between Silicon Cages. *Angew. Chem. Int. Ed.* **2012,** *51*, 2340–2344.
100. Lee, V. Ya; Yokoyama, T.; Takanashi, K.; Sekiguchi, A. Pentasilatricyclo[2.1.0.0^{2,5}]pentane and Its Anion. *Chem. Eur. J.* **2009,** *15*, 8401–8404.
101. Wiberg, N.; Niedermayer, W.; Polborn, K.; Nöth, H.; Knizek, J.; Fenske, D., et al. *Sterically Overloaded Silanes, Silylenes and Disilenes With Supersilyl Substituents* t*Bu*$_3$*Si*. In *Organosilicon Chemistry IV;* Wiley-VCH, 200093–97.
102. Bravo-Zhivotovskii, D.; Ruderfer, I.; Melamed, S.; Botoshansky, M.; Tumanskii, B.; Apeloig, Y. Nonsolvated, Aggregated 1,1-Dilithiosilane and the Derived Silyl Radicals. *Angew. Chem., Int. Ed.* **2005,** *44*, 739–743.
103. Rit, A.; Spaniol, T. P.; Okuda, J. Dinuclear Zinc Hydride Supported by an Anionic Bis(N-Heterocyclic Carbene) Ligand. *Chem. Asian J.* **2014,** *9*, 612–619.
104. Ring, M. A.; Ritter, D. M. Preparation and Reactions of Potassium Silyl. *J. Am. Chem. Soc.* **1961,** *83*, 802–805.
105. Lobreyer, T.; Oeler, J.; Sundermeyer, W. Über die verbesserte Darstellung von Silyl- und Germylkalium sowie die Synthese von Silylgermanen. *Chem. Ber.* **1991,** *124*, 2405–2410.
106. Chotard, J.-N.; Tang, W. S.; Raybaud, P.; Janot, R. Potassium Silanide ($KSiH_3$): A Reversible Hydrogen Storage Material. *Chem. Eur. J.* **2011,** *17*, 12302–12309.
107. Tang, W. S.; Chotard, J.-N.; Raybaud, P.; Janot, R. Enthalpy-Entropy Compensation Effect in Hydrogen Storage Materials: Striking Example of Alkali Silanides $MSiH_3$ (M = K, Rb, Cs). *J. Phys. Chem. C* **2014,** *118*, 3409–3419.
108. Leich, V.; Spaniol, T. P.; Okuda, J. Formation of α-[$KSiH_3$] by Hydrogenolysis of Potassium Triphenylsilyl. *Chem. Commun.* **2015,** *51*, 14772–14774.
109. Lobreyer, T.; Sundermeyer, W.; Oberhammer, H. Über dehydrierende Aufbaureaktionen zu silylsubstituierten Alkalimetallgermaniden, -stanniden und -phosphiden; Molekülstruktur von Neopentasilan. *Chem. Ber.* **1994,** *127*, 2111–2115.
110. Stueger, H.; Mitterfellner, T.; Fischer, R.; Walkner, C.; Patz, M.; Wieber, S. Selective Synthesis and Derivatization of Alkali Metal Silanides $MSi(SiH_3)_3$. *Chem. Eur. J.* **2012,** *18*, 7662–7664.
111. Lee, M. E.; Cho, H. M.; Ryu, M. S.; Kim, C. H.; Ando, W. A Stable Halosilylene at Room Temperature in THF Solution. *J. Am. Chem. Soc.* **2001,** *123*, 7732–7733.
112. Lee, M. E.; Cho, H. M.; Lim, Y. M.; Choi, J. K.; Park, C. H.; Jeong, S. E., et al. Syntheses and Reactivities of Stable Halosilylenoids, $(Tsi)X_2SiLi$ ($Tsi = C(SiMe_3)_3$, X = Br, Cl). *Chem. Eur. J.* **2004,** *10*, 377–381.
113. Lim, Y. M.; Cho, H. M.; Lee, M. E.; Baeck, K. K. A Stable Magnesium Bromosilylenoid: Transmetalation of a Lithium Bromosilylenoid by Magnesium Bromide. *Organometallics* **2006,** *25*, 4960–4964.
114. Lee, M. E.; Lim, Y. M.; Son, J. Y.; Seo, W. G. Stable Halodilithiosilanes: Synthesis and Reactivity of $TsiXSiLi_2$ (X = Br and Cl). *Chem. Lett.* **2008,** *37*, 680–681.
115. Fischer, R.; Baumgartner, J.; Kickelbick, G.; Marschner, C. The First Stable β-Fluorosilylanion. *J. Am. Chem. Soc.* **2003,** *125*, 3414–3415.
116. Zirngast, M.; Flock, M.; Baumgartner, J.; Marschner, C. Formation of Formal Disilene Fluoride Adducts. *J. Am. Chem. Soc.* **2008,** *130*, 17460–17470.
117. Molev, G.; Bravo-Zhivotovskii, D.; Karni, M.; Tumanskii, B.; Botoshansky, M.; Apeloig, Y. Synthesis, Molecular Structure, and Reactivity of the Isolable Silylenoid With a Tricoordinate Silicon. *J. Am. Chem. Soc.* **2006,** *128*, 2784–2785.

118. Tamao, K.; Kawachi, A.; Ito, Y. The First Stable Functional Silyl Anions: (Aminosilyl)lithiums. *J. Am. Chem. Soc.* **1992,** *114*, 3989–3990.
119. Tamao, K.; Kawachi, A.; Ohtani, H.; Ito, Y. Electronic Spectra of (Amino)(phenyl)disilanes. *J. Organomet. Chem.* **1994,** *473*, 29–34.
120. Kawachi, A.; Tamao, K. Structures of [(Amino)phenylsilyl]lithiums and Related Compounds in Solution and in the Solid State. *J. Am. Chem. Soc.* **2000,** *122*, 1919–1926.
121. Auer, D.; Kaupp, M.; Strohmann, C. "Unexpected" ^{29}Si NMR Chemical Shifts in Heteroatom-Substituted Silyllithium Compounds: A Quantum-Chemical Analysis. *Organometallics* **2004,** *23*, 3647–3655.
122. Zirngast, M.; Baumgartner, J.; Marschner, C. Preparation, Structure and Reactivity of $Et_2N(Me_3Si)_2SiK$. *Eur. J. Inorg. Chem.* **2008,** 1078–1087.
123. Popowski, E.; Rietz, I.; Fischer, C.; Erben, S.; Harloff, J.; Reinke, H. Reactions of Bis(trimethylsilyl) aminochlorosilanes With Lithium – N,Si- Migration of Trimethylsilylgroups or Self-condensation of Bis (trimethylsilyl) Aminosilyllithium Compounds. *Z. Anorg. Allg. Chem.* **2012,** *638*, 1187–1199.
124. Xiong, Y.; Yao, S.; Driess, M. Versatile Conversion of N-Heterocyclic Silylene to Silyl Metal Compounds by Insertion of Divalent Silicon into Metal–Carbon and Metal–Hydrogen Bonds. *Chem. Eur. J.* **2012,** *18*, 3316–3320.
125. Auer, G.; Oestreich, M. Silylzincation of Carbon–Carbon Multiple Bonds Revisited. *Chem. Commun.* **2006,** 311–313.
126. Armbruster, F.; Fernandez, I.; Breher, F. Syntheses, Structures, and Reactivity of Poly(pyrazolyl)silanes, -Disilanes, and the Ambidentate κ^1Si/κ^3N-Coordinating Tris(3,5-dimethylpyrazolyl)silanide Ligand [Si (3,5-Me_2pz)$_3$]-(MeTpsd). *Dalton. Trans.* **2009,** 5612–5626.
127. McNerney, B.; Whittlesey, B.; Krempner, C. Synthesis and Reactivity of New Pyrazolyl-Functionalized Potassium Silanides. *Eur. J. Inorg. Chem.* **2011,** 1699–1702.
128. Tamao, K.; Kawachi, A. The Chemistry of Silylenoids: Preparation and Reactivity of (Alkoxysilyl) lithium Compounds. *Angew. Chem. Int. Ed. Engl.* **1995,** *34*, 818–820.
129. Kawachi, A.; Tamao, K. Different Modes of Reaction of Monoalkoxy- and Dialkoxyphenylchlorosilanes With Lithium Metal: Selective Formation of (2-Alkoxydisilanyl)lithium vs (Dialkoxysilyl)lithium. *Organometallics* **1996,** *15*, 4653–4656.
130. Likhar, P. R.; Zirngast, M.; Baumgartner, J.; Marschner, C. Preparation and Structural Characterisation of Methoxybis(trimethylsilyl)silyl Potassium and Its Condensation Product. *Chem. Commun.* **2004,** 1764–1765.
131. Harloff, J.; Popowski, E.; Reinke, H. Substituted Lithium Trimethylsiloxysilanides LiSiRR' ($OSiMe_3$)—Investigations of Their Synthesis, Stability and Reactivity. *J. Organomet. Chem.* **2007,** *692*, 1421–1441.
132. Harloff, J.; Popowski, E. Reactions of Trimethylsiloxychlorosilanes With Lithium Metal—On the Mechanism of the Formation of Trimethylsiloxysilyllithium Compounds LiSiRR'($OSiMe_3$). *J. Organomet. Chem.* **2008,** *693*, 1283–1291.
133. Antolini, F.; Gehrhus, B.; Hitchcock, P. B.; Lappert, M. F. Crystalline Na-Si(NN) Derivatives [Si(NN) = Si{(NCH_2tBu)$_2C_6H_4$-1,2}]: the Silylenoid [Si(NN)OMe]$^-$, the Dianion [(NN)Si-Si(NN)]$^{2-}$, and the Radical Anion *c*-[Si(NN)]$_3^-$. *Chem. Commun.* **2005,** 5112–5114.
134. Kawachi, A.; Oishi, Y.; Kataoka, T.; Tamao, K. Preparation of Sulfur-Substituted Silyllithiums and Their Thermal Degradation to Silylenes. *Organometallics* **2004,** *23*, 2949–2955.
135. Fischer, R.; Frank, D.; Gaderbauer, W.; Kayser, C.; Mechtler, C.; Baumgartner, J., et al. α,ω-Oligosilyl Dianions and Their Application in the Synthesis of Homo- and Heterocyclosilanes. *Organometallics* **2003,** *22*, 3723–3731.
136. Mechtler, C.; Zirngast, M.; Gaderbauer, W.; Wallner, A.; Baumgartner, J.; Marschner, C. Synthesis of Oligosilyldi- and Trianions. *J. Organomet. Chem.* **2006,** *691*, 150–158.
137. Kira, M.; Kadowaki, T.; Yin, D.; Sakamoto, K.; Iwamoto, T.; Kabuto, C. Synthesis and Structure of Ethynylsilyllithiums. *Organometallics* **2007,** *26*, 4890–4895.

138. Markov, J.; Baumgartner, J.; Marschner, C.; Oehme, H.; Gross, T. *Vinyloligosilyl Anions—A New Class of Compounds*. In *Organosilicon Chemistry VI;* Wiley-VCH, 2005309−313.
139. Nakata, N.; Oikawa, T.; Matsumoto, T.; Kabe, Y.; Sekiguchi, A. Dilithium 1,4-Disilacyclohexa-2,5-diene-1,4-diide by the Reduction of 1,4-Disilabicyclo[2.2.0]hexa-2,5-diene: Synthesis and Characterization. *Organometallics* **2006,** *25*, 5850−5851.
140. Farwell, J. D.; Hitchcock, P. B.; Lappert, M. F. Sodium and Potassium 3-Sila-β-diketiminates Show New Coordination Modes. *Chem. Commun.* **2002,** 456−457.
141. Sekiguchi, A.; Ichinohe, M.; Yamaguchi, S. An Unexpected Reaction of Silacyclopropene To Form an Organodilithiosilane: Isolation and Characterization of a 1,1-Dilithiosilane Derivative. *J. Am. Chem. Soc.* **1999,** *121*, 10231−10232.
142. Ichinohe, M.; Arai, Y.; Sekiguchi, A.; Takagi, N.; Nagase, S. A New Approach to the Synthesis of Unsymmetrical Disilenes and Germasilene: Unusual ^{29}Si NMR Chemical Shifts and Regiospecific Methanol Addition. *Organometallics* **2001,** *20*, 4141−4143.
143. Igarashi, M.; Ichinohe, M.; Sekiguchi, A. Air-Stable Disilacyclopropene With a Si = C Bond and Its Conversion to Disilacyclopropenylium Ion: Silicon-Carbon Hybrid 2π-Electron Systems. *J. Am. Chem. Soc.* **2007,** *129*, 12660−12661.
144. Sekiguchi, A.; Izumi, R.; Lee, V. Ya; Ichinohe, M. The First Silastannene >Si = Sn. *J. Am. Chem. Soc.* **2002,** *124*, 14822−14823.
145. Nakata, N.; Izumi, R.; Lee, V. Ya; Ichinohe, M.; Sekiguchi, A. 1,3-Disila-2-gallata- and -indataallenic Anions $[>\text{Si-M-Si}<]^- \cdot \text{Li}^+$ (M = Ga, In): Compounds Featuring Double Bonds Between Elements of Groups 13 and 14. *J. Am. Chem. Soc.* **2004,** *126*, 5058−5059.
146. Nakata, N.; Fujita, T.; Sekiguchi, A. A Stable Schrock-Type Hafnium − Silylene Complex. *J. Am. Chem. Soc.* **2006,** *128*, 16024−16025.
147. Nakata, N.; Sekiguchi, A. A Stable Silaborene: Synthesis and Characterization. *J. Am. Chem. Soc.* **2006,** *128*, 422−423.
148. Lee, V. Ya; Kawai, M.; Sekiguchi, A.; Ranaivonjatovo, H.; Escudie, J. A "Push-Pull" Phosphasilene and Phosphagermene and Their Anion-Radicals. *Organometallics* **2009,** *28*, 4262−4265.
149. Lee, V. Ya; Aoki, S.; Kawai, M.; Meguro, T.; Sekiguchi, A. Stibasilene Sb = Si and Its Lighter Homologues: A Comparative Study. *J. Am. Chem. Soc.* **2014,** *136*, 6243−6246.
150. Kira, M.; Iwamoto, T.; Yin, D.; Maruyama, T.; Sakurai, H. Novel 1,2-Dilithiodisilanes Derived From Reduction of Stable Tetrakis(trialkylsilyl)disilenes With Lithium Metal. *Chem. Lett.* **2001,** 910−911.
151. Bravo-Zhivotovskii, D.; Ruderfer, I.; Melamed, S.; Botoshansky, M.; Schmidt, A.; Apeloig, Y. $[\{(t\text{Bu}_2\text{Me})_2\text{Si}\}_3\text{Li}_4]^{2-}$: An Aggregated Dianion of a 1,1-Dilithiosilane With a Unique Structural Motif. *Angew. Chem., Int. Ed.* **2006,** *45*, 4157−4159.
152. Bravo-Zhivotovskii, D.; Molev, G.; Kravchenko, V.; Botoshansky, M.; Schmidt, A.; Apeloig, Y. Novel Aggregation Motif of gem-Dilithiosilanes: Coaggregation of Two R_2SiLi_2 Molecules With Two RLi Molecules. *Organometallics* **2006,** *25*, 4719−4721.
153. Dobrovetsky, R.; Bravo-Zhivotovskii, D.; Tumanskii, B.; Botoshansky, M.; Apeloig, Y. Synthesis, Isolation, and Characterization of 1,1-DiGrignard and 1,1-Dizincio Silanes. *Angew. Chem., Int. Ed.* **2010,** *49*, 7086−7088.
154. Belzner, J.; Dehnert, U.; Stalke, D. Doubly Lithiated Oligosilanes: Synthesis and Structure of the First Vicinal Dilithiodisilane. *Angew. Chem., Int. Ed.* **1994,** *33*, 2450−2452.
155. Fischer, J.; Baumgartner, J.; Marschner, C. Silylgermylpotassium Compounds. *Organometallics* **2005,** *24*, 1263−1268.
156. Zirngast, M.; Baumgartner, J.; Marschner, C. Synthesis of Cyclic and Bicyclic Polysilanes of Variable Ring Sizes. *Organometallics* **2008,** *27*, 6472−6478.
157. Inoue, S.; Ichinohe, M.; Sekiguchi, A. Disilenyl Anions Derived From Reduction of Tetrakis(di-*tert*-butylmethylsilyl)disilene With Metal Naphthalenide Through a Disilene Dianion Intermediate: Synthesis and Characterization. *Chem. Lett.* **2005,** *34*, 1564−1565.

158. Inoue, S.; Ichinohe, M.; Sekiguchi, A. Isolable Silylene Anion Radical: Structural Characteristics in the Solid State and in Solution. *J. Am. Chem. Soc.* **2007,** *129*, 6096–6097.
159. Gaderbauer, W.; Balatoni, I.; Wagner, H.; Baumgartner, J.; Marschner, C. Synthesis and Structural Diversity of Oligosilanylzinc Compounds. *Dalton. Trans.* **2010,** *39*, 1598–1603.
160. Kyushin, S.; Kawai, H.; Matsumoto, H. (Trans-1,2,2,3,4,4-Hexa-*tert*-butyl-1,3-cyclotetrasilanediyl) dipotassium: Supramolecular Structure of the Silylpotassium-Benzene Complex. *Organometallics* **2004,** *23*, 311–313.
161. Fischer, J.; Gaderbauer, W.; Baumgartner, J.; Marschner, C. Group 14 Hetero- Mono- and Bicyclosilanes. *Heterocycles* **2006,** *67*, 507–510.
162. Lee, V. Ya; Takanashi, K.; Ichinohe, M.; Sekiguchi, A. The First Bicyclo[1.1.0]butane Dianion of Heavier Group 14 Elements. *Angew. Chem., Int. Ed.* **2004,** *43*, 6703–6705.
163. Takanashi, K.; Lee, V. Ya; Yokoyama, T.; Sekiguchi, A. Base-Free Molybdenum and Tungsten Bicyclic Silylene Complexes Stabilized by a Homoaromatic Contribution. *J. Am. Chem. Soc.* **2009,** *131*, 916–917.
164. Jarvie, A. W. P.; Winkler, H. J. S.; Peterson, D. J.; Gilman, H. Preparation and Characterization of Octaphenylcyclotetrasilane. *J. Am. Chem. Soc.* **1961,** *83*, 1921–1924.
165. Wagner, H.; Baumgartner, J.; Marschner, C. 1,1'-Oligosilylferrocene Compounds. *Organometallics* **2007,** *26*, 1762–1770.
166. Hengge, E.; Kollmann, G. Preparation of Cyclosilanes With a Silicon Side Chain. *J. Organomet. Chem.* **1975,** *92*, C43–C45.
167. Winkler, H. J. S.; Jarvie, A. W. P.; Peterson, D. J.; Gilman, H. Preparation and Characterization of Dodecaphenylcyclohexasilane. *J. Am. Chem. Soc.* **1961,** *83*, 4089–4093.
168. Choi, S.-B.; Boudjouk, P. A Novel α,ω Silyl Dianionic Salt. The Synthesis and Characterization of Remotely Connected Benzannulated Silole Monoanions. *J. Chem. Soc., Dalton Trans.* **2000,** 841–844.
169. Mechtler, C.; Marschner, C. Polysilyldianions—Synthesis and Reactivity. *Tetrahedron Lett.* **1999,** *40*, 7777–7778.
170. Marschner, C.; Mechtler, C. Cyclosilylanions From Dianionic Precursors. *Phosphorus, Sulfur, Silicon Relat. Elem.* **2001,** *169*, 193–196.
171. Blanton, J. R.; Diminnie, J. B.; Chen, T.; Wiltz, A. M.; Xue, Z. α,ω-Bis(potassiosilyl)alkanes. Synthesis and Characterization of $[K(18\text{-crown-}6)]_2[(Me_3Si)_2Si(CH_2)_nSi(SiMe_3)_2]$. *Organometallics* **2001,** *20*, 5542–5546.
172. Hong, J. H.; Boudjouk, P. A Stable Aromatic Species Containing Silicon. Synthesis and Characterization of the 1-*tert*-Butyl-2,3,4,5-tetraphenyl-1-silacyclopentadienide Anion. *J. Am. Chem. Soc.* **1993,** *115*, 5883–5884.
173. Choi, S.-B.; Boudjouk, P.; Wei, P. Aromatic Benzannulated Silole Dianions. The Dilithio and Disodio Salts of a Silaindenyl Dianion. *J. Am. Chem. Soc.* **1998,** *120*, 5814–5815.
174. Freeman, W. P.; Tilley, T. D.; Yap, G. P. A.; Rheingold, A. L. Silolyl Anions and Silole Dianions: Structure of $[K([18]crown\text{-}6)^+]_2[C_4Me_4Si^{2-}]$. *Angew. Chem. Int. Ed. Engl.* **1996,** *35*, 882–884.
175. Dysard, J. M.; Tilley, T. D. Synthesis and Reactivity of η^5-Silolyl, η^5-Germolyl, and η^5-Germole Dianion Complexes of Zirconium and Hafnium. *J. Am. Chem. Soc.* **2000,** *122*, 3097–3105.
176. Pietschnig, R.; West, R.; Powell, D. R. Reduction of Terphenyltrifluorosilanes: C–C Insertion Products and Possible Formation of a Disilyne. *Organometallics* **2000,** *19*, 2724–2729.
177. Liu, Y.; Stringfellow, T. C.; Ballweg, D.; Guzei, I. A.; West, R. Structure and Chemistry of 1-Silafluorenyl Dianion, Its Derivatives, and an Organosilicon Diradical Dianion. *J. Am. Chem. Soc.* **2002,** *124*, 49–57.
178. Hong, J.-H. Synthesis and NMR-Study of 1-Trimethylsilyl Substituted Silole Anion $[Ph_4C_4Si(SiMe_3)]^-\bullet[Li]^+$ and 3-Silolenide 2,5-Carbodianions $\{[Ph_4C_4Si(n\text{-}Bu)_2]^{2-}\bullet 2[Li]^+, [Ph4C4Si(t\text{-}Bu)_2]^{2-}\bullet 2[Li]^+\}$ via Silole Dianion $[Ph_4C_4Si]^{2-}\bullet 2[Li]^+$. *Molecules* **2013,** *18*, 10568–10579.
179. Lee, V. Ya; Takanashi, K.; Matsuno, T.; Ichinohe, M.; Sekiguchi, A. Cyclobutadiene Dianions Consisting of Heavier Group 14 Elements: Synthesis and Characterization. *J. Am. Chem. Soc.* **2004,** *126*, 4758–4759.

180. Lee, V. Ya; Kato, R.; Ichinohe, M.; Sekiguchi, A. The Heavy Analogue of CpLi: Lithium 1,2-Disila-3-Germacyclopentadienide, a 6π-Electron Aromatic System. *J. Am. Chem. Soc.* **2005,** *127*, 13142–13143.
181. Lee, V. Ya; Takanashi, K.; Kato, R.; Matsuno, T.; Ichinohe, M.; Sekiguchi, A. Heavy Analogues of the 6π-Electron Anionic Ring Systems: Cyclopentadienide Ion and Cyclobutadiene Dianion. *J. Organomet. Chem.* **2007,** *692*, 2800–2810.
182. Yasuda, H.; Lee, V. Ya; Sekiguchi, A. Si_3C_2-Rings: From a Nonconjugated Trisilacyclopentadiene to an Aromatic Trisilacyclopentadienide and Cyclic Disilenide. *J. Am. Chem. Soc.* **2009,** *131*, 6352–6353.
183. Bravo-Zhivotovskii, D.; Apeloig, Y.; Ovchinnikov, Y.; Igonin, V.; Struchkov, Y. T. A Novel Route to the Trisilacyclobutane Moiety. A Possible Silene-Disilene Reaction. *J. Organomet. Chem.* **1993,** *446*, 123–129.
184. Ohshita, J.; Masaoka, Y.; Ishikawa, M.; Takeuchi, T. Silicon-Carbon Unsaturated Compounds. 45. Reaction of Benzoyltris(trimethylsilyl)silane With Aryllithium Reagents. *Organometallics* **1993,** *12*, 876–879.
185. Ohshita, J.; Masaoka, Y.; Masaoka, S.; Ishikawa, M.; Tachibana, A.; Yano, T., et al. Silicon-Carbon Unsaturated Compounds. XLVIII. Synthesis and Reactions of Silicon Analogs of Lithium Enolates. *J. Organomet. Chem.* **1994,** *473*, 15–17.
186. Ohshita, J.; Masaoka, S.; Ishikawa, M. Oxidative Coupling of Lithium Silenolates: First Synthesis of Bis(acyl)-Substituted Polysilanes. *Organometallics* **1996,** *15*, 2198–2200.
187. Ohshita, J.; Sakurai, H.; Masaoka, S.; Tamai, M.; Kunai, A.; Ishikawa, M. Reactions of Lithium Silenolates With Benzophenone. *J. Organomet. Chem.* **2001,** *633*, 131–136.
188. Guliashvili, T.; El-Sayed, I.; Fischer, A.; Ottosson, H. The First Isolable 2-Silenolate. *Angew. Chem., Int. Ed.* **2003,** *42*, 1640–1642.
189. Ekloef, A. M.; Ottosson, H. Effects of Substituents and Counterions on the Structures of Silenolates: A Computational Investigation. *Tetrahedron* **2009,** *65*, 5521–5526.
190. Dobrovetsky, R.; Zborovsky, L.; Sheberla, D.; Botoshansky, M.; Bravo-Zhivotovskii, D.; Apeloig, Y. Isolation of Silenolates $(R_3Si)_2Si{:}C(OLi)Ad$ With a Doubly Bonded Silicon Atom. *Angew. Chem., Int. Ed.* **2010,** *49*, 4084–4087.
191. Haas, M.; Fischer, R.; Flock, M.; Mueller, S.; Rausch, M.; Saf, R., et al. Stable Silenolates and Brook-Type Silenes With Exocyclic Structures. *Organometallics* **2014,** *33*, 5956–5959.
192. Ohshita, J.; Kawamoto, H.; Kunai, A.; Ottosson, H. Formation of Acylsilenolates From Bis(acyl)trisilanes as the Silicon Analogues of Acylenolates. *Organometallics* **2010,** *29*, 4199–4202.
193. Zborovsky, L.; Dobrovetsky, R.; Botoshansky, M.; Bravo-Zhivotovskii, D.; Apeloig, Y. Synthesis of Silenyllithiums $Li(R'_3Si)Si{=}C(SiR_3)(1\text{-}Ad)$ via Transient Silyne–Silylidene Intermediates. *J. Am. Chem. Soc.* **2012,** *134*, 18229–18232.
194. Scheschkewitz, D. The Versatile Chemistry of Disilenides: Disila Analogues of Vinyl Anions as Synthons in Low-Valent Silicon Chemistry. *Chem. Lett.* **2011,** *40*, 2–11.
195. Scheschkewitz, D. A Silicon Analogue of Vinyllithium: Structural Characterization of a Disilenide. *Angew. Chem. Int. Ed.* **2004,** *43*, 2965–2967.
196. Ichinohe, M.; Sanuki, K.; Inoue, S.; Sekiguchi, A. Disilenyllithium From Tetrasila-1,3-Butadiene: A Silicon Analogue of a Vinyllithium. *Organometallics* **2004,** *23*, 3088–3090.
197. Weidenbruch, M.; Willms, S.; Saak, W.; Henkel, G. Hexaaryltetrasilabuta-1,3-Diene: A Molecule With Conjugated Si-Si Double Bonds. *Angew. Chem. Int. Ed. Engl.* **1997,** *36*, 2503–2504.
198. Abersfelder, K.; Gueclue, D.; Scheschkewitz, D. An Unsaturated α,ω-Dianionic oligosilane. *Angew. Chem., Int. Ed.* **2006,** *45*, 1643–1645.
199. Ichinohe, M.; Sanuki, K.; Inoue, S.; Sekiguchi, A. Tetrasila-1,3-butadiene and Its Transformation to Disilenyl Anions. *Silicon Chem.* **2007,** *3*, 111–116.
200. Sekiguchi, A.; Kinjo, R.; Ichinohe, M. A Stable Compound Containing a Silicon-Silicon Triple Bond. *Science* **2004,** *305*, 1755–1757.
201. Kinjo, R.; Ichinohe, M.; Sekiguchi, A. An Isolable Disilyne Anion Radical and a New Route to the Disilenide Ion upon Reduction of a Disilyne. *J. Am. Chem. Soc.* **2007,** *129*, 26–27.

202. Kira, M.; Ishida, S.; Iwamoto, T.; Kabuto, C. The First Isolable Dialkylsilylene. *J. Am. Chem. Soc.* **1999,** *121*, 9722–9723.
203. Iwamoto, T.; Kobayashi, M.; Uchiyama, K.; Sasaki, S.; Nagendran, S.; Isobe, H., et al. Anthryl-Substituted Trialkyldisilene Showing Distinct Intramolecular Charge-Transfer Transition. *J. Am. Chem. Soc.* **2009,** *131*, 3156–3157.
204. Inoue, S.; Ichinohe, M.; Sekiguchi, A. Conversion of a Disilenide Into A Silene: Silyl-Anion-Substituted Silene by a Sila-Peterson-Type Reaction From an sp^2-Type Silyl Anion. *Angew. Chem., Int. Ed.* **2007,** *46*, 3346–3348.
205. Tacke, R.; Pülm, M.; Wagner, B. Zwitterionic Pentacoordinate Silicon Compounds. *Adv. Organomet. Chem.* **1999,** *44*, 221–273.
206. Couzijn, E. P. A.; Slootweg, J. C.; Ehlers, A. W.; Lammertsma, K. Pentaorganosilicates. *Z. Anorg. Allg. Chem.* **2009,** *635*, 1273–1278.
207. Kost, D.; Kalikhman, I.; Krivonos, S.; Bertermann, R.; Burschka, C.; Neugebauer, R. E., et al. New Zwitterionic Pentacoordinate Silicates With $\mathit{Si}ONFC_2$, $\mathit{Si}ONF_2C$, and $\mathit{Si}O_2N_2C$ Frameworks: Synthesis, Structure, and Dynamic Stereochemistry. *Organometallics* **2000,** *19*, 1083–1095.
208. Tacke, R.; Mallak, M.; Willeke, R. Pentacoordination of Silicon by Four Covalent Si – S Bonds and One Covalent Si – C Bond. *Angew. Chem. Int. Ed.* **2001,** *40*, 2339–2341.
209. Tacke, R.; Bertermann, R.; Burschka, C.; Dragota, S.; Penka, M.; Richter, I. Spirocyclic Zwitterionic $\lambda^5\mathit{Si}$-Silicates With Two Bidentate Ligands Derived From α-Amino Acids or α-Hydroxycarboxylic Acids: Synthesis, Structure, and Stereodynamics. *J. Am. Chem. Soc.* **2004,** *126*, 14493–14505.
210. Seiler, O.; Büttner, M.; Penka, M.; Tacke, R. Zwitterionic $\lambda^5\mathit{Si}$-Silicates With an $\mathit{Si}S_4C$ or $\mathit{Si}S_2O_2C$ Skeleton, Compounds Containing a Square-Pyramidal or Trigonal-Bipyramidal Si-Coordination Polyhedron. *Organometallics* **2005,** *24*, 6059–6062.
211. Tacke, R.; Bertermann, R.; Burschka, C.; Dragota, S. A Zwitterionic Spirocyclic Pentacoordinate Silicon Compound Synthesized in Water by Si-O and Si-C Bond Cleavage. *Angew. Chem. Int. Ed.* **2005,** *44*, 5292–5295.
212. Richter, I.; Penka, M.; Tacke, R. The Hexacoordinate Silicate Dianions *mer*-Tris[glycolato(2 –)-O^1,O^2] silicate and *fac*-Tris[benzilato(2 –)-O^1,O^2]silicate: Syntheses and Structural Characterization. *Inorg. Chem.* **2002,** *41*, 3950–3955.
213. Tacke, R.; Penka, M.; Popp, F.; Richter, I. Bis[citrato(3 –)-O^1,O^3,O^6]silicate: A Dianionic Complex With Hexacoordinate Silicon(IV) and Two Tridentate Dioato(2 –)olato(1 –) Ligands. *Eur. J. Inorg. Chem.* **2002,** 1025–1028.
214. Theis, B.; Burschka, C.; Tacke, R. Optically Active Zwitterionic λ^5Si,λ^5Si'-Disilicates: Syntheses, Crystal Structures, and Behavior in Aqueous Solution. *Chem. Eur. J.* **2008,** *14*, 4618–4630.
215. Klüfers, P.; Kopp, F.; Vogt, M. Monosaccharides as Silicon Chelators: Pentacoordinate Bis(diolato) (phenyl)silicates With the *cis*-Furanose Isomers of Common Pentoses and Hexoses. *Chem. Eur. J.* **2004,** *10*, 4538–4545.
216. Lenormand, H.; Goddard, J.-P.; Fensterbank, L. Spirosilane Derivatives as Fluoride Sensors. *Org. Lett.* **2013,** *15*, 748–751.
217. Couzijn, E. P. A.; Schakel, M.; de Kanter, F. J. J.; Ehlers, A. W.; Lutz, M.; Spek, A. L., et al. Dynamic Configurational Isomerism of a Stable Pentaorganosilicate. *Angew. Chem. Int. Ed.* **2004,** *43*, 3440–3442.
218. Couzijn, E. P. A.; Ehlers, A. W.; Schakel, M.; Lammertsma, K. Electronic Structure and Stability of Pentaorganosilicates. *J. Am. Chem. Soc.* **2006,** *128*, 13634–13639.
219. Couzijn, E. P. A.; van den Engel, D. W. F.; Slootweg, J. C.; de Kanter, F. J. J.; Ehlers, A. W.; Schakel, M., et al. Configurationally Rigid Pentaorganosilicates. *J. Am. Chem. Soc.* **2009,** *131*, 3741–3751.
220. Corriu, R. J.; Guerin, C.; Henner, B.; Wang, Q. Pentacoordinate Hydridosilicates: Synthesis and Some Aspects of Their Reactivity. *Organometallics* **1991,** *10*, 2297–2303.
221. Rot, N.; Nijbacker, T.; Kroon, R.; de Kanter, F. J. J.; Bickelhaupt, F.; Lutz, M., et al. Introduction of Bulky Substituents at the Bridgehead Position of a 9-Silatriptycene: Pentacoordinate Hydridoorganylsilicates as Intermediates. *Organometallics* **2000,** *19*, 1319–1324.

222. Bearpark, M. J.; McGrady, G. S.; Prince, P. D.; Steed, J. W. The First Structurally Characterized Hypervalent Silicon Hydride: Unexpected Molecular Geometry and Si − H· · · K Interactions. *J. Am. Chem. Soc.* **2001,** *123*, 7736−7737.

223. Prince, P. D.; Bearpark, M. J.; McGrady, G. S.; Steed, J. W. Hypervalent Hydridosilicates: Synthesis, Structure and Hydride Bridging. *Dalton. Trans.* **2008,** 271−282.

224. Jochmann, P.; Davin, J. P.; Spaniol, T. P.; Maron, L.; Okuda, J. A Cationic Calcium Hydride Cluster Stabilized by Cyclen-Derived Macrocyclic N,N,N,N Ligands. *Angew. Chem. Int. Ed.* **2012,** *51*, 4452−4455.

225. Wei, B.; Li, H.; Yin, J.; Zhang, W.-X.; Xi, Z. Reaction of Dilithio Reagents With $PhSiH_3$: Formation of Siloles and 3-Silacyclopentenes. *J. Org. Chem.* **2015,** *80*, 8758−8762.

CHAPTER 8

Stable Silylenes and Their Transition Metal Complexes

Takeaki Iwamoto and Shintaro Ishida
Tohoku University, Sendai, Japan

Chapter Outline

Organosilicon Compounds.
DOI: http://dx.doi.org/10.1016/B978-0-12-801981-8.00008-3

8.0 List of Abbreviations

2-Ad 2-Adamantylidene
AIM Atoms-in-molecule
9-BBN 9-Borabicyclo[3.3.1]nonane
Cp η^5-Cyclopentadienyl
Cp* η^5-Pentamethylcyclopentadienyl
Cy Cyclohexyl
cod Cyclooctadiene
coe Cyclooctene
DFT Density functional theory
Dip 2,6-Diisopropylphenyl
DMAP 4-(Dimethylamino)pyridine
DMB 2,3-Dimethyl-1,3-butadiene
DME 1,2-Dimethoxyethane
Dsi Bis(trimethylsilyl)methyl
Hex Hexyl
Im*t*-Bu 1,3-Bis(*tert*-butyl)-1,3-dihydro-2*H*-imidazol-2-ylidene
ImDip 1,3-Bis(2,6-diisopropylphenyl)-1,3-dihydro-2*H*-imidazol-2-ylidene
Im*i*-Pr 1,3-Diisopropyl-1,3-dihydro-2*H*-imidazol-2-ylidene
ImMe 1,3,4,5-Tetramethyl-1,3-dihydro-2*H*-imidazol-2-ylidene
IR Infrared
***i*-Pr** Isopropyl
Me Methyl
Mes Mesityl
Mu Muonium

MO	Molecular orbital
Nac	β-Diketiminate
Nap	Naphthalene anion radical
NBO	Natural bond orbital
NHC	*N*-heterocyclic carbene
NICS	Nucleus independent chemical shift
NMR	Nuclear magnetic resonance
Np	Neopentyl (2,2-dimethylpropyl)
Ph	Phenyl
PPy	4-Pyrolidinopyridine
Pr	Propyl
***t*-Bu**	*tert*-Butyl
***t*-Amyl**	*tert*-Amyl (1,1-dimethylpropyl)
***t*-Oct**	*tert*-Octyl (1,1,3,3-tetramethylbutyl)
Tbb	4-*tert*-Butyl-2,6-bis[bis(trimethylsilyl)methyl]phenyl
Tbt	2,4,6-Tris[bis(trimethylsilyl)methyl]phenyl
THF	Tetrahydrofuran
Tip	2,4,6-Triisopropylphenyl
TPFPB$^-$	Tetrakis(pentafluorophenyl)borate
Tsi	Tris(trimethylsilyl)methyl
VT	Variable-temperature
XRD	X-ray diffraction
Xyl	2,6-Dimethylphenyl

8.1 Introduction

Silylenes (divalent silicon species, silicon analogs of carbenes) have been one of the most important low coordinate silicon species in modern silicon chemistry. Silylenes are usually singlet in the ground state and the divalent silicon center has two substituents and possesses lone pair electrons and vacant 3p orbital which serve as Lewis base and Lewis acid, respectively (Fig. 8.1), similar to singlet carbenes.

In the latter half of the 20th century, silylenes had been extensively investigated as reactive intermediates by various methods, such as chemical trapping reactions, spectroscopic observation at low temperatures, and theoretical calculations, etc., and the fundamental

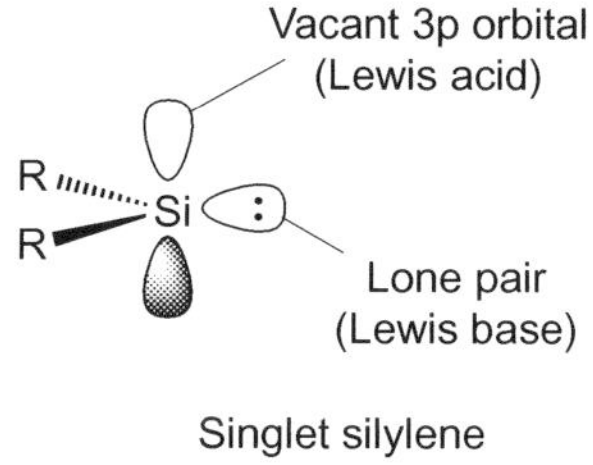

Figure 8.1
Electronic structure of the singlet silylene.

structure and reactivity of silylenes have been disclosed in detail.[1–5] From the middle of the 1980s, stable (isolable, persistent) silylenes stabilized electronically and kinetically by various ligands, such as **1a–7a** (Fig. 8.2), were reported by Jutzi,[6] Karsch,[7] West,[8] Lappert,[9] Okazaki and Tokitoh,[10] Kira.[11] In addition, transition metal-coordinated silylenes (silylene-transition metal complexes), such as **8–10**, have been also reported by Zybill,[12] Tilley,[13] Ogino.[14] The discovery of these stable silylenes made a significant impact into the development of the structural chemistry of silylenes. In 1998, Gaspar and West published a seminal review of silylenes (GW review in 1998).[1] In addition to the GW review, several comprehensive reviews of silylenes have been available.[15–30]

In the 21st century, the chemistry of stable silylenes has entered a new era and been dramatically developed. The notable important topics are: (1) The vast number of stable silylenes with the various functional substituents and the coordination numbers of the divalent silicon atoms and compounds with two or more silylene centers have been synthesized in addition to stable silylenes shown in Fig. 8.2, and their molecular structures and reactivity have been extensively studied. (2) Stable silylenes have been applied as synthons for new low coordinate silicon compounds. (3) Stable silylenes have been applied as ligands for transition metals and the resulting silylene complexes have been investigated as catalysts and/or precatalysts in several catalytic reactions. (4) Some of the functional silylenes activate small molecules, such as dihydrogen. (5) Ground-state triplet silylenes and excited-state singlet silylene have been successfully generated and their properties have

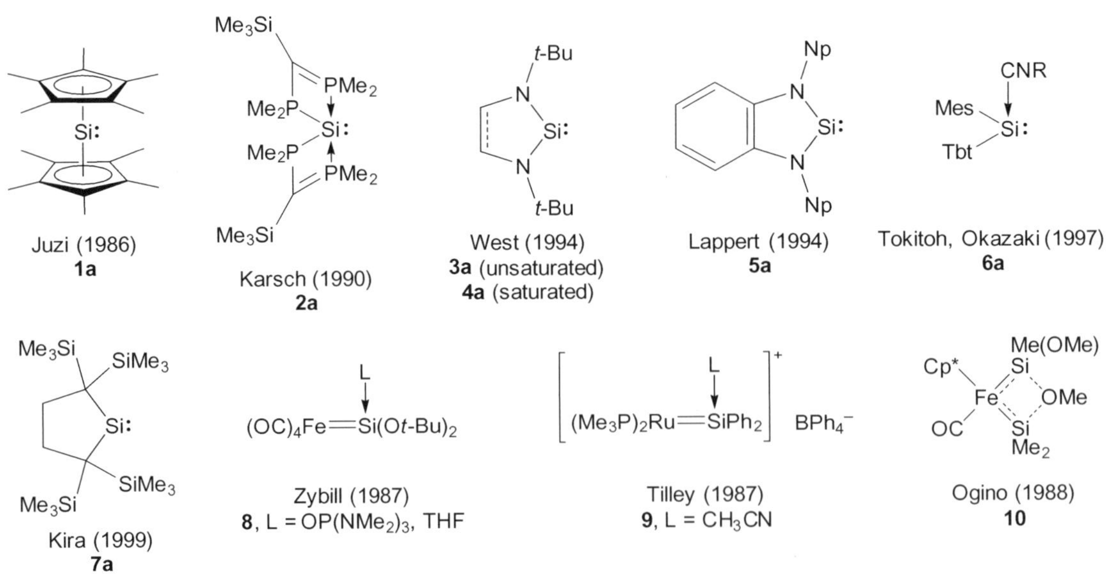

Figure 8.2
Trailblazing stable silylenes and silylene-transition metal complexes.

been disclosed. This dramatic progress in the chemistry of stable silylenes has been enabled by the diversity of ligands or substituents on the divalent silicon atoms. In addition to stable two-coordinate silylenes, various hypercoordinate silylenes bearing three or more ligands and/or substituents on the divalent silicon atom have been stabilized by coordination of internal bases like in **2a** and external bases like in **6a** and external Lewis acids.

This chapter will provide a brief overview of progress in the chemistry of stable (isolable and persistent) silylenes since the GW review in 1998, although various comprehensive reviews on stable silylenes[15–20,22–26] and their application for activation of small molecules and transformation of organic molecules[21,27–30] are available. Because the structure and reactivity of silylenes are considerably dependent on the ligands and substituents on the divalent silicon center, sections of this chapter are classified depending on the type of the ligands and substituents. Since comprehensive reviews on silylene transition-metal complexes[31–34] are available, this chapter will focus on only transition metal complexes derived from stable silylenes.

8.2 Cyclic Diaminosilylenes

8.2.1 Synthesis and Molecular Structures

The first stable two-coordinate silylene is cyclic diaminosilylene **3a** reported in 1994 by West and Denk (Scheme 8.1).[8] Diaminosilylene **4a** having a saturated ligand backbone is also synthesized. Silylene **3a** is thermally stable both in solution and in the solid state, whereas **4a** shows unique monomer–tetramer equilibrium, and both of the compounds are characterized by X-ray structural analysis, NMR (nuclear magnetic resonance), and UV-vis absorption spectroscopies.[3,35,36] Electrochemical investigation of **3a** and **4a** reveal that unsaturated silylene **3a** has lower oxidation potential than that of saturated **4a**, and **4a** was more readily reduced than **3a**.[37]

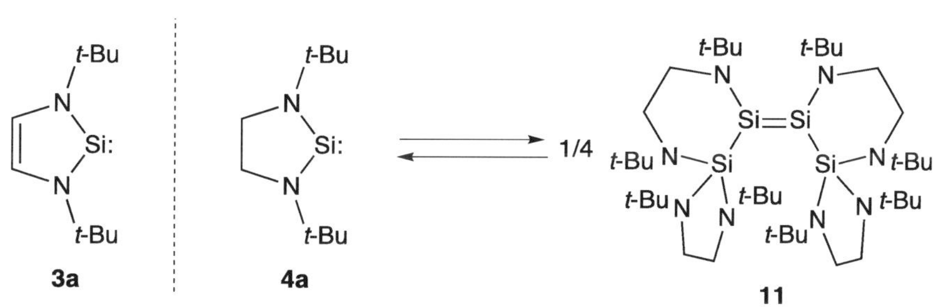

Scheme 8.1
Unsaturated and saturated cyclic diaminosilylenes.

Benzo- and pyrido-fused diaminosilylenes **5a–c** are also synthesized (Scheme 8.2).[9,38–40] Bis-diaminosilylene **5d** on biphenyl-scaffold is obtained as a pale yellow solid and was analyzed by XRD (X-ray diffraction) study.[41] The chemistry of stable cyclic diaminosilylenes are already well documented in several reviews,[15,17,20,22] thus this section mainly focuses on recent topics and stable silylene-based transition metal complexes.

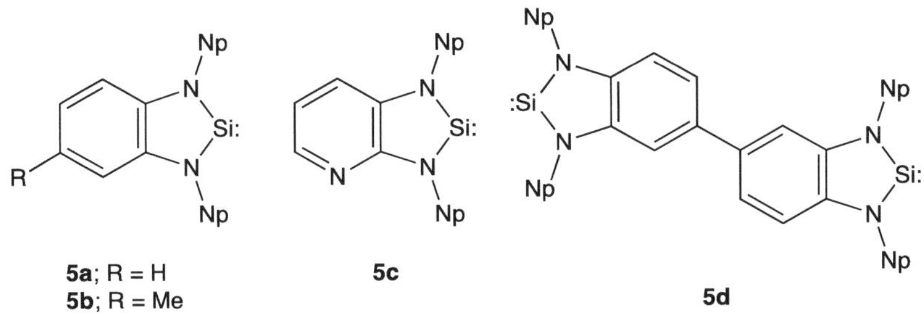

Scheme 8.2
Fused cyclic diaminosilylenes.

Recently, several new cyclic diaminosilylenes **4b–e** that are structurally similar to **4a** have been synthesized by the reaction of the corresponding dibromosilanes with KC_8 in a mixed solvent (THF:triethylamine = 5:1) for **4b** and in DME (1,2-Dimethoxyethane) for **4c–e** (Scheme 8.3).[42,43] In contrast to **4a**, silylenes **4b–e** show no oligomerization behavior. Synthesis of cyclic diaminosilylene **3b** bearing a bulky aryl group (2,6-diisopropylphenyl, Dip) on the nitrogen atom was originally reported in 1998[44] and its full characterization including X-ray structural analysis was conducted independently in 2009 and 2010.[45,46] Structurally similar *N*-aryl cyclic diaminosilylenes (**3c** and **3d**)[45,46] are also reported. New *N*-heterocyclic silylenes having alkyl groups (**3e** and **3f**) on the nitrogen atom have been prepared.[47]

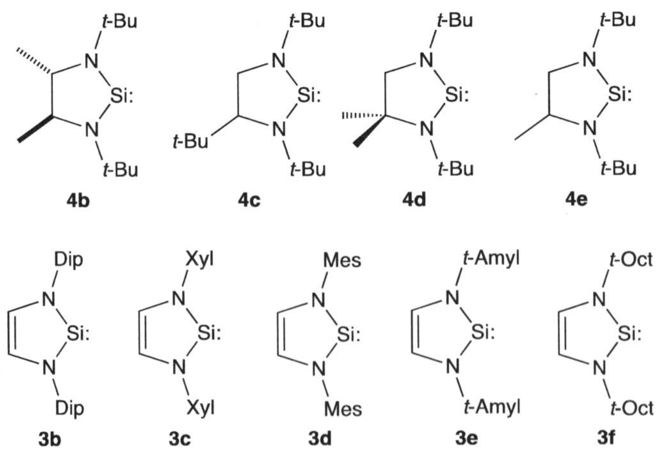

Scheme 8.3
Reported stable cyclic diaminosilylenes.

Cui and coworkers have reported dehydrochlorination of hydrochlorosilanes **12a** and **12d** using *N*-heterocyclic carbene (1,3-bis(*tert*-butyl)imidazol-2-ylidene, Im*t*-Bu) to afford **3a** and **3d** in good yields (Scheme 8.4). This nonmetallic synthetic method is an alternative route for silylenes.[48]

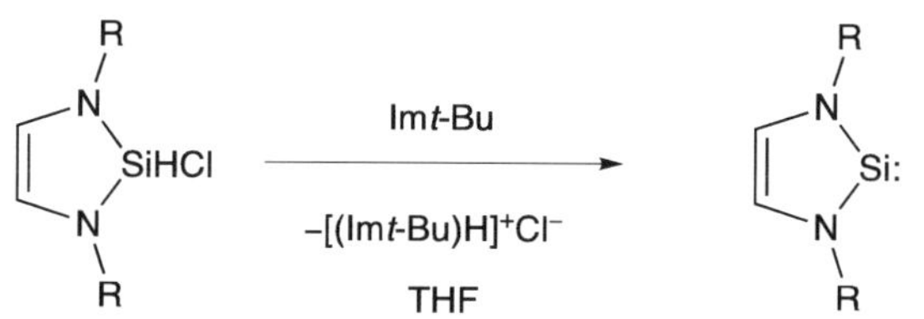

12a (R = *t*-Bu)
12d (R = Mes)

3a (R = *t*-Bu, 72%)
3d (R = Mes, 81%)

Scheme 8.4
Synthesis of silylenes by dehydrochlorination.

The δ_{Si} values of cyclic diaminosilylenes with unsaturated backbone appear in the range of 76–82 ppm, while those with saturated backbone are in the range of 119–141 ppm. Benzo or pyrido-fused cyclic diaminosilylenes show their ^{29}Si NMR signals in between those of silylenes bearing saturated and unsaturated backbone: δ_{Si} = 95–98 ppm. The N–Si bond lengths are 1.72–1.76 Å. The N–Si–N bond angles of these cyclic diaminosilylenes are similar to each other and close to the right angle: 87.5–92.0 degrees.

Cyclic bis(silylene) **13** was synthesized in 62% yield by the reduction of *N*-heterocyclic carbene (NHC)-coordinated silanimine **14** (Scheme 8.5).[49] This reaction would proceed via [2 + 2] dimerization of a silaisocyanide with elimination of NHC. The four-membered ring of **13** is essentially planar with the bond angles of N–Si–N (86.02(6) degrees) and Si–N–Si (94.02(9) degrees). The ^{29}Si NMR signal appeared at +183.3 ppm, being considerably downfield shifted compared to those of reported *N*-heterocyclic silylenes. The calculated NICS (nucleus independent chemical shift)(1) value of Si_2N_2 ring (+0.91) suggested some contribution of 4π electron antiaromatic character of **13**. Two unsaturated silicon atoms reacted as silylenes; reaction of **13** with two equiv of trimethylsilyl azide formed exocyclic bis(silanimine) **15** (Scheme 8.5). In addition, selective oxidation of **13** was also achieved by the reaction with N_2O to give monosilylene **16** in 62% yield together with **17** as a minor product (8%).[50] Oxidation of **13** with Me_3NO gave only **17**. This oxidation would proceed via formation of the corresponding Si = O species followed by hydrolysis. The silylene moiety and silanol moieties coexist in **16** and the ^{29}Si NMR spectrum shows two signals at +111.2 (N_2Si: unit) and −62.4 ppm ($Si(OH)_2$ unit).

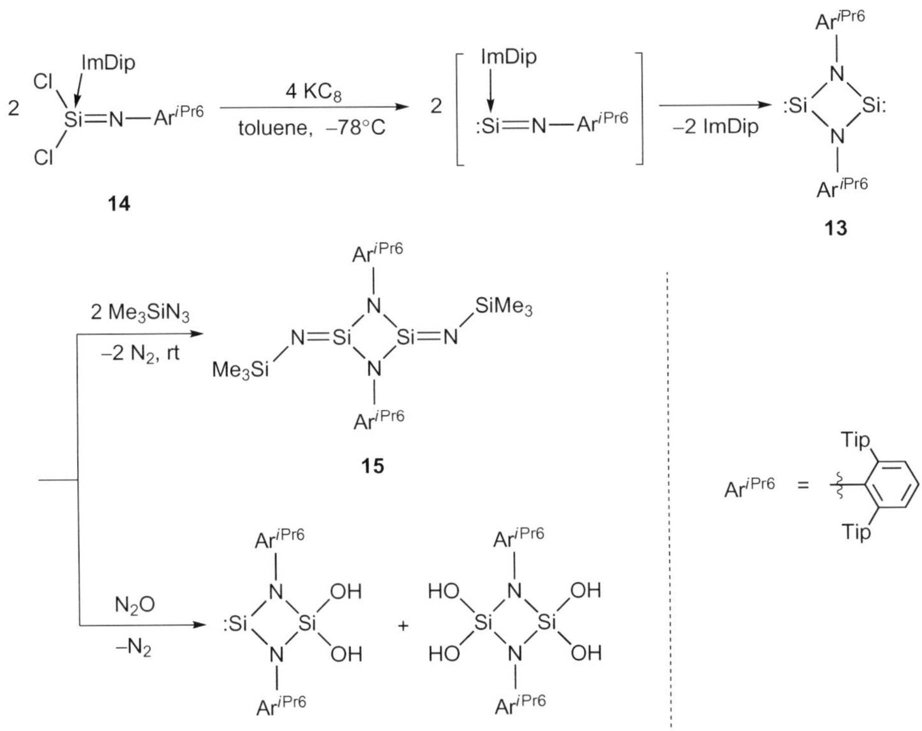

Scheme 8.5
Synthesis and reactions of cyclic bis(silylene) **13**.

8.2.2 Reactions With Haloalkanes and Halosilanes

Stable diaminosilylenes **3a** and **4a** react with various polyhaloalkanes and haloarenes to give 1-organo-2-halodisilanes as 2:1 adducts and organo(halo)silanes as 1:1 insertion products.[51–53] Selected examples for **3a** are illustrated in Scheme 8.6. Disilanes **18a–d** formed even with an excess amount of haloalkanes. Selectivity of the reactions depends on the reactants, and treatments with bulky haloalkanes, such as *tert*-butyl chloride and bis (trimethylsilyl)methyl chloride, leads to **19e** and **19f**. Silylene **3a** does not react with chlorobenzene but reacts with bromobenzene to give a mixture of **18g** and **19g**. No reaction proceeds upon the treatment of compound **19g** with **3a**. Reaction of **3a** with iodobenzene affords **19h** exclusively. Although several reaction mechanisms are proposed based on computational studies, radical chain reactions are more feasible.[54–56] The presence of CCl_3 radical, confirmed by matrix isolation study in the reaction of **3a** with CCl_4, and the presence of 1,1,2,2-tetrachloroethane, detected by ^1H and ^{13}C NMR spectroscopies during the reaction of **3a** with $CHCl_3$, also support the radical mechanism.[17]

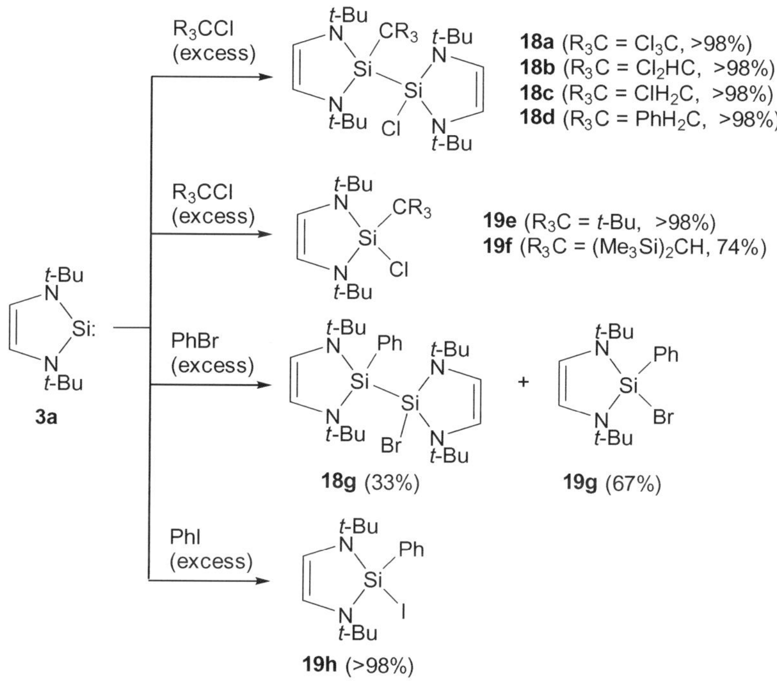

Scheme 8.6
Reactions of silylene **3a** with organic halides.

Holl and coworkers reported that C—H activation of solvents (cycloalkanes, ethers, and amines) occurs by mixing of **3a** and iodobenzene in these solvents.[57] Representative examples are displayed in Scheme 8.7. According to primary isotope kinetic effects for the reactions of **3a**, electron transfer from silylene to aryliodides was proposed at an initial stage, which facilitated homolysis of Ar—I bonds.

Benzo-fused cyclic diaminosilylene **5a** also reacts with various haloalkanes and selected examples are given in Scheme 8.8.[15,39,58] Silylene **5a** reacts with 1-chloropropane, dichloromethane, 2,2-dichloropropane, and chloroform to give 1-chloro-2-organodisilanes **21a—d** as primary 2:1 adducts even with an excess amount of chloroalkanes. Notably, reaction with bromomethylcyclopropane affords a mixture of bromodisilanes **22** and bromosilanes **23**. Formation of ring-opening products (**22b** and **23b**) as well as reactions of diaminosilylene **3a** with haloalkanes support the radical mechanism. Related reactions of **5a** with group 14 element halides are also reported.[59]

Disilanes **21a—d** are thermally labile, and they were converted to **24a—c** upon heating (Scheme 8.9). Thermolysis of **21d** results in the formation of a complex mixture. Disilanes **22a** and **22b** also decompose thermally into **23a** and **23b** with the elimination of **5a**.

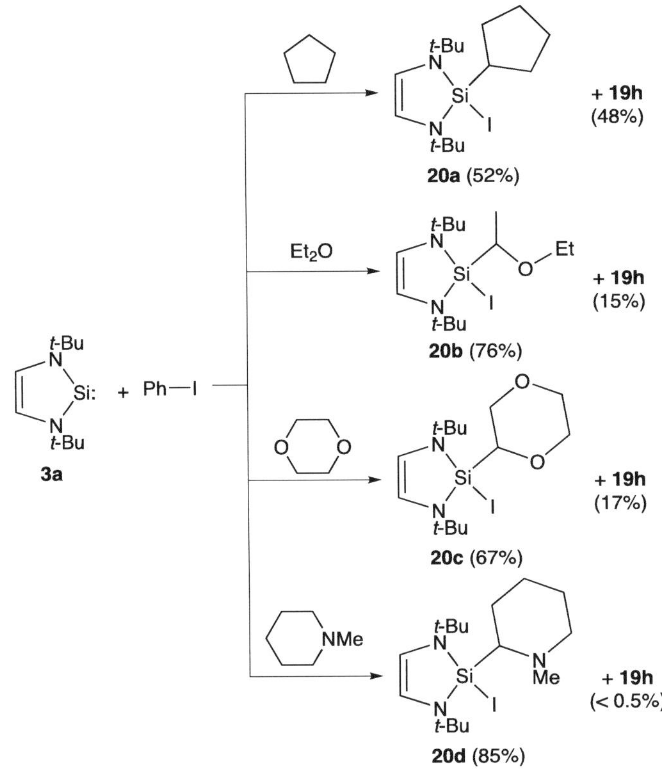

Scheme 8.7
C–H activation reactions of silylene **3a** in the presence of iodobenzene.

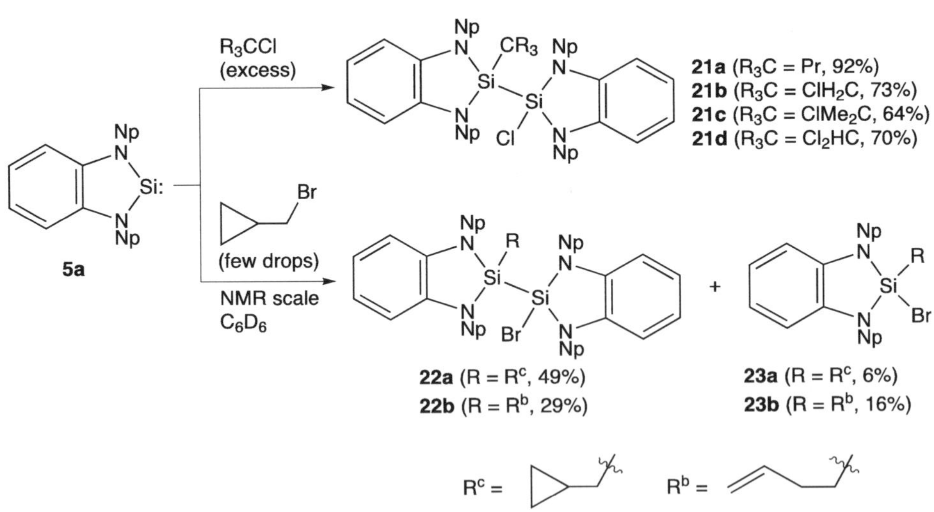

Scheme 8.8
Reactions of silylene **5a** with organic halides.

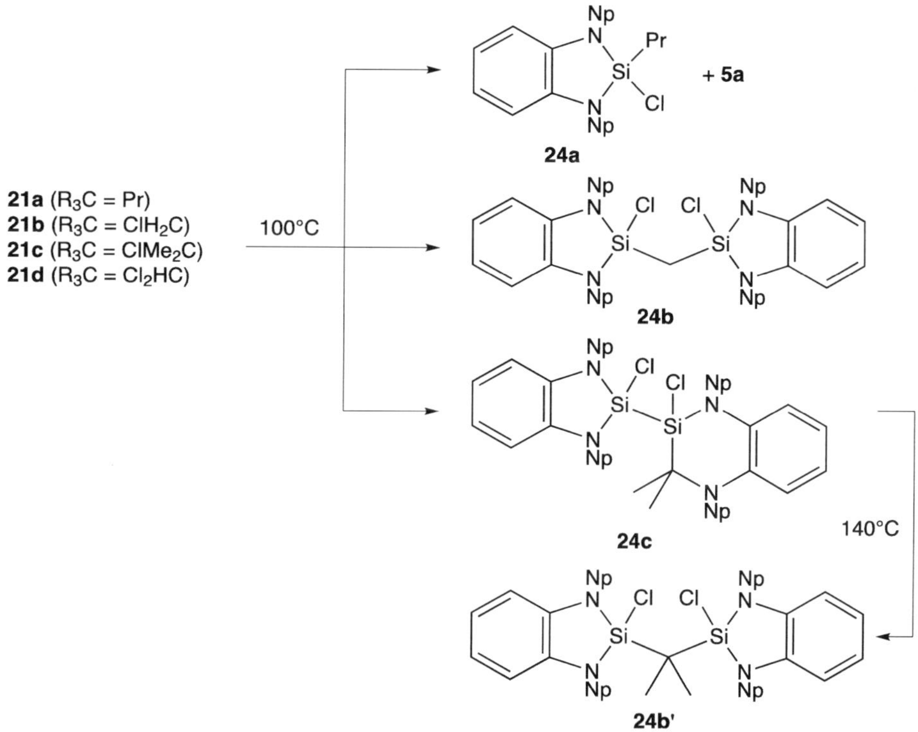

Scheme 8.9
Thermal reactions of silylene–organic halides adducts.

8.2.3 *Transition Metal Complexes and Related Metal Species of Stable Cyclic Diaminosilylenes*

Various transition metal complexes with cyclic diaminosilylene(s) as ligand(s) have been reported using mainly ligand exchange reactions. The coordination chemistry of the cyclic diaminosilylenes **3a**, **4a**, and **5a** is briefly summarized in this section.

8.2.3.1 Group 1 metals

Treatment of **4a** with Na/K alloy or KC_8 in THF affords dianionic disilane **25** (Scheme 8.10). Further reduction of **25** gives **26** accompanied by Si–Si bond cleavage.[60] Dianionic compounds **25** and **26** are characterized by NMR spectroscopy and trapping reactions with trimethylchlorosilane and EtOH. Reduction of unsaturated diaminosilylene **3a** results in the Si–N bond cleavage to form the corresponding diimine (*t*-BuN = CH–CH = N*t*-Bu).

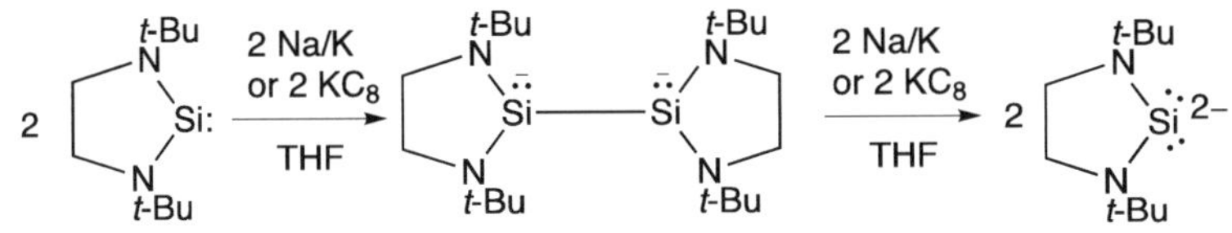

Scheme 8.10
Reactions of silylene **4a** with alikali metals.

Reduction of benzo-fused cyclic diaminosilylene **5a** with an excess amount of sodium also causes Si–Si bond formation to give 1,2-dianionic disilane **27** (Scheme 8.11). X-ray structural analysis of the salt recrystallized from THF reveals its silylsodium structure having two $[Na(thf)_2]^+$ units.[61] Similar reduction of **5a** and Na was conducted in a 3:1 molar ratio to give cyclotrisilane radical anion **28** that was isolated as a salt **28**$[Na(thf)_4]$. Since further reduction of **28** with excess sodium afforded **27**, anionic cyclotrisilane **28** is an intermediate to give **27**. The solvent-separated structure of the cyclotrisilane anion radical **28**$[Na(thf)_4]$ is determined by X-ray structural study and EPR parameters of **28** ($g = 2.0045$, $a(^{14}N) = 4.56$ G) closely resembles the reported values of structurally similar cyclotetrasilane radical anion.[62]

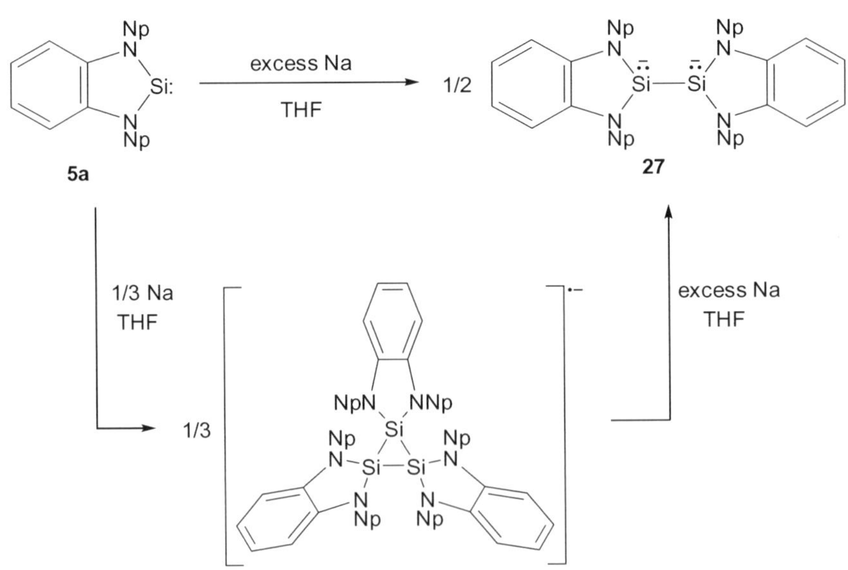

Scheme 8.11
Reactions of silylene **5a** with sodium.

Reactions of **5a** with **27** provide a mixture of cyclotrisilane dianion **29** and cyclotetrasilane dianion **30** (Scheme 8.12).[63] Treatment of a mixture of **29** and **30** with sodium leads to the reverse formation of **27**. XRD analysis exhibited that **29** and **30** existed as **29**$[Na(thf)_6]^+[Na(thf)_5]^+$ and **30**$[Na(thf)_6]^+[Na(thf)_5]^+$ salts, respectively.

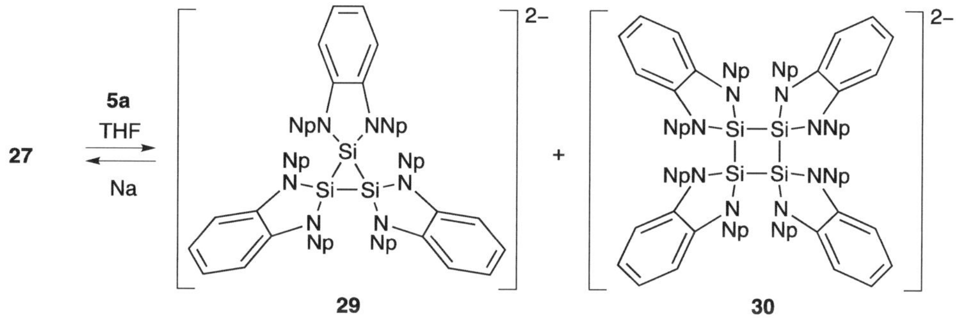

Scheme 8.12
Reaction of silylene **5a** with dianion **27**.

8.2.3.2 Group 2 metals

Very recently, alkaline earth element-silylene complexes were reported. Silylene **3a** reacts with Cp^*_2Ca, $Cp^*_2Sr(OEt_2)$, and Cp^*_2Ba to provide **31a–c** (Scheme 8.13).[64,65] Notably, complexes **31b** and **31c** decompose to diimine complexes and elemental silicon. According to several spectroscopic methods (inductively coupled plasma atomic emission spectroscopy, selected area electron diffraction, energy dispersive X-ray spectroscopy), the obtained elemental silicon is characterized as *allo*-silicon,[66] a rare allotrope of silicon. Similar thermolysis using the germanium analogs affords tetragonal germanium, an allotrope of elemental germanium.

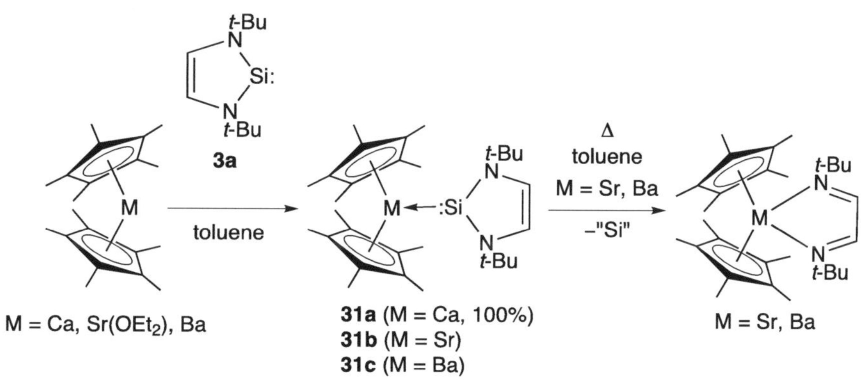

Scheme 8.13
Synthesis and reactions of group 2 metal complexes.

8.2.3.3 Group 6 metals

Irradiation of THF solution of group 6 hexacarbonyl complexes in the presence of silylene **3a** forms bis-silylene tetracarbonyl complexes **32a**, **32b**, and **32c** in 82, 84, 71%, respectively (Scheme 8.14).[67] Similar photochemical reactions using saturated silylene **4a** provided the corresponding complexes **32d–f**. The isolated yields of **4a**-coordinated complexes are low (26% for **32d**, 33% for **32e**, 36% for **32f**). In complexes **32a–f**, the 6-coordinate group 6 metal centers adopt a typical tetragonal bipyramidal geometry and the two silylene ligands located at trans-positions. Based on the comparison of the carbonyl stretching frequencies (υ_{CO}), the electronic character of **3a** and **4a** as ligands is roughly similar to that of triphenylphosphine, and saturated **4a** has increased π-acceptor character compared with unsaturated **3a**. Closely related tungsten carbonyl complex **32g** forms in 47% yield by the reaction of stable diaminosilylene **3c** with in situ generated $W(CO)_5(thf)$.[46]

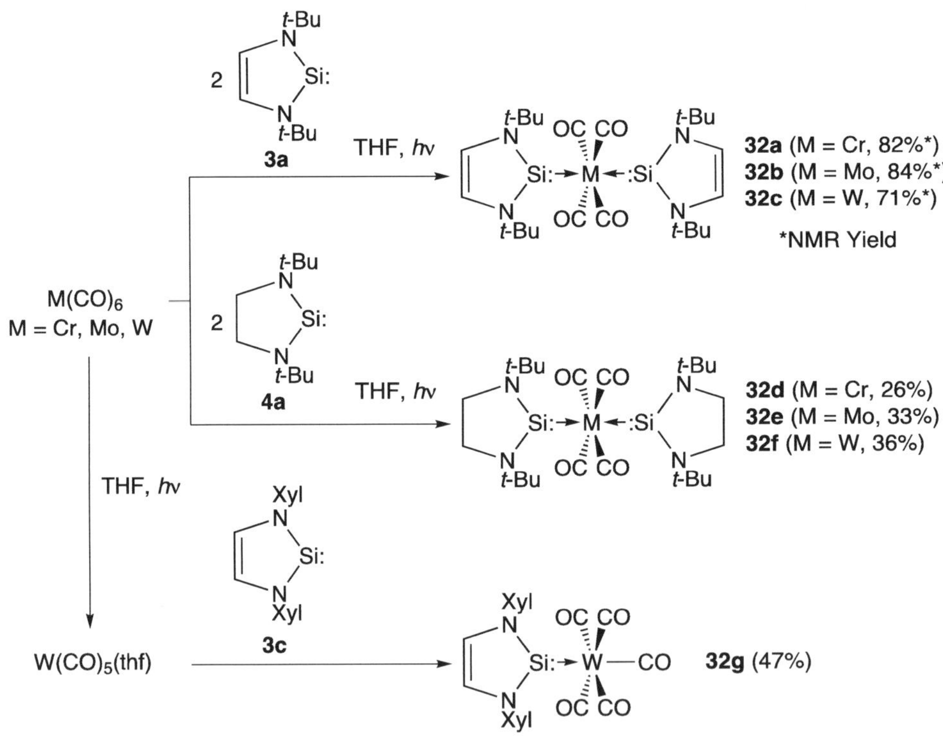

Scheme 8.14
Group 6 metal carbonyl complexes of cyclic diaminosilylenes.

Ligand exchange reaction of triethylphosphine-coordinated molybdenocene Cp_2MoPEt_3 with **3a** afforded deep-red crystals of **33** in 74% yield (Scheme 8.15). The M–H (M = Mo, W) bond insertion occurs at room temperature by the reaction of Cp_2MH_2 with **3a** to form molybdenum- and tungsten-silyl complexes **34a** and **34b** in 47 and 91% yields as yellow crystals.[68]

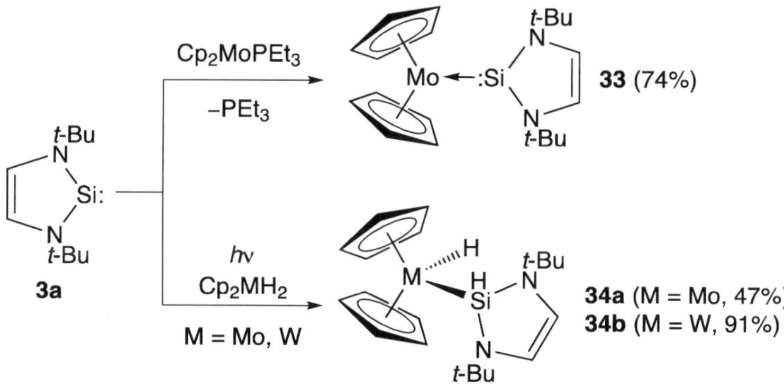

Scheme 8.15
Reactions of silylene **3a** with group 6 metallocenes.

8.2.3.4 Group 8 metals

Reactions of **3a** with group 8 metal carbonyl complexes $Fe_2(CO)_9$ and $Ru_3(CO)_{12}$ yield mono-silylene iron complex **35** and bis-silylene ruthenium complex **36**, respectively (Scheme 8.16).[3,67] No further reaction of **35** with excess **3a** proceeds. XRD analysis showed the iron atom of complex **35** adopted a distorted trigonal bipyramidal geometry and the silylene ligand located at the equatorial position. In ruthenium complex **36**, two crystallographically independent molecules existed in the asymmetric unit; one molecule has a roughly square-pyramidal structure with two **3a** ligands occupying basal positions and the other molecule adopts a trigonal bipyramidal geometry with two **3a** ligands locating at equatorial positions (Fig. 8.3). According to the NMR study, geometry change of complex **36** occurred rapidly in solution on the NMR timescale.

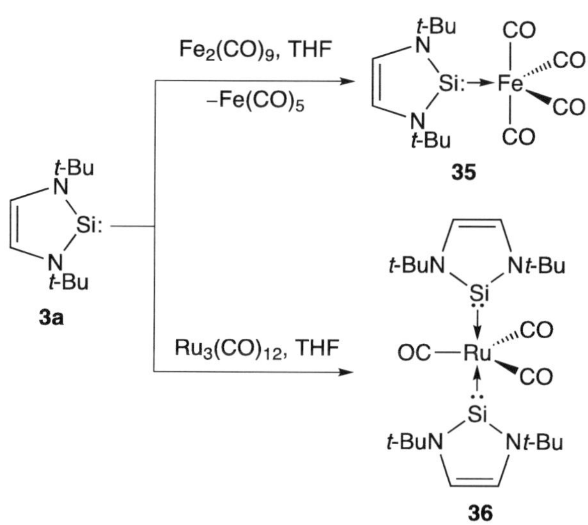

Scheme 8.16
Synthesis of group 8 metal carbonyl complexes with silylene ligand **3a**.

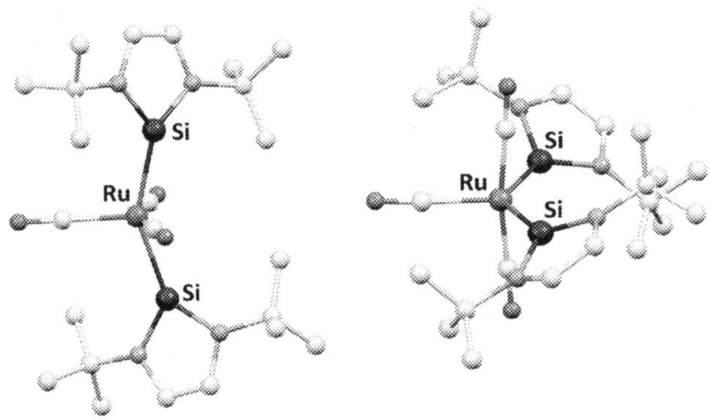

Figure 8.3
Molecular structures of **36** (CCDC: 157176).

Reaction of η^1-dinitrogen diruthenium complex **37** with four equivalents of **3a** at 50°C gives ruthenium(II)-silylene complex **38** in 81% yield with elimination of dinitrogen and a HCl adduct of **3a** (Scheme 8.17). Intramolecular C–H activation occurred on the bidentate phosphine ligand during the reaction.[69]

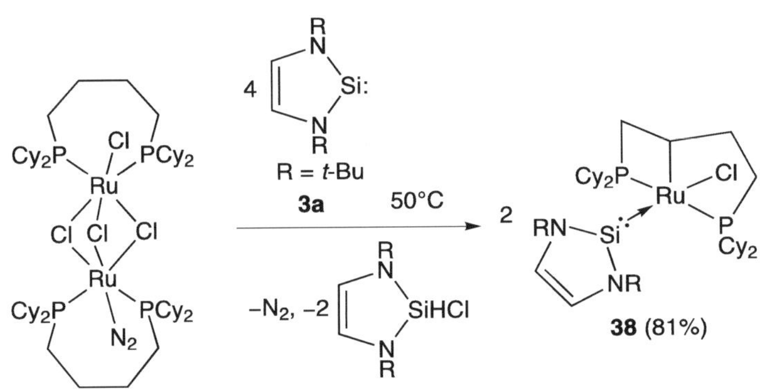

Scheme 8.17
Reaction of silylene **3a** with diruthenium complex **37**.

Silylene complex **38** is highly reactive because of its coordinative unsaturation and reactive silylene ligand. Hydrogenation of **38** in the presence of water provided chlorine- and hydrogen-bridged diruthenium complex **39** and a disiloxane (Scheme 8.18). Mechanistic survey including reaction with deuterium suggested that initial hydrolysis of **38** affords silyl complex **40a**, and the subsequent hydrogenolysis forms **40b** and silanol **40c**. Further reaction of **40b** with hydrogen gives **39**, while condensation of **40c** gives the corresponding disiloxane.

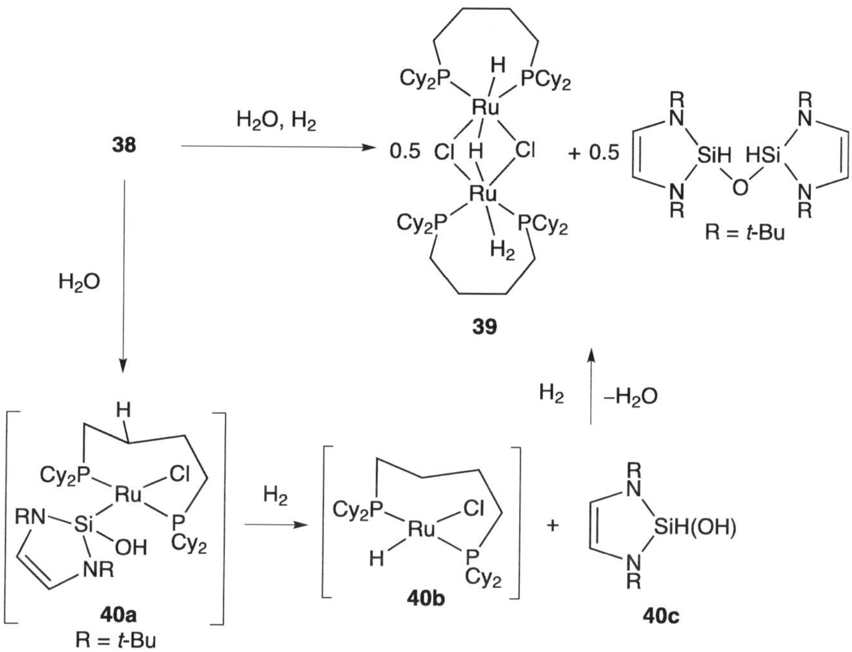

Scheme 8.18
Reactions of silylene complex **38**.

Reaction of silylene complex **38** with CO in toluene-d_8 afforded three isomeric Ru complexes **41a–c** and free silylene **3a** (Scheme 8.19).

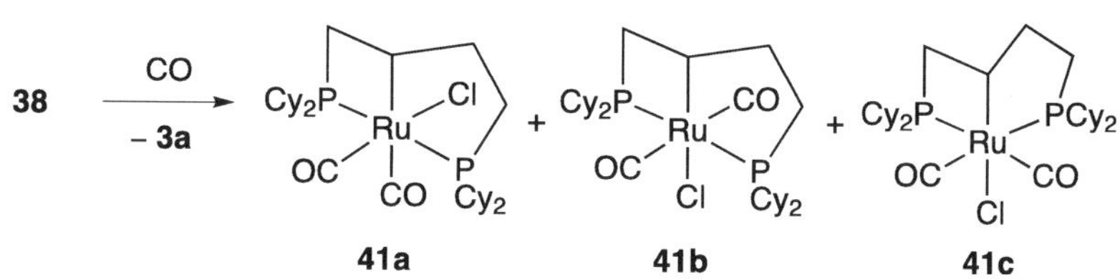

Scheme 8.19
Reaction of silylene complex **38** with CO.

Treatment of **3a** with a ruthenium η^2-dihydrogen complex affords **42** in 78% NMR yield (Scheme 8.20).[70] Complex **42** has a coordinatively unsaturated ruthenium center and one of *tert*-butyl groups of silylene ligand is close to the ruthenium atom. A similar reaction with Grubbs' first-generation complex [$RuCl_2$(=CHPh)(PCy_3)$_2$] gave a complex mixture with a trace amount of **42**. Reactions of **3a** with (Cp^*RuCl)$_4$ and [$Cp^*Ru(MeCN)_3$]$^+$[TfO]$^-$ give complexes **43** and **44** as deep-blue solid in 71% and dark yellow powder in 95% yield, respectively.[71]

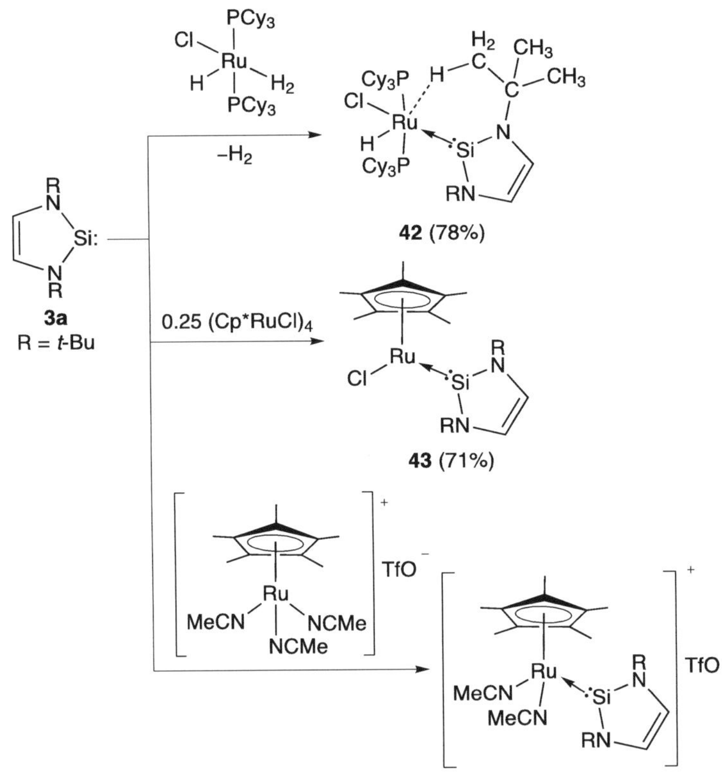

3a
R = *t*-Bu

42 (78%)

43 (71%)

44 (95%)

Scheme 8.20
Synthesis of ruthenium complexes **42**–**44**.

Compound **43** reacts with primary silanes (R′SiH$_3$, R′ = Ph, Hex) to provide diruthenium complexes **45a** and **45b** (Scheme 8.21). Since **43** is a 16-electron complex, Si–H oxidative addition of silanes to **43** would be an initial step of the formation mechanism of **45**.

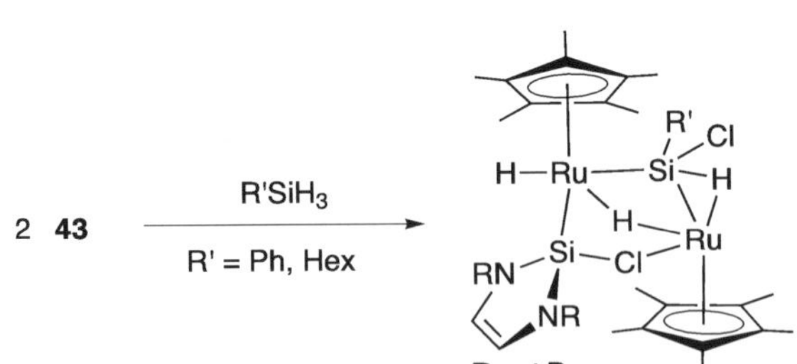

45a (R' = Ph, 88%)
45b (R' = Hex, 85%)

Scheme 8.21
Reaction of silylene complex **43** with hydrosilanes.

Thermolysis of cationic ruthenium complex **44** in THF results in the formation of dicationic diruthenium complex **46** featuring η^5-coordinating **3a** to one of the ruthenium atoms (Scheme 8.22).

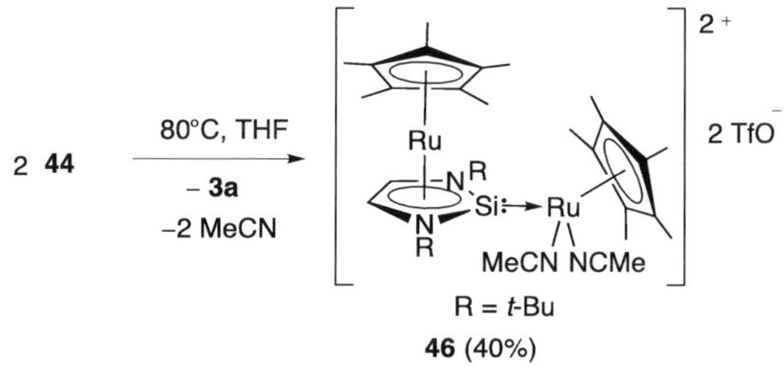

Scheme 8.22
Thermal reaction of silylene complex **44**.

A ruthenium(II) complex bearing silylene **5a** as a ligand (**47**) was synthesized as a red crystalline solid by the reaction of ruthenium(0) η^6-toluene complex **48** with hydrochlorosilane **49a** (Scheme 8.23).[72] The formation of **47** presumably proceeds by the oxidative addition of Si–H bond in **49a** to **48** and the subsequent 1,2-chlorine migration from silicon to ruthenium. Chlorosilyl complex **50** formed by the oxidative addition of Si–Cl bond in dichlorosilane **49b** to **48**.

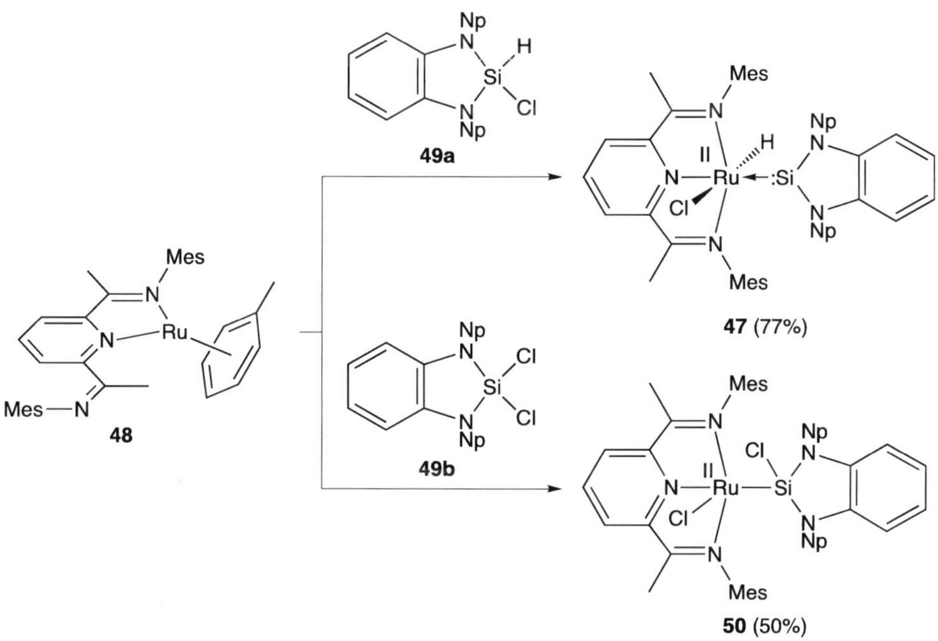

Scheme 8.23
Synthesis of silylene complex **47** and related reaction.

Ligation of **5a** to **48** and **52** provide ruthenium(0) and ruthenium(II) silylene complexes **51** and **53** (Scheme 8.24).[72] Silylene moiety can dissociate from **47** and **53**; treatment of silylene complex **47** or **53** with **48** under N_2 atmosphere in THF solution afford **51** and ruthenium(II)-THF complex $[N_3]RuClX(thf)$ ($[N_3] = 2,6\text{-}(MesN = CMe)_2C_5H_3N$, X = H, Cl). Similarly, chlorosilyl complex **50** reacts with **48** to provide **51**. XRD analysis showed that ruthenium(II) silylene complex **47** has a pseudo-octahedral geometry and hydride and chloride ligands sit at trans-positions. Five-coordinate ruthenium atom of **51** adopts an almost square pyramidal structure and the silylene moiety occupies the apical position. The Ru–Si bond lengths of Ru(II) silylene complexes **47** (2.304(1) Å) and **53** (2.3356(9) Å) are longer than those of five-coordinate Ru(0) complex **51** (2.227(2) and 2.232(2) Å).

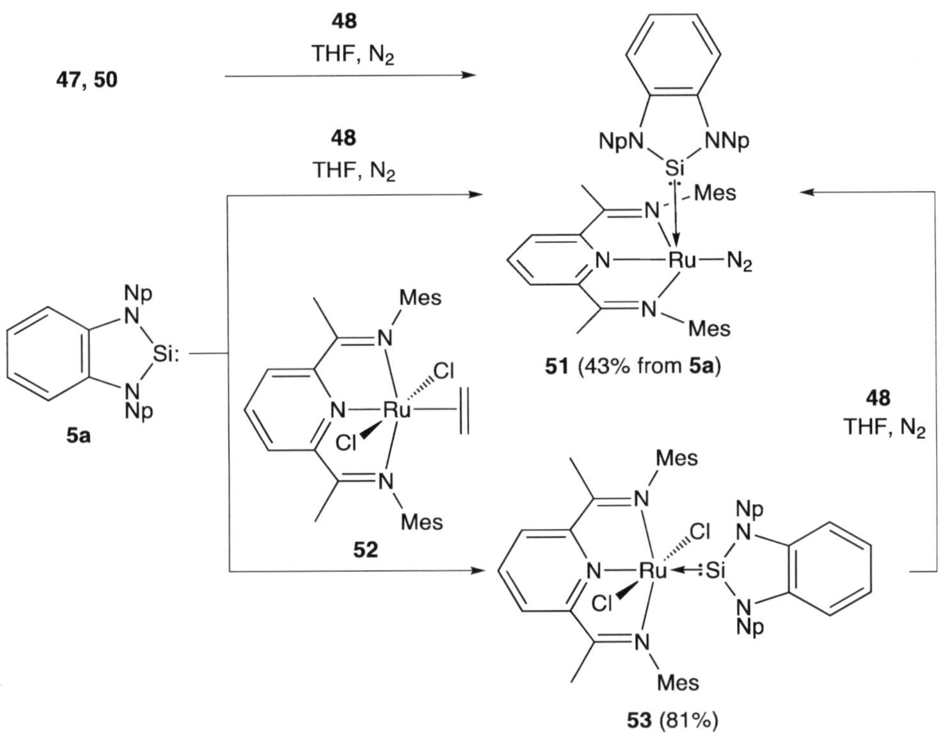

Scheme 8.24
Synthesis of ruthenium complex **51** with N_2 ligand.

8.2.3.5 Group 9 metals

Cationic rhodium-silylene complexes (**54a** and **54b**) are synthesized quantitatively by the ligand exchange reaction of $[Rh(cod)_2]^+[BAr^F{}_4]^-$ (Ar^F = 3,5-bis(trifluoromethyl)phenyl) with six equivalents of **3a** or **4a** in hexane (Scheme 8.25).[73] In contrast to tetrahedral geometry of four-coordinate Ni(0)-silylene complexes **64a** and **65** (vide infra), the cationic four-coordinate rhodium centers of **54a** and **54b** feature a square-planar geometry.

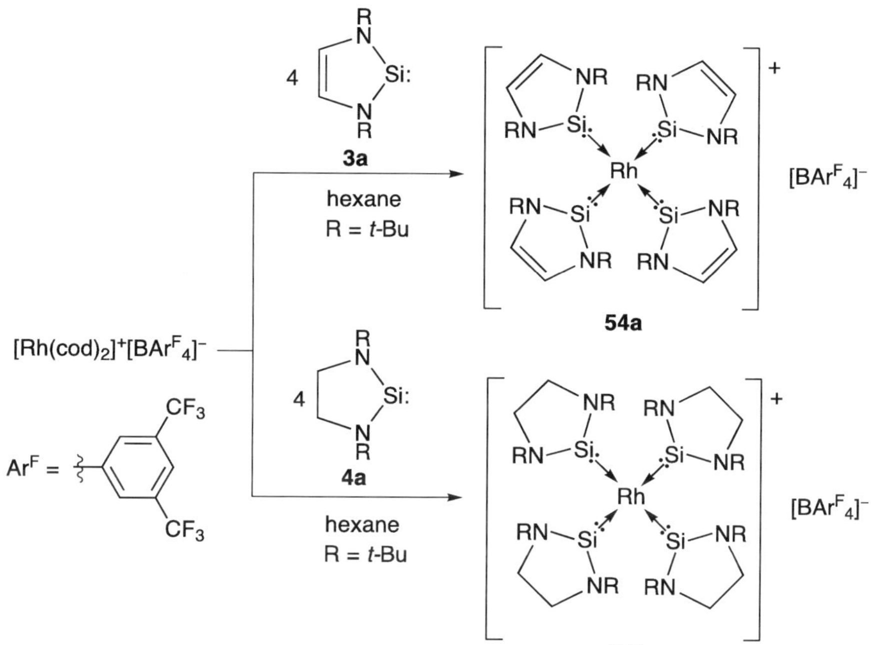

Scheme 8.25
Synthesis of rhodium-silylene complexes **54a** and **54b**.

8.2.3.6 Group 10 metals

Ligand exchange reactions of silylene **3a** with $Ni(CO)_4$ and $Ni(cod)_2$ give bis-silylene nickel(0) carbonyl complex **55a** and tri-silylene nickel(0) complex **55b** (Scheme 8.26).[74,75] Complex **55a** adopts a tetrahedral geometry, while the geometry of nickel atom in complex **55b** is trigonal planar. A structurally similar tri-silylene nickel complex **56** was also reported (Scheme 8.26).[75] The Ni–Si bond lengths of **55b** (av. 2.1515(13) Å) and **56** (av. 2.1577(8) Å) are almost identical. *N*-aryldiaminosilylenes **3b** and **3d** react with $Ni(cod)_2$ to afford complexes **57a** (Ar = Dip) and **57b** (Ar = Mes) in 64% and 53% yields, respectively (Scheme 8.26).[45] In these cases, different from the formation of **55b** and **56**, the third silylene coordination to nickel is suppressed probably due to steric reasons.

The reaction of $Pd[P(t\text{-}Bu)_3]_2$ with two equivalent of **4a** gives silylene-bridged dipalladium complex **58** as red crystals in 63% yield and its structure is confirmed by XRD (Scheme 8.27). Complex **58** is in an equilibrium with tetra-silylene complex **59**.[76] Reaction with four equivalent of **4a** forms **59** as a major product, but its isolation is prevented by the equilibrium with phosphine. Treatment of $PdMe_2(cod)$ with six-equivalent of **4a** also gives **59** in phosphine-free condition, however, **59** decomposes upon concentration.

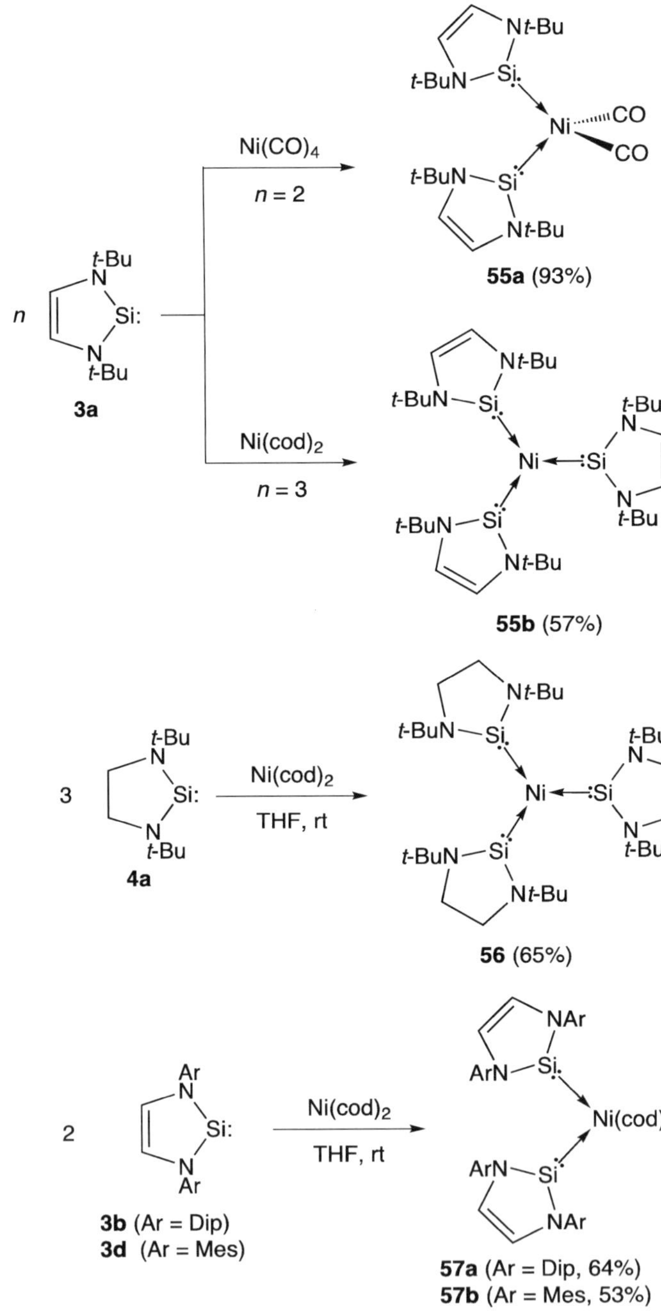

Scheme 8.26
Synthesis of nickel complexes of silylenes **3a**, **4a**, **3b**, and **3d**.

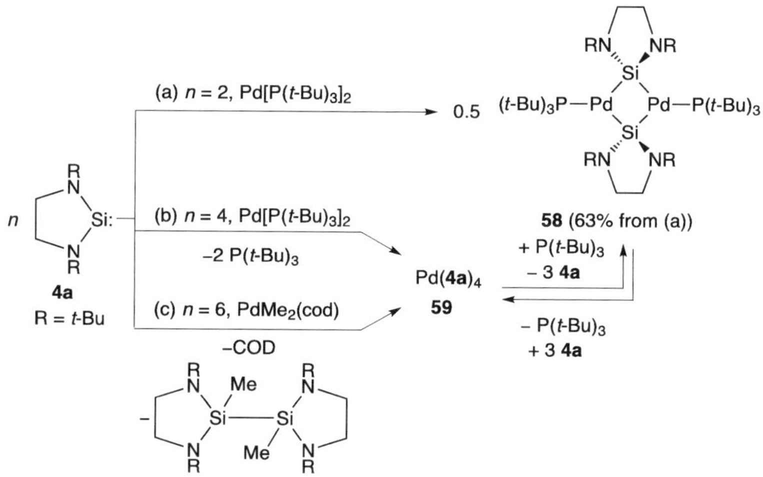

Scheme 8.27
Synthesis of palladium complexes with silylene **4a**.

Treatment of $Pd(PPh_3)_4$ with **3a** affords silylene-bridged dipalladium complex **60a** as dark-red crystalline solid in 50% yield (Scheme 8.28).[77] Two palladium atoms in **60a** adopt trigonal planar geometries with the Pd−Si bond lengths of 2.4154(4) and 2.3996(3) Å. This dimeric structure was preserved in solution based on the ^{29}Si NMR spectrum with the triplet signal of divalent silicon at 109.5 ppm due to two equivalent phosphorus nuclei ($^2J_{SiP}$ = 21.5 Hz). Compound **60a** has a catalytic activity for the Suzuki−Miyaura coupling reaction; coupling products of arylboronic acids with aryl bromides in DME are obtained in good yields. A representative example is shown in Scheme 8.29. η^3-Allyl palladium complex bearing silylene ligand (**61**) was synthesized in 54% yield by the reaction of **3a** and a half equivalent of η^3-allylpalladium chloride dimer.[78] Palladium complex **61** acts as a catalyst for the Heck reaction of aryl bromides with styrene (Scheme 8.29).

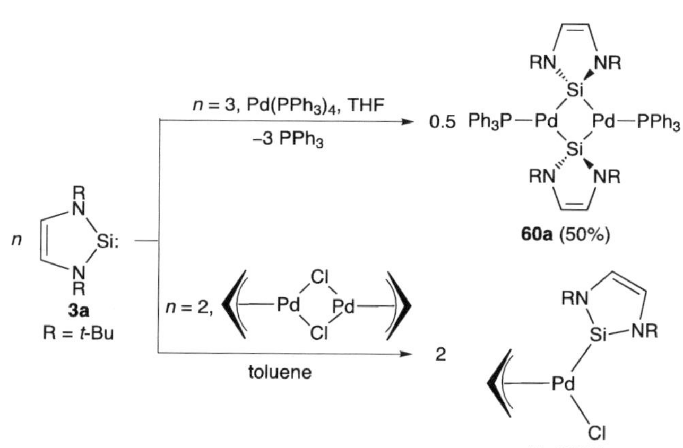

Scheme 8.28
Synthesis of palladium complexes with silylene **3a**.

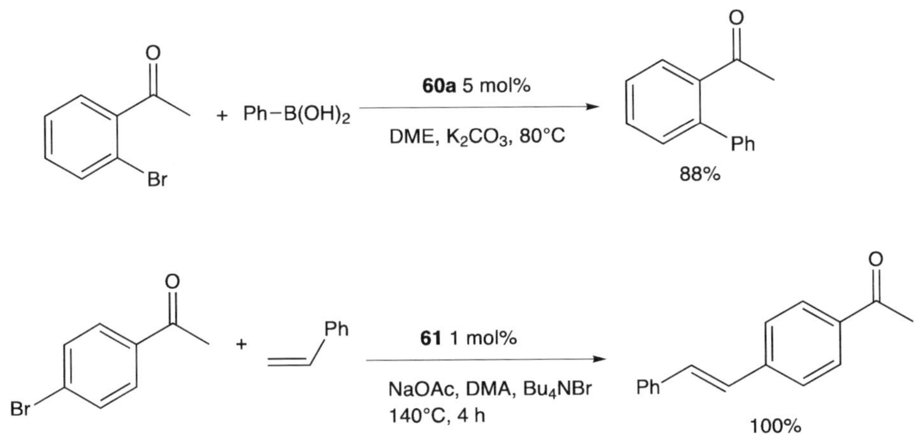

Scheme 8.29
Application of complexes **60a** and **61** for organic reactions.

Reaction of Pd[P(t-Bu)$_3$]$_2$ with three equivalents of **3a** forms palladium tri-silylene complex **62**.[76] However, silylene-bridged dipalladium complex **60b** is obtained upon the recrystallization process instead of **62** due to the equilibrium between the two species (Scheme 8.30). Isolation of **62** is achieved by employing phosphine-free condition; reaction of PdMe$_2$(cod) with **3a** affords **62** in 71% yield. In contrast to isolable nickel analog **55b**, compound **62** dissociates reversibly into **63** and free **3a** in solution. Ligand exchange reaction of bis(NHC) palladium complexes with **3a** provides **62**.[79] Reverse ligand exchange from **63** to palladium NHC-complex does not occur.

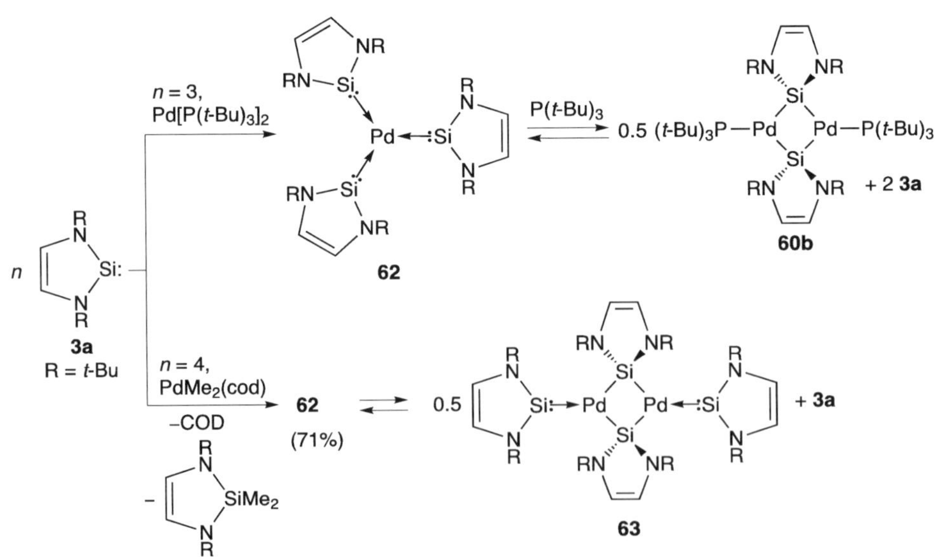

Scheme 8.30
Synthesis of silylene-palladium complexes **62** and **63**.

The coordination mode of **5a** is different from those of **3a** and **4a** due to the different steric environment. Three or four **5a** can coordinate to nickel(0) center to form four-coordinate nickel(0) complexes **64a** and **65** with tetrahedral geometry (Scheme 8.31).[80,81] In the reactions of **5a** with $NiCl_2(PPh_3)_2$, one of **5a** acts as a reducing reagent for nickel(II).

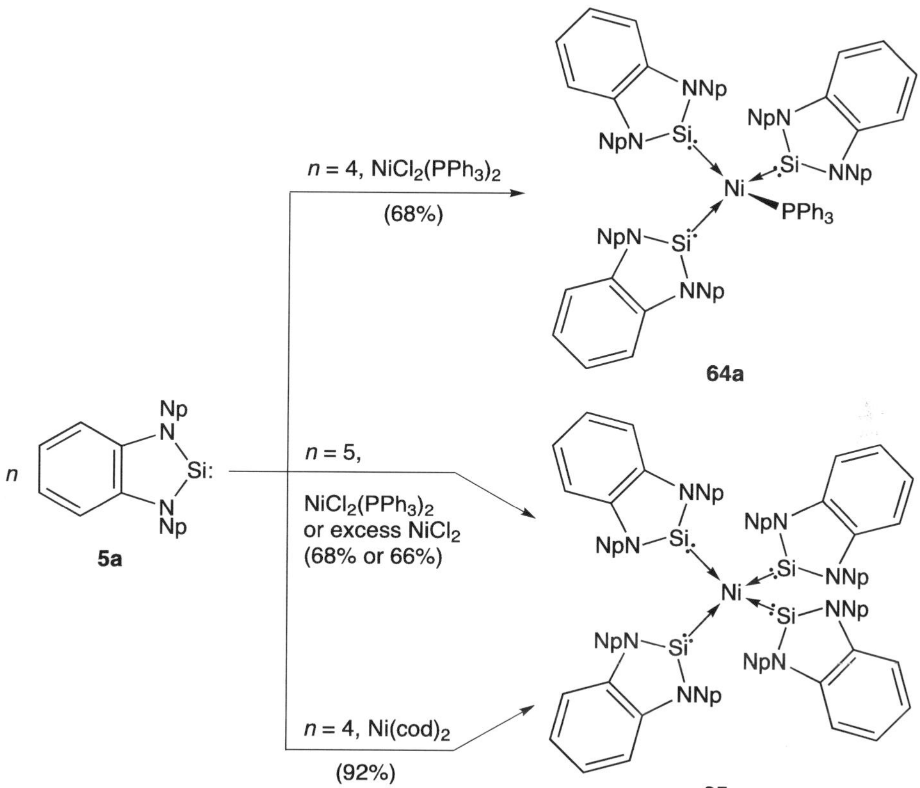

Scheme 8.31

Synthesis of silylene-nickel complexes **64a** and **65**.

Reactions of **5a** with $PdCl_2(PPh_3)_2$ and $PtCl_2(PPh_3)_2$ give palladium(II) and platinum(II) complexes having two chlorosilyl ligands and two silylene ligands, *trans*-**66a** and *trans*-**66b** (Scheme 8.32).[80,81] No metal(0) tetra-silylene complex similar to **64b** is observed in these reactions. Complexes *trans*-**66a** and *trans*-**66b** have square-planar tetracoordinate metal(II) centers. In complex *trans*-**66a**, Pd–Si(silylene) bond of 2.269(2) Å is shorter than Pd–Si(silyl) bond (2.437(2) Å). Similarly, bond of Pt–Si(silylene) of 2.266(1) Å in *trans*-**66b** is shorter than that of Pt–Si(silyl) (2.426(1) Å). In solution, equilibria between *cis–trans* isomers and a rotamer were suggested. Treatment of Pt $(PPh_3)_4$ with three equivalent of **5a** results in formation of **64b** having a tetrahedral geometry around the platinum(0) center.[81]

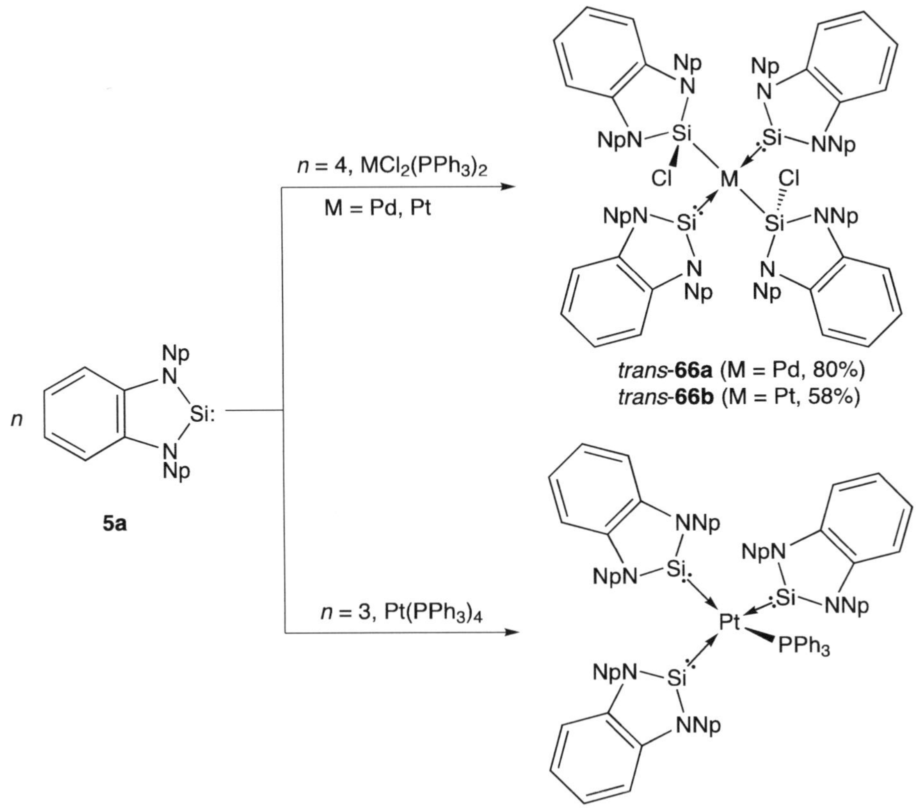

5a

trans-**66a** (M = Pd, 80%)
trans-**66b** (M = Pt, 58%)

64b

Scheme 8.32
Synthesis of group 10 metal complexes of silylene **5a**.

8.2.3.7 Group 11 metals

A silylene copper(I) complex (**67**) is prepared in 73% yield by the ligand exchange of CuI $(PPh_3)_3$ with **5a** (Scheme 8.33).[81] The copper atom of **67** adopts a distorted tetrahedral geometry with the significantly different two P–Cu–Si bond angles of 119.80(12) degrees and 108.70(13) degrees. The Cu–Si bond length is 2.289(4) Å.

8.2.3.8 Group 12 metals

Silylene group 12 complex is still elusive species. Oxidative addition of silylene **3a** across the O–Hg bond of bis(phosphoryl)mercury(II) compound affords bis(silyl)mercury(II) **68a** at −70°C (Scheme 8.34).[82] Compound **68a** decomposes at room temperature to silane **68b** and mercury metal even in the dark.

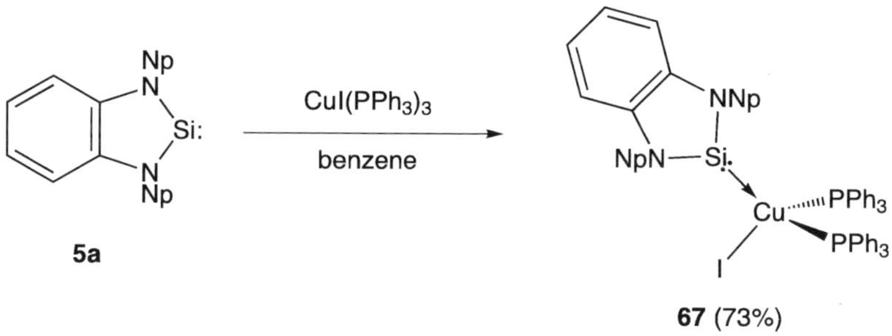

Scheme 8.33
Synthesis of the silylene–copper complex **67**.

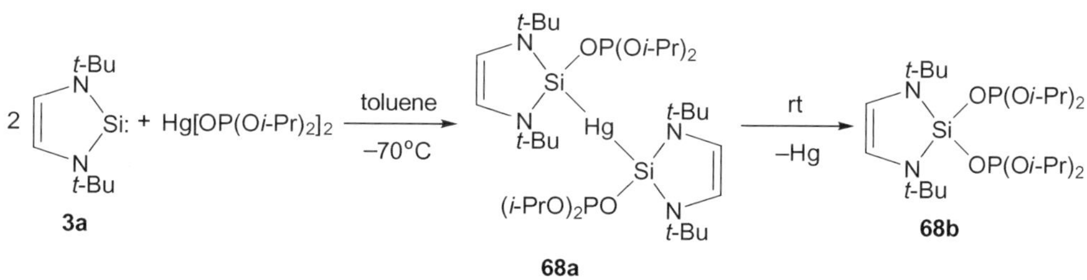

Scheme 8.34
Reaction of silylene **3a** with a mercury complex.

8.2.3.9 Other metals

Silylene **5a** coordinates to Cp_3M (M = Y and Yb) to form **69a** and **69b** (Scheme 8.35).[83] The central metal atoms of **69a** and **69b** adopt a distorted trigonal monopyramidal geometry and silylene ligands occupy the apical position. The Si–Y and Si–Yb bond lengths are 3.038(2) and 2.984(2) Å, respectively. In solution, **69a** and **69b** are in dissociative equilibrium with the starting materials.

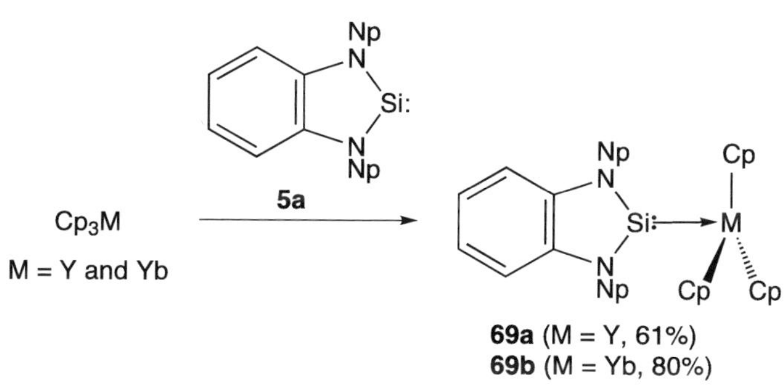

Scheme 8.35
Synthesis of complexes **69a** and **69b**.

Bis(pentamethylcyclopentadienyl)samarium (Cp^*_2Sm) reacts with **3a** in toluene to afford samarium(II)-silylene complex **70** as a purple solid in 90% yield (Scheme 8.36).[84] Ligand exchange of **70** in THF forms $Cp^*_2Sm(thf)_2$ and free **3a**. In the solid state, silylene moiety of **70** is bent relative to the Sm–Si bond direction; the N–Si–Sm bond angle of 118.2(1) degrees is much different from the other N–Si–Sm angle of 150.8(1) degrees. The unsymmetrical structure in the solid state would be due to filling the coordination site; one *tert*-butyl group is much closer (Sm...C; 3.396(4) Å) than the other *tert*-butyl group (4.741 (4) Å). The Sm–Si bond length of **70** is 3.1903(10) Å.

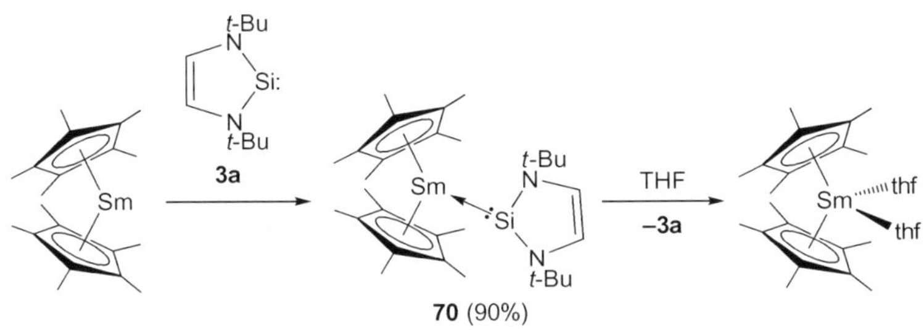

Scheme 8.36
Synthesis and reactions of silylene-samarium complex **70**.

8.2.4 Other Reactions

Muonium is a single-electron atom whose nucleus is the positive muon. Although muonium is chemically equivalent to a hydrogen atom, its mass is only one-ninth of the mass of a hydrogen atom. Muonium-coordinated silicon-centered radicals are generated by the exposure of **3a**, **3b**, **3e**, **3f**, and **4a** with muon beam, whose structures are discussed based on the muon hyperfine coupling constants obtained by transverse field muon spin rotation (μSR) spectroscopy (Scheme 8.37).[47,85–87] Addition of muonium to **3b** occurred at the

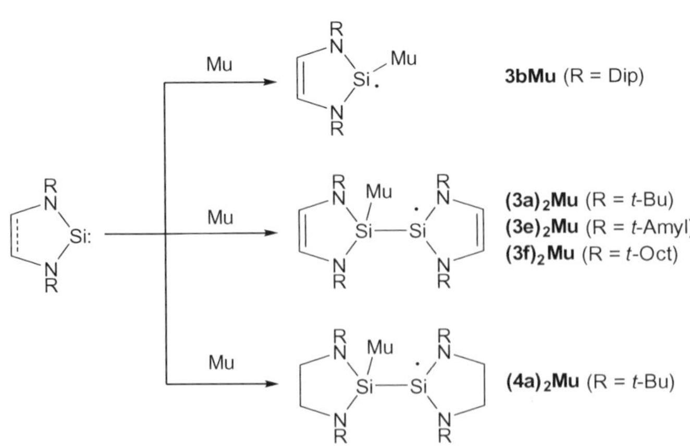

Scheme 8.37
Reactions of cyclic diaminosilylenes with muonium.

divalent silicon to give silyl radical **3bMu**, while muonium addition to *N*-alkyl diaminosilylenes (**3a**, **3e**, **3f**, **4a**) afforded the corresponding disilanyl radicals by the generation of muoniated silyl radicals similar to **3bMu** and the subsequent Si–Si bond formation with silylene.

8.3 Diaminosilylenes Derived From β-Diketiminate

8.3.1 Synthesis and Molecular Structures

In 2006, Driess et al. reported new diaminosilylenes with ylide-like structures derived from lithium β-diketiminate **71** (Scheme 8.38).[88] Reductive debromination of dibromosilane **72**, which was prepared by the reaction of lithium β-diketiminate **71** with $SiBr_4$, gave silylene **73** as yellow crystals in 77% yield. The ^{29}Si resonance due to two coordinate silicon nucleus appeared at +88.4 ppm. X-ray analysis showed a planar six-membered ring with apparent alternation of the endocyclic C–C bond lengths (Fig. 8.4). The Dip groups on the nitrogen atoms of the ligand are crucial for isolation of this silylene[89,90] because reduction of the dibromosilanes with cyclic diamino ligand where Dip groups were replaced with less bulky 2,6-dimethylphenyl or 2,6-diethylphenyl groups, with KC_8 did not lead to the corresponding silylenes, and reduction of dibromosilane where Dip groups were replaced with *t*-Bu groups, gave formal dimer of the corresponding silylene. Dissociation of Si = P bond of phosphasilene **74** also gave silylene **73**.[91] The PH unit in **74** was trapped by bulky *N*-heterocyclic carbene (Scheme 8.38).

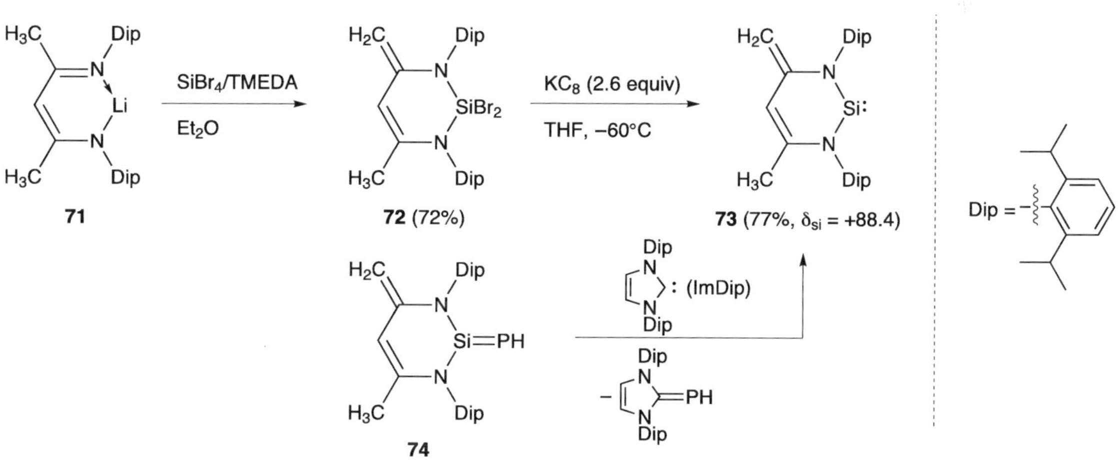

Scheme 8.38
Synthesis of ylide-like diaminosilylene **73**.

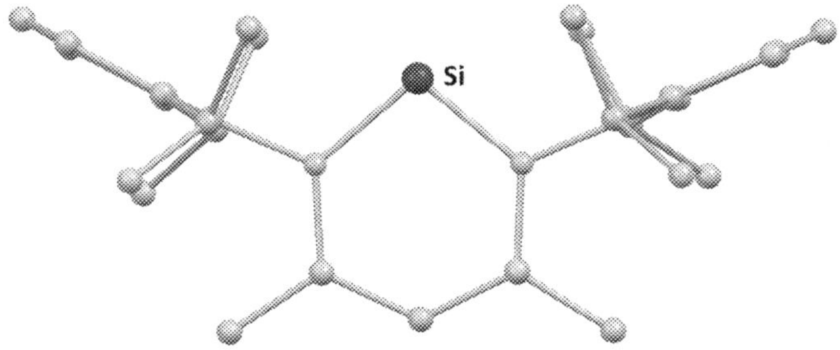

Figure 8.4
Molecular structure of silylene **73** (CCDC: 621936).

Other silylenes derived from **73** will be discussed in the next section.

8.3.2 Reactivity

8.3.2.1 Insertion reactions

Although silylene **73** does not insert into the H–H bond of dihydrogen,[92] it inserts into various single bonds. For instance, silylene **73** inserts into carbon–halogen and silicon–halogen (R–X) bonds of haloalkanes and halosilanes (Me–I, $PhCH_2$–Br, CH_2Cl–Cl, CH_2Cl–I, CH_2Br–Br, $CHCl_2$–Cl, CH_3CHCl_2–Cl, $SiHCl_2$–Cl, CH_3SiCl_2–Cl) to the corresponding insertion products **75** as major products (Scheme 8.39).[93] In the case of CH_2Br_2, $CRCl_3$ (R = H, CH_3), and $RSiCl_3$ (R = H, Cl), dibromosilane **72**, dichlorosilane **76a**, and monochlorosilane **76b** (Fig. 8.5), respectively, were formed as by-products, while no products with the Si–Si bond as found in the reaction of West's diaminosilylene with haloalkanes were observed (see Section 8.2.2). Bulky Me_2CCl_2, Ph_2SiCl_2, and $PhSiCl_3$ did not react with silylene **73**. The proposed mechanism for formation of these products **75** involves initial 1,4-addition of haloalkanes and halosilanes to give **75′** as a kinetic product due to its ylide-like structure (see next section), which was not observed experimentally, and its subsequent rearrangement to give the 1,1-addition thermodynamic product **75**, although direct formation of the 1,1-addition products was not ruled out. Interestingly, silylene **73** inserts into C–F bonds in C_6F_6, $CF_3C_6F_5$, 2,3,5,6-tetrafluoropyridine giving **77a–c**, while it inserts into C–H bond in pentafluorobenzene and 1,3,5-trifluorobenzene affording **77d** and **77e**. Insertion reaction of silylene **73** into the C–H bond of phosphorus ylide giving **78** was also reported.[94]

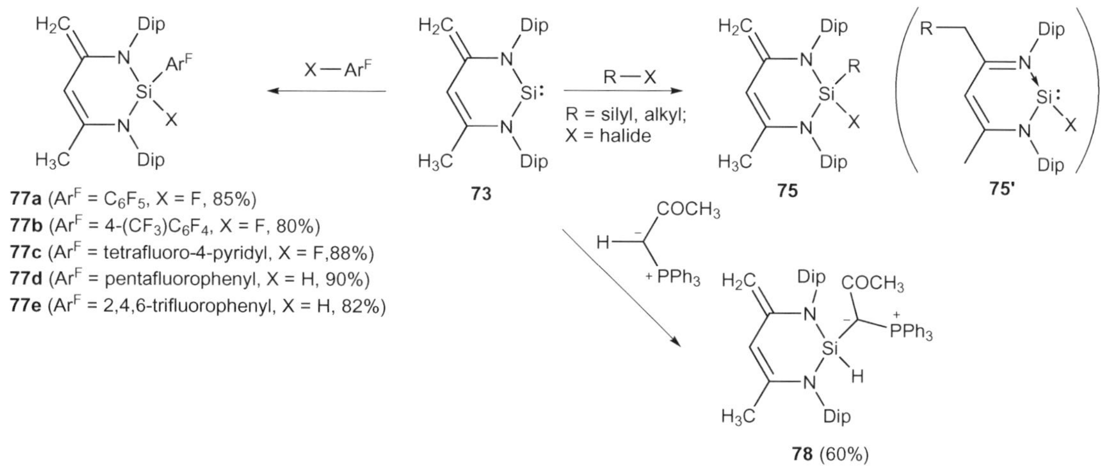

Scheme 8.39
Insertion of silylene **73** into R—X bonds.

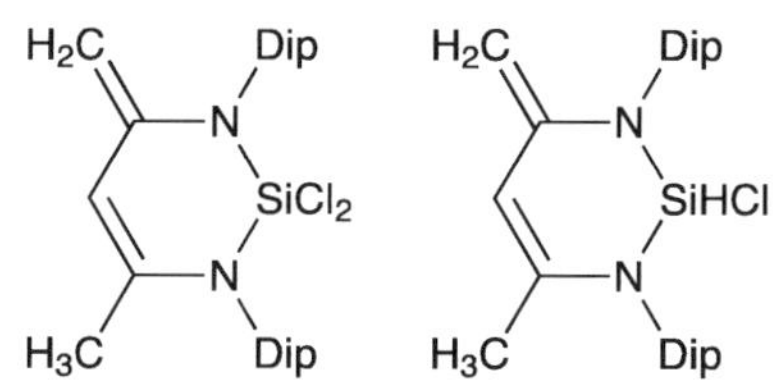

Figure 8.5
Products of reaction of **73** with chloromethanes and chlorosilanes.

Silylene **73** also inserts selectively into the terminal N—H bond in NH_3,[92] H_2NNH_2,[95] and H_2NNHMe[95] to give the corresponding adducts **79a—c**, respectively (Scheme 8.40). In contrast, **73** reacted with diphenylhydrazone to give dearomatized [1 + 4] cycloadduct **97** instead of the corresponding N—H insertion product (Scheme 8.44).[95] Similarly to N—H bond insertion reaction, reaction of **73** with an excess amount (20 equiv.) of PH_3 gave P-H insertion product **79d**.[96] On the other hand, reaction of **73** with one equivalent of AsH_3 gave blue crystals of arsasilene **80** in 48% yield. In solution, tautomerization between **80** and As—H insertion product **80′** was observed by NMR spectroscopy: the ratio of **80** to **80′** was 7:3 at room temperature.[96] Difference in stability between 1,1-addition product such as **79d** and **80′** and double bond activation product such as **80** was studied by the density functional theory (DFT) calculation of model compounds. Silylene **73** inserts into P—P bond in P_4 and the P—P bond in the resulting **81** to give **82**.[97]

Silylene **73** also inserts into Zn—C, Al—C, Al—H, and C—Li bonds to give the corresponding 1,1-adducts.[98]

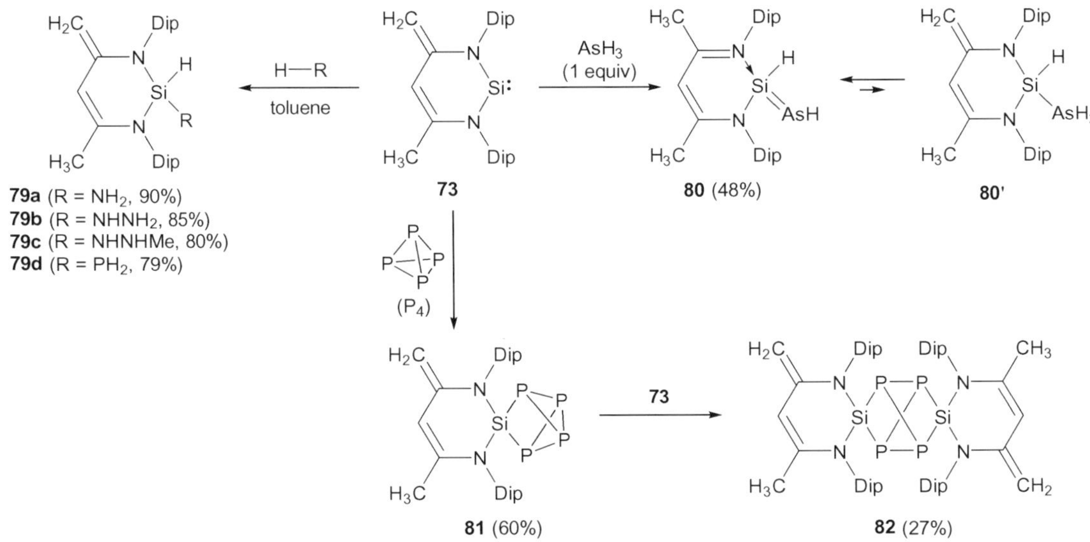

Scheme 8.40
Reactions of silylene **73** with H—N, H—P, H—As, and P—P bonds.

8.3.2.2 1,4-Addition reactions

Silylene **73** has a ylide-like (or zwitterionic) structure **73′** with nucleophilic terminal methylene carbon and electrophilic Si(II) centers which contributes to the following unique 1,4-addition reactions. The reaction of **73** with Me_3SiOTf ($OTf = OSO_2CF_3$) gave three-coordinate silylene **83a** and silyltriflate **83b** and the former underwent thermal isomerization to **83b** (Scheme 8.41).[88] The thermodynamic preference of **83b** over **83a** was supported by DFT calculations of model compounds. Reaction of **73** with C_6F_5OH and Ph_3SiOH gave 1:2 adducts **84a** and **84b**, which would be formed by the 1,4-addition of the phenol and the silanol followed by further addition of the phenol and silanol to the silylenes resulting from the 1,4-addition similar to **83a**.[99] Interestingly, mild reaction of **73** with H_2O gave a 2:1 adduct, siloxy silylene **85** instead of the desired hydroxysilylene **86**.[100] When $H_2O \cdot B(C_6F_5)_3$ was used for the hydrolysis of **73**, silaformamide-borane complex **87** was obtained in 67% yield.[100] The double bond character in **87** was supported by ^{18}O labeling experiment and theoretical calculations. Coordination of $B(C_6F_5)_3$ and imine nitrogen atom to oxygen and silicon atoms, respectively, would stabilize the silaformamide structure rather than hydroxysilylene structure like **86**. Ylide-like structure of **73** was also supported by its reactions with Lewis acid $B(C_6F_5)_3$ and Brønsted acid $[H(Et_2O)_2]^+[B(C_6F_5)_4]^-$ to give the corresponding cations **88** and **89**, respectively.[101] In contrast to reaction with H_2O, addition of H_2S to silylene **73** gave silathiaformamide **90** in 54% yield probably via intermediates **90′** and **90″** which were not observed.[102] Thermodynamic preference of **90** over **90′** and **90″** were supported by DFT calculations.

Reaction of **73** with HCl and NH_3BH_3 gave 1,1-adducts **76b** and **91**, respectively, probably via 1,4-adduct similar to **83a** followed by 1,4-migration.[103]

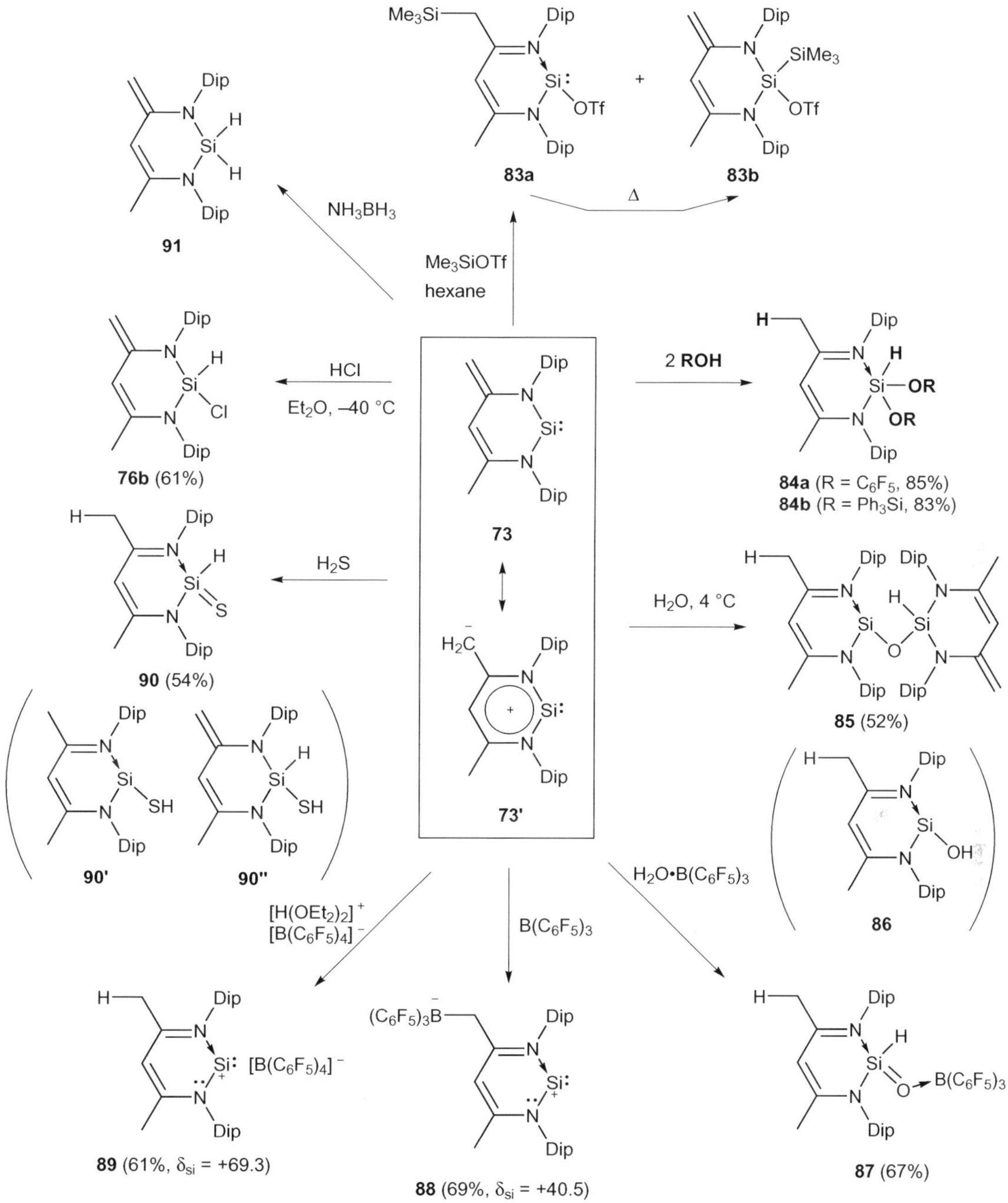

Scheme 8.41
Insertion reactions of silylene **73**.

8.3.2.3 Cycloadditions and related reactions

Cycloaddition reactions of silylene are known to give various kinds of cycloadducts. Ylide-like silylene **73** shows unique reactivity toward acetylenes (Scheme 8.42).[104] Reaction of silylene **73** with diphenylacetylene at room temperature gave the corresponding [1 + 2] cycloadduct **92a** quantitatively, while treatment of **73** with parent acetylene and phenylacetylene gave C–H insertion products **93a** and **93b** in good yields. When the reactions of **73** with acetylene and phenylacetylene were performed at low temperature (−78°C), [1 + 2] cycloadducts **92b** and **92c** were formed as a sole product. Interestingly, in the presence of a catalytic amount of [1 + 2] cycloadduct **92a–c**, [1 + 2] cycloadducts **92b** and **92c** were obtained as sole products even at room temperature, suggesting the autocatalytic reaction. The proposed mechanism for the autocatalytic cycle of **92a–c** involves the coordination of nucleophilic CH_2 moiety of the [1 + 2] cycloadduct to the terminal C–H moiety of acetylene followed by [1 + 2] addition of the coordinated acetylene to unreacted ylide-like silylene **73** (Scheme 8.43). Dibromosilane **72** also works as a catalyst for selective formation of [1 + 2] cycloadducts **92b** and **92c**. The DFT calculation of model compounds predicted that [1 + 2] cycloadducts are kinetic products, while C–H insertion products are thermodynamic products which are formed by 1,4-addition of terminal C–H bond of the acetylene to ylide-like silylene and the subsequent rearrangement similarly found in other 1,4-addition of silylene **73**.

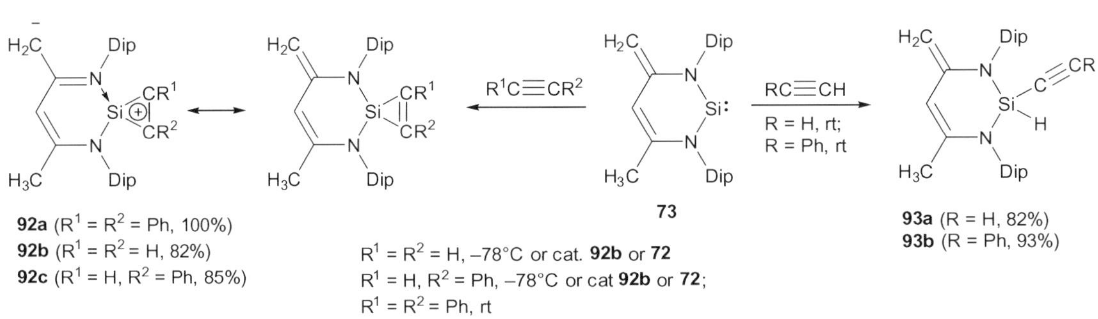

Scheme 8.42
Reactions of silylene **73** with acetylenes.

Ylide-like silylene **73** reacted with 2,3-dimethyl-1,3-butadiene (DMB), diisopropyl azodicarboxylate, and 3,5-di-*tert*-butyl-*o*-benzoquinone to give [1 + 4] cycloadducts **94**,[105] **95**,[106] and **96**[107] in high yields (Fig. 8.6). When 1,1,4,4-tetramethyl-1,3-butadiene was used, no reaction was observed probably due to steric reasons.[105]

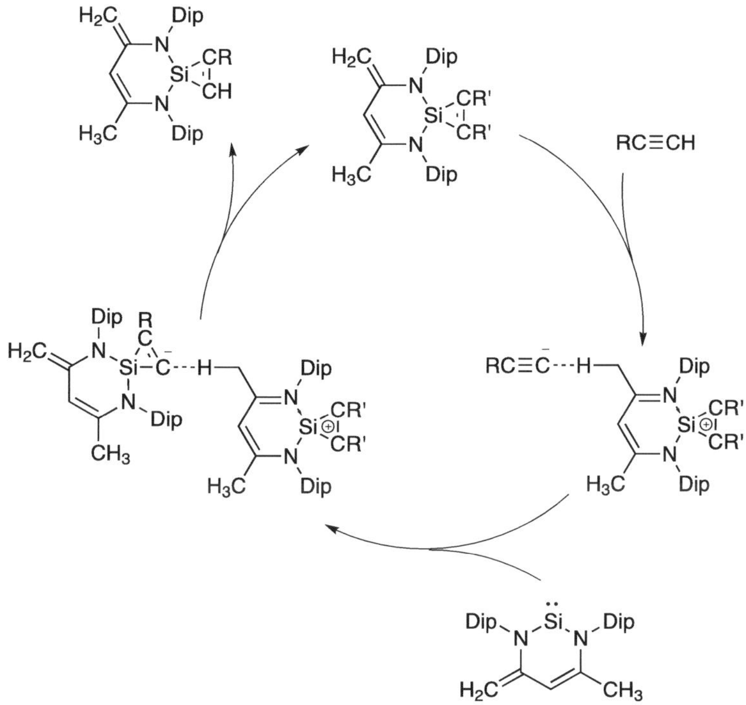

73

Scheme 8.43
Proposed mechanism of autocatalytic cycle of **92a–c**.

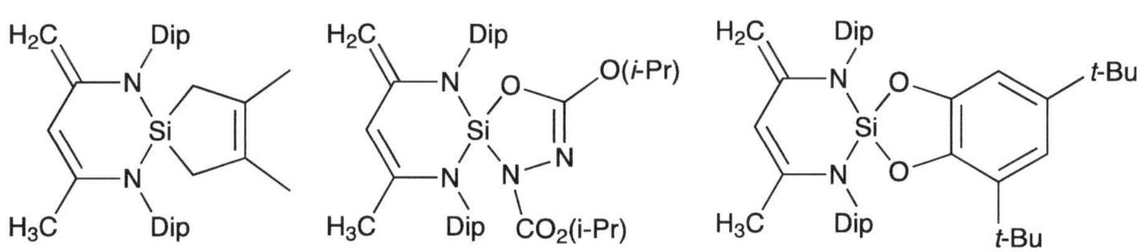

94 (90%) **95** (67%) **96** (78%)

Figure 8.6
Examples of cycloadducts of silylene **73**.

When diphenylhydrazone, benzophenone, and benzoylpyridine were used, dearomatized [1 + 4] cycloadducts **97**,[95] **98a**,[108] and **99**[106] were formed (Scheme 8.44). Dearomatized product **98a** undergoes thermal isomerization to give rearomatized product **98b**.[108] Addition of benzophenone giving **98a** and isomerization of **98a** to **98b** were stereoselective: only (R, S)-**98a** and (R*,R*)-**98b** were formed.[108] Similar dearomatized [1 + 4] cycloadduct and rearomatized product were also obtained when benzylideneacetone[108] and 4,4′-bis (dimethylamino)thiobenzophenone[109] were used.

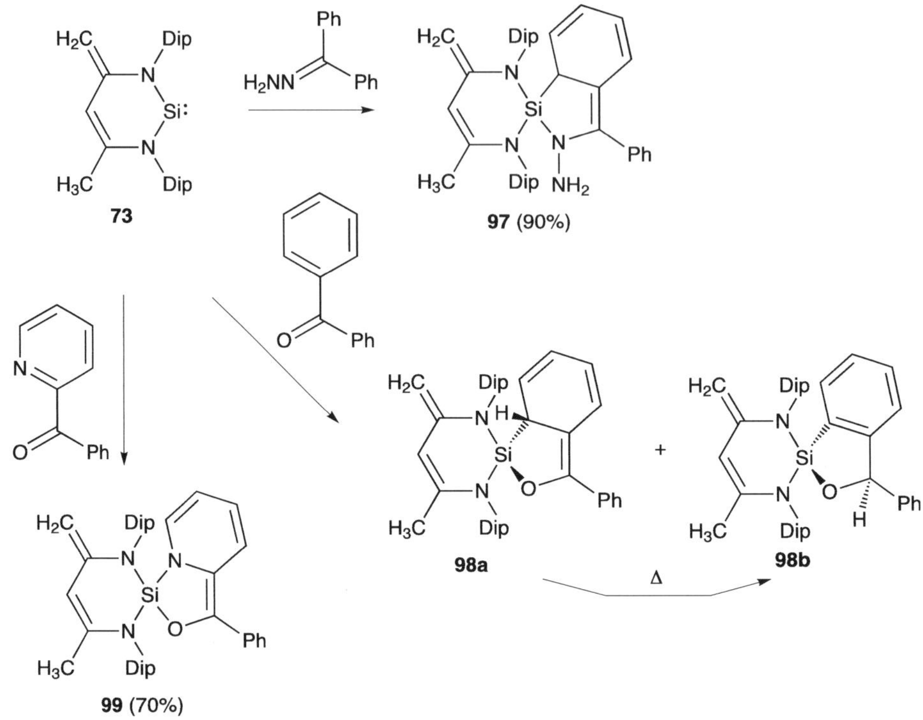

Scheme 8.44

[1 + 4] Cycloaddition of silylene **73** accompanied by dearomatization.

Interestingly, reaction of **73** with 1,2-diphenylhydrazine gave [1 + 4] cycloaddition product **100** resulting from elimination of H_2,[106] while reactions with hydrazines with carbonyl groups provide [1 + 4] cycloadduct resulting from double N–H bond activation (**101a–d**) (Scheme 8.45).[110]

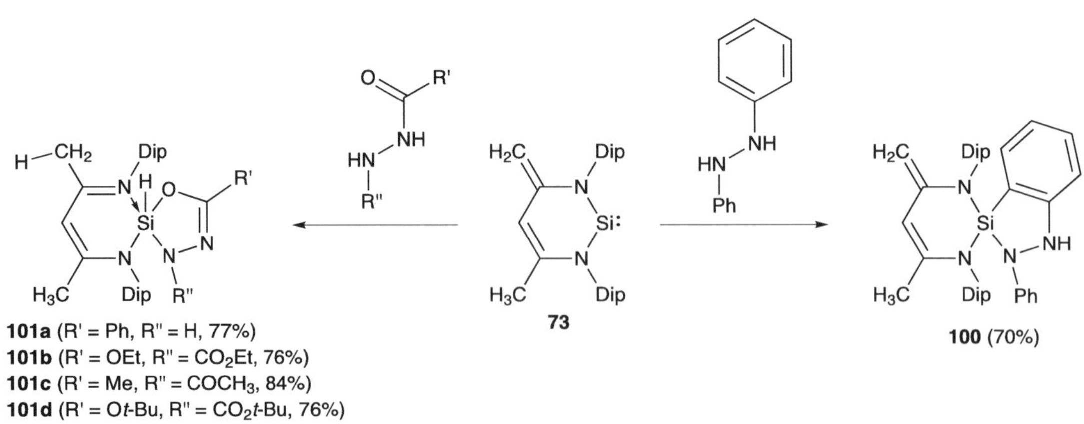

Scheme 8.45

[1 + 4] Cycloaddition of silylene **73** accompanied by elimination of H_2.

In the case of the reaction with acetophenone and acetylferrocene, ene products **102a**[105] and **102b**[107] were obtained instead of [1 + 4] cycloadducts like **98a** (Scheme 8.46). In the reaction of silylene **73** with 1,1,4,4-tetramethyl-2,3-diaza-1,3-butadiene, unusual [1 + 3] cycloadduct **103a** was obtained (Scheme 8.46).[105] Adduct **103a** is thermally unstable and it undergoes further isomerization to give **103b** in 71% through **103c**.

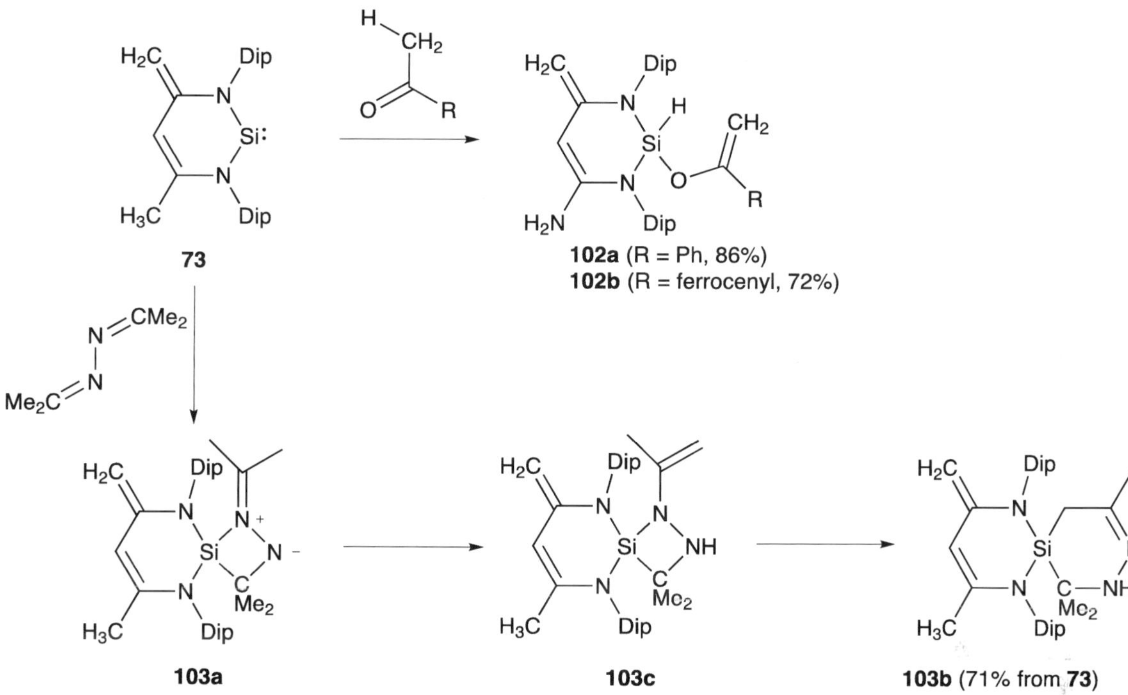

Scheme 8.46
Other reactions of silylene **73**.

8.3.2.4 Oxygen and chalcogen transfer reactions

Oxygen and chalcogen atom transfer to the divalent silicon center are fascinating routes to give Si = X doubly bonded compounds (X = O, S, Se, Te). Silylene **85** reacts with N_2O to give silanoic ester **104** as very thermally stable colorless crystals in 79% yield (Scheme 8.47).[111] X-ray analysis showed that **104** has a four-coordinate Si(=O) atom with the Si = O distance of 1.579(3) Å, which is slightly longer than that of **87** (1.552(2) Å).[100] Double bond character of the Si = O moiety in **104** was evidenced by the quantitative formation of 1,2-adduct of methanol to the Si = O bond. When O_2 was used for the reaction of **104**, 2,4-disila-1,3-disiletane **105** was formed. The corresponding silanoic thio-, seleno-, and telluroesters **106a**, **106b**, and **106c** were synthesized by the reaction of **85** with elemental chalcogens in good yields[112] similar to other stable silylenes.

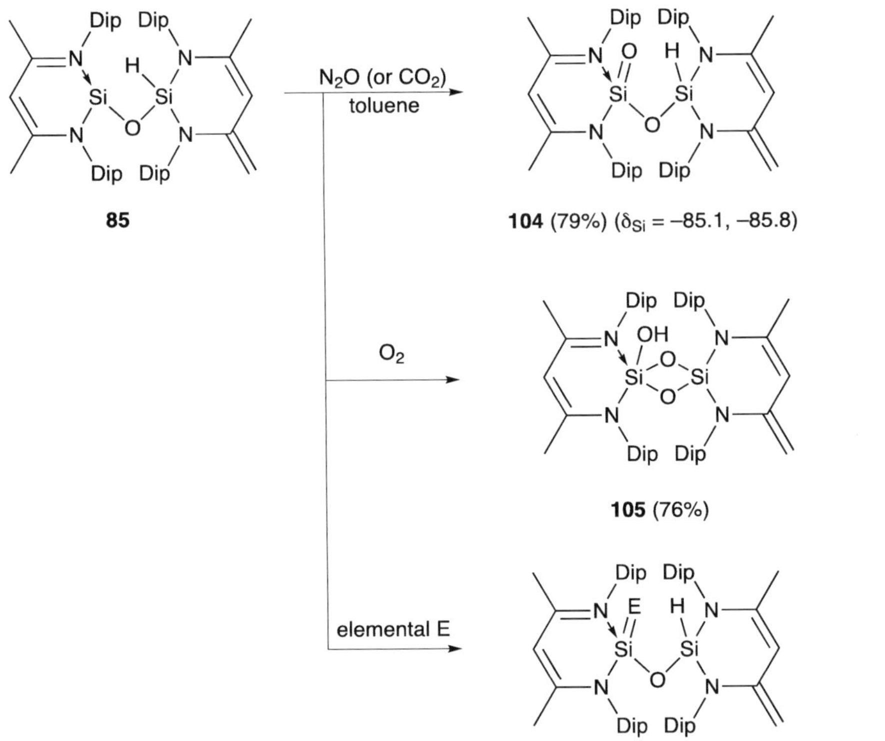

106a (E = S, 69%) (δ_{Si} = –40.7, –41.3)
106b (E = Se, 78%) (δ_{Si} = –38.4, –39.1)
106c (E = Te, 81%) (δ_{Si} = –51.8, –52.2)

Scheme 8.47
Reactions of silylene **85** with oxygen and chalcogen sources. Two ^{29}Si NMR chemical shifts (δ_{Si}) of sp^2-Si nuclei are due to two rotational isomers.

8.3.2.5 Coordination of Lewis bases

Ylide-like silylene **73** has an electrophilic silicon center thus enabling coordination of Lewis bases to Si(II) center similarly to other transient and persistent silylenes (for details, see Section 8.12).[113–118] For instance, reaction of **73** with tetramethylimidazol-2-ylidene (ImMe, 1,3,4,5-tetramethyl-1,3-dihydro-2*H*-imidazol-2-ylidene) in toluene at -60°C gave ImMe-coordinated silylene **107a** as yellow crystals in 86% yield (Scheme 8.48).[119] The ^{29}Si resonance of **107a** ($\delta_{Si} = -12.0$) is considerably upfield-shifted upon coordination of ImMe. Similarly Im*i*-Pr-coordinated silylene **107b** was also obtained.[120] X-ray analysis showed that ImMe coordinated to the Si(II) atom with the sum of the bond angles around the Si(II) atom of 295.5 degrees, and the Si(silylene)–C(ImMe) distance of 2.016(3) Å indicates a strong donor(ImMe)–acceptor(silylene) interactions in **107a** (Fig. 8.7). While compound **107b** is stable at room temperature, ImMe-coordinated silylene **107a** gradually undergoes rearrangement above −20°C to give carbene **108** in 75% yield.[121] Compound **108** works as a bidentate ligand featuring carbene and silylene moieties (see Section 8.3.3.4). Mechanisms involving activation of the C−H bond of ImMe at the electron rich butadiene moiety or Si(II) center in silylene **73** were proposed. Similar to the case of silylene **73** (Section 8.3.2.6), the addition of muonium occurred at the terminal methylene carbon and divalent silicon center to provide muoniated radicals **109a** and **109b**, the former of which was the major product.[122]

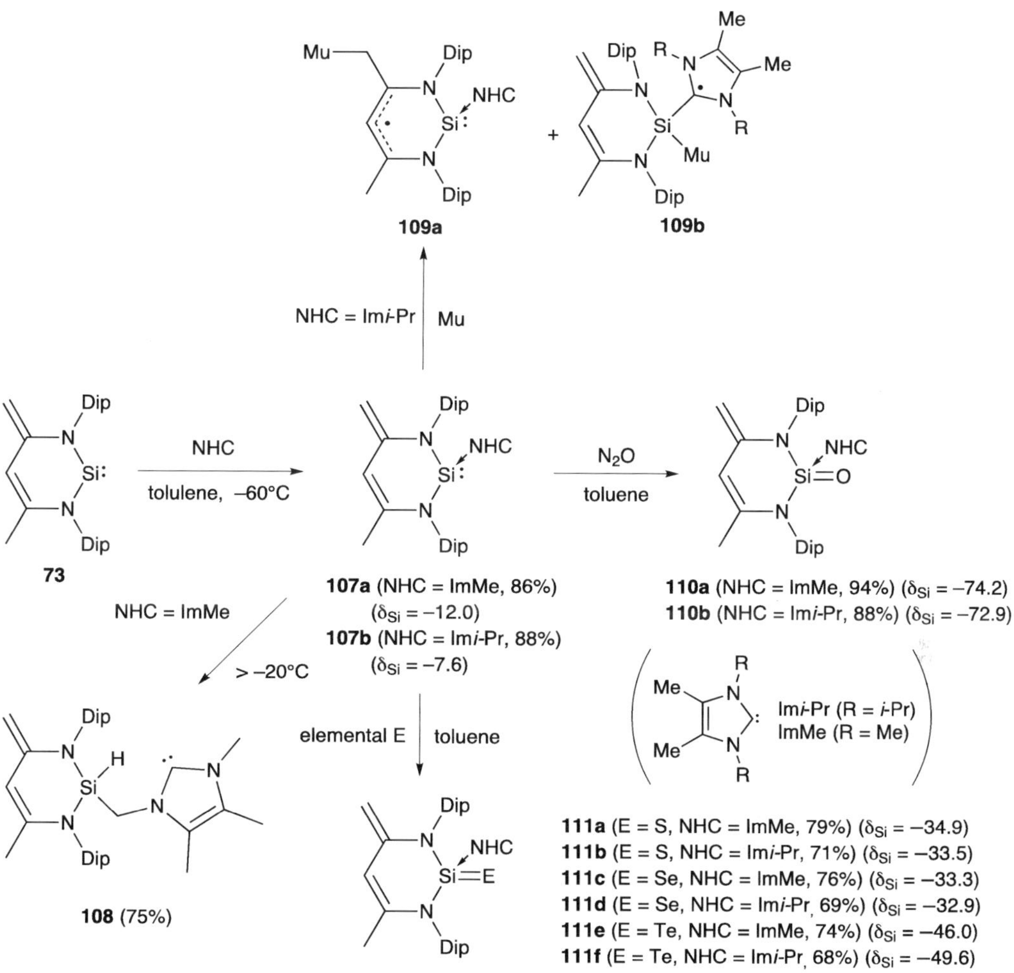

Scheme 8.48
Formation of Lewis-base adducts of silylene **73** and their reactions.

Coordination of NHC increased nucleophilicity of the Si(II) center; while silylene **73** does not react with N_2O, reaction of silylenes **107a** and **107b** with N_2O at −60°C proceeds to give NHC-coordinated silanones **110a**[119] and **110b**[120] in 94% and 88% yields. The Si−O

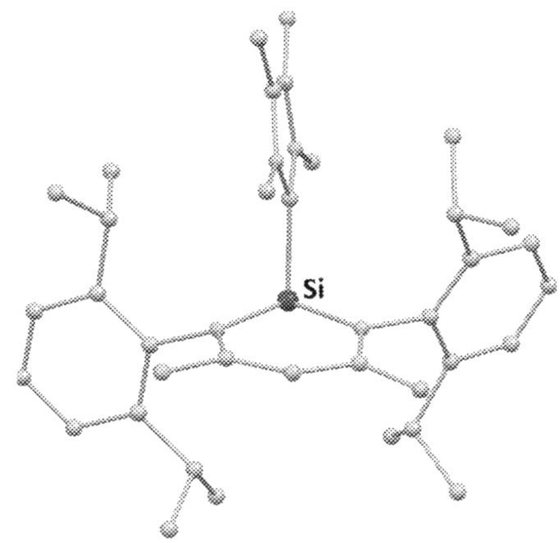

Figure 8.7
Molecular structure of **107a** (CCDC:744914).

distance of **110a** is 1.541(2) Å, while the Si(silylene)–C(ImMe) distance of 1.930(2) Å shortened compared to that of **107a** (2.016(3) Å), suggesting the contribution of an ylide-like resonance structure $(carbene)^{+}$–Si–O^{-}. Similarly treatment of **107a** and **107b** with elemental chalcogens provide the corresponding silanechalcogenones **111a–f** in high yields. NHC-free silylene **73** does not react with elemental Se and Te, while reaction of **73** with S_8 gave a complex mixture.[120]

8.3.2.6 Miscellaneous reactions

The reaction of **73** with diphenyldiazomethane gave a 1:2 adduct, diiminylsilane **112** in a moderate yield (Scheme 8.49).[123] The proposed mechanism involves formation of methylenediazasilacyclopropane **113** and subsequent addition of another Ph_2CN_2: silylene **73** does not react with $Ph_2C=N-N=CPh_2$ even in boiling toluene for several hours. Silylene **73** reacted with bulky arylazide giving silanimine **114**[124] similarly to the reaction of bulky silylenes with azides, while reaction of **73** with less bulky trimethylsilyl azide gave silatetrazoline **115**[123] similar to the reactions of West's NHSi.[36,43] While silylene **73** is inert toward organic nitriles such as acetonitrile and *t*-BuCN, reaction of **73** with cyclohexylisocyanide gave a mixture of **116** (32%) and 1:3 adduct **117** (41%).[123] Multistep mechanisms involving initial formation of silylene isocyanide complex similar to the reaction of other silylenes with isocyanides was proposed for formation of **116** and **117**. Addition of muonium occurred at the terminal methylene carbon and divalent silicon center to provide muoniated radicals **118** as a major product and **119** as a minor product.[122]

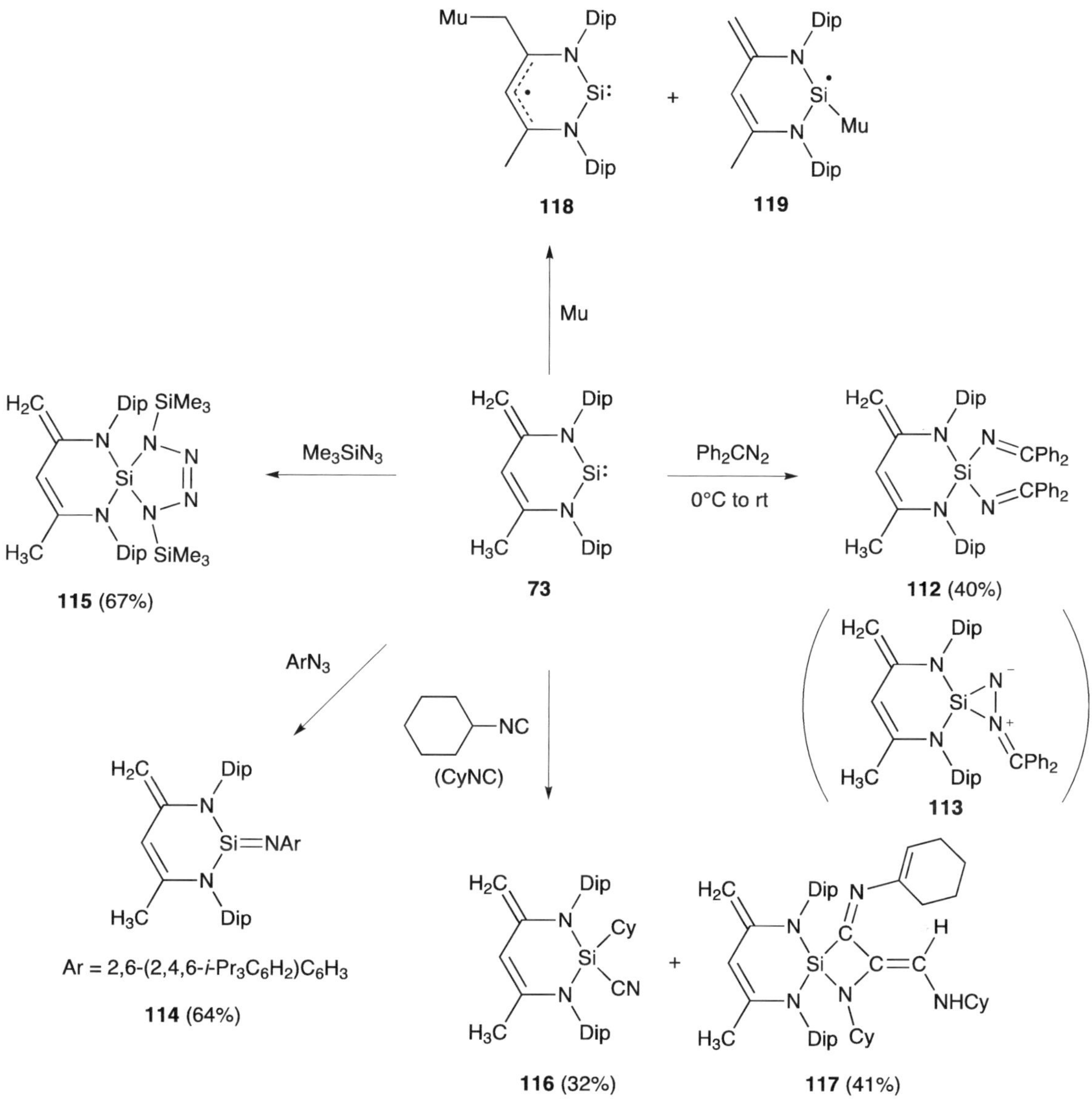

Scheme 8.49
Miscellaneous reactions of silylene **73**.

8.3.3 Transition Metal Complexes and Related Compounds

Silylene **73** works as a strong σ donor (Lewis base) with strong π-acceptor ability toward transition metal centers because of its ylide-like resonance structures. Late transition metal complexes of silylene **73** have been reported.

8.3.3.1 Group 9 metal complexes

Ylide-like silylene **73** reacted with Cp^*IrH_4 in toluene to give Ir–H insertion product **120** in 93% yield (Scheme 8.50).[125] The ^{29}Si resonance of **120** at −7.7 ppm is considerably upfield-shifted compared to that in the free silylene **73** (88.4 ppm) and the ^{1}H NMR spectrum of **120** showed a singlet signal at −15.58 ppm for the three hydrido ligands on the Ir metal. The Ir–Si distance of 2.3293(11) Å is within the typical range of values for other iridium–silyl complexes. The exocyclic methylene moiety of complex **120** still has nucleophilicity, which is demonstrated by the reaction of **120** with $B(C_6F_5)_3$ to provide **121** in 72% yield. Silyl complex **120** is not thermally stable; it undergoes thermal isomerization at room temperature for 24 h to generate an H-migration product, silylene iridium(III) complex **122** almost quantitatively.[125] The presence of NH_3 makes the reaction faster (2 h), which suggests intermolecular H transfer. The very short Si–Ir distance of **122** of 2.2328 (9) Å suggests the multiple bond character due to π-back bonding from Ir(III) center to silylene ligands.

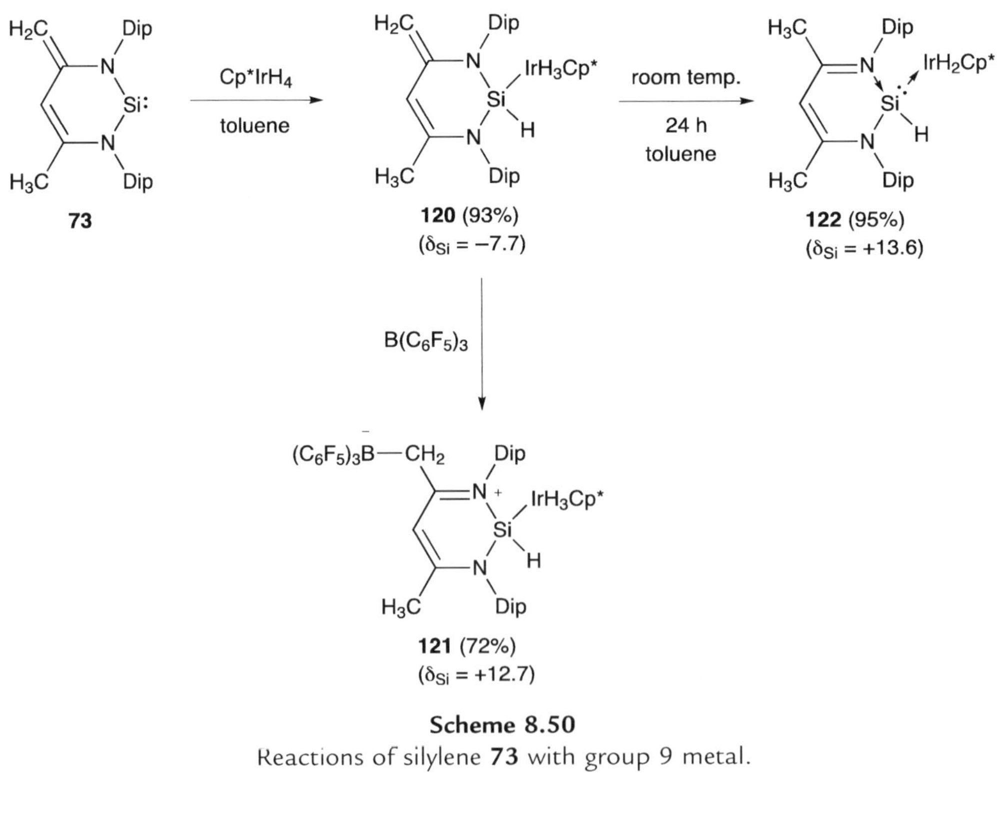

Scheme 8.50
Reactions of silylene **73** with group 9 metal.

8.3.3.2 Group 10 metal complexes

Ylide-like silylene **73** reacted with [Ni(cod)$_2$] complex in the presence of arenes giving Ni(0)(η^6-arene) complexes **123b–d** (Scheme 8.51).[126] These complexes are stable for weeks at ambient temperatures. More electron-rich arene ligands are thermodynamically favored: in C_6D_6 solution, the sample of complex **123b** equilibrates with Ni(0)(η^6-C_6D_6) complexes **123a-*d*$_6$** ($t_{1/2}$ = 18 min) faster than **123c** ($t_{1/2}$ = 166 min) or **123d** (22 days). The proposed mechanism for the equilibrium involves intermediates [(**73**)Ni(η^4-arene)(η^2-arene)] complexes. The bonding between the Ni(0) and Si(II) center was explained by the balanced σ,π-acid–base synergism between a strong σ donation from ylide-like silylene **73** to Ni(0)(arene) moiety and double π-back donation from $3d_{xz}$ and $3d_{yz}$ orbitals of Ni(0) to $3p_x$ and $3p_y$ orbitals of silicon. X-ray analysis of complexes **123b**, **123d**, and **123a** exhibited that complexation of Ni(η^6-arene) to the silylene moiety did not alter metric parameters of silylene moiety significantly.

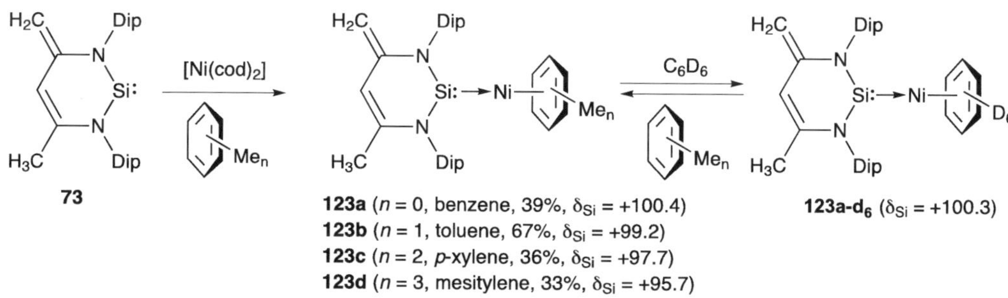

Scheme 8.51
Reactions of silylene **73** with Ni(cod)$_2$.

Complex **123b** is still electrophilic; reaction of **123b** with B(C_6F_5)$_3$ gave complex **124** in high yield (Scheme 8.52). Strong π-back donation from Ni to Si in **123** and **124** was supported by very short Si–Ni distances (2.0597(10), 2.060(1), 2.0647(6), and 2.0369(9) Å for **123b**,[126] **123d**,[102] **123a**,[102] and **124**,[126] respectively). The toluene ligand in **123b** was easily replaced by three carbonyl ligands and Ni(CO)$_3$ complex **125** was formed (Scheme 8.52).[127]

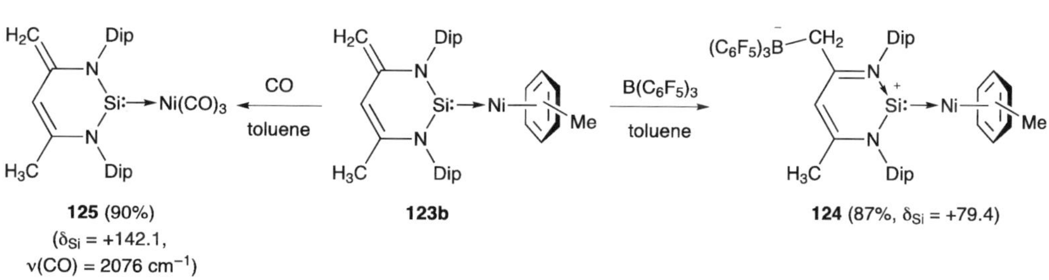

Scheme 8.52
Reactions of silylene complex **123b**.

Cyclic ligand of **125** still have a nucleophilicity toward Lewis acids. $[H(OEt_2)]^+[B(C_6F_5)_4]^-$ and $B(C_6F_5)_3$ added to the terminal methylene group of **125** to give complexes **126a** and **126b**,[127] respectively, while reactions of **125** with H_2O, TfOH, HCl, H_2S, NH_3, NH_2(*i*-Pr, isopropyl), and NH_2NH(*i*-Pr) gave 1,4-adducts **127a**, **127b**,[127] **127c**,[103] **127d–g**[102] (Scheme 8.53) rather than 1,2-adducts that have been often observed in the reactions of other silylene complexes. Formation of **127a** and **127d** is in contrast to the reaction of free silylene **73** with H_2O and H_2S giving siloxysilylene **85**[100] and silathioformamide **90**[102] (Scheme 8.41) probably because the complexation prevented further insertion of Si(II) center to O–H bond of the resulting H_2O adduct **127a** and migration of methyl proton in the ligand to the Si center was prevented by complexation. Functionalization of the ligand in **125** by addition of Lewis acids dramatically tune the σ-donor and π-acceptor abilities of the silylene ligand, which was disclosed by comparison of the carbonyl stretching frequencies (ν(CO)) of these $Ni(CO)_3$ complexes.[102,127] Complexes with zwitterionic or neutral ligands (Lewis-acid adduct of complex **125**), **126a** and **126b** exhibit high ν(CO) values (2095 and 2098 cm^{-1}) comparable to that of $Ni(CO)_3$ complex with a weak σ-donor and strong π-acceptor PF_3 ligand (2111 cm^{-1}).[127] In contrast, complexes **127e**, **127f**, and **127a** with a three-coordinate Si(II) center (1,4-adducts of complex **125**) show low ν(CO) values (2046, 2048, and 2050 cm^{-1}) akin to that of $Ni(CO)_3$ complex with a strong σ-donor and weak π-acceptor NHC ligand (2054 cm^{-1}). It should be noted that the range of ν(CO) values observed for these complexes (52 cm^{-1}) is as large as the difference between tri(*tert*-butyl)phosphine and trifluorophosphine (55 cm^{-1}; 2056 cm^{-1} for $(t\text{-}Bu_3P)Ni(CO)_3$ and 2111 cm^{-1} for $(F_3P)Ni(CO)_3$.[102]

Treatment of silylene complex **125** with NH_3BH_3 or **127c** with $Li[BEt_3H]$ gave hydridosilylene complex **128** (Scheme 8.53).[103] The Si–Ni distance of **128** determined by X-ray analysis is 2.2524(8) Å and ^1H NMR resonance of hydrido proton is 6.15 ppm. Hydridosilylene complex **128** underwent facile hydrosilylation reaction with diarylalkynes (Scheme 8.53).[103] Treatment of **128** with diphenylacetylene afforded only the *cis* addition product **129a**. When (*p*-tolyl)phenylacetylene was used, a mixture of **129b** and **129c** was obtained. Because excess CO suppressed the reaction, dissociation of CO from **128** and coordination of acetylenes to the Ni center is involved in the rate controlling step, which is supported by DFT calculation of model compounds.

8.3.3.3 Group 11 metal complexes

Silylene **73** is also useful for introducing silyl groups with various functionality under moderate conditions. Thus, silylene **73** inserted into the Cu–X (X = O*t*-Bu, OH, H, OC_6F_5) bonds of NHC stabilized Cu(I)X complexes to provide the corresponding silyl complexes **130a–d** in good yields (Scheme 8.54).[128] Silyl copper complexes **130a–d** are very efficient in the reduction of CO_2 (1 atm) to CO and the final products are siloxy complexes **131a–d**.

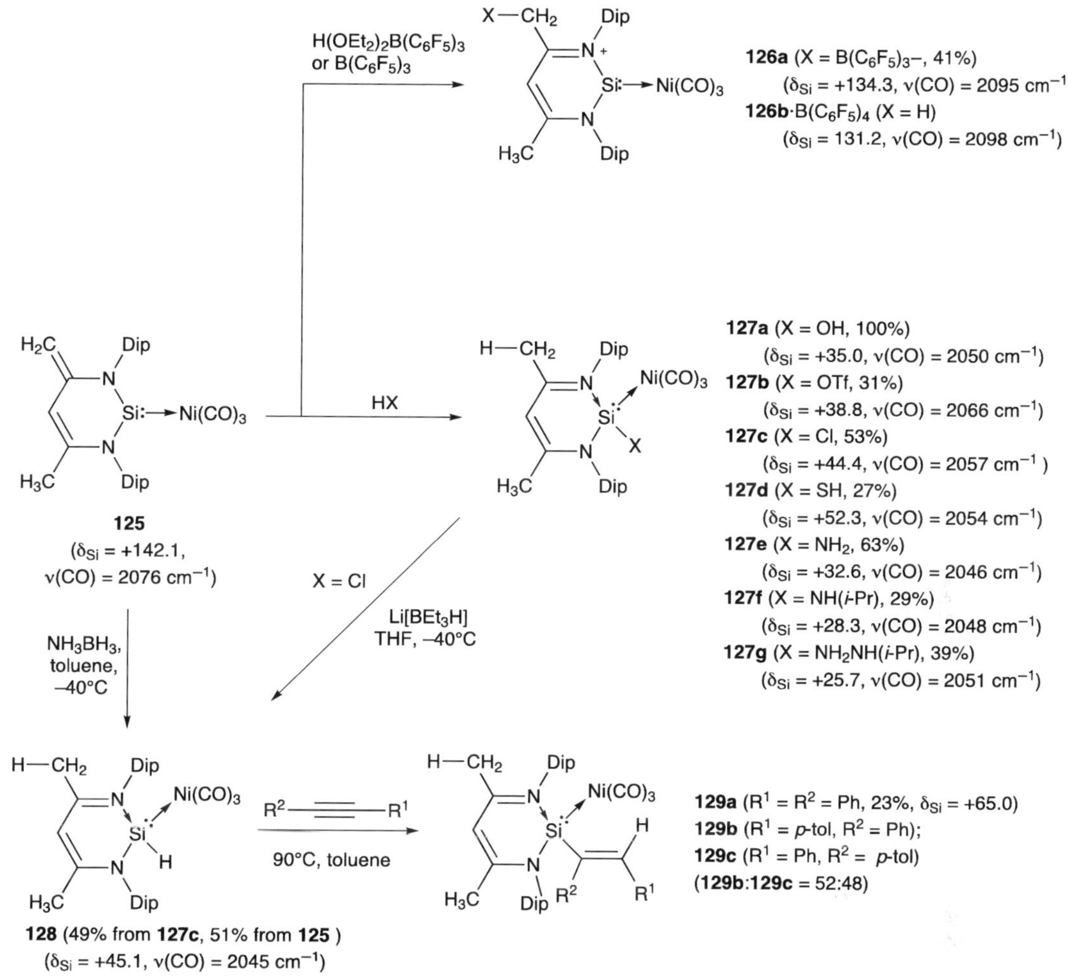

Scheme 8.53
Reactions of **125** and related complexes.

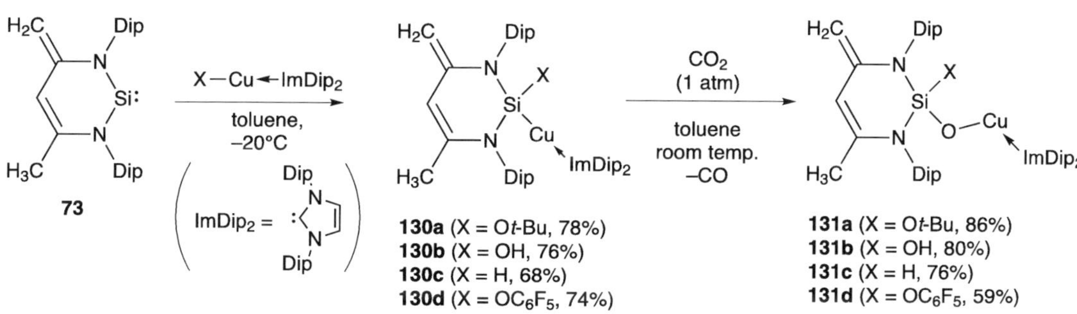

Scheme 8.54
Reactions of **73** with Cu complexes.

8.3.3.4 Bidentate ligands featuring silylene and carbene moieties

Carbene **108**, which is obtained from the reaction of silylene **73** with ImMe (Section 8.3.2.5), serves as a bidentate ligand featuring carbene and silylene moieties by 1,4-hydrogen migration. Reaction of **108** with $NiBr_2$(dme) in toluene gave novel $NiBr_2$ complex **132** in 82% yield (Scheme 8.55).[129] The reduction of **132** with KC_8 in the presence of dmpe gave dmpe-coordinate Ni(0) complex **133**, while the reduction of **132** with K[BEt_3H] in the presence of carbon monoxide afforded dicarbonyl Ni(0) complex **134**, respectively. The Ni–Si and Ni–C(carbene) distances in **134** (2.2131(13) and 1.947(4) Å) are longer than those in **133** (2.1740(18) and 1.918(7) Å), which is likely due to the stronger π acidity of CO than phosphine ligands resulting in decreasing of the π back donation. It should be noted that the IR (infrared) stretching vibrations for the CO ligands of **134** (1952 and 1887 cm^{-1}) are shorter compared to those in bisphosphine Ni complex [(dmpm)$Ni(CO)_2]_2$ (1991 and 1927 cm^{-1}) (dmpm = bis(dimethylphosphino)methane)[130] and bis-NHC Ni complex $(NHC^{Mes})_2Ni(CO)_2$ (2050 and 1877 cm^{-1}) (NHC^{Mes} = $(CHNMes)_2C$),[131] suggesting NHC–NHSi ligand in **134** is a stronger σ-donor than two phosphine or NHC ligands. Complex **132** efficiently catalyzes the Kumada–Corriu type cross-coupling reactions of 4-MeC_6H_4MgCl with various aryl halides having alkyl, CF_3, Me_2N groups at the para position, and heteroaryl halides. In contrast to the formation of the Ni(0) complexes, reduction of **132** with KC_8 in the presence of PMe_3 gave unexpected silylbromo nickel complex **135**, which undergoes further reduction to provide hydrosilyl-NHC Ni(0) **136**. Dissociation of PMe_3 ligand from **136** giving **137** was observed in solution.

8.4 Acyclic Heteroatom-Substituted Silylenes

8.4.1 Acyclic Diaminosilylenes

In contrast to cyclic diaminosilylenes, acyclic diaminosilylenes have been known as labile species. Bis(diisopropylamino)silylene (i-$Pr_2N)_2$Si: (**138**) can be generated by the reduction of the corresponding dichlorosilane (i-$Pr_2N)_2SiCl_2$ with alkali metal in the presence of trapping reagents (as solvents) under the refluxing conditions, which is confirmed by the formation of the corresponding trapping products.[132–134] Notably, ligand scrambling occurred during the reduction using deuterium-labeled diaminodichlorosilane, which strongly indicates the existence of ligand-bridged dimer **139a**. Photolysis of silacyclopropene **140** at room temperature can also generate **138** under milder conditions and the silylene **138** was observed by ^{1}H NMR and UV-vis spectroscopies and its generation was confirmed by trapping reactions with triethylvinylsilane and 2,3-dimethyl-1,3-butadiene (Scheme 8.56).[135] Under these conditions, no scrambling of substituents occurs and **138** dimerizes gradually to form disilene **139b**. Deuterium labeled experiments

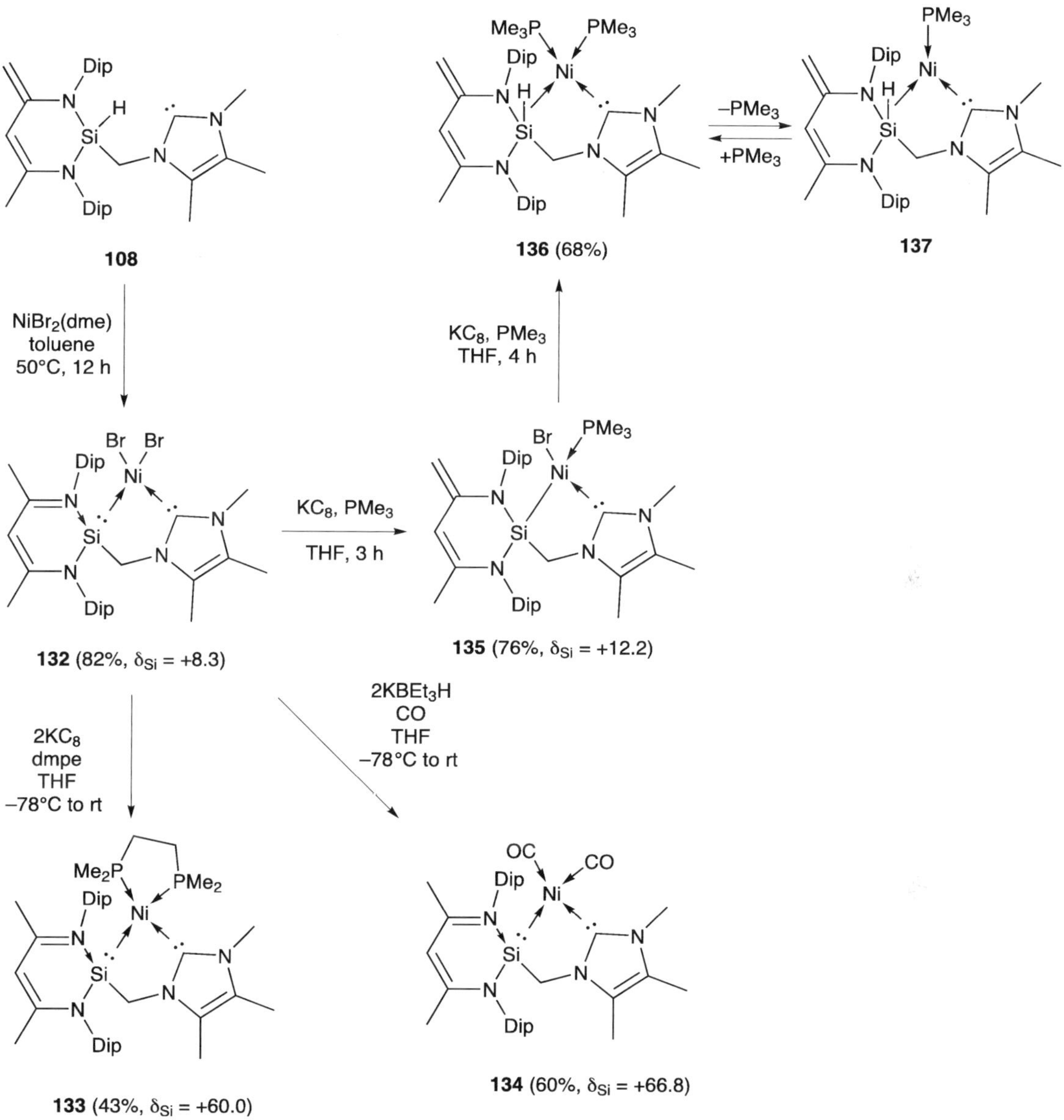

Scheme 8.55

Synthesis and reactivity of Ni complex **132** with bidentate ligand featuring both silylene and carbene moieties.

indicate that the bridged dimer **139a** forms only at higher temperature (for instance, benzene refluxing condition) but does not form at room temperature. Under the reaction conditions (THF, rt), **138** dimerizes reversibly to the corresponding disilene **139b** This unique behavior is also supported by computational studies.[136]

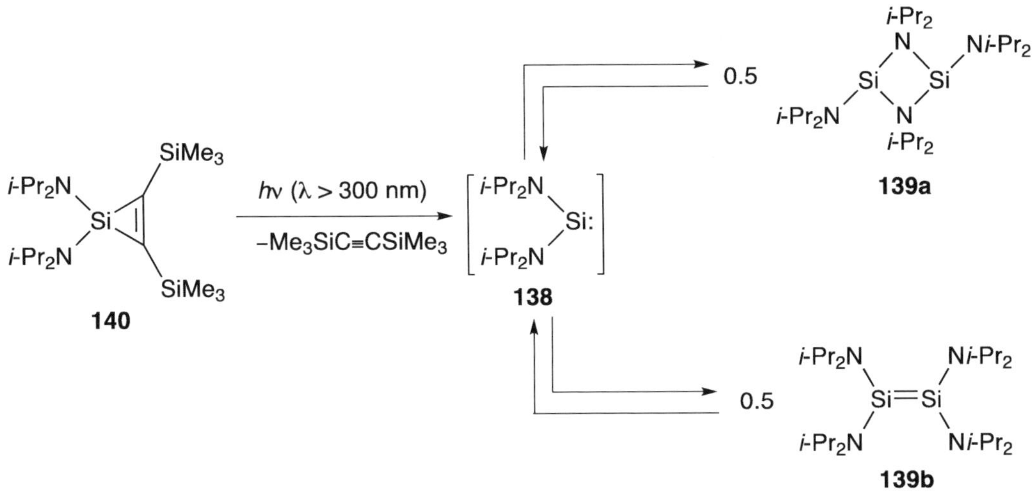

Scheme 8.56
Generation of acyclic diaminosilylene **138**.

Bis[bis(trimethylsilyl)amino]silylene **141** was generated in the reduction of diaminodibromosilane **142** with KC_8, and persistent over 12 h at $-20°C$, but it decomposed at room temperature (Scheme 8.57).[137] Silylene **141** was trapped with methanol and phenol to give the corresponding adducts **143a** and **143b**. The ^{29}Si NMR signal of divalent silicon of **141** was +223.9 ppm, which is much downfield shifted compared to the cyclic diaminosilylenes (90−110 ppm). Theoretical studies indicated that the interplanar angle Θ between SiNSi and NSi(II)N planes representing the angle between a lone pair orbital of the neighboring nitrogen and vacant 3p orbital of divalent silicon strongly correlated to ^{29}Si NMR chemical shift (δ_{Si} = +100 ppm for Θ = 0 degrees to +400 ppm for Θ = 90 degrees).

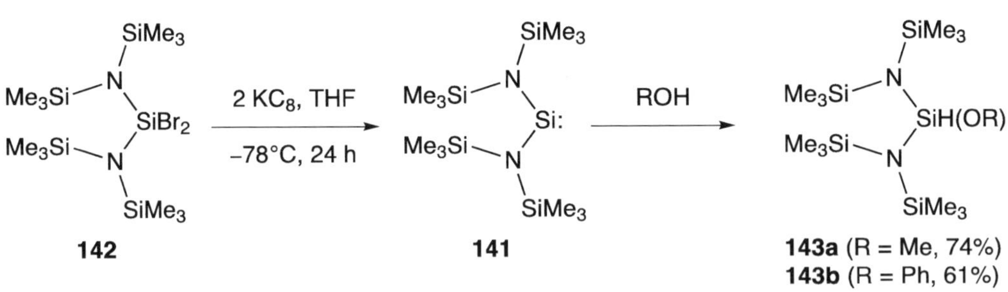

Scheme 8.57
Synthesis and reactions of acyclic diaminosilylene **141**.

Quite recently, Jones and coworkers have reported diaminosilylene **144** as a stable acyclic diaminosilylene (Scheme 8.58).[138] Nucleophilic reaction of very bulky amide **145** with an equimolar amount of the NHC-coordinated dichlorosilylene (Im*i*-Pr → $SiCl_2$) afforded **144** as a pale yellow crystalline solid in 51% yield after the removal of Im*i*-Pr by precipitation as insoluble complexes with $SiBr_4$ or CO_2. The ^{29}Si NMR spectrum of **144** showed the signal of divalent silicon at 204.6 ppm. The N–Si–N bond angle determined by XRD analysis was 110.94(5) degrees. Silylene **144** did not react with H_2 even at 40 atm and +60°C, while it was readily oxidized to form **146** probably via siladioxirane.

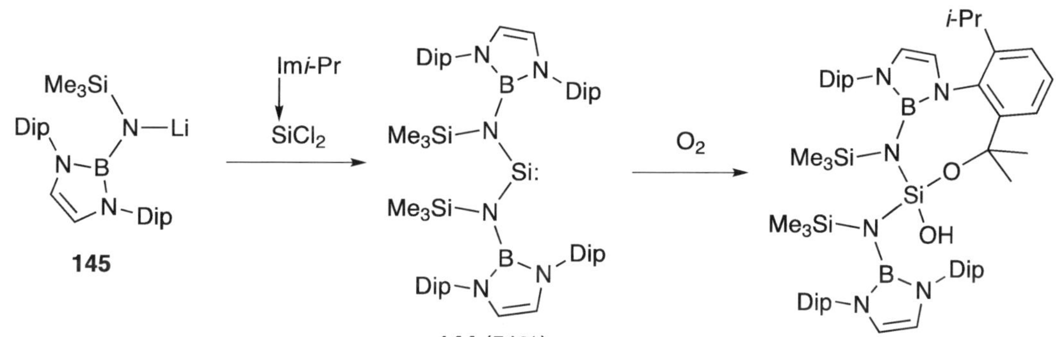

Scheme 8.58
Synthesis and oxidation of silylene **144**.

8.4.2 Amino(boryl)silylenes

Acyclic amino(boryl)silylene **147** was synthesized as a red solid by the reaction of **148** with two equiv of boryllithium **149**[139] (Scheme 8.59).[140] Boryllithium **149** serves as a bulky nucleophile and a reducing reagent in this reaction. In the ^{29}Si NMR spectrum, divalent silicon atom of **147** resonated at very low field, +439.7 ppm. The B–Si–N bond angle of 109.7(1) degrees is wider than those of cyclic dicoordinate stable silylenes (~90 degrees). DFT calculations showed that HOMO and LUMO were a lone pair orbital and vacant 3p orbital, respectively with no remarkable interactions with the neighboring boryl and amino substituents. Silylene **147** is stable even at 130°C in the solid state, whereas it isomerizes to **150** in solution at 50°C by the intramolecular C–H insertion of the divalent silicon atom. Notably, silylene **147** reacted with hydrogen gas in C_6D_6 solution at room temperature to give dihydrosilane **151**. Since mono-deuterio derivative **151-d$_1$** was obtained as a sole product by using HD in the reaction, the concerted reaction was proposed in the hydrogenation of **147**. The calculated $\Delta G^{\ddagger}$ value for the addition of hydrogen molecule to silylene **147** (+97.2 kJ mol^{-1}) is consistent with the experimental observations and the calculated singlet–triplet energy gap for **147**. The calculated transition state shows that the H_2 molecule approaches side-on to the silylene, which is also predicted for parent systems,[141] indicating donation of electron density from HOMO of H_2 into LUMO of **147**. This behavior is similar to that of transition metals toward H_2 rather than the nucleophilic activation of H_2 by carbenes.[142]

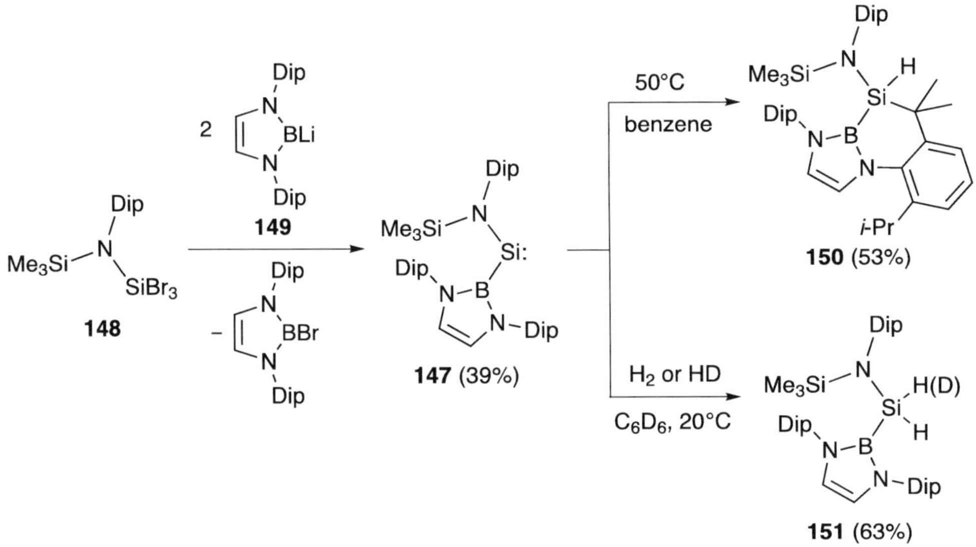

Scheme 8.59
Synthesis and reactions of silylene **147**.

8.4.3 *Amino(silyl)silylene*

Treatment of tribromosilane **152** with bulky silylpotassium reagents provided a purple crystalline solid of **153** in 51% isolated yield (Scheme 8.60).[143] The ^{29}Si NMR signals of the divalent silicon atom were observed at +438 and +467 ppm due to two rotamers. The Si–Si(II)–N bond angle(116.91(5) degrees) is in the wide end of those of dicoordinate stable silylenes. Similar to amino(boryl)silylene **147**, silylene **153** undergoes thermal intramolecular C–H insertion and room temperature H_2 activation to give **154** and **155**, respectively.

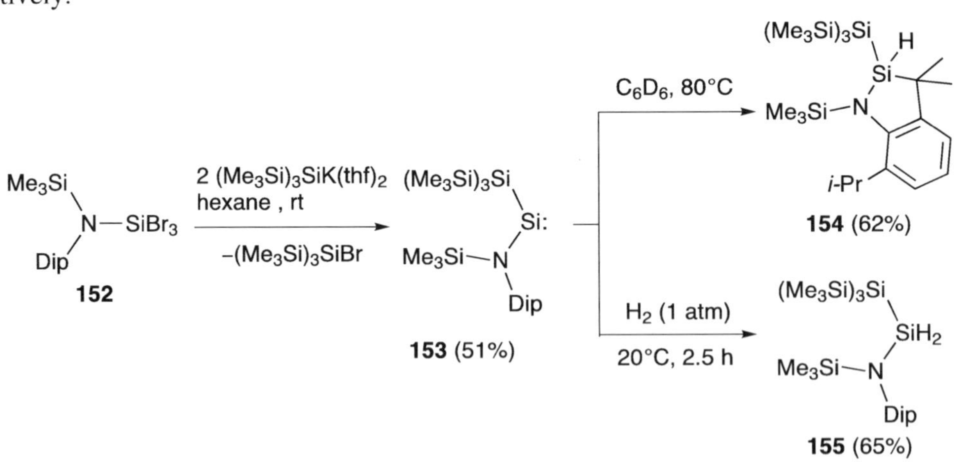

Scheme 8.60
Synthesis and reactions of silylene **153**.

Silylene **153** underwent cycloaddition with alkynes, DMB, and ethylene to give **156–158** (Scheme 8.61).[144,145] Notably, further reaction of silacyclopropane **158** to provide new silacyclopropane **159** proceeded via migratory insertion into Si–Si bond, which was confirmed by using ethylene-d_4 as a second reagent.

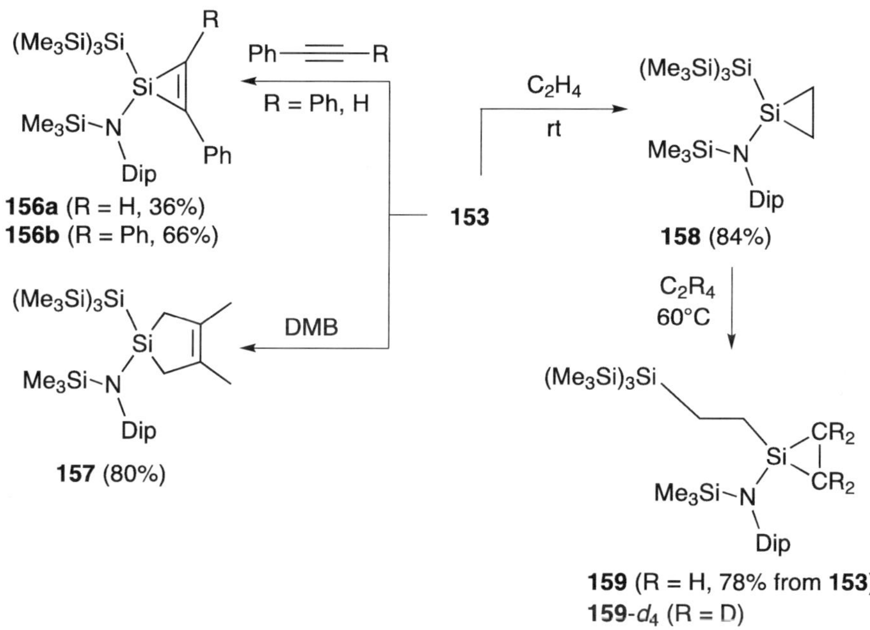

Scheme 8.61
Reactions of silylene **153** with alkenes and alkynes.

8.4.4 Di(arylthio)silylenes

In 2012, Power et al. have reported acyclic stable silylene **160a** bearing bulky arylthio groups SAr^{Me6} by the reaction of dibromosilane **161a** with Mg(I) complex $(NacMesMg)_2$[146–148] (Scheme 8.62).[149] Silylene **160a** has a small S – Si – S angle of 90.52 (2) degrees and its divalent ^{29}Si nucleus resonates at +285.5 ppm. The n→3p transition band appeared at 382 nm as a shoulder ($\varepsilon = 8300\ M^{-1}\ cm^{-1}$). Silylene **160a** reacted with MeI to form **162** but did not react with hydrogen gas.

Structurally similar bis(arylthio)silylenes **160b** and **160c** were also obtained by the reaction of dibromosilanes **161b** and **161c** with Rieke's magnesium and catalytic amount of anthracene in THF.[150] The ^{29}Si NMR chemical shifts of silylene moiety and the longest absorption maxima are as follows: 270.4 ppm and 405 nm ($\varepsilon = 3700\ M^{-1}\ cm^{-1}$)

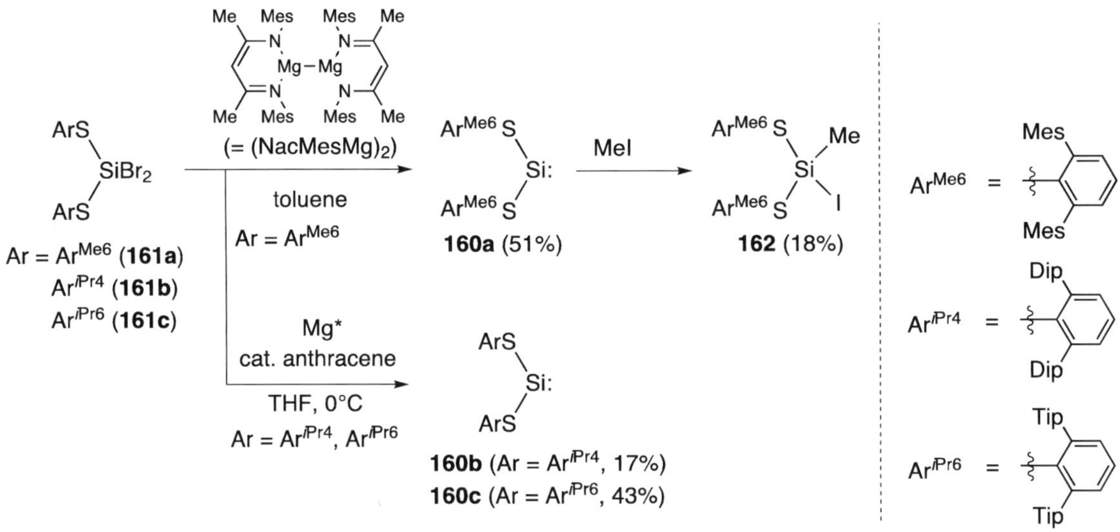

Scheme 8.62
Synthesis and reactions of silylenes **160**.

for **160b** and 270.9 ppm and 411 nm ($\varepsilon = 4400\ M^{-1}\ cm^{-1}$) for **160c**. The ArS–Si–SAr bond angles of these silylenes become acute with increasing the bulkiness of the substituents (90.52(2) degrees for **160a**, 85.08(5) degrees for **160b**, 84.8(1) degrees for **160c**). Since discrepancy between the optimized structures by DFT calculations and X-ray structures was improved with dispersion correction, dispersion forces contribute to such small bond angles.

Interestingly, reversible [1 + 2] cycloaddition of **160b** with ethylene proceeded to give silacyclopropane **163b** (Scheme 8.63).[151] Similar cycloaddition proceeded with less bulky silylene **160a** and ethylene or norbornadiene to afford **163c** and **163d**, but dissociation of **163c** and **163d** to the corresponding silylene and alkene was not observed. No cycloaddition of **160b** with norbornadiene proceeded. These results suggest that steric effect would play a crucial role in the reversible complexation of ethylene with silylene.

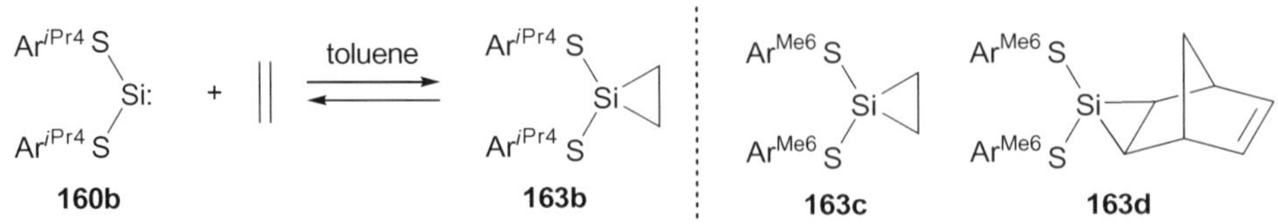

Scheme 8.63
Reversible reaction of silylenes **160** with alkenes.

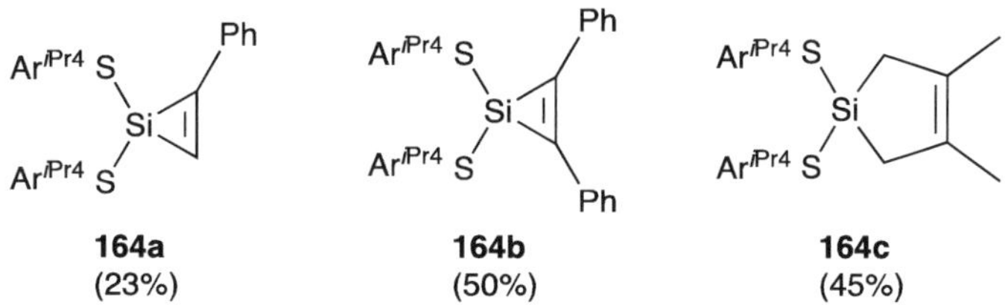

Figure 8.8
Cycloadducts of silylene **160b**.

Reactions of **160b** with phenylacetylene, diphenylacetylene, and 2,3-dimethyl-1,3-butadiene afford cycloadducts **164a–c** in moderate yields (Fig. 8.8).[152]

8.4.5 Persistent Diarylsilylenes

Although isolable diarylsilylenes are still elusive, persistent diarylsilylene **165** is generated by thermal dissociation ($>60°C$) of tetraaryldisilene (**166**) bearing bulky aryl groups (Scheme 8.64).[153–155] Despite the lack of spectroscopic observation, existence of silylene **165** in solution is confirmed by various reactions, and representative examples (insertion and cycloaddition) of the reactions of **165** to give **167a–d** are displayed in Scheme 8.64. Silylene-isocyanide complexes **6a–c** are formed by the reaction of **165** with the

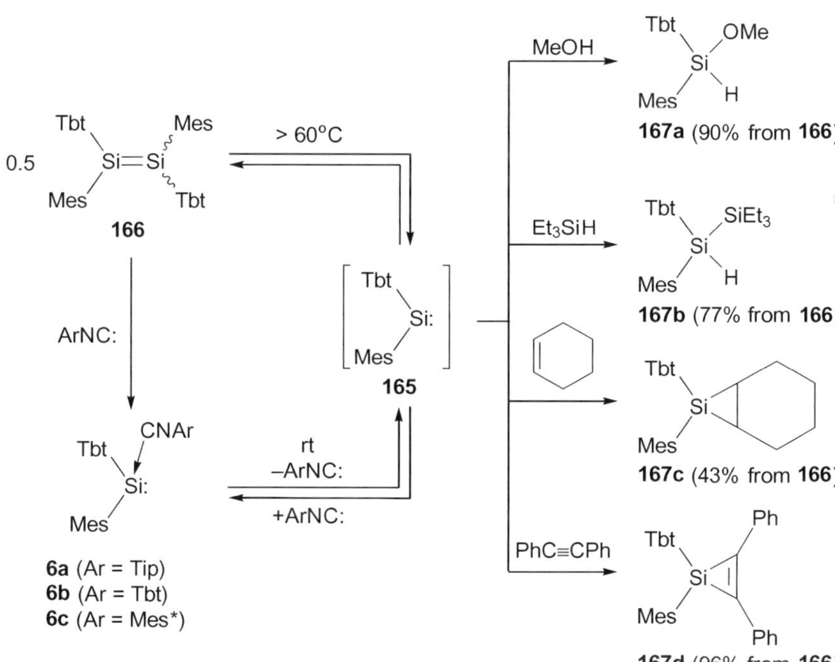

Scheme 8.64
Generation and various reactions of diarylsilylene **165**.

corresponding isocyanides.[10,114] Compounds **6a–c** constitute an important class of molecules because they are the first stable silylene-isocyanide complexes and a pioneering work of the external Lewis base stabilized silylenes (see also Section 8.12).[18] Compounds **6a–c** act as a synthetic equivalent of silylenes; diarylsilylene **165** is regenerated even at room temperature by the thermal dissociation of **6a–c** in solution.

Silylene **165** is a key precursor for unique silacycles (Scheme 8.65). [1 + 2] Cycloaddition of **165** with pivalonitrile and *tert*-butylphosphaalkyne proceeded to give the three-membered ring compounds **168a** and **168b** in excellent yields.[156] Treatment of **165** with isoprene gave vinylsilacyclopropane **169a** exclusively.[157] Compound **169a** isomerizes thermally into **170a** at 160°C for 12 days. Reaction of **165** (from silylene isocyanide complex **6c**) with 2,3-dimethyl-1,3-butadiene at room temperature gave a mixture of [1 + 2] and [1 + 4] cycloadducts (**169b** and **170b**) in 8:3 ratio. Ring expansion of **169b** to **170b** occurs at elevated temperature (100°C, 5 h), thus there is direct [1 + 4] cycloaddition that competes with [1 + 2] cycloaddition.

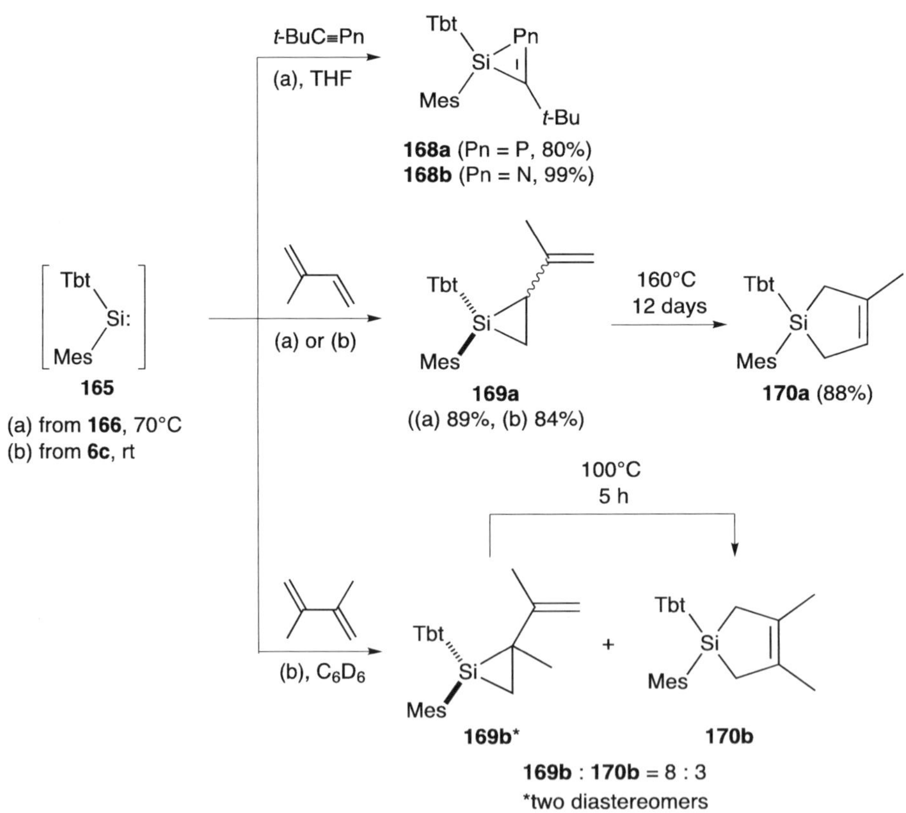

Scheme 8.65
Cycloaddition reactions of silylene **165**.

Silylene **165** reacts with benzene and pyridine derivatives to furnish dearomatized products (Scheme 8.66).[155,158,159] When **166** was heated in benzene solution or in THF solution of naphthalene, dearomatized bicyclic compounds **171** (a mixture of *E*, *Z*- isomers) and **172** were formed in 58% and 81% yields, respectively. Their molecular structures indicate that initial [1 + 2] cycloaddition and the following ring expansion to silepin operate in the formation mechanism of **171**. Thermal reaction of **165** (from **166**) with pyridine and DMAP (4-dimethylaminopyridine) provided 2:2 adduct **173** and azasilepin (1:1 adduct) **174** in 30% and 29%, respectively. Because azasilepin **174** was obtained in the case of DMAP, **173** would be formed by second silylene addition of an azasilepin and the subsequent attack of a second pyridine molecule on the strained silacyclopropane ring.

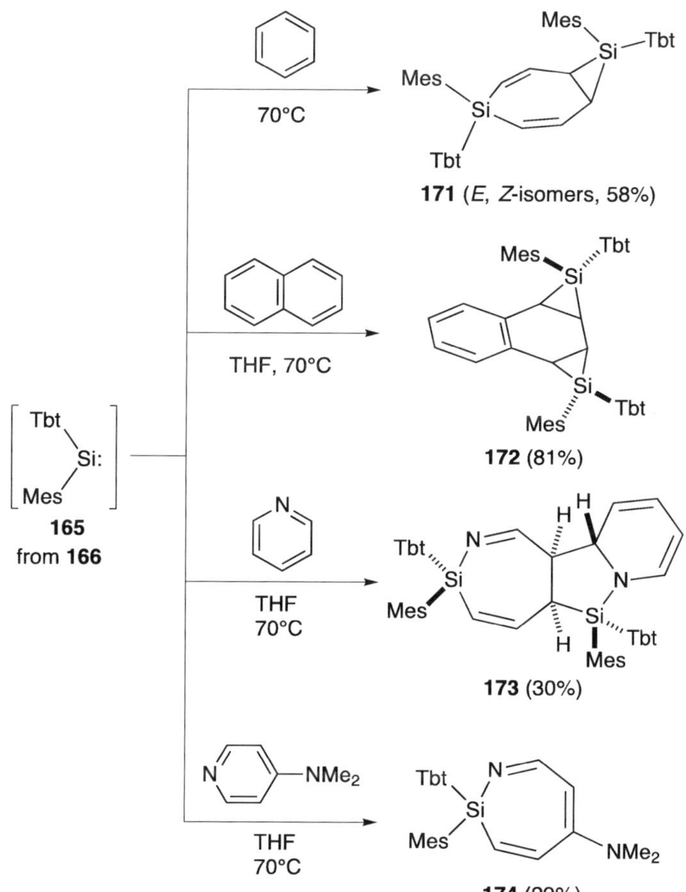

Scheme 8.66
Reactions of silylene **165** with aromatic compounds.

1,2-Dibromodisilene **175**[160] exists in a disilene form in solution as evidenced by VT-NMR and UV-vis spectroscopies, however, **175** reacts as bromosilylene **176** with various reagents, such as alcohols, ketones, hydrosilanes, alkynes, alkenes, and dienes (Scheme 8.67). These results apparently suggest the dissociation equilibrium between **175** and **176** in solution.[161] Notably, the reactivity of **175** as **176** is accelerated by LiBr; completion of the reaction of **175** with cyclohexene in DME to give silacyclopropane **177** is much faster in the presence of LiBr. In addition, thermal isomerization of **175** without LiBr afforded **178** as disilene isomer, while thermal isomerization with LiBr gave **179** as intramolecular C–H insertion product of silylene **176**. Facile generation of silylene-LiBr adduct [Ar(Br)SiLiBr, **180**] from **176** and LiBr may affect the observed reactivity of disilene **175**. Compound **180** is a bromosilylenoid, and related topics are treated in Section 8.4.8.

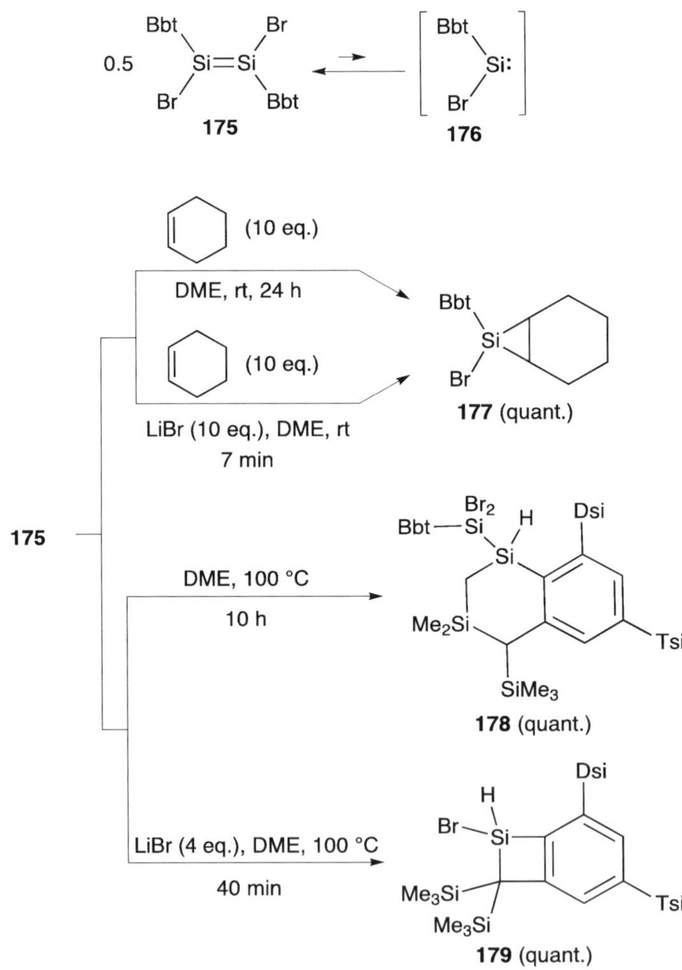

Scheme 8.67
Reactions of arylbromosilylene **176** generated from its dimer **175**.

Treatment of **175** with $Pt(PCy_3)_2$ provided bromosilylene-platinum complex **181** in 56% yield (Scheme 8.68).[162] Similar reaction with $Pd(PCy_3)_2$ gave a complex mixture. According to XRD study, platinum atom in **181** adopts trigonal planar geometry and the Si–Pt bond of 2.2076(15) Å is shorter than those of silylplatinum complexes (2.224–2.451 Å). Silylene silicon atom has a T-shape like geometry with the Pt–Si–C angle of 151.86(17) degrees and the Pt–Si–Br angle of 109.88(6) degrees, which would be attributed to steric repulsion between Bbt and cyclohexyl groups. In ^{29}Si NMR spectrum of **181** in C_6D_6 solution, unsaturated silicon atom appeared as a triplet with satellites ($^{2}J_{SiP} = 137$ Hz, $^{1}J_{SiPt} = 3660$ Hz) at +298.1 ppm.

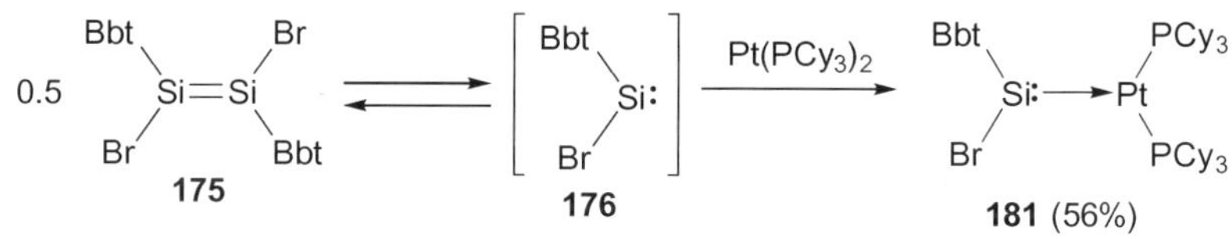

Scheme 8.68
Synthesis of platinum complex of silylene **176**.

8.4.6 Disilylsilylene Anion Radicals

Although disilylsilylenes are still elusive, their anion radicals have been reported. As shown in Scheme 8.69, silylene anion radicals **182a** and **182b** are synthesized in 56% yields by the reaction of disilene **183**[163] with 2.2 equiv. of MNap (M = Li, Na) and the subsequent addition of crown ethers.[164] In these reactions, the homolytic Si–Si bond cleavage of 1,2-dimetalladisilanes (**184a** and **184b**)[165] occurred by the addition of crown ethers. XRD analysis of **182a** revealed its silylene anion radical structure; the Si···Li distance of 6.7 Å is much

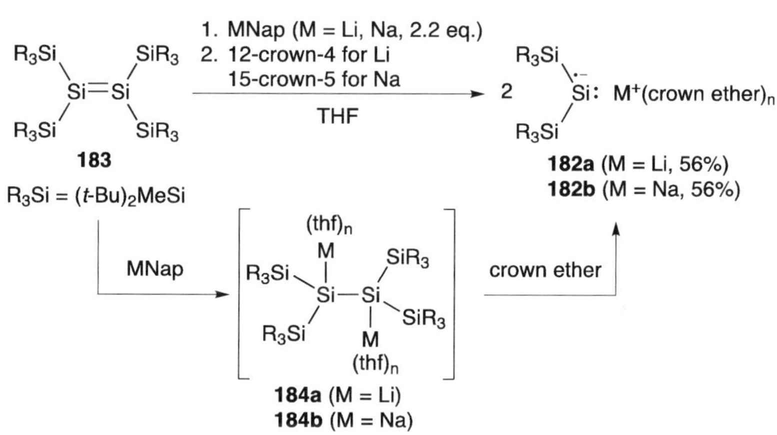

Scheme 8.69
Synthesis of silylene anion radicals **182**.

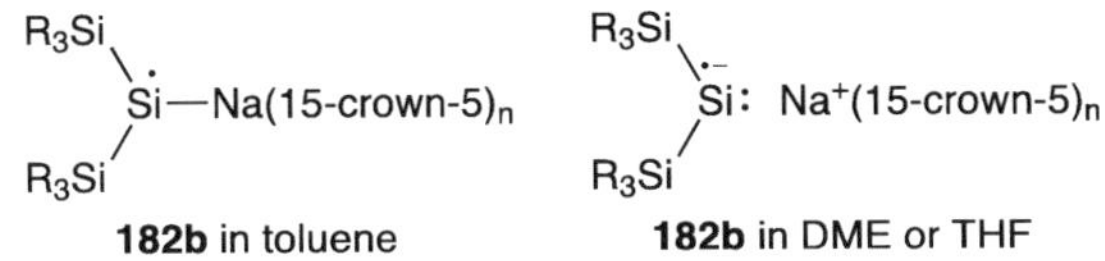

Figure 8.9
Structures of **182b** in various solvents.

larger than the sum of the van der Waals radii of Si and Li atoms. EPR studies demonstrated that structure of **182b** is dependent on the solvents; a contact ion pair (sodiosilyl radical having Si–Na bond) forms in toluene, while a solvent separated ion pair similar to XRD structure of **182a** forms in THF or DME solution (Fig. 8.9). In EPR spectrum of **182b** measured in toluene, a characteristic quartet owing to hyperfine splitting with sodium-23 (natural abundance 100%, $I = 3/2$) with a hyperfine coupling constant of 0.19 mT is observed.

8.4.7 Metallosilylenes

Two metallosilylenes that have transition metal fragments as substituents have been reported. Feriosilylene **497** is described in Section 8.13, while cationic chromiosilylene is described in this section. When cationic chromium silylidyne complex **185** was exposed to CO in fluorobenzene solution, cationic chromiosilylene **186** was obtained in 68% yield (Scheme 8.70).[166] In **186**, the Cr–Si bond lengths of 2.393(2) Å is close to the average length of Cr–Si single bonds (2.399 Å)[166] and the Cr–Si–C bond angle is 116.2(1)

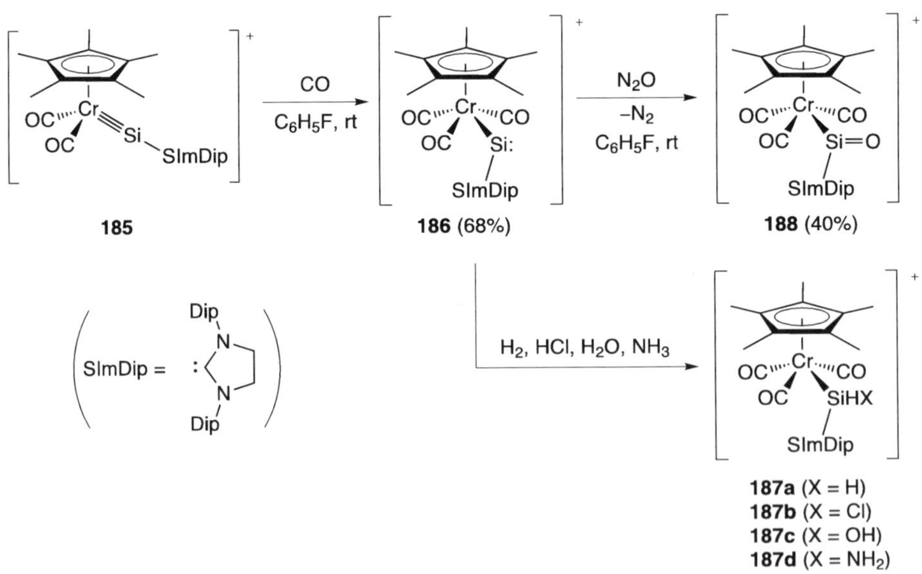

Scheme 8.70
Synthesis and reactions of chromiosilylene **186**. Counter anion $[BAr^F_4]^-$ (Ar^F = 3,5-bis (trifluoromethyl)phenyl) is omitted everywhere.

Type-I typical silyl anions	Type-II non-α-eliminative silylenoids	Type-III α-eliminative silylenoids
• nucleophile	• electrophile • nucleophile	• electrophile • nucleophile δ silylene source
Example: $(R_2N)_nR_{3-n}SiLi$ $n = 1,2$	Example: $(RO)_nR_{3-n}SiLi$ $n = 1,2$	Example: $(RS)_nR_{3-n}SiLi$ $n = 1,2$

Figure 8.10
Three types of silylenoids based on their reactivity.

degrees. Theoretical studies also supported the silylene character of **186**. Chromiosilylene **186** shows an extraordinary downfield shifted ^{29}Si NMR signal of +828.6 ppm and n→3p (HOMO→LUMO) transition band appeared at 724 nm ($\varepsilon = 391\ M^{-1}\ cm^{-1}$). Silylene **186** reacted with H_2, HCl, H_2O and NH_3 to give new silylchromium complexes **187a–d** as silylene H–X insertion products. Notably, exposure of **186** with N_2O in fluorobenzene afforded metallosilanone **188** in 40% yield as the first three-coordinate silanone species having trigonal planar silicon atom.[167]

8.4.8 Silylenoids

Silylenoids are functionalized silylmetals having electronegative leaving groups (typically, heteroatom or halogen substituent) on the anionic silicon atom (R_2SiMX, M = metal, X = leaving group), and are important class of intermediates in various silicon transformations.[168–171] Silylenoids are discussed in this section because they can generate the corresponding free silylenes by α-elimination. The character of silylenoids significantly depends on the electronegative heteroatom substituent. Kawachi and Tamao proposed three categories of silylenoids in the view of their reactivity (Fig. 8.10).[172] Type-I silylenoids act as only silyl anions and do not show electrophilic properties. Type-II silylenoids have both nucleophilic and electrophilic properties, but α-elimination does not occur. Type-III silylenoids are ambiphilic, and can be a silylene source by α-elimination.

Amino-substituted silyllithiums are typical type-I silylenoids; they behave only as nucleophilic silylanions and do not show electrophilic character.[173,174] Alkoxysilyllithiums represent type-II silylenoids; they show ambiphilic character and readily undergo self-condensation to form β-alkoxydisilanyllithiums. Typical reactions are shown in Scheme 8.71.[175–178] Alkoxysilyllithium **189** is synthesized by the Sn–Li exchange of silylstannane **190** with butyllithium. Self-condensation of **189** occurs upon elevated temperatures to give disilanyllithium **191a**. Self-condensation is supressed by addition of 12-crown-4, and the following reaction with Me_3SiCl leads to **192b**. Nucleophilic substitution of **189** with BuLi proceeds to give butylsilyllithium **191c**. Siloxysilyllithiums usually display a reactivity similar to alkoxysilyllithiums, but siloxysilyllithiums bearing bulky substituents on the silicon atom (**193a** and **193b**) do not form disilanyllithium and react with 2,3-dimethyl-1,3-butadiene to form the corresponding silacyclopentenes **194a** and **194b** in low yields, suggesting partial generation of silylene (Scheme 8.72).[179]

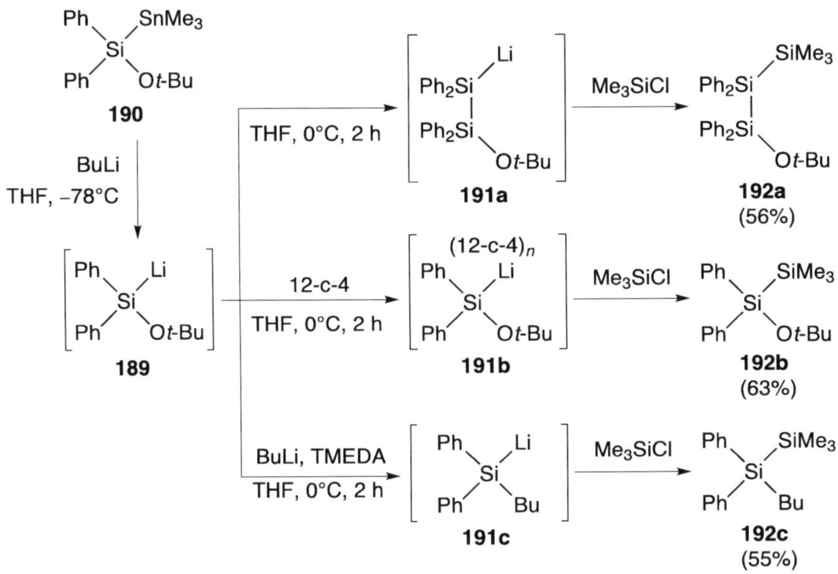

Scheme 8.71
Generation and reactions of silylenoid **189**.

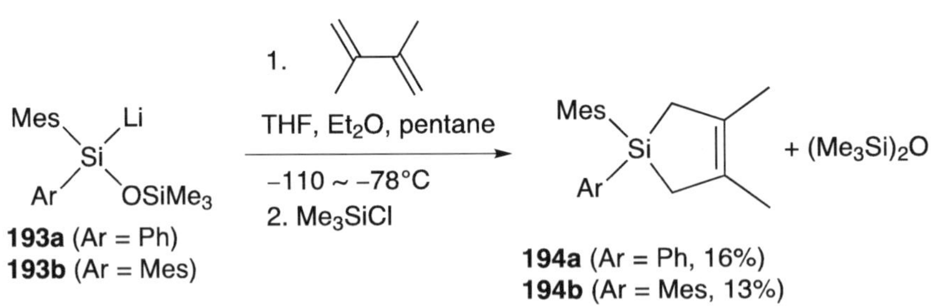

Scheme 8.72
Reactions of **193**.

Alkylthio-substituted silyllithiums are found to be type-III silylenoids. Compounds **195a** and **195b** are obtained by transmetalation of silylstannane **196a** with BuLi in THF at −78°C (Scheme 8.73) or metalation of **196b** with lithium naphthalenide (Scheme 8.74). Interestingly, in contrast to the above-mentioned functionalized silyllithiums, **195a** and

195b survive only at low temperatures and decompose into the corresponding sulfide (MesSLi) and silylenes **195a**′ and **195b**′ (Schemes 8.73 and 8.74) by α-elimination. Silylenes **195a**′ and **195b**′ can be trapped by triethylsilane, DMB, and diphenylacetylene to form the corresponding trapping products **197**–**200**.[172]

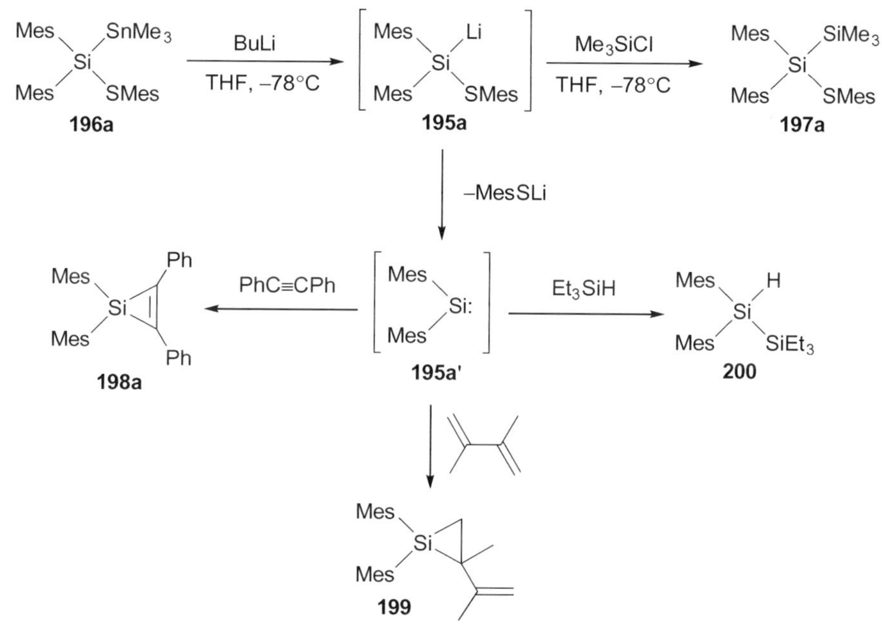

Scheme 8.73
Generation and reactions of silylenoid **195a**.

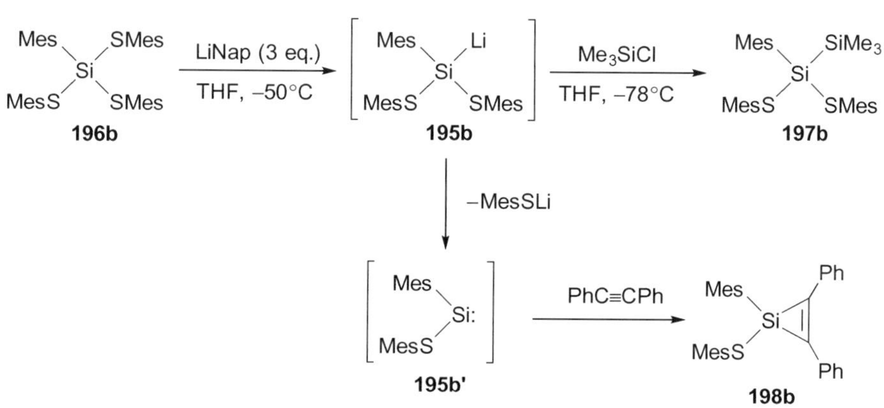

Scheme 8.74
Generation and reactions of silylenoid **195b**.

Halosilylenoids (R_2SiMX, M = metal, X = halogen) have attracted special attention as type-III silylenoids showing ambiphilicity and facile salt elimination character. Weiss and Oehme report the generation of trichlorolithiosilane $LiSiCl_3$ by bromine–lithium exchange reaction of $BrSiCl_3$ with Mes*Li, which is confirmed by the ^{29}Si NMR signal of $LiSiCl_3$ (30.9 ppm) and trapping reaction using DCl.[180]

Lee et al. have reported that reduction of tribromosilane **201Br** bearing bulky alkyl group, tris(trimethylsilyl)methyl group (Tsi), with two equiv of LiNap in THF provides bromosilylenoid **202Br** (Scheme 8.75).[181] Chloro derivative **202Cl** is also obtained in a similar mannar.[182] The ^{29}Si NMR spectra (+106 ppm at −50°C for **202Br** and +87 ppm at −70°C for **202Cl**) and DFT calculations[183] suggest that free silylene, THF coordinated silylene, and halogen-bridging dimer are excluded as the candidates for the structure of **202X** (X = Cl, Br), and $TsiSiX_2Li$ structure with X-Li interaction would be reasonable in solution. Thermal stability of the silylenoids depends on the halogen and metals. Bromosilylenoid **202Br** is stable at room temperature, whereas chlorosilylenoid **202Cl** slowly decomposes above −30°C. Potassium derivative $TsiSiBr_2K$ (**203Br**) can be generated by the reaction of **201Br** with KC_8 in THF,[182] which decomposes even at −40°C. Compound **203Br** was characterized by ^{29}Si NMR spectrum (+70 ppm at −60°C).

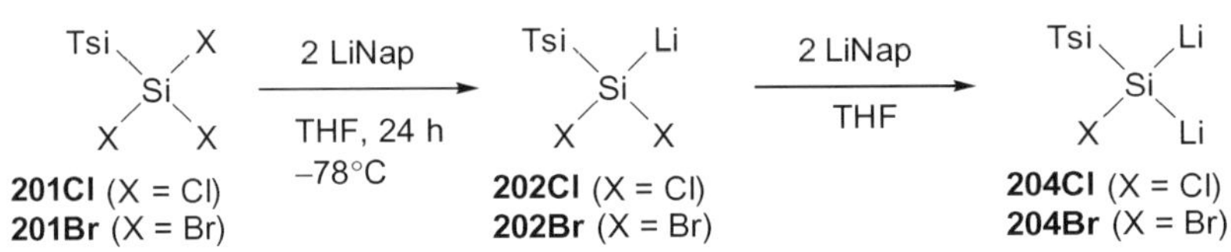

Scheme 8.75
Generation and further reduction of silylenoids **202Cl** and **202Br**.

Further reduction of **202X** (X = Cl, Br) using two equiv. of LiNap gives room temperature stable dilithiosilanes **204X** (Scheme 8.75).[184] Formation of **204X** is confirmed by ^{29}Si NMR signal of the central silicon (−106 ppm for **204Cl** and −110 ppm for **204Br** in THF at −70°C) and trapping experiments using MeOH(D). Compound **204Br** reacts with electrophiles, such as *i*-PrOH, Me_3SiBr, EtBr, and 1,4-dibromobutane, as shown in Scheme 8.76. Reactions of **204Br** with naphthalene and anthracene gave **205** and **206a** as the formal cycloadducts of silylene (Tsi(Li)Si:). Compound **206a** isomerizes to **206b** by the anionic silyl migration.[185]

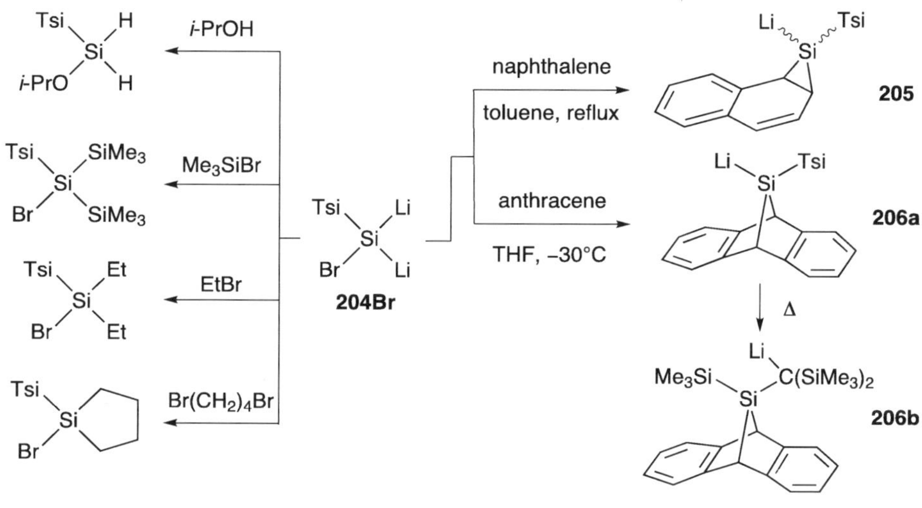

Scheme 8.76
Reactions of silylenoid **204Br**.

Nucleophilic substitution of **202Br** with MesLi affords arylsilylenoid **207a**, which shows ^{29}Si NMR signal at +88 ppm at −70°C and decomposes slowly at −10°C (Scheme 8.77).[186] Notably, addition of $MgBr_2$ to **207a** increases its thermal stability because of complexation of **207a** with $MgBr_2$. Complex **207a($MgBr_2$)** is also obtained by the reactions of **202Br** with MesMgBr, and by the reaction of the complex **202Br($MgBr_2$)**, which was obtained from **202Br** with $MgBr_2$, with MesLi or MesMgBr. Mild heating (45°C) of **207a($MgBr_2$)** in THF solution provides **207b** irreversibly by elimination of LiBr salt. Complex **207a($MgBr_2$)** and **207b** are stable at room temperature, and their ^{29}Si NMR signals appear at +87 ppm at −70°C for **207a($MgBr_2$)** and +140 ppm at 0°C for **207b**, respectively. DFT calculations propose bromosilyllithium or bromosilylmagnesium bromide structures having tetracoordinate silicon atoms with metal-bromine interactions within the observed silylenoids. Silylenoid **202Br** can also convert to new silylenoid **208** by the treatment with ArLi (Ar = 2-(Me_2NCH_2)C_6H_4.[187] Compound **208** is thermally stable even at 110°C. The ^{29}Si NMR signal (+126 ppm at room temperature) and several trapping products by the reactions with water and alcohols support silylenoid character of **208**. Intramolecular coordination of nitrogen to silicon atom and loss of Si−Br bond in structure of **208** was proposed by computational studies.[188]

Representative reactions of halosilylenoids **202X** are described in Schemes 8.78 and 8.79. Bromosilylenoid **202Br** underwent formal [1 + 2] cycloaddition of silylene Tsi(Br)Si: with alkenes to give silacyclopropanes **209a−d.**[189] Bromosilylenoid **202Br** also reacts with phenylisocyanate or phenyl isothiocyanate to give four membered ring compounds **210a** or **210b.**[190] Generation of Tsi(Br)Si = O or Tsi(Br)Si = S species that result from oxygen (sulfur) transfer from isocyanate or isothiocyanate to silicon and their subsequent dimerization were proposed as a possible mechanism.

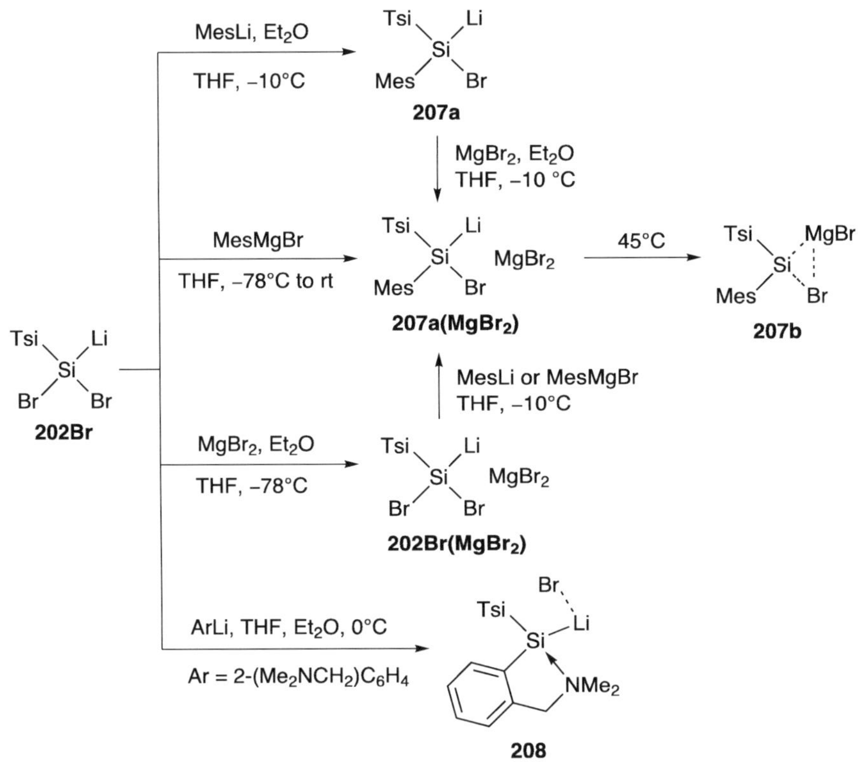

Scheme 8.77
Reactions of silylenoid **202Br** with nucleophiles.

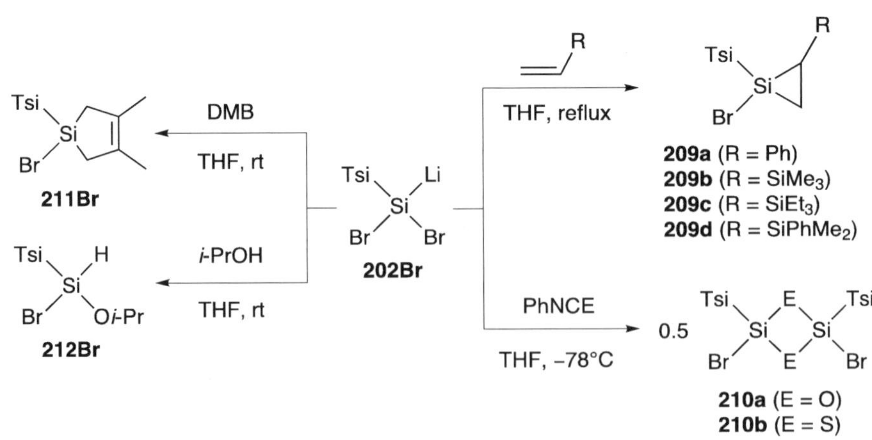

Scheme 8.78
Reactions of silylenoid **202Br**.

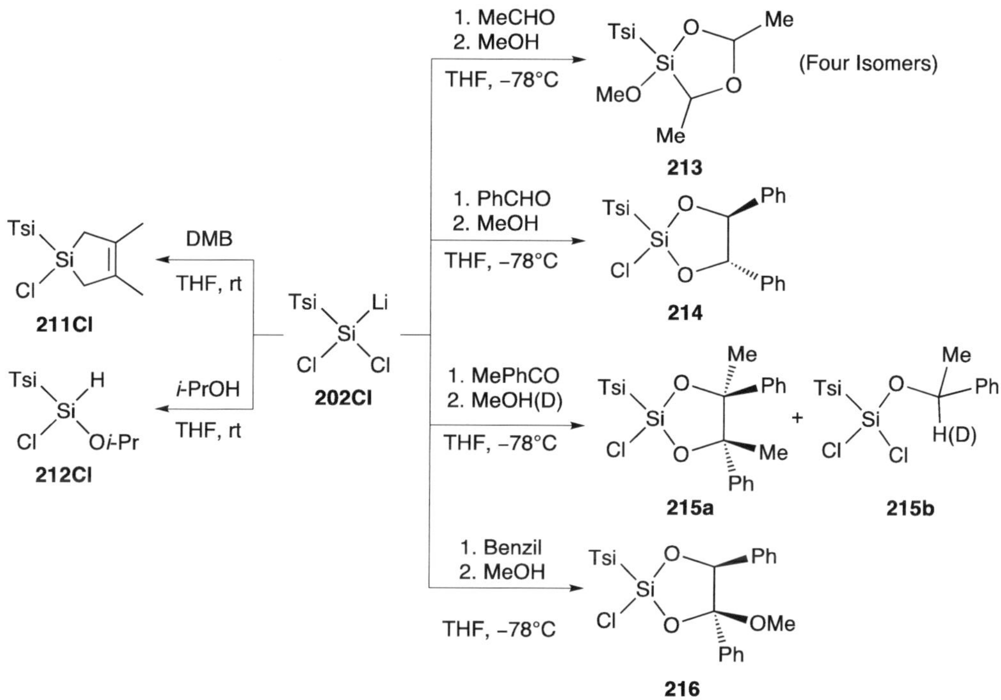

Scheme 8.79
Reactions of silylenoid **202Cl**.

Reactions of **202Cl** with DMB and *i*-PrOH gave expected products **211Cl** and **212Cl** similar to those of **202Br** (Scheme 8.79).[182] Reactions of **202Cl** with aldehydes provide 1:2 adducts **213** and **214**.[191] DFT calculations proposed these reactions proceed by the formation of silaoxiranes as formal [1 + 2] cycloadducts of silylene with aldehydes and the subsequent nucleophilic attack of the second aldehydes to silicon atom on the silaoxirane. Reactions of **202Cl** with acetophenone and benzil afford silaheterocyclic compounds **215a** and **216**, respectively.[192] Formation of **215b** as a by-product suggests stepwise cyclization involving Brook rearrangement of silyl group.

Apeloig et al. have synthesized isolable fluorosilylenoid **217**.[193] Silylenoid **217** was obtained in 40% yield by the bromine-lithium exchange reaction of bis(silyl) bromofluorosilane **218** with bulky silyllithium *t*-Bu_2MeSiLi in THF (Scheme 8.80). A striking feature of X-ray structure of **217** is considerably pyramidalized three-coordinate silicon atom (the sum of the bond angles: 307.6 degrees) (Fig. 8.11). The Si···Li distance is 3.21 Å, while the Si–F and Li–F bonds are 1.698(3) Å and 1.773(7) Å, respectively. Calculations for bonding character of **217** suggested highly polalized Si–F covalent bond. The ^{29}Si NMR signal of the three-coordinate silicon of **217** appeared at +107 ppm, which was reproduced by the theoretical calculation (+102 ppm). This confirms that **217** keeps three-coordinate silylenoid structure even in solution.

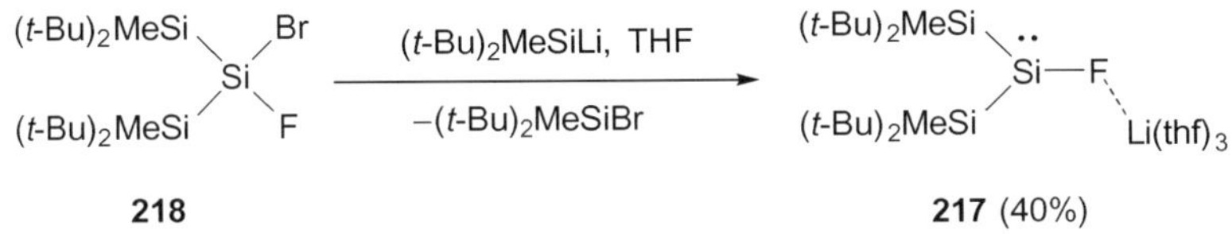

Scheme 8.80
Synthesis of silylenoid **217**.

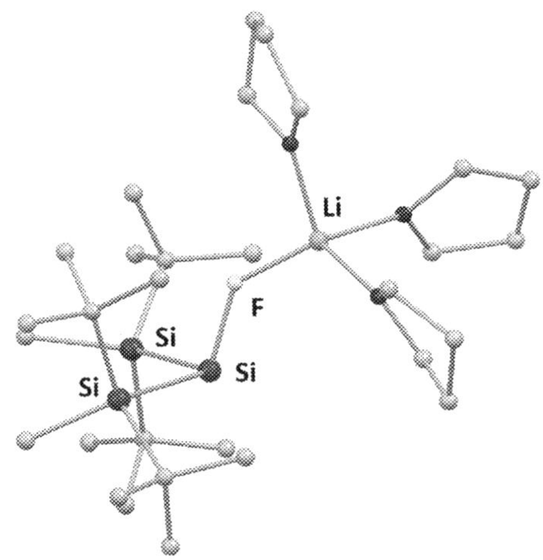

Figure 8.11
Molecular structure of **217** (CCDC: 288145).

Reactions of **217** with nucleophiles and an electrophile, dimerization to form disilene, and metalation to give lithiosilyl and sodiosilyl radicals were investigated in details (Scheme 8.81).

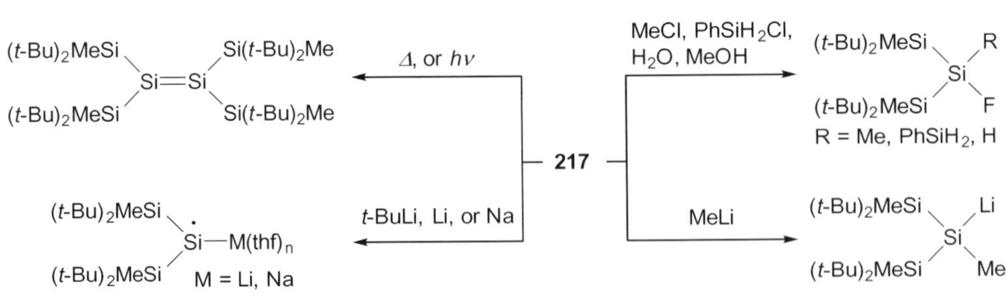

Scheme 8.81
Reactions of silylenoid **217**.

8.5 Dialkylsilylenes and Carbocyclic Silylenes

8.5.1 Synthesis and Molecular Structures

In 1999, Kira et al. have reported isolable dialkylsilylene **7a** synthesized by the reaction of dibromosilane **219** with two equiv of KC_8 in THF at $-50°C$ (Scheme 8.82).[11] Silylene **7a** is an orange crystalline solid and its structure is confirmed by NMR, MS spectroscopies and X-ray analysis. The ^{29}Si NMR signal of divalent silicon of **7a** appears at +567.4 ppm in C_6D_6 and a distinct n→3p (HOMO→LUMO) UV transition band is observed at 440 nm in hexane. Silylene **7a** is storable in the solid state even at room temperature but it isomerizes to silene **220** via intramolecular 1,2-silyl migration with the following kinetic parameters: $\Delta H^{\ddagger} = 18.49 \pm 0.43$ kcal mol^{-1} and $\Delta S^{\ddagger} = -20.32 \pm 1.35$ cal mol^{-1} K^{-1}.[194]

The fundamental reactivity of **7a** such as reactions with alkyl chlorides,[195] chlorosilanes and hydrosilanes,[196] aminoxy radicals,[197] alkenes and alkynes,[198] ketones,[199–201] carbon dioxide,[202] imines,[203] nitriles,[204] and diazocarbonyl compounds[205] are investigated. Since several reviews on the chemistry of **7a** are already available,[206–210] the selected topics including photochemistry and transition metal complexes are discussed in the following sections.

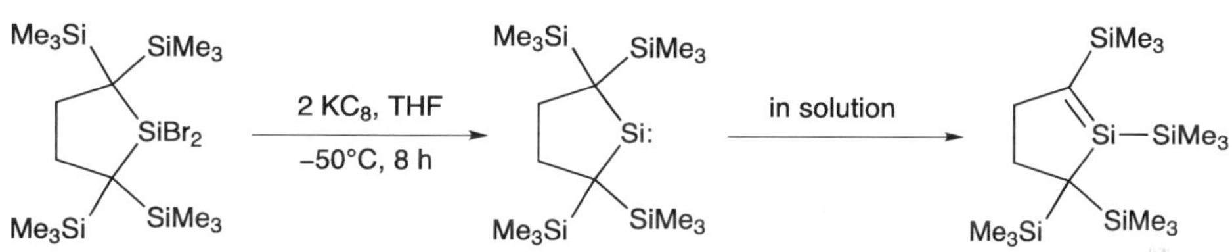

219 **7a** (75%) **220**

Scheme 8.82
Synthesis and thermal isomerization of silylene **7a**.

Structurally modified dialkylsilylene **7b** was also synthesized. Silylene **7b** does not form the corresponding silene via 1,2-silyl migration, but it dimerizes reversibly to disilene $(\mathbf{7b})_2$ in low temperature (Scheme 8.83).[211] Silylene **7b** exists almost exclusively at room temperature and considerable amount of disilene $(\mathbf{7b})_2$ is observed by VT-NMR (^{1}H, ^{13}C, and ^{29}Si NMR). The ^{29}Si NMR signals of divalent silicon in **7b** and unsaturated silicon in $(\mathbf{7b})_2$ appeared at +539.0 ppm and +123.3 ppm, respectively. VT-UV-vis spectroscopy also supports the equilibrium. Both **7b** and $(\mathbf{7b})_2$ were isolated as yellow and orange crystals and structurally characterized. Conformational change around 1,3-disilaindane moieties enable the dimerization of **7b**.

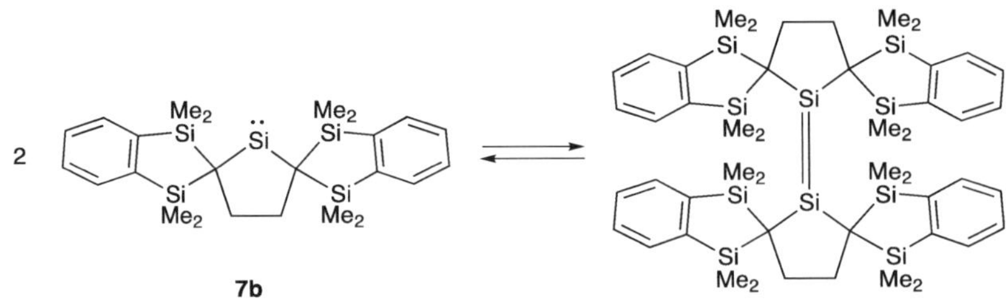

7b

(7b)$_2$

Scheme 8.83

Equilibrium between the silylene **7b** and its dimer.

More bulky dialkylsilylene **7c** having four *i*-PrMe$_2$Si instead of Me$_3$Si groups can be generated as an intermediate by the reduction of the corresponding dibromosilane with KC$_8$, but **7c** has poor crystallinity and undergoes facile 1,2-silyl migration to give a cyclic silene (the half-life of **7c** is only 20 min at 20°C), which prevents its isolation.[212] The existence of **7c** was confirmed by the intense yellow color of the resulting solution, ^{29}Si NMR spectroscopy (divalent silicon nuclei: +567.3 ppm in C$_6$D$_6$), and trapping reaction with H$_2$O to form **221**. Silanol **221** can be applied for preparation of persistent dialkylsilanone **222a** (Scheme 8.84). Oxidation of **7a** by N$_2$O or O$_3$ generates silanone **222b**, which survives only in Xe or Ar matrices at extremely low temperatures (< 27K).[213]

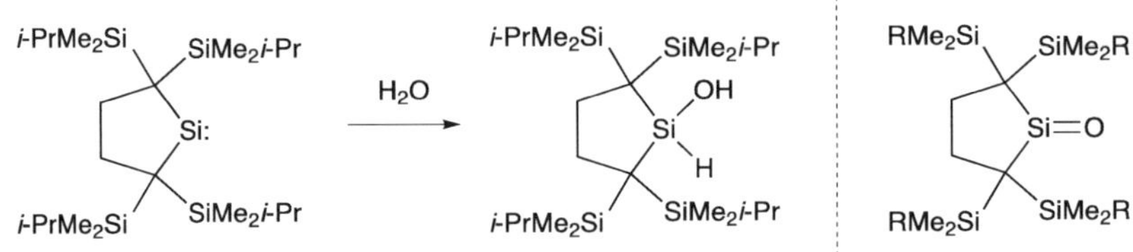

7c **221** **222a** (R = *i*-Pr) **222b** (R = Me)

Scheme 8.84

Reaction of persistent silylene **7c**.

Driess, Inoue et al. have reported carbocyclic silylene **223a** (R = Ph) and **223b** (R = *m*-Tol) having phosphorus ylide moiety.[214] Reduction of dibromosilanes **224** with excess amount of KC$_8$ (3–4 equiv in THF for **224a**, 3 equiv in DME for **224b**) provided the corresponding silylenes (Scheme 8.85). Their structures were confirmed by low-field shifted ^{29}Si NMR signals (+213.3 ppm, triplet, $^2J_{SiP}$ = 38.5 Hz for **223a**, +212.4 ppm, triplet, $^2J_{SiP}$ = 38.2 Hz for **223b**). Theoretical studies suggested the contribution of charge separated and 6π aromatic resonance structures (**223′** and **223″**) due to the ylide units, which stabilized electronically the silylene moiety. An independent theoretical study by Apeloig and Karni also disclosed electronic character of phosphorus-ylide containing cyclic silylenes.[215] Reaction of **223** with di-*t*-butyl-ortho-benzoquinone derivative gave the corresponding [1 + 4] cycloadducts.

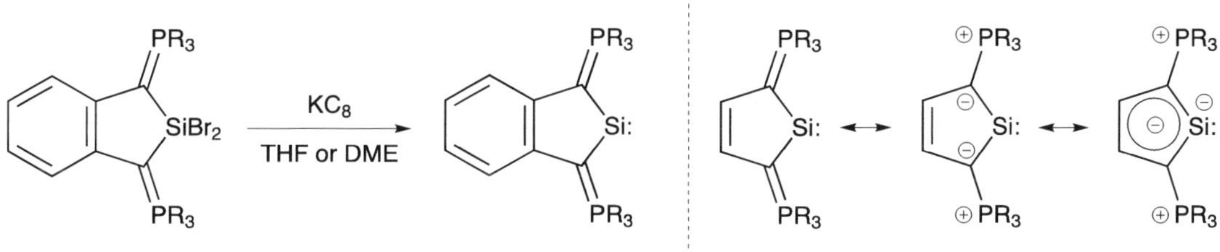

Scheme 8.85
Synthesis and canonical structures of silylenes **223**.

Oestreich et al. have reported generation of silylene **225** having all-carbon substituents by the photolysis of trisilane **226**, which is evidenced by typical trapping reactions to give **227a–c** (Scheme 8.86). [216] In this type of silylene, α-spirocyclopropyl groups are theoretically predicted to stabilize divalent silicon by donation of σ(C–C) orbital of cyclopropane rings to vacant 3p(Si) orbital.[217]

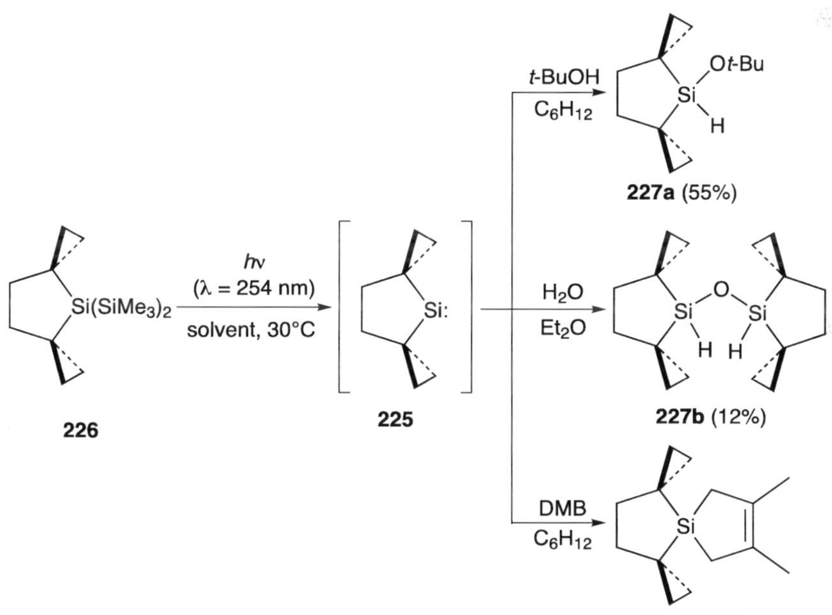

Scheme 8.86
Photochemical generation of silylene **225**.

8.5.2 Photochemical Cycloadditions to Aromatic Compounds

Photochemical reactions of **7a** with benzenes afford silepins **228–230** (Scheme 8.87).[218] These reactions seem to proceed via the formation of the [1 + 2] cycloadducts

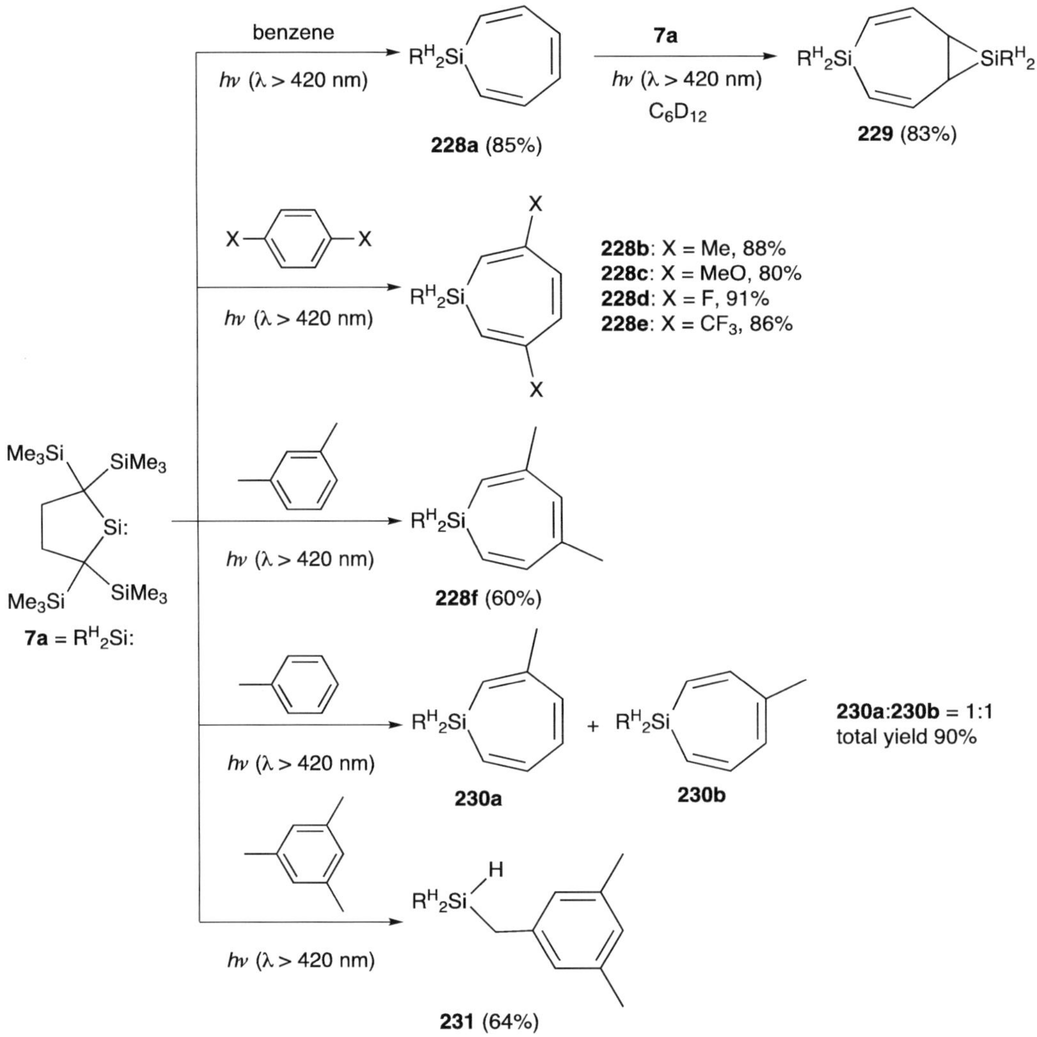

Scheme 8.87
Reactions of photochemically excited silylene **7a** with aromatic compounds.

(silanorcaradienes) and the subsequent ring expansion. The photochemical [1 + 2] cycloaddition of **7a** with silepin **228a** proceeds to give **229**. When irradiation of **7a** was conducted in mesitylene, hydrosilane **231** forms via the benzylic C–H bond insertion. In the case of naphthalene, two types of silacyclopropanes **232** and **233** are formed by photochemical [1 + 2] cycloaddition (Scheme 8.88). Compound **232** releases silylene **7a** and naphthalene by thermal retro [1 + 2] cycloaddition but **233** does not. In contrast to the previously reported dearomative cycloaddition of silylene with aromatic compounds,[155,158,159] the reactions shown in Scheme 8.87 do not proceed thermally. The singlet excited state of **7a** (**7a***) possessing diradical character is responsible for these dearomative cycloadditions, which was confirmed by theoretical investigation, fluorescent quenching experiments, and stereospecific photochemical [1 + 2] cycloaddition of **7a** with olefins.[219] Initially, two successive (stepwise) radical-like additions of **7a*** to C = C π bond were suggested, but recent theoretical study proposed one-step cycloaddition of **7a*** with benzene via a conical intersection between S1 and S0 potential energy surfaces.[220]

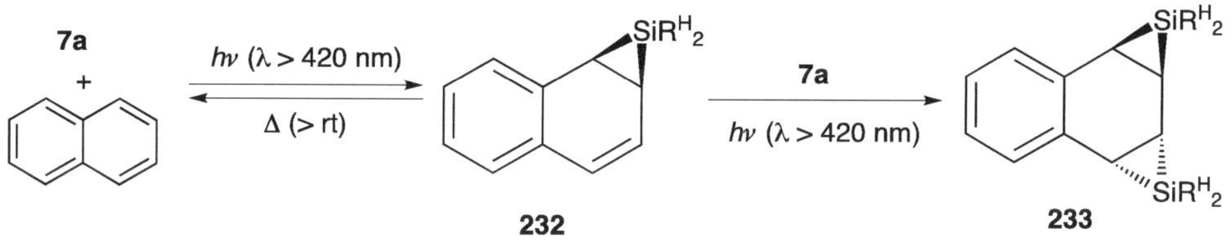

Scheme 8.88
Photochemical reaction of silylene **7a** with naphthalene.

Cycloaddition-ring expansion can be applied for the synthesis of bicyclic [4]dendralene **234a** and heptafulvene **234b** by the photochemical reactions of **7a** with azulene and guaiazulene (Scheme 8.89).[221] Thermal regioselective insertion of **7a** into nickel(II) norcorrole affords silicon-incorporated porphyrinoids (**235** and **236**) showing unique near-IR absorption (Scheme 8.90).[222]

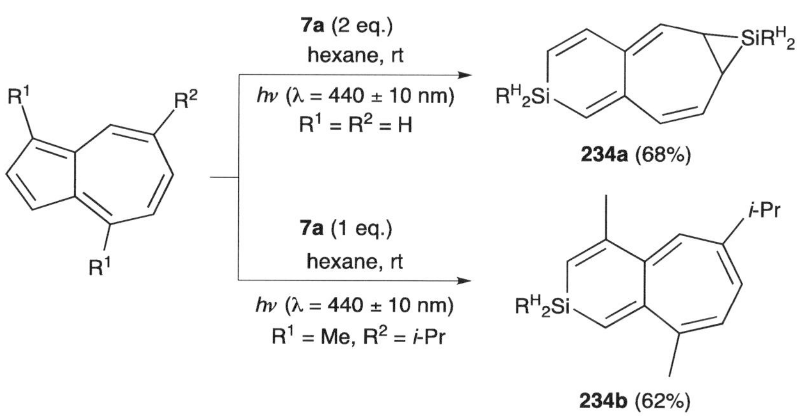

Scheme 8.89
Photochemical reactions of silylene **7a** with azulenes.

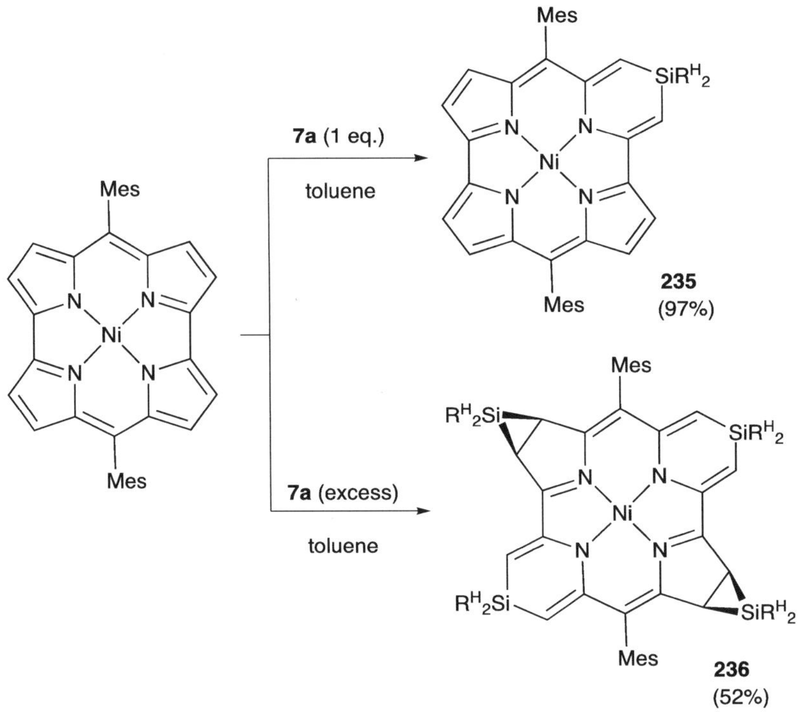

Scheme 8.90
Reaction of silylene **7a** with norcorrole.

8.5.3 Reactions With C–X and Si–X (X = H, Halogen) Bonds

Dialkylsilylene **7a** reacts with carbon tetrachloride and chloroform to give dichlorosilane **237** (Scheme 8.91).[195] A trace amount of hexachloroethane and tetrachloroethane are detected in these reactions, suggesting radical mechanism. Insertion of **7a** into Me–I bond of iodomethane proceeds to give **238**. Reactions of **7a** with an excess amount of dichloromethane and chloromethylcyclopropane afford 2:1 adducts (**239** and **240**) even in excess amount of haloalkanes. Ambiphilic character of **7a** would play an important role in the formation of **239** and **240**; one **7a** activates C–Cl bond as a Lewis acid and the second **7a** acts as a nucleophile to attack carbon atoms.

Silylene **7a** inserts into Si–Cl and Si–H bonds of various chlorosilanes and hydrosilanes (Scheme 8.92).[196,223] Notably, selective insertion proceeds in hydro(chloro)silanes; reaction of **7a** with dichlorosilane gives Si–Cl insertion product **241c**, whereas similar reaction with dimethylchlorosilane affords Si–H insertion product **242a**.

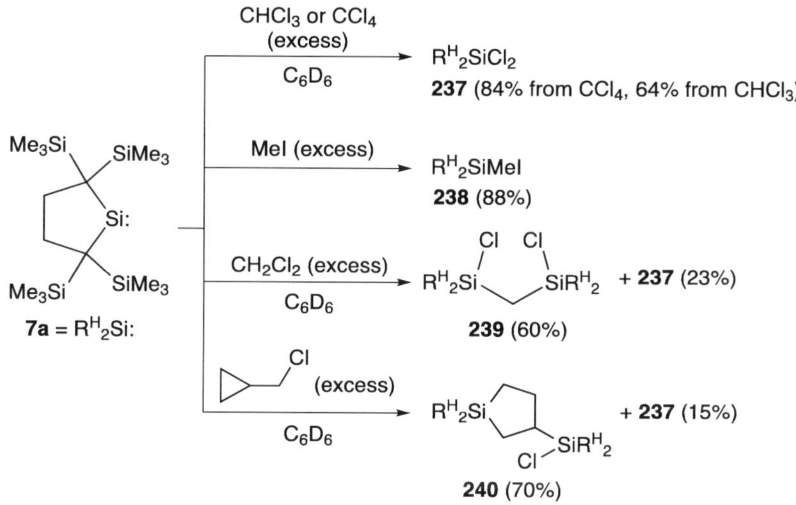

Scheme 8.91
Reactions of silylene **7a** with haloalkenes.

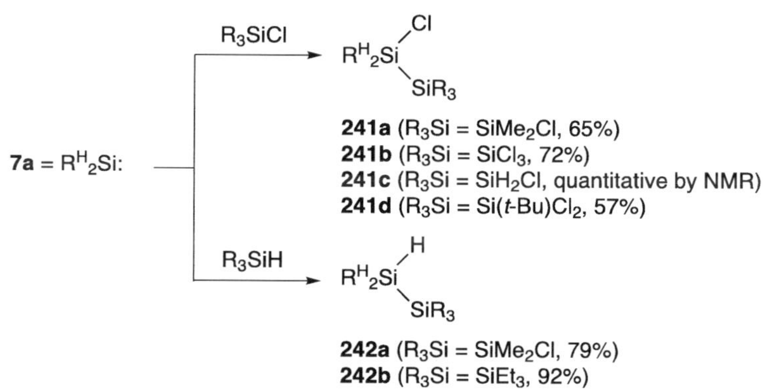

Scheme 8.92
Reactions of **7a** with chlorosilanes and hydrosilanes.

Theoretical studies proposed concerted Si—H and Si—Cl insertion mechanisms. These studies also pointed out the importance of two types of the orbital interactions in the Si···Si···X (X = H, Cl) plane during the insertion process (Scheme 8.93): (a) vacant 3p(Si) orbital of silylene and nonbonding orbital of X (n(X)) or σ (Si—X) orbital, (b) nonbonding n(Si) orbital of silylene and antibonding σ^*(Si—Y) orbital, where Y = in-plane substituent.[224–226] In-plane auxiliary substituents Y are important because effective in-plane orbital interactions accelerate insertion reactions. Out-of-plane substituents (R^1—R^4) mainly provide an environment of the reaction site.

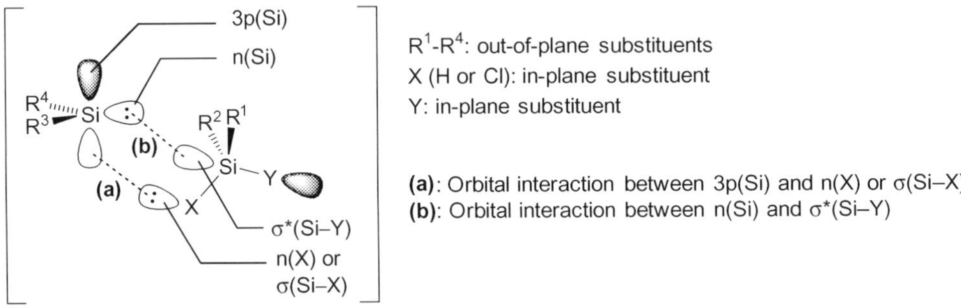

Scheme 8.93
Schematic representation of important orbital interactions of silylene insertion into Si–X bond.

8.5.4 Transition Metal Complexes and Related Metal Species

Several dialkylsilylene-transition metal complexes have been synthesized by the ligand exchange reactions of group 10 phosphine or cyclooctadiene complexes with isolable dialkylsilylene **7a**. When **7a** is treated with two equiv of $(Cy_3P)_2Pd$, dinuclear palladium complex **243** forms as dark blue-purple crystals in 24% yield (Scheme 8.94).[227] This reaction proceeds by the stepwise ligand exchange reactions via monosilylene monopalladium complex **244a**. Complex **243** is a mono(μ-silylene) dipalladium complex lacking bridging ligands (Pd–L–Si). Related palladium and platinum complexes **244–246** are also obtained by the stepwise ligand exchange reactions as shown in Scheme 8.94.[228] Planar three-coordinate silicon and metals (group 10) are found in the structures of **246a** and **246b**. The interplane angles between the C–Si–C and P–M–P planes are nearly right angles (88.36 degrees for **246a**, 87.59 degrees for **246b**). Reaction of **7a** with a half equivalent of $(Cy_3P)_2Pd$ gives bis(dialkylsilylene)palladium complex **247** as dark red crystalline solid in 40% yield.[229] Complex **247** adopts a linear structure and palladium atom is coordinated by two silylene units. The Si–Pd bond lengths of in **247** (2.263(1) and 2.260 (1) Å) are in the range of the reported bis(silylene)palladium complexes.[81] Two silylene moieties are almost perpendicular to each other. DFT calculations suggest that linear structure of **247** is due to severe steric repulsion between silylene units; the optimized structure of model compound $(Me_2Si)_2Pd$ has considerable bending Si–Pd–Si angle of 116.7 degrees and this angle increases with increasing the size of the silylene units. Complexes **243, 244a**, and **247** exhibit low-field shifted ^{29}Si NMR signals of divalent silicon at +303.5 ppm with $^2J_{SiP} = 78.5$ Hz (**243**), +413.5 ppm with $^2J_{SiP} = 132.3$ Hz (**244a**), and +447.7 ppm (**247**), respectively.

Dialkylsilylene-nickel complexes **248a** and **248b** are obtained by the reaction of **7a** with Ni $(cod)_2$ in benzene or toluene and the following ligand exchange reactions to give **246c** (Scheme 8.95).[228] The structure of **246c** resembles those of **246a** and **246b** (the interplanar angle between the C–Si–C and P–Ni–P planes is 87.32 degrees).

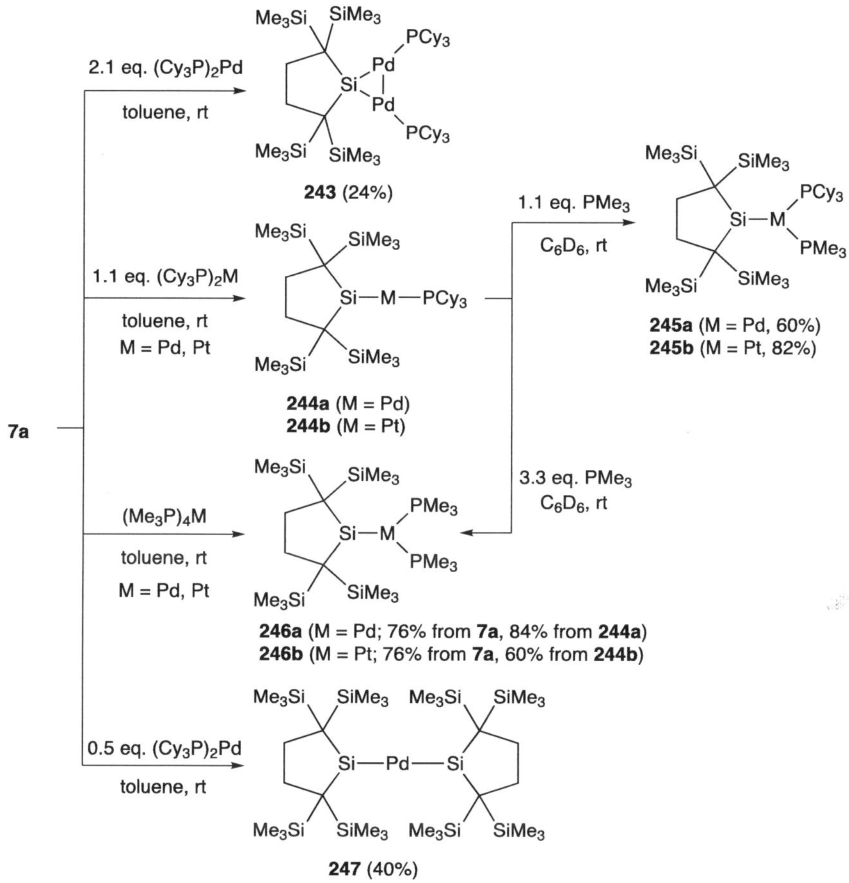

Scheme 8.94
Synthesis of palladium and platinum complexes of silylene **7a**.

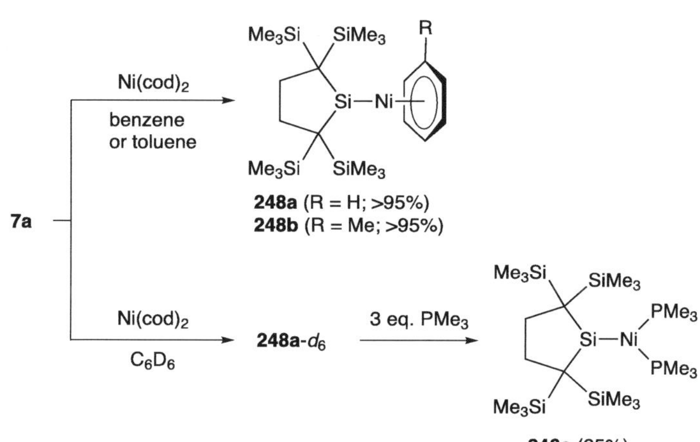

Scheme 8.95
Synthesis of nickel complexes of silylene **7a**.

Interestingly, complex **243** reacts with molecular hydrogen to give dihydrosilane **249** in 85% yield together with $(Cy_3P)_2Pd$ and palladium metal (Scheme 8.96). Since **7a** does not react with hydrogen under the same condition, oxidative addition of H_2 to **243** followed by 1,2-hydrogen shift from Pd to Si would occur as the initial step.[227] Reaction of **247** with molecular hydrogen proceeds immediately to give bis(hydrosilane)palladium complex **250** as orange crystals in 84% yield, which is the first formal 12-valence electron palladium complex.[229] Further exposure of **250** to H_2 results in the formation of dihydrosilane **249** in 97% yield.

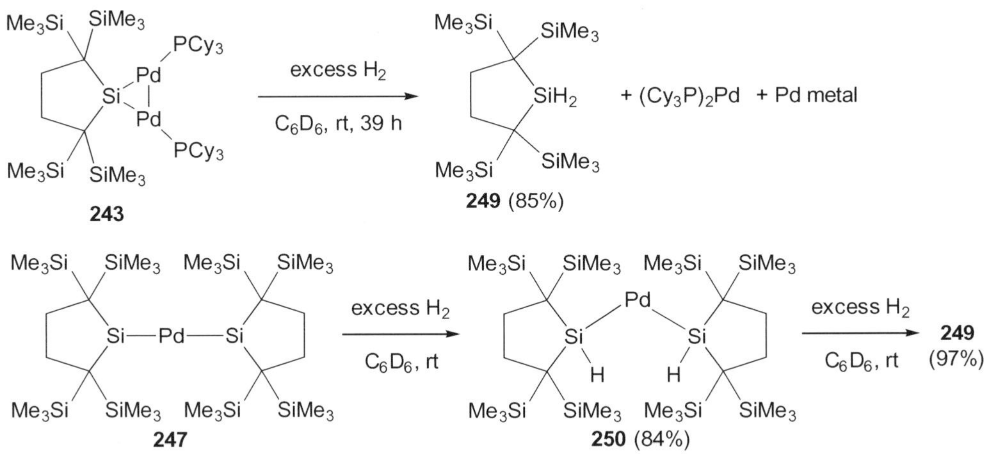

Scheme 8.96
Reactions of silylene complexes **243** and **247**.

Cationic complexes of dialkylsilylene and group 11 metals that are isoelectronic neutral dialkylsilylene-group 10 complexes are also synthesized. Reactions of **7a** with Cu^+TPFPB^- and $[Ag(benzene)_3]^+TPFPB^-$ afforded silylene copper complex $\mathbf{251a}^+TPFPB^-$ and silver complex $\mathbf{251b}^+TPFPB^-$ in 26% and 47% yields, respectively (Scheme 8.97).[230] These complexes exist as solvent separated ion pairs in the solid state and adopt linear structures with two-coordinate Cu and Ag centers. Two silylene units are almost perpendicular to each other; the interplanar angle between two C–Si(silylene)–C planes are 71.0 degrees for $\mathbf{251a}^+$ and 86.2 degrees for $\mathbf{251b}^+$. These structural characteristics are similar to those of the palladium complex **247**. Computational studies suggest that the reason for the linear dicoordinate structure of $\mathbf{251a}^+$ is not of steric origin, as in the palladium complex but is due to the intrinsic nature of cationic bis(silylene)copper. Weak π back-donation in $\mathbf{251a}^+$ and $\mathbf{251b}^+$ caused by the low-lying filled d orbitals of Cu(I) and Ag(I) cations is responsible for their intrinsic linear structures.

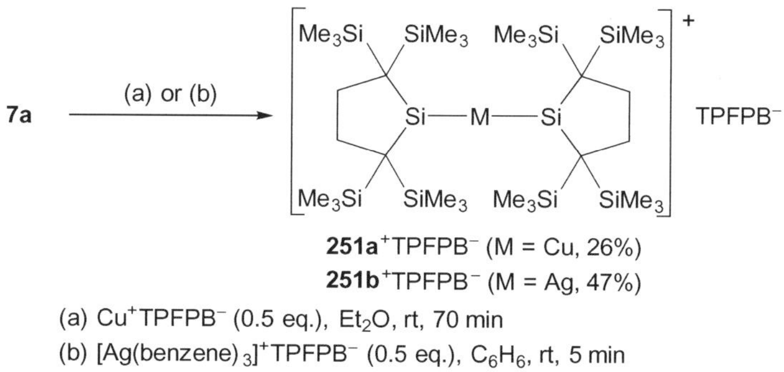

Scheme 8.97
Synthesis of group 11 metal complexes with silylene **7a**.

Reduction of **7a** with alkali metals (Na, K, Rb, Cs) or lithium (in the presence of 4,4′-di-*tert*-butylbiphenyl) in DME solution at −60°C generates the corresponding silylene anion radical (**252**), formation of which being confirmed by EPR spectroscopy (Scheme 8.98).[231] EPR parameters of **252** at −60°C are as follows: *g*-value = 2.0077, $a(^{29}\text{Si}$, silylene center) = 2.99 mT, $a(^{29}\text{Si}, \text{SiMe}_3)$ = 1.30 and 1.66 mT. This anion radical decomposes with a half-life of ca. 20 min at 25°C. In this context, an isolable silylene anion radical synthesized by Si−Si bond cleavage of 1,2-dimetalladisilane has been reported (Section 8.4.6).[164]

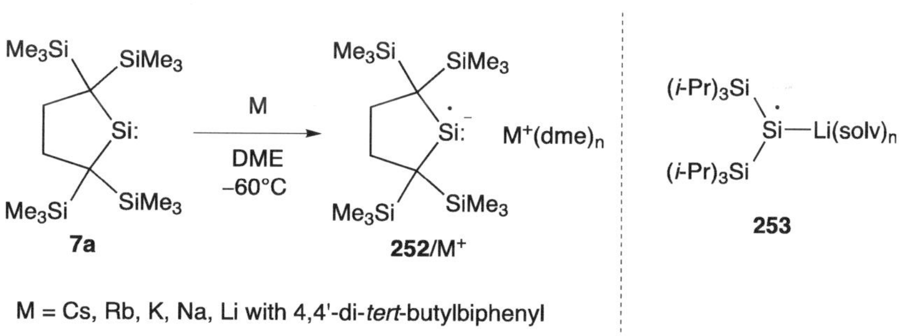

Scheme 8.98
Reduction of silylene **7a** with alkali metals.

No observation of hyperfine coupling constants owing to group-1 metals suggests **252**/M$^+$ exists as a solvent-separated ion pair in solution, which is in sharp contrast to lithiosilyl radical **253** showing hyperfine coupling constants owing to Si−^{7}Li bond in its EPR spectrum.[232] Similar reductions of stable cyclic diaminosilylenes result in the formation of 1,2-dianionic disilanes by the formal dimerization of the silylene anion radicals (Section 8.2.3.1).

8.6 Triplet Silylenes

In contrast to carbenes, silylenes are usually in the singlet ground state because of large singlet-triplet energy difference (ΔE_{ST}) that is closely related to the energy difference between nonbonding orbital (usually HOMO) and vacant 3p orbital (usually LUMO) on the divalent silicon. The ΔE_{ST} value reduces by the introduction of electropositive substituents because these substituents raise the energy level of the nonbonding orbital of silylene. Bulky substituents also reduce ΔE_{ST} value because they cause widening of the bond angle at the divalent silicon to raise the energy level of nonbonding orbital.[233–235] Consequently, silylenes bearing bulky silyl groups and/or electropositive metals as substituents are good candidates as the triplet ground state silylenes. Bis(tri-*tert*-butylsilyl)silylene **254** was generated by the photolysis of the corresponding silacyclopropenes **255a** and **255b** in methylcyclohexane glass matrix at 77K (Scheme 8.99).[236] Similarly, lithio- and potassio(tri-*tert*-butylsilyl)silylene **256a** and **256b** can be generated by photolysis of **257** in 2-methyltetrahydrofuran at 14K.[237] Their triplet ground states are proved by EPR study: characteristic signals are observed at 845 mT for **254**, 790 mT for **256a**, and 780 mT for **256b**, and their intensities are inversely proportional to the measurement temperatures. Considerably large zero-field splitting (ZFS) parameters D of $1.5 \sim 1.6\ cm^{-1}$ are found in the observed EPR spectra employing the free electron g-value, which indicate wide Si–Si–M (M = Si, Li, K) bond angles in accord with the expected molecular structures. Silylene **254** undergoes intramolecular C–H insertion to form a disilacyclobutane. Recent computational studies have pointed out that the intramolecular C–H insertion is not a diagnostic reaction of the triplet silylene; thermally accessible singlet silylene inserts into C–H bond much faster than the triplet silylene.[238]

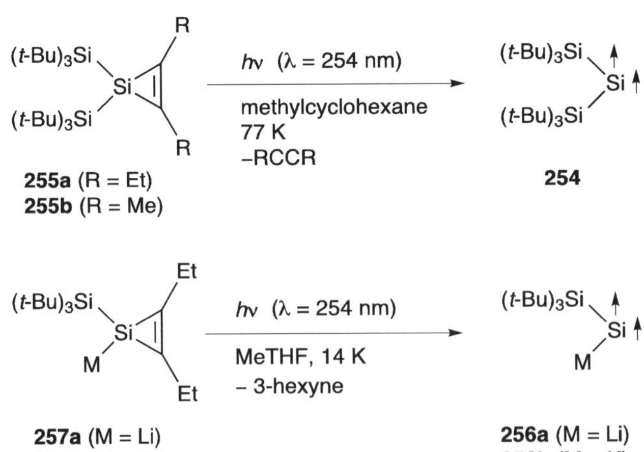

Scheme 8.99
Generation of triplet silylenes.

8.7 Monosilylenes With N,N-di(tert-butyl)amidinato Ligands

8.7.1 Synthesis and Molecular Structures

Since 2006, various functionalized silylenes having three-coordinate silicon atom with bulky amidinato ligands have been synthesized. Although reviews on functionalized silylenes are available recently by Roesky et al.,[22,239] the recent progress is briefly summarized in this section.

The first example of three-coordinate silylene with amidinato ligands **258** was reported by Roesky et al.[240] The reaction of di(*t*-butyl)carbodiimide with PhLi in Et_2O and subsequent addition of silicon tetrachloride gave five-coordinate trichlorosilane **259** in 47% yield (Scheme 8.100). The reduction of **259** with two equivalents of potassium metal in THF gave the first stable chlorosilylene **258** in 10% yield as colorless crystals. Alternatively, silylene **258** was synthesized in higher yields by dehydrochlorination of dichlorohydridosilane **260** with $LiN(SiMe_3)_2$ or NHC in toluene[241] or the reaction of di(*t*-butyl)carbodiimide with PhLi and subsequent addition of hexachlorodisilane.[242] The ^{29}Si NMR resonance due to three-coordinate silicon nucleus appeared at +14.6 ppm. XRD analysis of **258** showed that the silicon center was coordinated by two nitrogen atoms of the amidinato ligand and chlorine atom with the considerably small sum of bond angles of 260.73 degrees. The four-membered ring is planar and the two C–N distances (C–N bond and C = N bond) being the same (1.333(2) Å) suggesting a delocalized N–C–N moiety. The N–Si–N angle is very acute (68.35(8) degrees) and the Si–N distances are 1.870(2) Å. These metric parameters are well reproduced by DFT calculations at the B3LYP/ 6–31 + G(2pd,2df) level. The NBO (natural bond orbital) analysis shows that the Si–N bonds are formed by 85% s-orbital of nitrogen atom and pure p-orbital of silicon atom, and

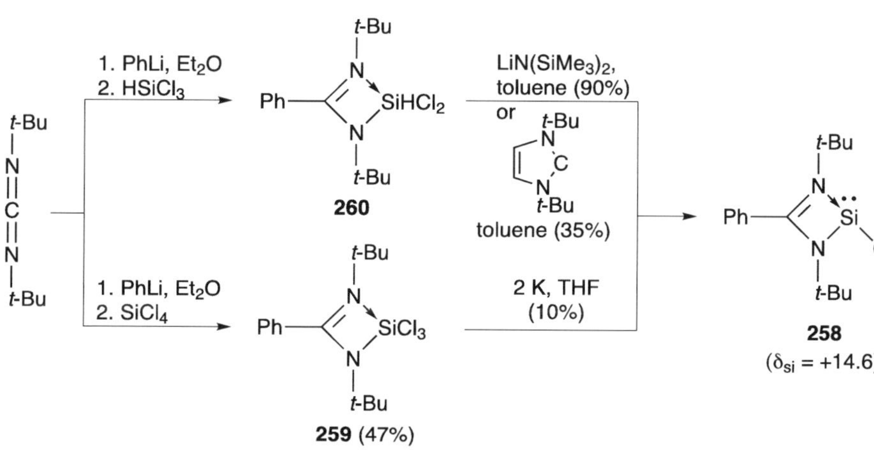

Scheme 8.100
Synthesis of amidinatochlorosilylene **258**.

the highly acute N–Si–N angle results from this s-p overlap. The corresponding bromosilylene **261** was prepared by the reaction of bis(silylene) **262** with bromine (see Scheme 8.135, next Section 8.8). The structural characteristics of **261** is close to those of **258.**[243]

The Ph group on the carbon atom and *t*-Bu on the nitrogen atoms are crucial for synthesis of **258**.[244] When *t*-Bu was replaced by $SiMe_3$, reduction of the corresponding trichlorosilane $[PhC(NSiMe_3)_2]SiCl_3$ did not give the corresponding silylene but disproportionation products. When Ph group on the carbon atom was replaced by alkyl groups R, such as *i*-Pr and cyclohexyl, and *t*-Bu groups on the nitrogen atoms were replaced by less hindered alkyl groups R′, such as Me and *i*-Pr groups, reduction of the corresponding trichlorsilane $[RC(NR')_2]SiCl_3$ gave a complex mixture.

Treatment of **260** with two or four equivalent of KC_8 gave silylsilylenes **263a** or **263b**, respectively (Scheme 8.101).[245] Although hydridosilylene **264** was not observed, generation of **264** and subsequent 1,3-addition of Si-H moiety of **264** to unreacted chlorosilylene **258** or hydridosilylene **264** was proposed for the mechanism of the formation of **263a** and **263b**. The mechanism of the formation of **263a** was supported by the reaction of chlorosilylene **258** with $K^+[BH(i\text{-}Bu)_3]^-$ giving **263a** in 33% yield. When two equivalents of $KN(SiMe_3)_2$ were used for the reaction of **260**, aminosilylene **265** was obtained.[246]

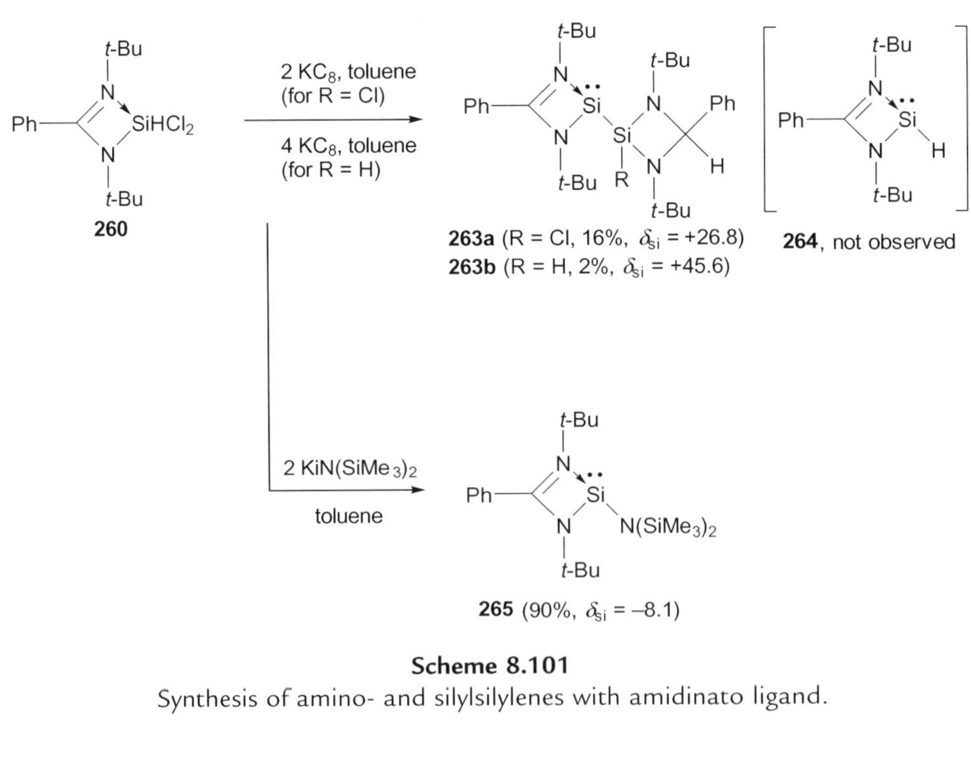

Scheme 8.101
Synthesis of amino- and silylsilylenes with amidinato ligand.

Various heteroatom-substituted three-coordinate silylenes **265**,[246,247] **266a–d**,[244,247] **266e–h**,[247] **266i**,[248] **266j**[64] were synthesized by introduction of heteroatom substituents to **259** and the subsequent reductive dechlorination of the adducts **267a–d**[244] (Scheme 8.102) or metathesis reaction of chlorosilylene **258** with the corresponding metal amides, alkoxides, or phosphides (Scheme 8.103).[247,248] Alkyl-substituted three-coordinate silylenes **266k** and **266l** were also obtained by the metathesis of chlorosilylene **258** with alkyl anions (Scheme 8.103).[249]

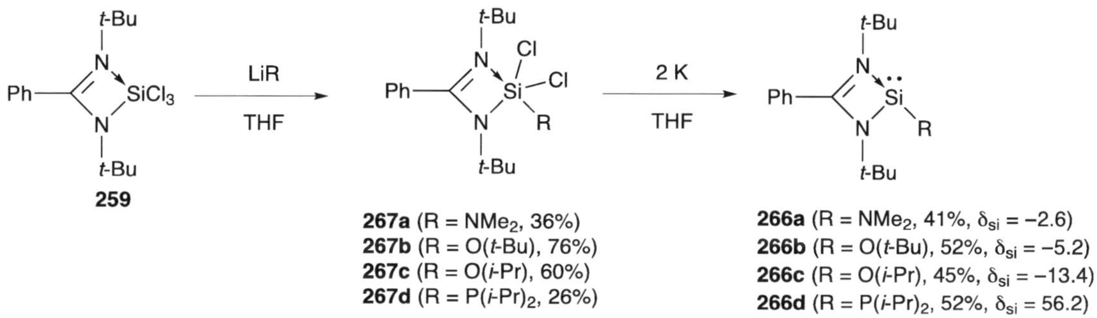

Scheme 8.102
Synthesis of amidinatosilylenes via reductive dechlorination of substituted dichlorosilanes.

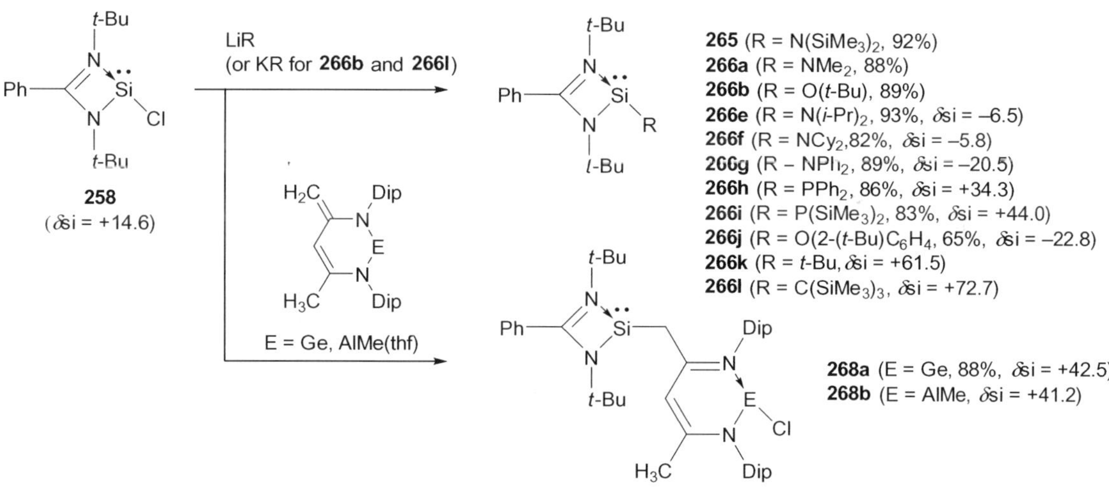

Scheme 8.103
Synthesis of amidinatosilylenes via metathesis of **258**.

Interestingly, reaction of **258** with ylide-like germylene gave *bifunctional* germylenesilylene **268a** ($\delta_{Si} = 42.5$) in 88% yield resulting from nucleophilic addition of **258** to electrophilic CH_2 moiety in the ylide-like germylene followed by migration of Cl (Scheme 8.103).[250] The corresponding alane-silylene **268b** ($\delta_{Si} = 41.2$) was also obtained by the similar reaction.[250]

8.7.2 Reactivity

8.7.2.1 Insertion reactions

Silylene **258** regioselectively inserts into the C–F bonds (Scheme 8.104). Reaction of **258** with hexafluorobenzene, octafluorotoluene, pentafluoropyridine, and pentafluorobenzene gave the corresponding C–F insertion products **269a–d**.[251] Formation of **269d** is in contrast to the reaction of diaminosilylene **73** with pentafluorobenzene and 1,3,5-trifluorobenzene giving C–H insertion products instead of C–F insertion products (Scheme 8.39). Selective formation of **269d** is explained by effective stabilization of the pentacoordinate structure of **269d** by electronegative F group rather than H atom. Similar C–F insertion of silylene **266b** with pentafluoropyridine was reported.[252] In the reaction of silylene **258** with $(F_3C)_2C=NPh$, N–F insertion occurred to give **270** rather than [1 + 2] cycloaddition.[253] Insertion reactions of **258** into S–S bond in diphenyldisulfide giving **271**[254] and N–H bond in diphenylhydrazone giving **272**[255] were also observed. Similarly to other silylenes, chlorosilylene **258** inserted into C–Cl bond of dichloromethane to give 1:2 adduct **273**. Even when an excess amount of dichloromethane was used, the 1:2 adduct **273** was obtained, which suggested that the reaction proceeded via chlorosilylene-dichloromethane complex as an intermediate.[256]

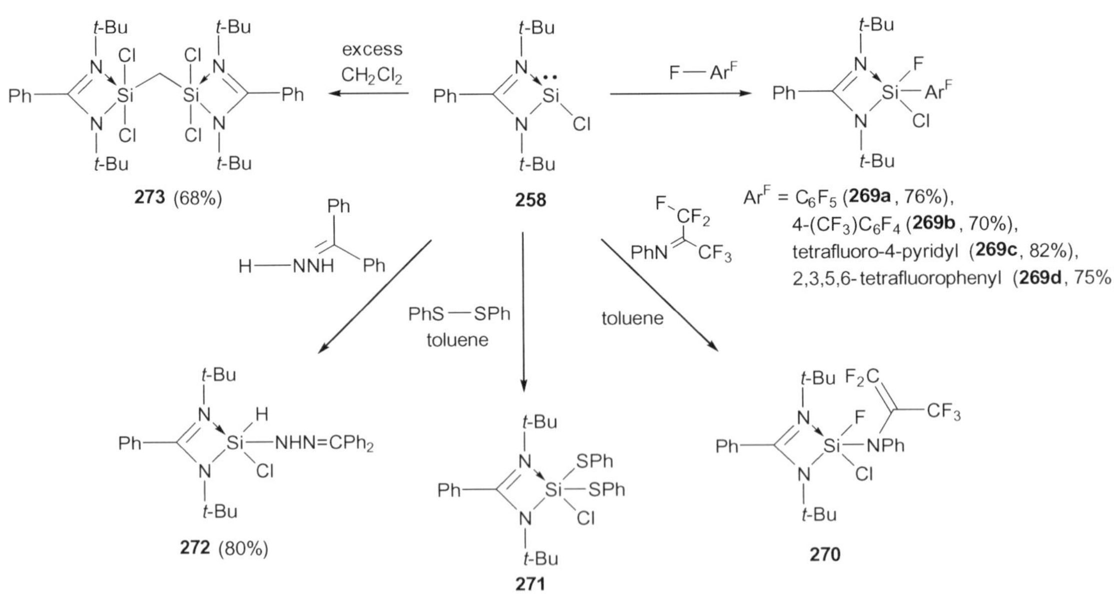

Scheme 8.104
Insertion reactions of chlorosilylene **258**.

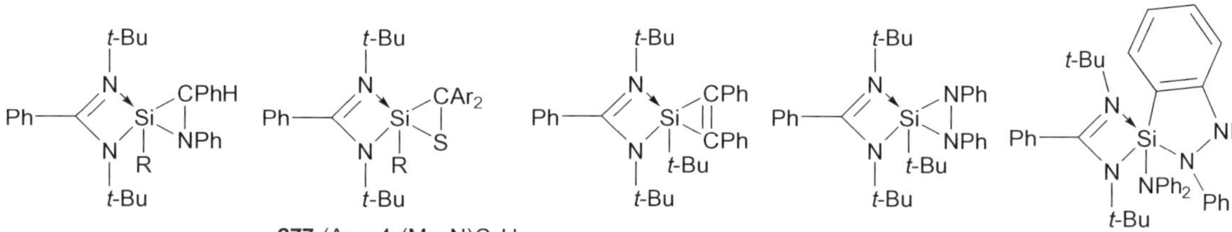

276a (R = t-Bu)
276b (R = NPh_2)

277 (Ar = 4-(Me_2N)C_6H_4, R = t-Bu)

283

285

286

Figure 8.12
Cycloadducts of **266g** and **266k**.

8.7.2.2 Cycloaddition reactions

Similar to other silylenes, silylene **258** undergoes [1 + 2] cycloaddition with benzophenone to give five-coordinate silaoxirane **274** (Scheme 8.105). Compound **274** showed a distorted square pyramidal geometry with the chlorine atom at the apical site.[257] Similar [1 + 2] cycloadditions were observed in the reactions of chlorosilylene **258** with 2-adamantanone,[107] *N*-benzylideneaniline giving **275**,[255] and the reactions of *t*-butylsilylene **266k** and *N*,*N*-diphenylaminosilylene **266g** with *N*-benzylideneaniline forming **276a**[258] and **276b**,[259] respectively (Fig. 8.12), the reaction of *t*-butylsilylene **266k** with 4,4′-bis(dimethylamino) thiobenzophenone giving **277**,[260] and the reaction of NPh_2-substituted silylene **266g** with diphenylacetylene.[261] In the case of benzyl, [1 + 4] cycloadduct **278** was obtained (Scheme 8.105).[262] Similar [1 + 4] cycloaddition was found in the reaction of chlorosilylene **258**, alkylsilylene **266l**, or NPh_2-substituted silylene **266g** with 3,5-di(*tert*-butyl)-*o*-benzoquinone, or Ar–N = CH–CH = N–Ar (Ar = 2,6-diisopropylphenyl)[254] giving [1 + 4] cycloadduct (for instance, **279** for the reaction with chlorosilylene **258** in Scheme 8.105),[107,260,261] and with cyclooctatetraene giving amidinato ligand-opened product **280**.[254] No reaction of chlorosilylene **258** with 1,5-cyclooctadiene and cyclooctene proceeded.

In the reaction of **258** with diphenylcetylene, disilacyclobutene **281** was obtained instead of [1 + 2] cycloadduct, silacyclopropene **282** (Scheme 8.106). Proposed mechanism for formation of **281** involves initial [1 + 2] cycloaddition giving **282** followed by insertion of **258** into Si–C bond in **282**.[241] In the case of more bulky *t*-Bu-substituted silylene **266k**, [1 + 2] cycloadduct **283** (Fig. 8.12) was obtained probably due to suppression of further addition of silylene **266k** by bulky *t*-butyl group.[258] Interestingly, the reaction mode of amidinatosilylenes with diazobenzene depends on the substituents on the three-coordinate silylene center. Reaction of cholorosilylene **258** with diazobenzene gave a bicyclic product in which the N–N bond of the diazobenzene remained intact (**284**) and silylene center inserted into C(aryl)–H bond (Scheme 8.106).[263] In contrast, more bulky *t*-butyl-substituted silylene **266k** reacted with diazobenzene to afford [1 + 2] cycloadduct **285** [258] (Fig. 8.12), while Ph_2N-substituted silylene **266g** reacted with azobenzene to provide diazasilacyclopentane **286** (Fig. 8.12).[259] Mechanisms for formation of **284** and **286** were not disclosed.

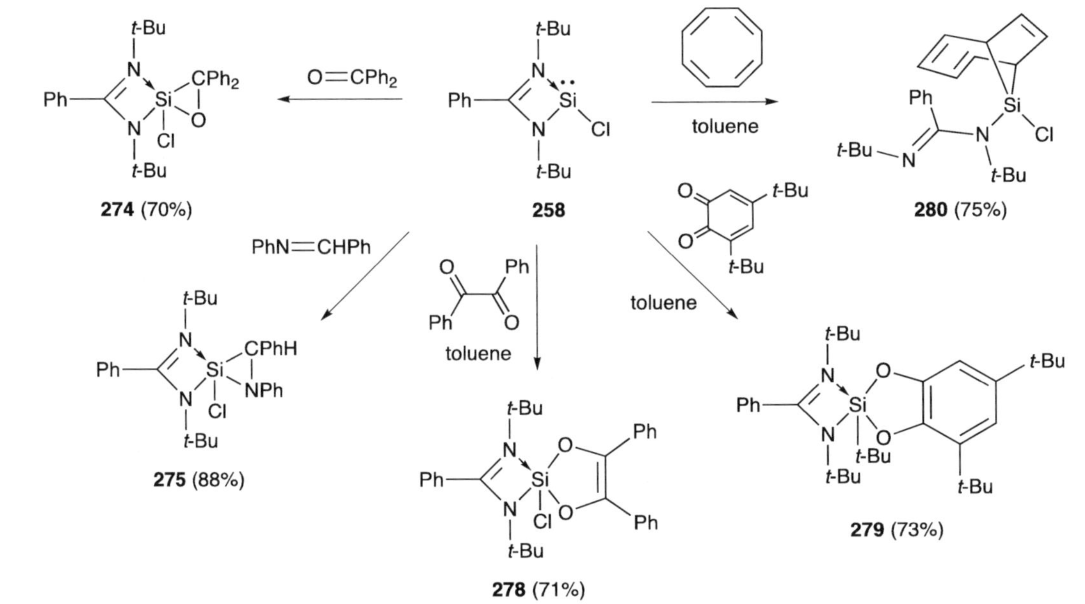

Scheme 8.105
Cycloaddition of chlorosilylene **258**.

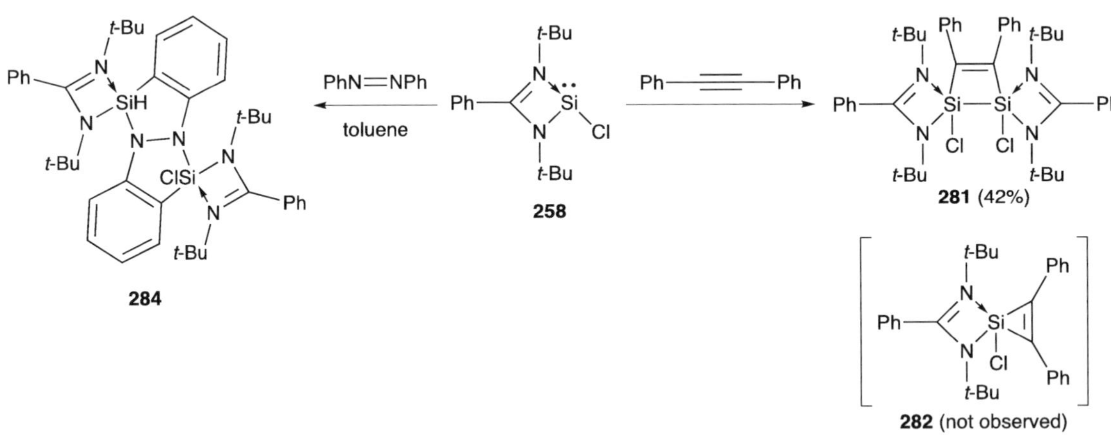

Scheme 8.106
Reactions of chlorosilylene **258**.

8.7.2.3 Oxygenation

Reaction of **258** with N_2O gave cyclotrisiloxane **287**,[264] while that with *t*-butylisocyanate or trimethylamine oxide gave cyclodisioxiane **288** (Scheme 8.107).[254] Similar cyclodisiloxiane **289a**,[249] **289b–e**[265] were obtained by the reaction of the corresponding silylenes with N_2O.

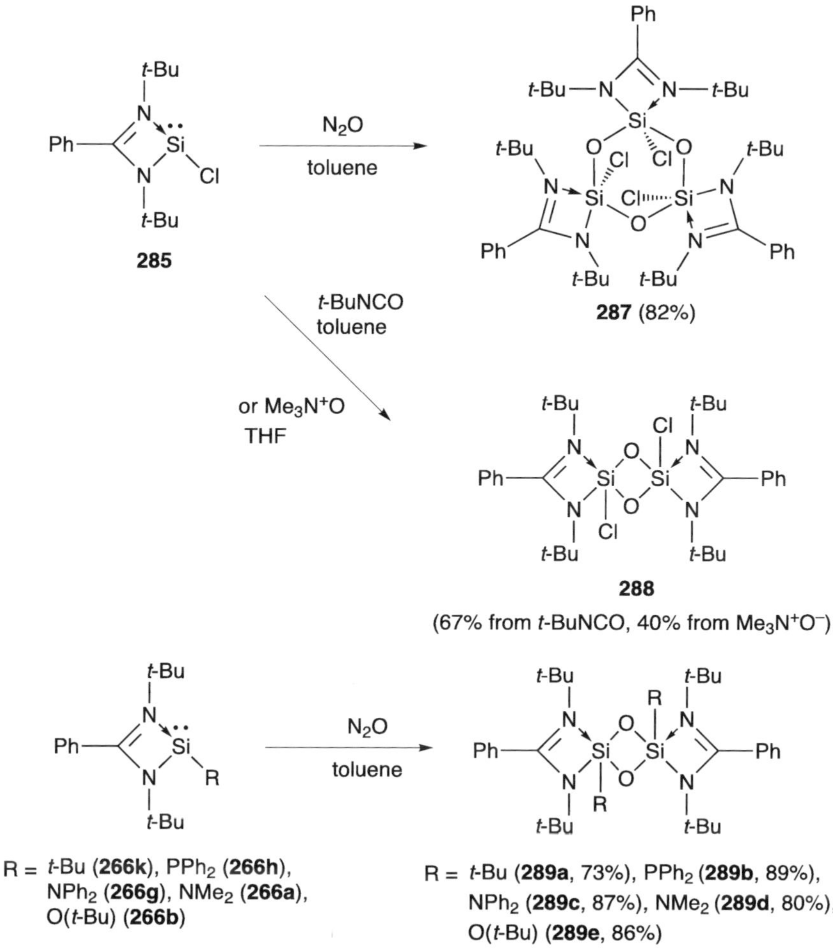

Scheme 8.107
Reactions of chlorosilylene **258** with oxygen sources.

8.7.2.4 Reaction with Lewis acids and muonium

Chlorosilylene **258** reacted with boranes to give the corresponding adducts **290a,b**[264] and **290c**.[266] (Scheme 8.108). Reaction of aminosilylene **265** with dioxane complex of dichlorogermylene in toluene gave silylene-coordinated dichlorogermylene **291**, which undergoes reductive dechlorination with KC_8 giving silylene-coordinate $Ge(0)_2$ species **292** (Scheme 8.109).[267] The detailed theoretical analysis for **292** suggested the π-back bonding from germanium to silylene ligand based on the shape of HOMOs and results of Quantum Theory of the Atom in Molecule (QTAIM) analysis, which are different in the corresponding $NHC \rightarrow Ge(0)_2 \leftarrow NHC$ species. Interestingly, muonioum adds to the carbon in the amidinato ligand to give the corresponding silyl radical **293**.[268]

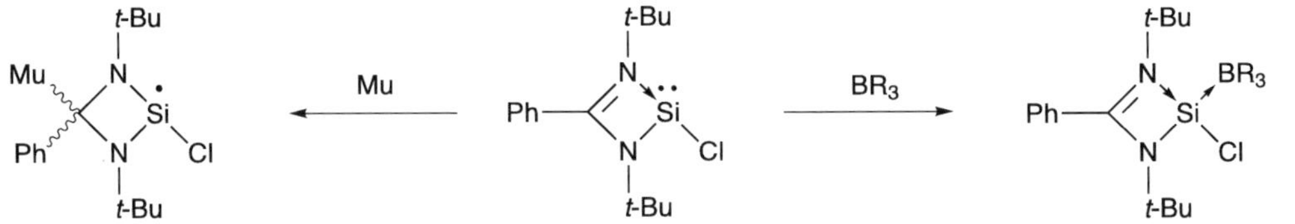

293 258 290a ($BR_3 = B(C_6F_5)_3$) 290b (BR_3 = 9-BBN) 290c ($BR_3 = BH_3$, 80%)

Scheme 8.108
Reactions of chlorosilylene **258** with Lewis acids and muonium.

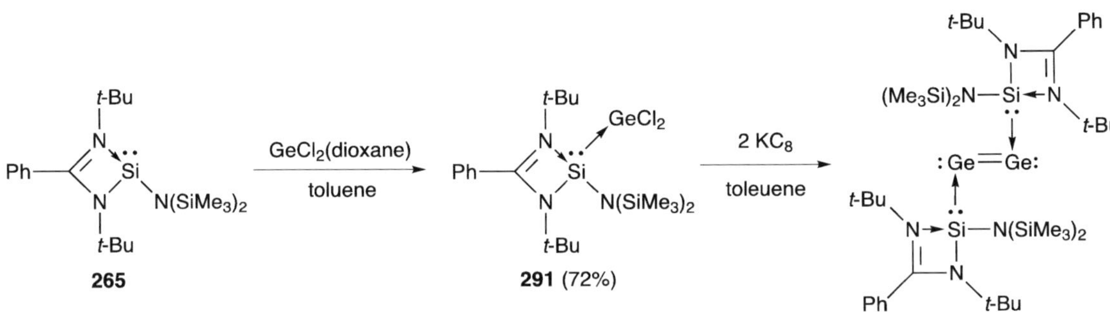

265 291 (72%) 292 (16%)

Scheme 8.109
Reactions of aminosilylene **265** with Lewis acid.

8.7.2.5 Substitution

As shown in Scheme 8.103, substitution reactions on the silicon atom of chlorosilylene **258** occurred to give the corresponding substituted silylenes upon treatment of **258** with nucleophile RLi (or RK). With organozinc compounds, such as $Zn(C_5Me_5)_2$, $ZnPh_2$, and $ZnEt_2$, the corresponding C_5Me_5, Ph, and Et-substituted silylenes were obtained as zinc complexes (**293**, **294**, and **295**, as shown in Scheme 8.110)[269]. The Zn–Si distances of **293**, **294**, and **295** are 2.3750(9), 2.4171(7), and 2.416(2) Å, respectively, which are either comparable to or longer than that of $(t\text{-}Bu_3Si)_2Zn$ (2.384(1) Å).[270] The short Zn–Si distance (2.3750(9) Å) and rather long Zn–Cl bond (2.3148(9) of **293** suggest the contribution of the ion pair resonance structure **293′** with a silyl cation character and a strong Zn–Si bond.

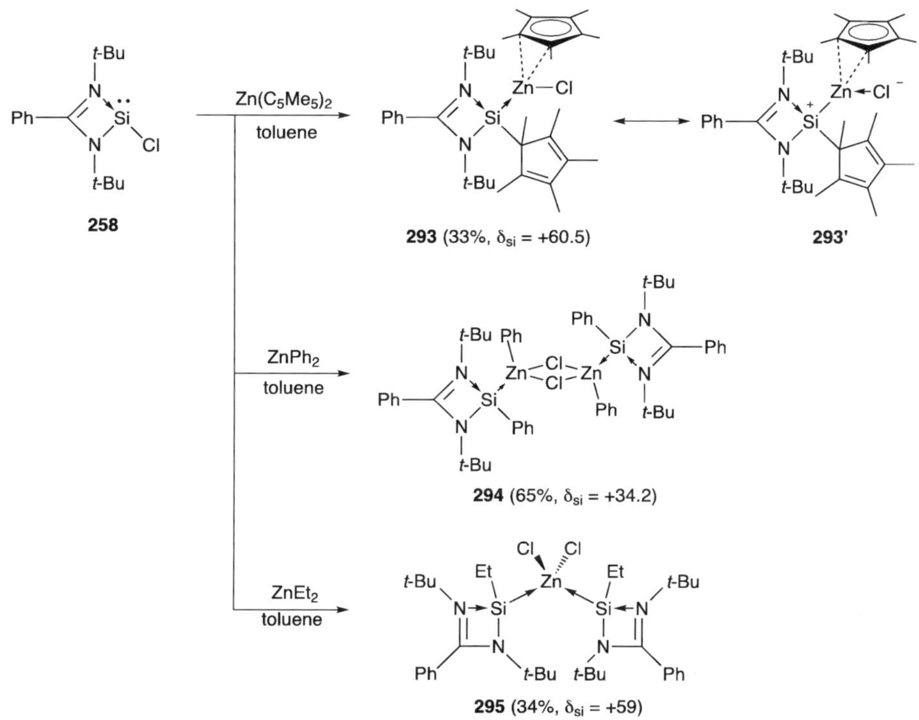

Scheme 8.110
Reactions of chlorosilylene **258** with organozinc compounds.

Borane adduct **290c** was reduced by $K^+[(sec\text{-Bu})_3BH]^-$ to give monohydridosilylene **296** (Scheme 8.111).[266] The high resolution single crystal X-ray analysis of **296** and the topological analysis of the resulting electron density based on Bader's QTAIM indicates that the central silicon atom has covalent interactions toward boron and hydrogen atoms and donor-acceptor interactions toward nitrogen atoms, as shown in a canonical structure **296′**. The Si–Cl moiety of **290c** was selectively fluorinated by β-diketiminatolead (II) monofluoride (**297**) to give monofluorosilylene **298** in 60% yield. [271]

8.7.2.6 As precursors of multiply-bonded silicon compounds

Reactions of silylenes with azide gave various products including Si = N double bonds. *N*, *N*-bis(trimethylsilyl)aminosilylene **265** reacted with trimethylsilyl azide to give *N*,*N*-bis (trimethylsilyl)aminosilanimine **299a** (Scheme 8.112). Various silanimines **299b–e** were obtained from **265**, **266g**, and **266a** with the corresponding azides.[272] Similarly, *N*-arylsilanimines were obtained from chlorosilylene **258**.[124] When adamantyl azide was used for the reaction of *N*,*N*-bis(trimethylsilyl)silylene **265**, silanimine **300** which would result from 1,3-migration of one trimethylsilyl group was obtained.[124]

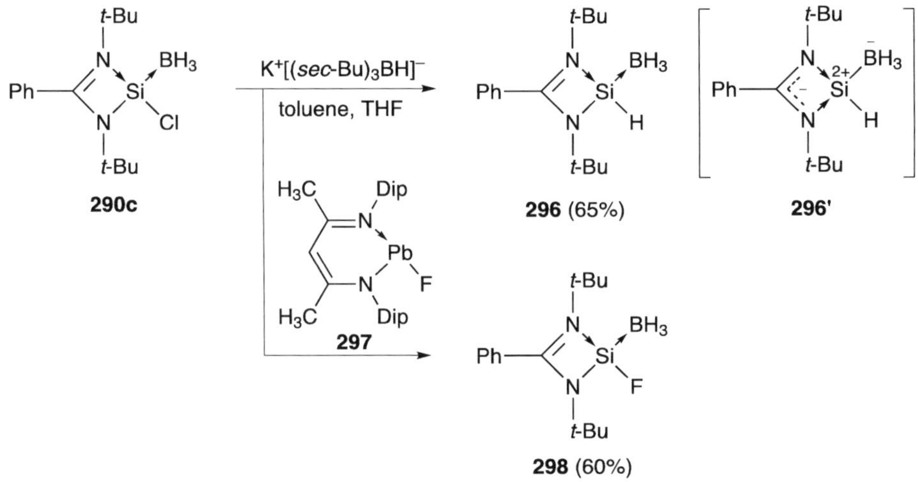

Scheme 8.111
Substitution reactions of **290c**.

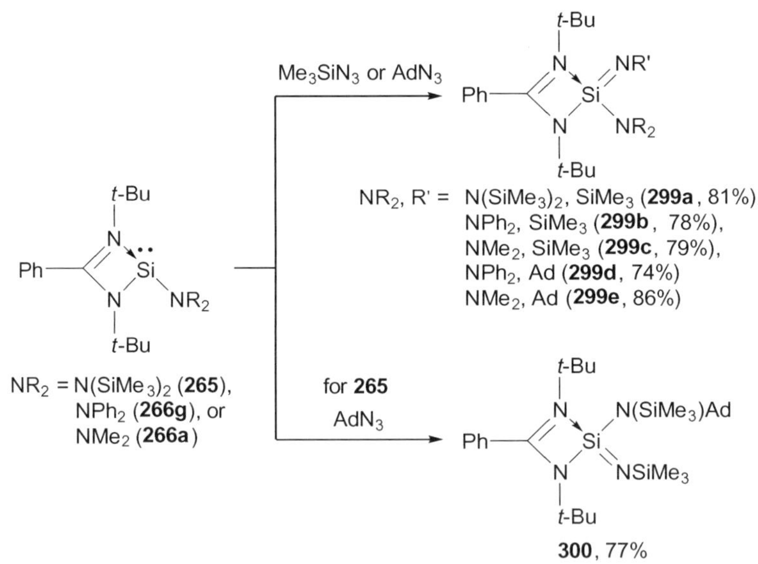

Scheme 8.112
Reactions of silylenes with azides.

Reaction of chlorosilylene **258** with *N,N*-diarylcarbodiimide gave silanimine **301** (Scheme 8.113).[254] In the reaction of chlorosilylene **258** with trimethylsilyldiazomethane, formal dimer of silanimine (**302**) resulting from the addition of the diazomethane to **258** was obtained.[256]

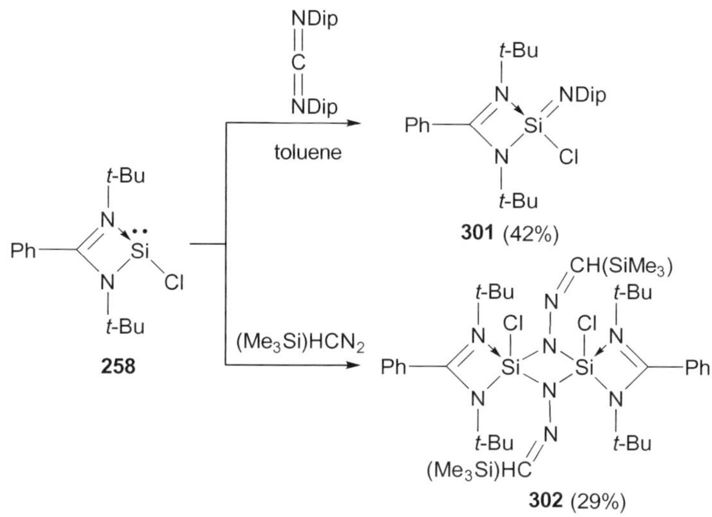

Scheme 8.113
Reactions of silylene **258** with a carbodiimide and a diazomethane.

In the reaction of NPh_2-substuted silylene **266g** with PhNCS, sulfur transfer reaction occurred to give the corresponding silanethione **303** in 82% yield (Scheme 8.114).[261] Similar chalcogen transfer reactions (S, Se, Te) giving the corresponding silanethione, silaneselone, and silanetellone were observed for $(Me_3Si)_2N$-substituted silylene **265**.[273]

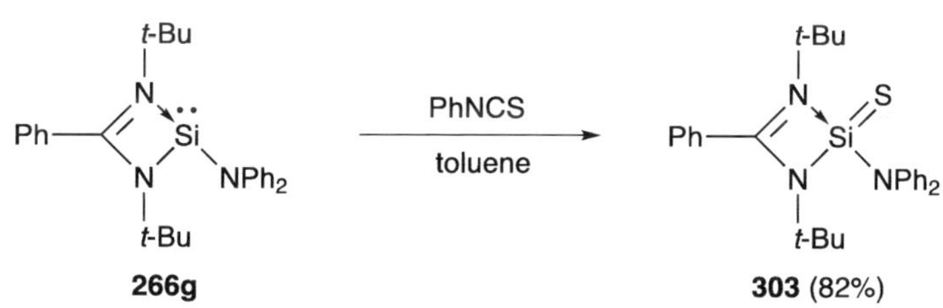

Scheme 8.114
Reaction of silylene **266g** with PhNCS.

Cholosilylene **258** reacted with $H_2O \cdot B(C_6F_5)_3$ in the presence of ImDip (1,3-Bis(2,6-diisopropylphenyl)-1,3-dihydro-2*H*-imidazol-2-ylidene) to give a Lewis acid- and Lewis base-stabilized Si = O species **304** (Scheme 8.115).[274] Although detailed mechanism for formation of **304** remains open, Lewis acid coordinated silaformaldehyde **305a** and silacarboxylic acid **305b** were suggested as possible intermediates (shown in parentheses in Scheme 8.115) and ImDip working as a scavenger of HCl. Similar to other stable silylenes, treatment of **258** with S_8 in toluene gave the corresponding three-coordinate silanethione **306**,[275] while reduction of **258** with four equivalents of KC_8 in the presence of S_8 gave a dimer of potassium siladithiocarboxylate **307**.[276]

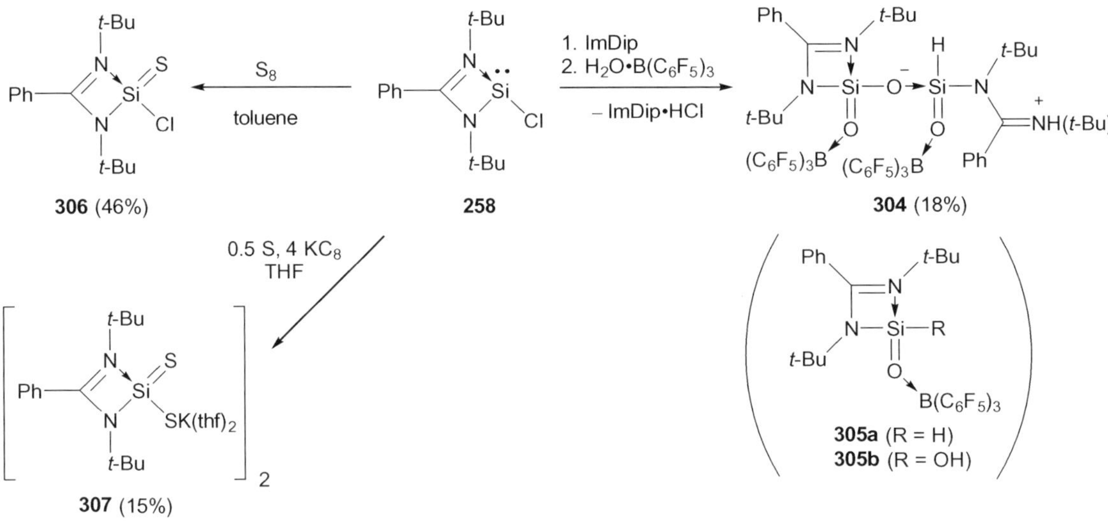

Scheme 8.115
Reaction of silylene **258** with group 16 element compounds.

Bis(trimethylsilyl)phosphinosilylene **266i** undergoes facile 1,2-silyl migration at 100°C to give the corresponding Si = P doubly-bonded compound, phosphasilene **308** (Scheme 8.116).[248] In the solid state, **308** adopts an *E*-geometry and the Si = P distance of **308** (2.095(3) Å) is much shorter than that of **266i** (2.2838(12) Å). The ^{29}Si resonance due to doubly-bonded silicon nucleus appeared at +40.5 ppm (^1J(Si,P) = 191.4 Hz). Interestingly, treatment of **266i** with Ph_3PCl_2, a typical desilylation reagent, gave a head-to-tail dimer of phosphasilyne (Si≡P), **309**,[248] which is also obtained by the reaction of silylene **258** with P_4.[277] The Si_2P_2 core of **309** is planar and diamond-shaped with the Si-P distances of 2.1701(12) and 2.1717(11) Å, which are intermediate between Si = P double bond in **308** (2.095(3) Å) and Si-P single bond in **308** (2.2838(12) Å). Theoretical analysis suggested that ylide-like **309′** is a main contributor to the electronic structure of **309**. The negative NICS calculated for **309** (NICS(1) = −2.57, NICS(0) = −6.01) indicate that **309** has a somewhat aromatic character in contrast to antiaromatic cyclobutadiene. Interestingly, bis(trimethylsilyl)phosphinosilylene **266i** reacted with pivaloyl chloride to give pivaloylphosphasilene **310** (Scheme 8.116).[278] DFT calculation predicted that **310** is formed by the initial insertion of the silicon center into the C−Cl bond of pivaloyl chloride followed by 1,2-elimination of Me_3SiCl at the rate controlling step.

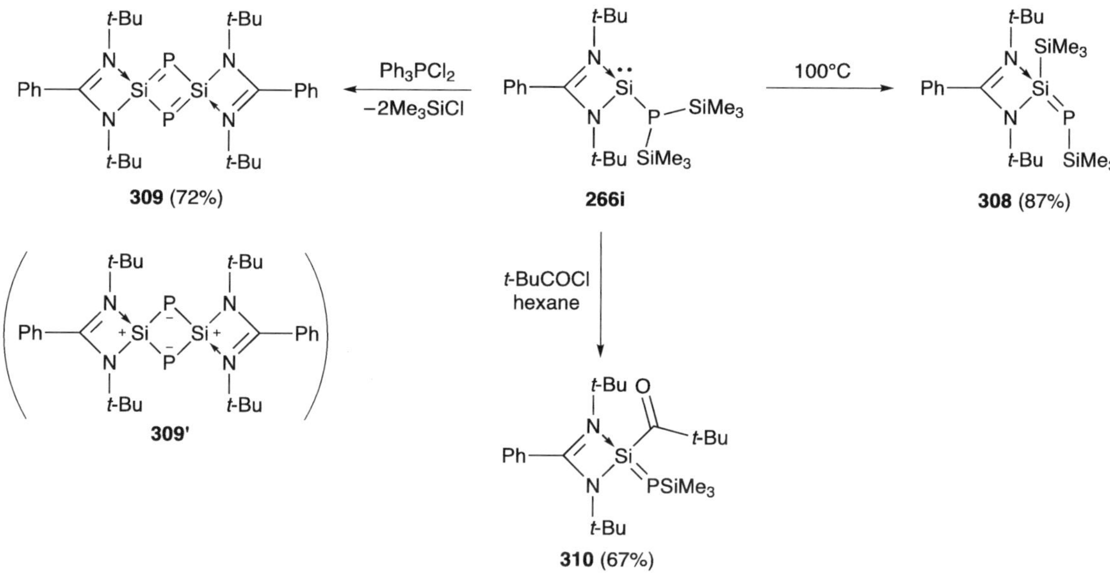

Scheme 8.116
Reaction of phosphinosilylene **266i** giving Si = P species.

Reaction of chlorosilylene **258** with AdC≡P gave a CSi_2P cycle **311** with zwitterionic rather than cyclobutadiene character which was proposed on the basis of structural characteristics (C-Si and P-Si distances that are in-between the corresponding single and double bonds) and theoretical calculations (Scheme 8.117).[277] Similar zwitterionioc Si_2P_2 cycle **309** was obtained by treatment of **258** with P_4 (Scheme 8.117).[277,279] Compounds **311** and **309** were also obtained by the reaction of bissilylene **262** with AdC≡P and P_4, respectively.[277] Interestingly, reaction of **258** with a smaller phophaalkyne MeC≡P gave a cationic CSi_3P cycle **312**.[280] The chloride ion in **312** is far from cationic Si (6.32 Å) indicating no interaction between the Cl and Si atoms. XRD analysis exhibited the planar CSi_3P cycle. Two Si–C bonds and two Si–P bonds are between the typical single and double bonds, while the Si–Si bond is in the range of the typical Si–Si single bond distances. The proposed mechanism for formation of **312** involves an initial formation of a CSi_2P cycle like **311** followed by further addition of silylene **258**.

Interestingly, reaction of **258** with Cp^*ZrCl_3 gave base-stabilized tetrasilacyclobutadiene dication **313** as yellow crystals in 29% yield (Scheme 8.117).[281] The Si_4 core of **313** is planar rhombic with the endocyclic Si–Si–Si angles of 98.85(7) and 81.15(7) degrees, and the endocyclic Si–Si distances of 2.321(2) and 2.331(2) Å. The cationic Si centers and silylene centers are coordinated by amidinato ligands and chlorosilylene ligand **258**,

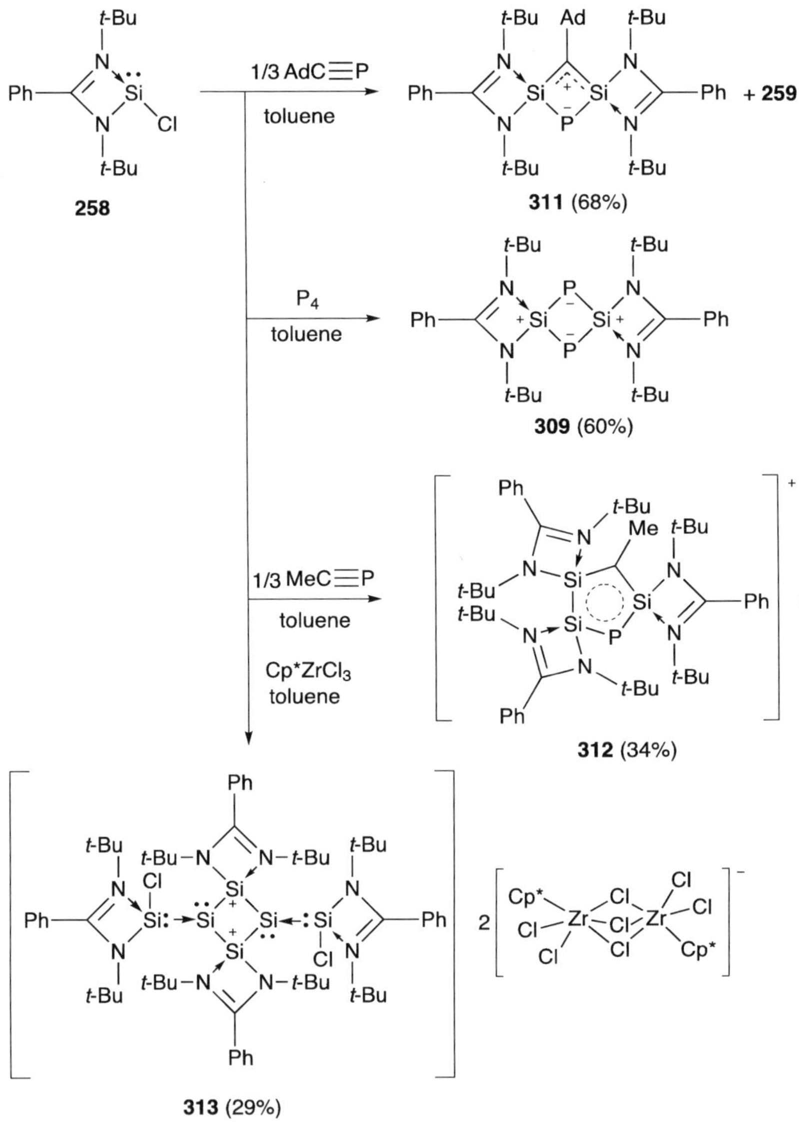

Scheme 8.117

Reaction of cholorosilylene **258** giving double-bonded silicon compounds.

respectively. Three ^{29}Si NMR resonances appeared at −128.7 (ring $Si(Si)_3$), +7.7 (silylene $Si(Cl)$), +53.4 (ring $Si(Si_2N_2)$) ppm. Theoretical analysis indicates that compound **313** is best described by the charge localized resonance structure featuring two cationic centers and two lone pair centers at the Si_4 core with some contribution of the π-electron delocalized resonance form. The NICS of **313** (NICS(1) = −3.8, NICS(0) = −4.7 ppm) indicate the compound **313** has an aromatic ring current.

Bis(trimethylsilyl)amino-substituted silylene **265** reacted with adamantylphosphaalkyne (AdC≡P) to give cyclic silene **314** resulting from migration of methyl group from silicon to phosphorus atom (Scheme 8.118)[246] in contrast to the reaction of chlorosilylene **258** with AdC≡P (Scheme 8.117), while **265** does not react with diphenylacetylene.[246] The proposed mechanism for formation of **314** involves an attack of Si(II) center of **265** to the electron deficient carbon atom of AdC≡P and an attack of the terminal phosphorus of AdC≡P to silicon atom of $SiMe_3$ group on nitrogen atom accompanied by methyl migration.[246]

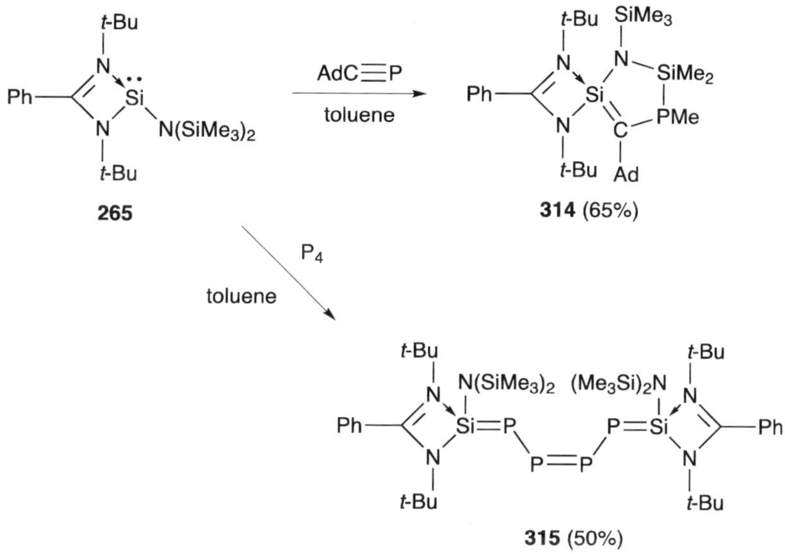

Scheme 8.118
Reaction of aminosilylene **265** giving double-bonded silicon compounds.

Aminosilylene **265** also reacted with 0.5 equiv of P_4 to give adduct **315** having a (*Z*)-diphosphene and two phosphoasilene moieties (Scheme 8.118)[279,282]. The structural parameters and ^{31}P NMR resonances are consistent with the presence of Si = P–P = P–P = Si moiety in **315**.

8.7.2.7 Miscellaneous reactions

Amidinato(chlorosilyl)silylene **263a** show unusual reactivity.[275] In the reactions with azobenzene and diphenylacetylene, diazadisilacyclobutane **316** and disilacyclobutene **317** were obtained (Scheme 8.119). Formation of these products was explained by the initial formation of [1 + 2] cycloadduct followed by 1,2-chlorine shift accompanied by ring expansion. Reaction of **263a** with $DipN_3$ gave **318**. Initial formation of silatetrazoline intermediate, as often observed for other silylenes, followed by 1,2-silyl shift with elimination of N_2, or formation of silanimine followed by insertion of nitrene from $DipN_3$ were proposed for formation mechanism of **318**. While reaction of **263a** with S_8 gave a 2:1

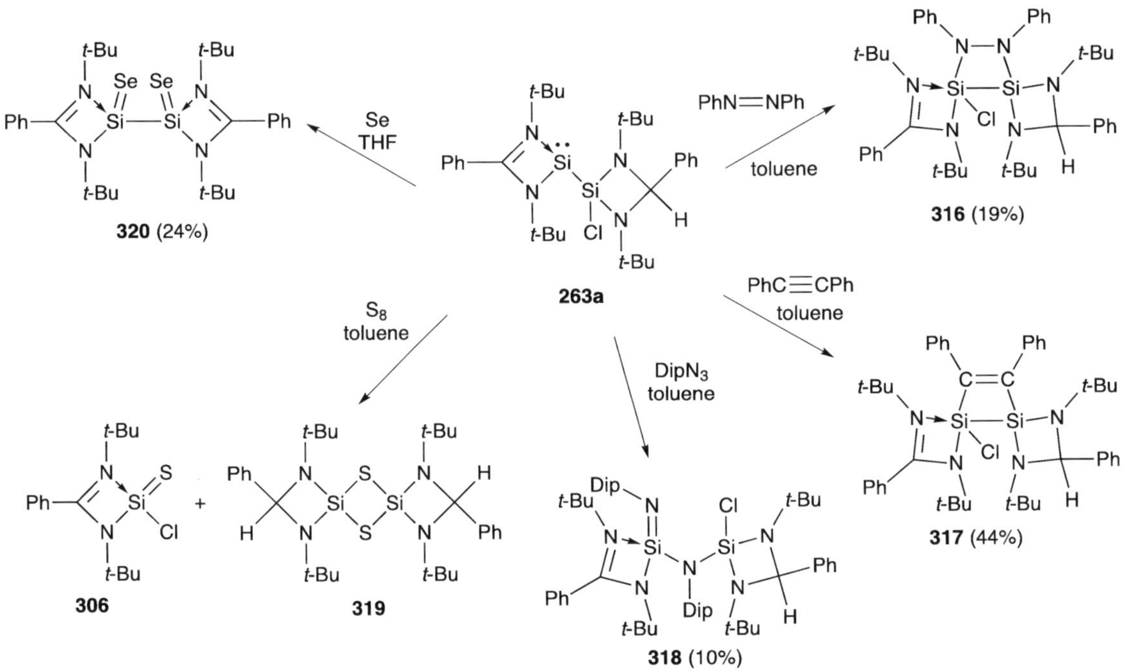

Scheme 8.119
Reactions of (chlorosilyl)silylene **263a**.

mixture of chlorosilanethione **306** and dithiodisiletane **319**, reaction with Se gave bis (silaneselone) **320**. In these reactions, cleavage of Si(silylene)-Si(Cl) bond occurred.

8.7.3 Transition Metal Complexes and Related Compounds From Monosilylenes

Three-coordinate silylenes can serve as ligands for transition metals similar to diaminosilylenes (NHSi) (Section 8.2.3).[67]

8.7.3.1 Group 2 metals

In contrast to transition metal-silylene complexes, complex of silylenes with s-block elements are still quite rare. The reaction of phenoxysilylene **266j** ($\delta_{Si} = -22.8$) with $Cp^{*}{}_2Ca$ gave calcium complex **321** ($\delta_{Si} = -13.7$) as colorless crystals in 99% yield (Scheme 8.120).[64] The Si–Ca distance in **321** (3.2732(5) Å) was much longer than the sum of the single bond covalent radii for Ca (1.71 Å) and Si (1.16 Å)[283] and Wiberg bond index (WBI) for Si–Ca of **321** is 0.47. Theoretical calculation showed a donor-acceptor interaction between Si and Ca atoms. Similar calcium complex was also obtained by the reaction of NHSi with $Cp^{*}{}_2Ca$ (Section 8.2.3.2).

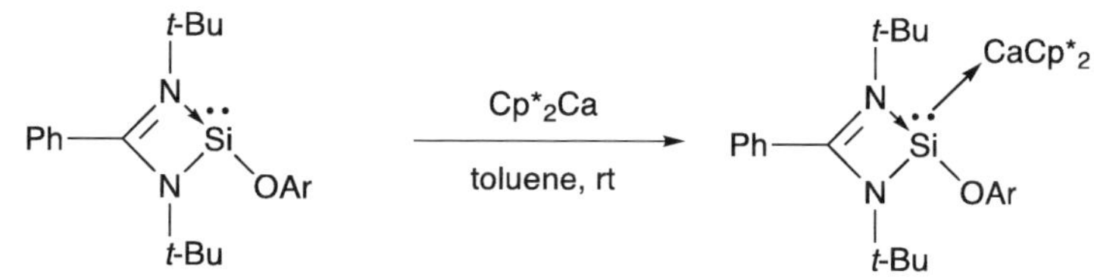

266j (Ar = 2-(*tert*-Bu)C_6H_4, δ_{Si} = −22.8)

321 (99%, δ_{Si} = −13.7)

Scheme 8.120
Synthesis of silylene-group 2 metal complex.

8.7.3.2 Group 4 metals

Bis(chlorosilylene) titanocene complex **322a** was prepared as thermally robust but extremely air-sensitive red-brown crystals by the ligand exchange reaction of $Cp_2Ti(PMe_3)_2$ with chlorosilylene **258** (Scheme 8.121).[284] The Cl moieties on the silylene ligands in **322a** are easily converted to Me and H groups by treatment of **322a** with MeLi and $LiBHEt_3$, respectively. The ^{29}Si resonance due to Si(II) center are downfield-shifted in the order **322a** (δ_{Si} = 120.6), **322c** (153.2), **322b** (189.2), which are consistent with the order of Hammett constants σ_p (0.23 (Cl), 0 (H), −0.17 (Me)). XRD analysis discloses that complexes **322a** and **322b** have considerably short Si–Ti distances (2.486(1) Å (**322a**) and 2.515(1) Å (**322b**)) among the silyl titanium complexes suggesting multiple bond character between Ti and Si. Theoretical calculations of model compounds for **322a–c** predicted that the Si–Ti bonds were considerably polarized (NPA charge: −1.2 ~ −1.3 (Ti), 1.2 ~ 1.6 (Si)), and two σ-type orbitals and one π-type orbital appeared to be delocalized over the Si–Ti–Si moiety, supporting the multiple bond character.

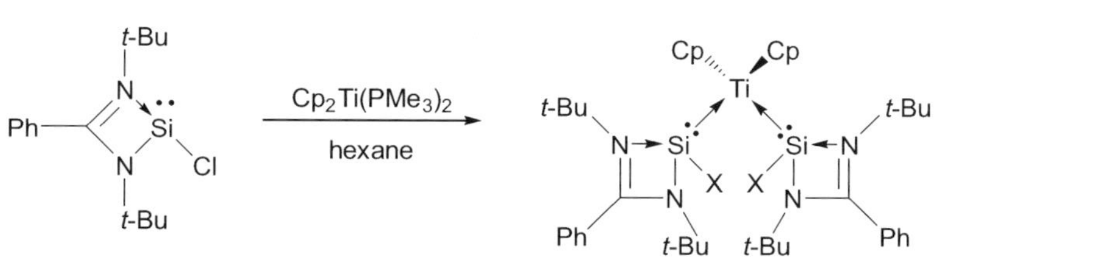

258

322a (X = Cl, 67%, δ_{Si} = +120.6)
322b (X = Me, 57%, δ_{Si} = +189.2)
322c (X = H, 40%, δ_{Si} = +153.2)

Scheme 8.121
Synthesis of silylene-group 4 metal complexes.

8.7.3.3 Group 5 metals

Mononuclear vanadium complex **333** (with silylene ligand **258**) was synthesized by ligand exchange reaction of the corresponding carbonyl complex (Scheme 8.122).[294] The ^{29}Si NMR and ^{51}V NMR resonances of **333** appear at +116.1 ppm and −1520 ppm, respectively. The Si−V distance (2.3866(5) Å) was shorter than those of silyl-vanadium complexes (2.56 Å) indicating significant π-back bonding contribution.

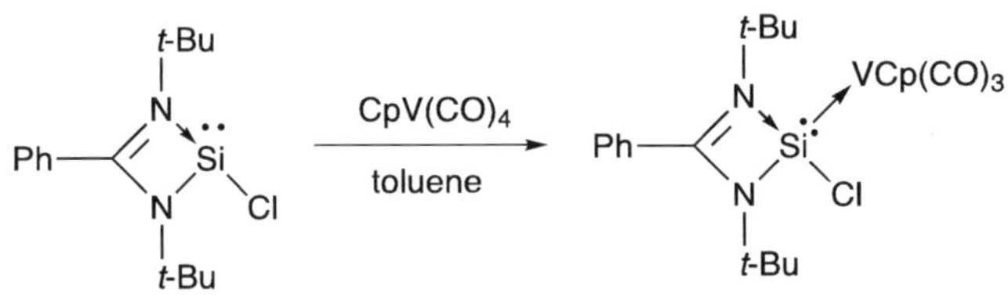

258 **333** (71%, δ_{Si} = +116.1)

Scheme 8.122
Synthesis of silylene-group 5 metal carbonyl complex **333**.

8.7.3.4 Group 6 metals

Group 6 metal carbonyl complexes **323a−c**[285] were synthesized by the reaction of chlorosilylene **258** with the corresponding metal carbonyl complexes (Scheme 8.123). Similar to other complexes, downfield-shifted ^{29}Si NMR resonances were observed for **323a−c** (+92.3 (**323a**), +72.6 (**323b**), +53.0 (**323c**) ppm).[285] The corresponding fluorosilylenes **324a−c** which were obtained by treatment of chlorosilylenes **323a−c** with Me_3SnF[285] exhibited similar downfield-shifted ^{29}Si resonances (+74.0 (**324a**), +56.7 (**324b**), +41.8 (**324c**) ppm) and Si−M distances; the Si−M distances are 2.3458(7) Å (Si−Cr), 2.4550(14) Å (Si−Mo), 2.5086(11) Å (Si−W) for chlorosilylenes **323a**, **323b**, and **323c**, respectively, while those are 2.3398(4) Å (Si−Cr) and 2.4990(8) Å (Si−W) for fluorosilylenes **324a** and **324c**. The ^{19}F NMR signals of fluorosilenes **324a−c** appeared at around −90 ppm.[285]

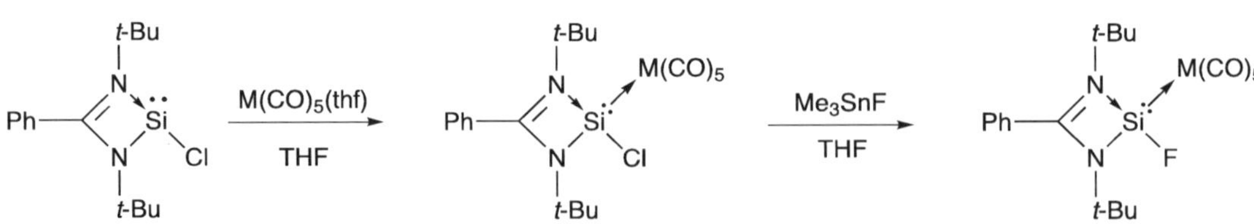

258

323a (M = Cr, 91%, δ_{Si} = +92.3)
323b (M = Mo, 84%, δ_{Si} = +72.6)
323c (M = W, 85%, δ_{Si} = +53.0)

324a (M = Cr, 84%, δ_{Si} = +74.0)
324b (M = Mo, 7%, δ_{Si} = +56.7)
324c (M = W, 88%, δ_{Si} = +41.8)

Scheme 8.123
Synthesis of silylene-group 6 metal carbonyl complexes.

Similar tungsten complex **325** was obtained by the reaction of bis(trimethylsilyl) phosphinosilylene **266i** with $W(CO)_5(thf)$ or thermal isomerization of phosphasilene-tungsten complex **326** (Scheme 8.124).[286]

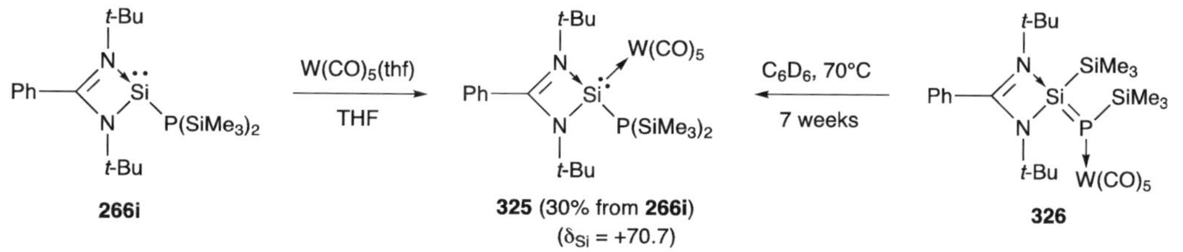

Scheme 8.124
Synthesis of silylene-group 6 metal carbonyl complex **325**.

8.7.3.5 Group 7 metals

When $Mn_2(CO)_{10}$ and $Re_2(CO)_{10}$ were used for the reaction with **258**, the cationic manganese complex with two silylene ligands (**327**)[287] and the rhenium complex with three silylene ligands (**328**)[287] were obtained, respectively (Scheme 8.125). In complex **327**, two silylene ligands were located in trans position with the Si–Mn–Si angle of 178.30(3) degrees. The Si–Mn distances are 2.2816(8) and 2.2789(8) Å. The ^{29}Si NMR spectrum of **327** shows a single signal at +92.5 ppm.[287] A similar cationic manganese complex with Ph_2N-substituted silylene **266g** was also obtained.[288] Complex **328** exists as a meridional isomer with two shorter Si–Re distances of 2.4384(18) and 2.4354(17) Å and one longer Si–Re distance of 2.4928(17) Å. Accordingly, the ^{29}Si NMR spectrum of **328** shows two signals at +40.3 and +47.4 ppm.[287] In the reaction of Ph_2N-substituted silylene **266g** with $Re_2(CO)_{10}$, a cationic bissilylene rhenium complex similar to **327** was obtained due to the bulky NPh_2 substituents.[288]

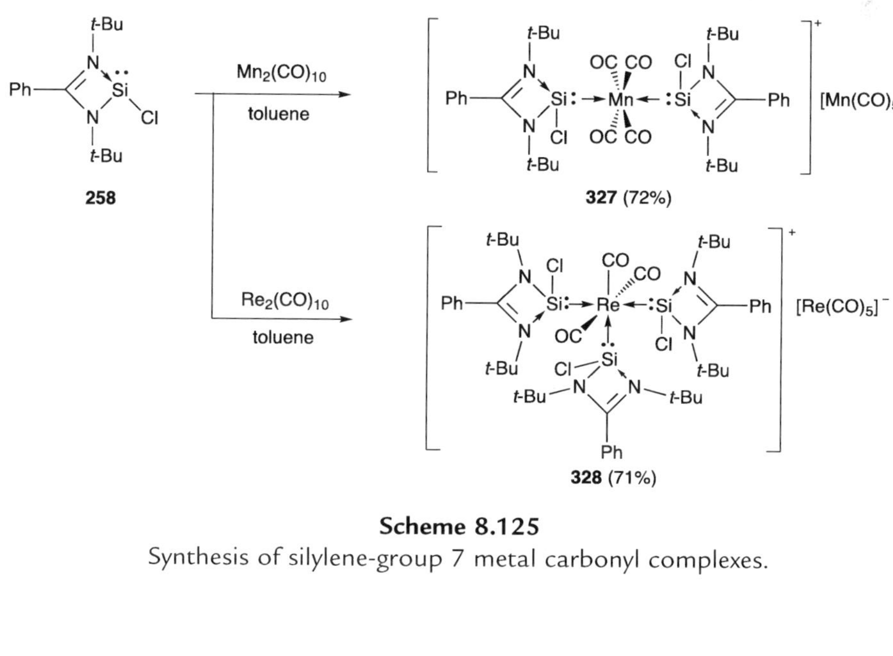

Scheme 8.125
Synthesis of silylene-group 7 metal carbonyl complexes.

8.7.3.6 Group 8 metals

Reaction of alkoxysilylene **266b** with $Fe_2(CO)_9$ gave iron complex **329a** in 72% yield (Scheme 8.126).[289] Similar iron-carbonyl complexes with hydridosilylene **264** ligand bearing other aryl and alkyl substituents on the nitrogen atoms of amidinato moiety **329b–d** were synthesized by the reaction of the corresponding dichlorohydridosilanes **330a–c** with $K_2[Fe(CO)_4]$,[290] although hydridosilylene **264** is not persistent silylene (Scheme 8.101).[245] XRD study of **329a**[289] and **329c**[290] exhibited that the geometry around the iron center is trigonal bipyramidal and the silylene ligand coordinating to the iron center occupies the apical position similar to $[(t\text{-BuO})_2\{(Me_2N)_3PO\}Si]Fe(CO)_4$ (**VIII**)[291] but in contrast to $[Si[N(t\text{-Bu})CH]_2Fe(CO)_4$ (2.196 Å)[3,5] in which the silylene ligand occupies the equatorial position. The Fe–Si distances of **329a** and **329c** (2.237(7) and 2.234(1) Å) are slightly longer than those of $[Si[N(t\text{-Bu})CH]_2Fe(CO)_4$ (**35**, 2.196 Å)[3,67] but shorter than that of $[(t\text{-BuO})_2\{(Me_2N)_3PO\}Si]Fe(CO)_4$ (**VIII**, 2.289(2) Å).[291] The CO stretching frequency observed in the IR spectra of **329a–c** are consistent with the geometry where the silylene ligand is at the apical position as was observed in the XRD analysis. Complexes **329b–d** show no catalytic activity for hydrosilylation of ketones.[290]

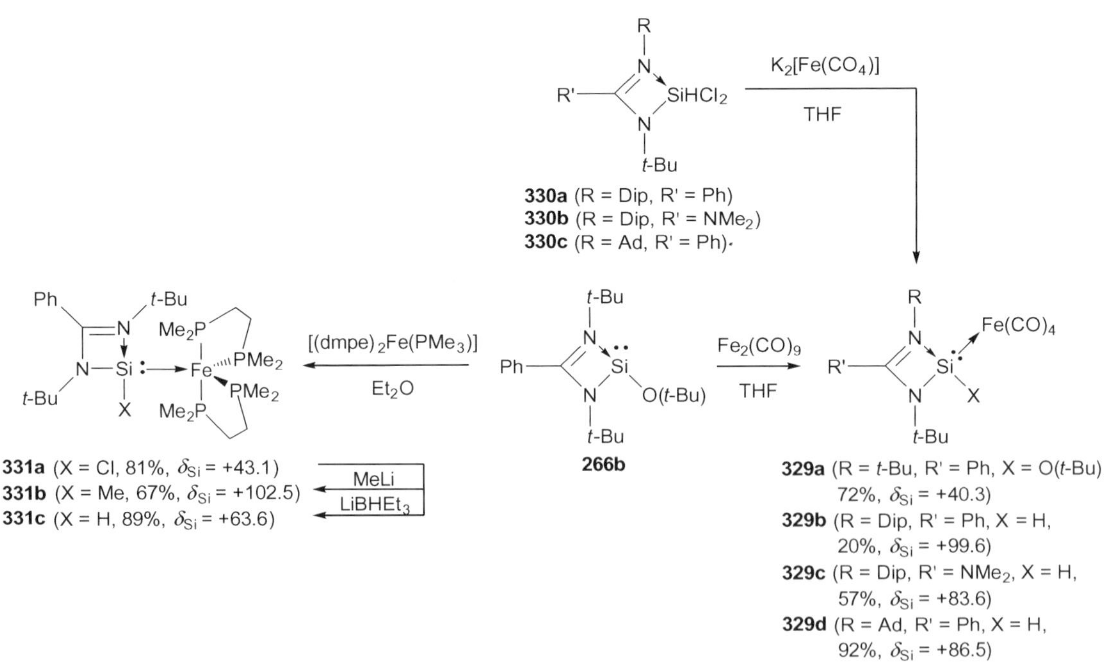

Scheme 8.126
Synthesis of silylene-iron complexes.

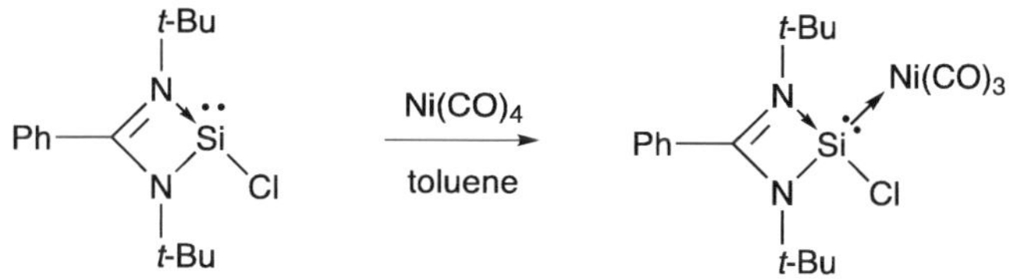

258

338 (55%)

Scheme 8.129
Synthesis of silylene-nickel complex.

8.7.3.8 Group 10 metals

Nickel carbonyl complex **338**[296] was synthesized by the reaction with chlorosilylene **258** with the corresponding metal carbonyl complex (Scheme 8.129). The downfield shifted ^{29}Si NMR resonance (δ_{Si} = +62.7 in **338**) (compared to free silylene **258** (δ_{Si} = +14.2 ppm)[240]), short Si–Ni distance (2.2111(8) Å) [compared to the sum of covalent radii of Si (1.11 Å) and Ni (1.24 Å)], and short Ni–C (carbonyl, averaged) distance (1.796(2) Å) (relative to that found in $Ni(CO)_4$ (1.817 Å)[297]) were observed.

Inoue et al. have found that bis(trimethylsilyl)phosphinosilylene **266i** reacted with $Ni(cod)_2$ (cod = cyclooctadiene) in toluene to give bis(silylene) Ni(cod) complex **339** as a major product.[298] Complex **339** was also formed by the reaction of phosphasilene **308**, which is a thermal isomerization product of **266i** (Scheme 8.130), [248] with $Ni(cod)_2$ in 73% yield. The Ni center in **339** is coordinated by two silylene ligands and one cod ligand with the Si–Ni–Si angle of 79.80(6) degrees. The Si–Ni distances (2.201(2) and 2.202(1) Å) are in-between those in the structurally similar bis(silylene) Ni(cod) complex **356** (2.1908(7) and 2.1969(7) Å, in Scheme 8.138)[299] and bis(silylene) complex $[(CHNt\text{-}Bu)_2Si]_2Ni(CO)_2$ **55a** (2.207(2) and 2.216(2) Å).[74]

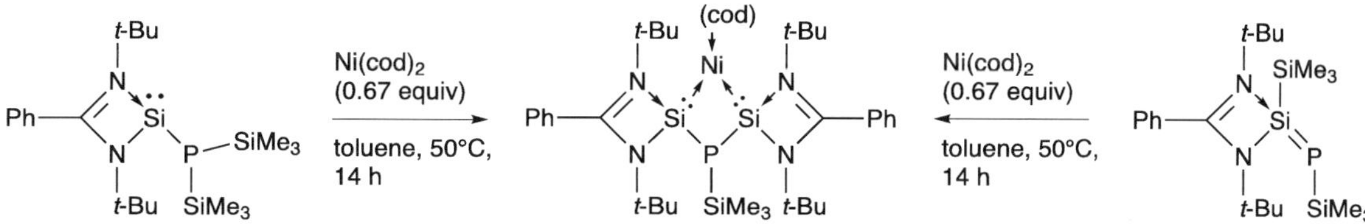

266i

339 (73% from 308)
(δ_{Si} = +51.5)

308

Scheme 8.130
Synthesis of bissilylene-nickel complex **339**.

8.7.3.9 Group 11 metals

Reaction of **258** with cationic copper complex [Cu(CH_3CN)(TMEDA)][OTf] gave mono (cholorosilylene) copper(I) complex **340a** (δ_{Si} = +32.9) in 80% yield (Scheme 8.131).[300] Similar ligand exchange reactions gave the corresponding Cu(I) complexes with aminosilylene and alkoxysilylene ligands **340b** (δ_{Si} = +18.3) and **340c** (δ_{Si} = +5.4) in 66% and 72% yields. The Si–Cu distances (2.1716(12) (**340a**), 2.1982 (12) (**340b**), 2.2003(6) Å (**340c**)) are within the short limits of Si–Cu bonds. XRD analysis revealed that complexes **340b** and **340c** adopt a trigonal planar geometry around the Cu(I) centers, while chlorosilylene complex **340a** has a pyramidalized geometry around the Cu(I) center with the sum of the bond angles of 348.72 degrees probably due to the coordination of OTf anion to the Cu(I) center with the distance of 2.312 Å. The increased Lewis acidity of the Cu(I) center in **340a** is rationalized by the effect of the electronegativity of Cl substituent on the Si(II) center.

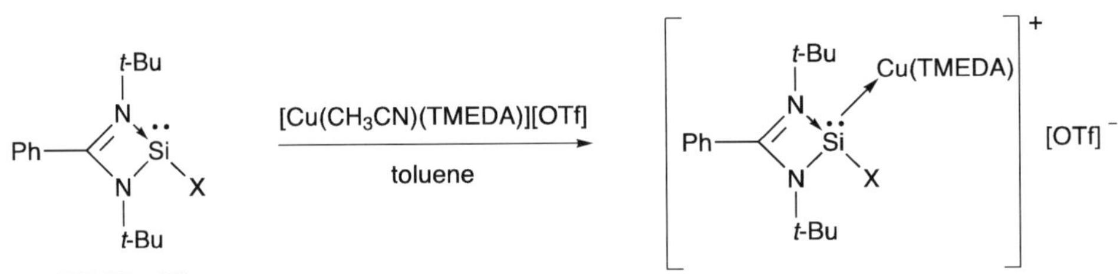

258 (X = Cl)
266a (X = NMe$_2$)
266b (X = O(*t*-Bu))

340a (X = Cl, 80%, δ_{Si} = +32.9)
340b (X = NMe$_2$, 66%, δ_{Si} = +18.3)
340c (X = O(*t*-Bu), 72%, δ_{Si} = +5.4)

Scheme 8.131
Synthesis of silylene–Cu complexes.

8.8 1,2-Bis(silylene) With Amidinato Ligands

When **259** was reduced by three equivalents of KC_8, intramolecularly base-stabilized bis (silylene) **262** bearing Si–Si bond was isolated as orange-red crystals in 5% yield (Scheme 8.132).[301] The ^{29}Si NMR resonance appeared at low-field (+75.7 ppm in tetrahydrofuran-d_8, +76.3 ppm in benzene-d_6). The central Si–Si bond distance in **262** (2.413(2) Å) is comparable to that of Robinson's NHC-stabilized $Si(0)_2$ (2.393 Å)[116] and within the range of typical Si–Si single bonds. Compound **262** adopts a gauche-bent conformation, as shown in Fig. 8.13. The calculated small chemical shielding anisotropy of **262** suggests no multiple bond character in the central Si–Si bond and QTAIM analysis shows the existence of lone pair electrons on each silicon atom.[243]

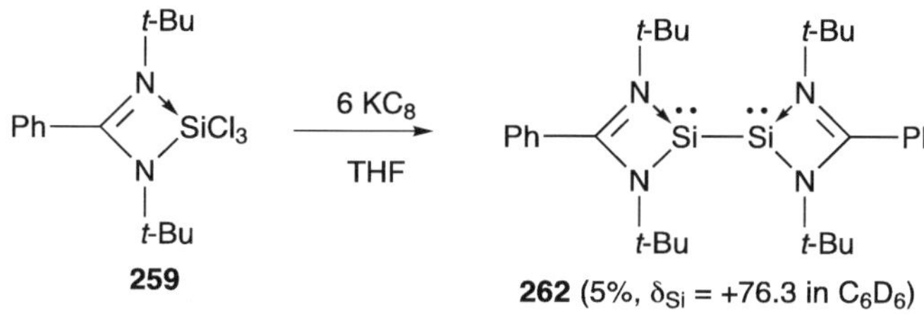

Scheme 8.132
Synthesis of bis(silylene) **262**.

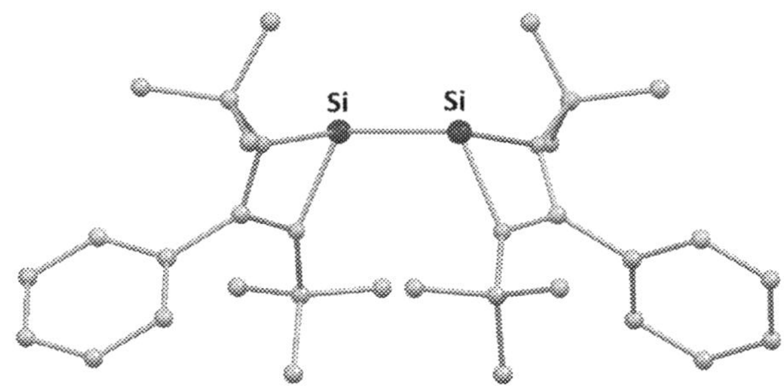

Figure 8.13
Molecular structure of **262** (CCDC: 734289).

Bis(silylene) **262** reacted with benzophenone in THF to give adduct **341** (Scheme 8.133).[302] When the reaction was performed in deuteriated THF (THF-d_8), **341**-d_2 was formed in which the benzylic protons were replaced by deuteriums. This result indicates that the benzylic protons in **341** come from THF. As a mechanism for formation of **341**, initial [1 + 2] cycloaddition between each silylene center and one C = O bond of benzophenone followed by ring expansion was proposed. In contrast, reaction of **262** with benzil gives [1 + 4] cycloadduct **342** whose Si–Si bond remains intact.[262] Similar [1 + 4] cycloaddition was found in the reaction of **262** with cyclooctatetraene giving **343**.[303]

Bis(silylene) **262** reacts with two equivalents of diphenylacetylene to give 1,4-disilabenzene **344** in 30% yield (Scheme 8.134).[304,305] This result is in contrast to the reaction of 1,2-disilyl- and 1,2-diaryldisilyne with acetylenes giving 1,2-disilabenzene derivatives.[306–309] Interestingly, reaction of **262** with phenylacetylene gave selectively new bis(silylene), *cis*-1,2-disilylenylethene **345**, instead of the corresponding 1,4-disilabenzene.[305] Although the mechanism for formation of **344** has been controversial,[304,305] higher stability of **345** compared to *cis*-[LSi{C(Ph) = C(Ph)SiL}, which is a possible initial product of **262** with diphenylacetylene, is proposed as a reason for the lack of further reaction of **345** with phenylacetylene giving 1,4-disilabenzene derivative occurs.[305] No reactions were observed

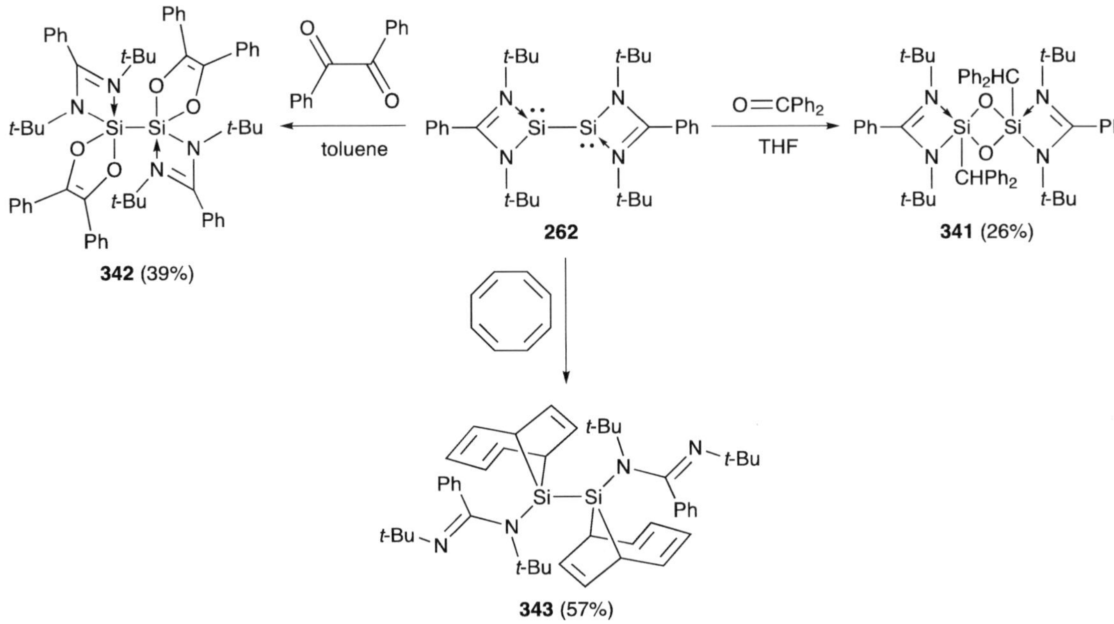

Scheme 8.133
Reactions of bis(silylene) **262** with unsaturated organic substrates.

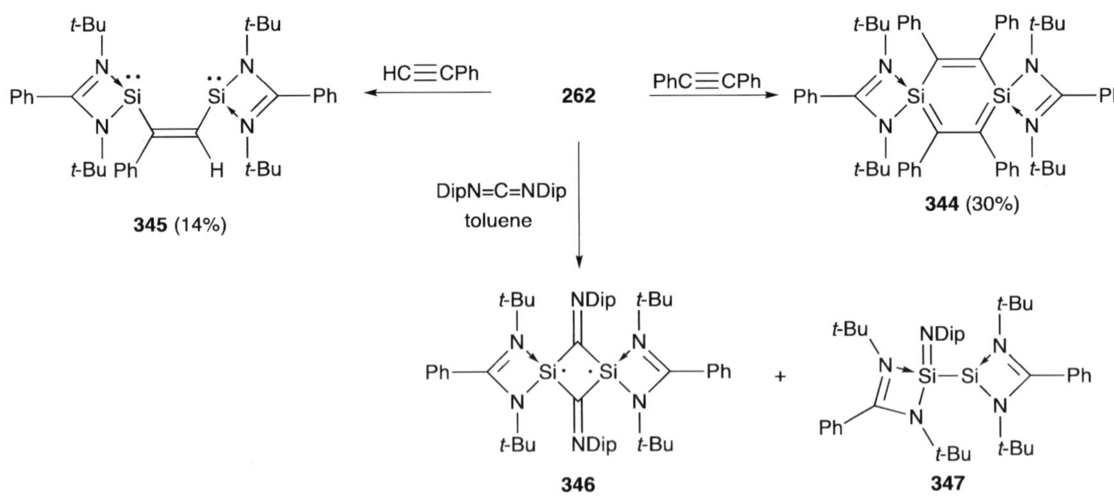

Scheme 8.134
Reactions of bis(silylene) **262** with unsaturated organic substrates giving new unsaturated silicon compounds.

when trimethylsilylacetylene and bis(trimethylsilyl)acetylene were used for the reaction with **262**.[305] The ^{29}Si NMR resonances assignable to tricoordinate silicon nuclei of bis (silylene) **345** appear at +15.1 and +29.5 ppm. Interestingly, treatment of DipN = C = NDip gave cyclic singlet biradical **346** and silanimine **347** in the ratio of 1:2.[310] Pure **346** was obtained as yellow-brown crystals in 26% yield. The proposed mechanism for formation of **346** and **347** involved the initial [1 + 2] cycloaddition of one of the C = N bonds in DipN = C = NDip to form Si–C–N cycles, further elimination of nitrene DipN:, ring expansion to form 1,3-disilabicyclo[1.1.0]butane-2,4-diimine, and finally homolytic cleavage of the Si–Si bond to give **346**. The nitrene DipN: can add to the silylene moiety of **262** to give **347**. Reactions of **262** with AdC≡P and P_4 gave CSi_2P cycle **311** and Si_2P_2 cycle **309** (Scheme 8.117 in Section 8.7.2.6).[277]

Oxidation of **262** in the presence of one-oxygen donor N_2O proceeded to give **348** (Scheme 8.135).[302] The proposed mechanism for formation of **348** involves insertion of oxygen atom from N_2O to Si–Si bond of **262** and subsequent oxidation of each lone pair at silicon giving the corresponding Si = O, which undergoes dimerization to give **348**. In the reaction with Se, bis(silaneselone) **320** was obtained.[275] The central Si–Si bond was cleaved by Br_2 to give bromosilylene **261**.[243] Interestingly, bis(silylene) **262** reacted with one equivalent of *N*-trimethylsilyl-4-dimethylaminopyridinium triflate and two equivalents of DMAP to form silyliumylidene cation **349** stabilized by amidinato and DMAP ligands.[311] Cation **349** was found to work as a σ-donor ligand for tungsten and rhodium complexes.[312]

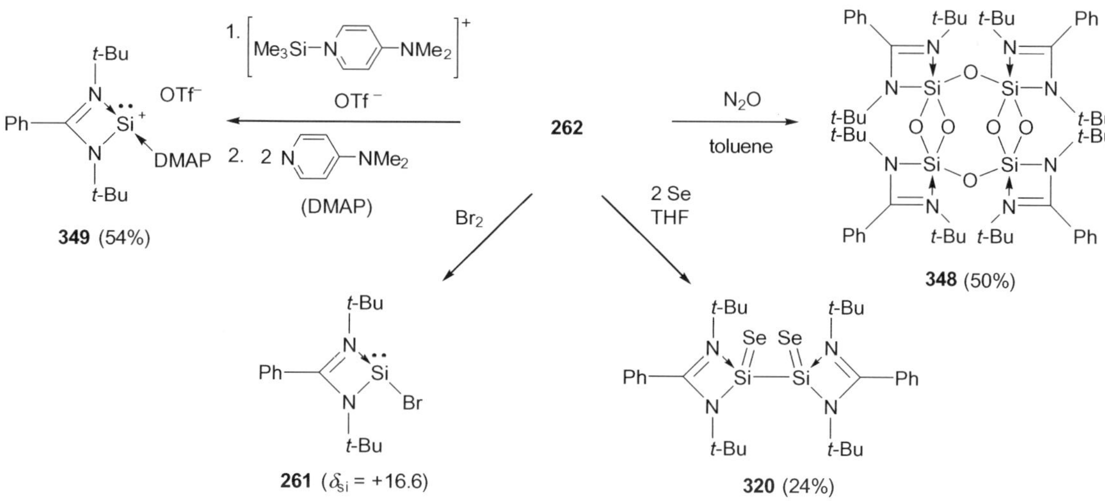

Scheme 8.135
Oxidation reactions of bis(silylene) **262**.

More interestingly, So et al. found that reaction of bis(silylene) **262** with digermylene **350** or chlorogermylene **351** with amidinato ligands gave zwitterionic germatrisilacyclobutadiene derivative coordinated by three amidinato ligands and one amidinatogermylene or amidinatosilylene ligand, **352a** and **352b**, as dark red crystals (Scheme 8.136).[313] Compound **352b** was also formed as a major product when an excess amount of **262** was used for the reaction with **350**. So et al. also found that the reduction of **262** with KC_8 in the presence of $Dip(Me_3Si)NSiCl_3$ afforded tetrasilacyclobutadiene derivative **353** as brown crystals in 5% yield.[314] The Si_4 ring in **353** is planar and rhombic. DFT calculations of model compound suggest the delocalization of two π, two σ, and two lone-pair electrons over the Si_4 ring in **353**.

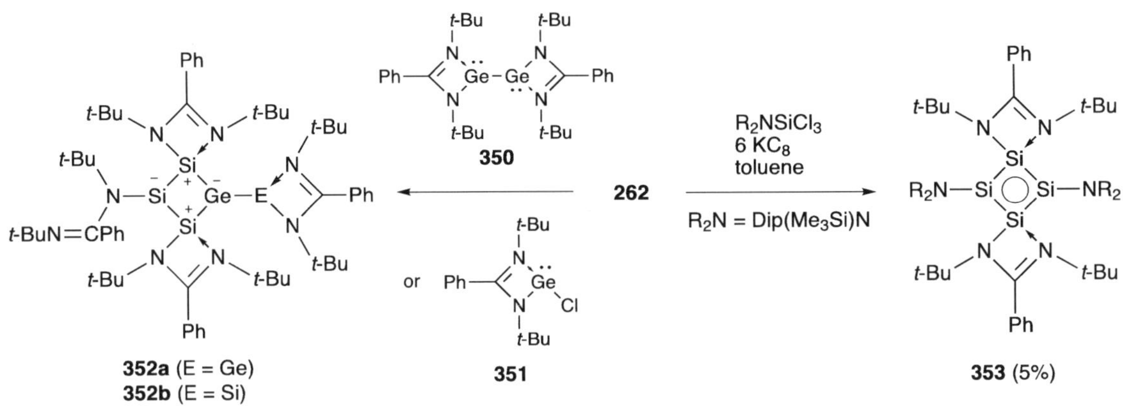

Scheme 8.136
Other reactions of bis(silylene) **262**.

8.9 Bis(silylene) With Amidinato Ligands Connected by Spacers

Oxygen-bridged bis(silylene) **354** was synthesized in two steps (Scheme 8.137).[299] Reaction of lithium amidinate with tetrachlorodisiloxane gave disiloxane with two amidinato ligands **355**. Dehydrochlorination of **355** with $LiN(SiMe_3)_2$ gave **354** in 76% yield as yellow crystals. The ^{29}Si resonance due to the Si(II) nuclei appeared at −16.1 ppm that is close to those of alkoxysilylenes **266b** (−5.2) and **266c** (−13.4).[244] XRD analysis showed that the two silylene centers adopt a remarkably pyramidalized geometry with the sum of bond angles of 277.69 and 275.34 degrees. The Si–O–Si moiety is considerably bent with the angle of 159.88(15) degrees and the Si–O distances of 1.641(2) and 1.652(2) Å.

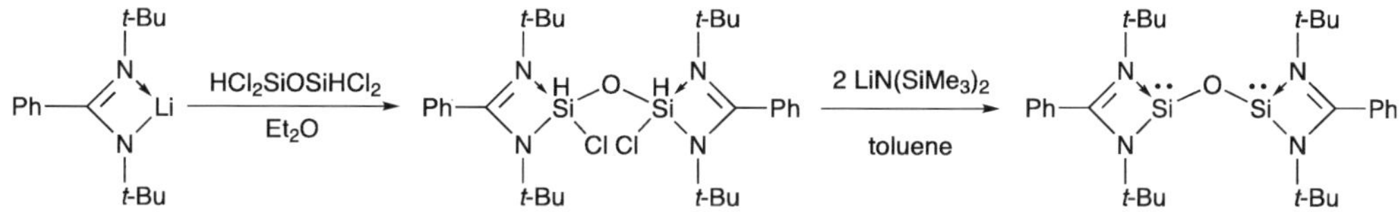

355 (53%) **354** (76%, δ_{Si} = −16.1)

Scheme 8.137
Synthesis of bis(silylene) **354**.

Oxygen-bridged bis(silylene) **354** works as a bidentate ligand (Scheme 8.138). Reaction of **354** with $Ni(cod)_2$ gave red crystals of nickel complex **356** (δ_{Si} = +32.8 ppm) in 91% yield.[299] The Si–O–Si angle of **356** of 93.44(8) degrees is acute compared to free bis (silylene) **354** (159.88(15) degrees) and the Si–O distances of 1.7011 (15) and 1.7081(17) Å are slightly longer than those of **354** (1.641(2) and 1.652(2) Å). The Si–Ni–Si angle of 68.90(3) degrees is also acute. The Si–Ni distances of 2.1908(7) and 2.1969(7) Å are within the range typical for the NHSi–nickel complexes. Complex **356** was applied as a precatalyst for cross-coupling reactions of aryl halides with organozinc and organomagnesium reagents. Interestingly, bis(silylene) **354** reacted with $[Cu(CH_3CN)_4]$ [OTf] giving cyclic 8-membered ring dinuclear complex **357**.[300] In the solid state, complex **357** exhibited a dicationic metallacyclooctane ring with Cu, Si, and O atoms. NMR spectra of **357** in CD_2Cl_2 suggest the existence of two species in solution, the solvent-free dication $[Cu_2\{\eta^1{:}\eta^1\text{-}\mathbf{354}\}_2]^{2+}$ and solvent-coordinated dication $[Cu_2\{\eta^1{:}\eta^1\text{-}\mathbf{354}\}_2]^{2+}(OTf^-)_n$ (n = 1,2). The ^{29}Si NMR resonances appeared at +3.3 and +7.7 ppm.

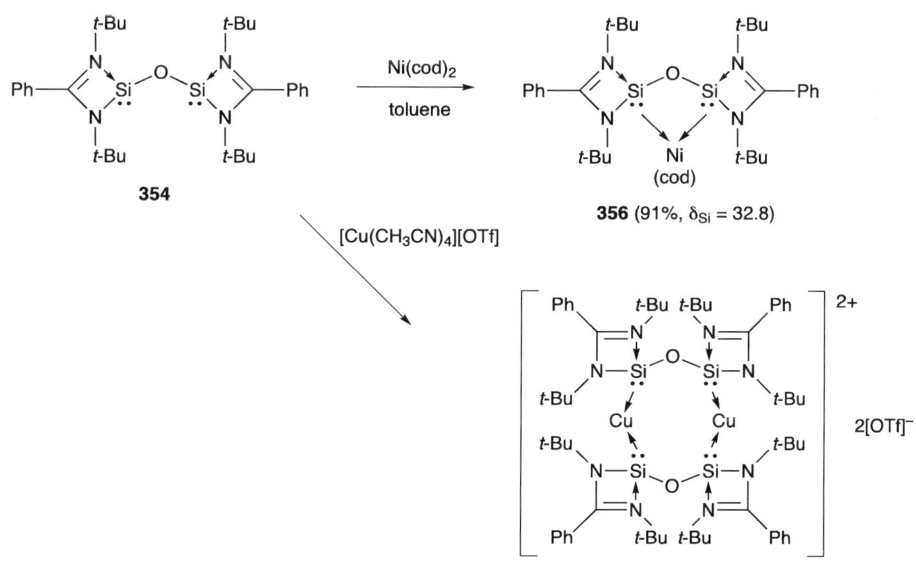

354

356 (91%, δ_{Si} = 32.8)

357 (52%, δ_{Si} = 3.3, 7.7 (two conformers))

Scheme 8.138
Synthesis of transition metal complexes of bis(silylene) **354**.

Ferrocene-connected bis(silylene) **358** is also available. Reaction of chlorosilylene **258** with 1,1′-dilithioferrocene gave 1,1′-bis(silylene)ferrocene **358** (δ_{Si} = +43.3) in 70% yield (Scheme 8.139).[315] Reaction of **358** with CpCo precursors provided bis(silylene) Co(I) complex **359** (δ_{Si} = +82.0) in 30% yield, while reaction with two molar equivalents of [CpCo(CO)$_2$] gave bis(silylene)-Co complex **360** (δ_{Si} = +85.7) in 87% yield. XRD analysis reveals that the Co–Si distance of **359** is 2.1252(14) Å and it is close to that of **334** (2.1143(4) Å).

Interestingly, complex **359** works as a precatalyst for [2 + 2 + 2] cycloaddition reactions (Scheme 8.140).[315] Reaction of phenylacetylene in the presence of 2.5 mol% of complex **359** in toluene gave triphenylbenzenes **361a** and **361b** in 72 and 28% yields, while the reaction in acetonitrile resulted in formation of methyldiphenylpyridines **362a** and **362b** in 39% and 14% yields, respectively.

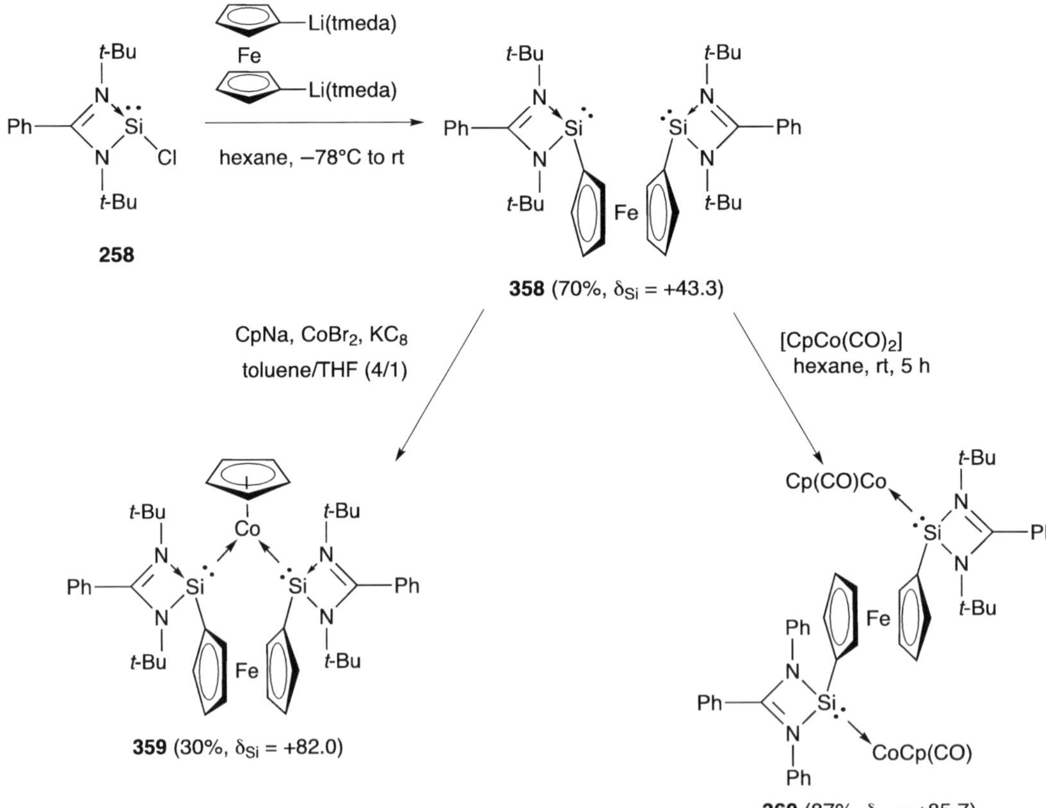

Scheme 8.139
Synthesis of ferrocene-connected bis(silylene) **358** and its transition metal complexes.

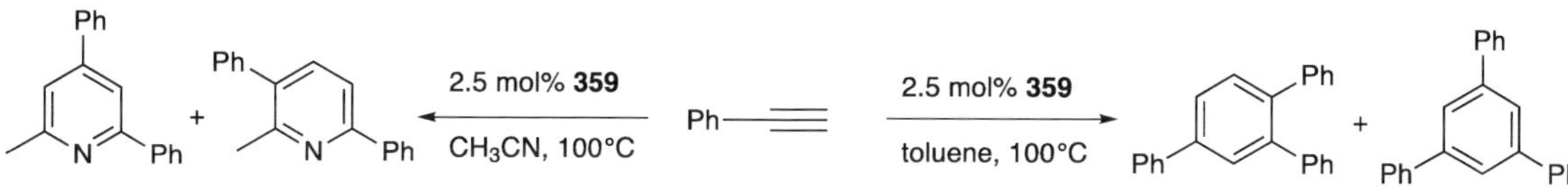

362a (39%) **362b** (14%) **361a** (72%) **361b** (28%)

Scheme 8.140
[2 + 2 + 2] Cycloaddition reactions catalyzed by complex **359**.

Bis(silylene) **363** with a resorcinolate connector was synthesized by the metathesis reaction of two molar equivalent of chlorosilylene **258** and dilithium resorcinolate and characterized by NMR spectroscopy and XRD analysis (Scheme 8.141).[316] Treatment of **363** with Pd $(PPh_3)_4$ gave unexpected bis(silylene)(silyl)phenylpalladium(II) **364** as a sole product in 81% yield. Complex **364** has a distorted square-planar geometry with two silylene moieties, phenyl, and hydridosilyl groups. In this complex, one Si–N bond was cleaved due to the migration of hydride to silyl center and one silylene unit remains uncoordinated. The Pd–Si (silylene) distances (2.3271(12) and 2.3038(11) Å) are shorter than the Pd–Si(–H) distance (2.3561(12) Å). Complex **364** exhibited four different ^{29}Si resonances at −8.7 (free silylene), +39.7 (Si-H), 62.3 and 65.8 (silylene→Pd), and a weak Si–H stretching vibration band at 2135 cm^{-1} in the IR spectrum. The proposed mechanism involves initial insertion of Pd metal into H–C(Ph) bond followed by 1,2-migration of hydride to one of the silylene centers and further coordination of one of the silylene moiety of another bis (silylene) **363**, which was supported by theoretical calculations of model compounds.

Reaction of **363** (SiCSi) with $[IrCl(coe)_2]_2$ (coe = cyclooctene) gave novel pincer-type iridium complex **365** ([(SiCSi)IrHCl(coe)]) in high yield (Scheme 8.141).[317] The corresponding digermylene complex **365′** (Fig. 8.14) was also synthesized in a similar manner. The Ir–Si distances of 2.305(1) Å and 2.301(1) Å are between the average Ir(III)–Si(IV) distances (about 2.39 Å) for octahedral silyliridium complexes and the distance for non-σ-donor stabilized Ir(III)-silylene complex (about 2.24 Å). Similarly, Ir and Rh complexes **366** and **367** were prepared in high yields (Scheme 8.141), although reaction of **363** with Vaska's complex $[IrCl(CO)(PPh_3)_2]$ gave a complex mixture.[317] Interestingly, the C = C distance and the ^{13}C NMR resonance of olefinic carbons in the coe ligand is decreased and downfield shifted in the order **365** (1.409(9) Å, 55.8 ppm), **365′** (n/a, 65.1 ppm), **368** (1.35(1) Å, 81.6 ppm, in Fig. 8.14), which indicate that the ligand donor strength decreases in the order: SiCSi (in **363**) > GeCGe (in **365′**) > *i*-PrN-PCP (in **368**) > *t*-Bu-PCP (in **369**) (Fig. 8.14; absence of the coe ligand coordinated to Ir center in **369** may suggest a less electron rich center due to weaker donor strength of *t*-Bu-PCP).

Ir complex **365** was applied for catalytic C–H borylation of arenes (benzene, toluene, and xylene) with pinacol borane (HBPin) (Scheme 8.142).[317] Interestingly, the presence of coe accelerates the reaction of benzene and pinacol borane (90% (with coe) and 53% (without

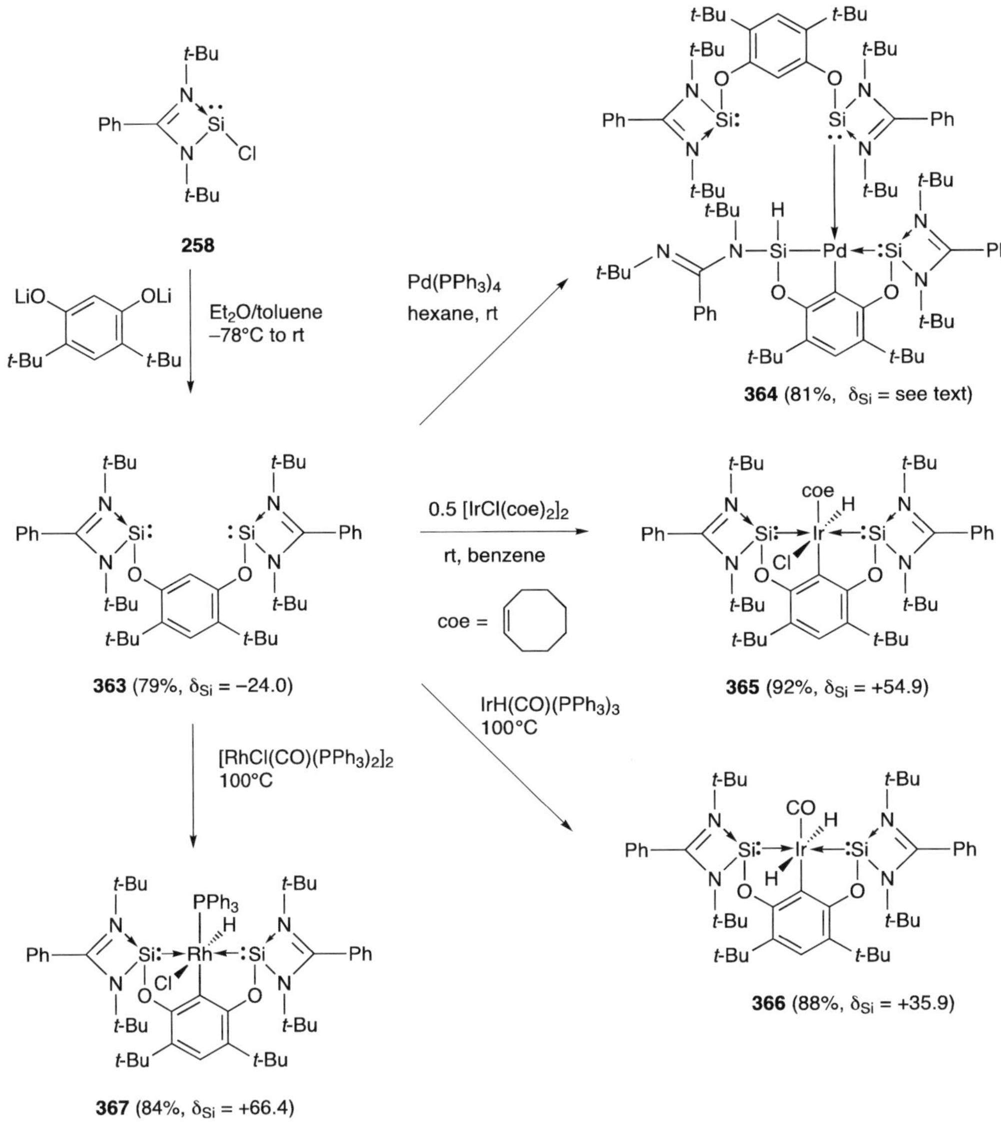

Scheme 8.141
Synthesis of bis(silylene) **363** and its transition metal complexes.

coe)). In the proposed mechanism, the following steps could be involved: oxidative addition of HBPin to Ir(I) center in [(SiCSi)Ir], C–H activation of the arene forming ArBPin and Ir (III)$(H)_2$ species, and regeneration of Ir(I) via (a) direct release of H_2 or (b) hydrogenation of coe. The stronger donor property of SiCSi (**363**) ligand may increase the energy barrier for the reductive elimination and release of H_2, and make coe working as an effective H_2 acceptor.

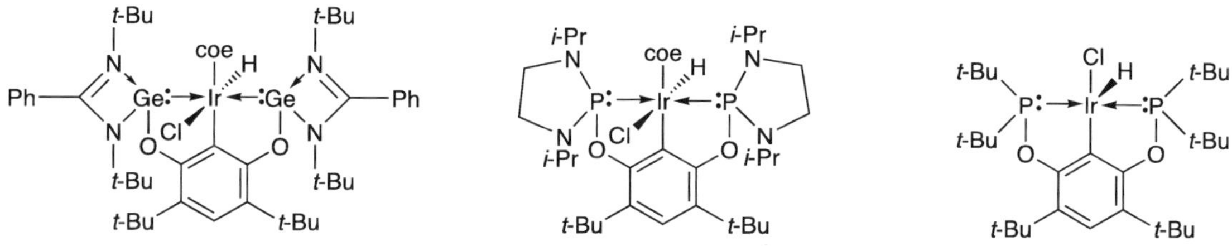

365'
[(GeCGe)IrHCl(coe)]

368
[(*i*-PrN-PCP)IrHCl(coe)]

369
[(*t*-Bu-PCP)IrHCl]

Figure 8.14
Complexes related to complex **365**.

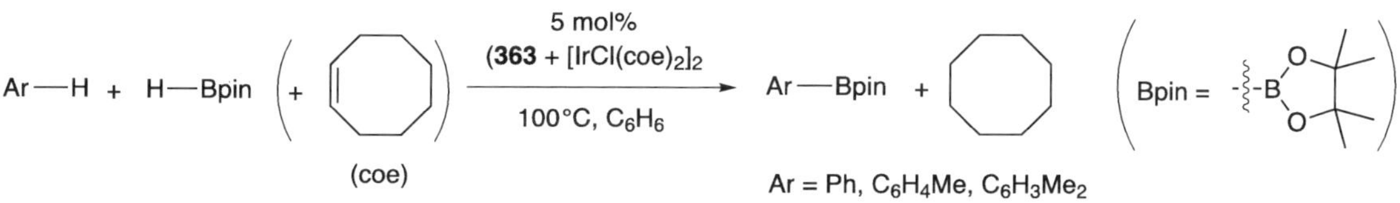

Scheme 8.142
C—H borylation of arenes with pinacol borane catalyzed by complex **365**.

Pincer-type Ni complexes **370a** (X = Br) and **370b** (X = I) were also synthesized by the reaction of **363** with NiX_2(dme) complex in the presence of excess NEt_3 (Scheme 8.143).[318] Complex **370a** serves as a catalyst for the Sonogashira reaction of (*E*)-1-iodo-1-octene with phenyl acetylene. The reaction proceeds with 5 mol% of **370a** and an excess of the halide to give the coupling product in a moderate yield (39%) (Scheme 8.144). The detailed analysis of the reaction disclosed that the catalytic cycle can involve complex **371** as an intermediate (Scheme 8.145) which was isolated and characterized by XRD analysis and MS spectroscopy. In complex **371** (X = Br), Cu metal interacted with one of the Si(II) centers and a carbon atom of the acetylide ligand.

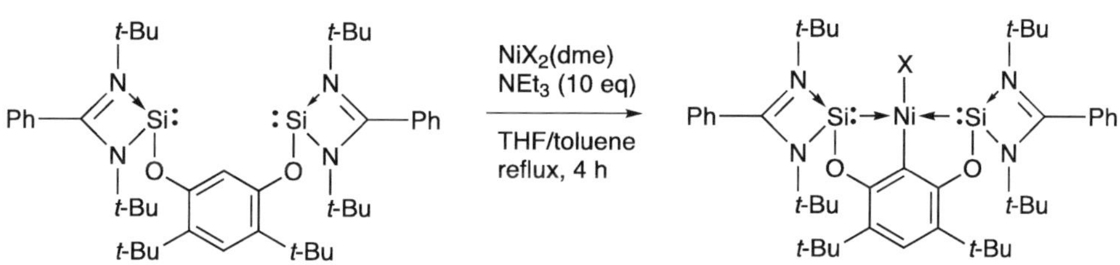

363 (= SiCSi)

370a (X = Br, 70%, δ_{Si} = +20.2)
370b (X = I, 66%, δ_{Si} = +30.7)

Scheme 8.143
Synthesis of the Ni complexes of bis(silylene) **363**.

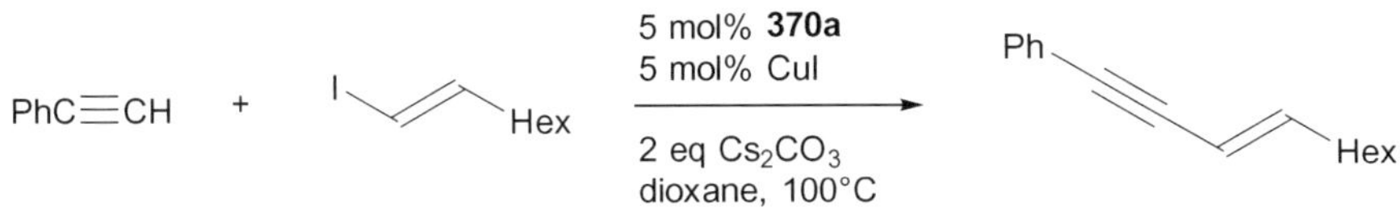

Scheme 8.144
Sonogashira reaction catalyzed by complex **370a**.

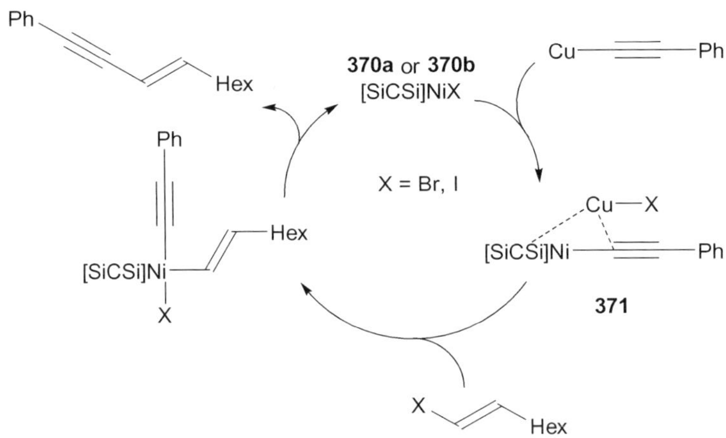

Scheme 8.145
Proposed mechanism for Sonogashira reaction catalyzed by complex **370**.

The SiNSi-type pincer ligand bearing two amidinato silylene moieties **372** (=SiNSi) was obtained in 90% yield as yellow crystals by the reaction of 2,6-*N*,*N*-di(ethylamino)pyridine (**373**) with 2 equiv of BuLi followed by addition of 2 equiv of chlorosilylene **258** (Scheme 8.146).[319] The ^{29}Si NMR spectrum suggests the existence of two rotational conformers ($\delta_{Si} = -14.9$ (major); -13.8 and -17.1 (minor)). Reaction of **372** with $FeCl_2 \cdot (thf)_{1.5}$ gave [SiNSi]$FeCl_2$ (**374**) in 73% yield. Interestingly, the N(pyridine)–Fe distance of 3.304 Å determined by XRD analysis indicates that complex **374** is better described as κ^2*Si*,*Si'*-[SiNSi]$FeCl_2$. XRD analysis also exhibits that complex **374** has a distorted tetrahedral coordination sphere, and the magnetic moment determined in solution ($\mu_{eff} = 4.6\ \mu B$) corresponds to a high-spin $S = 2$ ground state, which is consistent with the observed ^{57}Fe Mössbauer spectrum (isomer shift $\delta = 0.73$ mm s^{-1}, quadrupole splitting $\Delta E_q = 3.06$ mm s^{-1}).

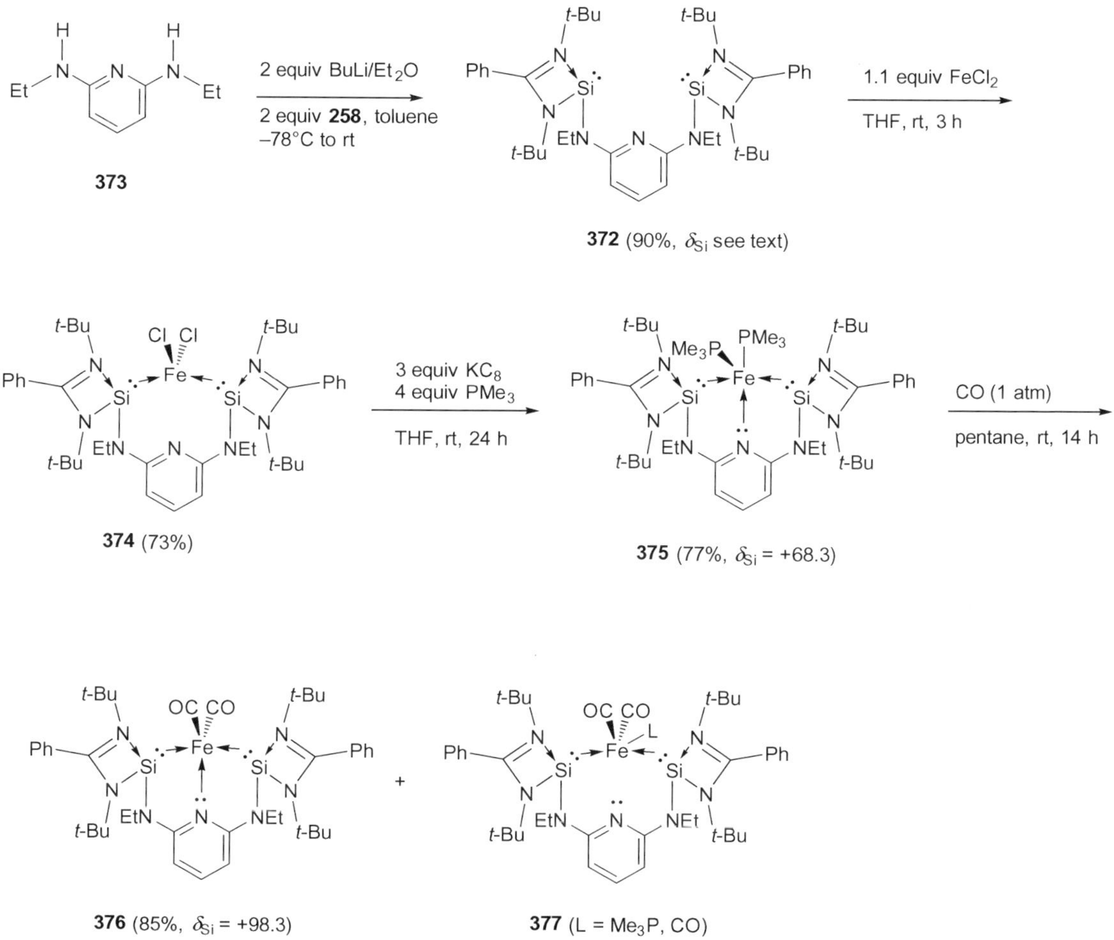

Scheme 8.146
Synthesis of bis(silylene) **372** and its transition metal complexes.

Reduction of **374** with excess KC_8 in the presence of excess PMe_3 gave diamagnetic Fe(0) complex **375** as a dark purple solid in 77% yield (Scheme 8.146).[319] Complex **375** was also obtained directly by the reaction of **372** with $Fe(PMe_3)_4$ in 48% yield. Interestingly, XRD analysis exhibited that the iron center has a slightly distorted pseudo-square pyramidal geometry with one of PMe_3 group occupying the apical position. The ^{31}P NMR spectrum of **375** showed two broad signals at +7.2 and +20.8 ppm, while the ^{29}Si NMR resonance was observed at +68.3 ppm as a doublet of doublets (^2J(Si–P) = 22.4 and 91.9 Hz), which is consistent with the structure in the solid state. The variable-temperature NMR spectra showed that the C_s symmetry was kept between 213 and 343K indicating that **375** is the first example of the configurationally stable pseudo-square pyramidal Fe(0) complex. The parameters of the ^{57}Fe Mössbauer spectrum of **375** at 77K (isomer shift $\delta = 0.24(1)$

mm s^{-1}, quadrupole splitting $\Delta E_q = 1.66(1)$ mm s^{-1}) are in the range expected for Fe(0) complexes with five-coordinate geometry and close to that of $(depe)_2Fe(N_2)$ with a five-coordinate geometry ($\delta = 0.24$ mm s^{-1}, $\Delta E_q = 2.14$ mm s^{-1}, depe = bis(diethylphosphino) ethane). Comparison of the ^{57}Fe Mössbauer parameters with reported values of the related Fe(II) and Fe(0) complexes indicate that ligand **372** has the innocent nature, which is supported by theoretical calculations.

Reaction of **375** with CO gave CO complexes **376** and **377**. In contrast to **375**, complex **376** has a C_{2v} symmetry which was evidenced by NMR spectroscopy and XRD analysis (Scheme 8.146).[319] Complex **375** can catalyze the hydrosilylation of 4-substituted-phenyl actophenone **378** with triethoxysilane (Scheme 8.147) (R (yield) = H (93%), OMe (>99%), Me (82%), Br (>99%) and CF_3 (95%)).

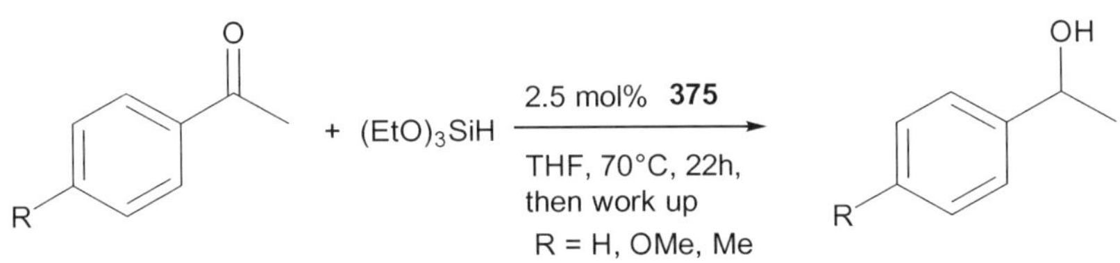

Scheme 8.147
Hydrosilylation of acetophenones catalyzed by **375**.

8.10 Stable Silylenes With N,N-(diisopropyl)amidinato and Guanidinato Ligands

8.10.1 Synthesis and Structure

In 2012, Tacke et al. reported new hypercoordinate silylene with two amidinato ligands **380a** synthesized by the reductive dehydrochlorination of six-coordinate silicon precursor **381a** (Scheme 8.148).[320,321] XRD analysis revealed that silylene **380a** has a distorted pseudo-tetrahedral geometry around the Si atom with the lone pair as the fourth ligand (Fig. 8.15). Different coordination mode of the amidinato ligands was observed: the Si–N distances in the bidentate ligand (1.9065(12) and 1.9026(12) Å) are much longer than that in the monodentate ligand (1.7885(12) Å). In contrast, ^{1}H, ^{13}C, and ^{14}N NMR spectra indicate that all isopropyl groups and nitrogen atoms are equivalent suggesting a rapid exchange of four nitrogen sites of amidinato ligands involving a four-coordinate silicon(II) species with two bidentate amidinato ligands (**380a′**) in solution (Scheme 8.149). Detailed DFT calculations revealed that four-coordinate species **380a′** is a local minimum and is favored by $\Delta G(298K) = 27.6$ kJ mol^{-1} over three-coordinate species **380a** at the ZORA-BLYP-D3(BJ)/TZVP level of theory by using COSMO to estimate the effects of solvation.[321] Effective orbital overlap of LUMO + 1 of LSi^+ and HOMO − 1 of L^- units in a bidentate fashion (L = amidinato) rather than a monodentate fashion would be responsible for the four-coordinate geometry. The ^{29}Si resonance due to the central silicon nucleus appeared at −15.4 ppm in the solid state, while that at −31.4 ppm in C_6D_6. The ^{29}Si NMR chemical shift of **380a** in the solid state (−15.4 ppm) well agrees with that calculated for three-coordinate species **380a** (−19.7 ppm) at the ZORA-SAOP/ET-pVQZ//ZORA-BLYP-D3(BJ)/TZVP level of theory, while in C_6D_6 solution (−31.4 ppm) is reproduced by benzene-coordinated four-coordinate species **380a′** · C_6H_6 (−49.5 ppm) and **380a′** · $2C_6H_6$ (−43.0 ppm) rather than benzene-free **380a′** (−62.7 ppm), which also supports the contribution of four-coordinate species **380a′** in solution.[321]

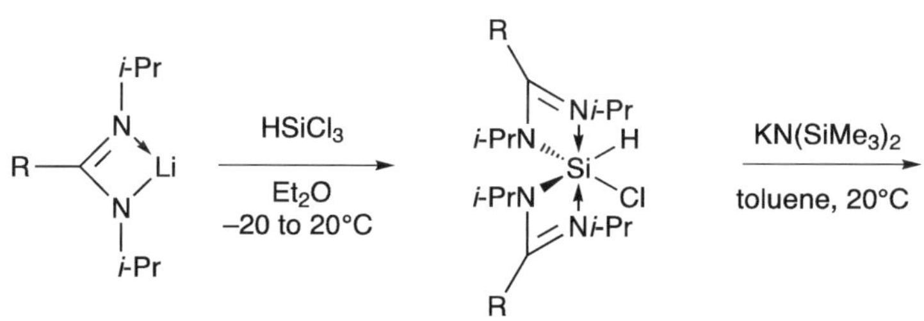

381a (70%, R = Ph)
381b (94%, R = *i*-Pr_2N)

380a (86%, R = Ph, δ_{Si} = −31.4 (C_6D_6), −15.4 (solid))
380b (95%, R = *i*-Pr_2N, δ_{Si} = −26.0 (toluene-d_8), −25.6 (solid))

Scheme 8.148
Synthesis of silylenes **380a,b**.

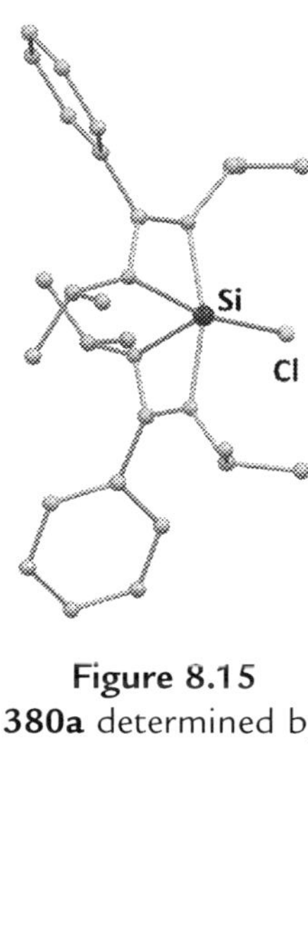

Figure 8.15
Molecular strucuture of silylene **380a** determined by XRD analysis (CCDC: 876289).

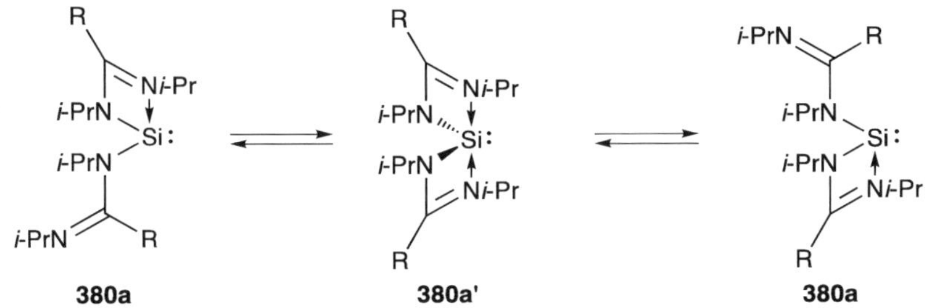

Scheme 8.149
Fluxional behavior of **380a**.

Silylene coordinated by a guanidinato ligand (**380b**), where Ph group in **380a** is replaced by i-Pr_2N group, is also synthesized by a method similar to that for **380a**.[322] Silylene **380b** shows structural characteristics similar to those of **380a**. No remarkable difference in ^{29}Si NMR resonances in solution and in the solid state was observed suggesting no significant contribution of a four-coordinate silicon(II) species with two bidentate guanidinato ligands in solution in contrast to **380a**. Detailed DFT calculation of **380b** showed that although four-coordinate species **380b′** is a local minimum, it is less stable than the three-coordinate species **380b** both in benzene solution and in the gas phase at the ZORA-BLYP-D3(BJ)/TZVP level of theory.[323] Effective orbital overlap of LUMO of LSi^+ and HOMO of L^- units in a monodentate fashion rather than a bidentate fashion (L = guanidinato) would be responsible for the three-coordinate geometry.

Silylene coordinated by more bulky guanidinato ligand (**382**) was synthesized according to a procedure similar to that of amidinato-coordinated silylenes (Scheme 8.150).[322]

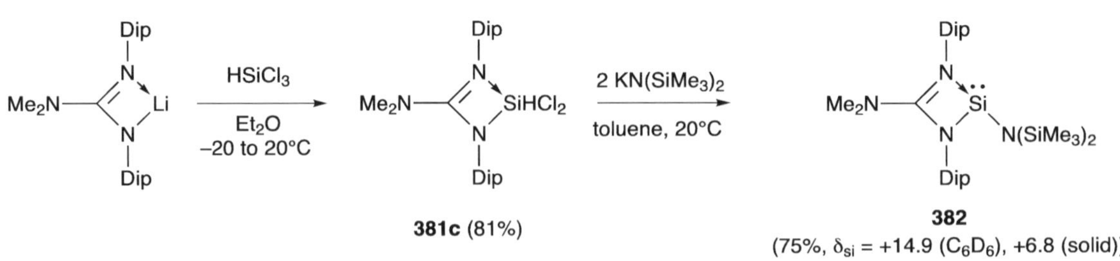

Scheme 8.150
Synthesis of silylene **382**.

8.10.2 Reactivity

8.10.2.1 Insertion reaction

Similarly to other stable silylenes, silylene **380a** inserts into various bonds, such as I–I,[324] PhSe–SePh,[321] C_6F_5–F[321], to give the corresponding insertion products with six-coordinate silicon atom **383a**, **383b**, and **383c** (Scheme 8.151). In the case of guanidinato-coordinate silylene **382**, reaction with PhEEPh gave silanimines **384a–c** (Scheme 8.152), which probably resulted from formation of insertion product followed by elimination of Me_3SiEPh. [325] Interestingly, silylene **380a** inserts into the C–H bond of 1,2-dicarba-*closo*-dodecaborane (1,2-$C_2B_{10}H_{12}$) to give **385** in good yield (Scheme 8.151) in contrast to NHC destroying the carborane cage of 1,2-$C_2B_{10}H_{12}$.[326] Reaction of **380a** with SO_2 in hexane gave an adduct with *O,O′*-sulfito moiety **386a**, while reaction with liquid SO_2 gave adducts with *O,O′*-dithionito ligand **387** (*cis*:*trans* = 64:36).[327] The mechanism for formation of **386a** involved initial formation of intermediate **380a** · SO_2, followed by elimination of SO giving a Si = O species **389a** (see Scheme 8.154) whose structure is similar to **389b–d**, and further addition of SO_2. In the case of **387**, further addition of SO_2 to the initial adduct **380a** · SO_2 would occur to give **387** before elimination of SO from **380a** · SO_2. Similar adduct **386b** was obtained by the reaction of **380a** with CO_2.[322]

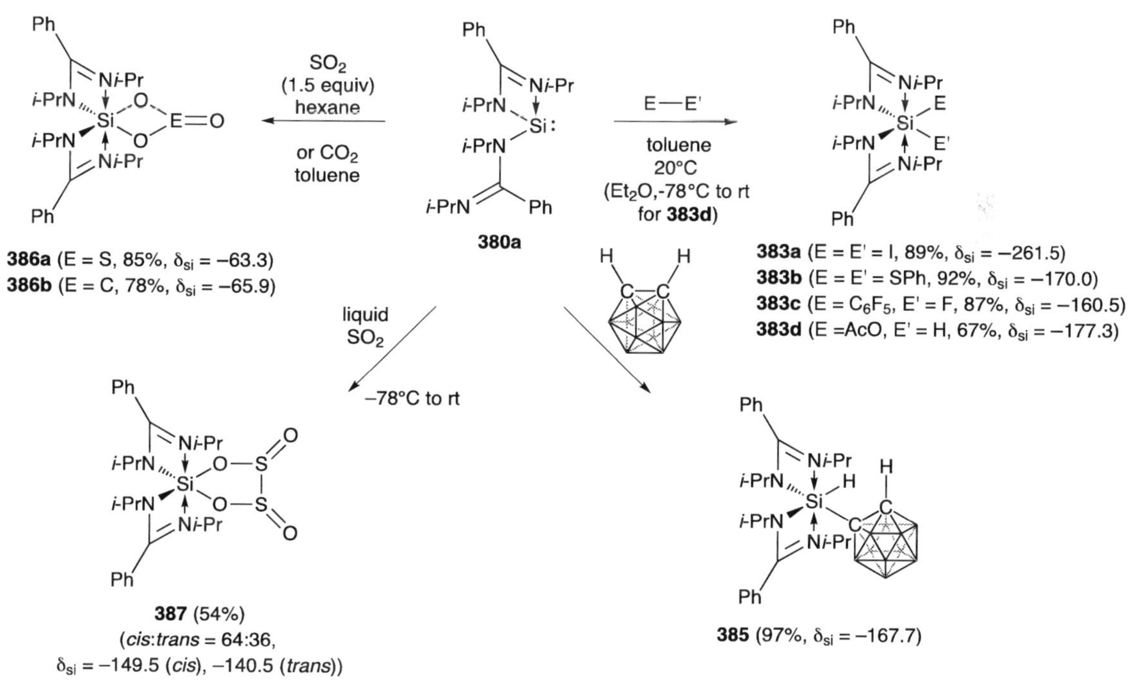

Scheme 8.151
Insertion reactions of silylene **380a**.

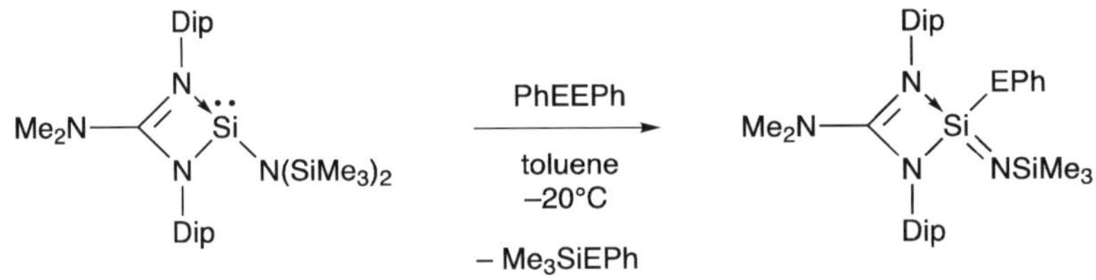

382

384a (X = S, 71%, δ_{Si} = −59.6)
384b (X = Se, 82%, δ_{Si} = −64.2)
384c (X = Te, 48%, δ_{Si} = −78.9)

Scheme 8.152
Insertion reactions of silylene **382**.

8.10.2.2 Reaction with Lewis acids and Brønsted acids

Reaction of **380a** with BR_3 (R = Et, Ph) gave the corresponding BR_3 adducts **388a** and **388b** in good yields (Scheme 8.153).[328] Both **388a** and **388b** adopt a distorted trigonal bipyramidal geometry with the (angles N(axial)−Si−N(axial) = 148.12(9) degrees and 150.22(8) degrees for **388a** and **388b** (similar to **383a**), and the Si−B distances of 2.076(4) Å (**388a**) and 2.067(3) Å (**388b**) are within the range of the reported Si−B distances for silylene-borane adducts (1.96 − 2.11 Å). The slightly stronger Si−B donor−acceptor bond of **388b** than that of **388a** was explained by DFT calculations.

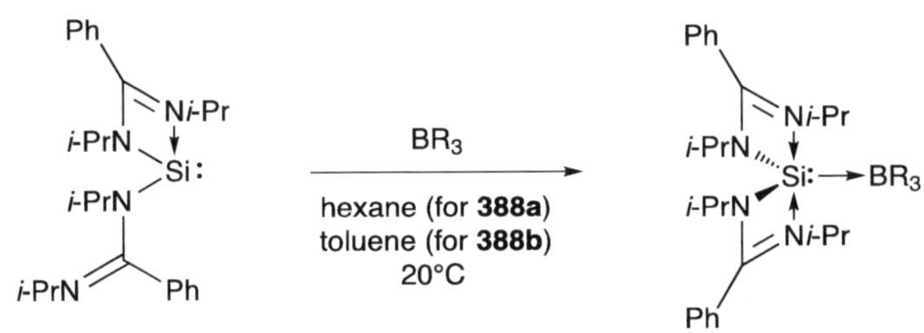

380a

388a (R = Et, 75%, δ_{Si} = −23.3)
388b (R = Ph, 82%, δ_{Si} = −40.3)

Scheme 8.153
Reactions of silylene **380a** with Lewis acids.

Reaction of **380a** with acetic acid gave the corresponding O−H insertion product **383d** (see Scheme 8.151 in Section 8.10.2.1).[329] The proposed mechanism for formation of **383d** involves formation of protonated silylene **380a** · H^+ with a five coordinate silicon as found in **398a** and **398b** (see Scheme 8.157 in Section 8.10.2.5), followed by addition of acetate, which was supported by quantitative formation of **383d** by the reaction of **398b** with potassium acetate in THF-d_8.

8.10.2.3 Reaction with chalcogens and N_2O

Similarly to other silylenes, silylene **380a** reacted with elemental sulfur, selenium, and tellurium to afford the corresponding silanechalcogenones with a five coordinate silicon center **389b–d** (Scheme 8.154).[324] Similar reactions giving four-coordinate silanechacogenones were observed for guanidinato-coordinate silylenes **380b** and **382**.[323] In the case of the reaction with N_2O, **390** (as a dimer of **389a**) was obtained in good yield.[322] In the case of guanidinato-coordinate silylene **382**, silanimine **391** was obtained probably resulting from the formation of silanone similar to **389a** followed by migration of $SiMe_3$ group from nitrogen atom to oxygen atom (Scheme 8.155),[325] while reaction of **380b** with N_2O gave disiladioxetane similar to **390**.[323]

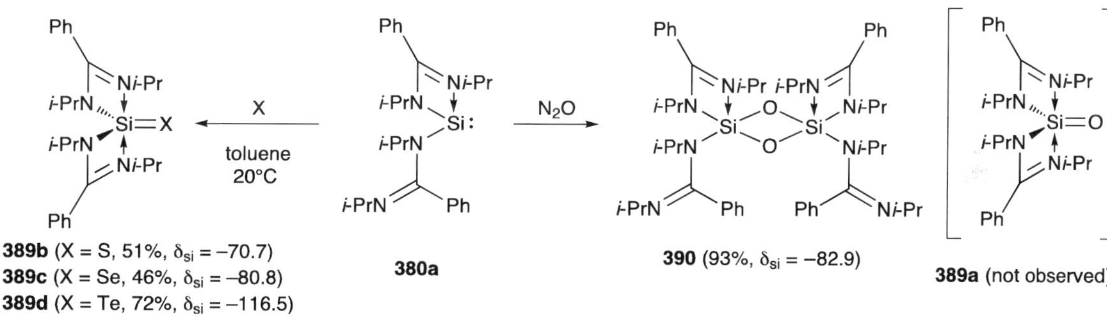

Scheme 8.154
Reactions of silylene **380a** with group 16 element compounds.

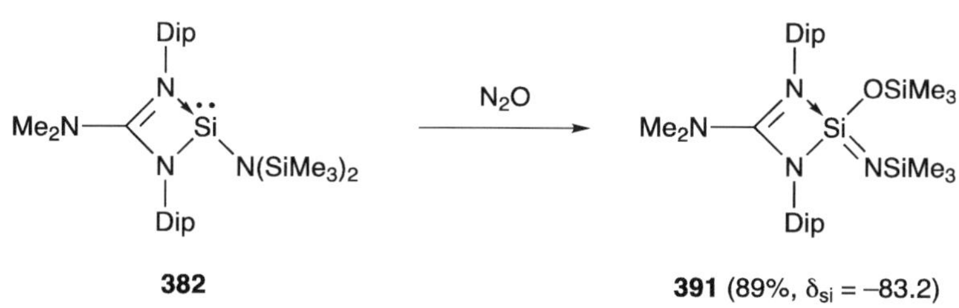

Scheme 8.155
Reactions of silylene **382** with N_2O.

8.10.2.4 Cycloaddition

Similar to other silylenes, **380a** undergoes cycloaddition reactions. Silylene **380a** reacted with various dienes, such as 2,3-dimethyl-1,3-butadiene, 1,4-diphenyl-1,3-butadiene, 2,3-dibenzyl-1,3-butadiene, cyclohexa-1,3-diene, cyclooctatetraene to give the corresponding [1 + 4] cycloadducts **392** (six-coordinate) and/or **393** (four-coordinate), except for the parent 1,3-diene giving bicyclic adduct **394** (Scheme 8.156).[330] Similarly, silylene **380a** reacted with 1,2-diketone, such as 3,5-di-*tert*-butyl-1,2-benzoquionone and 1,2-diphenylethane-1,2-dione gave the corresponding [1 + 4] cycloadduct (2,4-dioxa-3-silacyclopent-1-ene). The coordination number of the silicon atoms in the resulting cycloadduct is dependent on the substituents on the substrate and the aggregation state (in solution or in the solid state).[330]

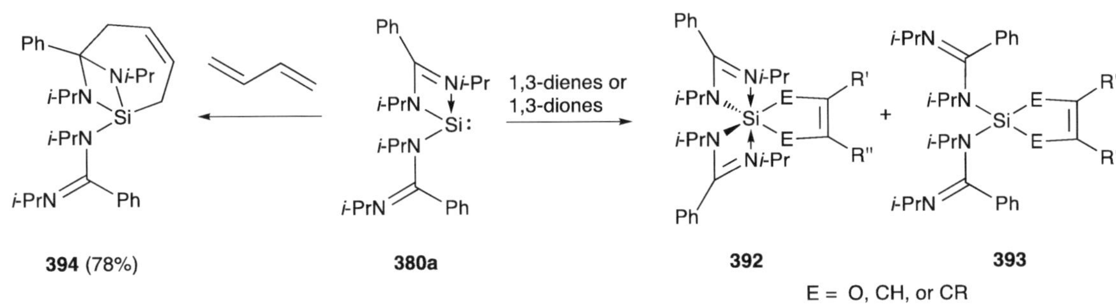

Scheme 8.156
Cycloaddition reactions of silylene **380a**.

8.10.2.5 Reactions with metal complexes

Silylene **380a** works as a four-coordinate silylene ligand for transition metals, while silylene **380b** serves as three- or four-coordinate silylene ligand. Electron donation from silicon to transition metal would reduce the electron density at the silicon atom and increase its Lewis acidity, which is responsible for the six-coordinate mode.

For instance, reaction of **380a** with $M(CO)_5$ (M = Cr, Mo, W) gave the corresponding group 6 metal carbonyl complexes **395a** (M = Cr, 76%),[331] **395b** (M = Mo, 89%),[331] and **395c** (M = W, 92%) (Scheme 8.157).[320] Similarly, reaction of **380a** with $Fe(CO)_5$ afforded complex **396a** (65%).[331] Interestingly, these complexes have a five-coordinate silicon with a strongly distorted trigonal bipyramidal geometry with one nitrogen atom of each of the two amidinato ligands occupying the axial positions (N(axial)–Si–N(axial) = 144.64(7) degrees (**395a**), 145.44(6) degrees (**395b**), 145.97(11) degrees (**395c**), 154.64(6) degrees (**396a**)).[320,331]

The iron complex **396a** has a distorted trigonal bipyramidal geometry with the silylene ligand occupying an equatorial site. The Si–M distances (2.4181(1) Å (**395a**), 2.5784(6) Å (**395b**), 2.5803(9) Å (**395c**), 2.3175(6) Å (**396a**)) are slightly longer than those reported for Si–M single bond distances. The theoretical study suggests that the Si–M bond in these complexes is a single bond resulting from an overlap of an orbital of Si with high s-character and an orbital of metal with high p-character. The ^{29}Si NMR resonances of **395a–c** and **396a** in solution (−22.3 ppm (**395a**) to +6.9 ppm (**395b**)) are slightly

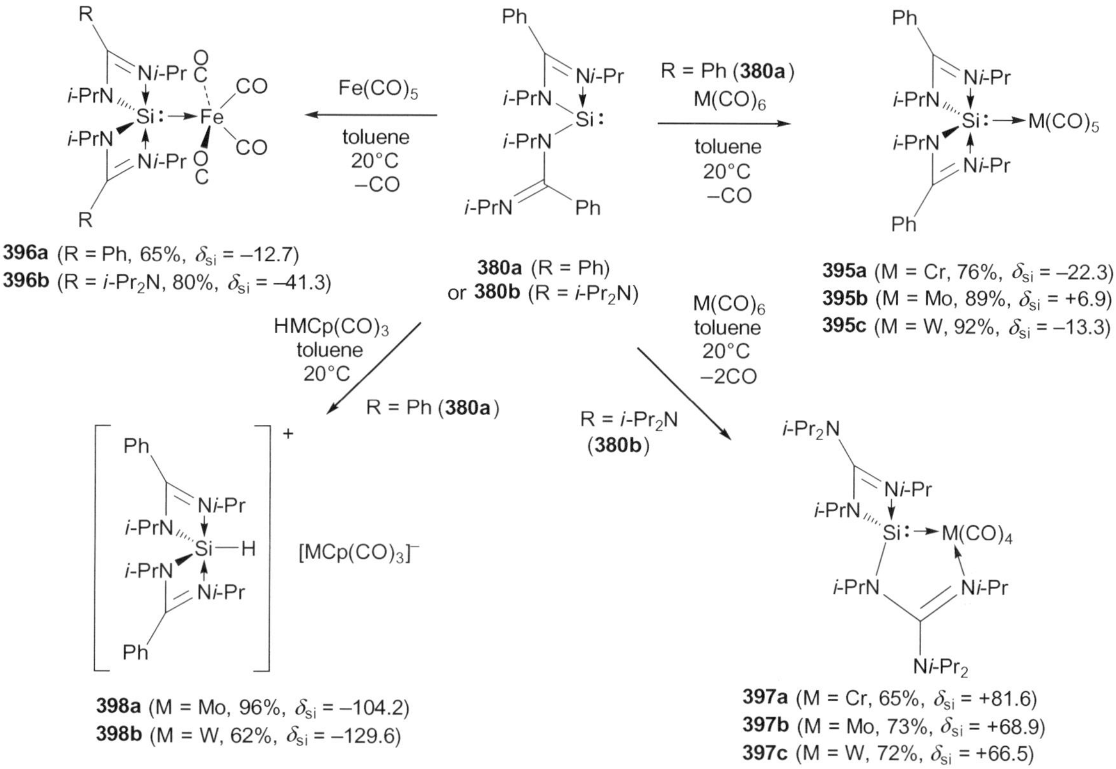

Scheme 8.157
Synthesis of transition metal complexes of silylenes **380a,b**.

downfield shifted compared to that of free silylene **380a** (−31.4 ppm). The ^{29}Si NMR chemical shifts for these complexes in solution are very similar to those observed in the solid state suggesting these complexes have a five-coordinate silicon as observed by XRD analysis.

It should be noted that reaction of guanidinato-coordinate silylene **380b** with M $(CO)_5$ (M = Cr, Mo, W) gave new complexes whose structures are totally different from **395a**–**c**.[332] In this case, spirocyclic complexes with $MSiN_2C$ ring **397a** (M = Cr), **397b** (M = Mo), **397c** (M = W) were obtained accompanied by elimination of two CO ligands from the metal center (Scheme 8.157). The Si–M distances (2.3400(8) Å (**397a**), 2.4877(7) Å (**397b**), 2.4899(8) Å (**397c**)) are slightly shorter than those for **395a**, **395b**, and **395c**, which is consistent with the reduced coordination number of silicon atom. In contrast, reaction of **380b** with $Fe(CO)_5$ gave **396b** structurally similar to **396a** (five-coordinate silylene ligands). The reasons for the observed different coordination modes of the amidinato and guanidinato ligands are still unclear.

Reaction of **380a** with metal hydrido complexes $HMCp(CO)_3$ (M = Mo, W) gave the ionic complexes **398a** (M = Mo, 96%) and **398b** (M = W, 62%) with a five-coordinate Si(IV) cation $[$**380a**$\cdot H]^+$ and a $[MCp(CO_3)]^-$ counter anion.[329] Interestingly, this reaction involves proton transfer from the transition metal to silicon atom formally associated with a redox process (Si(II) to Si(IV) and H^+ to H^-).

8.11 Phosphine-Supported Silylenes

8.11.1 Synthesis and Molecular Structure

Recently, Kato, and Baceiredo et al. have reported the chemistry of phosphine coordinated functionalized silylenes. Reduction of **399** with magnesium in THF affords phosphine coordinated silylene **400a** in 61% yield (Scheme 8.158).[333] Silylene **400a** is isolated as a diastereomeric pair owing to the pyramidalized three-coordinated silicon. The ^{29}Si NMR spectrum of **400a** shows two sets of doublets at −17.7 ppm ($^1J_{PSi}$ = 157 Hz) and −13.9 ppm ($^1J_{PSi}$ = 141 Hz), whose coupling constants due to ^{31}P nucleus are considerably larger than that of $[(Me_3Si)_4P]^+$ cation ($^1J_{PSi}$ = 1.4 Hz).[334] XRD analysis of **400a** shows that the P–Si bond length of 2.304(5) Å is comparable to that of $[(Me_3Si)_4P]^+$ cation (2.31 Å) and unsaturated silicon atom is remarkably pyramidalized with the sum of the bond angles around the silicon atom of 298.8 degrees. These spectroscopic and structural features indicate that **400a** possesses considerable phosphonium sila-ylide character (**400a′**).

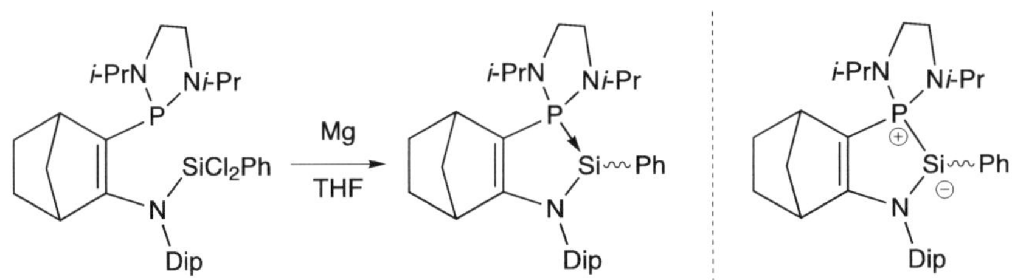

Scheme 8.158
Synthesis of silylene **400a**.

Reduction of dichlorosilanes **401a** and **401b** with magnesium in THF affords phosphine-coordinated hydridosilylenes **402a** and **402b** in 70% and 79% yields, respectively (Scheme 8.159).[335] These silylenes exist as a diastereomeric pair and **402a** shows two doublet signals at −44.8 ($^1J_{SiP}$ = 143.2 Hz) and −38.0 ($^1J_{SiP}$ = 140.1 Hz) ppm in the ^{29}Si NMR spectrum, which are due to silylene centers of the diastereomers. The ^1H NMR spectrum of **402a** showed two doublets accompanied by satellites due to the coupling with ^{31}P and ^{29}Si nuclei at 5.99 ppm ($^2J_{PH}$ = 3.3 Hz, $^1J_{SiH}$ = 85.1 Hz) and 5.76 ppm

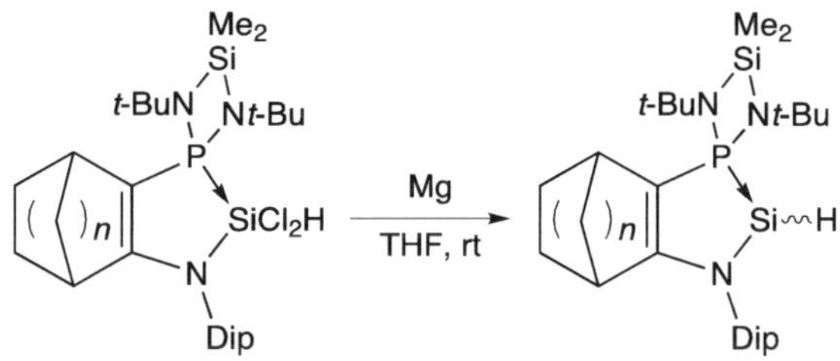

401a (n = 1)
401b (n = 2)

402a (n = 1, 70%)
402b (n = 2, 79%)

Scheme 8.159
Synthesis of silylenes **402**.

($^2J_{PH}$ = 2.9 Hz, $^1J_{SiH}$ = 85.7 Hz). In the molecular structure of **402b**, the Si···P distance of 2.318(15) Å indicates the intramolecular coordination and the unsaturated silicon atom adopts trigonal-pyramidal geometry with the sum of the bond angles of 275 degrees.

Phosphorus-supported chlorosilylene **403a** is synthesized by reduction of trichlorosilane **404a** with magnesium in THF (Scheme 8.160).[336]

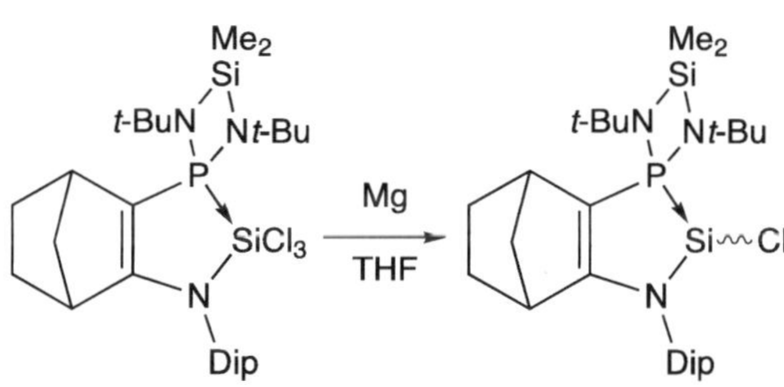

404a

403a (67%)

Scheme 8.160
Synthesis of silylene **403a**.

Other phosphorus-supported silylenes derived from hydridosilylene **402a** and chlorosilylene **403a** will be discussed in the next section.

8.11.2 Reactivity

Silylene **400a** featuring considerable phosphonium sila-ylide character has unique reactivity. Treatment of **400a** with one equivalent of mesitylaldehyde in ether gives phosphine oxide-coordinated silene **405** in 61% yield (Scheme 8.161). This reaction is regarded as a silicon version of Wittig reaction (sila-Wittig reaction).[333] Monitoring of the reaction at low-temperature revealed that five-coordinate silaoxirane **406** that resulted from a [1 + 2] cycloaddition of **400a** with mesitylaldehyde was observed as an intermediate.[337] With increasing the monitoring temperature ($> -30°C$), **406** isomerizes to **405**. The molecular structure of silaoxirane **406** was unequivocally determined by single crystal XRD analysis. Theoretical studies suggested that silaoxirane is an initial complex resulting from an almost barrierless [1 + 2] cycloaddition of silylene across the C = O double bond. Then, 1,2-oxygen migration from silicon to phosphorus of silaoxirane occurs to give a sila-Wittig product **405**.

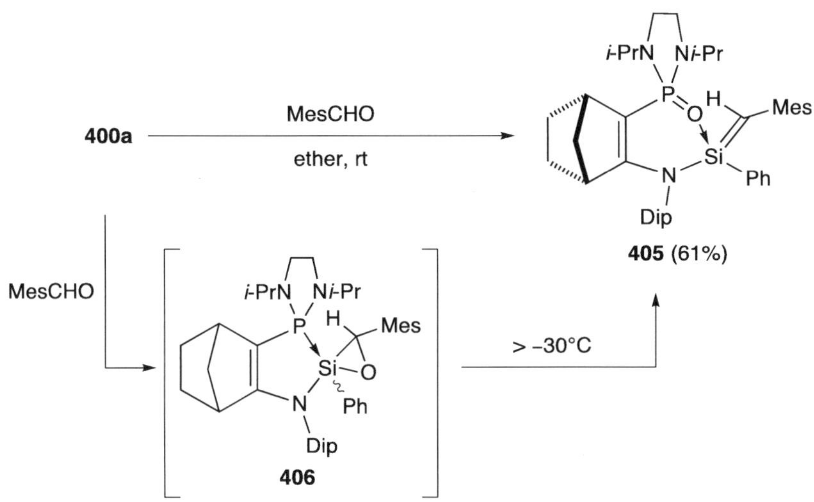

Scheme 8.161
Reaction of silylene **400a** with mesitylaldehyde.

Reactions of **400a** with alkynes furnish tricyclic phosphines **407a–c** (Scheme 8.162).[338] In the case of phenylacetylene, the reaction proceeds regioselectively to give only **408c**. Since silacyclopropene **408a** was observed by NMR spectroscopy as an intermediate and its structure was revealed by XRD analysis, [1 + 2] cycloaddition of silylene **400a** with alkynes to afford **408a–c** is the initial step of the reaction. Because configurationally stable phosphines **407a–c** formed as a single product from a diastereomeric pair of **400a**, a dynamic kinetic resolution occurred during the transformation of **408** to **407** probably due to the fluxional five-coordinate silicon of **408**. Tricyclic phosphines **407a–c** are air- and thermally-stable and show strong nucleophilic character. Moreover, enantiopure **407a** and **407b** are obtained from commercially available chiral *endo*-norborneol as a starting material.

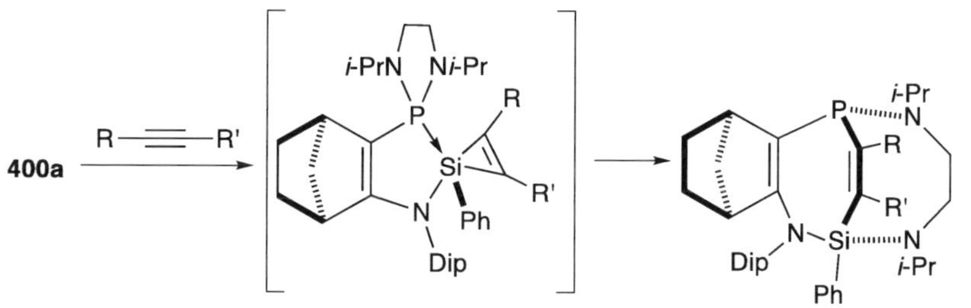

408a (R = R' = Et)
408b (R = R' = Ph)
408c (R = Ph, R' = H)

407a (89%, R = R' = Et)
407b (83%, R = R' = Ph)
407c (32%, R = Ph, R' = H)

Scheme 8.162
Reaction of silylene **400a** with acetylenes.

Reaction of **400a** with carbon dioxide gave tricyclic phosphine **409** in 32% yield accompanied by elimination of CO (Scheme 8.163).[339] Since single diastereomer of **409** formed from diastereomeric pair of **400a**, a kinetic resolution similar to the case of the formation of **407a–c** occurs during the reaction. Formation of silaketene **410** via sila-Wittig reaction, elimination of CO from **410**, and migration of one of the amino groups from phosphorus to silicon in **410′** were proposed as key steps. The presented transition-metal free reduction of CO_2 to form CO is noteworthy.

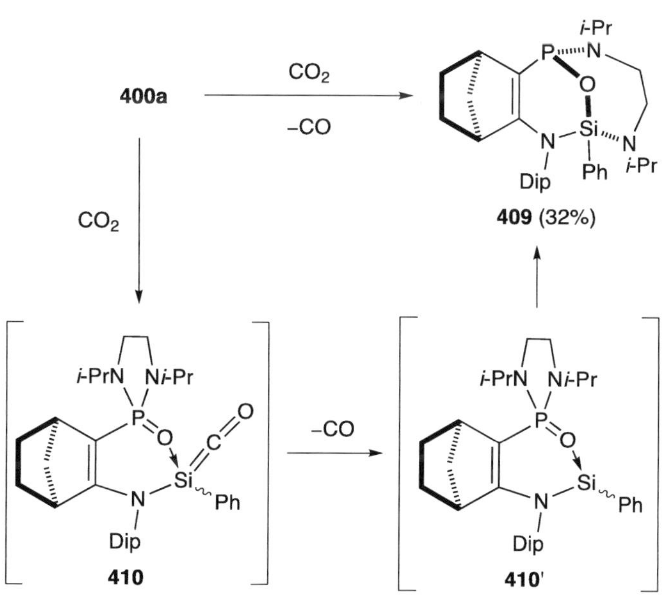

Scheme 8.163
Reaction of **400a** with CO_2.

Reaction of silylene **400a** with ethylene proceeded in a [1 + 2] cycloaddition manner to afford silacyclopropane **411a** (Scheme 8.164).[340] This reaction is reversible at room temperature and the ratio of **400a/411a** depends on the temperature and the pressure of ethylene. Equilibrium constant $K_{\mathbf{400a/411a}}$ = [**411a**]/[ethylene][**400a**] was estimated to be 1.15 mol L^{-1}, at 2.5 bar of ethylene, in THF-d_8, at −25°C. This equilibrium is sensitive to the substituents on the phosphorus atom; structurally similar **400b** shows similar reversible [1 + 2] cycloaddition with ethylene, but formation of **411b** is less favorable ($K_{\mathbf{400b/411b}} = 1.42 \times 10^{-1}$ mol L^{-1} at the same conditions). **400c** generated in-situ by the reductive dechlorination of the corresponding dichlorosilane with magnesium reacts with ethylene to give **411c** in 80% yield, but no reverse reaction of **411c** is observed in toluene even at 100°C.

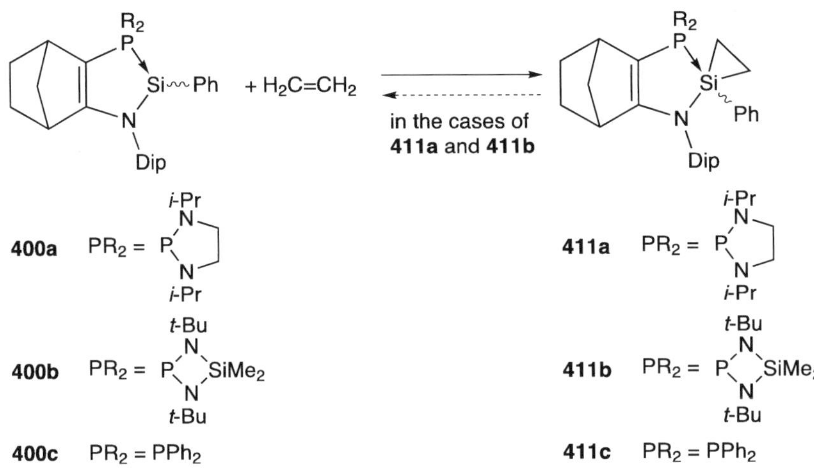

Scheme 8.164
Reactions of silylenes **400** with ethylene.

Hydridosilylene **402a** demonstrates diverse reactivity. Hydrosilylation of cyclopentene with **402a** proceeded without any metal catalysts and cyclopentyl silylene **412** (diastereomeric pair) forms in 40% isolated yield (Scheme 8.165).[335] Since reaction of **402a** with vinyltrimethylsilane provides silacyclopropane **413** and the subsequent thermolysis provides a mixture of hydrosilylation products (**414a** and **414b**), silacyclopropane was proposed as an intermediate of the observed hydrosilylation by hydridosilylene.

A [1 + 2] cycloaddition of hydridosilylene **402a** with diphenylacetylene at room temperature gives **415** (Scheme 8.166). Compound **415** isomerizes thermally to a new silylene, nitrogen-coordinated silacycloprop-1-ylidene **416** in 92% yield.[341] This isomerization involves a decrease in the silicon valency from Si(IV) to Si(II). Computational studies show that **415** and **416** are close in energy; **415** is slightly more stable by 3.2 kcal mol^{-1} than **416** at the B3LYP/6−31 G(d,p) level, which would be related to the unusual decrease in the silicon valency.

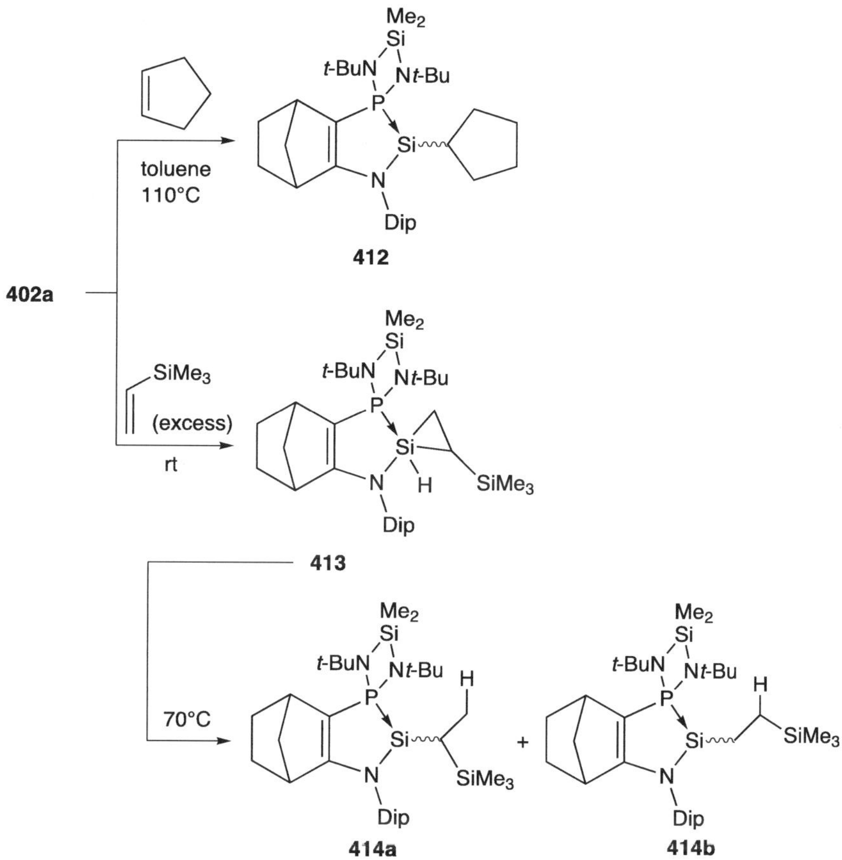

Scheme 8.165
Reaction of silylene **402a** with alkenes.

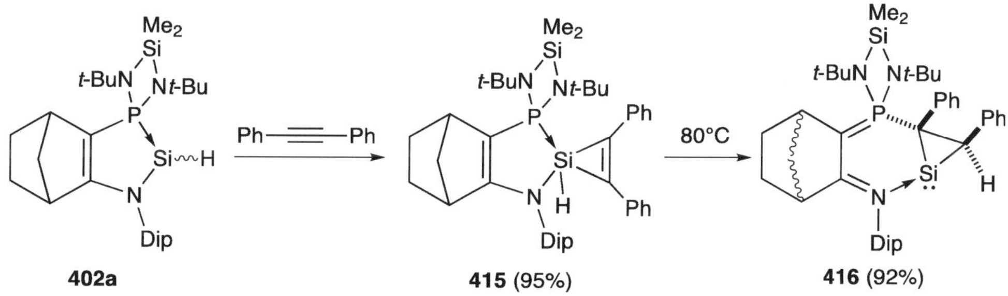

Scheme 8.166
Reaction of silylene **402a** with diphenylacetylene.

Silacyclopropylidene **416** reacts with BH_3 to afford the corresponding adduct (**417**) and with ethylvinylketone to give **418** probably via [1 + 4] cycloaddition and ring-opening of silacyclopropane unit (Scheme 8.167).[341] Irradiation ($\lambda = 300$ nm) or heating (150°C) of a toluene solution of **416** results in the formation of a stable silene with strong π-donating substituents (**419**).[342]

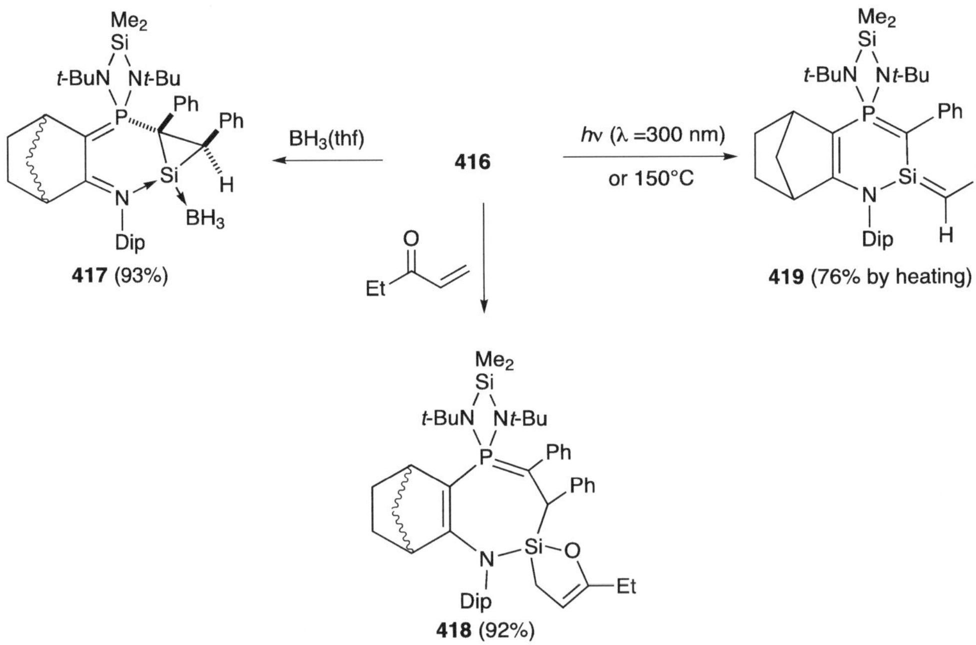

Scheme 8.167
Reactions of silylene **416**.

Silylene **416** leads to $R_2Si = O$ (silanone) species as shown in Scheme 8.168. Oxidation of **416** by N_2O affords nitrogen-coordinated silacyclopropan-1-one **420** in 85% yield.[343] X-ray diffractional study showed a short Si—O bond length of 1.5467(6) Å and an unusually long C—C bond in silacyclopropane unit of 1.6671(9) Å. Topological analysis of electron density map using diffraction data demonstrates the contribution of oxyallyl-like canonical structure (**420′**) to the overall structure. Oxidation of **416** with O_2 gives sila-β-lactone **421** in 51% yield.[344] Molecular structure of **421** reveals the short exocyclic Si—O bond of 1.539(2) Å and the long endocyclic Si—O bond of 1.691(2) Å, displaying the sila-lactone structure. Reaction of **416** with benzaldehyde in the presence of a catalytic amount of magnesium salt as a Lewis acid causes the formation of nitrogen-stabilized silacyclobutanone **422**.[345] A [1 + 2] cycloaddition to form oxasilaspiropentane followed by Lewis-acid catalyzed rearrangement to silacyclobutanone **422** was proposed as a possible reaction mechanism. The Si—O bond length of 1.544(2) Å determined by XRD analysis is reasonable for the base-supported silanone.

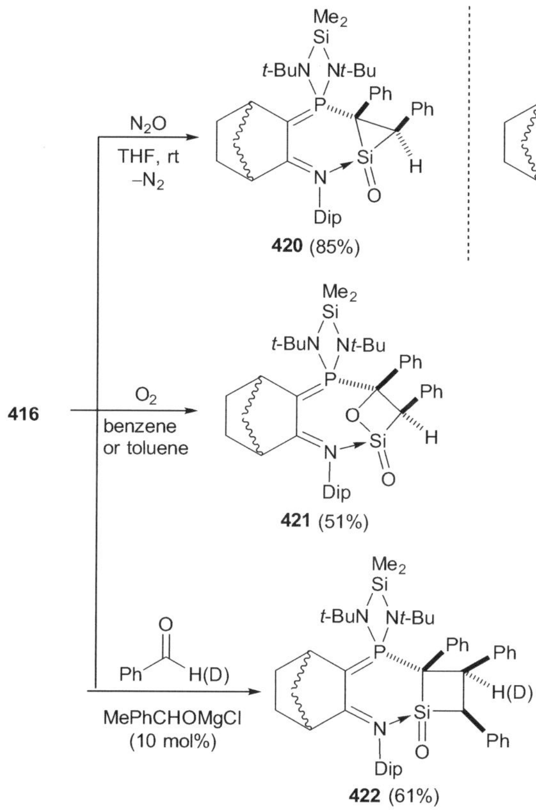

Scheme 8.168
Reactions of silylene **416** giving Si = O species.

Chlorosilylene **403a** is converted to a pair of diastereomers of diazomethylsilylene **423** as orange solid in 42% yield (Scheme 8.169). In the ^{29}Si NMR spectroscopy, one of the unsaturated silicon atoms appears at −20 ppm with coupling constants due to the ^{31}P nuclei of $^{1}J_{SiP} = 204$ Hz and $^{2}J_{SiP} = 240$ Hz. Interestingly, silylene **423** is converted to the first isolable silyne (silicon-carbon triply bonded species) **424**; irradiation of **423** (λ = 300 nm) in THF at −60°C resulted in complete consumption of **423** and formation of a dark-red solution of **424**.[336]

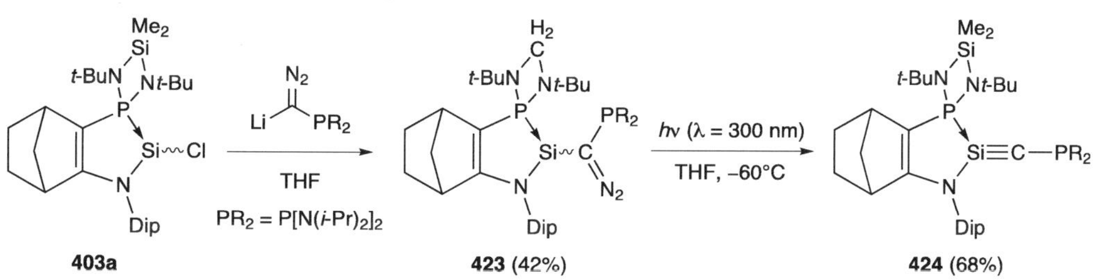

Scheme 8.169
Synthesis of stable silyne **424** from silylene **403a**.

Reactions of chlorosilylenes **403a** and **403b** with Me_3SnLi form stannylsilylenes **425a** and **425b** in 81% and 85% yields (Scheme 8.170).[346] Migratory insertion of stannyl group of **425a** slowly occurred with ethylene at room temperature to form **426a**. Isolated **426a** is stable at room temperature but a reverse reaction is observed at 85°C and **425a** is recovered. Reaction of silylene **425b** with ethylene to give **426b** is much faster than that of **425a** and completes within 15 min at room temperature. Reverse reaction of **426b** was observed at a lower temperature of 25°C. Computational studies found that the migratory insertion to form **426** proceeds via [1 + 2] cycloaddition of **425** with ethylene followed by 1,2-migration of stannyl group from silicon to carbon, similarly to the reactions of hydridosilylene **402a** with olefins.[335]

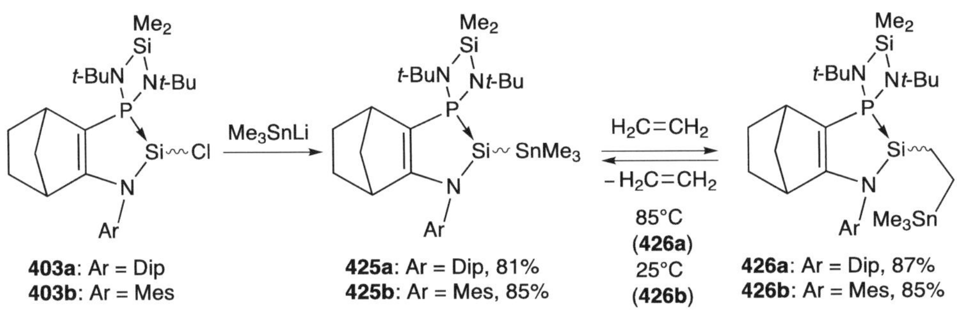

Scheme 8.170
Synthesis of stannylsilylenes **425** and their reaction with ethylene.

Similar reversible migratory insertion was observed in the reactions of **425a** with substituted styrenes at 45 °C to provide regioselectively **427** (Scheme 8.171).[346] The insertion was favorable for electron-deficient styrenes as linear Hammett correlation between equilibrium constants and *para*-substituent constants (σ_p) with positive reaction constant (ρ) was displayed.[347] These results are in accord with nucleophilic character of phosphine-coordinated silylene **425**.

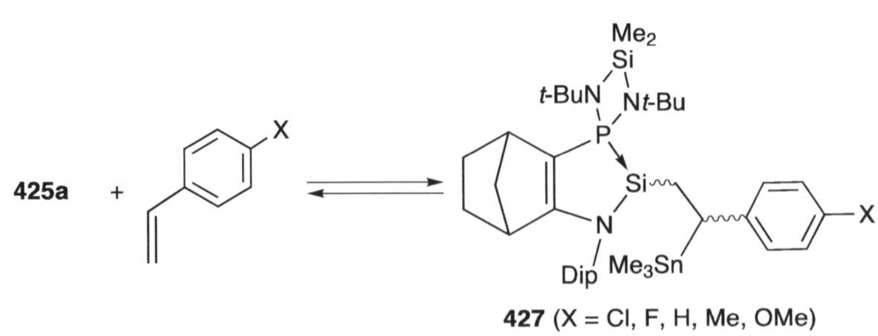

Scheme 8.171
Reaction of silylene **425a** with styrene derivatives.

8.11.3 Bis(silylenes)

Reduction of trichlorosilane **428** with three equivalents of lithium in THF affords bis (silylene)-bis(phosphine) adduct **429** as red crystals in 13% yield (Scheme 8.172).[339] In the ^{29}Si NMR spectra, unsaturated silicon atom of **429** appears as a doublet of doublets due to the coupling with the two ^{31}P nuclei at −18.5 ppm ($^{1}J_{SiP} = 192.1$ Hz, $^{2}J_{SiP} = 60.1$ Hz), suggesting P−Si−Si−P connectivity. The Si-Si bond length of **429** is 2.3306(12) Å. These experimental properties and DFT studies indicate the major contribution of polarized resonance form **429′**, rather than Si−Si triple bond to its electronic structure. Compound **429** immediately reacts with four equivalents of CO_2 to form **430** in 78% yield, accompanied by elimination of CO.

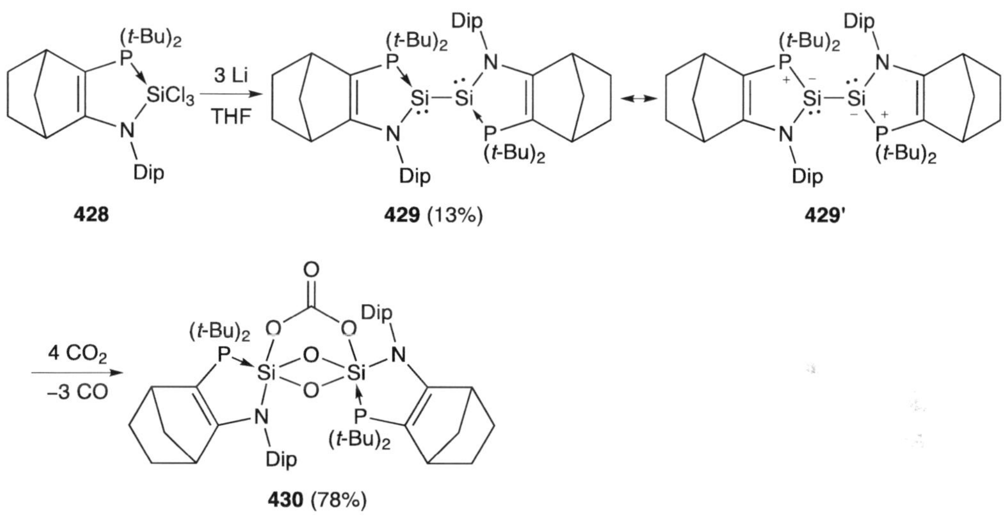

Scheme 8.172
Synthesis of bis(silylene) **429** and its reaction with CO_2.

Hydridosilylene **402a** does not dimerize, but hydridosilylene **431** that has a smaller substituent on the phosphorous atom undergoes reversible dimerization by Si(II)−H bond insertion to form **432** (Scheme 8.173(A)).[348] Similarly, in contrast to stable chlorosilylene **403a**, chlorosilylene **403b** that has a smaller substituent on the nitrogen atom is in an equilibrium with Si(II)−Cl insertion dimer **433** (Scheme 8.173(B)).

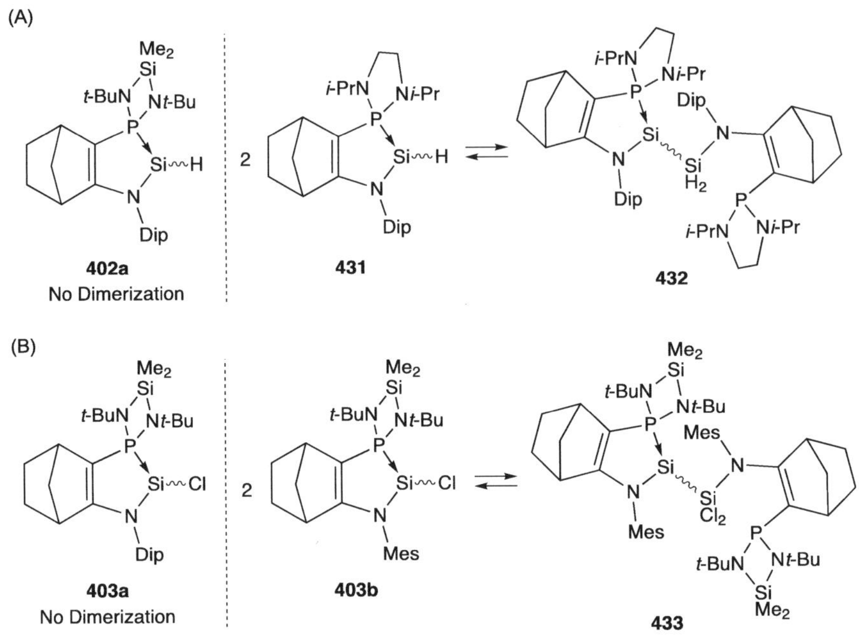

Scheme 8.173
Dimerization of silylenes **431** and **403b** generating new silylenes.

8.12 Intermolecularly Lewis Base-Stabilized Silylenes and Bis(silylenes)

8.12.1 Base-Stabilized Diarylsilylenes

As trailblazing work of silylenes stabilized by coordination of external bases, Tokitoh and Okazaki et al. have reported isocyanide-stabilized diarylsilylenes **6a**–**c** prepared by the reaction of transient diarylsilylene **165** generated by the dissociation of Si = Si double bond of the disilene **166** (Section 8.4.5) with isocyanides in 1997 (Scheme 8.174).[10,114] Although XRD analysis of **6a**–**c** has not been performed, DFT calculation of a model compound for

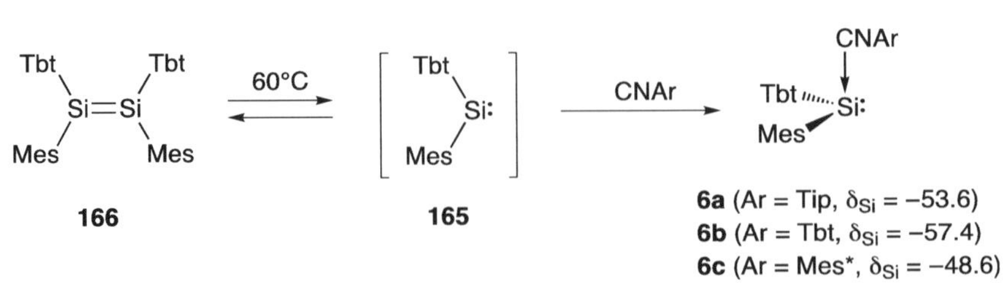

Scheme 8.174
Synthesis of isocyanide-stabilized diarylsilylenes.

6a–**c** exhibited a pyramidalized three-coordinate silicon center. Complexes **6a-c** reacts as diarylsilylene **165**. Similar to other transient silylenes, insertion of **165** into Si–H, C–H, C–C, C–I, C = C, O–H, H–Cl, B–B, B–H, B bonds, as well as [1 + 2] and [1 + 4] cyclizations, were reported.[10,114] Details of chemistry of **165** were summarized in a comprehensive review[18] and Section 8.4.5.

8.12.2 Base-Stabilized Halosilylenes

In 2008, Robinson et al. synthesized NHC stabilized bis(silylene) **434** as orange-red crystals by the reduction of ImDip-coordinated $SiCl_4$ (**435a**) with four equivalents of KC_8 in hexane accompanied by formation of $Si(0)_2$ species **436** as a minor product, the latter is obtained as a major product when **435a** was reduced with six equivalents of KC_8 in THF (Scheme 8.175).[116] XRD analysis showed that the Si–Si distance (2.393(3) Å) in **434** is within the Si–Si single bonds and the geometry around the Si atoms is considerably pyramidalized with the sum of the bond angles of 308.0 degrees (mean) (Fig. 8.16). The mean Si–C(ImDip) distance of 1.934(6) Å is comparable to that the starting material **435a** (1.928(2) Å). NBO analysis predicted that the model compound for **434** has a Si–Si bond orbital and lone pair orbitals on each silicon atom. These structural characteristics exhibited that **434** has a bonding situation as a carbene-coordinated bis (silylene).

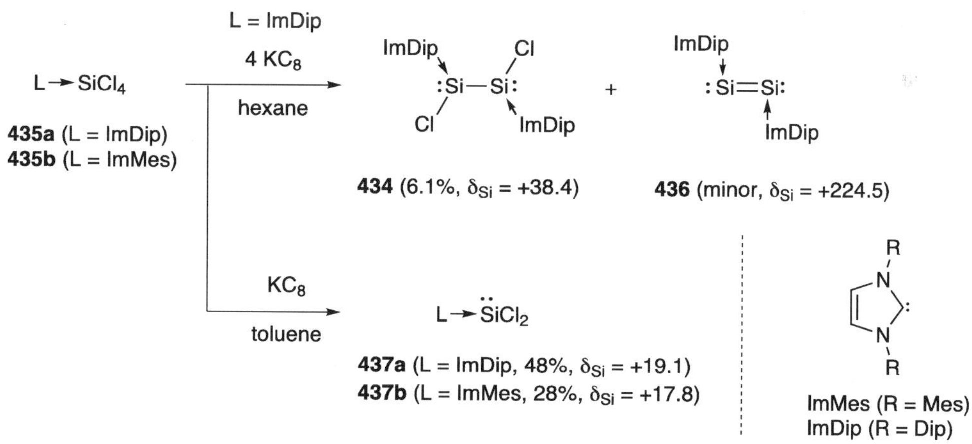

Scheme 8.175
Synthesis of halosilylenes **434**, **437a**, and **437b**.

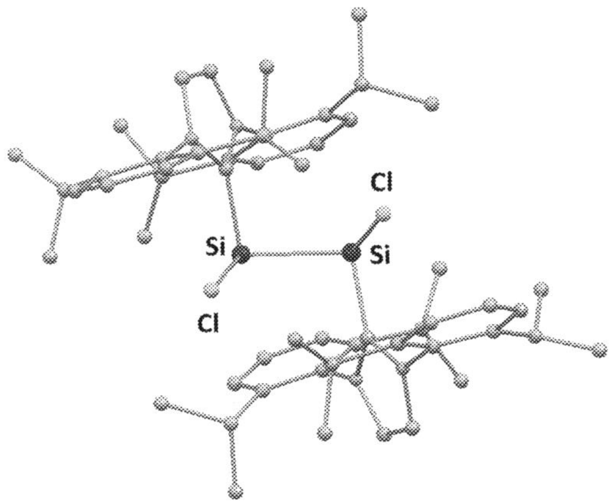

Figure 8.16
Molecular structure of **434** (CCDC: 686708).

In 2009, Roesky, Stalke et al. and Filippou et al. independently reported new tricoodinate Si (II) species stabilized by external coordination of NHC. Roesky, Stalke et al. reported that NHC-stabilized dichlorosilylenes **437a** and **437c** were synthesized by the reaction of two equivalents of ImDip or SImDip with trichlorosilane (Scheme 8.176)[117] or reductive dechlorination of tetrachlorosilane stabilized by ImDip (**435a**) or 1,3-bis(2,4,6-trimethylphenyl)imidazol-2-ylidene (ImMes) (**435b**) (Schcmc 8.175).[117] In the former reaction, hexachlorodisilane instead of trichlorosilane can also be used.[242] Filippou et al. have prepared dichlorosilylene with SImDip **437c** by similar method although **437c** was not obtained in a pure form due to its low solubility (Scheme 8.176).[166] Filippou et al. reported NHC-stabilized dibromosilylenes **438a**[118] and **438c**[166] synthesized by the reduction of **439a** and **439c**, which were prepared by the reaction of $SiBr_4$ with ImDip or SImDip, with two equivalents of KC_8 (Scheme 8.177). In a similar manner, diiodosilylene **440a** was obtained as yellow crystals.[349]

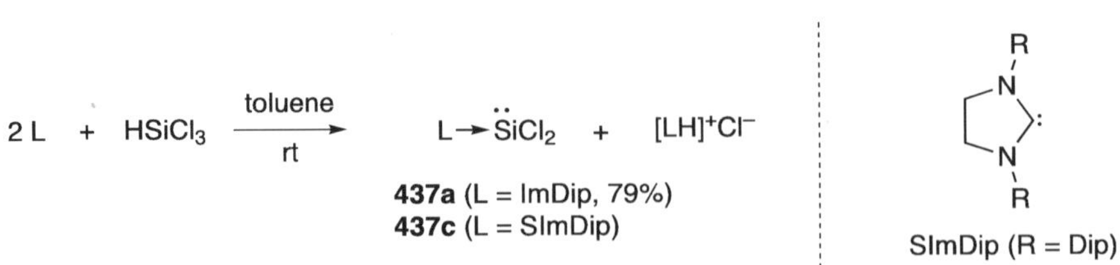

Scheme 8.176
Alternative synthesis of halosilylenes **437a** and **437c**.

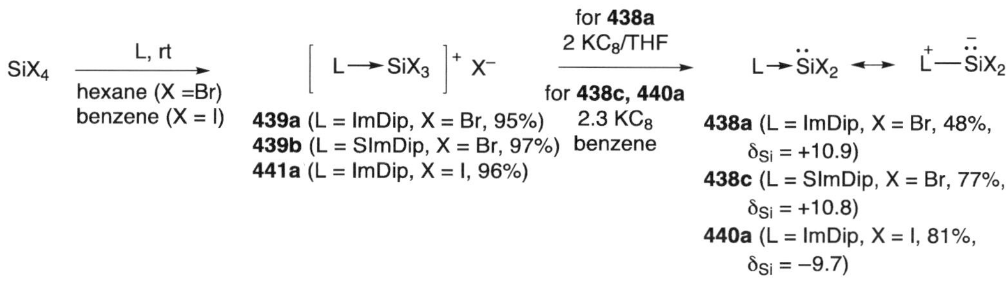

Scheme 8.177
Synthesis of halosilylenes **438a, 438c, and 440a**.

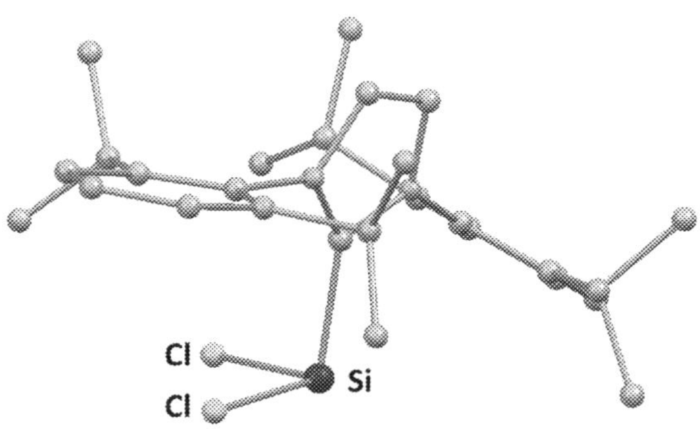

Figure 8.17
Molecular structure of **437a** (CCDC: 720236).

X-ray analysis of **437a** exhibited a highly pyramidalized geometry around the three-coordinate silicon atom with the sum of the bond angles of 289.7 degrees (Fig. 8.17).[117] The $Si(Cl_2)$–C(ImDip) distance is 1.985(2) Å. Bader's Atoms-in-Molecule (AIM) analysis suggested existence of lone-pair electrons on the three-coordinate silicon atoms. NHC-stabilized dibromosilylenes **438a** and **438c** also show similar structural characteristics.[118,166] The $Si(Br_2)$–C(carbene) distances of 1.989(3) Å (**438a**) and 2.007(5) Å (**438c**) and the sums of the bond angles around the Si atom of 292.7 degrees (**438a**) and 290(1) degrees (**438c**) suggesting that silicon atom uses mainly p-orbitals for the bonding. NBO analysis of **438a** predicted that the lone pair orbital existed on the $Si(Br_2)$ atom and the natural orbital of $Si(Br_2)$–C(ImDip) was highly filled (1.96) σ-orbital with donor-acceptor character resulting from the overlap of a filled $sp^{1.36}$ orbital of C(ImDip) (79.9%) and empty p-orbital of the Si atom (20.1%), supporting the bonding picture as a silylene-carbene adduct. The ^{29}Si NMR resonances due to three-coordinate silicon nuclei appear at +19.1, +17.8, +10.9, and +10.8 ppm for **437a**,[117] **437b**,[117] **438a**,[118] and **438c**,[166] respectively.

In 2010, Filippou et al. reported synthesis NHC-coordinated arylchlorosilylenes **442a,b** by dehydrochlorination of the corresponding dichlorohydridosilanes with two equivalents of ImMe (Scheme 8.178).[350] Both silylenes **442a,b** show structural characteristics similar to other NHC-coordinated halosilylenes: highly pyramidalized three-coordinate Si center (the sum of the bond angles: 299.2 degrees for **442b**), long bonds Si–C(carbene) (1.963(2) Å), Si–C(aryl) (1.937(2) Å), Si–Cl (2.1836(8) Å) for **442b**, compared to typical Si–C and Si–Cl distances of four-coordinate silicon compounds. These parameters suggest the silicon atom uses mainly p-orbitals for bonding to the substituents in **442b**, which is supported by the values of the ^1J(Si–C) coupling constants for **442a** (35 Hz (Si–C(carbene)), 48 Hz (Si–C(aryl)) which are smaller than the typical values for the Si–C(sp^2) bonds in silanes (64–70 Hz).

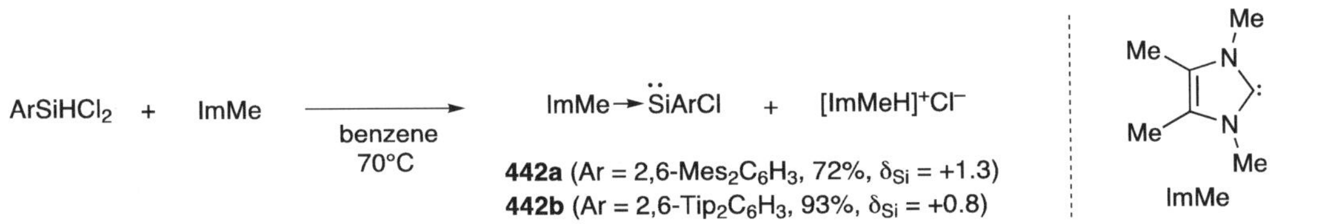

Scheme 8.178
Synthesis of arylhalosilylenes **442a,b**.

As another example of base-coordinated arylhalosilylenes, Matsuo, Tamao, et al. generated arylbromosilylenes **443a,b** by dissociation of the corresponding disilene **444** (a formal dimer of the silylene) observed as persistent species **445a,b** by coordination of 4-pyrrolidinopyridine (PPy) (Scheme 8.179).[351] The ^{29}Si NMR resonance of Si(II) nuclei appeared at +59.3 and 63.1 ppm for **445a** and **445b**, respectively. Bromosilylene **445a** was isolated as orange crystals and its structure was determined by XRD analysis. The coordination of PPy to the silicon center results in a trigonal-pyramidal geometry with the sum of the bond angles of 303.6 degrees and the Si–N(PPy) distance of 1.939(2) Å.

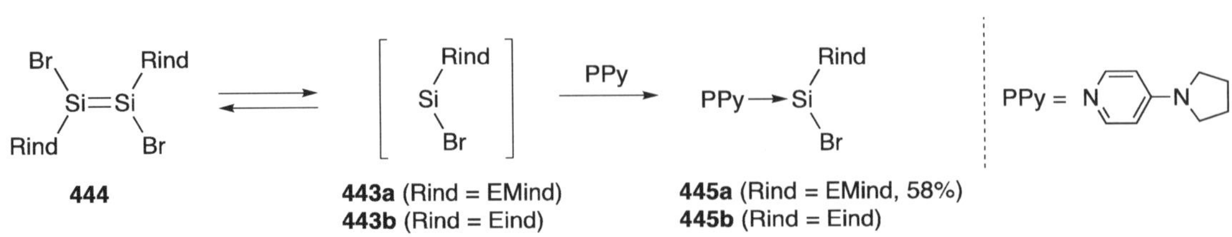

Scheme 8.179
Synthesis of arylhalosilylenes **445a,b**.

In 2015, Filippou et al. reported SImDip-coordinated bromosilylsilylene **446** obtained by the reaction of dibromosilylene **438c** with silylene TbbBrSi: generated by thermolysis of disilene TbbBrSi = SiBrTbb (Scheme 8.180A) or reaction of **438c** with LiTbb (Scheme 8.180B).[352] The proposed mechanism for formation of **446** involves generation of SImDip-coordinated TbbBrSi-$SiBr_2$ adduct **447** followed by bromine migration. The three-coordinate silicon center is highly pyramidalized with the sum of bond angles of 287.4 degrees and the Si–C(carbene) distance of 1.978(3) Å comparable to those of related carbene-coordinated bromosilylenes.

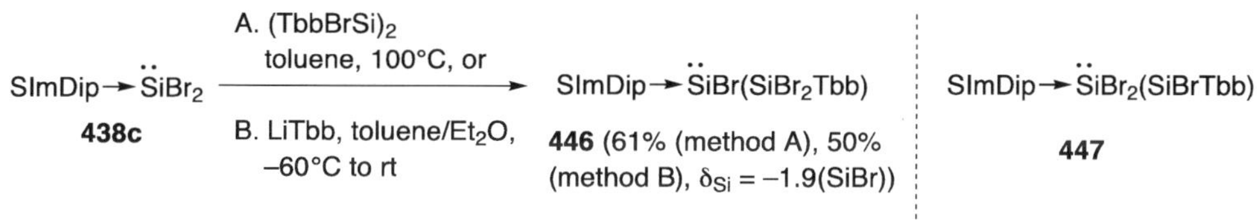

Scheme 8.180
Synthesis of halosilylsilylene **446**.

In 2011, Cui et al. reported NHC-stabilized aminochlorosilylene **448** bearing bulky Dip(Me_3Si)N group prepared by dehydrochlorination of aminodichlorosilane **449** with two equivalents of Im*i*-Pr in THF or reaction of 1,2-diaminotrichlorodisilane **450** with three equivalents of Im*i*-Pr in refluxing benzene (Scheme 8.181).[353] During latter reaction, disproportionation of **450** to aminochlorosilylene [Dip(Me_3Si)N]ClSi and **449** was proposed. Similar to other NHC-coordinated silylenes, highly pyramidalized geometry around three-coordinate silicon (the sum of the bond anlges of 305.2 degrees) and considerably long Si–C(carbene) distance (2.0023(19) Å) were observed.

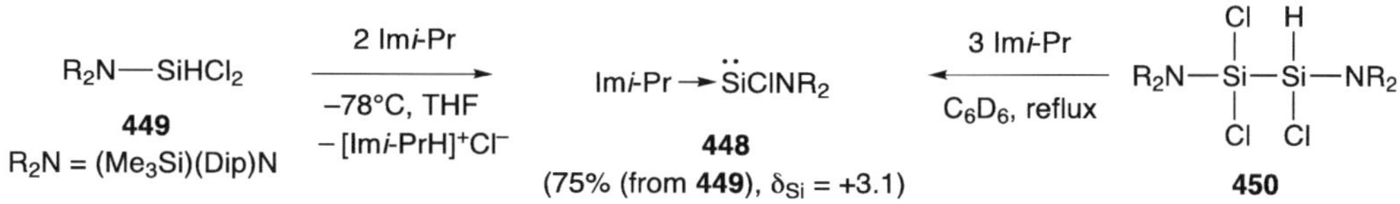

Scheme 8.181
Synthesis of amino(chloro)silylene **448**.

Another example of NHC-stabilized aminosilylene **451** was prepared by the metathesis of dichlorosilylene **437a** with lithium amide (Scheme 8.182).[354] Treatment of **451** with $LiBH_4$ gave the corresponding push–pull hydridosilylene (see Section 8.12.7).[354]

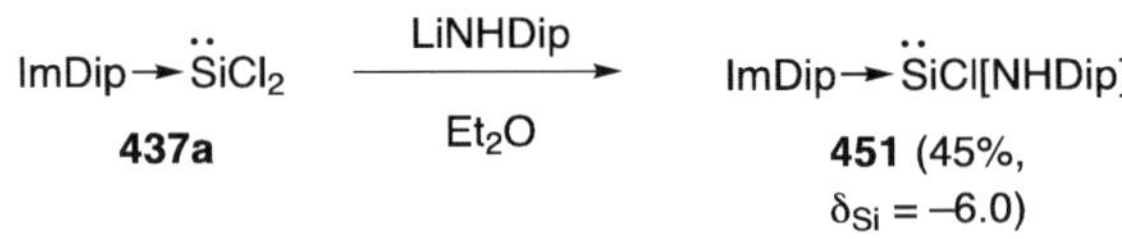

Scheme 8.182
Synthesis of aminohalosilylene **451**.

Silylene **437a** serves as a σ donor toward Lewis acids giving push–pull silylenes. For details, see Section 8.12.7.

Silylene **437a** undergoes cyclization reaction with various unsaturated compounds similar to other silylenes. For instance, **437a** reacted with excess diphenylacetylene to give trisilacyclopentene **452** as a colorless solid in 68% yield (Scheme 8.183), suggesting **437a** reacts as $SiCl_2$.[117] This is also supported by the reaction of **437a** with diarylketones giving the corresponding [1 + 2] cycloadducts **453a** and **453b**,[257] with *N*,*N*-dimesityl-1,4-diazabutadiene giving **454**,[355] and with benzyl giving [1 + 4] cycloadduct **455**.[257] Similarly

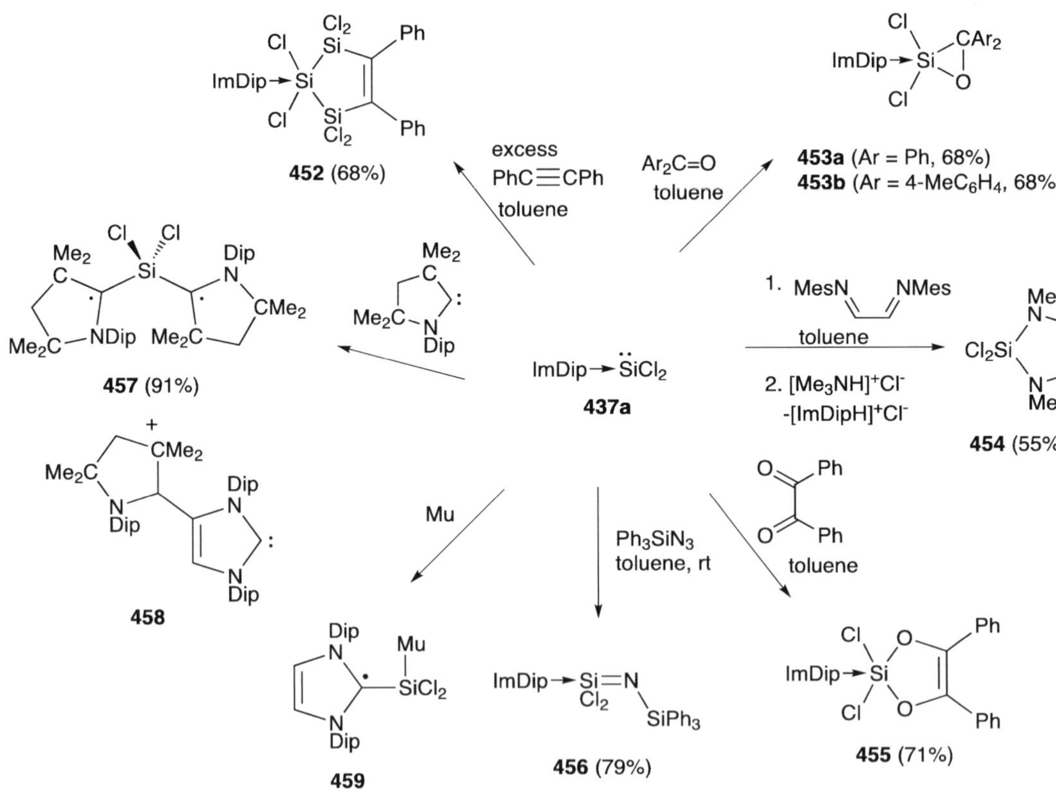

Scheme 8.183
Reactions of **437a**.

to other stable silylenes, silylene **437a** reacted with azide Ph_3SiN_3 giving the corresponding silanimine **456** in good yield.[124] When stronger Lewis base was treated with **437a**, Im*i*-Pr was replaced: cyclic alkylaminocarbene (CAAC) converts silylene **437a** to diradical **457** in 91% accompanied by formation of C–H insertion product of CAAC **458**.[356] In the solid state, diradical **457** have two different polymorphs: major polymorph with a diamagnetic closed-shell electronic configuration and minor polymorph showing paramagnetic contribution. Similar to other stable silylenes, muonium added to the divalent silicon center of **437a** to provide muoniated carbon-centered radical **459**.[268]

NHC-stabilized halosilylenes are good precursors for novel unsaturated silicon compounds. Silicon version of vinylidene is an important class of unsaturated silicon(II) species and an isomer of silicon-containing triply bonded compounds. Free disilavinylidene and related silenylidene are still unknown but NHC (SImDip) coordination allows their isolation in recent years as well as NHC-stabilized disilicon(0) (:Si = Si:)[116] and silagermenylidene (Tip_2Si = Ge:).[357] NHC-coordinated disilavinylidene **460** was isolated as a bright-red solid by the reduction of SImDip coordinated bromosilylene **446** with KC_8 (Scheme 8.184).[352] In **460**, the Si–Si bond length of 2.167(2) Å is suggestive of the double bond character. Two ^{29}Si NMR signals due to unsaturated silicon nuclei were observed at +86.0 ppm (Br side) and +34.6 ppm (SImDip side). Theoretical studies indicated that charge-separated canonical form **460′** is a major resonance hybrid.

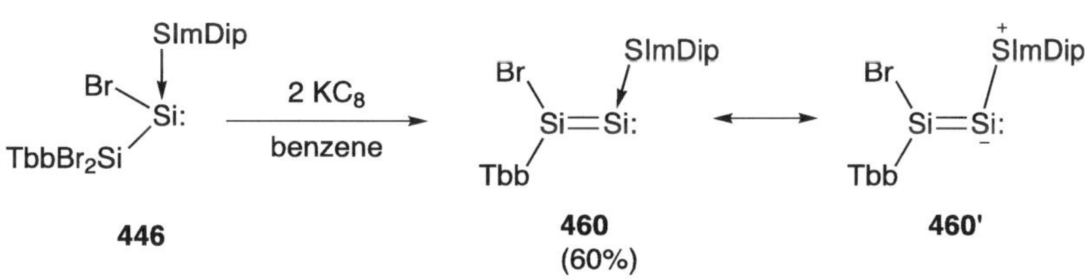

Scheme 8.184
Reduction of **446**.

Phosphasilenylidene, a heavier congener of isocyanide (RNC), is also obtained in NHC-stabilized form **461**. Treatment of **437a** with lithium silylphosphide LiPMes*($SiMe_3$) gave carbene-coordinated phosphasilenylidene **461** as bright orange crystals (Scheme 8.185).[358] Because **462** was observed during the reaction, **461** was formed by initial methathesis of **437a** and the lithium silylphosphide followed by elimination of chlorotrimethylsilane. **461** shows very downfield shifted ^{29}Si NMR signal at +267.3 ppm with $^1J_{PSi}$ = 170.4 Hz and small bond angle (C(ImDip)–Si–P) of 96.90(6) degrees. Computational studies suggest that charge-separated form **461′** is its major canonical structure.

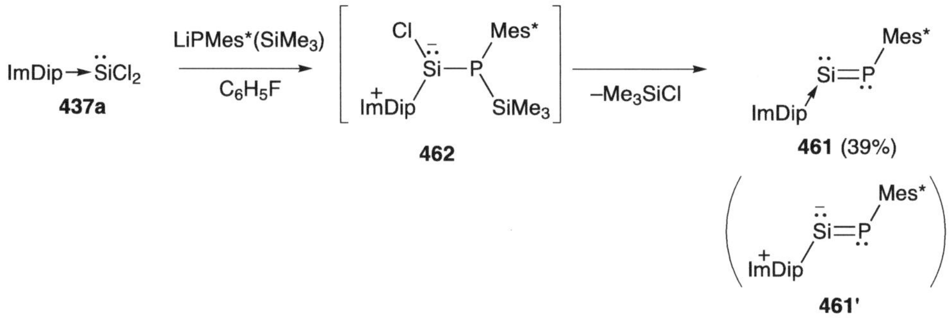

Scheme 8.185
Reaction of **437a** with lithium phosphide.

The iodides in the NHC-stabilized diiodosilylene **440a** are easily replaced by other NHC (Scheme 8.186).[349] Treatment of **440a** with 2.4 equivalents of carbene ImMe gave a dicationic Si(II) species **463** where ImDip and two I atoms are replaced by ImMe ligands. When more bulky carbene Im*i*-Pr was used, one I group was replaced to give NHC adduct of the iodosilyliumylidene cation **464**.

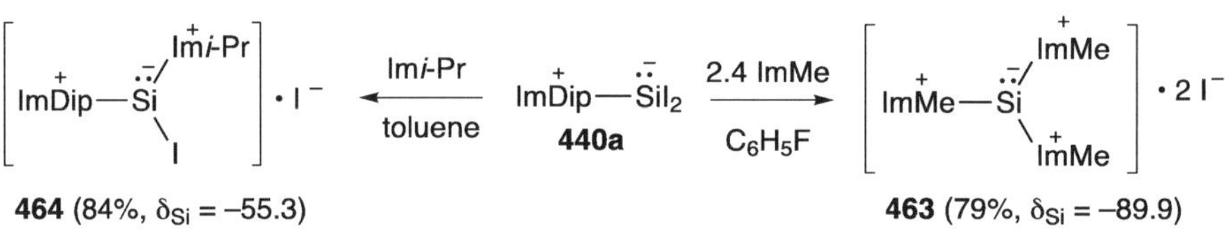

Scheme 8.186
Reaction of **440a** with carbenes.

Silylene **437a** behaves as a σ-donating ligand for transition metals. Silylene **437a** reacted with $Co_2(CO)_8$ to give cationic bis(silylene) cobalt complex **465** in 74% yield (Scheme 8.187).[359] X-ray analysis disclosed that **465** adopted a trigonal bipyramidal geometry with two silylene ligands occupying apical positions: the Si–Co distances are 2.2278(13) and 2.2276(12) Å, and the Si–Co–Si is 176.78(5) degrees. The IR spectrum of **465** measured in Nujol mull exhibited three bands due to carbonyl groups at 2052 (very weak), 1994 (strong), and 1969 (strong) cm^{-1}, which frequencies were lower than those of analogous complex containing $[Co(CO)_3(PR_3)_2]^+$. These results indicate that $SiCl_2$(ImDip) is a stronger electron donor than phosphine ligands. **437a** also reacted with $Ni(CO)_4$ to give bis(silylene) nickel complex **466** (Scheme 8.187).[296] The downfield shift of the ^{29}Si NMR resonance of **466** (δ_{Si} = +43.2) compared to **437a** (δ_{Si} = +19.1) was observed similarly to other transition metal complexes of silylene **437a**. X-ray analysis

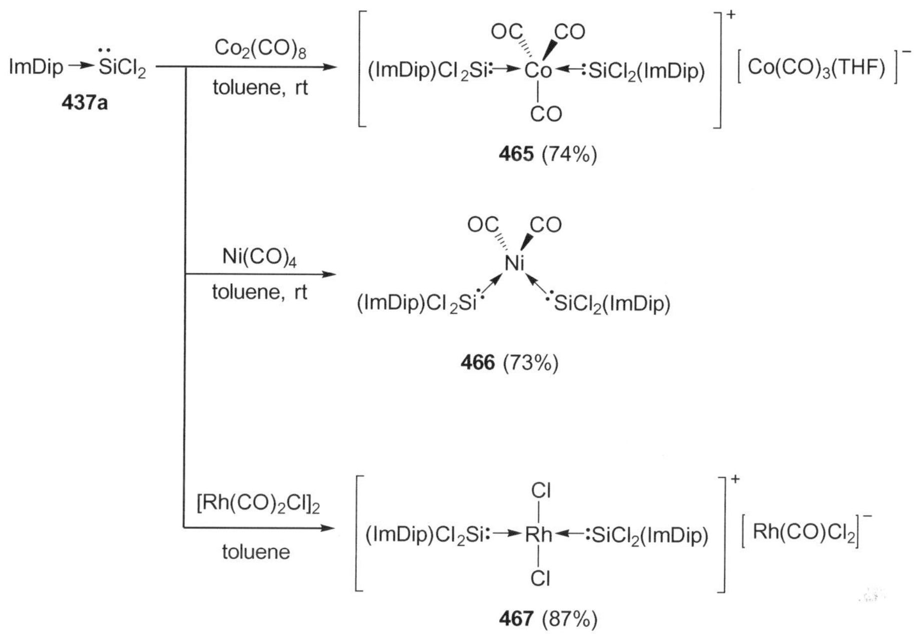

Scheme 8.187
Synthesis of transition metal complexes of silylene **437a**.

exhibits that **466** has a distorted tetrahedral geometry around the Ni center with two silylene **437a** and two carbonyl ligands. The Si–Ni distances of 2.1955(9) and 2.1854(8) Å in **466** are shorter than the sum of covalent radii of silicon (1.11 Å) and nickel (1.24 Å) suggesting π-back bonding within the Si–Ni bonds. The average Ni–C bond length of **466** (1.764(2) Å) is much shorter than that found in $Ni(CO)_4$ (1.817(2) Å).[297] **437a** also reacted with $[Rh(CO)_2Cl]_2$ to give bis(silylene) rhodium complex **467** as orange crystals in 87% yield (Scheme 8.187).[360]

Halosilylenes with SImDip ligand **438c** and **437c** react with $Li[CpCr(CO)_3]$ complex to give the corresponding halosilylidyne complexes **468** and **469** in moderate yield as the first examples of base-coordinated silylidyne complexes (Scheme 8.188).[166] XRD analysis showed the following structural characteristics: (1) planar three-coordinate silicon (the sum of bond angles: 359.9 degrees (**468**), 360.0 degrees (**469**)) and (2) considerably short Si–Cr bonds (2.1618(9) Å (**468**), 2.1603(7) Å (**469**)). The Si–Cr distances are much shorter than those for base-coordinated silylene complexes $[(LR_2Si)Cr(CO)_5]$ (2.33–2.52 Å). These structural characteristics suggest double bond character between Si and Cr atoms, which is supported by theoretical calculations (NBOs analysis). The ^{29}Si NMR resonance appeared at +95.1 (**468**) and +113.6 (**469**) ppm, that is considerably low-field shifted compared to those of starting materials **438c** (+10.9 ppm) and **437c** (+19.1 ppm).

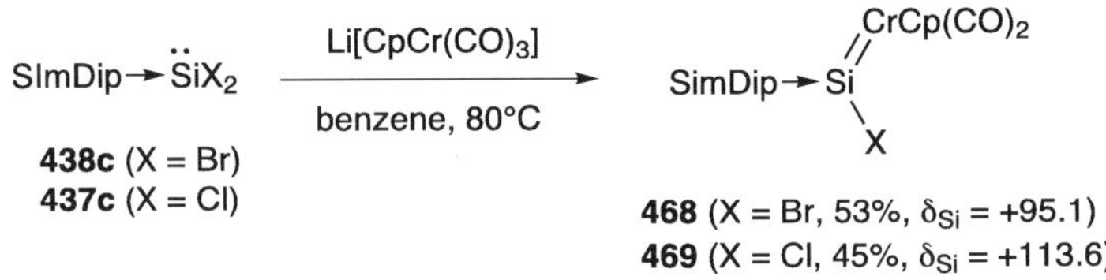

Scheme 8.188
Reaction of **438c** and **437c** with Li[CpCr(CO)$_3$] complex.

8.12.3 Base-Stabilized Hydridosilylenes

Inoue et al. have reported NHC-cooridinated monohydridosilylsilylene **470** prepared by dehydrochlorination reaction as another example of hydridosilylene (Scheme 8.189).[361] In the ^{1}H NMR spectrum, the signal due to proton bound to the silylene center appeared at 3.17 ppm. The ^{29}Si NMR resonance due to Si(II) nuclei was observed at −137.8 ppm. The ^{1}J(Si-H) coupling constant is 101.3 Hz, while ν(Si−H) is found at 1984 cm^{-1} as a sharp band. XRD analysis revealed the highly pyramidalized three-coordinate silicon atom with the sum of the bond angles of 290.84 degrees and the Si(silylene)−C(carbene) distance is 1.942(3) Å.

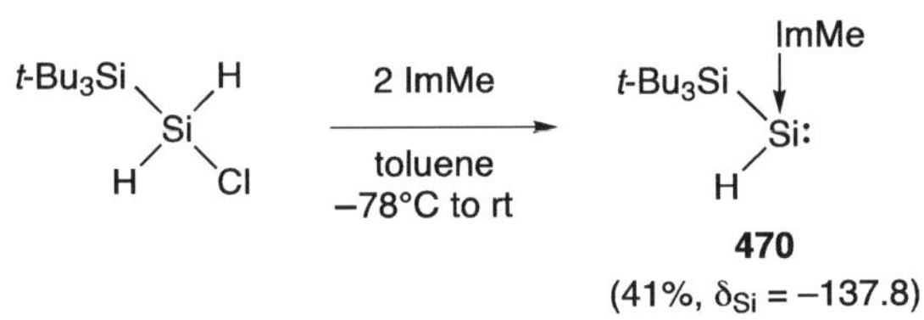

Scheme 8.189
Synthesis of hydridosilylene **470**.

As shown in Scheme 8.190, NHC-coordinated hydridosilylene **470** reacted with diphenylacetylene to afford silole **471**, while **470** reacted with phenylacetylene to provide 1-alkynyl-1-alkenylsilane **472** similarly to the formation of **478** from **476b** (vide infra, see Scheme 8.194).[362] Proposed mechanism for formation of **471** and **472** involves initial nucleophilic attack of silylenes to sp-carbon forming the corresponding vinyl anion. In the case the diphenylacetylene, the resulting vinyl anion attacks on the sp-carbon of another diphenylacetylene followed by cyclization and elimination of ImMe to provide **471**. Hydridosilylene **470** reacted with Ni(cod)$_2$ to afford 1,2-dihydrosilylene-nickel complex **473** showing metallacyclopropane character. As a possible mechanism for formation of **473**, stepwise coordination of **470** to the Ni center followed by stepwise 1,2-migration of ImMe from Si to Ni to give **473** was proposed and supported by DFT calculations.[361]

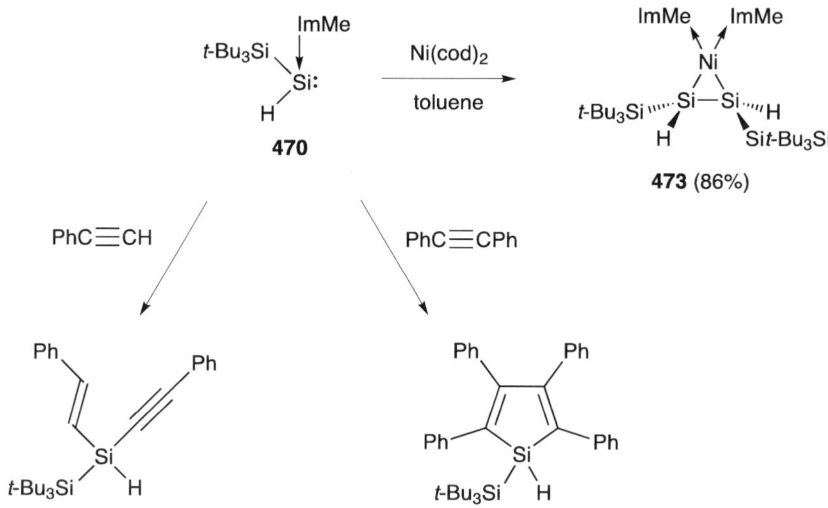

Scheme 8.190
Reactions of **470**.

8.12.4 Base-Stabilized Bis(silyl)silylenes

Sekiguchi et al. have synthesized NHC-coordinated bis(silyl)silylenes **474a** ($R_3Si = t\text{-}Bu_3Si$) and **474b** ($R_3Si = t\text{-}Bu_2MeSi$) in moderate yield by the reduction of the corresponding dibromosilane in the presence of ImMe (Scheme 8.191).[363] XRD analysis revealed the pyramidalized three-coordinate silicon with the sum of the bond angles of 344.3 degrees and the Si–C(carbene) distance is 1.933(4) Å. The ^{29}Si NMR resonance due to three-coordinate silicon nuclei appeared at −128.9 ppm, which is much upfield-shifted compared to those of NHC-coordinate silylenes and comparable to that for monohydridosilylsilylene **470** (Scheme 8.189) due to electropositive silyl substituents. The UV-Vis absorption spectrum of **474a** show a very weak absorption band assignable to $n \rightarrow \pi^*$ transition at 588 nm.

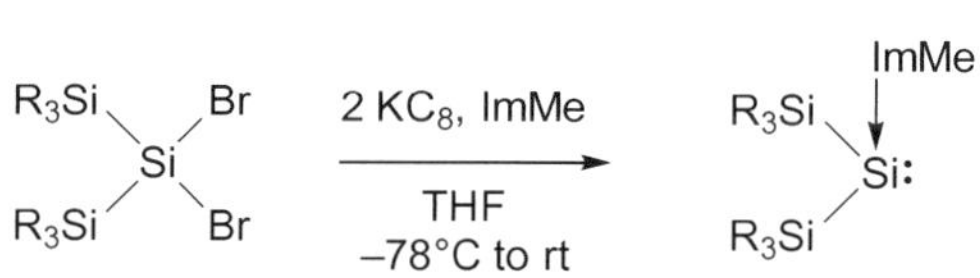

474a ($R_3Si = t\text{-}Bu_3Si$, 43%, $\delta_{Si} = -128.9$)
474b ($R_3Si = (t\text{-}Bu)_2MeSi$, 48%, $\delta_{Si} = -132.3$)

Scheme 8.191
Synthesis of bis(silyl)silylenes **474a** and **474b**.

Sekiguchi et al. have shown facile redox reactions of NHC-coordinated bis(silyl)silylene **474a** (Scheme 8.192).[363] Treatment of **474a** with $Ph_3C^+[BAr_4]^-$ (Ar = C_6F_4-4-$SiMe_2$*t*-Bu) affords the corresponding radical-cation **475** as yellow crystals in 86% yield. XRD analysis revealed that upon oxidation the geometry around the three-coordinate silicon center became almost planar (the sum of the bond angles is 359.8 degrees) and the Si–Si–Si angle was widened from 134.81(6) degrees (**474a**) to 144.96(4) degrees (**475**). The Si–Si–Si plane and NHC plane are nearly perpendicular to each other. The EPR spectrum of **475** exhibited a quintet signal ($g = 2.00466$) due to coupling with two ^{14}N nuclei ($I = 1$) with a hyperfine coupling constant (hfcc) $a(^{14}N) = 0.26$ mT. The signal is accompanied by the satellite signals due to ^{29}Si nuclei. The hfcc $a(^{29}Si)$ of 7.16 mT is smaller than those of trialkylsilyl radicals with silicon center with a pyramidalized geometry, which is consistent with the planar geometry around the silicon center. The EPR and X-ray analysis data, and DFT calculations indicate that **475** has a nearly sp^2-hybridized silicon center on which most of the positive charge is localized.

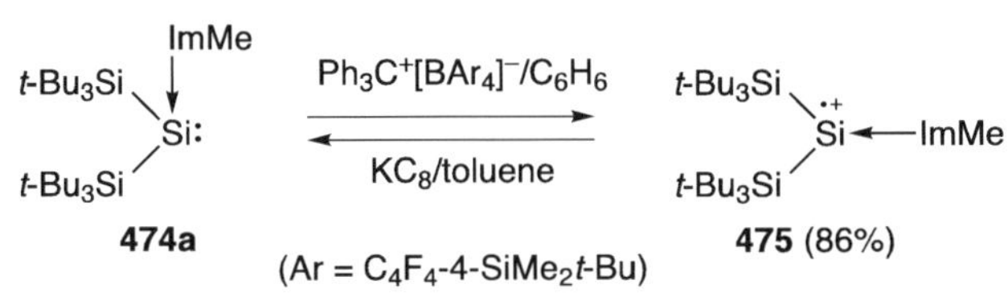

Scheme 8.192
Reactions of **474a**.

8.12.5 Base-Stabilized Carbocyclic Silylenes

In 2010, Cui et al. reported NHC-stabilized 1-silacyclopentadienylidenes **476a** ($\delta_{Si} = -48.6$) and **476b** ($\delta_{Si} = -43.6$) synthesized by the dehydrochlorination of the corresponding chlorotetraphenylsilole **477** with the corresponding NHC (Scheme 8.193).[364] The three-coordinate silicon center of **476b** is highly pyramidalized with the sum of the bond angles of 302.6 degrees. The Si–C(carbene) distance in **476b** (1.926(3) Å) is

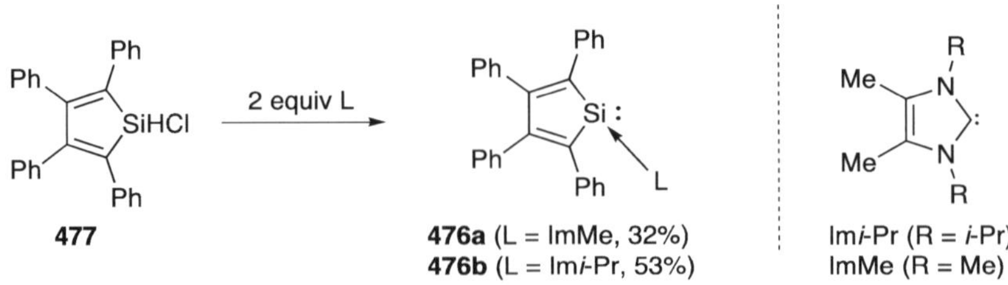

Scheme 8.193
Synthesis of carbocyclic silylenes **476a,b**.

considerably shorter than those in other NHC-coordinated silylenes but still longer than typical Si–C single bond (1.87 Å), suggesting highly electrophilic character of the free 1-silacyclopentadienylidene. The NBO analysis indicates that the Si–C(carbene) bond in **476b** can be described as a polarized donor-acceptor bond formed by the donation of the lone-pair of carbene to the vacant p-orbital of the Si atom.

Interestingly, silylene **476b** reacted with 2 equivalents of phenylacetylene to give 1-alkynyl-1-alkenylsilole **478** instead of typical [1 + 2] cycloadduct (Scheme 8.194).[364] The possible mechanism for formation of **478** involves that the initial nucleophilic attack of the **476b** to phenylacetylene to provide an alkenyl anion, which abstracts the terminal proton from the second phenylacetylene molecule followed by the attack of the resulting alkynyl anion to the silicon atom accompanied by release of NHC. Silylene **476b** reacted with aldehydes to give C = O bond cleavage products. Reaction of **476b** with *t*-BuCHO gave zwitterionic product **479**.[365] Formation of **480** and **481a–c** in the presence of $AlCl_3$ suggested the initial formation of the corresponding bicyclic silanone and ring expansion to give six-membered ring silanone. Although the initial step for the reaction of **476b** with aldehydes is unclear, initial formation of silaepoxide resulting from formal [1 + 2] cycloaddition followed by rearrangement giving the bicyclic silanone was proposed.

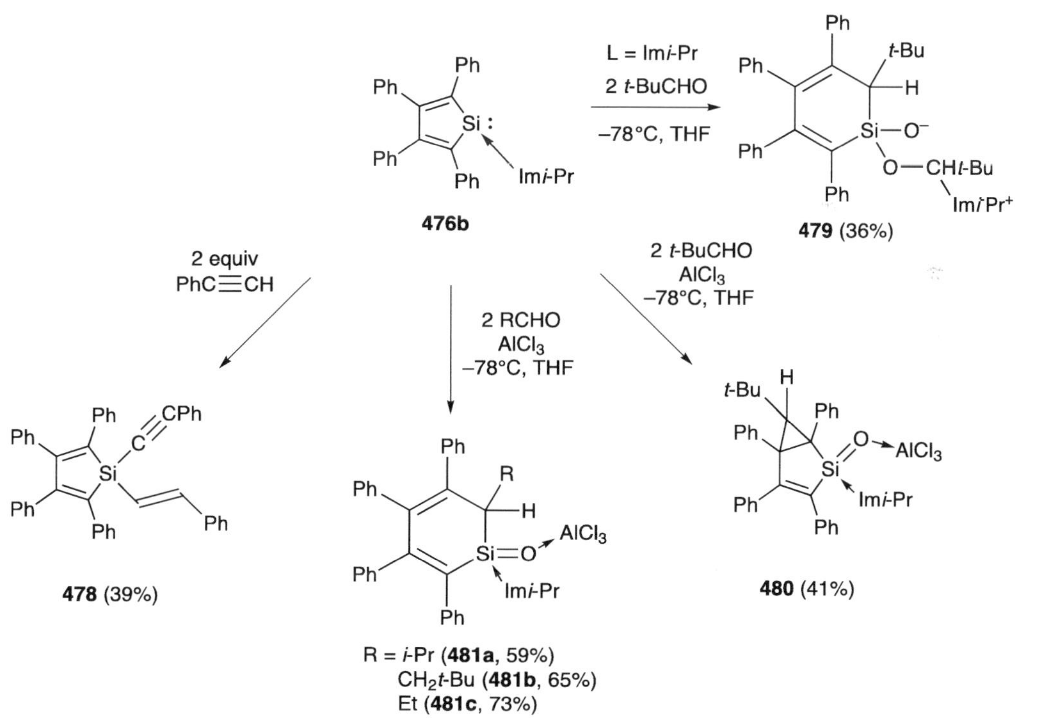

Scheme 8.194
Reactions of **476b**.

8.12.6 Base-Stabilized Diaminosilylene

Driess et al. have reported reactions of stable ylide-like diaminosilylene **73** with carbenes giving carbene-coordinate silylenes. For details, see Section 8.3.2.5.

8.12.7 Base- and Acid-Stabilized (Push–Pull) Silylenes

Robinson et al. reported unique route to Lewis base and Lewis acid-coordinated parent silylenes **482** and **483** as the first "push–pull" parent silylenes prepared from NHC-coordinated disilicon(0) **436** and BH_3(thf) in the presence of ImDip (Scheme 8.195).[366] In the absence of ImDip, only **482** was obtained in high yield (72%). **482** has two silylene units, SiH_2 coordinated by ImDip and silylborane, and borylhydridosilylene coordinated ImDip and B_3H_7, while **483** has ImDip and BH_3 coordinated silylene unit in the three-membered ring. Both compounds **482** and **483** were isolated as colorless crystals. X-ray analysis of **482** showed that the (ImDip)H_2Si:$\rightarrow$$BH_2$, H_2B–SiH(ImDip)(B_3H_7) and Si:$\rightarrow$ B_3H_7 distances are 1.934(4), 1.944(4) and 1.938(3) Å. The dative bond character for (ImDip)H_2Si:$\rightarrow$$BH_2$ and Si:$\rightarrow$$B_3H_7$ bonds was supported by the natural charge (ca. +0.93 for Si and −0.97 for B) DFT calculations.

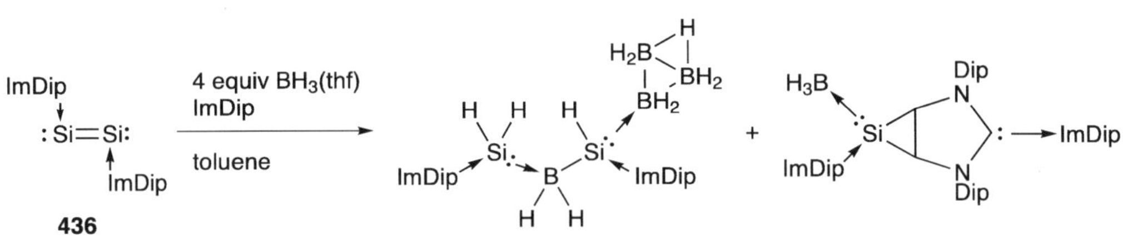

Scheme 8.195
Synthesis of "push–pull" silylenes **482** and **483**.

Reaction of **437a** with $B(C_6F_5)_3$ gave silylene-borane adduct **484** in 65% yield as colorless crystals (Scheme 8.196).[367] Upon coordination of the borane, the ^{29}Si NMR resonance of the central silicon nucleus (δ_{Si} = −53.2) is 72.3 ppm upfield-shifted from that of **437a** (δ_{Si} = +19.1). X-ray analysis showed that **484** has a tetrahedral geometry around the Si atom with the C→Si and averaged Si–Cl bond distances of 1.965(5) and 2.06195(19) Å, which are slightly shorter and noticeably shorter than those of **437a** (1.985(4) and 2.1664 (16) Å). The Si→B distance of 2.1135(6) Å is in the range of typical Si–B distances in silylboranes. The QTAIM analysis suggested ionic and covalent characters of C→Si and Si→B bonds, respectively.[367] Similar borane adduct **485** was also obtained by the reaction of **437a** with $LiBH_4$ or BH_3(thf).[368] Although structural characteristics found in **485** are

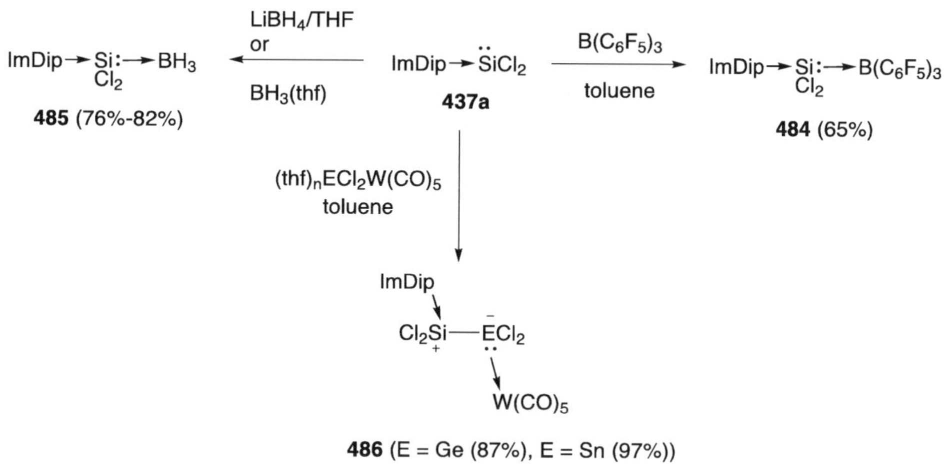

Scheme 8.196
Synthesis of "push–pull" silylenes from **437a**.

similar to those in **484**,[367] Si–B distance (1.965(2) Å) was shorter than that of **484**[367] and the ^{29}Si NMR resonance of the central silicon nucleus ($\delta_{Si} = +30.7$) was downfield-shifted compared to that of **484** ($\delta_{Si} = -53.2$)[367] and **437a** ($\delta_{Si} = +19.1$).[117] Reaction of **437a** with $(thf)_nECl_2 \rightarrow W(CO)_5$ complex (E = Ge, Sn) gave the corresponding tetrachlorogermasilene and tetrachlorstannasilene complexes **486**, which are good precursors for ImDip and W $(CO)_5$-coordinated parent germasilene and parent stannasilene.[369]

Mild reduction of **485** with $LiAlH_4$ in toluene-ether mixture gave the corresponding dihydridosilylene **487** in 55% yield (Scheme 8.197).[360] The NBO analysis of model compound $ImMe_2SiH_2BH_3$ ($ImMe_2 = [(HCNMe)_2C:]$) indicated the presence of polar dative C(carbene)–Si interaction and a relatively nonpolar Si–B single bond. When **487** was treated with $W(CO)_5(thf)$, the BH_3 group on silicon atom in **487** was replaced by $W(CO)_5$ to give the corresponding tungsten complex **488** in 66% yield.[360] The ν(CO) stretching frequency of the CO group trans to SiH_2 (2044 cm^{-1}) is slightly lower than the related stretching frequencies in Im*i*-Pr · EH_2 · $W(CO)_5$ (2047 cm^{-1} for E = Ge and Sn) suggesting that ImDip · SiH_2 is a marginally stronger donor than ImDip · EH_2 · $W(CO)_5$ (E = Ge, Sn).[370]

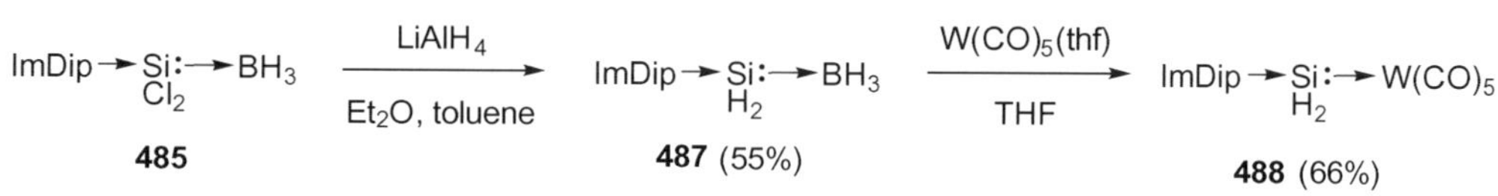

Scheme 8.197
Reactions of **485**.

Treatment of **451** with $LiBH_4$ gave the corresponding "push–pull" hydridosilylene coordinated additionally by BH_3 (**489**) (Scheme 8.198).[354]

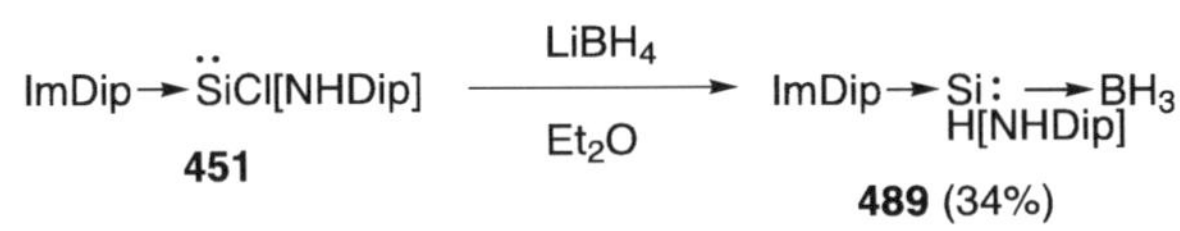

Scheme 8.198
Synthesis of "push–pull" silylene **489**.

8.13 Decamethylsilicocene and Its Derivatives

Decamethylsilicocene **1a** was synthesized by the reduction of dichlorosilane **490** with metal naphthalenides (MNap, M = Li, Na, K)[6] or decamethylsamarocenes ($Cp^*{}_2Sm$, $Cp^*{}_2Sm(THF)_2$) (Scheme 8.199).[371] Since decamethylsilicocene **1a** is a milestone of the silylene chemistry, its structural and spectroscopic character have been extensively studied and well established.[372] This section focused on pentamethylcyclopentadienyl Si(II) cation, Cp^*Si^+ (**491a**), and related recent topics of Si(II) species derived from **491a**. A recent review on this topic is also available.[373]

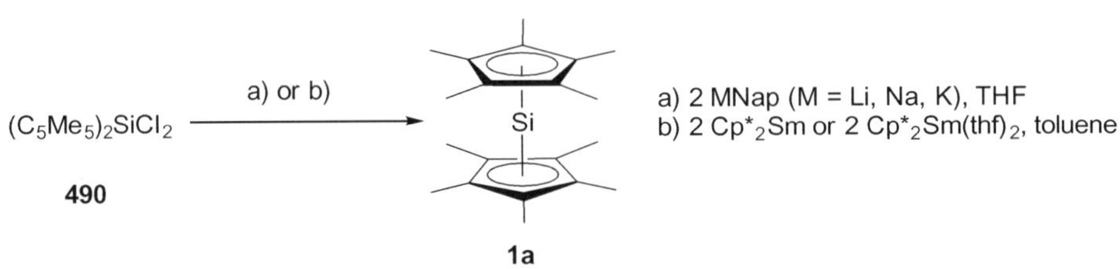

Scheme 8.199
Synthesis of silicocene **1a**.

One of the Cp^* ligands can be removed by protonated pentamethylcyclopentadiene to form Cp^*Si^+ (**491a**) salt (Scheme 8.200).[374] X-ray structural analysis, NMR spectroscopy, and theoretical calculations exhibit that cation **491a** is a π-complex of the Si(II) η^5-coordinated to the Cp^* ligand and can be regarded as a Cp^* derivative of silyliumylidene cation (HSi^+).

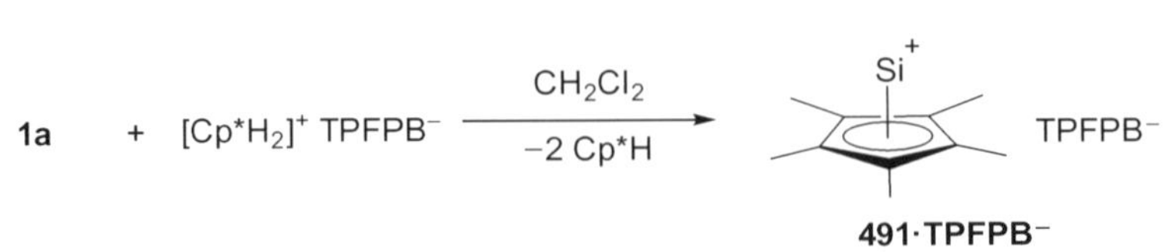

Scheme 8.200
Elimination of a Cp^* ligand from **1a** giving cation **491a**·**TPFPB**$^-$.

Silicocenes with different substitution pattern on the cyclopentadienyl ligands, and related unsaturated silicon compounds, are derived from the reactions of **491a · TPFPB⁻** with nucleophiles (Scheme 8.201).[375] Reactions of **491a · TPFPB⁻** with lithium cyclopentadienides gave new silicocene derivatives **1b–f**. Selective removal of Cp* ligand of **1b** by proton transfer reagent produced new Si(II) cation **491b**. The reaction of **491a · TPFPB⁻** with 1,2,4-trimethylcyclopentadienide provided a mixture of three silicocenes (**1a**, **1d**, **1e** with an approximate ratio of 1:8:1). The smallest silicocene **1f** is thermally labile and decomposes above −30°C. These structures are confirmed by the characteristic upfield shifted ^{29}Si NMR signals of central silicon atoms (−310 ~ −420 ppm). The ^{1}H and ^{13}C NMR signals indicate facile ring rotation of η^5-cyclopentadienyl groups in silicocenes and/or haptotropic shifts.

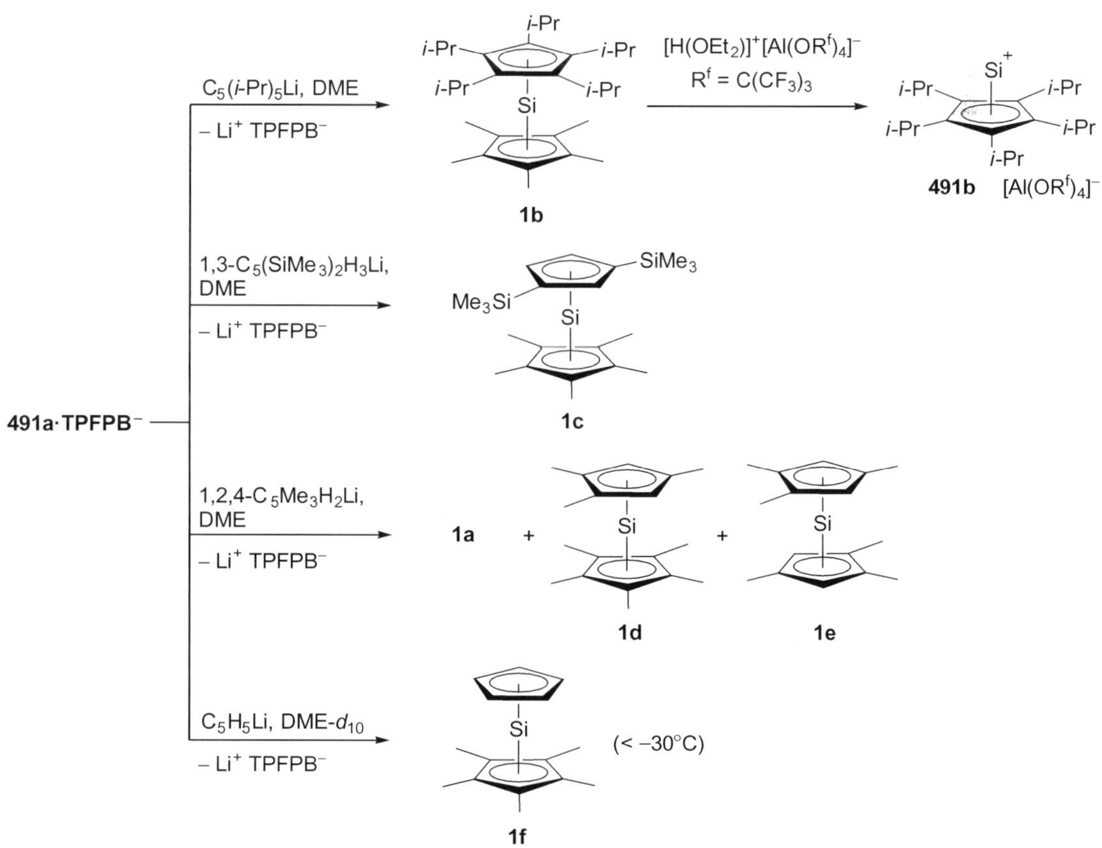

Scheme 8.201
Reactions of **491a · TPFPB⁻** with lithium cyclopentadienides.

Reaction of **491a · TPFPB**$^-$ salt with $LiN(SiMe_3)_2$ in DME at −78°C gave new silylene **492** in 20%–50% yield (yield based on the amount of disilene **493**, vide infra) (Scheme 8.202).[374,376] Temperature is crucial in this reaction; when the reaction was conducted at −40°C, compound **494** formed as a formal O–Me bond insertion product of **492** with DME. Silylene **492** underwent cycloaddition with 2,3-dimethyl-1,3-butadiene to give **495** quantitatively.

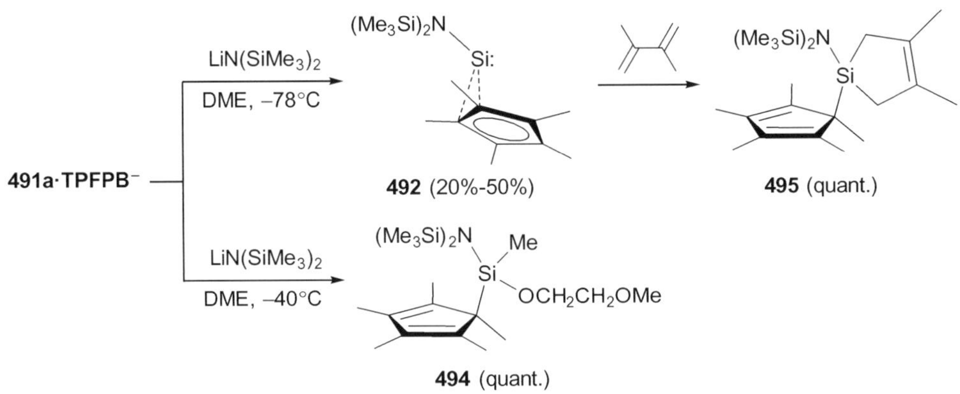

Scheme 8.202
Reactions of **491a · TPFPB**$^-$ with lithium bis(trimethylsilyl)amide.

Alternatively, η^1-Cp*$SiHCl_2$ with two equivalents of $KN(SiMe_3)_2$ in toluene afforded **492** in an improved yield (68% as disilene **493**) (Scheme 8.203).[377] Potassium amide acts as both base and nucleophile. The corresponding sodium amide was less effective (31% as disilene **493**), and lithium amide gave tricyclic compound **496** in 55% yield instead of **492**.

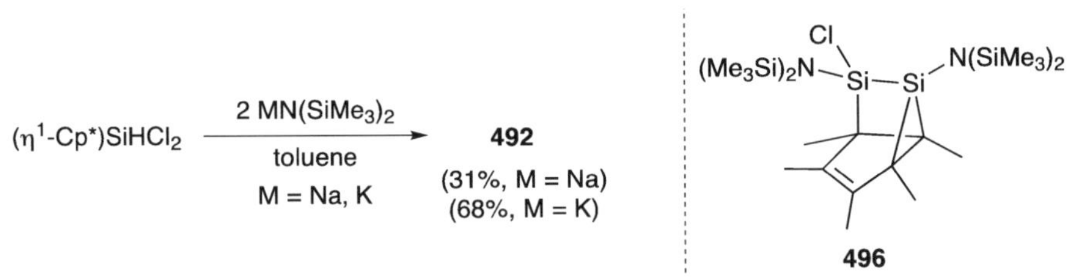

Scheme 8.203
Alternative synthesis of **492**.

Silylene **492** was isolated as a colorless viscous oil and its unsaturated ^{29}Si nucleus resonated at −10.2 ppm in C_6D_6. Five methyl groups and sp^2-ring carbon nuclei in Cp* ligand of **492** are equivalent in NMR spectra (^1H NMR; 1.90 ppm for methyl protons, ^{13}C NMR; 10.9 and 120.4 ppm for Me and sp^2-C). These characteristics of NMR spectra indicate the fast fluxional behavior of **492** on the NMR timescale. Theoretical study suggested the optimized structure of **492** with η^2-coordination of Cp* ligand and facile haptotropic shift, which is consistent with the observed NMR spectra. Interestingly, it dimerizes upon crystallization to form disilene **493** as deep-yellow crystals. Compound **493** dissociates to **492** by dissolving in solvents (Scheme 8.204). This "phase-dependent transformation"[378–382] is different from equilibrium between silylenes and the corresponding disilenes. Change of Cp* hapticity (η^2- to η^1-) upon dimerization would play an important role in this phenomenon. Reactions of **492** with N_2O and elemental sulfur afforded 1,3-dioxa-2,4-disilacyclobutane and 1,3-dithia-2,4-disilacyclobutane derivatives with η^1-Cp* ligands.[383]

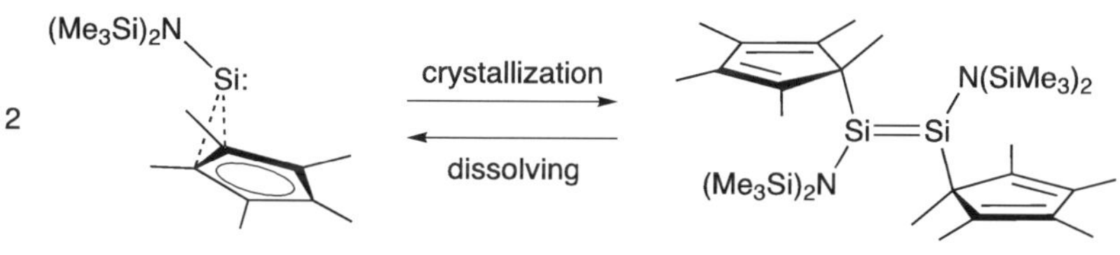

492 **493**

Scheme 8.204
Dimerization of **492**.

Treatment of **491a · TPFPB⁻** with $Na^+[Fe(CO)_2Cp^*]^-$ in a mixed solvent of hexane and Et_2O (50:1) provided ferriosilylene **497** in 48% yield (Scheme 8.205).[384] XRD study of **497** revealed that one of the Cp* ligands is coordinated to Si(II) center in η^3-fashion. The Fe–Si bond of 2.3677(6) Å is longer than that of silyliron complex $Cp^*(CO)_2Fe(SiH_3)$ (2.287(2) Å).[385] The ^{29}Si NMR spectrum displayed Si(II) signal of **497** at +316.7 ppm. Thermal intramolecular C–H bond insertion of silylene **497** proceeded quantitatively to give **498** with a (η^1-Cp*)–Si bond.

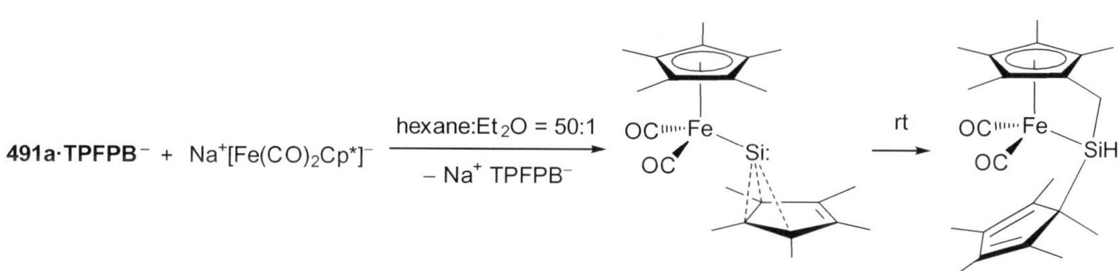

497 (48%) **498** (100%)

Scheme 8.205
Synthesis of ferriosilylene **497** from **491a · TPFPB⁻**.

Silylene **499** was synthesized by the reaction of **491a · TPFPB**$^-$ with a bulky lithium reagent and isolated as colorless crystals in 57% yield (Scheme 8.206).[386] XRD analysis revealed its structure with η^2-coordinated Cp* ligand to Si(II) center. The Si–N bond in **499** of 1.691(5) Å is shorter than that of aminosilylenes (1.696(3) – 1.774(4) Å),[9,41–43,45,88,138,140] suggesting stronger electron donation from nitrogen to silicon atom. The ^{29}Si NMR signal of divalent silicon in **499** appeared at −43.8 ppm. DFT calculations including population analysis suggest that the iminosilylene character plays a dominant role in **499**. Treatment of **499** with $B(C_6F_5)_3$ gave adduct **500** in 89% yield. The NMR spectra, XRD analysis, and theoretical study indicate that **500** has a significant 1-sila-2-azaallene character.

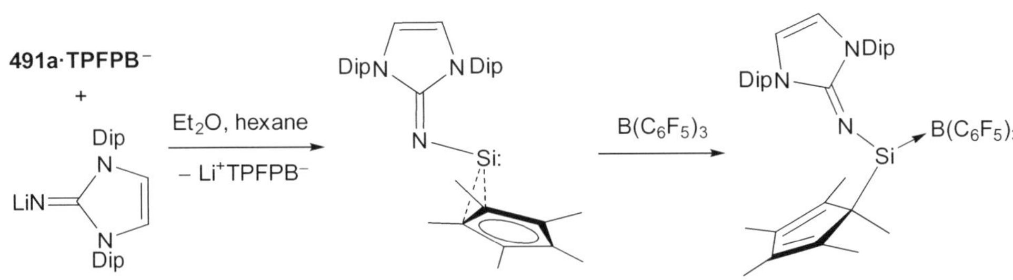

499 (57%)

500 (89%)

Scheme 8.206
Synthesis of silylene **499**.

Reaction of **491a · TPFPB**$^-$ with a bulky aryllithium ($Ar^{iPr6}Li \cdot Et_2O$) led to the first stable arylsilylene **501** as yellow crystals in 81% yield (Scheme 8.207).[387] Silylene **501** was fully characterized. XRD study revealed its structure with η^3-coordination of Cp* ligand. The ^{29}Si NMR signal owing to Si(II) atom was observed at 51.6 ppm.

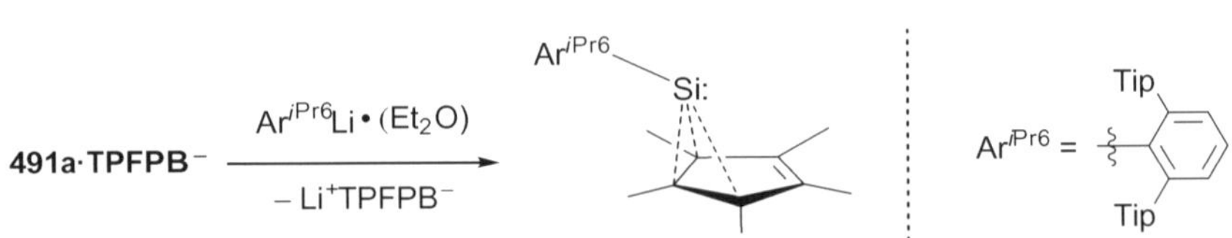

501 (81%)

Scheme 8.207
Synthesis of silylene **501**.

8.14 Conclusions

Chemistry of silylenes was dramatically developed since the Gaspar and West's review in 1998. Isolation of the stable silylenes with various substituents and coordination modes show that the electronic structures and reactivity of silylenes become tunable to some extent. Synthesis and reactivity of transition metal complexes with stable silylene ligands, especially *N*-heterocyclic silylenes, demonstrated that the stable silylenes have great potential as useful ligands for transition metal catalysts, similarly to the *N*-heterocyclic carbenes. Although this chapter mainly focused on the stable silylenes, various experimental and theoretical studies of transient silylenes and their transition metal complexes have also demonstrated the dramatic progress in the chemistry of silylenes. New findings of synthetic methods, structures, and diverse reactivity of stable silylenes should contribute to further development of not only new functional silylenes but also new silylene-based catalysts and low-coordinate species of the heavy group 14 and other main group elements.

References

1. Gaspar, P. P.; West, R. In The Chemistry of Organic Silicon Compounds; Rappoport, Z., Apeloig, Y., Eds.; Vol. 2; John Wiley & Sons, Ltd: Chichester, UK, 1998; pp 2463–2568.
2. Weidenbruch, M. Silylenes and Disilenes: Examples of Low Coordinated Silicon Compounds. *Coord. Chem. Rev.* **1994,** *130*, 275–300.
3. West, R.; Denk, M. Stable Silylenes: Synthesis, Structure, Reactions. *Pure Appl. Chem.* **1996,** *68*, 785–788.
4. Tokitoh, N.; Okazaki, R. Recent Topics in the Chemistry of Heavier Congeners of Carbenes. *Coord. Chem. Rev.* **2000,** *210*, 251–277.
5. Haaf, M.; Schmedake, T. A.; West, R. Stable Silylenes. *Acc. Chem. Res.* **2000,** *33*, 704–714.
6. Jutzi, P.; Kanne, D.; Krüger, C. Decamethylsilicocene - Synthesis and Structure. *Angew. Chem. Int. Ed. Engl.* **1986,** *25*, 164–264.
7. Karsch, H. H.; Keller, U.; Gamper, S.; Müller, G. $Si[(Me_2P)_2C(SiMe_3)]_2$, A Stable σ-Bonded Compound Containing Divalent Silicon. *Angew. Chem. Int. Ed. Engl.* **1990,** *29*, 295–296.
8. Denk, M.; Lennon, R.; Hayashi, R.; West, R.; Belyakov, A. V.; Verne, H. P., et al. Synthesis and Structure of A Stable Silylene. *J. Am. Chem. Soc.* **1994,** *116*, 2691–2692.
9. Gehrhus, B.; Lappert, M. F.; Heinicke, J.; Boese, R.; Bläser, D. Synthesis, Structures and Reactions of New Thermally Stable Silylenes. *J. Chem. Soc., Chem. Commun.* **1995,** 1931–1932.
10. Takeda, N.; Suzuki, H.; Tokitoh, N.; Okazaki, R.; Nagase, S. Reaction of a Sterically Hindered Silylene With Isocyanides: The First Stable Silylene – Lewis Base Complexes. *J. Am. Chem. Soc.* **1997,** *119*, 1456–1457.
11. Kira, M.; Ishida, S.; Iwamoto, T.; Kabuto, C. The First Isolable Dialkylsilylene. *J. Am. Chem. Soc.* **1999,** *121*, 9722–9723.
12. Zybill, C.; Müller, G. Synthesis and Structure of $[(OC)_4Fe = Si(OtBu)_2 \cdot HMPT]$, A Donor-Stabilized Silanediyl ("Silylene") Complex. *Angew. Chem. Int. Ed. Engl.* **1987,** *26*, 669–670.
13. Straus, D. A.; Tilley, T. D.; Rheingold, A. L.; Geib, S. J. Preparation, Characterization, and X-Ray Crystal Structure of An Acetonitrile-Complexed Ruthenium Silylene. *J. Am. Chem. Soc.* **1987,** *109*, 5872–5873.
14. Ueno, K.; Tobita, H.; Shimoi, M.; Ogino, H. Synthesis, Characterization, and X-Ray Crystal Structure of A Donor-Stabilized Bis(silylene)iron Complex. *J. Am. Chem. Soc.* **1988,** *110*, 4092–4093.

15. Gehrhus, B.; Lappert, M. F. Chemistry of Thermally Stable Bis(amino)silylenes. *J. Organomet. Chem.* **2001,** *617–618*, 209–223.
16. Tokitoh, N.; Ando, W. Silylenes (and Germylenes, Stannylenes, Plumbylenes). In *Reactive Intermediate Chemistry;* Moss, R. A., Platz, M. S., Jones, M., Eds.; John Wiley & Sons, Inc: Hoboken, NJ, 2004; pp 651–715.
17. Hill, N. J.; West, R. Recent Developments in the Chemistry of Stable Silylenes. *J. Organomet. Chem.* **2004,** *689*, 4165–4183.
18. Takeda, N.; Tokitoh, N. A Bulky Silylene Generated Under Mild Conditions: Its Application to the Synthesis of Organosilicon Compounds. *Synlett* **2007,** 2483–2491.
19. Mizuhata, Y.; Sasamori, T.; Tokitoh, N. Stable Heavier Carbene Analogues. *Chem. Rev.* **2009,** *109*, 3479–3511.
20. Mandal, S. K.; Roesky, H. W. Interstellar Molecules: Guides for New Chemistry. *Chem. Commun.* **2010,** *46*, 6016–6041.
21. Yao, S.; Xiong, Y.; Driess, M. Zwitterionic and Donor-Stabilized *N*-Heterocyclic Silylenes (NHSis) for Metal-Free Activation of Small Molecules. *Organometallics* **2011,** *30*, 1748–1767.
22. Sen, S. S.; Khan, S.; Samuel, P. P.; Roesky, H. W. Chemistry of Functionalized Silylenes. *Chem. Sci.* **2012,** *3*, 659–682.
23. Sen, S. S.; Khan, S.; Nagendran, S.; Roesky, H. W. Interconnected Bis-Silylenes: A New Dimension in Organosilicon Chemistry. *Acc. Chem. Res.* **2012,** *45*, 578–587.
24. Roesky, H. W. Chemistry of Low Valent Silicon. *J. Organomet. Chem.* **2013,** *730*, 57–62.
25. Ghadwal, R. S.; Azhakar, R.; Roesky, H. W. Dichlorosilylene: A High Temperature Transient Species to An Indispensable Building Block. *Acc. Chem. Res.* **2013,** *46*, 444–456.
26. Tokitoh, N.; Sasamori, T. Low-Coordinate Main Group Compounds – Group 14 (Si, Ge). In *Comprehensive Inorganic Chemistry II;* Reedijk, J., Poeppelmeier, K., Eds.; . *2nd ed.* Elsevier, 2013; pp 567–577.
27. Blom, B.; Stoelzel, M.; Driess, M. New Vistas in *N*-Heterocyclic Silylene (NHSi) Transition-Metal Coordination Chemistry: Syntheses, Structures and Reactivity Towards Activation of Small Molecules. *Chem. Eur. J.* **2013,** *19*, 40–62.
28. Blom, B.; Driess, M. Recent Advances in Silylene Chemistry: Small Molecule Activation En-Route Towards Metal-Free Catalysis. In *Structure and Bonding. Functional Molecular Silicon Compounds;* Scheschkewitz, D., Ed.; , 156; Springer: Cham, Switzerland, 2014; pp 85–123.
29. Blom, B.; Gallego, D.; Driess, M. *N*-Heterocyclic Silylene Complexes in Catalysis: New Frontiers in an Emerging Field. *Inorg. Chem. Front.* **2014,** *1*, 134–148.
30. Inoue, S. The Next Generation of Silylene Ligands for Better Catalysts. In *Discovering the Future of Molecular Sciences;* Pignataro, B., Ed.; Wiley-VCH Verlag GmbH & Co. KGaA, 2014; pp 243–273.
31. Ogino, H. Synthesis of Silylene and Silyl(silylene)metal Complexes. *Chem. Rec.* **2002,** *2*, 291–306.
32. Okazaki, M.; Tobita, H.; Ogino, H. Reactivity of Silylene Complexes. *Dalton Trans.* **2003,** 493–506.
33. Waterman, R.; Hayes, P. G.; Tilley, T. D. Synthetic Development and Chemical Reactivity of Transition-Metal Silylene Complexes. *Acc. Chem. Res.* **2007,** *40*, 712–719.
34. Álvarez-Rodríguez, L.; Cabeza, J. A.; García-Álvarez, P.; Polo, D. The Transition-Metal Chemistry of Amidinatosilylenes, -Germylenes and -Stannylenes. *Coord. Chem. Rev.* **2015,** *300*, 1–28.
35. Schmedake, T. A.; Haaf, M.; Apeloig, Y.; Mueller, T.; Bukalov, S.; West, R. Reversible Transformation Between a Diaminosilylene and a Novel Disilene. *J. Am. Chem. Soc.* **1999,** *121*, 9479–9480.
36. Hill, N. J.; Moser, D. F.; Guzei, I. A.; West, R. Reactions of Stable Silylenes With Organic Azides. *Organometallics* **2005,** *24*, 3346–3349.
37. Dhiman, A.; Mueller, T.; West, R.; Becker, J. Y. Electrochemistry and Computations of Stable Silylenes and Germylenes. *Organometallics* **2004,** *23*, 5689–5693.
38. Denk, M.; Green, J. C.; Metzler, N.; Wagner, M. Electronic Structure of a Stable Silylene: Photoelectron Spectra and Theoretical Calculations of Si(NRCHCHNR), $Si(NRCH_2CH_2NR)$ and $SiH_2(NRCHCHNR)$. *J. Chem. Soc., Dalton Trans.* **1994,** 2405–2410.

39. Gehrhus, B.; Hitchcock, P. B.; Lappert, M. F.; Heinicke, J.; Boese, R.; Bläser, D. Synthesis, Structures and Oxidative Addition Reactions of New Thermally Stable Silylenes; Crystal Structures of $[(CH_2{}^tBu)\}_2C_6H_4\text{-}1,2]$ and $[((CH_2{}^tBu)\}_2C_6H_4\text{-}1,2)(\mu\text{-}E)]_2$ (E = Se or Te). *J. Organomet. Chem.* **1996,** *521*, 211–220.
40. Heinicke, J.; Oprea, A.; Kindermann, M. K.; Karpati, T.; Nyulászi, L.; Veszprémi, T. Unsymmetrical Carbene Homologues: Isolable Pyrido[b]-1,3,$^2\lambda^2$-diazasilole, -Germole and -Stannole and Quantum-Chemical Comparison With Unstable Pyrido[c] Isomers. *Chem. Eur. J.* **1998,** *4*, 541–545.
41. Gehrhus, B.; Hitchcock, P. B.; Lappert, M. F. Synthesis of a Stable Biphenyl-bis(carbene) and -bis (silylene) $[M[(NCH_2Bu^t)_2C_6H_3\text{-}3,4]]_2$ (M = C or Si). *Z. Anorg. Allg. Chem.* **2005,** *631*, 1383–1386.
42. Li, W.; Hill, N. J.; Tomasik, A. C.; Bikzhanova, G.; West, R. A New Monomeric Saturated *N*-Heterocyclic Silylene as a Racemic Mixture. *Organometallics* **2006,** *25*, 3802–3805.
43. Tomasik, A. C.; Mitra, A.; West, R. Synthesis and Reactivity of Three New *N*-Heterocyclic Silylenes. *Organometallics* **2009,** *28*, 378–381.
44. Karsch, H. H.; Schlüter, P. A.; Bienlein, F.; Herker, M.; Witt, E.; Sladek, A., et al. Eine neue Methode zur Darstellung von Silaheterocyclen durch Cycloaddition im System Heterobutadien/$HSiCl_3$/NR_3. *Z. Anorg. Allg. Chem.* **1998,** *624*, 295–309.
45. Kong, L.; Zhang, J.; Song, H.; Cui, C. *N*-Aryl Substituted Heterocyclic Silylenes. *Dalton Trans.* **2009,** 5444–5446.
46. Zark, P.; Schaefer, A.; Mitra, A.; Haase, D.; Saak, W.; West, R., et al. Synthesis and Reactivity of *N*-Aryl Substituted *N*-Heterocyclic Silylenes. *J. Organomet. Chem.* **2010,** *695*, 398–408.
47. Mitra, A.; Brodovitch, J.-C.; Krempner, C.; Percival, P. W.; Vyas, P.; West, R. A Silyl Radical Formed by Muonium Addition to a Silylene. *Angew. Chem. Int. Ed.* **2010,** *49*, 2893–2895.
48. Cui, H.; Shao, Y.; Li, X.; Kong, L.; Cui, C. Dehydrochlorination to silylenes by N-heterocyclic carbenes. *Organometallics.* **2009,** *28*, 5191–5195.
49. Ghadwal, R. S.; Roesky, H. W.; Propper, K.; Dittrich, B.; Klein, S.; Frenking, G. A Dimer of Silaisonitrile With Two-Coordinate Silicon Atoms. *Angew. Chem. Int. Ed. Engl.* **2011,** *50*, 5374–5378.
50. Ghadwal, R. S.; Azhakar, R.; Proepper, K.; Dittrich, B.; John, M. Selective Functionalization of a Bis-Silylene. *Chem. Commun.* **2013,** *49*, 5987–5989.
51. Moser, D. F.; Bosse, T.; Olson, J.; Moser, J. L.; Guzei, I. A.; West, R. Halophilic Reactions of a Stable Silylene With Chloro and Bromocarbons. *J. Am. Chem. Soc.* **2002,** *124*, 4186–4187.
52. Moser, D. F.; Naka, A.; Guzei, I. A.; Mueller, T.; West, R. Formation of Disilanes in the Reaction of Stable Silylenes With Halocarbons. *J. Am. Chem. Soc.* **2005,** *127*, 14730–14738.
53. Ishida, S.; Nishinaga, T.; West, R.; Komatsu, K. Generation and Aromaticity of 2-Silaimidazolium Ion, A New Pi-Conjugated Silylium Ion. *Chem. Commun.* **2005,** 778–780.
54. Su, M.-D. Theoretical Study of Halophilic Reactions of Stable Silylenes With Chloro- and Bromocarbons. *J. Am. Chem. Soc.* **2003,** *125*, 1714–1715.
55. Su, M.-D. Density Functional Study of the Mechanism of Halophilic Reactions of Some Silylenes. *Chem. Phys. Lett.* **2003,** *374*, 385–391.
56. Joo, H.; McKee, M. L. Computational Study of the "Stable" Bis(amino)silylene Reaction With Halomethanes. A Radical or Concerted Mechanism? *J. Phys. Chem. A.* **2005,** *109*, 3728–3738.
57. Walker, R. H.; Miller, K. A.; Scott, S. L.; Cygan, Z. T.; Bartolin, J. M.; Kampf, J. W., et al. Silylene- and Germylene-Mediated C-H Activation: Reaction With Alkanes, Ethers, and Amines. *Organometallics* **2009,** *28*, 2744–2755.
58. Delawar, M.; Gehrhus, B.; Hitchcock, P. B. Reactions of a Stable Silylene With Halocarbons. *Dalton Trans.* **2005,** 2945–2953.
59. Gehrhus, B.; Hitchcock, P. B.; Jansen, H. The Stable Silylene $Si[(NCH_2Bu^t)_2C_6H_4\text{-}1,2]$: Reactions With Group 14 Element Halides. *J. Organomet. Chem.* **2006,** *691*, 811–816.
60. West, R.; Schmedake, T. A.; Haaf, M.; Becker, J.; Mueller, T. A Disilanyl Dianion and a Silyl Dianion From a Stable Silylene. *Chem. Lett.* **2001,** 68–69.

61. Antolini, F.; Gehrhus, B.; Hitchcock, P. B.; Lappert, M. F. Crystalline Na-Si(NN) Derivatives [Si(NN) = Si {(NCH$_2$*t*Bu)$_2$C$_6$H$_4$-1,2}]: The Silylenoid [Si(NN)OMe]$^-$, the Dianion [(NN)Si-Si(NN)]$^{2-}$, and the Radical Anion *c*-[Si(NN)]$_3^-$. *Chem. Commun.* **2005,** 5112–5114.
62. Gehrhus, B.; Hitchcock, P. B.; Zhang, L. An Isolable Radical Anion and Dianion of a Cyclotetrasilane: Synthesis and Structure of [Si{1,2-(NEt)$_2$C$_6$H$_4$}]$_4^{-\bullet}$ and [Si{1,2-(NEt)$_2$C$_6$H$_4$}]$_4^{2-}$. *Angew. Chem. Int. Ed.* **2004,** *43*, 1124–1126.
63. Gehrhus, B.; Hitchcock, P. B.; Pongtavornpinyo, R.; Zhang, L. Insights into the Making of a Stable Silylene. *Dalton Trans.* **2006,** 1847–1857.
64. Blom, B.; Klatt, G.; Gallego, D.; Tan, G.; Driess, M. Unprecedented Silicon(II)→Calcium Complexes With *N*-Heterocyclic Silylenes. *Dalton Trans.* **2015,** *44*, 639–644.
65. Blom, B.; Said, A.; Menezes, P. W.; Tan, G.; Driess, M.; Szilvasi, T., et al. Alkaline-Earth-Metal-Induced Liberation of Rare Allotropes of Elemental Silicon and Germanium From *N*-Heterocyclic Metallylenes. *Inorg. Chem.* **2015,** *54*, 8840–8848.
66. Von Schnering, H.-G.; Schwarz, M.; Nesper, R. The Lithium Sodium Silicide Li$_3$NaSi$_6$ and the Formation of Allo-Silicon. *J. Less-Common Met.* **1988,** *137*, 297–310.
67. Schmedake, T. A.; Haaf, M.; Paradise, B. J.; Millevolte, A. J.; Powell, D. R.; West, R. Electronic and Steric Properties of Stable Silylene Ligands in Metal(0) Carbonyl Complexes. *J. Organomet. Chem.* **2001,** *636*, 17–25.
68. Petri, S. H. A.; Eikenberg, D.; Neumann, B.; Stammler, H.-G.; Jutzi, P. Reaction of Molybdenocene and Tungstenocene Derivatives With the Divalent Silicon Species SiN^{t}BuCHCHN^{t}Bu and (C$_5$Me$_5$)$_2$Si. *Organometallics* **1999,** *18*, 2615–2618.
69. Amoroso, D.; Haaf, M.; Yap, G. P. A.; West, R.; Fogg, D. E. A Stable Silylene in a Reactive Environment: Synthesis, Reactivity, and Silicon Extrusion Chemistry of a Coordinatively Unsaturated Ruthenium Silylene Complex Containing Chloride and η^3-P-C-P Ligands. *Organometallics* **2002,** *21*, 534–540.
70. Cade, I. A.; Hill, A. F.; Kaempfe, A.; Wagler, J. Five-Coordinate Hydrido-Ruthenium(II) Complexes Featuring *N*-Heterocyclic Silylene and Carbene Ligands. *Organometallics* **2010,** *29*, 4012–4017.
71. Dysard, J. M.; Tilley, T. D. Synthesis and Reactivity of (Pentamethylcyclopentadienyl)ruthenium Complexes of the Stable Neutral Silylene Si(^{t}BuNCH = CHN^{t}Bu). *Organometallics* **2000,** *19*, 4726–4732.
72. Yoo, H.; Carroll, P. J.; Berry, D. H. Synthesis and Structure of Ruthenium-Silylene Complexes: Activation of Si-Cl Bonds in *N*-Heterocyclic Silanes. *J. Am. Chem. Soc.* **2006,** *128*, 6038–6039.
73. Neumann, E.; Pfaltz, A. Synthesis and Characterization of Cationic Rhodium Complexes With Stable Silylenes. *Organometallics* **2005,** *24*, 2008–2011.
74. Denk, M.; Hayashi, R. K.; West, R. Silylene Complexes From a Stable Silylene and Metal Carbonyls: Synthesis and Structure of [Ni{(Bu^{t}NCH = CHNBu^t)Si}$_2$(CO)$_2$], a Donor-Free Bis-Silylene Complex. *J. Chem. Soc., Chem. Commun.* **1994,** 33–34.
75. Schmedake, T. A.; Haaf, M.; Paradise, B. J.; Powell, D.; West, R. Two Trigonal Ni(silylene)$_3$ Complexes. *Organometallics* **2000,** *19*, 3263–3265.
76. Herrmann, W. A.; Harter, P.; Gstottmayr, C. W. K.; Bielert, F.; Seeboth, N.; Sirsch, P. Complexation of *N*-Heterocyclic Silylenes to Pd(0): Formation of Highly Labile Homoleptic Complexes Pd[Si(^{t}BuNCH$_2$CH$_2$N^{t}Bu)]$_4$ and Pd[Si(^{t}BuNCHCHN^{t}Bu)]$_3$. *J. Organomet. Chem.* **2002,** *649*, 141–146.
77. Fürstner, A.; Krause, H.; Lehmann, C. W. Preparation, Structure and Catalytic Properties of a Binuclear Pd(0) Complex With Bridging Silylene Ligands. *Chem. Commun.* **2001,** 2372.
78. Zhang, M.; Liu, X.; Shi, C.; Ren, C.; Ding, Y.; Roesky, H. W. The Synthesis of (η^3-C$_3$H$_5$)Pd{Si[N(^{t}Bu)CH]$_2$}Cl and the Catalytic Property for Heck Reaction. *Z. Anorg. Allg. Chem.* **2008,** *634*, 1755–1758.
79. Zeller, A.; Bielert, F.; Haerter, P.; Herrmann, W. A.; Strassner, T. Replacement of *N*-Heterocyclic Carbenes by *N*-Heterocyclic Silylenes at a Pd(0) Center: Experiment and Theory. *J. Organomet. Chem.* **2005,** *690*, 3292–3299.

80. Gehrhus, B.; Hitchcock, P. B.; Lappert, M. F.; Maciejewski, H. Silylenenickel(0) or Silyl(silylene)platinum (II) Complexes by Reaction of $Si[(NCH_2Bu^t)_2C_6H_4\text{-}1,2]$ With $[NiCl_2(PPh_3)_2]$, $[Ni(cod)_2]$, or $[PtCl_2(PPh_3)_2]$. *Organometallics* **1998,** *17*, 5599–5601.
81. Avent, A. G.; Gehrhus, B.; Hitchcock, P. B.; Lappert, M. F.; Maciejewski, H. Synthesis and Characterization of Bis(amino)silylene-nickel(0), -Palladium(II), -Platinum(0), -Platinum(II) and Copper(I) Complexes. *J. Organomet. Chem.* **2003,** *686*, 321–331.
82. Sheberla, D.; Tumanskii, B.; Tomasik, A. C.; Mitra, A.; Hill, N. J.; West, R., et al. Different Electronic Structure of Phosphonyl Radical Adducts of *N*-Heterocyclic Carbenes, Silylenes and Germylenes: EPR Spectroscopic Study and DFT Calculations. *Chem. Sci.* **2010,** *1*, 234–241.
83. Cai, X.; Gehrhus, B.; Hitchcock, P. B.; Lappert, M. F. Reactions of the Stable Bis(amino)silylene $Si[\{N(CH_2{}^tBu)\}_2C_6H_4\text{-}1,2]$ With Group 3 or Lanthanide Metal Organic Compounds. Crystal Structures of $[Ln(\eta^5\text{-}C_5H_5)_3Si\{[N(CH_2{}^tBu)]_2C_6H_4\text{-}1,2\}]\cdot C_7H_8$ (Ln = Y or Yb). *Can. J. Chem.* **2000,** *78*, 1484–1490.
84. Evans, W. J.; Perotti, J. M.; Ziller, J. W.; Moser, D. F.; West, R. Evaluation of a Silylene Divalent Lanthanide Interaction in the Metallocene Complex $(C_5Me_5)_2Sm[SiN^tBuCHCHN^tBu]$. *Organometallics* **2003,** *22*, 1160–1163.
85. McCollum, B. M.; Brodovitch, J.-C.; Clyburne, J. A. C.; Mitra, A.; Percival, P. W.; Tomasik, A., et al. Reaction of Stable *N*-Heterocyclic Silylenes and Germylenes With Muonium. *Chem. Eur. J.* **2009,** *15*, 8409–8412.
86. McCollum, B. M.; Brodovitch, J.-C.; Clyburne, J. A. C.; Percival, P. W.; West, R. Detection of a Secondary Muoniated Radical. *Phys. B* **2009,** *404*, 940–942.
87. West, R.; Percival, P. W. Organosilicon Compounds Meet Subatomic Physics: Muon Spin Resonance. *Dalton Trans.* **2010,** *39*, 9209–9216.
88. Driess, M.; Yao, S.; Brym, M.; Van Wuellen, C.; Lentz, D. A New Type of *N*-Heterocyclic Silylene With Ambivalent Reactivity. *J. Am. Chem. Soc.* **2006,** *128*, 9628–9629.
89. Xiong, Y.; Yao, S.; Driess, M. Twice Silicon-Induced C–H Activation and Tautomerisation of a Beta-Diketiminato Ligand and Formation of New Types of *N*-Heterocyclic Silanes. *Dalton Trans.* **2009,** 421–423.
90. Xiong, Y.; Yao, S.; Driess, M. A Zwitterionic Silylene as Reactive Intermediate and Its Unusual Dimerization to an *N*-Heterobicyclic Disilane. *Chem. Asian J.* **2009,** *4*, 1323–1328.
91. Hansen, K.; Szilvasi, T.; Blom, B.; Inoue, S.; Epping, J.; Driess, M. A Fragile Zwitterionic Phosphasilene as a Transfer agent of the Elusive Parent Phosphinidene (:PH). *J. Am. Chem. Soc.* **2013,** *135*, 11795–11798.
92. Jana, A.; Schulzke, C.; Roesky, H. W. Oxidative Addition of Ammonia at a Silicon(II) Center and an Unprecedented Hydrogenation Reaction of Compounds With Low-Valent Group 14 Elements Using Ammonia Borane. *J. Am. Chem. Soc.* **2009,** *131*, 4600–4601.
93. Xiong, Y.; Yao, S.; Driess, M. Reactivity of a Zwitterionic Stable Silylene Toward Halosilanes and Haloalkanes. *Organometallics* **2009,** *28*, 1927–1933.
94. Azhakar, R.; Sarish, S. P.; Roesky, H. W.; Hey, J.; Stalke, D. Regiospecific C–H Bond Activation: Reactivity Study of *N*-Heterocyclic Silylene Toward Ambidentate Phosphorus Ylide. *Organometallics* **2011,** *30*, 2897–2900.
95. Jana, A.; Roesky, H. W.; Schulzke, C.; Samuel, P. P. Insertion Reaction of a Silylene Into a N–H Bond of Hydrazine and a [1 + 4] Cycloaddition With Diphenyl Hydrazone. *Organometallics* **2009,** *28*, 6574–6577.
96. Prasang, C.; Stoelzel, M.; Inoue, S.; Meltzer, A.; Driess, M. Metal-Free Activation of EH_3 (E = P, As) by an Ylide-Like Silylene and Formation of a Donor-Stabilized Arsasilene With a HSi = AsH Subunit. *Angew. Chem. Int. Ed. Engl.* **2010,** *49*, 10002–10005.
97. Xiong, Y.; Yao, S.; Brym, M.; Driess, M. Consecutive Insertion of a Silylene into the P_4 Tetrahedron: Facile Access to Strained SiP_4 and Si_2P_4 Cage Compounds. *Angew. Chem. Int. Ed.* **2007,** *46*, 4511–4513.
98. Xiong, Y.; Yao, S.; Driess, M. Versatile Conversion of *N*-Heterocyclic Silylene to Silyl Metal Compounds by Insertion of Divalent Silicon into Metal-Carbon and Metal-Hydrogen Bonds. *Chem. Eur. J.* **2012,** *18*, 3316–3320.

99. Azhakar, R.; Ghadwal, R. S.; Roesky, H. W.; Granitzka, M.; Stalke, D. Reactivity Studies of a Stable *N*-Heterocyclic Silylene With Triphenylsilanol and Pentafluorophenol. *Organometallics* **2012,** *31*, 5506–5510.
100. Yao, S.; Brym, M.; van Wuellen, C.; Driess, M. From a Stable Silylene to a Mixed-Valent Disiloxane and an Isolable Silaformamide-Borane Complex With Considerable Silicon-Oxygen Double-Bond Character. *Angew. Chem. Int. Ed.* **2007,** *46*, 4159–4162.
101. Driess, M.; Yao, S.; Brym, M.; Van Wuellen, C. Low-Valent Silicon Cations With Two-Coordinate Silicon and Aromatic Character. *Angew. Chem. Int. Ed.* **2006,** *45*, 6730–6733.
102. Meltzer, A.; Inoue, S.; Praesang, C.; Driess, M. Steering S–H and N–H Bond Activation by a Stable *N*-Heterocyclic Silylene: Different Addition of H_2S, NH_3, and Organoamines on a Silicon(II) Ligand Versus Its Si(II)$\rightarrow$Ni$(CO)_3$ Complex. *J. Am. Chem. Soc.* **2010,** *132*, 3038–3046.
103. Stoelzel, M.; Praesang, C.; Inoue, S.; Enthaler, S.; Driess, M. Hydrosilylation of Alkynes by Ni$(CO)_3$-Stabilized Silicon(II) Hydride. *Angew. Chem. Int. Ed.* **2012,** *51*, 399–403.
104. Yao, S.; van Wuellen, C.; Sun, X.-Y.; Driess, M. Dichotomic Reactivity of a Stable Silylene Toward Terminal Alkynes: Facile C–H Bond Insertion Versus Autocatalytic Formation of Silacycloprop-3-Ene. *Angew. Chem. Int. Ed.* **2008,** *47*, 3250–3253.
105. Xiong, Y.; Yao, S.; Driess, M. Unusual [3 + 1] Cycloaddition of a Stable Silylene With a 2,3-Diazabuta-1,3-Diene Versus [4 + 1] Cycloaddition Toward a Buta-1,3-Diene. *Organometallics* **2010,** *29*, 987–990.
106. Azhakar, R.; Sarish, S. P.; Tavcar, G.; Roesky, H. W.; Hey, J.; Stalke, D., et al. Formation of Silicon Centered Spirocyclic Compounds: Reaction of *N*-Heterocyclic Stable Silylene With Benzoylpyridine, Diisopropyl Azodicarboxylate, and 1,2-Diphenylhydrazine. *Inorg. Chem.* **2011,** *50*, 3028–3036.
107. Azhakar, R.; Ghadwal, R. S.; Roesky, H. W.; Hey, J.; Stalke, D. Reactions of Stable *N*-Heterocyclic Silylenes With Ketones and 3,5-di-*tert*-butyl-*o*-benzoquinone. *Organometallics* **2011,** *30*, 3853–3858.
108. Xiong, Y.; Yao, S.; Driess, M. Versatile Reactivity of a Zwitterionic Isolable Silylene Toward Ketones: Silicon-Mediated, Regio- and Stereoselective C–H Activation. *Chem. Eur. J.* **2009,** *15*, 5545–5551.
109. Azhakar, R.; Ghadwal, R. S.; Roesky, H. W.; Mata, R. A.; Wolf, H.; Herbst-Irmer, R., et al. Reaction of *N*-Heterocyclic Silylenes With Thioketone: Formation of Silicon-Sulfur Three (Si-C-S)- and Five (Si-C-C-C-S)-Membered Ring Systems. *Chem. Eur. J.* **2013,** *19*, 3715–3720.
110. Azhakar, R.; Ghadwal, R. S.; Roesky, H. W.; Hey, J.; Stalke, D. Double N–H Bond Activation of N,N′-Bis-Substituted Hydrazines With Stable *N*-Heterocyclic Silylene. *Dalton Trans.* **2012,** *41*, 1529–1533.
111. Yao, S.; Xiong, Y.; Brym, M.; Driess, M. An Isolable Silanoic Ester by Oxygenation of a Stable Silylene. *J. Am. Chem. Soc.* **2007,** *129*, 7268–7269.
112. Yao, S.; Xiong, Y.; Brym, M.; Driess, M. A Series of Isolable Silanoic Thio-, Seleno-, and Telluroesters (LSi(=X)OR) With Donor-Supported Si = X Double Bonds (L = beta-diketiminate; X = S, Se, Te). *Chem. Asian J.* **2008,** *3*, 113–118.
113. Boesveld, W. M.; Gehrhus, B.; Hitchcock, P. B.; Lappert, M. F.; Schleyer, P. v. R. A Crystalline Carbene–Silylene Adduct 1,2-$C_6H_4[N(R)]_2C$-Si$[N(R)]_2C_6H_4$-1,2 (R = CH_2Bu^t); Synthesis, Structure and Bonding in Model Compounds. *Chem. Commun.* **1999,** 755–756.
114. Takeda, N.; Kajiwara, T.; Suzuki, H.; Okazaki, R.; Tokitoh, N. Synthesis and Properties of the First Stable Silylene—Isocyanide Complexes. *Chem. Eur. J.* **2003,** *9*, 3530–3543.
115. Abe, T.; Iwamoto, T.; Kabuto, C.; Kira, M. Synthesis, Structure, and Bonding of Stable Dialkylsilaketenimines. *J. Am. Chem. Soc.* **2006,** *128*, 4228–4229.
116. Wang, Y.; Xie, Y.; Wei, P.; King, R. B.; Schaefer, H. F., III; Schleyer, P. v. R.; Robinson, G. H. A Stable Silicon(0) Compound With a Si = Si Double Bond. *Science* **2008,** *321*, 1069–1071.
117. Ghadwal, R. S.; Roesky, H. W.; Merkel, S.; Henn, J.; Stalke, D. Lewis Base Stabilized Dichlorosilylene. *Angew. Chem. Int. Ed.* **2009,** *48*, 5683–5686.
118. Filippou, A. C.; Chernov, O.; Schnakenburg, G. $SiBr_2$(Idipp): A Stable *N*-Heterocyclic Carbene Adduct of Dibromosilylene. *Angew. Chem. Int. Ed.* **2009,** *48*, 5687–5690.
119. Xiong, Y.; Yao, S.; Driess, M. An Isolable NHC-Supported Silanone. *J. Am. Chem. Soc.* **2009,** *131*, 7562–7563.
120. Yao, S.; Xiong, Y.; Driess, M. *N*-Heterocyclic Carbene (NHC)-Stabilized Silanechalcogenones: NHC$\rightarrow$Si(R_2) = E (E = O, S, Se, Te). *Chem. Eur. J.* **2010,** *16*, 1281–1288.

121. Xiong, Y.; Yao, S.; Driess, M. Synthesis and Rearrangement of stable NHC→Silylene Adducts and Their Unique Reactivity Towards Cyclohexylisocyanide. *Chem. Asian J.* **2010,** *5*, 322–327.
122. Percival, P. W.; McCollum, B. M.; Brodovitch, J.-C.; Driess, M.; Mitra, A.; Mozafari, M., et al. Dual Reactivity of a Stable Zwitterionic *N*-Heterocyclic Silylene and Its Carbene Complex Probed With Muonium. *Organometallics* **2012,** *31*, 2709–2714.
123. Xiong, Y.; Yao, S.; Driess, M. Striking Reactivity of a Stable, Zwitterionic Silylene Towards Substituted Diazomethanes, Azides, and Isocyanides. *Chem. Eur. J.* **2009,** *15*, 8542–8547.
124. Samuel, P. P.; Azhakar, R.; Ghadwal, R. S.; Sen, S. S.; Roesky, H. W.; Granitzka, M., et al. Stable Silaimines With Three- and Four-Coordinate Silicon Atoms. *Inorg. Chem.* **2012,** *51*, 11049–11054.
125. Jungton, A.-K.; Meltzer, A.; Praesang, C.; Braun, T.; Driess, M.; Penner, A. Addition of $[(\eta^5\text{-}C_5Me_5)IrH_4]$ to a Zwitterionic Silylene: Stepwise Formation of Iridium(V)-Silyl and Iridium(III)-Silylene Complexes. *Dalton Trans.* **2010,** *39*, 5436–5438.
126. Meltzer, A.; Praesang, C.; Milsmann, C.; Driess, M. The Striking Stabilization of $Ni^0(\eta^6\text{-arene})$ Complexes by an Ylide-Like Silylene Ligand. *Angew. Chem. Int. Ed.* **2009,** *48*, 3170–3173.
127. Meltzer, A.; Prasang, C.; Driess, M. Diketiminate Silicon(II) and Related NHSi Ligands Generated in the Coordination Sphere of Nickel(0). *J. Am. Chem. Soc.* **2009,** *131*, 7232–7233.
128. Tan, G.; Blom, B.; Gallego, D.; Irran, E.; Driess, M. Elucidating the Effect of the Nucleophilicity of the Silyl Group in the Reduction of CO_2 to CO Mediated by Silyl-Copper(I) Complexes. *Chem. Eur. J.* **2014,** *20*, 9400–9408.
129. Tan, G.; Enthaler, S.; Inoue, S.; Blom, B.; Driess, M. Synthesis of Mixed Silylene-Carbene Chelate Ligands From *N*-Heterocyclic Silylcarbenes Mediated by Nickel. *Angew. Chem. Int. Ed.* **2015,** *54*, 2214–2218.
130. Poerschke, K. R.; Tsay, Y. H.; Krueger, C. Structure of $[\mu\text{-}(CH_3)_2PCH_2P(CH_3)_2]_2Ni_2(CO)_4$. *Inorg. Chem.* **1986,** *25*, 2097–2099.
131. Scott, N. M.; Clavier, H.; Mahjoor, P.; Stevens, E. D.; Nolan, S. P. Synthetic, Structural, and Thermochemical Studies of *N*-Heterocyclic Carbene (NHC) and Tertiary Phosphine Ligands in the $[(L)_2Ni(CO)_2]$ ($L = PR_3$, NHC) System. *Organometallics* **2008,** *27*, 3181–3186.
132. Huppmann, F.; Maringgele, W.; Kottke, T.; Meller, A. Reaktionen Subvalenter Verbindungen Des Siliciums Mit Benzol Und Mit Biphenyl. *J. Organomet. Chem.* **1992,** *434*, 35–44.
133. Huppmann, F.; Noltemeyer, M.; Meller, A. Reaktionen Subvalenter Verbindungen des Siliciums mit alkylierten Aromaten. *J. Organomet. Chem.* **1994,** *483*, 217–228.
134. Sakamoto, K.; Tsutsui, S.; Sakurai, H.; Kira, M. Generation and Trapping of Bis(dialkylamino)silylenes: Experimental Evidence for Bridged Structure of Diaminosilylene Dimers. *Bull. Chem. Soc. Jpn.* **1997,** *70*, 253–260.
135. Tsutsui, S.; Sakamoto, K.; Kira, M. Bis(diisopropylamino)silylene and Its Dimer. *J. Am. Chem. Soc.* **1998,** *120*, 9955–9956.
136. Takahashi, M.; Tsutsui, S.; Sakamoto, K.; Kira, M.; Müller, T.; Apeloig, Y. Dimers of Diaminosilylenes: Doubly Bonded or Bridged? The Dimers of $(i\text{-}Pr_2N)_2Si$. *J. Am. Chem. Soc.* **2001,** *123*, 347–348.
137. Lee, G.-H.; West, R.; Mueller, T. Bis[bis(trimethylsilyl)amino]silylene, an Unstable Divalent Silicon Compound. *J. Am. Chem. Soc.* **2003,** *125*, 8114–8115.
138. Hadlington, T. J.; Abdalla, J. A.; Tirfoin, R.; Aldridge, S.; Jones, C. Stabilization of a Two-Coordinate, Acyclic Diaminosilylene (ADASi): Completion of the Series of Isolable Diaminotetrylenes:$E(NR_2)_2$ (E = group 14 element). *Chem. Commun.* **2016,** *52*, 1717–1720.
139. Segawa, Y.; Yamashita, M.; Nozaki, K. Boryllithium: Isolation, Characterization, and Reactivity as a Boryl Anion. *Science* **2006,** *314*, 113–115.
140. Protchenko, A. V.; Birjkumar, K. H.; Dange, D.; Schwarz, A. D.; Vidovic, D.; Jones, C., et al. A Stable Two-Coordinate Acyclic Silylene. *J. Am. Chem. Soc.* **2012,** *134*, 6500–6503.
141. Wang, Y.; Ma, J. Silylenes and Germylenes: The Activation of H–H Bond in Hydrogen Molecule. *J. Organomet. Chem.* **2009,** *694*, 2567–2575.
142. Frey, G. D.; Lavallo, V.; Donnadieu, B.; Schoeller, W. W.; Bertrand, G. Facile Splitting of Hydrogen and Ammonia by Nucleophilic Activation at a Single Carbon Center. *Science* **2007,** *316*, 439–441.

143. Protchenko, A. V.; Schwarz, A. D.; Blake, M. P.; Jones, C.; Kaltsoyannis, N.; Mountford, P., et al. A Generic One-Pot Route to Acyclic Two-Coordinate Silylenes From Silicon(IV) Precursors: Synthesis and Structural Characterization of a Silylsilylene. *Angew. Chem. Int. Ed.* **2013,** *52*, 568–571.
144. Protchenko, A. V.; Blake, M. P.; Schwarz, A. D.; Jones, C.; Mountford, P.; Aldridge, S. Reactivity of Boryl- and Silyl-Substituted Carbenoids Toward Alkynes: Insertion and Cycloaddition Chemistry. *Organometallics* **2015,** *34*, 2126–2129.
145. Wendel, D.; Eisenreich, W.; Jandl, C.; Pöthig, A.; Rieger, B. Reactivity of an Acyclic Silylsilylene Toward Ethylene: Migratory Insertion into the Si–Si Bond. *Organometallics* **2016,** *35*, 1–4.
146. Green, S. P.; Jones, C.; Stasch, A. Stable Magnesium(I) Compounds With Mg–Mg Bonds. *Science* **2007,** *318*, 1754–1757.
147. Bonyhady, S. J.; Jones, C.; Nembenna, S.; Stasch, A.; Edwards, A. J.; McIntyre, G. J. Beta-Diketiminate-Stabilized Magnesium(I) Dimers and Magnesium(II) Hydride Complexes: Synthesis, Characterization, Adduct Formation, and Reactivity Studies. *Chem. Eur. J.* **2010,** *16*, 938–955.
148. Stasch, A.; Jones, C. Stable Dimeric Magnesium(I) Compounds: From Chemical Landmarks to Versatile Reagents. *Dalton Trans.* **2011,** *40*, 5659–5672.
149. Rekken, B. D.; Brown, T. M.; Fettinger, J. C.; Tuononen, H. M.; Power, P. P. Isolation of a Stable, Acyclic, Two-Coordinate Silylene. *J. Am. Chem. Soc.* **2012,** *134*, 6504–6507.
150. Rekken, B. D.; Brown, T. M.; Fettinger, J. C.; Lips, F.; Tuononen, H. M.; Herber, R. H., et al. Dispersion Forces and Counterintuitive Steric Effects in Main Group Molecules: Heavier Group 14 (Si-Pb) Dichalcogenolate Carbene Analogues With Sub-90 Degrees Interligand Bond Angles. *J. Am. Chem. Soc.* **2013,** *135*, 10134–10148.
151. Lips, F.; Fettinger, J. C.; Mansikkamaki, A.; Tuononen, H. M.; Power, P. P. Reversible Complexation of Ethylene by a Silylene Under Ambient Conditions. *J. Am. Chem. Soc.* **2014,** *136*, 634–637.
152. Lips, F.; Mansikkamaki, A.; Fettinger, J. C.; Tuononen, H. M.; Power, P. P. Reactions of Alkenes and Alkynes With an Acyclic Silylene and Heavier Tetrylenes Under Ambient Conditions. *Organometallics* **2014,** *33*, 6253–6258.
153. Tokitoh, N.; Suzuki, H.; Okazaki, R.; Ogawa, K. Synthesis, Structure, and Reactivity of Extremely Hindered Disilenes: The First Example of Thermal Dissociation of a Disilene into a Silylene. *J. Am. Chem. Soc.* **1993,** *115*, 10428–10429.
154. Suzuki, H.; Tokitoh, N.; Okazaki, R.; Harada, J.; Ogawa, K.; Tomoda, S., et al. Synthesis and Structures of Extremely Hindered and Stable Disilenes. *Organometallics* **1995,** *14*, 1016–1022.
155. Suzuki, H.; Tokitoh, N.; Okazaki, R. Thermal Dissociation of Disilenes into Silylenes. *Bull. Chem. Soc. Jpn.* **1995,** *68*, 2471–2481.
156. Tokitoh, N.; Suzuki, H.; Takeda, N.; Kajiwara, T.; Sasamori, T.; Okazaki, R. Stable 2H-Azasilirene and 2H-Phosphasilirene: Addition Reaction of an Overcrowded Silylene to a Nitrile and a Phosphaalkyne. *Silicon Chem.* **2004,** *1*, 313–319.
157. Takeda, N.; Tokitoh, N.; Okazaki, R. A Novel Reaction Mode in the Cycloaddition of Thermally Generated Silylenes With Conjugated Dienes. *Chem. Lett.* **2000,** 622–623.
158. Suzuki, H.; Tokitoh, N.; Okazaki, R. A Novel Reactivity of a Silylene: The First Examples of [1 + 2] Cycloaddition With Aromatic Compounds. *J. Am. Chem. Soc.* **1994,** *116*, 11572–11573.
159. Mizuhata, Y.; Sato, T.; Tokitoh, N. Reactions of an Overcrowded Silylene With Pyridines: Formation of a Novel 2H-1,2-Azasilepine and Its Further Cycloaddition. *Heterocycles* **2012,** *84*, 413–418.
160. Sasamori, T.; Hironaka, K.; Sugiyama, Y.; Takagi, N.; Nagase, S.; Hosoi, Y., et al. Synthesis and Reactions of a Stable 1,2-Diaryl-1,2-Dibromodisilene: A Precursor for Substituted Disilenes and a 1,2-Diaryldisilyne. *J. Am. Chem. Soc.* **2008,** *130*, 13856–13857.
161. Han, J. S.; Sasamori, T.; Mizuhata, Y.; Tokitoh, N. Evidence for LiBr-Assisted Generation of a Silylene From a 1,2-Diaryl-1,2-Dibromodisilene. *Chem. Asian J.* **2012,** *7*, 298–300.
162. Agou, T.; Sasamori, T.; Tokitoh, N. Synthesis of an Arylbromosilylene-Platinum Complex by Using a 1,2-Dibromodisilene as a Silylene Source. *Organometallics* **2012,** *31*, 1150–1154.
163. Sekiguchi, A.; Inoue, S.; Ichinohe, M.; Arai, Y. Isolable Anion Radical of Blue Disilene $({}^{t}Bu_2MeSi)_2Si = Si(SiMe{}^{t}Bu_2)_2$ Formed Upon One-Electron Reduction: Synthesis and Characterization. *J. Am. Chem. Soc.* **2004,** *126*, 9626–9629.

164. Inoue, S.; Ichinohe, M.; Sekiguchi, A. Isolable Silylene Anion Radical: Structural Characteristics in the Solid State and in Solution. *J. Am. Chem. Soc.* **2007,** *129*, 6096–6097.
165. Inoue, S.; Ichinohe, M.; Sekiguchi, A. Disilenyl Anions Derived From Reduction of Tetrakis(di-*tert*-butylmethylsilyl)disilene With Metal Naphthalenide Through a Disilene Dianion Intermediate: Synthesis and Characterization. *Chem. Lett.* **2005,** *34*, 1564–1565.
166. Filippou, A. C.; Chernov, O.; Schnakenburg, G. Chromium-Silicon Multiple Bonds: The Chemistry of Terminal *N*-Heterocyclic-Carbene-Stabilized Halosilylidyne Ligands. *Chem. Eur. J.* **2011,** *17*, 13574–13583.
167. Filippou, A. C.; Baars, B.; Chernov, O.; Lebedev, Y. N.; Schnakenburg, G. Silicon-Oxygen Double Bonds: A Stable Silanone With a Trigonal-Planar Coordinated Silicon Center. *Angew. Chem. Int. Ed.* **2014,** *53*, 565–570.
168. Wiberg, N.; Niedermayer, W. Silylene R*XSi (R* = Si^tBu_3; X = H, Me, Ph, Hal, R*): Bildung und Reaktionen. *J. Organomet. Chem.* **2001,** *628*, 57–64.
169. Wiberg, N.; Niedermayer, W.; Nöth, H.; Warchhold, M. Alkalimetalldisupersilylsilanide $MSiXR^*_2$ (R* = Si^tBu_3; X = H, Me, Ph, Hal): Synthesen, Charakterisierung, Strukturen, Reaktionen. *J. Organomet. Chem.* **2001,** *628*, 46–56.
170. Wiberg, N.; Niedermayer, W.; Fischer, G.; Nöth, H.; Suter, M. On the way to a Disilyne RSi≡SiR, II. Synthesis, Structure and Dehalogenation of the Disilene RClSi = SiClR [R = $(^tBu_3Si)_2MeSi$]. *Eur. J. Inorg. Chem.* **2002,** *2002*, 1066–1070.
171. Fischer, R.; Baumgartner, J.; Kickelbick, G.; Marschner, C. The First Stable Beta-Fluorosilylanion. *J. Am. Chem. Soc.* **2003,** *125*, 3414–3415.
172. Kawachi, A.; Oishi, Y.; Kataoka, T.; Tamao, K. Preparation of Sulfur-Substituted Silyllithiums and Their Thermal Degradation to Silylenes. *Organometallics* **2004,** *23*, 2949–2955.
173. Tamao, K.; Kawachi, A.; Ito, Y. The First Stable Functional Silyl Anions: (Aminosilyl)lithiums. *J. Am. Chem. Soc.* **1992,** *114*, 3989–3990.
174. Tamao, K.; Kawachi, A.; Ito, Y. Coupling of (Amino)alkylchlorosilanes With Lithium: New Access to Symmetrical Di- and Tetrafunctional Alkyldisilanes. *Organometallics* **1993,** *12*, 580–582.
175. Tamao, K.; Kawachi, A. The Chemistry of Silylenoids: Preparation and Reactivity of (Alkoxysilyl) lithium Compounds. *Angew. Chem. Int. Ed. Engl.* **1995,** *34*, 818–820.
176. Kawachi, A.; Tamao, K. Different Modes of Reaction of Monoalkoxy- and Dialkoxyphenylchlorosilanes With Lithium Metal: Selective formation of (2-Alkoxydisilanyl)lithium vs (Dialkoxysilyl)lithium. *Organometallics* **1996,** *15*, 4653–4656.
177. Tamao, K.; Kawachi, A. Reduction of Phenylchlorosilanes With Lithium 1-(Dimethylamino) naphthalenide: A New Access to Functionalized Silyllithiums. *Organometallics* **1995,** *14*, 3108–3111.
178. Tamao, K.; Kawachi, A.; Asahara, M.; Toshimitsu, A. Recent Developments in Silicon Interelement Linkage: The Case of Functionalized Silyllithium, Silylenoid and Sila-Ylide. *Pure Appl. Chem.* **1999,** *71*, 393–400.
179. Harloff, J.; Popowski, E.; Reinke, H. Substituted Lithiumtrimethylsiloxysilanides LiSiRR'$(OSiMe_3)$ – Investigations of Their Synthesis, Stability and Reactivity. *J. Organomet. Chem.* **2007,** *692*, 1421–1441.
180. Oehme, H.; Weiss, H. Reaction of 2,4,6-tri-*t*-butylphenyllithium With Bromotrichlorosilane. Generation of trichlorosilyllithium, $LiSiCl_3$. *J. Organomet. Chem.* **1987,** *319*, C16–C18.
181. Lee, M. E.; Cho, H. M.; Ryu, M. S.; Kim, C. H.; Ando, W. A Stable Halosilylene At Room Temperature in THF Solution. *J. Am. Chem. Soc.* **2001,** *123*, 7732–7733.
182. Lee, M. E.; Cho, H. M.; Lim, Y. M.; Choi, J. K.; Park, C. H.; Jeong, S. E., et al. Syntheses and Reactivities of Stable Halosilylenoids, $(Tsi)X_2SiLi$ (Tsi = $C(SiMe_3)_3$, X = Br, Cl). *Chem. Eur. J.* **2004,** *10*, 377–381.
183. Flock, M.; Dransfeld, A. The Solution Structure of $(Me_3Si)_3CSiBr$—An ab Initio/NMR Study. *Chem. Eur. J.* **2003,** *9*, 3320–3323.
184. Lee, M. E.; Lim, Y. M.; Son, J. Y.; Seo, W. G. Stable Halodilithiosilanes: Synthesis and Reactivity of $TsiXSiLi_2$ (X = Br and Cl). *Chem. Lett.* **2008,** *37*, 680–681.
185. Lim, Y. M.; Lee, M. E.; Lee, J.; Do, Y. Reactivity of Bromodilithiosilane to Naphthalene and Anthracene. *Organometallics* **2008,** *27*, 6375–6378.

186. Lim, Y. M.; Cho, H. M.; Lee, M. E.; Baeck, K. K. A Stable Magnesium Bromosilylenoid: Transmetalation of a Lithium Bromosilylenoid by Magnesium Bromide. *Organometallics* **2006,** *25*, 4960–4964.
187. Cho, H. M.; Lim, Y. M.; Lee, B. W.; Park, S. J.; Lee, M. E. Synthesis and Reactivity of Stable Trisyl (aminomethylphenyl)silylenoid. *J. Organomet. Chem.* **2011,** *696*, 2665–2668.
188. Qi, Y.; Mu, Z.; Feng, D. Theoretical Studies on the Pentacoordinate Silylenoid $PhCH_2(NH_2)CH_3SiLiF$. *Struct. Chem.* **2009,** *21*, 123–129.
189. Cho, H. M.; Bok, K.; Park, S. H.; Lim, Y. M.; Lee, M. E.; Choi, M.-G., et al. A New Synthetic Route for Silacyclopropanes: Reactions of a Bromosilylenoid With Olefins. *Organometallics* **2012,** *31*, 5227–5230.
190. Song, J. H.; Park, S. H.; Cho, H. M.; Lee, M. E. Reactivity of Bromosilylenoid With Heterocumulenes. *J. Organomet. Chem.* **2015,** *799-800*, 128–131.
191. Lim, Y. M.; Park, C. H.; Yoon, S. J.; Cho, H. M.; Lee, M. E.; Baeck, K. K. New Synthetic Routes for Silaheterocycles: Reactions of a Chlorosilylenoid With Aldehydes. *Organometallics* **2010,** *29*, 1355–1361.
192. Cho, H. M.; Lim, Y. M.; Lee, M. E. Reactivities of Chlorotrisylsilylenoid With Ketones. *Dalton Trans.* **2010,** *39*, 9232–9234.
193. Molev, G.; Bravo-Zhivotovskii, D.; Karni, M.; Tumanskii, B.; Botoshansky, M.; Apeloig, Y. Synthesis, Molecular Structure, and Reactivity of the Isolable Silylenoid With a Tricoordinate Silicon. *J. Am. Chem. Soc.* **2006,** *128*, 2784–2785.
194. Ishida, S.; Iwamoto, T.; Kira, M. Isomerization of an Isolable Silylene into a Silene via 1,2-Silyl Migration. *Organometallics* **2009,** *28*, 919–921.
195. Ishida, S.; Iwamoto, T.; Kabuto, C.; Kira, M. Unexpected Reactions of an Isolable Dialkylsilylene With Haloalkanes. *Chem. Lett.* **2001,** 1102–1103.
196. Ishida, S.; Iwamoto, T.; Kabuto, C.; Kira, M. Insertion of a Stable Dialkylsilylene into Silicon-Chlorine Bonds. *Silicon Chem.* **2003,** *2*, 137–140.
197. Iwamoto, T.; Masuda, H.; Ishida, S.; Kabuto, C.; Kira, M. Addition of Stable Nitroxide Radical to Stable Divalent Compounds of Heavier Group 14 Elements. *J. Am. Chem. Soc.* **2003,** *125*, 9300–9301.
198. Ishida, S.; Iwamoto, T.; Kira, M. Addition of a Stable Dialkylsilylene to Carbon-Carbon Unsaturated Bonds. *Heteroat. Chem.* **2011,** *22*, 432–437.
199. Ishida, S.; Iwamoto, T.; Kira, M. Reactions of an Isolable Dialkylsilylene With Ketones. *Organometallics* **2010,** *29*, 5526–5534.
200. Ishida, S.; Uchida, K.; Onodera, T.; Oikawa, H.; Kira, M.; Iwamoto, T. Siloxy-Substituted Cyclopentadiene Showing Aggregation-Enhanced Emission: An Application of Cycloaddition of Isolable Dialkylsilylene. *Organometallics* **2012,** *31*, 5983–5985.
201. Ishida, S.; Uchida, K.; Iwamoto, T. Unexpected Formation of Highly Strained 2,5-Dioxa-3-Silabicyclo [2.1.0]pentane by the Reaction of Isolable Dialkylsilylene With Benzil. *Heteroat. Chem.* **2014,** *25*, 348–353.
202. Liu, X.; Xiao, X.-Q.; Xu, Z.; Yang, X.; Li, Z.; Dong, Z., et al. Reactions of an Isolable Dialkylsilylene With Carbon Dioxide and Related Heterocumulenes. *Organometallics* **2014,** *33*, 5434–5439.
203. Chen, W.; Wang, L.; Li, Z.; Lin, A.; Lai, G.; Xiao, X., et al. Diverse Reactivity of an Isolable Dialkylsilylene Toward Imines. *Dalton Trans.* **2013,** *42*, 1872–1878.
204. Wang, L.; Chen, W.; Li, Z.; Xiao, X.-Q.; Lai, G.; Liu, X., et al. Reactions of an Isolable Dialkylsilylene With Aromatic Nitriles Providing a New Type of Heterosilole. *Chem. Commun.* **2013,** *49*, 9776–9778.
205. Dong, Z.; Xiao, X. Q.; Li, Z.; Lu, Q.; Lai, G.; Kira, M. Elusive 2H-1,2-Oxasiletes Through Reactions of an Isolable Dialkylsilylene With Diazocarbonyl Compounds. *Org. Biomol. Chem.* **2015,** *13*, 9471–9476.
206. Kira, M. Isolable Silylene, Disilenes, Trisilaallene, and Related Compounds. *J. Organomet. Chem.* **2004,** *689*, 4475–4488.
207. Kira, M.; Ishida, S.; Iwamoto, T. Comparative Chemistry of Isolable Divalent Compounds of Silicon, Germanium, and Tin. *Chem. Rec.* **2004,** *4*, 243–253.
208. Kira, M.; Iwamoto, T.; Ishida, S. A Helmeted Dialkylsilylene. *Bull. Chem. Soc. Jpn.* **2007,** *80*, 258–275.
209. Kira, M. An Isolable Dialkylsilylene and Its Derivatives. A Step Toward Comprehension of Heavy Unsaturated Bonds. *Chem. Commun.* **2010,** *46*, 2893–2903.

210. Kira, M. Reactions of a Stable Dialkylsilylene and Their Mechanisms. *J. Chem. Sci.* **2012,** *124*, 1205–1215.
211. Abe, T.; Tanaka, R.; Ishida, S.; Kira, M.; Iwamoto, T. New Isolable Dialkylsilylene and Its Isolable Dimer That Equilibrate in Solution. *J. Am. Chem. Soc.* **2012,** *134*, 20029–20032.
212. Ishida, S.; Abe, T.; Hirakawa, F.; Kosai, T.; Sato, K.; Kira, M., et al. Persistent Dialkylsilanone Generated by Dehydrobromination of Dialkylbromosilanol. *Chem. Eur. J.* **2015,** *21*, 15100–15103.
213. Linden, M. M.; Reisenauer, H. P.; Gerbig, D.; Karni, M.; Schafer, A.; Muller, T., et al. Preparation of a Silanone Through Oxygen Atom Transfer to a Stable Cyclic Silylene. *Angew. Chem. Int. Ed.* **2015,** *54*, 12404–12409.
214. Asay, M.; Inoue, S.; Driess, M. Aromatic Ylide-Stabilized Carbocyclic Silylene. *Angew. Chem. Int. Ed.* **2011,** *50*, 9589–9592.
215. Karni, M.; Apeloig, Y. Conjugation, Resonance, and Stability in *N*-Heterocyclic Silylenes and in Phosphorus Ylide Substituted Silylenes. *Organometallics* **2012,** *31*, 2403–2415.
216. Redies, K. M.; Fallon, T.; Oestreich, M. En Route to Stable All-Carbon-Substituted Silylenes: Synthesis and Reactivity of a Bis(α-spirocyclopropyl)silylene. *Organometallics* **2014,** *33*, 3235–3238.
217. Momeni, M. R.; Shakib, F. A.; Azizi, Z. New Generation of Dialkylsilylenes With Stabilities Comparable to Diaminosilylenes: A theoretical Study. *J. Phys. Chem. A* **2011,** *115*, 10550–10555.
218. Kira, M.; Ishida, S.; Iwamoto, T.; Kabuto, C. Excited-State Reactions of an Isolable Silylene With Aromatic Compounds. *J. Am. Chem. Soc.* **2002,** *124*, 3830–3831.
219. Kira, M.; Ishida, S.; Iwamoto, T.; de Meijere, A.; Fukitsuka, M.; Ito, O. The Singlet Excited State of a Stable Dialkylsilylene is Responsible for Its Photoreactions. *Angew. Chem. Int. Ed.* **2004,** *43*, 4510–4512.
220. Su, M.-D. A Theoretical Investigation of Photochemical Reactions of an Isolable Silylene With Benzene. *Chem. Eur. J.* **2014,** *20*, 9419–9423.
221. Kosai, T.; Ishida, S.; Iwamoto, T. Transformation of Azulenes to Bicyclic [4]dendralene and Heptafulvene Derivatives via Photochemical Cycloaddition of Dialkylsilylene. *Chem. Commun.* **2015,** *51*, 10707–10709.
222. Fukuoka, T.; Uchida, K.; Sung, Y. M.; Shin, J.-Y.; Ishida, S.; Lim, J. M., et al. Near-IR Absorbing Nickel (II) Porphyrinoids Prepared by Regioselective Insertion of Silylenes into Antiaromatic Nickel(II) Norcorrole. *Angew. Chem. Int. Ed.* **2014,** *53*, 1506–1509.
223. Iwamoto, T.; Kobayashi, M.; Uchiyama, K.; Sasaki, S.; Nagendran, S.; Isobe, H., et al. Anthryl-Substituted Trialkyldisilene Showing Distinct Intramolecular Charge-Transfer Transition. *J. Am. Chem. Soc.* **2009,** *131*, 3156–3157.
224. Xu, Z.; Jin, J.; Li, Z.; Qiu, H.; Jiang, J.; Lai, G., et al. Remarkable Substituent Effects on the Activation Energy of Silylene Insertion into Silicon-Chlorine Bonds. *Chem. Eur. J.* **2009,** *15*, 8605–8612.
225. Lai, G.; Xu, Z.; Li, Z.; Jiang, J.; Kira, M.; Qiu, H. Stereoelectronic Substituent Effects on Silylene Insertion into the Si–Cl Bond. *Organometallics* **2009,** *28*, 3591–3593.
226. Xu, Z.; Jin, J.; Zhang, H.; Li, Z.; Jiang, J.; Lai, G., et al. Insertion of Silylenes into Si–H and Si–Cl Bonds. Comparison of Mechanism and Substituent Effects. *Organometallics* **2011,** *30*, 3311–3317.
227. Watanabe, C.; Iwamoto, T.; Kabuto, C.; Kira, M. Dinuclear Palladium Complex With a Bridging Dialkylsilylene Ligand. *Chem. Lett.* **2007,** *36*, 284–285.
228. Watanabe, C.; Inagawa, Y.; Iwamoto, T.; Kira, M. Synthesis and Structures of (Dialkylsilylene)bis (phosphine)-nickel, Palladium, and Platinum Complexes and (η^6-arene)(dialkylsilylene)nickel Complexes. *Dalton Trans.* **2010,** *39*, 9414–9420.
229. Watanabe, C.; Iwamoto, T.; Kabuto, C.; Kira, M. Fourteen-Electron Bis(dialkylsilylene)palladium and Twelve-Electron Bis(dialkylsilyl)palladium Complexes. *Angew. Chem. Int. Ed.* **2008,** *47*, 5386–5389.
230. Inagawa, Y.; Ishida, S.; Iwamoto, T. Two-Coordinate Dialkylsilylene-Coinage Metal Complexes. *Chem. Lett.* **2014,** *43*, 1665–1667.
231. Ishida, S.; Iwamoto, T.; Kira, M. Radical Anion of Isolable Dialkylsilylene. *J. Am. Chem. Soc.* **2003,** *125*, 3212–3213.

232. Bravo-Zhivotovskii, D.; Yuzefovich, M.; Sigal, N.; Korogodsky, G.; Klinkhammer, K.; Tumanskii, B., et al. The Synthesis of the First Compound With Li-Si-Hg Bonding: [{Li($i$$Pr_3$$Si)_2$Si}$_2$Hg]—A Source for the [Li($i$$Pr_3$$Si)_2$Si] Radical. *Angew. Chem. Int. Ed.* **2002,** *41*, 649–651.
233. Holthausen, M. C.; Koch, W.; Apeloig, Y. Theory Predicts Triplet Ground-State Organic Silylenes. *J. Am. Chem. Soc.* **1999,** *121*, 2623–2624.
234. Yoshida, M.; Tamaoki, N. DFT Study on Triplet Ground State Silylenes Revisited: The Quest for the Triplet Silylene Must Go On. *Organometallics* **2002,** *21*, 2587–2589.
235. Apeloig, Y.; Paunez, R.; Karni, M.; West, R.; Steiner, W.; Chapman, D. Why Is Methylene a Ground State Triplet While Silylene Is a Ground State Singlet? *Organometallics* **2003,** *22*, 3250–3256.
236. Sekiguchi, A.; Tanaka, T.; Ichinohe, M.; Akiyama, K.; Tero-Kubota, S. Bis(tri-*tert*-butylsilyl)silylene: Triplet Ground State Silylene. *J. Am. Chem. Soc.* **2003,** *125*, 4962–4963.
237. Sekiguchi, A.; Tanaka, T.; Ichinohe, M.; Akiyama, K.; Gaspar, P. P. Tri-*tert*-butylsilylsilylenes With Alkali Metal Substituents (^{t}Bu_3Si)SiM (M = Li, K): Electronically and Sterically Accessible Triplet Ground States. *J. Am. Chem. Soc.* **2008,** *130*, 426–427.
238. Kosa, M.; Karni, M.; Apeloig, Y. Were Reactions of Triplet Silylenes Observed? *J. Am. Chem. Soc.* **2013,** *135*, 9032–9040.
239. Nagendran, S.; Roesky, H. W. The Chemistry of Aluminum(I), Silicon(II), and Germanium(II). *Organometallics* **2008,** *27*, 457–492.
240. So, C.-W.; Roesky, H. W.; Magull, J.; Oswald, R. B. Synthesis and Characterization of [PhC(N^t$Bu)_2$] SiCl: A Stable Monomeric Chlorosilylene. *Angew. Chem. Int. Ed.* **2006,** *45*, 3948–3950.
241. Sen, S. S.; Roesky, H. W.; Stern, D.; Henn, J.; Stalke, D. High Yield Access to Silylene RSiCl (R = PhC (N^t$Bu)_2$) and Its Reactivity Toward Alkyne: Synthesis of Stable Disilacyclobutene. *J. Am. Chem. Soc.* **2010,** *132*, 1123–1126.
242. Ghadwal, R. S.; Proepper, K.; Dittrich, B.; Jones, P. G.; Roesky, H. W. Neutral Pentacoordinate Silicon Fluorides Derived From Amidinate, Guanidinate, and Triazapentadienate Ligands and Base-Induced Disproportionation of Si_2Cl_6 to Stable Silylenes. *Inorg. Chem.* **2011,** *50*, 358–364.
243. Yeong, H.-X.; Lau, K.-C.; Xi, H.-W.; Hwa Lim, K.; So, C.-W. Reactivity of a Disilylene [{PhC(NBu^t)} $Si]_2$ Toward Bromine: Synthesis and Characterization of a Stable Monomeric Bromosilylene. *Inorg. Chem.* **2010,** *49*, 371–373.
244. So, C.-W.; Roesky, H. W.; Gurubasavaraj, P. M.; Oswald, R. B.; Gamer, M. T.; Jones, P. G., et al. Synthesis and Structures of Heteroleptic Silylenes. *J. Am. Chem. Soc.* **2007,** *129*, 12049–12054.
245. Zhang, S.-H.; Yeong, H.-X.; Xi, H.-W.; Lim, K. H.; So, C.-W. Hydrosilylation of a Silicon(II) Hydride: Synthesis and Characterization of a Remarkable Silylsilylene. *Chem. Eur. J.* **2010,** *16*, 10250–10254.
246. Sen, S. S.; Hey, J.; Herbst-Irmer, R.; Roesky, H. W.; Stalke, D. Striking Stability of a Substituted Silicon (II) Bis(trimethylsilyl)amide and the Facile Si-Me Bond Cleavage Without a Transition Metal Catalyst. *J. Am. Chem. Soc.* **2011,** *133*, 12311–12316.
247. Azhakar, R.; Ghadwal, R. S.; Roesky, H. W.; Wolf, H.; Stalke, D. Facile Access to the Functionalized *N*-Donor Stabilized Silylenes PhC(N^t$Bu)_2$SiX (X = PPh_2, NPh_2, NCy_2, N^iPr_2, NMe_2, N($SiMe_3)_2$, O^{t}Bu). *Organometallics* **2012,** *31*, 4588–4592.
248. Inoue, S.; Wang, W.; Prasang, C.; Asay, M.; Irran, E.; Driess, M. An Ylide-Like Phosphasilene and Striking Formation of a 4π-Electron, Resonance-Stabilized 2,4-Disila-1,3-Diphosphacyclobutadiene. *J. Am. Chem. Soc.* **2011,** *133*, 2868–2871.
249. Azhakar, R.; Ghadwal, R. S.; Roesky, H. W.; Wolf, H.; Stalke, D. A Debut for Base Stabilized Monoalkylsilylenes. *Chem. Commun.* **2012,** *48*, 4561–4563.
250. Sarish, S. P.; Sen, S. S.; Roesky, H. W.; Objartel, I.; Stalke, D. Elegant Approach to Spacer Arranged Silagermylene and Bis(germylene) Compounds. *Chem. Commun.* **2011,** *47*, 7206–7208.
251. Jana, A.; Samuel, P. P.; Tavcar, G.; Roesky, H. W.; Schulzke, C. Selective Aromatic C–F and C–H Bond Activation With Silylenes of Different Coordinate Silicon. *J. Am. Chem. Soc.* **2010,** *132*, 10164–10170.
252. Samuel, P. P.; Singh, A. P.; Sarish, S. P.; Matussek, J.; Objartel, I.; Roesky, H. W., et al. Oxidative Addition Versus Substitution Reactions of Group 14 Dialkylamino Metalylenes With Pentafluoropyridine. *Inorg. Chem.* **2013,** *52*, 1544–1549.

253. Azhakar, R.; Roesky, H. W.; Wolf, H.; Stalke, D. Metal Free and Selective Activation of one C–F Bond in a Bound CF_3 Group. *Chem. Commun.* **2013,** *49*, 1841–1843.
254. Khan, S.; Sen, S. S.; Kratzert, D.; Tavcar, G.; Roesky, H. W.; Stalke, D. Synthesis of Stable Silicon Heterocycles by Reaction of Organic Substrates With a Chlorosilylene [PhC($N^tBu)_2$SiCl]. *Chem. Eur. J.* **2011,** *17*, 4283–4290.
255. Sarish, S. P.; Jana, A.; Roesky, H. W.; Samuel, P. P.; Abad Andrade, C. E.; Dittrich, B., et al. Synthesis of a Lewis Base Stabilized Dimer of *N*-Substituted Hydrosilahydrazone and a Silaaziridine. *Organometallics* **2011,** *30*, 912–916.
256. Sen, S. S.; Hey, J.; Kratzert, D.; Roesky, H. W.; Stalke, D. A Remarkable End-On Activation of Diazoalkane and Cleavage of Both C–Cl Bonds of Dichloromethane With a Silylene to a Single Product With Five-Coordinate Silicon Atoms. *Organometallics* **2012,** *31*, 435–439.
257. Ghadwal, R. S.; Sen, S. S.; Roesky, H. W.; Granitzka, M.; Kratzert, D.; Merkel, S., et al. Convenient Access to Monosilicon Epoxides With Pentacoordinate Silicon. *Angew. Chem. Int. Ed.* **2010,** *49*, 3952–3955.
258. Azhakar, R.; Roesky, H. W.; Ghadwal, R. S.; Holstein, J. J.; Dittrich, B. An Access to Base-Stabilized Three-Membered Silicon Heterocycles. *Dalton Trans.* **2012,** *41*, 9601–9603.
259. Azhakar, R.; Roesky, H. W.; Wolf, H.; Stalke, D. Reactivity of Stable Heteroleptic Silylene PhC ($N^tBu)_2SiNPh_2$ Toward Diazobenzene and *N*-Benzylideneaniline. *Organometallics* **2012,** *31*, 8608–8612.
260. Azhakar, R.; Roesky, H. W.; Holstein, J. J.; Dittrich, B. Reaction of the Silylene PhC($N^tBu)_2$SitBu With 4,4′-Bis(dimethylamino)thiobenzophenone and Treatment of the Silylene PhC($N^tBu)_2SiC(SiMe_3)_3$ With 3,5-di-*tert*-butyl-*o*-benzoquinone. *Eur. J. Inorg. Chem.* **2013,** *2013*, 2777–2781.
261. Azhakar, R.; Roesky, H. W.; Wolf, H.; Stalke, D. On the Reactivity of the Silylene PhC($N^tBu)_2SiNPh_2$ Toward Organic Substrates. *Z. Anorg. Allg. Chem.* **2013,** *639*, 934–938.
262. Tavcar, G.; Sen, S. S.; Roesky, H. W.; Hey, J.; Kratzert, D.; Stalke, D. Reactions of a Bis-Silylene (LSi-SiL, L = PhC($N^tBu)_2$) and a Heteroleptic Chloro Silylene (LSiCl) With Benzil: Formation of Bis (siladioxolene) and Monosiladioxolene Analogue With Five-Coordinate Silicon Atoms in Both Ring Systems. *Organometallics* **2010,** *29*, 3930–3935.
263. Khan, S.; Sen, S. S.; Michel, R.; Kratzert, D.; Roesky, H. W.; Stalke, D. Formation of a Unsymmetrical Ring System via C–H Bond Activation of Diazobenzene by Stable *N*-Heterocyclic Chlorosilylene (PhC ($N^tBu)_2$SiCl). *Organometallics* **2011,** *30*, 2643–2645.
264. Jana, A.; Azhakar, R.; Sarish, S. P.; Samuel, P. P.; Roesky, H. W.; Schulzke, C., et al. Reactions of Stable Amidinate Chlorosilylene and [1 + 4]-Oxidative Addition of *N*-Heterocyclic Silylene With *N*-Benzylideneaniline. *Eur. J. Inorg. Chem.* **2011,** *2011*, 5006–5013.
265. Azhakar, R.; Proepper, K.; Dittrich, B.; Roesky, H. W. Reactivity Studies of Heteroleptic Silylenes With N_2O. *Organometallics* **2012,** *31*, 7586–7590.
266. Jana, A.; Leusser, D.; Objartel, I.; Roesky, H. W.; Stalke, D. A Stable Silicon(II) Monohydride. *Dalton Trans.* **2011,** *40*, 5458–5463.
267. Shan, Y.-L.; Yim, W.-L.; So, C.-W. An *N*-Heterocyclic Silylene-Stabilized Digermanium(0) Complex. *Angew. Chem. Int. Ed.* **2014,** *53*, 13155–13158.
268. Percival, P. W.; Brodovitch, J.-C.; Mozafari, M.; Mitra, A.; West, R.; Ghadwal, R. S., et al. Free Radical Reactivity of Mono- and Dichlorosilylene With Muonium. *Chem. Eur. J.* **2011,** *17*, 11970–11973.
269. Schaefer, S.; Koeppe, R.; Gamer, M. T.; Roesky, P. W. Zinc-Silylene Complexes. *Chem. Commun.* **2014,** *50*, 11401–11403.
270. Wiberg, N.; Amelunxen, K.; Lerner, H. W.; Nöth, H.; Appel, A.; Knizek, J., et al. Disupersilylmetals ($^tBu_3Si)_2$M and Supersilylmetal Halides $^tBu_3$SiMX With M = Zn, Cd, Hg: Syntheses, Properties, Structures. *Z. Anorg. Allg. Chem.* **1997,** *623*, 1861–1870.
271. Jana, A.; Sarish, S. P.; Roesky, H. W.; Leusser, D.; Objartel, I.; Stalke, D. Pentafluoropyridine as a Fluorinating Reagent for Preparing a Hydrocarbon Soluble β-Diketiminatolead(II) Monofluoride. *Chem. Commun.* **2011,** *47*, 5434–5436.
272. Azhakar, R.; Roesky, H. W.; Holstein, J. J.; Proepper, K.; Dittrich, B. Reactivity Studies of Heteroleptic Silylenes PhC($N^tBu)_2$SiX (X = NPh_2, NMe_2) Toward Selected Azides. *Organometallics* **2013,** *32*, 358–361.

273. Chan, Y.-C.; Li, Y.; Ganguly, R.; So, C.-W. Acyclic Amido-Containing Silanechalcogenones. *Eur. J. Inorg. Chem.* **2015,** *2015*, 3821–3824.
274. Ghadwal, R. S.; Azhakar, R.; Roesky, H. W.; Proepper, K.; Dittrich, B.; Klein, S., et al. Donor-Acceptor-Stabilized Silicon Analogue of an Acid Anhydride. *J. Am. Chem. Soc.* **2011,** *133*, 17552–17555.
275. Zhang, S.-H.; Yeong, H.-X.; So, C.-W. Reactivity of a Silylsilylene Bearing a Functionalized Diaminochlorosilyl Substituent. *Chem. Eur. J.* **2011,** *17*, 3490–3499.
276. Zhang, S.-H.; Xi, H.-W.; Lim, K. H.; So, C.-W. Synthesis and Characterization of an Amidinate-Stabilized Siladithiocarboxylate and Its Germanium(II) Complex. *Organometallics* **2011,** *30*, 3686–3689.
277. Sen, S. S.; Khan, S.; Roesky, H. W.; Kratzert, D.; Meindl, K.; Henn, J., et al. Zwitterionic Si-C-Si-P and Si-P-Si-P Four-Membered Rings With Two-Coordinate Phosphorus Atoms. *Angew. Chem. Int. Ed.* **2011,** *50*, 2322–2325.
278. Breit, N. C.; Szilvasi, T.; Inoue, S. New Route to Access an Acyl-Functionalized Phosphasilene and a Four-Membered Si-P-C-O Heterocycle. *Chem. Eur. J.* **2014,** *20*, 9312–9318.
279. Khan, S.; Sen, S. S.; Roesky, H. W. Activation of Phosphorus by Group 14 Elements in Low Oxidation States. *Chem. Commun.* **2012,** *48*, 2169–2179.
280. Sen, S. S.; Hey, J.; Eckhardt, M.; Herbst-Irmer, R.; Maedl, E.; Mata, R. A., et al. A Stable Cation of a CSi_3P Five-Membered Ring With a Weakly Coordinating Chloride Anion. *Angew. Chem. Int. Ed.* **2011,** *50*, 12510–12513.
281. Inoue, S.; Epping, J. D.; Irran, E.; Driess, M. Formation of a Donor-Stabilized Tetrasilacyclobutadiene Dication by a Lewis Acid Assisted Reaction of an *N*-Heterocyclic Chloro Silylene. *J. Am. Chem. Soc.* **2011,** *133*, 8514–8517.
282. Khan, S.; Michel, R.; Sen, S. S.; Roesky, H. W.; Stalke, D. A P_4 Chain and Cage From Silylene-Activated White Phosphorus. *Angew. Chem. Int. Ed.* **2011,** *50*, 11786–11789.
283. Pyykko, P.; Atsumi, M. Molecular Single-Bond Covalent Radii for Elements 1-118. *Chem. Eur. J.* **2009,** *15*, 186–197.
284. Blom, B.; Driess, M.; Gallego, D.; Inoue, S. Facile Access to Silicon-Functionalized Bis-Silylene Titanium(II) Complexes. *Chem. Eur. J.* **2012,** *18*, 13355–13360.
285. Azhakar, R.; Ghadwal, R. S.; Roesky, H. W.; Wolf, H.; Stalke, D. Stabilization of Low Valent Silicon Fluorides in the Coordination Sphere of Transition Metals. *J. Am. Chem. Soc.* **2012,** *134*, 2423–2428.
286. Breit, N. C.; Szilvasi, T.; Inoue, S. Facile Rotation Around a Silicon-Phosphorus Double Bond Enabled Through Coordination to Tungsten. *Chem. Commun.* **2015,** *51*, 11272–11275.
287. Azhakar, R.; Sarish, S. P.; Roesky, H. W.; Hey, J.; Stalke, D. Syntheses of Group 7 Metal Carbonyl Complexes With a Stable *N*-Heterocyclic Chlorosilylene. *Inorg. Chem.* **2011,** *50*, 5039–5043.
288. Azhakar, R.; Roesky, H. W.; Holstein, J. J.; Dittrich, B. The Group 7 Metal Carbonyl Complexes From a Stable Heteroleptic Silylene $PhC(N^tBu)_2SiNPh_2$. *Dalton Trans.* **2012,** *41*, 12096–12100.
289. Yang, W.; Fu, H.; Wang, H.; Chen, M.; Ding, Y.; Roesky, H. W., et al. A Base-Stabilized Silylene With a Tricoordinate Silicon Atom as a Ligand for a Metal Complex. *Inorg. Chem.* **2009,** *48*, 5058–5060.
290. Blom, B.; Pohl, M.; Tan, G.; Gallego, D.; Driess, M. From Unsymmetrically Substituted Benzamidinato and Guanidinato Dichlorohydridosilanes to Novel Hydrido *N*-Heterocyclic Silylene Iron Complexes. *Organometallics* **2014,** *33*, 5272–5282.
291. Zybill, C.; Mueller, G. Synthesis and Structure of the Donor-Stabilized Silylene (Silanediyl) Complexes $(t\text{-}C_4H_9O)_2Si = Cr(CO)_5 \cdot HMPT$ and $(t\text{-}C_4H_9O)_2Si = Fe(CO)_4 \cdot HMPT$. *Organometallics* **1988,** *7*, 1368–1372.
292. Blom, B.; Enthaler, S.; Inoue, S.; Irran, E.; Driess, M. Electron-Rich *N*-Heterocyclic Silylene (NHSi)-Iron Complexes: Synthesis, Structures, and Catalytic Ability of an Isolable Hydridosilylene-Iron Complex. *J. Am. Chem. Soc.* **2013,** *135*, 6703–6713.
293. Wilson, S. T.; Coville, N. J.; Shapely, J. R.; Osborn, J. A. Rapid Intramolecular Rearrangements in Pentacoordinate Transition Metal Compounds. V. Coupling of Olefin Rotation and Berry Pseudorotation in Tetracarbonyliron-Olefin Complexes. *J. Am. Chem. Soc.* **1974,** *96*, 4038–4040.

294. Azhakar, R.; Ghadwal, R. S.; Roesky, H. W.; Hey, J.; Stalke, D. Facile Access to Transition-Metal-Carbonyl Complexes With an Amidinate-Stabilized Chlorosilylene Ligand. *Chem. Asian J.* **2012,** *7*, 528–533.
295. Azhakar, R.; Ghadwal, R. S.; Roesky, H. W.; Hey, J.; Krause, L.; Stalke, D. Mixed Valence η^6-Arene Cobalt(i) and Cobalt(ii) Compound. *Dalton Trans.* **2013,** *42*, 10277–10281.
296. Tavcar, G.; Sen, S. S.; Azhakar, R.; Thorn, A.; Roesky, H. W. Facile Syntheses of Silylene Nickel Carbonyl Complexes From Lewis Base Stabilized Chlorosilylenes. *Inorg. Chem.* **2010,** *49*, 10199–10202.
297. Braga, D.; Grepioni, F.; Orpen, A. G. Nickel Carbonyl [$Ni(CO)_4$] and Iron Carbonyl [$Fe(CO)_5$]: Molecular Structures in the Solid State. *Organometallics* **1993,** *12*, 1481–1483.
298. Breit, N. C.; Szilvasi, T.; Suzuki, T.; Gallego, D.; Inoue, S. From a Zwitterionic Phosphasilene to Base Stabilized Silyliumylidene-Phosphide and Bis(silylene) Complexes. *J. Am. Chem. Soc.* **2013,** *135*, 17958–17968.
299. Wang, W.; Inoue, S.; Yao, S.; Driess, M. An Isolable Bis-Silylene Oxide ("Disilylenoxane") and Its Metal Coordination. *J. Am. Chem. Soc.* **2010,** *132*, 15890–15892.
300. Tan, G.; Blom, B.; Gallego, D.; Driess, M. Facile Access to Mono- and Dinuclear Heteroleptic *N*-Heterocyclic Silylene Copper Complexes. *Organometallics* **2014,** *33*, 363–369.
301. Sen, S. S.; Jana, A.; Roesky, H. W.; Schulzke, C. A Remarkable Base-Stabilized Bis(silylene) With a Silicon(I)-Silicon(I) Bond. *Angew. Chem. Int. Ed.* **2009,** *48*, 8536–8538.
302. Sen, S. S.; Tavcar, G.; Roesky, H. W.; Kratzert, D.; Hey, J.; Stalke, D. Synthesis of a Stable Four-Membered Si_2O_2 Ring and a Dimer With Two Four-Membered Si_2O_2 Rings Bridged by Two Oxygen Atoms, With Five-Coordinate Silicon Atoms in both Ring Systems. *Organometallics* **2010,** *29*, 2343–2347.
303. Sen, S. S.; Khan, S.; Kratzert, D.; Roesky, H. W.; Stalke, D. Reaction of a Base-Stabilized Bis(silylene) $[PhC(N^tBu)_2Si]_2$ With Cyclooctatetraene Without Cleavage of the Si–Si Bond. *Eur. J. Inorg. Chem.* **2011,** 1370–1373.
304. Sen, S. S.; Roesky, H. W.; Meindl, K.; Stern, D.; Henn, J.; Stueckl, A. C., et al. Synthesis, Structure, and Theoretical Investigation of Amidinato Supported 1,4-Disilabenzene. *Chem. Commun.* **2010,** *46*, 5873–5875.
305. Yeong, H.-X.; Xi, H.-W.; Lim, K. H.; So, C.-W. Synthesis and Characterization of an Amidinate-Stabilized *cis*-1,2-Disilylenylethene [*cis*-LSi{C(Ph) = C(H)}SiL] and a Singlet Delocalized Biradicaloid $[LSi(\mu^2\text{-}C_2Ph_2)_2SiL]$ ($L = PhC(N^tBu)_2$). *Chem. Eur. J.* **2010,** *16*, 12956–12961.
306. Kinjo, R.; Ichinohe, M.; Sekiguchi, A.; Takagi, N.; Sumimoto, M.; Nagase, S. Reactivity of a Disilyne RSi≡SiR ($R = Si^iPr[CH(SiMe_3)_2]_2$) Toward π-Bonds: Stereospecific Addition and a New Route to an Isolable 1,2-Disilabenzene. *J. Am. Chem. Soc.* **2007,** *129*, 7766–7767.
307. Takeuchi, K.; Ichinohe, M.; Sekiguchi, A. Reactivity of the Disilyne RSi≡SiR ($R = Si^iPr[CH(SiMe_3)_2]_2$) Toward Silylcyanide: Two Pathways to Form the Bis-Adduct [$RSiSiR(CNSiMe_3)_2$] With Some Silaketenimine Character and a 1,4-Diaza-2,3-Disilabenzene Analogue. *J. Am. Chem. Soc.* **2008,** *130*, 16848–16849.
308. Han, J. S.; Sasamori, T.; Mizuhata, Y.; Tokitoh, N. Reactivity of an Aryl-Substituted Silicon-Silicon Triple Bond: 1,2-Disilabenzenes From the Reactions of a 1,2-Diaryldisilyne With Alkynes. *Dalton Trans.* **2010,** *39*, 9238–9240.
309. Takeuchi, K.; Ichinohe, M.; Sekiguchi, A.; Guo, J.-D.; Nagase, S. Reactivity of the Disilyne RSiSiR ($R = Si^iPr[CH(SiMe_3)_2]_2$) Toward Bis(silylcyanide) Forming a 1,4-Diaza-2,3-Disilabenzene Analog. *J. Phys. Org. Chem.* **2009,** 390–394.
310. Zhang, S.-H.; Xi, H.-W.; Lim, K. H.; Meng, Q.; Huang, M.-B.; So, C.-W. Synthesis and Characterization of a Singlet Delocalized 2,4-Diimino-1,3-Disilacyclobutanediyl and a Silylenylsilaimine. *Chem. Eur. J.* **2012,** *18*, 4258–4263.
311. Yeong, H.-X.; Xi, H.-W.; Li, Y.; Lim, K. H.; So, C.-W. A Silyliumylidene Cation Stabilized by an Amidinate Ligand and 4-Dimethylaminopyridine. *Chem. Eur. J.* **2013,** *19*, 11786–11790.

312. Yeong, H.-X.; Li, Y.; So, C.-W. A Base-Stabilized Silyliumylidene Cation as a Ligand for Rhodium and Tungsten Complexes. *Organometallics* **2014,** *33*, 3646–3648.
313. Yeong, H.-X.; Zhang, S.-H.; Xi, H.-W.; Guo, J.-D.; Lim, K. H.; Nagase, S., et al. An Amidinate-Stabilized Germatrisilacyclobutadiene Ylide. *Chem. Eur. J.* **2012,** *18*, 2685–2691.
314. Zhang, S.-H.; Xi, H.-W.; Lim, K. H.; So, C.-W. An Extensive n,π,σ-Electron Delocalized Si_4 Ring. *Angew. Chem. Int. Ed.* **2013,** *52*, 12364–12367.
315. Wang, W.; Inoue, S.; Enthaler, S.; Driess, M. Bis(silylenyl)- and Bis(germylenyl)-Substituted Ferrocenes: Synthesis, Structure, and Catalytic Applications of Bidentate Silicon(II)-Cobalt Complexes. *Angew. Chem. Int. Ed.* **2012,** *51*, 6167–6171.
316. Wang, W.; Inoue, S.; Irran, E.; Driess, M. Synthesis and Unexpected Coordination of a Silicon(II)-Based SiCSi Pincerlike Arene to Palladium. *Angew. Chem. Int. Ed.* **2012,** *51*, 3691–3694.
317. Brueck, A.; Gallego, D.; Wang, W.; Irran, E.; Driess, M.; Hartwig, J. F. Pushing the σ-Donor Strength in Iridium Pincer Complexes: Bis(silylene) and Bis(germylene) Ligands are Stronger Donors than Bis(phosphorus(III)) Ligands. *Angew. Chem. Int. Ed.* **2012,** *51*, 11478–11482.
318. Gallego, D.; Brueck, A.; Irran, E.; Meier, F.; Kaupp, M.; Driess, M., et al. From Bis(silylene) and Bis(germylene) Pincer-Type Nickel(II) Complexes to Isolable Intermediates of the Nickel-Catalyzed Sonogashira Cross-Coupling Reaction. *J. Am. Chem. Soc.* **2013,** *135*, 15617–15626.
319. Gallego, D.; Inoue, S.; Blom, B.; Driess, M. Highly Electron-Rich Pincer-Type Iron Complexes Bearing Innocent Bis(metallylene)pyridine Ligands: Syntheses, Structures, and Catalytic activity. *Organometallics* **2014,** *33*, 6885–6897.
320. Junold, K.; Baus, J. A.; Burschka, C.; Tacke, R. Bis[N,N′-Diisopropylbenzamidinato(-)]silicon(II): A Silicon(II) Compound With Both a Bidentate and a Monodentate Amidinato Ligand. *Angew. Chem. Int. Ed.* **2012,** *51*, 7020–7023.
321. Junold, K.; Nutz, M.; Baus, J. A.; Burschka, C.; Guerra, C. F.; Bickelhaupt, F. M., et al. The Donor-Stabilized Silylene Bis[N,N′-Diisopropylbenzamidinato(-)]silicon(II): Synthesis, Electronic Structure, and Reactivity. *Chem. Eur. J.* **2014,** *20*, 9319–9329.
322. Mück, F. M.; Junold, K.; Baus, J. A.; Burschka, C.; Tacke, R. Donor-Stabilized Silylenes With Guanidinato Ligands. *Eur. J. Inorg. Chem.* **2013,** *2013*, 5821–5825.
323. Mück, F. M.; Kloss, D.; Baus, J. A.; Burschka, C.; Bertermann, R.; Poater, J., et al. Stable Four-Coordinate Guanidinatosilicon(IV) Complexes With SiN_3El Skeletons (El = S, Se, Te) and Si = El Double Bonds. *Chem. Eur. J.* **2015,** *21*, 14011–14021.
324. Junold, K.; Baus, J. A.; Burschka, C.; Auerhammer, D.; Tacke, R. Stable Five-Coordinate Silicon(IV) Complexes With SiN_4 X Skeletons (X = S, Se, Te) and Si = X Double Bonds. *Chem. Eur. J.* **2012,** *18*, 16288–16291.
325. Mück, F. M.; Ulmer, A.; Baus, J. A.; Burschka, C.; Tacke, R. Thermally Stable Four-Coordinate Silicon(IV) Complexes With an Si = N Double Bond and an SiN_3El Skeleton (El = O, S, Se, Te). *Eur. J. Inorg. Chem.* **2015,** *2015*, 1860–1864.
326. Junold, K.; Baus, J. A.; Burschka, C.; Finze, M.; Tacke, R. Selective C–H Bond Activation of 1,2-Dicarba-Closo-Dodecaborane by the Donor-Stabilized Silylene Bis[N,N′-Diisopropylbenzamidinato(-)]silicon(II). *Eur. J. Inorg. Chem.* **2014,** *2014*, 5099–5102.
327. Junold, K.; Mueck, F. M.; Kupper, C.; Baus, J. A.; Burschka, C.; Tacke, R. Activation of Sulfur Dioxide by Bis[N,N′-Diisopropylbenzamidinato(-)]silicon(II): Synthesis of Neutral Six-Coordinate Silicon(IV) Complexes With Chelating *O,O′*-Sulfito or *O,O′*-Dithionito Ligands. *Chem. Eur. J.* **2014,** *20*, 12781–12785.
328. Junold, K.; Baus, J. A.; Burschka, C.; Fonseca Guerra, C.; Bickelhaupt, F. M.; Tacke, R. Bis[N,N′-Diisopropylbenzamidinato(-)]silicon(II): Lewis Acid/Base Reactions With Triorganylboranes. *Chem. Eur. J.* **2014,** *20*, 12411–12415.
329. Junold, K.; Sinner, K.; Baus, J. A.; Burschka, C.; Fonseca Guerra, C.; Bickelhaupt, F. M., et al. Reactions of the Donor-Stabilized Silylene Bis[N,N′-Diisopropyl-Benzamidinato(-)]silicon(II) With Brønsted Acids. *Chem. Eur. J.* **2014,** *20*, 16462–16466.

330. Laskowski, N.; Junold, K.; Kupper, C.; Baus, J. A.; Burschka, C.; Tacke, R. Bis[N,N′-Diisopropylbenzamidinato(-)]silicon(II): Cycloaddition Reactions With Organic 1,3-Dienes and 1,2-Diketones. *Organometallics* **2014,** *33*, 6141–6148.

331. Junold, K.; Baus, J. A.; Burschka, C.; Vent-Schmidt, T.; Riedel, S.; Tacke, R. Five-Coordinate Silicon(II) Compounds With Si–M Bonds (M = Cr, Mo, W, Fe): Bis[N,N′-Diisopropylbenzamidinato(-)]silicon(II) as a Ligand in Transition-Metal Complexes. *Inorg. Chem.* **2013,** *52*, 11593–11599.

332. Mück, F. M.; Kloss, D.; Baus, J. A.; Burschka, C.; Tacke, R. Novel Transition-Metal (M = Cr, Mo, W, Fe) Carbonyl Complexes With Bis(guanidinato)silicon(II) Ligands. *Chem. Eur. J.* **2014,** *20*, 9620–9626.

333. Gau, D.; Kato, T.; Saffon-Merceron, N.; Cossio, F. P.; Baceiredo, A. Stable Phosphonium Sila-Ylide With Reactivity as a Sila-Wittig Reagent. *J. Am. Chem. Soc.* **2009,** *131*, 8762–8763.

334. Driess, M.; Barmeyer, R.; Monsé, C.; Merz, K. E. $(SiMe_3)_4^+$ Ions (E = P, As): Persilylated Phosphonium and Arsonium Ions. *Angew. Chem. Int. Ed.* **2001,** *40*, 2308–2310.

335. Rodriguez, R.; Gau, D.; Contie, Y.; Kato, T.; Saffon-Merceron, N.; Baceiredo, A. Synthesis of a Phosphine-Stabilized Silicon(II) Hydride and Its Addition to Olefins: A Catalyst-Free Hydrosilylation Reaction. *Angew. Chem. Int. Ed. Engl.* **2011,** *50*, 11492–11495.

336. Gau, D.; Kato, T.; Saffon-Merceron, N.; De Cozar, A.; Cossio, F. P.; Baceiredo, A. Synthesis and Structure of a Base-Stabilized C-Phosphino-Si-Amino Silyne. *Angew. Chem. Int. Ed.* **2010,** *49*, 6585–6588.

337. Gau, D.; Rodriguez, R.; Kato, T.; Saffon-Merceron, N.; Cossio, F. P.; Baceiredo, A. Nucleophilic Silylenoid Character of Stable Phosphonium Sila-Ylides. *Chem. Eur. J.* **2010,** *16*, 8255–8258.

338. Gau, D.; Rodriguez, R.; Kato, T.; Saffon-Merceron, N.; Baceiredo, A. Diastereoselective Synthesis of Bulky, Strongly Nucleophilic, and Configurationally Stable P-Stereogenic Tricyclic Phosphines. *J. Am. Chem. Soc.* **2010,** *132*, 12841–12843.

339. Gau, D.; Rodriguez, R.; Kato, T.; Saffon-Merceron, N.; de Cozar, A.; Cossio, F. P., et al. Synthesis of a Stable Disilyne Bisphosphine Adduct and Its Non-Metal-Mediated CO_2 Reduction to CO. *Angew. Chem. Int. Ed. Engl.* **2011,** *50*, 1092–1096.

340. Rodriguez, R.; Gau, D.; Kato, T.; Saffon-Merceron, N.; De Cozar, A.; Cossio, F. P., et al. Reversible Binding of Ethylene to Silylene-Phosphine Complexes at Room Temperature. *Angew. Chem. Int. Ed.* **2011,** *50*, 10414–10416.

341. Rodriguez, R.; Troadec, T.; Kato, T.; Saffon-Merceron, N.; Sotiropoulos, J.-M.; Baceiredo, A. Synthesis and Characterization of an Isolable Base-Stabilized Silacycloprop-1-Ylidene. *Angew. Chem. Int. Ed.* **2012,** *51*, 7158–7161.

342. Nakata, N.; Rodriguez, R.; Troadec, T.; Saffon-Merceron, N.; Sotiropoulos, J. M.; Baceiredo, A., et al. A stable Silene Substituted by Strong Pi-Donors at the Silicon Center. *Angew. Chem. Int. Ed. Engl.* **2013,** *52*, 10840–10844.

343. Rodriguez, R.; Troadec, T.; Gau, D.; Saffon-Merceron, N.; Hashizume, D.; Miqueu, K., et al. Synthesis of a Donor-Stabilized Silacyclopropan-1-One. *Angew. Chem. Int. Ed.* **2013,** *52*, 4426–4430.

344. Rodriguez, R.; Gau, D.; Troadec, T.; Saffon-Merceron, N.; Branchadell, V.; Baceiredo, A., et al. A Base-Stabilized Sila-β-Lactone and a Donor/Acceptor-Stabilized Silanoic Acid. *Angew. Chem. Int. Ed. Engl.* **2013,** *52*, 8980–8983.

345. Troadec, T.; Lopez Reyes, M.; Rodriguez, R.; Baceiredo, A.; Saffon-Merceron, N.; Branchadell, V., et al. Donor-Stabilized Silacyclobutanone: A Precursor of 1-Silaketene Via Retro-[2 + 2]-Cycloaddition Reaction at Room Temperature. *J. Am. Chem. Soc.* **2016,** *138*, 2965–2968.

346. Rodriguez, R.; Contie, Y.; Gau, D.; Saffon-Merceron, N.; Miqueu, K.; Sotiropoulos, J.-M., et al. Reversible Insertion of Unactivated Alkenes into Silicon(II)-Tin Bonds. *Angew. Chem. Int. Ed.* **2013,** *52*, 8437–8440.

347. Hansch, C.; Leo, A.; Taft, R. W. A Survey of Hammett Substituent Constants and Resonance and Field Parameters. *Chem. Rev.* **1991,** *91*, 165–195.

348. Rodriguez, R.; Contie, Y.; Mao, Y.; Saffon-Merceron, N.; Baceiredo, A.; Branchadell, V., et al. Reversible Dimerization of Phosphine-Stabilized Silylenes by Silylene Insertion into Si(II) -H and Si(II) -Cl σ-Bonds At Room Temperature. *Angew. Chem. Int. Ed.* **2015,** *54*, 15276–15279.
349. Filippou, A. C.; Lebedev, Y. N.; Chernov, O.; Strassmann, M.; Schnakenburg, G. Silicon(II) Coordination Chemistry: *N*-Heterocyclic Carbene Complexes of Si^{2+} and SiI^{+}. *Angew. Chem. Int. Ed.* **2013,** *52*, 6974–6978.
350. Filippou, A. C.; Chernov, O.; Blom, B.; Stumpf, K. W.; Schnakenburg, G. Stable *N*-Heterocyclic Carbene Adducts of Arylchlorosilylenes and Their Germanium Homologues. *Chem. Eur. J.* **2010,** *16*, 2866–2872.
351. Suzuki, K.; Matsuo, T.; Hashizume, D.; Tamao, K. Room-Temperature Dissociation of 1,2-Dibromodisilenes to Bromosilylenes. *J. Am. Chem. Soc.* **2011,** *133*, 19710–19713.
352. Ghana, P.; Arz, M. I.; Das, U.; Schnakenburg, G.; Filippou, A. C. Si = Si Double Bonds: Synthesis of an NHC-Stabilized Disilavinylidene. *Angew. Chem. Int. Ed.* **2015,** *54*, 9980–9985.
353. Cui, H.; Cui, C. Silylation of *N*-Heterocyclic Carbene With Aminochlorosilane and -Disilane: Dehydrohalogenation vs. Si–Si Bond Cleavage. *Dalton Trans.* **2011,** *40*, 11937–11940.
354. Al-Rafia, S. M.; McDonald, R.; Ferguson, M. J.; Rivard, E. Preparation of Stable Low-Oxidation-State Group 14 Element Amidohydrides and Hydride-Mediated Ring-Expansion Chemistry of *N*-Heterocyclic Carbenes. *Chem. Eur. J.* **2012,** *18*, 13810–13820.
355. Jana, A.; Tavcar, G.; Roesky, H. W.; Schulzke, C. Facile Synthesis of Dichlorosilane by Metathesis Reaction and Dehydrogenation of Dihydrogermane by a Frustrated Lewis Pair. *Dalton Trans.* **2010,** *39*, 6217–6220.
356. Mondal, K. C.; Roesky, H. W.; Schwarzer, M. C.; Frenking, G.; Tkach, I.; Wolf, H., et al. Conversion of a Singlet Silylene to a Stable Biradical. *Angew. Chem. Int. Ed.* **2013,** *52*, 1801–1805.
357. Jana, A.; Huch, V.; Scheschkewitz, D. NHC-Stabilized Silagermenylidene: A Heavier Analogue of Vinylidene. *Angew. Chem. Int. Ed.* **2013,** *52*, 12179–12182.
358. Geiss, D.; Arz, M. I.; Strassmann, M.; Schnakenburg, G.; Filippou, A. C. Si = P Double Bonds: Experimental and Theoretical Study of an NHC-Stabilized Phosphasilenylidene. *Angew. Chem. Int. Ed.* **2015,** *54*, 2739–2744.
359. Li, J.; Merkel, S.; Henn, J.; Meindl, K.; Doring, A.; Roesky, H. W., et al. Lewis-Base-Stabilized Dichlorosilylene: A Two-Electron σ-Donor Ligand. *Inorg. Chem.* **2010,** *49*, 775–777.
360. Al-Rafia, S. M. I.; Malcolm, A. C.; McDonald, R.; Ferguson, M. J.; Rivard, E. Efficient Generation of Stable Adducts of Si(II) Dihydride Using a Donor-Acceptor Approach. *Chem. Commun.* **2012,** *48*, 1308–1310.
361. Inoue, S.; Eisenhut, C. A Dihydrodisilene Transition Metal Complex From an *N*-Heterocyclic Carbene-Stabilized Silylene Monohydride. *J. Am. Chem. Soc.* **2013,** *135*, 18315–18318.
362. Eisenhut, C.; Szilvasi, T.; Breit, N. C.; Inoue, S. Reaction of an *N*-Heterocyclic Carbene-Stabilized Silicon(II) Monohydride With Alkynes: [2 + 2 + 1] Cycloaddition Versus Hydrogen Abstraction. *Chem. Eur. J.* **2015,** *21*, 1949–1954.
363. Tanaka, H.; Ichinohe, M.; Sekiguchi, A. An Isolable NHC-Stabilized Silylene Radical Cation: Synthesis and Structural Characterization. *J. Am. Chem. Soc.* **2012,** *134*, 5540–5543.
364. Gao, Y.; Zhang, J.; Hu, H.; Cui, C. Base-Stabilized 1-Silacyclopenta-2,4-Dienylidenes. *Organometallics* **2010,** *29*, 3063–3065.
365. Gao, Y.; Hu, H.; Cui, C. The Reactivity of a Silacyclopentadienylidene Towards Aldehydes: Silole Ring Expansion and the Formation of Base-Stabilized Silacyclohexadienones. *Chem. Eur. J.* **2011,** *17*, 8803–8806.
366. Abraham, M. Y.; Wang, Y.; Xie, Y.; Wei, P.; Schaefer, H. F.; Schleyer, P. v R., et al. Cleavage of Carbene-Stabilized Disilicon. *J. Am. Chem. Soc.* **2011,** *133*, 8874–8876.
367. Ghadwal, R. S.; Roesky, H. W.; Merkel, S.; Stalke, D. Ambiphilicity of Dichlorosilylene in a Single Molecule. *Chem. Eur. J.* **2010,** *16*, 85–88.

368. Azhakar, R.; Tavčar, G.; Roesky, H. W.; Hey, J.; Stalke, D. Facile Synthesis of a Rare Chlorosilylene-BH_3 Adduct. *Eur. J. Inorg. Chem.* **2011,** *2011*, 475–477.

369. Al-Rafia, S. M.; Malcolm, A. C.; McDonald, R.; Ferguson, M. J.; Rivard, E. Trapping the Parent Inorganic Ethylenes H_2SiGeH_2 and H_2SiSnH_2 in the Form of Stable Adducts At Ambient Temperature. *Angew. Chem. Int. Ed.* **2011,** *50*, 8354–8357.

370. Al-Rafia, S. M.; Malcolm, A. C.; Liew, S. K.; Ferguson, M. J.; Rivard, E. Stabilization of the Heavy Methylene Analogues, GeH_2 and SnH_2, Within the Coordination Sphere of a Transition Metal. *J. Am. Chem. Soc.* **2011,** *133*, 777–779.

371. Evans, W. J.; Ulibarri, T. A.; Jutzi, P. Relative Reactivity of Decamethylsilicocene and Decamethylsamarocene: Reduction of $(C_5Me_5)_2SiCl_2$ by Sm(II) Reagents. *Inorg. Chim. Acta* **1990,** *168*, 5–6.

372. Kühler, T.; Jutzi, P. Decamethylsilicocene: Synthesis, Structure, Bonding and Chemistry. *Adv. Organomet. Chem.* **2003,** *49*, 1–34.

373. Jutzi, P. The Pentamethylcyclopentadienylsilicon(II) Cation: Synthesis, Characterization, and Reactivity. *Chem. Eur. J.* **2014,** *20*, 9192–9207.

374. Jutzi, P.; Mix, A.; Rummel, B.; Schoeller, W. W.; Neumann, B.; Stammler, H. G. The $(Me_5C_5)Si^+$ Cation: A Stable Derivative of HSi^+. *Science* **2004,** *305*, 849–851.

375. Jutzi, P.; Mix, A.; Neumann, B.; Rummel, B.; Stammler, H. G. Novel π-Complexes of Divalent Silicon: Mixed Substituted Neutral Sandwich Compounds and the Half-Sandwich Cation $(^iPr_5C_5)Si^+$. *Chem. Commun.* **2006,** 3519–3521.

376. Jutzi, P.; Mix, A.; Neumann, B.; Rummel, B.; Schoeller, W. W.; Stammler, H.-G., et al. Reversible Transformation of a Stable Monomeric Silicon(II) Compound into a Stable Disilene by Phase Transfer: Experimental and Theoretical Studies of the System $\{[(Me_3Si)_2N](Me_5C_5)Si\}_n$ With $n = 1,2$. *J. Am. Chem. Soc.* **2009,** *131*, 12137–12143.

377. Khan, S.; Sen, S. S.; Roesky, H. W.; Kratzert, D.; Michel, R.; Stalke, D. One Pot Synthesis of Disilatricycloheptene Analogue and Jutzi's Disilene. *Inorg. Chem.* **2010,** *49*, 9689–9693.

378. Hinchley, S. L.; Morrison, C. A.; Rankin, D. W. H.; Macdonald, C. L. B.; Wiacek, R. J.; Cowley, A. H., et al. Persistent Phosphinyl Radicals From a Bulky Diphosphine: An Example of a Molecular Jack-in-the-Box. *Chem. Commun.* **2000,** 2045–2046.

379. Hinchley, S. L.; Morrison, C. A.; Rankin, D. W. H.; Macdonald, C. L. B.; Wiacek, R. J.; Voigt, A., et al. Spontaneous Generation of Stable Pnictinyl Radicals From "Jack-in-the-Box" Dipnictines: A Solid-State, Gas-Phase, and Theoretical Investigation sakof the origins of steric stabilization. *J. Am. Chem. Soc.* **2001,** *123*, 9045–9053.

380. Fjeldberg, T.; Haaland, A.; Schilling, B. E. R.; Lappert, M. F.; Thorne, A. J. Subvalent Group 4B Metal Alkyls and Amides. Part 8. Germanium and Tin Carbene Analogues MR_2 [M = Ge or Sn, R = CH$(SiMe_3)_2$]: Syntheses and Structures in the Gas Phase (Electron Diffraction); Molecular-Orbital Calculations for MH_2 and $GeMe_2$. *J. Chem. Soc., Dalton Trans.* **1986,** 1551–1556.

381. Goldberg, D. E.; Hitchcock, P. B.; Lappert, M. F.; Thomas, K. M.; Thorne, A. J.; Fjeldberg, T., et al. Subvalent Group 4B Metal Alkyls and Amides. Part 9. Germanium and Tin Alkene Analogues, the Dimetallenes M_2R_4[M = Ge or Sn, R = CH$(SiMe_3)_2$]: X-ray Structures, Molecular Orbital Calculations for M_2H_4, and Trends in the Series $M_2R'_4$[M = C, Si, Ge, or Sn; R′ = R, Ph, $C_6H_2Me_3$-2,4,6, or $C_6H_2Et_2$-2,6]. *J. Chem. Soc., Dalton Trans.* **1986,** 2387–2394.

382. Guo, J.-D.; Nagase, S.; Power, P. P. Dispersion Force Effects on the Dissociation of "Jack-in-the-Box" Diphosphanes and Diarsanes. *Organometallics* **2015,** *34*, 2028–2033.

383. Khan, S.; Michel, R.; Koley, D.; Roesky, H. W.; Stalke, D. Reactivity Studies of a Disilene With N_2O and Elemental Sulfur. *Inorg. Chem.* **2011,** *50*, 10878–10883.

384. Jutzi, P.; Leszczynska, K.; Mix, A.; Neumann, B.; Rummel, B.; Schoeller, W., et al. Synthesis and Characterization of the Ferrio-Substituted Silicon(II) Compound $Me_5C_5(CO)_2FeSiC_5Me_5$. *Organometallics* **2010,** *29*, 4759–4761.

385. Malisch, W.; Möller, S.; Fey, O.; Wekel, H.-U.; Pikl, R.; Posset, U., et al. Synthese und Reaktivität von Silicium-Übergangsmetallkomplexen. XXIX. Trihydridosilyl-Komplexe von Eisen und Ruthenium: Ein einfacher Zugang, Ligandaustauschreaktionen am Metall und schwingungsspektroskopische Analyse. *J. Organomet. Chem.* **1996,** *507*, 117–124.
386. Inoue, S.; Leszczynska, K. An Acyclic Imino-Substituted Silylene: Synthesis, Isolation, and Its Facile Conversion into a Zwitterionic Silaimine. *Angew. Chem. Int. Ed.* **2012,** *51*, 8589–8593.
387. Jutzi, P.; Leszczynska, K.; Neumann, B.; Schoeller, W. W.; Stammler, H.-G. [2,6-$(Trip)_2H_3C_6$](Cp^*)Si: A Stable Monomeric Arylsilicon(II) Compound. *Angew. Chem. Int. Ed.* **2009,** *48*, 2596–2599.

CHAPTER 9

Multiple Bonds to Silicon (Recent Advances in the Chemistry of Silicon Containing Multiple Bonds)

Antoine Baceiredo and Tsuyoshi Kato

University of Toulouse, Toulouse, France

Chapter Outline

9.0 List of Abbreviations

Ad	Adamantyl
Ant	Anthracenyl
Ar	Aryl
COD	1,5-Cyclooctadiene
Cp	Cyclopentadienyl
Cp*	Pentamethylcyclopentadienyl
Cy	Cyclohexyl
DBU	1,8-Diazabicyclo[5.4.0]undec-7-ene
Dipp	2,6-Di(*iso*-propyl)phenyl
DMAP	4-(Dimethylamino)pyridine
***i*Pr**	*iso*-Propyl
Me	Metyl
Mes	2,4,6-Trimethylphenyl
Mes*	2,4,6-Tri(*tert*-butyl)phenyl
NHC	*N*-Heterocyclic carbene

Organosilicon Compounds.
DOI: http://dx.doi.org/10.1016/B978-0-12-801981-8.00009-5

Np	Naphthyl
OTf	Trifluoromethanesulfonate
***o*Tol**	*ortho*-Methylphenyl
Ph	Phenyl
R	Alkyl
***t*Bu**	*tert*-Butyl
THF	Tetrahydrofurane
Tip	2,4,6-Tri(*iso*-propyl)phenyl
TMDA	*N,N,N,N*-Tetramethylethylenediamine
TMS	Trimethylsilyl
Xyl	Dimethylphenyl

9.1 General Introduction

Over the previous 40 years the understanding of bonding in multiply bonded compounds involving main group elements has developed greatly, and the coined "Double bond rule" has been disproved.[1] In this context the first reports of the stable disilene and silene derivatives featuring Si=Si and Si=C double bonds, respectively, by the groups of West and Brook in 1981, have been major advances in silicon chemistry (Scheme 9.1).[2] Since that, the chemistry of stable low-coordinate silicon compounds has experienced considerable development, as reflected in many review articles published in the meantime.[3–12]

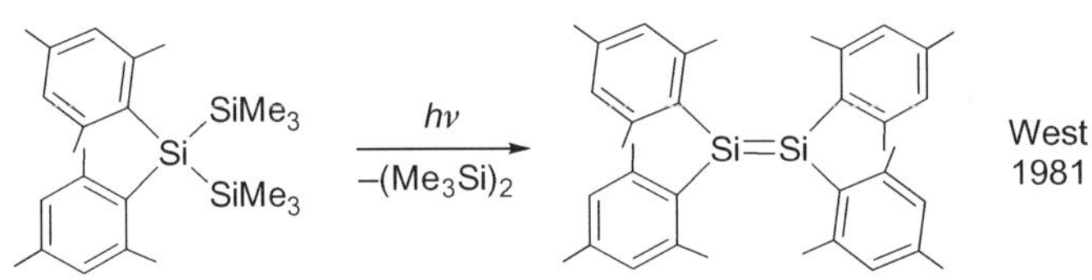

Emphasis in this chapter is placed mainly on the more recent developments of stable heteronuclear double and triple bond silicon species. For completeness, the numerous pioneering studies in this field as well as the studies of homonuclear disilenes and disilynes have not been neglected but rather briefly discussed, since they have already been summarized in numerous reviews. They include, in particular, a very recent review on disilene and disilyne chemistry from Iwamoto and Ishida in 2014[4g] as well as a book chapter on Si containing multiple bond compounds written by Lee and Sekiguchi in 2010.[3a] The literature has been covered mainly since 2000, although no claim for completeness is made, especially with regard to older publications.

9.2 Silicon Containing Double Bonds

9.2.1 Homonuclear Compounds (Si=Si Double Bond)

The synthesis of new stable disilenes is a constantly expanding field largely due to the great interest in the further reactivity of these species. Nowadays several synthetic methods for the preparation of disilenes are available, utilizing mainly two strategies: (1) starting from readily available low-oxidation state precursors such as silylenes (dimerization), or silynes (1,2-addition); and (2) reduction of higher oxidation state precursors (1,2-dihalodisilanes) (Scheme 9.1). In addition, the use of new silicon chemical tools such as disilenide ions **1**[13,14] as well as dilithiosilane **2**[15] allowed the synthesis of numerous types of unsymmetrical and functionalized disilenes.[16,17]

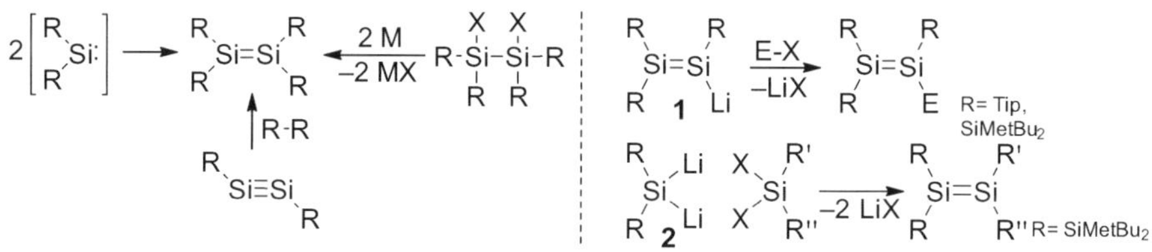

Scheme 9.1
Synthetic routes to stable disilene derivatives.

The amazing development of the disilene chemistry has led to the preparation of a large variety of functionalized derivatives taking advantage of the kinetic stabilization by bulky substituents. To date, various examples of metal-substituted disilenes (M = Li, Mg, Zr, Cu, Zn,...) are known,[13,14,18] as well as stable disilenes with heteroatom substituents (halogen,[19,20] amino,[21] phosphino,[20] boryl,[22] hydro-[22a] and 1,2-dihydrodisilenes).[23] Even structures featuring two cumulated Si=Si double bonds have been prepared. Indeed, the first trisilaallene derivative **3**, featuring a sp-hybridized central Si atom, was reported by the group of Kira in 2003 (Scheme 9.2).[24]

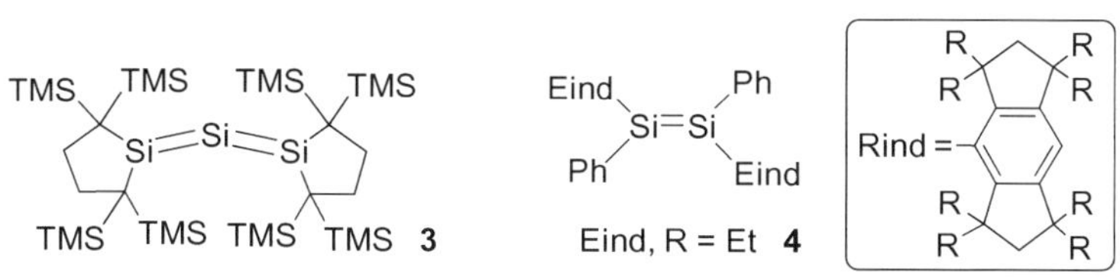

Scheme 9.2
Trisilaallene and π-conjugated disilene stabilized by the Rind-substituents.

Generally the use of extremely bulky substituents induces an important twist about the silicon–silicon double bond, which reduces the extent of the π-conjugation. This problem was solved by developing a new family of bulky groups based on a fused-ring octa-R-substituted *s*-hydrindacene skeleton (Rind groups).[25] Indeed, Tamao and coworkers have demonstrated that the corresponding Eind-substituted (R = Et) disilastilbene **4** is perfectly stable (air-stable in the solid state for more than one month), and presents an unusual planar geometry (Scheme 9.2).[26] In this case, the peculiar cavity produced by the Eind substituents effectively protect the reactive Si=Si fragment which is coplanar with the phenyl groups as demonstrated by photophysical data.

9.2.1.1 Disilene complexes

Olefin-transition metal complexes have been studied extensively as important reactive intermediates in various catalytic processes. In contrast, only a few examples of disilene–transition metal complexes are known so far. The first example of a Pt(0) η^2-disilene complex were reported by West et al. in 1989,[27] and the first structurally characterized was the related tungsten complex.[28] These results demonstrated that disilenes with small substituents can be isolated as stable transition metal complexes (Scheme 9.3).

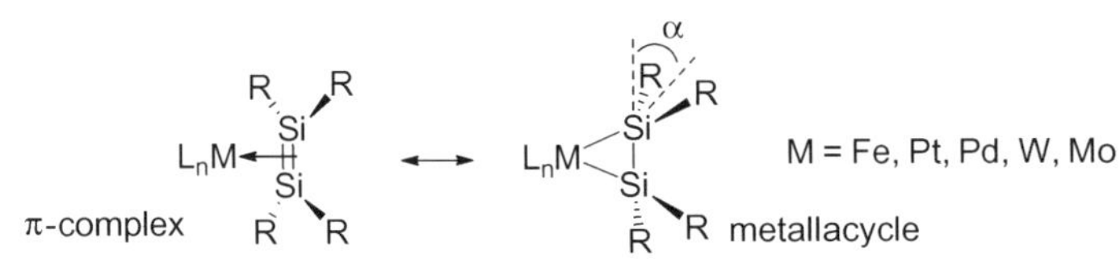

Scheme 9.3
Isolated mononuclear η^2-disilene complexes.

According to the Dewar–Chatt–Duncanson model,[29] the structure of these complexes can be depicted as a combination of two resonance hybrids: (1) a π-complex; and (2) a metallacycle (Scheme 9.3). The structure of several 16-electron Pt[30] and Pd[30b] η^2-disilene complexes have been determined by X-ray crystallography, and the structural parameters around the disilene moiety have shown that these complexes are better described as metallacyclopropanes rather than π-complexes. However, the nature of the metal center strongly influences the structure of the final disilene complex, and the related 14-electron disilene Pd complex presents an important π-complex character, which is demonstrated by the presence of two sp^2-hybridized silicon atoms.[30c]

In most of the disilene transition metal complexes reported, the disilene ligand is not stable as a free species, and must be prepared in situ. Mainly three synthetic approaches have been used to prepare such disilene complexes: (1) oxidative addition of 1,2-dihydrosilanes to an unsaturated platinum fragment[27]; (2) reductive dehalogenation of a precursor containing the X-M-Si-Si-X fragment (M = W, Mo)[28]; (3) treatment of 1,2-disilanyldianions with platinum and palladium dichlorides.[30] Complexes **6** involving an early transition metal (Hf, Ti, Zr) can be also easily prepared by reaction of 1,2-dipotassiodisilane **5** with the corresponding metal dichloride.[31] The hafnocene disilene complex presents some disilylene character, and the addition of one equivalent of trimethylphosphine leads to a more conventional metallacycle structure **7** (Scheme 9.4).

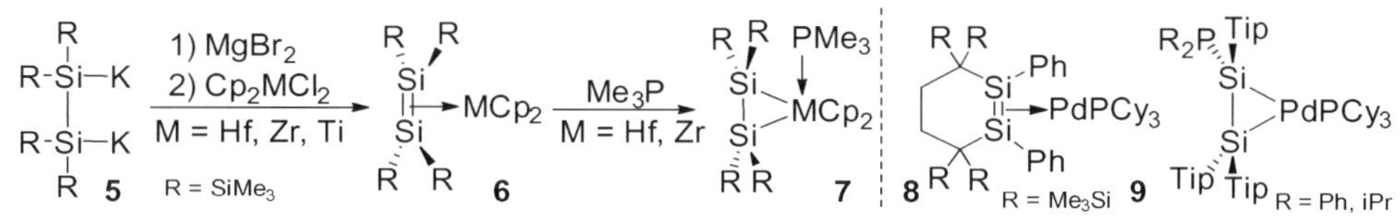

Scheme 9.4
Synthesis of group 4 metallacycle disilene complexes.

The first attempts to prepare disilene transition metal complexes from kinetically stabilized disilenes were reported only very recently.[32] Thus, a 14-electron Pd(0) complex featuring a cyclic disilene fragment **8** was obtained by reaction of a stable 1,2-disilacyclohexene with $(Cy_3P)_3Pd$. Structural data indicate that this complex **8** presents the strongest π-complex character among known disilene palladium complexes, which is in agreement with theoretical calculations.[32a] The original phosphino disilenes recently reported by Scheschkewitz and coworkers readily react with the same Pd(0) precursor affording the corresponding disilene complexes **9**.[20] This result demonstrated the preferential coordination of the Si=Si double bond rather than the phosphine fragment. Moreover, despite the coordination to a very electron-poor 12-electron metal fragment, an important degree of π-back-donation explains the substantial metallacyclopropane character of these complexes (Scheme 9.4).[20]

In contrast, the 1,2-dibromo-1,2-diaryldisilene **10** featuring very bulky aryl substituents reacts under mild conditions with 2 equiv. of $Pt(PCy_3)_2$ to give the corresponding aryl (bromo)silylene-Pt complex **11**, through the cleavage of the Si=Si double bound of the starting disilene (Scheme 9.5).[32b] A similar dissociation reaction into bromosilylene can be assisted by traces of LiBr,[33] and using one equiv. of *N*-heterocyclic carbene (NHC) the NHC-arylbromosilyne adduct can be isolated.[33b]

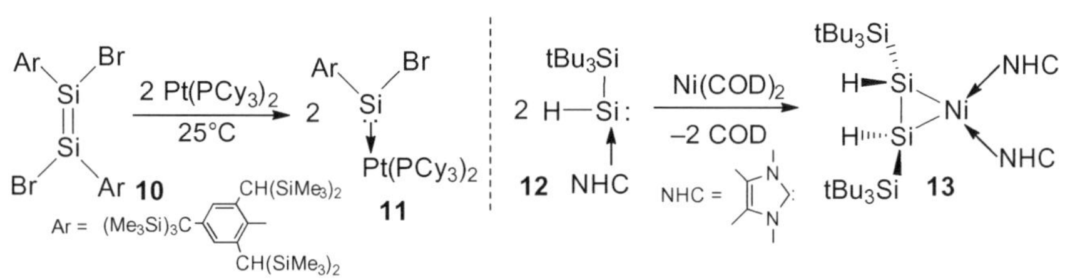

Scheme 9.5
Synthesis of aryl(bromo)silylene-Pt(0) and dihydro(disilene)-Ni(0) complexes.

Conversely, the new NHC-stabilized silylene hydride **12** immediately reacts with (cyclooctadiene)Ni(0) complex to give the first example of a dihydrodisilene nickel complex **13** in 86% yield (Scheme 9.5). Structural observations (large bent back angles) in addition to the very high-field chemical shift in ^{29}Si NMR spectroscopy ($\delta = -115.0$ ppm) clearly indicate that this complex presents an important metallacyclopropane character.[34]

9.2.1.2 Miscellaneous

Among the original structures featuring a disilene moiety, the structural isomers of hexasilabenzene (Si_6H_6), have been the subject of many investigations, and several isomers have been isolated: the hexasilaprismane **I**,[35] the chair-like Si_6 Siliconoid structure **II**,[36] and the [1.1.1]propellane-like structure **III**.[37] Recently, the missing hexasilabenzvalene structure **IV**, has been isolated thanks to the fusion of the hexasilabenzvalene moiety with cyclopentasilane rings and the presence of *tert*-butyl groups which bring about drastic changes in the relative energies of the Si_6R_6 isomers (Scheme 9.6).[38]

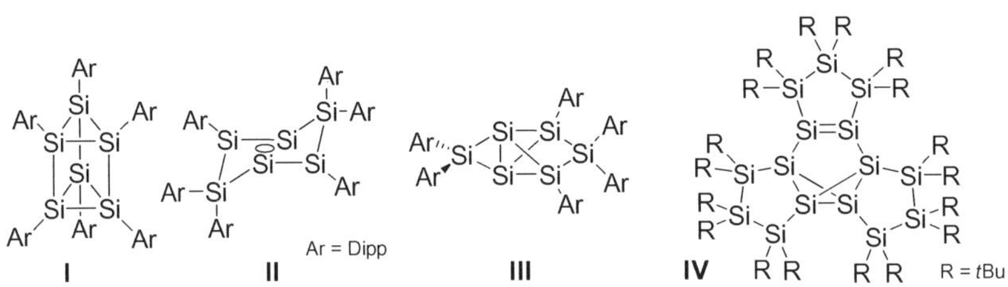

Scheme 9.6
Isolated hexasilabenzene (Si_6R_6) isomers.

We have seen that, in the case of 1,2-dibromodisilene **14**, the Si=Si double bond can be cleaved by adding one equiv. of NHC.[33b] Similarly, taking advantage of the Lewis base character of NHC-stabilized dibromosilylene, Filippou and coworkers have reported the preparation of the first NHC-stabilized bromo(silyl)silylene **15** by cleavage of the corresponding dibromodisilene **14**. Interestingly, the reduction of this (silyl)silylene with two equivalents of C_8K affords selectively the original NHC-stabilized disilavinylidene **16** (Scheme 9.7).[39]

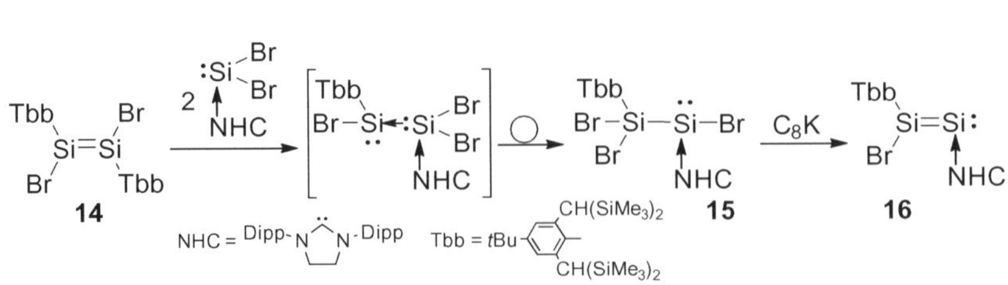

Scheme 9.7
Synthesis of an NHC-stabilized disilavinylidene.

A similar cleavage of the Si=Si double bond was recently observed by Robinson et al. in the case of the original NHC-stabilized disilicon(0) complex **17**.[40] Indeed, this disilene-like compound immediately reacts with $Fe(CO)_5$ at RT to give, first, an η^1-complex **18**, which then can react with a second equiv. of $Fe(CO)_5$ at 100°C leading to Si[μ-$Fe_2(CO)_6$](μ-CO)Si **19**. The formation of the latter complex involves the cleavage of the Si=Si double bond and the insertion of CO and $Fe_2(CO)_6$ into the Si_2 core (Scheme 9.8).[41]

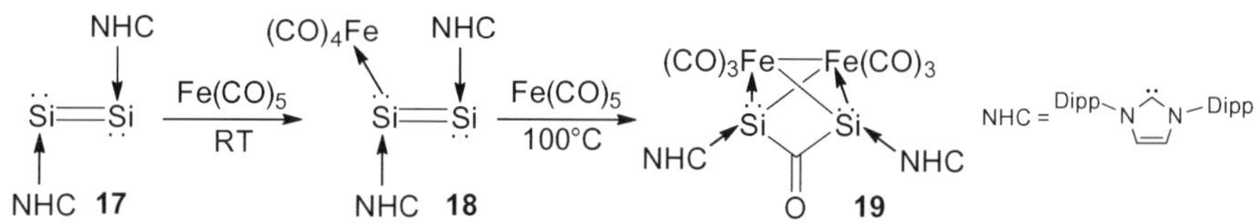

Scheme 9.8
Reactivity of carbene-stabilized disilicon(0) toward $Fe(CO)_5$.

Some of the isolated kinetically stabilized disilene structures present an extremely twisted geometry depending on the substituents, and particularly in the case of the overcrowded di- and tri-(*tert*-butyl)silyl groups: (tBu_3Si)(Ph)Si=Si(Ph)($SitBu_3$) (torsional angle = 85.6 degrees),[42a] ($tBu_2MeSi)_2Si$=$Si(SiMetBu_2)_2$ (torsional angle = 54.9 degrees).[42b] As a consequence of the remarkably twisted geometry around the Si=Si double bond, the latter disilene has an intense blue color, indicating a small HOMO-LUMO gap. Therefore the corresponding disilene anion radical can be readily generated by a thermally induced rotation around the Si=Si bond.[43] The corresponding triplet state has been characterized using EPR spectroscopy.

Baines and coworkers have demonstrated that the reactivity of tetramesityldisilene mirrors that of Si(100)-2 × 1 surface,[44] and therefore can serve as a reasonable molecular model for this surface. Indeed, nitromethane readily reacts either with silicon surfaces or tetramesityldisilene via an original 1,3-dipolar cycloaddition process affording, in the case of disilene, an unusual 1,3,2,4,5-dioxazadisila-ring system which then slowly isomerizes to the 1,4,2,3,5-dioxazadisilolidine (Scheme 9.9).[45] Similarly, the adducts obtained with nitriles, mainly 1,2,3-azadisiletines and their enamine tautomers, are structurally similar to those formed on the Si(100)-2 × 1 surface.[45b]

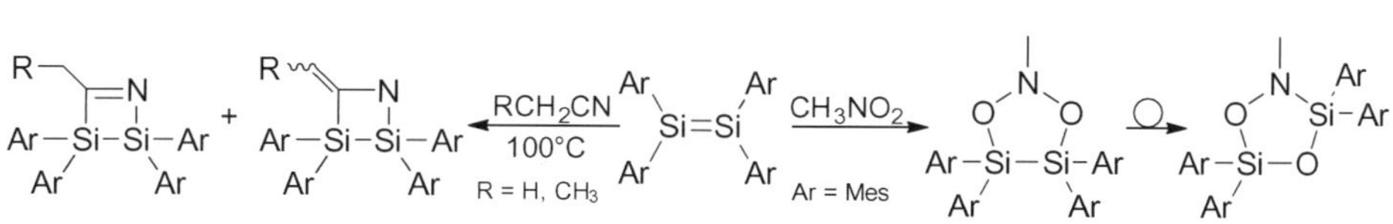

Scheme 9.9
Addition of nitromethane and nitriles to tetramesityldisilene.

9.2.2 Heteronuclear Compounds

9.2.2.1 Si=E13

9.2.2.1.1 Si=B, Si=Ga, Si=In

An early computational study predicted that the π-bonds of $H_2Si{=}BH$ and $H_2Si{=}AlH$ derivatives are particularly weak (27.1 and 14.1 kcal mol^{-1}, respectively) and thus, related molecules should be highly reactive.[46] Indeed, there are very few types of compounds with Si=E (E = B, Ga, In) and none with Si=Al double bonds that have been described to date.

The first attempt to synthesize a silaborene by the reaction of dilithiosilane with dichloromesitylborane in THF failed.[47] The reaction leads to the formation of a seven-membered cyclic borane **20**, presumably formed by the isomerization of transient silaborene complexed with THF (Scheme 9.10). However, the same reaction with a π-donating amino substituted dichloroborane cleanly affords the first stable silaborene **21**.[48] Its ^{29}Si NMR resonance is exceptionally high field-shifted (−128 ppm) for unsaturated silicon compounds. In contrast, the ^{11}B NMR spectrum shows a signal at 87.7 ppm, which is shifted downfield relative to amino-substituted boranes (58.8−59.8 ppm). The structure reveals a significant shortening of the Si=B bond (1.838 Å) relative to Si−B single bonds (2.038−2.125 Å).[49] The strong π-electron donation from the amino group to the boron center was clearly indicated by the essentially linear structure of Si−B−N fragment (179.9 degrees) and short B−N bond (1.370 Å). The availability of the vacant orbital on the boron center has been demonstrated by the reaction with lithium trimethylsilylacetylide which gives the corresponding silaborenide anion **22**.

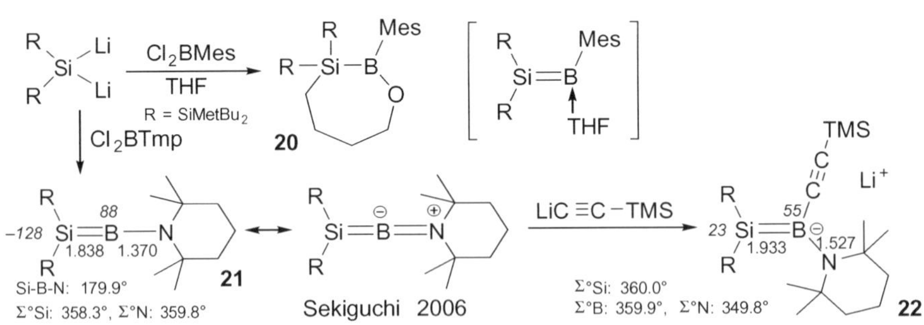

Scheme 9.10
Synthesis of the first stable silaborene.

Similarly, the reaction of $GaCl_3$ with two equiv. of dilithiosilane afforded the first stable Si=Ga double-bonded species, 1,3-disila-2-gallataallenic anion **23** (Scheme 9.11).[50] The ^{29}Si nuclei is also strongly shielded in this molecule (−80 ppm) similarly to silaborene **21**. The disilagallataallenic anion **23** presents a bent structure (Si−Ga−Si: 161.6 degrees) and two pyramidalized silicon centers (Σ°Si = 343.0 degrees). The same synthetic method allowed to prepare the first stable 1,3-disila-2-indatataallenic anion **24**, which presents geometry and spectroscopic properties similar to those of the gallium analog[50] (Scheme 9.11).

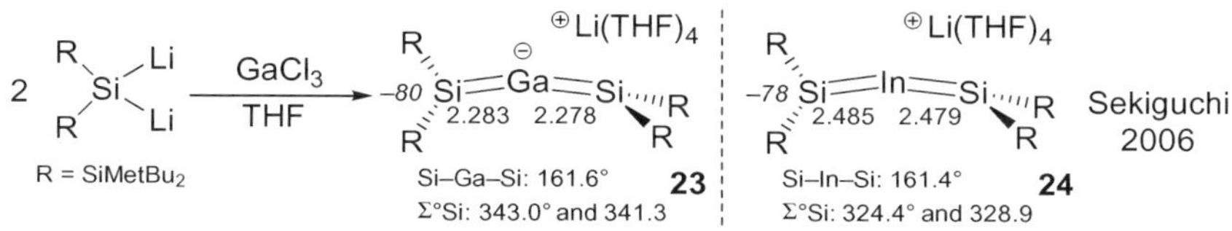

Scheme 9.11
Synthesis of the first stable Si═Ga and Si═In derivatives.

9.2.2.2 Si═E14

9.2.2.2.1 Si═C

Prehistory In 1981, the first stable silene **25** was synthesized by Brook by photochemical rearrangement of tris(trimethylsilyl)acylsilane (Scheme 9.12).[51] Silene **25** with bulky substituents presents a significantly shorter Si═C bond (1.764 Å) than typical single bonds (1.87–1.91 Å) but it is somewhat elongated relative to the theoretically predicted one ($H_2Si{=}CH_2$, 1.718 Å). Two years later, Wiberg reported the second stable silene **26** obtained by lithiation of a fluorosilyl bromomethane and subsequent elimination of lithium fluoride.[52] The resulting *Si*-alkyl-*C*-silyl silene **26** with π-accepting silyl substituents on the C atom shows a much shorter Si═C bond (1.702 Å) and low-field shifted ^{29}Si NMR signal (144.2 ppm) than that of Brook's silene **25** (41.8 ppm).[53] More recently, Apeloig has reported a new synthetic method via a Sila-Peterson-type reaction for preparing stable *Si*-silyl-*C*-alkyl silenes **27**.[54] It is worth noting that all the silenes characterized in the solid state present a planar geometry in contrast to disilenes which generally present a trans-bent structure.

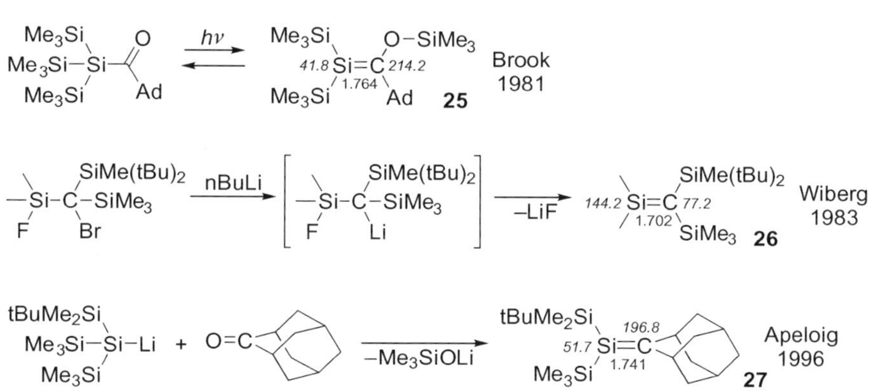

Scheme 9.12
Synthesis of stable silenes.

Since the discovery of these three synthetic methods reported before 2000, the chemistry of silenes has been extensively studied and several synthetic methods to access various other stable silenes have been developed.[5]

Base-adducts of silenes Since silicon is much more electropositive than carbon (Pauling electronegativities: 1.9(Si) versus 2.5(C)), Si=C bond in silenes is strongly polarized toward carbon atom. As a consequence, the silicon center presents a strong Lewis acid character. Indeed, Wiberg's silenes **26** form stable adducts **26-D** with various Lewis bases (D)[55] including weakly coordinating THF (Scheme 9.13).[56] The Lewis base coordination results in enhanced nucleophilic (basic) character at the silene carbon atom. For example, the pyridine complexes of silenes **28** isomerize at RT to 2-pyridyl silanes via deprotonation of pyridine in the 2-position by the silene carbon.[53b]

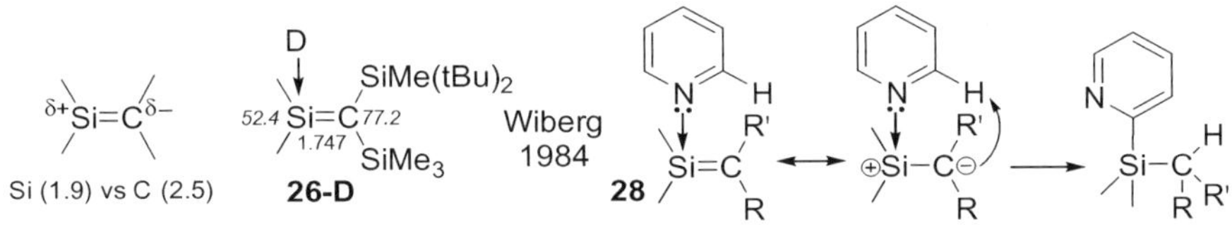

Scheme 9.13
Donor-stabilized silenes and isomerization reactions.

Silenes intramolecularly-stabilized show an enhanced stability. Several syntheses of such silenes have been described. The first one was achieved by the reaction of tris (trimethylsilyl)chloromethysilane **29** with two equiv. of aryl lithium,[57] which gives various silenes **30** having a different substitution pattern on the silicon center (Scheme 9.14).[58]

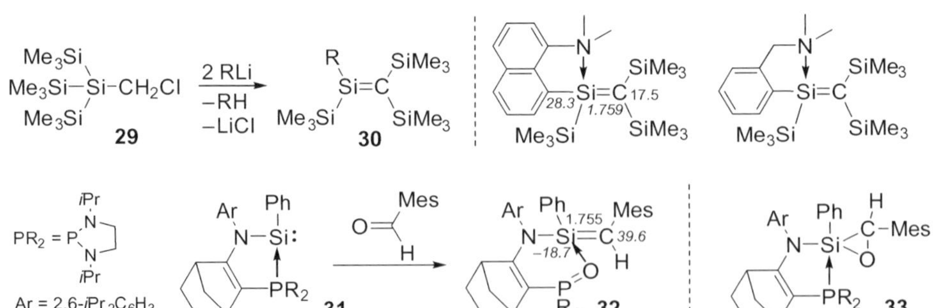

Scheme 9.14
Synthesis of stable silenes with an intramolecularly coordinating ligand.

Kato and Baceiredo described the first sila-Wittig reaction between the stable sila-phosphonium ylide **31** (phosphine-stabilized silylene) and an aldehyde which affords a silene **32** stabilized by an intramolecular coordination of phosphine oxide.[59] The experimental and theoretical studies on the reaction mechanism indicated that the reaction starts with [1 + 2] cycloaddition between silaylide and the aldehyde to form a silaoxirane intermediate **33**. The following reactions (formation of a transient silaoxaphosphetane and retro-[2 + 2] cycloaddition reaction) to form the silene require high thermal activation energies.[60] These results suggest that the sila-Wittig reaction for acyclic silaphosphonium ylides would be difficult to achieve. Indeed, if the phosphine ligand is not fixed on the silicon center by structurally rigid bridge, the dissociation of phosphine ligand from the silaoxirane intermediate **33** would easily occur (Scheme 9.14).

C-donor substituted silenes Intrinsic polarization of silenes (canonical structure **A**) can be significantly reduced by the presence π-donating substituents on carbon (canonical structure **B**) (Scheme 9.15). Such a phenomenon leading to an unexpectedly long Si=C bond in Brook's silene **25** was well studied by theoretical calculations.[61] This effect, so-called "inversed polarity," is known to significantly alter the electronic and structural properties of silenes as well as their reactivity and to increase their stability. Particularly, in 2002, Kira prepared a silene **34** with a strong "inversed polarity," which is attributed to the aromatic cyclopropenium fragment (Scheme 9.15).[62] Indeed, the silene **34** shows an exceptionally high field ^{29}Si NMR signal (−73.7 ppm). Although theoretical calculations predict a pyramidalized silicon center with a localized lone pair on it, X-ray structure revealed a planar geometry. Authors explained this discrepancy by the packing forces effects in the crystal. In addition, its reaction with methanol proceeded with the regioselectivity opposite to that of all other silenes including Brook's silene **25**.[51a,63] leading to the corresponding methoxy-substituted cyclopropene (Scheme 9.15).

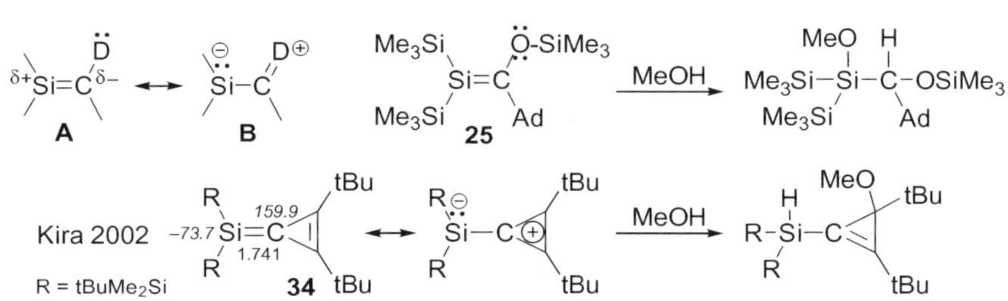

Scheme 9.15
Silenes with inversed polarity and their reactions with MeOH.

Scheschkewitz and Sekiguchi successfully synthesized several four-membered cyclic Brook-type silenes **35** by reaction of a disilenide anion with acylchlorides (Scheme 9.16).[64] These silenes feature a strongly pyramidalized silene silicon center (Σ°Si = 342.2 degrees), and the UV/Vis absorption is slightly red-shifted (351 nm) compared to that of acyclic one (340 nm). However, the Si=C distance (1.775 Å) is close to that of Brook's silene **25** (1.764 Å) and the Si–O bond (1.416 Å) is even slightly longer (1.400 Å for **35**). Authors propose that the strong pyramidalization of the silicon center is mainly due to the cyclic constitution of silenes rather than the π-donation of substituents. Of particular interest, these silenes present an enhanced stability and do not react with MeOH.

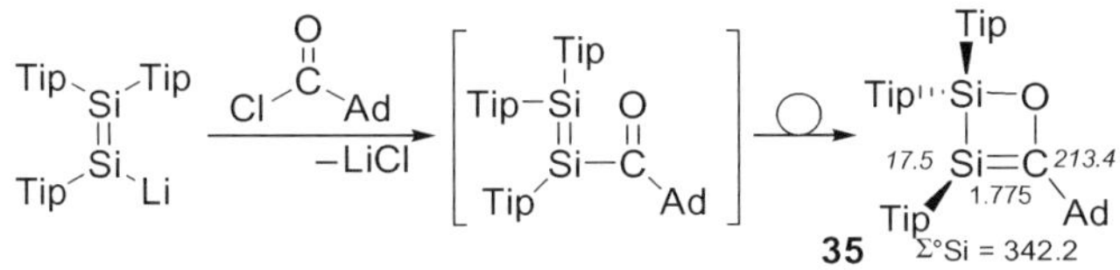

Scheme 9.16
Synthesis of a four-membered ring cyclic Brook-type silenes.

Scheschkewitz et al. also reported a unique reaction of an unsymmetrically substituted disilene **36** with an isocyanide which leads to the formation of a sila-enamine derivative **37** after cleavage of Si=Si bond by isocyanide insertion (Scheme 9.17).[65] The "inversed polarity" of the silene due to the strong π-donation by the amino substituent was clearly indicated by ^{29}Si NMR signal observed for the silene silicon atom at significantly high field (−4.4 ppm) as well as by a slightly pyramidalized silicon center (Σ°Si = 355.9 degrees). Similarly, an extended conjugated system with two silene units **38** was also synthesized from the corresponding *p*-phenylene bridged bis-disilene.

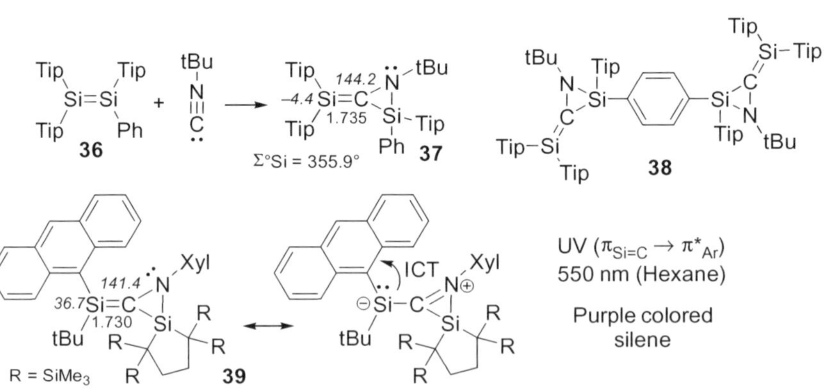

Scheme 9.17
Silenes with strongly π-donating amino groups on C atom.

Using the same synthetic strategy, Iwamoto et al. obtained *Si*-anthracene substituted sila-enamine **39** (Scheme 9.17).[66] This molecule shows UV/Vis absorption bands at 400 nm [π(Si=C)→π*(Si=C)] and at 550 nm. The latter was attributed to an intramolecular charge transfer (ICT) interaction between π(Si=C) and π*(anthryl). They explained the easy ICT in this molecule by its high HOMO energy level ($\pi_{Si=C}$) due to the strong electron donation from the amino group.

As donor-substituted silenes, silaenolate anions can also be considered. Although there had been several reports concerning 2-silenolate anions from the groups of Bravo-Zhivotovskii,[67] Apeloig,[68] Ishikawa and Ohshita,[69] they are moderately stable at low temperature, and at RT they undergo degradation within few hours. The first stable and isolable "potassium salt" of 2-silaenolate anion **40** was synthesized by Ottoson et al. in 2003 by the reaction of tris(trimethylsilyl)acylsilane with a potassium *tert*-butoxide (Scheme 9.18).[70] The most prominent features of the molecule are a markedly pyramidalized silicon center (Σ°Si = 317 degrees) and a long Si–C bond (1.926 Å) corresponding to a Si–C single bond, as well as a short C=O bond (1.245 Å) typical for a double bond. These data clearly indicate that this molecule **40** is dominated by keto form rather than enol form. Indeed, the exceptionally high-field ^{29}Si NMR chemical shift (−78.7 ppm) demonstrates the negatively charged silicon center ("inversed polarity"). Moreover, the signal shifts to even higher field (−93.8 ppm) in the presence of 18-crown-6 which increases the ionic character of the potassium silenolate **40**.

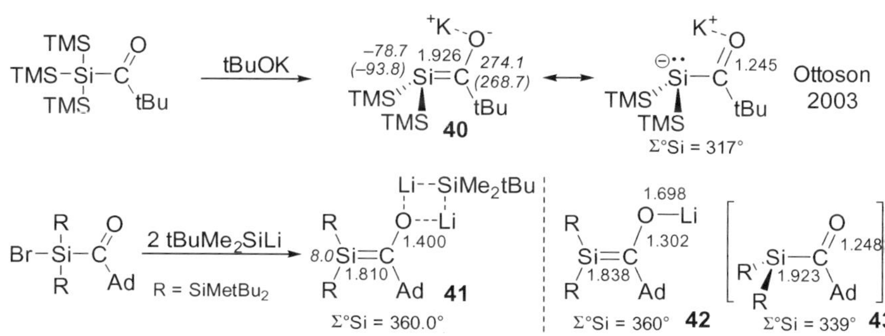

Scheme 9.18
Synthesis of silaenolate anions.

In marked contrast, the related lithium salt **41** exhibits an enol type structure (Scheme 9.18),[71] as indicated by a planar geometry of the silene fragment (Σ°Si = 360 degrees) and a short Si=C bond (1.810 Å). The differences between potassium and lithium salts are mainly due to the stronger covalent bond character of O–Li bond than that of O–K bond. Actually, computational studies clearly demonstrated that the elongation of O–Li distance (increasing ionic character, **42** → **43**) in a step-by-step manner induces the proportional elongation of Si=C bond and the pyramidalization of silicon center, which indicates the increasing keto form character of the silenolate.

Interestingly, Stueger et al. demonstrated that the coordination mode of potassium cation in 2-silaenolate anion strongly depends on the nature of acyl substituent.[72] Indeed, although, in the case of *C*-alkyl silaenolate ion **44**, the potassium cation coordinates on the silicon and oxygen atoms, it interacts with oxygen atom and aryl group in the *C*-aryl substituted one **45** (Scheme 9.19). As a consequence, the aryl derivative **45** presents a shorter Si=C bond and less pyramidalized silicon center, indicating an enhanced enol character. The reactivity of these 2-sila-enolates also reflects their different coordination modes. Indeed, the reaction of the alkyl-substituted one **44** with iPr_3SiCl takes place at the silicon center affording a silylketone **46**, whereas the *O*-silylation reaction was observed for the aryl-substituted **45** leading to the corresponding silene **47** (Scheme 9.19).

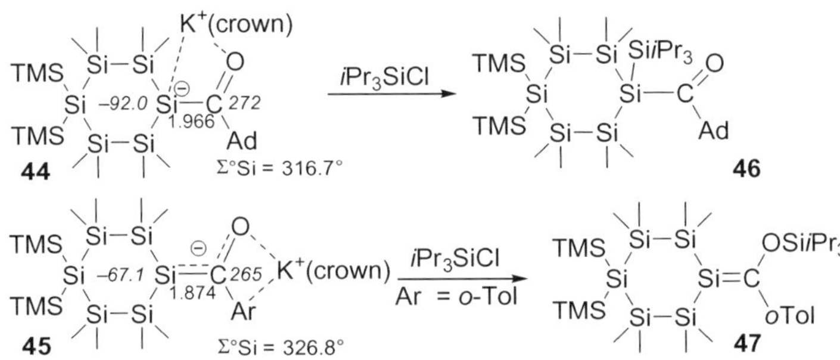

Scheme 9.19

Different reactivity of C-alkyl and -aryl substituted silaenolate anions toward iPr_3SiCl.

Scheschkewitz and Sekiguchi reported a unique reaction of cyclotrisilenes **48** with CO, which generates a short-lived intermediate such as bicyclobutanone **49**/oxyallyl species **49′** (Scheme 9.20).[73] This intermediate **49** was successfully trapped by methanol affording the first stable 2-hydroxy silene (2-silenol) **50**, which is stable up to 100°C and decomposes at 120°C without any evidence of keto-enol tautomerization. They explained the surprising stability of the enol form relative to the keto form by steric reasons.

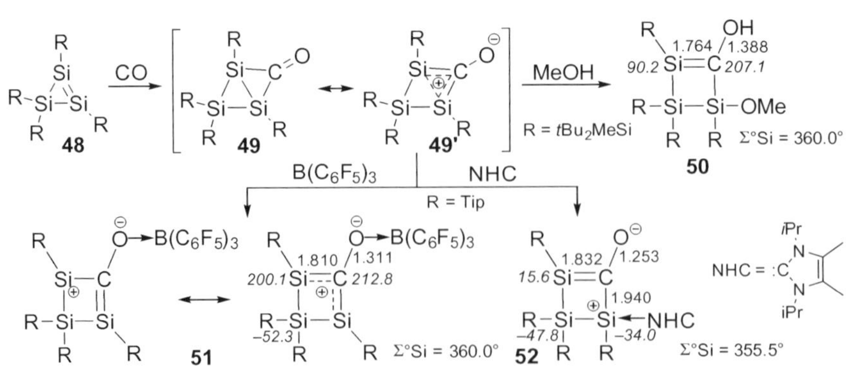

Scheme 9.20

Generation of 1,3-disilaoxaallyl intermediate and its trapping reactions.

The intermediate was also trapped either by a Lewis acid ($B(C_6F_5)_3$) or a Lewis base (NHC) to give the corresponding donor–acceptor adducts **51** and **52** (Scheme 9.20).[74] The borane adduct **51** presents a 1,3-disilaallyl cation type structure with two identical and relatively long SiC bonds (1.810 Å) and a ^{29}Si NMR chemical shift at very low field (200.1 ppm). The NHC adduct **52** can be regarded as a 2-silaenolate type zwitterionic species, intermediate between keto and enol forms, indicated by a moderately long Si=C bond (1.832 Å) and a pyramidalized silicon center (Σ°Si = 355.5 degrees) as well as a high-field ^{29}Si NMR chemical sshift (15.6 ppm).

Si-*donor substituted silenes* In contrast to various types of available silenes with donating substituents on the carbon atom, those with donating substituents on the silicon center **53** are very rare (Scheme 9.21). In the latter case, an enhanced polarization of silene can be expected. However, according to a recent study by Kato and Baceiredo, the situation seems not to be so simple.[75]

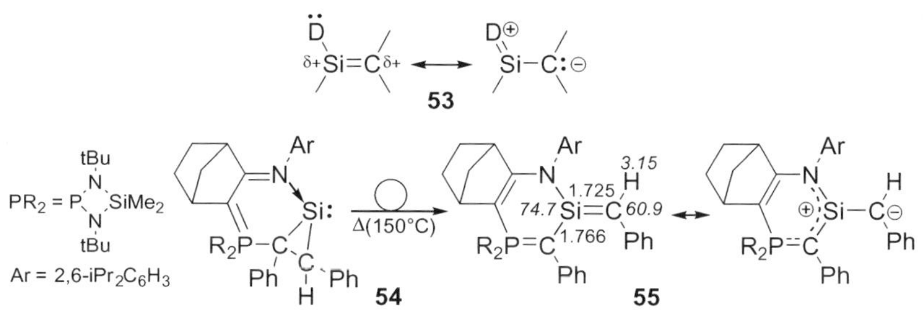

53 **54** **55**

Scheme 9.21

Synthesis of a stable silene with strongly π-donating substituents on Si atom.

Silene **55** with two strongly π-donating substituents, amino and phosphonium ylide groups, on the silicon center was obtained by thermolysis of the corresponding donor-stabilized silacyclopropylidene **54** (Scheme 9.21).[76] Contrary to the case of *C*-donor-substituted silenes with an elongated Si=C bond, the Si=C bond in **55** remains quite short (1.725 Å) and the value is closer to that observed for Wiberg's silene **26** (1.704 Å) and shorter than that of Brook's one **25** (1.764 Å). Enhanced polarization of Si=C moiety toward carbon atom is suggested by significantly high-field chemical shifts in ^{13}C- and ^{1}H NMR spectra observed for the vinylic CH group (60.9 and 3.15 ppm, respectively). Contrary to these experimental observations, the NBO charge analysis indicates that the charge distribution within the silene fragment in **55** (q_C: −1.08, q_{Si}: +1.83) is almost the same as those for the silene **56-Ph** without donating substituents (q_C: −0.96, q_{Si}: +1.60, Scheme 9.22).

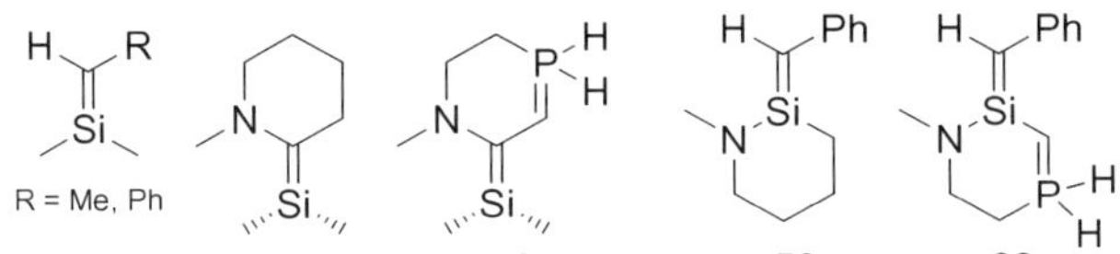

56 57 58 59 60

	Si=C (Å)	q_C[a]	q_{Si}[a]	E_{HOMO} (eV)	$\Delta E_{HOMO\text{-}LUMO}$ (eV)
56-Me	1.709[b]	–0.96	1.54	–5.35	5.67
57	1.816[c]	–0.28	1.14	–4.32	3.91
58	1.914[c]	–0.19	0.93	–3.83	3.09
56-Ph	1.719[b]	–0.96	1.60	–5.16	4.51
59	1.718[b]	–1.04	1.79	–4.77	4.54
60	1.721[b]	–1.07	1.84	–4.19	4.40
55	1.727[b]	–1.08	1.83	–4.11	4.20[d]

Scheme 9.22

Calculated Si=C bond length (M06/6–31 G** level) of silenes, and their properties (NBO charges of silene fragment, HOMO energy level and HOMO-LUMO gap): [a] NBO charges of silene fragment; [b] silene fragment is essentially planar; [c] Si center is pyramidalized $\Sigma°$Si = 334.0(6) (in **57**) and 308.2(7) (in **58**) degrees; [d] $\Delta E_{(HOMO\text{-}LUMO+2)}$.

In addition, calculations also demonstrated that the effects of π-donating substituents on silene fragment are totally different depending on their positions (Si or C atom). Indeed, one or two π-donors, such as amino- and phosphonium ylide groups, on the carbon atom considerably alter both structural and electronic properties of the corresponding silenes (**57**, **58** in Scheme 9.22). This C-substitution with π-donating substituents induces a significantly elongated Si=C bond, with a pyramidalized silicon center. Besides, this effect results in a significant diminution of polarity of Si=C fragment and a decrease of HOMO-LUMO gap (**57**, **58** in Scheme 9.22). In marked contrast, the introduction of the same type of substituents on silicon atom does not induce detectable changes in the silene function (**59**, **60**). Indeed, in the case of silenes (**59**, **60**), there are no significant alterations neither in their geometry nor in the charge distributions (q_C and q_{Si}) compared to those of **56-Ph**. It affects only the HOMO energy level which is significantly higher, but the HOMO-LUMO gap remains approximately the same.

1-Silaallenes The first stable 1-silaallene **61** was synthesized by West in 1993 by intramolecular addition of the intermediate aryllithium to fluorosilyl substituted alkyne followed by lithium fluoride elimination (Scheme 9.23).[77] Silaallene **61** presents a slightly bent structure (Si–C–C = 173.5 degrees), a short Si=C bond (1.704 Å) as well as a planar geometry around the silicon center. Several years later, the same group reported a simpler synthetic method involving an intermolecular addition of alkyl- or aryl-lithium on the fluorosilylalkyne derivative **62** (Scheme 9.23).[78]

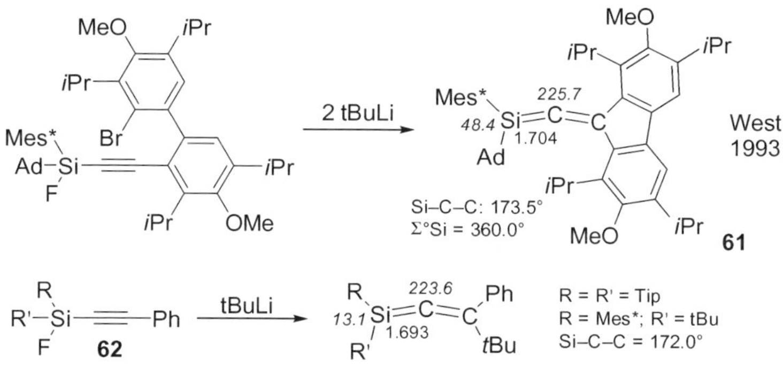

Scheme 9.23
Synthesis of 1-silaallenes.

An attempt to synthesize the still elusive silyne derivative has been made by the double addition-elimination reaction of a bulky dichlorosilylalkyne **63** with two equivalents of *t*BuLi (Scheme 9.24).[79] However, the reaction did not give the desired silyne **64** but a 1-silaallene **65**. In fact, the second nucleophilic attack did not occur at the terminal carbon of the chloro-silaallene intermediate but at the silicon center substituting the chlorine atom to give the silaallene **65**. Theoretical calculations revealed that the formation of 1-silaallene **66** is thermodynamically favored over the silyne isomer **67** ($\Delta E_{\text{silaallene-silyne}} = 40.2$ kcal mol^{-1}).

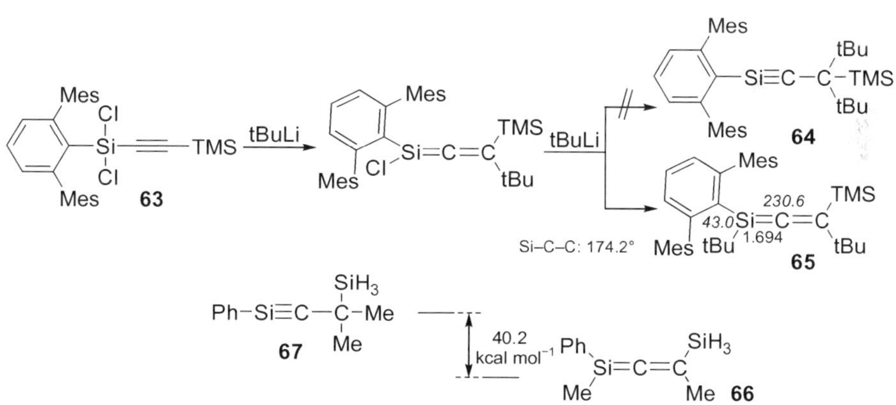

Scheme 9.24
Synthesis of 1-silaallene from a dichlorosilylalkyne derivative.

Kira et al. successfully synthesized the first example of a stable dialkyl 1-silaketenimine **68**, and characterized it in solution and in the solid state (Scheme 9.25).[80] In contrast to the previously reported Tokitoh's stable diarylketeneimine **69**[81] which can be regarded rather as a silylene complexed with an isocyanide ligand, the dialkyl derivative **68** exhibits a strong allenic character with a Si—C bond (1.786 Å) which is only slightly longer than those of silenes (1.693–1.764 Å). It also shows a significantly elongated C—N bond (1.210 Å) relative to that of the corresponding arylisocyanide (c.1.160 Å for **70**). The C—N—C moiety (130.7 degrees) is considerably bent.

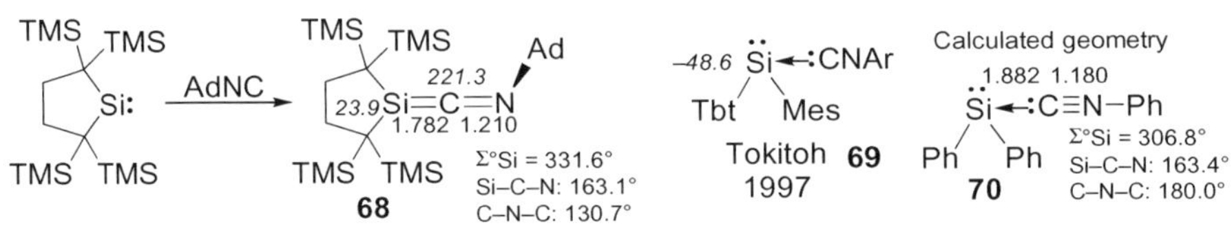

Scheme 9.25
Synthesis of stable 1-silaketeneimines.

Sekiguchi et al. described the reactivity of stable disilyne **71** with silylcyanide- and alkylisocyanide (Scheme 9.26). In both cases, the reactions lead to the formation of bis-adducts **72**[82] and **74**.[83] In the case of silylcyanide, the reaction also affords a 1,4-diazo-2,3-disilabenzene derivative **73** as byproduct. Bis-adducts (**72**, **74**) show moderate silaketenimine character with relatively short Si—C bonds (1.809–1.826 Å) and elongated C=N bonds (1.185 Å). These structural data are intermediate between the values observed for dialkyl derivatives (ketenimine) and diaryl compounds (silylene-isocyanide complex). Of particular interest, the N-alkyl bis-silaketenimine derivative **74** is not stable at RT and it decomposes into several products including the first stable disilene **75** with strong π-accepting nitrile substituents (Scheme 9.26).

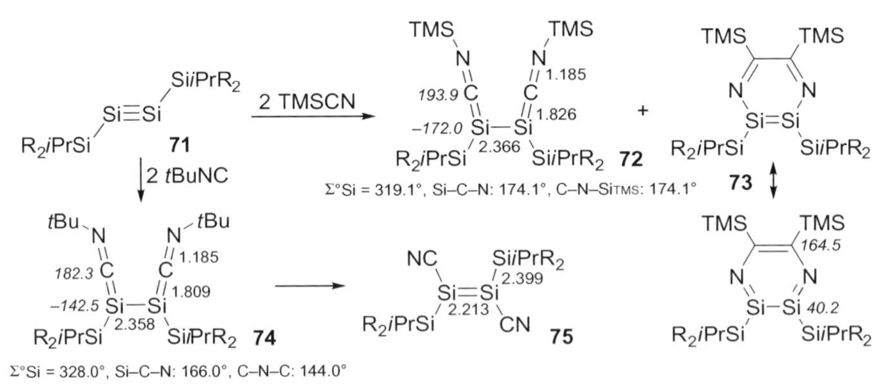

Scheme 9.26
Synthesis of bis-silaketeneimines.

2,3-Disila-1,3-butadienes Coordination of strongly σ-donating ligands such as NHCs or cyclic amino alkyl carbenes (CAACs) significantly modifies the electronic and geometric properties of disilynes. Indeed, Robinson et al. clearly demonstrated that a formal dichlorodisilyne complex **76** with two NHC ligands does not present any Si–Si multiple-bond character (long Si–Si bond: 2.393 Å) but shows two strongly pyramidalized silicon centers with a lone pair on each of them (Scheme 9.27).[40] The Si–C_{NHC} distances (1.939 Å) correspond to those of Si–C single bond. The disilyne complex **76** can be thus regarded as a NHC stabilized 1,2-bis-silylene. In contrast, a similar dichlorodisilyne complex **77** with stronger σ-donating and π-accepting CAAC ligands exhibits an enhanced butadiene character which was indicated by much shorter Si–C bonds (1.826 Å) and less pyramidalized silicon centers (Σ°Si = 328.0 degrees for **77** and 306.7 degrees for **76**).[84]

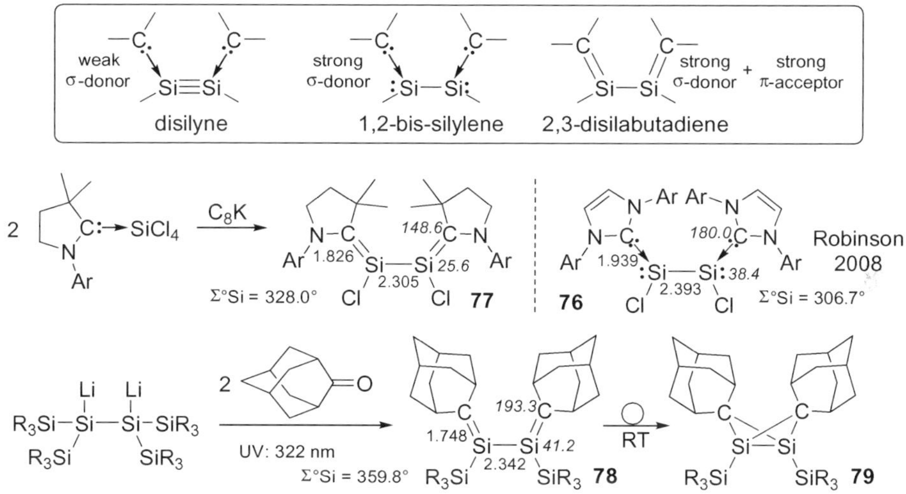

Scheme 9.27
Synthesis of 2,3-disila-1,3-butadienes.

Iwamoto et al. have obtained a 2,3-disila-1,3-butadiene **78** with little electronic perturbation by double sila-Peterson olefination between a dilithiodisilane and two equiv. of 2-adamantanone (Scheme 9.27).[85] The butadienic character of **78** was shown by short Si=C distances within the typical range for silenes (1.692–1.764 Å) as well as essentially planar geometry of two silenes moieties (Σ°Si = 359.8 and Σ°C = 360.0 degrees). The longest absorption band maximum at 377 nm in UV/VIS spectra is considerably red-shifted compared with that of silenes with a similar structure (322 nm) owing to the conjugation of the two Si=C bonds. This disilabutadiene is not stable at RT and undergoes thermal isomerization to 1,3-disilabicyclo[1.1.0]butane **79** (Scheme 9.27), and the presence of an isobestic point in the UV/VIS absorption spectra suggests that the process is unimolecular.

Small cyclic silenes Very few silenes incorporated into a small cyclic structure are known to date. Only one example of three-membered cyclic silene **80** has been synthesized by Sekiguchi et al. by the reaction of dilithiosilane with 1-adamantylcarbonyl chloride

(Scheme 9.28).[86] Due to the extreme steric protection provided by bulky substituents, the disilacyclopropene **80** is stable in air and can be purified by HPLC (*t*BuOMe/MeOH). The longest wavelength absorption maximum of **80** ($\pi-\pi^*$) in the UV/VIS spectra is observed at 394 nm ($\varepsilon = 1800$) in hexane, being the most red-shifted among other known silenes including 2,3-disilabutadiene (377 nm). They explained this important red-shift by a significantly raised HOMO energy level due to σ(Si–Si)–π(Si=C) interaction. They have used this disilacyclopropene as a precursor for the synthesis of a stable disilacyclopropenium cation **81** (Scheme 9.28).

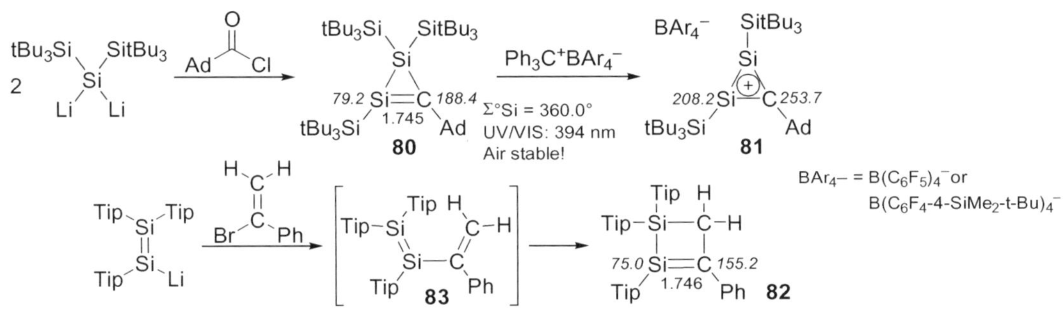

Scheme 9.28
Synthesis of three- and four-membered cyclic silenes.

A four-membered cyclic silene (disilacyclobutene) **82** has also been synthesized by Scheschkewitz and Sekiguchi.[87] The reaction of disilenide anion with a vinylbromide, probably generates a transient 1,2-disila-1,3-butadiene intermediate **83**, which readily isomerizes to disilacyclobutene (Scheme 9.28).

Metallated silenes The lithium disilenide **84** developed by Scheschkewitz has been well demonstrated to be powerful synthetic tool in organosilicon chemistry.[13,88] By analogy, we can also expect the same usefulness for the mono-silicon analogs, silenide anions **85**. However, only two types of these metallated-silenide derivatives have been described to date.

The first one is the *Si*-mercury-substituted silene **87**, which was synthesized by a Brook rearrangement of mercury-substituted acylsilane **86** generated by mixing bis(lithiosilyl) mercury and two equiv. of 1-adamantylcarbonyl chloride (Scheme 9.29).[89] The Brook rearrangement is related to the size of silyl substituents (SiR_3). Indeed, although the rearrangement of mercury acylsilane **86** with bulky silyl substituents ($Si\textit{i}Pr_3$) proceeds at RT, the use of a smaller substituent ($SiMe_2\textit{t}Bu$) prevents the thermal isomerization even at 200°C. To date, no reactivity of the mercury-substituted silene **87** has been reported.

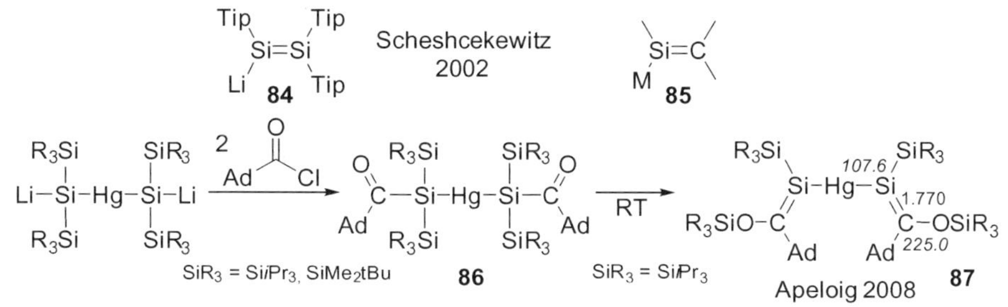

Scheme 9.29
Synthesis of *Si*-mercury substituted silene.

More recently, Apeloig et al. reported the synthesis of the first stable lithium-substituted silene **89** from the bis-metallated lithium-sila-enolate-lithium-silane **88**, after elimination of one equiv. of lithium siloxide (Scheme 9.30).[90] They postulated that the reaction starts with an 1,2-elimination of lithium siloxide to generate a transient silyne **90** which isomerizes into the thermodynamically more stable silylidene intermediate **91** ($\Delta E_{\text{silylidene-silyne}} = -12.8\ \text{kcal mol}^{-1}$). The silylidene intermediate **91** subsequently reacts with the lithium silane to give the experimentally obtained silenyllithium **89**. The substituent patterns of the silenyllithium **89** obtained using two lithium silanes with different substituents ($R_2R'SiLi = Me\textit{t}Bu_2SiLi$ or $Me_2\textit{t}BuSiLi$) demonstrated that the reaction proceeds through the migration of one of $Me\textit{t}Bu_2Si$ groups on silicon atom to the adjacent carbon atom (silyne–silylidene isomerization **90** → **91**) followed by the addition of lithium–silane on the silicon center of silylidene **91**. The ^{29}Si NMR chemical shift of the doubly bonded Si atom appears at 243.9 ppm, strongly deshielded compared to the corresponding neutral silenes (51.7 ppm). The chemical shift is comparable to that of disilenide anion **84** (328 ppm). To date, only their reactivity toward H_2O is reported (Scheme 9.30).

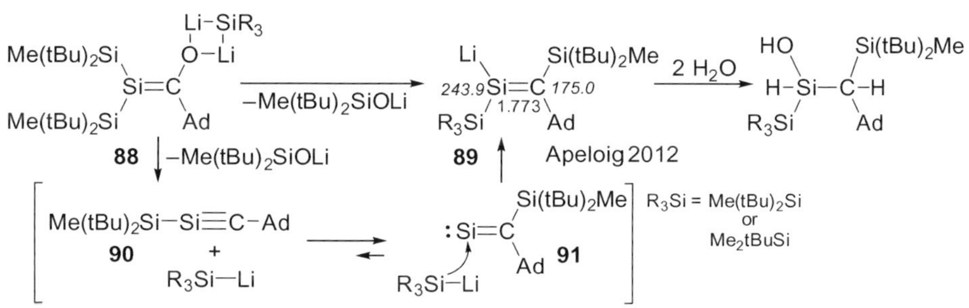

Scheme 9.30
Synthesis of stable lithium silenides.

2-Silaallenes 2-Silaallenes **92** with a sp-hybridized silicon atom are extremely rare. The trisilaallene **93** described by Kira in 2000[24] and related 1,3-digermasilaallene[91] are almost unique examples (Scheme 9.31). In contrast with 1-silaallenes, the chemistry of 2-silaallenes **92** is still in its infancy. This is principally owing to their high reactivity due to the presence of a sp-hyridized silicon center as well as the strongly electron deficient character resulting from the two cumulated silene functions on the same silicon atom. Indeed, So et al. synthesized a 2-silaallene **94** stabilized by intramolecular coordination of donating phosphine thioxides on the central silicon atom (Scheme 9.31).[92] According to its X-ray structure, the strong coordination of the ligands on the silicon center significantly elongates Si=C bonds (1.810 Å) compared with other donor-stabilized silenes (1.747 − 1.759 Å). Attempt to synthesize a 1,2-disilaallene via a sila-Peterson reaction between the disilenide anion and 2-adamantanone has failed due to the subsequent attack of the eliminated siloxide anion on the central sp-hybridized silicon atom of the generated disilaallene **95** giving a 1,2-disilaallyl lithium **96** (Scheme 9.31).[93] In this molecule, the negative charge is localized on the terminal silicon center, as suggested by its strong pyramidalization (Σ°Si = 337.1 degrees) and short Si=C bond (1.743 Å).

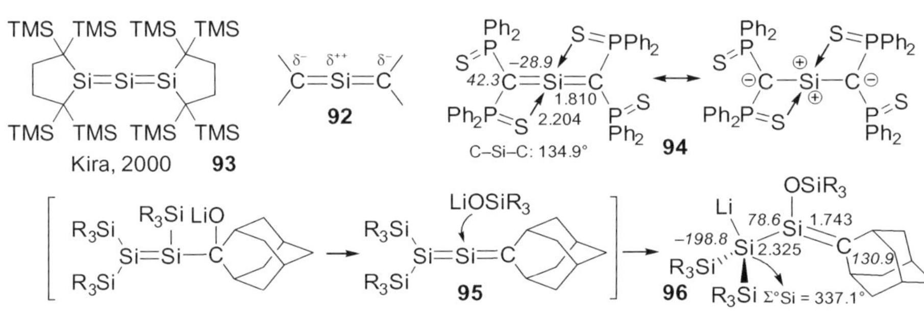

Scheme 9.31
Stable donor-stabilized 2-silaallene and an attempt to synthesize 1,2-disilaallene.

As already mentioned, π-donor substituted silenes **97** with inverted polarization present an enhanced stability. The same stabilization technique can be adopted for 2-silaallenes **98** (Scheme 9.32). In this case, the central silicon exhibits a formal dianionic character.

The first synthesis of such a molecule has been realized by Roesky's group. They reduced a unique CAAC (cyclic amino-alkyl carbene) stabilized dichlorosilylene **99** with a diradical character[94] by C_8K to prepare the corresponding 2-silaallene **100** (Scheme 9.32).[95]

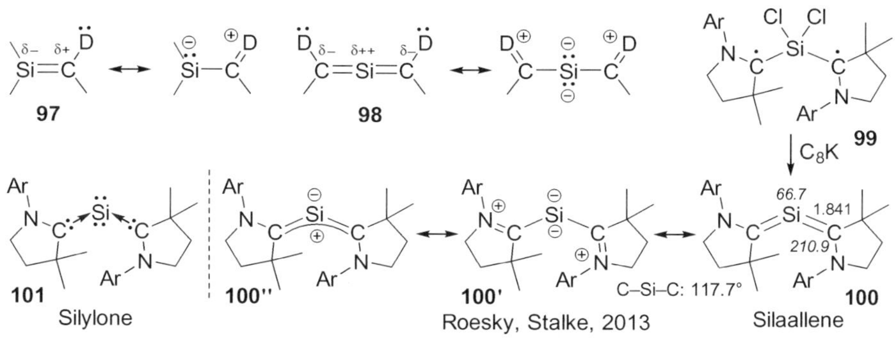

Scheme 9.32
Stabilization of 2-silaallenes by π-donation of substituents on C atoms and synthesis of the stable silylone with CAAC ligands with some allene character.

Prominent features of the molecule are strongly bent structure (C–Si–C: 117.7 degrees) and considerably elongated Si=C bonds (1.841 Å), suggesting a significant "inverted polarization" character due to the strong π-electron donation from the amino groups of CAAC fragments. Theoretical calculations indicate that the HOMO-1 is an in-plane σ-lone pair orbital at Si and the HOMO is a π-orbital extending over the C–Si–C fragment. This might suggest the π-electron delocalization of an on-plane lone pair at Si toward C atoms (allyl cation type π-conjugation) which is in agreement with moderately short Si–C bonds. Thus the 2-silaallene **100** can be represented as a combination of allene **100** and dianionic silicon species (**100′**, **100″**). This molecule can also be regarded as an atomic silicon complex stabilized by two carbene ligands, so-called "silylone" **101** (Scheme 9.32). The experimental charge density analysis of the 2-silaallene **100** indicated the presence of residual charge densities around the central silicon corresponding to the two nonbonding electrons at the silicon center.[96]

As expected, the related compound **102** with cyclic diamino carbene ligands (NHC) presents a much less allenic character (Scheme 9.33).[97] Indeed, the cyclic silylone **102** with a bidentate NHC ligand reported by Driess exhibits an enormous upfield shift of the ^{29}Si NMR signal (−83.3 ppm) relative to that observed for the silylone **100** with CAAC ligands (+66.7 ppm). This strong shielding of ^{29}Si nuclei (inversed polarization effect) probably originates from poor back-donation from the silicon atom to NHC ligands with a weak π-accepting character. This was strongly supported by calculations which indicate much less positive charge found at the Si atom of NHC-complex **103** (+0.191) than that of the CAAC-complex **104** (+0.351).

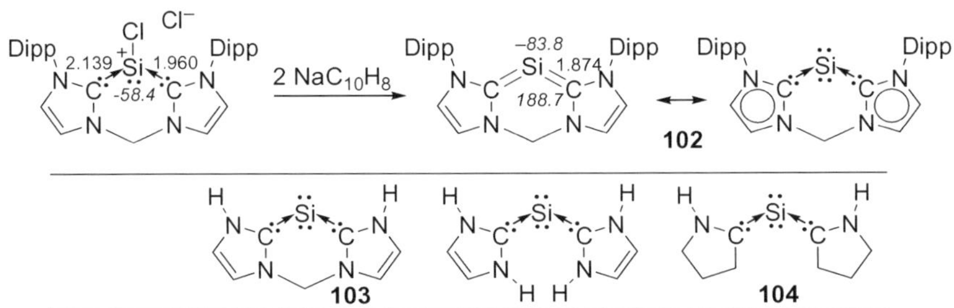

	103		104
C–Si–C [°]	88.2	94.7	100.0
Si–C [Å]	1.880	1.877	1.861
$\sigma(^{29}Si)$ [ppm]	–69.2	–67.2	+22.5
NBO charge at Si	+0.191	+0.163	+0.351

Calculated at B3LYP/6-31G(d) level of theory. GIAO/B3LYP/6-311G(d) [NHC]: 6-311G(3d) [Si]

Scheme 9.33
Stable silylone with NHC ligands.

Silene complexes Although the chemistry of metal complexes containing reactive silicon-based fragments appears to be highly relevant to a number of catalytic processes, very few types of complexes with silene ligands are known to date.

The first stable and isolable η^2-silene complex of transition metal **105** was synthesized by Tilley's group by the intramolecular oxidative addition of Si–H to Ru(II) complex (Scheme 9.34).[98] This reaction is also regarded as a β-hydride elimination of alkylsilane complex **106**. The same synthetic strategy was adopted for the synthesis of a stable η^2-silene-Ir(II) complex **107**.[99] For both complexes, signals of the silene moiety in ^{13}C- and ^{29}Si-NMR significantly shifted to high field (−29.0 ppm (C) and 6.1 ppm (Si) for Ru, −33.4 ppm (C) and −20.8 ppm (Si) for Ir) compared with those of silenes (50–144 ppm (C) and

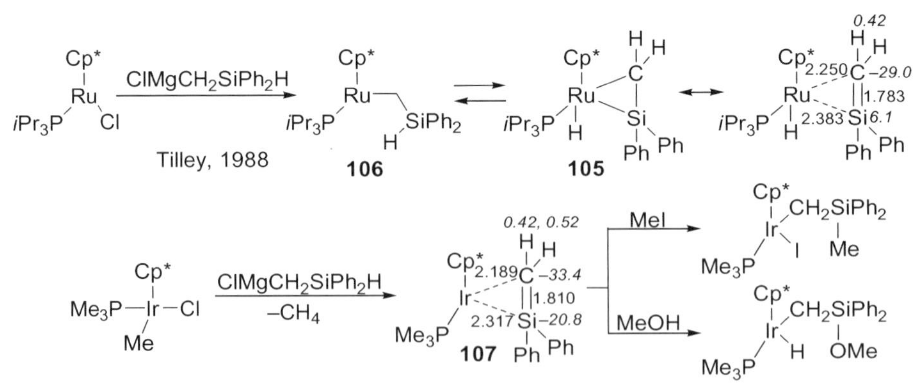

Scheme 9.34
Synthesis of the first η^2-silene complexes via β-hydride elimination.

77–197 ppm (Si)). The Si═C bond (1.783 Å for Ru and 1.810 Å for Ir) is shorter than normal Si–C single bonds (1.87–1.91 Å), which reflects a partial double bond character. Reactions of the Ir complex **107** with MeI and MeOH result in clean Ir–Si bond cleavages to afford the corresponding silylmethyl or the silylmethoxy complexes (Scheme 9.34).

In addition, the formation of η^2-silene complex **108** via β-hydride transfer proceeding through C–H activation of an alkylsilane complex has also been described (Scheme 9.35). In some cases, these processes are reversible under mild conditions and allow to functionalize aliphatic C–H bonds of silanes **109**. This process is considered to be directly related to the catalytic dehydropolymerization of silanes reported by Berry's group in the early 1990s.[100] The direct observation of intramolecular activation of aliphatic C–H in the 16-electron trimethylsilyl Ru complex **110** reversibly transforming into the corresponding η^2-silene complex **111** at RT has been achieved by Berry's group.[101] Although silene complex **111** could not be isolated by crystallization, its treatment with $CHCl_3$ led to the formation of an isolable Ru complex **112** with an agostic Si-H interaction (Scheme 9.35).[102] The partial η^2-silene complex character is also indicated by the short Si–C distance (1.788 Å), which is only slightly longer than those observed for silenes (1.704–1.764 Å). This complex **112** can be seen as a snapshot of the β-hydrogen transfer process of the complex transforming into a silene complex **112′**.

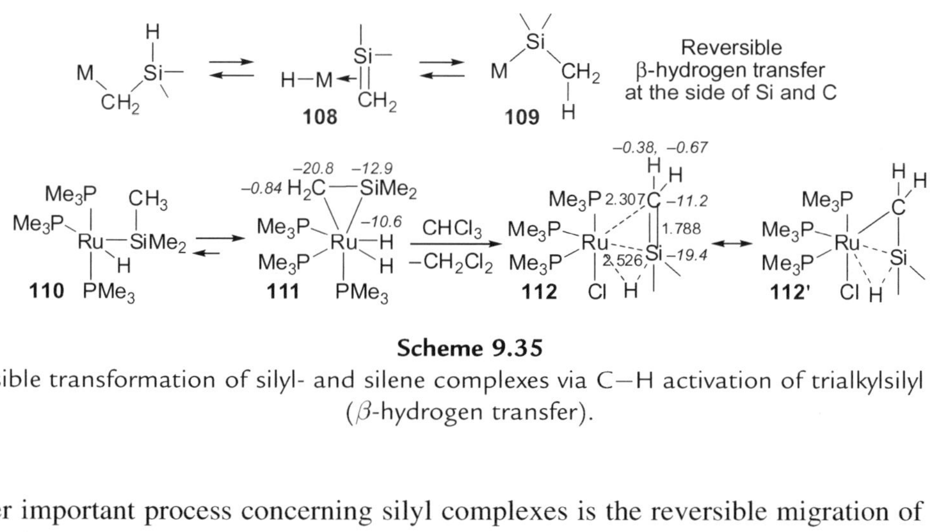

Scheme 9.35
Reversible transformation of silyl- and silene complexes via C–H activation of trialkylsilyl groups (β-hydrogen transfer).

Another important process concerning silyl complexes is the reversible migration of one substituent of the silyl group to the metal center (α-migration) leading to a silylene complex (Scheme 9.36). An original silylene-silene interconversion within the coordination sphere of transition metal via this process (α-methyl migration) as well as the β-hydride migration via C–H bond activation was observed for a cationic silylene complex of iridium **113** (Scheme 9.36).[103] Understanding of these two processes is particularly important for developing catalytic transformations of silanes.

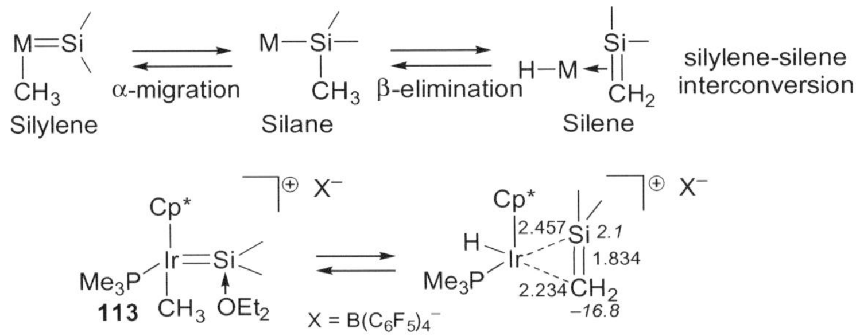

Scheme 9.36
Reversible silylene-silene transformation on transition metal via α-methyl migration and β-hydrogen transfer.

As another synthetic method for silene complexes, reductive dechlorination of $Cp_2W(Cl)(CH_2SiMe_2Cl)$ **114** has also been reported (Scheme 9.37).[104] The resulting complex **115** readily reacts with MeOH via cleavage of the W–Si bond. In contrast, the reaction with trimethylsilane results in the M–C bond cleavage.

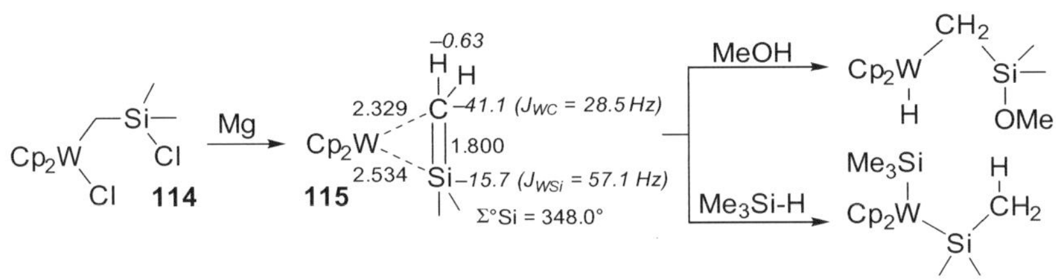

Scheme 9.37
Reductive dechlorination of $Cp_2W(Cl)(CH_2SiMe_2Cl)$.

The first synthesis of silene complexes by direct reaction between a silene and a metal precursor was reported by Apeloig et al. Indeed, they synthesized the η^2-silene Pt(0) complex **117** by the reaction of a transient silene generated in situ with $Pt(PCy_3)_2$ (Scheme 9.38).[105]

As already mentioned, the stable silene **55** substituted by strongly σ-donating substituents on the silicon atom presents classical silene-like properties as well as a high HOMO energy level. This silene **55** is an excellent ligand for transition metals, and the reactions with several metal precursors such as Karstedt's Pt(0) complex or $Ni(COD)_2$ result in the formation of the corresponding silene complexes **118** and **119**, respectively (Scheme 9.38).[75] This direct method should allow to synthesize various silene complexes. Of particular interest, the Pt-silene complex **118** shows an enhanced catalytic activity for the hydrosilylation reaction of alkenes, compared with that observed for the classical Karstedt's catalyst **120** (Scheme 9.39).[106]

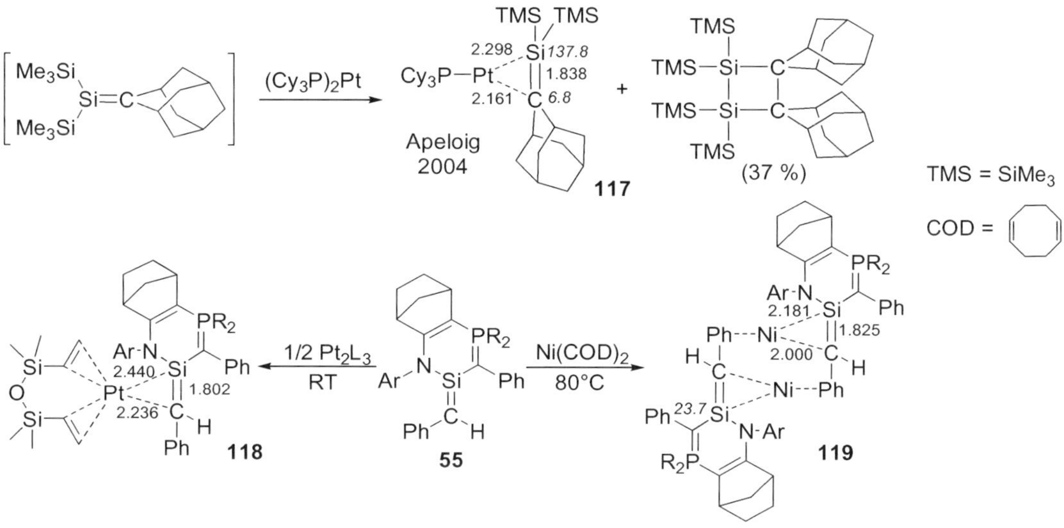

Scheme 9.38
Synthesis of silene complexes by direct reaction of silenes with metallic precursors.

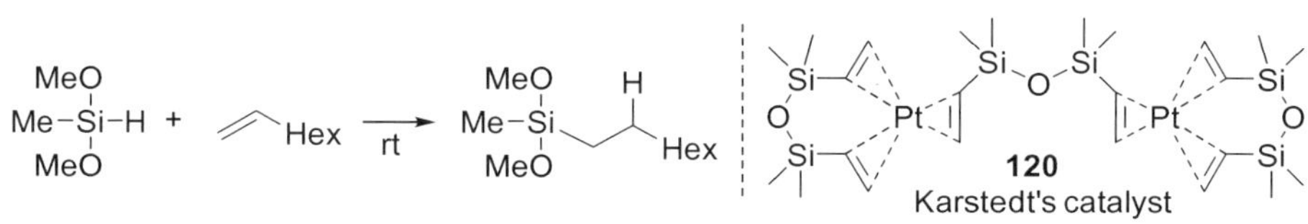

118	Karstedt's catalyst 120
92% in 10 min	84% in 180 min

Scheme 9.39
Hydrosilylation reaction catalyzed by Pt(0)-silene complex 118.

9.2.2.2.2 Si=Ge

The first stable and isolable silagermene, 2-disilagermirene **121**, has been synthesized by Lee, Sekiguchi et al. by the photochemical isomerization of 1-disilagermirene **122** (Scheme 9.40).[107] The endocyclic double-bonded silicon atom exhibits a ^{29}Si NMR downfield resonance at 100.7 ppm and the endocyclic tetravalent Si atom has an upfield resonance at −120 ppm. This isomerization (**122** → **121**) can also be performed under thermal conditions, and at 120°C an equilibrium mixture of 1-disilagermirene **122** (2%) and 2-disilagermirene **121** (98%) was obtained. This result implies the small difference in energy between both isomers (3 kcal mol^{-1}), which is in agreement with the calculated one (2.3 kcal mol^{-1}, Scheme 9.40). The 2-disilagermirene **121** readily reacts with a phenylacetylene to give a heavier cyclopentadiene analog **123** with Si=Ge and C=C double bonds.[108] The Si=Ge double bond length is 2.250 Å, which is intermediate between Si═Si (2.138–2.289 Å) and Ge═Ge (2.213–2.450 Å) double bond lengths. The reaction probably proceeds through a formal [2 + 2]-cycloaddition of the 2-disilagermirene **121** with the acetylene to form a bicyclic compound **124** which isomerizes to a transient cyclopentadiene **125** with Ge═C and Si═C double bonds. This intermediate **125** subsequently isomerizes to the final product. The isomerization from **125′** to **123′** was calculated to be thermodynamically favored by 14.3 kcal mol^{-1} (Scheme 9.40).[109]

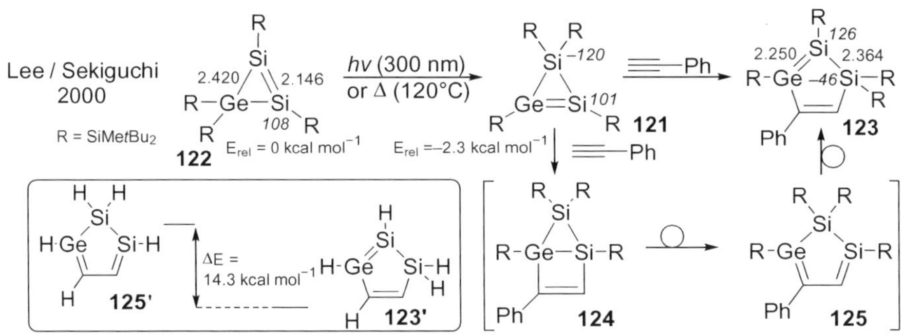

Scheme 9.40
Synthesis of the first stable compounds with a Si═Ge bond.

The first stable allenes with two cumulated Si = Ge double bonds such as 1,3-digermasilaallene **126**[91] and 1,3-disilagermaallene **127**[110] have been synthesized by Kira's group (Scheme 9.41). The former **126** is synthesized by reduction of cyclic trichlorosilyl-chlorogermane by KC_8 and the latter **127** was prepared by the reduction of a mixture of stable cyclic dialkylsilylene and $GeCl_2$-dioxane complex with KC_8. Both compounds present a strongly bent structure (Ge–Si–Ge = 125.7 degrees and Si–Ge–Si = 132.4 degrees) and pyramidalized Ge and Si centers (Σ°Ge = 354 degrees for **126**, Σ°Si═349.3 degrees for **127**). In both cases, the Si = Ge bonds (2.269 Å for **126**, 2.237 Å for **127**) are comparable with that of silagermene **123** (2.250 Å).

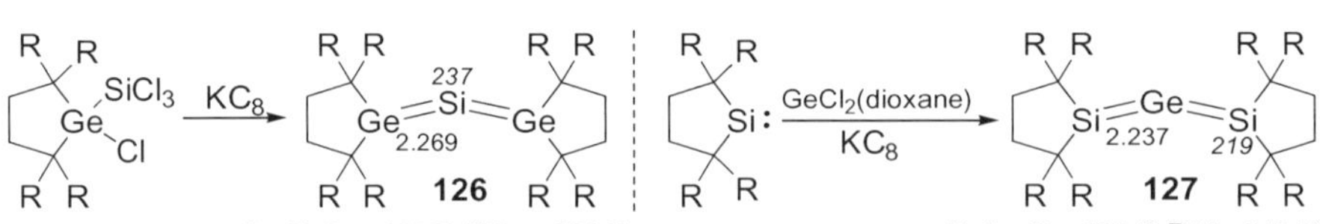

Scheme 9.41
Synthesis of allenes with Si═Ge double bonds.

The first silagermylidene **128** has been isolated by Scheschkewitz et al. as a stable NHC-complex (Scheme 9.42).[111] The silagermylidene **128** was obtained by the reduction of a mixture of a dichlorosilane and a NHC-stabilized dichlorogermylene with 4 equiv. of $LiC_{10}H_6$ The Si═Ge bond length (2.252 Å) is similar to other Si═Ge containing compounds. As expected, the angle Si–Ge–C_{NHC} is quite small (98.9 degrees) and Ge–C_{NHC} distance (2.047 Å) corresponds to a Ge–C single bond. The silagermylidene **128** readily reacts with phenylacetylene, via a formal [2 + 2]-cycloaddition, to give a four-membered cyclic base-stabilized germylene **129**.

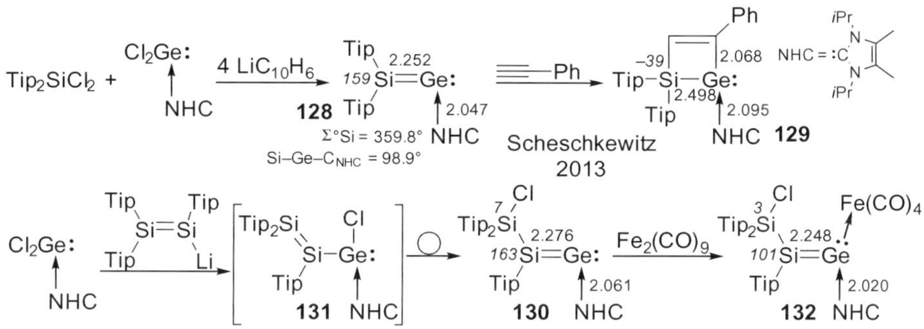

Scheme 9.42
Synthesis and reactivity of NHC-stabilized silagermylidenes.

A similar NHC-stabilized silagermylidene **130** was obtained, in one step, by the reaction of NHC-stabilized dichlorogermylene with a lithium disilenide (Scheme 9.42).[112] The reaction probably produces a disilene-substituted chloro-germylene **131** which isomerizes to the silagermylidene **130**. The coordination ability of the silagermylidene has been demonstrated by the synthesis of the corresponding $Fe(CO)_4$ complex **132**.

9.2.2.2.3 Si=Sn

Only one example of stable silastannene **133** has been reported, by Lee and Sekiguchi et al. in 2002 (Scheme 9.43). It was synthesized by the reaction of dilithiosilane with diaryldichlorostannane.[113] Silastannene **133** features strongly polarized Si = Sn bond toward Si atom ($Si^{\delta-} = Sn^{\delta+}$, NPA charge − 0.54 on Si and +1.40 on Sn), which is demonstrated

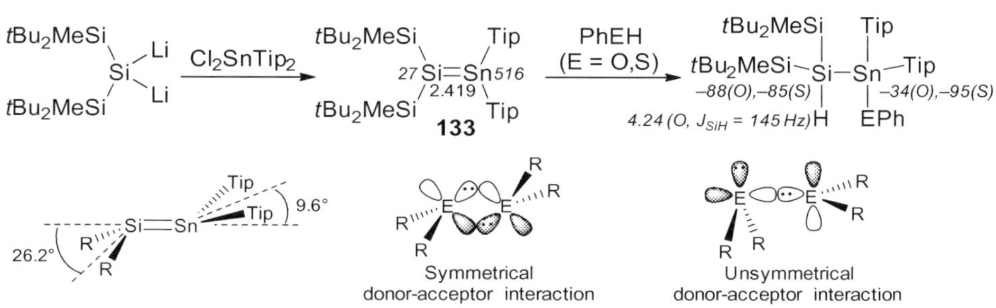

Scheme 9.43
Synthesis of the first stable silastannene.

by deshielded sp^2-Sn nuclei (516.7 ppm) as well as significantly shielded sp^2-Si nuclei (27.4 ppm). The Si=Sn bond length is intermediate between Si=Si (2.138–2.289 Å)[3] and Sn=Sn (2.59–3.087 Å)[3] bond lengths, and considerably shorter than Si–Sn single bond (2.60 Å).[114] The molecule is remarkably twisted by 34.6 degrees. In general, disilenes prefer to have a planar or near-planar geometry, whereas distannenes tend to have a highly pronounced *trans*-bent geometry. The silastannene shows a completely opposite tendency and presents a quite unusual *trans*-bent structure with a larger bonding angle at the Si atom (26.2 degrees) than that at the Sn atom (9.6 degrees). This was rationalized in terms of the unconventional unsymmetrical donor–acceptor interaction mode, promoted by the $Si^{\delta-} = Sn^{\delta+}$ double bond polarization. The regioselectivity of the addition of PhEH (E = O, S) agrees well with the polarity of the Si=Sn bond (Scheme 9.43).

9.2.2.3 Si=E15

9.2.2.3.1 Si=N

Prehistory The first synthesis of stable and isolable silaimines (**134**, **135**) was realized independently by two German groups (Klingebiel and Wiberg) in 1986 (Scheme 9.44). Klingebiel's group synthesized a kinetically stabilized silaimine **134** using bulky substituents on Si and N atoms by LiCl elimination of the lithiated aminochlorosilane at 80°C under vacuum (0.01 mbar) and simultaneously isolated it by distillation.[115] Although the structural analysis of the silaimine failed, its strongly low-field shifted ^{29}Si NMR resonance (66.3 ppm) is characteristic for an unsaturated silicon compound.

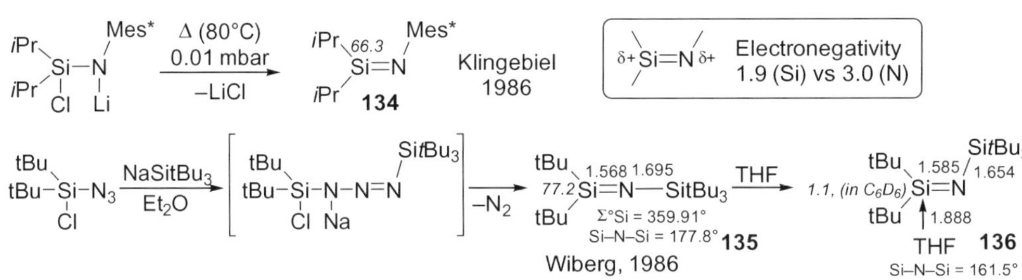

Scheme 9.44
Synthesis of the first stable silaimines.

Wiberg's groups demonstrated that the reaction of chlorosilylazide and sodium tris(*tert*-butylsilyl)silanide leads to the formation of a stable silaimine **135** which could be isolated as crystalline material (Scheme 9.44).[116] Its X-ray structure shows a planar geometry around the silicon center (Σ°Si = 359.9 degrees) and a significantly shorter Si=N bond (1.568 Å) than the neighboring N–SitBu_3 single bond (1.695 Å). The Si=N–Si fragment is essentially linear (177.8 degrees), which is consistent with the results obtained by theoretical calculations for the model compound (H_2Si=N–SiH_3: 175.6 degrees).

The linear geometry of the silaimine **135** is attributed to the interactions between the lone pair on nitrogen atom and the two silicon centers rather than steric effect.[117] The calculations also indicated only a small energy difference between the bent and linear form of the silaimine ($\Delta E_{linear\text{-}bent} = 6.0$ kcal mol^{-1} for $H_2Si{=}NH$). Not surprisingly, because of the significant difference in electronegativity between Si (1.9) and N (3.0), silaimine presents a strong Lewis acidic character at silicon, much stronger than that of silenes. This was clearly shown by the formation of a stable complex **136** with a molecule of coordinated THF in spite of the strong kinetic protection around the silicon atom. As a consequence, the signal in ^{29}Si NMR shifts to higher field (1.1 ppm) relative to the base-free silaimine (77 ppm) and the Si=N bond is slightly elongated (1.585 Å).

Base-stabilized silaimines Before the isolation of the base-free silaimines **135**, Wiberg's group had already reported in 1985 the isolation of THF-silaimine **136** complex (Scheme 9.45).[118] Of particular interest, complex **136** is highly reactive, and an ene-reaction with propene was observed at RT in the absence of any catalyst. It also undergoes a [2 + 2] cycloaddition reaction with an electron rich olefin ($CH_2{=}CHOMe$). The reaction with N_2O affords oligosiloxanes **137** and trialkylsilylazide **138**. This reaction probably proceeds through the transient formation of [3 + 2] cycloadduct **139** and its decomposition into the silyl azide and a highly reactive silanone which immediately oligomerizes. As expected, the reaction with H_2O gives the corresponding 1,2-adduct **140** with a hydroxyl group on the electrophilic silicon center.

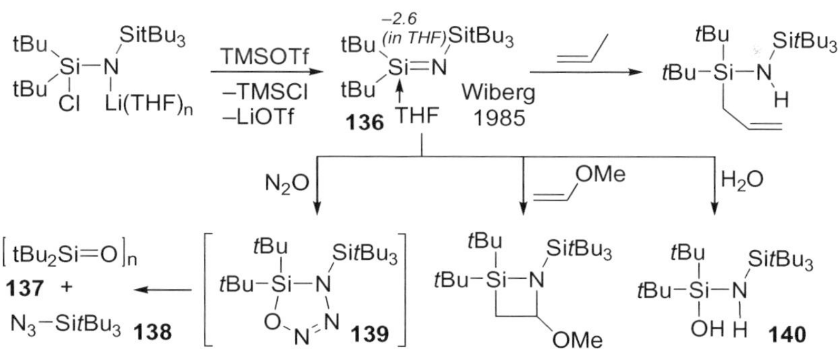

Scheme 9.45
Reactivity of THF-silaimine complex **136**.

Similarly to the case of silenes (Section 9.2.2.2.1), the pyridine complex of silaimine **141** also isomerizes under mild conditions by insertion into the C–H bond of the pyridine at the 2-position (Scheme 9.46).[119] The amine adduct of silaimine **142** is also labile and the amine ligand under vacuum easily dissociates to generate a free silaimine which readily dimerizes to give a head-to-tail dimer **143**.[120] This amine-complex also isomerizes in C_6D_6 solution at RT by insertion of the silaimine fragment into the sp^3-C—H bond of the amine ligand to afford an aminosilane **144**.

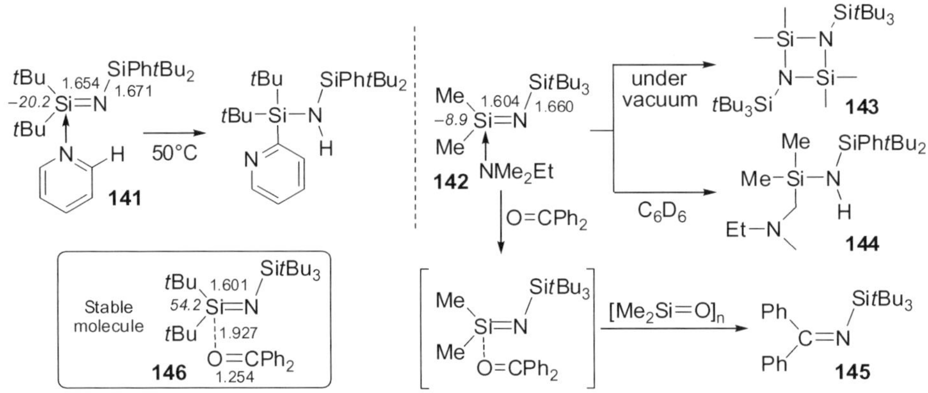

Scheme 9.46
Isomerization and reactions of base-stabilized silaimines.

Interestingly, the amine-adduct **142** readily reacts with diphenylketone, *via* sila-Wittig type reaction, to give the corresponding imine **145** and polydimethylsiloxanes.[121] The reaction of a more bulky silaimine with the diphenylketone stops at the formation of a stable donor—acceptor complex **146**, which is supposed to be an intermediate of the reaction.[121]

Different synthetic methods Since 2000, several other synthetic methods have been developed for stable silaimines. As a unique synthesis, Klingebiel et al. reported a reversible [3 + 2] cycloaddition of silatetraazoline **147** (Scheme 9.47). Indeed, the cycloadduct **147** (silaimine + silylazide) decomposes under mild conditions (40°C) into the corresponding silaimine and azide.[122] Using this technique, they synthesized a *N*-chlorosilyl-silaimine **148** in which the chlorine atom is in between the two silicon centers via a [1,3]sigmatropic migration. Indeed, the molecule is symmetric at the NMR timescale and only one signal at 3.0 ppm for the two silicon atoms was observed in ^{29}Si NMR.

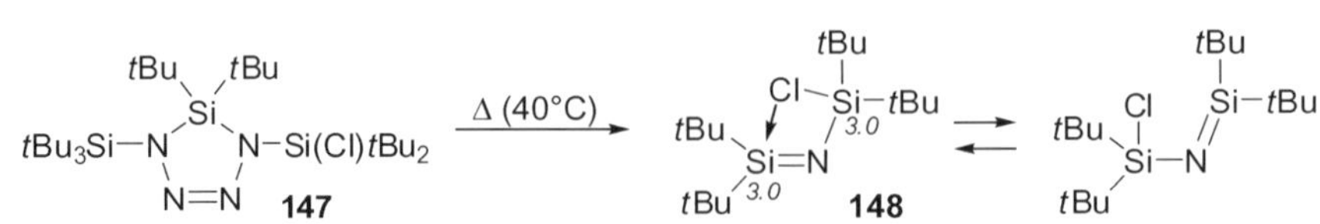

Scheme 9.47
Retro-[3 + 2] cycloaddition of silatetraazoline.

Stable silylene + N_3R Denk and West et al. have described the reaction of stable cyclic diaminosilylene **149** (NHSi) with an azide which generates a silaimine (Scheme 9.48).[123] Although the small trimethylsilyl azide subsequently reacts with the generated silaimine to give the corresponding 1,2-adduct **150**, the reaction with the bulky triphenylmethylazide in THF selectively affords a stable THF complex of silaimine **151**.

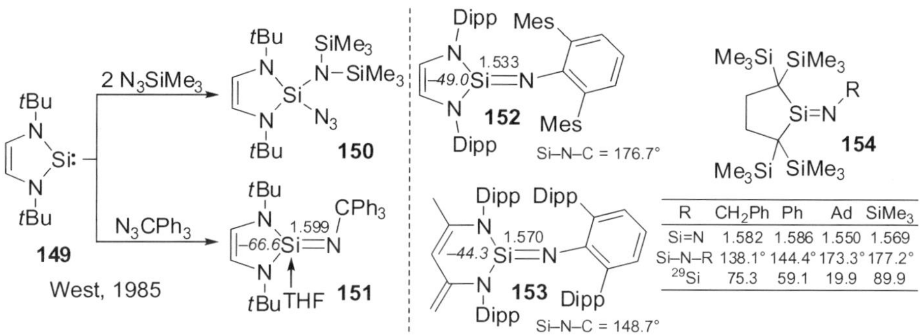

Scheme 9.48

Synthesis of various silaimines by the reaction of different stable silylenes with azides.

Among the available synthetic methods, the latter appeared to be very important as various base-free silaimines have been synthesized starting from different stable silylenes. Particularly, two stable diamino-silaimines (**152**, **153**) with bulky substituents were synthesized (Scheme 9.48). They are perfectly stable at RT[124] and show considerably high-field shifted ^{29}Si NMR resonances (−49 and − 44 ppm) compared with those observed for the previously reported alkyl-substituted silaimines (66−77 ppm). Stable silaimines **154** derived from Kira/Iwamoto's cyclic dialkylsilylene are well protected by the bulky substituent system and tolerates various kind of substituents on the nitrogen atom, including relatively small benzyl and phenyl groups.[125]

Donor-stabilized silylenes + N_3R The same synthetic methodology appeared to be appropriate for the synthesis of silaimines **155** and **156** starting from base-stabilized silylenes such as silylenes with a bulky amidine ligand as well as the NHC-stabilized dichlorosilylene (Scheme 9.49).[124b,126,127]

The additional kinetic stabilization provided by the bulky NHC ligand allows the isolation of the reactive *Si*-dichloro-silaimine **156**. Starting from this dichlorosilaimine **156**, Roesky et al. successfully synthesized the first stable dimer of sila-isonitrile **157** (Scheme 9.49).[128] This four-membered cyclic molecule **157** with 4π-electrons with antiaromatic character (NICS(0) = +5.01, NICS(1) = +0.91) presents two divalent silicon atoms and it reacts with azides to afford a bis-silaimine **158**.

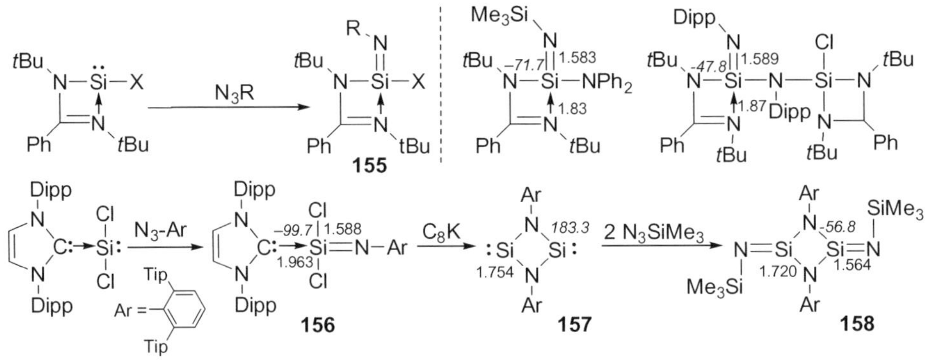

Scheme 9.49
Reactions of base-stabilized silylenes with N_3R.

D-stabilized silylene + RN═C═NR and others Roesky et al. also described that the same types of donor-stabilized silaimines **159** and **160** can also be synthesized by reaction of donor-stabilized silylenes with a carbodiimide, acting as an aryl nitrene precursor (Scheme 9.50).[126,129]

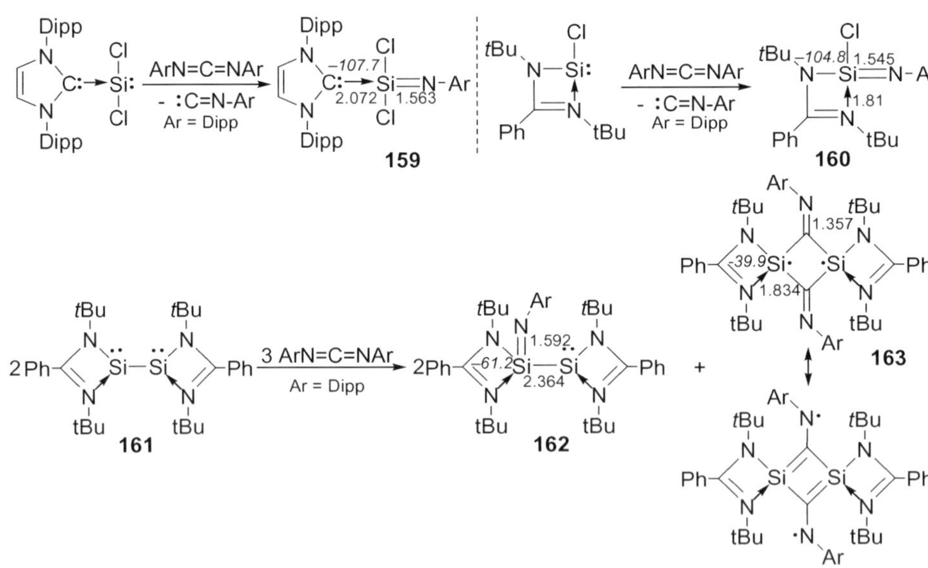

Scheme 9.50
Reaction of donor-stabilized silylenes with carbodiimides.

A 1,2-bis-silylene **161** with amidine ligands also reacts with a carbodiimide to afford a base-stabilized silaimine **162** featuring an intact silylene center (Scheme 9.50).[130] More interestingly this reaction also generates a four-membered heterocyclic singlet biradical **163** which might result from the reaction of the bis-silylene with the generated isonitrile as byproduct of the former reaction.

Tacke et al. described an isomerization of a transient silanone **165**, generated by the reaction of a bis(trimethylsilyl)amino-silylene **164** with a N_2O, via 1,3-migration of trimethylsilyl group from N to O atom to afford a base-stabilized silaimine **166** (Scheme 9.51).[131] The same type of isomerization has also been observed for a bis (trimethylsilyl)amino silaimine to give **167**.[124b]

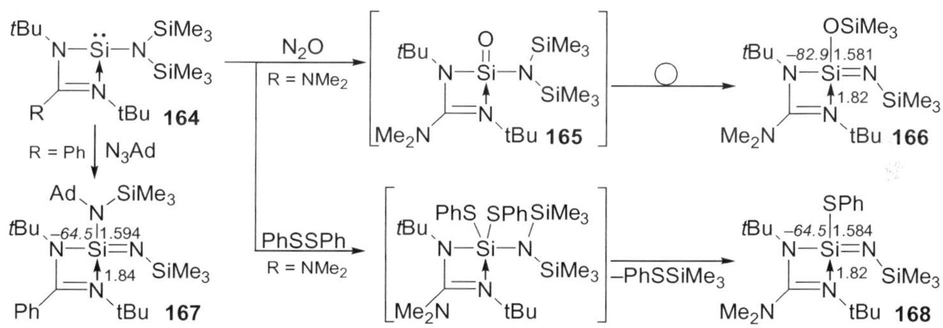

Scheme 9.51
Synthesis of silaimines by the reaction of bis(trimethylsilyl)amino-substituted silylene with various reagents.

Tacke et al. also reported the reaction of the silylene **164** with diphenyldisulfide which affords a thiophenoxy substituted silaimine **168** (Scheme 9.51).[131] The reaction probably proceeds via oxidative addition of PhSSPh to the silylene followed by an elimination of $PhSSiMe_3$ to give the silaimine **168**.

As a unique synthesis of silaimines and its application in organic synthesis, Cui et al. reported the reaction of NHC-stabilized chloroaminosilylene **169**, easily prepared from the corresponding dichlorosilane derivative in the presence of NHCs,[132] with electrophilic alkynes, which results in the regio- and stereo-selective formation of a silaimine substituted olefins **170** (Scheme 9.52).[133] The reaction probably starts with nucleophilic attack of the silylene **169** on the terminal carbon atom of the alkyne followed by migration of trimethylsilyl group from nitrogen to the negatively charged vinylic carbon to form the silaimine. Treatment of the obtained chlorovinylsilaimines **170** with an excess of methanol gives a cis-1,2-disilylalkene **171** with two different silyl groups (TMS and $Si(OMe)_3$). The reaction can be considered as a formal bis-silylation of an alkyne without metallic catalysts.

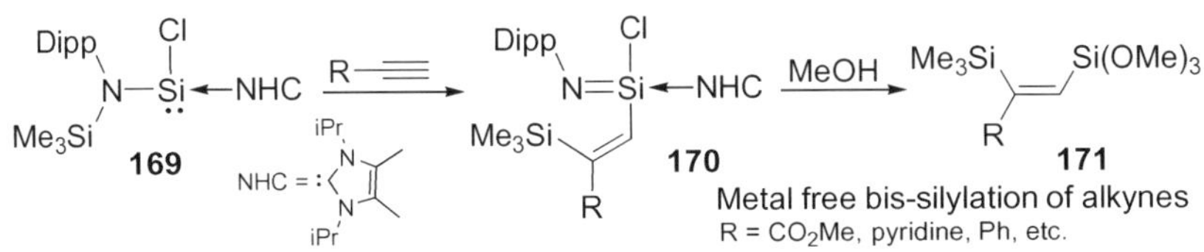

Scheme 9.52
Metal free bis-silylation of activated alkynes.

9.2.2.3.2 Si=P

Prehistory The first moderately stable phosphasilene **172** was synthesized by 1,2-elimination of LiCl from the lithiated chlorosilylphosphane by Bickelhaupt et al. in 1984 (Scheme 9.53).[134] The phosphasilene **172** presents characteristic low-field shifted ^{29}Si- and ^{31}P NMR resonances (76.7 ppm and 136.0 ppm, respectively) with a large phosphorus-silicon coupling constant (149 Hz). However, this synthetic route suffered from an inevitable side reaction, the substitution reaction of chloride by *n*BuLi to give **173**. The same group solved the problem by the reaction of dichlorosilane derivative with $ArPH_2$ in the presence of two equivalents of *n*BuLi, which proceeds more selectively and allows to generate various phosphasilenes **174**.[135] Nevertheless, the reactions are not sufficiently clean and all the phosphasilenes prepared have only been characterized in solution by ^{29}Si- and ^{31}P NMR spectroscopy.

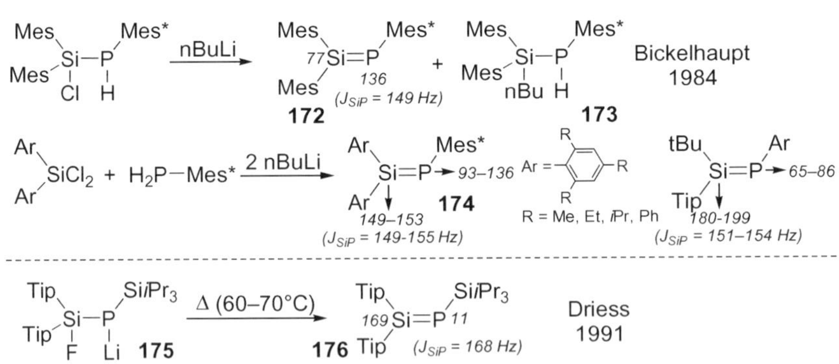

Scheme 9.53
Synthesis of the first stable phosphasilenes.

In the meantime, Driess et al. reported the synthesis of a stable *P*-silyl phosphasilene **176**, by thermal LiF elimination from lithiated fluorosilylphosphane **175** at 60°C, which was fully characterized in solution (Scheme 9.53).[136] However, it took about 10 years, from the first characterization of persistent phosphasilenes in solution, to successfully synthesize the first isolable phosphasilene **178** by Niecke.[137] The synthesis was achieved by the reaction of stable 1,3-diphospha-2-silaallyl anion **177**[138] with a chlorodiphenylphosphine, and isolated as yellow crystals (Scheme 9.54). The X-ray diffraction analysis discloses a Si=P bond length of 2.092 Å, which is characteristically shorter than the neighboring Si–P single bond (2.254 Å), and the silicon center is slightly pyramidalized (Σ°Si = 356.8 degrees).

Several years later, Driess et al. also isolated a *P*-silyl phosphasilene **179** presenting a planar geometry around the tricoordinate silicon center (Σ°Si = 359.9 degrees) and a slightly shorter Si=P bond (2.062 Å) than that of Niecke's one (Scheme 9.54).[139] This value is very close to that predicted by theoretical calculations (2.04–2.06 Å).[46,140] It was also shown by calculations that the silyl group on the phosphorus atom strengthens the Si=P π-bond.[141]

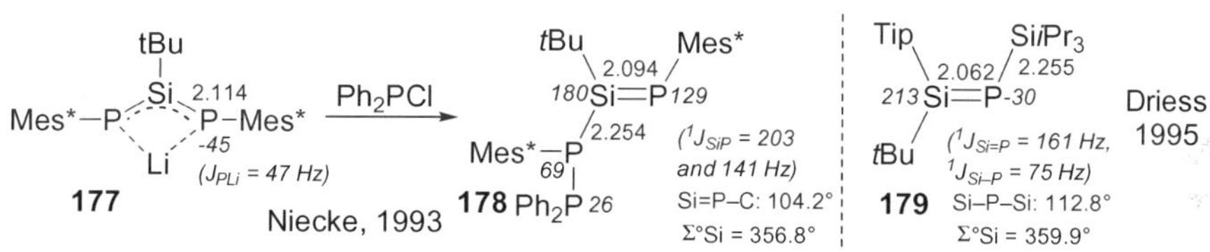

Scheme 9.54
Synthesis of first isolable phosphasilenes.

Half parent phosphasilene The first important achievement in this domain since 2000 has been made by Driess's group in 2006. Indeed they reported the first synthesis of surprisingly stable "half parent" phosphasilene **180** with a terminal P–H bond as a mixture of *cis*- and *trans*-isomers (*Z*:*E* = 1:1.5).[142a] Although the deprotonation on the phosphorus atom in **180** by strong bases such as *n*BuLi was unsuccessful, its metalation with Me_2Zn in the presence of TMEDA (tetramethylethylenediamine) proceeds cleanly at −78°C to afford the *P*-zinciophosphasilene **181**. The plumbylene derivative **182** has also been prepared[142b] (Scheme 9.55).

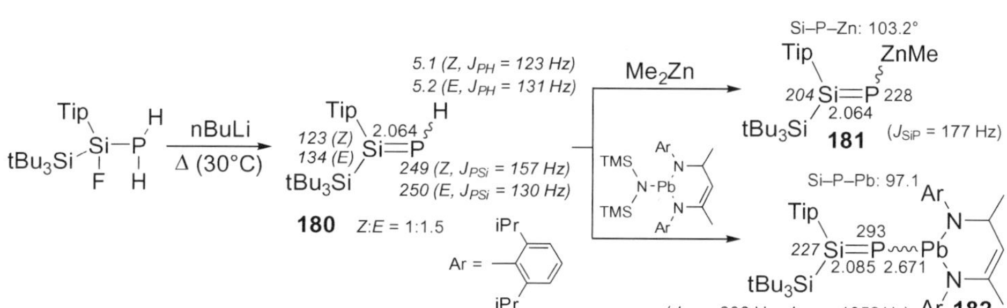

Scheme 9.55
"Half" parent phosphasilene and its P-functionalization.

The same group also reported another "half parent" phosphasilene **183** with a highly polarized Si=P π-bond toward P atom due to the unique zwitterionic cyclic diaminosilylene fragment (Scheme 9.56).[143] The strong polarization was clearly indicated by the dramatic high field shift of the ^{31}P NMR resonance (−294 ppm) and by NBO analysis (73.5% on P atom). In contrast to the previous stable "half parent" phosphasilene **180**, this molecule **183** is thermally labile and it decomposes even at RT liberating the cyclic silylene **184** and polyphosphane of PH as red-brown solid, suggesting an easy dissociation of "parent phosphinidene (PH)." Indeed, they demonstrated a facile transfer of PH fragment to a NHC, which takes place at RT to give the corresponding phosphaalkene **185**. Recently the nitrogen analog was reported, but "half parent" silaimines **186** are only transient species.

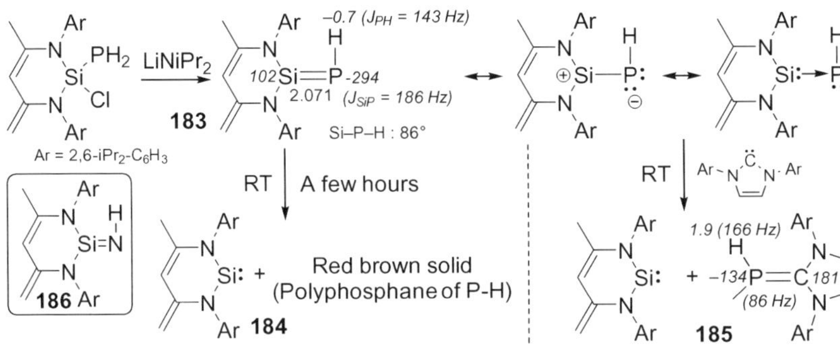

Scheme 9.56

Si-Diamino "Half parent" phosphasilene and its reaction as a "parent phosphinidene PH" reagent.

In contrast, the addition of strongly σ-donating but poorly π-accepting ligands such as DMAP (*N,N*-dimetylaminopyridine) or alkyl-substituted NHC on the phosphasilene does not induce the PH transfer (Scheme 9.57).[144] Instead they coordinate on the silicon center to form stable donor−acceptor complexes **187** and **188**, respectively. In spite of an enhanced stability at RT, PH-transfer toward aryl-substituted NHC takes place at 90°C. It is worth noting that, in contrast to silenes and silaimines, phosphasilenes are poorly electrophilic due to the small electronegativity difference between silicon (1.9) and phosphorus (2.1). Indeed, there are few examples of such donor−acceptor complexes. The formation of a stable complex in this case is probably due to the electronegative amino substituents on the silicon center as well as the strong nucleophilic character of the ligands.

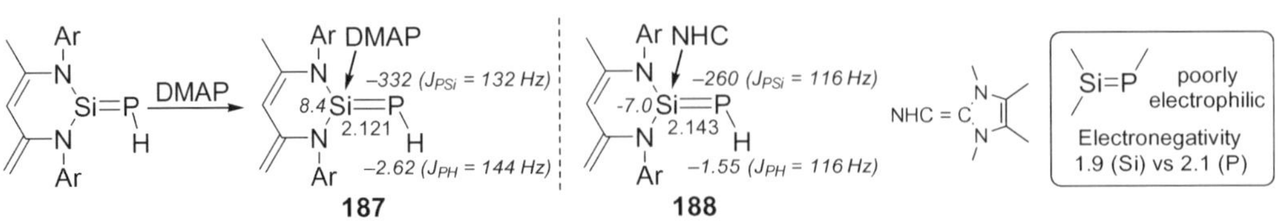

Scheme 9.57

Synthesis of donor-stabilized "half parent" phosphasilenes.

Air stable phosphasilene The introduction of fused ring bulky aryl groups (Eind), developed by Tamao/Matsuo, as substituents, significantly increases the stability of the corresponding phosphasilenes (Scheme 9.58).[145] Indeed, phosphasilenes **189** with two Eind groups on Si and P atoms are stable even under air. The introduction of larger aromatic substituents (Ph → Np → Ant) on the silicon center causes gradual red-shift of the absorption peaks in UV/VIS spectrum (385 → 430 nm). A *p*-phenylene bridged bis-phosphasilene **190** with an extended π-conjugation presents a further red-shifted absorption band (449 nm).

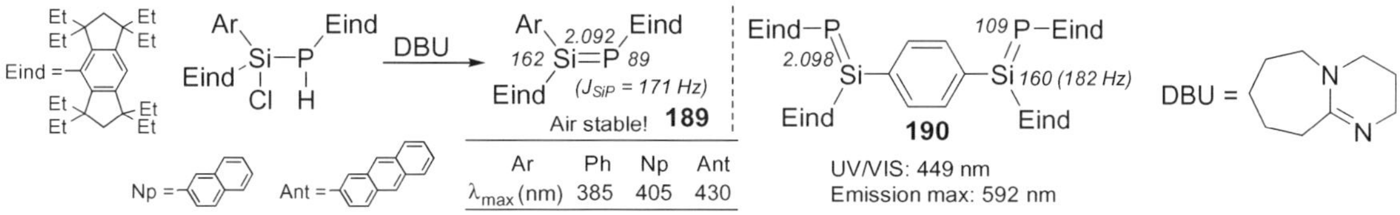

Ar	Ph	Np	Ant
λ_{max} (nm)	385	405	430

Scheme 9.58
Synthesis of air stable phosphasilenes.

New synthetic methods All stable phosphasilenes reported in early stage have been synthesized by H–X elimination (X = halogen) from the corresponding halogenosilylphosphine in the presence of a strong base (*n*BuLi or LDA). Since several years, new silicon based reagents allowed to develop several new synthetic methods.

2-phospha-1,3-disilaallyl fluoride derivative **191** undergoes a 1,3-sigmatropic shift of fluorine atom under mild conditions (40°C).[146] Similarly, a diaminophosphino disilene **192** also isomerizes via a 1,3-migration of an amino group on the phosphorus toward silicon atom to give the corresponding phosphasilene **193** as a mixture of *cis*- and *trans*-isomers (E:Z = 95:5).[147] The isomerization of the smaller model compound (Ph instead of Tip group) was calculated to be thermodynamically favored by 28.8 kcal mol^{-1}. Due to the presence of the electronegative π-donating amino group on the phosphorus atom, the polarization of Si = P fragment is inverted (Si: −0.095, P: +0.169). The addition of *tert*-butylisonitrile induces the 1,2-migration of the second amino group on phosphorus to the adjacent silicon atom to afford a 1-phoshaketenimine **194** (Scheme 9.59).

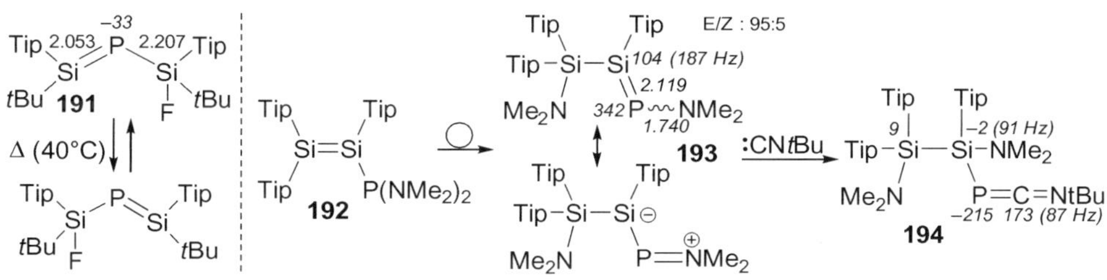

Scheme 9.59
Synthesis of a phosphasilene by isomerization of phosphinodisilene and its reaction with an isonitrile.

The stable dilithiosilane, developed by Sekiguchi's group, readily reacts with a bulky dichlorophosphine to afford a stable phosphasilene **195** (Scheme 9.60).[148] Interestingly, it is important to note that a dilithiophosphine (Li_2PTip **196**) does not react at all with bulky dichlorosilanes such as $Cl_2Si(tBu)(Tip)$.[136b]

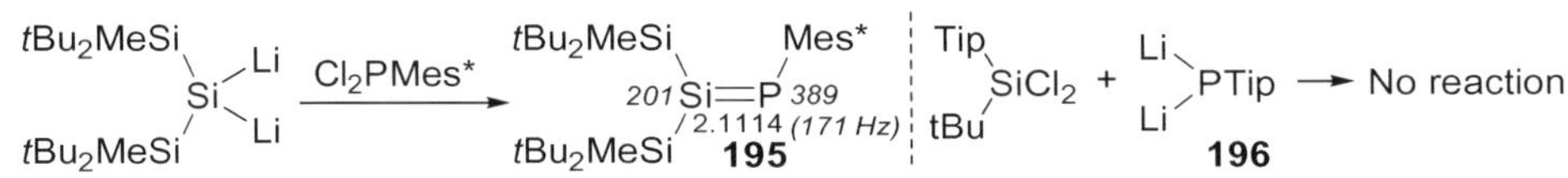

Scheme 9.60
Reaction of a dilithiosilane with a dichlorophosphine.

Recently, Scheschkewitz and Goicoechea demonstrated that using 2-phosphaethynolate anion **197** (PCO^-), a phosphide fragment (P^-) can be introduced into a cyclotrisilene **198** under photolytic conditions (Scheme 9.61). After loss of carbon monoxide, and formal ring expansion, the original four membered cyclic 1-phospha-2,3-disilaallyl anion **199** was obtained.[149] The terminal ^{29}Si- (−40 ppm) and ^{31}P- nucleus (−57 ppm in THF, −94 ppm in toluene) are strongly shielded whereas the central ^{29}Si nucleus is strongly deshielded (193 ppm), which is a typical NMR pattern for allyl anions.

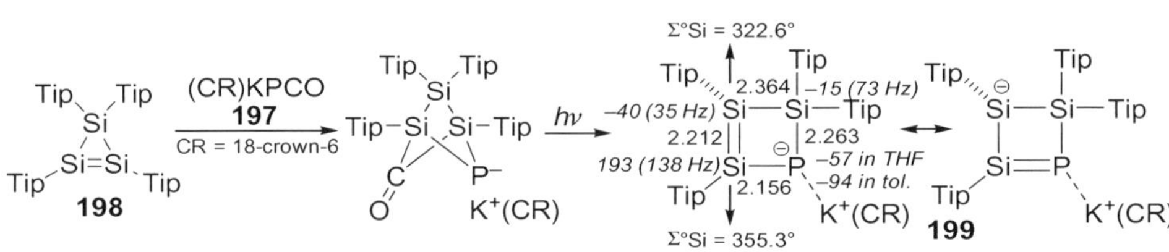

Scheme 9.61
Reaction of the stable cyclotrisilenes with K–P=C=O.

Mono- and diphospha-cyclobutadienes **201** and **202** stabilized by donating ligands have been synthesized by the reaction of cyclic amidine-stabilized chlorosilylene **200** with a phosphaalkyne and white phosphorus (P_4) respectively (Scheme 9.62).[150] The chlorosilylene **200** also behaves as a reducing agent in this reaction. The same results were obtained starting from the 1,2-disilylene derivative **203**. The diphospha-cyclobutadiene **202** has also been synthesized by the treatment of a bis(trimethysilyl)phosphino-silylene **204** with a dichlorophosphonium derivative.[151] In both compounds, the strong coordination of donating ligands on the silicon atoms causes strong polarization of the Si=P bonds toward P atoms, as indicated by high field ^{31}P resonances (**201**: −243 ppm, **202**: −166 ppm), and the loss of anti-aromatic character (NICS(1) = −4.14 for **201** and −2.56 for **202**).

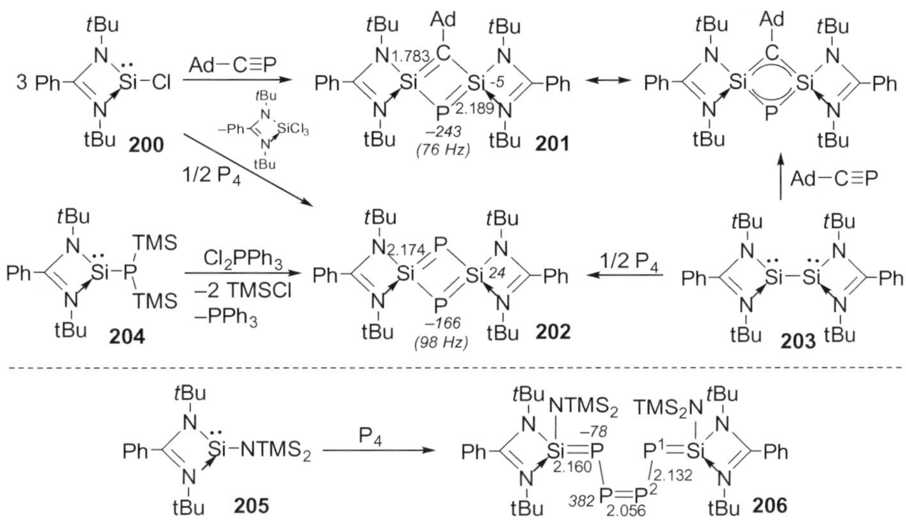

Scheme 9.62
Synthesis of mono- and diphospha-cyclobutadienes as well as of 2,3,4,5-tetraphospha-1,3,5-hexatriene.

In contrast to the case of chlorosilylene, the reaction of the related aminosilylene **205** with P_4 leads to the formation of a unique 2,3,4,5-tetraphospha-1,3,5-hexatriene derivative **206** as an *E*-isomer (Scheme 9.62).[152] $P^2{=}P^2$ bond length (2.056 Å) is in the typical range of double bonds, while the P1–P2 distance (2.132 Å) is halfway between a single (2.21 Å) and double bond (1.954–2.044 Å), suggesting an enhanced electron delocalization in the π-conjugated system due to the strong polarization of Si=P bonds coordinated by a donating ligand.

The bis(trimethysilyl)phosphino-silylene **204** is thermally labile, and isomerizes at 100°C by a 1,2-migration of a trimethylsilyl group to the adjacent Si atom to give the corresponding base stabilized phosphasilene **207** (Scheme 9.63).[151] Similarly, an

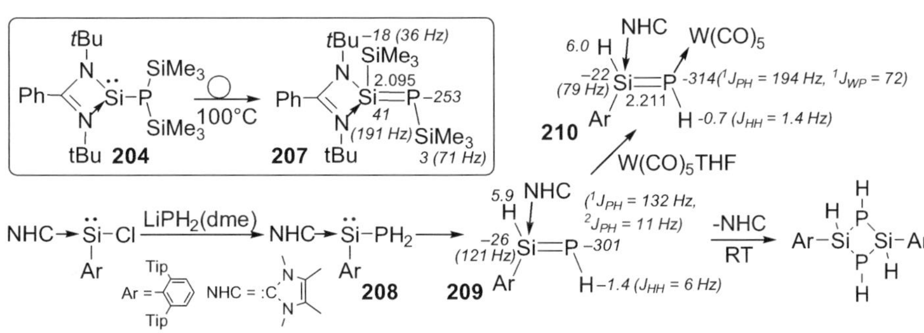

Scheme 9.63
Phosphinosilylene–Phosphasilene conversion and the synthesis of the first persistent 1,2-dihydrophosphasilene.

NHC-stabilized silylene **208** with a PH_2 substituent also undergoes a similar isomerization affording the first persistent 1,2-dihydrophosphasilene **209**.[153] Although it slowly dimerizes at room temperature, the addition of $W(CO)_5$ leads to a donor–acceptor complex **210** which is perfectly stable at RT.

Roesky et al. recently synthesized stable dichlorophosphasilene complexes with a NHC ligand such as cyclic (amino)(alkyl)carbene (cAAC) (**211**) or NHC (**212**, Scheme 9.64).[154] Of particular interest, crystals of the cAAC- (**211**) and NHC-complexes (**212**) are blue and red, respectively, and they present considerably red-shifted absorption bands (665 nm for **211**, 475 nm for **212**) in UV/VIS spectrum compared to those observed for other phosphasilenes (335–430 nm). This was explained by a particular HOMO-LUMO excitation process via ICT transition. Indeed, the HOMO is located on $\pi_{Si=P}$ orbital, while the LUMO is located at the carbene moiety ($\pi^*_{carbene}$). The LUMO is lower in energy for cAAC-complex **211** than that of NHC-complex **212** by 0.78 eV, which explains the enhanced red-shift observed for the cAAC-complex **211**.

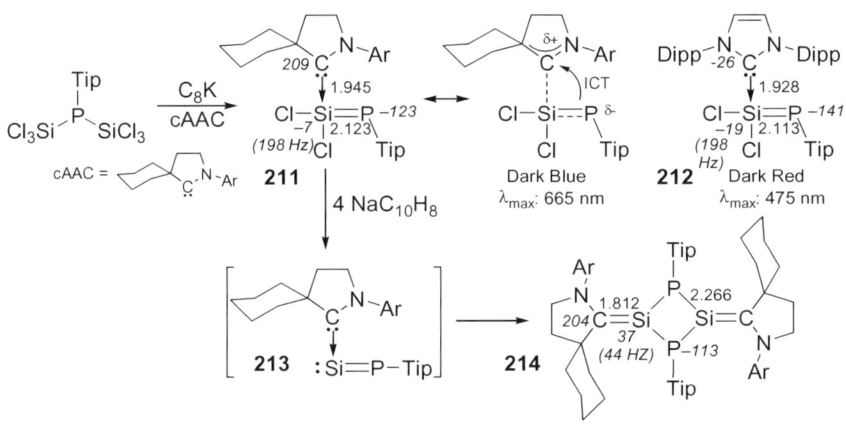

Scheme 9.64

Synthesis of NHC- and CAAC-stabilized dichlorophosphasilenes and reductive dechlorination reaction.

The attempt to synthesize a stable phosphasilenylidene derivative **213** (heavier analog of isonitrile) by the reductive dechlorination of the dichlorophosphasilene complexes **211** using $NaC_{10}H_8$ failed and the reaction resulted in the formation of a head-to-tail dimer **214** (Scheme 9.64).[155] Contrary to this result, Filippou et al. described that a labile NHC-stabilized phosphinosilylene **215** readily eliminates TMSCl at $-10°C$, leading to the formation of the first stable phosphasilenylidene **216** (Scheme 9.65).[156] The prominent features of **216** are a strongly bent structure (C_{NHC}–Si–P: 96.9 degrees, Si–P–C_{Mes^*}: 95.4 degrees) and a considerably low-field shifted ^{29}Si- and ^{31}P NMR resonances (267 ppm and 402 ppm, respectively).

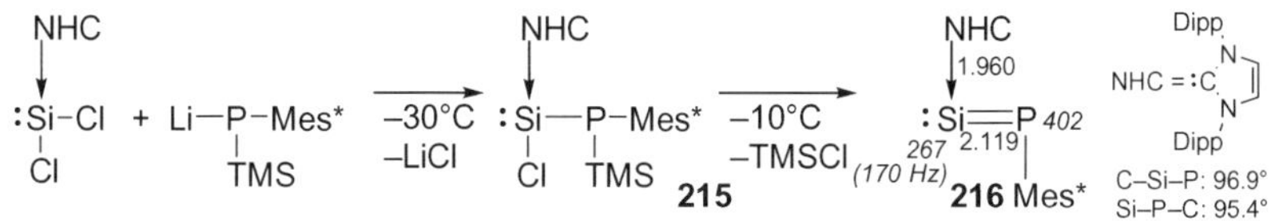

Scheme 9.65

Synthesis of the first stable heavier analog of isonitrile: NHC-stabilized phosphasilenylidene.

In contrast to the classical carbon centered phosphonium ylides which are important chemical synthetic tools (Wittig reagent),[157] their silicon analogs, phosphonium sila-ylides are highly reactive species and fragile molecules due to the weak P→Si interaction.[158] Kato/Baceiredo's group stabilized such a molecule **217** incorporating the fragile ylidic P = Si bond into a cyclic structure (Scheme 9.66).[59] The X-ray diffraction analysis revealed a long P–Si bond (2.30 Å) like a single bond and a strongly pyramidalized silicon center, indicating localized electron pair on the silicon atom without back-donation toward the phosphonium fragment (absence of ylidic type P═Si double bond). Nevertheless, theoretical calculations demonstrated that the inversion barrier around the silicon center, through its planarization, is quite small (19.2 kcal mol^{-1}). Of particular interest, this molecule **217** behaves not only as a phosphonium sila-ylide[59,159] but also as a base-stabilized silylene.[60,160]

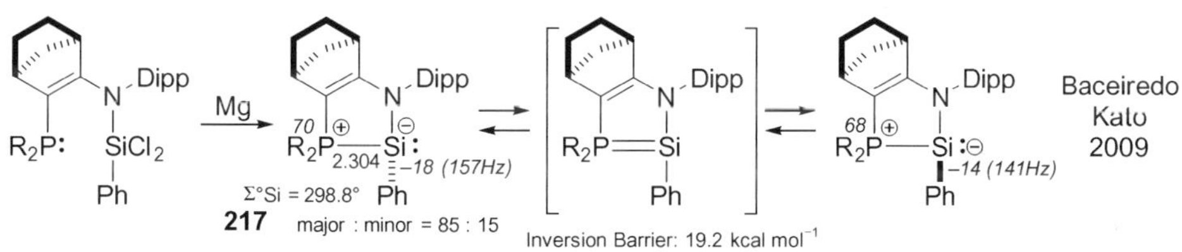

Scheme 9.66

Synthesis of the first stable phosphonium sila-ylide.

9.2.2.3.3 Si═As, Si═Sb

Until Lee, Sekiguchi et al. reported the third type of stable arsasilene in 2014, only Driess's group had contributed to the chemistry of stable arsasilenes. They reported in 1991 the first synthesis of stable arsasilene **218** by thermal LiF elimination from the corresponding fluorosilyl-lithioarsane at 80°C (Scheme 9.67).[139,161] Several years later, they isolated an arsasilene **219** in the solid state. It presents a significantly low-field shifted ^{29}Si NMR resonance (228 ppm) and a significantly shorter Si═As (2.165 Å) bond than adjacent single As–Si bond (2.369 Å). The arsasilene **218** readily reacts with benzophenone, via a pseudo-Wittig reaction, to give an arsaalkene **220**. In the presence of two equivalents of isocyanide an original four-membered ring **221** featuring an arsaalkene fragment was

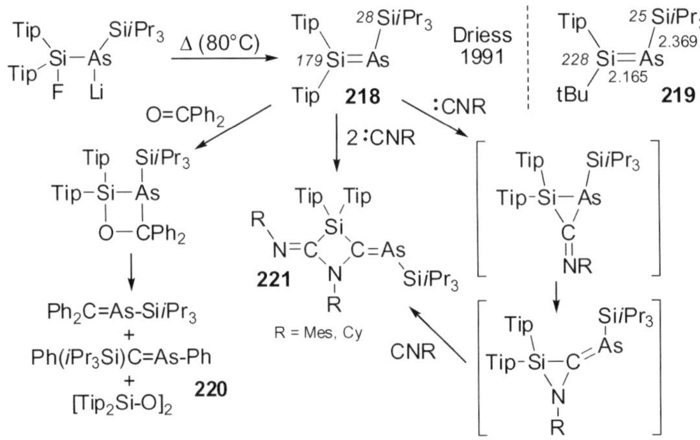

Scheme 9.67
Synthesis and reactivity of the first stable arsasilenes.

obtained.[162] This reaction probably proceeds through the transient formation of a silaaziridine with an exocyclic arsaalkene function (similarly to the case of the disilenes (Section 9.2.2.2.1)), which subsequently reacts with a second equivalent of isocyanide to give the final arsaalkene.

As already mentioned, another stable arsasilene **222** has been synthesized by Lee, Sekiguchi et al. by the coupling of a dilithiosilane with a dichloroarsane (Scheme 9.68).[163] The Si=As bond (2.216 Å) is longer than that observed for the Driess's arsasilene (2.165 Å) probably due to the steric reason. The ^{29}Si NMR chemical shift (214 ppm) is similar to those observed for the Driess's ones (179–228 ppm).

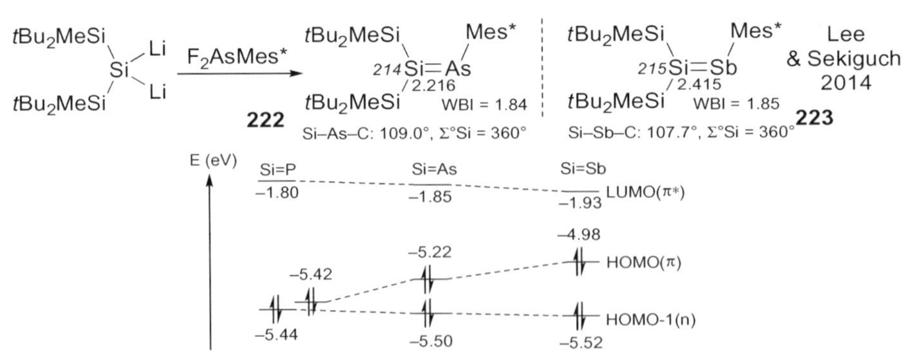

Scheme 9.68
Synthesis of a stable arsasilene and the first stable stibasilene.

The first stable stibasilene **223** has also been synthesized by the same method using F_2SbMes^* as starting material (Scheme 9.68).[163] This molecule exhibits a Si=Sb bond of 2.415 Å, which is notably shorter than the sum of the covalent radii of Sb and Si atoms,[164] and a trigonal planar geometry around the sp^2-silicon center ($\Sigma°Si = 360$ degrees), suggesting an important double bond character (calculated Wiberg bond index: 1.85). Theoretical calculations demonstrated that, from phosphasilene to arsasilene, and to stibasilene, the HOMO energy level ($\pi_{Si=E}$) remarkably increases, in contrast to the energy level of HOMO-1 which progressively decreases (Scheme 9.68). Such trend can be well rationalized given the greater s-character of the lone pair MO (62% for Si=P, 68% for Si=As, 75% for Si=Sb).

A base stabilized arsasilene **225** with a small H–Si=As–H subunit has also been prepared by Driess et al., by the reaction of an *N*-heterocyclic silylene with AsH_3 (Scheme 9.69).[165] The arsasilene **225** was isolated as deep blue crystals. It exhibits a Si=As bond (2.218 Å) as short as that of a base free arsasilene **222** (2.216 Å) and a high-field shifted ^{29}Si NMR resonance (17.6 ppm) due to the base coordination on the silicon atom. The As–H proton signal appears at remarkably high field (−2.22 ppm), which is presumably attributed to a relativistic effect (spin-orbit coupling). The intense blue color of the arsasilene is due to its HOMO ($\pi_{Si=As}$) → LUMO (π^*_{ligand}) transition ($\lambda_{max} = 690$ nm). In benzene solution a prototropic tautomerism occurs leading to an equilibrium between both isomers (silylarsane **224**: arsasilene **225** = 3:7, at RT). Calculations indicate that arsasilene **225** is only slightly more stable than silylarsane **224** by 5.7 kcal mol^{-1}.

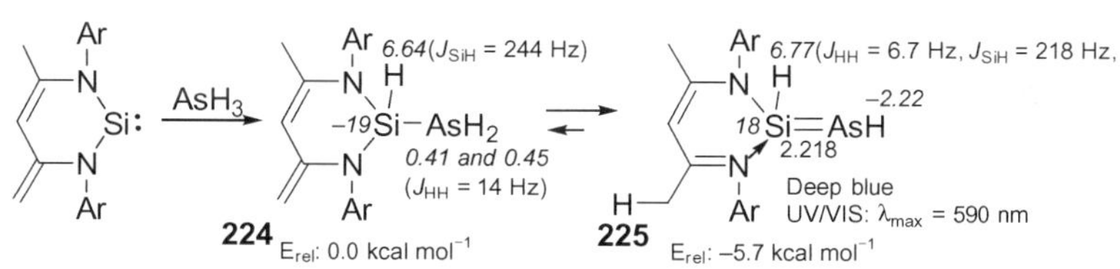

Scheme 9.69
Synthesis of a stable donor-stabilized arsasilene.

9.2.2.4 Si=E16

9.2.2.4.1 Si=O

Introduction The carbonyl group is stable, ubiquitous, and without any doubt, one of the most important functional groups in organic chemistry. In marked contrast, its heavier silicon analogs, silanones, are highly reactive species due to the weak and strongly polarized Si–O π-bond resulting from significant electronegativity difference (O–Si: 1.6) compared with that of organic carbonyl function (O–C: 1.0). This accounts for a very strong tendency to give thermodynamically stable polysiloxanes that are now among the most important building blocks for organic-inorganic hybrid polymers.[166] Although transient silanones in noble gas matrix or in gas phase have been characterized by several methods such as IR spectroscopy,[167] their chemistry, from the practical and synthetic points of view, has been underdeveloped compared to the organic ones. The synthesis of the stable silanone is known as a "Kipping's dream."

The first stable silanones The situation was dramatically improved since the discovery of an efficient stabilization method by Driess's group in 2007, by forming a silanone complex with a donor ligand (**II**) or with donor–acceptor ligands (**I**, Scheme 9.70). This technique allowed to synthesize several silanone derivatives that are stable and easy-to-manipulate.[9a] Indeed, the first stable and isolable silanone derivative such as a donor–acceptor stabilized sila-formamide **226**, has been synthesized by reaction of the Driess's stable cyclic diaminosilylene with a hydrated tris(pentafluorophenyl)borane (Scheme 9.71).[168] This air-stable silanone **226** is characterized by a ^{29}Si NMR signal at higher field (−61.5 ppm) than free silanones (170 ppm for **271**, 129 ppm for **273**, *vide infra*), which is in agreement with the complexation of the silanone fragment. In contrast, the Si=O bond length is significantly shorter than Si–O single bonds (1.87 Å) and is only marginally longer than the value calculated for the parent silaformamide (1.537 Å). In addition, the IR absorption band was observed at 1165 cm^{-1} for Si=O bond, which is shifted to lower frequencies compared to matrix isolated silanones (about 1200 cm^{-1}) but is far above frequencies typical for Si–O single bonds (800–900 cm^{-1}). These results suggest that the multiple bond character of Si=O bond is conserved even after the stabilization by coordination of donor–acceptor ligands.

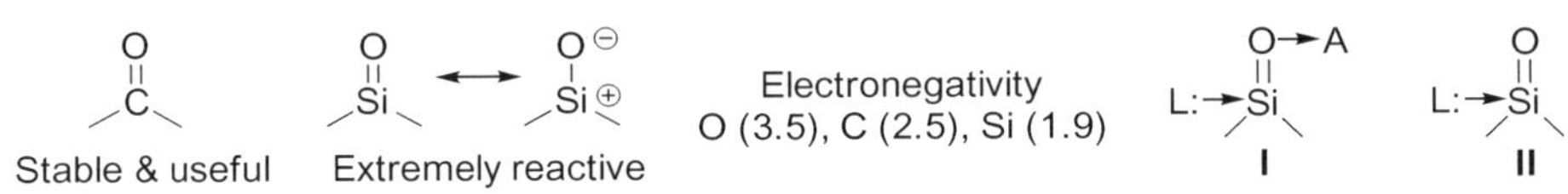

Scheme 9.70
Differences between ketones and silanones and stabilization of silanones by coordination of ligands.

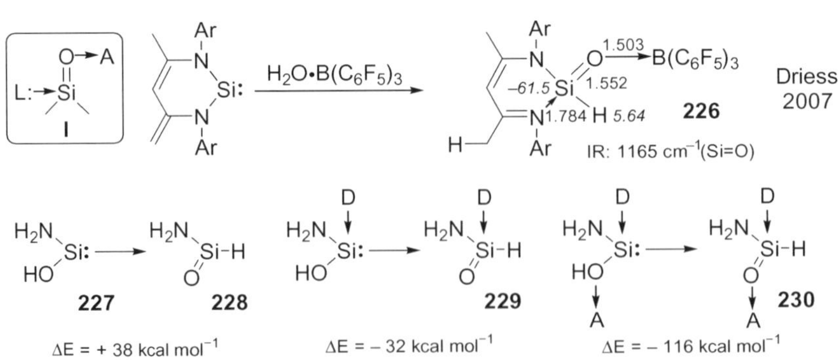

Scheme 9.71
Synthesis of the first donor–acceptor stabilized silanone.

Theoretical calculations for the model compound ($HOSiNH_2$) demonstrated that the coordination of ligands inverses the relative stability between both isomers, silaformamide **228** and hydroxysilylene **227** (Scheme 9.71). Indeed, the silanone form is much more stable, both in the case of donor-stabilized **229** ($\Delta E = -31.7$ kcal mol^{-1}) and donor–acceptor-stabilized formamides **230** ($\Delta E = -115.5$ kcal mol^{-1}).

Just after, Driess's group has also reported the first stable donor-supported silanone, a sila-ester **232** stabilized by intramolecular coordination of an imine ligand (Scheme 9.72). This sila-ester derivative **232** was directly obtained by oxidation of the corresponding siloxy-silylene **231** by N_2O or CO_2.[169] This derivative **232** presents a similar ^{29}Si chemical shift (−55 ppm) and Si=O bond length (1.579 Å) to those observed for the donor–acceptor stabilized formamide **226** (−61.5 ppm and 1.552 Å).

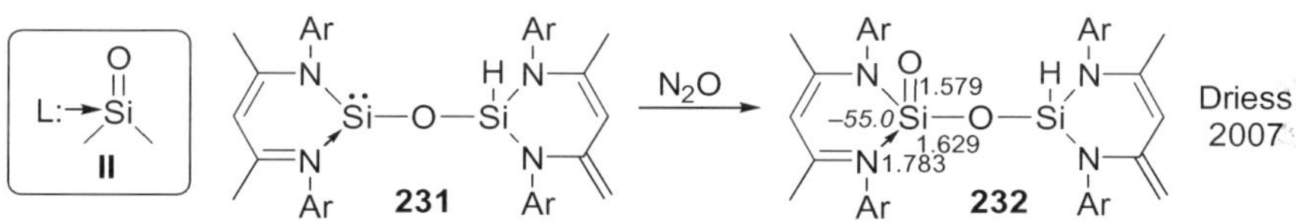

Scheme 9.72
Synthesis of the first donor-stabilized silanone.

Base-stabilized silanones The oxidation of donor-stabilized silylenes **233** appeared to be a straight way for the preparation of various silanones (Scheme 9.73). Indeed, the first silanone (silaurea) stabilized by intermolecular coordination of NHC ligand **234** was also obtained from the corresponding NHC stabilized silylene **233**.[170] The Si=O bond is short (1.541 Å) and the particularly long interatomic distance between Si and C_{NHC} (1.930 Å) indicates its dative bond character. The shorter Si–C_{NHC} bond distance than the value observed for the silylene-NHC complex (2.016 Å) clearly indicates the enhanced electrophilic character of silicon center.

It was known that the reaction of silylenes with O_2 leads to the transient formation of siladioxiranes **235**. Although these species are particularly interesting as new silicon-based oxidizing agents, they are generally unstable transient species and readily isomerize to the corresponding sila-esters **236** (Scheme 9.73).[167e,171] Driess et al. have successfully synthesized and isolated such an elusive siladioxirane **237** as a complex with a NHC ligand.[172] This dioxirane derivative **237** is stable at low temperature (−20°C), but slowly undergoes an oxygen transfer to the NHC ligand at RT to afford a silanone adduct with a cyclic urea **238**, which is stable under inert conditions.

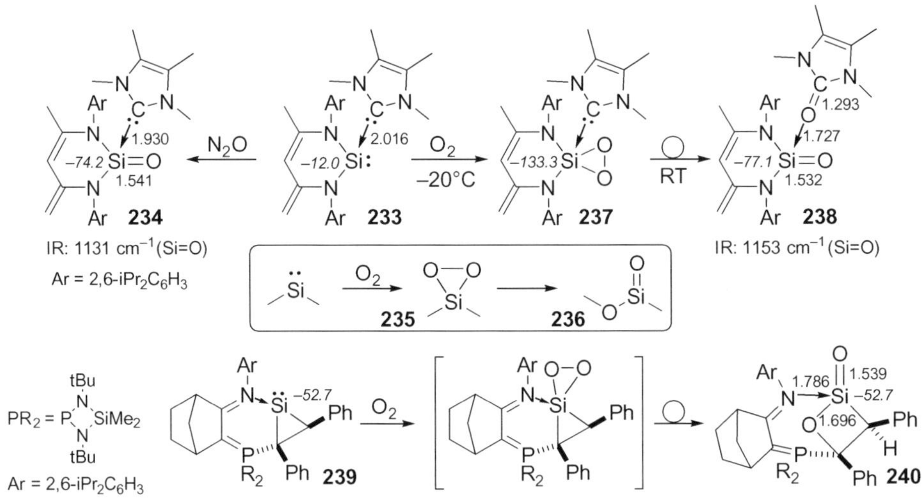

Scheme 9.73
Synthesis of donor-stabilized silanones by oxidation of the corresponding donor-stabilized silylenes.

Similarly the donor-stabilized three-membered cyclic silylene (sila-cyclopropylidene) **239** reacts with O_2 affording the corresponding cyclic sila-ester **240** which is the first example of a sila-β-lactone derivative (Scheme 9.73).[173]

The same synthetic methodology was also applied for the synthesis of a stable cationic silanone **241** by Inoue et al. (Scheme 9.74). Indeed, the oxidation of a cationic silylene stabilized by two NHC ligands by CO_2 affords a stable silicon analog of the acylium ion complex.[174]

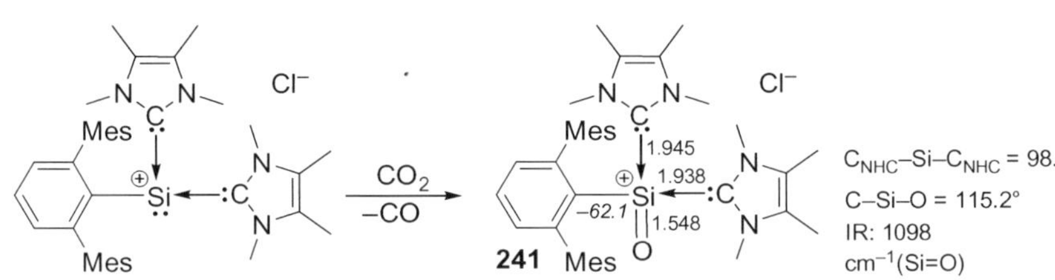

Scheme 9.74
Stable silicon analog of acylium ion stabilized by NHC ligands.

The silanone complex **242** stabilized a less nucleophilic *N*,*N*-dimethyl-4-amino-pyridine (DMAP), which presents much higher reactivity, has also been reported by the group of Driess (Scheme 9.75).[175] This molecule **242** reacts even with poorly reactive NH_3 to give a sila-hemiaminal **243**, in equilibrium with its tautomer, the silanoic amide **244**,[176] which is thermodynamically favored ($\Delta E_{\text{amido-hemiaminal}} = -3.1$ kcal mol^{-1} (gas phase), -5.2 kcal mol^{-1} (toluene)). These two isomers reversibly form a complex **245** in which they are connected with each other by Si–OH ··· O–Si hydrogen bond.

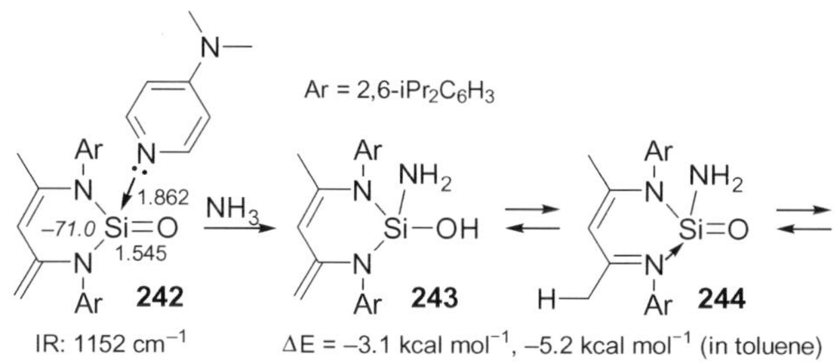

Scheme 9.75
Reaction of DMAP-stabilized silanone with NH_3.

Small cyclic base-stabilized silanones Cyclopropanones **246** are highly reactive ketones that display unusual properties arising from the incorporation of the carbonyl group into a strained three-membered ring. Particularly, substituted cyclopropanones may undergo ring-opening to form oxyallyl zwitterions **247** (Scheme 9.76),[177] which can behave as 1,3-dipoles. Inherently reactive, they have been implicated as intermediates in a number of organic reactions,[178] although there are very few reports on the direct experimental observation of the oxyallyl intermediate **247**.[179] The bicyclobutanone **248**, with an increased ring strain, presents a unique hybrid structure between cyclopropanone **248** and oxyallyl **248′** with a long endocyclic C–C bond.[180]

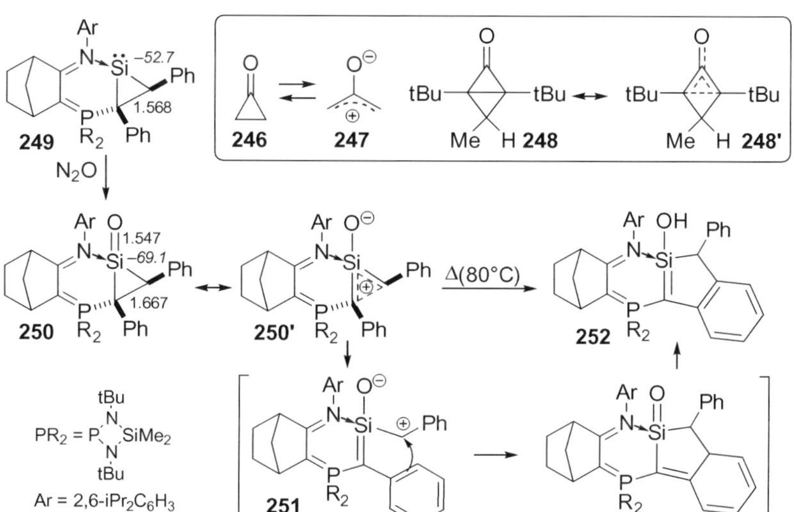

Scheme 9.76
Synthesis of a silacyclopropanone and its isomerization.

Recently the Kato/Baceiredo group successfully synthesized a stable sila-analog of cyclopropanone **250** as a complex with donating ligand (Scheme 9.76).[181] This molecule **250** presents a significantly elongated C–C bond (1.667 Å) and shorten Si–C bonds (1.918 and 1.836 Å) in the three-membered ring compared with those observed for the corresponding silylene **249** (1.568 Å for C–C, 1.993 and 1.920 Å for Si–C). This feature was explained by a negative hyperconjugation involving the strong Lewis acidity of silanone function, in spite of the coordination of the donating ligand, resulting in the original silacyclopropanone-oxyallyl hybrid structure **250′**. This silacyclopropanone **250** isomerizes under mild conditions, probably via the formation of oxaallyl intermediate **251** which undergoes a cyclization by intramolecular Frieldel-Craft reaction, to give the base-stabilized 1-silenol **252**. Similar isomerizations of organic cyclopropanones have already been reported.[182]

Donor–acceptor-stabilized silanones The donor–acceptor ligand system provides a more efficient thermodynamic and kinetic stabilization than that of only donating ligands. The coordination of two ligands on the Si and the O atoms of silanone fragment protects it well from both sides of the highly reactive function.

Roesky et al. demonstrated well this concept with the synthesis of small sila-formyl chloride **253** (O=SiHCl) as a stable complex with a bulky donating NHC ligand and an accepting $B(C_6F_5)_3$ ligand (Scheme 9.77).[183] Although the stabilization energy by coordination of acceptor ligand is smaller than that by donor ligand, both of them are

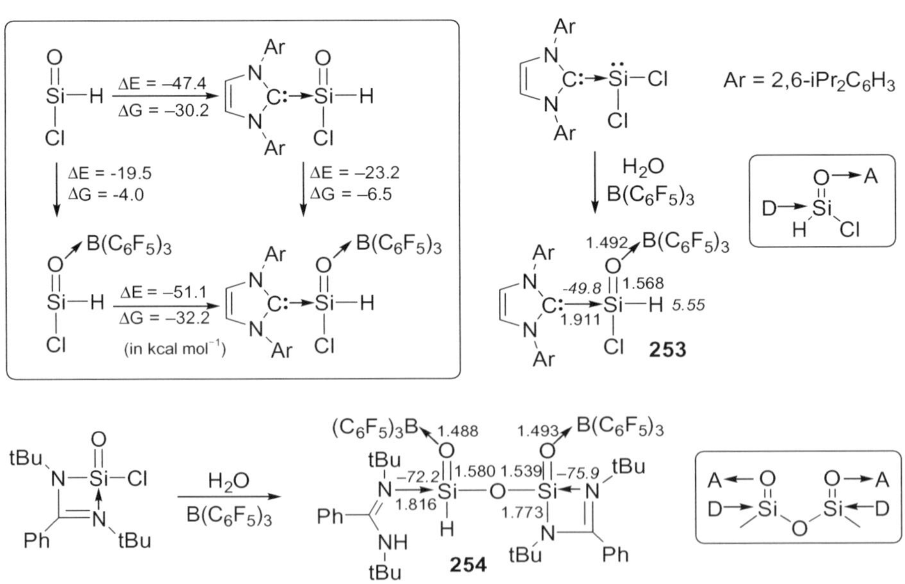

Scheme 9.77
Stabilization of small silanones by a donor–acceptor ligand system.

energetically favored. The total stabilization energy provided by coordination of donor–acceptor ligands is considerably large ($\Delta E = -70.6$ kcal mol^{-1}, $\Delta G = -36.4$ kcal mol^{-1}). Using this efficient stabilization technique, the same group also synthesized the first stable sila-acid anhydride **254** derivative by the selective hydrolysis of donor-stabilized mono-chlorosilanone in the presence of $B(C_5F_6)_3$.[184]

A more striking example of the use of the donor–acceptor ligand system has been reported by Driess et al. They have demonstrated that an attempt to synthesize donor-stabilized silanoic acid **255** by selective hydrolysis of the DMAP stabilized silanone failed because of an immediate degradation producing silica (SiO_2) and two free ligands (Scheme 9.78).[185] In contrast, the same reaction in the presence of a Lewis acid such as $B(C_5F_6)_3$ leads to the formation of persistent donor–acceptor stabilized silanoic acid **256**, which is enough stable to be fully characterized in solution. Nevertheless, it slowly decomposes eliminating silica at RT.

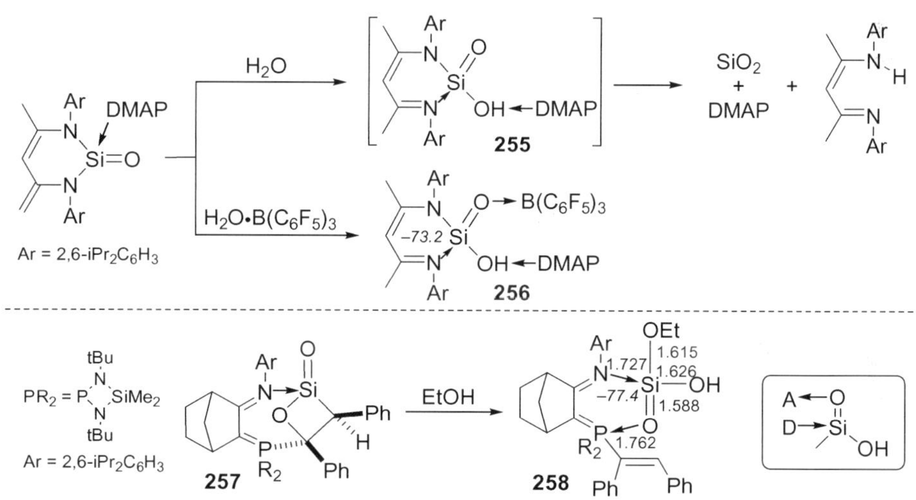

Scheme 9.78
Donor–acceptor stabilized silanoic acids.

The first isolable silanoic acid derivative **258** was synthesized as a donor–acceptor complex by Kato/Baceiredo group by the reaction of donor-stabilized sila-β-lactone **257** with ethanol (Scheme 9.78).[173] The reaction probably proceeds through the first formation of a 1,2-adduct of alcohol, followed by a retro-[2 + 2] cycloaddition reaction of the oxasiletane ring with a pentacoordinate silicon center similarly to Peterson olefination.[186] The silanone function of the acid **258** is stabilized by coordination of a unique ambiphilic iminophosphorane ligand with an imine group as donating site and a phosphorane group as accepting site. The chemical shift in ^{29}Si NMR for the silanone function (−77.4 ppm) is similar to the value observed for Driess's silanoic acid derivative (−73.2 ppm). The signal in ^{31}P NMR appearing at significantly high field (−57.5 ppm) compared to the precursor (51.3 ppm) is consistent with the presence of a pentacoordinate phosphorus center due to the coordination of silanone oxygen. The silicon–oxygen bond distance for the silanone function (1.588 Å) is shorter than others (1.615 and 1.626 Å), suggesting a remaining multiple bond character. The efficient stabilization of silanoic acid **258** by coordination of the donor–acceptor iminophosphorane ligand on the silanone function was supported by DFT calculations revealing the strong exergonic nature of the reaction ($\Delta G = -44.2$ kcal mol^{-1}).

Base-stabilized silanone complex of transition metals As shown by the formation of various donor–acceptor complexes of silanones, donor-stabilized silanones with an enhanced polarization should be potential ligands for metals. The first attempt by Driess et al. to synthesize such metallic complexes **260** using $Zn(OAc)_2$ failed due to its high reactivity (Scheme 9.79).[187] Indeed, the silanone **259** inserts into the polarized Zn–OAc bond to afford a zinc siloxide complex **260** with covalent O–Zn bonds. In contrast, the base-stabilized silanone **259** forms stable complexes with dimethylzinc (**261**) and trimethylaluminium (**262**). The Zn–O distance (2.023 Å) in **261** is significantly longer than those of **260** (1.896 and 1.913 Å), indicating the dative nature of the $Si^+–O^- \rightarrow Zn$ interaction. In the case of the Al-complex **262**, the O–Al distance (1.804 Å) is considerably shorter than the corresponding value observed for ketone-alkylaluminum complexes (1.90–2.07 Å),[188] indicating a stronger Al–O interaction probably due to the highly polarized nature of the base-stabilized silanone.

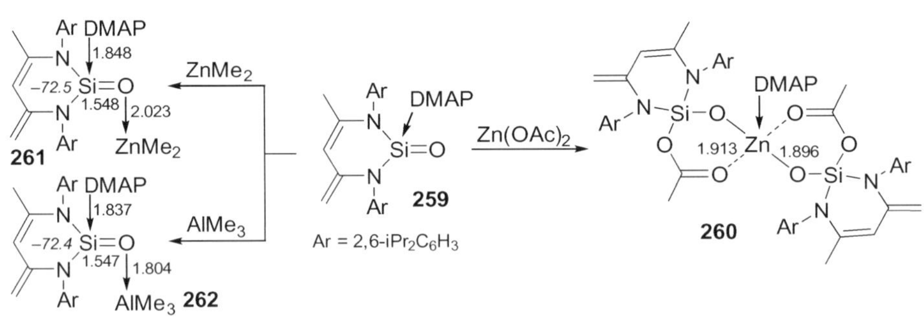

Scheme 9.79
Metal complexes with a donor-stabilized silanone ligand.

The first transition metal complex **264** featuring a base-stabilized silanone ligand has been reported by Ueno et al. by oxidization of a dimesitylsilylene-W(II) complex **263**[189] with pyridine oxide in the presence of DMAP (Scheme 9.80).[190] The structure of the complex **264** shows a η^1-coodination of the silanone on the metallic center via oxygen atom, indicated by a long Si–W distance (3.639 Å) and a wide Si–O–W angle (155.3 degrees). The W–O bond (2.165 Å) is only slightly shorter than those of the η^1-coordinated ketones to tungsten (2.21–2.39 Å).

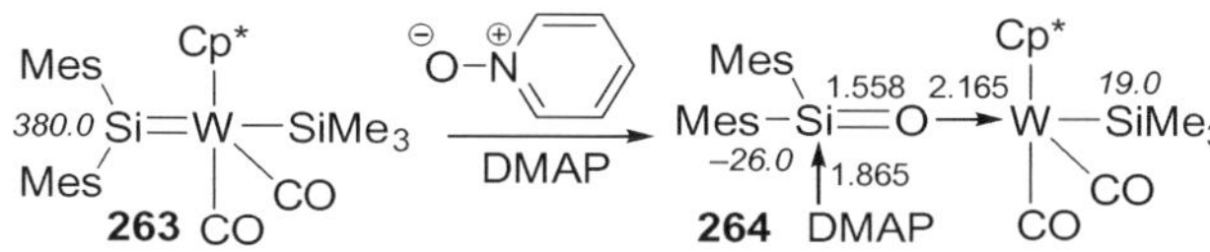

Scheme 9.80
The first transition metal complex of donor-stabilized silanone.

Silicon oxide complexes In contrast to stable molecular CO_2 with two cumulated C=O double bonds, SiO_2 is an extremely reactive molecule and exists in nature as extremely stable polymeric solid (silica, quarts and sands) constituted of only Si–O σ-bonds. Stabilization of the small silicon oxides overcoming enormous thermodynamic driving force to form such polymers is a great challenge. The stabilization technique of silanones by coordination of donating ligands appeared to be an efficient method for the isolation of small silicon oxides. Indeed, Robinson et al. realized the oxidation of stable disilicon(0) complex **265**[40] by two different oxidants, O_2 and N_2O, which leads, in each case, to the formation of different NHC-stabilized silicon oxide complexes with different compositions, such as Si_2O_4 **266** (SiO_2 dimer) and Si_2O_3 **267** (SiO_2 + SiO), respectively (Scheme 9.81).[191] Both compounds present two Si=O functions stabilized by NHC ligands. IR absorption bands for Si=O functions are in the same region (1147 cm^{-1} for Si_2O_4 **266** and 1092 cm^{-1} for Si_2O_3 **267**) as the base-stabilized silanones (1100–1200 cm^{-1}). Wiberg bond indices of these two silicon oxides for Si=O bonds (1.10 for **266** and 1.05 for **267**) are almost double compared with those for Si–O_{bridge} bonds (0.57, 0.59), which indicate their double bond character.

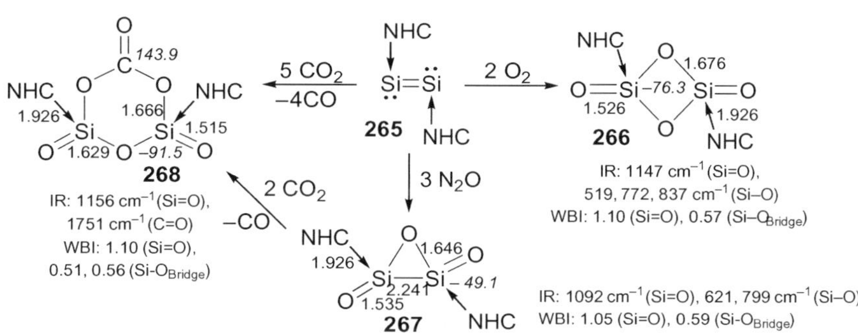

Scheme 9.81
Synthesis of NHC-stabilized silicon oxides as well as silicon and carbon mixed oxides.

Of particular interest, the oxidation of the disilicon(0) complex with five equivalents of CO_2 produces a six-membered cyclic silicon-carbon mixed oxide complex **268** with two NHC ligands, and elimination of four equiv. of CO (Scheme 9.81).[192] The Si=O bond (1.515 Å) is slightly shorter compared to other NHC-silanone complexes (1.526–1.568 Å), and is as short as that of free $Me_2Si{=}O$ found by rotational spectroscopy (1.515 Å).[193] Although the formal composition of the mixed oxide **268** is constituted of a Si_2O_4 and a CO_2, the reaction of Si_2O_4 **266** with CO_2 results in an immediate decomposition giving NHC–CO_2 complex as the only characterized byproduct. Instead, the reaction of Si_2O_3 complex **267** with one equivalent of CO_2 cleanly produces the same mixed oxide **268**, and elimination of CO.

Three coordinate silanones Filippou et al. have recently reported the isolation of the first RT stable silanone **271** with a three coordinate silicon center, taking advantage of the electronic stabilization provided by a transition metal fragment (Scheme 9.82).[194] The synthesis was achieved by the oxidation of the first stable metallosilylene **270**, derived from the corresponding cationic chromium silylidyne complex salt **269**, by N_2O.

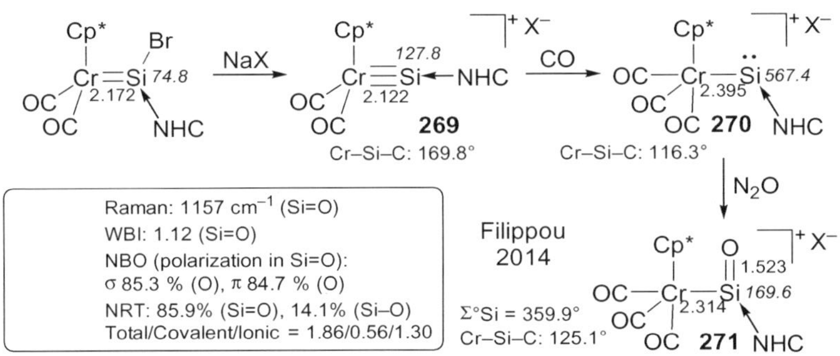

Scheme 9.82
Synthesis of stable metallasilanone with a three coordinate silicon center.

Silanone **271** was isolated as thermally stable bright yellow crystals (40% yield). The structure reveals a trigonal planar geometry around the Si center ($\Sigma°$Si = 359.9 degrees) and a short Si=O bond (1.523 Å) which is slightly shorter than Si = O bonds in base-stabilized silanones (1.531–1.579 Å), but within the range of the values found for the NHC stabilized silicon oxides (1.515–1.535 Å). The Cr–Si bond (2.314 Å) is shorter than that of the precursor (2.395 Å) due to the significantly increased s-character of the silicon hybrid orbital used for the bond ($sp^{12.0} \rightarrow sp$) after oxidation. As expected, the complex **271** features a distinctive low-field shifted ^{29}Si NMR signal at 169.6 ppm, and a $\nu_{(Si=O)}$ absorption band in the Raman spectra at 1157 cm^{-1}, only at slightly lower frequency than that of $Me_2Si{=}O$ (1204 cm^{-1}).[195] The HOMO corresponds to the $\pi_{Si=O}$ orbital and the

LUMO is the $\pi^*_{Si=O}$ orbital with a large contribution of silicon. The complex **271** reacts instantaneously with H_2O to give the dihydroxysilyl complex. This reaction occurs rapidly even under argon atmosphere of a glove box containing 1 ppm of H_2O.

A persistent dialkylsilanone **273** has also been recently synthesized by Iwamoto et al. by dehydrobromination of dialkylbromosilanol **272** using a bulky strong base such as tris(trimethylsilyl)silyl potassium, and has been characterized in solution at −80°C (Scheme 9.83).[196] Similarly to the previous metallasilanone **271**, the ^{29}Si NMR signal for the tricoordinate silicon atom appears at low-field (128.7 ppm). In situ IR spectra shows an absorption band for Si=O at 1150 cm^{-1}, which is also close to the value obtained for the metallasilanone **271** (1157 cm^{-1}). The silanone **273** reacts with a mesitylenenitrile *N*-oxide to give a [3 + 2] cycloadduct **274**. The silanone **273** is stable up to −80°C and, above this temperature, it isomerizes to the corresponding silyl silenol ether **275** via 1,3-silyl migration from C to O atom. The synthesis of this silanone **273** is strongly dependent on the nature of the base used, and in the case of *t*BuLi the formation of a stable lithium bromosiloxide **276** was observed, which does not undergo dehydrobromination. The moderately short Si–O bond (1.587 Å), as well as elongated Si–Br bond (2.359 Å) relative to the typical bond length (2.24 Å), suggest some silenoid character.

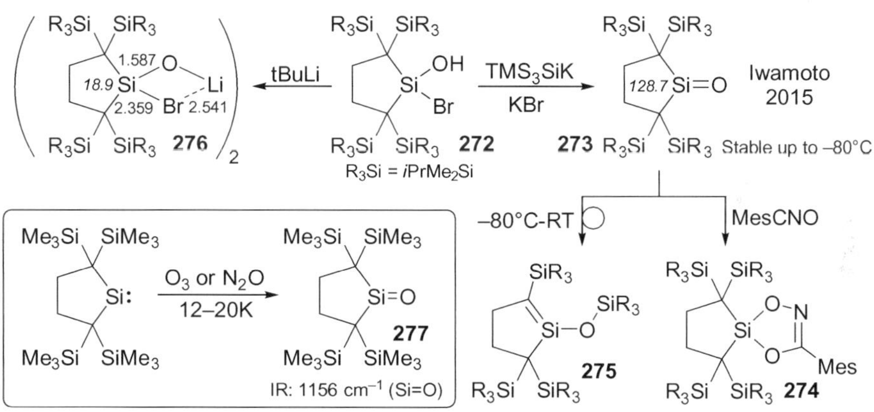

Scheme 9.83
Synthesis and reactivity of the first persistent dialkyl silanone.

The same type of dialkylsilanone **277** was also synthesized in argon matrix by oxidation of Kira's stable silylene by N_2O or O_3 (Scheme 9.83) and was characterized by IR spectroscopy ($\nu_{Si=O} = 1156$ cm^{-1}).[197]

9.2.2.4.2 Si=S

The energies of σ- and π-bonds in a C=O double bond are almost equal to each other (93.6 and 95.3 kcal mol^{-1} respectively, Scheme 9.84). This is the main reason why the

	Calculated bond energy (kcal mol^{-1})			
	$H_2C{=}O$	$H_2C{=}S$	$H_2Si{=}O$	$H_2Si{=}S$
σ	93.6	73.0	119.7	81.6
π	95.3	54.6	58.5	47.0
σ+π	188.9	127.6	178.2	128.6

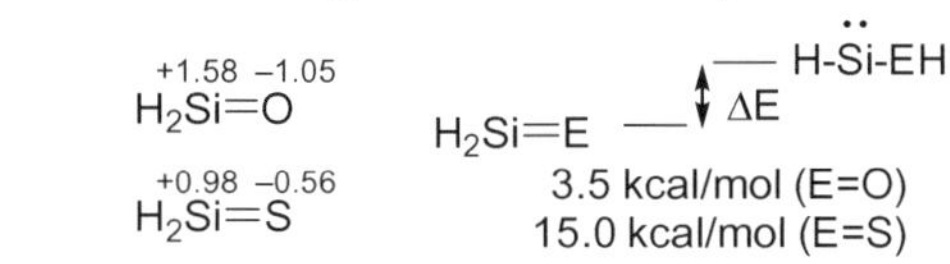

Scheme 9.84
Comparison of calculated physical properties of silanones and silanethiones (B3LYP/TZ(d,p) level).

addition–elimination reaction of carbonyl compounds (esters, acyl halides, . . .) easily occurs. In contrast, in the case of the heavier silicon analog (silanone), although the total double bond energies of Si=O bond is similar to that of C=O bond (178.2 vs 188.9 kcal mol^{-1}), σ-bond energy (119.7 kcal mol^{-1}) is much greater than that of π-bond energy (58.5 kcal mol^{-1}).[198] Thus silanones have a strong tendency to polymerize to form polysiloxanes with only Si–O single bonds. Thioketones and their silicon analogs (silanethiones) show the same trend, whereas the difference between σ- and π-bond energies is much smaller. Although the π-bond energy of silanethiones is smaller than that of silanones, Okazaki and Tokitoh have predicted that the double bond of silanethiones is kinetically more stable due to a weaker polarization than that of silanones.[199] In addition, $H_2Si{=}S$ is calculated to be 15 kcal mol^{-1} more stable than its divalent isomer (H–Si–SH), though $H_2Si{=}O$ is only 3.5 kcal mol^{-1} more stable than H–Si–OH. For these reasons, Kudo and Nagase concluded that stable silanethiones should be easier to be synthesized than silanones.[199]

The first stable and isolable silanethione **278** was synthesized by Corriu et al. in 1989 as a complex with an intramolecularly coordinating amine ligand by the reaction of the corresponding pentacoordinate silane with S_8 (Scheme 9.85).[200] Due to the coordination of the ligand on the silicon atom, it presents a ^{29}Si NMR resonance at 22.3 ppm, at higher-field relative to that expected for unsaturated silicon compounds and a slightly elongated Si=S bond (2.013 Å) compared to the values predicted by calculations (1.945 Å). Nevertheless, the Si=S bond is considerably shorter than Si–S single bond (2.16 Å).

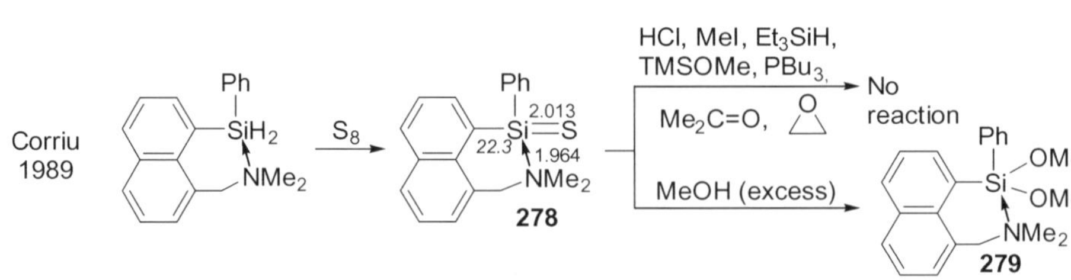

Scheme 9.85
Synthesis of the first donor-stabilized silanethione.

The Si–N bond is longer than typical Si–N σ-bonds (1.76 Å), in agreement with a dative bond character. The stabilization of Si=S bond by the donor ligand is very efficient, inducing an unexpectedly low reactivity. Indeed, the silanethione **278** is inert toward various nucleophilic and electrophilic reagents. It undergoes a methanolysis in the presence of a large excess of MeOH to give dimethoxysilane **279**.[200]

There are only two stable base-free silanethiones known to date. The first example is the diarylsilanethione **280**, reported by Okazaki et al. in 1994, which was synthesized by the reaction of tetrathiasilolane derivative with three equiv. of triphenylphosphine (Scheme 9.86).[198,201] This silanethione **280** with two extremely bulky aryl groups is isolable as yellow crystals. As expected, it presents a low-field shifted ^{29}Si NMR resonance (166.6 ppm) and a shorter Si=S bond (1.948 Å) than that observed for the Corriu's base-stabilized silanethione (2.013 Å). Despite its high thermal stability (mp = 185–189°C), it readily reacts with various reagents, such as 2,3-dimethyl-1,3-butadiene, mesitylenenitrile *N*-oxide, as well as phenylisothiocyanate to give the corresponding [4 + 2]-, [3 + 2]-, and [2 + 2]-cycloadducts (Scheme 9.86).

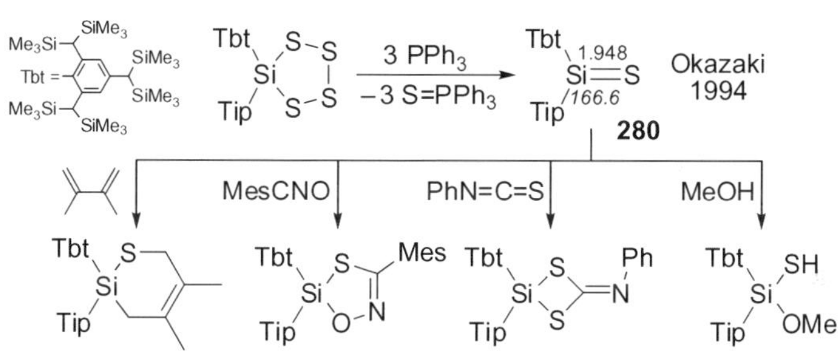

Scheme 9.86
Synthesis and reactivity of the first base-free silanethione.

Kira's group also reported the synthesis of a stable dialkyl silanethione **281** by thiolation of the stable cyclic dialkylsilylene using triphenylphosphine thioxide (Scheme 9.87).[202] Although the Si=S bond length (1.958 Å) is similar to that of the Okazaki's diarylsilanethione **280**, the ^{29}Si NMR chemical shift appears at a much lower field (216.8 ppm), which is probably indicative of the less perturbed nature of Si=S double bond. Due to the efficient kinetic stabilization, **281** is poorly reactive, and no reaction was observed with a butadiene even at high temperature (100°C). Instead, under these conditions, it slowly dimerizes to give a 1,3-dithiadisiletane **282** (4%), which clearly shows greater stability of **281** than the corresponding silanone **283** which is only a transient species and undergoes a spontaneous dimerization.[203]

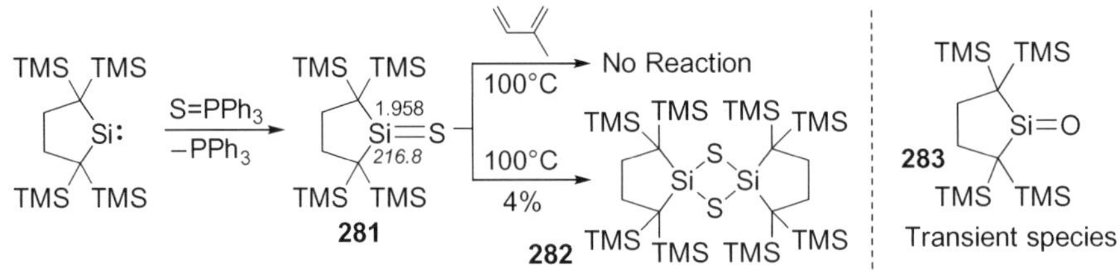

Scheme 9.87
The synthesis of a stable dialkylsilanethione.

In contrast to the reaction of free silylenes with S_8 resulting in the formation of tetrathiasilolane **284**, the same reaction with base-stabilized silylenes affords the corresponding silanethione **285** (Scheme 9.88). Due to simplicity of the method, various base-stabilized silanethiones (**286**, **287**) have been synthesized starting from different base-stabilized silylenes.[127b,170b] Silathionium complexes (**288**[204] and **289**[205]) have also been synthesized from the corresponding base-stabilized silyliumylidene cations. The first pentacoordinated silanethione **291** has also been obtained by the thiolation of amidine substituted silylene **290**.[206] Coordination of an additional donating ligand on the silicon center after oxidation of silylene **290** suggests an enhanced electrophilic character of silanethione **291** relative to silylene **290**.

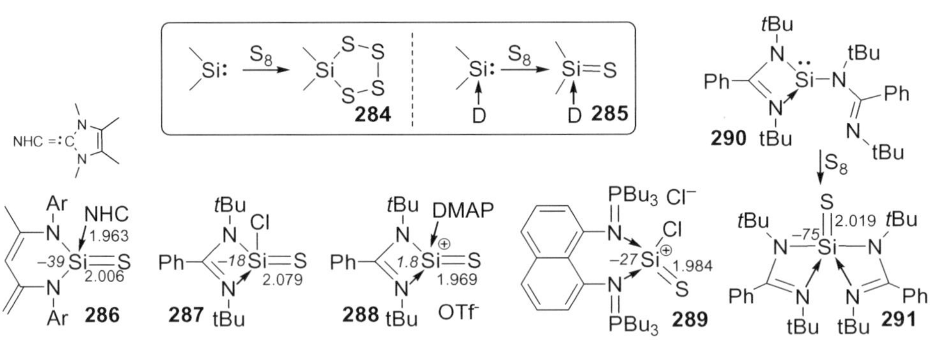

Scheme 9.88
Synthesis of base-stabilized silanethiones.

Alternatively, the base-stabilized silanethiones **292** and **293** with an amidine ligand have also been synthesized by thiolation of the corresponding silylene using a phenylthioisocyanate[207] and by the reaction of dichlorosilane derivative with two equivalents of potassium,[208] respectively (Scheme 9.89). The mechanism of the latter reaction is still unclear.

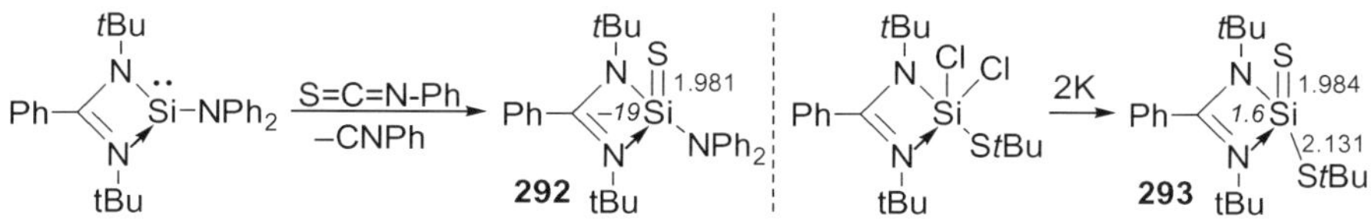

Scheme 9.89
Synthesis of base-stabilized silanethiones by other methods.

Driess's stable *N*-heterocyclic silylene **294** with a ligand system presenting unique zwitterionic properties readily reacts with H_2S to give a donor-stabilized silathioformamide **295** (Scheme 9.90).[209] The same reaction with H_2O results in the formation of a hydroxysilylene isomer **296** rather than silaformamide.[168] These results are consistent with theoretical calculations which predicted the different relative stability of the two divalent and tetravalent silicon isomers (H–Si–ES and $H_2Si{=}E$) between silanone and silanethione. The "half-parent" phosphasilene **297** with the same ligand system on the silicon atom undergoes a similar reaction with H_2S, affording a PH_2-substituted silanethione **298**.[210] The same synthetic strategy has also been employed to obtain the first stable silathiocarboxylic acid **299** as an acid-base adduct.[185,211]

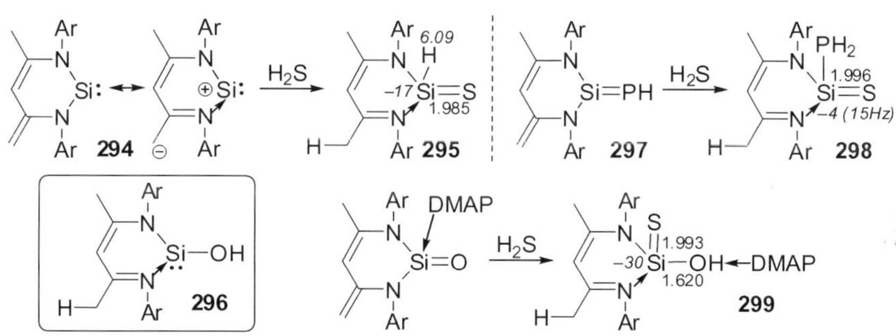

Scheme 9.90
The reaction of zwitterionic type *N*-heterocylic silylene and related compounds with H_2S.

In contrast to stable CO_2 and CS_2 as monomers, the molecular SiS_2 is an extremely reactive species and undergoes spontaneous polymerization. Such a molecule has only been detected in an inert gas matrix at very low temperature.[212] Driess's group has successfully stabilized such a highly reactive monomeric SiS_2 forming a complex **300** with a bulky and strongly donating bidentate NHC ligand (Scheme 9.91).[213] The X-ray structure of complex **301** with an additional $GaCl_3$ ligand on the S atom shows strongly elongated Si=S bonds, especially the one with a S atom coordinated by $GaCl_3$ (2.262 and 2.006 Å). Theoretical calculations indicated strongly polarized Si=S bonds (toward S atom) with a moderate multiple bonding character enabled by a negative hyperconjugation (no bonding/double bond resonance).

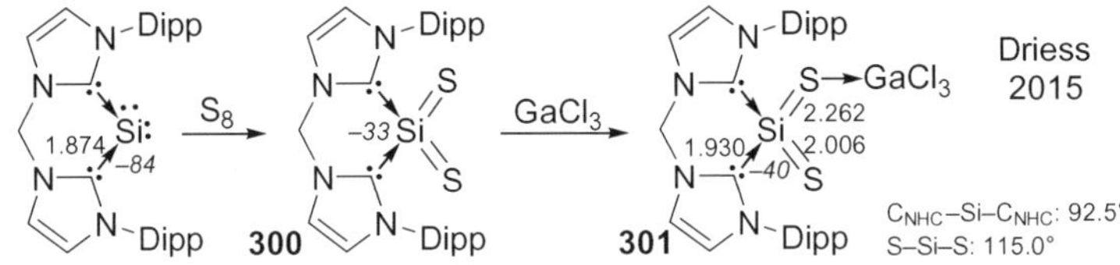

Scheme 9.91
Synthesis of NHC-stabilized SiS_2.

9.2.2.4.3 Si=Se and Si=Te

Similarly to the case of silanethiones, there are only two types of bulky substituent systems which can kinetically stabilize selenium and tellurium analogs of silanones (silaneselone and silanetellone). The first stable diaryl silaneselone **303** and -tellone **305** were reported by Okazaki/Tokitoh in 2002, and were synthesized by the reaction of the diselanasilirane **302** with triphenylphosphine, and the treatment of dilithiosilane **304** with tellurium dichloride, respectively (Scheme 9.92).[214] Both products present characteristic signals at significantly low-field ^{29}Si NMR (174 ppm for **303** and 171 ppm for **305**), ^{77}Se NMR (635 ppm), and ^{125}Te NMR (731 ppm), which is in good agreement with their double bond character.

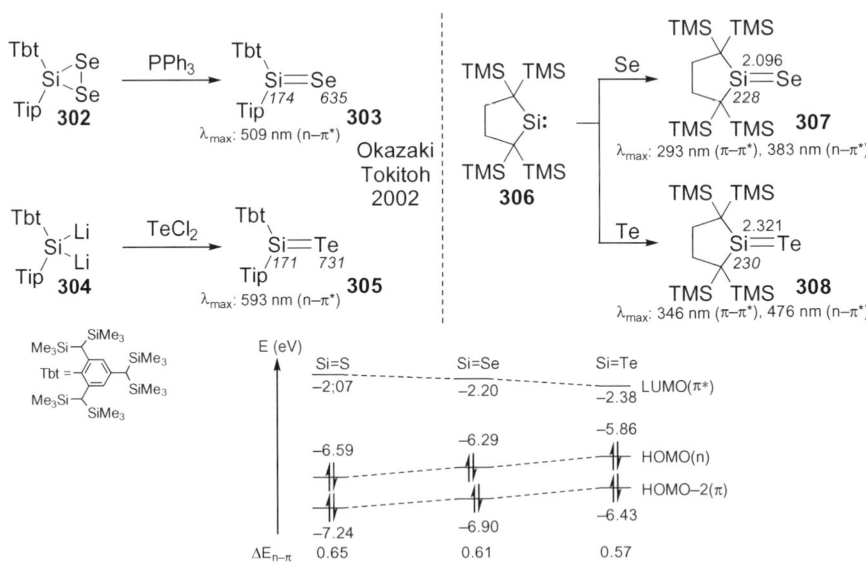

Scheme 9.92
The synthesis of the first stable silaneselone and silanetellone.

Several years later, Kira et al. also synthesized a stable dialkyl silaneselone **307** and silanetellone **308** by the direct reaction of dialkylsilylene **306** with elemental Se and Te respectively (Scheme 9.92).[215] The Si=Se and Si=Te bonds of both products

(Si=Se: 2.096 Å, Si=Te: 2.321 Å) are significantly shorter than the corresponding Si-E single bond lengths (Si–Se 2.27 Å, Si–Te: 2.52 A). They also present a planar geometry around silicon centers (Σ°Si = 360 degrees). The $n \rightarrow \pi^*$ transition bands in UV-Vis spectra of silaneselone **307** (383 nm) and silanetellone **308** (476 nm) are significantly blue-shifted than the corresponding bands of diaryl ones (509 nm for **303** and 593 nm for **305**), suggesting less perturbation of $\pi_{Si=E}$-bond orbitals by the alkyl substituents. Of particular interest, $\Delta\nu(\pi\pi^* - n\pi^*)$ values are in the same range among silanethione **281**, -selone **307** and–tellone **308** (11430, 11090, 10620 cm^{-1} respectively), which indicates that the energy gap between n- and π-orbitals are similar for all products (Scheme 9.92). This result is quite different from that observed for the Si=E (E = P, As, Sb) compounds (Section 9.2.2.3.2, Scheme 9.67), which shows an increasing energy levels of HOMO (π-orbital) and decreasing energy levels of lone pair orbital (n-orbital) on going from P to As to Sb.

Similarly to the case of sulfur analogs, the best way to synthesize base-stabilized silaneselones and silanetellones is the direct reaction of base-stabilized silylenes with elemental Se and Te (Scheme 9.93). Using this technique, NHC-stabilized silaneselone **309** and silanetellone **310**,[170b] silanoic seleno – **311** and telluroesters **312**[216] as well as pentacoordinate ones (**313** and **314**)[206] have been synthesized. A stable 1,2-di(silaneselone) **315** has also recently been reported by So et al.[127b] In all cases, ^{77}Se (–323 to –485 ppm) and ^{125}Te NMR (–983 to –1208 ppm) resonances are significantly high-field shifted relative to those of donor-free diarylsilaneselone (+636 ppm) and silanetellone (+1143 ppm), probably due to the strong $Si^{+}=E^{-}$ polarization induced by the ligand coordination on the silicon atom.

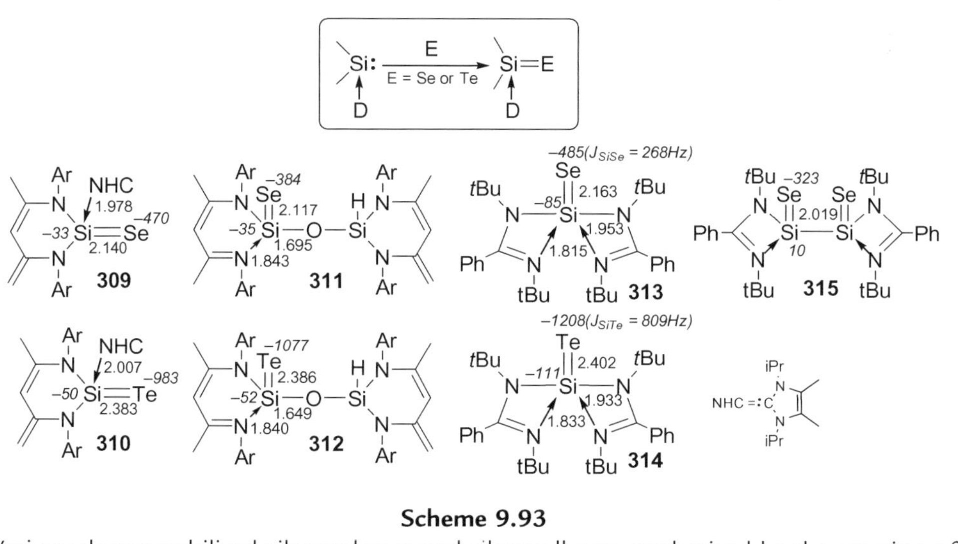

Scheme 9.93
Various donor-stabilized silaneselones and silanetellones synthesized by the reaction of base-stabilized silylenes with elemental Se and Te.

As a unique synthetic method, West et al. reported the hydrolysis of diselenadisiletane **316** affording the corresponding donor-stabilized dimeric di(silaselenone) **317** (Scheme 9.94).[217] The reaction with *t*BuOH leads to the formation of a *tert*-butyl selenosilacarbamate **318**. The selenosilanoic acid **319** also appeared to be accessible from the reaction of the corresponding donor-stabilized silanone with a dilithium selenide (Li_2Se) followed by protonation with two equiv. of trimethylammonium chloride.[211]

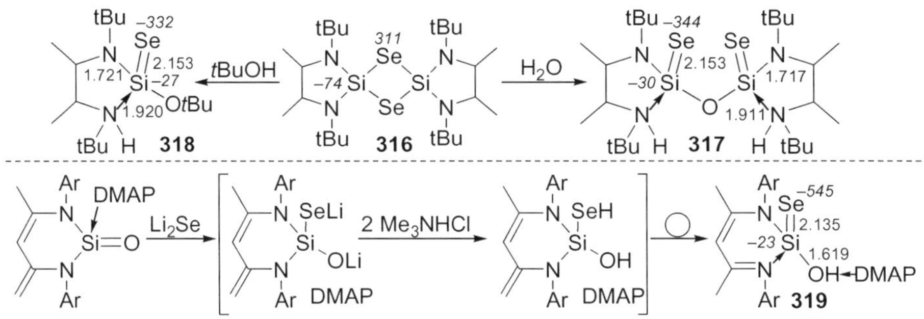

Scheme 9.94
Unique synthesis of stable silaneselone derivatives.

9.3 Silicon Containing Triple Bonds

9.3.1 Si≡Si Triple Bond

Among the most significant discoveries in the area of the stable low-coordinate silicon compounds is probably the synthesis of the first disilyne derivatives featuring Si–Si triple bonds (Scheme 9.95).[221,222b,218] The first example of a heavier alkyne analog was the lead derivative isolated by Power in 2000,[219] and since then the remaining heavier group 14 element alkyne analogs have been successfully isolated.[220] Here, we will discuss the generation and characterization of disilynes, and some of their further chemistry.

R–Si≡Si–R'	SiSi (Å)	SiSi-R (°)	^{29}Si NMR (ppm)	Refs.
R = R' = Si*i*Pr$[CH(SiMe_3)_2]_2$	2.0622	137.44	89.9	221
R = Si*i*Pr$[CH(SiMe_3)_2]_2$; R' = $SiCH_2$*t*Bu$[CH(SiMe_3)_2]_2$	2.0569	138.78	62.6–106.3	223
R = R' = Bbt	2.108	133.0	18.7	225
R = R' = R_S	2.0863	132.05	31.8	224

Bbt =
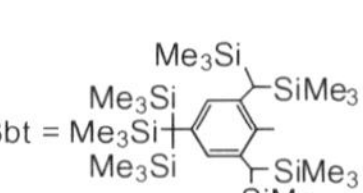

Rs =
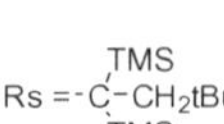

Scheme 9.95
Isolated and characterized disilynes derivatives.

By using extremely bulky ligands, independently, Sekiguchi,[221] Wiberg,[222b] and their coworkers reported the synthesis of the first stable disilynes in 2004 by reductive dehalogenation of tetrahalodisilanes or 1,2-dihalodisilenes. Several other disilynes have also been reported.[223–226]

Although most of the disilynes are perfectly stable in the solid state, they readily rearrange in solution. The two disilyldisilynes (**320** and **322**), symmetrical and unsymmetrical, slowly isomerize to give cyclotrisilene structures (**321** and **323**), via migration of one substituent,[223] while in the case of dialkyl(disilyne) **324** and diaryldisilyne C–H-insertion reactions were observed leading to bis(silacyclopropane) **325** and bis(silacyclobutene) **326** derivatives (Scheme 9.96).[224,225b]

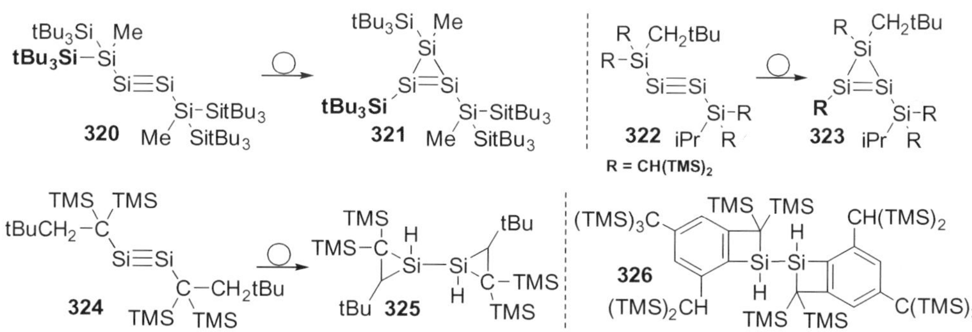

Scheme 9.96
Thermal isomerization of disilynes.

All disilynes have essentially planar, *trans*-bent core structures with substantial deviation from linearity, and feature an important triple bond character (Scheme 9.96). Calculated molecular orbitals (MOs) show the presence of two nondegenerate highest occupied $\pi_{(Si\text{-}Si)}$ orbitals (out-of-plane: HOMO, and in-plane: HOMO-1).[222a] The bending of the linear geometry in these compounds has been viewed as arising from a second-order Jahn–Teller effect mainly involving the antibonding $\sigma^*_{(Si-Si)}$ and the in-plane $\pi_{(Si-Si)}$ orbitals.[221,224,220] The LUMO is particularly low-lying in energy and as a consequence, the formation of Lewis base bis-adducts **327** by the addition of two donating ligands is strongly exothermic.[40,225] A few stable disilyne bis-adducts have been synthesized, using strongly σ-donating groups, NHCs **328**,[40] isocyanides **329**,[82] imines **330**,[226a,d] and phosphines **331**[226e] (Scheme 9.97). The coordination of the Lewis bases at the two silicon centers induces highly polarized 1,2-diylidic structures with very long Si–Si-single bonds. As a consequence, each silicon center in these bis-adducts exhibit a lone pair of electrons, and they behave mostly as bis-silylene derivatives.

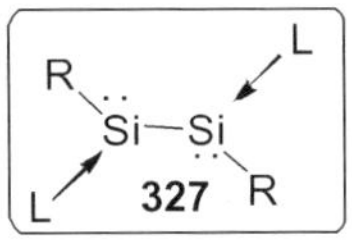

	R = Cl L = NHC **328**	R = Si(iPr)CH(TMS)$_2$ L = TMSNC: **329**	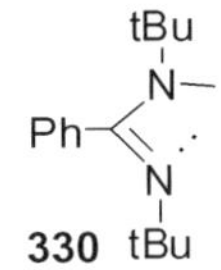 **330**	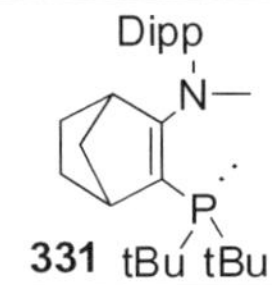 **331**
SiSi (Å)	2.393	2.369	2.413	2.331
Refs.	40	82	226a,d	226e

Scheme 9.97
Base-stabilized disilyne derivatives.

The reactivity of these compounds mostly involves the cleavage of the Si–Si bond, and in the case of the amidinate-stabilized disilyne **330**, a clean reaction was observed with bromine to give a bromosilylene **332**.[226b] The same compound treated with benzophenone in THF at RT overnight leads to the formation of a four-membered Si_2O_2 ring **333** with pentacoordinate silicon atoms, while the reaction with two equivalents of a diketone affords a bis(siladioxolene) derivative **334** involving two [1 + 4] cycloaddition reactions at each silylene center (Scheme 9.98).[226d]

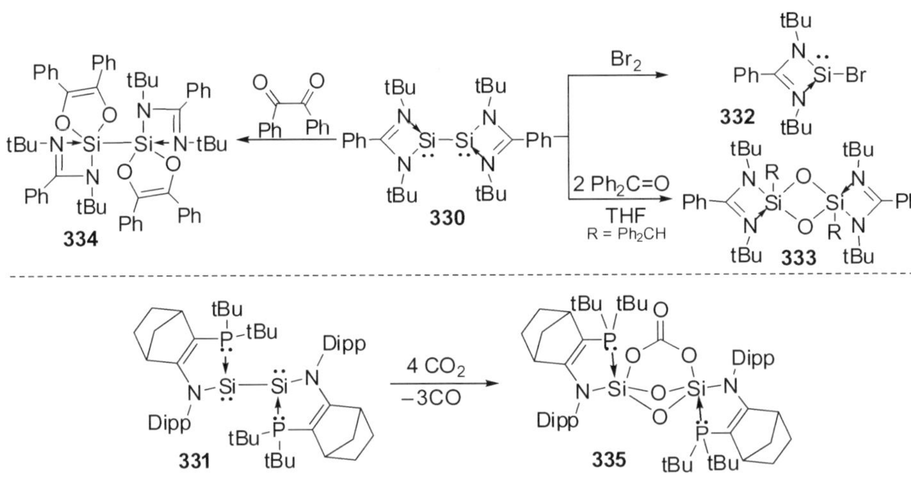

Scheme 9.98
Reactivity of Amidinate- and Phosphine-stabilized disilynes.

The phosphine-stabilized disilyne **331** is stable at RT under inert atmosphere conditions for weeks, however it readily reacts with carbon dioxide, under mild conditions, affording an aminosilicate **335** featuring two pentacoordinate silicon atoms (Scheme 9.98).[226e] These new low-valent silicon derivatives featuring an interconnected bis-silylene system have demonstrated a high potential as new building blocks in organosilicon chemistry, which has been recently reviewed.[226]

In contrast, the chemistry of nonbase-stabilized disilynes is strongly related to their important silicon-silicon triple bond character (bond order of 2.618 in the case of Sekiguchi's disilyne).[221] These species have been involved in many cycloaddition reactions with unsaturated compounds, and some key reactions are shown in Schemes 9.99 and 9.100. Sekiguchi's disilyne **320** reacts under mild conditions with *cis*- and *trans*-2-butenes to give stereospecifically the corresponding disilacyclobutenes (**321**, **322**) in good yields (Scheme 9.99).[227b] Based on theoretical calculations, the reaction proceeds via concerted [2 + 1] cycloaddition starting with the interaction between in-plane LUMO (π^*in) of the disilyne and the HOMO of 2-butene. The reaction generates a silacyclopropyl(silylene) intermediate **323** which isomerizes by a 1,2-insertion of silylene (typical for singlet silylenes and carbenes) to give the final 1,2-disilacyclobutene **321, 322**.

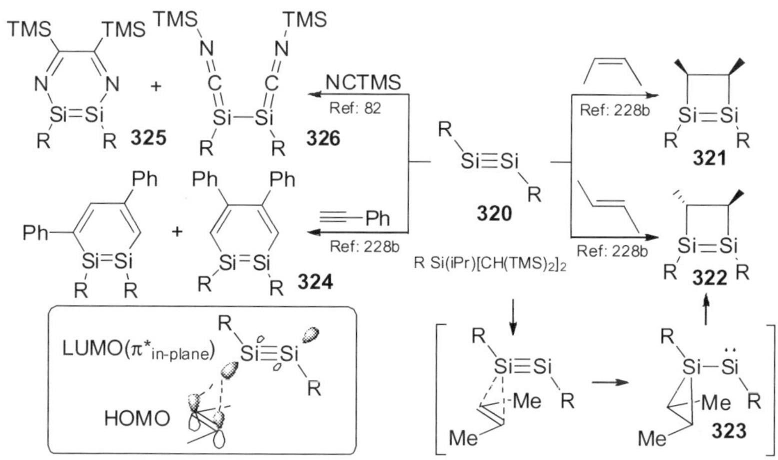

Scheme 9.99
Examples of cycloaddition reactions with Sekiguchi's disilyldisilyne.

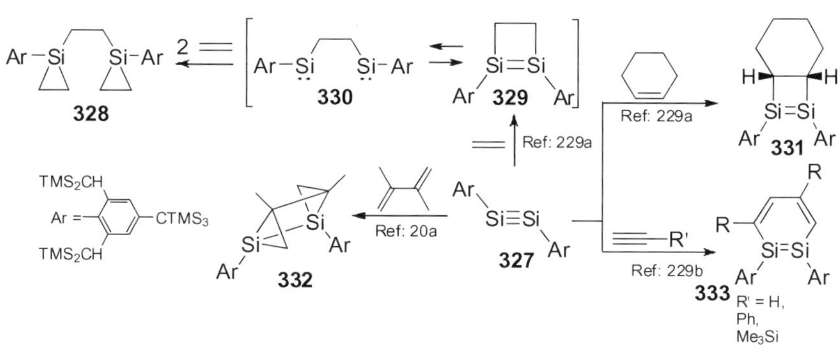

Scheme 9.100
Examples of cycloaddition reactions with Tokitoh's diaryldisilyne.

In the case of Tokitoh's diaryldisilyne **327**, the reaction with 3 equiv. of ethylene leads to an original ethylene-bridged bis(silacyclopropane) derivative **328** (Scheme 9.100). The reaction probably involves the first formation of a disilacyclobutene intermediate **329**, which dissociate into two fragments of silylenes (**330**). Such an intermediate can be isolated by reacting 1 equiv. of cyclohexene, and in this case the corresponding fused bicyclic disilacyclobutene **331** was isolated. Moreover, a unique reactivity was observed with dimethylbutadiene, which gives, in moderate yields, an *anti*-tricyclodisilahexane structure **332**.[228] Both types of disilynes (**320**, **327**) also react with alkynes or nitrile to give the corresponding disilabenzene (**324**, **333**) and 1,4-diaza-2,3-disilabenzene derivatives (**325**), and the reader can find more details in a recent review on this topic.[12]

Along with the formation of diaza(disilabenzene) **325** by the reaction of symmetrical disilyl (disilyne) **320** with trimethylsilylcyanide, some traces of bis(silaketeneimine) **326** were observed, probably due to the equilibrium cyanide-isocyanide (Scheme 9.99).[82] Interestingly, the use of 2 equiv. of alkylisocyanide led to the quantitative formation of this type of bis-adduct, which can be spectroscopically characterized below $-30°C$.[83] These disilyne-alkylisocyanide bis-adducts are thermally labile compounds and readily decompose at RT to give an original 1,2-dicyanodisilene (21%) and 1,2-dicyanodisilane via C–N bond cleavage. Therefore, disilyne **320** can be used as precursor of functionalized disilenes, and amino- and boryl-substituted disilenes (**334**, **335**) can be readily obtained by addition of amines[21c,d] and hydroboranes, respectively (Scheme 9.101).[22a]

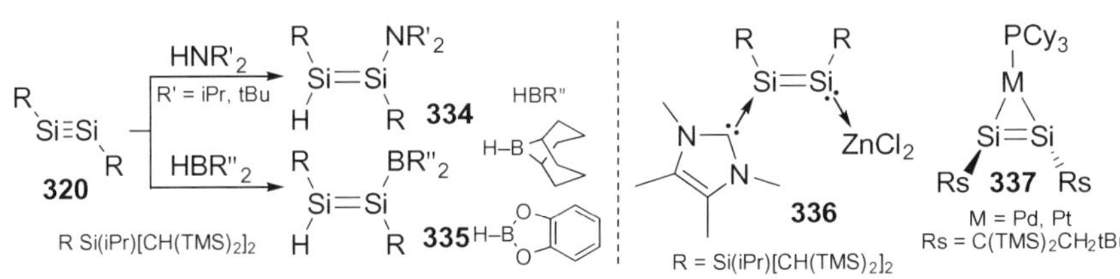

Scheme 9.101

Reaction of Sekiguchi's disilyne with alkylisocyanides, amines and hydroboranes and its use as ligand for transition metals.

As predicted by theoretical studies,[229] disilynes should be unique ligands for transition metals. However, only two experimental studies have been reported to date. The first one concerns a η^1-(NHC-coordinated disilyne) zinc complex **336**, and more recently Pd and Pt η^2-disilyne complexes **337** have been isolated (Scheme 9.101).[230]

9.3.2 Si≡E Triple Bond

In contrast to the chemistry of stable disilynes with a homonuclear Si≡Si triple bond, that of alkyne analogs with a heteronuclear triple bond (Si≡E, E = 14- and 15-group elements) is still underdeveloped. Indeed, the base-stabilized silyne reported by Kato/Baceiredo's team in 2010 is the only known example of an isolable heteronuclear silicon containing triple bond species to date.[231] As related compounds, vinylidene isomers of germasilyne (**128**, Scheme 9.41 in Section 9.2.2.2.2)[111] and of phosphasilyne (**216**, Scheme 9.64 in Section 9.2.2.3.2)[156] have also recently been isolated as stable complexes with a NHC ligand (Scheme 9.102).

One major obstacle for their synthesis is the extremely favored isomerization into the corresponding silavinylidene isomers (−Si≡E14−,[232,233] − Si≡E15[234]). This tendency is particularly strong in the case of silynes (−Si≡C−) and silanitriles (−SiN) compared to the heavier analogs (Scheme 9.102). Indeed, for the heavier silanitrile analogs, the triple-bonded species (−Si≡E15) are more stable than the silavinylidene isomers (−E15 = Si:).

E_{14}	$H_2Si{=}E_{14}$:	$HSi{\equiv}E_{14}H$	:$Si{=}E_{14}H_2$	E_{15}	$E_{15}{\equiv}SiH$	$HE_{15}{=}Si$:
C	48.0	0.0	−41.8	N	0.0	−67.5
Ge	−9.2	0.0	2.1	P	0.0	1.4
				As	0.0	9.3
				Sb	0.0	17.6

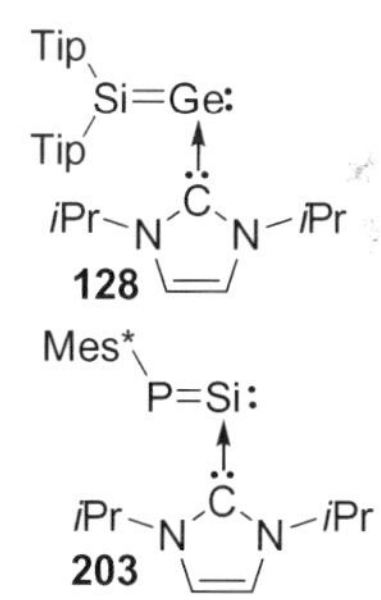

128
203

Scheme 9.102
Relative stability of triple-bonded species and their isomers.

In the case of silynes (−Si≡C−), several computational studies predict a *trans*-bent geometry, similar to disilynes, and the linear structure is not a minimum on the potential energy surfaces.[233−235] It was also predicted that the order of stability between both isomers, silyne and silavinylidene, can be reversed either by introducing on silicon a more electronegative substituent, such as F ($\Delta E_{\text{vinylidene-silyne}} = -6.4$ kcal mol^{-1}),[233,235] or a strongly π-donating substituent, such as an amino group ($\Delta E_{\text{vinylidene-silyne}} = -16.8$ kcal mol^{-1}),[235] or by using bulky substituents.[233,235] Nevertheless, donor-free silynes are still elusive molecules.

The first persistent and isolable silyne **339** (stable up to −30°C) was obtained as a complex with an intramolecular coordination of a phosphine ligand on the silicon center by photolysis of the corresponding diazo-precursor **338** at −60°C (Scheme 9.103). The silyne derivative presents a very short silicon−carbon bond (1.667 Å), which is significantly shorter than Si=C double bonds (1.70−1.77 Å, Section 9.2.2.2.1),[3b,5j] and is in the range predicted for Si≡C triple bonds (1.63−1.67 Å).[233,235] The geometry around the central carbon is essentially linear (Si−C−P^2: 178.2 degrees) in contrast to the strongly bent silicon center (N-Si−C: 128.5 degrees), which is also pyramidalized ($\Sigma_{Si\alpha}$: 345 degrees).

These data are in agreement with important silyne character (Canonical structure A). Of particular interest, the geometry of silyne is completely different from that of its germanium analog **340**, which presents a phosphinocarbene type structure[235] with a long Ge–C bond (1.887 Å) which is between a single (1.98 Å)[236] and a double (1.80 Å) bond[3b] as well as a very short C–P (1.549 Å) bond with a double bond character (Scheme 9.103).

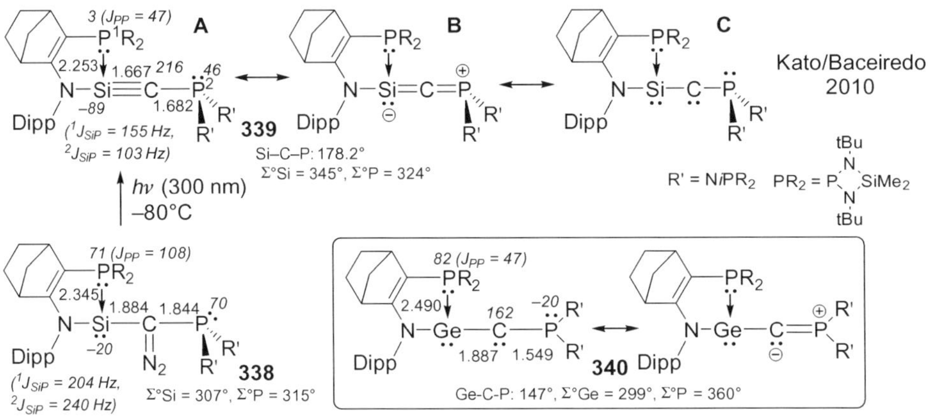

Scheme 9.103
Synthesis of the first isolable base-stabilized silyne.

The silicon–carbon triple bond of the silyne is perturbed by the phosphine coordination on the Si center as well as by the π- delocalization of the lone pair on the phosphine substituent to the π^*_{SiC} orbital (3-center-4-electron system), suggested by the shorten C–P^2 distance (1.682 Å) compared with that observed for the diazo precursor (1.844 Å). Probably due to these electronic effects (canonical structure B) and hypervalency of silicon center (λ^5-Si atom), it exhibits a strongly high-field shifted ^{29}Si NMR resonance (–89.4 ppm, $^2J_{SiP}$ = 103.0 Hz, $^1J_{SiP}$ = 155.4 Hz). In the ^{13}C NMR spectrum, the silyne carbon appears at 216 ppm, similarly to those observed for silaallene derivatives (214–268 ppm).[3b,5j] In addition, the calculated Wiberg bond order for the Si–C fragment (1.687) is much smaller than that calculated for the Si–Si triple bond (2.618).[221] This is probably due to the 3-center–4-electron system of the Si–C–P fragment (canonical structure **B)** as well as the highly polarized Si–C π-bonds due to the phosphine coordination on the silicon center, suggesting a certain carbenic character (Canonical structure C, Scheme 9.103).

Indeed, at RT, this base-stabilized silyne **339** readily transforms into a phosphaalkene derivative **340** via the 1,2-migration of a diisopropylamino group from the phosphorus to the central carbon atom, which is a typical rearrangement for singlet carbenes (Scheme 9.104).[237] Its carbenic character was also highlighted by the coupling reaction with *tert*-butyl-isocyanide, a well-known carbene trapping agent, to cleanly afford the corresponding keteneimine **341**.

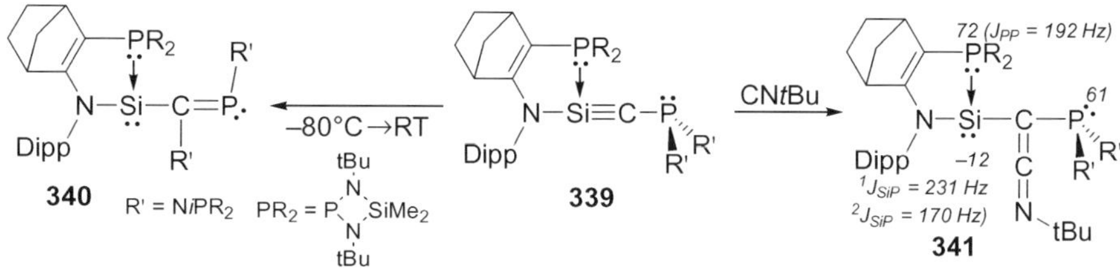

Scheme 9.104
Isomerization of silyne and reaction with an isonitrile.

9.4 Conclusion

Since 2000, significant progress in the synthesis of silicon containing double and triple-bonded species has been made. In particular, one of the most important events is the first synthesis of stable and isolable disilyne achieved by Sekiguchi et al. in 2004, which opened the way to study the silicon-containing triple-bonded species. However, any other stable silicon containing triple-bonded species remain elusive except for the recently reported isolable donor-stabilized silyne. Therefore, the chemistry of mono-silicon triple-bonded species is still completely undeveloped in contrast to the chemistry of silicon-containing double bonded derivatives in which almost all the series of compounds are now available. The further development of such fundamental chemistry is expected in the next decade, which should enable the development of original materials with new applications taking advantage of all the accumulated knowledge, particularly in the stabilization of silicon-based unsaturated compounds.

References

1. (a) Pitzer, K. S. Repulsive Forces in Relation to Bond Energies, Distances and Other Properties. *J. Am. Chem. Soc.* **1948,** *70*, 2140.
   (b) Mulliken, R. S. Overlap Integrals and Chemical Binding. *J. Am. Chem. Soc.* **1950,** *72*, 4493.
   (c) Jutzi, P. New Element-Carbon (p−p)π Bonds. *Angew. Chem. Int. Ed.* **1975,** *14*, 232.
2. West, R.; Fink, M. J.; Michl, J. Tetramesityldisilene, a Stable Compound Containing a Silicon-Silicon Double Bond. *Science* **1981,** *214*, 1343.
3. Reviews on heavier analogues of alkenes and alkynes:
   (a) Lee, V. Ya; Sekiguchi, A. Heavy Analogs of Alkenes, 1,3-Dienes, Allenes and Alkynes: Multiply Bonded Derivatives of Si, Ge, Sn and Pb. In *Organometallic Compounds of Low-Coordinate Si, Ge, and Pb*; , John Wiley & Sons, Ltd: Chichester, 2010; p 199, Chapter 5.
   (b) Power, P. P. π-Bonding and the Lone Pair Effect in Multiple Bonds Between Heavier Main Group Elements. *Chem. Rev.* **1999,** *99*, 3463.
   (c) Fischer, R. C.; Power, P. P. π-Bonding and the Lone Pair Effect in Multiple Bonds Involving Heavier Main Group Elements: Developments in the New Millennium. *Chem. Rev.* **2010,** *110*, 3877.
4. Reviews on disilenes:
   (a) West, R. Chemistry of the Silicon−Silicon Double Bond. *Angew. Chem. Int. Ed.* **1987**, 26, 1201.
   (b) Raabe, G.; Michl, J. Multiple Bond to Silicon. *Chem. Rev.* **1985,** *85*, 419.

(c) Raabe, G.; Michl, J. Multiple Bonds to Silicon. In *The Chemistry of Organic Silicon Compounds;* Patai, S., Rappoport, Z., Eds.; John Wiley & Sons, Ltd: Chichester, 1989; p 1015. Chapter 17.
(d) Okazaki, R.; West, R. Chemistry of Stable Disilenes. *Adv. Organomet. Chem.* **1996,** *39*, 231.
(e) Weidenbruch, M. Recent Advances in the Chemistry of Silicon–Silicon Multiple Bonds. In *The Chemistry of Organic Silicon Compounds;* Rappoport, Z., Apeloig, Y., Eds.; Vol. 3; John Wiley & Sons, Ltd: Chichester, 2001; p 391 Chapter 5.
(f) Kira, M.; Iwamoto, T. Progress in the Chemistry of Stable Disilenes. *Adv. Organomet. Chem.* **2006,** *54*, 73.
(g) Iwamoto, T.; Ishida, S. Multiple Bonds With Silicon: Recent Advances in Synthesis, Structure, and Functions of Stable Disilenes. In *Functional Molecular Silicon Coumounds II;* Scheschkewitz, D., Ed.; Springer: Switzerland, 2014; p 125.

5. Reviews on silenes:
(a) Schaefer, H. F., III The Silicon–Carbon Double Bond: A Healthy Rivalry Between Theory and Experiment. *Acc. Chem. Res.* **1982**, *15*, 283.
(b) Brook, A. G.; Baines, K. M. Silenes. *Adv. Organomet. Chem.* **1986,** *25*, 1.
(c) Grev, R. S. Structure and Bonding in the Parent Hydrides and Multiply Bonded Silicon and Germanium Compounds: From Mh_n to $R_2M{=}M'R_2$ and $RM{\equiv}M'R$. *Adv. Organomet. Chem.* **1991,** *33*, 125.
(d) Brook, A. G.; Brook, M. A. The Chemistry of Silenes. *Adv. Organomet. Chem.* **1996,** *39*, 71.
(e) Müller, T.; Ziche, W.; Auner, N. Silicon–Carbon and Silicon–Nitrogen Multiply Bonded Compounds. In *The Chemistry of Organic Silicon Compounds;* Rappoport, Z., Apeloig, Y., Eds.; Vol. 2; John Wiley & Sons, Ltd.: Chichester, 1998; p 867. Chapter 16.
(f) Escudié, J.; Couret, C., et al. Silenes >Si=C<, Germenes >Ge=C< and Stannenes >Sn=C<: The French Contribution. *Coord. Chem. Rev.* **1998,** *178*, 565.
(g) Morkin, T. L.; Leigh, W. J. Substituent Effects on the Reactivity of the Silicon–Carbon Double Bond. *Acc. Chem. Res.* **2001,** *34*, 129.
(h) Gusel'nikov, L. E. Hetero π-Systems from 2 + 2 Cycloreversions. Part 1. Gusel'nikov–Flowers Route to Silenes and Origination of the Chemistry of Doubly Bonded Silicon. *Coord. Chem. Rev.* **2003,** *244*, 149.
(i) Lee, V. Ya; Sekiguchi, A. Heteronuclear Heavy Alkenes $E = E'$ (E, E' = Group 14 Elements): Germasilenes, Silastannenes, Germastannenes...Next Stop? *Organometallics* **2004,** *23*, 2822.
(j) Ottosson, H.; Eklof, A. M. Silenes: Connectors Between Classical Alkenes and Nonclassical Heavy Alkenes. *Coord. Chem. Rev.* **2008,** *252*, 1287.

6. Reviews on applications of silenes and of disilenes in organic synthesis:
(a) Ottosson, H.; Steel, P. G. Silylenes, Silenes, and Disilenes: Novel Silicon-Based Reagents for Organic Synthesis? *Chem. Eur. J.* **2006**, *12*, 1576.
(b) Milnes, K. K.; Pavelka, L. C.; Baines, K. M. Cycloaddition of Carbonyl Compounds and Alkynes to (Di)silenes and (Di)germenes: Reactivity and Mechanism. *Chem. Soc. Rev.* **2016,** *45*, 1019.

7. Reviews on silaimines:
(a) Walter, S.; Klingebiel, U. From Aminosilane to Iminosilane. *Coord. Chem. Rev.* **1994**, *130*, 481.
(b) Hemme, I.; Klingebiel, U. Iminosilanes and Related Compounds—Synthesis and Reactions. *Adv. Organomet. Chem.* **1996,** *39*, 159.
(c) Neugebauer, P.; Jaschke, B.; Klingebiel, U. Recent Developments in the Chemistry of Compounds With Silicon–Nitrogen Bonds. In *The Chemistry of Organic Silicon Compounds;* Rappoport, Z., Apeloig, Y., Eds.; Vol. 3; John Wiley & Sons, Ltd.: Chichester, 2001; p 429. Chapter 6.
(d) Klingebiel, U.; Matthes, C. From Lithiumhalosilylamide to Small and Large Ring Compounds, Iminosilenes and Iminosilanes. *J. Organomet. Chem.* **2007,** *692*, 2633.

8. Reviews on phospha- and arsasilenes:
(a) Driess, M. Some Aspects of the Chemistry of Silylidene-Phosphanes and -Arsanes. *Coord. Chem. Rev.***1996**, *145*, 1.
(b) Driess, M. Silicon–Phosphorus and Silicon–Arsenic Multiple Bonds. *Adv. Organomet. Chem.* **1996,** *39*, 193.

9. Reviews on silanones and silanethions:
   (a) Xiong, Y.; Yao, S.; Driess, M. Chemical Tricks To Stabilize Silanones and Their Heavier Homologues With E═O Bonds (E═Si–Pb): From Elusive Species to Isolable Building Blocks. *Angew. Chem. Int. Ed.* **2013**, 52, 4302.
   (b) Tokitoh, N.; Okazaki, R. Recent Advances in the Chemistry of Silicon–Heteroatom Multiple Bonds. In *The Chemistry of Organic Silicon Compounds;* Rappoport, Z., Apeloig, Y., Eds.; Vol. 2; John Wiley & Sons, Ltd.: Chichester, 2001; p 1063. Chapter 17.
10. Reviews on 1,3-dienes:
   (a) Willms, S.; Weidenbruch, M. Hexaaryltetrasilabuta-1, 3-Diene: Synthesis and Some Reactions. In *Organosilicon Chemistry IV: From Molecules to Materials*; Auner, N., Weis, J., Eds.; Wiley-VCH: Weinheim, 2000; p 117.
   (b) Weidenbruch, M. Some Recent Advances in the Chemistry of Silicon and Its Homologues in Low Coordination States. *J. Organomet. Chem.* **2002,** *646*, 39.
   (c) Weidenbruch, M. A Terasilabuta-1,3-Diene and Related Compounds With Conjugated Multiple Bonds. In *Silicon Chemistry;* Jutzi, P., Schubert, U., Eds.; Wiley-VCH: Weinheim, 2003; p 100.
   (d) Weidenbruch, M. From Silylenes and Disilenes to a Tetrasilabuta-1,3-Diene and Related Compounds. In *Organosilicon Chemistry V: From Molecules to Materials;* Auner, N., Weis, J., Eds.; Wiley-VCH: Weinheim, 2003; p 114.
11. Reviews on heavier analogues of allenes:
   (a) Escudié, J.; Ranaivonjatovo, H.; Rigon, L. Heavy Allenes and Cumulenes E = C = E′ and E = C = C = E′ (E = P, As, Si, Ge, Sn; E′ = C, N, P, As, O, S). *Chem. Rev.* **2000**, *100*, 3639.
   (b) Eichler, B.; West, R. Chemistry of Group 14 Heteroallenes. *Adv. Organomet. Chem.* **2001,** *46*, 1.
   (c) Escudié, J.; Ranaivonjatovo, H. Group 14 and 15 Heteroallenes E═C═C and E═C═E′. *Organometallics* **2007,** *26*, 1542.
12. Reviews on heavier analogues of alkynes:
   (a) Power, P. P. Silicon, Germanium, Tin and Lead Analogues of Acetylenes. *Chem. Commun.* **2003**, 2091.
   (b) Power, P. P. Bonding and Reactivity of Heavier Group 14 Element Alkyne Analogues. *Organometallics* **2007,** *26*, 4362.
   (c) Sekiguchi, A. Disilyne With a Silicon–Silicon Triple Bond: A New Entry to Multiple Bond Chemistry. *Pure Appl. Chem.* **2008,** *80*, 447.
13. Scheschkewitz, D. A Silicon Analogue of Vinyllithium: Structural Characterization of a Disilenide. *Angew. Chem. Int. Ed.* **2004,** *43*, 2965.
14. Ichinohe, M.; Sanuki, K.; Inoue, S.; Sekiguchi, A. Disilenyllithium From Tetrasila-1,3-Butadiene: A Silicon Analogue of a Vinyllithium. *Organometallics* **2004,** *23*, 3088.
15. (a) Sekiguchi, A.; Ichinohe, M.; Yamaguchi, S. An Unexpected Reaction of Silacyclopropene to Form an Organodilithiosilane: Isolation and Characterization of a 1,1-Dilithiosilane Derivative. *J. Am. Chem. Soc.* **1999,** *121*, 10231.
   (b) Ichinohe, M.; Arai, Y.; Sekiguchi, A.; Nagase, S. A New Approach to the Synthesis of Unsymmetrical Disilenes and Germasilene: Unusual ^{29}Si NMR Chemical Shifts and Regiospecific Methanol Addition. *Organometallics* **2001,** *20*, 4141.
16. Scheschkewitz, D. Anionic Reagents With Silicon-Containing Double Bonds. *Chem. Eur. J.* **2009,** *15*, 2476.
17. (a) Tanaka, H.; Inoue, S.; Ichinohe, M.; Driess, M.; Sekiguchi, A. Synthesis and Striking Reactivity of an Isolable Tetrasilyl-Substituted Trisilaallene. *Organometallics* **2011,** *30*, 3475.
   (b) Lee, V. Ya; Sekiguchi, A. Novel Organometallic Reagents: Geminal Dianionic Derivatives of the Heavy Group 14 Elements. *Inorg. Chem.* **2011,** *50*, 12303.
18. (a) Cowley, M. J.; Abersfelder, K.; White, A. J.; Majumdar, M.; Scheschkewitz, D. Transmetallation Reactions of a Lithium Disilenide. *Chem. Commun.* **2012,** *48*, 6595.
   (b) Nguyen, T.-I.; Scheschkewitz, D. Activation of a Si═Si bond by η^1-Coordination to a Transition Metal. *J. Am. Chem. Soc.* **2005,** *127*, 10174.
   (c) Abersfelde, K., et al. An Unsaturated α,ω-Dianionic Oligosilane. *Angew. Chem. Int. Ed.* **2006,** *45*, 1643.

(d) Leszczynska, K.; Abersfelder, K.; Mix, A.; Neumann, B.; Stammler, H. G.; Cowley, M. J.; Jutzi, P.; Scheschkewitz, D. Reversible Base Coordination to a Disilene. *Angew. Chem. Int. Ed.* **2012,** *51*, 6785.
(e) Iwamoto, T.; Kobayashi, M.; Uchiyama, K.; Sasaki, S.; Nagendran, S.; Isobe, H.; Kira, M. Anthryl-Substituted Trialkyldisilene Showing Distinct Intramolecular Charge-Tranfer Transition. *J. Am. Chem. Soc.* **2009,** *131*, 3156.
(f) Ichinohe, M.; Sanuki, K.; Inoue, S.; Sekiguchi, A. Tetrasila-1,3-Butadiene and Its Transformation to Disilenyl Anions.. *Silicon Chem.* **2007,** *3*, 111.
(g) Inoue, S.; Ichinohe, M.; Sekiguchi, A. Disilenyl Anions Derived From Reduction of Tetrakis(di-tert-butylmethylsilyl) Disilene With Metal Naphthalenide Through a Disilene Dianion Intermediate: Synthesis and Characterization. *Chem. Lett.* **2005,** *34*, 1564.
(h) Kira, M.; Iwamoto, T.; Yin, D.; Maruyama, T.; Sakurai, H. Novel 1,2-Dilithiodisilanes Derived From Reduction of Stable Tetrakis(trialkylsilyl)disilene With Lithium Metals. *Chem. Lett.* **2001,** *31*, 910.

19. (a) Wiberg, N.; Niedmayer, W.; Fischer, G.; Nöth, H.; Suter, M. Synthesis, Structure and Dehalogenation of the Disilene RClSi=SiClR [R = $(tBu_3Si)_2MeSi$]. *Eur. J. Inorg. Chem.* **2002,** 1066.
(b) Sasamori, T.; Hironaka, K., et al. Synthesis and Reactions of a Stable 1,2-Diaryl-1,2-Dibromodisilene: A Precursor for Substituted Disilenes and A 1,2-Diaryldisilyne. *J. Am. Chem. Soc.* **2008,** *120*, 13856.
(c) Suzuki, K.; Matsuo, T.; Hashizume, D.; Tamao, K. Room-Temperature Dissociation of 1,2-Dibromodisilenes to Bromosilylenes. *J. Am. Chem. Soc.* **2011,** *133*, 19710.

20. Hartmann, M.; Haji-Abdi, A.; Abersfelder, K.; Haycock, P. R.; White, A. J. P.; Scheschkewitz, D. Synthesis, Characterisation and Complexation of Phosphino Disilenes. *Dalton Trans.* **2010,** *39*, 9288.

21. (a) Jutzi, P.; Mix, A.; Rummel, B.; Schoeller, W. W.; Neumann, B.; Stammler, H. G. The $(Me_5C_5)Si^+$ Cation: A Stable Derivative of HSi^+. *Science* **2004,** *305*, 849.
(b) Jutzi, P., et al. Reversible Transformation of a Stable Monomeric Silicon(II) Compound into a Stable Disilene by Phase Transfer: Experimental and Theoretical Studies of the System $\{[(Me_3Si)_2N](Me_5C_5)Si\}n$ With n = 1,2. *J. Am. Chem. Soc.* **2009,** *131*, 12137.
(c) Takeuchi, K.; Ikoshi, M.; Ichinohe, M.; Sekiguchi, A. Addition of Amines and Hydroborane to the Disilyne RSi≡SiR (R = Si*i*Pr[CH(SiMe$_3$)$_2$]$_2$) Giving Amino- and Boryl-Substituted Disilenes. *J. Am. Chem. Soc.* **2010,** *132*, 930.
(d) Takeuchi, K.; Ikoshi, M.; Ichinohe, M.; Sekiguchi, A. Silicon Version of Enamines: Amino-Substituted Disilenes by the Reactions of the Disilyne RSi≡SiR (R = Si*i*Pr[CH(SiMe$_3$)$_2$]$_2$) With Amines. *J. Organomet. Chem.* **2011,** *696*, 1156.

22. (a) Takeuchi, K.; Ichinohe, M.; Sekiguchi, A. Hydroboration of Disilyne RSi≡SiR (R = Si*i*Pr[CH(SiMe$_3$)$_2$]$_2$), Giving Boryl-Substituted Disilenes. *Organometallics* **2011,** *30*, 2044.
(b) Inoue, S.; Ichinohe, M.; Sekiguchi, A. A Stable Boryl-Substituted Disilene From the Reaction of an sp^2-Type Silyl Anion With Haloboranes: Synthesis and Characterization. *Chem. Lett.* **2008,** *30*, 1044.

23. Agou, T.; Sugiyama, Y.; Sasamori, T.; Sakai, H.; Furukawa, Y.; Takagi, N., et al. Synthesis of Kinetically Stabilized 1,2-Dihydrodisilenes. *J. Am. Chem. Soc.* **2012,** *134*, 4120.

24. (a) Ishida, S.; Iwamoto, T.; Kabuto, C.; Kira, M. A Stable Silicon-Based Allene Analogue With a Formally sp-Hybridized Silicon Atom. *Nature* **2003,** *421*, 725.
(b) Iwamoto, T.; Abe, T.; Ishida, S., et al. Reactions of Trisilaallene and 2-Germadisilaallene With Various Reagents. *J. Organomet. Chem.* **2007,** *692*, 263.

25. Matsuo, T.; Kobayashi, M.; Tamao, K. π-Conjugated Disilenes Stabilized by Fused-Ring Bulky "Rind" Groups. *Dalton Trans.* **2010,** *39*, 9203.

26. Fukazawa, A.; Li, Y.; Yamaguchi, S.; Tsuji, H.; Tamao, K. Coplanar Oligo(*p*-Phenylenedisilenylene)s Based on the Octaethyl-Substituted s-Hydrindacenyl Groups. *J. Am. Chem. Soc.* **2007,** *129*, 1416.

27. Pham, E. K.; West, R. Synthesis and Characterization of the First Transition-Metal η^2-Disilene Complexes. *J. Am. Chem. Soc.* **1989,** *111*, 7667.

28. Berry, D. H.; Chey, J. H.; Zipin, H. S.; Carroll, P. J. Disilene Complexes of Molybdenum and Tungsten. *J. Am. Chem. Soc.* **1990,** *112*, 452.

29. (a) Dewar, M. J. S. A Review of the π-Complex Theory. *Bulletin de la Socie´te´ Chimique de France* **1951,** *18*, C71.
(b) Chatt, J.; Duncanson, L. A. Olefin Coordination Compounds. Part III. Infra-red Spectra and Structures: Attempted Preparation of Acetylene Complexes. *J. Chem. Soc.* **1953,** 2939.

30. (a) Hashimoto, H.; Sekiguchi, Y.; Iwamoto, T.; Kabuto, C.; Kira, M. Synthesis and X-ray Structure of a Platinum η^2-Disilene Complex. *Organometallics* **2002,** *21*, 454.
(b) Hashimoto, H.; Sekiguchi, Yu, et al. Comparison of Structures Between Platinum and Palladium Complexes of a Tetrasilyldisilene. *Can. J. Chem.* **2003,** *81*, 1241.
(c) Kira, M.; Sekiguchi, Y.; Iwamoto, T.; Kabuto, C. 14-Electron Disilene Palladium Complex Having Strong π-Complex Character. *J. Am. Chem. Soc.* **2004,** *126*, 12778.
(d) Hashimoto, H.; Suzuki, K.; Setaka, W.; Kabuto, C.; Kira, M. Iron Complexes of (E)- and (Z)-1,2-Dichlorodisilenes. *J. Am. Chem. Soc.* **2004,** *126*, 13628.
(e) Iwamoto, T.; Sekiguchi, Y.; Yoshida, N.; Kabuto, C.; Kira, M. Ligand Dependence of π-Complex Character in Disilene-Palladium Complexes. *Dalton Trans.* **2006,** 177.

31. (a) Fischer, R.; Zirngast, M.; Flock, M.; Baumgartner, J.; Marschner, C. Synthesis of a Hafnocene Disilene Complex. *J. Am. Chem. Soc.* **2005,** *127*, 70.
(b) Zirngast, M.; Flock, M., et al. Group 4 Metallocene Complexes of Disilenes, Digermenes and a Silagermene. *J. Am. Chem. Soc.* **2009,** *131*, 15952.

32. (a) Abe, T.; Iwamoto, T.; Kira, M. A Stable 1,2-Disilacyclohexene and Is 14-Electron Palladium(0) Complex. *J. Am. Chem. Soc.* **2010,** *132*, 5008.
(b) Agou, T.; Sasamori, T.; Tokitoh, N. Synthesis of an Arylbromosilylene−Platinum Complex by Using a 1,2-Dibromodisilene as a Silylene Source. *Organometallics* **2012,** *31*, 1150.

33. (a) Han, J. S.; Sasamori, T.; Mizuhata, Y.; Tokitoh, N. Evidence for LiBr-Assisted Generation of a Silylene From a 1,2-Diaryl-1,2-Dibromodisilene. *Chem. Asian J.* **2012,** *7*, 298.
(b) Agou, T.; Hayakawa, N.; Sasamori, T., et al. Reactions of Diaryldibromodisilenes With *N*-Heterocyclic Carbenes: Formation of Formal bis-NHC Adducts of Silyliumylidene Cations. *Chem. Eur. J.* **2014,** *20*, 9246.

34. Inoue, S.; Eisenhut, C. A Dihydrosilene Transition Metal Complex From an *N*-Heterocyclic Carbene-Stabilized Silylene Monohydride. *J. Am. Chem. Soc.* **2003,** *135*, 18315.

35. (a) Sekiguchi, A.; Yatabe, T.; Kabuto, C.; Sakurai, H. The "Missing" Hexasilaprismane: Synthesis, X-ray Analysis, and Photochemical Reaction. *J. Am. Chem. Soc.* **1993,** *115*, 5853.
(b) Abersfelder, K.; Russel, A.; Rzepa, H. S., et al. Contraction and Expansion of the Silicon Scaffold of stable Si_6R_6 Isomers. *J. Am. Chem. Soc.* **2012,** *134*, 16008.

36. Abersfelder, K.; White, A. J. P.; Rzepa, H. S.; Scheschkewitz, D. A Tricyclic Aromatic Isomer of Hexasilabenzene. *Science* **2010,** *327*, 564.

37. Abersfelder, K.; White, A. J. P.; Berger, R. J. F.; Rzepa, H. S.; Scheschkewitz, D. A Stable Derivative of the Global Minimu on the Si_6H_6 Potential Energy Surface. *Angew. Chem. Int. Ed.* **2011,** *50*, 7936.

38. Tsurusaki, A.; Iizuka, C.; Otsuka, K.; Kyushin, S. Cyclopentasilane-Fused Hexasilabenzvalene. *J. Am. Chem. Soc.* **2013,** *135*, 16340.

39. Ghana, P.; Arz, M. I.; Das, U.; Schnakenburg, G.; Filippou, A. C. Si=Si double bonds: synthesis af an NHC-stabilized disilavinylidene. *Angew. Chem. Int. Ed.* **2015,** *54*, 9980.

40. Wang, Y.; Xie, Y.; Wei, P.; King, R. B.; Schaefer, H. F., III; Schleyer, Pv. R., et al. A Stable Silicon(0) Compound With a Si=Si Double Bond. *Science* **2008,** *321*, 1069.

41. Hickox, H. P.; Wang, Y.; Xie, Y.; Chen, M.; Wei, P.; Schaefer, H. F., III, et al. Transition Metal-Mediated Cleavage of Si=Si Double Bond. *Angew. Chem. Int. Ed.* **2015,** *54*, 10267.

42. (a) Wiberg, N.; Niedermayer, W.; Polborn, K. Disilenes R*RSi=SiRR* (R* = Si*t*Bu_3) With Silicon-Bound Me and Ph Groups R: Formation, Identification, Thermolysis, Structure. *Z. Anorg. Allg. Chem.* **2002,** *628*, 1045.

(b) Sekiguchi, A.; Inoue, S.; Ichinobe, M., et al. Isolable Anion Radical of Blue Disilene $(tBu_2MeSi)_2Si{=}Si(SiMetBu_2)_2$ Formed Upon One-Electron Reduction: Synthesis and Characterization. *J. Am. Chem. Soc.* **2004,** *126*, 9626.

43. Kostenko, A.; Tumanskii, B.; Karni, M.; Inoue, S.; Ichinohe, M.; Sekiguchi, A., et al. Observation of Thermally Accessible Triplet State Resulting From Rotation Around a Main-Group Bond. *Angew. Chem. Int. Ed.* **2015,** *54*, 12144.

44. (a) Hardwick, J. A.; Pavelka, L. C.; Baines, K. M. The Addition of Amides to Group 14 (Di)metallenes. *Dalton Trans.* **2012,** *41*, 609.

(b) Buriak, J. M. Organometallic Chemistry on Silicon and Germanium Surfaces. *Chem. Rev.* **2002,** *102*, 1271.

45. (a) Tashkandi, N. Y.; Parsons, F.; Guo, J.; Baines, K. M. Addition of Nitromethane to a Disilene and a Digermene: Comparison to Surface Reactivity and Facile Formation of 1,3,2-Dioxazolidines. *Angew. Chem. Int. Ed.* **2015,** *54*, 1612.

(b) Hardwick, J. A.; Baines, K. M. The Addition of Nitriles to Tetramesityldisilene: A Comparison of the Reactivity Between Surface and Molecular Disilenes. *Chem. Eur. J.* **2015,** *21*, 2480.

46. Schleyer, P. v. R.; Kost, D. A Comparison of the Energies of Double Bonds of Second-Row Elements With Carbon and Silicon. *J. Am. Chem. Soc.* **1988,** *110*, 2105.

47. Nakata, N.; Izumi, R.; Lee, V. Ya; Ichinohe, M.; Sekiguchi, A. Reaction of Dilithiosilane R_2SiLi_2 and Dilithiogeramane R_2GeLi_2 ($R{=}SiMetBu_2$) With $MesBCl_2$ (Mes = 2,4,6-Trimethylphenyl): Evidence for the Foramtion of Silaborene $R_2Si{=}BMes$ and Germaborene $R_2Ge = BMes$. *Chem. Lett.* **2005,** *34*, 582.

48. Nakata, N.; Sekiguchi, A. A Stable Silaborene: Synthesis and Characterization. *J. Am. Chem. Soc.* **2006,** *128*, 422.

49. Kajiwara, T.; Takeda, N.; Sasamori, T.; Tokitoh, N. Unprecedented Insertion Reaction of a Silylene into a B−B Bond and Generation of a Novel Borylsilyl Anion by Boron−Metal Exchange Reaction of the Resultant Diborylsilane. *Chem. Commun.* **2004,** 2218.

50. (a) Nakata, N.; Izumi, R.; Lee, V. Ya; Ichinohe, M.; Sekiguchi, A. 1,3-Disila-2-gallata- and -Indataallenic Anions [> Si-M-Si <]-, Li^+ (M = Ga, In): Compounds Featuring Double Bonds Between Elements of Groups 13 and 14. *J. Am. Chem. Soc.* **2004,** *126*, 5058.

(b) Hsiao, J., et al. Theoretical Insight into the Mechanism of [3 + 2] Cycloaddition Reactions of 1,3-Disila-2-group-13 Atomic Anions $[>Si{=}M{=}Si<]^-$ (M = B, Al, Ga, In, and Tl). *Organometallics* **2007,** *26*, 4432.

51. (a) Brook, A. G.; Abdesaken, F.; Gutekunst, B.; Gutekunst, G.; Kallury, R. K. A Solid Silaethene: Isolation and Characterization. *J. Chem. Soc., Chem. Commun.* **1981,** 191.

(b) Brook, A. G.; Nyburg, S. C., et al. Stable Solid Silaethylenes. *J. Am. Chem. Soc.* **1982,** *104*, 5667.

52. Wiberg, N.; Wagner, G. A Novel Stable Silaethene. *Angew. Chem. Int. Ed.* **1983,** *22*, 1005.

53. (a) Wiberg, N.; Wagner, G.; Müller, G. Isolation and Structure of a Stable Molecule Containing a Silicon-Carbon Double Bond. *Angew. Chem. Int. Ed.* **1985,** *24*, 229.

(b) Wiberg, N.; Wagner, G.; Riede, J.; Müller, G. Structure of the Stable Silaethene $Me_2Si{=}C(SiMe_3)(SiMetBu_2)$. *Organometallics* **1987,** *6*, 35.

54. Apeloig, Y.; Bendikov, M.; Yuzefovich, M.; Nakash, M.; Bravo-Zhivotovskii, D.; Bläser, D., et al. Novel Stable Silenes via a Sila-Peterson-Type Reaction. Molecular Structure and Reactivity. *J. Am. Chem. Soc.* **1996,** *118*, 12228.

55. Wiberg, N.; Joo, K.-S.; Polborn, K. Structure of a Silaethene Amine Adduct. *Chem. Ber.* **1993,** *126*, 67.

56. Wiberg, N.; Wagner, G.; Miiller, G.; Riede, J. Unsaturated Silicon Compounds. *J. Organomet. Chem.* **1984,** *271*, 381.

57. (a) Mickoleit, M.; Schmohl, K.; Kempe, R.; Oehme, H. Reaction of Dichlorometyltris(trimethylsilyl)silane With Organolithium Reagents: Synthesis of an Intramolecularly Donor-Stabilized Silene. *Angew. Chem. Int. Ed.* **2000,** *39*, 1610.

(b) Mickoleit, M.; Kempe, R.; Oehme, H. Synthesis and Characterization of Novel Silenes Stabilized by Intramolecular Coordination of a Dialkylamino Group. *Chem. Eur. J.* **2001,** *7*, 987.

58. (a) Pötter, M.; Bäumer, U.; Mickoleit, M.; Kempe, R.; Oehme, H. Intramolecularly Donor-Stabilized Silenes Part 3. Synthesis and Structure of 1-[2,6-Bis(dimethylaminomethyl)phenyl]-1,2,2-tris(trimethylsilyl)silene. *J. Organomet. Chem.* **2001,** *621*, 261.
(b) Schmohl, K.; Reinke, H., et al. Reaction of the (Dichloromethyl)oligosilanes $R(Me_3Si)_2Si\text{-}CHCl_2$ (R = Me, Ph, Me_3Si) With Organolithium Reagents and the Synthesis of Novel Kinetically Stabilized Silenes. *Eur. J. Inorg. Chem.* **2001,** 481.
(c) Mickoleit, M.; Schmohl, K.; Michalik, M.; Oehme, H. The Synthesis of Stable Intramolecularly Dialkylamino-Coordinated Silenes and Their Methyl Iodide Induced Isomerization into Cyclic Aminosilanes. *Eur. J. Inorg. Chem.* **2004,** 1538.
(d) Bäumer, U.; Reinke, H.; Oehme, H. Intramolecularly Donor-Stabilized Silenes: Part 6. The Synthesis of 1-[2,6-Bis(dimethylaminomethyl)phenyl]silenes and Their Reaction With Aromatic Aldehydes. *J. Organomet. Chem.* **2006,** *691*, 229.
59. Gau, D.; Kato, T.; Saffon-Merceron, N.; Cossío, F. P.; Baceiredo, A. Stable Phosphonium Sila-Ylide With Reactivity as a Sila-Wittig Reagent. *J. Am. Chem. Soc.* **2009,** *131*, 8762.
60. Gau, D.; Rodriguez, R.; Kato, T.; Saffon-Merceron, N.; Cossío, F. P.; Baceiredo, A. Nucleophilic Silylenoid Character of Stable Phosphonium Sila-Ylides. *Chem. Eur. J.* **2010,** *16*, 8255.
61. Apeloig, Y.; Karni, M. Substituent Effects on the Carbon-Silicon Double Bond. Monosubstituted Silenes. *J. Am. Chem. Soc.* **1984,** *106*, 6676.
62. Sakamoto, K.; Ogasawara, J.; Kon, Y.; Sunagawa, T.; Kabuto, C.; Kira, M. The First Isolable 4-Silatriafulvene. *Angew. Chem. Int. Ed.* **2002,** *41*, 1402.
63. Brook, A. G.; Safa, K. D.; Lickiss, P. D.; Baines, K. M. Photoisomerization of a Highly Substituted Silene. *J. Am. Chem. Soc.* **1985,** *107*, 4338.
64. Bejan, I.; Güclü, D.; Inoue, S.; Ichinohe, M.; Sekiguchi, A.; Scheschkewitz, D. Stable Cyclic Silenes From Reaction of Disilenides With Carboxylic Acid Chlorides. *Angew. Chem. Int. Ed.* **2007,** *46*, 3349.
65. Majumdar, M.; Huch, V.; Bejan, I.; Meltzer, A.; Scheschkewitz, D. Reversible, Complete Cleavage of Si=Si Double Bonds by Isocyanide Insertion. *Angew. Chem. Int. Ed.* **2013,** *52*, 3516.
66. Iwamoto, T.; Ohnishi, N.; Akasaka, N.; Ohno, K.; Ishida, S. Anthryl-Substituted 3-Silylene-2-Silaaziridine Obtained by Isomerization of Disilacyclopropanimine: An Exocyclic Silene Showing a Distinct Intramolecular Charge Transfer Transition. *J. Am. Chem. Soc.* **2013,** *135*, 10606.
67. Biltueva, I. S.; Bravo-Zhivotovskii, D. A.; Kalikhman, I. D.; Vitkovskii, V. Y.; Shevchenko, S. G.; Vyazankin, N. S., et al. Sila- and Germa-Enolate Anions. The Reaction of Acylgermanes With Triethylgermyllithium. *J. Organomet. Chem.* **1989,** *368*, 163.
68. Bravo-Zhivotovskii, D.; Apeloig, Y.; Ovchinnikov, Y.; Igonin, V.; Struchkov, Y. T. A Novel Route to the Trisilacyclobutane Moiety. A Possible Silene-Disilene Reaction. *J. Organomet. Chem.* **1993,** *446*, 123.
69. (a) Ohshita, J.; Masaoka, Y.; Masaoka, S.; Ishikawa, M.; Tachibana, A.; Yano, T., et al. Silicon-Carbon Unsaturated Compounds. XLVIII. Synthesis and Reactions of Silicon Analogs of Lithium Enolates. *J. Organomet. Chem.* **1994,** *473*, 15.
(b) Ohshita, J.; Masaoka, S.; Masaoka, Y.; Hasebe, H.; Ishikawa, M.; Tachibana, A.; Yano, T.; Yamabe, T. Silicon-Carbon Unsaturated Compounds. 55. Synthesis and Reactions of Lithium Silenolates, Silicon Analogs of Lithium Enolates. *Organometallics* **1996,** *15*, 3136.
(c) Ohshita, J.; Masaoka, S.; Ishikawa, M. Oxidative Coupling of Lithium Silenolates: First Synthesis of Bis(acyl)-substituted Polysilanes. *Organometallics* **1996,** *15*, 2198.
(d) Ohshita, J.; Masaoka, S.; Morimoto, Y.; Sano, M.; Ishikawa, M. Silicon-Carbon Unsaturated Compounds. 60. Reactions of Lithium Silenolates With Dienes. *Organometallics* **1997,** *16*, 1123.
(e) Ohshita, J.; Sakurai, H.; Tokunaga, Y.; Kunai, A. Formation and Reactions of Lithium Ester Silenolates: Silicon Analogues of Lithium Ester Enolates. *Organometallics* **1999,** *18*, 4545.
(f) Ohshita, J.; Sakurai, H.; Masaoka, S.; Tamai, M.; Kunai, A.; Ishikawa, M. Reactions of Lithium Silenolates With Benzophenone. *J. Organomet. Chem.* **2001,** *633*, 131.
(g) Ohshita, J.; Tamai, M.; Sakurai, H.; Kunai, A. Reactions of Lithium Silenolates With Acetylenes. Formation and characterization of 2-siladienes. *Organometallics* **2001,** *20*, 1065.

70. Guliashvili, T.; El-Sayed, I.; Fischer, A.; Ottosson, H. The First Isolable 2-Silenolate. *Angew. Chem. Int. Ed.* **2003,** *42*, 1640.
71. Dobrovetsky, R.; Zborovsky, L.; Sheberla, D.; Botoshansky, M.; Bravo-Zhivotovskii, D.; Apeloig, Y. Isolation of Silenolates $(R_3Si)_2Si{=}C(OLi)Ad$ With a Doubly Bonded Silicon Atom. *Angew. Chem. Int. Ed.* **2010,** *49*, 4084.
72. Haas, M.; Fischer, R.; Flock, M.; Mueller, S.; Rausch, M.; Saf, R., et al. Stable Silenolates and Brook-Type Silenes With Exocyclic Structures. *Organometallics* **2014,** *33*, 5956.
73. Cowley, M. J.; Ohmori, Y.; Huch, V.; Ichinohe, M.; Sekiguchi, A.; Scheschkewitz, D. Carbonylation of Cyclotrisilenes. *Angew. Chem. Int. Ed.* **2013,** *52*, 13247.
74. Cowley, M. J.; Huch, V.; Scheschkewitz, D. Donor−Acceptor Adducts of a 1,3-Disila-2-oxyallyl zwitterion. *Chem. Eur. J.* **2014,** *20*, 9221.
75. Nakata, N.; Rodriguez, R.; Troadec, T.; Saffon-Merceron, N.; Sotiropoulos, J.-M.; Baceiredo, A., et al. A Stable Silene Substituted by Strong π-Donors at the Silicon Center. *Angew. Chem. Int. Ed.* **2013,** *52*, 10840.
76. Rodriguez, R.; Troadec, T.; Kato, T.; Saffon-Merceron, N.; Sotiropoulos, J.-M.; Baceiredo, A. Synthesis and Characterization of an Isolable Base-Stabilized Silacycloprop-1-ylidene. *Angew. Chem. Int. Ed.* **2012,** *51*, 7158.
77. Miracle, G. E.; Ball, J. L.; Powell, D. R.; West, R. The First Stable 1-Silaallene. *J. Am. Chem. Soc.* **1993,** *115*, 11598.
78. Trommer, M.; Miracle, G. E.; Eichler, B. E.; Powell, D. R.; West, R. Synthesis and Reactivity of Several Stable 1-Silaallenes. *Organometallics* **1997,** *16*, 5737.
79. Spirk, S.; Belaj, F.; Albering, J. H.; Pietschnig, R. Formation of a Silylated 1-Silaallene via an Intermediate 1-Chloro-1-Silaallene. *Organometallics* **2010,** *29*, 2981.
80. Abe, T.; Iwamoto, T.; Kabuto, C.; Kira, M. Synthesis, Structure, and Bonding of Stable Dialkylsilaketenimines. *J. Am. Chem. Soc.* **2006,** *128*, 4228.
81. (a) Takeda, N.; Suzuki, H.; Tokitoh, N.; Okazaki, R.; Nagase, S. Reaction of a Sterically Hindered Silylene With Isocyanides: The First Stable Silylene-Lewis Base Complexes. *J. Am. Chem. Soc.* **1997,** *119*, 1456.
    (b) Takeda, N.; Kajiwara, T., et al. Synthesis and Properties of the First Stable Silylene-Isocyanide Complexes. *Chem. Eur. J.* **2003,** *9*, 3530.
82. Takeuchi, K.; Ichinohe, M.; Sekiguchi, A. Reactivity of the Disilyne RSi≡SiR (R = Si*i*Pr[$CH(SiMe_3)_2$]) Toward Silylcyanide: Two Pathways to Form the Bis-Adduct [$RSiSiR(CNSiMe_3)_2$] With Some Silaketeneimine Character and a 1,4-Diaza-2,3-Disilabenezene Analogue. *J. Am. Chem. Soc.* **2008,** *130*, 16848.
83. Takeuchi, K.; Ichinohe, M.; Sekiguchi, A. A New Disilene With π-Accepting Groups From the Reaction of Disilyne RSi≡SiR (R = Si*i*Pr[$CH(SiMe_3)_2$]) With Isocyanides. *J. Am. Chem. Soc.* **2012,** *134*, 2954.
84. Mondal, K. C.; Roesky, H. W.; Dittrich, B.; Holzmann, N.; Hermann, M.; Frenking, G., et al. Formation of a 1,4-Diamino-2,3-disila-1,3-butadiene Derivative. *J. Am. Chem. Soc.* **2013,** *135*, 15990.
85. Motomatsu, D.; Ishida, S.; Ohno, K.; Iwamoto, T. Isolable 2,3-Disila-1,3-butadiene From a Double Sila-Peterson Reaction. *Chem. Eur. J.* **2014,** *20*, 9424.
86. Igarashi, M.; Ichinohe, M.; Sekiguchi, A. Air-Stable Disilacyclopropene With a Si=C Bond and Its Conversion to Disilacyclopropenylium Ion: Silicon-Carbon Hybrid 2-Electron Systems. *J. Am. Chem. Soc.* **2007,** *129*, 12660.
87. Bejan, I.; Inoue, S.; Ichinohe, M.; Sekiguchi, A.; Scheschkewitz, D. 1,2-Disilacyclobut-2-enes: Donor-Free Four-Membered Cyclic Silenes From Reaction of Disilenides With Vinylbromides. *Chem. Eur. J.* **2008,** *14*, 7119.
88. Review on disilenide anions. Scheschkewitz, D. The Versatile Chemistry of Disilenides: Disila Analogues of Vinyl Anions as Synthons in Low-Valent Silicon Chemistry. *Chem. Lett.* **2011,** *40*, 2.
89. Bravo-Zhivotovskii, D.; Dobrovetsky, R.; Nemirovsky, D.; Molev, V.; Bendikov, M.; Molev, G., et al. The Synthesis and Isolation of a Metal-Substituted Bis-Silene. *Angew. Chem. Int. Ed.* **2008,** *47*, 4343.

90. Zborovsky, L.; Dobrovetsky, R.; Botoshansky, M.; Bravo-Zhivotovskii, D.; Apeloig, Y. Synthesis of Silenyllithiums Li(R'$_3$Si)Si=C(SiR$_3$)(1-Ad) via Transient Silyne−Silylidene Intermediates. *J. Am. Chem. Soc.* **2012,** *134*, 18229.
91. Iwamoto, T.; Masuda, H.; Kabuto, C.; Kira, M. Trigermaallene and 1,3-Digermasilaallene. *Organometallics* **2005,** *24*, 197.
92. Yang, Y.-F.; Foo, C.; Xi, H.-W.; Li, Y.; Lim, K. H.; So, C.-W. A Base-Stabilized 2-Silaallene. *Organometallics* **2013,** *32*, 2267.
93. Inoue, S.; Ichinohe, M.; Sekiguchi, A. Conversion of a Disilenide into a Silene: Silyl-Anion-Substituted Silene by a Sila-Peterson-Type Reaction From an sp2-Type Silyl Anion. *Angew. Chem. Int. Ed.* **2007,** *46*, 3346.
94. Mondal, K. C.; Roesky, H. W.; Schwarzer, M. C.; Frenking, G.; Tkach, I.; Wolf, H., et al. Conversion of a Singlet Silylene to a Stable Biradical. *Angew. Chem. Int. Ed.* **2013,** *52*, 1801.
95. Mondal, K. C.; Roesky, H. W.; Schwarzer, M. C.; Frenking, G.; Niepötter, B.; Wolf, H., et al. A Stable Singlet Biradicaloid Siladicarbene: (L:)$_2$Si. *Angew. Chem. Int. Ed.* **2013,** *52*, 2963.
96. Niepötter, B.; Herbst-Irmer, R.; Kratzert, D.; Samuel, P. P.; Mondal, K. C.; Roesky, H. W., et al. Experimental Charge Density Study of a Silylone. *Angew. Chem. Int. Ed.* **2014,** *53*, 2766.
97. Xiong, Y.; Yao, S.; Inoue, S.; Epping, J. D.; Driess, M. A Cyclic Silylone ("Siladicarbene") With an Electron-Rich Silicon(0) Atom. *Angew. Chem. Int. Ed.* **2013,** *52*, 7147.
98. Campion, B. K.; Heyn, R. H.; Tilley, T. D. Preparation, Isolation, and Characterization of Transition-Metal η^2-silene complexes. X-ray crystal structure of (η^5-C$_5$Me$_5$)[P(*i*-Pr)$_3$]Ru(H)(η^2-CH$_2$SiPh$_2$). *J. Am. Chem. Soc.* **1988,** *110*, 7558.
99. Campion, B. K.; Heyn, R. H.; Tilley, T. D. A Stable η^2-Silene Complex of Iridium: (η^5-C$_5$Me$_5$) (PMe$_3$)Ir (η^2-CH$_2$=SiPh$_2$). *J. Am. Chem. Soc.* **1990,** *112*, 4079.
100. Procopio, L. J.; Berry, D. H. Dehydrogenative Coupling of Trialkylsilanes Mediated by Ruthenium Phosphine Complexes: Catalytic Synthesis of Carbosilanes. *J. Am. Chem. Soc.* **1991,** *113*, 4039.
101. Dioumaev, V. K.; Plössl, K.; Carroll, P. J.; Berry, D. H. Formation and Interconversion of Ruthenium-Silene and 16-Electron Ruthenium Silyl Complexes. *J. Am. Chem. Soc.* **1999,** *121*, 8391.
102. Dioumaev, V. K.; Carroll, P. J.; Berry, D. H. Tandem β-C-H activation/Si-H elimination reactions: stabilization of C-H activation products by β-agostic Si-H interactions. *Angew. Chem. Int. Ed.* **2003,** *42*, 3947.
103. Klei, S. R.; Tilley, T. D.; Bergman, R. G. An Observable Silene/Silylene Rearrangement in a Cationic Iridium Complex. *Organometallics* **2001,** *20*, 3220.
104. Koloski, T. S.; Carroll, P. J.; Berry, D. H. Structure and Reactivity of Cp$_2$W(η^2-Me$_2$Si=CH$_2$), a Tungsten Silene Complex. *J. Am. Chem. Soc.* **1990,** *112*, 6405.
105. Bravo-Zhivotovskii, D.; Peleg-Vasserman, H.; Kosa, M.; Molev, G.; Botoshanskii, M.; Apeloig, Y. The Direct Synthesis of a Silene-Organometallic Complex. *Angew. Chem. Int. Ed.* **2004,** *43*, 745.
106. (a)Nakata, N.; Rodriguez, R.; Kato, T.; Saint Jalmes, L.; Marrot, S.; Baceiredo, A.; patent WO2015004396.
    (b) Troadec, T., et al. Silacyclopropylideneplatinum(0) Complex as a Robust and Efficient Hydrosilylation Catalyst. *Inorg. Chem.* **2016,** *55*, 8234.
107. (a) Lee, V. Ya; Ichinohe, M.; Sekiguchi, A.; Takagi, N.; Nagase, S. The First Three-Membered Unsaturated Rings Consisting of Different Heavier Group 14 Elements: 1-Disilagermirene With a Si=Si Double Bond and Its Isomerization to a 2-Disilagermirene With a Si=Ge Double Bond. *J. Am. Chem. Soc.* **2000,** *122*, 9034.
    (b) Lee, V. Ya; Ichinohe, M., et al. Disilagermirenes: Heavy Cyclopropenes of Si and Ge Atoms. *J. Organomet. Chem.* **2003,** *685*, 168.
108. Lee, V. Ya; Ichinohe, M.; Sekiguchi, A. The First Metalladiene of Group 14 Elements With a Silole-Type Structure With Si=Ge and C = C Double Bonds. *J. Am. Chem. Soc.* **2000,** *122*, 12604.
109. Lee, V. Ya; Ichinohe, M.; Sekiguchi, A. Reaction of Disilagermirenes With Phenylacetylene: From a Germasilene -Ge=Si- to a Metalladiene of the Type −Si=Ge-C=C-. *J. Organomet. Chem.* **2001,** *636*, 41.

110. Iwamoto, T.; Abe, T.; Kabuto, C.; Kira, M. A Missing Allene of Heavy Group 14 Elements: 2-Germadisilaallene. *Chem. Commun.* **2005,** 5190.

111. Jana, A.; Huch, V.; Scheschkewitz, D. NHC-Stabilized Silagermenylidene: A Heavier Analogue of Vinylidene. *Angew. Chem. Int. Ed.* **2013,** *52*, 12179.

112. Jana, A.; Majumdar, M.; Huch, V.; Zimmer, M.; Scheschkewitz, D. NHC-Coordinated Silagermenylidene Functionalized in Allylic Position and Its Behaviour as a Ligand. *Dalton Trans.* **2014,** *43*, 5175.

113. Sekiguchi, A.; Izumi, R.; Lee, V. Ya; Ichinohe, M. The First Silastannene >Si=Sn<: A New Doubly-Bonded System of Heavier Group 14 Elements. *J. Am. Chem. Soc.* **2002,** *124*, 14822.

114. Mackay, K. M. In *The Chemistry of Organic Germanium, Tin and Lead Compounds;* Patai, S., Ed.; John Wiley & Sons Ltd: New York, 1995; Chapter 2.

115. Hesse, M.; Klingebiel, U. Diisopropyl(2,4,6-tri-tert-butylphenylimino)silane - a Very Stable, Free Silaimine. *Angew. Chem. Int. Ed.* **1986,** *25*, 649.

116. Wiberg, N.; Schurz, K.; Reber, G.; Müller, G. Preparation and Structure of a Stable Molecule Containing a Silicon Nitrogen Double Bond and of Its Tetrahydrofuran Adduct. *J. Chem. Soc., Chem. Commun.* **1986,** 591.

117. Schleyer, P. v. R.; Stout, P. O. On the Nature of the Si=N Double Bond and the Ease of Bending at Nitrogen. *J. Chem. Soc., Chem. Commun.* **1986,** 1373.

118. Wiberg, N.; Schurz, K.; Fiischer, G. Isolation of the Stable Silaketimine $t\text{Bu}_2\text{Si{=}N-Si}t\text{Bu}_3$. *Angew. Chem. Int. Ed.* **1985,** *24*, 1053.

119. Niesmann, J.; Klingebiel, U.; Schäfer, M.; Boese, R. Free and Coordinated Iminosilanes: Synthesis and Structures. *Organometallics* **1998,** *17*, 947.

120. Lerner, H.-W.; Bolte, M.; Schurz, K.; Wiberg, N.; Baum, G.; Fenske, D., et al. Chemical and Structural Features of NMe_2Et Adducts of the Silanimines $\text{Me}_2\text{Si{=}NSi}t\text{Bu}_3$ and $t\text{Bu}_2\text{Si{=}NSiCl}t\text{Bu}_2$. *Eur. J. Inorg. Chem.* **2006,** 4998.

121. Wiberg, N.; Schurz, K.; Müller, G.; Riede, J. "Wittig Reactions" of Unsaturated Silicon Compounds: Isolation and Structure of Possible Intermediates. *Angew. Chem. Int. Ed.* **1988,** *27*, 935.

122. Lerner, H.-W.; Wiberg, N.; Bats, J. W. Das Chlorsilylsubstituierte Silanimin $t\text{Bu}_2\text{Si{=}N-SiCl}t\text{Bu}_2$. *J. Organomet. Chem.* **2005,** *690*, 3898.

123. Denk, M.; Hayashi, R. K.; West, R. Reaction of a Stable Silylene With Covalent Azides: A New Synthesis for Silaimines. *J. Am. Chem. Soc.* **1994,** *116*, 10813.

124. (a) Kong, L.; Cui, C. Synthesis and Reactivity of a Base-Free *N*-Heterocyclic Silanimine. *Organometallics* **2010,** *29*, 5738.
(b) Samuel, P. P.; Azhakar, R.; Ghadwal, R. S.; Sen, S. S.; Roesky, H. W., et al. Stable Silaimines With Three- and Four-Coordinate Silicon Atoms. *Inorg. Chem.* **2012,** *51*, 11049.

125. Iwamoto, T.; Ohnishi, N.; Gui, Z.; Ishida, S.; Isobe, H.; Maeda, S., et al. Synthesis and Structure of Stable Base-Free Dialkylsilanimines. *New J. Chem.* **2010,** *34*, 1637.

126. Ghadwal, R. S.; Roesky, H. W.; Schulzke, C.; Granitzk, M. *N*-Heterocyclic Carbene Stabilized Dichlorosilaimine $\text{IPr}\cdot\text{Cl}_2\text{Si{=}NR}$. *Organometallics* **2010,** *29*, 6329.

127. (a) Azhakar, R.; Roesky, H. W.; Holstein, J. J.; Pröpper, K.; Dittrich, B. Reactivity Studies of Heteroleptic Silylenes $\text{PhC(N}t\text{Bu)}_2\text{SiX}$ ($\text{X} = \text{NPh}_2, \text{NMe}_2$) Toward Selected Azides. *Organometallics* **2013,** *32*, 358.
(b) Zhang, S.-H.; Yeong, H.-X., et al. Reactivity of a Silylsilylene Bearing a Functionalized Diaminochlorosilyl Substituent. *Chem. Eur. J.* **2011,** *17*, 3490.

128. Ghadwal, R. S.; Roesky, H. W.; Pröpper, K.; Dittrich, B.; Klein, S.; Frenking, G. A Dimer of Silaisonitrile With Two-Coordinate Silicon Atoms. *Angew. Chem. Int. Ed.* **2011,** *50*, 5374.

129. Khan, S.; Sen, S. S.; Kratzert, D.; Tavcar, G.; Roesky, H. W.; Stalke, D. Synthesis of Stable Silicon Heterocycles by Reaction of Organic Substrates With a Chlorosilylene $[\text{PhC(N}t\text{Bu)}_2\text{SiCl}]$. *Chem. Eur. J.* **2011,** *17*, 4283.

130. Zhang, S.-H.; Xi, H.-W.; Lim, K. H.; Meng, Q.; Huang, M.-B.; So, C.-W. Synthesis and Characterization of a Singlet Delocalized 2,4-Diimino-1,3-disilacyclobutanediyl and a Silylenylsilaimine. *Chem. Eur. J.* **2012,** *18*, 4258.

131. Mück, F. M.; Ulmer, A.; Baus, J. A.; Burschka, C.; Tacke, R. Thermally Stable Four-Coordinate Silicon (IV) Complexes With an Si=N Double Bond and an SiN_3El Skeleton (El = O, S, Se, Te). *Eur. J. Inorg. Chem.* **2015,** 1860.

132. (a) Cui, H.; Cui, C. Silylation of *N*-Heterocyclic Carbene With Aminochlorosilane and -Disilane: Dehydrohalogenation vs. Si–Si Bond Cleavage. *Dalton Trans.* **2011,** 11937.
(b) Cui, H.; Cui, C. Synthesis of a Base-Stabilized 1-Hydrosilanimine via NHC-Mediated Dehydrohalogenation of Hydrochlorosilane. *Chem. Asian J.* **2011,** *6*, 1138.

133. Cui, H.; Ma, B.; Cui, C. Metal-Free, Stereospecific Bis-Silylation of Functionalized Alkynes With NHC-Supported Silylaminosilylene. *Organometallics* **2012,** *31*, 7339.

134. Smit, C. N.; Lock, F. M.; Bickelhaupt, F. 2,4,6-Tri-*tert*-butylphenylphosphene-dimesitylsilene; The First Phosphasilaalkene. *Tetrahedron. Lett.* **1984,** *25*, 3011.

135. (a) Smit, C. N.; Bickelhaupt, F. Phosphasilenes: Synthesis and Spectroscopic Characterization. *Organometallics* **1987,** *6*, 1156.
(b) van den Winkel, Y.; Bastiaans, H. M. M.; Bickelhaupt, F. Phosphasilene Synthesis and Reactivity: An Improved Route to 1-(2,4,6-Tri-*tert*-butylphenyl)-2-*tert*-butyl-2-(2,4,6-tri-isopropylphenyl) phosphasilene. *J. Organomet. Chem.* **1991,** *405*, 183.

136. Driess, M. 1,2,3-Triphospha-4-silabicyclo[1.1.0]butanes From Activated, Stable Phosphasilenes and White Phosphorus. *Angew. Chem. Int. Ed.* **1991,** *30*, 1022.

137. Niecke, E.; Klein, E.; Niegev, M. Synthesis, Structure, and Reactivity of a 1,3-Diphospha-2-silaallyl Anion. *Angew. Chem. Int. Ed.* **1989,** *28*, 751.

138. Bender, H. R. G.; Niecke, E.; Nieger, M. The First X-ray Structure of a Phosphasilene: 1,3,4-Triphospha-2-sila-1-butene. *J. Am. Chem. Soc.* **1993,** *115*, 3314.

139. Driess, M.; Rell, S.; Pritzkow, H. First Structural Characterization of Silicon-Arsenic and Silicon-Phosphorus Multiple Bonds in Silylated Silylidene-Arsanes and -Phosphanes: X-ray Structure of a Tellura-Arsasilirane Derivative. *J. Chem. Soc., Chem. Commun.* **1995,** 253.

140. (a) Dykema, K. J.; Truong, T. N.; Gordon, M. Studies of Silicon-Phosphorus Bonding. *J. Am. Chem. Soc.* **1985,** *107*, 4535.
(b) Lee, L.-J.; Boggs, J. E.; Cowley, A. H. Phosphasilene ($HP{=}SiH_2$) and Its Valence Isomers. A Theoretical Study. *J. Chem. Soc., Chem. Commun.* **1985,** 773.

141. Driess, M.; Janoschek, R. A Comparison of Silylidene-Amines, -Phosphanes and -Arsanes: Syntheses and Quantum Chemical Calculations. *J. Mol. Struct.: THEOCHEM* **1994,** *313*, 129.

142. (a) Driess, M.; Block, S.; Brym, M.; Gamer, M. T. Synthesis of A "Half"-Parent Phosphasilene $R_2Si{=}PH$ and Its Metalation to the Corresponding P-Zinciophosphasilene [$R_2Si = PM$]. *Angew. Chem. Int. Ed.* **2006,** *45*, 2293.
(b) Yao, S.; Block, S.; Brym, M., et al. A New Type of Heteroleptic Complex of Divalent Lead and Synthesis of the P-Plumbyleniophosphasilene, $R_2Si{=}P$-Pb(L): (L = β-diketiminate). *Chem. Commun.* **2007,** 3844.

143. Hansen, K.; Szilvasi, T.; Blom, B.; Inoue, S.; Epping, J.; Driess, M. A Fragile Zwitterionic Phosphasilene as a Transfer Agent of the Elusive Parent Phosphinidene (:PH). *J. Am. Chem. Soc.* **2013,** *135*, 11795.

144. Hansen, K.; Szilvási, T.; Blom, B.; Irran, E.; Driess, M. A Donor-Stabilized Zwitterionic "Half-Parent" Phosphasilene and Its Unusual Reactivity Towards Small Molecules. *Chem. Eur. J.* **1947,** *2014*, 20.

145. Li, B.; Matsuo, T.; Hashizume, D.; Fueno, H.; Tanaka, K.; Tamao, K. π-Conjugated Phosphasilenes Stabilized by Fused-Ring Bulky Groups. *J. Am. Chem. Soc.* **2009,** *131*, 13222.

146. Driess, M.; Rell, S.; Pritzkow, H.; Janoschek, R. $R_2Si = P\text{-}SiR_2F$: 1,3-Sigmatropic Migration of Fluorine in a 2-Phospha-1,3-disilaallyl Derivative Capable of Conjugation and Its Conversion to Phosphadisilacyclopropanes. *Angew. Chem. Int. Ed.* **1997,** *36*, 1326.

147. Willmes, P.; Cowley, M. J.; Hartmann, M.; Zimmer, M.; Huch, V.; Scheschkewitz, D. From Disilene (Si=Si) to Phosphasilene (Si=P) and Phosphacumulene (P=C=N). *Angew. Chem. Int. Ed.* **2014,** *53*, 2216.
148. Lee, V. Ya; Kawai, M.; Sekiguchi, A.; Ranaivonjatovo, H.; Escudié, J. A. "Push – Pull" Phosphasilene and Phosphagermene and Their Anion-Radicals. *Organometallics* **2009,** *28*, 4262.
149. Robinson, T. P.; Cowley, M. J.; Scheschkewitz, D.; Goicoechea, J. M. Phosphide Delivery to a Cyclotrisilene. *Angew. Chem. Int. Ed.* **2015,** *54*, 683.
150. Sen, S. S.; Khan, S.; Roesky, H. W.; Kratzert, D.; Meindl, K.; Henn, J., et al. Zwitterionic Si-C-Si-P and Si-P-Si-P Four-Membered Rings With Two-Coordinate Phosphorus Atoms. *Angew. Chem. Int. Ed.* **2011,** *50*, 2322.
151. Inoue, S.; Wang, W.; Präsang, C.; Asay, M.; Irran, E.; Driess, M. An Ylide-like Phosphasilene and Striking Formation of a 4-Electron, Resonance-Stabilized 2,4-Disila-1,3-diphosphacyclobutadiene. *J. Am. Chem. Soc.* **2011,** *133*, 2868.
152. Khan, S.; Michel, R.; Sen, S. S.; Roesky, H. W.; Stalke, D. A P_4 Chain and Cage From Silylene-Activated White Phosphorus. *Angew. Chem. Int. Ed.* **2011,** *50*, 11786.
153. Xiong, Y.; Yao, S.; Müller, R.; Kaupp, M.; Driess, M. From Silylone to an Isolable Monomeric Silicon Disulfide Complex. *Angew. Chem. Int. Ed.* **2015,** *54*, 10254.
154. Roy, S.; Stollberg, P.; Herbst-Irmer, R.; Stalke, D.; Andrada, D. M.; Frenking, G., et al. Carbene Supported Dimer of Heavier Ketenimine Analogue With P and Si Atoms. *J. Am. Chem. Soc.* **2015,** *137*, 150.
155. Roy, S.; Dittrich, B.; Monda, T.; Koley, D.; Stuückl, A. C.; Schwederski, B., et al. Carbene-Dichlorosilylene Stabilized Phosphinidenes Exhibiting Strong Intramolecular Charge Transfer Transition. *J. Am. Chem. Soc.* **2015,** *137*, 6180.
156. Geiß, D. I; Arz, M. I.; Straßmann, M.; Schnakenburg, G.; Filippou, A. C. Si=P Double Bonds: Experimental and Theoretical Study of an NHC-Stabilized Phosphasilenylidene. *Angew. Chem. Int. Ed.* **2015,** *54*, 2739.
157. Hoffmann, R. W. Wittig and His Accomplishments: Still Relevant Beyond His 100th Birthday. *Angew. Chem. Int. Ed.* **2001,** *40*, 1411.
158. (a) Ando, W.; Sekiguch, A.; Hagiwara, K.; Sakakibara, A.; Yoshida, H. Lewis Base Adducts to Diorganosilylenes. *Organometallics* **1988,** *7*, 558.
(b) Gillette, G. R.; Noren, G. H., et al. Lewis Base Adducts to Diorganosilylenes. *Organometallics* **1989,** *8*, 487.
159. Gau, D.; Rodriguez, R.; Kato, T.; Saffon-Merceron, N.; Baceiredo, A. Diastereoselective Synthesis of Bulky, Strongly Nucleophilic, and Configurationally Stable P-Stereogenic Tricyclic Phosphines. *J. Am. Chem. Soc.* **2010,** *132*, 12841.
160. (a) Rodriguez, R.; Gau, D.; Kato, T.; Saffon-Merceron, N.; De Cózar, A.; Cossío, F. P. Baceiredo, Reversible Binding of Ethylene to Silylene–Phosphine Complexes at Room Temperature. *Angew. Chem. Int. Ed.* **2011,** *50*, 10414.
(b) Rodriguez, R., et al. Synthesis of a Phosphine-Stabilized Silicon(II) Hydride and Its Addition to Olefins: A Catalyst-Free Hydrosilylation Reaction. *Angew. Chem. Int. Ed.* **2011,** *50*, 11492.
(c) Rodriguez, R.; Contie, Y.; Gau, D.; Saffon-Merceron, N.; Miqueu, K.; Sotiropoulos, J.-M.; Baceiredo, A.; Kato, T. Reversible Insertion of Unactivated Alkenes into Silicon(II)-Tin Bonds. *Angew. Chem. Int. Ed.* **2013,** *52*, 8437.
(d) Rodriguez, R.; Contie, Y.; Mao, Y.; Saffon-Merceron, N.; Baceiredo, A.; Branchadell, V.; Kato, T. Reversible Dimerization of Phosphine-Stabilized Silylenes by Silylene Insertion into Si^{II}–H and Si^{II}–Cl σ-Bonds at Room Temperature. *Angew. Chem. Int. Ed.* **2015,** *54*, 15276.
161. Driess, M.; Pritzkow, H.; Rell, S.; Winkler, U. Synthesis and Unusual Reactivity of Compounds Containing Silicon-Phosphorus and Silicon-Arsenic Double Bonds: New Silylidenephosphanes and -Arsanes of the Type $R_2Si{=}E(SiR_3)$ (E = P, As). *Organometallics* **1996,** *15*, 1845.

162. Driess, M.; Pritzkow, H. An Arsanylidensilane ("Arsasilene") and Its Derivatization With Tellurium and Benzophenone. *Angew. Chem. Int. Ed.* **1992,** *31*, 316.
163. Lee, V. Ya; Aoki, S.; Kawai, M.; Meguro, T.; Sekiguchi, A. Stibasilene Sb=Si and Its Lighter Homologues: A Comparative Study. *J. Am. Chem. Soc.* **2014,** *136*, 6243.
164. Pyykkö, P.; Atsumi, M. Molecular Single-Bond Covalent Radii for Elements 1-118. *Chem. Eur. J.* **2009,** *15*, 186.
165. Präsang, C.; Stoelzel, M.; Inoue, S.; Meltzer, A.; Driess, M. Metal-Free Activation of EH_3 (E = P, As) by an Ylide-like Silylene and Formation of a Donor-Stabilized Arsasilene With a HSi=AsH Subunit. *Angew. Chem. Int. Ed.* **2010,** *49*, 10002.
166. Brook, M. A. *Silicon in Organic, Organometallic, and Polymer Chemistry;* WileyVCH: Weinheim, 2000.
167. (a) Withnall, R.; Andrews, L. Infrared Spectroscopic Evidence for Silicon-Oxygen Double Bonds: Silanone and the Silanoic and Silicic Acid Molecules. *J. Am. Chem. Soc.* **1985,** *107*, 2567.
(b) Withnall, R.; Andrews, L. Infrared Spectroscopic Evidence for Silicon-Oxygen Double Bonds: Methyl- and Dimethylsilanones in Solid Argon. *J. Am. Chem. Soc.* **1986,** *108*, 8118.
(c) Withnall, R.; Andrews, L. Matrix Reactions of Methylsilanes and Oxygen Atoms. *J. Phys. Chem.* **1988,** *92*, 594.
(d) Schnöckel, H. Matrix Reactions of SiO. IR-Spectroscopic Identification of the Molecules SiO_2 and $OsiCl_2$. *Zeitschrift für Anorganische und Allgemeine Chemie* **1980,** *460*, 37.
(e) Schnöckel, H. Matrix Reaction von SiO mit F_2. IR-Spektroskopischer Nachweis Von Molekularem $OSiF_2$. *J. Mol. Struct.* **1980,** *65*, 115.
(f) Khabashesku, V. N.; Kerzina, Z. A.; Baskir, E. G., et al. Direct Spectroscopic Study of Unstable Molecules With Silicon Oxygen Multiple Bonds: Low Temperature Matrix Stabilization of $(CH_3)_2Si{=}O$ and $(CD_3)_2Si{=}O$ in the Gas Phase. *J. Organomet. Chem.* **1988,** *347*, 277.
(g) Khabashesku, V. N.; Kerzina, Z. A.; Kudin, K. N.; Nefedov, O. M. Matrix Isolation Infrared and Density Functional Theoretical Studies of Organic Silanones, $(CH_3O)_2Si{=}O$ and $(C_6H_5)_2Si{=}O$. *J. Organomet. Chem.* **1998,** *566*, 45.
(h) Khabashesku, V. N.; Kudin, K. N.; Margrave, J. L.; Fredin, L. Matrix Isolation Infrared and Density Functional Theoretical Studies of Cryogenic Reactions of Silicon Atoms With Dimethyl Ether and Methanol: New Route to Generation and Stabilization of Transient Silanones. *J. Organomet. Chem.* **2000,** *595*, 248.
(i) Bornemann, H.; Sander, W. Oxidation of Methyl(phenyl)silylenes-Synthesis of a Dioxasilirane. *J. Am. Chem. Soc.* **2000,** *122*, 6727.
(j) Karni, M.; Apeloig, Y.; Kapp, J.; Schleyer, P. v. R. Theoretical Aspects of Compounds Containing Si, Ge, Sn and Pb. In *The Chemistry of Organic Silicon Compounds;* Rappoport, Z., Apeloig, Y., Eds.; Vol. 3; John Wiley & Sons, Ltd.: Chichester, 2001; p 1. Chapter 1.
168. Yao, S.; Brym, M.; van Wüllen, C.; Driess, M. From s Stable Silylene to a Mixed-Valent Disiloxane and an Isolable Silaformamide-Borane Complex With Considerable Silicon-Oxygen Double-Bond Character. *Angew. Chem. Int. Ed.* **2007,** *46*, 4159.
169. Yao, S.; Xiong, Y.; Brym, M.; Driess, M. An Isolable Silanoic Ester by Oxygenation of a Stable Silylene. *J. Am. Chem. Soc.* **2007,** *129*, 7268.
170. (a) Xiong, Y.; Yao, S.; Driess, M. An Isolable NHC-Supported Silanone. *J. Am. Chem. Soc.* **2009,** *131*, 7562.
(b) Yao, S.; Xiong, Y.; Driess, M. *N*-Heterocyclic Carbene (NHC)-Stabilized Silanechalcogenones: NHC- > $Si(R_2)$ = E (E = O, S, Se, Te). *Chem. Eur. J.* **2010,** *16*, 1281.
(c) Yao, S.; Xiong, Y., et al. Synthesis and Rearrangement of Stable NHC- > Silylene Adducts and Their Unique Reactivity Towards Cyclohexylisocyanide. *Chem. Asian J.* **2010,** *5*, 322.
171. (a) Becerra, R.; Bowes, S.-J.; Ogden, J. S.; Cannady, J. P.; Adamovic, I.; Gordon, M. S., et al. Time-Resolved Gas-Phase Kinetic and Quantum Chemical Studies of the Reaction of Silylene With Oxygen. *Phys. Chem. Chem. Phys.* **2005,** *7*, 2900.

(b) Patyk, A.; Sander, W.; Gauss, J.; Cremer, D. Dimethyldioxasilirane. *Angew. Chem. Int. Ed.* **1989,** *28*, 898.

172. Xiong, Y.; Yao, S.; Müller, R.; Kaupp, M.; Driess, M. From Silicon(II)-Based Dioxygen Activation to Adducts of Elusive Dioxasiliranes and Sila-Ureas Stable At Room Temperature. *Nat. Chem.* **2010,** *2*, 577.
173. Rodriguez, R.; Gau, D.; Troadec, T.; Saffon-Merceron, N.; Branchadell, V.; Baceiredo, A., et al. A Base-Stabilized Sila—Lactone and a Donor/Acceptor-Stabilized Silanoic Acid. *Angew. Chem. Int. Ed.* **2013,** *52*, 8980.
174. Ahmad, S. U.; Szilvasi, T.; Irran, E.; Inoue, S. An NHC-Stabilized Silicon Analogue of Acylium Ion: Synthesis, Structure, Reactivity, and Theoretical Studies. *J. Am. Chem. Soc.* **2015,** *137*, 5828.
175. Xiong, Y.; Yao, S.; Irran, E.; Driess, M. From a Stable Silanone Complex to Isolable, Donor-Supported Silicoxonium Halides $[LSi(dmap) = O\text{-}SiMe_3]^+X^-$. *Chem. Eur. J.* **2011,** *17*, 11274.
176. Xiong, Y.; Yao, S.; Müller, R.; Kaupp, M.; Driess, M. Activation of Ammonia by a Si=O Double Bond and Formation of a Unique Pair of Sila-Hemiaminal and Silanoic Amide Tautomers. *J. Am. Chem. Soc.* **2010,** *132*, 6912.
177. (a) Cordes, M. H. J.; Berson, J. A. Thermal Interconversion of a Pair of Diastereomeric Cyclopropanones. An Upper Limit for a Cyclopropanone-Oxyallyl Energy Separation. *J. Am. Chem. Soc.* **1992,** *114*, 11010.
(b) Ichino, T.; Villan, S. M.; Gianola, A. J.; Goebbert, D. J.; Velarde, L., et al. The Lowest Singlet and Triplet States of the Oxyallyl Diradical. *Angew. Chem. Int. Ed.* **2009,** *48*, 8509.
178. (a) Noyori, R. Organic Syntheses via the Polybromo Ketone-Iron Carbonyl Reaction. *Acc. Chem. Res.* **1979,** *12*, 61.
(b) Takaya, H.; Makino, S.; Hayakawa, Y.; Noyori, R. Carbon-Carbon Bond Formation Promoted by Transition Metal Carbonyls. 17. Reactions of Polybromo Ketones With 1,3-Dienes in the Presence of Iron Carbonyls. New $3 + 4 \rightarrow 7$ Cyclocoupling Reaction Forming 4-Cycloheptenones. *J. Am. Chem. Soc.* **1978,** *100*, 1765.
(c) Hayakawa, Y.; Yokoyama, K.; Noyori, R. Carbon-Carbon Bond Formation Promoted by Transition Metal Carbonyls. 20. Iron Carbonyl Promoted Reaction of α,α'-dibromo Ketones and Aromatic Olefins Leading to 3-Arylcyclopentanones. The [3 + 2] Cycloaddition Involving an Allylic Cation. *J. Am. Chem. Soc.* **1978,** *100*, 1791.
(d) Hayakawa, Y., et al. Carbon-Carbon Bond Formation Promoted by Transition Metal Carbonyls. 19. General Synthesis of Tropane Alkaloids via the Polybromo Ketone-Iron Carbonyl Reaction. *J. Am. Chem. Soc.* **1978,** *100*, 1786.
(e) Lee, K.; Cha, J. K. A New Approach to Phorbol by [4 + 3] Oxyallyl Cycloaddition and Intramolecular Heck Reaction. *Org. Lett.* **1999,** *1*, 523.
179. Hirano, T.; Kumagai, T.; Miyashi, T.; Akiyama, K.; Ikegama, Y. Novel Photoreactions of Benzhydrylidenequadricyclane and Quadricyclanone: A New Route to Trimethylenemethane and Oxyallyl Derivatives. *J. Org. Chem.* **1991,** *56*, 1907.
180. Bhargava, S.; Hou, J.; Parvez, M.; Sorensen, T. S. Preparation and Crystal Structure of Two Bicyclo [1.1.0]butan-2-ones: A Hybrid Oxyallyl−Cyclopropanone Motif. *J. Am. Chem. Soc.* **2005,** *127*, 3704.
181. Rodriguez, R.; Troadec, T.; Gau, D.; Saffon-Merceron, N.; Hashizume, D.; Miqueu, K., et al. Synthesis of a Donor-Stabilized Silacyclopropan-1-one. *Angew. Chem. Int. Ed.* **2013,** *52*, 4426.
182. Moiseev, A. G.; Abe, M.; Danilov, E. O.; Neckers, D. C. First Direct Detection of 2,3-Dimethyl-2,3-diphenylcyclopropanone. *J. Org. Chem.* **2007,** *72*, 2777.
183. Ghadwal, R. S.; Azhakar, R.; Roesky, H. W.; Pröpper, K.; Dittrich, B.; Goedecke, C., et al. Donor−Acceptor Stabilized Silaformyl Chloride. *Chem. Commun.* **2012,** *48*, 8186.
184. Ghadwal, R. S.; Azhakar, R.; Roesky, H. W.; Pröpper, K.; Dittrich, B.; Klein, S., et al. Donor−Acceptor-Stabilized Silicon Analogue of an Acid Anhydride. *J. Am. Chem. Soc.* **2011,** *133*, 17552.
185. Xiong, Y.; Yao, S.; Driess, M. Silicon Analogues of Carboxylic Acids: Synthesis of Isolable Silanoic Acids by Donor−Acceptor Stabilization. *Angew. Chem. Int. Ed.* **2010,** *49*, 6642.

186. Peterson, D. J. Carbonyl Olefination Reaction Using Silyl-Substituted Organometallic Compounds. *J. Org. Chem.* **1968,** *33*, 780.
187. Xiong, Y.; Yao, S.; Driess, M. Coordination of a Si = O Subunit to Metals: Complexes of Donor-Stabilized Silanone Featuring a Terminal Si = O → M Coordination (M = Zn, Al). *Dalton Trans.* **2010,** *39*, 9282.
188. (a) Branch, C. S.; Poppel, L. C.; Bott, S. G.; Barron, A. R. Molecular Structure of $(t\text{Bu})_3\text{Al}[\text{O}{=}\text{C}(\text{OPh})_2]$. *J. Chem. Crystallogr.* **1999,** *29*, 993 and references cited therein.
   (b) Power, M. B.; Bott, S. C.; Atwood, J. L., et al. π-Face Selectivity of Coordinated Ketones to Nucleophilic Additions: The Importance of Aluminum-Oxygen π-Bonding. *J. Am. Chem. Soc.* **1990,** *112*, 3446.
   (c) van Vliet, M. R. P.; van Koten, G.; Rotteveel, M. A.; Schrap, M.; Vrieze, K.; Kojic-Prodic, B.; Spek, A. L.; Duisenberg, A. J. M. Reactivity of 1-Aza-4-oxa-1,3-butadienes (α-Imino Ketones) Toward Triorganoaluminum Reagents. 1. X-ray Crystal Structures of $[(\text{AlMe}_3)_2\{\sigma,\sigma\text{-N,O-(MeN:C(Ph)C(Ph):O)}\}]$ With a N,O-Bridge-Bonded α-Imino Ketone, and of Cyclic $\text{Me}_2\text{Al}(\textit{tert}\text{-Bu})\text{N:CHC}(\text{Me})_2\text{OAlMe}_3$, A Carbonyl Alkylated Product Which is O-Coordinated to AlMe_3. *Organometallics* **1986,** *5*, 1389.
189. Muraoka, T.; Nakamura, T.; Nakamura, A.; Ueno, K. Reactions of (Silyl)(silylene)tungsten and -Molybdenum Complexes With Sulfur Reagents. *Organometallics* **2010,** *29*, 6624.
190. Muraoka, T.; Abe, K.; Haga, Y.; Nakamura, T.; Ueno, K. Synthesis of a Base-Stabilized Silanone-Coordinated Complex by Oxygenation of a (Silyl)(silylene)tungsten Complex. *J. Am. Chem. Soc.* **2011,** *133*, 15365.
191. Wang, Y.; Chen, M.; Xie, Y.; Wei, P.; Schaefer, H. F., III; Schleyer, P. v. R., et al. Stabilization of elusive silicon oxides. *Nat. Chem.* **2015,** *7*, 509.
192. Wang, Y.; Chen, M.; Xie, Y.; Wei, P.; Schaefer, H. F., III; Robinso, G. H. Stabilization of Silicon−Carbon Mixed Oxides. *J. Am. Chem. Soc.* **2015,** *137*, 8396.
193. Bogey, M.; Delcroix, B.; Walters, A.; Guillemin, J.-C. Experimentally Determined Structure of H_2SiO by Rotational Spectroscopy and Isotopic Substitution. *J. Mol. Spectrosc.* **1996,** *175*, 421.
194. Filippou, A. C.; Baars, B.; Chernov, O.; Lebedev, Y. N.; Schnakenburg, G. Silicon−Oxygen Double Bonds: A Stable Silanone With a Trigonal-Planar Coordinated Silicon Center. *Angew. Chem. Int. Ed.* **2014,** *53*, 565.
195. Arrington, C. A.; West, R.; Michl, J. On the Proposed Thermal Interconversion of Matrix-Isolated Dimethylsilylene and 1-Methylsilene: Their Reactions With Oxygen Atom Donors. *J. Am. Chem. Soc.* **1983,** *105*, 6176.
196. Ishida, S.; Abe, T.; Hirakawa, F.; Kosai, T.; Sato, K.; Kira, M., et al. Persistent Dialkylsilanone Generated by Dehydrobromination of Dialkylbromosilanol. *Chem. Eur. J.* **2015,** *21*, 15100.
197. Linden, M. M.; Reisenauer, H. P.; Gerbig, D.; Karni, M.; Schäfer, A. T. M.; Apeloig, Y., et al. Preparation of a Silanone Through Oxygen Atom Transfer to a Stable Cyclic Silylene. *Angew. Chem. Int. Ed.* **2015,** *54*, 12404.
198. Suzuki, H.; Tokitoh, N.; Okazaki, R.; Nagase, S.; Goto, M. Synthesis, Structure, and Reactivity of the First Kinetically Stabilized Silanethione. *J. Am. Chem. Soc.* **1998,** *120*, 11096.
199. Kudo, T.; Nagase, S. Theoretical Study of Silanethione ($H_2Si{=}S$) in the Ground, Excited, and Protonated States. Comparison With Silanone ($H_2Si{=}O$). *Organometallics* **1986,** *5*, 1207.
200. Arya, P.; Boye, J.; Carré, F.; Corriu, R.; Lanneau, G.; Lapasset, J., et al. Formation and Reactivity of Silicon−Sulfur and Silicon−Selenium Double Bonds. The First X-Ray Structure of a Silanethione. *Angew. Chem. Int. Ed.* **1989,** *28*, 1016.
201. Suzuki, H.; Tokitoh, N.; Nagase, S.; Okazaki, R. Formation and Reactivity of Silicon−Sulfur and Silicon−Selenium Double Bonds. The First X-Ray Structure of a Silanethione. *J. Am. Chem. Soc.* **1994,** *116*, 11578.
202. Iwamoto, T.; Sato, K.; Ishida, S.; Kabuto, C.; Kira, M. Synthesis, Properties, and Reactions of a Series of Stable Dialkyl-Substituted Silicon−Chalcogen Doubly Bonded Compounds. *J. Am. Chem. Soc.* **2006,** *128*, 16914.

203. Iwamoto, T.; Masuda, H.; Ishida, S.; Kabuto, C.; Kira, M. Addition of Stable Nitroxide Radical to Stable Divalent Compounds of Heavier Group 14 Elements. *J. Am. Chem. Soc.* **2003,** *125*, 9300.
204. Yeong, H.-X.; Xi, H.-W.; Li, Y.; Lim, K. H.; So, C.-W. A Silyliumylidene Cation Stabilized by an Amidinate Ligand and 4-Dimethylaminopyridine. *Chem. Eur. J.* **2013,** *19*, 11786.
205. Xiong, Y.; Yao, S.; Inoue, S.; Irran, E.; Driess, M. The Elusive Silyliumylidene $[ClSiD]^+$ and Silathionium $[ClSi{=}S]^+$ Cations Stabilized by Bis(Iminophosphorane) Chelate Ligand. *Angew. Chem. Int. Ed.* **2012,** *51*, 10074.
206. Junold, K.; Baus, J. A.; Burschka, C.; Auerhammer, D.; Tacke, R. Stable Five-Coordinate Silicon(IV) Complexes With SiN4X Skeletons (X = S, Se, Te) and Si=X Double Bonds. *Chem. Eur. J.* **2012,** *18*, 16288.
207. Azhakar, R.; Roesky, H. W.; Wolf, H.; Stalke, D. On the Reactivity of the Silylene $PhC(N\textit{t}Bu)_2SiNPh_2$ Toward Organic Substrates. *Z. Anorg. Allg. Chem.* **2013,** *639*, 934.
208. So, C.-W.; Roesky, H. W.; Oswald, R. B.; Pala, A.; Jones, P. G. Synthesis and Characterization of $[\{PhC(N\textit{t}Bu)_2\}Si(S)S\textit{t}Bu]$: A Silicon Thioester Analogue With the Si(=S)-S-Skeleton. *Dalton Trans.* **2007,** 5241.
209. Meltzer, A.; Inoue, S.; Präsang, C.; Driess, M. Steering S-H and N-H Bond Activation by a Stable N-Heterocyclic Silylene: Different Addition of H_2S, NH_3, and Organoamines on a Silicon(II) Ligand versus Its $Si(II) \rightarrow Ni(CO)_3$ Complex. *J. Am. Chem. Soc.* **2010,** *132*, 3038.
210. Hansen, K.; Szilvási, T.; Blom, B.; Irran, E.; Driess, M. A Donor-Stabilized Zwitterionic "Half-Parent" Phosphasilene and Its Unusual Reactivity Towards Small Molecules. *Chem. Eur. J.* **1947,** *2014*, 20.
211. Tan, G.; Xiong, Y.; Inoue, S.; Enthaler, S.; Blom, B.; Epping, J. D., et al. *Chem. Commun.* **2013,** *49*, 5595.
212. (a) Schnöckel, H.; Köppe, R. Matrix IR Spectrum and ab Initio SCF Calculations of Molecular SiS_2. *J. Am. Chem. Soc.* **1989,** *111*, 4583.
(b) Friesen, M.; Schnöckel, H. l Raman Spectrum and Bonding of Matrix Isolated, Molecular Silicon Disulfide. *Z. Anorg. Allg. Chem.* **1999,** *625*, 1097.
213. Xiong, Y.; Yao, S.; Müller, R.; Kaupp, M.; Driess, M. From Silylone to an Isolable Monomeric Silicon Disulfide Complex. *Angew. Chem. Int. Ed.* **2015,** *54*, 10254.
214. Tokitoh, N.; Sadahiro, T.; Hatano, K.; Sasaki, T.; Takeda, N.; Okazaki, R. Synthesis of Kinetically Stabilized Silaneselone and Silanetellone. *Chem. Lett.* **2002,** *31*, 34.
215. Iwamoto, T.; Sato, K.; Ishida, S.; Kabuto, C.; Kira, M. Synthesis, Properties, and Reactions of a Series of Stable Dialkyl-Substituted Silicon-Chalcogen Doubly Bonded Compounds. *J. Am. Chem. Soc.* **2006,** *128*, 16914.
216. Yao, S.; Xiong, Y.; Brym, M.; Driess, M. A Series of Isolable Silanoic Thio-, Seleno-, and Telluroesters (LSi(=X)OR) With Donor-Supported Si=X Double Bonds (L = β-Diketiminate; X = S, Se, Te). *Chem. Asian. J.* **2008,** *3*, 113.
217. Mitra, A.; Wojcik, J. P.; Lecoanet, D.; Müller, T.; West, R. A Bis(silaselenone) With Two Donor-Stabilized Si=Se Bonds From an Unexpected Stereoconvergent Hydrolysis of a Disilenadisiletane. *Angew. Chem. Int. Ed.* **2009,** *48*, 4069.
218. See ref. 3 (a).;
(b) Sekiguchi, A.; Ichinohe, M.; Kinjo, R. The Chemistry of Disilyne With a Genuine Si−Si Triple Bond: Synthesis, Structure, and Reactivity. *Bull. Chem. Soc. Jpn.* **2006**, *79*, 825.
(c) Power, P. P. Bonding and Reactivity of Heavier Group 14 Element Alkyne Analogues. *Organometallics* **2007,** *26*, 4362.
(d) Sekiguchi, A. Disilyne With A Silicon-Silicon Triple Bond: A New Entry to Multiple Bond Chemistry. *Pure Appl. Chem.* **2008,** *80*, 447.
(e) Power, P. P. Main-Group Elements as Transition Metals. *Nature* **2010,** *463*, 171.
(f) Asay, M.; Sekiguchi, A. *Bull. Chem. Soc. Jpn.* **2012,** *85*, 1245.
219. Pu, L.; Twamley, B.; Power, P. P. Synthesis and Characterization of $2,6\text{-}Trip_2H_3C_6PbPbC_6H_3\text{-}2,6\text{-}Trip_2$ (Trip = $C_6H_2\text{-}2,4,6\text{-}\textit{i}\text{-}Pr_3$): A Stable Heavier Group 14 Element Analogue of an Alkyne. *J. Am. Chem. Soc.* **2000,** *122*, 3524.

220. Jung, Y.; Brynda, M.; Power, P. P.; Head-Gordon, M. Ab Initio Quantum Chemistry Calculations on the Electronic Structure of Heavier Alkyne Congeners: Diradical Character and Reactivity. *J. Am. Chem. Soc.* **2006,** *128*, 7185.

221. Sekiguchi, A.; Kinjo, R.; Ichinohe, M. A Stable Compound Containing a Silicon-Silicon Triple Bond. *Science* **2004,** *305*, 1755.

222. (a) Takagi, N.; Nagase, S. Theoretical Study of an Isolable Compound With a Short Silicon-Silicon Triple Bond, $(tBu_3Si)_2MeSiSi{\equiv}SiSiMe(SitBu_3)_2$. *Eur. J. Inorg. Chem.* **2002,** 2775.
(b) Wiberg, N.; Vasisht, S. K.; Fischer, G.; Mayer, Z. A Relatively Stable Disilyne RSi≡SiR (R = SiMe $(SitBu_3)_2$). *Z. Anorg. Allg. Chem.* **2004,** *630*, 1823.

223. Murata, Y.; Ichinohe, M.; Sekiguchi, A. Unsymmetrically Substituted Disilyne $Dsi_2iPrSi\text{-}Si{\equiv}Si\text{-}SiNpDsi_2$ ($Np = CH_2tBu$): Synthesis and Characterization. *J. Am. Chem. Soc.* **2010,** *132*, 16768.

224. (a) Sasamori, T.; Hironaka, K.; Sugiyama, Y.; Takagi, N.; Nagase, S.; Hosoi, Y., et al. Synthesis and Reactivity of a Stable 1,2-Diaryl-1,2-dibromodisilene: A Precursor for Substituted Disilenes and a 1,2-Diaryldisilyne. *J. Am. Chem. Soc.* **2008,** *130*, 13856.
(b) Sasamori, T.; Han, J. S.; Hironaka, K.; Takagi, N.; Nagase, S.; Tokitoh, N. Synthesis and Structure of Stable 1,2-Diaryldisilyne. *Pure Appl. Chem.* **2010,** *82*, 603.

225. (a) Sen, S. S.; Jana, A.; Roesky, H. W.; Schulzke, C. A Remarkable Base-Stabilized Bis(silylene) With a Silicon(I)-Silicon(I) Bond. *Angew. Chem. Int. Ed.* **2009,** *48*, 8536.
(b) Yeong, H.-X.; Lau, K. C.; Xi, H.-W., et al. Reactivity of a Disilylene $[\{PhC(NtBu)_2\}Si]_2$ Toward Bromine: Synthesis and Characterization of a Stable Monomeric Bromosilylene. *Inorg. Chem.* **2010,** *49*, 371.
(c) Sen, S. S.; Tavcar, G.; Roesky, H. W. Synthesis of a Stable Four-Membered Si_2O_2 Ring and a Dimer With Two Four-Membered Si_2O_2 Rings Bridged by Two Oxygen Atoms, With Five-Coordinate Silicona Atoms in Both Ring Systems. *Organometallics* **2010,** *29*, 2343.
(d) Tavcar, G.; Sen, S. S.; Roesky, H. W.; Hey, J.; Kratzert, D.; Stalke, D. Reactions of a Bis-Silylene (LSi-SiL, L = $PhC(NtBu)_2$) and a Heteroleptic Chloro Silylene (LSiCl) With Benzil: Formation of Bis(siladioxolene) and Monosiladioxolene Analogue With Five-Coordinate Silicon Atoms in Both Ring Systems. *Organometallics* **2010,** *29*, 3930.
(e) Gau, D.; Rodriguez, R.; Kato, T.; Saffon-Merceron, N.; de Cozar, A.; Cossio, F. P.; Baceiredo, A. Synthesis of a Stable Disilyne Bisphosphine Adduct and Its Non-Metal-Mediated CO_2 Reduction to CO. *Angew. Chem. Int. Ed.* **2011,** *50*, 1092.

226. Sen, S. S.; Khan, S.; Nagendran, S.; Roesky, H. W. Interconnected Bis-Silylenes: A New Dimension in Organosilicon Chemistry. *Acc. Chem. Res.* **2012,** *45*, 578.

227. (a) Sekiguchi, A.; Ichinohe, M.; Kinjo, R. The Chemistry of Disilyne With a Guenuine Si-Si Triple Bond: Synthesis, Structure, and Reactivity. *Bull. Chem. Soc. Jpn.* **2006,** *79*, 825.
(b) Kinjo, R.; Ichinohe, M.; Sekiguchi, A.; Takagi, N., et al. Reactivity of a Disilyne RSi≡SiR. ($R = SiiPr[CH(SiMe_3)_2]_2$) Toward π-Bonds: Stereospecific Addition and a New Route to an Isolable 1,2-Disilabenzene. *J. Am. Chem. Soc.* **2007,** *129*, 7766.

228. (a) Han, J. S.; Sasamori, T.; Mizuhata, Y.; Tokitoh, N. Reactivity of an Aryl-Substituted Silicon-Silicon Triple Bond: Reactions of a 1,2-Diaryldisilyne With Alkenes. *J. Am. Chem. Soc.* **2010,** *132*, 2546.
(b) Han, J. S.; Sasamori, T.; Mizuhata, Y., et al. Reactivity of an Aryl-Substituted Silico-Silicon Triple Bond: 1,2-Disilabenzenes From the Reactions of a 1,2-Diaryldisilyne With Alkynes. *Dalton Trans.* **2010,** *39*, 9238.

229. (a) Stegmann, R.; Frenking, G. Theoretical Studies of Organometallic Compounds. 15. Si_2H_2 and $CSiH_2$ Isomers as Ligands in High-Valent Transition Metal Complexes. *Organometallics* **1995,** *14*, 5308.
(b) Kuramoto, Y.; Sawai, N.; Fujiwara, Y.; Sumimoto, M.; Nakao, Y., et al. Stabilization of Vinylidene-Type and Acetylene-Type Si_2H_2 Species by Coordination With Rhodium(I) and Platinium(0) Complexes. Theoretical Proposals. *Organometallics* **2005,** *24*, 3655.

230. (a) Yamaguchi, T.; Sekiguchi, A.; Driess, M. An N-Heterocyclic carbene—Disilyne Complex and its Reactivity Toward $ZnCl_2$. *J. Am. Chem. Soc.* **2010,** *132*, 14061.

(b) Ishida, S.; Sugawara, R.; Misawa, Y.; Iwamoto, T. Palladium and Platinum η^2-Disilyne Complexes Bearing an Isolable Dialkyldisilyne as a Ligand. *Angew. Chem. Int. Ed.* **2013,** *52*, 12869–12873.

231. Gau, D.; Kato, T.; Saffon-Merceron, N.; De Cózar, A.; Cossío, F. P.; Baceiredo, A. Synthesis and Reactivity of a Base-Stabilized C-Phosphino-Si-Amino Silyne. *Angew. Chem. Int. Ed.* **2010,** *49*, 6585.

232. Danovich, D.; Ogliaro, F.; Karni, M.; Apeloig, Y.; Cooper, D. L.; Shaik, S. Silynes (RC≡SiR′) and Disilynes (RSi≡SiR′): Why Are Less Bonds Worth Energetically More? *Angew. Chem. Int. Ed.* **2001,** *40*, 4023.

233. Boone, A. J.; Magers, D. H.; Leszczyński, J. Searches on the Potential Energy Hypersurfaces of $GeCH_2$, $GeSiH_2$, and Ge_2H_2. *Int. J. Quantum. Chem.* **1998,** *70*, 925.

234. Devarajan, D.; Frenking, G. Are They Linear, Bent, or Cyclic? Quantum Chemical Investigation of the Heavier Group 14 and Group 15 Homologues of HCN and HNC. *Chem. Asian. J.* **2012,** *7*, 1296.

235. (a) Bourissou, D.; Guerret, O.; Gabbaï, F. P.; Bertrand, G. Stable Carbenes. *Chem. Rev.* **2000,** *100*, 39.

(b) Canac, Y.; Soleilhavoup, M.; Conejero, S., et al. Stable non-*N*-Heterocyclic Carbenes (non-NHC): Recent Progress. *J. Organomet. Chem.* **2004,** *689*, 3857.

(c) Vignolle, J.; Cattoën, X.; Bourissou, D. Stable Noncyclic Carbenes. *Chem. Rev.* **2009,** *109*, 3333.

(d) Kato, T.; Maerten, E.; Baceiredo, A. Non-NHCs Stable Singlet Carbene Ligands. In *Transition Metal Complexes of Neutral η^1-Carbon Ligands. Topics in Organometallic Chemistry;* Chauvin, R., Canac, Y., Eds.; 30; Springer-Verlag, 2010; pp 131–147.

236. Baines, K. M.; Stibbs, W. G. The Molecular Structure of Organogermanium Compounds. *Coord. Chem. Rev.* **1995,** *145*, 157.

237. (a) Hoffmann, R.; Zeiss, J. D.; Dine, G. W. V. The Electronic Structure of Methylenes. *J. Am. Chem. Soc.* **1968,** *90*, 1485.

(b) Nickon, A. New Perspectives on Carbene Rearrangements: Migratory Aptitudes, Bystander Assistance, and Geminal Efficiency. *Acc. Chem. Res.* **1993,** *26*, 84.

(c) Storer, J. W.; Hook, K. N. Origin of Anomalous Kinetic Parameters in Carbene 1,2-Shifts by Direct Dynamics. *J. Am. Chem. Soc.* **1993,** *115*, 10426.

(d) Sander, W.; Bucher, G.; Wierlacher, S. Carbenes in Matrixes: Spectroscopy, Structure, and Reactivity. *Chem. Rev.* **1993,** *93*, 1583.

CHAPTER 10

Silaaromatics and Related Compounds

Yoshiyuki Mizuhata and Norihiro Tokitoh

Kyoto University, Kyoto, Japan

Chapter Outline

10.1 Introduction

10.1.1 Aromatic Compounds

Aromatic compounds as represented by benzene and naphthalene are a group of compounds, which occupy a very important position in organic chemistry. A detailed commentary of the aromatic compounds and aromaticity is beyond the scope of this chapter, but here are the important elements.[1–4]

Organosilicon Compounds.
DOI: http://dx.doi.org/10.1016/B978-0-12-801981-8.00010-1

An aromatic compound generally refers to a compound having $[4n+2]\pi$ electrons with cyclic conjugated structure and is particularly stable (Hückel's rule). Aromatic compounds exhibit distinctive properties in their structures, reactivity, and magnetic properties besides stability, and are called comprehensively "having aromaticity." Meanwhile, cyclic conjugated structures having 4n number of π electrons are destabilized and show "anti-aromaticity." On the one hand, cyclic conjugated structures with these electronic states in the case of nonplanar molecules are not subject to any of the stabilization and destabilization above, and such compounds are called "nonaromatic" compounds.

Aromatic compounds exhibit the following features specifically:

- Structural features: Equivalency of bond lengths are observed. The C–C bond lengths in the aromatic ring take values between those of the single and double bonds, and in particular, the lengths of the six bonds in benzene are equal. This is described by "no bond alternation." Bond alternation can be observed depending on the substitution pattern and the ring structure.
- Chemical behavior: The reactivity of the double bond in the aromatic ring is reduced compared to the usual ones. They are also prone to the electrophilic substitution reaction rather than the addition reaction.
- Magnetic properties: Aromatic compounds show characteristic magnetic effects such as magnetic anisotropy and diamagnetism by the ring current effect of π electrons on the aromatic ring.

While aromaticity is estimated by these features, it cannot be clearly quantified and strictly defined. There have been many proposals from both sides of the experimental and theoretical studies to measure aromaticity, but they are still insufficient. Therefore, when discussing aromaticity, it is necessary to study carefully from various viewpoints to evaluate comprehensively the characteristics described above.

The ambiguity of the aromatic concept gives a fluctuation (degree of freedom) for the definition of the aromatic compounds themselves, and the present definition is expanded to a "cyclic arrangement of atoms stabilized by the resonance effect of electrons (presence of a bond on the cyclic arrangement is not absolutely essential)" from that following Hückel's rule. For instance, "Möbius Aromaticity"[5–7] (π electron systems having the same topology as the so-called "Möbius strip" are stabilized when they have a $4n\pi$ electron system on the contrary to Hückel's rule) and "homoaromaticity"[8] (interaction of π electron systems cleaved by saturated atoms) are referred to as "nonclassical aromatic compounds." While aromatic compounds are very common, the chemistry of aromaticity leaves room for discussion.

10.1.2 Heteroaromatic Compounds

Heteroaromatic compounds, i.e., compounds containing heteroatoms partially or fully instead of the carbon atom of the aromatic hydrocarbon, have been actively studied for a long time from the viewpoints of characteristic differences from aromatic hydrocarbons and the applicable scope of aromaticity. In fact, nitrogen-containing aromatic compounds, such as pyridine and quinoline, and aromatic compounds containing a Group 16 element, such as furan and thiophene, have been deployed in numerous research, and they have been revealed as aromatic compounds. Thus, there have been numerous examples of synthesis and isolation of the aromatic compounds containing Groups 13, 15, and 16 elements including heavier analogs, and their structures and properties become apparent.

Meanwhile, the synthesis of silicon-containing aromatic compounds in the form of the replacement of carbon to silicon, an element of the same Group 14 as carbon, had been difficult due to their high reactivity, despite a number of experimental and theoretical studies over a long period. This is because the double bond between silicon and carbon described in their resonance structure is very labile.[9–11]

In the theoretical calculations, silabenzene takes a planar and π-electron delocalized structure, and its aromaticity was predicted as 70%–85% of benzene.[12] The experimental verification had been waited for.

10.2 Neutral Silaaromatic Compounds

10.2.1 Syntheses of Neutral Silaaromatic Compounds

Generation of silatoluene **1**, the first neutral silicon-containing aromatic compound, was reported in 1977.[13] The generation of **1** was confirmed by the formation of [4 + 2] adduct **3** in the dehydrochlorination reaction of chlorosilanes **2** with base in the presence of acetylene. In response to this discovery, the generation method of **1** and silabenzene **5** by FVP (Flash Vacuum Pyrolysis) of allylsilanes **4** has been developed,[14] and their photoelectron spectra in the gas phase were measured in 1980.[15,16] Around the same time, the ultraviolet and infrared spectra of **1** and **5** were measured by capture in low-temperature argon matrix,[17,18] and their spectroscopic properties were revealed. However, they can exist only under very low temperature, and compound **5** decomposes to oligomer at higher than 80 K. This high reactivity marks a stark contrast to the high thermal stability of phosphabenzene, which has a third row element, phosphorus, as well as silicon, that does not change at all at room temperature under an inert atmosphere.

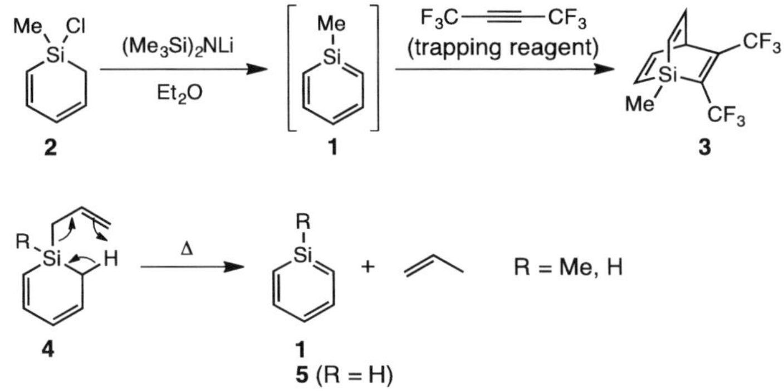

As a method to stabilize species having high tendency to oligomerize, there is a steric protection of the active site by the introduction of bulky substituents (kinetic stabilization) to suppress the intermolecular reaction. Based on this concept, the syntheses of sila- and germabenzenes **6**[19,20] and silabenzenes **7**[21] by introducing bulky substituents were attempted. However, they were still highly reactive at room temperature, and dimerized or decomposed to complicated mixtures. It was difficult to elucidate their detailed properties.

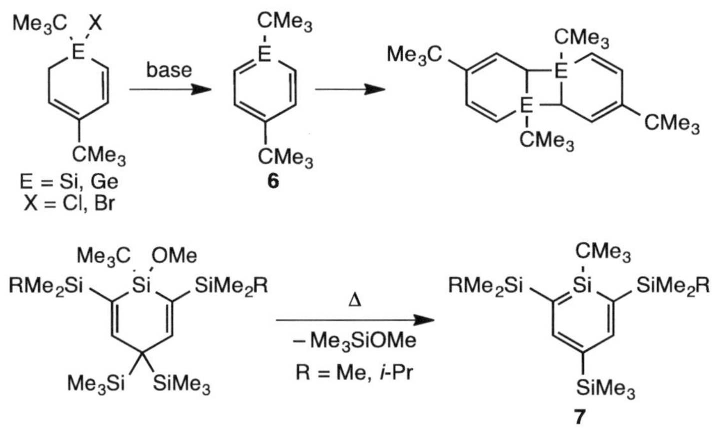

Tokitoh and coworkers have succeeded in the synthesis and isolation of a series of neutral silicon-containing aromatic compounds **8**–**12** including silabenzene **9** as compounds stable even at room temperature by the introduction of a bulky aryl substituent, Tbt or Bbt, onto the silicon atom.[22–31] Basically, their syntheses were achieved by the dehydrohalogenation reactions of halosilanes **14** with lithium diisopropylamide (LDA). In the case of 9-silaanthracene **11**, halogen atoms (Cl, Br) did not work well as a leaving group, but silyl triflate **15** was a good precursor.[31] Although details will be described later, the verification of their structures and characters showed that they have aromatic character comparable to the parent hydrocarbons. Similarly, the synthesis of germanium analogs has been achieved,[32–34] and also in 2006 the synthesis and isolation of the 2-stannanaphthalene having a tin atom at the 2-position of naphthalene was reported.[35,36] These results showed that the concept of aromaticity could be applicable even in the systems including a heavier Group 14 element. 4,4′-Disila-1,1′-biphenyl **13**, whose silabenzene rings were directly connected, was synthesized in 2010.[37]

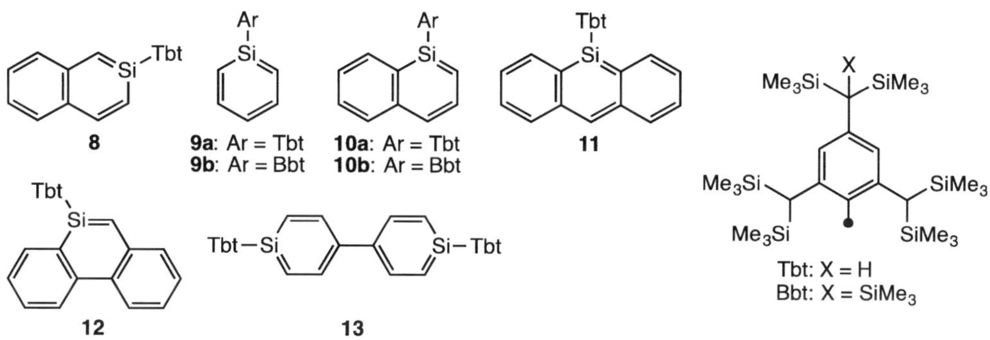

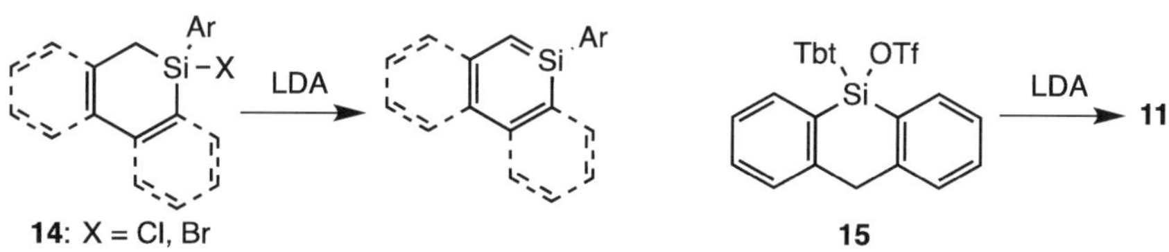

10.2.2 Structures and Properties of Neutral Silaaromatic Compounds

As for compounds **8**–**13** except for **10**, their structures have been revealed by the X-ray crystallographic analysis. As for their structural features of the aromatic compound as described above, the planarity of the ring and the equivalence of the bond lengths should be noted. With respect to the planarity of the ring skeleton, all compounds have a nearly perfect planar structure. The sums of the bond angles around the silicon are all 360 degrees. In addition, the bond lengths clearly reflect the bond alternation of parent hydrocarbon. In the case of silabenzene, whose parent hydrocarbon (benzene) does not indicate a bond alternation, two of Si–C and four of the C–C bond lengths in the ring are substantially equal respectively. In systems such as 2-silanaphthalene, whose parent compound exhibits bond alternation, the differences of bond lengths can be observed. On the other hand, the Si–C bond lengths are within the range of the single and double bond lengths. The structural features indicate that delocalization of π electrons, i.e., $[4n + 2]\pi$ electronic structure, is retained even in the silicon-containing aromatic compounds.

The ^{1}H and ^{13}C NMR signals assignable to the ring protons and carbons are both observed in the usual aromatic region. These results indicated that the ring current effect is expressed in heavier analogs of aromatic compounds. Moreover, ^{29}Si nucleus is possible and effective to NMR measurement. Previous studies indicated that the chemical shifts of the central silicon of low-coordinated silicon compounds are known to be significantly downfield shifted as compared to those of conventional four-coordinated compounds. Compounds **8**–**13** give a signal to the 87–94 ppm at low magnetic field region.

In the case of 4,4′-disila-1,1′-biphenyl **13**, the X-ray crystallographic study revealed a twisted structure with a dihedral angle of ca. 41 degrees. In the UV-vis spectrum in hexane, not only a red shift but also a sixfold increase in absorbance of the longest and broadened absorption maximum was observed, indicating free rotation around the central C–C bond and possible π-electronic conjugation through the single bond between the two silaaromatic rings.

10.2.3 Reactivity of Neutral Silaaromatic Compounds

While the results supporting the high aromaticity of neutral silicon-containing aromatic compounds as mentioned above are obtained, the Si=C bond still maintains high reactivity and is easily prone to 1,2- and 1,4-addition reactions.[22,38–41] For instance, the reactivity of silabenzene **9a** was summarized. Reaction sites are different according to the skeleton.

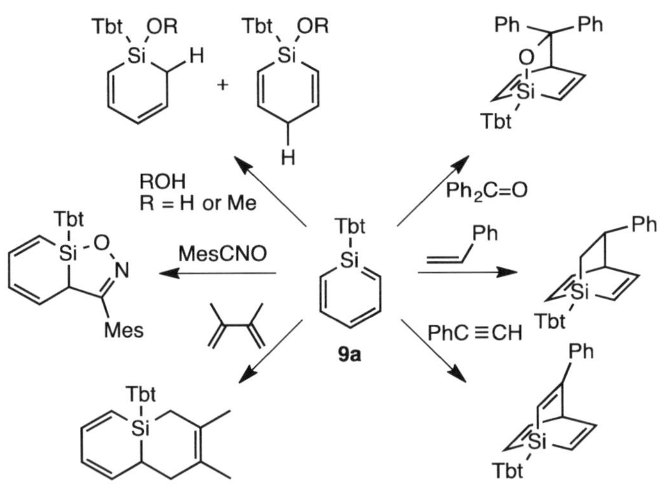

One of the characteristic reactions of aromatic compounds is coordination to various transition metals, giving η^6-arene complexes. In the case of silabenzene **9a**, ligand exchange reaction with Group 6 transition metals (Cr, Mo) provided novel η^6-arene complexes **16**.[42] Similar reactions proceed in the case of germabenzene,[43,44] and its W complex can be obtained in addition to the Cr and Mo complexes. These results showed that metallabenzenes behave similarly to the parent hydrocarbons in view of reactivity. In their IR spectra, complexes **16** show the absorption bands of carbonyl stretching at a region similar to the corresponding η^6-benzene complexes, suggesting their coordination ability as an arene ligand as well as that of benzene.

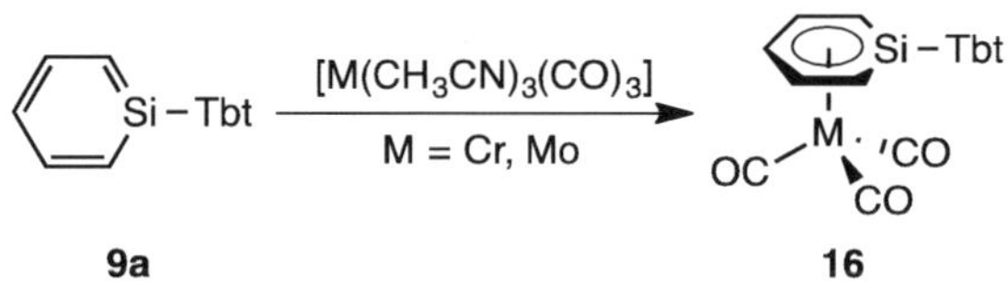

There are two other synthetic methods not depending on the ligand exchange reaction of isolated metallabenzene, and complexes **17** and **18** were synthesized from the corresponding hydrosilanes **19** and **20**, respectively.[45] In complex **17**, a large contribution of the η^1,η^5 type **17′** was suggested from the results of the NMR measurements. In disilabenzene complex **18**, the ruthenanorbornadiene structure was revealed by X-ray crystal structure analysis, but the contribution of the disila-Dewar benzene complex structure **18′** was pointed out from the short distance between two silicon atoms.

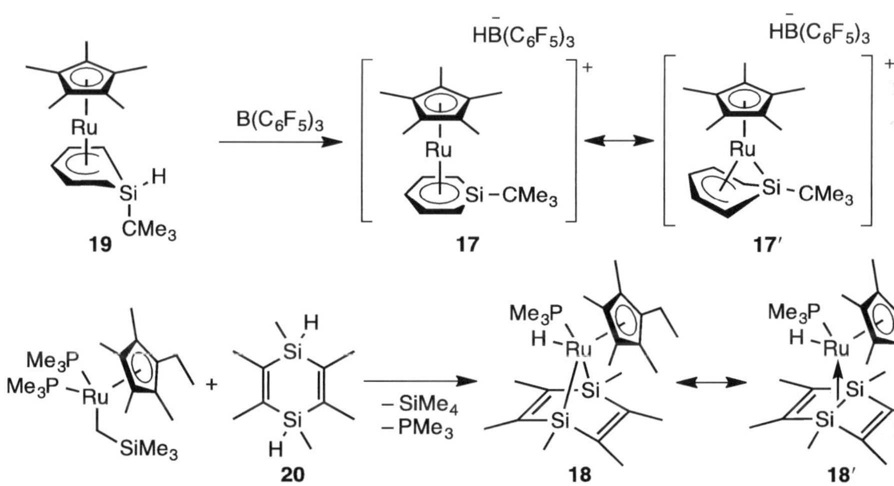

10.2.4 Disilaaromatic Compounds Derived From the Silicon—Silicon Triple-Bonded Compounds

In 2004, Sekiguchi and coworkers succeeded in the synthesis and isolation of the silicon—silicon triple-bond compound "disilyne" **21** as a stable compound for the first time, and revealed a number of its interesting properties.[46,47] Above all, its reactivity is very interesting and it has been reported to be a good precursor of various asymmetric disilenes and low-coordinated silicon compounds, whose synthesis had been difficult heretofore. In particular, its reaction with phenyl acetylene afforded 1:2 adducts, the mixture of 1,2-disilabenzenes **22** and **23**.[48] These are the first aromatic compounds containing two silicon atoms in a molecule.

Meanwhile, Tokitoh and coworkers have also succeeded in the synthesis of the carbon-substituted disilyne **24** by using Bbt groups as protecting groups.[49] Compound **24** gave the corresponding 1,2-disilabenzenes **25**–**27** by reactions with alkynes similarly to **21**.[50] Acetylene gas is available in addition to terminal acetylenes such as phenyl- and trimethylsilylacetylenes, giving compound **27** with only hydrogen substituents on the carbon backbone. In systems where isomers can occur, the formation of 4,5-disubstituted isomer has not been confirmed. In 2015, the germanium analog, 1,2-digermabenzene, was successfully synthesized and isolated via the reaction of a germanium–germanium triple-bond compound, digermyne, with 2 equiv of acetylene. Even though its molecular structure showed a *trans*-bent geometry for the Ge–Ge moiety, theoretical calculations suggested substantial levels of aromaticity.[51]

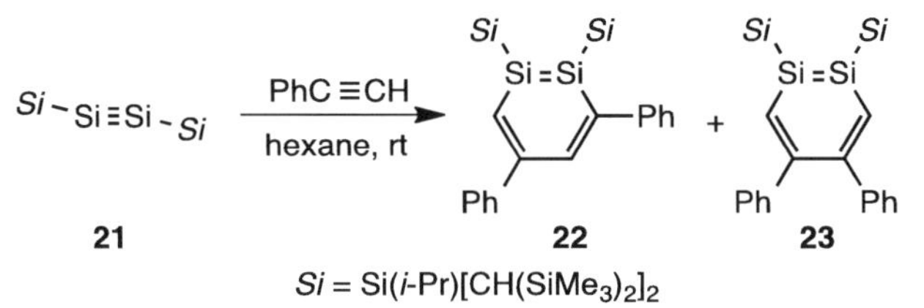

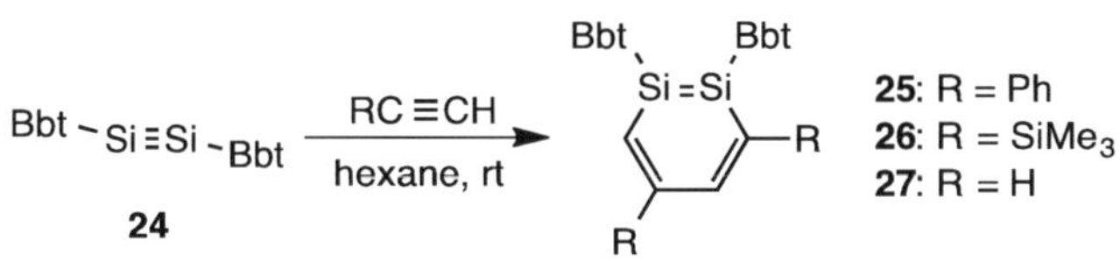

The results of X-ray crystallographic analysis of these disilabenzenes showed planar Si_2C_4 ring and not only equivalence of C–C bond lengths but also that lengths of Si–Si and Si–C were in the middle of those of single and double bonds. They have the characteristic π-electron delocalized structure similar to monosilabenzene. The most striking structural difference between silyl- and aryl-substituted derivatives, **23** and **27**, is the torsional angle *R*–Si–Si–*R* (*R*: substituent). While the torsional angle *C*(Bbt)–Si–Si–*C*(Bbt) in **27** is 45.6 degrees, compound **23** shows almost planar geometries at the two cyclic Si atoms [*Si*–Si–Si–*Si* = 13.1(2) degrees]. This is due to the electropositive silyl substituents, larger s-orbital at the central silicon atoms and smaller difference in the size of the s- and p-orbitals. That is, the extent of the s/p-hybridization is greater in the case of silyl-substituted derivative **23**.

In their ^{29}Si NMR spectra, the signals assignable to the silicon atoms in the disilabenzene ring were observed at 99.4/106.8 (**22**), 99.2 (**23**), 55.0/61.7 (**25**), 64.1/65.4 (**26**), and 57.1 (**27**) ppm. The signals of **22** and **23** bearing silyl-substituents, were observed at a lower magnetic field than those of **25**–**27**, for the reasons of structural differences described above. Meanwhile, the values of **25**–**27** were clearly upfield shifted as compared with 90 ppm of the mono-silabenzene **9b** having the same aryl substituents.

In addition, 2,3-disilapyrazines **28** and **29** were obtained by the reactions of disilyne **21** and silylcyanide.[52] Each bond length in the six-membered rings is an intermediate value of those of single and double bonds from the results of X-ray crystallographic analysis, suggesting the contribution of delocalized 6π electrons.

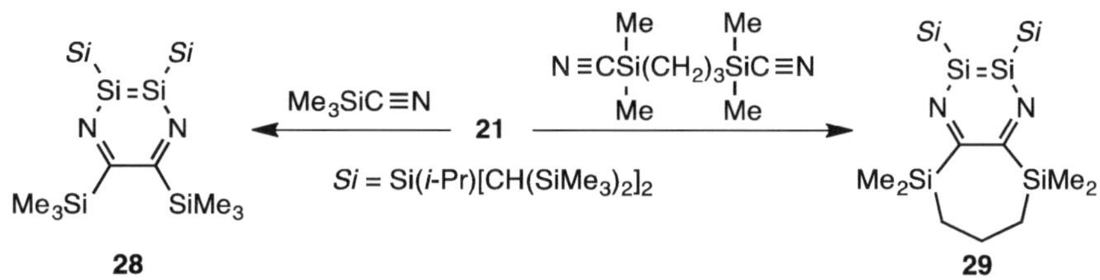

10.2.5 Valence Isomers

Under the photoreaction conditions, benzene (C_6H_6) is known to give various valence isomers such as benzvalene and Dewar benzene.[53,54] As for the silicon-containing aromatic compounds, photoreaction of silabenzene ($HSiC_5H_5$) (320 nm) gave sila-Dewar benzene observed by the IR spectrum in cryogenic conditions of Ar matrix at 10 K.[55] On the other hand, the photo irradiation of isolated silabenzene **9a** by the high-pressure mercury lamp (290–350 nm) gave silabenzvalene **30** rather than the corresponding sila-Dewar benzene.[26] Compound **30** was relatively stable in solution at room temperature, and gave a characteristic signal at −71.6 ppm in the ^{29}Si NMR. Compound **30** readily reacted with moisture in the air, giving a compound **31**.

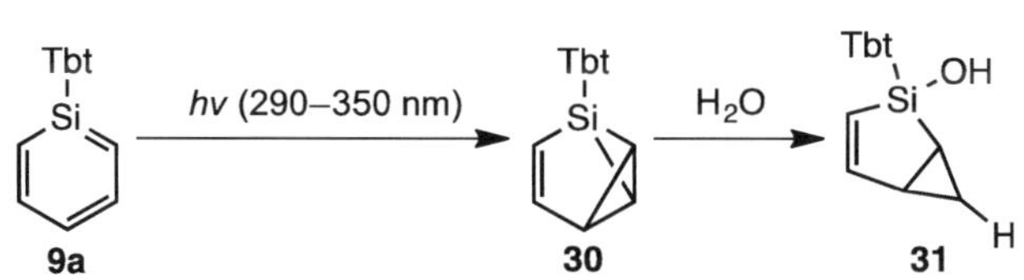

Also, photoirradiation of 9-silaanthracene **11** with light of 300–500 nm at room temperature afforded 9-sila-9,10-Dewar anthracene **32**.[56] Compound **32** returns to the 9-silaanthracene **11** with a relatively short half-life (16 min at 20°C in C_6D_6; 5 min at 10°C in hexane). In the ^{29}Si and ^{13}C NMR of **32**, signals derived from the central silicon and the 10-position carbon were observed at 1.4 and 48.6 ppm, respectively.

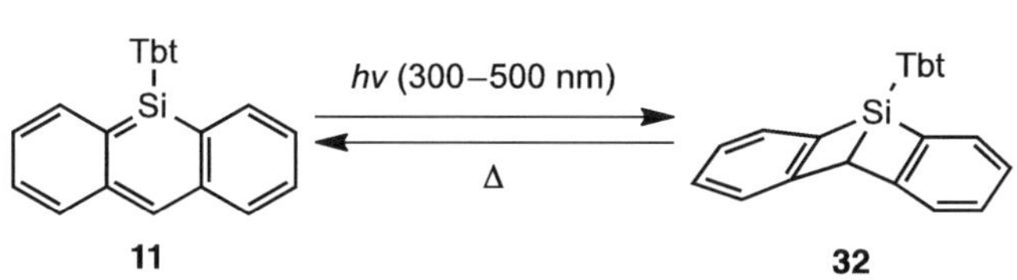

There are many examples of the silicon analogs of valence isomers such as compounds **33–35** directly prepared without going through the silicon-containing aromatic compounds.[57–60] In addition, hexasilaprismane **36** was synthesized, and its valence isomerization to hexasila-Dewar benzene **37** under photoirradiation at low temperature has been suggested.[61]

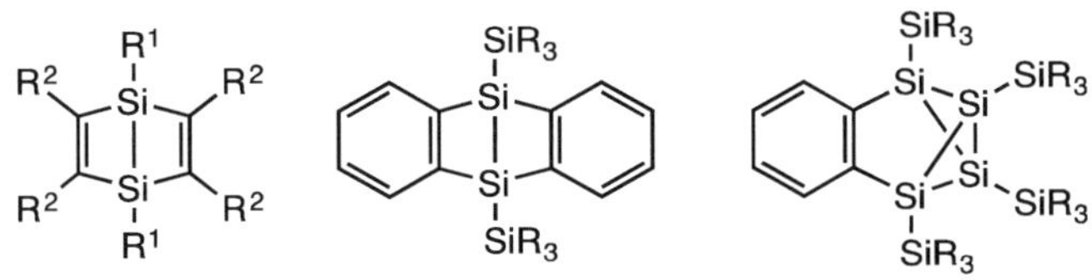

33 **34** **35**

R^1 = Me, R^2 = $SiMe_3$
R^1 = $SiMe(t\text{-}Bu)_2$, R^2 = Et

SiR_3 = $SiMe(t\text{-}Bu)_2$

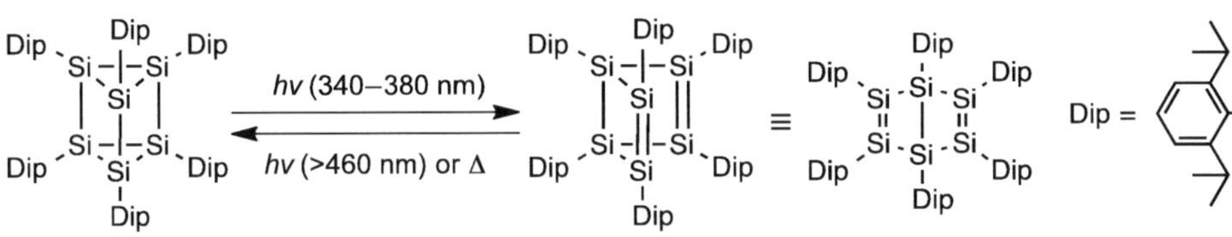

36 **37**

In 2013, hexasilabenzvalene **38** was synthesized by the reduction of tetrachlorocyclopentasilane **39** as a green powder.[62]

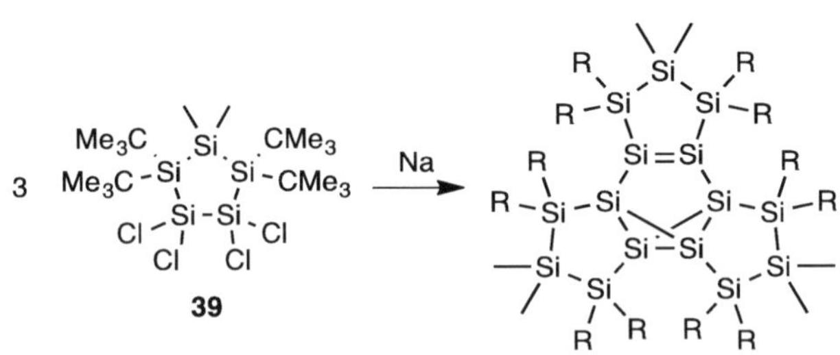

39

38 R = CMe_3

Not a valence isomer, but tricyclic isomer of hexasilabenzene **39** was synthesized in 2010.[63] Compound **39** was considered to have the electronic states depicted in resonance structures **39a** and **39b**, on which 6π electrons were delocalized in the central four-membered ring moiety.

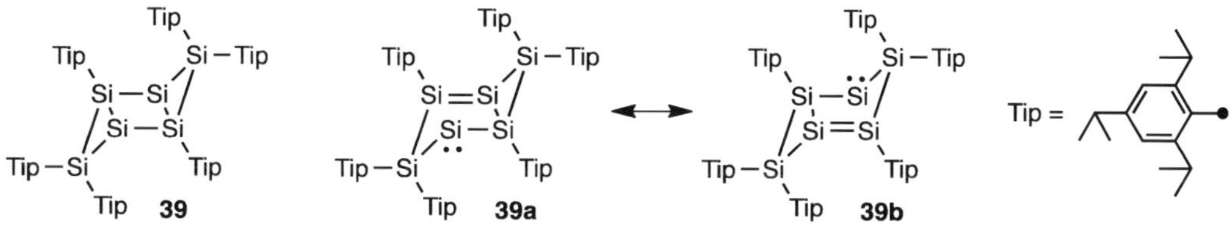

The isolation of the Si_4Ge_2 analog **40** was also achieved.[64] Compounds **39** and **40** showed unique rearrangement giving compounds **42** and **43**, respectively.[65] The attempted synthesis of the analog **41** having Si_4Sn_2 core resulted in the formation of **44**, suggesting the initial formation of **41**. In 2014, an all-tin-substituted derivative having an Sn_6 core of **44** was achieved by the reaction of $TipSnH_3$ with *N*-heterocyclic carbene.[66]

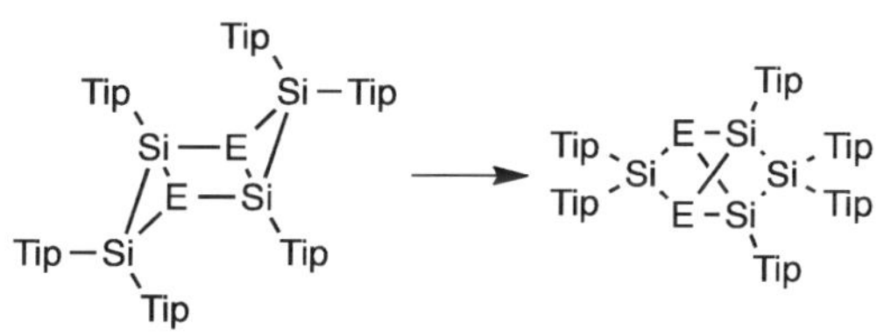

39 (E = Si)
40 (E =Ge)
41 (E = Sn): not detected

42 (E = Si)
43 (E =Ge)
44 (E = Sn)

10.3 Cationic and Anionic Silaaromatic Compounds

While the synthesis and isolation of the neutral silicon-containing aromatic compounds as described in Section 10.2 is extremely difficult, a number of silicon-containing aromatic compounds with an electric charge are known.

10.3.1 Three-Membered Ring Compounds

Cyclopropenyl cation is an aromatic compound of the minimum-membered ring, corresponding to the case of $n = 0$ of $4n + 2$ in the Hückel's rule. Its silicon analogs, **45** and **46**, have been synthesized by the desilylation reactions of the corresponding cyclic disilene **47** and silene **48**, respectively.[67,68] Germanium analog of **45** has also been reported.[69] As for **45**, the results of X-ray crystallographic analysis showed that the lengths of three Si–Si bonds in the three-membered ring were approximately equal to 2.211(3)–2.221(3) Å. In the ^{29}Si NMR, the signals of the three-membered ring were observed at 284.6 and 288.1 ppm (**45**) (intensity ratio of 2:1, due to the differences of the substituent) and 208.2 ppm (**46**), quite a low magnetic field region.

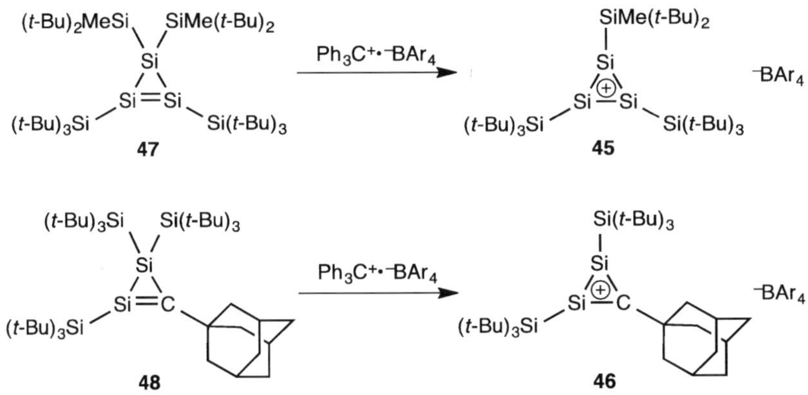

$^{-}BAr_4$ = $^{-}B(C_6F_5)_4$ (for **45**, **46**)
$^{-}B[C_6F_4\text{-}4\text{-}SiMe_2(t\text{-}Bu)]_4$ (for **45**, **46**)
$^{-}B(C_6F_4\text{-}4\text{-}H)_4$ (for **45**)

10.3.2 Four-Membered Ring Compounds

In the compounds four-membered-ring of carbons, cyclobutadiene dication of 2π electron system and cyclobutadiene dianion of 6π electron system exhibit aromaticity according to Hückel's rule. However, their chemical stability is reduced due to the electrostatic repulsion with two positive or negative charges.

As for cyclobutadiene dication, the generations and NMR observations of tetramethyl- and tetraphenyl-derivatives at low temperature have been reported.[70] There have been no reports of the silicon analog. In 2011, the donor-stabilized cyclo-Si_4 dication **49** having two silylene-like moieties was synthesized.[71] In the central four-membered ring of **49**, the charge-localized resonance structure with some contribution from the π-electron delocalized aromatic resonance form was suggested from the results of theoretical calculations.

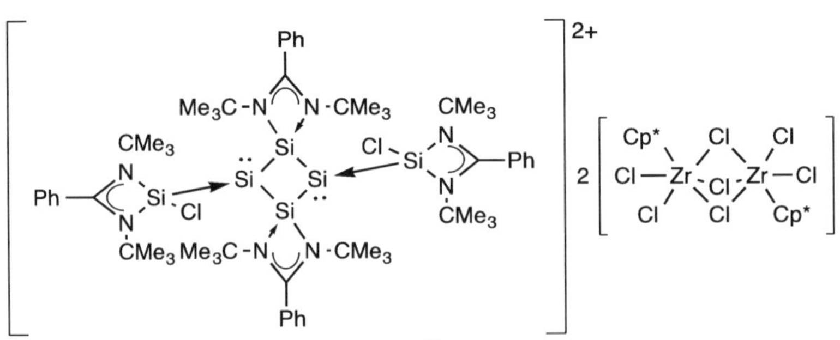

49

Cyclobutadiene dianion was synthesized and isolated as a stable compound in 2000 by the electronic perturbation of four trimethylsilyl groups.[72,73] The four-membered ring actually has a high planarity and delocalized electronic structure, indicating its aromaticity. The syntheses of the silicon (and germanium) analogs **50** and **51** were also achieved with a similar substitution pattern.[74] As opposed to the carbon system, compounds **50** and **51** have a nonplanar structure with bond alternation and were regarded as nonaromatic compounds. However, theoretical calculations predicted the large effect of the countercation for the intramolecular electrostatic repulsion including carbon analogs.[75] Therefore, there is a room for the synthesis of heavier analogs of cyclobutadiene dianion as an aromatic compound by means of a suitable countercation selected.

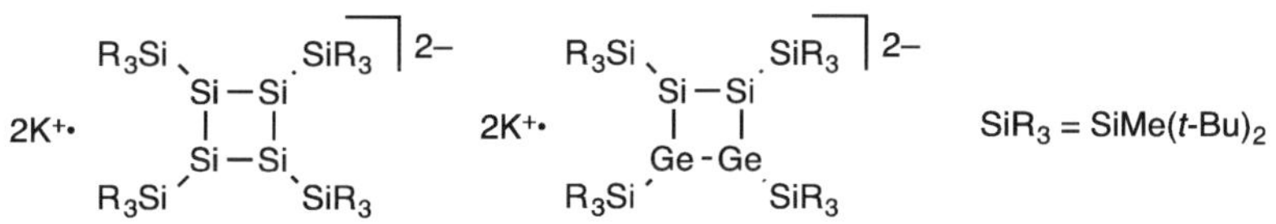

50 **51**

Meanwhile, neutral cyclobutadiene is a molecule with 4π electron system and has been extensively studied in view of the anti-aromaticity. While unsubstituted cyclobutadiene is extremely reactive, the observation at low temperature and synthetic and structural analysis studies of stable substituted derivatives showed their rectangular structure with clear bond alternation.

Similarly, a number of the generation of silicon analogs have been reported, mostly resulting in the conversion into other chemical species by the intramolecular reaction or oligomerization.[76,77] In 2011, tetrasilacyclobutadiene **52** was synthesized and isolated, showing a planar diamond structure without bond alternation.[78] The results of solid state NMR suggested a contribution of an alternately charge-separated structure **52**′.

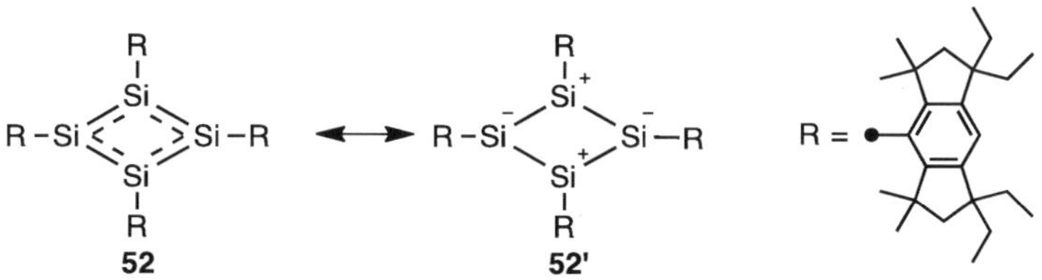

52 **52'**

With respect to their transition metal complexes, compounds **53**–**56** have been reported, and their structures are revealed.[79–86] The transition metal complexes of parent cyclobutadiene become 6π electron system by two-electron donation from the metal, showing no bond alternation, i.e., π-electron delocalization and aromaticity. Similarly to the carbon analogs, silicon ones have a generally planar four-membered ring structure without bond alternation.

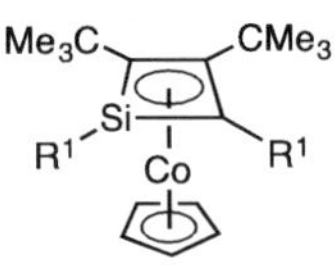

53

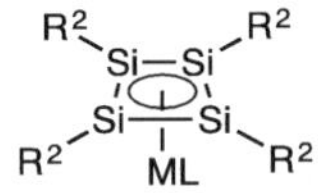

54

ML = $Co(CO)_2/K(diglyme)_2(thf)$
$Co(CO)_2/K([2.2.2]cryptand)$
CoCp
$Ru(CO)_3$
$RuCp^*/K(thf)_2$
$Fe(CO)_3$

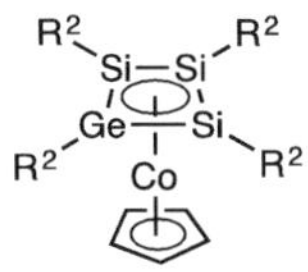

55

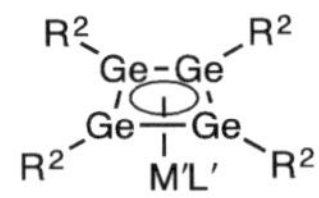

56

M′L′ = CoCp
$Fe(CO)_3$

$R^1 = SiMe_2(t\text{-Bu})$
$R^2 = SiMe(t\text{-Bu})_2$

10.3.3 Five-Membered Ring Compounds

Cyclopentadienyl anion is frequently used as a ligand in organometallic complexes and aromatic compounds of 6π electron system. The conjugated acid, cyclopentadiene, indicates pKa of 16 and is abnormally strongly acidic hydrocarbon by the driving force of aromatization to cyclopentadienyl anion. Indeed, cyclopentadienyl anion has a high (80%–90%) aromaticity of benzene.

Studies on the silicon analogs have been reported since the early stages of organosilicon chemistry, and the first example of the synthesis tracks back to 1958.[87] However, the detailed discussion including its aromaticity has come to be recognized since the 1990s. So far, the various silole (=1-silacyclopenta-2,4-diene) analogs of monoanions **57**–**62** and dianionic species **63**–**66** have been synthesized and isolated, and their aromaticity has been discussed.[88–92] The derivatives having sodium as a countercation have also been synthesized with respect to **63** and **64**.[93]

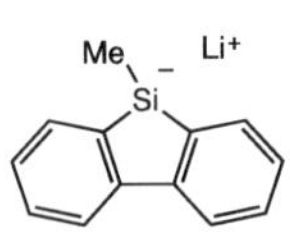

57

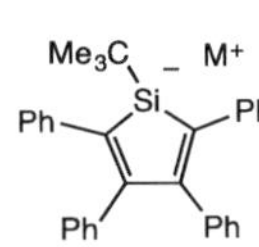

M = Li, Na

58

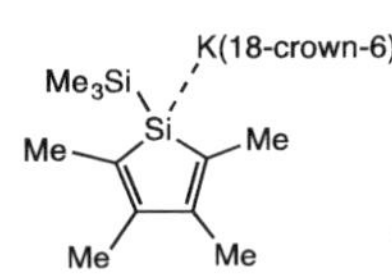

59

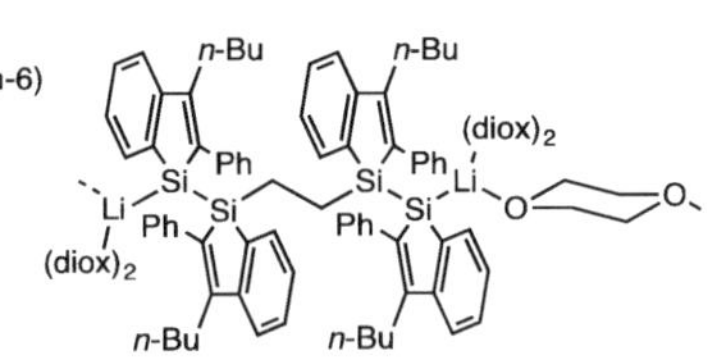

60

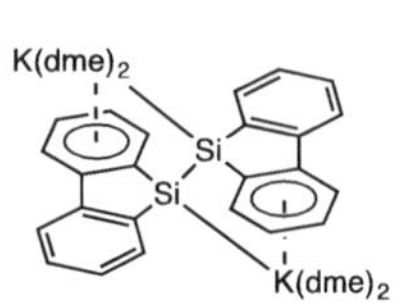

61

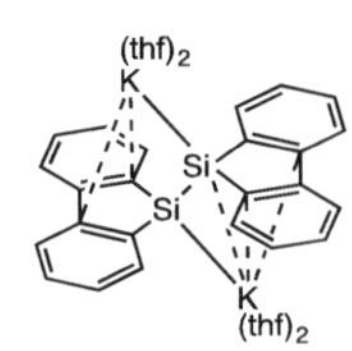

62

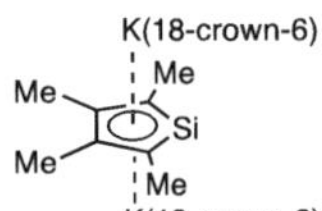

63

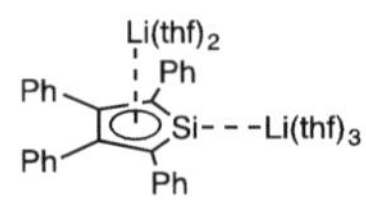

64

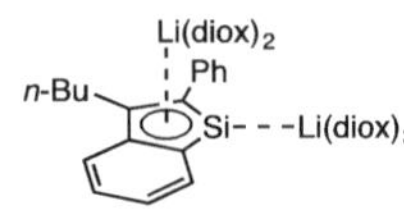

65

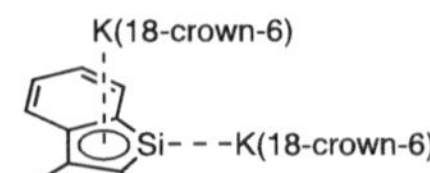

66

Interestingly, monoanion species **59–62** by X-ray crystallographic analysis have a structure with localized anionic charge on the silicon atom unlike the carbon systems, and the pyramidal structure around the silicon atom and clear bond alternation were observed. The results of NMR measurement in addition to the features above indicated their nonaromatic character. On the other hand, two countercations in the dianion species coordinated to the ring as η^5,η^1 or η^5,η^5 fashions, and the lengths of four C–C and two Si–C bonds in the ring were almost equal, respectively. These are the characteristic properties of an aromatic compound. In addition, verification of its properties by theoretical calculations also have been performed widely, showing that the structures having a η^5,η^5-coordination are 21 kcal mol^{-1} more stable as compared to those having η^5,η^1-coordination. Experimentally, both structures were observed in germanium analogs depending on the difference of the recrystallization temperature, and also only η^5,η^5-type in tin analogs.[94] The η^5,η^1-type structure in the silicon analogs is considered to be due to the effect of the solvation and packing force in the crystalline state. Notably, the synthesis of a lead analog was achieved in 2010.[95] One of the countercations was coordinated to the ring in η^5-fashion, and another existed as a separated ion pair solvated by DME.

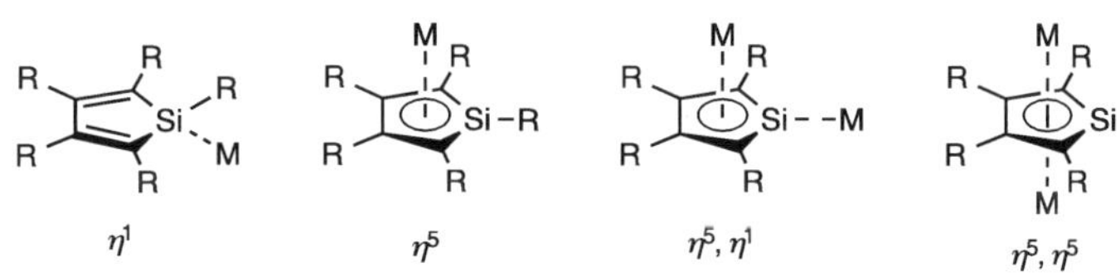

On the contrary to the charge localization in the monoanion species, π electrons delocalized structure has been reported in the transition metal complex **67**.[96]

Cyclopentadienyl anion analogs containing plural silicon (and germanium) atoms have been actively studied, and complexes **68** and **69** were synthesized and isolated as stable compounds.[97,98] Unlike silole monoanions, they have an aromaticity even in the case of monoanion. However, **69** takes a structure with localized charge (**70**) in a polar solvent, regarded as a nonaromatic compound. In addition, their transition metal complexes **71** and **72** have been synthesized and fully characterized.[80,99–101]

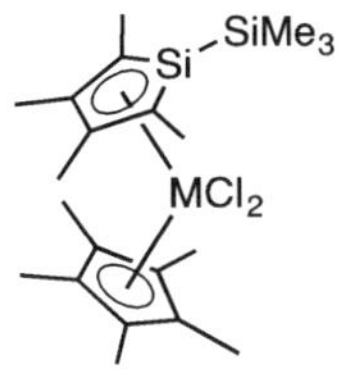

M = Zr, Hf

67

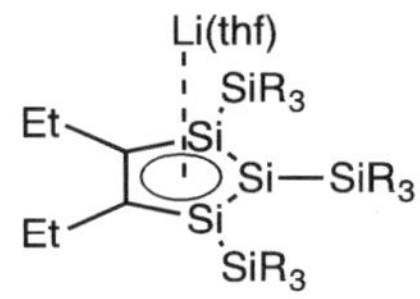

68

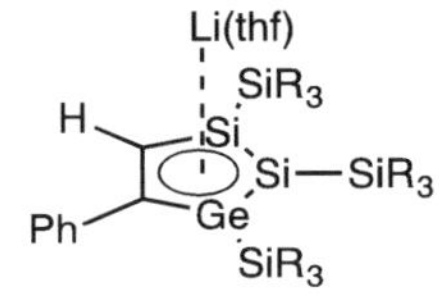

69

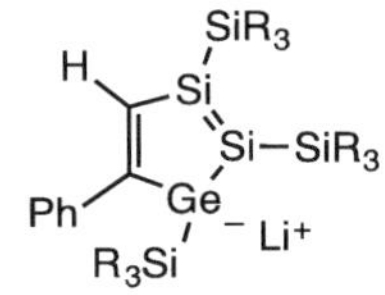

70

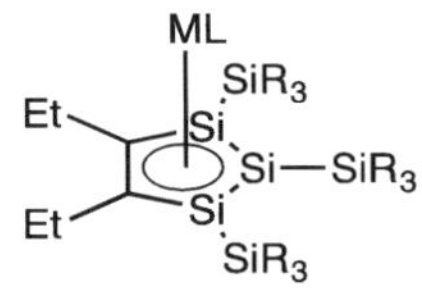

ML = Rh(CO)$_2$
RuCp*

71

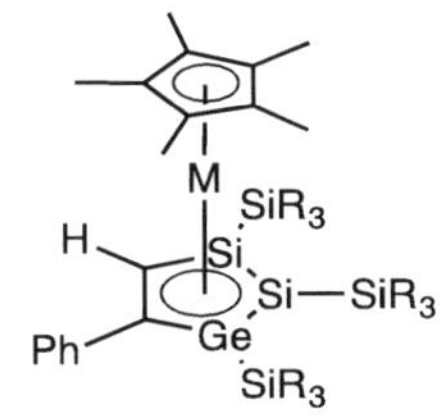

72 (M = Fe, Ru)

SiR$_3$ = SiMe(*t*-Bu)$_2$

10.3.4 Seven-Membered Ring Compounds

The cationic species given by hydride abstraction from cyclohepta-1,3,5-triene is called tropylium ion. Tropylium ion has a 6π electron system as well as benzene and shows aromaticity. Its existence is widely known: It becomes a ligand for transition metals and is observed in the mass analysis as a stable fragment of a compound having a benzyl ($-CH_2-Ph$) moiety (rearrangement of benzyl cation to the more stable tropylium ion).

As for the silicon analogs, the generation and the NMR observations of **73** at low temperature were reported in 2000.[102] Compound **73** is a very thermally unstable, and undergoes chlorine abstraction from solvent (dichloromethane) at over −50°C. In the ^{29}Si NMR, the signal derived from the central silicon was observed at a low magnetic field region, 142.9 ppm. Similarly, the signals from the ring carbons in the ^{13}C NMR were downfield shifted, suggesting the delocalization of positive charge on the entire ring. So far, there are no other reports.

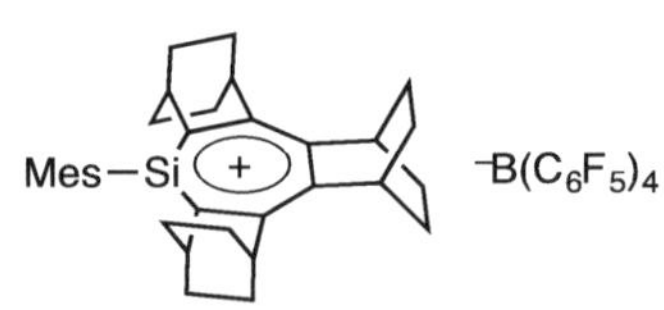

73

10.4 Other Silaaromatic Compounds

A divalent species of the silicon, silylene, is a highly reactive compound. Five-membered ring silylenes **74** and **75** containing nitrogen atoms have unique stability and are known to be the first examples of isolable silylenes.[103] These are regarded as the silicon analogs of NHCs (*N*-heterocyclic carbenes), which have shown remarkable development in recent years. While compound **74**, consisting of saturated bonds, gradually decomposes at room temperature in the crystalline state, compound **75**, having a double bond, does not decompose even by heating of the toluene solution at 150°C. The high thermal stability of **75** as compared to **74** is understood to be due to the expression of 6π aromaticity contributed by the resonance structures as shown in **75a–c**.

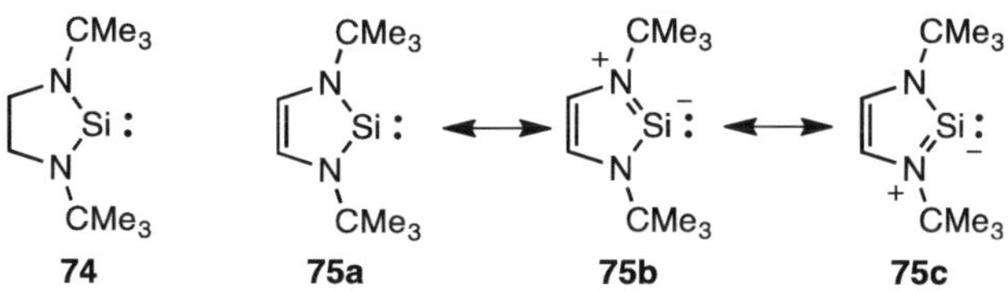

As a related compound, the generation of 2-silaimidazolium ion **76** was reported in 2005.[104] Compound **76** is stable at temperatures below −10°C. In its ^{29}Si NMR spectrum, the signal derived from the cyclic silicon atom was observed at 53.0 (in CD_2Cl_2) and 53.2 (in toluene-d_8) ppm. Although the spectroscopic features along with the results of the theoretical calculations suggested the presence of aromaticity in 2-silaimidazolium cation, its extent is smaller than that of the imidazolium cation.

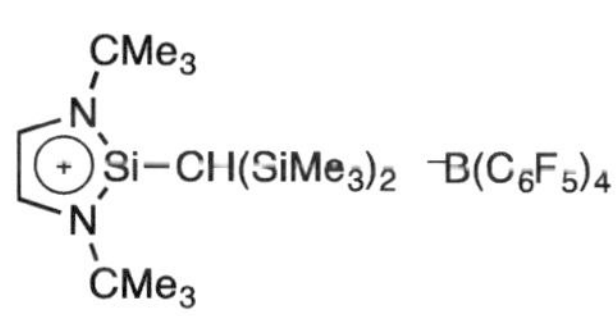

76

As a nonclassical aromatic compound described in Section 10.1, homo aromatic compound **77** has been reported.[105] The bond lengths of Si1–Si2 and Si2–Si3 are approximately equal (2.24 Å) to each other, and the distance of Si1···Si3 [2.692 (2) Å] is larger as compared to the general Si–Si single-bond length (about 2.4 Å) but still indicative of interaction. These experimental facts and the results of theoretical calculations indicated the 2π-electron system on the three-membered ring shown in the figure.

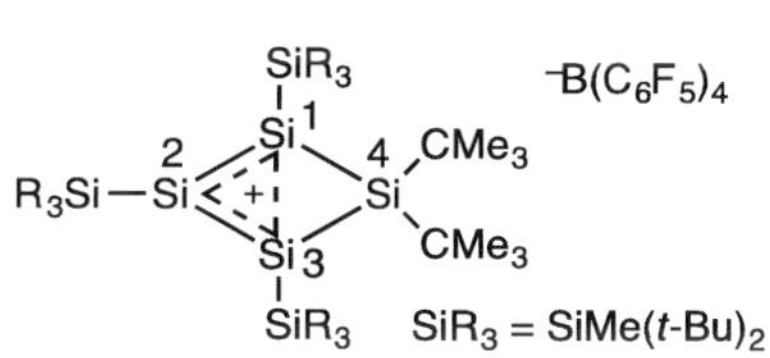

77

10.5 Summary and Outlook

By the recent remarkable progress as described thus far, various silicon-containing aromatic compounds have been synthesized and isolated, and their properties have been revealed. The foundations for the systematic understanding of their nature including aromaticity are now ready. The compounds that exceed the boundaries of traditional understanding for aromatic compounds are appearing, and further developments with future and important findings will be able to answer the question "what is the aromatic stabilization?"

As for neutral silicon-containing aromatic compounds, the introduction of up to two silicon atoms has been achieved as the approach to polysilabenzene. Hexasilabenzene, the ultimate silicon analog of benzene, is predicted to have not a planar but a chair-formed bent structure. The synthetic study of missing polysilabenzenes including hexasilabenzene is really desired together with the interplay with theoretical calculations.

References

1. Minkin, V. I.; Glukhovtsev, M. N.; Simkin, B. Ya *Aromaticity and Antiaromaticity: Electronic and Structural Aspects;* John Wiley & Sons, Inc: New York, 1994.
2. Schleyer, P. R. Introduction: Aromaticity. *Chem. Rev.* **2001,** *101* (5), 1115–1118.
3. Balaban, A. T.; Schleyer, P. R.; Rzepa, H. S. Crocker, Not Armit and Robinson, Begat the Six Aromatic Electrons. *Chem. Rev.* **2005,** *105* (10), 3436–3447.
4. Schleyer, P. R. Introduction: Delocalization–Pi and Sigma. *Chem. Rev.* **2005,** *105* (10), 3433–3435.
5. Rzepa, H. S. Möbius Aromaticity and Delocalization. *Chem. Rev.* **2005,** *105* (10), 3697–3715.
6. Jux, N. The Porphyrin Twist: Hückel and Möbius Aromaticity. *Angew. Chem. Int. Ed.* **2008,** *47* (14), 2543–2546.
7. Yoon, Z. S.; Osuka, A.; Kim, D. Möbius Aromaticity and Antiaromaticity in Expanded Porphyrins. *Nat. Chem.* **2009,** *1* (2), 113–122.
8. Williams, R. V. Homoaromaticity. *Chem. Rev.* **2001,** *101* (5), 1185–1204.
9. Ottosson, H.; Steel, P. G. Silylenes, Silenes, and Disilenes: Novel Silicon-Based Reagents for Organic Synthesis?. *Chem. Eur. J.* **2006,** *12* (6), 1576–1585.
10. Raabe, G.; Michl, J. Multiple Bonding to Silicon. *Chem. Rev.* **1985,** *85* (5), 419–509.
11. Lee, V. Ya; Sekiguchi, A. Heteronuclear Heavy Alkenes EE′ (E, E′ = Group 14 Elements): Germasilenes, Silastannenes, Germastannenes... Next Stop? *Organometallics* **2004,** *23* (12), 2822–2834.
12. Baldridge, K. K.; Uzan, O.; Martin, J. M. L. The Silabenzenes: Structure, Properties, and Aromaticity. *Organometallics* **2000,** *19* (11), 1477–1487.
13. Barton, T. J.; Banasiak, D. S. Evidence for the Generation and Trapping of a Silabenzene. *J. Am. Chem. Soc.* **1977,** *99* (15), 5199–5200.
14. Barton, T. J.; Burns, G. T. Unambiguous Generation and Trapping of a Silabenzene. *J. Am. Chem. Soc.* **1978,** *100* (16), 5246–6246.
15. Bock, H.; Bowling, R. A.; Solouki, B.; Barton, T. J.; Burns, G. T. Analysis and Optimization of Gas-Phase Reactions. 13. Silatoluene. *J. Am. Chem. Soc.* **1980,** *102* (1), 429–430.
16. Solouki, B.; Rosmus, P.; Bock, H.; Maier, G. Short-Path Pyrolsis: Silabenzene. *Angew. Chem. Int. Ed. Engl.* **1980,** *19* (1), 51–52.

17. Kreil, C. L.; Chapman, O. L.; Burns, G. T.; Barton, T. J. Spectroscopic Observation of Matrix-Isolated Silatoluene. *J. Am. Chem. Soc.* **1980,** *102* (2), 841–842.
18. Maier, G.; Mihm, G.; Reisenauer, H. P. Silabenzene. *Angew. Chem. Int. Ed. Engl.* **1980,** *19* (1), 52–53.
19. Märkl, G.; Hofmeister, P. 1, 4-Di-Tert-Butylsilabenzol; Erzeugung Und Abfangreaktionen. *Angew. Chem.* **1979,** *91* (10), 863–864.
20. Märkl, V. G.; Rudnick, D. 1.4-Di-Tert.butyl-Germabenzol-Erzeugung Und Abfangreaktionen Germabenzol- Und 1.2-Dihydrogermabenzol-Dimere. *Tetrahedron Lett.* **1980,** *21* (15), 1405–1408.
21. Jutzi, P.; Meyer, M.; Reisenauer, H. P.; Maier, G. Zur Kinetischen Stabilisierung von Silabenzol: Matrix-Isolierung Und Charakterisierung von 2,4,6-Tris(trialkylsilyl)-Substituiertem 1-Tert-Butyl-1-Silabenzol. *Chem. Ber.* **1989,** *122* (7), 1227–1230.
22. Tokitoh, N. New Progress in the Chemistry of Stable Metallaaromatic Compounds of Heavier Group 14 Elements. *Acc. Chem. Res.* **2004,** *37* (2), 86.
23. Tokitoh, N.; Wakita, K.; Okazaki, R.; Nagase, S.; Schleyer, P. R., et al. A Stable Neutral Silaaromatic Compound, 2-{2,4,6-Tris[bis(trimethylsilyl)methyl]phenyl}- 2-Silanaphthalene. *J. Am. Chem. Soc.* **1997,** *119* (29), 6951–6952.
24. Wakita, K.; Tokitoh, N.; Okazaki, R.; Nagase, S.; Schleyer, P. R, et al. Synthesis of Stable 2-Silanaphthalenes and Their Aromaticity. *J. Am. Chem. Soc.* **1999,** *121* (49), 11336–11344.
25. Wakita, K.; Tokitoh, N.; Okazaki, R.; Nagase, S. Synthesis and Properties of an Overcrowded Silabenzene Stable at Ambient Temperature. *Angew. Chem. Int. Ed.* **2000,** *39* (3), 634–636.
26. Wakita, K.; Tokitoh, N.; Okazaki, R.; Takagi, N.; Nagase, S. Crystal Structure of a Stable Silabenzene and Its Photochemical Valence Isomerization into the Corresponding Silabenzvalene. *J. Am. Chem. Soc.* **2000,** *122* (23), 5648–5649.
27. Tokitoh, N.; Wakita, K.; Matsumoto, T.; Sasamori, T.; Okazaki, R.; Takagi, N., et al. The Chemistry of Stable Silabenzenes. *J. Chinese Chem. Soc.* **2008,** *55* (3), 487–507.
28. Takeda, N.; Shinohara, A.; Tokitoh, N. Synthesis and Properties of the First 1-Silanaphthalene. *Organometallics* **2002,** *21* (20), 4024–4026.
29. Shinohara, A.; Takeda, N.; Sasamori, T.; Tokitoh, N. Synthesis of Kinetically Stabilized 1-Silanaphthalenes and Their Properties. *Bull. Chem. Soc. Jpn.* **2005,** *78* (6), 977–987.
30. Tokitoh, N.; Shinohara, A.; Matsumoto, T.; Sasamori, T.; Takeda, N.; Furukawa, Y. Synthesis and Properties of a Kinetically Stabilized 9-Silaphenanthrene. *Organometallics* **2007,** *26* (16), 4048–4053.
31. Takeda, N.; Shinohara, A.; Tokitoh, N. The First Stable 9-Silaanthracene. *Organometallics* **2002,** *21* (2), 256–258.
32. Nakata, N.; Takeda, N.; Tokitoh, N. Synthesis and Structure of a Kinetically Stabilized 2-Germanaphthalene: The First Stable Neutral Germaaromatic Compound. *Organometallics* **2001,** *20* (9), 5507–5509.
33. Nakata, N.; Takeda, N.; Tokitoh, N. Synthesis and Properties of the First Stable Germabenzene. *J. Am. Chem. Soc.* **2002,** *124* (24), 6914–6920.
34. Sasamori, T.; Inamura, K.; Hoshino, W.; Nakata, N.; Mizuhata, Y.; Watanabe, Y., et al. Synthesis and Characterization of Two Isomers of 14π-Electron Germaaromatics: Kinetically Stabilized 9-Germaanthracene and 9-Germaphenanthrene. *Organometallics* **2006,** *25* (15), 3533–3536.
35. Mizuhata, Y.; Sasamori, T.; Takeda, N.; Tokitoh, N. A Stable Neutral Stannaaromatic Compound: Synthesis, Structure and Complexation of a Kinetically Stabilized 2-Stannanaphthalene. *J. Am. Chem. Soc.* **2006,** *128* (4), 1050–1051.
36. Mizuhata, Y.; Sasamori, T.; Nagahora, N.; Watanabe, Y.; Furukawa, Y.; Tokitoh, N. Synthesis and Properties of Stable 2-Metallanaphthalenes of Heavier Group 14 Elements. *Dalton Trans.* **2008,** *No. 33*, 4409.
37. Tanabe, Y.; Mizuhata, Y.; Tokitoh, N. Synthesis and Structure of a Heavier Congener of Biphenyl: 1,1′-Disila-4,4′-Biphenyl. *Organometallics* **2010,** *29* (4), 721–723.

38. Nakata, N.; Takeda, N.; Tokitoh, N. Reactions of 2-Germanaphthalene With Elemental Sulfur and Selenium: Synthesis of Novel Cyclic Polychalcogenides Containing a Germanium, Trichalcogenagermolanes. *Chem. Lett.* **2002,** *94* (8), 818–918.
39. Nakata, N.; Takeda, N.; Tokitoh, N. Reaction of a Stable Germabenzene With Chalcogens: Synthesis and Structure of a Novel Germanium Analog of Pentathiepane, 1,2,3,4,5,6-Pentathiagermepane. *J. Organomet. Chem.* **2003,** *672* (1–2), 66–71.
40. Mizuhata, Y.; Shinohara, A.; Tanabe, Y.; Tokitoh, N. Cycloaddition Reactions of Kinetically Stabilized 9-Silaanthrancene With Polycyclic Aromatic Hydrocarbons. *Chem. Eur. J.* **2009,** *15* (34), 8405–8408.
41. Mizuhata, Y.; Tokitoh, N. Reactions of a Kinetically Stabilized 2-Stannanaphthalene With Elemental Sulfur and Selenium: Synthesis of Novel Cyclic Chalcogenides Containing a Tin Atom. *J. Sulfur Chem.* **2009,** *30* (3–4), 385–392.
42. Shinohara, A.; Takeda, N.; Sasamori, T.; Matsumoto, T.; Tokitoh, N. Synthesis and Properties of η^6-Silabenzene – $M(CO)_3$ Complexes (M = Cr, Mo). *Organometallics* **2005,** *24* (25), 6141–6146.
43. Nakata, N.; Takeda, N.; Tokitoh, N. η^6-Germabenzene Complexes of Chromium and Molybdenum. *Angew. Chem. Int. Ed.* **2003,** *42* (1), 115–117.
44. Tokitoh, N.; Nakata, N.; Shinohara, A.; Takeda, N.; Sasamori, T. Coordination Chemistry of a Kinetically Stabilized Germabenzene: Syntheses and Properties of Stable η^6-Germabenzene Complexes Coordinated to Transition Metals. *Chem. Eur. J.* **2007,** *13* (6), 1856–1862.
45. Dysard, J. M.; Tilley, T. D.; Woo, T. K. Silabenzene and Disilabenzene Complexes of Ruthenium. *Organometallics* **2001,** *20* (6), 1195–1203.
46. Sekiguchi, A.; Kinjo, R.; Ichinohe, M. A Stable Compound Containing a Silicon-Silicon Triple Bond. *Science* **2004,** *305* (5691), 1755–1757.
47. Sekiguchi, A.; Ichinohe, M.; Kinjo, R. The Chemistry of Disilyne With a Genuine Si–Si Triple Bond: Synthesis, Structure, and Reactivity. *Bull. Chem. Soc. Jpn.* **2006,** *79* (6), 825–832.
48. Kinjo, R.; Ichinohe, M.; Sekiguchi, A.; Takagi, N.; Sumimoto, M.; Nagase, S. Reactivity of a Disilyne $RSi{\equiv}SiR$ ($R = Si\mathit{i}Pr[CH(SiMe_3)_2]_2$) Toward π-Bonds: Stereospecific Addition and a New Route to an Isolable 1,2-Disilabenzene. *J. Am. Chem. Soc.* **2007,** *129* (25), 7766–7767.
49. Sasamori, T.; Hironaka, K.; Sugiyama, Y.; Takagi, N.; Nagase, S.; Hosoi, Y., et al. Synthesis and Reactions of a Stable 1,2-Diaryl-1,2-Dibromodisilene: A Precursor for Substituted Disilenes and a 1,2-Diaryldisilyne. *J. Am. Chem. Soc.* **2008,** *130* (42), 13856–13857.
50. Han, J. S. S.; Sasamori, T.; Mizuhata, Y.; Tokitoh, N. Reactivity of an Aryl-Substituted Silicon–silicon Triple Bond: 1,2-Disilabenzenes From the Reactions of a 1,2-Diaryldisilyne With Alkynes. *Dalton Trans.* **2010,** *39* (39), 9238.
51. Sasamori, T.; Sugahara, T.; Agou, T.; Guo, J.-D.; Nagase, S.; Streubel, R., et al. Synthesis and Characterization of a 1,2-Digermabenzene. *Organometallics* **2015,** *34* (11), 2106–2109.
52. Takeuchi, K.; Ichinohe, M.; Sekiguchi, A. Reactivity of the Disilyne $RSi{\equiv}SiR$ ($R = Si\mathit{i}Pr[CH(SiMe_3)_2]_2$) toward Silylcyanide: Two Pathways To Form the Bis-Adduct $[RSiSiR(CNSiMe_3)_2]$ With Some Silaketenimine Character and a 1,4-Diaza-2,3-Disilabenzene Analogue. *J. Am. Chem. Soc.* **2008,** *130* (50), 16848–16849.
53. Scott, L. T.; Jones, M. Rearrangements and Interconversions of Compounds of the Formula $(CH)_n$. *Chem. Rev.* **1972,** *72* (2), 181–202.
54. Bryce-Smith, D.; Gilbert, A. The Organic Photochemistry of Benzene—I. *Tetrahedron* **1976,** *32* (12), 1309–1326.
55. Maier, G.; Mihm, G.; Baumgärtner, R. O. W.; Reisenauer, H. P. Hetero-π-Systeme, 7. Silabenzol. *Chem. Ber.* **1984,** *117* (7), 2337–2350.
56. Shinohara, A.; Takeda, N.; Tokitoh, N. Photochemical and Thermal Reactions of a Kinetically Stabilized 9-Silaanthracene: The First Spectroscopic Observation of a 9,10-Dewar-9-Silaanthracene Isomer. *J. Am. Chem. Soc.* **2003,** *125* (36), 10804–10805.
57. Takanashi, K.; Lee, V. Ya; Ichinohe, M.; Sekiguchi, A. 1,2,5,6-Tetrasilabenzobenzvalene: A Valence Isomer of 1,2,3,4-Tetrasilanaphthalene. *Chem. Lett.* **2007,** *36* (9), 1158–1159.

58. Nakata, N.; Masubuchi, S.; Sekiguchi, A. Spiro[2,5-Dibenzo-1,4-Disilacyclohexa-2,5-Diene-1,7′-2′,5′-Dibenzo-1′,4′,7′-Trisilanorbornadiene]: Unusual Dimerization of Silyl-Substituted 9,10-Disila(Dewar Anthracene). *Phosphorus, Sulfur, Silicon Relat. Elem.* **2010,** *185* (5–6), 957–963.
59. Kabe, Y.; Ohkubo, K.; Ishikawa, H.; Ando, W. 1,4-Disila(Dewar-Benzene) and 1,4-Disilabenzene: Valence Isomerization of Bis(alkylsilacyclopropenyl)s. *J. Am. Chem. Soc.* **2000,** *122* (15), 3775–3776.
60. Oikawa, T.; Nakata, N.; Matsumoto, T.; Kabe, Y.; Sekiguchi, A. Photochemical Reaction of Silyl-Substituted 1,4-Disila(dewar-Benzene) With Isocyanide and Phenylacetylene. *Heteroat. Chem* **2008,** *19* (1), 87–92.
61. Sekiguchi, A.; Yatabe, T.; Kabuto, C.; Sakurai, H. The "Missing" Hexasilaprismane: Synthesis, X-Ray Analysis, and Photochemical Reactions. *J. Am. Chem. Soc.* **1993,** *115*, 5853–5854.
62. Tsurusaki, A.; Iizuka, C.; Otsuka, K.; Kyushin, S. Cyclopentasilane-Fused Hexasilabenzvalene. *J. Am. Chem. Soc.* **2013,** *135* (44), 16340–16343.
63. Abersfelder, K.; White, A. J. P.; Rzepa, H. S.; Scheschkewitz, D. A Tricyclic Aromatic Isomer of Hexasilabenzene. *Science* **2010,** *327* (5965), 564–566.
64. Jana, A.; Huch, V.; Repisky, M.; Berger, R. J. F.; Scheschkewitz, D. Dismutational and Global-Minimum Isomers of Heavier 1,4-Dimetallatetrasilabenzenes of Group 14. *Angew. Chem. Int. Ed.* **2014,** *53* (13), 3514–3518.
65. Abersfelder, K.; White, A. J. P.; Berger, R. J. F.; Rzepa, H. S.; Scheschkewitz, D. A Stable Derivative of the Global Minimum on the Si_6H_6 Potential Energy Surface. *Angew. Chem. Int. Ed.* **2011,** *50* (34), 7936–7939.
66. Sindlinger, C. P.; Wesemann, L. Hydrogen Abstraction From Organotin Di- and Trihydrides by N-Heterocyclic Carbenes: A New Method for the Preparation of NHC Adducts to Tin(II) Species and Observation of an Isomer of a Hexastannabenzene Derivative [R_6Sn_6]. *Chem. Sci.* **2014,** *5* (7), 2739–2746.
67. Ichinohe, M.; Igarashi, M.; Sanuki, K.; Sekiguchi, A. Cyclotrisilenylium Ion: The Persilaaromatic Compound. *J. Am. Chem. Soc.* **2005,** *127* (28), 9978–9979.
68. Igarashi, M.; Ichinohe, M.; Sekiguchi, A. Air-Stable Disilacyclopropene With a SiC Bond and Its Conversion to Disilacyclopropenylium Ion: Silicon – Carbon Hybrid 2π-Electron Systems. *J. Am. Chem. Soc.* **2007,** *129* (42), 12660–12661.
69. Sekiguchi, A.; Tsukamoto, M.; Ichinohe, M. A Free Cyclotrigermenium Cation With a 2π-Electron System. *Science* **1997,** *275* (5296), 60–61.
70. Olah, G. A.; Staral, J. S. Novel Aromatic Systems. 4. Cyclobutadiene Dications. *J. Am. Chem. Soc.* **1976,** *98* (20), 6290–6304.
71. Inoue, S.; Epping, J. D.; Irran, E.; Driess, M. Formation of a Donor-Stabilized Tetrasilacyclobutadiene Dication by a Lewis Acid Assisted Reaction of an *N*-Heterocyclic Chloro Silylene. *J. Am. Chem. Soc.* **2011,** *133* (22), 8514–8517.
72. Sekiguchi, A.; Matsuo, T.; Watanabe, H. Synthesis and Characterization of a Cyclobutadiene Dianion Dilithium Salt: Evidence for Aromaticity. *J. Am. Chem. Soc.* **2000,** *122* (23), 5652–5653.
73. Matsuo, T.; Sekiguchi, A. Cyclobutadiene Dianion. *Bull. Chem. Soc. Jpn.* **2004,** *77* (2), 211–226.
74. Lee, V. Ya; Takanashi, K.; Matsuno, T.; Ichinohe, M.; Sekiguchi, A. Cyclobutadiene Dianions Consisting of Heavier Group 14 Elements: Synthesis and Characterization. *J. Am. Chem. Soc.* **2004,** *126* (15), 4758–4759.
75. Shainyan, B. A.; Sekiguchi, A. Computational Study of Tetrasilylcyclobutadiene Dianion and Its Dilithium Salt. 6e–6c Three-Dimensional Aromaticity. *J. Mol. Struct. THEOCHEM* **2005,** *728* (1–3), 1–5.
76. Puranik, D. B.; Fink, M. J. The Thermal Isomerization of a Silacyclobutadiene to a Cyclopropenylsilylene: Evidence for a Stable Silylene in Fluid Solution. *J. Am. Chem. Soc.* **1989,** *111* (15), 5951–5952.
77. Maier, G.; Reisenauer, H. P.; Jung, J.; Pacl, H.; Egenolf, H. C_3H_4Si Species: Generation and Matrix-Spectroscopic Identification of Some Silacyclobutadiene Isomers. *Eur. J. Org. Chem.* **1998,** *1998* (7), 1297–1305.
78. Suzuki, K.; Matsuo, T.; Hashizume, D.; Fueno, H.; Tanaka, K.; Tamao, K. A Planar Rhombic Charge-Separated Tetrasilacyclobutadiene. *Science* **2011,** *331* (6022), 1306–1309.

79. Kon, Y.; Sakamoto, K.; Kabuto, C.; Kira, M. A Cobalt Silacyclobutadiene Complex. *Organometallics* **2005,** *24* (7), 1407−1409.
80. Lee, V. Ya; Sekiguchi, A. Cyclic Polyenes of Heavy Group 14 Elements: New Generation Ligands for Transition-Metal Complexes. *Chem. Soc. Rev.* **2008,** *37* (8), 1652.
81. Takanashi, K.; Lee, V. Ya; Matsuno, T.; Ichinohe, M.; Sekiguchi, A. Tetrasilacyclobutadiene $(tBu_2MeSi)_4Si_4$: A New Ligand for Transition-Metal Complexes. *J. Am. Chem. Soc.* **2005,** *127* (16), 5768−5769.
82. Takanashi, K.; Lee, V. Ya; Ichinohe, M.; Sekiguchi, A. A (Tetrasilacyclobutadiene)tricarbonyliron Complex $[\{\eta^4\text{-}(tBu_2MeSi)_4Si_4\}Fe(CO)_3]$: The Silicon Cousin of Pettit's (Cyclobutadiene)tricarbonyliron Complex $[(\eta^4\text{-}H_4C_4)Fe(CO)_3]$. *Angew. Chem. Int. Ed.* **2006,** *45* (20), 3269−3272.
83. Takanashi, K.; Lee, V. Ya; Ichinohe, M.; Sekiguchi, A. (η^5-Cyclopentadienyl)(η^4-Tetrasila- and η^4-Trisilagermacyclobutadiene)cobalt: Sandwich Complexes Featuring Heavy Cyclobutadiene Ligands. *Eur. J. Inorg. Chem.* **2007,** *2007* (35), 5471−5474.
84. Takanashi, K.; Lee, V. Ya; Sekiguchi, A. Tetrasilacyclobutadiene and Cyclobutadiene Tricarbonylruthenium Complexes: $[\eta^4\text{-}(tBu_2MeSi)_4Si_4]Ru(CO)_3$ and $[\eta^4\text{-}(Me_3Si)_4C_4]Ru(CO)_3$. *Organometallics* **2009,** *28* (4), 1248−1251.
85. Lee, V. Ya; Takanashi, K.; Sekiguchi, A. A Two-and-a-Half-Layer Sandwich: Potassium Salt of Anionic (η^4-tetrasilacyclobutadiene)(η^5-Cyclopentadienyl)ruthenium. *Dalton Trans.* **2010,** *39* (39), 9229−9231.
86. Lee, V. Ya; Ito, Y.; Yasuda, H.; Takanashi, K.; Sekiguchi, A. From Tetragermacyclobutene to Tetragermacyclobutadiene Dianion to Tetragermacyclobutadiene Transition Metal Complexes. *J. Am. Chem. Soc.* **2011,** *133* (13), 5103−5108.
87. Gilman, H.; Gorsich, R. D. Cyclic Organosilicon Compounds. II. Reactions Involving Certain Functional and Related Dibenzosilole Compounds. *J. Am. Chem. Soc.* **1958,** *80* (13), 3243−3246.
88. Liu, Y.; Stringfellow, T. C.; Ballweg, D.; Guzei, I. A.; West, R. Structure and Chemistry of 1-Silafluorenyl Dianion, Its Derivatives, and an Organosilicon Diradical Dianion. *J. Am. Chem. Soc.* **2002,** *124* (1), 49−57.
89. Choi, S.; Boudjouk, P.; Wei, P. Aromatic Benzannulated Silole Dianions. The Dilithio and Disodio Salts of a Silaindenyl Dianion. *J. Am. Chem. Soc.* **1998,** *120*, 5814−5815.
90. Choi, S.-B.; Boudjouk, P. A Novel α,ω Silyl Dianionic Salt. The Synthesis and Characterization of Remotely Connected Benzannulated Silole Monoanions. *J. Chem. Soc., Dalton Trans.* **2000,** *6*, 841−844.
91. West, R.; Sohn, H.; Bankwitz, U.; Calabrese, J.; Apeloig, Y.; Mueller, T. Dilithium Derivative of Tetraphenylsilole: An $\eta^1-\eta^5$ Dilithium Structure. *J. Am. Chem. Soc.* **1995,** *117* (46), 11608−11609.
92. Freeman, W. P.; Tilley, T. D.; Yap, G. P. A.; Rheingold, A. L. Silolyl Anions and Silole Dianions: Structure of $[K([18]crown\text{-}6)^+]_2[C_4Me_4Si^{2-}]$. *Angew. Chem. Int. Ed. Engl.* **1996,** *35* (8), 882−884.
93. Saito, M.; Yoshioka, M. The Anions and Dianions of Group 14 Metalloles. *Coord. Chem. Rev.* **2005,** *249* (7−8), 765−780.
94. Saito, M.; Haga, R.; Yoshioka, M.; Ishimura, K.; Nagase, S. The Aromaticity of the Stannole Dianion. *Angew. Chem. Int. Ed.* **2005,** *44* (40), 6553−6556.
95. Saito, M.; Sakaguchi, M.; Tajima, T.; Ishimura, K.; Nagase, S.; Hada, M. Dilithioplumbole: A Lead-Bearing Aromatic Cyclopentadienyl Analog. *Science* **2010,** *328* (5976), 339−342.
96. Dysard, J. M.; Tilley, T. D. Synthesis and Reactivity of η^5-Silolyl, η^5-Germolyl, and η^5-Germole Dianion Complexes of Zirconium and Hafnium. *J. Am. Chem. Soc.* **2000,** *122* (13), 3097−3105.
97. Yasuda, H.; Lee, V. Ya; Sekiguchi, A. Si_3C_2-Rings: From a Nonconjugated Trisilacyclopentadiene to an Aromatic Trisilacyclopentadienide and Cyclic Disilenide. *J. Am. Chem. Soc.* **2009,** *131* (18), 6352−6353.
98. Lee, V. Ya; Kato, R.; Ichinohe, M.; Sekiguchi, A. The Heavy Analogue of CpLi: Lithium 1,2-Disila-3-Germacyclopentadienide, a 6π-Electron Aromatic System. *J. Am. Chem. Soc.* **2005,** *127* (38), 13142−13143.
99. Lee, V. Ya; Kato, R.; Sekiguchi, A.; Krapp, A.; Frenking, G. Heavy Ferrocene: A Sandwich Complex Containing Si and Ge Atoms. *J. Am. Chem. Soc.* **2007,** *129* (34), 10340−10341.

100. Yasuda, H.; Lee, V. Ya; Sekiguchi, A. η^5-1,2,3-Trisilacyclopentadienyl—A Ligand for Transition Metal Complexes: Rhodium Half-Sandwich and Ruthenium Sandwich. *J. Am. Chem. Soc.* **2009,** *131* (29), 9902–9903.
101. Lee, V. Ya; Kato, R.; Sekiguchi, A. Heavy Metallocenes of the Group 8 Metals: Ferrocene and Ruthenocene Derivatives. *Bull. Chem. Soc. Jpn.* **2013,** *86* (12), 1466–1471.
102. Nishinaga, T.; Izukawa, Y.; Komatsu, K. The First Cyclic π-Conjugated Silylium Ion: The Silatropylium Ion Annelated With Rigid σ-Frameworks. *J. Am. Chem. Soc.* **2000,** *122* (38), 9312–9313.
103. Haaf, M.; Schmedake, T. A.; West, R. Stable Silylenes. *Acc. Chem. Res.* **2000,** *33* (10), 704–714.
104. Ishida, S.; Nishinaga, T.; West, R.; Komatsu, K. Generation and Aromaticity of 2-Silaimidazolium Ion, a New π-Conjugated Silylium Ion. *Chem. Commun.* **2005,** *No. 6*, 778–780.
105. Sekiguchi, A.; Matsuno, T.; Ichinohe, M. The Homocyclotrisilenylium Ion: A Free Silyl Cation in the Condensed Phase. *J. Am. Chem. Soc.* **2000,** *122* (45), 11250–11251.

SECTION 1.3

Organosilicon Compounds of Hypercoordinate Silicon

CHAPTER 11

Penta- and Hexacoordinated Silicon(IV) Compounds

Naokazu Kano

University of Tokyo, Tokyo, Japan

Chapter Outline

Organosilicon Compounds.
DOI: http://dx.doi.org/10.1016/B978-0-12-801981-8.00011-3

11.1 Introduction

A silicon atom in organosilicon compounds is mostly bounded to four atoms in the same manner as that in crystals of elemental silicon. On one hand, this structural motif resembles saturated aliphatic carbon analogs among a variety of organic compounds. On the other hand, silicon has a capacity for increase in the coordination number from four to five and six in some cases, making a clear difference from the organic compounds, which cannot expand the coordination number to greater than four. The organosilicon compounds containing pentacoordinated and hexacoordinated silicon atoms show interesting structures, reactivities, and properties, which are different from common tetracoordinated organosilicon compounds. These compounds are often called *hypercoordinate* silicon compounds because of the expansion of their coordination number. Although the term "hypervalent" has also been used to designate these compounds, it has been under criticism for a long time.[1] There have been several excellent reviews on the highly coordinated organosilicon compounds to date. Fundamental properties, structural aspects, and reactivities were summarized in these prior reviews.[2–11] Applications of stable and transient hypercoordinate organosilicon compounds to organic synthesis were summarized in other outstanding review papers.[12,13] In this section, the author focuses on only the publication from year 2000 to survey recent advances in the chemistry of stable hypercoordinate organosilicon compounds. Synthesis, structures, reactivities, and properties of the isolated hypercoordinate silicon compounds are the main issues of this section rather than theoretical studies on these compounds, although theoretical calculations, promoted by advances in computer technology, have contributed to disclose reasons for their stability and unique structures of the hypercoordinate states.[14–24]

11.2 Hypercoordinate Silicon Compounds Bearing Silicon–Halogen Bonds

Silicon–fluorine and silicon–chlorine bonds are often found in stable hypercoordinate silicon compounds because their electron-withdrawing ability and small atomic size help for the stabilization of the expanded coordination states. On one hand, the silicon–chlorine bond sometimes derives from the starting compounds such as $SiCl_4$ and/or chloride ion which adds to tetracoordinated silanes.[25] On the other hand, the silicon–fluorine bond in hypercoordinate silicon compounds can be formed by nucleophilic addition reactions of fluoride ion to Lewis acidic silicon compounds,[26,27] conversion of a silicon–chlorine bond with a fluorinating reagent,[28,29] and insertion of a donor-stabilized silylene into a carbon–fluorine bond.[30,31] For a peculiar example of combination of the first and second methods, the reaction of pentafluoroethyllithium with $SiCl_4$ gave the pentacoordinated difluorosilicate and hexacoordinated trifluorosilicate by fluoride abstraction of a high Lewis acidic intermediary tris(pentafluoroethyl)silane derivative (Fig. 11.1).[32,33] Silicates bearing

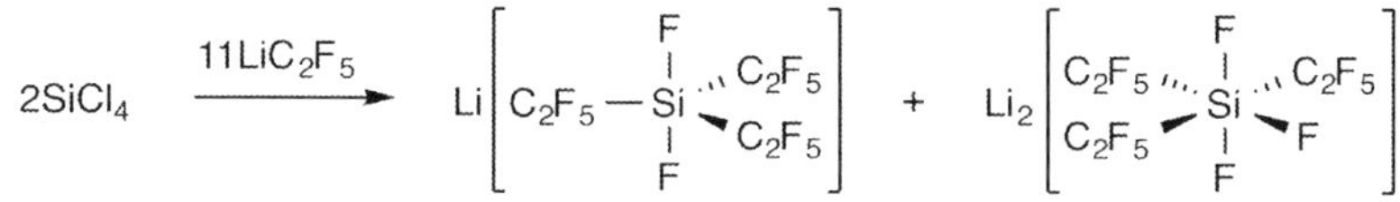

Figure 11.1
Synthesis of pentacoordinated difluorosilicate and hexacoordinated trifluorosilicate.

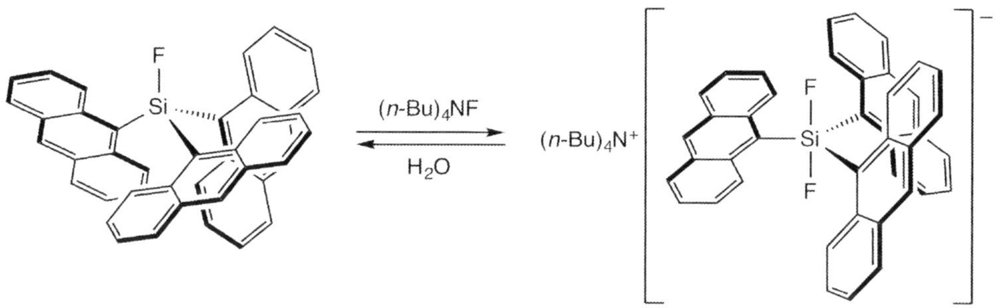

Figure 11.2
Interconversion between tri(9-anthryl)fluorosilane and the corresponding difluorosilicate.

up to five fluorine atoms are often used as counter anions for crystallographic characterization of some cationic species.[34–38]

The changes of the structure upon the Si–F bond formation process accompanying the increase of coordination number of silicon have been applied to development of fluorescent fluoride ion sensors. Tri(9-anthryl)fluorosilane reacts reversibly with a fluoride ion to form the corresponding difluorosilicate, to cause intense visible fluorescence emission (Fig. 11.2).[39,40] The fluorescent quantum yield increases considerably from 0.033 to 0.64 with about 20 nm hypsochromic shift of the emission maxima upon the addition of fluoride ion. The degree of through-space interaction between the anthryl groups changed due to the pentacoordination, and this change was the origin of the optical changes.

Likewise, transformation of the cationic di(9-anthryl)fluorosilane bearing dimethylsulfonium moiety to the difluorosilicate by binding to fluoride ion induced a blue shift of the anthryl-based absorption bands (Fig. 11.3).[41] The sulfonium moiety enhanced the fluoride ion affinity. The binding to the fluoride anion was assisted via inductive and Coulombic effects based on "onium-based strategy." Furthermore, the fluoride ion was engaged in a bonding donor–acceptor interaction involving a fluorine lone pair as a donor and a carbon–sulfur σ*-orbital as the acceptor. Similar assistance for the fluoride ion binding was found in *o*-(fluorosilyl)(dimesitylboryl)benzenes.[42]

Tetracoordinated spirosilanes bearing two sets of chelate ligands showed high affinity toward fluoride ion,[43] and capture of fluoride ion caused the shift of the maximum emission

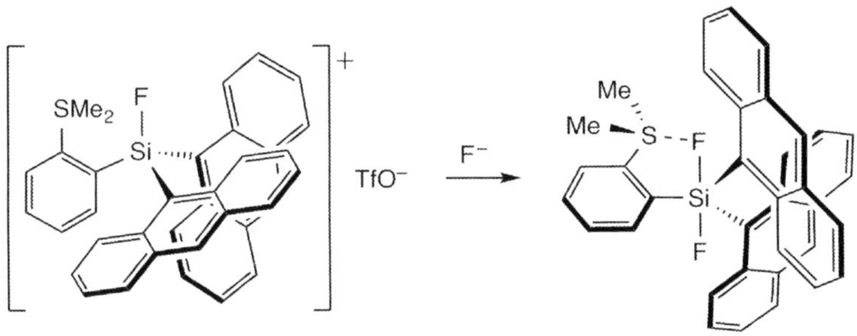

Figure 11.3
Fluoride ion binding of the cationic di(9-anthryl)fluorosilane bearing dimethylsulfonium moiety.

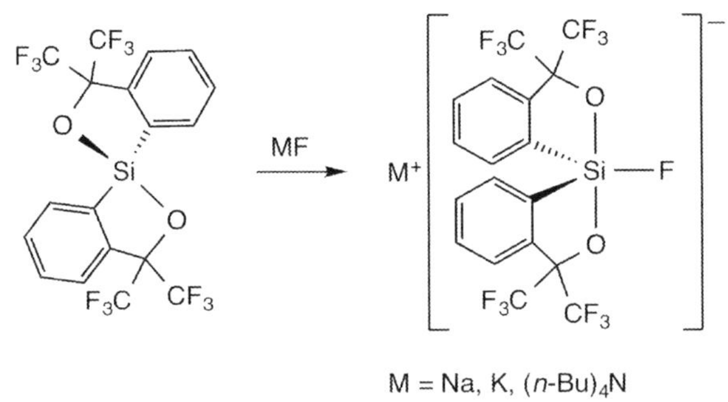

Figure 11.4
Fluoride ion binding of the spirosilane to form the fluorosilicate.

of fluorescence from 290 to 311 nm in dichloromethane (Fig. 11.4).[44] Formation of the fluorosilicate is quick and selective, even in the presence of other naturally abundant anions such as phosphate.

The structural change upon the hypercoordination associated with Si–F bond formation was applied to formation of a molecular gear system. Interconversion between bis(4-methyl-9-triptycyl)difluorosilane and the corresponding trifluorosilicate plays a role of molecular gear system (Fig. 11.5).[45] Rapid gear rotation of a meshed gear of the silane is declutched by the hypercoordination forming the silicate, which rotates around the Si–C bonds without correlation. This silicate is also interesting as an example of pentacoordinated silicates against the Muetterties rule, which describes the apicophilicity of a group in relation to electronegativity.[3]

Tetrabutylammonium triphenyldifluorosilicate (TBAT) was found to be a versatile reagent with numerous synthetic applications (Fig. 11.6).[46] Despite there being some similar difluoro(triorgano)silicates used as fluorinating reagents, such as tris(dimethylamino)sulfonium (trimethylsilyl)difluorosilicate (TASF) and its derivatives,[47–50] TBAT is one of the ideal fluoride sources because it can be obtained as

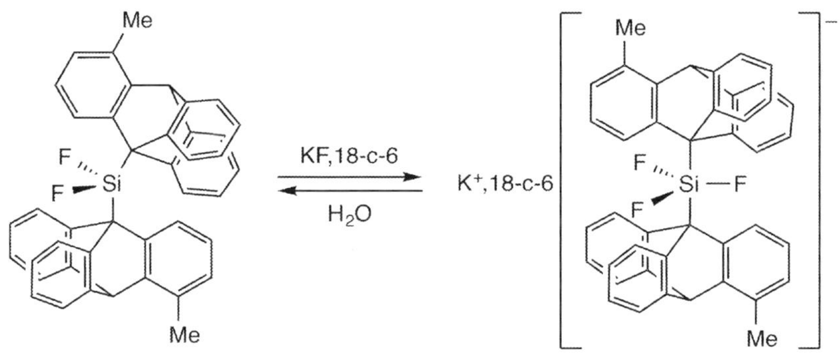

Figure 11.5
A molecular gear system using interconversion between bis(4-methyl-9-triptycyl)difluorosilane and the corresponding trifluorosilicate.

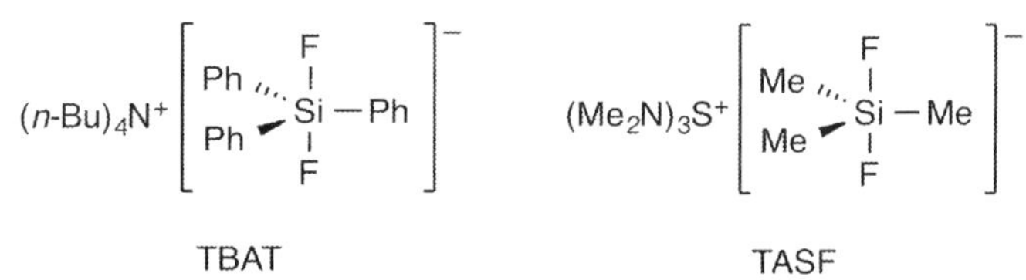

Figure 11.6
Fluorinating reagents, TBAT and TASF.

anhydrous and nonhygroscopic crystalline solid, which is soluble to most organic solvents. In addition to the application to fluorination reactions,[51] TBAT is utilized for the intramolecular addition reactions of allylsilane to 2,3-dihydro-4-pyrridones[52] and the palladium-catalyzed arylation reactions of allylic benzoates[53] and aryl halides.[54]

11.3 Hypercoordinate Silicon Compounds Bearing Silicon–Nitrogen Bonds

11.3.1 Pyridine- and N-Containing Heterocycles Ligands

Silicon can be penta- and hexacoordinated by additional coordination of a highly Lewis basic donor, if it exists in the reaction medium, to a highly Lewis acidic silicon moiety. For example, each dichlorosilane and trichlorosilane forms the 1:2 adducts with pyridines (Fig. 11.7).[55–57] Two molecules of pyridines coordinate to the silicon of $HSiCl_3$ through intermolecular N–Si interaction. The Si–N bond lengths range between 1.975(3) and 1.989 (3) Å in the crystalline state. The adducts are stable in the solid state, but tend to dismutate to the pyridine adducts of H_2SiCl_2 and $SiCl_4$ upon treatment with polar solvents or heating. The dismutation reactions can be prevented by substitution of one chlorine atom of

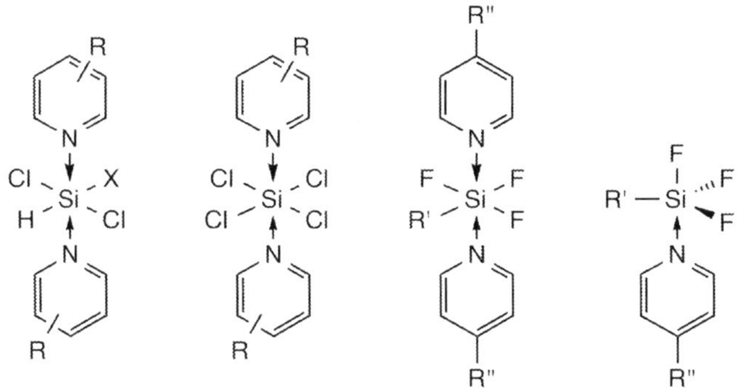

Figure 11.7
Penta- and hexacoordinated halosilanes coordinated by pyridines.

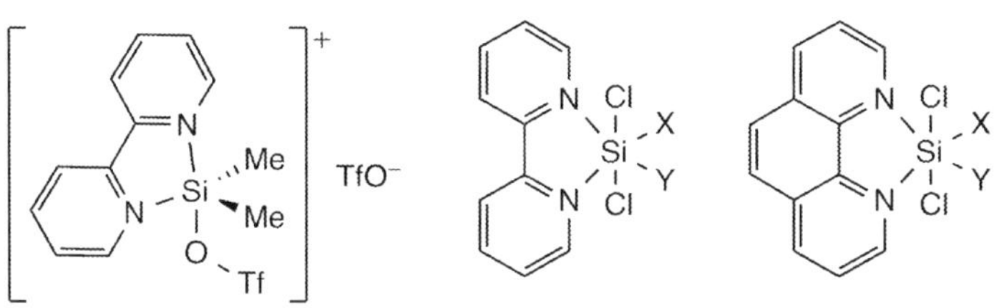

Figure 11.8
Penta- and hexacoordinated silicon-bipyridine complexes.

trichlorosilane with a methyl group. Trifluoro(pentafluorophenyl)silane and 4-methoxypyridine also form the 1:2 adduct at room temperature in a highly concentrated solution.[58] In contrast, trifluoro(phenylethynyl)silane, trifluorohexylsilane, and trifluorophenylsilane form the corresponding 1:1 pyridine-adducts featuring pentacoordinated silicon atoms in solution at room temperature.[58,59] Binding constant for the formation of the pentacoordinated trifluorophenylsilane—4-methoxypyridine complex is $550 \pm 100\ M^{-1}$ at 25°C in benzene. Both trifluoro(phenylethynyl)silane and trifluorophenylsilane equilibrate with the corresponding hexacoordinated 1:2 pyridine complexes, respectively, at low temperatures.

Hypercoordinate silicon compounds coordinated by a bidentate *N,N′*-chelating 2,2′-bipyridine (bipy) ligand are thermodynamically more stable, in general, than those coordinated by two pyridine ligands. Actually, treatment of $Me_2Si(OTf)_2$ with bipy gave the pentacoordinated cationic silicon complex, $[Me_2Si(bipy)(OTf)][OTf]$, in which one of the triflates is coordinating to the silicon in a distorted trigonal bipyramidal (TBP) geometry (Fig. 11.8).[60] It contrasts with the formation of the tetracoordinated dicationic silicon

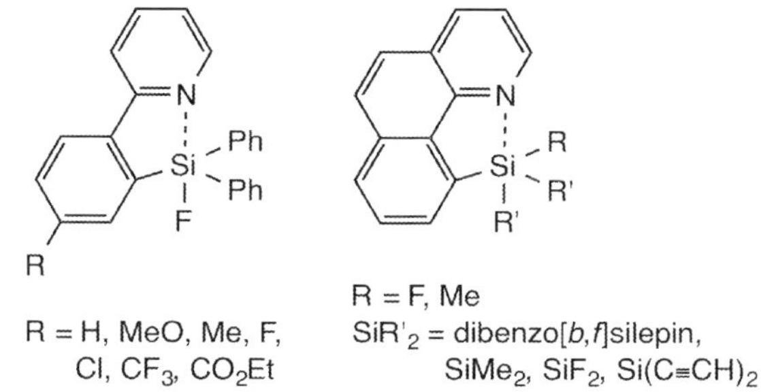

Figure 11.9
Pentacoordinated 2-(2-pyridinyl)phenylsilanes and benzo[*h*]quinolin-10-ylsilanes.

complex, $[Me_2Si(dmap)_2][OTf]_2$, with two equivalents of 4-(dimethylamino)pyridine (dmap). The less steric demands of the essentially planar bipy ligand bring the additional coordination of an oxygen of a triflate. Chlorosilanes such as H_2SiCl_2, $HSiCl_3$, and $RSiCl_3$ (R = Me, Ph) also form the isolable adducts with 2,2′-bipyridine.[61] In the octahedral geometry around the hexacoordinated silicon center, the carbon substituents and hydrogen atoms are exclusively *trans*-disposed to the *N*-donor atoms of bipy ligand. The resembling complexes with 1,10-phenanthroline (phen) show the same tendency.

2-Phenylpyridines form both a Si–C covalent bond and a Si–N dative bond in the fluorosilanes, making the silicon atoms pentacoordinated (Fig. 11.9). The 2-phenylpyridines are straightforwardly introduced onto the silicon atom by *ortho*-selective C–H silylation with fluorodiphenylsilane and Ir(acac)(cod) catalyst.[62] Alternatively, the C–H fluorosilylation of 2-phenylpyridines with a palladium catalyst $[Pd(MeCN)_4](BF_4)_2$ was used for the synthesis of the pentacoordinated 2-(2-pyridyl)phenylsilanes, which exhibit fluorescence.[63] Benzo[*h*]quinoline is also used for the *C,N*-bidentate ligand to form the neutral pentacoordinated silicon compounds that also fluoresce.[64,65]

Besides bipyridine, 2-phenylpyridines, and benzo[*h*]quinolone ligands, the pyridines with an anchoring side arm at the 2-position serve hypercoordinate silicon compounds as bidentate ligands to form a chelate ring. For the side arms to bind pyridines to the silicon atom, $C(SiMe_3)_2$,[66] $CPh(SiMe_3)$, and $NSiMe_3$[67] groups as well as sulfur atom[68] were used to furnish the hypercoordinate sphere to silicon (Fig. 11.10). Despite small N–Si–E bite angles [69.95(9)–74.69(14) degrees] in the four-membered chelate rings (E = C, N, S), the pyridine moieties intramolecularly interact with the silicon atom through short enough Si–N bond lengths between 1.893(3) and 2.072(4) Å.

Moreover, a dianionic tridentate ligand including pyridine is useful for the hypercoordination of silicon. The pentacoordinated silicon compounds derived from *N* -(2-pyridylmethyl)salicylamide[69] and 2-{[(pyridin-2-yl)methyl]amino}benzenethiol[70] show TBP geometry, while those derived from 2,6-pyridinebis(1,1-diarylmethanol)[71] show

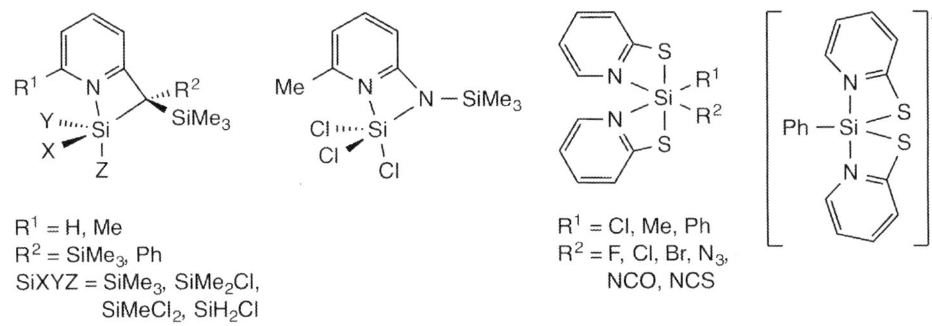

Figure 11.10
Penta- and hexacoordinated silicon compounds coordinated by pyridines bearing an anchoring side arm.

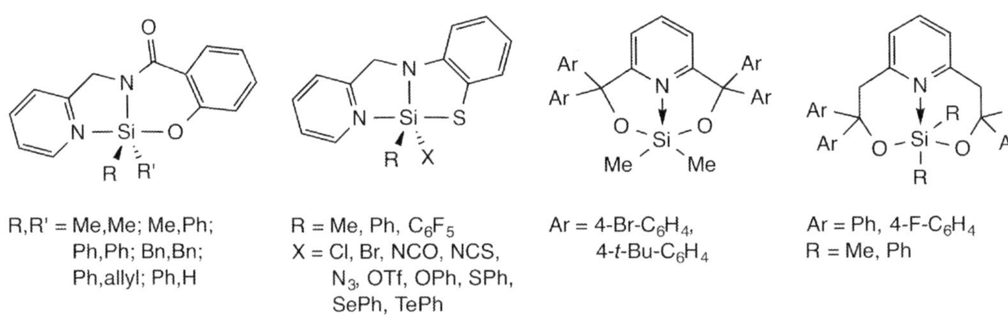

Figure 11.11
Penta- and hexacoordinated silicon compounds bearing tridentate pyridine-containing ligands.

square pyramidal (SP) geometry (Fig. 11.11). The intramolecular N–Si coordination, which ranges between 1.950(8) and 2.069(1) Å, constitutes two fused five-membered rings. Increase of the chelate ring-size from five to six leads to a change in the conformation from SP to TBP, resulting in lengthening of the intramolecular N–Si distances [2.570(4)–2.665(2) Å].[72]

Nitrogen-containing heterocyclic compounds other than pyridine can be used for coordination to silicon if they are incorporated with an anchoring ligand. 1-Methyl-2-mercaptoimidazole (methimazole),[73] 8-oxyquinoline,[74] 8-thioquinoline,[75] and bis{1,3,4,6,7,8-hexahydro-2*H*-pyrimido[1,2*a*]pyrimidinyl}methane,[76,77] were used for the coordination to the silicon (Fig. 11.12). In the silacycloalkanes bearing two 8-quinolinate ligands, the N–Si coordination attenuates with increasing the ring size.[74] As a result, whereas the silacyclobutane and silacyclopentane feature the hexacoordinated silicon atom, the silacyclohexane and silacycloheptane show an equilibrium between the tetra- and hexacoordinated states of the silicon.

The bidentate coordination of the nitrogen-containing heterocycle ligands is expanded to tridentate coordination for some hypercoordinate silicon compounds. Pentacoordination of

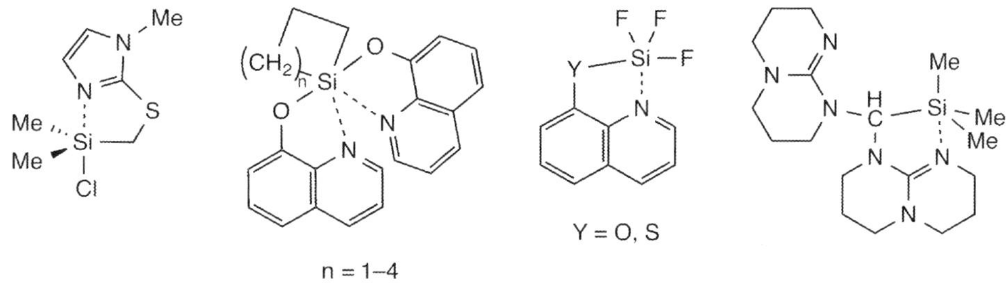

Figure 11.12
Penta- and hexacoordinated silicon compounds chelated by bidentate ligands including nitrogen-containing heterocycles.

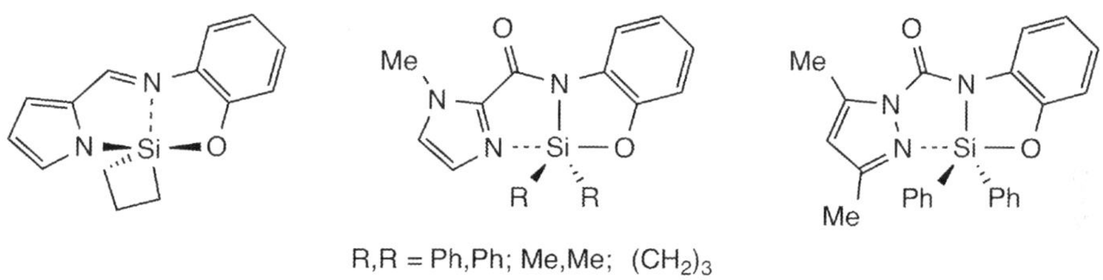

Figure 11.13
Pentacoordinated silicon compounds chelated by bidentate ligands including nitrogen-containing heterocycles.

silicon is achieved by using the *N,N′,O* tridentate ligands including pyrrole,[78] *N*-methylimidazole,[79,80] and 3,5-dimethylpyrazole (Fig. 11.13).[79] In the latter two compounds, the amide-nitrogen is covalently bound to silicon as an anionic ligand.

11.3.2 Imine Ligands

In the previous examples, a lone pair of nitrogen atom of *N*-containing heterocycles is utilized for the coordination. However, the donor nitrogen atom is not always involved in heterocycles for hypercoordination to silicon. The nitrogen atom of an acyclic imine is also used for hypercoordination to silicon in many compounds. For example, both *N*-methyl-*o*-oxyacetophenoneiminato[81] and β-enaminoiminato[82] ligands are found to be effective for the hypercoordination (Fig. 11.14).

Imines bearing an additional coordination site can easily be introduced on the imine-nitrogen by treatment of the corresponding ketone and aldehyde precursors with functionalized amines. Accordingly, a variety of tridentate imine ligands have been synthesized and applied to achieve hypercoordination to silicon. Dianionic tridentate *O,N,O′*-ligands bearing an imine moiety, which is conjugated with alkene[83–86] and benzene,[87,88] are often used for stabilization of hypercoordinate silicon compounds

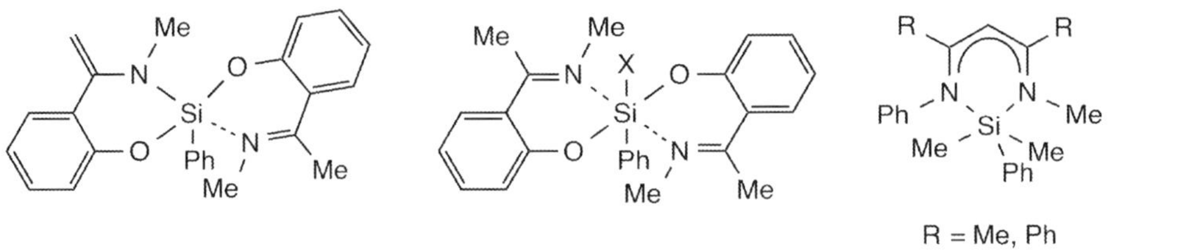

Figure 11.14
Penta- and hexacoordinated silicon compounds chelated by bidentate imine-containing ligands.

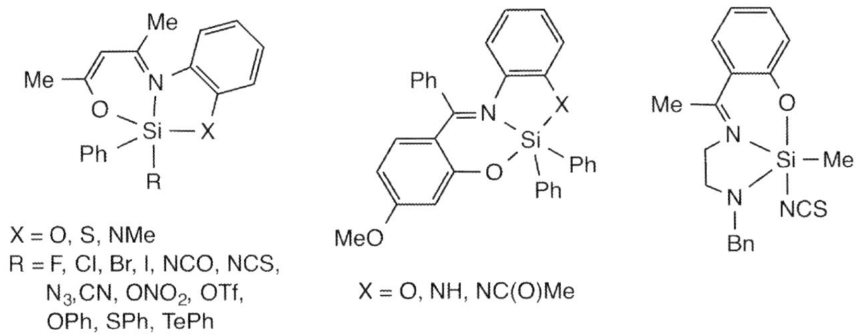

Figure 11.15
Pentacoordinated silicon compounds chelated by tridentate imine-containing ligands.

(Fig. 11.15). Two other coordination sites in the pentacoordinated silicon compounds are occupied by phenyl, halogen, pseudohalogen, chalcogenophenyl, and so on. The analogous dianionic tridentate *O,N,S*-ligand instead of the *O,N,O′*-ligands similarly stabilizes the pentacoordinated silicon compounds, resulting in formation of a silicon–sulfur bond.[89–91] The *O,N,N′*-ligands are also combined with halogen, pseudohalogen, and chalcogenophenyl groups to accomplish the hypercoordinate states.[84–86,92]

Combination of one tridentate and one bidentate ligand on the same silicon atom effectively stabilize the hypercoordination state. The bidentate ligands such as propylene,[93] 8-quinolinolato,[94] 2-oxidobenzoate,[95] and β-diketonato ligands occupy two coordination sites of penta- or hexacoordinated silicon compounds (Fig. 11.16).[96,97] Synthesis of hexacoordinated silicon compounds with a CF_3-containing β-diketonato ligand such as hexafluoroacetylacetonato (hfacac) was attempted, but migration of a nitrogen or sulfur to a CF_3-bound carbon atom proceeded to form the pentacoordinated silicon compounds coordinated by the resulting tetradentate ligand.

Some tridentate ligands containing an asymmetric carbon atom are found to stabilize pentacoordinated silicon compounds.[98,99] An enantiomeric pure pentacoordinated silicon compound, which showed optical rotation, was synthesized starting from L-phenylalanine

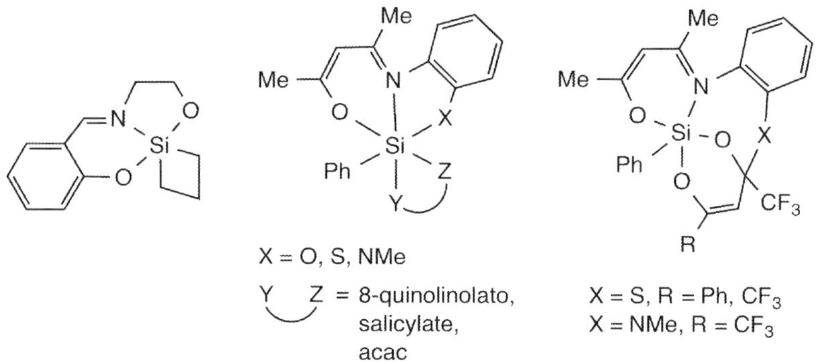

Figure 11.16
Penta- and hexacoordinated silicon compounds chelated by both bi- and tridentate ligands, and its derivatives.

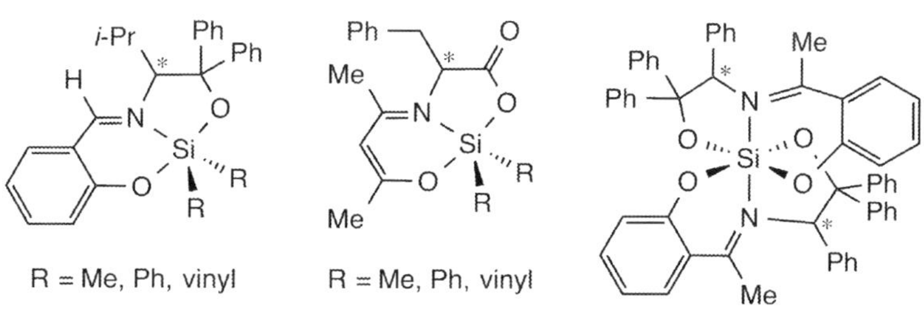

Figure 11.17
Chiral penta- and hexacoordinated silicon compounds chelated by tridentate ligands containing an asymmetric carbon atom.

(Fig. 11.17). Introduction of two sets of another dianionic tridentate chiral ligand on the silicon atom is an easy access to chiral hexacoordinated silicon compounds.

Neutral hexacoordinated silicon compounds are available by using twofold coordination of the identical achiral tridentate ligands (Fig. 11.18).[83,87,93] Although each tridentate ligand occupies three meridional positions in most cases, tridentate *N*-(2-carbamidophenyl)imine ligands, which form both five- and seven-membered rings by the coordination to the silicon center, occupy three facial positions.[100]

11.3.3 Amidinato Ligands

There have been many hypercoordinate silicon compounds bearing amidinato ligands, which coordinate to silicon in a bidentate fashion. Equimolar reactions of the lithium amidinato reagents, prepared by addition of alkyl- or aryllithium to carbodiimides, with tetra- and trichlorosilane gave the corresponding pentacoordinated tri- and dichlorosilanes, respectively (Fig. 11.19).[28,29,101–104] The bulkiness around the silicon atoms can easily be

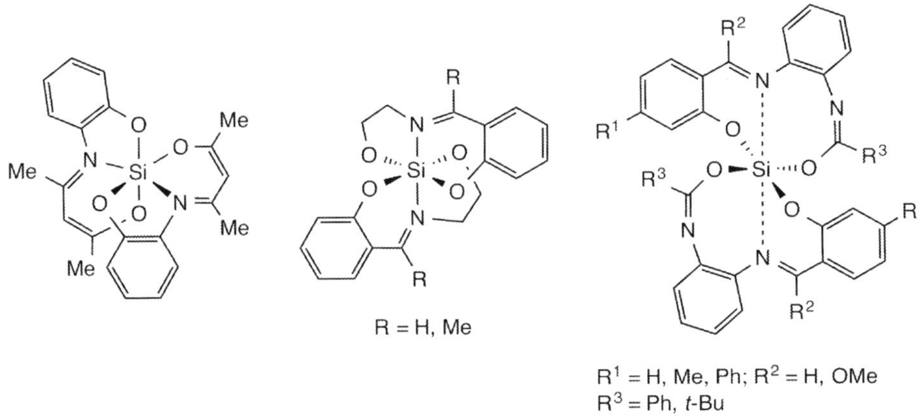

Figure 11.18
Hexacoordinated silicon compounds chelated by two tridentate ligands containing an imine moiety.

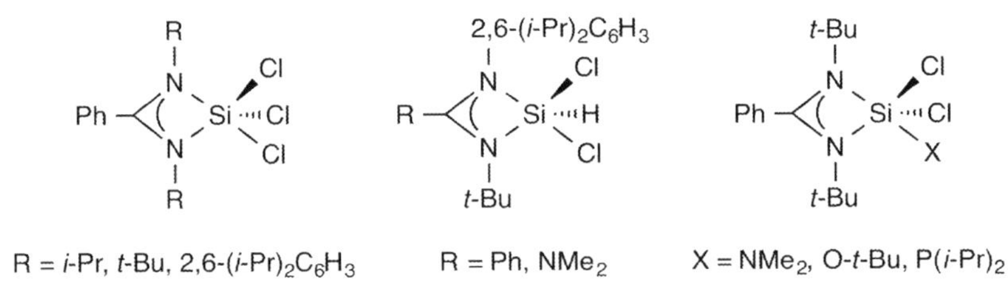

Figure 11.19
Pentacoordinated silicon compounds chelated by amidinato ligands.

tuned by using a different carbodiimide precursor having another substituent on the nitrogen atoms.[103] The crystal structure of the trichlorosilane exhibits a distorted TBP geometry around the silicon atom showing different Si–Cl bond lengths (2.0711(14)–2.1449(14) Å).[105] Two nitrogen atoms of the amidinato ligand are unsymmetrically bound to the silicon center, featuring different Si–N bond lengths (1.780(3) and 1.931(3) Å) and an acute bite angle (70.14(12) degrees). The dichlorosilanes chelated by the amidinato ligand show similar structural parameters.[103,106] Substitution reactions of the (amidinato)trichlorosilanes leads to formation of the amino-, alkoxy-, and phosphino-substituted derivatives.[107]

Reactions of the lithium amidinato reagents with tetrachlorosilane, trichlorosilane, and tetrabromosilane in molar ratio 2:1 afforded the corresponding hexacoordinated silanes bearing two sets of the amidinato ligands (Fig. 11.20).[106,108,109] Both two chloro ligands of the bis(amidinato)dichlorosilanes can be replaced with other halogen, pseudohalogen, and benzene-1,2-dichalcogenolato ligands, which present *cis* configuration in a strongly distorted octahedron geometry.[108] These configurations are stable in the solution states, as confirmed

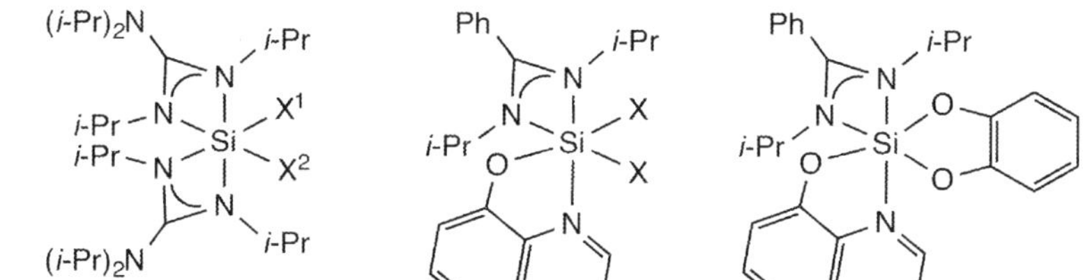

$X^1 = X^2$ = F, Cl, Br, CN, N_3, NCO, NCS
X^1 = H, X^2 = Cl

X = F, Cl, Br, CN, N_3, NCO, NCS

Figure 11.20
Hexacoordinated silicon compounds chelated by amidinato ligands.

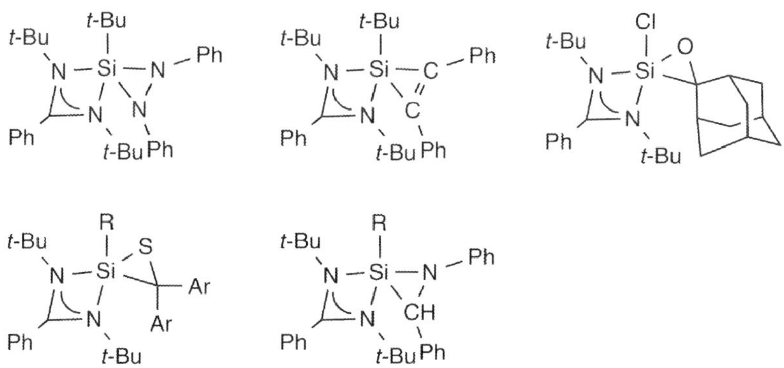

R = Cl, *t*-Bu
Ar = 4-Me_2N-C_6H_4

R = *t*-Bu, NPh_2

Figure 11.21
Three-membered ring compounds containing a pentacoordinated silicon atom chelated by an amidinato ligand.

by the variable-temperature (VT) nuclear magnetic resonance (NMR) study. However, the difluorosilane derivative exchanged two nitrogen sites of the amidinato ligands. One or two more bidentate ligands in addition to the amidinato ligands were introduced to the silicon center to stabilize the hexacoordinated state. For example, the amidinato, 8-quinolinolato, and catecholato ligands were adequately introduced to form a hexacoordinated silicon complex.[29] The VT-NMR experiments showed exchange processes of two nitrogen sites of the amidinato ligand and two binding sites of the 8-quinolinolato ligand in solution.

Moreover, a variety of hypercoordinate silicon compounds bearing amidinato ligands are available by addition reactions of tricoordinated donor-stabilized silylenes (see Chapter 8: Stable Silylenes and Their Transition Metal Complexes). Some of the products have remarkable three-membered ring systems such as the SiN_2,[110] SiC_2, SiCO,[111] SiCS,[112,113] and SiCN,[110,114] which contains a pentacoordinated silicon atom (Fig. 11.21). Deviation

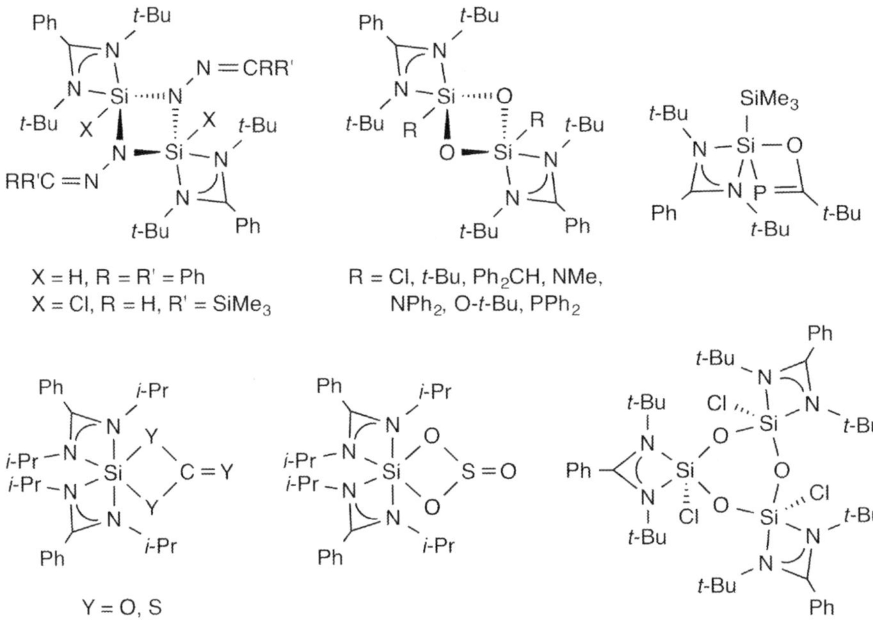

Figure 11.22
Four-membered ring compounds containing penta- and hexacoordinated silicon atoms chelated by amidinato ligands.

from TBP to SP structures is remarkable in these three-membered ring compounds. The silicon atoms of the silirene and siladiaziridine are nearly in a SP geometry.

Furthermore, there are some four-membered ring compounds containing one or two pentacoordinated silicon atoms. The silicon atoms in a distorted TBP geometry are incorporated in almost planar Si_2O_2,[115–117] Si_2N_2,[118,119] and SiCOP[120] four-membered rings (Fig. 11.22). Hexacoordinated silicon atom is found in the $SiCS_2$,[121] $SiCO_2$,[122] and SiO_2S[123] four-membered rings, and two chalcogen atoms are inevitably situated in the *cis* positions around the silicon atom. Furthermore, the ring size has been expanded from four to six, and the Si_3O_3 ring including three pentacoordinated silicon atoms is situated in a slightly distorted planar configuration.[124]

11.3.4 Salen Ligands

Salen ligand, in which two salicylaldimine moieties are bridged by ethylene linker, contributes to stabilize hexacoordinated silicon compounds as a tetradentate ligand (Fig. 11.23). Insufficient electron-withdrawing ability of the *trans*-group occasionally causes dissociation of a leaving group such as chloride from the silicon atom, resulting in formation of the corresponding cationic pentacoordinated silicon complex.[125] If there is a

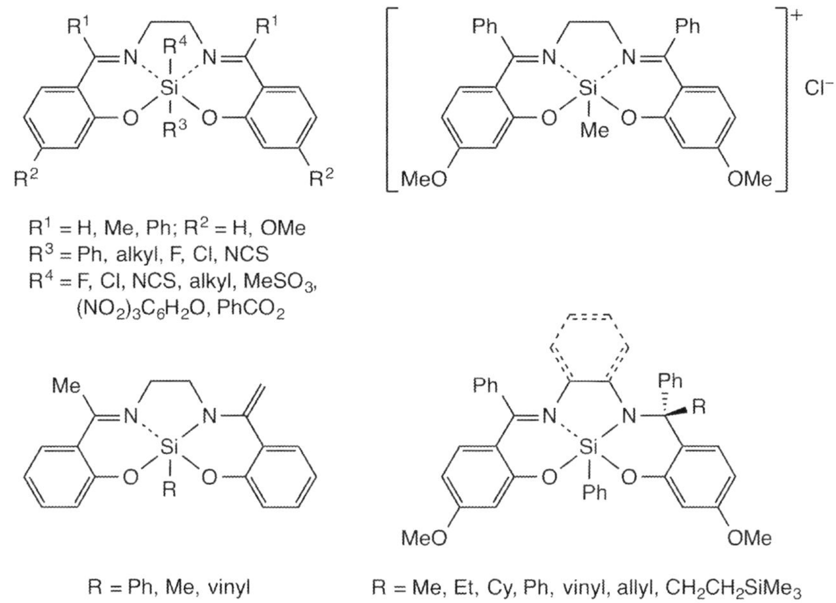

Figure 11.23
Hypercoordinate silicon compounds chelated by salen-type ligands and derivatives.

methyl group on the imine carbon of the salen ligand, a pentacoordinated silicon complex bearing a trianionic enamine-functionalized ligand is generated by the imine–enamine tautomerization, which is enhanced by a base included in the reaction medium.[126] Treatment of the enamine complexes with Brønsted acids caused regeneration of the hexacoordinated silicon–salen complexes by protonation of the enamine moiety and addition of the conjugate base of the acids to the silicon.[126,127] In the hexacoordinated silicon complexes, a silicon–carbon(alkyl) bond was cleaved upon UV light irradiation. The Si–C bond cleavage was accompanied by rearrangement of the alkyl group to an imine carbon, forming a pentacoordinated silicon complex coordinated by a trianionic tetradentate amide-imine ligand.[128–130] Such photo-driven Si–C bond cleavage reactivity was also found in similar silicon–salphen complexes.[131]

In the pentacoordinated silicon complexes bearing the salen or salen-derived ligand, the silicon atom is always situated in a distorted TBP coordination sphere with a nitrogen and oxygen atom in the apical positions. In the hexacoordinated silicon–salen complexes, the salen ligand surrounds the silicon atom and two other groups are usually *trans*-situated in the crystal, whereas two other isomers are observed in solution.[132] Only the silacyclobutane complex bearing the salen-type ligand shows the all-*cis* arrangement of the O, N, and C atoms in the crystal because the four-membered ring forces the salen-type ligand to this torsion.[133]

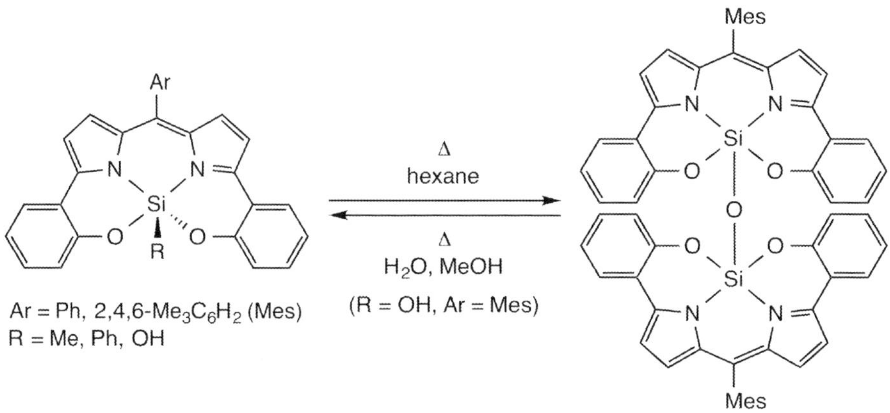

Figure 11.24
Pentacoordinated silicon compounds chelated by a dipyrrin-based ligand.

Another tetradentate open-chain N_2O_2-type ligand, which features a dipyrrin moiety like boron-dipyrromethene (BODIPY), forms pentacoordinated silicon complexes (Fig. 11.24).[134,135] The colored dipyrrin–silicon complexes emit intense red or near-IR fluorescence (λ_{FL} 626–662 nm) with their fluorescent quantum yields up to 0.81. The optical properties change reversibly upon conversion of the silanol to the disiloxane derivative.

11.3.5 Phthalocyanine and Porphyrin Ligands

Macrocyclic tetradentate ligands such as phthalocyanine and porphyrins are suitable to stabilize a hexacoordinated state of silicon through four polar Si–N bonds. As a matter of fact, phthalocyanine surrounds a silicon atom to form various stable hexacoordinated silicon complexes (SiPc) (Fig. 11.25).[136] The silicon atom has an octahedral geometry in the center of the phthalocyanine ring and two other groups are forced to occupy the *trans* positions in the SiPc. The SiPc derivatives with electron-withdrawing axial ligands showed slight redshift of the Q bands and relatively lower fluorescence quantum yields compared to the SiPc derivatives with electron-donating axial ligands.[137] The nature of the axial ligands on the SiPc has an effect on the aggregation tendency and low solubility, both of which restricts their application. Solubility in common organic solvents and water is improved by employment of axial ligands with a long alkyl chain and quaternized ammonium moieties.[138–143] These ligands on the silicon also prevent the aggregation of the SiPcs. Photoresponsive property of some SiPc derivatives are promising for organic photovoltaic devices,[144,145] photocatalysts for hydrogen generation,[146] and photodynamic therapy agents.[147] Combination of the SiPc framework with siloxane linkages makes the cofacial, oxygen-bridged SiPc oligomers in addition to its monomer.[148,149] In these frameworks,

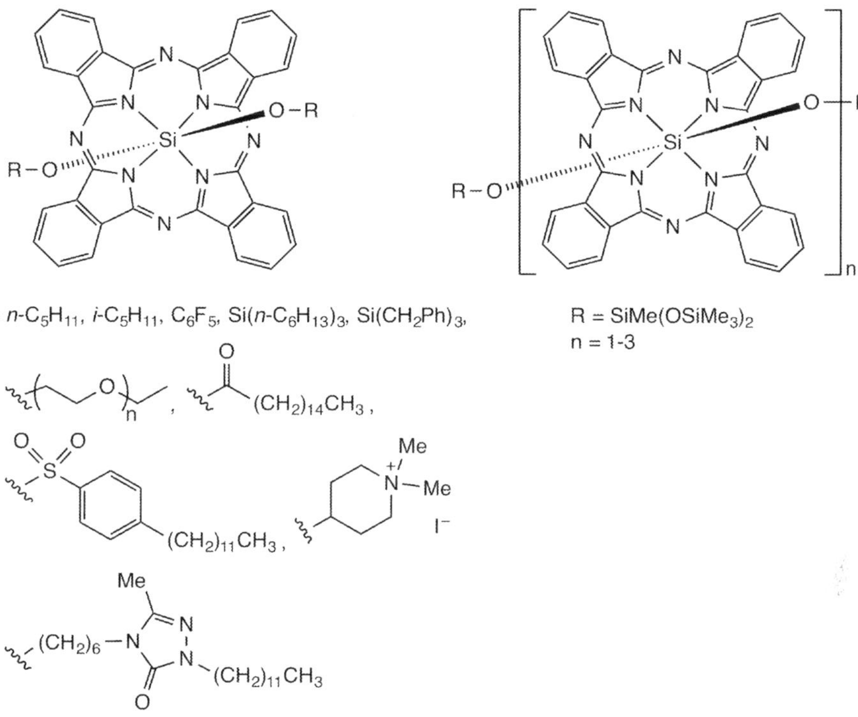

Figure 11.25
Hexacoordinated silicon compounds chelated by phthalocyanine ligand.

hexacoordinated silicon atoms are aligned on the line of the axial axis with each interatomic Si−Si distance of 3.314−3.330 Å.

Similarly, porphyrins constitute a framework of hexacoordinated silicon complexes as a macrocyclic tetradentate ligand together with two axial ligands. In the tetraphenylporphyrinato (TPP) silicon complexes with both a phenyl and methyl group (Fig. 11.26), the silicon−carbon(methyl) bond is selectively cleaved upon irradiation and chemical oxidation, giving new phenyl substituted silicon porphyrins.[150] More complicated *meso*-aryl [28]hexaphyrin was used to form pentacoordinated silicon complexes, which exhibited Möbius aromaticity.[151,152] The UV/vis/near-IR absorption spectra of the silicon complexes showed red-shifted Soret-like band and enhanced Q-like bands compared with those of the free ligand.

11.3.6 Amine Ligands

In the aforementioned hypercoordinate silicon compounds shown so far, a nitrogen atom coordinating to silicon is incorporated in a π-bond or aromatic ring in the ligand. However, tertiary amines effectively coordinate to the silicon atom as well. Especially, if the nitrogen

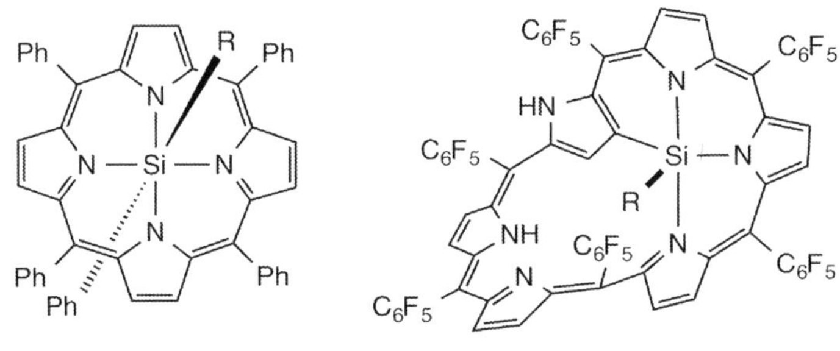

Figure 11.26
Hypercoordinate silicon compounds chelated by tetraphenylporphyrin and *meso*-aryl [28] hexaphyrin ligands.

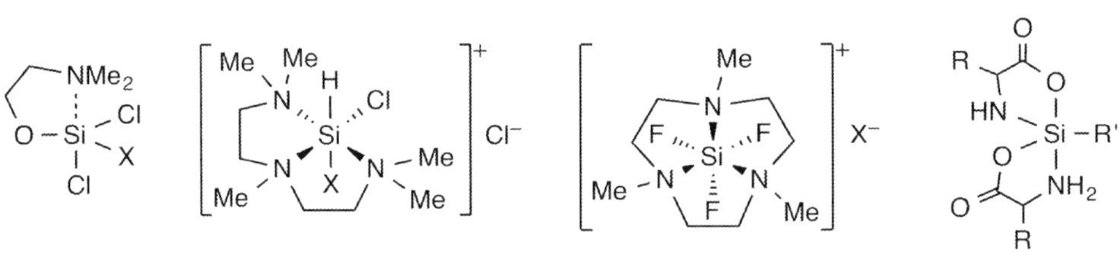

Figure 11.27
Hypercoordinate silicon compounds chelated by amine-containing ligands.

atom is integrated into a chelate ligand to form a five-membered ring by the coordination, it effectively stabilizes the hypercoordinate state. Both $Cl_2HSiOCH_2CH_2NMe_2$ and $Cl_3SiOCH_2CH_2NMe_2$ adopt five-membered chelate ring structures involving pentacoordinated silicon atoms through short intramolecular Si–N dative bonds [2.037(2) Å, 2.060(2) Å] in the crystal (Fig. 11.27).[153] Nevertheless, the dative bonds are labile in solution and dissociate upon increasing the temperature. Instead of the labile bidentate ligand, *N,N,N′,N″,N″*-pentamethyldiethylenetriamine (pmdeta) was used for firm coordination toward silicon to be hexacoordinated with a static framework.[154] Since the number of coordinating sites increase, the hypercoordinate structures are stabilized much more. The triaza macrocycle *N,N′,N″*-trimethyl-1,4,7-triazacyclononane (Me_3tacn) also stabilizes the cationic hexacoordinated silicon.[155] In addition to the tertiary amines, primary amine fragments are also used for the hypercoordination. Actually, some natural and unnatural α-amino acids such as (*S*)-alanine, (*S*)-phenylalanine, and (*S*)-*tert*-leucine act as mono- and dianionic bidentate *O,N*-ligands for the stable pentacoordinated silicon compounds.[156]

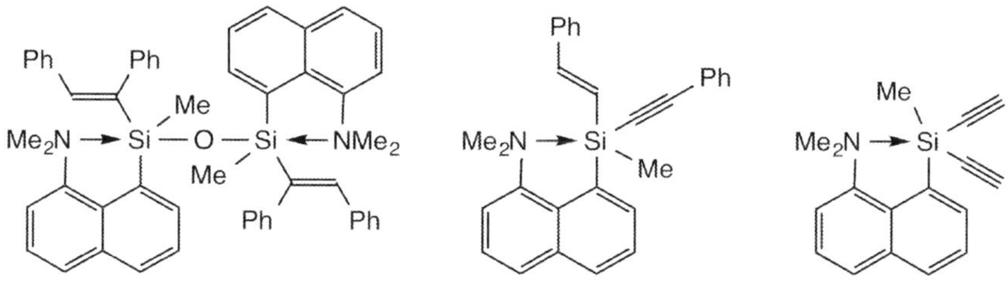

Figure 11.28
Hypercoordinate silicon compounds chelated by 8-(dimethylamino)naphthyl ligands.

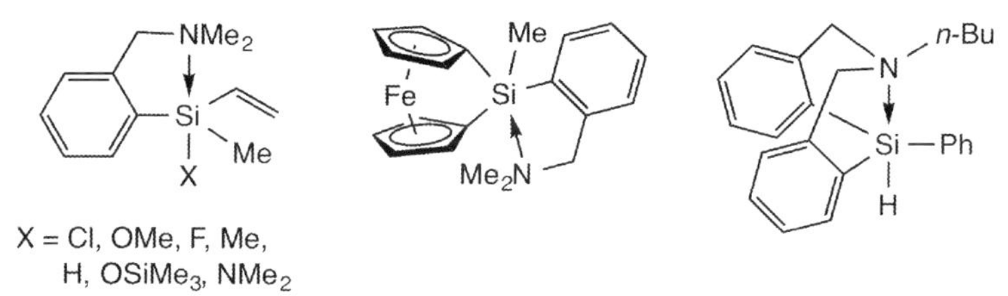

Figure 11.29
Hypercoordinate silicon compounds chelated by tertiary amine-containing ligands.

Pentacoordinated 2-(dimethylaminomethyl)phenylsilane and 8-(dimethylamino)naphthylsilane derivatives have been studied since the early stages of hypercoordinate silicon chemistry, and they were still utilized for some applications in recent years. For example, disiloxanes and ethynylsilanes are pentacoordinated by using the rigid 8-(dimethylamino)naphthyl ligand (Fig. 11.28).[157–159] 2-(Dimethylaminomethyl)phenyl ligand is useful for stabilization of pentacoordinated vinylsilanes, which are promising precursors for generation of a silene,[160,161] and a structurally modified silicon-bridged [1]ferrocenophane (Fig. 11.29).[162] Two sets of the 2-(dialkylaminomethyl)phenyl ligand are unified to form a tridentate ligand, which coordinates to the silicon through a transannular dative bond in addition to two-fold covalent Si–C bonds.[163,164] In the phenylsilane derivative, the hydrogen and nitrogen atoms occupy the apical positions, and three carbon atoms occupy the equatorial positions. The pentacoordinated hydrosilane is useful for the end-capping synthesis of [2]rotaxanes.[165,166]

The intramolecular coordination of an amine moiety to silicon forming a five-membered ring effectively works for the hypercoordination as shown above. Can this coordination form a three-membered ring, if electrophilicity of the silicon atom is high enough? (Dimethylaminomethyl)trifluorosilane, which might have formed an intramolecular Si–N dative bond, did not form such a three-membered ring, but formed the dimer which features two Si–N dative bonds in the crystalline state (Fig. 11.30).[167] In contrast, trifluorosilylhydrazines showed strong intramolecular Si–N β-donor–acceptor bonds,

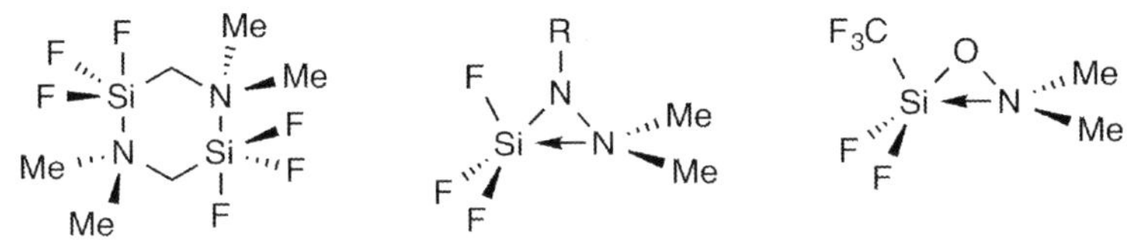

R = Me, $SiMe_3$, $SnMe_3$

Figure 11.30
Three-membered ring compounds containing a pentacoordinated silicon atom and the dimer.

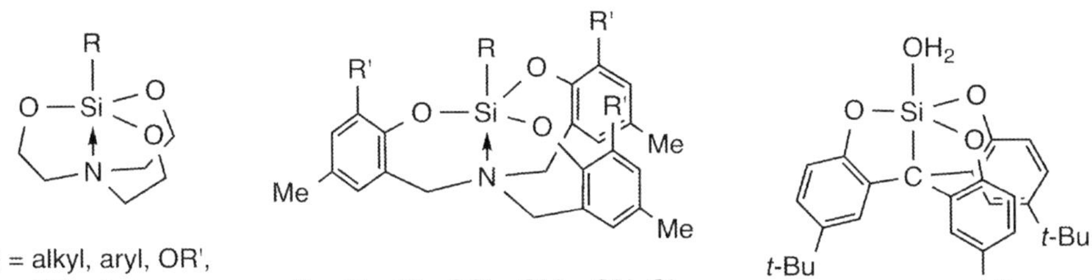

R = alkyl, aryl, OR', NR'_2, halogen, $Si(SiMe_3)_3$

R = Me, Ph, CCl_3, OMe, OH, Cl
R' = Me, *t*-Bu

Figure 11.31
Silatranes and the six-membered ring versions.

ranging from 2.102(1) to 2.510(6) Å, to form a three-membered ring both in the crystal and gas-phase.[168] In a *Si*-trifluoromethylated silylhydroxylamine derivative, acute SiON bond angle of 74.1 degrees and the short Si–N distance of 1.904(2) Å were observed.[169] The Si–N bond could not be described as a classical dative bond, but as a dative bond without covalent and classical charge-transfer contributions, while increasing the electrostatic component.

11.3.7 Silatranes

Silatranes are series of molecules that have three five-membered rings with a transannular dative bond between nitrogen and silicon atoms in tetra- and pentacoordinated states, respectively (Fig. 11.31). A good point of the silatranes is the capability of modification of their structures. Not only the substituent R at the opposite side of the nitrogen–silicon dative bond, but also the ring size of the fused rings[170–172] and the element to form the transannular bond[173] can be modified to revise various properties. After publication of an outstanding review by Voronkov,[174] another excellent critical review paper was published some years ago.[175] A variety of aspects of silatranes such as their synthesis, structure, reactivity, and applications to medical uses were well summarized in these reviews. Many papers related to the silatranes have been published in due course. Especially, the publications reported the synthesis of new silatranyl derivatives bearing various functional groups R on the 1-position (silicon atom) such as methylaminomethyl,[176] aminopropyl,[177] *N*-(2-aminoethyl)-3-aminopropyl, 3-amidopropyl,[178] propylammonium chloride,[179] triethylammonium propyldithiocarbamate,[180] *N*-ethylimidazole,[181] *N*-ethylpyrazole,[182] isothiocyanato,[183] *N*-propylimines,[184] *N*-propylimidazole,[185,186] *N*-propylbenzimidazoles,[187] *N*-propyltriazoles[188–190], *N*-propylisoquinolines,[191] *N*-propylphthalimide,[192] *N*-methylacetamides,[193] *N*-methylpyrrole,[194] *N*-propyl-*N*′-[(*p*-phenyldiazenyl)phenyl]urea,[195] and tris(trimethylsilyl)silyl[196] moieties. Some transition metal complexes containing a silatrane moiety in the ligand have been reported for cobalt complexes.[197–199]

1-Arylsilatranes were also studied both experimentally and theoretically.[200–204] The study was facilitated by efficient synthetic methods such as rhodium(I)-catalyzed cross-coupling reactions of 1-hydrosilatrane with iodoarene[205] and iridium(I)-catalyzed aromatic C–H silylation of arenes with 1-hydrosilatrane.[206] Many of the derivatives synthesized from 1-(γ-aminopropyl)silatranes were more enthusiastically studied than the 1-arylsilatranes. This is because of their high biological activities against tumor cancer cells with low cytotoxicity,[207] suppressive activity on the expression of Hepatitis B virus,[208] and plant growth-regulating activity.[209,210] Because of their toxicity and biological activity, which are quite rarely found in organosilicon compounds, the silatranes have been studied for a long time. The antitumor activities of the silatranes are well summarized in another review paper.[211]

In addition to the point of views of the biological applications, the silatranes are useful compounds for several other applications. They have been applied to Hiyama cross-coupling reactions,[212–214] grafting to carbon interfaces,[215] functionalization of tips of atomic force microscopy and mica surfaces,[216–220] nematic liquid crystals,[221] and ion-conductive matrices.[222]

11.3.8 Hydrazone and Azine Ligands

Some hypercoordinate silicon compounds equilibrate with other compounds bearing the silicon atom in a different coordination number by elimination or addition of a ligand in a solution state, if the two compounds have similar thermodynamic stabilities. For example, neutral hexacoordinated silicon compounds bearing two identical *N*,*O*-chelate hydrazone ligands, showing two N–Si dative bonds, are equilibrated with the corresponding cationic pentacoordinated compounds (Fig. 11.32).[223–225] The ionization, the dissociation of the Si–Cl bond, is reversible in most cases, if there is neither steric hindrance on the silicon atom nor strong electron-withdrawing groups on the chelate ligand, which stabilizes both states before and after the ionization.[226–228] The reversibility makes good contrast with the cationic pentacoordinated silicon compounds bearing similar *N*,*O*-chelate ligands.[229] The extent of dissociation can be changed by replacement of the leaving group. The equilibrium position is shifted by decreasing temperature and increasing polarity of the solvents.[230]

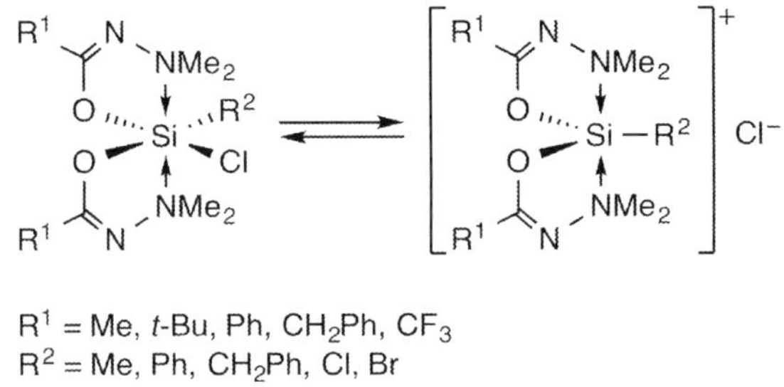

Figure 11.32
Equilibrium between hexa- and pentacoordination states of the silicon compounds chelated by hydrazone ligands.

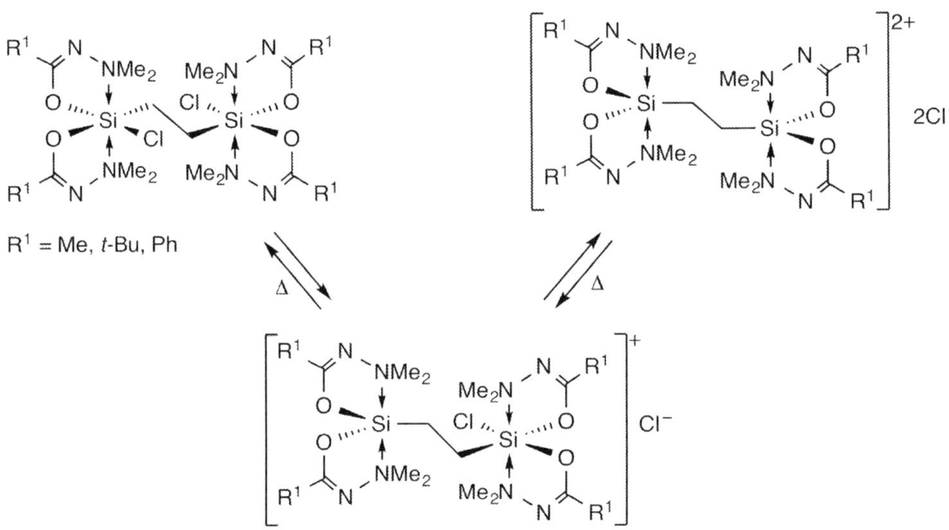

Figure 11.33
Equilibrium between hexa- and pentacoordination states of the silicon compounds bearing two silicon atoms bridged by ethylene linker.

Thus, the coordination states of the hypercoordinate silicon compounds are affected by the external reaction conditions in this system.

Bridging two hexacoordinated silicon moieties with ethylene linker showed greater tendency to ionize (Fig. 11.33). The shift of the equilibrium was indicated to be predominated by the steric factor.[231] The dissociation–recombination reactions constitute ligand exchange in these neutral hexacoordinated complexes, and the mechanism is closely related to the S_N1 type mechanism in carbon compounds.

The azine ligands containing *N*-isopropylideneimino moiety are more powerful donors than the hydrazine ligands. Utilization of the azine ligands leads to *trans*-configuration of two

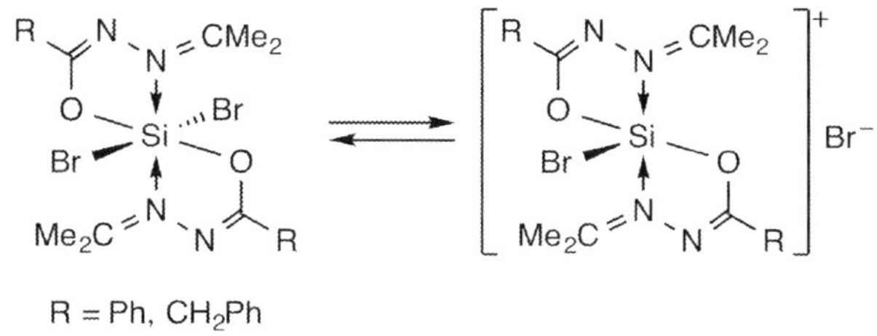

Figure 11.34
Equilibrium between hexa- and pentacoordination states of the silicon compounds chelated by azine ligands.

halogen atoms and enhanced the ionization (Fig. 11.34).[232–234] Moreover, adoption of the chelate ligands containing *N*-triphenylphosphinimino moiety caused complete ionization without hexacoordination.[235,236] Heating of the pentacoordinated silicon compounds bearing the former azine ligands caused an intramolecular aldol-type condensation of the imine moieties instead of the hexacoordination.[237–241]

11.3.9 Azobenzene Ligands

As the structures of silicon compounds are fairly affected or even determined by the coordination number of the silicon atom, the change of the coordination number of the silicon atom by any method leads to the control of the structure of the silicon compounds. In some 2-(phenylazo)phenylsilanes, the azobenzene moiety serves as both photoresponsible moiety and coordination site. The photoirradiation of fluorosilanes bearing the *E*-azobenzene moiety as a ligand caused the reversible *E–Z* isomerization of the azobenzenes in good yields and induced the coordination number changes (Fig. 11.35).[242] The yellow-colored *E*-isomers have a pentacoordinated silicon atom coordinated by a nitrogen atom of the azo moiety, whereas the red-colored *Z*-isomers have a tetracoordinated silicon atom in the absence of the N–Si interaction both in the solution and crystalline states. The coordination numbers of the silicon atoms can be distinguished by the colors.[243] The *E*-2,2′-bis(trifluorosilyl)azobenzene, where two nitrogen atoms are used for the intramolecular N–Si coordination, underwent both the photoisomerization and measurable fluorescence emission at room temperature.[244] After the photoisomerization to the corresponding *Z*-isomer, where the silicon atom is tetracoordinated, it does not fluoresce.

The trifluorosilane derivative can be converted to the corresponding tetrafluorodisiloxane with two pentacoordinated silicon atoms by hydrolysis[245] and to the tetrafluorosilicate with a hexacoordinated silicon atom by fluorination.[242,246] The photoirradiation of the tetrafluorosilicate caused the reversible *E–Z* isomerization almost quantitatively in both directions, and induced complete change of the configuration around the silicon atom between distorted octahedral and TBP structures (Fig. 11.36).

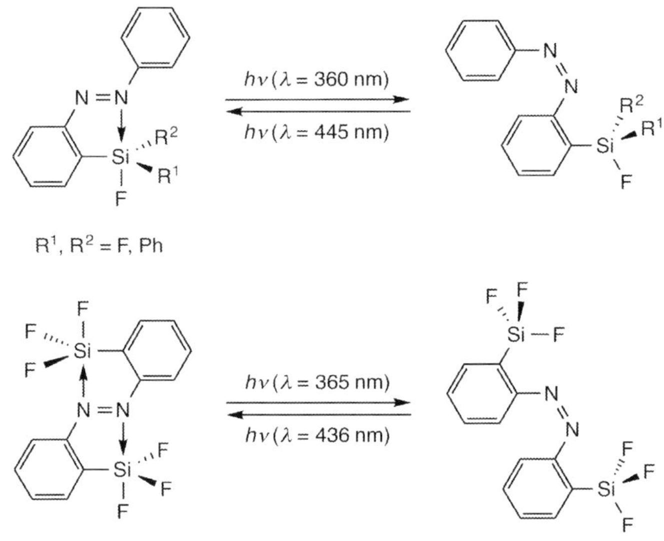

Figure 11.35
Photoswitching of the coordination number of the silicon compounds coordinated by azobenzene ligands.

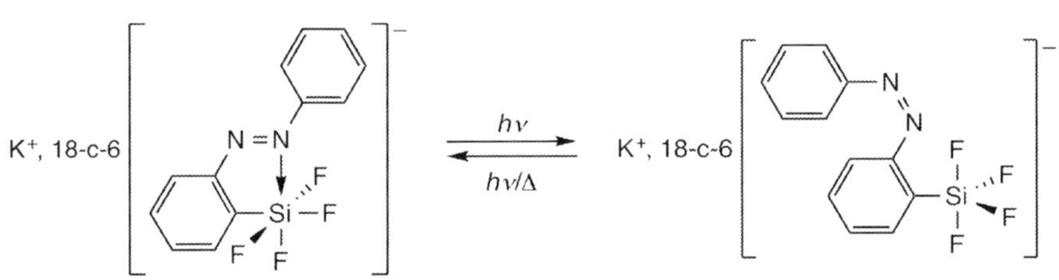

Figure 11.36
Photoswitching of the coordination number of the silicate coordinated by azobenzene ligand.

Such photoswitching of the coordination number was applied to the control of the reactivity of the silicon compounds. For instance, the allyldifluorosilane bearing the *E*-azobenzene moiety as a ligand quickly reacted with potassium fluoride in the presence of crown ether to proceed intramolecular N–Si allyl migration, while it resulted in no reaction after the photoisomerization to the *Z*-isomer (Fig. 11.37).[247,248] The difference of the reactivity definitely depends on the coordination number of the silicon atom. The hydrosilane derivatives also showed the photoinduced difference in the reactivity.[249,250]

11.4 Hypercoordinate Organosilicon Compounds Bearing Silicon–Oxygen Bonds

Many pentacoordinated organosilicon compounds have silicon–oxygen bonds as well as silicon–nitrogen bonds. Both covalent bonds of alkoxy groups to the silicon center and

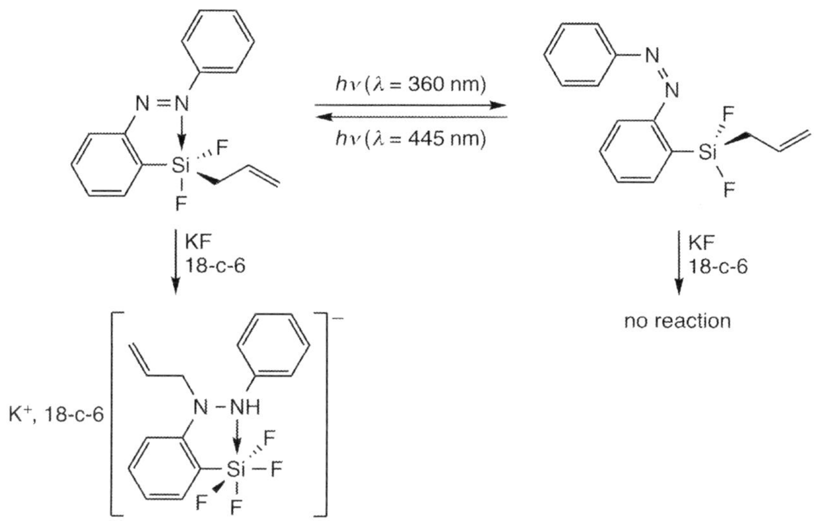

Figure 11.37
Photocontrol of the reactivity of the allylsilane chelated by azobenzene ligand.

dative bonds of oxygen-containing functional groups to the silicon atom are found in many stable hypercoordinate silicon compounds. Such examples are illustrated below.

11.4.1 Amide and Imide Ligands

There are several studies on the hypercoordinate, especially pentacoordinated, silicon compounds bearing C,O-chelate ligands featuring an amide moiety on the silicon atom. Most of the ligands form a five-membered chelate ring by the intramolecular coordination of the amide oxygen atom to the silicon atom, which is in the distorted TBP environment (Fig. 11.38). Some lactams,[251–253] such as 2-pyridone,[254] 1(2*H*)-isoquinolinone,[255] and 2(1*H*)-quinolinone,[256] are also used as the ligands by anchoring them through a methylene group. In the trifluorosilanes chelated by the $-CH_2N(Me)C(=O)R$ ligand, where R is Me, OMe, and CF_3, the increase in the Si–F(apical) bond length corresponds to the increase in the C = O bond length and decrease in the O–Si distance.[257] The Si–O distances are shortened as the C–O–Si angles become larger, showing mutual effects of the structural parameters.[258] As the geometry around the silicon atom gets closer to the TBP structure, the axial O–Si–F fragment changes from a dative bond of the $n_O-\sigma^*_{Si-F}$ interaction to a three-centered four-electron bond.[259]

Such structural changes of the silicon compounds with an anchoring amide ligand are suitable for study on structures of intermediary species in the S_N2 reaction at silicon by a carbonyl oxygen because the geometry around the silicon atom may change consecutively from tetrahedral to TBP structures in correlation to the progress of the Si–O bond formation (Fig. 11.39).[256] Afterward, the silicon–halogen(apical) bond in the halosilanes

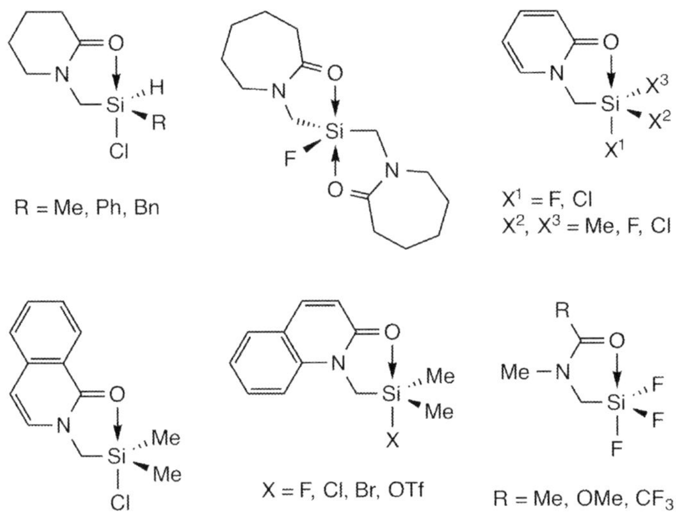

Figure 11.38
Pentacoordinated silicon compounds chelated by amide ligands.

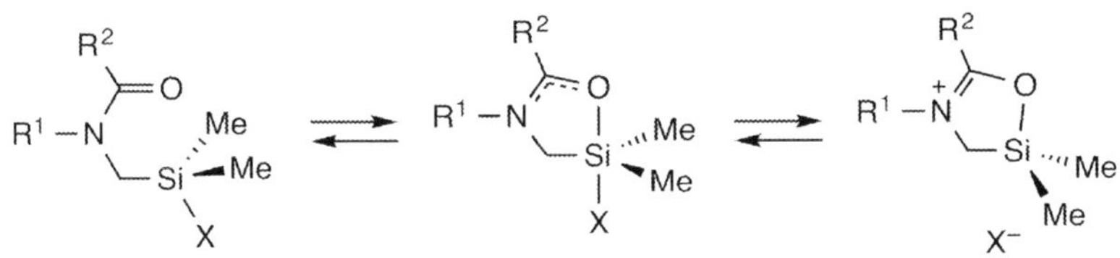

Figure 11.39
Structural changes in the S_N2 reaction of the silicon compound with the intramolecular carbonyl oxygen atom of the amide ligand.

elongates and finally cleaves along with shortening of the Si–O bond. In fact, the Si–O bond lengths in 1-(halodimethylsilylmethyl)-2-quinolinones change in the range from 1.745(3) to 2.065(1) Å depending on the leaving ability of the halogens. The axial halogens, which are recognized as a leaving group in the S_N2 reaction profiles, play an important role for the geometry around the silicon atom.[256,260] In addition, the extent of pentacoordination of the silicon atom, showing approximation of the Si-geometry to the TBP structure, increases as the number of electron-withdrawing chlorine atoms on the silicon atom increases by replacement of methyl groups.[254]

Such elimination of a leaving group is also related to the reaction mechanism of permutation of the ligands on the silicon atom. The permutation mechanism was studied in the positional exchange of equatorial ligands on the silicon center.[261–264] In the pentacoordinated silicon compounds bearing chelating amide ligands, the permutation was caused by the initial dissociation of a leaving group from the silicon, rotation, and association of the leaving group again, rather than the Berry's pseudorotation without the dissociation of halogen.[261,262]

The pentacoordinated chlorosilanes are susceptible to hydrolysis. The hydrolysis under basic conditions leads to formation of the corresponding disiloxanes, in which two pentacoordinated silicon atoms are bridged by an oxygen atom (Fig. 11.40).[265,266] In the hydrolysis under neutral conditions, coordination of a water molecule to the silicon atom proceeds, but the reaction occasionally stops before dimerization in some cases. As a result, protonated silanols containing a pentacoordinated silicon atom can be isolated, showing a wide range of $Si{-}OH_2$ bond lengths (1.878(4)–1.977(8) Å) (Fig. 11.41).[255,267–270] Deprotonation does not occur under neutral conditions because of their relatively weak acidity, but counteranions may interact directly, or via another water molecule, with one of the protons. The silanols can be recognized as a silyl cation stabilized by the water molecule through a dative bond.[271] Treatment of the silanols with aqueous $NaHCO_3$ achieves conversion to the corresponding dicationic disiloxanes including two pentacoordinated silicon atoms.[267]

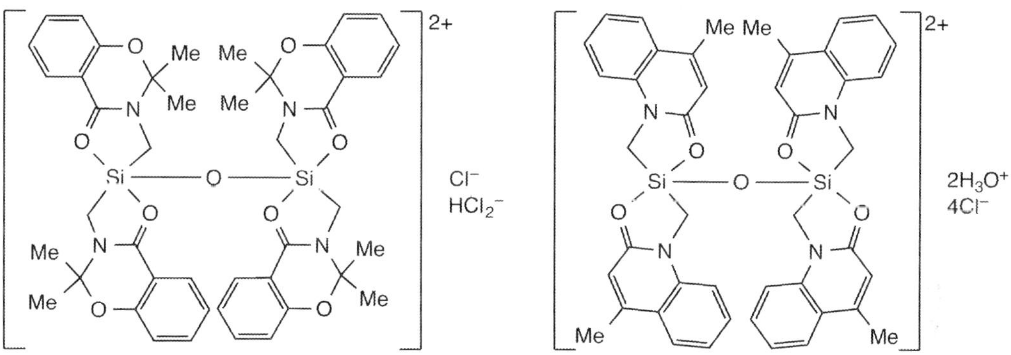

Figure 11.40
Pentacoordinated disiloxanes chelated by amide ligands.

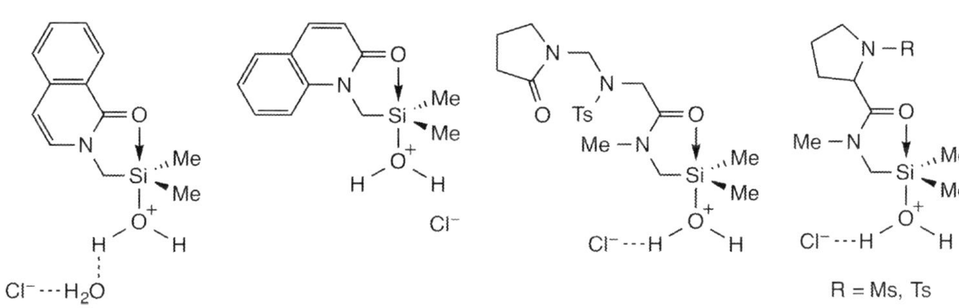

Figure 11.41
Pentacoordinated protonated silanols chelated by amide ligands.

Incorporation of the amide-chelated silicon atoms in polymers changes the conformation and properties of the tetracoordinated silicon-based polymers. While treatment of a cyclic pentacoordinated sila[1]ferrocenophane yielded the cyclic dimer selectively,[272] the ring-opening polymerization initiated by a silyl cation catalyst afforded poly(ferrocenylsilane), which included pentacoordinated silicon atoms in the polymer backbone (the weight-average molar mass M_w 9.0×10^3) (Fig. 11.42).[273] The UV absorption spectrum of the pentacoordinated polymer showed an appreciable difference in the near IR region from that of the tetracoordinated polymer.

Imides, as well as amides, can coordinate to the silicon atom. As compared to isostructural amide analogs, compounds of the imide ligand show a longer Si—O bond and a shorter apical Si—halogen bond in the pentacoordinated halosilanes.[274,275] The structural differences are attributed to the reduced basicity of oxygen atoms in the imide than in the amide as a result of competing n,π-conjugation of the nitrogen lone pair with two C=O bonds. Increasing the number of methylene units in cyclic derivatives and changing to an acyclic structure makes the Si—O bond somewhat shorter both in the amide and imide compounds because of the attenuated steric strain. The imide compounds may show alternating coordination of silicon to each oxygen atom. Indeed, trifluorosilanes with the anchored phthalimide and succinimide ligands showed the fast bridge-flipping (pendulum) effect even at low temperature in the solution state (Fig. 11.43).[276–279]

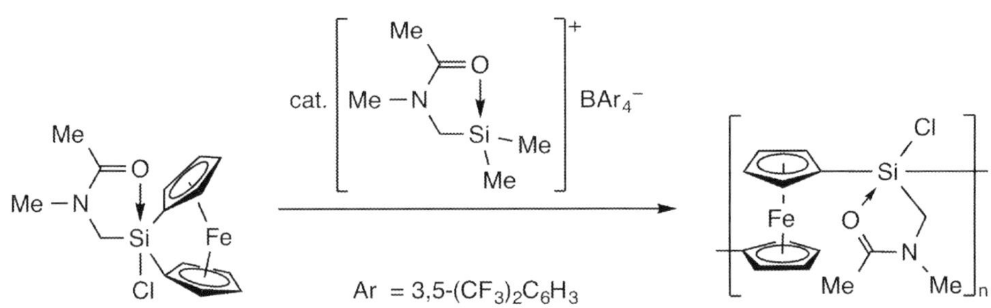

Figure 11.42
Formation of poly(ferrocenylsilane) including pentacoordinated silicon atoms in the polymer backbone by ring-opening polymerization.

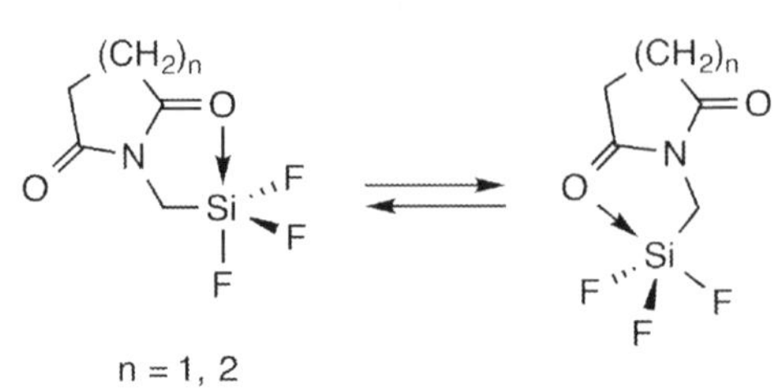

Figure 11.43
Bridge flipping of pentacoordinated trifluorosilane chelated by imide ligands.

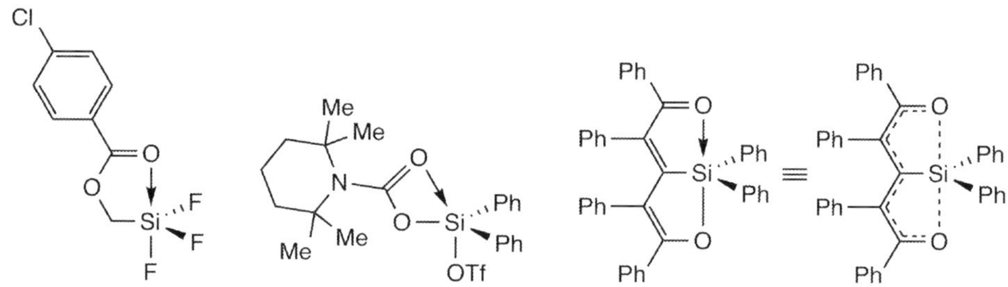

Figure 11.44
Pentacoordinated silicon compounds chelated by ester, carbamate, and vinylogous β-diketonato ligands.

11.4.2 Ester, Carbamate, and Ketone Ligands

Carbonyl oxygen atoms in some esters also contribute to form pentacoordinated states of silicon if the electrophilicity of the silicon atom is high enough. For example, both the silylmethyl ester[280] and silyl carbamate[281] present the pentacoordinated states of silicon (Fig. 11.44). The O–Si–C bond angles differ by 18 degrees between the trifluorosilylmethyl ester and the silyl carbamate, reflecting the different ring size. In these cases, the carbonyl C=O bonds keep high or some degree of double-bond character. It is worth mentioning that the silyl carbamate was synthesized by cooperative activation of CO_2 by $Ph_2Si(OTf)_2$ and 2,2,6,6-tetramethylpiperidine.

Some ketones also coordinate to the silicon center provided that the C=O double bond is delocalized over the ligand to support the coordination of the oxygen. For instance, the tridentate vinylogous β-diketonato ligand coordinates to the silicon atom through the Si–C and two Si–O bonds in the silicon complex, which shows bilaterally symmetrical structure as a result of electron delocalization over the ligand (Fig. 11.44).[282] This silicon complex is notable for its bright-orange fluorescence emission (emission maximum λ_{em} 554 nm, fluorescence quantum yield Φ_F 0.13) under white light.

2-Acylpyrroles were applied to pyrrole-functionalized mono-anionic bidentate *O,N*-ligands for hypercoordinate silicon complexes, in which the coordinating oxygen belongs to the ketone moiety (Fig. 11.45).[283,284] Twofold coordination of the ligand achieves hexacoordination of the silicon atom. The L_2SiX_2 complexes, where L and X are the bidentate 2-acylpyrrolide ligand and a monodentate ligand, respectively, exhibit three of five possible configurations of the octahedral geometry around the silicon atom in the crystalline states. That is, the L_2SiCl_2 and $L_2Si(OTf)_2$ complexes showed all-*trans* configuration, while the L_2SiF_2 and L_2SiPh_2 complexes showed *N,N-trans* and *O,O-trans* configurations, respectively.

11.4.3 Phosphoramide and Phosphonate Ligands

Hexamethylphosphoramide (HMPA), which is a strong Lewis base, coordinates to the highly Lewis acidic silicon compounds through the oxygen atom. Both

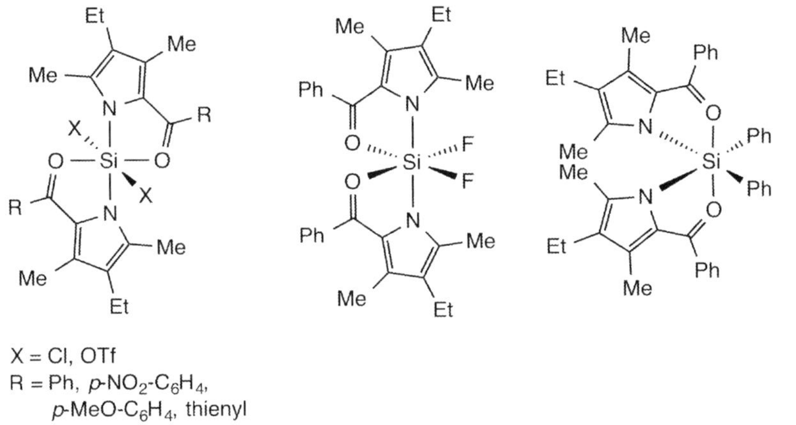

Figure 11.45
Hexacoordinated silicon compounds chelated by 2-acylpyrrolide ligands.

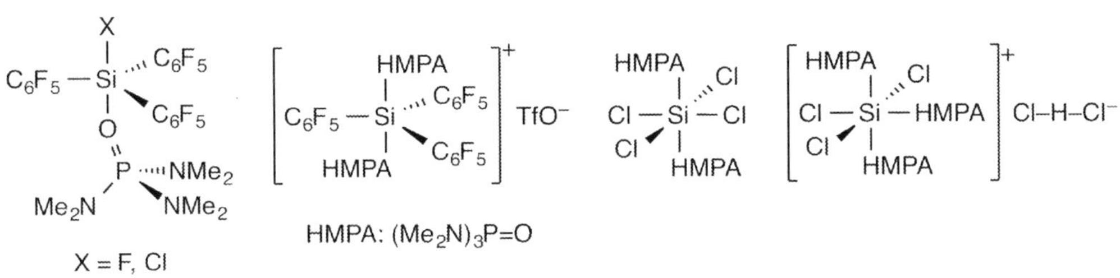

Figure 11.46
Penta- and hexacoordinated silicon compounds coordinated by HMPA.

tris(pentafluorophenyl)silyl fluoride and chloride formed the corresponding neutral 1:1 complexes with HMPA, while tris(pentafluorophenyl)silyl triflate formed the cationic 1:2 complex accompanied by dissociation of the triflate (Fig. 11.46).[285] In all cases, the silicon atom is in TBP geometry with apical Si–O bonds (1.795(1)–1.871(2) Å). The Si–O–P angles in the complexes are close to linear (165.16(9)–177.27(8) degrees). The HMPA and $Si(C_6F_5)_3$ moieties are in an unfavorable eclipsed conformation despite the high steric crowding of silicon center. The stability of the HMPA complexes is associated with strong electron-deficiency of tris(pentafluorophenyl)silyl part. In contrast, tetrachlorosilane and HMPA yielded neutral 1:2 and cationic 1:3 adducts, which were in equilibrium in solution (Fig. 11.46).[286] The ratio depends on the HMPA concentration. The isolable hexacoordinated neutral *trans*-2HMPA · $SiCl_4$ and cationic [*mer*-3HMPA · $SiCl_3]^+[HCl_2]^-$ complexes were characterized crystallographically. This investigation helps characterization of reaction intermediates in Lewis base catalyzed aldol addition reactions in the presence of $SiCl_4$ and *N*,*N*-diisopropylethylamine[287].

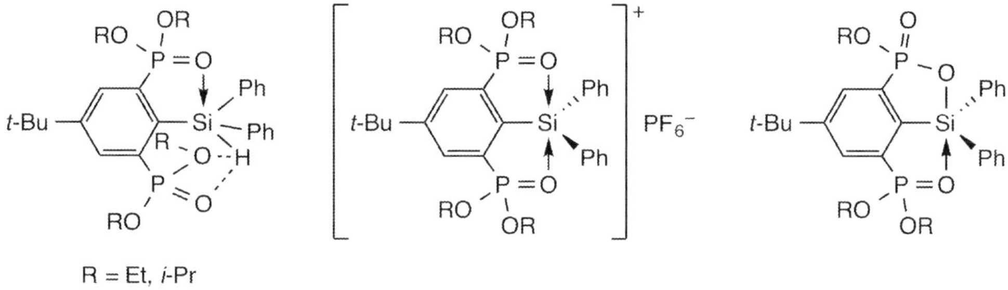

Figure 11.47
Pentacoordinated silicon compounds chelated by phosphonate ester ligands.

Considering the highly polar bond character of the P=O bond in phosphonic acid derivatives, the P=O groups potentially coordinate intramolecularly to the silicon atom. In the hydrodiphenylsilanes incorporated with twofold dialkyl phosphonate moiety on the aryl substituent, one P=O group of a phosphonate ester interacts with the silicon atom (Si–O, 2.7378(18)–2.918(2) Å) (Fig. 11.47).[288,289] The intramolecular interaction produces a monocapped-tetrahedral configuration around the pentacoordinated silicon center. Intramolecular Si–H···O hydrogen bonding interferes with additional Si–O interaction. These silanes were converted to the intramolecularly coordinated triarylsilylium by a reaction with triphenylcarbenium hexafluorophosphate or a photoreaction with tungsten hexacarbonyl.[288,290] The coordination geometry around the silicon atom in the silyl cation is described by TBP geometry with three carbon atoms in equatorial positions and two oxygen atoms in apical positions. The intramolecular Si–O distances in the silylium ion range from 1.9225(14) to 1.9312(14) Å (R = *i*-Pr), which are notably shorter than the closest Si–O distances in the triphenylsilyl and trimethylsilyl derivatives (Si–O for $SiPh_3$, 3.091(2); for $SiMe_3$, 3.2849(13) Å) because of the high electrophilicity of the silylium ion.[291,292] The intramolecularly coordinated triorganosilylium ion was hydrolyzed to give the corresponding neutral dealkylated product, benzoxaphosphasilole, where the silicon atom was still pentacoordinated by an intramolecular O–Si coordination.[288,293,294]

11.4.4 Diolato Ligands

There are many anionic hypercoordinate silicon compounds, namely, silicates, containing Si–O bonds. Since dianionic bidentate ligands forming a five-membered ring are suitable chelators for silicon, two sets of catecholato ligands are often used for stabilization of the silicates. The resulting spirocycle together with another substituent shows SP structure in many cases. Actually, tris(dimethylamino)sulfonium bis(perfluorocatecholato)fluorosilicate has the relatively narrow ranges of both F–Si–O bond angles (99.8–105.0 degrees) and Si–O bond lengths (1.72–1.75 Å), exhibiting a distorted SP structure of the Si-polyhedron (Fig. 11.48).[295] On the other hand, potassium

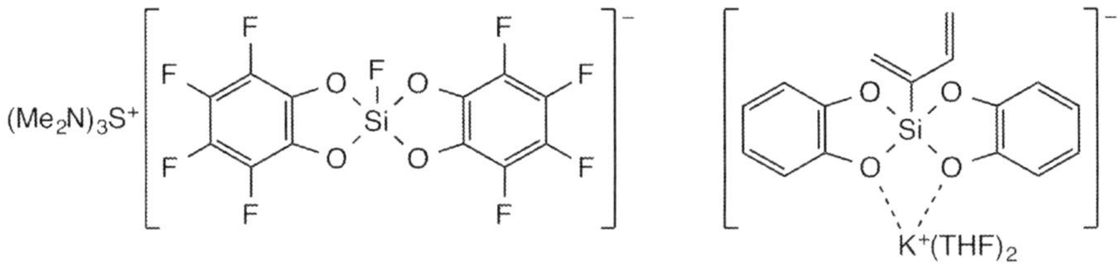

Figure 11.48
Pentacoordinated silicon compounds chelated by catecholato ligands.

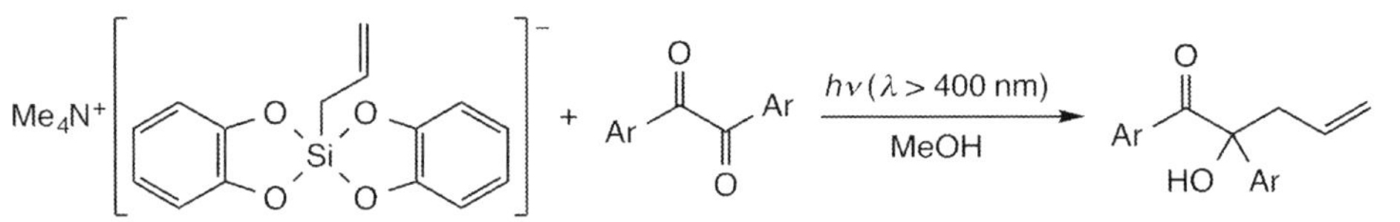

Figure 11.49
Photoinduced allylation of diketones with the pentacoordinated allyl silicate chelated by catecholato ligands.

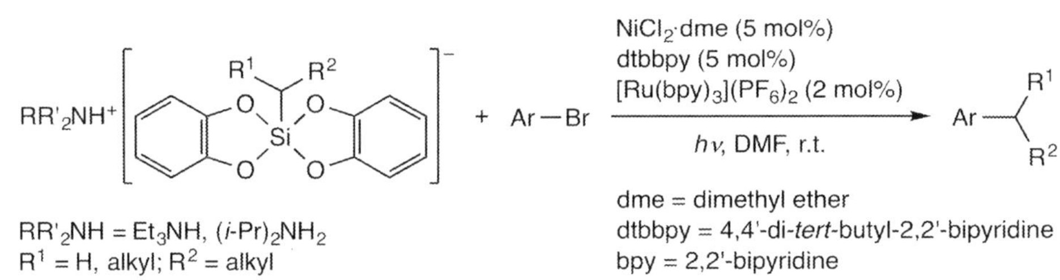

Figure 11.50
Photoredox/nickel dual-catalyzed cross-coupling reactions of alkyl silicates with aryl bromides.

bis(catecholato)(1,3-butadien-2-yl)silicate, which demonstrates intramolecular O–K interactions, has wider ranges of bond angles (98.6–116.9 degrees) and the Si–O bond lengths (1.72–1.77 Å), showing intermediate of TBP and SP structures.[213]

The bis(catecholato)silicates are useful for their applicability to organic synthesis. The allyl silicates undergo photoinduced allylation reactions with 1,2-diketones, aromatic ketones, and dicyanoarenes (Fig. 11.49).[296–299] The generation of allyl radical by photoinduced oxidation of the silicates plays a crucial role in the reactions. Other various 1° and 2° alkyl radicals can be generated from the alkyl silicates upon visible-light irradiation in the presence of photocatalysts such as iridium and ruthenium catalysts.[300,301] The photocatalytic generation of the alkyl radicals were merged with the nickel-catalyzed cross-coupling reactions. As a consequence, the photoredox/nickel cross-coupling dual catalysis served for the C(sp^2)–C(sp^3) cross-coupling reactions between various alkyl silicates and aryl bromides (Fig. 11.50).

The triethylammonium aryl silicates have been utilized, without requiring the visible light irradiation, for the palladium catalyzed Hiyama cross-coupling reactions with aryl triflates and aryl bromides by addition of TBAF[302] or employing microwave irradiation,[303] respectively. This reactivity of the C(sp^2)–C(sp^2) cross-coupling reactions were combined with the Diels–Alder reactions. The 1,3-dien-2-yl silicate showed much higher reactivity than the Danishefsky's diene. The cycloadducts in the Diels–Alder reaction can be converted to various aryl substituted cycloadducts via the palladium-catalyzed cross-coupling reactions (Fig. 11.51).

Although silicates are always accompanied by their counter cations, an ammonium is merged with a part of a substituent in zwitterionic silicates.[304] The zwitterionic silicates present excellent crystallinity, which provides advantages for both their isolation and structural characterization.[305] The zwitterionic silicates were rationally synthesized by using (dialkylamino)alkylsilanes as precursors. In the precursors, trimethoxysilyl,[306–308] dimethoxy(phenyl)silyl,[306,309] and dimethoxy(methyl)silyl[310] groups were used for the silyl groups. Treatment of the (dialkylamino)alkylsilanes with 1,2-diols,[307–310] α-hydroxyacetic acids,[311,312] and hydroxamic acids[306] caused both protonation of the amine and substitution of the silane, resulting in formation of the zwitterionic pentacoordinated silicates with the spirocyclic structures and an SiO_4C skeleton (Fig. 11.52). This procedure is advantageous

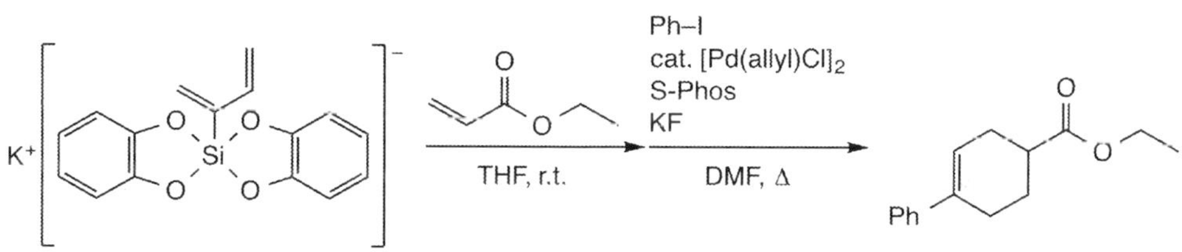

Figure 11.51
Connection of three components by the Diels–Alder reaction and the cross-coupling reaction starting from the 1,3-dien-2-yl silicate.

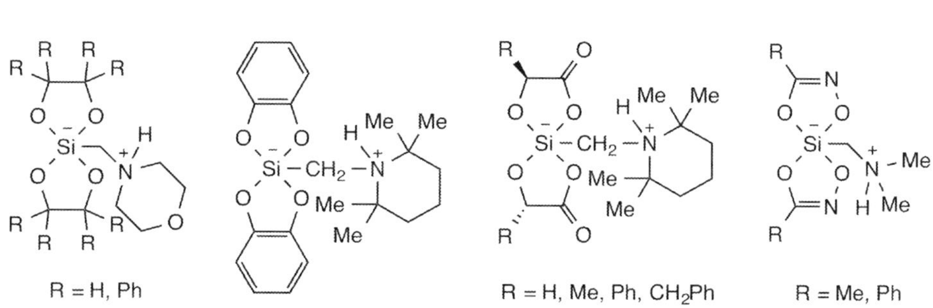

Figure 11.52
Pentacoordinated zwitterionic silicates including ammonium moiety in the ligands with an SiO_4C skeleton.

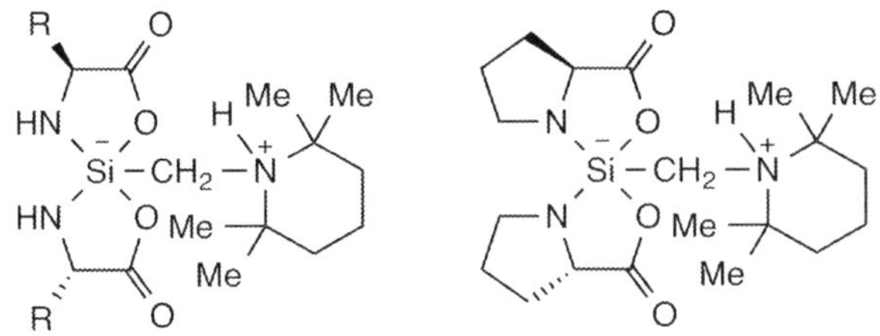

R = H, Me, CH_2Ph, *i*-Pr, *t*-Bu

Figure 11.53
Pentacoordinated zwitterionic silicates including ammonium moiety in the ligands with an SiO_2N_2C skeleton.

because the released side products are only methanol together with benzene or methane originating from the substituents on the silicon atom in the precursors. An application of this method was proposed for the solid-phase synthesis of aromatic compounds in a traceless fashion.[313] The zwitterionic silicates showed distorted TBP structures in most cases. The alkyl group occupies the equatorial position, and four oxygen atoms occupy the resulting two equatorial and two apical positions. The oxygen atoms bounded to the carbonyl groups and nitrogen atom in the derivatives prepared from α-hydroxyacetic acids and hydroxamic acids, respectively, occupy the apical positions rather than the equatorial positions. An intramolecular N–H···O hydrogen bond between the ammonium hydrogen atom and one of two apical oxygen atoms is observed in many cases.

Some α-amino acids can be used for the *O,N*-bidentate ligands for the zwitterionic pentacoordinated silicates with an SiO_2N_2C skeleton.[305,314] The silicates derived from (*S*)-alanine and (*S*)-phenylalanine were isolated as diastereomerically and enantiomerically pure crystalline compounds (Fig. 11.53). Upon dissolution of the crystals in a solvent, (Λ)/(Δ)-epimerization led to equilibrium mixtures of the (Λ,*S*,*S*)/(Δ,*S*,*S*)-diastereomers. In contrast, the silicates derived from (*S*)-valine, (*S*)-*tert*-leucine, and (*S*)-proline remained as diastereo- and enantiomerically pure in solution between −100 and 23°C without epimerization.

In general, the zwitterionic silicates are rather sensitive to moisture. For example, hydrolysis of the bis(ethane-1,2-diolato)silicate gave the corresponding octasilsesquioxane accompanied by formation of ethylene glycol.[309] However, the zwitterionic silicate containing *meso*-oxolane-3,4-diolato and morpholinio group is fairly resistant to hydrolysis (Fig. 11.54).[308,315] Some of the bis(silicate)s bearing two pentacoordinated silicon atoms bridged by two tartrate ligands showed a remarkable kinetic stability in aqueous solution.[316–318]

Threefold chelation of the catecholato ligands on the silicon atom forms the dianionic hexacoordinated silicate (Fig. 11.55). They should be paired with a couple of monovalent

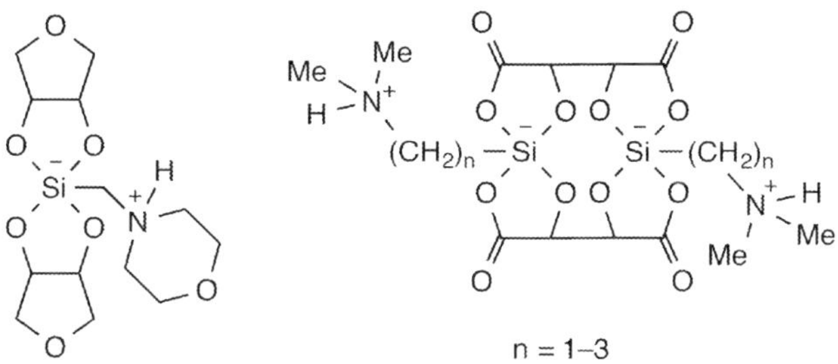

Figure 11.54
Hydrolysis-resistant pentacoordinated zwitterionic silicate and bis(silicate)s.

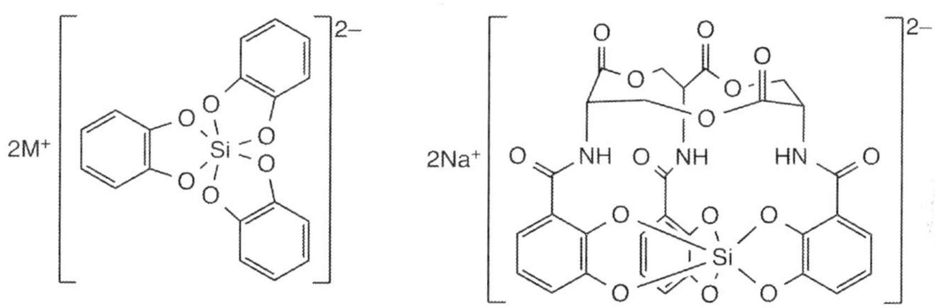

Figure 11.55
Hexacoordinated tris(catecholato)silicates and silicon complex of enterobactin.

cations such as potassium and tetraalkylammonium or a divalent cation. A phosphatrane-derived phosphonium[319] and transition metal ions[320] were used, but rarely, for the countercations. The hexacoordinated silicates show an octahedral structure surrounded by three ligands.[321–323] The structural motif shows good resemblance to the siderophore enterobactin, indicating its capability of binding silicon. Actually, its silicon complex was synthesized.[324,325] Surprisingly, the enterobactin silicon complex was found and isolated from the endophytic *Streptomyces* sp. occurring in *Piper guinensis* roots.[326]

11.5 Hypercoordinate Organosilicon Compounds Bearing Silicon–Sulfur Bonds

The sulfur-bound silicates are less common because of hygroscopic property of the silicon–sulfur bond and relatively low electronegativity of sulfur (electronegativity χ 2.58), which is rather unfavorable for stabilization of the hypercoordinate state. There are some hypercoordinate silicon compounds bearing silicon–sulfur bonds.[327–331] Certain sulfur-bound hypercoordinate silicon compounds have already appeared in previous pages because such hypercoordinate silicon compounds have the silicon–nitrogen or silicon–oxygen bonds together as a part of bidentate and tridentate ligands.

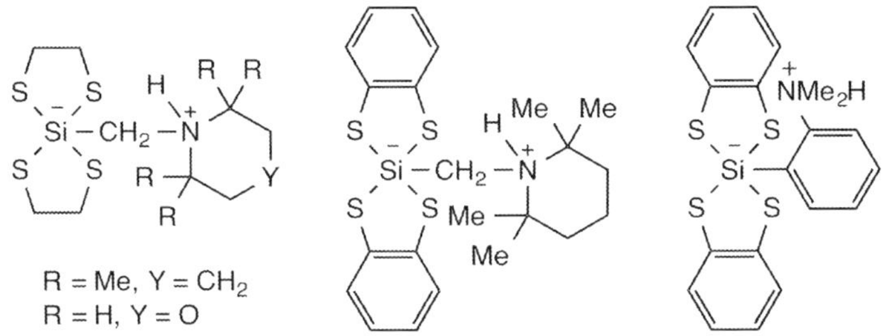

Figure 11.56
Pentacoordinated zwitterionic silicates including ammonium moiety in the ligands with an SiS_4C skeleton.

The zwitterionic pentacoordinated silicates with the silicon bonding to four sulfur atoms and one carbon atom were synthesized by the reactions of [(dialkylamino)methyl]trihydrosilanes with ethane-1,2-dithiol and benzene-1,2-dithiol, accompanied by extrusion of hydrogen gas (Fig. 11.56).[332,333] For the latter compound, the selenium analog was also synthesized.[334] They showed slightly distorted TBP structures. In contrast, the similar zwitterionic silicate derived from [2-(dimethylamino)phenyl]trihydrosilane and benzene-1,2-dithiol showed slightly distorted SP polyhedron of Si-coordination, with four sulfur atoms occupying the four basal sites and the aryl *ipso*-carbon atom occupying the apical site.[335] These Si–S bonds (2.2251(6)–2.2558(6) Å) are longer than the equatorial Si–S bonds (2.1588(9)–2.1728(9) Å) and shorter than the apical Si–S bonds (2.2871(7)–2.3475(8) Å) in the above-mentioned zwitterionic silicates, which all contain a TBP Si-coordination polyhedron. The structural difference between these silicates arises probably from difference in the hybridization, sp^2 or sp^3, of the carbon atom that is bound to the silicon atom. The detailed bonding analysis of the zwitterionic silicate indicates that all five bonds in the SiS_4C framework in TBP geometry are significantly and almost equally polarized toward the ligand atoms, stressing the necessity of polar bonds to stabilize such a hypercoordinate silicon center.[333]

11.6 Hypercoordinate Organosilicon Compounds Bearing Silicon–Silicon Bonds

A silicon–silicon bond is a crucial and fundamental backbone of oligosilanes and polysilanes. The involvement of one or more highly coordinated silicon atoms would change the structure of the framework and, fascinatingly, induce some deviations in their properties. However, the electronic properties and bulkiness of the silyl group, which inevitably appears in disilanes and oligosilanes, destabilize highly coordinated silicon to which the silyl groups bind. As a result, a silicon–silicon bond of the intermediary disilanes containing one or two hypercoordinate silicon atoms was sometimes cleaved.[336,337] Thus, synthesis of stable hypercoordinate Si–Si bonded compounds was a challenging topic.

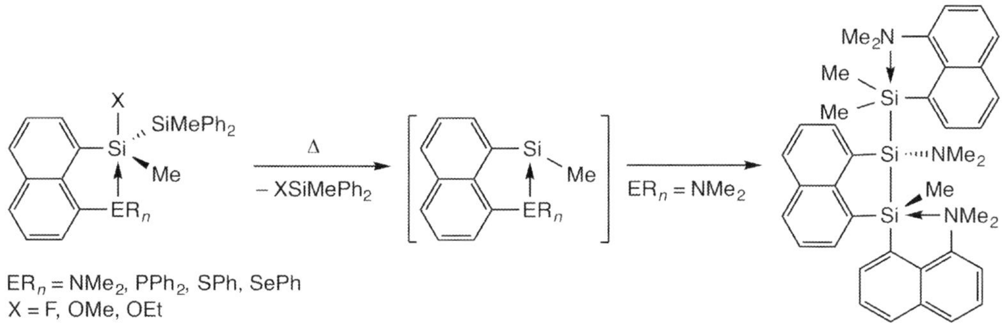

Figure 11.57
Pentacoordinated disilanes chelated by naphthalene-containing bidentate ligands and conversion to a trisilane via the silylene intermediate.

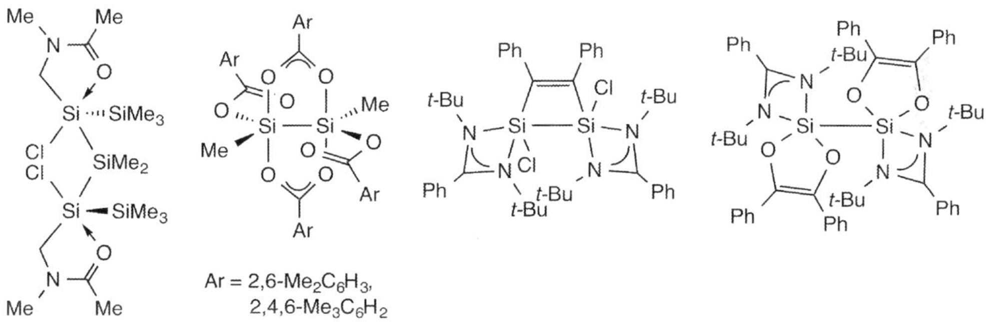

Figure 11.58
Pentasilane and disilanes containing two pentacoordinated silicon atoms.

The pentacoordinated disilanes bearing 8-dimethylamino-, 8-diphenylphosphino-, 8-methylthio-, and 8-methylselenonaphthyl groups, were synthesized as stable compounds (Fig. 11.57).[338–341] Thermolysis of the (8-dimethylaminonaphthyl)disilane leads to formation of the trisilane.[158,342] Trapping reactions of the intermediary nitrogen-coordinated silylene, which was formed by the α-elimination of fluorosilane or ethoxysilane, results in formation of various types of adducts.[158,343] The reactivity of α-elimination of fluorosilane or alkoxysilane under thermal conditions is common to the pentacoordinated disilanes bearing a naphthyl group furnished with a coordinating heteroatom at the 8-position.[338–341]

The incorporation of pentacoordinated silicon atoms in oligosilanes can fix the conformation of the silicon chains because silicon chains have been locked by inhibition of the rotation around the Si–Si single bonds to pentacoordinated silicon atoms. In the pentasilane including two pentacoordinated silicon atoms, the fixation provides somewhat different photophysical properties because of the unusual locking of the silicon chains into the all-*transoid* conformation (Fig. 11.58).[344]

The pentacoordinated states of both two silicon atoms of a disilane could be achieved by the bridging with a suitable ligand. Two of four sets of arylcarboxylate ligands bridge the two silicon atoms in the disilane and occupy four apical positions of the two silicon atoms (Fig. 11.58). Four other atoms on the silicon atoms at the equatorial site showed coplanarity with the silicon atoms.[345] However, the labile oxygen–silicon dative bonds dissociate in solution in this compound. In another case, two pentacoordinated silicon atoms are incorporated into the almost planar four-membered ring and found to be stable also in the solution.[346,347] Nevertheless, the four-membered ring is dispensable for the stability. Actually, the bis-silylene bearing the same amidinato ligands reacts with benzyl affording a similar compound that contains two 1,3-dioxa-2-silacyclopent-4-ene rings that were directly linked by a Si–Si bond (2.3628(7) Å).[348]

Reactions of 3,5-dimethylpyrazolyllithium with hexachlorodisilane gave a mixture including three disilanes bridged by the 3,5-dimethylpyrazolyl ligands (Fig. 11.59).[349] Change of the precursor to (3,5-diphenylpyrazolyl)trimethylsilane gave the tetrachlorodisilanes more selectively.[350] The silicon–silicon bonds are in the range of 2.281(3)–2.307(2) Å. Terminal and bridging pyrazolyl ligands exchange in solution, as suggested by the NMR spectra.

In the disilanes above, they show symmetric structures. By contrast, the trisilane bearing two adjacent hexacoordinated silicon atoms with four sets of an asymmetrically bridging ligand was synthesized by the reaction of 1-methyl-3-trimethylsilylimidazoline-2-thione with hexachlorodisilane (Fig. 11.60).[351] The paddlewheel shape with a Si–Si bond length of 2.387(1) Å was confirmed. Natural bond orbital (NBO) analysis showed that the Si–Si bond is mainly a polar bond of a donor–acceptor model as indicated by the canonical structures in Fig. 11.60. The first and second two canonical structures are more important contributions. Such a polar bonding property is different from that of a Si–Si linkage between two hexacoordinated silicon atoms in phthalocyanine dimers, which were spectroscopically characterized.[352–354]

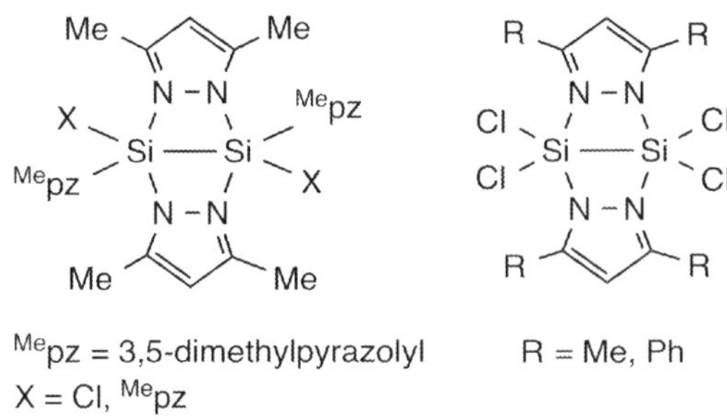

Figure 11.59
Pentacoordinated disilanes bridged by 3,5-dimethylpyrazolyl ligands.

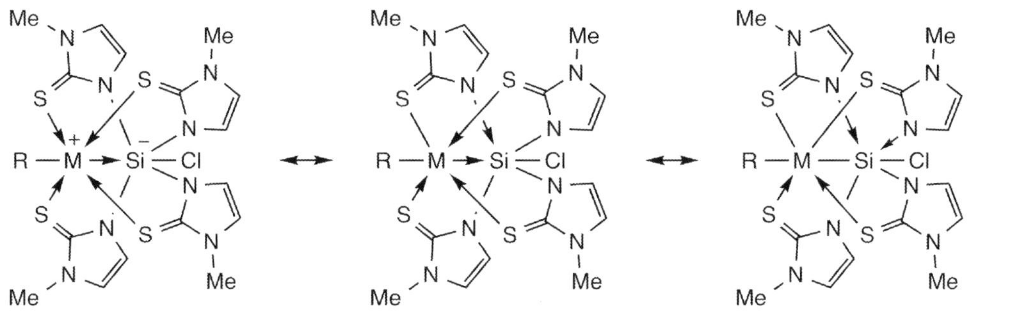

Figure 11.60
Hexacoordinated silicon compounds containing homo- and heteronuclear bonds between two group 14 elements.

Heteronuclear bonds between two hexacoordinated group 14 elements, Si–Ge and Si–Sn bonds, are also available by using the same *N,S*-bidentate ligand.[355] These complexes showed major contributions of a dative bond ($M^{II} \rightarrow Si^{IV}$) (middle in Fig. 11.60) and a covalent bond ($M^{III}-Si^{III}$) (right in Fig. 11.60) for silicon–germanium and silicon–tin bonds, respectively, suggesting dependence of the bond character on the group 14 element combination.

All the above examples of the Si–Si bonded compounds are neutral. The dianionic Si–Si bonded compounds linking two pentacoordinated silicon atoms, disilicate, bearing two sets of the C,O-bidentate ligand was synthesized by reductive coupling reaction of the corresponding tetracoordinated spirosilane with lithium (Fig. 11.61).[356] Although the disilicate shows high stability against hydrolysis, the Si–Si bond can be cleaved by oxidation with 2,3-dichloro-5,6-dicyanobenzoquinone (DDQ) under basic conditions, making this bond formation/cleavage reversible.[357] The disilicate is protonated by treatment with hydrochloric acid to afford the corresponding silylsilicate and disilane.[358] In both compounds, silicon atoms keep pentacoordinated because of intramolecular silicon–oxygen coordination from the resulting hydroxy moiety. Treatment of a base, butyllithium or lithium hydride, caused deprotonation, making the transformations reversible. These transformations caused changes in the conformation along the silicon–silicon bond. Reflecting the changes in the charge, the oxidation potential and UV/vis absorption maximum of the silylsilicate are both intermediate between those of the disilicate and the disilane.

Although either chelating or bridging ligands were used for the stabilization of the hypercoordinate state of the silicon in the disilanes and oligosilanes above, these ligands are

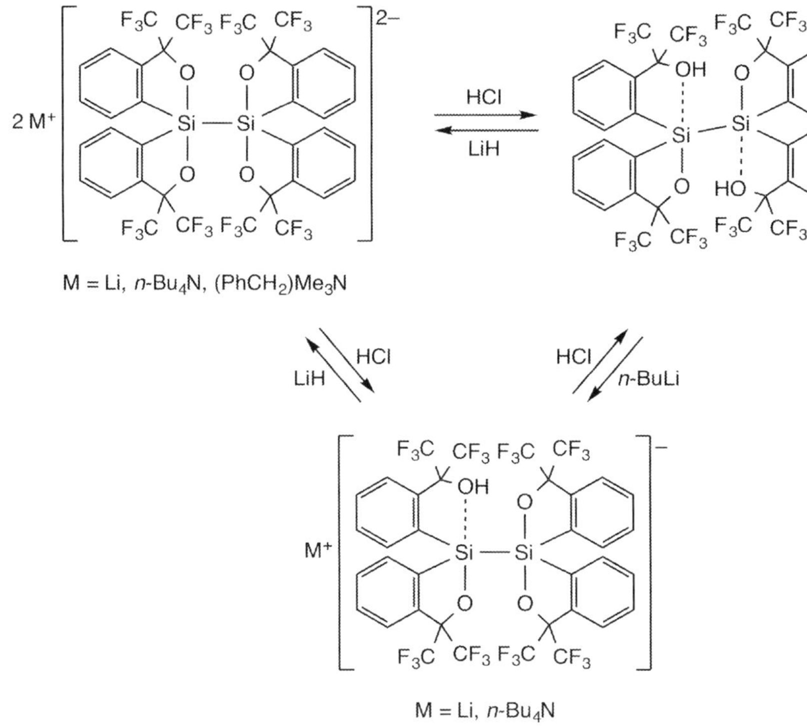

Figure 11.61
Interconversion among pentacoordinated Si–Si bonded compounds by acid–base reactions.

not necessarily essential. The heptachlorinated di(silyl)silicate with a radical cation of tetrakis(dimethylamino)ethylene as a counter cation was synthesized via a disproportionation reaction of chloromethyldisilanes and the electron-rich alkene (Fig. 11.62).[359] In other cases, pentacoordinated di(silyl)silicate, tri(silyl)silicates, and hexacoordinated di(silyl)silicate were obtained and isolated by simply mixing hexachlorodisilane with tetrabutylammonium chloride below −40°C (Fig. 11.62).[360] Release of intermediary $SiCl_3^-$ was induced by the attack of chloride to hexachlorodisilane and its addition to neutral perchlorinated disilane and trisilane are the key steps in the formation of the perchlorinated oligosilylsilicates.

11.7 Hypercoordinate Organosilicon Compounds Bearing Silicon–Carbon Bonds

Silicon–carbon bonds are common in organosilicon compounds including hypercoordinate compounds. Only the carbene-coordinating hypercoordinate silicon compounds and pentacoordinated silicon compounds bearing all five hydrocarbon ligands are described here.

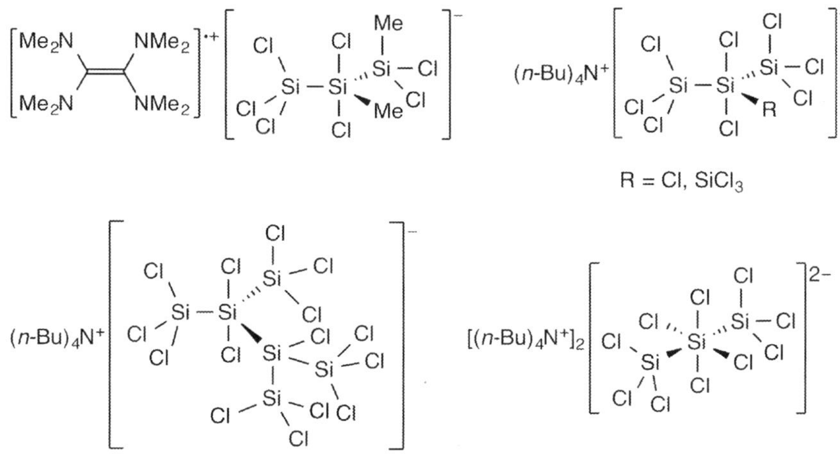

Figure 11.62
Penta- and hexacoordinated silicates bearing perchlorinated silyl groups.

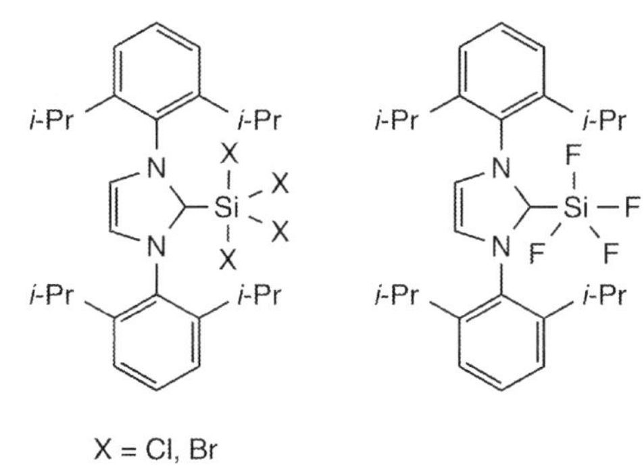

Figure 11.63
Pentacoordinated tetrahalosilanes coordinated by NHC ligands.

11.7.1 N-Heterocyclic Carbene Ligands

N-Heterocyclic carbenes (NHCs) have become essential ligands for the stabilization of unusual oxidation states and coordination modes of both transition metals and main group elements in a couple of decades. The NHCs have also been applied to the stabilization of the highly coordinated organosilicon species. After the synthesis of stable pentacoordinated tetrachlorosilane coordinated by an NHC ligand (Fig. 11.63),[361] studies on the pentacoordinated organosilicon compounds have started to search for their application. The synthesis of NHC–$SiCl_4$ naturally expanded to the synthesis of the $SiBr_4$, SiF_4, and Ph_2SiCl_2 analogs.[362–364] All the neutral mono-NHC adducts of silicon tetrahalides show TBP structure around the silicon atom. The NHC ligand in the NHC-adducts of $SiCl_4$ and $SiBr_4$ was at an equatorial position regardless of the bulkiness of the NHCs, but it resides at

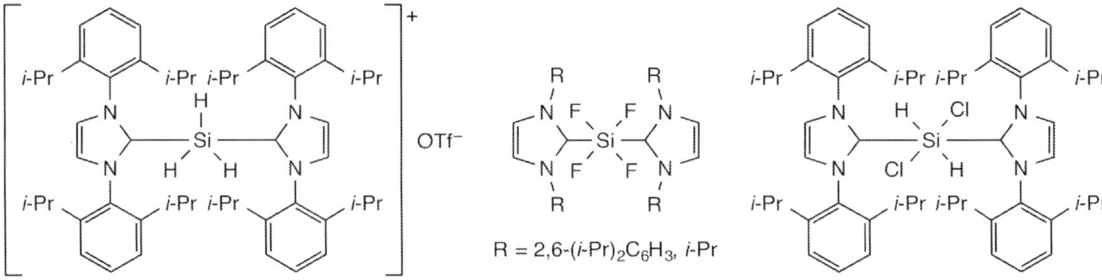

Figure 11.64
Penta- and hexacoordinated silicon compounds coordinated by two NHC ligands.

an apical position in the sterically demanding NHC–SiF_4. Such the difference of the NHC ligand position between the chloride and fluoride matches with the calculated apicophilicity.[365] Different steric repulsion and hydrogen bonding of these two halogens induced the differences between the apicophilicity of the carbene in this case. The bonding analysis using density functional theory (DFT) calculations on NHC–$SiCl_4$ demonstrates that the electrostatic attraction is a key for their silicon–carbon(carbene) bond, which has a significant ionic character to compose the stabilized carbene complexes.[366] Namely, the silicon–carbene bond cannot be described in terms of donor–acceptor interactions depicted in the Dewar–Chatt–Duncanson model.

Twofold coordination of the NHC ligand is available by the addition of a sterically demanding NHC to H_3SiOTf, forming a cationic pentacoordinated organosilicon compound (Fig. 11.64).[367] The product, $[(NHC)_2SiH_3]^+(OTf)^-$, represents the bis-NHC adduct of the parent silylium cation. In this compound, both NHC ligands occupy the apical positions to avoid the steric repulsion. However, attempted formation of the mono-NHC adduct of Ph_3SiH, Ph_2SiH_2 and $PhSiH_3$ resulted in insertion of the silanes into the C–N bond of NHCs.[368,369] Hexacoordinated adducts of SiF_4 and H_2SiCl_2 bearing two sets of NHCs were synthesized similarly under different reaction conditions or with an increased stoichiometry of the NHCs.[362,365,370] All the hexacoordinated bis-NHC adducts of the silanes show an octahedral geometrical environment around the silicon center with the NHC ligands in *trans* position.

Some NHC-$SiCl_4$ adducts were found to function as a precursor for the NHC-stabilized diatomic silicon and dichlorosilylene (see Chapter 9: Multiple Bonds to Silicon (Recent Advances in the Chemistry of Silicon Containing Multiple Bonds) and Chapter 10: Silaaromatics and Related Compounds).[361,371] On the way to search for further utilization of the NHC-$SiCl_4$ adduct, the 1,3-dimethylimidazolidin-2-ylidene adduct of silicon tetrachloride was applied to carbene-transfer reactions.[372,373] The stable adduct can transfer the saturated NHC fragment to the stronger Lewis acids such as $(C_2F_5)_2SiCl_2$ and BF_3, and even to $RPCl_2$ (R = Cl, Ph, Me) (Fig. 11.65). Moreover, a reaction of the NHC–$SiCl_4$ with

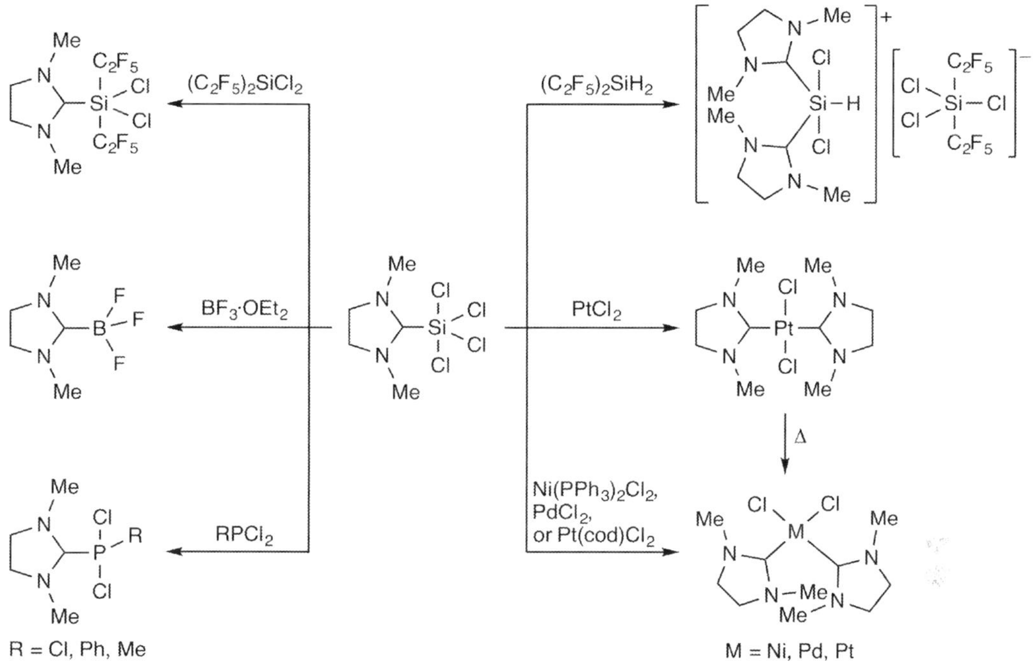

Figure 11.65
Carbene transfer reactions of pentacoordinated NHC-Si complex.

another strong Lewis acid, bis(pentafluoroethyl)dihydrosilane, gave the salt $[(NHC)_2SiCl_2H]^+[(C_2F_5)_2SiCl_3]^-$ containing each pentacoordinated silicon atom in the cation and anion.[374] The reaction mechanism is considered to proceed by the sequential chloride/hydride metathesis, chloride abstraction, and carbene transfer.

The carbene-transfer is pertinent to synthesis of the transition metal carbene complexes. A reaction of $Ni(PPh_3)_2Cl_2$ and $PdCl_2$ in THF gave the bis-NHC transition metal complexes *cis*-$[(NHC)_2NiCl_2]$ and *cis*-$[(NHC)_2PdCl_2]$, respectively.[372] In contrast, a reaction of two equivalents of NHC with $PtCl_2$ gave exclusively *trans*-$[(NHC)_2PtCl_2]$, while addition of $[Pt(cod)Cl_2]$ gave the *cis* product.[375] The stereoselective synthesis of these two isomers enabled direct observation of a thermally induced uncommon *trans*–*cis* isomerization of the bis-NHC platinum complex.

The carbene transfer reactivity of a pentacoordinated silicon compound bearing an unsaturated NHC ligand is deployed to use as a matrix to protect the moisture-sensitive NHCs with keeping reactivity almost identical to that of the free NHCs.[376] Complexation of moisture-sensitive NHCs with oligosiloxanes and polydimethylsiloxane (PDMS) forms the respective pentacoordinated organosilicon compounds, which can be handled and stored in the air without any special precautions. The NHC–PDMS complex facilitates some organic

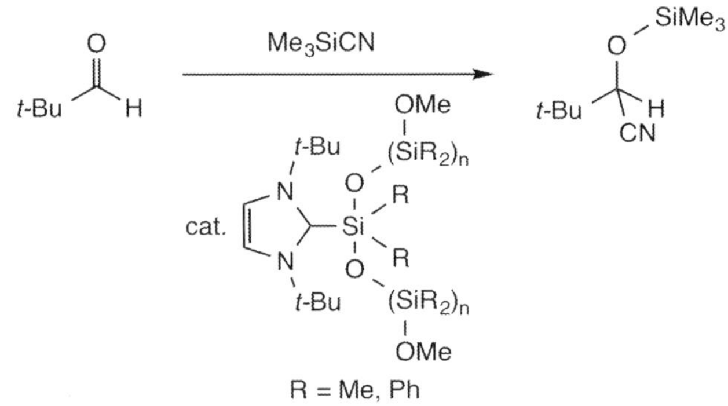

Figure 11.66
Cyanosilylation of an aldehyde with trimethylsilyl cyanide using the NHC–PDMS catalyst.

reactions catalyzed by NHCs such as cyanosilylation of an aldehyde, transesterification of an ester, and polymerization of octamethylcyclotetrasiloxane (Fig. 11.66).

11.7.2 Five Hydrocarbon Ligands on the Silicon

Stable pentacoordinated organosilicates usually require two or more electron-withdrawing ligands that are more electronegative than carbon, such as F, Cl, NR_2, and OR, to restrain elimination of a ligand from the silicon center. If not, they usually appear as the reactive intermediates. Thus, the silicates with five hydrocarbon ligands on the silicon center tend to release a substituent. Such a heterolytic Si–C bond dissociation property is utilized for the application of bis(trifluoromethyl)trimethylsilicate as a trifluoromethylation reagent.[377–379]

2,2′-Biphenyldiyl was used as a ligand to obtain stable pentaorganosilicates.[380] The incorporation of a silicon atom to form 9*H*,9-silafluorene skeleton is beneficial for the stabilization of the silicates because the biphenyldiyl groups are electron-withdrawing, uncongested, and conformationally preventing dissociation.[381] The lithium methylsilicate was synthesized by the reaction of 5,5′-spirobi[9*H*-9-silafluorene] with lithium metal in DME (Fig. 11.67).[382] Alternatively, it was rationally synthesized by the reaction of the spirosilane with methyllithium in THF.[383] Cation exchange gave the corresponding tetraalkylammonium pentaorganosilicates as stable solids with high melting point. The crystal structure of the tetrabutylammonium methylsilicate, which resonates at −105.2 ppm in the ^{29}Si NMR spectrum in DMF, shows a slightly distorted TBP structure with the methyl group in an equatorial position. Reflecting the bond properties, the apical Si–C bonds (2.011(2)–2.018(2) Å) are meaningfully longer than equatorial bonds (1.924(2)–1.948(2) Å). The pentaorganosilicate reacted with methanol to cleave a silicon–carbon(aryl) bond and protonate the carbon atom.

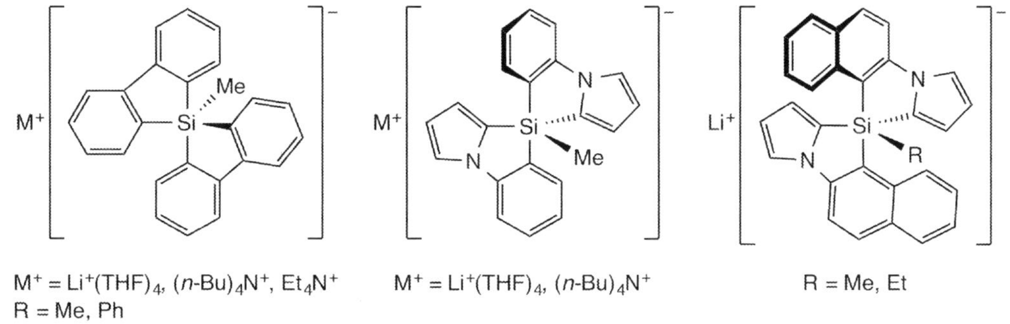

Figure 11.67
Pentaorganosilicates.

The exceptional stability of the silicates launched a study on configurational isomerism of the pentaorganosilicates with two bidentate 1-phenylpyrrole-2,2′-diyl ligands. The air-stable silicate could adopt three configurations, in which the orientation of the bidentate substituents differ.[384] NMR study showed the configuration with two phenyl moieties on the apical positions and two pyrrole moieties on the equatorial positions in the TBP structure to be predominant configuration in solution at −50°C in accordance with the X-ray crystal structure. Further NMR studies and calculations revealed that the silicate was in dynamic equilibrium with two other isomers by the stereomutation. The energy differences between them are estimated to be 1.8 and 2.6 kcal mol^{-1}. A permutational barrier was estimated to be 15.0 kcal mol^{-1}.

With the assistance of the DFT calculations, two principle mechanisms have been identified for the stereomutation of silicates bearing five hydrocarbon ligands: Berry's pseudorotation and threefold cyclic permutation.[385] The study leads to design of ethyl- and methylbis([2]naphthylpyrrol-2,1′-diyl)silicates which do not cause any stereomutation. Increasing the π-electron density and ortho substitution were found to be the key for increasing the barrier heights for the interconversion.[386] Despite the possibility of three enantiomeric pairs, both derivatives showed only single propeller-like conformation that did not epimerize at silicon at room temperature. Such a result is interesting for the application of these pentaorganosilicates as chiral cocatalysts or chiral ionic liquids.

11.8 Hypercoordinate Organosilicon Compounds Bearing Silicon–Boron Bonds

Pentacoordinated organosilicon compounds with a silicon–boron bond are quite rare. One obstacle is relatively low electronegativity of boron (χ, 2.04), which destabilizes the pentacoordinated state of the silicon, and another is lack of a suitable synthetic method and an appropriate precursor. A donor-stabilized tricoordinated silylene bearing amidinato ligands is

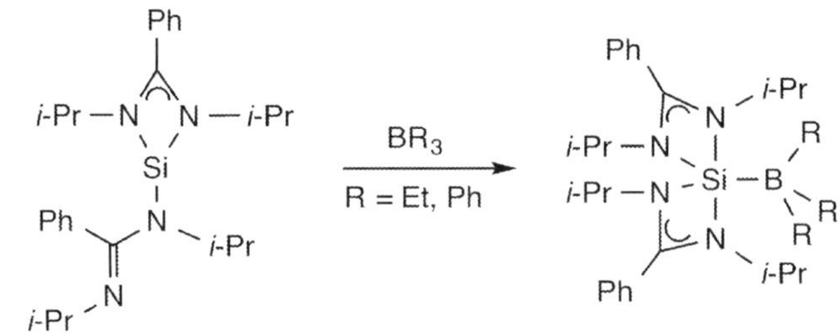

Figure 11.68
Formation of the pentacoordinated silicon compounds bearing a Si—B bond.

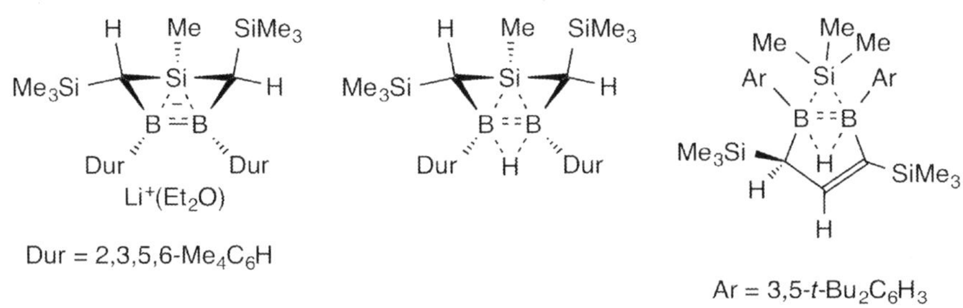

Figure 11.69
Pentacoordinated silicon compounds bearing 3c—2e B—Si—B bonds.

a good precursor for such the silicon—boron bonded compounds and reacts with triethylborane and triphenylborane giving the silylene-borane adducts, in which another nitrogen coordinates to the silicon center (Fig. 11.68).[387] In the distorted TBP geometry around the pentacoordinated silicon atom, the boron atom occupies an equatorial position showing normal Si—B bond lengths (2.067(3)—2.077(3) Å). The ^{11}B and ^{29}Si NMR spectra suggest rapid exchange of four nitrogen sites in solution. The adducts are recognized as Lewis acid/base complexes stabilized by HOMO—LUMO interactions between the silylene and borane. Some extent of dissociation of the silicon—boron bond in the triethylborane adduct was noticed in the solution at high temperature, but not in the triphenylborane adduct.

The creation of other types of pentacoordinated silicon species with a silicon—boron bond has been achieved by utilization of the three-centered two-electron (3c—2e) B—Si—B bond. This ionic compound, lithium (*deloc*-1,3,4)-1-sila-3,4-diboracyclopentane-1-ide, showed a strongly distorted five-membered ring with short transannular Si · · · B distances (2.079(3)—2.08(3) Å) in the crystal (Fig. 11.69).[388] The neutral bishomoaromatic compounds featuring the 3c—2e B—Si—B bond were also synthesized.[389,390] The neutral compounds exhibit both the B—Si—B and B—H—B bridging structures in the crystal. The Si · · · B distances (2.110(2)—2.322(3) Å) in the neutral compounds showed considerable elongation from those in the aforementioned anionic compound. The 3c—2e B—Si—B bond achieves the

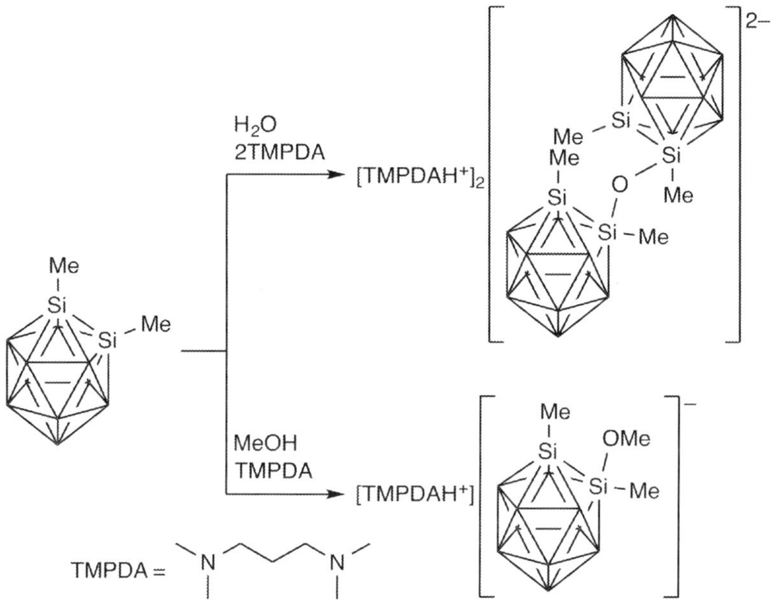

Figure 11.70
Reactions of the *o*-silaborane with water and methanol.

homoaromatic stabilization. The homoaromatic stabilization energies for bis-methylene bridged model compounds of anionic and neutral species are estimated to be 79.9 and 32.5 kcal mol^{-1}, respectively, demonstrating strong homoaromaticity. The bis-methylene bridge is not essential to the 3c–2e bond as seen in the third example, where a trimethylsilyl group bridges two boron atoms without the assistance of methylene groups.

Utilization of the 3c–2e B–Si–B bond for hypercoordination of silicon is expanded to incorporate the silicon atom in a boron cluster. Both of the two carbon atoms of *o*-carborane are swapped with hexacoordinated silicon atoms in 1,2-dimethyl-1,2-disila-*closo*-dodecaborane(12), *o*-silaborane.[391] The *o*-silaborane reacts with water and methanol in the presence of a diamine to yield the dianionic oxygen-bridged dimer and the anionic methoxy derivative, respectively (Fig. 11.70).[392] In the cluster opening reactions, the silicon atom kept hexacoordinated. The Si–B distances ranged from 2.0 to 2.3 Å in both products. In solution, their NMR spectra indicated a dynamic behavior involving oxygen-migration from one silicon atom to another at room temperature.

11.9 Hypercoordinate Organosilicon Compounds Bearing Silicon–Phosphorus Bonds

Despite many examples of penta- and hexacoordinated organosilicon compounds with a nitrogen–silicon dative bond, those with a phosphorus donor are scarcely studied.[393]

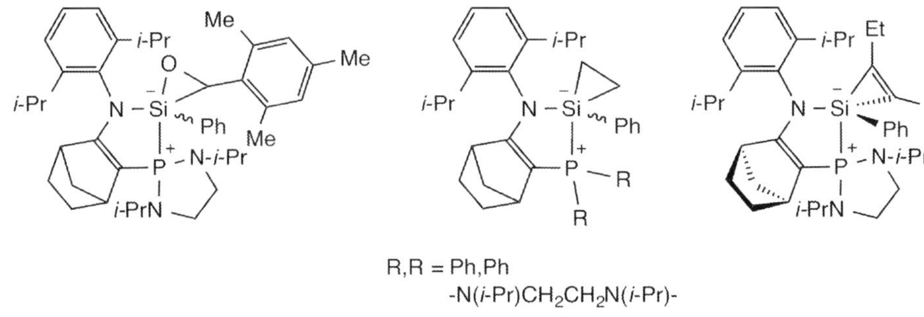

Figure 11.71
Three-membered ring compounds containing a pentacoordinated silicon atom.

Oxasilacyclopropane,[394] silacyclopropane,[395] and silacyclopropene,[396] all of which contain a bond between a pentacoordinated silicon and phosphorus atom, were synthesized by [2 + 1] cycloaddition reactions of a stable phosphonium sila–ylide as a precursor (Fig. 11.71). The P–Si bond in the oxasilacyclopropane is evidenced by the large P–Si coupling constants (J (P,Si) = 87.7 and 71.3 Hz) in the NMR spectra of a pair of the diastereomers. In the crystal, it presents a strongly deformed TBP geometry with the phosphorus and oxygen atoms in apical positions. Its P–Si bond (2.4905(13) Å) is much shorter than those of the silacyclopropane (3.273(1) Å) and silacyclopropene (3.2324(7) Å). The oxasilacyclopropane and the silacyclopropene are thermally unstable and treated only at low temperature. In contrast, the silacyclopropane is stable in the solid state at room temperature although it is returned to the sila–ylide above −30°C in solution upon decreasing the ethylene pressure.

11.10 Hypercoordinate Compounds Bearing Silicon–Metal Bonds

The interaction of a silicon atom with transition metals results in the formation of a penta- or hexacoordinated silicon moiety in several transition metal complexes. A detailed description is given in the section of transition metal complexes of silicon (Chapter 2: Transition Metal Complexes of Silicon (Excluding Silylene Complexes) and Chapter 8: Stable Silylenes and Their Transition Metal Complexes). Transition metal complexes with a penta- or hexacoordinated silicon ligand that play a role of a Z-type ligand are discussed below.

A diphosphine-silane forms the gold complex manifesting an interaction between the gold(I) and silicon (Fig. 11.72).[397] The interaction is indicated by relatively close Au–Si distance (3.090(2) Å), noticeable elongation of the Si–F bond (1.635(3) Å) in the *trans* position, and slightly distorted TBP structure around the silicon atom. The silicon plays a role of a Lewis acid to accept a σ-electron pair from gold, showing the donor–acceptor interaction. Because the overlap between the occupied donating d orbital at gold and the

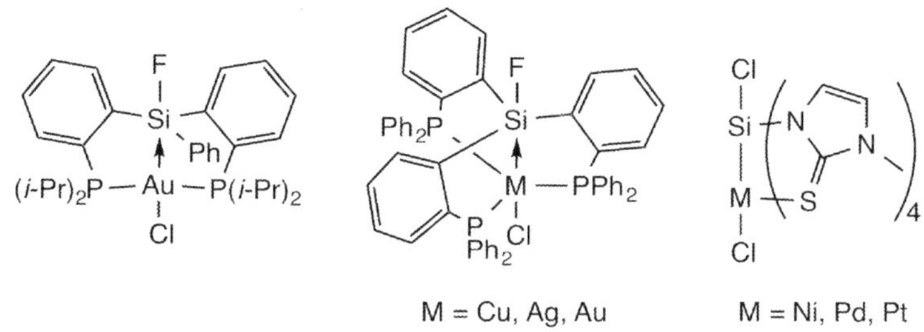

Figure 11.72
Transition metal complexes featuring a bond between silicon and transition metal atoms.

low-lying accepting orbital centered at silicon (σ^*_{Si-F}) is significant for the Au→Si interaction, it is maximized when the fluorine atom is located trans to the donor gold atom. The *cis* form, in which the fluorine atom is *cis* to gold, was observed as a minor contribution in solution, while it lacked the Au–Si interaction.[398]

A tetradentate tripodal ligand was formed by increasing a number of the phosphine moieties from two to three and was applied for the coordination to group 11 metal chlorides to synthesize a series of metallasilatranes.[399] The Au–Si distance (3.223(2) Å) is slightly longer than that with the diphosphine scaffold because of differences in the flexibility of the system and electron-donating ability of the phosphine. Comparison of structural parameters showed an increasing tendency of the Lewis basicity of the group 11 metals going down a group leading to the stronger M→Si interactions.

Formation of the hexacoordinated state of a silicon atom is available by coordination from a transition metal to silicon in some transition metal complexes. Reactions of chloro[tris(methimazolyl)]silane with $[MCl_2(PPh_3)_2]$ (M = Ni, Pd, Pt) furnished the corresponding cage-like metallasilatranes (Fig. 11.72).[400–402] In the paddlewheel-shaped complexes, the silicon atoms, forming short M–Si axis (Si–Ni, 2.598(1) Å; Si–Pd, 2.527(2)–2.569(1); Si–Pt, 2.447(3)–2.469(2) Å), are almost in plane with the four surrounding nitrogen atoms. The strong M–Si bonds could be realized by a tilt of the methimazolyl groups, furnishing the C_4-symmetric structures. NBO and electron density analyses confirmed the M–Si dative bonds for nickel and palladium complexes, whereas the Pt analog exhibits marked covalent contributions.

In a palladium complex of tetra(7-azaindolyl-1-yl)silane, two ligands coordinated to the palladium atom, which showed weak Si···Pd interaction (3.31 Å) (Fig. 11.73).[403] Upon heating at 150°C, two other dangling ligands coordinate to palladium accompanying migration of chlorine from the metal to silicon to give the corresponding paddlewheel complex. The addition of $GaCl_3$ as a chloride scavenger and recrystallization in acetonitrile afforded a cationic complex with a similar structural motif.

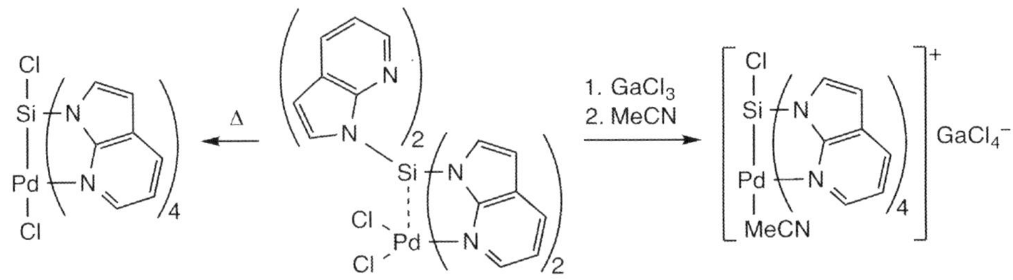

Figure 11.73
Formation of hexacoordinated silicon compounds bearing a Si–Pd bond.

11.11 Conclusions

Various types of penta- and hexacoordinated silicon compounds, some of which could not be available before 2000, have been synthesized as discussed in this chapter. Previously, most of the stable highly coordinated silicon compounds were those featuring halogen, nitrogen, oxygen, and carbon ligands. Nowadays, rational ligand design has enabled the stabilization of the compounds bearing a bond to a silicon center from boron, phosphorus, sulfur, transition metals, and another silicon. Utilization of a variety of ligands also made it possible to incorporate of a pentacoordinated silicon atom into unique heterocyclic structures, such as strained three-membered rings. Most studies on the hypercoordinate silicon compounds have been limited to elucidation of structures, spectral properties and fundamental reactivities, but some of the compounds were found to be applicable to functional materials, such as ion sensors and luminescent materials. Moreover, some NHC-coordinated silicon compounds were applied to carbene transfer reactions and as the catalysts for cyanosilylation. Studies on hypercoordinate silicon compounds in the future may shift from fundamental research to application of the hypercoordinate silicon compounds to development of functional materials and organic reactions for practical usage. Because silatranes have been extensively investigated for their toxicity, bioactivity of other hypercoordinate silicon compounds might be a promising direction to be developed in the future.

References

1. Gillespie, R. J.; Silvi, B. The Octet Rule and Hypervalence: Two Misunderstood Concepts. *Coord. Chem. Rev.* **2002,** *233*, 53–62.
2. Corriu, R. J. P.; Young, J. C. Hypervalent Silicon Compounds. In *The Chemistry of Organic Silicon Compounds;* Patai, S., Rappoport, Z., Eds.; Wiley: Chichester, UK, 1989; pp 1241–1288.
3. Holmes, R. R. The Stereochemistry of Nucleophilic-Substitution at Tetracoordinated Silicon. *Chem. Rev.* **1990,** *90* (1), 17–31.
4. Corriu, R. J. P.; Young, J. C. Hypervalent Silicon Compounds. In *The Silicon-Heteroatom Bond;* Patai, S., Rappoport, Z., Eds.; Wiley: Chichester, UK, 1991; pp 1–66.

5. Chuit, C.; Corriu, R. J. P.; Reye, C.; Young, J. C. Reactivity of Penta- and Hexacoordinate Silicon Compounds and Their Role as Reaction Intermediates. *Chem. Rev.* **1993,** *93* (4), 1371–1448.
6. Corriu, R. Reactivity of Penta- and Hexacoordinated Silicon Species. In *Organosilicon Chemistry I: From Molecules to Materials;* Auner, N., Weis, J., Eds.; Wiley-VCH: Weinheim, Germany, 1994; pp 157–163.
7. Bassindale, A. R.; Glynn, S. G.; Jiang, J.; Parker, D. J.; Turtle, R.; Taylor, P. G., et al. Recent Explorations of the Chemistry of Pentacoordinate Silicon. In *Organosilicon Chemistry II: From Molecules to Materials;* Auner, N., Weis, J., Eds.; Wiley-VCH: Weinheim, Germany, 1996; pp 411–425.
8. Belzner, J.; Schaer, D. Highly Coordinated Silicon Compounds—Hydrazino Groups as Intramolecular Donors. In *Organosilicon Chemistry II: From Molecules to Materials;* Auner, N., Weis, J., Eds.; Wiley-VCH: Weinheim, Germany, 1996; pp 459–465.
9. Holmes, R. R. Comparison of Phosphorus and Silicon: Hypervalency, Stereochemistry, and Reactivity. *Chem. Rev.* **1996,** *96* (3), 927–950.
10. Kost, D.; Kalikhman, I. Hypervalent Silicon Compounds. In *Chemistry of Organic Silicon Compounds. Vol. 2, Pt. 2;* Rappoport, Z., Apeloig, Y., Eds.; Wiley: Chichester, UK, 1998; pp 1339–1445.
11. Chuit, C.; Corriu, R. J. P.; Reye, C. Structure and Reactivity of Hypercoordinate Silicon Species. In *Chemistry of Hypervalent Compounds;* Akiba K-y, Ed.; Wiley-VCH: New York, NY, 1999; pp 81–146.
12. Rendler, S.; Oestreich, M. Hypervalent Silicon as a Reactive Site in Selective Bond-Forming Processes. *Synthesis* **2005,** *11*, 1727–1747.
13. Benaglia, M.; Guizzetti, S.; Pignataro, L. Stereoselective Reactions Involving Hypervalent Silicate Complexes. *Coord. Chem. Rev.* **2008,** *252* (5–7), 492–512.
14. Alkorta, I.; Rozas, I.; Elguero, J. Molecular Complexes Between Silicon Derivatives and Electron-Rich Groups. *J. Phys. Chem. A.* **2001,** *105* (4), 743–749.
15. Fleischer, H. Molecular "Floppyness" and the Lewis Acidity of Silanes: A Density Functional Theory Study. *Eur. J. Inorg. Chem.* **2001,** (2), 393–404.
16. Molina, J. M.; Dobado, J. A. The Three-Center-Four-Electron (3c-4e) Bond Nature Revisited. An Atoms-In-Molecules Theory (AIM) and ELF Study. *Theor. Chem. Acc.* **2001,** *105* (4–5), 328–337.
17. Timoshkin, A. Y.; Davydova, E. I.; Sevastianova, T. N.; Suvorov, A. V.; Schaefer, H. F. Relationship Between the Energy of Donor-Acceptor Bond and the Reorganization Energy in Molecular Complexes. *Int. J. Quantum. Chem.* **2002,** *88* (4), 436–440.
18. Denmark, S. E.; Beutner, G. L.; Wynn, T.; Eastgate, M. D. Lewis Base Activation of Lewis Acids: Catalytic, Enantioselective Addition of Silyl Ketene Acetals to Aldehydes. *J. Am. Chem. Soc.* **2005,** *127* (11), 3774–3789.
19. Davydova, E. I.; Timoshkin, A. Y.; Sevastianova, T. N.; Suvorov, A. V.; Frenking, G. Lewis Acidity of Group 14 Tetrahalides in Gas Phase. *J. Mol. Struct.:THEOCHEM* **2006,** *767* (1–3), 103–111.
20. Pierrefixe, S. C.; Fonseca Guerra, C.; Bickelhaupt, F. M. Hypervalent Silicon Versus Carbon: Ball-in-a-Box Model. *Chem. Eur. J.* **2008,** *14* (3), 819–828.
21. Bento, A. P.; Bickelhaupt, F. M. Frontside Versus Backside S_N2 Substitution at Group 14 Atoms: Origin of Reaction Barriers and Reasons for Their Absence. *Chem. Asian. J.* **2008,** *3* (10), 1783–1792.
22. Marchaj, M.; Freza, S.; Skurski, P. Why are SiX_5^- and GeX_5^- (X = F, Cl) Stable But Not CF_5^- and CCl_5^-? *J. Phys. Chem. A.* **2012,** *116* (8), 1966–1973.
23. Getmanskii, I. V.; Minyaev, R. M.; Minkin, V. I. Effect of Counterions on Hypervalent Interactions in Prereaction Complexes of S_N2 Reactions of Bisphenoid Compounds of the Second and Third Period Elements. *Russ. Chem. Bull.* **2012,** *61* (11), 2036–2048.
24. Grabowski, S. J. Tetrel Bond-σ-Hole Bond as a Preliminary Stage of the S_N2 Reaction. *Phys. Chem. Chem. Phys.* **2014,** *16* (5), 1824–1834.
25. Steinhauer, S.; Bottcher, T.; Schwarze, N.; Neumann, B.; Stammler, H. G.; Hoge, B. Synthesis of Chlorosilicates. *Angew. Chem. Int. Ed.* **2014,** *53* (48), 13269–13772.
26. Mąkosza, M.; Bujok, R. Simple Synthesis of Tetrabutylammonium and Potassium Salts of Difluorotriorganostannate and Silicate Anions. *Synlett* **2004,** *2*, 371–373.

27. Bolli, C.; Gellhaar, J.; Jenne, C.; Kessler, M.; Scherer, H.; Seeger, H., et al. Bis(triphenyl-λ^5-phosphanylidene)ammonium Fluoride: A Reactive Fluoride Source to Access the Hypervalent Silicates $[Me_nSiF_{5-n}]^-$ (n = 0–3). *Dalton Trans.* **2014,** *43* (11), 4326–4334.
28. Ghadwal, R. S.; Pröpper, K.; Dittrich, B.; Jones, P. G.; Roesky, H. W. Neutral Pentacoordinate Silicon Fluorides Derived From Amidinate, Guanidinate, and Triazapentadienate Ligands and Base-Induced Disproportionation of Si_2Cl_6 to Stable Silylenes. *Inorg. Chem.* **2011,** *50* (1), 358–364.
29. Junold, K.; Burschka, C.; Bertermann, R.; Tacke, R. Novel Neutral Hexacoordinate Benzamidinatosilicon (IV) Complexes With SiN_3OF_2, SiN_3OCl_2, SiN_3OBr_2, SiN_5O and SiN_3O_3 Skeletons. *Dalton Trans.* **2011,** *40* (38), 9844–9857.
30. Jana, A.; Samuel, P. P.; Tavcar, G.; Roesky, H. W.; Schulzke, C. Selective Aromatic C–F and C–H Bond Activation With Silylenes of Different Coordinate Silicon. *J. Am. Chem. Soc.* **2010,** *132* (29), 10164–10170.
31. Samuel, P. P.; Singh, A. P.; Sarish, S. P.; Matussek, J.; Objartel, I.; Roesky, H. W., et al. Oxidative Addition Versus Substitution Reactions of Group 14 Dialkylamino Metalylenes With Pentafluoropyridine. *Inorg. Chem.* **2013,** *52* (3), 1544–1549.
32. Steinhauer, S.; Stammler, H. G.; Neumann, B.; Ignat'ev, N.; Hoge, B. Synthesis of Five- and Six-Coordinate Tris(pentafluoroethyl)fluorosilicates. *Angew. Chem. Int. Ed.* **2014,** *53* (2), 562–564.
33. Steinhauer, S.; Bader, J.; Stammler, H. G.; Ignat'ev, N.; Hoge, B. Synthesis of Tris- and Tetrakis (pentafluoroethyl)silanes. *Angew. Chem. Int. Ed.* **2014,** *53* (20), 5206–5209.
34. Kolomeitsev, A. A.; Bissky, G.; Barten, J.; Kalinovich, N.; Lork, E.; Röschenthaler, G.-V. Hexaalkylguanidinium and 2-(Dialkylamino)-1,3-dimethylimidazolinium Trimethyldifluorosiliconates and Perfluoroalkoxides. Accidental Isolation and Molecular Structure of $[C(NMe_2)_3]^+F^- \cdot 6CH_2Cl_2$. *Inorg. Chem.* **2002,** *41* (23), 6118–6124.
35. Driess, M.; Merz, K.; Monsé, C. Synthesis of the First Fluoro(phosphanyl)- and Diphosphanyl-Stannanes and Surprising Formation of $[P(SnMe_3)_4]^+SiF_5^-$. *Chem. Commun.* **2003,** *20*, 2608–2609.
36. Henrich, M.; Marhold, A.; Kolomeitsev, A. A.; Kalinovich, N.; Röschenthaler, G.-V. A New Family of Delocalized Lipophilic Cations. *Tetrahedron Lett.* **2003,** *44* (31), 5795–5798.
37. Gorol, M.; Mösch-Zanetti, N. C.; Roesky, H. W.; Noltemeyer, M.; Schmidt, H.-G. Synthesis of a Novel Organoiridium(I) Fluoro Complex. *Eur. J. Inorg. Chem.* **2004,** (13), 2678–2682.
38. Spirk, S.; Belaj, F.; Nieger, M.; Köfeler, H.; Rechberger, G. N.; Pietschnig, R. Exploring the Anion-Cation Interaction in *m*-Terphenyltetrafluorosilicates by Using Multinuclear NMR Spectroscopy, X-ray Diffraction, and ICR-FT-MS. *Chem. Eur. J.* **2009,** *15* (37), 9521–9529.
39. Yamaguchi, S.; Akiyama, S.; Tamao, K. Photophysical Properties Changes Caused by Hypercoordination of Organosilicon Compounds: From Trianthrylfluorosilane to Trianthryldifluorosilicate. *J. Am. Chem. Soc.* **2000,** *122* (28), 6793–6794.
40. Yamaguchi, S.; Akiyama, S.; Tamao, K. A New Approach to Photophysical Properties Control of Main Group Element π-Electron Compounds Based on the Coordination Number Change. *J. Organomet. Chem.* **2002,** *652* (1–2), 3–9.
41. Kim, Y.; Kim, M.; Gabbai, F. P. Synthesis and Anion Affinity of a Bidendate Sulfonium Fluorosilane Lewis Acid. *Org. Lett.* **2010,** *12* (3), 600–602.
42. Kawachi, A.; Tani, A.; Shimada, J.; Yamamoto, Y. Synthesis of B/Si Bidentate Lewis acids, *o*-(Fluorosilyl)(dimesitylboryl)benzenes, and Their Fluoride Ion Affinity. *J. Am. Chem. Soc.* **2008,** *130* (13), 4222–4223.
43. Lenormand, H.; Corcé, V.; Sorin, G.; Chhun, C.; Chamoreau, L. M.; Krim, L., et al. Versatile Access to Martin's Spirosilanes and Their Hypervalent Derivatives. *J. Org. Chem.* **2015,** *80* (6), 3280–3288.
44. Lenormand, H.; Goddard, J. P.; Fensterbank, L. Spirosilane Derivatives as Fluoride Sensors. *Org. Lett.* **2013,** *15* (4), 748–751.
45. Setaka, W.; Nirengi, T.; Kabuto, C.; Kira, M. Introduction of Clutch Function into a Molecular Gear System by Silane-Silicate Interconversion. *J. Am. Chem. Soc.* **2008,** *130* (47), 15762–15763.

46. Handy, C. J.; Lam, Y. F.; DeShong, P. On the Synthesis and NMR Analysis of Tetrabutylammonium Triphenyldifluorosilicate. *J. Org. Chem.* **2000,** *65* (11), 3542–3543.
47. Omotowa, B. A.; Shreeve, J. M. Difluoroorganometalate-Assisted Generation of Perfluorocarbanions From Trimethylsilyl Synthons and Their Interactions With Perfluoroaryl Compounds. *Organometallics* **2000,** *19* (14), 2664–2670.
48. Kvicala, J.; Mysik, P.; Paleta, O. Tetrabutylammonium Butyldifluorodimethylsilicate and Difluorodimethylphenylsilicate, New Nucleophilic Fluorinating Reagents. *Synlett* **2001,** *4*, 547–549.
49. Wessel, J.; Behrens, U.; Lork, E.; Borrmann, T.; Stohrer, W.-D.; Mews, R. Tris(dimethylamino) oxosulfonium Difluorotrimethylsilicate, $(Me_2N)_3SO^+Me_3SiF_2^-$ (TAOS Fluoride). *Inorg. Chem.* **2002,** *41* (18), 4715–4721.
50. Mąkosza, M.; Surowiec, M. Oxidative Nucleophilic Substitution of Hydrogen in Nitroarenes by Silyl Enol Ethers. *Tetrahedron* **2003,** *59* (33), 6261–6266.
51. Hirai, M.; Gabbai, F. P. Squeezing Fluoride Out of Water With a Neutral Bidentate Antimony(V) Lewis Acid. *Angew. Chem. Int. Ed.* **2015,** *54* (4), 1205–1209.
52. Furman, B.; Dziedzic, M. Tetrabutylammonium Triphenyldifluorosilicate (TBAT) Initiated Intramolecular Addition of Allylsilanes to 2,3-Dihydro-4-pyridones. A Novel Route for the Stereoselective Construction of Indolizidine Systems. *Tetrahedron Lett.* **2003,** *44* (35), 6629–6632.
53. Hoke, M. E.; Brescia, M. R.; Bogaczyk, S.; DeShong, P.; King, B. W.; Crimmins, M. T. Regioselectivity in the Palladium-Catalyzed Addition of Carbon Nucleophiles to Carbocyclic Derivatives. *J. Org. Chem.* **2002,** *67* (2), 327–335.
54. Penso, M. Biaryl Formation: Palladium Catalyzed Cross-Coupling Reactions Between Hypervalent Silicon Reagents and Aryl Halides. *J. Mol. Catal. A: Chem.* **2003,** *204–205*, 177–185.
55. Fester, G. W.; Wagler, J.; Brendler, E.; Kroke, E. Stable Trichlorosilane-Pyridine Adducts. *Eur. J. Inorg. Chem.* **2008,** (32), 5020–5023.
56. Fester, G. W.; Wagler, J.; Brendler, E.; Bohme, U.; Roewer, G.; Kroke, E. Octahedral Adducts of Dichlorosilane With Substituted Pyridines: Synthesis, Reactivity and a Comparison of Their Structures and ^{29}Si NMR Chemical Shifts. *Chem. Eur. J.* **2008,** *14* (10), 3164–3176.
57. Fester, G. W.; Wagler, J.; Brendler, E.; Bohme, U.; Gerlach, D.; Kroke, E. Octahedral $HSiCl_3$ and $HSiCl_2Me$ Adducts With Pyridines. *J. Am. Chem. Soc.* **2009,** *131* (19), 6855–6864.
58. Nakash, M.; Gut, D.; Goldvaser, M. Formation of Hypervalent Complexes of Trifluorosilanes With Pyridine and With 4-Methoxypyridine, Through Intermolecular Silicon···Nitrogen Interactions. *Inorg. Chem.* **2005,** *44* (4), 1023–1030.
59. Nakash, M.; Goldvaser, M. Formation of Hypervalent Complexes of $PhCCSiF_3$ With Pyridine Through Intermolecular Silicon···Nitrogen Interaction. *J. Am. Chem. Soc.* **2004,** *126* (11), 3436–3437.
60. Robertson, A. P.; Friedmann, J. N.; Jenkins, H. A.; Burford, N. Exploring Structural Trends for Complexes of $Me_2E(OSO_2CF_3)_2$ (E = Si, Ge, Sn) With Pyridine Derivatives. *Chem. Commun.* **2014,** *50* (59), 7979–7981.
61. Fester, G. W.; Eckstein, J.; Gerlach, D.; Wagler, J.; Brendler, E.; Kroke, E. Reactions of Hydridochlorosilanes With 2,2′-Bipyridine and 1,10-Phenanthroline: Complexation Versus Dismutation and Metal-Catalyst-Free 1,4-Hydrosilylation. *Inorg. Chem.* **2010,** *49* (6), 2667–2673.
62. Wakaki, T.; Kanai, M.; Kuninobu, Y. Iridium-Catalyzed Ortho-Selective C-H Silylation of Aromatic Compounds Directed Toward the Synthesis of π-Conjugated Molecules With Lewis Acid-Base Interaction. *Org. Lett.* **2015,** *17* (7), 1758–1761.
63. Xiao, Q.; Meng, X.; Kanai, M.; Kuninobu, Y. Palladium-Catalyzed C-H Fluorosilylation of 2-Phenylpyridines: Synthesis of Silafluorene Equivalents. *Angew. Chem. Int. Ed.* **2014,** *53* (12), 3168–3172.
64. Tokoro, Y.; Yeo, H.; Tanaka, K.; Chujo, Y. Synthesis of Benzo[*h*]quinoline-Based Neutral Pentacoordinate Organosilicon Complexes. *Chem. Commun.* **2012,** *48* (68), 8541–8543.
65. Tokoro, Y.; Tanaka, K.; Chujo, Y. Synthesis of Dibenzo[*b,f*]silepins With a Benzoquinolyl Ligand. *Org. Lett.* **2013,** *15* (10), 2366–2369.

66. van den Ancker, T. R.; Raston, C. L.; Skelton, B. W.; White, A. H. Hypervalent Silicon via Intramolecular Coordination in a Four-Membered Ring in Complexes of Substituted Pyridyl Ligands. *Organometallics* **2000,** *19* (22), 4437–4444.
67. Jones, C.; Junk, P. C.; Leary, S. G.; Smithies, N. A.; Steed, J. W. Investigations Into the Small-Bite *N*-Functionalised Amido Chemistry of Silicon and Tin(IV). *Inorg. Chem. Commun.* **2002,** *5* (7), 533–536.
68. Baus, J. A.; Burschka, C.; Bertermann, R.; Guerra, C. F.; Bickelhaupt, F. M.; Tacke, R. Neutral Six-Coordinate and Cationic Five-Coordinate Silicon(IV) Complexes With Two Bidentate Monoanionic N,S-Pyridine-2-thiolato(-) Ligands. *Inorg. Chem.* **2013,** *52* (18), 10664–10676.
69. Schwarz, D.; Brendler, E.; Kroke, E.; Wagler, J. Pentacoordinate Silicon Complexes With *N*-(2-pyridylmethyl)salicylamide as a Dianionic (*ONN'*) Tridentate Chelator. *Z. Anorg. Allg. Chem.* **2012,** *638* (11), 1768–1775.
70. Kobelt, C.; Burschka, C.; Bertermann, R.; Fonseca Guerra, C.; Bickelhaupt, F. M.; Tacke, R. Synthesis and Structural Characterisation of Neutral Pentacoordinate Silicon(IV) Complexes With a Tridentate Dianionic N,N,S Chelate Ligand. *Dalton Trans.* **2012,** *41* (7), 2148–2162.
71. Gómez, E.; Santes, V.; de la Luz, V.; Farfán, N. Synthesis and Structure of Pentacoordinated Monoorganosilane Derivatives of Pyridine Ligands. *J. Organomet. Chem.* **2001,** *622* (1–2), 54–60.
72. Gómez, E.; Hernandez, Z.; Alvarez-Toledano, C.; Toscano, R. A.; Santes, V.; Sharma, P. Multinuclear NMR and X-Ray Diffraction Study of Pentacoordinated Siloxane Structures Derived of Pyridine Diols. *J. Organomet. Chem.* **2002,** *648* (1–2), 280–287.
73. Brendler, E.; Heine, T.; Hill, A. F.; Wagler, J. A Pentacoordinate Chlorotrimethylsilane Derivative: A Very Polar Snapshot of a Nucleophilic Substitution and Its Influence on ^{29}Si Solid State NMR Properties. *Z. Anorg. Allg. Chem.* **2009,** *635* (9–10), 1300–1305.
74. Brendler, E.; Wächtler, E.; Wagler, J. Hypercoordinate Silacycloalkanes: Step-by-Step Tuning of N→Si Interactions. *Organometallics* **2009,** *28* (18), 5459–5465.
75. Aksamentova, T. N.; Voronkov, M. G.; Trofimova, O. M.; Chipanina, N. N.; Bolgova, Y. I.; Chernov, N. F., et al. Unusual Lability of Intramolecular Coordination Bond O→Si *N*-(Trifluorosilylmethyl) succinimide. *J. Struct. Chem.* **2005,** *46* (2), 248–252.
76. Maaranen, J.; Andell, O. S.; Vanne, T.; Mutikainen, I. Dimethyl-, Diphenyl- and Cyclotetramethylene Silylchloride Derivatives of Guanidine, Their Synthesis and Structure. Formation of Pentacoordinated Silicon Compounds and an Unexpected Diionic Disila-Diguanidinium Dichloride. *J. Organomet. Chem.* **2006,** *691* (1–2), 240–246.
77. Coles, M. P.; Khalaf, M. S.; Hitchcock, P. B. A New Aliphatic *N,C,N'*-Pincer Ligand With Pendant Guanidine Groups. *Inorg. Chim. Acta.* **2014,** *422*, 228–234.
78. Gerlach, D.; Brendler, E.; Heine, T.; Wagler, J. Dianion of Pyrrole-2-*N*-(*o*-hydroxyphenyl)carbaldimine as an Interesting Tridentate (ONN) Ligand System in Hypercoordinate Silicon Complexes. *Organometallics* **2007,** *26* (1), 234–240.
79. Wagler, J.; Hill, A. F. Ring Opening of Organosilicon-Substituted Benzoxazolinone: A Convenient Route to Chelating Ureato and Carbamido Ligands. *Organometallics* **2008,** *27* (24), 6579–6586.
80. Gericke, R.; Gerlach, D.; Wagler, J. Ring-Strain-Formation Lewis Acidity? A Pentacoordinate Silacyclobutane Comprising Exclusively Equatorial Si–C Bonds. *Organometallics* **2009,** *28* (23), 6831–6834.
81. Wagler, J.; Bohme, U.; Brendler, E.; Roewer, G. Equilibrium Between Tetra-, Penta-, and Hexacoordinate Imine and Enamine Chelates of Silicon: Crystal Structure and Variable-Temperature NMR. *Organometallics* **2005,** *24* (6), 1348–1350.
82. González-Nogal, A. M.; Calle, M.; Cuadrado, P. Reactions of Lithium Silylcuprates With Pyrazolium and Indazolium Salts. *Eur. J. Org. Chem.* **2007,** *2007* (36), 6089–6096.
83. Seiler, O.; Burschka, C.; Metz, S.; Penka, M.; Tacke, R. Synthesis and Structural Characterization of Neutral Higher-Coordinate Silicon(IV) Complexes With Tridentate Dianionic Chelate Ligands. *Chem. Eur. J.* **2005,** *11* (24), 7379–7386.
84. Theis, B.; Metz, S.; Back, F.; Burschka, C.; Tacke, R. Neutral Pentacoordinate Silicon(IV) Complexes With $SiClO_2NC$, $SiClON_2C$, SiO_3NC, or SiO_2N_2C Skeletons. *Z. Anorg. Allg. Chem.* **2009,** *635* (9–10), 1306–1312.

85. Weiß, J.; Theis, B.; Metz, S.; Burschka, C.; Guerra, C. F.; Bickelhaupt, F. M., et al. Neutral Pentacoordinate Halogeno- and Pseudohalogenosilicon(IV) Complexes With a Tridentate Dianionic *O,N,O* or *N,N,O* Ligand: Synthesis and Structural Characterization in the Solid State and in Solution. *Eur. J. Inorg. Chem.* **2012,** (19), 3216−3228.
86. Weiß, J.; Theis, B.; Baus, J. A.; Burschka, C.; Bertermann, R.; Tacke, R. Neutral Pentacoordinate Silicon (IV) Complexes With a Tridentate Dianionic *O,N,O* or *N,N,O* Ligand: Synthesis and Structural Characterization in the Solid State and in Solution. *Z. Anorg. Allg. Chem.* **2014,** *640* (2), 300−309.
87. Böhme, U.; Günther, B. Five and Six-Coordinate Silicon Complexes With an *O,N,O′*-Chelating Ligand Derived From *o*-Hydroxyacetophenone−*N*-(2-hydroxyethyl)imine. *Inorg. Chem. Commun.* **2007,** *10* (4), 482−484.
88. Lippe, K.; Gerlach, D.; Kroke, E.; Wagler, J. *N*-(*o*-Aminophenyl)-2-oxy-4-methoxybenzophenoneimine − Si-Chelation by a Tridentate ONN Ligand System Versus Benzimidazoline Formation. *Inorg. Chem. Commun.* **2008,** *11* (5), 492−496.
89. Metz, S.; Burschka, C.; Platte, D.; Tacke, R. Pentacoordination of Silicon by Five Different Ligand Atoms: Neutral Silicon(IV) Complexes With SiClSONC and SiISONC Skeletons. *Angew. Chem. Int. Ed.* **2007,** *46* (37), 7006−7009.
90. Metz, S.; Burschka, C.; Tacke, R. A Neutral Pentacoordinate Silicon(IV) Complex With a Monodentate Nitrato Ligand: Synthesis and Structural Characterization. *Organometallics* **2008,** *27* (22), 6032−6034.
91. Metz, S.; Theis, B.; Burschka, C.; Tacke, R. Neutral Pentacoordinate Halogeno- and Pseudohalogenosilicon(IV) Complexes With an SiSONCX Skeleton (X = F, Cl, Br, I, N, C): Synthesis and Structural Characterization in the Solid State and in Solution. *Chem. Eur. J.* **2010,** *16* (23), 6844−6856.
92. González-García, G.; Gutiérrez, J. A.; Cota, S.; Metz, S.; Bertermann, R.; Burschka, C., et al. Synthesis and Structural Characterization of Novel Neutral Higher-Coordinate Silicon(IV) Complexes With $\mathit{Si}ON_3C$ and $\mathit{Si}ON_4C$ Skeletons. *Z. Anorg. Allg. Chem.* **2008,** *634* (8), 1281−1286.
93. Paul, L. E. H.; Foehn, I. C.; Schwarzer, A.; Brendler, E.; Böhme, U. Salicylaldehyde-(2-hydroxyethyl) imine—A Flexible Ligand for Group 13 and 14 Elements. *Inorg. Chim. Acta.* **2014,** *423*, 268−280.
94. Weiß, J.; Baus, J. A.; Burschka, C.; Tacke, R. Neutral Hexacoordinate Silicon(IV) Complexes With a Tridentate Dianionic *O,N,X* Ligand (X = O, N, S), Bidentate Monoanionic *X,N* Ligand (X = O, S), and Phenyl Ligand. Compounds With a $\mathit{Si}O_3N_2C$, $\mathit{Si}SO_2N_2C$, $\mathit{Si}O_2N_3C$, $\mathit{Si}SON_3C$, or $\mathit{Si}S_2ON_2C$ Skeleton. *Eur. J. Inorg. Chem.* **2014,** (14), 2449−2455.
95. Seiler, O.; Burschka, C.; Fenske, T.; Troegel, D.; Tacke, R. Neutral Hexa- and Pentacoordinate Silicon (IV) Complexes With SiO_6 and SiO_4N Skeletons. *Inorg. Chem.* **2007,** *46* (13), 5419−5424.
96. Metz, S.; Burschka, C.; Tacke, R. Neutral Hexacoordinate Silicon(IV) Complexes With an $\mathit{Si}SO_3NC$ Skeleton and a Neutral Pentacoordinate Silicon(IV) Complex Containing a Trianionic Tetradentate *O,N,O,O* Ligand. *Organometallics* **2009,** *28* (7), 2311−2317.
97. Weiß, J.; Sinner, K.; Baus, J. A.; Burschka, C.; Tacke, R. Neutral Hexacoordinate Silicon(IV) Complexes With a $\mathit{Si}O_4NC$ or $\mathit{Si}O_3N_2C$ Skeleton and Neutral Pentacoordinate Silicon(IV) Complexes Containing a Trianionic Tetradentate *O,N,O,O* Ligand. *Eur. J. Inorg. Chem.* **2014,** (3), 475−483.
98. Böhme, U.; Wiesner, S.; Günther, B. Easy Access to Chiral Penta- and Hexacoordinate Silicon Compounds. *Inorg. Chem. Commun.* **2006,** *9* (8), 806−809.
99. Böhme, U.; Fels, S. A New Type of Chiral Pentacoordinated Silicon Compounds With Azomethine Ligands Made From Acetylacetone and Amino Acids. *Inorg. Chim. Acta.* **2013,** *406*, 251−255.
100. Kämpfe, A.; Kroke, E.; Wagler, J. Hypercoordinate Silicon Complexes of (O,N,N' vs. O,N,O') Schiff Base Type *N*-(2-Carbamidophenyl)imines: Examples of Exclusively O-Silylated Carbamides. *Eur. J. Inorg. Chem.* **2009,** (8), 1027−1035.
101. Sen, S. S.; Jana, A.; Roesky, H. W.; Schulzke, C. A Remarkable Base-Stabilized Bis(silylene) With a Silicon(I)−Silicon(I) Bond. *Angew. Chem. Int. Ed.* **2009,** *48* (45), 8536−8538.

102. Jones, C.; Bonyhady, S. J.; Holzmann, N.; Frenking, G.; Stasch, A. Preparation, Characterization, and Theoretical Analysis of Group 14 Element(I) Dimers: A Case Study of Magnesium(I) Compounds as Reducing Agents in Inorganic Synthesis. *Inorg. Chem.* **2011,** *50* (24), 12315–12325.
103. Blom, B.; Pohl, M.; Tan, G.; Gallego, D.; Driess, M. From Unsymmetrically Substituted Benzamidinato and Guanidinato Dichlorohydridosilanes to Novel Hydrido N-Heterocyclic Silylene Iron Complexes. *Organometallics* **2014,** *33* (19), 5272–5282.
104. Tacke, R.; Kobelt, C.; Baus, J. A.; Bertermann, R.; Burschka, C. Synthesis, Structure and Reactivity of a Donor-Stabilised Silylene With a Bulky Bidentate Benzamidinato Ligand. *Dalton Trans.* **2015,** *44* (33), 14959–14974.
105. Lv, L.-D.; Li, J.-J.; Yang, W.; Ren, C.-X.; Ding, Y.-Q. Trichlorido(*N,N'*-di-*tert*-butylbenzamidinato-κ^2*N,N'*)silicon. *Acta Crystallogr., Sect. E: Struct. Rep. Online* **2008,** *64* (5), o870.
106. Mück, F. M.; Junold, K.; Baus, J. A.; Burschka, C.; Tacke, R. Donor-Stabilized Silylenes With Guanidinato Ligands. *Eur. J. Inorg. Chem.* **2013,** *2013* (34), 5821–5825.
107. So, C.-W.; Roesky, H. W.; Gurubasavaraj, P. M.; Oswald, R. B.; Gamer, M. T.; Jones, P. G., et al. Synthesis and Structures of Heteroleptic Silylenes. *J. Am. Chem. Soc.* **2007,** *129* (39), 12049–12054.
108. Junold, K.; Burschka, C.; Bertermann, R.; Tacke, R. Novel Neutral Hexacoordinate Silicon(IV) Complexes With Two Bidentate Monoanionic Benzamidinato Ligands. *Dalton Trans.* **2010,** *39* (39), 9401–9413.
109. Junold, K.; Baus, J. A.; Burschka, C.; Tacke, R. Bis[*N,N'*-diisopropylbenzamidinato(−)]silicon(II): A Silicon(II) Compound With Both a Bidentate and a Monodentate Amidinato Ligand. *Angew. Chem. Int. Ed.* **2012,** *51* (28), 7020–7023.
110. Azhakar, R.; Roesky, H. W.; Ghadwal, R. S.; Holstein, J. J.; Dittrich, B. An Access to Base-Stabilized Three-Membered Silicon Heterocycles. *Dalton Trans.* **2012,** *41* (32), 9601–9603.
111. Azhakar, R.; Ghadwal, R. S.; Roesky, H. W.; Hey, J.; Stalke, D. Reactions of Stable *N*-Heterocyclic Silylenes With Ketones and 3,5-Di-*tert*-butyl-*o*-benzoquinone. *Organometallics* **2011,** *30* (14), 3853–3858.
112. Azhakar, R.; Roesky, H. W.; Holstein, J. J.; Dittrich, B. Reaction of the Silylene PhC(N*t*Bu)$_2$Si*t*Bu With 4,4′-Bis(dimethylamino)thiobenzophenone and Treatment of the Silylene PhC(N*t*Bu)$_2$SiC(SiMe$_3$)$_3$ With 3,5-Di-*tert*-butyl-*o*-benzoquinone. *Eur. J. Inorg. Chem.* **2013,** (15), 2777–2781.
113. Azhakar, R.; Ghadwal, R. S.; Roesky, H. W.; Mata, R. A.; Wolf, H.; Herbst-Irmer, R., et al. Reaction of *N*-Heterocyclic Silylenes With Thioketone: Formation of Silicon–Sulfur Three (Si-C-S)- and Five (Si-C-C-C-S)-Membered Ring Systems. *Chem. Eur. J.* **2013,** *19* (11), 3715–3720.
114. Azhakar, R.; Roesky, H. W.; Wolf, H.; Stalke, D. Reactivity of Stable Heteroleptic Silylene PhC (N*t*Bu)$_2$SiNPh$_2$ Toward Diazobenzene and *N*-Benzylideneaniline. *Organometallics* **2012,** *31* (24), 8608–8612.
115. Khan, S.; Sen, S. S.; Kratzert, D.; Tavcar, G.; Roesky, H. W.; Stalke, D. Synthesis of Stable Silicon Heterocycles by Reaction of Organic Substrates With a Chlorosilylene [PhC(N*t*Bu)$_2$SiCl]. *Chem. Eur. J.* **2011,** *17* (15), 4283–4290.
116. Azhakar, R.; Proepper, K.; Dittrich, B.; Roesky, H. W. Reactivity Studies of Heteroleptic Silylenes With N$_2$O. *Organometallics* **2012,** *31* (21), 7586–7590.
117. Azhakar, R.; Ghadwal, R. S.; Roesky, H. W.; Wolf, H.; Stalke, D. A Debut for Base Stabilized Monoalkylsilylenes. *Chem. Commun.* **2012,** *48* (38), 4561–4563.
118. Sarish, S. P.; Jana, A.; Roesky, H. W.; Samuel, P. P.; Abad Andrade, C. E.; Dittrich, B., et al. Synthesis of a Lewis Base Stabilized Dimer of N-Substituted Hydrosila Hydrazone and a Silaaziridine. *Organometallics* **2011,** *30* (4), 912–916.
119. Sen, S. S.; Hey, J.; Kratzert, D.; Roesky, H. W.; Stalke, D. A Remarkable End-On Activation of Diazoalkane and Cleavage of Both C–Cl Bonds of Dichloromethane With a Silylene to a Single Product With Five-Coordinate Silicon Atoms. *Organometallics* **2012,** *31* (1), 435–439.
120. Breit, N. C.; Szilvasi, T.; Inoue, S. New Route to Access an Acyl-Functionalized Phosphasilene and a Four-Membered Si-P-C-O Heterocycle. *Chem. Eur. J.* **2014,** *20* (30), 9312–9318.

121. Mueck, F. M.; Baus, J. A.; Nutz, M.; Burschka, C.; Poater, J.; Bickelhaupt, F. M., et al. Reactivity of the Donor-Stabilized Silylenes [*i*PrNC(Ph)N*i*Pr]$_2$Si and [*i*PrNC(NiPr$_2$)N*i*Pr]$_2$Si: Activation of CO_2 and CS_2. *Chem. Eur. J.* **2015,** *21* (46), 16665–16672.
122. Junold, K.; Nutz, M.; Baus, J. A.; Burschka, C.; Guerra, C. F.; Bickelhaupt, F. M., et al. The Donor-Stabilized Silylene Bis[*N,N'*-diisopropylbenzamidinato(−)]silicon(II): Synthesis, Electronic Structure, and Reactivity. *Chem. Eur. J.* **2014,** *20* (30), 9319–9329.
123. Junold, K.; Mueck, F. M.; Kupper, C.; Baus, J. A.; Burschka, C.; Tacke, R. Activation of Sulfur Dioxide by Bis[*N,N'*-diisopropylbenzamidinato(−)]silicon(II): Synthesis of Neutral Six-Coordinate Silicon(IV) Complexes With Chelating *O,O'*-Sulfito or *O,O'*-Dithionito Ligands. *Chem. Eur. J.* **2014,** *20* (40), 12781–12785.
124. Jana, A.; Azhakar, R.; Sarish, S. P.; Samuel, P. P.; Roesky, H. W.; Schulzke, C., et al. Reactions of Stable Amidinate Chlorosilylene and [1 + 4]-Oxidative Addition of *N*-Heterocyclic Silylene With *N*-Benzylideneaniline. *Eur. J. Inorg. Chem.* **2011,** (32), 5006–5013.
125. Wagler, J.; Bohme, U.; Brendler, E.; Roewer, G. First X-ray Structure of a Cationic Silicon Complex With Salen-Type Ligand: An Unusual Compound With Two Different Si-N Dative Bonds. *Z. Naturforsch., B.* **2004,** *59* (11–12), 1348–1352.
126. Wagler, J.; Böhme, U.; Roewer, G. Silicon-Enamine Complexes: Pentacoordinate Silicon Compounds. *Angew. Chem. Int. Ed.* **2002,** *41* (10), 1732–1734.
127. Wagler, J.; Böhme, U.; Brendler, E.; Thomas, B.; Goutal, S.; Mayr, H., et al. Switching Between Penta- and Hexacoordination With Salen-Silicon-Complexes. *Inorg. Chim. Acta.* **2005,** *358* (14), 4270–4286.
128. Wagler, J.; Doert, T.; Roewer, G. Synthesis of Amines From Imines in the Coordination Sphere of Silicon-Surprising Photo-Rearrangement of Hexacoordinate Organosilanes. *Angew. Chem. Int. Ed.* **2004,** *43* (18), 2441–2444.
129. Wagler, J.; Roewer, G. Syntheses of Allyl- and 3-Silylpropyl-Substituted Salen-Like Tetradentate Ligands via Hypercoordinate Silicon Complexes. *Z. Naturforsch., B.* **2006,** *61* (11), 1406–1412.
130. Wagler, J.; Roewer, G.; Gerlach, D. Photo-Driven Si-C Bond Cleavage in Hexacoordinate Silicon Complexes. *Z. Anorg. Allg. Chem.* **2009,** *635* (9–10), 1279–1287.
131. Lippe, K.; Gerlach, D.; Kroke, E.; Wagler, J. Hypercoordinate Organosilicon Complexes of an *ONN'O'* Chelating Ligand: Regio- and Diastereoselectivity of Rearrangement Reactions in Si–Salphen Systems. *Organometallics* **2009,** *28* (2), 621–629.
132. González-García, G.; Gutiérrez, J. A.; Álvarez, E.; López, J. A.; Jiménez-Halla, J. O. C. Synthesis, Stereoisomerism and Crystal Structures of Neutral Hexacoordinate Silicon(IV) Complexes With *Salen-O, N,N,O* and Thiocyanato-*N* Ligands. *Inorg. Chim. Acta.* **2015,** *428*, 93–99.
133. Wagler, J.; Böhme, U.; Brendler, E.; Blaurock, S.; Roewer, G. Novel Hexacoordinate Diorganosilanes With Salen-Type Ligands: Molecular Structure Versus ^{29}Si NMR Chemical Shifts. *Z. Anorg. Allg. Chem.* **2005,** *631* (13–14), 2907–2913.
134. Sakamoto, N.; Ikeda, C.; Yamamura, M.; Nabeshima, T. Structural Interconversion and Regulation of Optical Properties of Stable Hypercoordinate Dipyrrin-Silicon Complexes. *J. Am. Chem. Soc.* **2011,** *133* (13), 4726–4729.
135. Yamamura, M.; Albrecht, M.; Albrecht, M.; Nishimura, Y.; Arai, T.; Nabeshima, T. Red/Near-Infrared Luminescence Tuning of Group-14 Element Complexes of Dipyrrins Based on a Central Atom. *Inorg. Chem.* **2014,** *53* (3), 1355–1360.
136. Yang, Y. Hexacoordinate Bonding and Aromaticity in Silicon Phthalocyanine. *J. Phys. Chem. A.* **2010,** *114* (50), 13257–13267.
137. Chen, X.; Ma, D.; Cai, K.; Pan, S.; Wang, Y.; Zhang, T., et al. Effect of Axial Ligands on the Photophysical Properties of New Silicon(IV) Phthalocyanines. *J. Coord. Chem.* **2015,** *68* (4), 732–740.
138. Biyiklioglu, Z. Electrochemical and Aggregation Properties of Newly Synthesized Dendritic Axially Morpholine-Disubstituted Silicon Phthalocyanine, Mono-Substituted Subphthalocyanine and Their Quaternized Derivatives. *Inorg. Chem. Commun.* **2015,** *55*, 60–64.

139. Biyiklioglu, Z.; Baş, H. Synthesis and Electrochemistry of Non-Aggregated Axially Disubstituted Silicon Phthalocyanines Bearing Benzoxazin Substituents. *Inorg. Chim. Acta.* **2015,** *427*, 293–298.
140. Biyiklioglu, Z.; Bekircan, O. Synthesis and Electrochemical Properties of Axially Disubstituted Silicon Phthalocyanine and Peripherally Tetra Substituted Manganese(III) Phthalocyanine Bearing 1,2,4-Triazole Substituents. *Synth. Met.* **2015,** *200*, 148–155.
141. Baş, H.; Biyiklioglu, Z. Non-Aggregated Axially Naphthoxazin Group Substituted Silicon Phthalocyanines: Synthesis and Electrochemistry. *J. Organomet. Chem.* **2015,** *791*, 238–243.
142. Çakır, V.; Göksel, M.; Durmuş, M.; Biyiklioglu, Z. Synthesis and Photophysicochemical Properties of Novel Water Soluble Phthalocyanines. *Dyes Pigm.* **2016,** *125*, 414–425.
143. Bandera, Y.; Burdette, M. K.; Shetzline, J. A.; Jenkins, R.; Creager, S. E.; Foulger, S. H. Synthesis of Water Soluble Axially Disubstituted Silicon (IV) Phthalocyanines With Alkyne & Azide Functionality. *Dyes Pigm.* **2016,** *125*, 72–79.
144. Lessard, B. H.; White, R. T.; Al-Amar, M.; Plint, T.; Castrucci, J. S.; Josey, D. S., et al. Assessing the Potential Roles of Silicon and Germanium Phthalocyanines in Planar Heterojunction Organic Photovoltaic Devices and How Pentafluoro Phenoxylation Can Enhance $\pi-\pi$ Interactions and Device Performance. *ACS Appl. Mater. Interfaces* **2015,** *7* (9), 5076–5088.
145. Xu, H.; Ohkita, H.; Tamai, Y.; Benten, H.; Ito, S. Interface Engineering for Ternary Blend Polymer Solar Cells With a Heterostructured Near-IR Dye. *Adv. Mater.* **2015,** *27* (39), 5868–5874.
146. Huang, J.; Wu, Y.; Wang, D.; Ma, Y.; Yue, Z.; Lu, Y., et al. Silicon Phthalocyanine Covalently Functionalized *N*-Doped Ultrasmall Reduced Graphene Oxide Decorated With Pt Nanoparticles for Hydrogen Evolution From Water. *ACS Appl. Mater. Interfaces* **2015,** *7* (6), 3732–3741.
147. Zhao, Z.; Chan, P. S.; Li, H.; Wong, K. L.; Wong, R. N.; Mak, N. K., et al. Highly Selective Mitochondria-Targeting Amphiphilic Silicon(IV) Phthalocyanines With Axially Ligated Rhodamine B for Photodynamic Therapy. *Inorg. Chem.* **2012,** *51* (2), 812–821.
148. Yang, Y.; Samas, B.; Kennedy, V. O.; Macikenas, D.; Chaloux, B. L.; Miller, J. A., et al. Long, Directional Interactions in Cofacial Silicon Phthalocyanine Oligomers. *J. Phys. Chem. A.* **2011,** *115* (45), 12474–12485.
149. Oniwa, K.; Shimizu, S.; Shiina, Y.; Fukuda, T.; Kobayashi, N. A. μ-Oxo Hetero Dimer of Silicon Phthalocyanine and Naphthalocyanine. *Chem. Commun.* **2013,** *49* (75), 8341–8343.
150. Ishida, S.; Yoshimura, K.; Matsumoto, H.; Kyushin, S. Selective Si–C Bond Cleavage on a Diorganosilicon Porphyrin Complex Bearing Different Axial Ligands. *Chem. Lett.* **2009,** *38* (4), 362–363.
151. Ishida, S.; Tanaka, T.; Lim, J. M.; Kim, D.; Osuka, A. Si^{IV} Incorporation Into a [28]hexaphyrin That Triggered Formation of Möbius Aromatic Molecules. *Chem. Eur. J.* **2014,** *20* (27), 8274–8278.
152. Ishida, S.; Osuka, A. *meso*-Aryl [28]Hexaphyrin Silicon Complexes Bearing Various Si-Substituents and 1,16-Dihydrohexaphyrin Bis-Chlorosilicon Complex. *Chem. Asian. J.* **2015,** *10* (10), 2200–2206.
153. Hagemann, M.; Mix, A.; Berger, R. J.; Pape, T.; Mitzel, N. W. Strong Intramolecular Si–N Interactions in the Chlorosilanes $Cl_{3-n}H_nSiOCH_2CH_2NMe_2$ (n = 1–3). *Inorg. Chem.* **2008,** *47* (22), 10554–10564.
154. Kim, B.-K.; Choi, S.-B.; Kloos, S. D.; Boudjouk, P. Synthesis and Characterization of New Cationic Hexacoordinate Silanes. *Inorg. Chem.* **2000,** *39* (4), 728–731.
155. Cheng, F.; Hector, A. L.; Levason, W.; Reid, G.; Webster, M.; Zhang, W. Preparation and Structure of the Unique Silicon(IV) Cation $[SiF_3(Me_3tacn)]^+$. *Chem. Commun.* **2009,** *11*, 1334–1336.
156. Cota, S.; Beyer, M.; Bertermann, R.; Burschka, C.; Gotz, K.; Kaupp, M., et al. Neutral Penta- and Hexacoordinate Silicon(IV) Complexes Containing Two Bidentate Ligands Derived From the α-Amino Acids (*S*)-Alanine, (*S*)-Phenylalanine, and (*S*)-*tert*-leucine. *Chem. Eur. J.* **2010,** *16* (22), 6582–6589.
157. Tamao, K.; Asahara, M.; Saeki, T.; Feng, S. Y.; Kawachi, A.; Toshimitsu, A. Nucleophilic Addition of a Divalent Silicon Species Bearing the 8-Dimethylamino-1-naphthyl Group to Diphenylacetylenes to Form the 1-Amino-1-silaphenalene Skeleton. *Chem. Lett.* **2000,** *6*, 660–661.

158. Tamao, K.; Asahara, M.; Saeki, T.; Toshimitsu, A. Enhanced Nucleophilicity of Ambiphilic Silylene and Silylenoid Bearing 8-(Dimethylamino)-1-naphthyl Group. *J. Organomet. Chem.* **2000,** *600* (1–2), 118–123.
159. Bushuk, S. B.; Carré, F. H.; Guy, D. M. H.; Douglas, W. E.; Kalvinkovskya, Y. A.; Klapshina, L. G., et al. Hypercoordinate Silicon and Phosphorus Acetylene Compounds: Crystal Structure Determinations and Fluorescence Spectroscopic Study. *Polyhedron* **2004,** *23* (17), 2615–2623.
160. Cho, H. M.; Lee, M. E.; Kim, C. H. Facile Thermal Elimination of LiOR (R = Me, Ph) in the Reaction of [*o*-(*N*,*N*-Dimethylaminomethyl)phenyl]- organoxymethylvinylsilane With *t*-BuLi. *Organometallics* **2002,** *21* (9), 2017–2019.
161. Cho, H. M.; Jang, J. I.; Kim, C. H.; Lee, S.-G.; Lee, M. E. Syntheses and NMR Study of [(2-Dimethylaminomethyl)phenyl]vinylsilane Derivatives and [(2-Dimethylaminomethyl)phenyl]ethylsilane Derivatives. *J. Organomet. Chem.* **2003,** *679* (1), 43–47.
162. Jakle, F.; Vejzovic, E.; Power-Billard, K. N.; MacLachlan, M. J.; Lough, A. J.; Manners, I. Synthesis and Structure of a Hypercoordinate Silicon-Bridged [1]ferrocenophane. *Organometallics* **2000,** *19* (15), 2826–2828.
163. Domoto, Y.; Saruhashi, K.; Fukushima, A.; Sase, S.; Goto, K.; Kawashima, T. Synthesis and Properties of Pentacoordinated Phenoxysilane and Carboxysilanes With Intramolecular Nitrogen–Silicon Coordination. *Phosphorus, Sulfur Silicon Relat. Elem.* **2010,** *185* (5–6), 1221–1229.
164. Saruhashi, K.; Goto, K.; Kawashima, T. Structural Characterization of a Pentacoordinate Monohydrosilane With the Apical Si–H Bond. *Chem. Heterocycl. Compd.* **2001,** *37* (11), 1394–1395.
165. Domoto, Y.; Fukushima, A.; Kasuga, Y.; Sase, S.; Goto, K.; Kawashima, T. Catalyst-Free Syntheses of [2]Rotaxanes Utilizing a Pentacoordinated Hydrosilane as an End-Capping Agent. *Org. Lett.* **2010,** *12* (11), 2586–2589.
166. Domoto, Y.; Sase, S.; Goto, K. Efficient End-Capping Synthesis of Neutral Donor-Acceptor [2]rotaxanes Under Additive-Free and Mild Conditions. *Chem. Eur. J.* **2014,** *20* (48), 15998–16005.
167. Mitzel, N. W.; Vojinovic, K.; Foerster, T.; Robertson, H. E.; Borisenko, K. B.; Rankin, D. W. (Dimethylaminomethyl)trifluorosilane, $Me_2NCH_2SiF_3$—A Model for the α-Effect in Aminomethylsilanes. *Chem. Eur. J.* **2005,** *11* (17), 5114–5125.
168. Vojinovic, K.; McLachlan, L. J.; Hinchley, S. L.; Rankin, D. W.; Mitzel, N. W. Strong Intramolecular Secondary Si···N Bonds in Trifluorosilylhydrazines. *Chem. Eur. J.* **2004,** *10* (12), 3033–3042.
169. Mitzel, N. W.; Vojinovic, K.; Frohlich, R.; Foerster, T.; Robertson, H. E.; Borisenko, K. B., et al. Three-Membered Ring or Open Chain Molecule–$(F_3C)F_2SiONMe_2$ a Model for the α-Effect in Silicon Chemistry. *J. Am. Chem. Soc.* **2005,** *127* (39), 13705–13713.
170. Chandrasekaran, A.; Day, R. O.; Holmes, R. R. A New Class of Silatranes: Structure and Dynamic NMR Behavior. *J. Am. Chem. Soc.* **2000,** *122* (6), 1066–1072.
171. Timosheva, N. V.; Chandrasekaran, A.; Day, R. O.; Holmes, R. R. Synthesis of Trioxy- and Tetraoxysilatranes With All Six-Membered Rings. Structure and Dynamic NMR Behavior. *Organometallics* **2000,** *19* (26), 5614–5622.
172. Timosheva, N. V.; Chandrasekaran, A.; Day, R. O.; Holmes, R. R. Synthesis of Silatranes With All Six-Membered Rings. Influence of Steric Effect on Structure and NMR Behavior. *Organometallics* **2001,** *20* (11), 2331–2337.
173. Kobayashi, J.; Kawaguchi, K.; Kawashima, T. Water-Coordinated Neutral Silane Complex: A Frozen Intermediate of Hydrolysis of Alkoxysilanes. *J. Am. Chem. Soc.* **2004,** *126* (50), 16318–16319.
174. Pestunovich, V.; Kirpichenko, S.; Voronkov, M. Silatranes and Their Tricyclic Analogs. *In The Chemistry of Organic Silicon Compounds: John Wiley & Sons, Ltd* **2001,** 1447–1537.
175. Puri, J. K.; Singh, R.; Chahal, V. K. Silatranes: A Review on Their Synthesis, Structure, Reactivity and Applications. *Chem. Soc. Rev.* **2011,** *40* (3), 1791–1840.
176. Sterkhova, I. V.; Lazarev, I. M.; Smirnov, V. I.; Lazareva, N. F. 1-(Methylaminomethyl)silatrane: Synthesis, Characterization and Reactivity. *J. Organomet. Chem.* **2015,** *775*, 27–32.

177. Dumitriu, A. M.-C.; Cazacu, M.; Shova, S.; Turta, C.; Simionescu, B. C. Synthesis and Structural Characterization of 1-(3-Aminopropyl)silatrane and Some New Derivatives. *Polyhedron* **2012,** *33* (1), 119–126.
178. Singh, G.; Rani, S.; Saroa, A.; Promila.; Arora, A.; Choquesillo-Lazarte, D. Amide-Tethered Organosilatranes: Syntheses, Structural Characterization and Photophysical Properties. *Inorg. Chim. Acta.* **2015,** *433*, 78–91.
179. Voronkov, M. G.; Zel'bst, È. A.; Fundamensky, V. S.; Gurzhiy, V. V.; Bolgova, Y. I.; Trofimova, O. M. Crystal and Molecular Structure of 1-(3-Ammoniopropyl)silatrane Chloride. *J. Struct. Chem.* **2014,** *55* (2), 370–373.
180. Singh, R.; Puri, J. K.; Chahal, V. K.; Sharma, R. P.; Kohli, S.; Kant, R., et al. Synthesis, Characterization, Reactivity, and Antibacterial Studies of Triethylammonium-3-silatranylpropyldithiocarbamate, Spectroscopic, and Quantum Mechanical Studies of 3-(Silatranyl)propylammonium Chloride. *Synth. React. Inorg., Met.-Org., Nano-Met. Chem.* **2012,** *42* (6), 823–832.
181. Sheludyakov, V. D.; Kuz'mina, N. E.; Abramkin, A. M.; Korlyukov, A. A.; Arkhipov, D. E.; Loginov, S. V., et al. Synthesis and Structure of *N*-(2-silatranylethyl)imidazoles. *Russ. J. Gen. Chem.* **2012,** *81* (12), 2468–2477.
182. Sheludyakov, V. D.; Kuz'mina, N. E.; Abramkin, A. M.; Korlyukov, A. A.; Arkhipov, D. E.; Loginov, S. V., et al. Synthesis and Structure of *N*-(2-silatranylethyl)pyrazoles. *Russ. J. Gen. Chem.* **2012,** *81* (12), 2478–2486.
183. Singh, G.; Saroa, A.; Garg, M.; Sharma, R. P.; Gubanov, A. I.; Smolentsev, A. I. 1-Isothiocyanatosilatrane Derived From Trisisopropanolamine: Synthesis, Characterization, Reactivity and Theoretical Studies. *J. Organomet. Chem.* **2012,** *719*, 21–25.
184. Singh, R.; Mutneja, R.; Kaur, V.; Wagler, J.; Kroke, E. Derivatization of 3-Aminopropylsilatrane to Introduce Azomethine Linkage in the Axial Chain: Synthesis, Characterization and Structural Studies. *J. Organomet. Chem.* **2013,** *724*, 186–191.
185. Singh, G.; Girdhar, S.; Sharma, R. P.; Starynowicz, P.; Singh, B. Carbofunctional Silatrane Possessing Imidazole Moiety: Synthesis, Characterization, and Antibacterial Studies. *Phosphorus, Sulfur Silicon Relat. Elem.* **2014,** *189* (11), 1732–1745.
186. Singh, G.; Girdhar, S.; Khullar, S.; Mandal, S. K. Imidazolyl-Substituted Silatranes Derived From Triethanolamine and Tris(isopropanol)amine: Syntheses and Structural Characterization. *J. Coord. Chem.* **2015,** *68* (5), 875–894.
187. Singh, G.; Girdhar, S.; Khullar, S.; Mandal, S. K. 1,3-Diazolyl Functionalized Organopropylsilatranes: Synthesis and Structural Characterization. *Inorg. Chim. Acta.* **2014,** *413*, 203–207.
188. Singh, G.; Mangat, S. S.; Singh, J.; Arora, A.; Garg, M. Synthesis of Novel 1,2,3-Triazole Based Silatranes via "Click Silylation". *J. Organomet. Chem.* **2014,** *769*, 124–129.
189. Singh, G.; Mangat, S. S.; Sharma, H.; Singh, J.; Arora, A.; Pannu, A. P. S., et al. Design and Syntheses of Novel Fluorescent Organosilicon-Based Chemosensors Through Click Silylation: Detection of Biogenic Amines. *RSC Adv.* **2014,** *4* (69), 36834–36844.
190. Singh, G.; Arora, A.; Mangat, S. S.; Singh, J.; Chaudhary, S.; Kaur, N., et al. Synthesis and Characterization of Modified Schiff Base Silatranes (MSBS) via 'Click Silylation'. *J. Mol. Struct.* **2015,** *1079*, 173–181.
191. Zablotskaya, A.; Segal, I.; Belyakov, S.; Lukevics, E. Silyl Modification of Biologically Active Compounds. 11. Synthesis, Physico-Chemical and Biological Evaluation of *N*-(Trialkoxysilylalkyl) tetrahydro(iso,silaiso) Quinoline Derivatives. *Appl. Organomet. Chem.* **2006,** *20* (2), 149–154.
192. Singh, G.; Saroa, A.; Girdhar, S.; Rani, S.; Sahoo, S.; Choquesillo-Lazarte, D. Synthesis, Characterization, Electronic Absorption and Antimicrobial Studies of *N*-(silatranylpropyl)phthalimide Derived From Phthalic Anhydride. *Inorg. Chim. Acta.* **2015,** *427*, 232–239.
193. Pukhalskaya, V. G.; Kramarova, E. P.; Kozaeva, L. P.; Korlyukov, A. A.; Shipov, A. G.; Bylikin, S. Y., et al. Synthesis, Structure and Muscarinic Agonist Activity of Substituted *N*-(silatran-1-ylmethyl) acetamides. *Appl. Organomet. Chem.* **2010,** *24* (3), 162–168.

194. Voronkov, M. G.; Trofimova, O. M.; Turchaninov, V. K.; Zel'bst, E. A.; Bolgova, Y. I.; Belyaeva, V. V., et al. *N*-Silatranylmethyl Derivatives of Pyrrole, Indole, and Carbazole. *Russ. J. Org. Chem.* **2003,** *39* (10), 1458–1466.

195. Singh, G.; Saroa, A.; Girdhar, S.; Rani, S.; Choquesillo-Lazarte, D.; Sahoo, S. C. Incorporation of Azo Group at Axial Position of Silatranes: Synthesis, Characterization and Antimicrobial Activity. *Appl. Organomet. Chem.* **2015,** *29* (8), 549–555.

196. Aghazadeh Meshgi, M.; Baumgartner, J.; Marschner, C. Oligosilanylsilatranes. *Organometallics* **2015,** *34* (15), 3721–3731.

197. Vasiliev, A. D.; Zel'bst, E. A.; Soldatenko, A. S.; Bolgova, Y. I.; Trofimova, O. M.; Voronkov, M. G. Synthesis and Molecular Structure of Cobalt(II) Complex With 1-(1-Silatranylmethyl)-1,2,4-triazole. *Struct. Chem.* **2012,** *24* (4), 1001–1005.

198. Voronkov, M. G.; Zel'bst, É. A.; Vasiliev, A. D.; Soldatenko, A. S.; Bolgova, Y. I.; Trofimova, O. M. Crystal and Molecular Structure of the Complex of *N*-(1-silatranylmethyl)-3,5-dimethylpyrazole With Cobalt Dichloride. *J. Struct. Chem.* **2012,** *53* (3), 614–617.

199. Voronkov, M. G.; Zel'bst, E. A.; Vasiliev, A. D.; Soldatenko, A. S.; Bolgova, Y. I.; Trofimova, O. M. Crystal and Molecular Structure of the Complex of 2-(1-silatranylmethylthio)-4,5-benzo-1,3-thiazole With $CoCl_2$ and MeCN. *J. Struct. Chem.* **2012,** *53* (1), 193–196.

200. Lazareva, N. F.; Pestunovich, V. A.; Albanov, A. I.; Vesnin, M. G.; Voronkov, M. G. Trans-silylation of Silatranes and 1,2-Dimethyl-2-Azasilatran-3-one by Si-Substituted Trimethoxysilanes. *Tetrahedron Lett.* **2000,** *41* (24), 4823–4826.

201. Wagner, G.; Herrmann, R.; Pedersen, B.; Scherer, W. Synthesis and Structure of Chiral Silatranes Derived From Terpenes. *Z. Naturforsch., B: Chem. Sci.* **2001,** *56* (1), 25–38.

202. Voronkov, M. G.; Egorochkin, A. N.; Kuznetsova, O. V. Polarizability Effect in Silatranes and Related Compounds. *J. Organomet. Chem.* **2006,** *691* (1–2), 159–164.

203. Voronkov, M. G.; Albanov, A. I.; Grebneva, E. A.; Trofimova, O. M.; Chernov, N. F.; Chipanina, N. N. Reaction of Phenyltrifluorosilane and Phenyl(hydrocarbyl)difluorosilanes With Bis(2-hydroxyethyl)-and Tris(2-hydroxyethyl)amine and Their *N*-Methyl and *O*-Trimethylsilyl Derivatives. A Novel Route to Quasisilatranes. *Russ. J. Gen. Chem.* **2006,** *76* (12), 1854–1859.

204. Kovács, I.; Matern, E.; Sattler, E.; Anson, C. E.; Párkányi, L. The Synthesis, Crystal Structures and NMR Spectroscopic Investigation of 3,7,10-Trimethylsilatranes and Carbasilatranes. *J. Organomet. Chem.* **2009,** *694* (1), 14–20.

205. Murata, M.; Yamasaki, H.; Uogishi, K.; Watanabe, S.; Masuda, Y. Synthesis of Organosilatranes via Rhodium(I)-Catalyzed Silylation of Organic Iodides With Hydrosilatrane. *Synthesis* **2007,** *19*, 2944–2946.

206. Ishiyama, T.; Saiki, T.; Kishida, E.; Sasaki, I.; Ito, H.; Miyaura, N. Aromatic C–H Silylation of Arenes With 1-Hydrosilatrane Catalyzed by an Iridium(I)/2,9-dimethylphenanthroline (dmphen) Complex. *Org. Biomol. Chem.* **2013,** *11* (47), 8162–8165.

207. Ping, G.; Yue-Wu, W.; Xin-Tong, L.; Xiao-Lu, Q.; Le-Ping, H.; Zi-Xin, X., et al. Synthesis, Antitumor Activity, and SAR of *N*-Substituted γ-Aminopropylsilatrane Derivatives. *Phosphorus, Sulfur Silicon Relat. Elem.* **2014,** *189* (4), 511–518.

208. Han, A. Y.; Li, L. N.; Qing, K. Y.; Qi, X. L.; Hou, L. P.; Luo, X. T., et al. Synthesis and Biological Evaluation of Nucleoside Analogues Than Contain Silatrane on the Basis of the Structure of Acyclovir (ACV) as Novel Inhibitors of Hepatitis B Virus (HBV). *Bioorg. Med. Chem. Lett.* **2013,** *23* (5), 1310–1314.

209. Lin, Y.; Song, B.; Han, A.; Hu, S.; Ye, F.; Xie, Z. Synthesis of γ-Arylmethylene-aminopropyl-3,7,10-trimethyl-silatrane Derivatives and Their Activities of Regulating Plant Growth. *Phosphorus, Sulfur Silicon Relat. Elem.* **2011,** *186* (2), 298–303.

210. Xie, Z.-x; Chen, L.-f; Wang, Y.-w; Song, X.-q; Qi, X.-l; Guo, P., et al. Synthesis and Stimulation of Seed Germination of γ-Aminopropyl Silatrane Derivatives. *Phytochem. Lett.* **2014,** *8*, 202–206.

211. Voronkov, M. G.; Baryshok, V. P. Antitumor Activity of Silatranes (Review). *Pharm. Chem. J.* **2004,** *38* (1), 3–9.

212. Riggleman, S.; DeShong, P. Application of Silicon-Based Cross-Coupling Technology to Triflates. *J. Org. Chem.* **2003,** *68* (21), 8106–8109.
213. Pidaparthi, R. R.; Welker, M. E.; Day, C. S.; Wright, M. W. Preparation of 2-Trialkylsiloxy-Substituted 1,3-Dienes and Their Diels-Alder/Cross-Coupling Reactions. *Org. Lett.* **2007,** *9* (9), 1623–1626.
214. Pidaparthi, R. R.; Junker, C. S.; Welker, M. E.; Day, C. S.; Wright, M. W. Preparation of 2-Silicon-Substituted 1,3-Dienes and Their Diels-Alder/Cross-Coupling Reactions. *J. Org. Chem.* **2009,** *74* (21), 8290–8297.
215. Peureux, C.; Jouikov, V. Covalent Grafting of Silatranes to Carbon Interfaces. *Chem. Eur. J.* **2014,** *20* (30), 9290–9294.
216. Riener, C. K.; Stroh, C. M.; Ebner, A.; Klampfl, C.; Gall, A. A.; Romanin, C., et al. Simple Test System for Single Molecule Recognition Force Microscopy. *Anal. Chim. Acta.* **2003,** *479* (1), 59–75.
217. Shlyakhtenko, L. S.; Gall, A. A.; Filonov, A.; Cerovac, Z.; Lushnikov, A.; Lyubchenko, Y. L. Silatrane-Based Surface Chemistry for Immobilization of DNA, Protein-DNA Complexes and Other Biological Materials. *Ultramicroscopy* **2003,** *97* (1–4), 279–287.
218. Lyubchenko, Y. L.; Shlyakhtenko, L. S. AFM for Analysis of Structure and Dynamics of DNA and Protein-DNA Complexes. *Methods* **2009,** *47* (3), 206–213.
219. Lyubchenko, Y. L.; Shlyakhtenko, L. S.; Gall, A. A. Atomic Force Microscopy Imaging and Probing of DNA, Proteins, and Protein DNA Complexes: Silatrane Surface Chemistry. *Methods Mol. Biol.* **2009,** *543* (DNA–Protein Interactions), 337–351.
220. Senapati, S.; Manna, S.; Lindsay, S.; Zhang, P. Application of Catalyst-Free Click Reactions in Attaching Affinity Molecules to Tips of Atomic Force Microscopy for Detection of Protein Biomarkers. *Langmuir* **2013,** *29* (47), 14622–14630.
221. Alvarez, R.; Mehl, G. H. Nematic Tribenzosilatranes. *Liq. Cryst.* **2005,** *32* (4), 469–476.
222. Mizumo, T.; Fujita, R.; Ohno, H.; Ohshita, J. Lithium Ion Conduction in Silatrane Matrices. *Chem. Lett* **2011,** *40* (8), 798–800.
223. Kingston, V.; Gostevskii, B.; Kalikhman, I.; Kost, D. Equilibrium Between Neutral Hexacoordinate Silicon Complexes and Ionic Pentacoordinate Siliconium Salts Through Fast Dissociation–Recombination of the Si–Cl Bond. *Chem. Commun.* **2001,** (14), 1272–1273.
224. Kalikhman, I.; Girshberg, O.; Lameyer, L.; Stalke, D.; Kost, D. Tautomeric Equilibrium Between Penta- and Hexacoordinate Silicon Chelates. A Chloride Bridge Between Two Pentacoordinate Silicons. *J. Am. Chem. Soc.* **2001,** *123* (20), 4709–4716.
225. Kost, D.; Kingston, V.; Gostevskii, B.; Ellern, A.; Stalke, D.; Walfort, B., et al. Donor-Stabilized Silyl Cations. 3. Ionic Dissociation of Hexacoordinate Silicon Complexes to Pentacoordinate Siliconium Salts Driven by Ion Solvation. *Organometallics* **2002,** *21* (11), 2293–2305.
226. Kost, D.; Gostevskii, B.; Kocher, N.; Stalke, D.; Kalikhman, I. An Unexpected, Sterically Driven, Methyl Halide Elimination in Pentacoordinate Siliconium Halide Salts: Silicon Complexes With Equatorial Nitrogen Coordination. *Angew. Chem. Int. Ed.* **2003,** *42* (9), 1023–1026.
227. Gostevskii, B.; Adear, K.; Sivaramakrishna, A.; Silbert, G.; Stalke, D.; Kocher, N., et al. Neutral and Ionic Dissociation Patterns in Hexacoordinate Silicon Chelates: A Model Nucleophilic Substitution at Pentacoordinate Silicon. *Chem. Commun.* **2004,** *14*, 1644–1645.
228. Gostevskii, B.; Silbert, G.; Adear, K.; Sivaramakrishna, A.; Stalke, D.; Deuerlein, S., et al. Donor-Stabilized Silyl Cations. 9. Two Dissociation Patterns of Hexacoordinate Silicon Complexes: A Model Nucleophilic Substitution at Pentacoordinate Silicon. *Organometallics* **2005,** *24* (12), 2913–2920.
229. Kalikhman, I.; Krivonos, S.; Lameyer, L.; Stalke, D.; Kost, D. Pentacoordinate Silyl Cations Stabilized by Coordination With Oxygen Donors: Crystal Structure, Charge Distribution, and Stereodynamics. *Organometallics* **2001,** *20* (6), 1053–1055.
230. Jouikov, V.; Gostevskii, B.; Kalikhrnan, I.; Kost, D. Donor-Stabilized Silyl Cations. 12. Electrochemical Evidence for Ionic Dissociation of Hexacoordinate Silicon Complexes. *Organometallics* **2007,** *26* (24), 5791–5798.

231. Kalikhman, I.; Kingston, V.; Gostevskii, B.; Pestunovich, V.; Stalke, D.; Walfort, B., et al. Donor-Stabilized Silyl Cations. 5. Comparison Between Mono- and Binuclear Siliconium Chelates. *Organometallics* **2002,** *21* (21), 4468–4474.
232. Kalikhman, I.; Gostevskii, B.; Girshberg, O.; Krivonos, S.; Kost, D. Donor-Stabilized Silyl Cations 4: *N*-Isopropylidene Hydrazides, Novel Bidentate Ligands for Penta- and Hexacoordinate Silicon Chelates. *Organometallics* **2002,** *21* (12), 2551–2554.
233. Kalikhman, I.; Gostevskii, B.; Girshberg, O.; Sivaramakrishna, A.; Kocher, N.; Stalke, D., et al. Donor-Stabilized Silyl Cations. *J. Organomet. Chem.* **2003,** *686* (1–2), 202–214.
234. Janardan, S.; Suman, P.; Sivaramakrishna, A.; Vijayakrishna, K. Donor-Stabilized Hypercoordinated Silicon(IV) Chelates With Cyclohexylideneimino-N Ligand Group: Role of Substituents on Ionization. *Polyhedron* **2015,** *85*, 34–40.
235. Sivaramakrishna, A.; Kalikhman, I.; Kertsnus, E.; Korlyukov, A. A.; Kost, D. Donor-Stabilized Silyl Cations. 10. Pentacoordinate Siliconium-Ion Salts With a Triphenylphosphinimino-*N* Ligand Group: Two-Bond P–N–Si Coupling as a Measure for Coordination Strength. *Organometallics* **2006,** *25* (15), 3665–3669.
236. Suman, P.; Janardan, S.; Lone, M. Y.; Jha, P.; Vijayakrishna, K.; Sivaramakrishna, A. Role of *N*-Donor Groups on the Stability of Hydrazide Based Hypercoordinate Silicon(IV) Complexes: Theoretical and Experimental Perceptions. *Polyhedron* **2015,** *99*, 266–274.
237. Gostevskii, B.; Pestunovich, V.; Kalikhman, I.; Sivaramakrishna, A.; Kocher, N.; Deuerlein, S., et al. Donor-Stabilized Silyl Cations. 8. Carbon-Carbon Bond Formation Through a Novel Interchelate Molecular Rearrangement in Pentacoordinate Siliconium-Ion Salts. *Organometallics* **2004,** *23* (19), 4346–4348.
238. Gostevskii, B.; Kalikhman, I.; Tessier, C. A.; Panzner, M. J.; Youngs, W. J.; Kost, D. Hexacoordinate Complexes of Silacyclobutane: Spontaneous Ring Opening and Rearrangement. *Organometallics* **2005,** *24* (24), 5786–5788.
239. Kalikhman, I.; Gostevskii, B.; Kertsnus, E.; Botoshansky, M.; Tessier, C. A.; Youngs, W. J., et al. Competitive Molecular Rearrangements in Hexacoordinate Cyano-Silicon Dichelates. *Organometallics* **2007,** *26* (10), 2652–2658.
240. Kalikhman, I.; Gostevskii, B.; Kertsnus, E.; Deuerlein, S.; Stalke, D.; Botoshansky, M., et al. Competing Reactions of Hypercoordinate Silicon Dichelates. *J. Phys. Org. Chem.* **2008,** *21* (12), 1029–1034.
241. Kertsnus-Banchik, E.; Kalikhman, I.; Gostevskii, B.; Deutsch, Z.; Botoshansky, M.; Kost, D. Hydride Migration From Silicon to an Adjacent Unsaturated Imino Carbon: Intramolecular Hydrosilylation. *Organometallics* **2008,** *27* (20), 5285–5294.
242. Kano, N.; Komatsu, F.; Yamamura, M.; Kawashima, T. Reversible Photoswitching of the Coordination Numbers of Silicon in Organosilicon Compounds Bearing a 2-(Phenylazo)phenyl Group. *J. Am. Chem. Soc.* **2006,** *128* (21), 7097–7109.
243. Yamamura, M.; Kano, N.; Kawashima, T.; Matsumoto, T.; Harada, J.; Ogawa, K. Crucial Role of N · · · Si Interactions in the Solid-State Coloration of Disilylazobenzenes. *J. Org. Chem.* **2008,** *73* (21), 8244–8249.
244. Kano, N.; Yamamura, M.; Kawashima, T. 2,2'-Disilylazobenzenes Featuring Double Intramolecular Nitrogen···Silicon Coordination: A Photoisomerizable Fluorophore. *Dalton Trans.* **2015,** *44* (37), 16256–16265.
245. Kano, N.; Yamamura, M.; Komatsu, F.; Kawashima, T. Synthesis, Crystal Structure, and Photoreaction of a Disiloxane Bearing Two 2-(Phenylazo)phenyl Groups. *J. Organomet. Chem.* **2003,** *686* (1–2), 192–197.
246. Kano, N.; Komatsu, F.; Kawashima, T. Reversible Photocontrol of the Coordination Number of Silicon in a Tetrafluorosilicate Bearing a 2-(Phenylazo)phenyl Group. *J. Am. Chem. Soc.* **2001,** *123* (43), 10778–10779.
247. Kano, N.; Yamamura, M.; Kawashima, T. Reactivity Control of an Allylsilane Bearing a 2-(Phenylazo)phenyl Group by Photoswitching of the Coordination Number of Silicon. *J. Am. Chem. Soc.* **2004,** *126* (20), 6250–6251.

248. Yamamura, M.; Kano, N.; Kawashima, T. Intramolecular Allylation of the Azo Group of 2-(Allylsilyl) azobenzenes and Its Photocontrol. *J. Organomet. Chem.* **2007,** *692* (1–3), 313–325.
249. Yamamura, M.; Kano, N.; Kawashima, T. Photoswitching of the Reactivity Involving Hydrosilylation of a 1,1,3,3-Tetrahydrodisiloxane Bearing Two Azo Groups. *Tetrahedron Lett.* **2007,** *48* (23), 4033–4036.
250. Yamamura, M.; Kano, N.; Kawashima, T. Synthesis, Structure, and Reactivity of Pentacoordinate Hydrosilanes Bearing a 2-(Phenylazo)phenyl Group. *Z. Anorg. Allg. Chem.* **2009,** *635* (9–10), 1295–1299.
251. Kramarova, E. P.; Pogozhikh, S. A.; Shipov, A. G.; Negrebetsky, V. V.; Tandura, S. N.; Shumskii, A. N., et al. Bis(O–Si)-chelate Fluorobis(2-oxohexahydroazepinomethyl)silylium Tetrafluoroborate, A Cation-Anionic Complex Containing a Silylium Ion Stabilized by O→Si Coordination. *Russ. Chem. Bull.* **2001,** *50* (2), 331–332.
252. Zamyshlyaeva, O. A.; Shipov, A. G.; Kramarova, E. P.; Negrebetsky, V. V.; Shumskii, A. N.; Tandura, S. N., et al. Synthesis and Some Chemical Properties of 2,2-Dimethyl-4-R-1-Oxa-4-Aza-2-Sila-5-Benzocycloheptanones. *Chem. Heterocycl. Compd.* **2002,** *38* (1), 116–127.
253. Pestunovich, V. A.; Kirpichenko, S. V.; Lazareva, Ny. F.; Albanov, A. I.; Voronkov, M. G. Pentacoordinate Hydrochlorosilanes With Lactamomethyl Ligand. *J. Organomet. Chem.* **2007,** *692* (11), 2160–2167.
254. Bassindale, A. R.; Borbaruah, M.; Glynn, S. J.; Parker, D. J.; Taylor, P. G. Modelling Nucleophilic Substitution at Silicon Using Hypervalent Silicon Compounds Based on Di and Tri Halosilanes. *J. Organomet. Chem.* **2000,** *606* (2), 125–131.
255. Bassindale, A. R.; Parker, D. J.; Taylor, P. G.; Auner, N.; Herrschaft, B. The X-ray Structure of a Protonated Hypervalent Silanol. *Chem. Commun.* **2000,** *7*, 565–566.
256. Bassindale, A. R.; Parker, D. J.; Taylor, P. G.; Auner, N.; Herrschaft, B. Modelling S_N2 Nucleophilic Substitution at Silicon by Structural Correlation With X-ray Crystallography and NMR Spectroscopy. *J. Organomet. Chem.* **2003,** *667* (1–2), 66–72.
257. Voronkov, M. G.; Zel'bst, E. A.; Kashaev, A. A.; Katkevich, Y. V.; Bruskov, V. A.; Lazareva, N. F., et al. Molecular Structure of (O→Si)methyl *N*-Methyl-*N*-(trifluorosilylmethyl)carbamate in Crystal and in Solution. *Dokl. Chem.* **2002,** *386* (4–6), 265–268.
258. Aksamentova, T. N.; Chipanina, N. N.; Voronkov, M. G.; Trofimova, O. M.; Zelbst, E. A.; Turchaninov, V. K. Directivity of the Intramolecular O→Si Bond in Pentacoordinate Organosilicone Compounds. *Russ. J. Gen. Chem.* **2010,** *79* (11), 2328–2338.
259. Korlyukov, A. A.; Lyssenko, K. A.; Antipin, M. Y. Chemical Bonding in Isolated Molecules and Crystals of Zwitterionic Pentacoordinate Silicon Chelates. *Russ. Chem. Bull.* **2002,** *51* (8), 1423–1432.
260. Bassindale, A. R.; Glynn, S. J.; Taylor, P. G.; Auner, N.; Herrschaft, B. Modelling Nucleophilic Substitution at Silicon Using Hypervalent Silicon Compounds Based on Urea Ligands. *J. Organomet. Chem.* **2001,** *619* (1–2), 132–140.
261. Bassindale, A. R.; Baukov, Y. I.; Taylor, P. G.; Negrebetsky, V. V. Proton Catalysis of Nucleophilic Substitution at Pentacoordinate Silicon. *J. Organomet. Chem.* **2002,** *655* (1–2), 1–6.
262. Negrebetsky, V. V.; Taylor, P. G.; Kramarova, E. P.; Bylikin, S. Y.; Belavin, I. Y.; Shipov, A. G., et al. Structures and Stereochemical Non-Rigidity of Si-Substituted *N*-(Dimethylsilylmethyl)- and *N*-(Methylphenylsilylmethyl)amides and -Lactams. *J. Organomet. Chem.* **2006,** *691* (19), 3976–3988.
263. Negrebetsky, V. V.; Taylor, P. G.; Kramarova, E. P.; Shipov, A. G.; Pogozhikh, S. A.; Ovchinnikov, Y. E., et al. Synthesis, Structure and Dynamic Stereochemistry of (O→Si)-Chelate *N*-(Trifluorosilylmethyl)-[*N*-(*S*)-(1-Phenylethyl)]acetamide and 1-(Trifluorosilylmethyl)-2-Oxoperhydroazepine: Retention of the O→Si Coordination in the Adduct With KF and 18-Crown-6. *J. Organomet. Chem.* **2008,** *693* (7), 1309–1320.
264. Bassindale, A. R.; Parker, D. J.; Taylor, P. G.; Turtle, R. A Kinetic and Mechanistic Study of Nucleophilic Substitution at a Pentacoordinated Silicon Atom. *Z. Anorg. Allg. Chem.* **2009,** *635* (9–10), 1288–1294.
265. Shipov, A. G.; Kramarova, E. P.; Pogozhikh, S. A.; Negrebetskii, V. V.; Smirnova, L. S.; Artamkina, O. B., et al. Synthesis and Molecular and Crystal Structures of Mono- and Bis-Chelate Hypercoordinate

Silicon Compounds Containing the C,O-Chelating 2,2-Dimethyl-4-oxo-2,3-dihydro-1,3-oxazin-3-ylmethyl Ligand. *Russ. Chem. Bull.* **2007,** *56* (3), 461–474.

266. Shipov, A. G.; Korlyukov, A. A.; Kramarova, E. P.; Arkhipov, D. E.; Bylikin, S. Y.; Hunze, F., et al. Donor-Stabilized Five-Coordinate Cationic Chelate Silicon Compounds With Two (O→Si)-Coordinating Ligands. *Russ. J. Gen. Chem.* **2012,** *81* (12), 2412–2427.
267. Korlyukov, A. A.; Pogozhikh, S. A.; Ovchinnikov, Y. E.; Lyssenko, K. A.; Antipin, M. Y.; Shipov, A. G., et al. (Amidomethyl)dimethylsilanol Hydrohalides: Synthesis, NMR and IR Studies. Characteristic Features of the Electronic Structure From High-Resolution X-ray Study and Quantum Chemical Calculation. *J. Organomet. Chem.* **2006,** *691* (19), 3962–3975.
268. Nikolin, A. A.; Arkhipov, D. E.; Shipov, A. G.; Kramarova, E. P.; Koval'chuk, N. A.; Korlyukov, A. A., et al. Pentacoordinated Chlorosilanes With *C,O*-Chelate Ligands Derived From *N*-Methyl-*N'*-Organosulfonyl-Prolinamides. *Chem. Heterocycl. Compd.* **2012,** *47* (12), 1565–1583.
269. Shipov, A. G.; Kramarova, E. P.; Murasheva, T. P.; Korlyukov, A. A.; Pogozhikh, S. A.; Tarasenko, S. A., et al. Scheme of Hydrolysis of Five-Coordinate Chlorosilanes by X-Ray Diffraction Data. *Russ. J. Gen. Chem.* **2012,** *81* (12), 2428–2439.
270. Shipov, A. G.; Kramarova, E. P.; Fang, H.; Arkhipov, D. E.; Nikolin, A. A.; Bylikin, S. Y., et al. 1-Organosulfonyl-2-sila-5-piperazinones: Synthesis, Molecular and Crystal Structure, and Chemical Transformations Into 2-Aminoacid Derivatives. *J. Organomet. Chem.* **2013,** *741-742*, 114–121.
271. Pestunovich, V. A.; Albanov, A. I.; Pogozhikh, S. A.; Ovchinnikov, Y. E.; Lazareva, N. F. *N*-[Hydroxy (dimethyl)silylmethyl]-*N,N'*-Propylene Urea Hydrochloride: Synthesis and Structure. *J. Mol. Struct.* **2012,** *1024*, 136–141.
272. Bao, M.; Hatanaka, Y.; Shimada, S. The First Synthesis and X-ray Structure of [1.1]Silaferrocenophane Containing Pentacoordinate Silicon Moieties. *Chem. Lett.* **2004,** *33* (5), 520–521.
273. Hatanaka, Y.; Okada, S.; Minami, T.; Goto, M.; Shimada, K. Synthesis, X-ray Structure, and Ring-Opening Polymerization of Pentacoordinate Silicon-Bridged [1]Ferrocenophane. *Organometallics* **2005,** *24* (6), 1053–1055.
274. Pogozhikh, S. A.; Ovchinnikov, Y. E.; Kramarova, E. P.; Negrebetskii, V. V.; Shipov, A. G.; Albanov, A. I., et al. Synthesis, Structure, and Spatial Arrangement of Si-Substituted *N*-(Dimethylsilylmethyl) imides. *Russ. J. Gen. Chem.* **2004,** *74* (10), 1501–1507.
275. Sohail, M.; Panisch, R.; Bowden, A.; Bassindale, A. R.; Taylor, P. G.; Korlyukov, A. A., et al. Pentacoordinate Silicon Complexes With Dynamic Motion Resembling a Pendulum on the S_N2 Reaction Pathway. *Dalton Trans.* **2013,** *42* (30), 10971–10981.
276. Voronkov, M. G.; Zel'bst, É. A.; Kashaev, A. A.; Katkevich, Y. V.; Fundamenskii, V. S.; Bolgova, Y. I., et al. Crystal and Molecular Structure of *N*-(Trifluorosilylmethyl)phthalimide. *Dokl. Chem.* **2003,** *393* (4–6), 275–278.
277. Voronkov, M. G.; Chipanina, N. N.; Shainyan, B. A.; Bolgova, Y. I.; Trofimova, O. M.; Chernov, N. F., et al. Intramolecular O→Si Coordination in *N*-(Trifluorosilylmethyl)phthalimide and Its Analogues. *Dokl. Chem.* **2004,** *394* (1–3), 4–7.
278. Voronkov, M. G.; Zel'bst, É. A.; Katkevich, Y. V.; Kashaev, A. A.; Fundamenskii, V. S.; Bolgova, Y. I., et al. Stereoelectronic Structure and Alternating Intramolecular Coordination O→Si in *N*-(Trifluorosilylmethyl)succinimide. *Dokl. Chem.* **2005,** *402* (1–3), 77–80.
279. Voronkov, M. G.; Trofimova, O. M.; Chernov, N. F.; Bolgova, Y. I.; Albanov, A. I.; Chipanina, N. N., et al. Pentacoordinate Organosilicon Compounds—Molecular Structure and Stereodynamic Behavior of (Trifluorosilylmethyl)succinimide and (Trifluorosilylmethyl)glutarimide. *Heteroat. Chem.* **2006,** *17* (6), 567–571.
280. Zel'bst, É. A.; Korlyukov, A. A.; Voronkov, M. G. Crystal and Molecular Structure of 4-Chloro (benzoyloxymethyl)trifluorosilane at 120 K. *J. Struct. Chem.* **2013,** *54* (1), 274–276.
281. Weicker, S. A.; Stephan, D. W. Activation of Carbon Dioxide by Silyl Triflate-Based Frustrated Lewis Pairs. *Chem. Eur. J.* **2015,** *21* (37), 13027–13034.

282. Liu, Y. X.; Steiner, S. A.; Spahn, C. W.; Guzei, I. A.; Toulokhonova, I. S.; West, R. A Stable, Highly Fluorescent Vinylogous β-Diketonate Silicon Complex. *Organometallics* **2007,** *26* (6), 1306–1307.
283. Kampfe, A.; Brendler, E.; Kroke, E.; Wagler, J. 2-Acylpyrroles as Mono-Anionic *O,N*-Chelating Ligands in Silicon Coordination Chemistry. *Chem. Eur. J.* **2014,** *20* (30), 9409–9418.
284. Kämpfe, A.; Brendler, E.; Kroke, E.; Wagler, J. Tp*Cu(I)-CN-SiL$_2$-NC-Cu(I)Tp*—A Hexacoordinate Si-Complex as Connector for Redox Active Metals *via* π-Conjugated Ligands. *Dalton Trans.* **2015,** *44* (10), 4744–4750.
285. Dilman, A. D.; Levin, V. V.; Korlyukov, A. A.; Belyakov, P. A.; Struchkova, M. I.; Antipin, M. Y., et al. Complexation of Tris(pentafluorophenyl)silanes With Neutral Lewis Bases. *J. Organomet. Chem.* **2008,** *693* (6), 1005–1019.
286. Denmark, S. E.; Eklov, B. M. Neutral and Cationic Phosphoramide Adducts of Silicon Tetrachloride: Synthesis and Characterization of Their Solution and Solid-State Structures. *Chem. Eur. J.* **2008,** *14* (1), 234–239.
287. Denmark, S. E.; Eklov, B. M.; Yao, P. J.; Eastgate, M. D. On the Mechanism of Lewis Base Catalyzed Aldol Addition Reactions: Kinetic and Spectroscopic Investigations Using Rapid-Injection NMR. *J. Am. Chem. Soc.* **2009,** *131* (33), 11770–11787.
288. Peveling, K.; Schürmann, M.; Ludwig, R.; Jurkschat, K. From Intramolecularly [4 + 1]- and [4 + 2]-Coordinated Tri- and Tetraorganosilanes to Hypercoordinated Benzoxasilaphospholes. *Organometallics* **2001,** *20* (22), 4654–4663.
289. Dannappel, K.; Schürmann, M.; Costisella, B.; Jurkschat, K. Restricted Rotation About the P – C Bond in the Triorganosilane {4-*t*-Bu-2,6-[P(O)(O-*i*-Pr)$_2$]$_2$C$_6$H$_2$}SiPh$_2$H: Identification of a Novel Rotamer and Its Conversion to the Siliconium Salt {4-*t*-Bu-2,6-[P(O)(O-*i*-Pr)$_2$]$_2$C$_6$H$_2$}SiPh$_2^+$PF$_6$. *Organometallics* **2005,** *24* (5), 1031–1034.
290. Dannappel, K.; Nienhaus, R.; Schürmann, M.; Costisella, B.; Jurkschat, K. The UV-Light Initiated Reaction of Organosilanes With Tungsten Hexacarbonyl: Formation of an Organosilylene Complex and Organosilylium Salts. *Z. Anorg. Allg. Chem.* **2009,** *635* (13–14), 2126–2134.
291. Peveling, K.; Schürmann, M.; Jurkschat, K. The Molecular Structure of {2,6[P(O)(O-*i*-Pr)$_2$]$_2$-4-*t*-BuC$_6$H$_2$} SiPh$_3$. *Main Group Met. Chem.* **2001,** *24* (4), 251–252.
292. Wagner, M.; Lutter, M.; Dietz, C.; Prosenc, M. H.; Jurkschat, K. On the Reactivity of RSnCl and RSiMe$_3${R = 4-*t*Bu-2,6-[P(O)(OiPr)$_2$]$_2$C$_6$H$_2$} Towards BF$_3\cdot$OEt$_2$: Competing Lewis Acidities. *Eur. J. Inorg. Chem.* **2015,** *2015* (12), 2152–2158.
293. Peveling, K.; Henn, M.; Löw, C.; Mehring, M.; Schürmann, M.; Costisella, B., et al. Mechanistic Studies on the Cyclization of Organosilicon and Organotin Compounds Containing the O,C,O-Coordinating Pincer-Type Ligand {4-*t*-Bu-2,6-[P(O)(OR)$_2$]$_2$C$_6$H$_2$}$^-$ (R = *i*-Pr, Et): Phosphorus (*P*OC)-Versus Carbon (PO*C*)-Attack. *Organometallics* **2004,** *23* (7), 1501–1508.
294. Peveling, K.; Dannappel, K.; Schürmann, M.; Costisella, B.; Jurkschat, K. Structure-Directing Properties of Lithium Chloride in Supramolecular {4-*t*-Bu-2,6-[P(O)(OEt)$_2$]$_2$C$_6$H$_2$}SiH$_2$Ph · LiCl · 2H$_2$O. Intermolecular P = O→Li Versus Intramolecular P = O→(H)Si Coordination. *Organometallics* **2006,** *25* (2), 368–374.
295. Liberman-Martin, A. L.; Bergman, R. G.; Tilley, T. D. Lewis Acidity of Bis(perfluorocatecholato)silane: Aldehyde Hydrosilylation Catalyzed by a Neutral Silicon Compound. *J. Am. Chem. Soc.* **2015,** *137* (16), 5328–5331.
296. Nishigaichi, Y.; Suzuki, A.; Saito, T.; Takuwa, A. First Examples of Hypervalent Enhancement of Photoallylation by Allylsilicon Compounds via Photoinduced Electron Transfer. *Tetrahedron Lett.* **2005,** *46* (31), 5149–5151.
297. Nishigaichi, Y.; Suzuki, A.; Takuwa, A. Remarkable Enhancement of Photo-Allylation of Aromatic Carbonyl Compounds With a Hypervalent Allylsilicon Reagent by Donor Molecules. *Tetrahedron Lett.* **2007,** *48* (2), 211–214.
298. Matsuoka, D.; Nishigaichi, Y. Photosubstitution of Dicyanoarenes by Hypervalent Allylsilicon Reagents via Photoinduced Electron Transfer. *Chem Lett.* **2014,** *43* (4), 559–561.

299. Matsuoka, D.; Nishigaichi, Y. Allyl-Transfer Reaction From Photoexcited Hypervalent Allylsilicon Reagent Toward Dicyanobenzenes. *Chem Lett.* **2015,** *44* (2), 163–165.
300. Corce, V.; Chamoreau, L. M.; Derat, E.; Goddard, J. P.; Ollivier, C.; Fensterbank, L. Silicates as Latent Alkyl Radical Precursors: Visible-Light Photocatalytic Oxidation of Hypervalent Bis-Catecholato Silicon Compounds. *Angew. Chem. Int. Ed.* **2015,** *54* (39), 11414–11418.
301. Jouffroy, M.; Primer, D. N.; Molander, G. A. Base-Free Photoredox/Nickel Dual-Catalytic Cross-Coupling of Ammonium Alkylsilicates. *J. Am. Chem. Soc.* **2016,** *138*, 475–478.
302. Seganish, W. M.; DeShong, P. Preparation and Palladium-Catalyzed Cross-Coupling of Aryl Triethylammonium Bis(catechol) Silicates With Aryl Triflates. *J. Org. Chem.* **2004,** *69* (4), 1137–1143.
303. Seganish, W. M.; DeShong, P. Palladium-Catalyzed Cross-Coupling of Aryl Triethylammonium Bis (catechol) Silicates With Aryl Bromides Using Microwave Irradiation. *Org. Lett.* **2004,** *6* (23), 4379–4381.
304. Tacke, R.; Pülm, M.; Wagner, B. Zwitterionic Pentacoordinate Silicon Compounds. In *Advances in Organometallic Chemistry;* West, R., Hill, A. F., Eds.; Volume 44; Academic Press, 1999; pp 221–273.
305. Tacke, R.; Bertermann, R.; Burschka, C.; Dragota, S.; Penka, M.; Richter, I. Spirocyclic Zwitterionic λ^5*Si*-Silicates With Two Bidentate Ligands Derived From α-Amino Acids or α-Hydroxycarboxylic Acids: Synthesis, Structure, and Stereodynamics. *J. Am. Chem. Soc.* **2004,** *126* (44), 14493–14505.
306. Tacke, R.; Bertermann, R.; Biller, A.; Dannappel, O.; Penka, M.; Pülm, M., et al. Zwitterionic Spirocyclic λ^5*Si*-Silicates With Two Bidentate Acetohydroximato(2−) or Benzohydroximato(2−) Ligands: Synthesis, Structure, and Dynamic Stereochemistry. *Z. Anorg. Allg. Chem.* **2000,** *626* (5), 1159–1173.
307. Tacke, R.; Burschka, C.; Heermann, J.; Richter, I.; Wagner, B.; Willeke, R. Macrocyclic Siloxanes With Two Exocyclictrans-Silanol and Two Exocyclictrans-Amino Functions—Stereoselective Syntheses and Structures. *Eur. J. Inorg. Chem.* **2001,** (9), 2211–2215.
308. Tacke, R.; Bertermann, R.; Burschka, C.; Dragota, S. A Zwitterionic Spirocyclic Pentacoordinate Silicon Compound Synthesized in Water by Si−O and Si−C Bond Cleavage. *Angew. Chem. Int. Ed.* **2005,** *44* (33), 5292–5295.
309. Richter, I.; Burschka, C.; Tacke, R. Octakis[(2,2,6,6-Tetramethylpiperidino)methyl]octasilsesquioxane: Synthesis and Crystal Structure Analysis of a New Aminoorganyl-Functionalized Octasilsesquioxane. *J. Organomet. Chem.* **2002,** *646* (1–2), 200–203.
310. Richter, I.; Penka, M.; Tacke, R. [Benzene-1,2-diolato(2−)][benzene-1,2-diolato(1−)]methyl[(2,2,6,6-tetramethylpiperidinio)methyl]silicate: Isolation, Structural Characterization, and Thermally Induced Methane Elimination. *Organometallics* **2002,** *21* (14), 3050–3053.
311. Belyakov, S.; Erchak, N.; Zheimote, S.; Fleisher, M.; Lukevics, E. Structural Investigation of 1-(*N,N*-dimethylaminomethyl)-spirobi(3-oxo-2,5-dioxa-1-silacyclo-pentan)ates. *Chem. Heterocycl. Compd.* **2008,** *44* (12), 1520–1526.
312. Erchak, N.; Belyakov, S.; Kalvinsh, I.; Pypowski, K.; Valbahs, E.; Lukevics, E. Two Polymorphic Modifications of 1-(*N*-Morpholiniomethyl)Spirobi(3-Oxo-2,5-Dioxa-1-Silacyclopentan)-Ate Hydrate. *Chem. Heterocycl. Compd.* **2009,** *45* (9), 1137–1143.
313. Tacke, R.; Ulmer, B.; Wagner, B.; Arlt, M. A Novel Silicon-Based Linkage and Cleavage Strategy for Solid-Phase Synthesis: Formation of Resin-Linked Zwitterionic Pentacoordinate Silicates as the Key Step and Release of the Target Molecules in a Traceless Fashion. *Organometallics* **2000,** *19* (25), 5297–5309.
314. Dragota, S.; Bertermann, R.; Burschka, C.; Penka, M.; Tacke, R. Diastereo- and Enantiomerically Pure Zwitterionic Spirocyclic λ^5*Si*-[(Ammonio)methyl]silicates With an SiO_2N_2C Skeleton Containing Two Bidentate Chelate Ligands Derived From α-Amino Acids. *Organometallics* **2005,** *24* (23), 5560–5568.
315. Kastele, X.; Klufers, P.; Tacke, R. Not Unhydrolyzed at pH 8. *Angew. Chem. Int. Ed.* **2006,** *45* (20), 3212–3214.
316. Theis, B.; Burschka, C.; Tacke, R. Optically Active Zwitterionic λ^5Si,λ^5Si′-Disilicates: Syntheses, Crystal Structures, and Behavior in Aqueous Solution. *Chem. Eur. J.* **2008,** *14* (15), 4618–4630.

317. Theis, B.; Weiss, J.; Lippert, W. P.; Bertermann, R.; Burschka, C.; Tacke, R. Zwitterionic and Anionic Multinuclear Pentacoordinate Silicon(IV) Complexes With Bridging (*R*,*R*)-tartrato(4−) Ligands and SiO_5 Skeletons: Synthesis and Reactivity in Aqueous Solution. *Chem. Eur. J.* **2012,** *18* (8), 2202−2206.
318. Shipov, A. G.; Korlyukov, A. A.; Arkhipov, D. E.; Kramarova, E. P.; Negrebetskii, V. V.; Bylikin, S. Y., et al. Synthesis and Structural Characterization of the Anionic Chelates of Hypercoordinate Silicon, The Derivatives of Glycolic and Tartaric Acids. *Mendeleev Commun.* **2010,** *20* (5), 273−274.
319. Kingston, J. V.; Verkade, J. G. $P[N(i\text{Bu})CH_2CH_2]_3N$: A Versatile Non-Ionic Base for the Synthesis of Higher Coordinate Silicates. *Inorg. Chem. Commun.* **2005,** *8* (7), 643−646.
320. Bindu, P.; Kingston, J. V.; Rao, M. N. S. Dianionic Silicate Derivatives of Transition Metals: Synthesis by A Facile Ion Exchange Route, Characterization and Thermolysis Studies. *Polyhedron* **2004,** *23* (4), 679−686.
321. Tacke, R.; Stewart, A.; Becht, J.; Burschka, C.; Richter, I. Di[(hydroxyalkyl)dimethylammonium] Tris [benzene-1,2-diolato(2−)]silicates and Their Germanium Analogs: Syntheses, Crystal Structure Analyses, and NMR Studies. *Can. J. Chem.* **2000,** *78* (11), 1380−1387.
322. Bindu, P.; Varghese, B.; Rao, M. N. S. Six Coordinate Tris(catecholato)silicates of Primary Amine Residues—Synthesis, Characterization, and Thermolysis Studies. X-Ray Structures of $[n\text{-}C_3H_7NH_3]_2[Si(C_6H_4O_2)_3]\cdot 1/2(C_6H_{14}N_2)$ and of a Bulky Secondary Ammonium Ion, $[(i\text{-}C_4H_9)_2NH_2]_2[Si(C_6H_4O_2)_3]\cdot H_2O$. *Phosphorus Sulfur Silicon Relat. Elem.* **2003,** *178* (11), 2373−2386.
323. Suvitha, A.; Varghese, B.; Rao, M. N. S. Bis(diisobutylammonium) Tris(naphthalene-2,3-diolato)silicate Acetonitrile Trisolvate. *Acta. Crystallogr. Sect. E Struct. Rep. Online* **2005,** *62* (1), o344−o346.
324. Schmiederer, T.; Rausch, S.; Valdebenito, M.; Mantri, Y.; Mosker, E.; Baramov, T., et al. The *E. coli* Siderophores Enterobactin and Salmochelin Form Six-Coordinate Silicon Complexes at Physiological pH. *Angew. Chem. Int. Ed.* **2011,** *50* (18), 4230−4233.
325. Baramov, T.; Keijzer, K.; Irran, E.; Mosker, E.; Baik, M. H.; Sussmuth, R. Synthesis and Structural Characterization of Hexacoordinate Silicon, Germanium, and Titanium Complexes of the *E. coli* Siderophore Enterobactin. *Chem. Eur. J.* **2013,** *19* (32), 10536−10542.
326. Kenla, T. J.; Tatong, M. D.; Talontsi, F. M.; Dittrich, B.; Frauendorf, H.; Laatsch, H. Si-Enterobactin From the Endophytic *Streptomyces* sp. KT-S1-B5−A Potential Silicon Transporter in Nature. *Chem. Commun.* **2013,** *49* (69), 7641−7643.
327. Willeke, R.; Tacke, R. Zwitterionic λ^5Si-Silicates With SiO_2S_2C Skeletons: Syntheses and Structural Characterization in the Solid State. *Z. Anorg. Allg. Chem.* **2001,** *627* (7), 1537−1541.
328. Naganuma, K.; Kawashima, T. Synthesis and Structure of the First Pentacoordinate Thiasiliranides. *J. Organomet. Chem.* **2002,** *643-644*, 504−507.
329. Troegel, D.; Burschka, C.; Riedel, S.; Kaupp, M.; Tacke, R. Unusual Silicon Coordination Polyhedra: Non-VSEPR Structures of Zwitterionic λ^5-Si Silicon(IV) Complexes With an SiS_2N_2C or SiS_2O_2C Skeleton. *Angew. Chem. Int. Ed.* **2007,** *46* (37), 7001−7005.
330. Bockholt, A.; Jutzi, P.; Mix, A.; Neumann, B.; Stammler, A.; Stammler, H.-G. Neutral and Cationic Silicon Species Containing Aryl-OCO- or Aryl-SCS-Type Pincer Ligands: Synthesis, Structure, Bonding, and Comparison With Aryl-NCN Systems. *Z. Anorg. Allg. Chem.* **2009,** *635* (9−10), 1326−1334.
331. Kalashnikova, N. A.; Bylikin, S. Y.; Korlyukov, A. A.; Shipov, A. G.; Baukov, Y. I.; Taylor, P. G., et al. Cationic Complexes of Silicon and Germanium With (*O*,*S*)-Chelate Ligands. *Dalton Trans.* **2012,** *41* (41), 12681−12682.
332. Tacke, R.; Mallak, M.; Willeke, R. Pentacoordination of Silicon by Four Covalent Si−S Bonds and One Covalent Si−C Bond. *Angew. Chem. Int. Ed.* **2001,** *40* (12), 2339−2341.
333. Bertermann, R.; Biller, A.; Kaupp, M.; Penka, M.; Seiler, O.; Tacke, R. New Zwitterionic Spirocyclic λ^5Si-Silicates With an SiX_4C Skeleton (X = S, O) Containing Two Ligands of the Dithiolato(2−) or Diolato(2−) Type: Synthesis, Structure, and Bonding Situation. *Organometallics* **2003,** *22* (20), 4104−4110.
334. Theis, B.; Metz, S.; Burschka, C.; Bertermann, R.; Maisch, S.; Tacke, R. Neutral Pentacoordinate Silicon (IV) Complexes With Silicon-Chalcogen (S, Se, Te) Bonds. *Chem. Eur. J.* **2009,** *15* (30), 7329−7338.

335. Seiler, O.; Buettner, M.; Penka, M.; Tacke, R. Zwitterionic λ^5Si-Silicates With an SiS_4C or SiS_2O_2C Skeleton, Compounds Containing a Square-Pyramidal or Trigonal-Bipyramidal Si-Coordination Polyhedron. *Organometallics* **2005,** *24* (24), 6059–6062.
336. Wagler, J.; Böhme, U.; Roewer, G. Activation of a Si–Si Bond by Hypercoordination: Cleavage of a Disilane and Formation of a Si–C Bond. *Organometallics* **2004,** *23* (25), 6066–6069.
337. Wagler, J. A Disilane With a Hypercoordinate Silicon Atom: Coordination of an Imine Ligand Versus Si–Si Bond Splitting. *Organometallics* **2007,** *26* (1), 155–159.
338. Toshimitsu, A.; Hirao, S.; Saeki, T.; Asahara, M.; Tamao, K. Preparation, Structure, and Reactivity of Pentacoordinate Disilanes Bearing an 8-Charcogeno-1-naphthyl Group and a Heteroatom on the Same Silicon Atom. *Heteroat. Chem.* **2001,** *12* (5), 392–397.
339. Toshimitsu, A.; Saeki, T.; Tamao, K. Phosphonium Sila-Ylide: Reaction Pathway Different From Ammonium Sila-Ylide but Similar to Phosphonium Ylide. *J. Am. Chem. Soc.* **2001,** *123* (37), 9210–9211.
340. Toshimitsu, A.; Saeki, T.; Tamao, K. Intramolecularly Base-Coordinated Silylenes That Behave as Sila-Ylides. *Phosphorus, Sulfur Silicon Relat. Elem.* **2002,** *177* (6–7), 1409–1412.
341. Saeki, T.; Toshimitsu, A.; Tamao, K. Generation of Intramolecular Sulfur- or Selenium-Coordinated Silylenes by the Thermolysis of Pentacoordinate Disilanes Bearing an 8-Thio- or 8-Seleno-1-naphthyl Group and a Methoxy Group on the Same Silicon Atom. *Organometallics* **2003,** *22* (16), 3299–3303.
342. Tamao, K.; Asahara, M.; Toshimitsu, A. Trimerization of a Divalent Silicon Species Bearing the 8-Dimethylamino-1-naphthyl Group Accompanied by the Migration of Methyl and Dimethylamino Groups. *Chem. Lett.* **2000,** *29* (6), 658–659.
343. Saeki, T.; Toshimitsu, A.; Tamao, K. Thermolysis of Bis(pentacoordinate) Silicon Compound Bearing Two 1-Fluorodisilanyl Units at 4- and 8-Positions on 1,5-Bis(dimethylamino)naphthalene Ring: Construction of Disilapyrene Skeleton and Cyclophane-Shaped Siloxane Dimer. *J. Organomet. Chem.* **2003,** *686* (1–2), 215–222.
344. El-Sayed, I.; Hatanaka, Y.; Onozawa, S.; Tanaka, M. Unusual Locking of Silicon Chains Into All-Transoid Conformation by Pentacoordinate Silicon Atoms. *J. Am. Chem. Soc.* **2001,** *123* (15), 3597–3598.
345. Kano, N.; Nakagawa, N.; Shinozaki, Y.; Kawashima, T.; Sato, Y.; Naruse, Y., et al. Disilanes Containing Two High-Coordinate Silicon Atoms Bridged by Carboxylate Ligands: Synthesis, Structure, and Dynamic Behavior. *Organometallics* **2005,** *24* (12), 2823–2826.
346. Sen, S. S.; Roesky, H. W.; Stern, D.; Henn, J.; Stalke, D. High Yield Access to Silylene RSiCl (R = PhC$(NtBu)_2$) and Its Reactivity Toward Alkyne: Synthesis of Stable Disilacyclobutene. *J. Am. Chem. Soc.* **2010,** *132* (3), 1123–1126.
347. Zhang, S.-H.; Yeong, H.-X.; So, C.-W. Reactivity of a Silylsilylene Bearing a Functionalized Diaminochlorosilyl Substituent. *Chem. Eur. J.* **2011,** *17* (12), 3490–3499.
348. Tavčar, G.; Sen, S. S.; Roesky, H. W.; Hey, J.; Kratzert, D.; Stalke, D. Reactions of a Bis-Silylene (LSi – SiL, L = PhC$(NtBu)_2$) and a Heteroleptic Chloro Silylene (LSiCl) With Benzil: Formation of Bis (siladioxolene) and Monosiladioxolene Analogue With Five-Coordinate Silicon Atoms in Both Ring Systems. *Organometallics* **2010,** *29* (17), 3930–3935.
349. Armbruster, F.; Fernández, I.; Breher, F. Syntheses, Structures, and Reactivity of Poly(pyrazolyl)silanes, -Disilanes, and the Ambidentate κ^1Si/κ^3N-Coordinating Tris(3,5-dimethylpyrazolyl)silanide Ligand $[Si(3,5\text{-}Me_2pz)_3]^-$ ((Me)Tpsd). *Dalton Trans.* **2009,** *29*, 5612–5626.
350. Bitto, F.; Kraushaar, K.; Böhme, U.; Brendler, E.; Wagler, J.; Kroke, E. Chlorosilanes and 3,5-Dimethylpyrazole: Multinuclear Complexes, Acetonitrile Insertion and ^{29}Si NMR Chemical-Shift Anisotropy Studies. *Eur. J. Inorg. Chem.* **2013,** *2013* (16), 2954–2962.
351. Wagler, J.; Brendler, E.; Heine, T.; Zhechkov, L. Disilicon Complexes With Two Hexacoordinate Si Atoms: Paddlewheel-Shaped Isomers With $(ClN_4)Si–Si(S_4Cl)$ and $(ClN_2S_2)Si–Si(S_2N_2Cl)$ Skeletons. *Chem. Eur. J.* **2013,** *19* (42), 14296–14303.
352. Li, J. Z.; Subramanian, L. R.; Hanach, M. Phthalocyanine Dimer With a Direct Si–Si Linkage. A Response to a Recent Work. *J. Porphyrins Phthalocyanines* **2000,** *4* (8), 739–741.

353. Kobayashi, N.; Furuya, F.; Yug, G.-C.; Wakita, H.; Yokomizo, M.; Ishikawa, N. Synthesis and Characterization of Phthalocyanines With Direct Si–Si Linkages. *Chem. Eur. J.* **2002,** *8* (6), 1474–1484.
354. Ranta, J.; Viljanen, R.; Lemmetyinen, H.; Efimov, A. A Novel, One-Pot Method for the Synthesis of Dimeric and Monomeric Phthalocyanine Silicon Complexes From Free-Base Phthalocyanines. *Dyes Pigm.* **2009,** *83* (3), 317–323.
355. Wagler, J.; Brendler, E.; Langer, T.; Pöttgen, R.; Heine, T.; Zhechkov, L. Ylenes in the $M^{II} \rightarrow Si^{IV}$ (M = Si, Ge, Sn) Coordination Mode. *Chem. Eur. J.* **2010,** *16* (45), 13429–13434.
356. Kano, N.; Miyake, H.; Sasaki, K.; Kawashima, T.; Mizorogi, N.; Nagase, S. Dianionic Species With a Bond Consisting of Two Pentacoordinated Silicon Atoms. *Nat. Chem.* **2010,** *2* (2), 112–116.
357. Kano, N.; Miyake, H.; Sasaki, K.; Kawashima, T. Oxidative Bond Cleavage of the Silicon–Silicon Bond of a Pentacoordinated Disilicate. *Heteroat. Chem.* **2015,** *26* (2), 183–186.
358. Kano, N.; Sasaki, K.; Miyake, H.; Kawashima, T. Synthesis and Isolation of a Silylsilicate Containing Two Pentacoordinated Silicon Atoms by Monoprotonation of a Disilicate and Monodeprotonation of a Disilane. *Organometallics* **2014,** *33* (9), 2358–2362.
359. Knopf, C.; Herzog, U.; Roewer, G.; Brendler, E.; Rheinwald, G.; Lang, H. Interactions of Chloromethyldisilanes With Tetrakis(dimethylamino)ethylene (TDAE), Formation of $[TDAE]^+$ $[Si_3Me_2Cl_7]^-$. *J. Organomet. Chem.* **2002,** *662* (1–2), 14–22.
360. Tillmann, J.; Meyer, L.; Schweizer, J. I.; Bolte, M.; Lerner, H. W.; Wagner, M., et al. Chloride-Induced Aufbau of Perchlorinated Cyclohexasilanes From Si_2Cl_6: A Mechanistic Scenario. *Chem. Eur. J.* **2014,** *20* (30), 9234–9239.
361. Wang, Y.; Xie, Y.; Wei, P.; King, R. B.; Schaefer, H. F., III; Schleyer, P. v. R., et al. A Stable Silicon(0) Compound With a Si=Si Double Bond. *Science* **2008,** *321* (5892), 1069–1071.
362. Ghadwal, R. S.; Sen, S. S.; Roesky, H. W.; Tavcar, G.; Merkel, S.; Stalke, D. Neutral Penta- and Hexacoordinate *N*-Heterocyclic Carbene Complexes Derived From SiX_4 (X = F, Br). *Organometallics* **2009,** *28* (21), 6374–6377.
363. Jana, A.; Omlor, I.; Huch, V.; Rzepa, H. S.; Scheschkewitz, D. *N*-Heterocyclic Carbene Coordinated Neutral and Cationic Heavier Cyclopropylidenes. *Angew. Chem. Int. Ed.* **2014,** *53* (37), 9953–9956.
364. Schneider, H.; Schmidt, D.; Radius, U. A Facile Route to Backbone-Tethered *N*-Heterocyclic Carbene (NHC) Ligands via NHC to aNHC Rearrangement in NHC Silicon Halide Adducts. *Chem. Eur. J.* **2015,** *21* (7), 2793–2797.
365. Holloczki, O.; Nyulaszi, L. Stability and Structure of Carbene-Derived Neutral Penta- and Hexacoordinate Silicon Complexes. *Organometallics* **2009,** *28* (14), 4159–4164.
366. Liu, Z. Chemical Bonding in Silicon-Carbene Complexes. *J. Phys. Chem. A* **2009,** *113* (22), 6410–6414.
367. Xiong, Y.; Yao, S.; Driess, M. Coordination of *N*-Heterocyclic Carbene to H_2SiX_2 (X = Cl, OTf) and H_3SiOTf (OTf = OSO_2CF_3): Synthesis of Donor-stabilized Parent Silylium Salts With Four- and Five-coordinate Silicon Atoms. *Z. Naturforsch., B: Chem. Sci.* **2013,** *68* (5–6), 445–452.
368. Schmidt, D.; Berthel, J. H.; Pietsch, S.; Radius, U. C–N Bond Cleavage and Ring Expansion of *N*-Heterocyclic Carbenes Using Hydrosilanes. *Angew. Chem. Int. Ed.* **2012,** *51* (35), 8881–8885.
369. Iversen, K. J.; Wilson, D. J.; Dutton, J. L. Activation of the C–N Bond of *N*-Heterocyclic Carbenes by Inorganic Elements. *Dalton Trans.* **2014,** *43* (34), 12820–12823.
370. Uhlemann, F.; Köppe, R.; Schnepf, A. Synthesis of Metastable $Si^{II}X_2$ Solutions (X = F, Cl). A Novel Binary Halide for Synthesis. *Z. Anorg. Allg. Chem.* **2014,** *640* (8–9), 1658–1664.
371. Ghadwal, R. S.; Roesky, H. W.; Merkel, S.; Henn, J.; Stalke, D. Lewis Base Stabilized Dichlorosilylene. *Angew. Chem. Int. Ed.* **2009,** *48* (31), 5683–5686.
372. Böttcher, T.; Bassil, B. S.; Zhechkov, L.; Heine, T.; Röschenthaler, G.-V. $(NHC^{Me})SiCl_4$: A Versatile Carbene Transfer Reagent—Synthesis From Silicochloroform. *Chem. Sci.* **2013,** *4* (1), 77–83.
373. Böttcher, T.; Steinhauer, S.; Lewis-Alleyne, L. C.; Neumann, B.; Stammler, H. G.; Bassil, B. S., et al. $NHC \rightarrow SiCl_4$: An Ambivalent Carbene-Transfer Reagent. *Chem. Eur. J.* **2015,** *21* (2), 893–899.

374. Böttcher, T.; Steinhauer, S.; Neumann, B.; Stammler, H. G.; Röschenthaler, G. V.; Hoge, B. Pentacoordinate Silicon(IV): Cationic, Anionic and Neutral Complexes Derived From the Reaction of NHC→$SiCl_4$ With Highly Lewis Acidic $(C_2F_5)_2SiH_2$. *Chem. Commun.* **2014,** *50* (47), 6204–6206.
375. Lewis-Alleyne, L. C.; Bassil, B. S.; Böttcher, T.; Röschenthaler, G. V. Selective Synthesis of *cis*- and *trans*-[$(NHC^{Me})_2PtCl_2$] and [$NHC^{Me}Pt(cod)Cl$][$NHC^{Me}PtCl_3$] Using $NHC^{Me}SiCl_4$. *Dalton Trans.* **2014,** *43* (42), 15700–15703.
376. Bonnette, F.; Kato, T.; Destarac, M.; Mignani, G.; Cossío, F. P.; Baceiredo, A. Encapsulated *N*-Heterocyclic Carbenes in Silicones Without Reactivity Modification. *Angew. Chem. Int. Ed.* **2007,** *46* (45), 8632–8635.
377. Maggiarosa, N.; Tyrra, W.; Naumann, D.; Kirij, N. V.; Yagupolskii, Y. L. $[Me_3Si(CF_3)F]^-$ and $[Me_3Si(CF_3)_2]^-$: Reactive Intermediates in Fluoride-Initiated Trifluoromethylation With Me_3SiCF_3—An NMR Study. *Angew. Chem. Int. Ed.* **1999,** *38* (15), 2252–2253.
378. Kolomeitsev, A.; Movchun, V.; Rusanov, E.; Bissky, G.; Lork, E.; Röschenthaler, G.-V., et al. Different Fluoride Anion Sources and (Trifluoromethyl)trimethylsilane: Molecular Structure of Tris(dimethylamino) sulfonium Bis(trifluoromethyl)trimethylsiliconate, the First Isolated Pentacoordinate Silicon Species With Five Si–C Bonds. *Chem. Commun.* **1999,** *11*, 1017–1018.
379. Prakash, G. K.; Wang, F.; Zhang, Z.; Haiges, R.; Rahm, M.; Christe, K. O., et al. Long-Lived Trifluoromethanide Anion: A Key Intermediate in Nucleophilic Trifluoromethylations. *Angew. Chem. Int. Ed.* **2014,** *53* (43), 11575–11578.
380. de Keijzer, A. H. J. F.; de Kanter, F. J. J.; Schakel, M.; Osinga, V. P.; Klumpp, G. W. In Search of Stable Lithium Pentaorganylsilicates; Special Role of Five Phenyl Ligands and of Ligands Containing the 1,4-(1,3-Butadienediyl) Unit. *J. Organomet. Chem.* **1997,** *548* (1), 29–32.
381. Couzijn, E. P. A.; Ehlers, A. W.; Schakel, M.; Lammertsma, K. Electronic Structure and Stability of Pentaorganosilicates. *J. Am. Chem. Soc.* **2006,** *128* (41), 13634–13639.
382. Ballweg, D.; Liu, Y.; Guzei, I. A.; West, R. Pentacoordinate Spirosilicate Anion, Bis(2,2′-biphenyldiyl) methylsilicate, Synthesized by the Lithium Cleavage of Dimethoxyethane. *Silicon Chem.* **2002,** *1* (1), 57–60.
383. Deerenberg, S.; Schakel, M.; de Keijzer, A. H. J. F.; Kranenburg, M.; Lutz, M.; Spek, A. L., et al. Tetraalkylammonium Pentaorganosilicates: The First Highly Stable Silicates With Five Hydrocarbon Ligands. *Chem. Commun.* **2002,** *4*, 348–349.
384. Couzijn, E. P. A.; Schakel, M.; de Kanter, F. J. J.; Ehlers, A. W.; Lutz, M.; Spek, A. L., et al. Dynamic Configurational Isomerism of A Stable Pentaorganosilicate. *Angew. Chem. Int. Ed.* **2004,** *43* (26), 3440–3442.
385. Couzijn, E. P. A.; Slootweg, J. C.; Ehlers, A. W.; Lammertsma, K. Stereomutation of Pentavalent Compounds: Validating the Berry Pseudorotation, Redressing Ugi's Turnstile Rotation, and Revealing the Two- and Three-Arm Turnstiles. *J. Am. Chem. Soc.* **2010,** *132* (51), 18127–18140.
386. Couzijn, E. P. A.; van den Engel, D. W. F.; Slootweg, J. C.; de Kanter, F. J. J.; Ehlers, A. W.; Schakel, M., et al. Configurationally Rigid Pentaorganosilicates. *J. Am. Chem. Soc.* **2009,** *131* (10), 3741–3751.
387. Junold, K.; Baus, J. A.; Burschka, C.; Fonseca Guerra, C.; Bickelhaupt, F. M.; Tacke, R. Bis[*N,N'*-diisopropylbenzamidinato(−)]silicon(II): Lewis Acid/Base Reactions With Triorganylboranes. *Chem. Eur. J.* **2014,** *20* (39), 12411–12415.
388. Scheschkewitz, D.; Hofmann, M.; Ghaffari, A.; Amseis, P.; Prasang, C.; Mesbah, W., et al. Very Strong Anionic Homoaromaticity in (*deloc*-1,3,4)-1-sila-3,4-diboracyclopentane-1-ides, The Importance of the Energy of the Reference System for Homoaromatic Stabilization Energies. *J. Organomet. Chem.* **2002,** *646* (1–2), 262–270.
389. Prasang, C.; Amseis, P.; Scheschkewitz, D.; Geiseler, G.; Massa, W.; Hofmann, M., et al. Strong Neutral Homoaromatics. *Angew. Chem. Int. Ed.* **2006,** *45* (40), 6745–6747.
390. Präsang, C.; Sahin, Y.; Hofmann, M.; Geiseler, G.; Massa, W.; Berndt, A. 1,2-Diboracyclopentanes Without Strong Donor Substituents: Synthesis, Reactions, and Computational Analysis. *Eur. J. Inorg. Chem.* **2008,** (32), 5046–5055.

391. Wesemann, L.; Englert, U.; Seyferth, D. First Reaction of 1,2-Dimethyl-*o*-Silaborane—Nucleophilic Degradation. *Angew. Chem. Int. Ed.* **1995,** *34* (20), 2236–2238.
392. Wesemann, L.; Trinkaus, M.; Ramjoie, Y.; Ganter, B.; Englert, U.; Müller, J. Adducts of *o*-Silaborane With Water and Methanol. *Eur. J. Inorg. Chem.* **2000,** *4*, 735–739.
393. Sidorkin, V. F.; Doronina, E. P. Cage Silaphosphanes With a P→Si Dative Bond. *Organometallics* **2009,** *28* (18), 5305–5315.
394. Gau, D.; Rodriguez, R.; Kato, T.; Saffon-Merceron, N.; Cossío, F. P.; Baceiredo, A. Nucleophilic Silylenoid Character of Stable Phosphonium Sila-Ylides. *Chem. Eur. J.* **2010,** *16* (28), 8255–8258.
395. Rodriguez, R.; Gau, D.; Kato, T.; Saffon-Merceron, N.; De Cózar, A.; Cossío, F. P., et al. Reversible Binding of Ethylene to Silylene-Phosphine Complexes at Room Temperature. *Angew. Chem. Int. Ed.* **2011,** *50* (44), 10414–10416.
396. Gau, D.; Rodriguez, R.; Kato, T.; Saffon-Merceron, N.; Baceiredo, A. Diastereoselective Synthesis of Bulky, Strongly Nucleophilic, and Configurationally Stable P-Stereogenic Tricyclic Phosphines. *J. Am. Chem. Soc.* **2010,** *132* (37), 12841–12843.
397. Gualco, P.; Lin, T. P.; Sircoglou, M.; Mercy, M.; Ladeira, S.; Bouhadir, G., et al. Gold-Silane and Gold-Stannane Complexes: Saturated Molecules as σ-Acceptor ligands. *Angew. Chem. Int. Ed.* **2009,** *48* (52), 9892–9895.
398. Gualco, P.; Mercy, M.; Ladeira, S.; Coppel, Y.; Maron, L.; Amgoune, A., et al. Hypervalent Silicon Compounds by Coordination of Diphosphine-Silanes to Gold. *Chem. Eur. J.* **2010,** *16* (35), 10808–10817.
399. Kameo, H.; Kawamoto, T.; Bourissou, D.; Sakaki, S.; Nakazawa, H. Evaluation of the σ-Donation From Group 11 Metals (Cu, Ag, Au) to Silane, Germane, and Stannane Based on the Experimental/Theoretical Systematic Approach. *Organometallics* **2015,** *34* (8), 1440–1448.
400. Wagler, J.; Brendler, E. Metallasilatranes: Palladium(II) and Platinum(II) as Lone-Pair Donors to Silicon (IV). *Angew. Chem. Int. Ed.* **2010,** *49* (3), 624–627.
401. Truflandier, L. A.; Brendler, E.; Wagler, J.; Autschbach, J. ^{29}Si DFT/NMR Observation of Spin-Orbit Effect in Metallasilatrane Sheds Some Light on the Strength of the Metal→Silicon Interaction. *Angew. Chem. Int. Ed.* **2011,** *50* (1), 255–259.
402. Autschbach, J.; Sutter, K.; Truflandier, L. A.; Brendler, E.; Wagler, J. Atomic Contributions From Spin-Orbit Coupling to ^{29}Si NMR Chemical Shifts in Metallasilatrane Complexes. *Chem. Eur. J.* **2012,** *18* (40), 12803–12813.
403. Wahlicht, S.; Brendler, E.; Heine, T.; Zhechkov, L.; Wagler, J. 7-Azaindol-1-yl(organo)silanes and Their $PdCl_2$ Complexes: Pd-Capped Tetrahedral Silicon Coordination Spheres and Paddlewheels With a Pd–Si Axis. *Organometallics* **2014,** *33* (10), 2479–2488.

Index

Note: Page numbers followed by "*f*" and "*t*" refer to figures and tables, respectively.

E

I

K

L

N

O

T